T4-AKE-918

The Structure of Probability Theory with Applications

McGRAW-HILL SERIES IN PROBABILITY AND STATISTICS

David Blackwell and Herbert Solomon, *Consulting Editors*

Bharucha-Reid Elements of the Theory of Markov Processes and Their Applications
Drake Fundamentals of Applied Probability Theory
Dubins and Savage How to Gamble If You Must
Ehrenfeld and Littauer Introduction to Statistical Methods
Graybill Introduction to Linear Statistical Models, Volume I
Jeffrey The Logic of Decision
Li Introduction to Experimental Statistics
Miller Simultaneous Statistical Inference
Mood and Graybill Introduction to the Theory of Statistics
Morrison Multivariate Statistical Methods
Pearce Biological Statistics: An Introduction
Pfeiffer Concepts of Probability Theory
Raj Sampling Theory
Thomasian The Structure of Probability Theory with Applications
Wadsworth and Bryan Introduction to Probability and Random Variables
Weiss Statistical Decision Theory
Wolf Elements of Probability and Statistics

The Structure of Probability Theory with Applications

Aram J. Thomasian

Professor of Statistics and
Electrical Engineering and Computer Science,
University of California, Berkeley

McGraw-Hill Book Company

New York San Francisco St. Louis Toronto London Sydney

THE STRUCTURE OF PROBABILITY THEORY WITH APPLICATIONS

Printed in the United States of America.

Library of Congress catalog card number: 69-12052

1234567890 MAMM 7654321069

64276

to my mother and father
for their love and assistance
over the years
and
to my wife and three sons
who for five years contributed
so many of my hours
which should have been theirs

Preface

Probability theory is twice favored by having a beautifully simple structure and by having an almost unlimited range of applicability. This book emphasises both the structure and the applicational flavor. No prior exposure to probability is presumed. Almost all of the book should be accessible after a standard two-year calculus course. The presentation is aimed at those who desire a foundation stronger than usual for applications, in any area, and at those who are interested in the theory itself.

The first half of the book, Chapters 1 to 17, is suitable for a one-semester first course in probability. These chapters contain a liberal sprinkling of *optional* sections, subsections, and examples, which are identified by a *star*. The second half of the book contains incidental topics (Chapter 18 to 21) plus enough material for a course on stochastic processes (Chapters 22 to 24). A table showing the interdependence of this material appears at the end of the preface.

The first 15 chapters require little use of calculus. However, the reader is expected gradually to adopt a more abstract point of view when the interpretability, conceptual clarity, and applicational power of the presented viewpoint are increasingly demonstrated. This book is more abstract than most probability books at this mathematical level, but the reward is great.

Many examples and exercises have a practical flavor and also demonstrate the necessity for a thorough grasp of fundamentals. Formal results are interpreted verbally, and many illustrations encourage the visualization which is so essential. The reader's probabilistic intuition is kept current, hence awe-inspiring surprises are rare. Intuitive explanations are kept outside the formal proofs, which may usually be skipped. However, proofs are not neglected, and explicit attention is given to the strategy behind most of the more difficult ones.

The presentation is definitely self-contained enough for self-study. At the end of the book is an extensive summary of all definitions and theorems, as well as the essence of the more significant examples and discussion. The reader need refer only to the summaries of previously studied chapters, unless he seeks details or proofs. Some students have used the summary as a preview before trying some exercises and reading the main text. Instructors may write to the publisher for a solutions manual.

THE STRUCTURE OF PROBABILITY THEORY

Chapter 21 exhibits the structure of probability theory, as systematized by Kolmogorov and as used in almost all advanced probability research in the last 30 years. Chapter 15 retains the power and conceptual clarity of this structure while ignoring certain technical complications (the measurability of sets and functions) which are beyond the mathematical level of most of this book. A comparison of the summaries of Chapters 15 and 21 will reveal the closeness of this approximation. *Chapter 15 essentially exhibits the structure of probability theory as used in this book.* We adopt this structure and simultaneously accept the constraint that everything that is not rigorous should be so indicated, and should require only technical modifications, as exemplified by the transition from Chapter 15 to Chapter 21. Such technicalities are in some senses quite minor. However, since they tend to dominate and obscure the underlying simplicity, the first half of the book provides a valuable perspective even for those who go on to a rigorous presentation.

COMMENTS ON CHAPTERS 1 TO 17

The first half of the book encourages an early partial assimilation of Chapter 15, and also properly relates the usual first-course topics to this foundation. This foundation evolves from the ideas of sample space, event, random variable as a function, induced distributions, etc. It is *not* sufficient, as has long been common, merely to mention these ideas without explicitly and convincingly relating the theory and examples to them. If this foundation is not laid early then later studies suffer. Even if a proper foundation is eventually acquired and assimilated, in retrospect the whole process is seen to entail much confusion and wasted time and effort. I had this frustrating experience about 15 years ago as a graduate student. Later as a teacher my continuous irritation at perpetuating this state of affairs was a major motivation for starting this book.

I believe that the unstarred portions of Chapters 1 to 17 provide a reasonable solution to the foregoing problem. This material has been used satisfactorily in a graduate course and in several intensive one-quarter courses (11 weeks) taken primarily by undergraduates having widely varying interests but only a calculus background. A semester course, or a two-quarter course, seems preferable. In these courses the definitions, results, interpretations, and applications were emphasized, but not the proofs. A small percentage of the students acquired quite a thorough grasp of the subject at this level, but for most students probability is too novel for this to occur in a single course. However, the crucial point is that all the students had their facts and thoughts embedded in a familiar framework which is capable of as much clarity and firmness as later studies might necessitate.

A BRIEF SURVEY OF THE CONTENTS

Chapters 1 to 5: mathematical preliminaries and combinatorics For most courses Chapters 1, 3, 4, and 5 will probably serve primarily as reference, after having been covered rapidly as a review. The material of Chapter 5 on absolutely convergent series is not really needed until Chapter 9.

Chapters 1 and 3, together with Section 2-1, provide a self-contained exposition of sets and functions, in a depth sufficient for this book. This language must be used in a book having the goals that this one has. It permits a more comprehensible and powerful theory even in the discrete case. This somewhat more abstract style cannot be avoided by the uninitiated student for very long without serious disadvantage. The rapid trend in mathematical education, from grade school on up, and in the presentation of applied mathematical subjects, is clearly in this direction. Acquiring this way of thinking through a course in probability theory has much to recommend it, since the concepts can constantly be interpreted in a variety of real-world examples.

The unstarred portions of Sections 2-2 and 2-3 contain the minimum combinatorial background required for the rest of the book, while additional standard results are derived in Section 2-4. For a first course in probability to yield a general perspective it should not be dominated by the fascinating subtleties peculiar to combinatorics. Furthermore, Section 17-4 shows that a first course provides useful tools for combinatorial problems.

Chapters 6 to 17: the structure and some basic facts and examples Chapter 6 introduces probability densities and measures in the discrete case, where the basic concepts involve the fewest technical complications. Theorem 2 (Section 6-4) derives the elementary properties of probability measures in such a way that they are available for the rest of the book. Chapter 7 does the same for independent events. Chapter 8 examines the universally popular binomial, Poisson, and Pascal densities, while Chapters 9 and 10 introduce means, variances, etc., for discrete densities.

In the decisive Chapter 11 random variables are introduced, and then the earlier theory is reviewed and incorporated into a structure close to that of Chapter 15. The earlier chapters make Chapter 11 digestible, but it usually requires repeated chewing. These earlier chapters were inserted after experience with a preliminary edition.

Chapters 12 and 13 apply the theory to a variety of interesting nontrivial discrete examples. The statements of the theorems and most of the discussion in Chapters 12 and 13 are valid in general, as are the proofs in Chapter 12 for the weak and strong laws of large numbers (WLLN and SLLN). The central-limit theorem (CLT) is stated, interpreted, applied, and made believable in Chapter 13, but its proof is postponed to Chapter 19.

Chapter 15 provides a rapid review of a theory which now has intuitional significance. Most readers find the slight increase in abstraction of Chapter 15 quite easy. Chapter 16 replaces the summations of the discrete case by integrations in the (absolutely) continuous case. Due attention is given to the special continuous densities having wide applicability. The exercises in Section 16-4 derive some of the distributions of importance in statistics, although the book does not contain any statistical theory. Chapter 16 is long, but much of it can be covered rapidly, since it is subsumed within the now familiar conceptual framework of Chapter 15.

Chapter 17 develops the properties of distribution functions. For several reasons distribution functions appear later than usual. They are useful only for special kinds of problems, and they do not lead to the conceptual clarity obtainable from the probability-measure emphasis of Chapter 15. All the useful special distributions (except the *singular* multivariate normal) have discrete or continuous densities which deserve direct examination. Furthermore, as is apparent from Chapter 22, distribution functions do not have the generality needed for stochastic processes and provided by probability measures.

The starred Sections 14-1, 15-5, and 16-6 are devoted to conditional expectations, and each section presumes the earlier ones. Some instructors will reasonably feel that conditional expectations are too important to be ignored, although they are not essential to the remainder of the book.

Chapters 18 to 21: incidental topics Chapter 18 contains two topics which should appear in a book of this kind. Chapters 19 to 21 (as well as Section 12-4 and the Addendum to Section 24-2) are for proof-oriented readers. Prior exposure to rigorous proofs is desirable for Chapters 20 and 21.

Section 18-1 develops covariance matrices in enough detail for Section 18-2 to give a transparent development of multivariate normal densities. Sections 18-3 and 18-4 state the properties of characteristic functions, without proof, and then illustrate their use by deriving central-limit theorems. Characteristic functions are Fourier transforms of probability distributions, and they are used extensively in many areas of probability. The material of Chapter 18 is basic to statistical theory and fields such as communications theory.

Chapter 19 contains the Lindeberg non–characteristic-function proof of the CLT. This proof has generality and it yields real insight; hence its neglect is surprising.

Chapter 20 illustrates the powerful truncation technique by generalizing the CLT and SLLN. In Section 20-2 the CLT is obtained for general triangular arrays under the Lindeberg condition by truncating the result from Chapter 19. In Section 20-3 the result from Section 12-4 is truncated in order to obtain Kolmogorov's SLLN. Also the asymptotic version of the SLLN is related to the equivalent approach of Sections 12-3 and 12-4.

Chapter 21 provides a substantial introduction to the rigorous construction of probability theory, so that no vague gap should be left between this book and a later course on measure theory, abstract integration, or advanced probability theory. It also yields relatively easy access to some technical terms which are appearing more frequently in applied books.

Chapters 22 to 24: stochastic processes This material probably forms the most stimulating sequel to the first half of the book. Chapter 22 gives a thorough introduction to the intuitive nature and mathematical formalization of continuous-parameter processes. In a sense Chapter 22 builds, on top of Chapter 15, a superstructure adequate for Chapters 23 and 24. Section 22-3 goes into more than usual detail concerning the crucial relationship between realizations and consistent families of distributions. Chapters 23 and 24 introduce the Poisson and Wiener processes by way of approximating experiments which reveal at the outset the properties to be anticipated. Explicit constructions are given for each of the Poisson type processes, and the emphasis is always on the nature of the realizations. This approach seems more transparent and closer to intuition than the common method of studying solutions to partial-differential equations. It is pleasing that these last three chapters make essential use of almost everything from the first 17 chapters.

Chapters 14, 22, 23, and 24 have been used for a one-quarter graduate course on stochastic processes. In a semester course, or by omitting Section 24-2, it should be possible to spend some time on Markov chains. The omission of Markov chains is probably the most serious omission from this text, even though the techniques for their analysis are quite different.

ACKNOWLEDGMENTS

I thank the many students whose enthusiasm and performance kept me writing. Both my departments provided a stimulating and relatively constraint-free environment. My indebtedness to many members of the statistics department goes back to the decisive instruction I received as a graduate student many years ago.

The statistics department provided considerable manuscript assistance. The many reviews obtained by McGraw-Hill were helpful, especially the final review by Alvin W. Drake. Charles Desoer never tired of urging me on.

Much of my comprehension of probability theory was acquired during the many rejuvenating summers that my research was supported by the Information Systems Branch of the Office of Naval Research. Without these summers the book might never have been written.

It is a pleasure to acknowledge that this book is based on the work of many researchers and authors, most of whom are unmentioned.

Aram J. Thomasian

INTERDEPENDENCY TABLE

Each section presumes all preceding unstarred portions of Chapters 1 to 17.

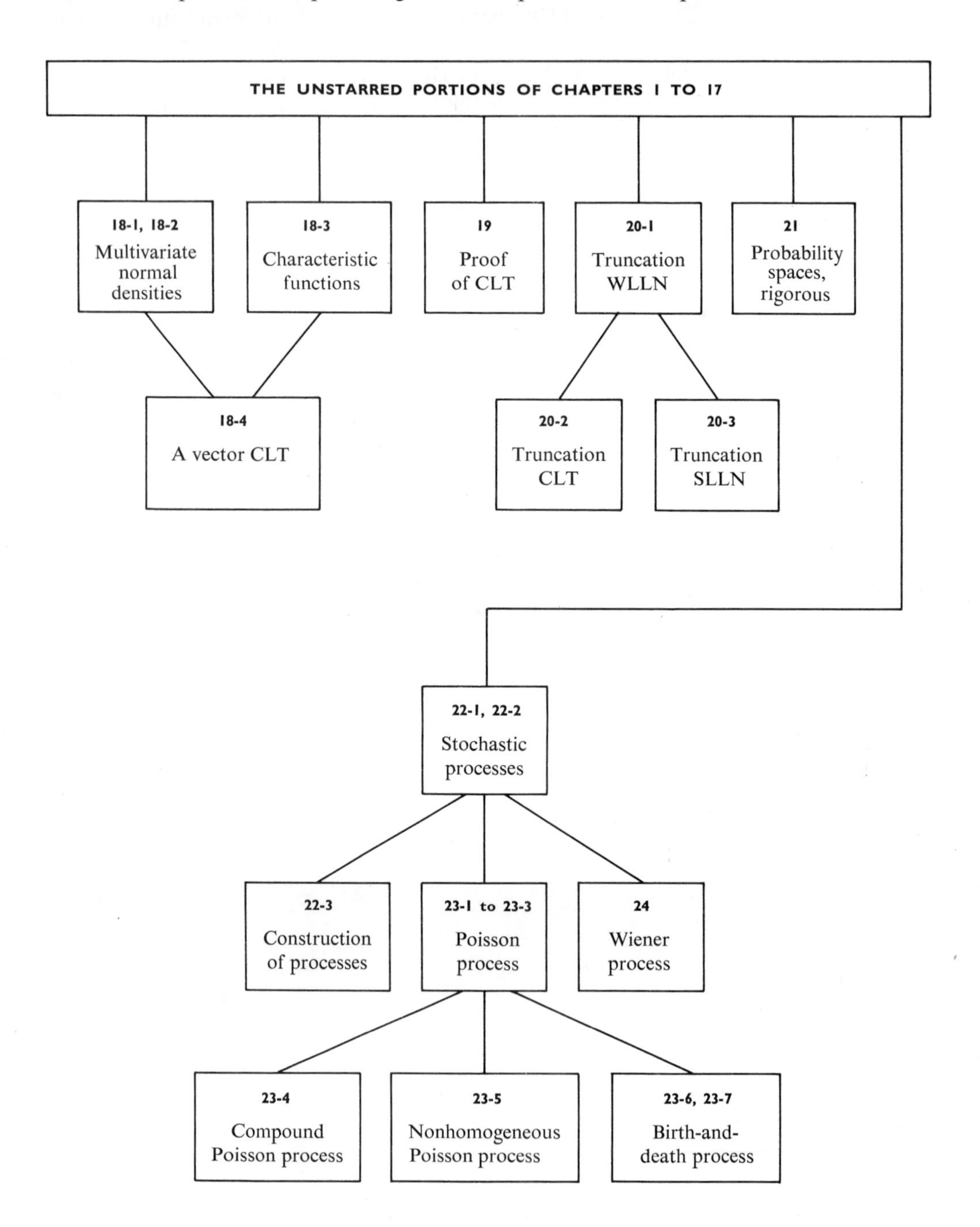

Contents

1
Sets

The mathematical concept of a *set* is basic. The more precise developments of this concept and its uses are lengthy. In this chapter we give a relatively informal discussion of the basic ideas and facts concerning sets. The many examples throughout the book should gradually make an uninitiated reader comfortable with elementary set theory as presented in this chapter.

1-1 SETS

A **set**[1] is a collection of objects which are called the *elements*, or *members*, or *points* of the set. Two sets are **equal** iff[2] they both consist of exactly the same members.

For example, $\{-1.5,8,H\}$ is a set with three elements, the numbers -1.5 and 8 and the letter H. Also, $\{-1.5,8,H\} = \{H,8,-1.5\}$, since both sets have exactly the same members. That is, as far as equality of sets is concerned the elements do *not* have an order. Also we assume that a set contains an element only once, so that the notation $\{x_1, x_2, \ldots, x_n\}$ means "the set of different elements among the objects $x_1, x_2, \ldots, x_n$." Thus $\{1,2,8\}$ and $\{2,1,8,8,1\}$ both have three elements and are actually equal.

Henceforth, unless stated otherwise, the notation $\{x_1, x_2, \ldots, x_k\}$ *for a set means that the* x_i *are all different.*

Thus $\{x_1,x_2,x_3\}$ represents a set with three elements.

A set is said to be **finite** iff it has a finite number of elements; otherwise it is said to be **infinite.** The set of all positive integers is an infinite set which is often denoted by $\{1, 2, \ldots\}$.

If π is a formula which is either true or false for every x then $\{x: \pi(x)\}$, read "the set of all x such that $\pi(x)$," denotes the set of all objects x which make $\pi(x)$ a true formula. Thus $\{x: x \text{ is a real number, and } x^2 < 4\}$ is a set whose elements are all those real numbers which are both less than 2 and greater than -2. The method of using a formula to define a set is our basic method for defining infinite sets and large finite sets.

[1] Defined terms are indicated by **boldface** type.
[2] **Iff** is an abbreviation of *if and only if*.

Clearly $\{x: \pi(x)\} = \{y: \pi(y)\}$. That is, the use of a symbol such as x merely helps to eliminate the need for such awkward descriptions as "the set of all objects which make π true." In the notation $\{x: \pi(x)\}$ we can think of the brace, $\{\ \}$, as an abbreviation for "the set of all" and the colon, :, as an abbreviation for "such that." In the expression $\{x: \pi(x)\}$ we often replace x by a symbol which helps to make the definition of the set more quickly comprehensible. For example, if in a particular discussion we are using the notation (m,n) for an ordered pair of positive integers, as defined in the next chapter, we might use $\{(m,n): m \leq n^2\}$ to denote the set of all ordered pairs of positive integers such that the first coordinate is less than or equal to the square of the second coordinate. This set may also be defined more lengthily as

$$\{x: x = (m,n), \text{ where } m \text{ and } n \text{ are positive integers satisfying } m \leq n^2\}.$$

Clearly $\{3,-4\} = \{x: x \text{ is a real number and } x^2 + x - 12 = 0\}$. Thus the same set can be defined in different ways, and it may be hard to find out whether two particular sets are equal. Also, each member of a set might be a rather complicated object. For example, each member of a particular set might be a continuous real-valued function defined for all real x satisfying $0 \leq x \leq 1$.

The preceding discussion should have made the concepts of set and equality of sets precise enough for the reader to continue on and increase his comprehension of these ideas as he gains experience with them. We have merely discussed the concept of set in informal terms and have not attempted a formal axiomatic approach to the idea.

1-2 SOME SET-THEORY TERMINOLOGY

Let us now introduce additional terminology relating to sets and then define some standard operations on sets. These definitions primarily involve the English language, elementary logic, and the general idea of a set. They can later be applied to any precisely defined sets of mathematical objects which we encounter.

We use $x \in A$ as an abbreviation for "x **belongs to** A," or "x is a member of A," or "x is an element of A," where x is some object and A is a set. Similarly $x \notin A$ means x **does not belong to** A. We can now restate the definition of equality for sets as follows: $A = B$ *iff for all* x,

$$x \in A \text{ iff } x \in B.$$

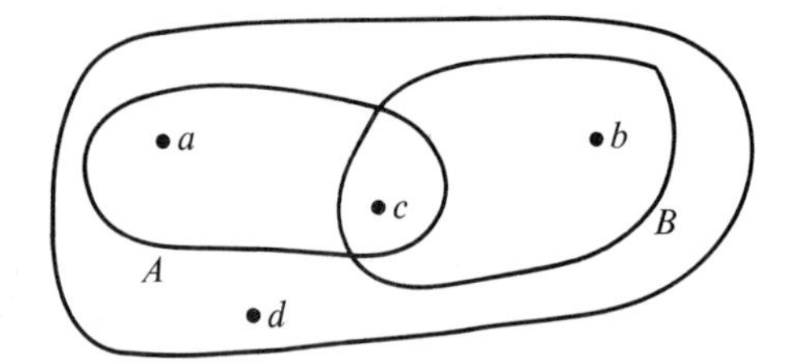

Fig. 1 $a \in A$, $a \notin B$; $b \notin A$, $b \in B$; $c \in A$, $c \in B$; $d \notin A$, $d \notin B$.

In considering sets a picture is often an aid to understanding. Many of the figures in this chapter illustrate set relationships as they might appear if the sets under discussion were sets of points in the plane—which they often are not. Such pictures in set theory are called *Venn diagrams* (after John Venn, 1834–1883). Although such pictures are an aid to visualization, it should always be kept in mind that the development has very general applicability.

If A and B are sets we say that A is a **subset** of B iff every member of A also belongs to B. We denote this relationship by $A \subset B$, or by $B \supset A$, and say that A **is contained in** B, or B contains A. Thus $A \subset B$ *iff for all* x, *if* $x \in A$ *then* $x \in B$. For example, $\{1,2\} \subset \{0,1,2\}$. Clearly $A \subset A$ for every set A. Also, $A = B$ iff $A \subset B$ and $A \supset B$; that is, $A = B$ iff each contains the other.

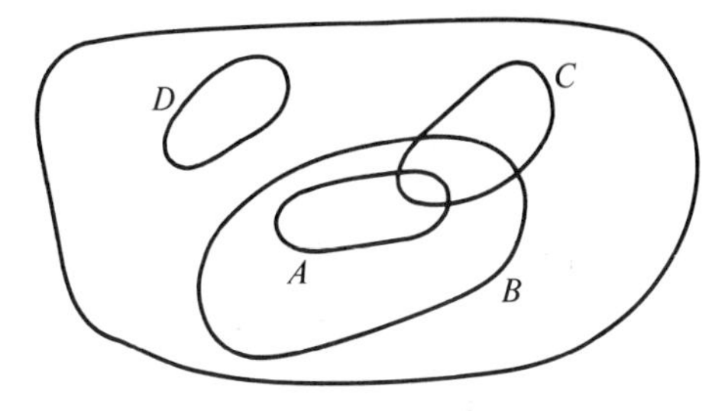

Fig. 2 $A \subset B$, but $A \not\subset C$, $A \not\subset D$, and $B \not\subset A$.

The assertion "A is a subset of B," does not rule out the possibility that A may be equal to B. In certain situations it may be desirable to rule out equality. This occurs so seldom in probability theory that we introduce a terminology for such situations, but not a symbolism. We say that A is a **proper subset** of B, or that A is properly contained in B, iff A is contained in B but is not equal to B. Clearly, this is equivalent to saying that every member of A also belongs to B but B has at least one member which does not belong to A.

A sequence $A_1, A_2, \ldots$ of sets is said to be **increasing** iff each set is contained in the next one, that is, iff $A_1 \subset A_2 \subset A_3 \subset \cdots$. A sequence of sets is said to be **decreasing** iff each set is contained in the preceding one, that is, iff $A_1 \supset A_2 \supset A_3 \cdots$.

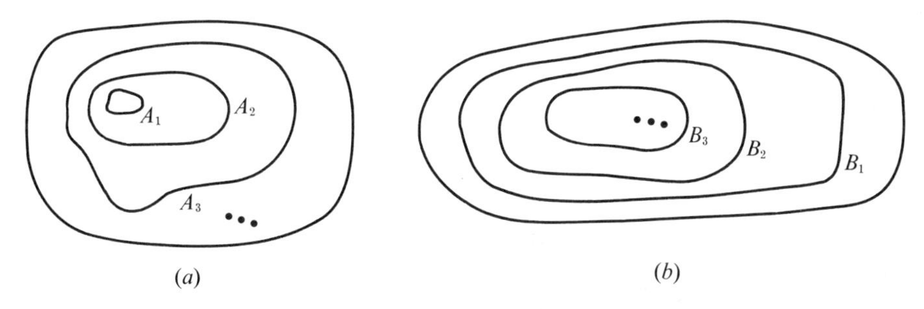

Fig. 3 (*a*) $A_1, A_2, \ldots$ is an increasing sequence of sets; (*b*) $B_1, B_2, \ldots$ is a decreasing sequence of sets.

The unique set which does not have any members is called the **empty set** and is denoted by φ. Naturally φ is considered to be a finite set rather than an infinite set. Clearly $\varphi \subset A$ is true for every set A. Also, $A \neq \varphi$ means that A is **nonempty;** that is,

A has at least one member. Thus $\{x: x \text{ is a real number and } x^2 \leq -2\} = \varphi$ and $\{x: x \text{ is a real number and } x^2 \leq 2\} \neq \varphi$.

A set may contain one and only one element. The *set* containing the one object x is considered different from the *object* x; that is, $\{x\}$ and x are different. This may be thought of intuitively as follows. Suppose we are considering 10 balls, numbered from 1 to 10, as objects. We might think of the set consisting of balls 1 and 3 as being a bag containing balls 1 and 3. Then "ball 3" is a different thing from "a bag with ball 3 inside." We will sometimes ignore this distinction in the interests of simplicity. The reader may wonder why a distinction was made if it can be ignored. The same kind of question can be raised more forcefully concerning some of the definitions and notations introduced later. Our justification is that we wish to keep the formal development reasonably precise so that this precision is available when needed, without losing the advantages of a more casual approach whenever such an approach is not likely to be misinterpreted.

Two sets are said to be **disjoint,** or mutually exclusive, iff they do not have a member in common. More generally sets $A_1, A_2, \ldots, A_n$ are disjoint iff no element belongs to more than one of $A_1, A_2, \ldots, A_n$. Similarly an infinite sequence of sets is disjoint iff no element belongs to more than one of the sets in the sequence.

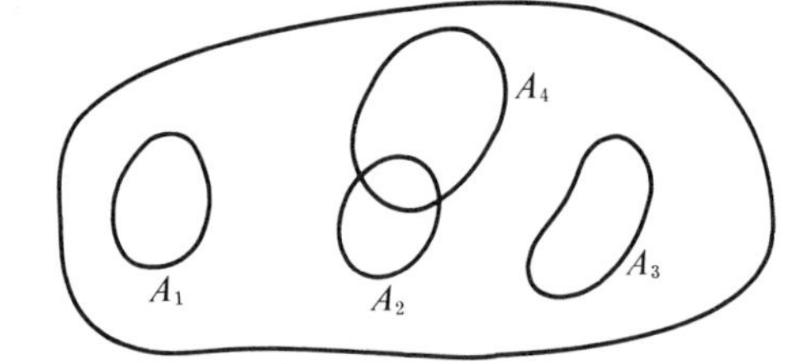

Fig. 4 A_1, A_2, A_3 are disjoint; A_1, A_2, A_3, A_4 are not disjoint.

Suppose we have three different sets A, B, and C and we form a new set $\mathscr{A} = \{A,B,C,\}$. Then $\mathscr{A}$ is a set which has three members, each of which is a set. In such a context it is customary to use words such as **collection** or **family** or **class** as synonyms for the word "set." We say that $\mathscr{A}$ is a class of sets rather than $\mathscr{A}$ is a set of sets. Thus a family of sets is said to be disjoint iff no object belongs to more than one set in the family.

1-3 OPERATIONS ON SETS

Let us now describe three fundamental techniques for constructing new sets from given sets. In many discussions some particular set Ω will be especially important, and many subsets of Ω will be considered. In such a context Ω may be called the *universal set* for the discussion. If Ω is such a set and A is a subset of Ω then the set of all members of Ω which do *not* belong to A is called the **complement of A with respect to Ω** and is denoted by A^c. Therefore, as depicted in Fig. 1,

$$A^c = \{x: x \in \Omega \quad \text{and} \quad x \notin A\}.$$

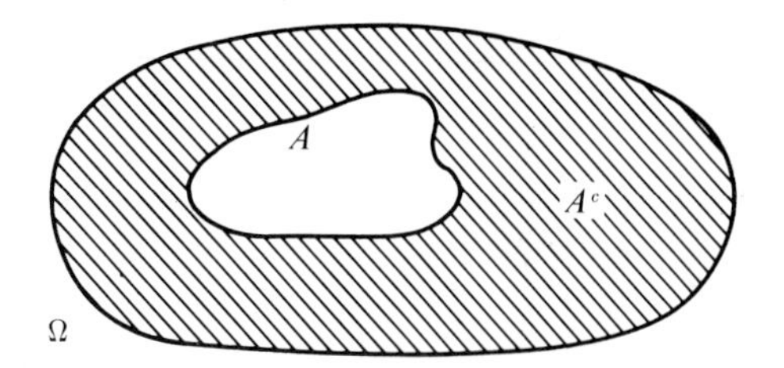

Fig. 1

Clearly A^c is not defined until Ω is specified. If the set Ω is clear from the context we may abbreviate to "the complement of A." Clearly $(A^c)^c = A$.

If A and B are sets we call the set of all objects which belong to *either A or B* the **union of** A **and** B. We denote this set by $A \cup B$, read "A union B." Therefore

$$A \cup B = \{\omega\colon \omega \in A \textit{ or } \omega \in B\}.$$

For example, $\{8,2,H\} \cup \{H,5,8\} = \{8,H,2,5\}$. Note that in mathematics "or" always means "or, or both." Thus the number 8 belongs to the union of the two sets $\{8,2,H\}$ and $\{H,5,8\}$. More generally, if $A_1, A_2, \ldots, A_n$ are sets their union is the set of all objects which belong to *at least one* of them; it is denoted by $A_1 \cup$

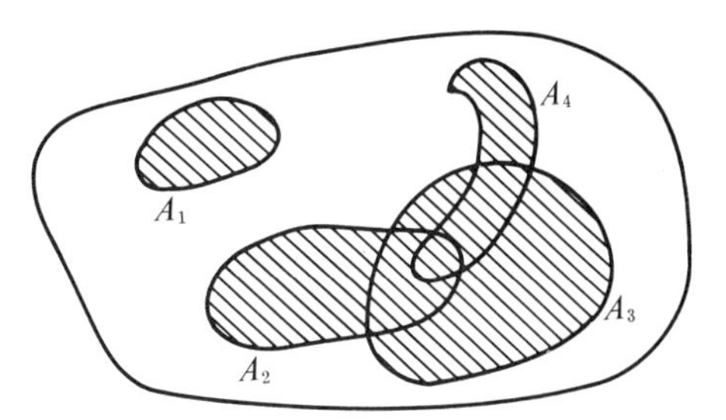

Fig. 2 $A_1 \cup A_2 \cup A_3 \cup A_4$ shaded.

$A_2 \cup \cdots \cup A_n$ or by $\bigcup_{k=1}^{n} A_k$. Similarly, if $A_1, A_2, \ldots$ is an infinite sequence of sets the union of these sets is the set of all objects which belong to at least one of them; it is denoted by $A_1 \cup A_2 \cup \cdots$ or by $\bigcup_{k=1}^{\infty} A_k$. Therefore, as depicted in Fig. 2,

$$\bigcup_{k=1}^{\infty} A_k = \{\omega\colon \textit{there exists} \text{ a positive integer } k \text{ such that } \omega \in A_k\}.$$

If A and B are sets we call the set of all objects which belong to both A and B the **intersection of** A **and** B. We denote it by $A \cap B$, read "A intersect B." Therefore

$$A \cap B = \{\omega\colon \omega \in A \textit{ and } \omega \in B\}.$$

We often abbreviate $A \cap B$ to AB. Note that there is no corresponding abbreviation for unions. More generally, if $A_1, A_2, \ldots, A_n$ are sets their intersection is the set of all objects which belong to *every one* of them; it is denoted by $A_1 \cap A_2 \cap \cdots \cap A_n$

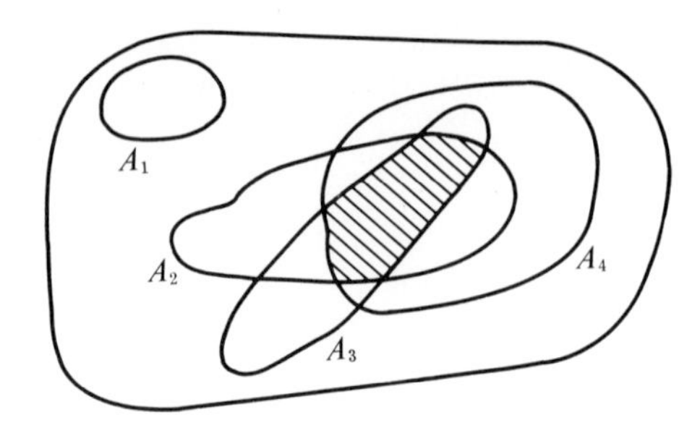

Fig. 3 $A_2A_3A_4$ shaded; $A_1A_2A_3A_4 = \varphi$.

or by $\bigcap_{k=1}^{n} A_k$ or by $A_1A_2 \cdots A_n$. Similarly if $A_1, A_2, \ldots$ is an infinite sequence of sets the intersection of these sets is the set of all objects which belong to every one of them; it is denoted by $A_1 \cap A_2 \cap \cdots$ or by $\bigcap_{k=1}^{\infty} A_k$ or by $A_1A_2 \cdots$. Therefore,

$$\bigcap_{k=1}^{\infty} A_k = \{\omega : \omega \in A_k \textit{ for every } \text{positive integer } k\}.$$

EXERCISES

1. Seven people labeled 1, 2, . . . , 7 are classified as in the table below. Let $\Omega = \{1, 2, \ldots, 7\}$ be the basic set with respect to which complements are taken. Let A be the set of married people, so that $A = \{1,2,4\}$.

Person	Height	Sex	Married
1	Tall	Male	Yes
2	Short	Female	Yes
3	Medium	Female	No
4	Short	Female	Yes
5	Medium	Male	No
6	Tall	Male	No
7	Tall	Female	No

(*a*) Exhibit B, the set of males.
(*b*) Are A and B disjoint?
(*c*) Exhibit C, the set of tall people.
(*d*) Exhibit $A \cup B$, the set of people who are either married or male.
(*e*) Exhibit $A \cup (B \cap C^c)$ and describe it verbally.
(*f*) Is AC a subset of B? Interpret the statement $AC \subset B$, assuming it to be true.

2. Sketch the following subsets of the plane: $A = \{(x,y): x + y \leq 1\}$, $B = \{(x,y): y \leq 2x + 2\}$, AB, $A \cup B$, $A \cup (B^c)$.

3. The subset A_n of the plane is defined by $A_n = \{(x,y): x^2 + y^2 \leq 1/n\}$. Describe the following sets: $A = \bigcup_{n=1}^{\infty} A_n$, $B = \bigcap_{n=1}^{\infty} A_n$, $C = A \cap B^c$.

4. Clearly $\{\varphi, \{H\}, \{3\}, \{H, 3\}\}$ is the class of all subsets of $\{H,3\}$. Find the class of all subsets of $\Omega = \{H,3,\alpha\}$.

1-4 SET IDENTITIES

An **identity** in the algebra of sets is an equality between sets which is *always* true regardless of which specific sets are used in the equation. If complements are involved the equation must naturally hold for every possible selection of a universal set. Both $(A \cup B) = (B \cup A)$ and $(A^c)^c = A$ are obviously identities. Very few set identities are used in this book. Almost all of those used are either obvious or appear in the present chapter.

Example 1 The equation $(A \cup B)^c = (A^c) \cap (B^c)$ is an identity which states that the complement of the union of two sets equals the intersection of their complements. It is interpreted pictorially in Fig. 1. The last picture in Fig. 1 is obtained from the previous two pictures by using the

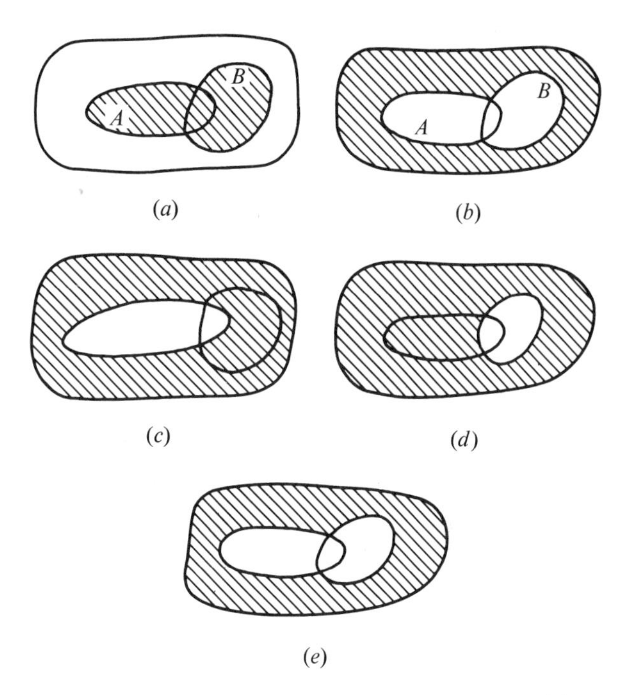

Fig. 1 (*a*) $A \cup B$ shaded; (*b*) $(A \cup B)^c$ shaded; (*c*) A^c shaded; (*d*) B^c shaded; (*e*) $(A^c) \cap (B^c)$ shaded.

definition of intersection, since a point belongs to $(A^c) \cap (B^c)$ iff it belongs to both A^c and B^c. This identity is obvious verbally if we observe that $(A \cup B)^c$ is the set of all points which do *not* belong to [either A or B], while $(A^c) \cap (B^c)$ is the set of all points which [do not belong to A] *and* [do not belong to B]. ■

In giving a more formal proof of an identity we prove that two sets which are defined in different ways are actually always equal, that is, that they are the *same set.* The following is a technique frequently employed to prove that $C = D$. First prove that anything which belongs to C must also belong to D, and then prove that anything which belongs to D must also belong to C; that is, first prove that C is a subset of D, and then prove that D is a subset of C. For example, let us give the first half of the proof that $C = D$ for the identity where $C = (A \cup B)^c$ and $D = (A^c) \cap (B^c)$. Let ω be an arbitrary member of C. We must show that ω belongs to D. Now, ω belongs to $(A \cup B)^c$, so ω does not belong to $(A \cup B)$. Thus ω does not belong to either A or B, so that ω must belong to both A^c and B^c. Thus ω belongs to the intersection of (A^c) and (B^c), which is D.

Example 2 Show that $A \cup (BC) = (A \cup B)C$ is *not* an identity. It is necessary only to somehow arrive at and exhibit one example for which the equation is false. Such an example is called a **counterexample.** Thus we can show that $(A) \cup (B \cap C) = (A \cup B) \cap (C)$ is not an identity by letting $A = \{1\}$, $B = \{2\}$, and $C = \{3\}$, so that $(A) \cup (B \cap C) = \{1\} \cup \varphi = \{1\}$, while $(A \cup B) \cap (C) = \{1,2\} \cap \{3\} = \varphi$. The fact that $(A) \cup (B \cap C) = (A \cup B) \cap (C)$ is not an identity is obvious from Fig.2. ■

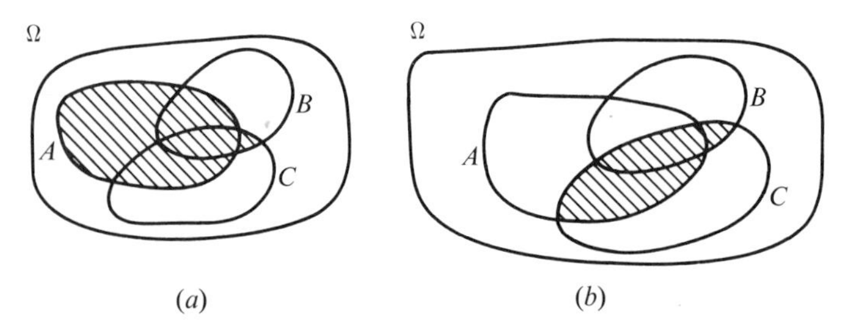

Fig. 2 (*a*) $A \cup (B \cap C)$ shaded: (*b*) $(A \cup B) \cap C$ shaded.

Example 3 The equation $(A \cap B)^c = (A^c) \cup (B^c)$ is an identity which states that the complement of the intersection of two sets equals the union of their complements. We now prove this identity. A point ω belongs to $(A \cap B)^c$ iff ω does not belong to $(A \cap B)$; that is, iff ω does not belong to both A and B, that is, iff ω belongs to at least one of A^c and B^c; that is, iff ω belongs to $(A^c) \cup (B^c)$. Thus a point belongs to the left-hand side of the equation iff it belongs to the right-hand side, and so the identity is proved. The identities of Examples 1 and 3 are called the **De Morgan laws.** ■

Example 4 The generalization of the De Morgan laws to infinite sequences of sets yields the two identities

$$(A_1 \cup A_2 \cup A_3 \cup \cdots)^c = (A_1{}^c) \cap (A_2{}^c) \cap (A_3{}^c) \cap \cdots$$

$$(A_1 \cap A_2 \cap A_3 \cap \cdots)^c = (A_1{}^c) \cup (A_2{}^c) \cup (A_3{}^c) \cup \cdots.$$

Thus the complement of a union equals the intersection of the complements, and the complement of an intersection equals the union of the complements. The first of these identities states that those objects which do not belong to at least one of $A_1, A_2, \ldots$ are precisely those objects which belong to every one of $A_1{}^c, A_2{}^c, \ldots$. The second identity states that those objects which do not belong to every one of $A_1, A_2, \ldots$ are precisely those objects which belong to at least one of $A_1{}^c, A_2{}^c, \ldots$. These verbalizations of the identities essentially constitute obvious proofs of them. These proofs also show that the corresponding equations for finite sequences of sets are also identities; that is,

$$(A_1 \cup A_2 \cup \cdots \cup A_n)^c = A_1{}^cA_2{}^c \cdots A_n{}^c \qquad (\bigcup_{k=1}^{n} A_k)^c = \bigcap_{k=1}^{n} A_k{}^c$$

$$(A_1A_2 \cdots A_n)^c = A_1{}^c \cup A_2{}^c \cup \cdots \cup A_n{}^c \qquad (\bigcap_{k=1}^{n} A_k)^c = \bigcup_{k=1}^{n} A_k{}^c. \quad \blacksquare$$

Example 5 If $A_1, A_2, \ldots$ is any sequence of sets then $A_1 \cup A_2 \cup \cdots = B_1 \cup B_2 \cup \cdots$, where the sets $B_1, B_2, \ldots$ are disjoint and B_n is that subset of A_n defined by

$$B_n = A_n \cap A_{n-1}^c \cap A_{n-2}^c \cap \cdots \cap A_2{}^c \cap A_1{}^c.$$

This identity permits the union of any sequence of sets to be expressed as the union of a sequence of disjoint sets. Clearly B_n is a subset of A_n, since B_n equals the intersection of A_n with another set. As indicated in Fig. 3, B_n consists of those points of A_n which do not belong

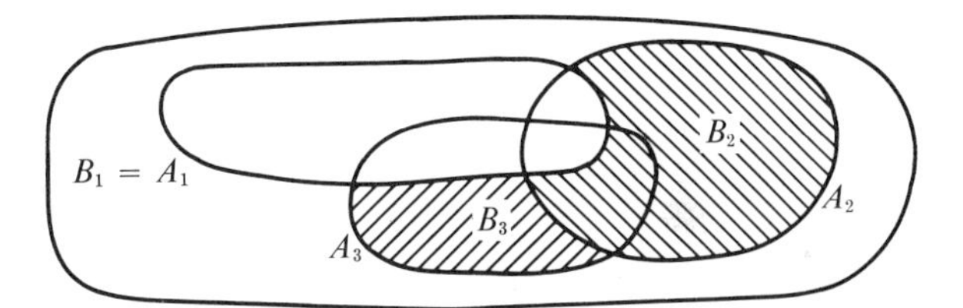

Fig. 3

to any of the earlier sets $A_1, A_2, \ldots, A_{n-1}$. Since B_n is contained in A_n, we know that $B_1 \cup B_2 \cup \cdots$ is contained in $A_1 \cup A_2 \cup \cdots$. Thus we need only show that $A_1 \cup A_2 \cup \cdots$ is contained in $B_1 \cup B_2 \cup \cdots$. If ω is a particular point belonging to $A_1 \cup A_2 \cup \cdots$ let n be the smallest integer such that $\omega \in A_n$. Clearly $\omega \in B_n$, since ω belongs to A_n but to none of $A_1, A_2, \ldots, A_{n-1}$. Thus ω belongs to $B_1 \cup B_2 \cup \cdots$, and the identity is proved. $\blacksquare$

In most of this book the manipulations with sets should, after a little experience, appear quite simple and natural. In a few places, such as the definition of Borel sets in Sec. 21-1, the uninitiated reader may be surprised that anyone could have confidence in the results of such general manipulations with sets. Far more complex set manipulations than are used in this book have been customary in mathematics for many years. Such manipulations permit a great deal of freedom and have not yet been found to introduce contradictions. A theory is contradictory if it contains a relation which can be proved to be both true and false. It should be pointed out, however, that some excessively loose set constructions, not used in normal mathematics, are easily shown to lead to contradictions. One such construction, for example, involves the set B of all sets which do not belong to themselves (is $B \in B$ true?).[1]

[1] For a deeper development of sets, functions, countability, real numbers, and much more, see, for example, Robert R. Stoll, "Set Theory and Logic," W. H. Freeman and Company, San Francisco, 1961, or Andrew M. Gleason, "Fundamentals of Abstract Analysis," Addison-Wesley, Reading, Mass., 1966. This latter work develops summability (pp. 200–209) in a fashion equivalent to Appendix 3. For detailed discussion and many more examples, at another level, Samuel Goldberg, "Probability," Prentice-Hall, Inc., Englewood Cliffs, N.J., 1960, presumes high school algebra and covers sets, combinatorics, and much of Chaps. 6 to 11 in the finite case.

EXERCISES

1. Most of the equations below are identities in the algebra of sets and are so obvious that they are true essentially by inspection. Find the ones which are not identities and give counterexamples for them. All sets are subsets of Ω, and complements are with respect to Ω. $\varphi =$ the empty set.

(*a*) $A \cup B = B \cup A$
(*b*) $\Omega \cap \varphi = \Omega$
(*c*) $A \cap B = B \cap A$
(*d*) $A \cup A^c = \Omega$
(*e*) $(A \cup B) \cup C = A \cup (B \cup C)$
(*f*) $A \cap A^c = \varphi$
(*g*) $(A \cap B) \cap C = A \cap (B \cap C)$
(*h*) $A \cup (A \cap B) = A$
(*i*) $A \cup \varphi = A$
(*j*) $A \cup B^c = A^c \cup B$
(*k*) $A \cap \Omega = A$
(*l*) $(A^c)^c = A$
(*m*) $\varphi^c = \Omega$
(*n*) $A \cap \varphi = \varphi$
(*o*) $\Omega^c = \varphi$
(*p*) $A \cup B = B \cap A$
(*q*) $A \cup A = A$
(*r*) $A \cup \Omega = \Omega$
(*s*) $A \cap A = A$
(*t*) $(A \cup B) \cap A = A$

2. Sketch Venn diagrams and shade areas to indicate the truth of the identities below (this was done in Fig. 1 for one identity):

(*a*) $A \cap (B \cup C) = (A \cap B) \cup (A \cap C)$
(*b*) $A \cup (B \cap C) = (A \cup B) \cap (A \cup C)$
(*c*) $(A \cap B)^c = (A^c) \cup (B^c)$

3. Prove that if $A_1, A_2, \ldots$ is a decreasing sequence of sets then $A_1 \cup A_2 \cup \cdots = A_1$. State and prove an analogous fact about $A_1 \cap A_2 \cap \cdots$ when $A_1, A_2, \ldots$ is an increasing sequence of sets.

4. Prove the identity $A\left(\bigcup_{n=1}^{\infty} B_n\right) = \bigcup_{n=1}^{\infty} (AB_n)$. Note that if $B_1, B_2, \ldots$ is a sequence of disjoint sets so is $AB_1, AB_2, \ldots$.

5. Let $A = \bigcup_{n=1}^{\infty} A_n$, where A_n is that subset of the plane defined by

$$A_n = \left\{(x,y): y \leq (-1)^n \frac{n}{n+1} x^3\right\}.$$

Make a sketch of A. Describe its boundaries and state whether or not they are included in A. Let B_3, B_4, and B_5 be defined as in Example 5, and sketch them.

2 Combinatorics

Combinatorics, or combinatorial analysis, is the study of problems in which a dominant concern is with the number of ways that different results can occur. It is a subtle and useful subject. Section 2-1 introduces the concept of cartesian product. This concept is useful in combinatorics and is fundamental to mathematics in general. Sections 2-2 and 2-3 are devoted to a few elementary and widely used combinatorial facts which in some form are probably already familiar to the reader. Section 2-4 derives several additional combinatorial results which are not needed in the remainder of the book.

2-1 CARTESIAN PRODUCTS

If n is a positive integer an **n-tuple** is an *ordered collection* of n elements, denoted by $(x_1, x_2, \ldots, x_n)$, where x_i is called the ith *coordinate*. An n-tuple $x = (x_1, x_2, \ldots, x_n)$ is equal to an r-tuple $y = (y_1, y_2, \ldots, y_r)$ iff $n = r$ and $x_1 = y_1, x_2 = y_2, \ldots, x_n = y_n$; that is, two tuples are equal iff they have the same number of coordinates and are equal coordinatewise. Thus $(3.2,H,1.5,-8) \neq (3.2,H,1.5) \neq (H,3.2,1.5)$. Different coordinates of a tuple may be equal, so that $(5,5,-2)$ is a legitimate 3-tuple. A 1-tuple is essentially a set with one member, but if $n \geq 2$ then $(x_1, x_2, \ldots, x_n) \neq \{x_1, x_2, \ldots, x_n\}$, since an n-tuple is ordered, while a set is not. We may call a 2-tuple an *ordered pair* and a 3-tuple an *ordered triple*, etc. If all the coordinates of an n-tuple are real numbers the n-tuple is essentially an n-dimensional vector, with real numbers for coordinates, where clearly $(1,3) \neq (3,1)$.

The **cartesian product** of two sets A and B is $\{(a,b)\colon a \in A \text{ and } b \in B\}$, denoted by $A \times B$; that is, $A \times B$ is the set of all ordered pairs such that the first coordinate belongs to A and the second coordinate belongs to B. More generally, for $n \geq 2$ the cartesian product of sets $A_1, A_2, \ldots, A_n$ is $\{(x_1, x_2, \ldots, x_n)\colon x_i \in A_i \text{ for all } i = 1, 2, \ldots, n\}$; that is, the cartesian product of $A_1, A_2, \ldots, A_n$ is the set of all n-tuples with ith coordinate belonging to A_i for all $i = 1, 2, \ldots, n$. This set is denoted by $A_1 \times A_2 \times \cdots \times A_n$ or by $\bigtimes_{k=1}^{n} A_k$. For example, an object is a member of $\{1, 2, \ldots\} \times \{\ldots, -1, 0, 1, 2, \ldots\} \times \{A, B, \ldots, Z\}$ iff it is an ordered triple (α,β,γ), with α a positive integer, β an integer, and γ a capital letter of the English alphabet. In the special case where all the sets $A_1, A_2, \ldots, A_n$ are equal to the same set A we denote the cartesian product by A^n. Thus A^n is the set of all n-tuples whose coordinates all belong to the set A. For example, $\{1,2,3,4\}^{38}$ is the set of all 38-tuples whose coordinates are integers from 1 to 4.

It is common to be casual with regard to extra parentheses in connection with cartesian products. For instance an element of $D \times C$ is (d,c), but if $D = A \times B$ then $d = (a,b)$, so that an element of $(A \times B) \times C$ is $((a,b),c)$, where $a \in A$, $b \in B$, and $c \in C$. Often we ignore this subtlety and make the convention that $(A \times B) \times C = A \times (B \times C) = A \times B \times C$, and we represent an element of this set by (a,b,c).

Unless stated otherwise, the letter R *will always denote the set of all* **real numbers,** *and* R^n *will denote the set of all* n-**tuples of real numbers.** *Thus* $R = R^1$.

A real number will often be referred to as **a real.** Clearly the notation R^n is consistent with our general notation for cartesian products; that is, $R^2 = R \times R$, $R^3 = R \times R \times R$, etc. An element, or point, of R^2 can be thought of as a point in the plane, or as the vector from the origin to that point. The same is true for R^3 and for R^n in general. We will sometimes use the convenient label "vector" for a point in R^n.

If A and B are contained in R then $A \times B$ can be depicted as a subset of R^2; that is, as shown in Fig. 1, the cartesian product of two sets of reals can be pictured in the plane. A member of R, R^2, R^3 can be thought of as a point on the real axis, in the plane, and in three-dimensional space, respectively. Members of R^n for $n \geq 4$

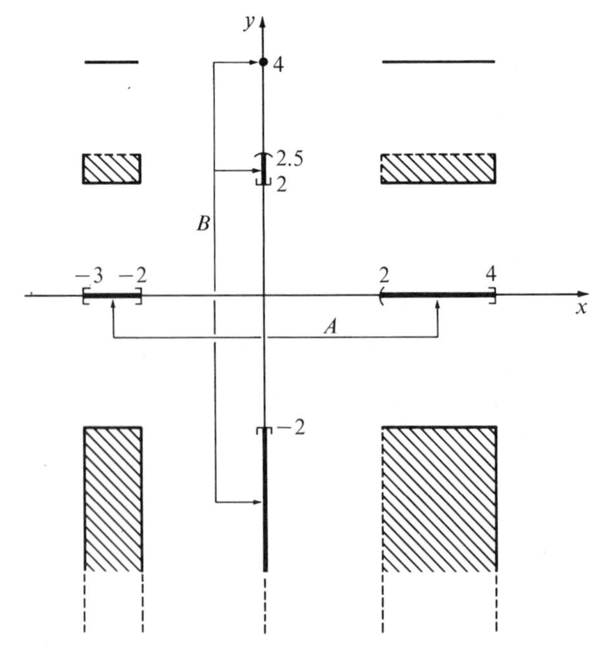

Fig. 1 $A = \{x: -3 \leq x \leq -2 \text{ or } 2 < x \leq 4\}$; $B = \{y: y \leq -2\} \cup \{y: 2 \leq y < 2.5\} \cup \{4\}$.

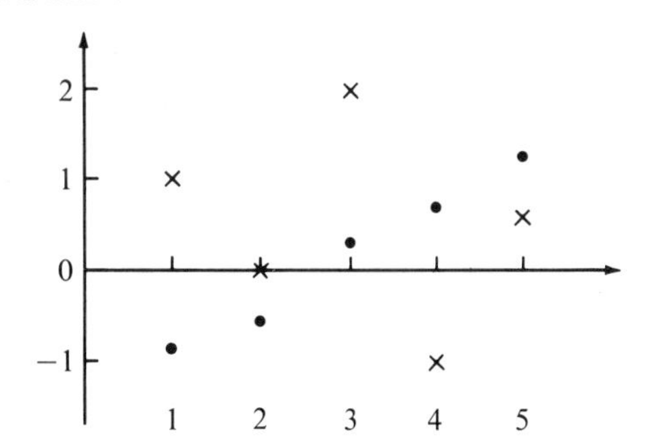

Fig. 2

are harder to visualize. A method of depicting a point in R^n is shown in Fig. 2 for the case $n = 5$. The member $(1,0,2,-1,.5)$ of R^5 is represented by a sequence of five crosses. Another point of R^5 is shown as a sequence of five dots. The horizontal spacing of these points will usually be made uniform. Occasionally the horizontal axis will be thought of as the time axis, and we may think of the ith coordinate x_i of $(x_1, x_2, \ldots, x_n)$ as obtained from an observation made at time i.

EXERCISES

1. Exhibit at least two different members of each of the following four cartesian products: $\{\alpha,\beta,\gamma\} \times \{\alpha,0,3.6,H\}, \{3\} \times \{4\} \times \{8,1\} \times \{3\}, \{-1,0,1,2\}^7, R^8$.

2. Let $A = \{-2,1,3\}$, $B = \{5,6,7\}$, $C = \{9\}$, $D = \{\alpha\colon \alpha \text{ is real and } -8 < \alpha \leq -3\}$, and $E = \{\beta\colon \beta \text{ is real and } 1.5 \leq \beta \leq 4\}$. Sketch pictures of each of the following sets:
(*a*) $A^2, B^2, A \times B, B \times A, A \times C, C \times A$
(*b*) $D^2, D \times E, C \times E, B \times E$
(*c*) $(B \cup D) \times E, D \times (D \cup E), E^3, E^2 \times B$

3. If the points (x,y) and (x',y') both belong to a cartesian product $A \times B$ prove that (x,y') and (x',y) also belong to $A \times B$. Conclude that a circular disk

$$\{(x,y)\colon x^2 + y^2 \leq r^2\} \qquad r \neq 0$$

in the plane cannot be represented as a cartesian product $A \times B$ of two sets A and B of reals.

4. List each of the following sets and state how many members each has:
(*a*) $\{0,1\}, \{0,1\}^2, \{0,1\}^3, \{\alpha,\beta\}^2, \{0,1\} \times \{\alpha,\beta\}$
(*b*) $\{1\} \times \{3\} \times \{2,1\} \times \{\alpha,\beta\}, \{1,2,3\}^2$

5. Give the number of members in each of the following sets: $\{1,2,3\}, \{1,2,3\}^2, \{1,2,3\}^3, \{1,2,3\}^4, \{1,2,3\}^n, \{1, 2, 3, \ldots, k\}^n$.

2-2 THE SIZE OF A CARTESIAN PRODUCT

The **size of a finite set** is the number of members which belong to it. Thus $\{-1.5,8,H\}$ has size 3. This section is largely devoted to the proof and applications of the following simple fact:

Theorem 1 If $A_1, A_2, \ldots, A_n$ are finite sets then the size of $A_1 \times A_2 \times \cdots \times A_n$ equals the product of the sizes of $A_1, A_2, \ldots, A_n$. In particular,

$$[\text{the size of } A^n] = [\text{the size of } A]^n. \quad \blacksquare$$

Example 1 Let $A = \{1,2,3,4,5,6\} \times \{H,T\} \times \{H,T\}$. From Theorem 1 we see that A has $6 \times 2 \times 2 = 24$ members. The member $(5,H,T)$ of A might be interpreted as follows: the roll of the die resulted in the upward face having *five dots*, and the first toss of the coin resulted in a *head*, and the second toss of the coin resulted in a *tail*. Thus there are 24 possible outcomes to the experiment of rolling a die and then tossing a coin twice. Note that the definition of equality for tuples is crucial to such a counting problem. That is, unlike a set, a 3-tuple is ordered, so $(5,H,T) \neq (5,T,H)$; hence each of these 3-tuples contributes to the size of A.

Similarly $\{H,T\}^5$ has $2^5 = 32$ members and one of them is (H,H,T,H,T). ■

Proof of Theorem 1 The theorem is certainly true if $n = 2$, since if $A = \{a_1, a_2, \ldots, a_r\}$ and $B = \{b_1, b_2, \ldots, b_s\}$ then $A \times B$ consists of the set of rs different elements in the rectangular array

$$\begin{array}{ccccc} (a_1,b_1) & (a_1,b_2) & (a_1,b_3) & \cdots & (a_1,b_s) \\ (a_2,b_1) & (a_2,b_2) & (a_2,b_3) & \cdots & (a_2,b_s) \\ \cdots & \cdots & \cdots & \cdots & \cdots \\ (a_r,b_1) & (a_r,b_2) & (a_r,b_3) & \cdots & (a_r,b_s). \end{array}$$

As usual, the notation $A = \{a_1, a_2, \ldots, a_r\}$ means that A has r *different* members. If we were to let $a_1 = a_2$ the top two rows of the array would be equal and the size of $A \times B$ would be less than rs. Now the size of $A_1 \times A_2 \times \cdots \times A_n$ equals the size of $(A_1 \times A_2 \times \cdots \times A_{n-1}) \times (A_n)$, and by the result for a pair of sets this equals the size of $(A_1 \times A_2 \times \cdots \times A_{n-1})$ times the size of A_n, so the result is true by induction. ■

As in the case of Theorem 1, we often use brief proofs depending on mathematical induction. At this point let us review the principle of induction and then give as an illustration a more formal proof of Theorem 1.

The principle of mathematical induction For every positive integer n let $S(n)$ be a sentence which is either true or false; that is, $S(n)$ may be true for some values of n but false for other values of n. *If*

1. $S(1)$ is true
2. For every positive integer n, if $S(n)$ is true for that particular n then $S(n + 1)$ is true

then $S(n)$ is true for every positive integer n. ■

We *assume* that the principle of induction is true. In axiomatic developments of the positive integers the principle of induction in some form is often assumed as an axiom.

For each $n = 1, 2, \ldots$ let $S(n)$ be a sentence which is either true or false, and let A be the set of positive integers for which it is true. Suppose we have proved that 1 belongs to A and that whenever an integer belongs to A the next higher integer also belongs to A. If n does not belong to A we do not assert anything about whether or not $n + 1$ belongs to A. Then the principle of induction permits us to draw the intuitively obvious conclusion that A equals the set of *all* positive integers. In somewhat more picturesque language, the principle of induction states that if A is some set of positive integers, and we imagine drawing a cross at precisely those integer points of the real axis which belong to A, and there is a cross at 1, and wherever there is a cross the next higher integer also has a cross, then every positive integer has a cross.

We now give *a more formal proof of Theorem 1.* Let $S(n)$ be the statement "if $A_1, A_2, \ldots, A_n$ are any n finite sets then the number of elements in $A_1 \times A_2 \times \cdots \times A_n$ is the product $N_1 N_2 \cdots N_n$, where N_k is the number of elements in A_k." Thus $S(3) =$ [if A_1, A_2, A_3 are any 3 finite sets . . .]. Now, $S(1)$ is trivially true. By exhibiting the rectangular array we showed that $S(2)$ is true. We will use the fact that $S(2)$ is true in proving part 2 of the induction principle. Assume that n is some particular integer for which $S(n)$ is true. We wish to prove that $S(n+1)$ is true. If $A_1, \ldots, A_n, A_{n+1}$ are $n+1$ finite sets then $A_1 \times A_2 \times \cdots \times A_n$ has $N_1 N_2 \cdots N_n$ members, since $S(n)$ is true. But by $S(2)$ we know that $(A_1 \times A_2 \times \cdots \times A_n) \times (A_{n+1})$ has $(N_1 N_2 \cdots N_n)(N_{n+1})$ members. However, $(A_1 \times A_2 \times \cdots \times A_n) \times (A_{n+1})$ is the same as $A_1 \times A_2 \times \cdots \times A_{n+1}$ except for some extra parentheses, so part 2 of the induction proof is completed. Hence from the principle of mathematical induction we know that $S(n)$ is true for every positive integer n. ■

Example 2 There are 5^{30} ways for an instructor to assign grades of A, B, C, D, and F to 30 students. In effect the instructor arbitrarily numbers the students from 1 to 30, selects a 30-tuple $(x_1, x_2, \ldots, x_{30})$ with all coordinates from $\{A,B,C,D,F\}$, and assigns grade x_k to student number k. ■

Exercises 1 *to* 5 *may be done before reading the remainder of this section.*

Example 3 From Theorem 1 we know that the size of $\{0,1\}^n$ is 2^n; that is, there are exactly 2^n different n-tuples with coordinates from $\{0,1\}$. In particular there are $2^3 = 8$ such 3-tuples: (0,0,0), (0,0,1), (0,1,0), (0,1,1), (1,0,0), (1,0,1), (1,1,0), (1,1,1). This simple result is used frequently. We now use it to prove that *any set of size n has exactly 2^n subsets.*

Let $A = \{x_1, x_2, \ldots, x_n\}$ be a set with exactly n members. For each subset of A we will show how to construct a corresponding n-tuple with coordinates from $\{0,1\}$. For example, to the subset $\{x_2,x_3,x_6\}$ we will correspond the n-tuple (0,1,1,0,0,1,0,0, . . . ,0,0). In constructing this correspondence we give the members of A an arbitrary ordering, that is, x_1 first, then x_2 second, etc. In general for the subset $\{x_i, x_j, x_k, \ldots, x_m\}$, with with $i < j < k < \cdots < m$, we correspond the n-tuple which has 1's for coordinates $i, j, k, \ldots, m$ and 0's for the other coordinates. Figure 1 exhibits this complete correspondence for a particular example.

Subsets of A	Corresponding 3-tuples
$\varphi =$ the empty set	(0,0,0)
$\{H\}$	(1,0,0)
$\{3\}$	(0,1,0)
$\{\alpha\}$	(0,0,1)
$\{H,3\}$	(1,1,0)
$\{H,\alpha\}$	(1,0,1)
$\{3,\alpha\}$	(0,1,1)
$\{H,3,\alpha\} = A$	(1,1,1)

Fig. 1 $A = \{H,3,\alpha\}$ has 3 members and $8 = 2^3$ subsets.

After a little thought it should be clear that by this technique a unique n-tuple has been assigned to each subset of A, and different subsets of A have different corresponding n-tuples. Thus instead of counting the number of subsets of A, we can count the number of n-tuples which arise from this technique. But clearly we obtain every one of the 2^n n-tuples with coordinates from $\{0,1\}$, so that A must have exactly 2^n subsets. In the terminology of Sec. 3-5, we have established a *one-to-one correspondence* between the class of all subsets of A and the set of n-tuples with coordinates from $\{0,1\}$. ∎

Example 4 At the end of a course there are 30 registered students. The instructor must select that subset of the students who are to receive a passing grade, with all the others receiving a failing grade. There are 2^{30} different subsets of students, so there are $2^{30} \doteq 10^{10}$ ways to grade them. The symbol $\doteq$ means **approximately equals** and does not have a precise definition. Naturally we include the case in which all students pass, as well as the case in which none pass. Note that our proof that a set of size n has 2^n subsets merely amounts to arbitrarily numbering the students and assigning a grade of 0 or 1 to each. ∎

Example 5 An urn contains k balls labeled $1, 2, \ldots, k$, and $n \geq 1$ balls are drawn from it with replacement. By **with replacement** we mean that a ball is drawn, its number x_1 is noted, and the ball is returned to the urn; then a second ball is drawn, its number x_2 is noted, and the ball is returned to the urn; etc. Thus it is theoretically possible to get the same ball on the first two draws, for example, in which case $x_1 = x_2$; hence *repeats are permitted in sampling with replacement*. The result of n such draws is an n-tuple $(x_1, x_2, \ldots, x_n)$, where x_i is the number of the ball obtained on the ith draw. x_i is one of the integers $1, 2, \ldots, k$. We may call $(x_1, x_2, \ldots, x_n)$ the resulting **sample**. From Theorem 1 *there are exactly k^n different possible samples obtainable from n draws with replacement from an urn with k balls.* Traditionally x_i is called the "ith sample" and $(x_1, x_2, \ldots, x_n)$ is referred to as a "sample of size n," but we avoid this terminology, which uses "size n" to refer to the number of coordinates rather than the size of a finite set. The preceding result can be thought of in the following way: there are k choices for x_1, and for each such choice there are k choices for x_2, so that there are $(k)(k) = k^2$ choices for (x_1,x_2); for each choice of (x_1,x_2) there are k choices for x_3, so that there are $(k^2)(k) = k^3$ choices for (x_1,x_2,x_3); etc. ∎

Example 6 An urn contains k balls labeled $1, 2, \ldots, k$, and n balls are drawn from it without replacement, where $1 \leq n \leq k$. By **without replacement** we mean that a ball is drawn, its number x_1 is noted, and the ball is *not* returned to the urn; then a second ball is drawn, its number x_2 is noted, and the ball is *not* returned to the urn; etc. Thus the number n of draws cannot exceed the number k of balls. Also, we cannot have $x_1 = x_2$; hence *repeats are not permitted in sampling without replacement*. The result of n such draws is an n-tuple $(x_1, x_2, \ldots, x_n)$ with no two coordinates equal, where each coordinate is one of the integers $1, 2, \ldots, k$. We may call $(x_1, x_2, \ldots, x_n)$ the resulting sample. Clearly if $n = 1$, so that there is only one draw, then there are k samples and they are $(1), (2), \ldots, (k)$. If there are a total of $n = 2$ draws there are $(k)(k-1)$ such samples and they are the 2-tuples which are not crossed out in Fig. 2, which has k rows and $k - 1$ acceptable 2-tuples in each row. Figure 2 can be thought of in the following way: There are k choices for x_1, and for each such choice there are $k - 1$ choices for x_2, so that there are $(k)(k-1)$ choices for (x_1,x_2). Similarly for each of the $(k)(k-1)$ choices for (x_1,x_2) there are $k - 2$ choices for x_3, so that there are $(k)(k-1)(k-2)$ choices

$$
\begin{array}{ccccccc}
\not(1,1) & (1,2) & \cdots & (1,i) & \cdots & (1,k) \\
(2,1) & \not(2,2) & \cdots & (2,i) & \cdots & (2,k) \\
\cdots & \cdots & \cdots & \cdots & \cdots & \cdots \\
(i,1) & (i,2) & \cdots & \not(i,i) & \cdots & (i,k) \\
\cdots & \cdots & \cdots & \cdots & \cdots & \cdots \\
(k,1) & (k,2) & \cdots & (k,i) & \cdots & \not(k,k)
\end{array}
$$

Fig. 2

for (x_1,x_2,x_3). Thus in general, *if* $1 \le n \le k$ *there are exactly* $(k)(k-1)(k-2)\cdots(k-n+1)$ *different possible samples obtainable from n draws without replacement from an urn with k balls.* Note that $(k)(k-1)\cdots(k-n+1)$ has n factors, starting at k and decreasing in integer steps. In particular if $n = k$ we see that there are $(k)(k-1)(k-2)\cdots(2)(1)$ such samples. We make the usual definitions that $0! = 1$ and $k! = (k)(k-1)(k-2)\cdots(2)(1)$ for any positive integer k, where $k!$ is called k **factorial**, or factorial k. Thus there are $k!$ samples obtainable from k draws without replacement from an urn with k balls, and these samples are just the $k!$ permutations of the integers $1, 2, \ldots, k$. For example, if $k = 3$ then (1,2,3), (1,3,2), (2,1,3), (2,3,1), (3,1,2), and (3,2,1) are the $3! = 6$ permutations of the integers 1, 2, and 3. ■

If A is a subset of the finite set Ω, and we think of each member of Ω as being, in some sense, equally likely, then we define the number $P(A)$ by

$$P(A) = \frac{[\text{the size of } A]}{[\text{the size of } \Omega]}.$$

We call this number $P(A)$ the **probability of** A **in** Ω, or more briefly, the probability of A. Thus $P(A)$ is the fraction of the points in Ω which belong to A. Clearly $P(\Omega) = 1$ and $0 \le P(A) \le 1$. *This is a temporary definition which will be generalized and discussed in Chap. 6.*

Example 7 A coin is tossed three times, so that a 3-tuple with each coordinate an H or a T is obtained. The assumption that each of these $2^3 = 8$ outcomes is equally likely corresponds quite well to normal coin tossing. Let A be the event that a head is obtained on the first toss. That is, a 3-tuple belongs to A iff its first coordinate is H. Thus $A = \{(H,H,H), (H,H,T), (H,T,H), (H,T,T)\}$, so that $P(A) = 4/8 = 1/2$, as it should. By $n \ge 1$ **fair coin tosses** we mean that each of the 2^n members of $\{H,T\}^n$ is assumed to be equally likely. ■

Example 8 Let Ω be the set of all k^n samples $(x_1, x_2, \ldots, x_n)$ obtainable from n draws *with* replacement from an urn with k balls. Assume that $1 \le n \le k$ and let A be the event that there are no repeats, that is, that no ball happens to be drawn twice. Thus A consists of those members $(x_1, x_2, \ldots, x_n)$ of Ω for which $x_i \ne x_j$ whenever $i \ne j$. From Examples 5 and 6 we have

$$P(A) = \frac{(k)(k-1)(k-2)\cdots(k-n+1)}{k^n}.$$

For example, if $k = 10$ and $n = 4$ then $P(A) = (10)(9)(8)(7)/10^4 = 0.504$ is the probability that no ball will be drawn twice if we perform the experiment of making four draws with replacement from an urn with 10 balls. That is, this event A should occur about half of the times that this experiment is performed.

For $k = 365$ and $n = 40$ we find after some computation that

$$P(A) = \frac{(365)(364)\cdots(326)}{365^{40}} \doteq .109.$$

Thus the probability is only .109 that of 40 people in a room, no two will have their birthdays on the same day of the year. That is, we arbitrarily number the people from 1 to 40 and define the 40-tuple $(x_1, x_2, \ldots, x_{40})$ by letting x_i be the birthday of the ith person. We *assume* that all 365^{40} members of Ω are equally likely. The value of only about 10 percent chance for 40 people may seem to be far too small, but surprising answers are not uncommon in probability theory. ■

EXERCISES

1. Each measurement of the temperature yields one of the integers $-32, -31, \ldots, 99, 100$. What is the number of different possible sequences of 1,000 temperature measurements?

2. Each of 100 items is tested and found to be defective or nondefective. How many possible outcomes are there?

3. Let $A_1 = A_3 = \{1, 2, \ldots, 10^4\}$, $A_2 = \{1, 2, \ldots, 10^3\}$, and $A = A_1 \times A_2 \times A_3$. Each element of A is interpreted as the description of the operations of a company for 1 day. The element (892,141,620) of A is interpreted as follows:

$892 worth of equipment was ordered by customers.
$141 worth of equipment was mailed to customers.
$620 worth of raw materials was ordered by the company.

(*a*) How many points has A?
(*b*) We are interested in describing the company's operations over a 20-day period. How many points has A^{20}?

4. Prove each of the following by induction on n:
(*a*) Equation (1) Appendix 1
(*b*) $1 + 4 + \cdots + n^2 = (1/6)n(n + 1)(2n + 1)$
(*c*) $(1 - y)^n \geq 1 - ny$ for all real y satisfying $y \leq 1$
(*d*) Equation 2 of Appendix 1

5. Prove by induction on n that the product of $n \geq 1$ factors, each of which is a sum of $k \geq 1$ numbers, equals the sum of the k^n products obtainable by selecting one number from each of the n factors. That is, if we start with an $n \times k$ matrix of numbers $(a_{i,j})$ and multiply together all row sums we obtain

$$\prod_{i=1}^{n}\left(\sum_{j=1}^{k} a_{i,j}\right) = \sum_{(\epsilon_1,\ldots,\epsilon_n)\in\Omega} a_{1,\epsilon_1}a_{2,\epsilon_2}a_{3,\epsilon_3}\cdots a_{n,\epsilon_n}$$

where Ω consists of the k^n different $(\epsilon_1, \epsilon_2, \ldots, \epsilon_n)$ and each ϵ_s is one of the integers $1, 2, \ldots, k$. The following alternate form is also standard:

$$\sum_{(\epsilon_1,\ldots,\epsilon_n)\in\Omega} = \sum_{\epsilon_1=1}^{k}\sum_{\epsilon_2=1}^{k}\cdots\sum_{\epsilon_n=1}^{k}$$

6. A company is considering building additional warehouses at new locations during the next year. There are 10 satisfactory locations and the company must decide how many and which ones to select. How many choices are there?

7. How many sets are there which consist of exactly two different letters from the set of 26 letters $\{A, B, C, \ldots, Z\}$ of the English alphabet?

8. Are there more samples obtainable in five draws from 10 objects with replacement than from 12 objects without replacement?

9. Let $\Omega = \{H,T\}^{100}$ be the set of possible outcomes from a sequence of 100 fair coin tosses. *An event is a subset of* Ω. Each of the following criteria defines an event (those outcomes which satisfy the criteria); find the probability of each event.
(*a*) The first two tosses are heads.
(*b*) Tosses 3 and 17 are heads.
(*c*) The first 50 tosses are heads.
(*d*) The first 25 tosses are heads *and* the next 25 tosses are tails.
(*e*) All even-numbered tosses result in heads.

10. Let $\Omega = \{1, 2, \ldots, 6\}^2$ correspond to the set of all outcomes obtainable by rolling a six-sided die twice. Assume that the outcomes are equally likely. What is the probability of $\{(x_1,x_2): x_1 + x_2 \le 4\}$?

11. What is the probability that no two of the next six people you meet will have birthdays during the same month of the year? Assume that each of the 12 months is an equally likely birthday month.

12. By **n random decimal digits** we mean that each member of $\{0, 1, 2, \ldots, 9\}^n$ is equally likely. What is the probability that four random decimal digits will all be different? For example, 3, 9, 0, 1 are all different, but 3, 9, 3, 1 are not.

13. An urn contains $n + 1 \ge 2$ balls, each labeled by a distinct real number $h_0 < h_1 < \cdots < h_n$. You make $n + 1$ drawings without replacement from this urn and so obtain some $(n + 1)$-tuple $(x_0, x_1, \ldots, x_n)$ which is one of the $(n + 1)!$ equally likely permutations of $(h_0, h_1, \ldots, h_n)$. Let B_k be the event that $x_0 = h_k$; that is, you obtained h_k on your first drawing. Show that

$$P(B_k) = \frac{1}{n+1} \qquad \text{for} \quad k = 0, 1, \ldots, n.$$

Thus, as is intuitively obvious, each ball has the same chance of being drawn first.

COMMENT: The extraneous numbers h_k were introduced so that this exercise can be given the following interpretation: Fix an integer $n \ge 1$ and let $h_0, \ldots, h_{n+1}$ be the heights of the next $n + 1$ people you will encounter; assume that the heights are distinct. Assume that the order in which you encounter these people can be any of the $(n + 1)!$ possible orders and that each is equally likely. Then B_k is the event that k of them will be shorter than the first person you encounter. Under these assumptions we see that for any $k = 0, 1, \ldots, n$ the probability is $1/(n + 1)$ that k of the next $n + 1$ people you encounter will be shorter than the first person you encounter. We shall return to this interpretation in Example 9 (Sec. 17-2).

14. *An Upper Bound for $P(A)$ of Example* 8. The bound derived in this paragraph is useful for this exercise. From inequality (9) of Appendix 1 we have

$$(1 - x_1)(1 - x_2) \cdots (1 - x_N) \le e^{-(x_1+x_2+\cdots+x_N)} \qquad \text{if each } x_i \le 1.$$

Applying this inequality to $P(A)$ in Example 8, we obtain

$$P(A) = (1)\left(1 - \frac{1}{k}\right)\left(1 - \frac{2}{k}\right) \cdots \left(1 - \frac{n-1}{k}\right) \le \exp -\left(\frac{1}{k} + \frac{2}{k} + \cdots + \frac{n-1}{k}\right)$$

From equation (1) of Appendix 1 we have $1 + 2 + \cdots + (n - 1) = (n - 1)n/2$, so

$$P(A) \le e^{(n-1)n/2k} \tag{1}$$

Inequality (1) *provides a simple upper bound on the probability $P(A)$ that there will be no repeats in n draws with replacement from k objects, for* $1 \le n \le k$. For example, if $k = 365$ and $n = 40$ a fair amount of computation is needed to find that $P(A) \doteq .109$; the use of inequality (1) requires less effort and provides the fairly sharp bound $P(A) \le e^{-2.137} = .118$.

(*a*) Show that the probability is small that no two of the next 184 people you meet will have birthdays on the same day of the year. The number 184 was chosen because $184 \doteq 365/2$ is about half the number of birthdays available.

(*b*) When a county clerk needs a citizen for jury duty he selects one at random with replacement from 10,000 eligible citizens. He needs 400 jurors in a certain year. Show that the probability is small that no citizen will be called twice for jury duty during the year.

(*c*) Show that the probability is small that 10 random decimal digits will all be different (see Exercise 12).

2-3 BINOMIAL COEFFICIENTS

This section considers combinational problems involving binomial coefficients. If k and n are any integers satisfying $0 \le k \le n$ we define the **binomial coefficient** $\binom{n}{k}$ by

$$\binom{n}{k} = \frac{n!}{[k!][(n-k)!]}.$$

Note that

$$\binom{n}{k} = \binom{n}{n-k} \qquad \binom{n}{0} = \binom{n}{n} = 1 \qquad \binom{n}{1} = \binom{n}{n-1} = n.$$

We first state and prove a basic *recursion formula for binomial coefficients:*

$$\binom{n+1}{k} = \binom{n}{k-1} + \binom{n}{k} \qquad \text{if } 1 \le k \le n.$$

The proof consists of the following simple manipulation:

$$\binom{n}{k-1} + \binom{n}{k} = \frac{n!}{(k-1)!\,(n-k)!}\left[\frac{1}{n-k+1} + \frac{1}{k}\right]$$

$$= \frac{n!}{(k-1)!\,(n-k)!}\,\frac{n+1}{k(n-k+1)} = \binom{n+1}{k}.$$

Clearly if we know all the binomial coefficients $\binom{n}{0}, \binom{n}{1}, \ldots, \binom{n}{n}$ for a certain n we can find all the binomial coefficients for $n+1$ from this recursion formula.

The recursion formula can be used to prove that the *Pascal triangle* of Fig. 1 generates the binomial coefficients. The construction of Fig. 1 is as follows. Rows 0

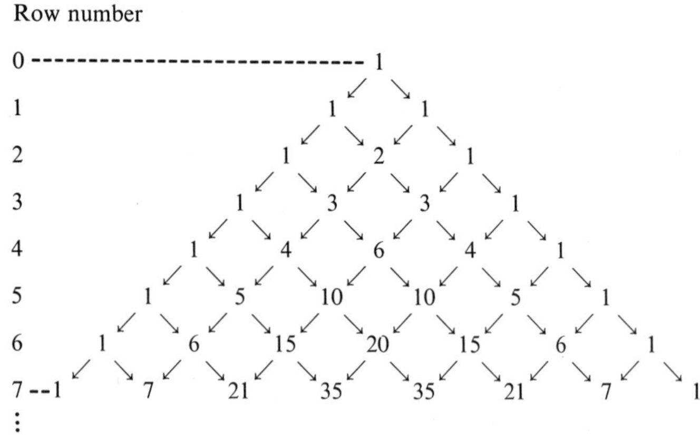

Fig. 1 Pascal triangle.

and 1 are made as shown. Then rows $n = 2, 3, 4, \ldots$ are constructed in that order, where the construction of the $(n+1)$st row involves the already available nth row. The $(n+1)$st row is constructed by making the two extreme entries 1, and every other entry in the $(n+1)$st row is obtained by adding the closest two entries in the nth row. In the next paragraph we easily prove that the Pascal triangle generates the binomial coefficients; that is, the row labeled n consists of the binomial coefficients $\binom{n}{0}$, $\binom{n}{1}, \ldots, \binom{n}{n}$. Note that the row labeled n has the integer n for its second entry, so that the row labels are already built into the Pascal triangle. Also the Pascal triangle, or equivalently the recursion formula, shows that each binomial coefficient is an integer since each is a 1 or the sum of two higher entries which are integers.

Let $A_{n,0}, A_{n,1}, \ldots, A_{n,n}$ stand for the $n+1$ numbers obtained by this construction as the nth row of the Pascal triangle. Let $S(n)$ be the statement "$A_{n,0} = \binom{n}{0}$ and $A_{n,1} = \binom{n}{1}$ and $A_{n,2} = \binom{n}{2}$ and $\cdots$ and $A_{n,n} = \binom{n}{n}$." Clearly $S(1)$ is true. If $S(n)$ is true for a particular n, then from the construction of the $(n+1)$st row,

$$\begin{array}{llllllll}
\textit{Row } n: & A_{n,0} & A_{n,1} & A_{n,2} \cdots A_{n,k-1} & & A_{n,k} \cdots A_{n,n-1} & A_{n,n} \\
 & \swarrow & & \searrow & & \swarrow & \searrow \\
\textit{Row } n+1: \; A_{n+1,0} & & & & A_{n+1,k} & & & A_{n+1,n+1}
\end{array}$$

we have $A_{n+1,k} = A_{n,k-1} + A_{n,k}$ for each k satisfying $1 \le k \le n$. But $S(n)$ is assumed to be true, so $A_{n+1,k} = \binom{n}{k-1} + \binom{n}{k}$. Hence from the recursion formula we have $A_{n+1,k} = \binom{n+1}{k}$ if $1 \le k \le n$, while from the construction we have $A_{n+1,0} = 1 = \binom{n+1}{0}$ and $A_{n+1,n+1} = 1 = \binom{n+1}{n+1}$. Thus $S(n+1)$ is true, so by induction $S(n)$ is true for all n; that is, $A_{n,k} = \binom{n}{k}$ whenever $0 \le k \le n$. ■

The same recursion formula can be used to prove the *binomial theorem.*

Binomial theorem For any reals x and y and any positive integer n

$$(x+y)^n = \sum_{k=0}^{n} \binom{n}{k} x^k y^{n-k}. \quad ■$$

Proof For any positive integer n let $S(n)$ be the statement "the above formula holds for every real x and y for this value of n." Clearly $S(1)$ is true. If $S(n)$ is true then

$$\begin{aligned}
(x+y)^{n+1} &= (x+y)(x+y)^n = x(x+y)^n + y(x+y)^n \\
&= x \sum_{k=0}^{n} \binom{n}{k} x^k y^{n-k} + y \sum_{k=0}^{n} \binom{n}{k} x^k y^{n-k}.
\end{aligned}$$

We now write this equation without using summation notation and then use the recursion formula to add terms placed one above the other. This results in

$$(x+y)^{n+1}= \binom{n}{0}xy^n + \binom{n}{1}x^2y^{n-1}+\cdots+\binom{n}{n-1}x^ny+\binom{n}{n}x^{n+1}$$

$$+\binom{n}{0}y^{n+1} + \binom{n}{1}xy^n + \binom{n}{2}x^2y^{n-1}+\cdots+ \binom{n}{n}x^ny$$

$$=\binom{n+1}{0}y^{n+1}+\binom{n+1}{1}xy^n+\binom{n+1}{2}x^2y^{n-1}+\cdots+\binom{n+1}{n}x^ny+\binom{n+1}{n+1}x^{n+1}$$

so that $S(n+1)$ is true. The theorem follows by induction. ■

From Theorem 1 (Sec. 2-2) there are exactly 2^n different n-tuples $(x_1, x_2, \ldots, x_n)$ with each coordinate a 0 or a 1. We now prove the following simple, useful theorem:

Theorem I If k and n are integers satisfying $0 \le k \le n$ then there are exactly $\binom{n}{k}$ different n-tuples $(x_1, x_2, \ldots, x_n)$ which have k coordinates equal to 1, and the remaining $n - k$ coordinates equal to 0. ■

The $\binom{5}{3} = 10$ different 5-tuples with 3 coordinates equal to 1 are listed in the right-hand column of Fig. 2.

Proof Clearly if $k = 0$ there is only $\binom{n}{0} = 1$ such n-tuple, and all its coordinates are 0. Similarly if $k = n$ there is only $\binom{n}{n} = 1$ such n-tuple, and all its coordinates are 1. Let $S(n)$ be the statement "for this n and all integers k satisfying $0 \le k \le n$ there are exactly $\binom{n}{k}$ different n-tuples $(x_1, x_2, \ldots, x_n)$ with k coordinates equal to 1, and the remaining $n - k$ coordinates equal to 0."

The first few of $S(1)$, $S(2)$, $S(3)$, . . . are true and should be checked by the reader. If $S(n)$ is true for a particular n we must show that $S(n + 1)$ is true. The proof is based on the recursion formula for the binomial coefficients. The cases $k = 0$ or $k = n + 1$ are already proved, so we take any k satisfying $1 \le k \le n$. If $S(n)$ is true there are $\binom{n}{k-1}$ n-tuples with coordinates from $\{0,1\}$ and having $(k - 1)$ 1's. Add a new 1 coordinate at the right of each of these, as shown in Fig. 2. Also add a new 0 coordinate to the right of each of the $\binom{n}{k}$ n-tuples having k "ones." We have thus obtained $\binom{n}{k-1} + \binom{n}{k} = \binom{n+1}{k}$ different $(n + 1)$-tuples with k "ones," and every such $(n + 1)$-tuple is clearly obtained by this process. Thus $S(n + 1)$ is true, so by induction $S(n)$ is true for all n. ■

	$n = 4$	$n + 1 = 5$	
$\binom{n}{k-1} = 6$	(0,0,1,1) →	(0,0,1,1,1)	
	(0,1,0,1) →	(0,1,0,1,1)	
	(0,1,1,0) →	(0,1,1,0,1)	
$k - 1 = 2$ "ones"	(1,1,0,0) →	(1,1,0,0,1)	
	(1,0,1,0) →	(1,0,1,0,1)	
	(1,0,0,1) →	(1,0,0,1,1)	$\binom{n+1}{k} = 10$
$\binom{n}{k} = 4$	(0,1,1,1) →	(0,1,1,1,0)	$k = 3$ "ones"
	(1,0,1,1)	(1,0,1,1,0)	
	(1,1,0,1) →	(1,1,0,1,0)	
$k = 3$ "ones"	(1,1,1,0) →	(1,1,1,0,0)	

Fig. 2

Example 1 In this example we prove and apply the fact that *a set of size n has exactly* $\binom{n}{k}$ *subsets of size k, where* $0 \leq k \leq n$. In Example 3 (Sec. 2-2) it was shown that a set of size n has a total of 2^n subsets. This was proved, as in Fig. 1 (Sec. 2-2), by corresponding to each subset such as $\{x_2,x_3,x_6\}$ a unique n-tuple $(0,1,1,0,0,1,0,\ldots,0)$ of zeros and ones. Clearly the subsets of size k correspond to n-tuples with exactly k ones. Thus from Theorem 1 there are $\binom{n}{k}$ subsets of size k, *as was to be proved.*

A set of size n has 2^n subsets, so that

$$\sum_{k=0}^{n} \text{[the number of subsets of size } k]$$

must equal 2^n. Thus the equation

$$\binom{n}{0} + \binom{n}{1} + \binom{n}{2} + \cdots + \binom{n}{n} = 2^n$$

must be an identity. This example has yielded a combinatorial proof and interpretation of this identity. The same identity can be proved by letting $x = y = 1$ in the binomial theorem.

If an urn has n balls numbered $1, 2, \ldots, n$ and we draw an unordered fistfull of k of them then there are $\binom{n}{k}$ possible outcomes. That is, there are $\binom{n}{k}$ possible subsets of size k which can be drawn.

Order is irrelevant in a card hand, so a poker hand is just a subset of size 5 from a card deck of size 52. Thus there are $\binom{52}{5} = 2{,}598{,}960$ different poker hands. ■

Example 2 We know that there are (100)(99)(98) samples obtainable from three draws without replacement from a population with 100 members. Thus if no person can die twice and no two of 100 people can die simultaneously then there are (100)(99)(98) possibilities for the first three deaths. That is, if we think of the people as labeled by the numbers $1, 2, \ldots, 100$ then (31,8,12) corresponds to person number 31 dying first, then person number 8, then person number 12. Now $(31,8,12) \neq (8,31,12)$, and we are considering these as different possibilities, since we are counting 3-tuples and tuples are always ordered. We might be interested only in which are the first three people to die, and not in the order of their deaths. If we wish to neglect order there are only $\binom{100}{3} = (100)(99)(98)/6$ unordered possibilities, since from Example 1 there are $\binom{n}{k}$ subsets of size k from a set with n members. That is, the six permutations of (31,8,12) are replaced by the unordered set {31,8,12}. Combinatorial problems can be tricky, and it is essential to define clearly what the various possibilities are and which of these possibilities are considered to be different before we count them. ■

***Example 3** In this example we use Theorem 1 to give a more transparent proof of the binomial theorem: For all real t_0 and t_1

$$(t_0 + t_1)^n = \sum_{k=0}^{n} \binom{n}{k} t_1^k t_0^{n-k}.$$

We first exhibit the expansions of $(t_0 + t_1)^n$ in the cases $n = 2$ and $n = 3$:

$$(t_0 + t_1)^2 = (t_0 + t_1)(t_0 + t_1) = t_0t_0 + t_0t_1 + t_1t_0 + t_1t_1$$

$$(t_0 + t_1)^3 = (t_0 + t_1)(t_0 + t_1)^2$$

$$= t_0t_0t_0 + t_0t_0t_1 + t_0t_1t_0 + t_0t_1t_1 + t_1t_0t_0 + t_1t_0t_1 + t_1t_1t_0 + t_1t_1t_1.$$

In general $(t_0 + t_1)^n$ equals the sum of 2^n terms, where each $(x_1, x_2, \ldots, x_n)$ belonging to $\{0,1\}^n$ contributes a term $t_{x_1}t_{x_2} \cdots t_{x_n}$ obtained by selecting t_{x_1} from the first of the n factors of $(t_0 + t_1)(t_0 + t_1) \cdots (t_0 + t_1)$, then selecting t_{x_2} from the second factor, etc. If $(x_1, x_2, \ldots, x_n)$ has exactly k coordinates equal to 1 then

$$t_{x_1}t_{x_2} \cdots t_{x_n} = (t_1)^k(t_0)^{n-k}.$$

Collecting these $\binom{n}{k}$ terms together yields $\binom{n}{k}(t_1)^k(t_0)^{n-k}$. Doing this for $k = 0, 1, \ldots, n$ yields the binomial theorem. This proof is related to Exercise 5 (Sec. 2-2). ■

Example 4 This example illustrates the inherent arbitrariness in constructing a model for a physical experiment.

Model α. Two pennies are considered to be distinguishable; for example, they may be numbered with scratches or have different mint dates. We can let $\Omega = \{H,T\}^2$ represent the set of possible outcomes obtainable by tossing the two pennies simultaneously; that is, the outcome is some (x_1,x_2), where x_i is H or T, depending on whether the ith penny came up a head or a tail. Thus Ω has four points and the event $A = \{(H,T), (T,H)\}$ that there is one head and one tail has two points, so $P(A) = 2/4 = 1/2$ *if* we assume that each of the four points is equally likely.

Model β. Two pennies considered to be indistinguishable are tossed simultaneously. Let Ω consist of three points ω_1, ω_2, ω_3, corresponding respectively to the outcomes both heads, one head and one tail, and both tails; that is, if we obtain the outcome one head and one tail we do not attempt to distinguish or determine which of the two pennies came up heads. Let $A = \{\omega_2\}$ consist of the single point for one head and one tail so that $P(A) = 1/3$ *if* we assume that each of the three points is equally likely.

Comments on the Two Models. Both models are worthy of consideration. Predictions based on model α agree quite well with *experimental coin tossing*, but this is not so for model β. Thus, for example, a *typical* long sequence of such paired-penny tossings yields the outcome one head and one tail about 50 percent of the time. Of course, this experimental result is unaffected by whether or not we happen during the experiment to distinguish which of the two pennies came up heads. In dealing with things like atoms and photons in statistical mechanics there are situations where "indistinguishable" types of models like β fit the facts, while "distinguishable" types of models like α do not. ■

Exercises 1 to 8 may be done before reading the remainder of this section.

***Example 5: Galton's quincunx** The purpose of this example is to analyze an idealized version of a clever device which now comes with probability-theory experimental kits for children. The device was described by F. Galton in his book "Natural Inheritance," 1889.

Imagine the plane of Fig. 3 to be a vertical board with a thin horizontal nail at each cross. The horizontal spacing between nails is d, and a nail in one row is directly above the midpoint of two nails in the row below. The distance between two essentially adjacent nails,

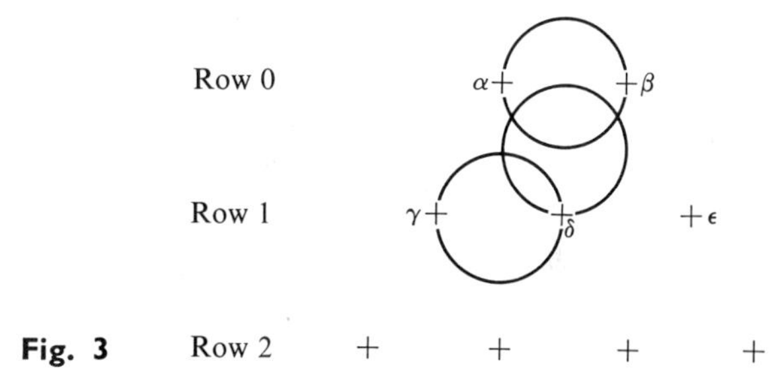

Fig. 3

say β and ϵ, in adjacent rows is d, but it could be larger. A pane of glass is mounted parallel to the board and at distance d in front of it. A steel ball of diameter d is dropped between nails α and β, and it just squeezes through and drops down to strike δ. The ball now has essentially an equal chance of rolling and slipping to the left or to the right. In Fig. 3 it is shown as moving to the left to just squeeze through between γ and δ. The ball continues in this way through, say, $N = 8$ rows and finally slips into a retaining channel. The device actually permits a number of balls, say $M = 100$, to follow such paths one after another and so yields a final distribution of balls in channels. This distribution typically has most of the balls in the more central channels, as shown in Fig. 4. We wish to prove that *the probability that a ball will come to rest in channel k is $\binom{N}{k}\frac{1}{2^N}$ if there are N rows and we use the labeling shown in Fig. 4*. In Sec. 8-2 this will be called the *binomial distribution with parameters N and $\frac{1}{2}$*.

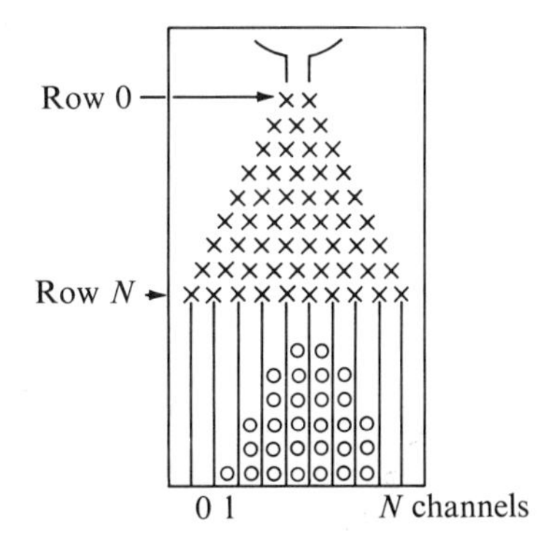

Fig. 4 A Galton quincunx.

We will consider each of the possible paths for a ball to the bottom to be equally likely and show that there are exactly $\binom{N}{k}$ different paths terminating in channel k. Then from Example 1 there are $\binom{N}{0} + \binom{N}{1} + \cdots + \binom{N}{N} = 2^N$ paths in all, and hence fraction $\binom{N}{k}2^{-N}$ of the paths terminate in channel k. The assumption that each path is equally likely is reasonable, since each choice of a right or left is assumed to be equally likely and should not depend on past choices.

Many devices of the kind described above actually have the spacing between adjacent horizontal nails much larger than the diameter of the ball. Such a design permits a ball to maintain a right or left velocity component when passing through a row of nails, so that each choice of a right or a left is *not* independent of the past history for the ball. Such a design typically results in more balls terminating in channels 0 and N than would be expected from our analysis. Also the distribution of balls in channels may be skewed to one side because nail δ is not precisely centered below nails α and β, or because the board is tilted slightly to right or left.

Consider the spaces between nails in the nth row as labeled $(n,0)$, $(n,1)$, $(n,2)$, . . . , (n,n) as shown in Fig. 5. Let $A_{n,k}$ be the number of different paths a ball can follow to reach

Row 0	$+\ (0, 0)\ +$
Row 1	$+\ (1, 0) + (1, 1)\ +$
Row 2	$+\ (2, 0) + (2, 1) + (2, 2)\ +$
$\vdots$	
Row n	$\cdots + (n, k-1) + (n, k) + \cdots$
Row $n+1$	$\cdots + (n+1, k) + \cdots$

Fig. 5

location (n,k). *We need only show that* $A_{n,k} = \binom{n}{k}$. Note that if the label (n,k) of Fig. 5 is replaced by the number $\binom{n}{k}$ we get the Pascal triangle of Fig. 1. Clearly $A_{n,0} = 1$ for all n, since to get to $(n,0)$ the ball must always take its left alternative. Similarly $A_{n,n} = 1$ for all n. Clearly if $1 \le k \le n$ then $A_{n+1,k} = A_{n,k-1} + A_{n,k}$, as indicated in Fig. 5, since the paths to $(n+1, k)$ are those which reach $(n, k-1)$ and take a right plus those which reach (n,k) and take a left. Let $S(n)$ be the statement "$A_{n,0} = \binom{n}{0}$ and $A_{n,1} = \binom{n}{1}$ and $\cdots$ and $A_{n,n} = \binom{n}{n}$." Now $S(1)$ is true, and if $S(n)$ is true for a particular n then $A_{n+1,k} = A_{n,k-1} + A_{n,k} = \binom{n}{k-1} + \binom{n}{k}$. From the recursion formula for binomial coefficients we have $A_{n+1,k} = \binom{n+1}{k}$, so $S(n+1)$ is true. Thus by induction $A_{n,k} = \binom{n}{k}$ whenever $0 \le k \le n$, *as was to be shown.*

The same result can be obtained somewhat differently. A path to the bottom can be described by an N-tuple $(x_1, x_2, \ldots, x_N)$, where each x_i belongs to $\{0,1\}$. That is, the path goes to the left just prior to the ith row iff $x_i = 0$, and to the right iff $x_i = 1$. The path $(0,0,0,1,1,0, \ldots ,1)$ makes three left choices, then two right choices, etc. Now $\{0,1\}^N$ has

2^N members, so there are 2^N paths. From Theorem 1 there are $\binom{N}{k}$ members of $\{0, 1\}^N$ which have k coordinates equal to 1, and these paths terminate in channel k. ■

***Example 6** By n fair coin tosses we mean that the outcome is some member $(H,H,T,\ldots,H)$ of the set $\{H,T\}^n$ of n-tuples, and each of these 2^n different n-tuples is equally likely. Naturally, for example, the fifth coordinate of an n-tuple is H iff the fifth toss resulted in a head. If n_H of the trials result in heads and n_T of them result in tails then $n = n_H + n_T$. In this example we will derive the following startlingly simple answer to a complicated problem:

Bertrand's ballot theorem (1887) If n fair coin tosses resulted in n_H heads and n_T tails with $n_H > n_T$ then the probability is

$$\frac{n_H}{n_H + n_T} - \frac{n_T}{n_H + n_T} = \frac{n_H - n_T}{n_H + n_T}$$

that for *every* $k = 1, 2, \ldots, n$ there were more heads than tails in the first k tosses, that is, that the number of heads exceeded the number of tails throughout the whole sequence of $n = n_H + n_T$ tosses. ■

Thus if a sequence of fair coin tosses resulted in 62 percent heads and 38 percent tails the probability is $.62 - .38 = .24$ that the number of heads exceeded the number of tails throughout the sequence of tosses. Equivalently if there are 48 more heads than tails in 200 tosses the probability is $48/200 = .24$ that heads led all the way. Similarly if only candidates H and T ran for election and candidate H received 62 percent of the votes the probability is .24 that candidate H led throughout the whole election. In this interpretation we assume that every possible sequence of n votes which leads to this same final election result is equally likely.

We now introduce notation which will permit us to state the result more analytically. Let n_H and n_T be any two nonnegative integers for which $n_H > n_T$ and let $n = n_H + n_T$. *These two numbers n_H and n_T will henceforth be considered to be fixed.* We will at times replace H and T by 1 and -1, respectively. Let $\epsilon = (\epsilon_1, \epsilon_2, \ldots, \epsilon_n)$ be any member of $\{1, -1\}^n$ and let Ω consist of those ϵ for which n_H of the coordinates of $\epsilon = (\epsilon_1, \epsilon_2, \ldots, \epsilon_n)$ equal 1; that is, Ω is the set of all possible sequences of n votes which lead to the same final numbers of n_H and n_T. Let A consist of those members $\epsilon = (\epsilon_1, \epsilon_2, \ldots, \epsilon_n)$ of Ω for which

$$\epsilon_1 + \epsilon_2 + \cdots + \epsilon_k \geq 1 \qquad \text{for all } k = 1, 2, \ldots, n.$$

Clearly $\epsilon_1 + \epsilon_2 + \cdots + \epsilon_k$ is the excess of heads over tails in the first k tosses. Thus $\epsilon = (\epsilon_1, \epsilon_2, \ldots, \epsilon_n)$ belongs to A iff every one of these n excesses is 1 or greater and $\epsilon_1 + \epsilon_2 + \cdots + \epsilon_n = n_H - n_T$. In this notation the ballot theorem states that

$$\frac{[\text{the size of } A]}{[\text{the size of } \Omega]} = \frac{n_H - n_T}{n_H + n_T} \tag{1}$$

since $P(A)$ is defined to equal the left-hand side of this equation. Note that the definitions of A, Ω, and $P(A)$ depend on the two fixed numbers n_H and n_T. In the terminology of Sec. 7-1, $P(A)$ is the conditional probability that heads led all the way, given that the final numbers of heads and tails were n_H and n_T, respectively. *We now prove equation* (1).

Let Ω^+ and Ω^- be the subsets of Ω for which $\epsilon_1 = 1$ and $\epsilon_1 = -1$, respectively. Clearly A is a subset of Ω^+, since the first toss must result in a head if the sequence is to belong to A.

Let B consist of those members of Ω^+ which do not belong to A. Clearly A and B are disjoint subsets of Ω^+ and their union equals Ω^+. Thus we have

$$[\text{the size of } \Omega^+] = [\text{the size of } A] + [\text{the size of } B]. \tag{2}$$

We will shortly prove the nonobvious fact that

$$[\text{the size of } B] = [\text{the size of } \Omega^-]. \tag{3}$$

We now show that the ballot theorem follows from equation (3). From equations (2) and (3) we obtain

$$\frac{[\text{the size of } A]}{[\text{the size of } \Omega]} = \frac{[\text{the size of } \Omega^+] - [\text{the size of } \Omega^-]}{[\text{the size of } \Omega]}. \tag{4}$$

Applying Theorem 1 to each of Ω, Ω^+, and Ω^- in equation (4) yields

$$\begin{aligned}
\frac{[\text{the size of } A]}{[\text{the size of } \Omega]} &= \frac{\binom{n-1}{n_H-1} - \binom{n-1}{n_H}}{\binom{n}{n_H}} \\
&= \frac{n_H!\,(n-n_H)!}{n!}\left[\frac{(n-1)!}{(n_H-1)!\,(n-n_H)!} - \frac{(n-1)!}{n_H!\,(n-n_H-1)!}\right] \\
&= \frac{1}{n}\Big[n_H - (n-n_H)\Big] = \frac{1}{n_H+n_T}\Big[n_H - n_T\Big].
\end{aligned}$$

Thus in order to prove the ballot theorem we need only prove equation (3), which we now proceed to do.

If we have an $\epsilon = (\epsilon_1, \epsilon_2, \ldots, \epsilon_n)$ in Ω and k satisfies $1 \le k \le n$ we can define

$$s_k(\epsilon) = \epsilon_1 + \epsilon_2 + \cdots + \epsilon_k$$

so that $s_k(\epsilon)$ is the excess of heads over tails *at time* k, that is, immediately after the kth toss. Each $\epsilon = (\epsilon_1, \epsilon_2, \ldots, \epsilon_n)$ determines a finite sequence $s_1(\epsilon), s_2(\epsilon), \ldots, s_n(\epsilon)$ of such excesses. We can plot $s_k(\epsilon)$ above k, and by joining adjacent points with straight lines we obtain a polygonal line, as in Fig. 6. Such a polygonal line starts at (1,1) or (1,−1), terminates at the

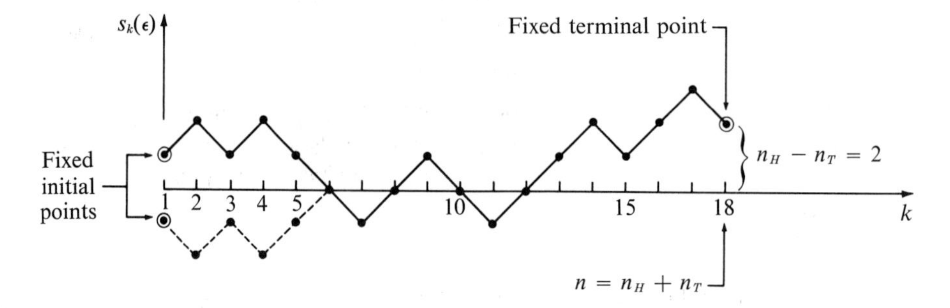

	1	2	3	4	5	6	7	8	9	10	11	12	13	14	15	16	17	18	
If	(1,	1,	−1,	1,	−1,	−1,	−1,	1,	1,	−1,	−1,	1,	1,	1,	−1,	1,	1,	−1)	$= (\epsilon_1, \cdots, \epsilon_n) = \epsilon$
then	(1,	2,	1,	2,	1,	0,	−1,	0,	1,	0,	−1,	0,	1,	2,	1,	2,	3,	2)	$= (s_1(\epsilon), \cdots, s_n(\epsilon))$.

Fig. 6 $n_H = 10$ and $n_T = 8$.

fixed point $(n, n_H - n_T)$, and has a constant slope of ± 1 between any two integers. Every ϵ in Ω has a unique corresponding polygonal line, and clearly every such polygonal line is obtained from one and only one ϵ in Ω. In the terminology of Sec. 3-5, there is a *one-to-one correspondence* between Ω and the set of such polygonal lines. The ballot theorem states that the fraction of these polygonal lines which never touch the abscissa equals the slope of the line from the origin to the fixed terminal point.

We wish to prove equation (3), where Ω^- is in effect the set of polygonal lines starting at $(1,-1)$ and B is the set of polygonal lines starting at $(1,1)$ and touching the abscissa at least once. If a line starting at $(1,1)$ does touch the abscissa then there is a first point at which it touches the abscissa, and by *reflecting* this initial portion in the abscissa, as in Fig. 6, we obtain a member of Ω^-. This reflection technique permits us to transform each member of B into a member of Ω^-. It should be apparent that this transformation yields each member of Ω^- once and only once, so that B and Ω^- have the same size. Thus equation (3), and hence the ballot theorem, is proved. [Exercise 7d (Sec. 8-2) generalizes the ballot theorem.] ■

EXERCISES

1. A bridge deck has 52 cards and a bridge hand has 13 cards. How many different bridge hands are there?

2. A company plans to build four additional warehouses at new locations during the next year. There are 10 satisfactory locations and the company must decide which four to select. How many choices are there?

3. Prove that the probability is $\frac{1}{2}$ that n fair coin tosses will result in an even number of heads. HINT: Let $x = -1$ and $y = 1$ in the binomial theorem.

4. In an extrasensory-perception experiment each of two people in different rooms selects three cards without replacement from a different deck of 52 cards. With the *assumption* that all possible outcomes are equally likely:

(*a*) What is the probability that they both select the same three cards?

(*b*) Would you expect them to select the same three cards in at least one of a proposed 1,000 independent repetitions of this experiment?

5. A number of applicants have satisfied the requirements for appointment as fireman. The fire department publishes their names in a list and announces that as vacancies occur they will be filled from this list, starting from the top. A city councilman charges that racial bias influenced the order in which the names were listed. The fire department agrees to use a lottery in order to fill each vacancy by random selection from among those remaining. After four vacancies have been filled the councilman charges bad faith because *every applicant selected, after the first, was higher on the original list than the preceding selection.* Assuming that the fire department operated in good faith, find the probability of this event, to three decimal places.

6. Any subset G of the set $\Omega_n = \{1, 2, \ldots, n\}^2$ of all ordered pairs of the integers $1, 2, \ldots, n$ may be called an *oriented graph with nodes* $1, 2, \ldots, n$. A member (i,j) of G may be interpreted as an oriented branch, or arrow, from i to j. The oriented branch (i,i) may be called a *self-loop*. The empty graph with nodes $1, 2, \ldots, n$ has no branches. Figure 7*a* depicts the oriented graph

$$G = \{(2,3), (3,2), (3,4), (4,4)\} \qquad \text{with nodes } 1, 2, 3, 4, 5.$$

This graph has four oriented branches.

(*a*) Given $n \geq 1$, how many such graphs G are there?

(*b*) How many such G have exactly k oriented branches?

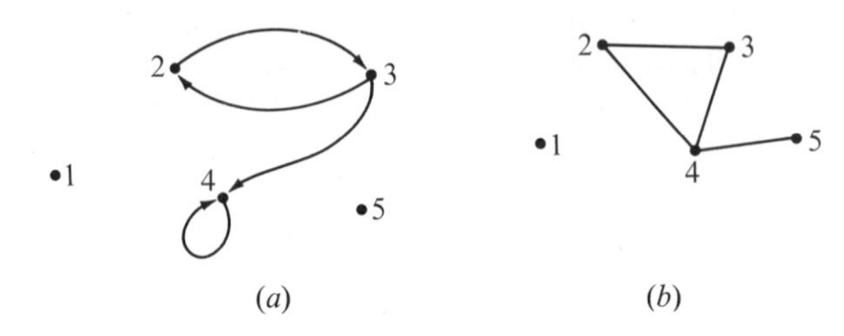

Fig. 7

7. Let A_n be the collection of all subsets of size 2 from the set $\{1, 2, \ldots, n\}$. Any subset G of A_n may be called an *unoriented graph with nodes* $1, 2, \ldots, n$ *and no self-loops*. If G is nonempty the member $\{i,j\}$ of G, with $i \neq j$, may be interpreted as an unoriented branch between node i and node j. Figure 7*b* depicts the unoriented graph

$$G = \{\{2,3\}, \{3,4\}, \{2,4\}, \{4,5\}\} \qquad \text{with nodes } 1, 2, 3, 4, 5$$

and no self-loops. This graph has four unoriented branches.
(*a*) Given $n \geq 2$, how many such graphs G are there?
(*b*) How many such G have exactly k unoriented branches?
8. How many 5-tuples $(r_1, r_2, \ldots, r_5)$ of nonnegative integers are there which satisfy the equation $r_1 + r_2 + \cdots + r_5 = 8$?

Exercise 9 depends on Example 5.

9. If there are $N = 8$ rows in the device of Example 5, find the probability that a ball will end up in one of the three center columns.

The remaining exercises depend on Example 6.

10. For which of the outcomes below does heads lead all the way, in the sense of Example 6?

(*a*) (H,H,T,T,H,H,T)
(*b*) (T,H,H,H,H,H)
(*c*) (H,T,H,H,H,H)
(*d*) (H,H,T,H,T,H,T,H,T,T)
(*e*) (H,H,T,H,T,H,T,H)

11. (*a*) If a sequence of fair coin tosses yields 78 percent heads, what is the probability that heads led all the way?
(*b*) If a sequence of 1,000 fair coin tosses yields 100 more tails than heads, what is the probability that tails led all the way?
12. Coffee beans are poured into a shute leading to a divider which sends about half the beans to each half of the divided hull of a river barge. It is found that when one side of the hull is full the other side is only 90 percent full. What is the probability that the full side was always heavier, from the very start of loading? What are your assumptions?
13. Using the notation introduced in the proof in Example 6, for the case $n_H = 4$ and $n_T = 2$ exhibit the polygonal lines as in Fig. 6 for the following:
(*a*) The set A
(*b*) The set B
(*c*) Two members of B and their corresponding members in Ω^-

*2-4 THE MULTINOMIAL AND HYPERGEOMETRIC FORMULAS

Three additional frequently used combinatorial formulas will be derived in this section. The first two can be interpreted in terms of the distributions of n balls among k

urns. A total of $n \geq 1$ balls labeled $1, 2, \ldots, n$ are distributed among $k \geq 1$ urns labeled $1, 2, \ldots, k$. A description of which balls are in which urns can be given by the n-tuple $x = (x_1, x_2, \ldots, x_n)$, where x_i is the number of the urn in which ball i is located. We call x the **location vector** and the x_i the *location numbers.* Two examples are exhibited in Fig. 1. From Theorem 1 (Sec. 2-2) *there are k^n different location vectors.*

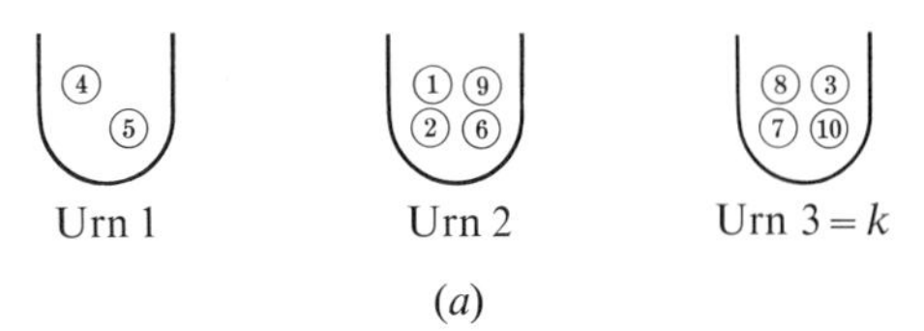

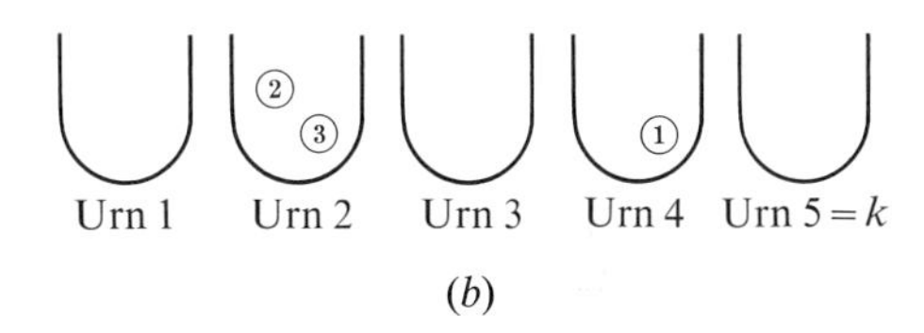

Fig. 1 (*a*) $n = 10$ balls, $k = 3$ urns, $x = (2,2,3,1,1,2,3,3,2,3)$ = location vector, and $r = (2,4,4)$ = occupancy vector; (*b*) $n = 3$ balls, $k = 5$ urns, $x = (4,2,2)$ = location vector, and $r = (0,2,0,1,0)$ = occupancy vector.

If we are interested in how many balls there are in each urn it is natural to introduce the **occupancy vector** $r = (r_1, r_2, \ldots, r_k)$, where r_j is the number of balls in urn number j. The r_j are called the *occupancy numbers.* Clearly $r_1 + r_2 + \cdots + r_k = n$ and r_j equals the number of coordinates of x which are equal to j. We wish to show that *there are exactly*

$$\binom{n+k-1}{n} = \binom{n+k-1}{k-1} \tag{1}$$

different occupancy vectors, that is, that *there are exactly* $\binom{n+k-1}{n}$ *different k-tuples* $(r_1, r_2, \ldots, r_k)$ *of nonnegative integers which satisfy the equation* $r_1 + r_2 + \cdots + r_k = n$. For the case $k = 3$ and $n = 5$, Fig. 2 lists these $\binom{5+3-1}{2} = 21$ different occupancy vectors. To an occupancy vector $r = (r_1, r_2, \ldots, r_k)$ we correspond a tuple r' of zeros and ones constructed as follows: the first $r_1 + 1$ coordinates

(5,0,0)	(0,1,4)	(0,2,3)
(0,5,0)	(1,0,4)	(3,1,1)
(0,0,5)	(3,2,0)	(1,3,1)
(4,1,0)	(3,0,2)	(1,1,3)
(4,0,1)	(0,3,2)	(2,2,1)
(0,4,1)	(2,3,0)	(2,1,2)
(1,4,0)	(2,0,3)	(1,2,2)

Fig. 2 The $\binom{5+3-1}{2} = 21$ different 3-tuples (r_1,r_2,r_3) of nonnegative integers satisfying $r_1 + r_2 + r_3 = 5$. $n = 5$ and $k = 3$.

of r' are r_1 zeros followed by a 1; the next $r_2 + 1$ coordinates of r' are r_2 zeros followed by a 1; and so on through $r_{k-1} + 1$. The last r_k coordinates of r' are r_k zeros.

For example, if

$r = (3, 2, 0, 4, \ldots, 3, 1)$ then

$$r' = (\underbrace{0, 0, 0}_{r_1}, 1, \underbrace{0, 0}_{r_2}, 1, \underbrace{}_{r_3} 1, \underbrace{0, 0, 0, 0}_{r_4}, 1, \ldots, 1, 0, 0, 0, 1, \underbrace{0}_{r_k}).$$

Thus each r has a corresponding r', which is just a different way of representing r, since r_j is the number of zeros between the $(j-1)$st and jth 1 in r'.

Such an r' can be any one of the $(n + k - 1)$-tuples having $k - 1$ coordinates equal to 1 and n coordinates equal to 0. Clearly each r' can be obtained from one and only one r, so the number of occupancy vectors equals the number of such r'. By Theorem 1 (Sec. 2-3) this is just $\binom{n+k-1}{k-1}$, so *the stated result concerning equation (1) is proved.*

Example 1 If we have a real-valued function $f(t_1, t_2, \ldots, t_k)$ of k real variables we can form a total of k^n different partial derivatives of order n. That is, given an n-tuple $(x_1, x_2, \ldots, x_n)$, where each x_i is one of the integers $1, 2, \ldots, k$, we can form

$$\frac{\partial^n f(t_1, t_2, \ldots, t_k)}{\partial t_{x_n} \partial t_{x_{n-1}} \cdots \partial t_{x_2} \partial t_{x_1}}$$

where we take the partial first with respect to t_{x_1}, then with respect to t_{x_2}, etc. Thus each location vector corresponds, at least formally, to such an nth-order partial derivative. Under quite general regularity conditions on the function f each such nth-order partial derivative will actually exist, *and* furthermore the resulting function will depend only on how many times we differentiate with respect to each variable. For example, we typically have

$$\frac{\partial^6 f(t_1,t_2,t_3)}{\partial t_1\, \partial t_3\, \partial t_1\, \partial t_3\, \partial t_2\, \partial t_1} = \frac{\partial^6 f(t_1,t_2,t_3)}{\partial^2 t_3\, \partial t_2\, \partial^3 t_1}.$$

In more general notation we have

$$\frac{\partial^n f(t_1, t_2, \ldots, t_k)}{\partial t_{x_n} \partial t_{x_{n-1}} \cdots \partial t_{x_2} \partial t_{x_1}} = \frac{\partial^n f(t_1, t_2, \ldots t_k)}{\partial_{t_k}^{r_k} \partial_{t_{k-1}}^{r_{k-1}} \cdots \partial_{t_2}^{r_2} \partial_{t_1}^{r_1}}$$

if $x = (x_1, x_2, \ldots, x_n)$ is a location vector and $r = (r_1, r_2, \ldots, r_k)$ is its corresponding occupancy vector. Thus all location vectors having the same occupancy vector result in the same function. Therefore we know from equation (1) that under general regularity conditions on the function $f(t_1, t_2, \ldots, t_k)$ there are only $\binom{n+k-1}{k-1}$ different functions which can be obtained by taking nth-order partial derivatives of f. In particular for $k = 3$ and $n = 5$ there are $3^5 = 243$ formally different fifth-order partial derivatives of $f(t_1,t_2,t_3)$, but each of them is equal to one of the 21 different fifth-order partials which correspond to the occupancy vectors listed in Fig. 2. ■

If $r_1, r_2, \ldots, r_k$ are nonnegative integers and $r_1 + r_2 + \cdots + r_k = n$ then

$$\frac{n!}{r_1!\, r_2! \cdots r_k!} \tag{2}$$

is called a **multinomial coefficient.** We wish to sketch a proof of the interpretation that *expression (2) equals the number of location vectors which have the same corresponding occupancy vector* $(r_1, r_2, \ldots, r_k)$; that is, there are exactly $n!/(r_1!\, r_2! \cdots r_k!)$ different ways of distributing n numbered balls among k numbered urns such that the jth urn has r_j balls for $j = 1, 2, \ldots, k$.

For $k = 2$ expression (2) reduces to the binomial coefficient $\binom{n}{r_1}$, and the interpretation follows from Theorem 1 (Sec. 2-3) if we number the two urns as 0 and 1 rather than as 1 and 2. Our proof imitates the induction proof of Theorem 1 (Sec. 2-3), which was based on the recursion formula for binomial coefficients. The generalization of that formula is the following *recursion formula for multinomial coefficients:* if $r_1 + r_2 + \cdots + r_k = n + 1$, where each $r_i \geq 1$, then

$$\frac{(n+1)!}{r_1!\, r_2! \cdots r_k!} = \sum_{j=1}^{k} \frac{n!}{r_1!\, r_2! \cdots r_{j-1}!\, (r_j - 1)!\, r_{j+1}! \cdots r_k!}.$$

The proof of this recursion formula is immediate if we use $(n + 1)! = n!\,(r_1 + r_2 + \cdots + r_k)$ in the left-hand side. If expression (2) has the stated interpretation in all cases for a particular n we extend to $n + 1$ as follows: Take any location vector $(x_1, x_2, \ldots, x_n)$ with occupancy vector $(r_1, r_2, \ldots, r_{j-1}, r_j - 1, r_{j+1}, \ldots, r_k)$, extend it to location vector $(x_1, x_2, \ldots, x_n, j)$ with occupancy vector $(r_1, r_2, \ldots, r_k)$ and use the recursion formula. *This concludes the abbreviated proof of the interpretation of expression (2).* Note that the right-hand side of the recursion formula can easily be interpreted as equaling the number of different distributions of n balls having the property that an $(n + 1)$st ball can be added to yield occupancy vector $(r_1, r_2, \ldots, r_k)$.

Example 2 If x_i is the output voltage from some apparatus at time i and each x_i must be one of three possible voltage levels v_1, v_2, v_3, then there are $100!/(22!)(38!)(40!)$ different sequences $(x_1, x_2, \ldots, x_{100})$ of $n = 100$ outputs having the property that voltage levels v_1, v_2, v_3 occurred 22, 38, and 40 percent of the time, respectively. That is, instead of considering a sequence such as $(v_3, v_3, v_1, v_2, v_3, \ldots, v_1)$, we can just as well consider the location vector $(3, 3, 1, 2, 3, \ldots, 1)$, so that ball 5 in urn 3 means that the voltage level at time 5 was v_3. ■

Let D balls be drawn *without replacement* from an urn which contains R red balls and B black balls. Then the probability is

$$\frac{\binom{R}{r}\binom{B}{D-r}}{\binom{R+B}{D}} \tag{3}$$

that exactly r of the D balls drawn will be red. Expression (3) considered as a function of r is called the **hypergeometric distribution.** We let R, B, D, and r be any nonnegative integers subject to the constraints

$$D \leq R + B \qquad r \leq D \qquad r \leq R \qquad D - B \leq r.$$

That is, the number D of draws cannot exceed the total number $R + B$ of balls in the urn; the number r of red balls drawn cannot exceed the total number D of balls drawn or the number R of red balls in the urn; and the number $D - r$ of black balls drawn cannot exceed the number B of black balls in the urn.

To prove (3) we number the red balls $1, 2, \ldots, R$ and the black balls $R + 1$, $R + 2, \ldots, R + B$. The outcome of D draws without replacement can be characterized by an $(R + B)$-tuple having D coordinates equal to 1 and the remaining coordinates equal to 0:

$$\begin{array}{l} 1, 2, 3, 4, 5, \ldots, R, R + 1, \ldots, R + B \\ (0, 1, 1, 0, 1, \ldots, 0, \ 1, \qquad \ldots, 0) \end{array}$$

Our interpretation of such a tuple is that its ith coordinate is 1 iff the ith ball was one of the D drawn. From Theorem 1 (Sec. 2-3) there are $\binom{R+B}{D}$ such tuples, so there are $\binom{R+B}{D}$ possible outcomes. That is, as in Example 1 (Sec. 2-3), there are $\binom{R+B}{D}$ subsets of size D from a set of $R + B$ objects. We assume that these outcomes are equally likely. For the sample drawn to contain exactly r red balls the first R coordinates of the tuple must have exactly r ones. Thus there are $\binom{R}{r}$ possible ways to select the first R coordinates, and for any such selection there are $\binom{B}{D-r}$ ways to select the last B coordinates to have $(D - r)$ ones. Hence there are $\binom{R}{r}\binom{B}{D-r}$ different outcomes which yield r red balls, so *the stated interpretation of expression* (3) *has been proved.*

Example 3 A sample of $D = 3$ items is drawn without replacement from a lot of $R + B = 10$ items. The items might be radio-transmitter tubes, or airplane propellers, or something else being subjected to quality control. A "red" item is defective, and a "black" item is nondefective. The number R of red items in the lot is unknown. The $D = 3$ items drawn are subjected to extensive and perhaps destructive tests, and the number r of them which are defective is determined. If $r = 0$ or 1 the lot of 10 items is accepted, and if $r = 2$ or 3 the lot of 10 items is rejected. Thus the probability p_R of accepting the lot if R items are defective is

$$p_R = \frac{\binom{R}{0}\binom{10-R}{3} + \binom{R}{1}\binom{10-R}{2}}{\binom{10}{3}}. \tag{4}$$

We find, for example, that $p_3 = .817$ and $p_6 = \frac{1}{3}$. Thus 30 percent defective lots are accepted about 82 percent of the time and 60 percent defective lots are accepted only about 33 percent of the time. ■

EXERCISES

1. State how many different functions can be obtained by taking sixth-order partial derivatives of $f(t_1,t_2,t_3) = (t_1)^3 e^{t_2+t_3 \cos t_1} + (t_1 t_2 \sin t_3)^{10}$.

2. How many ways can 30 essentially identical spare vacuum tubes be divided among a destroyer, a cruiser, and a battleship?

3. Let $n = 10$ and $k = 3$.
(*a*) How many different location vectors are there, and how many occupancy vectors are there?
(*b*) How many location vectors correspond to occupancy vector (2,4,4)?
4. How many ways can 40 sales districts be divided among four managers such that they have 10, 13, 8, and 9 sales districts, respectively? What is the significance of the word "respectively" in the previous sentence?
5. What is the coefficient of $t_1^{r_1}t_2^{r_2}\cdots t_k^{r_k}$ in the expansion of $(t_1 + t_2 + \cdots + t_k)^n$?
6. Eight blood samples are selected from 60 blood samples, of which six are cancerous. What is the probability that exactly two of the blood samples selected are cancerous?
7. Evaluate and interpret p_8, p_9, p_{10} as defined in equation (4) of Example 3.
8. What is the probability of obtaining exactly r red balls if D balls are drawn *with replacement* from an urn containing R red balls and B black balls?
9. In deriving the hypergeometric distribution (3) the balls were considered to be numbered. Another approach is to consider that R red and B black unnumbered balls are in an urn. All $R + B$ balls are drawn one after another, and an $(R + B)$-tuple of zeros and ones is formed by making the ith coordinate 1 iff the ith ball drawn is red. There are a total of $\binom{R+B}{R}$ such tuples. The first D balls drawn are considered to be the sample of D balls, so exactly r of the first D coordinates must be ones and $R - r$ of the remaining $R + B - D$ coordinates are ones. This approach yields

$$\frac{\binom{D}{r}\binom{R+B-D}{R-r}}{\binom{R+B}{R}} \tag{5}$$

as the probability of r reds among the first D drawn. Prove that (5) equals (3).
10. Suppose $R = 100$ animals (birds or fish) are caught and labeled (by tags or paint) and then let loose or mixed with B unlabeled animals of the same kind. It is desired to estimate B, which is unknown. A sample of $D = 200$ animals is later taken and found to contain $r = 10$ of the previously labeled animals. Thus 5 percent of the sample was found to be labeled, so it is not unreasonable to guess that about 5 percent of the whole population is labeled. That is, our guess $\hat{B}$ for the unknown B is defined by $R/(R + \hat{B}) = .05$, so $R + \hat{B} = 2{,}000$ is our guess for the unknown total population size and $\hat{B} = 1{,}900$ is our guess for the unlabeled population size.

Thus if the recapture data R, D, and r are known, and if $\hat{B}$ is defined in terms of them to be the largest integer for which $R/(R + \hat{B}) \geq r/D$, then $\hat{B}$ would seem to be a reasonable estimate for the unknown B. We wish to provide additional justification for using $\hat{B}$. For given, fixed R, D, and r define $f(B)$ to equal expression (3). Thus $f(B)$ is the probability of observing r reds in D draws when there are R reds in all, *assuming* that there are B black in all. We wish to show that $f(B)$ is maximized by setting $B = \hat{B}$, that is, that no value for B yields a higher probability for what was observed than does $B = \hat{B}$. Such an estimate $\hat{B}$ for the unknown parameter B is called a **maximum-likelihood estimate** in statistics. Verify that

$$\frac{f(B)}{f(B-1)} = 1 + \frac{D}{B-(D-r)}\left[\frac{R}{R+B} - \frac{r}{D}\right] \qquad \text{if } B > (D - r). \tag{6}$$

Clearly expression (6) is ≥ 1 iff [ ] ≥ 0. Thus $f(B)/f(B-1) \geq 1$ if $B \leq \hat{B}$ and $f(B)/f(B-1) < 1$ if $B > \hat{B}$; that is, f increases with B up to $B = \hat{B}$ and then decreases as B increases beyond $\hat{B}$.

3
Functions

This chapter consists of a brief but reasonably precise exposition of some of the more important concepts related to the fundamental mathematical concept of a function. The level of abstraction may be somewhat higher than the reader is accustomed to; but it permits a far greater conceptual clarity and generality in the development of probability theory.

3-1 FUNCTIONS

The basic idea of a function is that it assigns a *unique* value $f(x)$ to each x. In this book, as in almost all of mathematics, the word "function" means "single valued function." A function will sometimes be represented graphically as in Fig. 1, where for

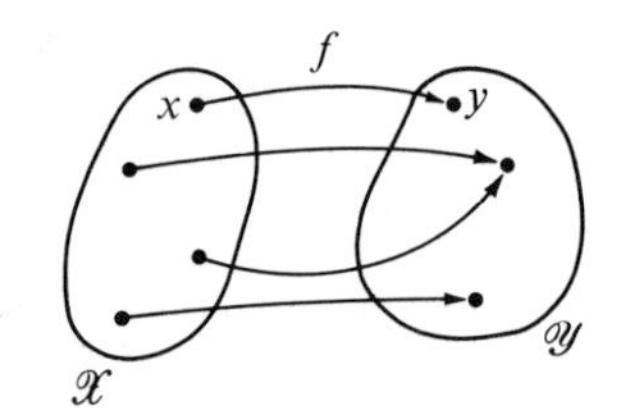

Fig. 1

each element x of a set $\mathscr{X}$ the function f assigns a unique value $y = f(x)$ in a set $\mathscr{Y}$. Thus to each x in $\mathscr{X}$ there is an arrow starting at that x and terminating in that point $y = f(x)$ which corresponds to that x. The property of single-valuedness is reflected in Fig. 1 by never having two arrows with the same starting point and different terminating points; that is, no two different arrows can have the same initial point. A function f can be thought of as a bundle or set of arrows, as indicated in Fig. 1. In this context the only relevant properties of an arrow are its initial and terminal points. Thus an arrow from x to y is equivalent to the ordered pair (x,y). Hence a function is essentially just a set of ordered pairs. The requirement of single-valuedness means that no two different ordered pairs can have the same first coordinate. We have thus been led to the *basic definition* that a **function** is a set of ordered pairs with the property that no two different ordered pairs in the set have the same first coordinate. Transformation, mapping, and correspondence are among the many synonyms for "function."

Several related definitions will be made before we consider examples of functions. The **domain** of a function is the set of all the first coordinates of the ordered pairs which belong to the function. A function is said to be defined **on** its domain. The **range** of a function is the set of all the second coordinates of the ordered pairs which belong to the function. Since a function is a set, we naturally say that two functions are **equal functions** iff they are equal as sets, that is, iff they both consist of the same set of ordered pairs. If (x,y) is an ordered pair which belongs to a function f another notation for y is $f(x)$; that is, $f(x)$ is the second coordinate of the ordered pair whose first coordinate is x. We say that $f(x)$ is the **value of** f **at** x, or f carries x into $f(x)$, or $f(x)$ is the **image of** x **under** f, or $f(x)$ is the value of f at the argument x.

Example 1 A simple example of a function is the set $f = \{(8,3), (2,-1), (\alpha,3), (\epsilon,\beta)\}$. For this example $f(8) = 3, f(2) = -1, f(\alpha) = 3$, and $f(\epsilon) = \beta$, the domain of f is the set $\{8,2,\alpha,\epsilon\}$, and the range of f is the set $\{3,-1,\beta\}$. Figure 2 depicts this function.

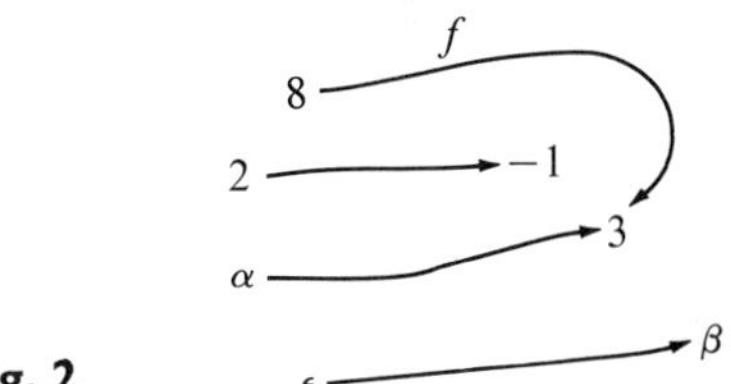

Fig. 2

Clearly the set $\{(8,3), (2,-1), (\alpha,3), (\epsilon,\beta), (2,5)\}$ is *not* a function, since the two different ordered pairs $(2,-1)$ and $(2,5)$ belong to it and have the same first coordinate. Also, the set $\{(8,3), (2,-1), \epsilon\}$ is *not* a function, since one of its members ϵ is not an ordered pair. ■

Example 2 A more typical example of a function is the set g of all ordered pairs (x,y) such that x and y are reals satisfying the inequality $-1 \leq x \leq 1$ and the equation $y = 3x^3 + 2x^2$. Thus $(\frac{1}{2},\frac{7}{8})$ belongs to g, so that $g(\frac{1}{2}) = \frac{7}{8}$. Neither $(2,8)$ nor $(.1,12)$ belongs to g. The function g is a subset of the plane and is shown as such in Fig. 3; that is, if (x,y) belongs to g, then

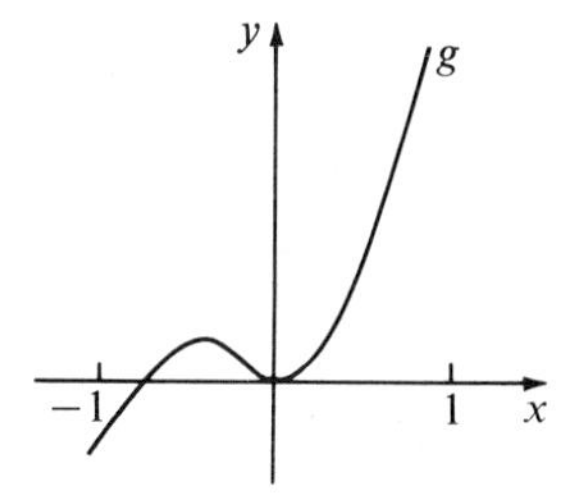

Fig. 3

it is plotted as a point in the plane in the usual way. The representation in Fig. 3 is essentially the usual one, with which the reader is familiar. Thus a function is nothing more than what might be called the *whole graph* of the function. The domain of g is the interval $-1 \leq x \leq 1$, which is the projection of g into the x axis. The range of g is the interval $-1 \leq x \leq 5$, which is the projection of g into the y axis. The property of single-valuedness corresponds to the fact that no vertical line intersects the graph at more than one point. Naturally this same function g can be defined as the set of all ordered pairs (t,α) such that t and α are reals satisfying the inequality $-1 \leq t \leq 1$ and the equation $\alpha = 3t^3 + 2t^2$. That is, the x and y, or the t and α, are just symbols which are useful in defining the function of Fig. 3. ■

Example 3 Let the function h be defined to consist of the set of all ordered pairs (x,y) of reals satisfying $0 \leq x \leq 1$ and $y = 3x^3 + 2x^2$. This function is plotted in Fig. 4, and clearly this graph is not the same as the graph of Fig. 3. ■

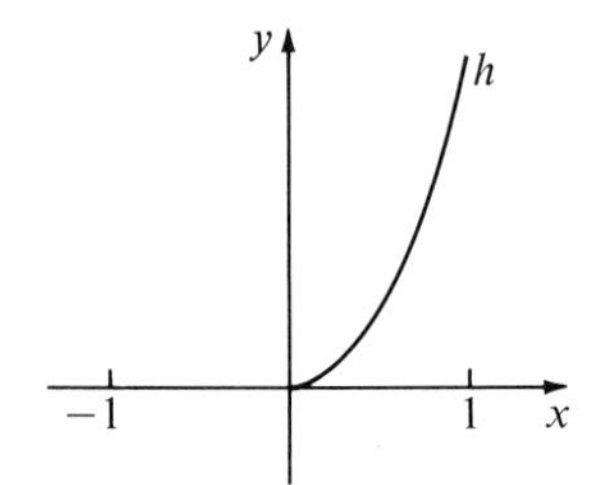

Fig. 4

The two functions g and h are *not* equal. These two examples illustrate the important fact that the domain of a function is determined by its definition. Stated more casually, two functions are equal iff they have the same domain and they both assign the same value to each member of this domain.

Example 4 Figure 5 shows still another function F, which is defined to be the set of all ordered pairs (z,w) such that $z = (x,y)$ is an ordered pair satisfying $0 \leq x \leq 1$ and $0 \leq y \leq 1$; and

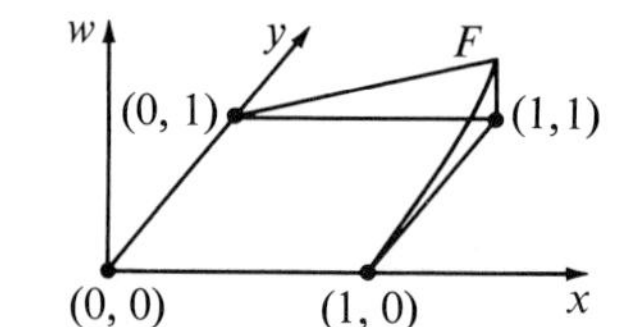

Fig. 5

x,y, and w satisfy the equation $w = xy^2$. The domain of F is a unit square and its range is the interval $0 \leq w \leq 1$. Thus a member of F has the form $((x,y), xy^2)$, where the first coordinate is a point (x,y) in the square and the second coordinate is xy^2, which is the height at $(x,y)$of the surface F above the xy plane. In more standard notation, F assigns the value $F(x,y) = xy^2$ to a point (x,y) in the unit square. ■

Example 5 The exponential function can be defined to be the set of all ordered pairs (x,y) of reals which satisfy

$$y = \sum_{k=0}^{\infty} \frac{x^k}{k!}.$$

It is also possible to define the exponential function as the set of all ordered pairs (x,y) of reals which satisfy

$$y = \lim_{n\to\infty} \left(1 + \frac{x}{n}\right)^n.$$

From elementary calculus we know that these two definitions do indeed define the same function; that is, they both assign the same value $f(x)$ to each real x. Thus these two *different* "rules" define the same function. Most explicit functions with which we deal will be defined by such fairly simple means as polynomials, Taylor series, integrals. However, we do not assume that a function has any particular amount of constructibility. It is often said that a function is a rule which assigns a unique value to each element of its domain. Whenever we use the word "rule" in such a context it is essentially equivalent to the word "function." ■

In the remainder of this book the notation and verbal treatment of functions will usually be more casual than is indicated by the definition of a function as a set of ordered pairs. However, we will always try to be both precise and consistent with the definitions which have been made.

We mention that the members of the domain and range of a function can be quite arbitrary objects. Also the domain of a function is part of its definition. Furthermore the rule assigning $f(x)$ to x may be very complex.

EXERCISES

1. The array below is used to define the set of all the different objects among $(x_1,y_1), \ldots, (x_n,y_n)$. If this set is to be a function, state whether two different rows can be:
(*a*) The same in both their left and right entries
(*b*) Different in both their left and right entries
(*c*) Different in their left entries but have the same right entries
(*d*) Different in their right entries but have the same left entries

$$\begin{array}{cc} x_1 & y_1 \\ x_2 & y_2 \\ \cdots & \cdots \\ x_n & y_n \end{array}$$

2. Sketch a picture like Fig. 2 for those of the sets below which are functions:
(*a*) $\{(a,b), (c,d), (e,e), (b,a)\}$
(*b*) $\{(a,a), (b,a), (a,c), (b,d)\}$
(*c*) $\{(\alpha,\beta), (\gamma,\delta), \epsilon\}$
(*d*) $\{(a,b), (c, \{d, \beta\})\}$
(*e*) $\{(a,b), (c,d), (a,b)\}$
(*f*) $\{(1,a), (2,b), (3,b)\}$
(*g*) $\{(a,1), (b,2), (b,3)\}$

3. Let D be any nonempty set. A function f defined on D and assigning to each d in D a *set* $f(d)$ is called an *indexed family of sets;* that is, the set $f(d)$ is assigned to the index d. Sometimes the set $f(d)$ is written as A_d and we say "the family of all sets A_d for d belonging to D." We denote the *union* of the sets in the family by $\bigcup_{d\in D} f(d)$. An object belongs to this union iff it belongs to $f(d)$ for *some* d in D. We denote the *intersection* of the sets in the family by $\bigcap_{d\in D} f(d)$. An object belongs to this intersection iff it belongs to $f(d)$ for *every* d in D. Note that if $D = \{1, 2, \ldots, N\}$ then

$$\bigcup_{d\in D} f(d) = f(1) \cup f(2) \cup \cdots \cup f(N), \quad \bigcap_{d\in D} f(d) = f(1) \cap f(2) \cap \cdots \cap f(N).$$

Similarly if $D = \{1, 2, \ldots\}$ we obtain the union and intersection of an infinite sequence of sets. This new notation is quite general; for example, we can let D equal the set R of all real numbers, and as shown in Sec. 4-3, the set R is uncountable. Prove the *generalized De Morgan laws*

$$\left[\bigcup_{d \in D} f(d)\right]^c = \bigcap_{d \in D} [f(d)]^c \qquad \left[\bigcap_{d \in D} f(d)\right]^c = \bigcup_{d \in D} [f(d)]^c$$

where each $f(d)$ is assumed to be a subset of the same set Ω with respect to which all complements are taken.

3-2 ADDITIONAL TERMINOLOGY CONCERNING FUNCTIONS

Recall that the domain of a function is the set of all arguments for which it is defined. A function is said to be *defined* on its domain. The range of a function is the set of all values which it takes on. Thus the range of a function f is the set of all y such that there exists at least one x belonging to the domain of f for which $y = f(x)$.

The notation $f: \mathscr{X} \to \mathscr{Y}$ means that f is a function whose **domain is** $\mathscr{X}$ and whose range is **contained in** $\mathscr{Y}$. Figure 1 illustrates this notation, which is common in mathematics and will be used frequently throughout this book. Thus $f: \mathscr{X} \to \mathscr{Y}$ means that f is a function which assigns to *each* x belonging to $\mathscr{X}$ a unique value $f(x)$ *belonging to* $\mathscr{Y}$. If $f: \mathscr{X} \to \mathscr{Y}$ we say that f is a function *on* $\mathscr{X}$ and **into** $\mathscr{Y}$. Figure 3 (Sec. 3-1)

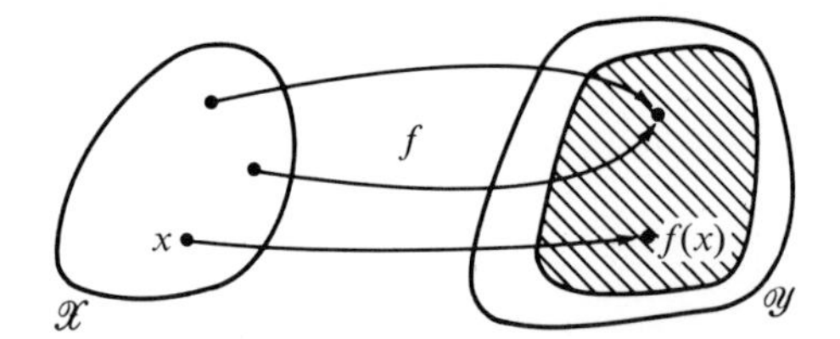

Fig. 1 The range of $f: \mathscr{X} \to \mathscr{Y}$ is shaded.

depicts a function g which is defined *on* the interval $-1 \leq x \leq 1$ and *into* the set R of all real numbers. The notation $f: \mathscr{X} \to \mathscr{Y}$ deemphasizes the role of the range of a function. One of the advantages of this is that, as in Fig. 3 (Sec. 3-1), it is often easy to define a function *into* a set $\mathscr{Y}$, while it may be troublesome to find its range.

A function is said to be a **function of a real variable** iff its domain is contained in the set R of all reals. The exponential function and the function of Fig. 3 (Sec. 3-1) are functions of a real variable. A function is said to be a **function of n real variables** iff its domain is contained in the set R^n of all n-tuples of reals. Such a function may also be called a **function with vector argument**. Figure 5 (Sec. 3-1) shows a function of two real variables. A function is said to be **real valued** iff its range is contained in R. Figures 4 and 5 (Sec. 3-1) depict real-valued functions. If there is a positive integer n such that the range of a function is contained in R^n the function is said to be **vector valued**. Thus the function $f: R^3 \to R^2$ defined by $f(x,y,z) = (e^x \cos y, z^2)$ is a vector-valued function of three real variables and it assigns the point $(e^x \cos y, z^2)$ in the plane to the point (x,y,z) in R^3. Clearly a function $f: \mathscr{X} \to R^n$ is equivalent to n functions $f_1, \ldots, f_n$ each on $\mathscr{X}$ and into R; that is, f can be written as $f = (f_1, \ldots, f_n)$ and the value of f at x is $f(x) = (f_1(x), \ldots, f_n(x))$, so that $f_i(x)$ is the ith component of $f(x)$.

A few physical interpretations of functions involving real numbers may be helpful. A curve in R^n, in parametric form, is a continuous function f on $0 \leq t \leq 1$ and into R^n; that is, f assigns the point $f(t)$ in R^n to the point t of the unit interval. For $n = 2$ we have a curve or trajectory in the plane. For $n = 3$ we can think of the trajectory of a particle in space as a function of time; that is, $f(t)$ is the location of the particle at time t. A function $p: R^3 \to R$ may be thought of as an electric charge density in space; that is, $p(x,y,z)$ is the charge density at (x,y,z); if $p(x)$ is always nonnegative then p can be interpreted as a mass density. A function $f: R^3 \to R^3$ may be thought of as a vector field in space; that is, the field at (x,y,z) is the vector $f(x,y,z)$. A function $f: R^4 \to R^3$ may be thought of as a vector field which varies as a function of time; that is, the field at (x,y,z) at time t is the vector $f(x,y,z,t)$.

Inequalities between real-valued functions with the same domain will often be used. If $f: \mathscr{X} \to R$ and $g: \mathscr{X} \to R$ then by $f \leq g$ we mean that $f(x) \leq g(x)$ for all x in $\mathscr{X}$. Thus $f \leq g$ is true if $\mathscr{X} = R^2$, $f(x,y) = y^2 + e^x \sin y$, and $g(x,y) = 2y^2 + e^x + 3$. For this example the stronger statement $f(x,y) < g(x,y)$ for all (x,y) in R^2 is true. We often ignore the possibility of such a strengthening. Frequently the added complexity in statement or proof required for such a strengthening is excessive. If $f: \mathscr{X} \to R$ and $g: \mathscr{X} \to R$ then by $|f| \leq g$ we mean that $|f(x)| \leq g(x)$ for all x in $\mathscr{X}$, and this is equivalent to $-g(x) \leq f(x) \leq g(x)$ for all x in $\mathscr{X}$.

3-3 COMPOSITION

Manipulations such as "substitute $y_1 = F(x_1,x_2,x_3)$ and $y_2 = H(x_1,x_2,x_3)$ into $g(y_1,y_2)$ to obtain $g(F(x_1,x_2,x_3), H(x_1,x_2,x_3))$" should be familiar to the reader. For such manipulations there is the concept of the composition of functions. If $f: \mathscr{X} \to \mathscr{Y}$ and $g: \mathscr{Y} \to \mathscr{Z}$ then the **composition of** g **applied after** f is the function $h: \mathscr{X} \to \mathscr{Z}$, defined to have the value $h(x) = g(f(x))$ at any point x of $\mathscr{X}$. The notations $g(f)$ and $g \circ f$ are common for the function h. Thus, as in Fig. 1, the value of h at x is obtained by

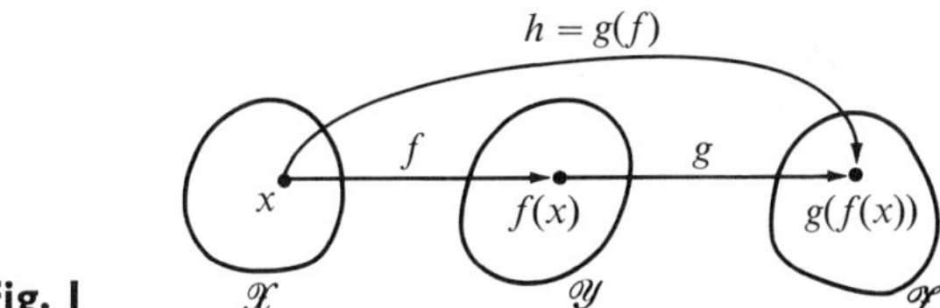

Fig. 1

first evaluating f at x and then evaluating g at $f(x)$. The function $g(f)$ should be thought of as a bundle of arrows going directly from $\mathscr{X}$ to $\mathscr{Z}$, even though it is initially defined via $\mathscr{Y}$.

Instead of saying "let $f: \mathscr{X} \to \mathscr{Y}$ and $g: \mathscr{Y} \to \mathscr{Z}$ and let $g(f)$ be the composition of g applied after f," we often use the picture below to define these relations:

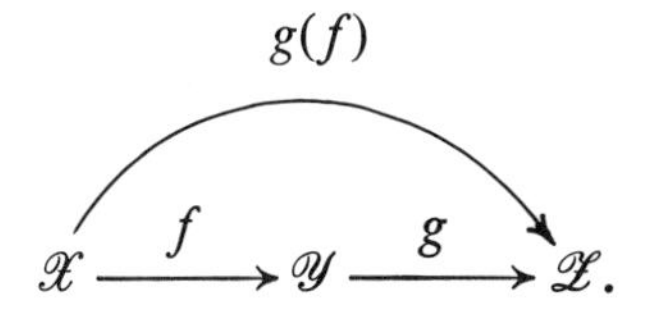

Example 1 As an example of composition, let $f\colon R^3 \to R^2$ be defined by $f(x_1,x_2,x_3) = (e^{x_1}\cos x_2, x_3 \sin x_1)$ and let $g\colon R^2 \to R$ be defined by $g(y_1,y_2) = y_1{}^2 + 2y_1y_2 + 4$. Then the value of $g(f)\colon R^3 \to R$ at (x_1,x_2,x_3) is

$$g(f(x_1,x_2,x_3)) = [e^{x_1}\cos x_2]^2 + 2[e^{x_1}\cos x_2][x_3 \sin x_1] + 4. \quad \blacksquare$$

Let $f\colon \mathscr{X} \to R$. If k is a positive integer the function $f^k\colon \mathscr{X} \to R$ is defined to have the value $[f(x)]^k$ at any x belonging to $\mathscr{X}$. This is an example of a composition where the function $g\colon R \to R$ has the value y^k at a real y. Similarly if $f\colon \mathscr{X} \to R$ is any real-valued function on some set $\mathscr{X}$ then the value of $af + b$ at x is $af(x) + b$, the value of e^f at x is $e^{f(x)}$, and the value of $|f|$ at x is $|f(x)|$.

Let $f_1, \ldots, f_n$ be n functions on $\mathscr{X}$ and into R. The value of $a_1f_1 + \cdots + a_nf_n$ at x is $a_1f_1(x) + \cdots + a_nf_n(x)$, the value of $f_1f_2\cdots f_n$ at x is $f_1(x)f_2(x)\cdots f_n(x)$, and the value of $\max\{f_1, \ldots, f_n\}$ at x is $\max\{f_1(x), \ldots, f_n(x)\}$. These are examples of compositions where $f\colon \mathscr{X} \to R^n$ is defined by $f(x) = (f_1(x), \ldots, f_n(x))$. For the middle case of the product of n functions we let $g\colon R^n \to R$ be defined by $g(y_1, \ldots, y_n) = y_1 \cdots y_n$. For the last case we let $g\colon R^n \to R$ be defined by $g(y_1, \ldots, y_n) = \max\{y_1, \ldots, y_n\}$, where $\max\{y_1, \ldots, y_n\}$ is defined to be the largest of the numbers $y_1, \ldots, y_n$; for example, $\max\{3.2,12,8,-6\} = 12$.

EXERCISES

1. Each of g, f, g_1 and f_1 is on R and into R, where $g(x) = \cos x$, $f(x) = x^4$, $g_1(x) = \cos x^2$, and $f_1(x) = x^2$. Is $g(f) = g_1(f_1)$?

2. Let $g\colon R^2 \to R$, where $g(x,y) = \sin^2 x + \cos^2 y$ for all (x,y) in R^2. Let $f\colon R \to R$, where $f(x) = e^{2x}$ for all x in R. Let $h\colon R \to R$, where $h(x) = \cos x$ for all x in R. Find the functions $f(h)$, $h(f)$, and $g(f, h)$ which are all defined on R. Note that $f(h) \neq h(f)$. Evaluate $g(f(\pi/2), h(\pi/2))$.

3. Using the functions graphed in Fig. 2, sketch the two functions $\frac{1}{2}f_1 + f_2 + f_3$ and $\max\{f_1,f_2,f_3\}$.

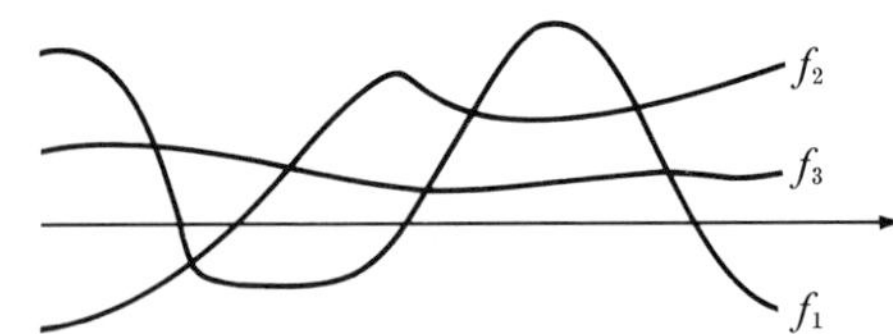

Fig. 2

4. Using the functions graphed in Fig. 3, sketch the functions $|f|$, f^2, and $-g$. Which of $|f| \le g$, $f^2 \le g$, and $\sin f \le g$ are true?

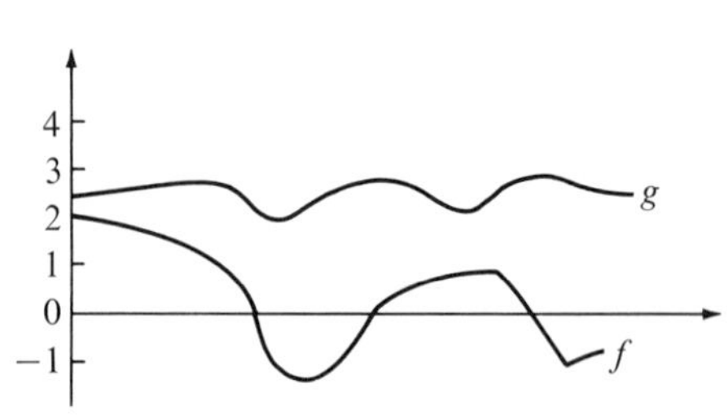

Fig. 3

3-4 INVERSE IMAGES

If $f: \mathscr{X} \to \mathscr{Y}$ and A is a subset of $\mathscr{Y}$, the **inverse image of** A **under** f is the set of all x belonging to $\mathscr{X}$ such that $f(x)$ belongs to A. The inverse image of A under f is denoted by $f^{-1}(A)$ and is shown in Fig. 1. Thus $f^{-1}(A)$ is the set of all x which are carried into A by f. By this definition we obtain for every subset A of $\mathscr{Y}$ a unique subset $f^{-1}(A)$ of $\mathscr{X}$.

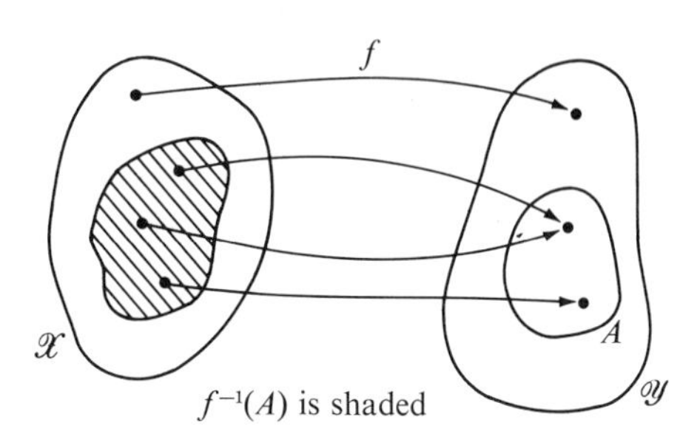

Fig. 1

We have defined a function f^{-1} such that an element of its domain is a subset of $\mathscr{Y}$ and an element of its range is a subset of $\mathscr{X}$. Also every subset of $\mathscr{Y}$ belongs to its domain.

Clearly $f^{-1}(\varphi) = \varphi$. Also, $f^{-1}(\mathscr{Y}) = \mathscr{X}$ if $f: \mathscr{X} \to \mathscr{Y}$; that is, the inverse image of the empty set is empty and the inverse image of all of $\mathscr{Y}$ is the whole domain $\mathscr{X}$.

Example 1 Let $f: \mathscr{X} \to R$ be defined as follows: the domain of f is the set of all reals greater than or equal to -2, and if x belongs to $\mathscr{X}$ then $f(x) = x^2 + 2$. This function is depicted in Fig. 2.

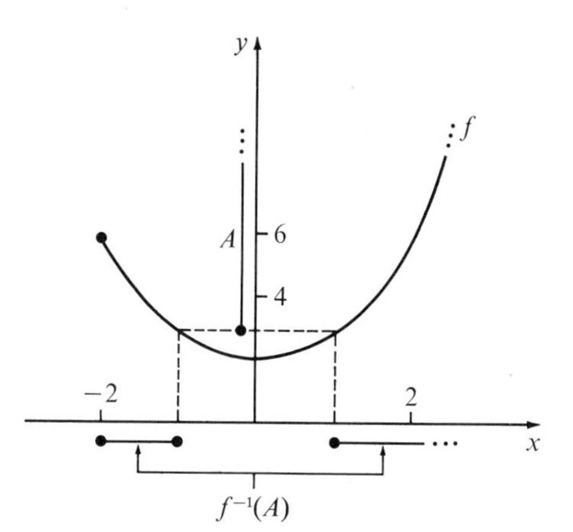

Fig. 2

If we let A be the set of all reals greater than or equal to 3 then $f^{-1}(A)$ is the union of the interval $-2 \leq x \leq -1$ with the set of all reals greater than or equal to 1. ■

An obvious property of f^{-1} is that $f^{-1}(A)$ is a subset of $f^{-1}(B)$ whenever A is a subset of B. This is depicted in Fig. 3.

In addition to preserving the property of being a subset, f^{-1} also preserves disjointness; that is, if A and B are disjoint so are $f^{-1}(A)$ and $f^{-1}(B)$, as shown in Fig. 4.

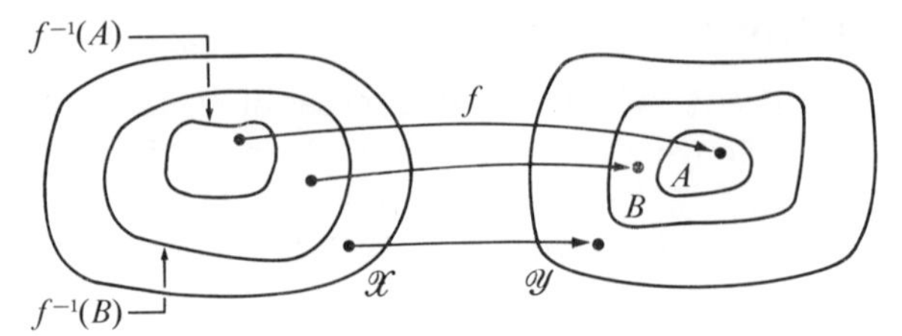

Fig. 3

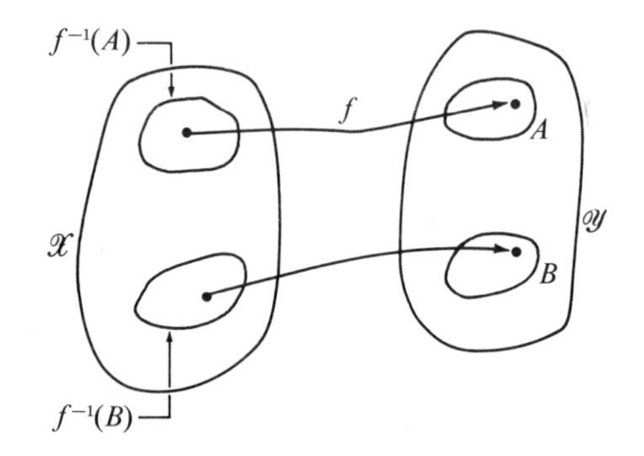

Fig. 4

The inverse image of a union is the union of the inverse images; that is, $f^{-1}(A \cup B) = f^{-1}(A) \cup f^{-1}(B)$, as in Fig. 5.

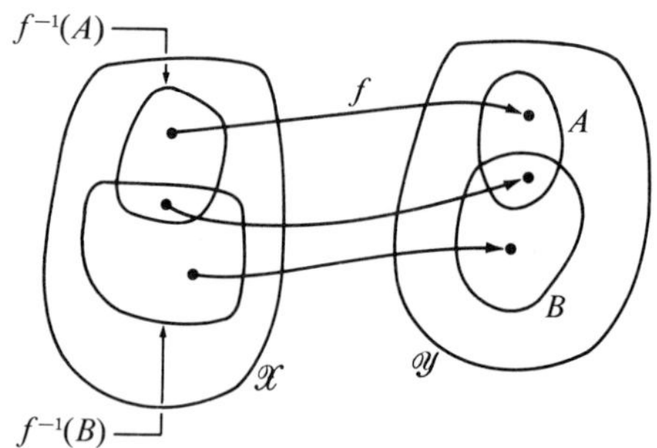

Fig. 5

If

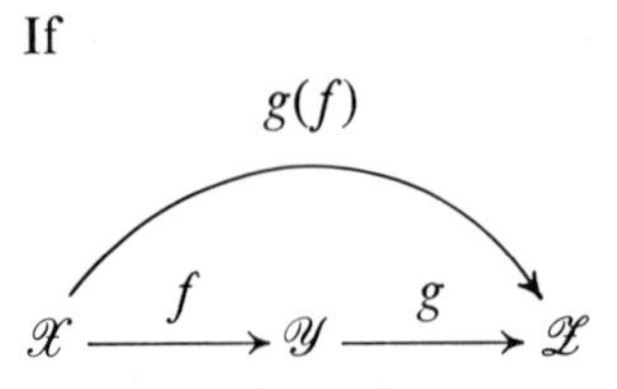

and A is a subset of $\mathscr{Z}$ then the inverse image of A under the function $g(f)$ is the set $f^{-1}(g^{-1}(A))$. That is, as shown in Fig. 6, we first find the set $g^{-1}(A)$ of all points of $\mathscr{Y}$ which are carried into A by g and then find the set of all points of $\mathscr{X}$ which are carried into $g^{-1}(A)$ by f.

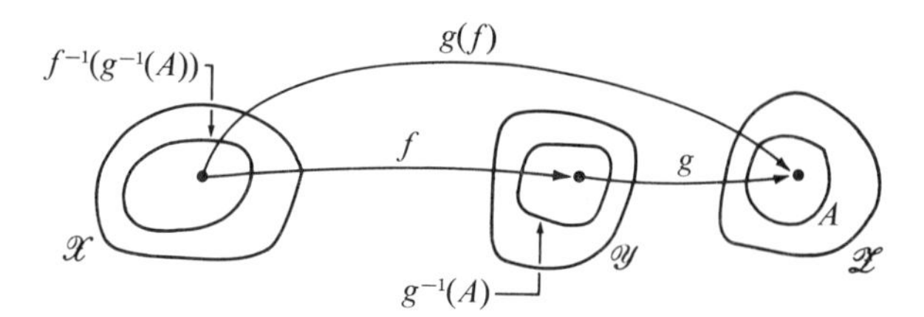

Fig. 6

Example 2 Let $g: R \to R$ be defined by $g(x) = x^3$ for all real x. Let the function f be as in Example 1 and let B be the set of all reals greater than or equal to 27. We wish to find the inverse image of B under $g(f)$. Now, $g^{-1}(B)$ is clearly the set of all reals greater than or equal to 3, and this happens to be the set A of Example 1; thus $f^{-1}(g^{-1}(B)) = f^{-1}(A)$. This set is shown in Fig. 2. ■

The concept f^{-1} which we have been considering is quite different from the concept of an inverse function as used in the statement "the logarithm function is the inverse function to the exponential function." This latter kind of inverse will be considered in the next section. The same notation f^{-1} is used in both cases, but the context should indicate which of the two kinds of inverse function is being considered.

EXERCISES

1. Let $f(i) = (h_i, w_i)$ equal the height and weight of person number i, where $f: \{1, 2, \ldots, n\} \to R^2$. Interpret the facts that f^{-1} preserves inclusion, disjointness, and unions.

2. The graph of a function $f: R \to R$ is shown in Fig. 7. Let A be the set of all reals greater than or equal to 2, and let B be the interval $1 \leq y \leq 3$, so that $A \cup B$ is the set of all reals

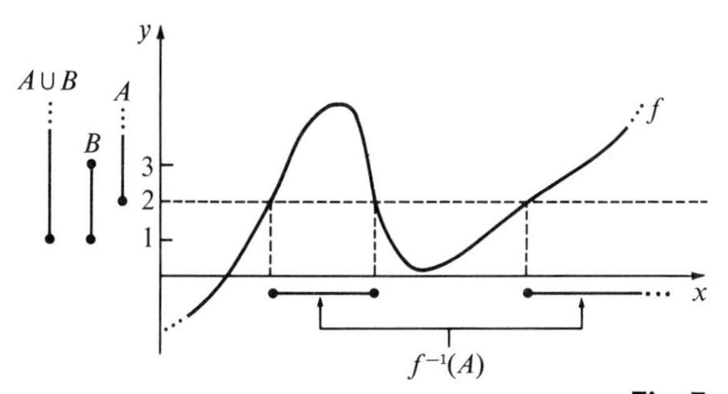

Fig. 7

greater than or equal to 1. Make a sketch showing the sets $f^{-1}(A)$, $f^{-1}(B)$, and $f^{-1}(A \cup B)$. Check that $f^{-1}(A \cup B) = f^{-1}(A) \cup f^{-1}(B)$.

3. Let $f: R^2 \to R^2$ be defined by $f(x,y) = (x^2, \sin y)$.

(*a*) Make a sketch of the range of f.

(*b*) Find $f^{-1}(A)$, where $A = \{(z,w): z \geq 0\}$ is the right half-plane.

(*c*) Find $f^{-1}(B)$, where $B = \{(z,w): w \geq 4\}$.

(*d*) Make a sketch of $f^{-1}(C)$, where

$$C = \left\{(z,w): 4 \leq z \leq 9, -\frac{1}{\sqrt{2}} \leq w \leq \frac{1}{\sqrt{2}}\right\}.$$

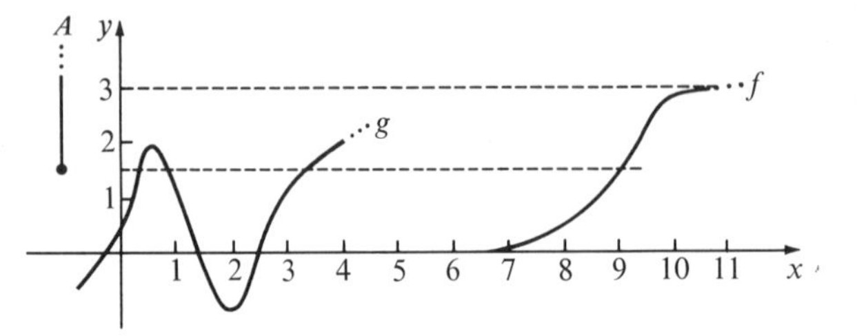

Fig. 8

4. Two functions, $f: R \to R$ and $g: R \to R$, are graphed in Fig. 8. Let A be the set of all reals greater than or equal to 1.5, let $B = g^{-1}(A)$, and let $C = f^{-1}(B)$, so that the inverse image of A under $g(f)$ equals C. Find B from g, sketch B along both axes, and find C from f. Sketch $g(f)$, find "the inverse image of A under $g(f)$" from $g(f)$, check that it equals C.

3-5 ONE-TO-ONE CORRESPONDENCE

Recall that the notation $f: \mathscr{X} \to \mathscr{Y}$ means that $\mathscr{X}$ *is* the domain of f and $\mathscr{Y}$ *contains* the range of f. We say f is *on* $\mathscr{X}$ and *into* $\mathscr{Y}$. If $\mathscr{Y}$ actually equals the range of f we say that f is **onto** $\mathscr{Y}$. The function of Fig. 3 (Sec. 3-1) is *onto* the interval $-1 \leq y \leq 5$. Equivalently $f: \mathscr{X} \to \mathscr{Y}$ is onto $\mathscr{Y}$ iff for every y belonging to $\mathscr{Y}$ there is *at least* one x belonging to $\mathscr{X}$ such that $y = f(x)$.

A function $f: \mathscr{X} \to \mathscr{Y}$ is said to be **one-to-one** iff for every y belonging to $\mathscr{Y}$ there is *at most* one x belonging to $\mathscr{X}$ for which $y = f(x)$; that is, each value is taken on at most once. For Figs. 1 to 5 (Sec. 3-1) only the function of Fig. 4 is one-to-one. A function $f: \mathscr{X} \to \mathscr{Y}$ is called a **one-to-one correspondence** between $\mathscr{X}$ and $\mathscr{Y}$ iff it is both one-to-one and onto $\mathscr{Y}$.

The four possibilities for onto and one-to-one are shown in Fig. 1, where in each case the range of f is shaded. In terms of such pictures f is *onto* $\mathscr{Y}$ iff every point of $\mathscr{Y}$

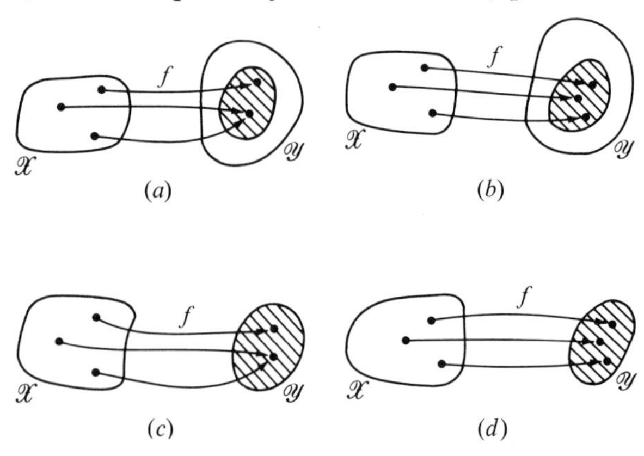

Fig. 1 The range of $f: \mathscr{X} \to \mathscr{Y}$ is shaded. The four possibilities for onto and one-to-one: (*a*) not onto, not one-to-one; (*b*) not onto, one-to-one; (*c*) onto, not one-to-one; (*d*) onto, one-to-one, a one-to-one correspondence.

has *at least* one arrow terminating on it, and f is *one-to-one* between $\mathscr{X}$ and $\mathscr{Y}$ iff every point of $\mathscr{Y}$ has *at most* one arrow terminating on it. Clearly a function $f: \mathscr{X} \to \mathscr{Y}$ is a one-to-one correspondence between $\mathscr{X}$ and $\mathscr{Y}$ iff for every y belonging to $\mathscr{Y}$ there is exactly one x belonging to $\mathscr{X}$ such that $y = f(x)$, that is, iff every point of $\mathscr{Y}$ has exactly one arrow terminating on it.

It is clear from the definition and from the last picture of Fig. 1 that a one-to-one correspondence effectively pairs up points of $\mathscr{X}$ and $\mathscr{Y}$ in an essentially symmetric fashion. Going back to the definition of a function as a set of ordered pairs, we see that a one-to-one correspondence f between $\mathscr{X}$ and $\mathscr{Y}$ is a set of ordered pairs (x,y) such that each x belonging to $\mathscr{X}$ is the first coordinate of exactly one ordered pair and each y belonging to $\mathscr{Y}$ is the second coordinate of exactly one ordered pair. It is precisely in this case that, as shown in Fig. 2, we can define an inverse function *on all of* $\mathscr{Y}$.

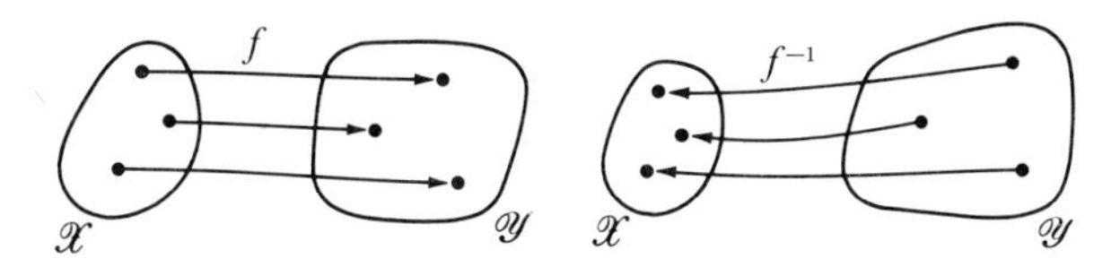

Fig. 2 The inverse function $f^{-1}: \mathscr{Y} \to \mathscr{X}$ of a one-to-one correspondence $f: \mathscr{X} \to \mathscr{Y}$.

If $f: \mathscr{X} \to \mathscr{Y}$ is a one-to-one correspondence between $\mathscr{X}$ and $\mathscr{Y}$ we can define the **inverse of f on $\mathscr{Y}$** as the function $f^{-1}: \mathscr{Y} \to \mathscr{X}$, where $f^{-1}(y)$ is the unique x such that $f(x) = y$. We see that f^{-1} is a one-to-one correspondence between $\mathscr{Y}$ *and* $\mathscr{X}$, and that $f^{-1}(f(x)) = x$ for all x in $\mathscr{X}$ and $f(f^{-1}(y)) = y$ for all y in $\mathscr{Y}$. As an example the exponential function graphed in Fig. 3 is a one-to-one correspondence between R and the set of all positive reals. Its inverse function is the logarithm function.

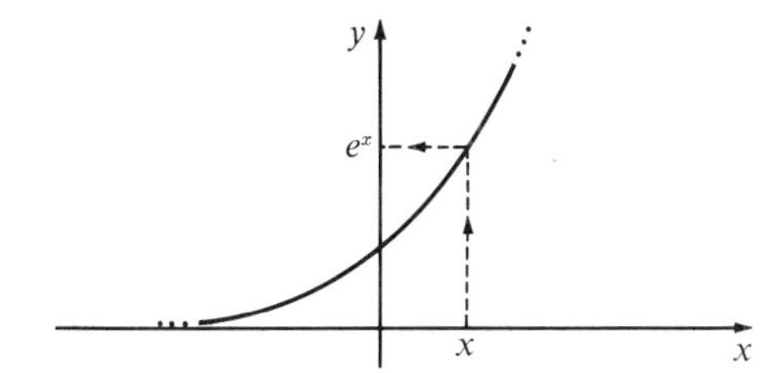

Fig. 3 The exponential function.

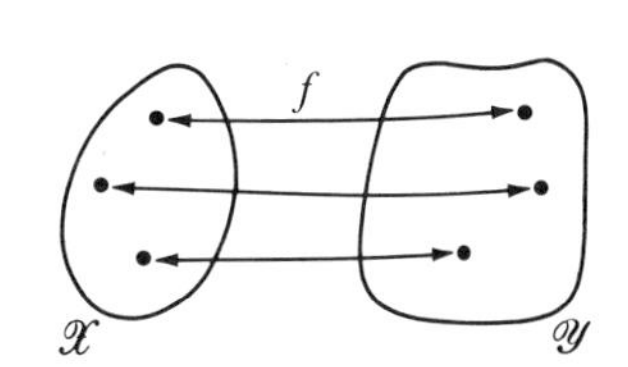

Fig. 4 A one-to-one correspondence.

The inverse function now being considered can be defined only for *certain* functions $f: \mathscr{X} \to \mathscr{Y}$, and it has *points* of $\mathscr{Y}$ as members of its domain. The inverse function considered in Sec. 3-4 can be defined for *any* function, and it has *subsets* of $\mathscr{Y}$ as members of its domain.

A one-to-one correspondence $f: \mathscr{X} \to \mathscr{Y}$ was defined in an asymmetric fashion with $\mathscr{X}$ for the domain. Usually our manipulations will be consistent with this definition. However, it is clear, particularly in terms of ordered pairs, that such a correspondence is essentially symmetric between $\mathscr{X}$ and $\mathscr{Y}$. If we wished we could use a notation such as $\mathscr{X} \overset{f}{\leftrightarrow} \mathscr{Y}$ and pictures such as Fig. 4 for a one-to-one correspondence.

EXERCISES

1. Each of the four functions $\sin x$, e^x, $3x^3 + 2x^2$, and x^3 is to be considered as having R for its domain. For each, state whether it is onto R and whether it is one-to-one.

2. The accepted *complete* mathematical meaning for the term "real-valued function" can be correctly inferred from the fact that "real-valued" is an adjective modifying the noun "function." That is, a "real-valued function" is *nothing more* than a function having the property of being real valued. The same comment cannot be made about the unfortunate accepted term "one-to-one correspondence." Why?

3. For each of the following cases exhibit a one-to-one correspondence $f\colon \mathscr{X} \to \mathscr{Y}$ between $\mathscr{X}$ and $\mathscr{Y}$:

(*a*) $\mathscr{X} = \{1,2,3,4\}$, $\mathscr{Y} = \{\alpha,\beta,\gamma,33.5\}$
(*b*) $\mathscr{X} = \{1, 2, 3, \ldots\}$, $\mathscr{Y} = \{6, 7, 8, \ldots\}$
(*c*) $\mathscr{X} = \{1, 2, 3, \ldots\}$, $\mathscr{Y} = \{2, 4, 6, 8, \ldots\}$
(*d*) $\mathscr{X} = \{x\colon x \text{ is real and } -2 \le x \le 3\}$, $\mathscr{Y} = \{y\colon y \text{ is real and } 5 \le y \le 6\}$
(*e*) $\mathscr{X} = R =$ [the set of all real numbers], $\mathscr{Y} = \{y\colon y \text{ is real and } y > 0\}$

4
Countability

The theory of cardinality was created by G. Cantor (1845–1918) and is fundamental to modern mathematics. A few of the simpler ideas and facts from this theory are needed in probability theory and will be developed in this chapter. The theory is typically referred to as the theory of cardinal numbers, or the theory of transfinite numbers, or the theory of sets. Cantor's "theory of sets" essentially presupposes the elementary ideas of set theory as developed in Chap. 1.

4-1 CARDINALITY

Any two sets $\mathscr{X}$ and $\mathscr{Y}$ are said to have the **same cardinality** iff there exists a one-to-one correspondence $f: \mathscr{X} \to \mathscr{Y}$ between $\mathscr{X}$ and $\mathscr{Y}$. We interpret "have the same cardinality" as meaning "have the same number of members." This definition is the decisive idea in the theory of cardinality.

The two sets $\mathscr{X} = \{1,2,3\}$ and $\mathscr{Y} = \{2, \epsilon, (\delta, \beta)\}$ have the same cardinality because the function $f: \mathscr{X} \to \mathscr{Y}$ defined by $f(1) = \epsilon, f(2) = (\delta,\beta)$, and $f(3) = 2$ is a one-to-one correspondence between $\mathscr{X}$ and $\mathscr{Y}$. This result is obvious, since both sets have three members. The idea of cardinality agrees with our usual ideas about the sizes of finite sets but gives us new insights into the sizes of infinite sets.

The property of having the same cardinality is *reflexive*, *symmetric*, and *transitive*. By reflexive we mean that any set $\mathscr{X}$ has the same cardinality as itself. This is true because the identity function on $\mathscr{X}$ is a one-to-one correspondence between $\mathscr{X}$ and $\mathscr{X}$. The **identity function on a set** $\mathscr{X}$ is the function $f: \mathscr{X} \to \mathscr{X}$ defined by $f(x) = x$ for all x in $\mathscr{X}$. Thus the identity function assigns the value x to a point x. The straight line through the origin in the plane and making a 45° angle with the x axis is the graph of the identity function on R. By symmetric we mean that if $\mathscr{X}$ and $\mathscr{Y}$ have the same cardinality then $\mathscr{Y}$ and $\mathscr{X}$ have the same cardinality. This is true because if $f: \mathscr{X} \to \mathscr{Y}$ is a one-to-one correspondence between $\mathscr{X}$ and $\mathscr{Y}$, then $f^{-1}: \mathscr{Y} \to \mathscr{X}$ is a one-to-one correspondence between $\mathscr{Y}$ and $\mathscr{X}$. By transitive we mean that if $\mathscr{X}$ and $\mathscr{Y}$ have the same cardinality and $\mathscr{Y}$ and $\mathscr{Z}$ have the same cardinality then $\mathscr{X}$ and $\mathscr{Z}$ have the same cardinality. This is true because the composition of two one-to-one correspondences is also one:

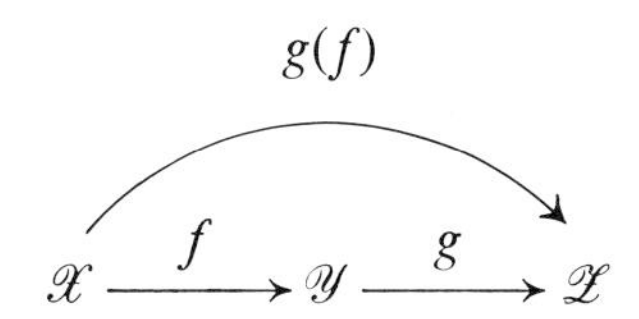

These three properties are certainly desirable for our interpretation that having the same cardinality means having the same number of elements. Thus $\mathscr{X}$ has the same number of elements as $\mathscr{X}$. *If* $\mathscr{X}$ has the same number of elements as $\mathscr{Y}$, *then* $\mathscr{Y}$ has the same number of elements as $\mathscr{X}$. *If* $\mathscr{X}$ has the same number of elements as $\mathscr{Y}$ *and if* $\mathscr{Y}$ has the same number of elements as $\mathscr{Z}$, *then* $\mathscr{X}$ has the same number of elements as $\mathscr{Z}$.

4-2 COUNTABLE SETS

Clearly a nonempty set $\mathscr{X}$ is finite iff there is a positive integer n such that the set $\mathscr{X}$ has the same cardinality as the set $\{1, 2, \ldots, n\}$. Recall that a set is said to be infinite iff it is not finite.

A set is said to be **denumerable** iff it has the same cardinality as **the set** $I = \{1, 2, \ldots\}$ **of all positive integers.** Thus a set $\mathscr{X}$ is denumerable iff there is a one-to-one correspondence $f: I \to \mathscr{X}$ between I and $\mathscr{X}$. In particular $I = \{1, 2, \ldots\}$ is denumerable. For a denumerable set $\mathscr{X}$ we may use the notation $x_n = f(n)$, so that, roughly speaking, a denumerable set $\mathscr{X}$ is one which can be written as an infinite sequence $x_1, x_2, \ldots$ of elements of $\mathscr{X}$ such that every element of $\mathscr{X}$ appears once and only once in the sequence. Such a function or sequence is called an **enumeration** of the denumerable set. Similarly if every member of a finite set $\mathscr{X}$ appears exactly once in a finite sequence $x_1, \ldots, x_n$, each of whose terms belong to $\mathscr{X}$, then $x_1, \ldots, x_n$ is called an enumeration of $\mathscr{X}$. Naturally two denumerable sets $\mathscr{X}$ and $\mathscr{Y}$ have the same cardinality, since we can compose the appropriate one-to-one correspondences:

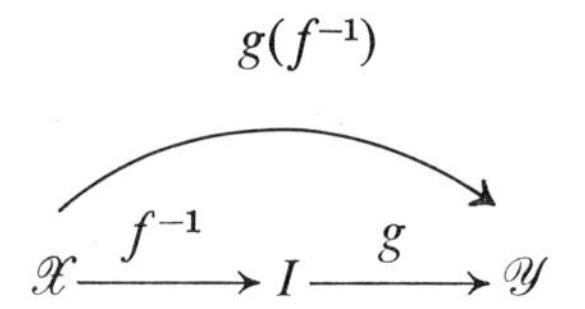

The denumerable sets are the smallest infinite sets in the sense that any infinite set has a denumerable subset. This is easily proved. Let $\mathscr{X}$ be an infinite set. Our task is to construct a function $f: I \to \mathscr{X}$ such that f is a one-to-one correspondence between I and the range of f. Select any member x_1 of $\mathscr{X}$ and define $f(1) = x_1$; then select any member x_2 of $\mathscr{X}$ such that $x_2 \neq x_1$ and define $f(2) = x_2$. After $x_1, \ldots, x_n$ have been decided on select any member x_{n+1} of $\mathscr{X}$ different from $x_1, \ldots, x_n$ and set $f(n + 1) = x_{n+1}$. This construction can always be carried out, since $\mathscr{X}$ has an infinite number of members, and so there are always additional members to select. Thus the desired function is defined inductively, and the statement is proved.

A set is said to be **countable** iff it is either finite or denumerable. Thus there are two kinds of countable sets, those which are finite and those which are infinite and denumerable. Countable sets are of fundamental importance in mathematics and probability theory. They are sometimes called **discrete** sets. Our most useful results from this chapter are that certain manipulations with countable sets give us new countable sets.

A set is said to be **uncountable** iff it is not countable. Thus a set is uncountable iff it is infinite but not denumerable. We will show in Sec. 4-3 that there are uncountable sets. Such sets are larger than the set I of positive integers in the sense that they cannot be placed in one-to-one correspondence with I. If uncountable sets did not exist, the theory of cardinality would be trivial.

Most of the foregoing comments and definitions concerning cardinality should seem reasonable, since they agree with our knowledge about the sizes of finite sets. A phenomenon which occurs only with infinite sets is that an infinite set can have the same cardinality as a proper subset of itself. For example, the set $\mathscr{X} = \{3, 4, 5, \ldots\}$ is denumerable even though it is a proper subset of $I = \{1, 2, 3, \ldots\}$. A one-to-one correspondence $f: I \to \mathscr{X}$ is defined by $f(n) = n + 2$ and is depicted in Fig. 1.

$$\begin{array}{cccccc} \mathscr{X} & 3 & 4 & 5 & 6 & \ldots \\ f\ \uparrow & \uparrow & \uparrow & \uparrow & \uparrow & \ldots \\ I & 1 & 2 & 3 & 4 & \ldots \end{array}$$

Fig. 1

$$\begin{array}{cccccc} \mathscr{Y} & 2 & 4 & 6 & 8 & \ldots \\ f\ \uparrow & \uparrow & \uparrow & \uparrow & \uparrow & \ldots \\ I & 1 & 2 & 3 & 4 & \ldots \end{array}$$

Fig. 2 $f(n) = 2n$.

Similarly the set $\mathscr{Y} = \{2, 4, 6, \ldots\}$ of all even integers is denumerable, as shown in Fig. 2. Also, the set of all integers is denumerable, as shown in Fig. 3.

$$\begin{array}{cccccccc} 0 & 1 & -1 & 2 & -2 & 3 & -3 & \ldots \\ \uparrow & \uparrow & \uparrow & \uparrow & \uparrow & \uparrow & \uparrow & \ldots \\ 1 & 2 & 3 & 4 & 5 & 6 & 7 & \ldots \end{array}$$

Fig. 3

The principal facts about countable sets needed for this book are collected into Theorem 1, which is stated, proved, and then used in examples.

Theorem 1

(*a*) A subset of a countable set is countable.
(*b*) If the domain of a function is countable its range is countable.
(*c*) The cartesian product of a finite number of countable sets is countable.
(*d*) The union of a finite or infinite sequence of countable sets is countable. ∎

Proof

(*a*) Let A be a subset of a countable set $\mathscr{X}$. If A is finite, then by definition A is countable. If A is infinite $\mathscr{X}$ must also be infinite, since a subset of a finite set is certainly finite. In this case $\mathscr{X}$ is denumerable, so we let $x_1, x_2, \ldots$ be an enumeration of $\mathscr{X}$. "Erase" any term in the sequence $x_1, x_2, \ldots$ if it does not belong to A. The remaining terms in their order give an enumeration of A so A is countable. ∎

For example, let $\mathscr{X}$ be the set of odd integers greater than 8 and let A consist of those members of $\mathscr{X}$ which are primes. A picture of the preceding proof is given in Fig. 4.

An enumeration of $\mathscr{X}$:	9, 11, 13, 15, 17, 19, 21, 23, 25, 27, ...
Resulting enumeration of A:	11, 13, 17, 19, 23, ...

Fig. 4

The preceding casual proof of Theorem 1*a* exhibits the main ideas of a more formal proof. We often prefer such a casual proof. A somewhat more formal proof of Theorem 1*a* can be given as follows: Let $f: I \to \mathscr{X}$ be a one-to-one correspondence between I and $\mathscr{X}$. Define $g(1)$ to equal the first member of A in the sequence $f(1), f(2), \ldots$. After $g(1), g(2), \ldots, g(n)$ have been defined, let $g(n+1)$ equal that member of A which is different from $g(1), g(2), \ldots, g(n)$ and appears earliest in the sequence $f(1), f(2), \ldots$. The function g defined inductively in this manner is a one-to-one correspondence between I and A.

(*b*) Let the domain $\mathscr{X}$ of the function $f: \mathscr{X} \to \mathscr{Y}$ be countable. If $\mathscr{X} = \{x_1, \ldots, x_n\}$ is finite the range of f is the set of different elements among $f(x_1), \ldots, f(x_n)$ and hence is finite. If $\mathscr{X}$ is denumerable it has an enumeration $x_1, x_2, \ldots$. In the sequence $f(x_1), f(x_2), \ldots$ erase any term which has appeared earlier in the sequence. The remaining terms in their order clearly form a finite or infinite sequence which is an enumeration of the range of f. ■

(*c*) Let $\mathscr{X}_1, \mathscr{X}_2, \ldots, \mathscr{X}_n$ be n countable sets. We wish to prove that $\mathscr{X}_1 \times \mathscr{X}_2 \times \cdots \times \mathscr{X}_n$ is countable. Let us assume that each $\mathscr{X}_k$ is denumerable, since the needed modifications to the proof are simple when some of the $\mathscr{X}_k$ are finite. We first show that $\mathscr{X}_1 \times \mathscr{X}_2$ is countable. Let $\alpha_1, \alpha_2, \ldots$ be an enumeration of $\mathscr{X}_1$ and $\beta_1, \beta_2, \ldots$ an enumeration of $\mathscr{X}_2$. The enumeration

$$(\alpha_1,\beta_1), (\alpha_1,\beta_2), (\alpha_2,\beta_1), (\alpha_3,\beta_1), (\alpha_2,\beta_2), (\alpha_1,\beta_3), \ldots$$

of $\mathscr{X}_1 \times \mathscr{X}_2$ results from following the arrows in Fig. 5. This technique is called the *Cauchy diagonal method*, or the first diagonal method.

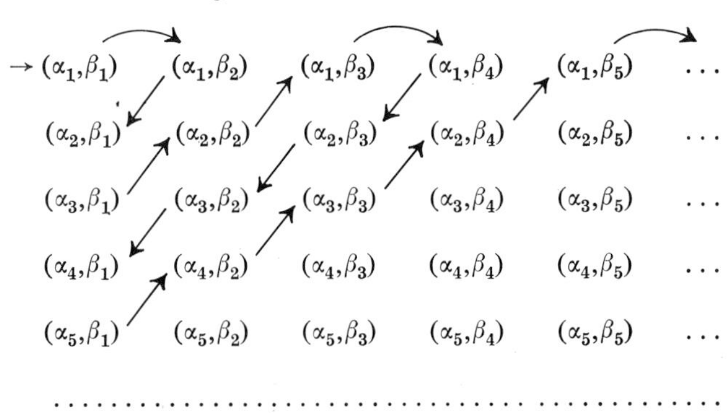

Fig. 5

Thus $\mathscr{X}_1 \times \mathscr{X}_2$ is countable. Now, $(\mathscr{X}_1 \times \mathscr{X}_2) \times (\mathscr{X}_3)$ is the cartesian product of two countable sets, and so it is countable. But $(\mathscr{X}_1 \times \mathscr{X}_2) \times (\mathscr{X}_3)$ is the same as $\mathscr{X}_1 \times \mathscr{X}_2 \times \mathscr{X}_3$ except for extra parentheses, so this later set is countable. Thus by induction Theorem 1*c* is proved. ■

(*d*) We prove the result in the case of an infinite sequence of denumerable sets. The modifications for some finiteness are simple. Let each of $\mathscr{X}_1, \mathscr{X}_2, \ldots$ be denumerable and let $\mathscr{X}$ be the union of these sets. We must show that $\mathscr{X}$ is denumerable.

Let[1] $x_{k1}, x_{k2}, \ldots$ be an enumeration of $\mathscr{X}_k$. Form the sequence $x_{11}, x_{12}, x_{21}, x_{31}, x_{22}, x_{13}, \ldots$ by following the arrows in Fig. 6.

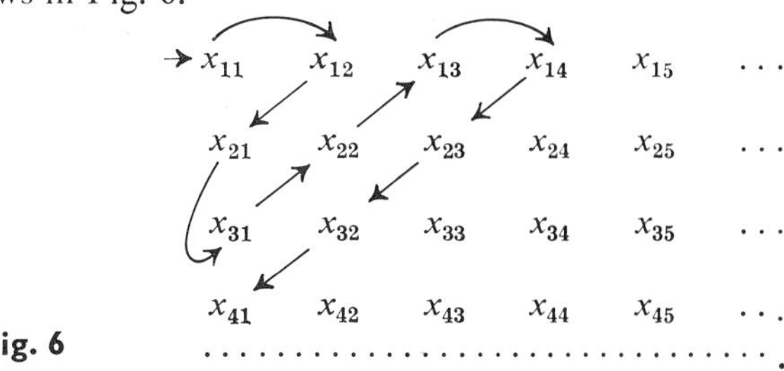

Fig. 6

Some of the $\mathscr{X}_k$ may have members in common, but by erasing those terms which have already appeared we get an enumeration of $\mathscr{X}$. Theorem 1*d* is proved, so all of Theorem 1 has been proved. ■

Example 1 By Theorem 1*c* the set $I \times I$ of ordered pairs (m,n) of positive integers is countable. This set is shown in Fig. 7. The Cauchy diagonal method of enumeration is easily visualized for this set. Incidentally, as shown in Exercise 1 (Sec. 4-3), *the cartesian product of a denumerable number of finite sets may be uncountable.* ■

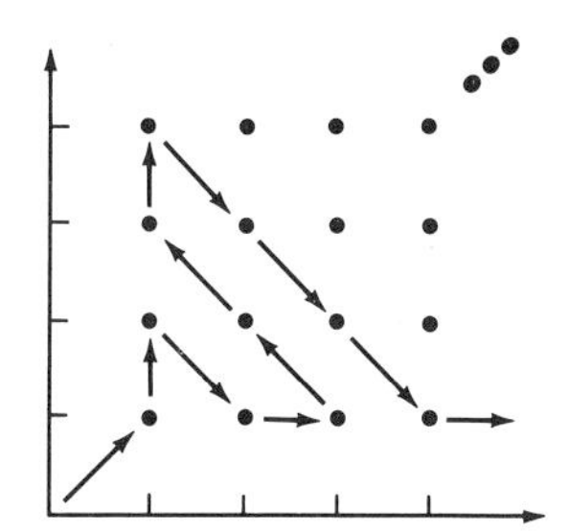

Fig. 7

Example 2 The set $\{-1,-2,-3,\ldots\}$ of all negative integers is obviously denumerable, so by Theorem 1*d* we know that the union of the three sets $\{-1, -2, \ldots\}$, $\{0\}$, and $\{1, 2, \ldots\}$ is denumerable. But this is just the set of all integers, which we saw earlier to be denumerable. ■

A **rational number** is just a real number which can be expressed as a ratio m/n of two integers. We consider $3/5$ and $6/10$ to be equal, so there are many representations for the same rational. A unique representation can be obtained by requiring that the numerator and denominator have no common factors. Rationals can also be described as those reals whose infinite decimal expansions eventually become repeating.

Example 3 We can define a real-valued function $f: I \times I \to R$ whose domain is the countable set $I \times I$ by $f(m,n) = m/n$. The range of this function is clearly the set of all positive rational numbers, which by Theorem 1*b* must be countable. Clearly *the set of all rationals*—positive, negative, or zero—*is countable.* The rationals are an example of a denumerable subset R' of the reals R with the property that for any member x of R there are members of R' which are arbitrarily close to x. An enumeration of the positive rationals cannot be made in the same

[1] The jth term of this sequence is $x_{k,j}$ but we omit the comma and abbreviate to x_{kj} when no confusion is likely. Similarly x_{43} in Fig. 6 is an abbreviation for $x_{4,3}$.

order in which the rationals appear in the real number axis, since for a given rational there is no "next larger rational." Thus any enumeration of the rationals must hop back and forth on the number axis, as illustrated in Exercise 1. ■

Example 4 For any integer $n \geq 2$ let $\mathscr{X}_n$ be the set of all n-tuples of integers. By Theorem 1*d* the set $\mathscr{X} = \mathscr{X}_2 \cup \mathscr{X}_3 \cup \cdots$ *of all tuples of integers is countable.* If $(a_1, a_2, \ldots, a_n)$ is a member of $\mathscr{X}$, we form the polynomial

$$a_n x^{n-1} + a_{n-1} x^{n-2} + \cdots + a_2 x + a_1$$

and let $A_{(a_1,\cdots,a_n)}$ be the set of all real roots of this polynomial. Since each of these sets is finite and there are only countably many of them, we know that their union is countable. This proves that *the set of all algebraic numbers is countable.* A real number x_0 is said to be **algebraic** iff there exists a polynomial $a_n x^{n-1} + a_{n-1} x^{n-2} + \cdots + a_2 x + a_1$ with *integer* coefficients which has x_0 as a root, that is, $a_n x_0^{n-1} + a_{n-1} x_0^{n-2} + \cdots + a_2 x_0 + a_1 = 0$. The rational number m/n is the root of the polynomial $nx + (-m)$, so every rational number is algebraic. However, many algebraic numbers, such as $\sqrt{2}$ which is a root of $x^2 - 2$, are not rational. ■

EXERCISES

1. The positive rationals can be enumerated by following the arrows in the array of Fig. 8

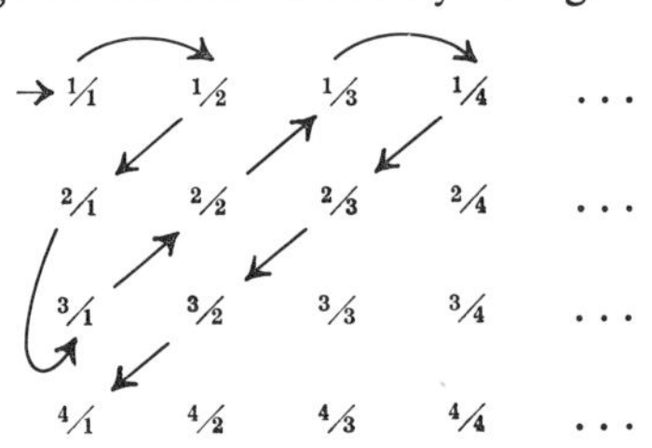

Fig. 8

to obtain the sequence $1/1, 1/2, 2/1, 3/1, 2/2, 1/3, 1/4, \ldots$ and then erasing terms which have appeared previously to obtain $1, 1/2, 2, 3, 1/3, 1/4, \ldots$. The enumeration can then be indicated on the real axis, as in Fig. 9, where for the kth term a_k of the sequence $a_1, a_2, \ldots$ we

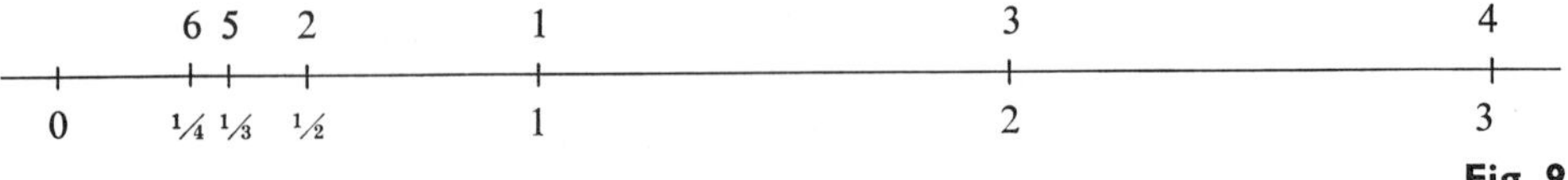

Fig. 9

write the number k above the point a_k on the real axis. Construct such a picture out to the appearance of $2/5$ in the enumeration.

2. If $r > 0$ and x and y are three reals we call the subset

$$\{(x',y'): (x' - x)^2 + (y' - y)^2 < r^2\}$$

of R^2 the *open disk* of radius r and center (x,y). We call this set a *rational open disk* iff r, x, and y are three rational numbers. Prove that the collection of all rational open disks is countable.

3. Let $F: R \to R$ satisfy $0 \le F(x) \le 1$ for all x. Assume that F is nondecreasing, that is, that $F(x) \le F(x')$ whenever $x < x'$. We say that F has a *positive jump* at x_0 iff

$$\lim_{\substack{\epsilon \to 0 \\ \epsilon > 0}} F(x_0 - \epsilon) < \lim_{\substack{\epsilon \to 0 \\ \epsilon > 0}} F(x_0 + \epsilon).$$

Prove that the set of reals at which F has a positive jump is countable.

4. Prove that if A is a countable subset of the domain $\mathscr{X}$ of a function $f: \mathscr{X} \to \mathscr{Y}$ then the set of all y such that $y = f(x)$ for some x in A is countable, that is, that the set of all images of points in A is countable.

5. Prove that if $\mathscr{X}$ is *any* infinite set then there is a one-to-one correspondence $f: \mathscr{X} \to \mathscr{X}'$ between $\mathscr{X}$ and some proper subset $\mathscr{X}'$ of $\mathscr{X}$. HINT: Let D be a denumerable subset of $\mathscr{X}$ and let $f(x) = x$ whenever x belongs to $\mathscr{X}$ but not to D. Now "appropriately" define $f(x)$ for x in D.

6. Prove that the addition of a countable set does not affect the cardinality of an infinite set; that is, if $\mathscr{X}$ is any infinite set and C is any countable set then $\mathscr{X}$ and $\mathscr{X} \cup C$ have the same cardinality. Note that this result implies in particular that if $a < b$ then adding one or both end points to the interval $a < x < b$ of reals does not change its cardinality. HINT: Start as in Exercise 5.

7. Let $A_n = \{x: x \text{ is real and } x \ge n\}$. It will be shown in Sec. 4-3 that every A_n is uncountable.

(*a*) Is the family $\{A_1, A_2, \ldots\}$ a countable family?

(*b*) Is $\bigcup_{n=1}^{\infty} A_n$ a countable set?

8. For this exercise let (a,b) denote the interval $a < x < b$ of reals. The three functions of Fig. 10 show that: the interval $(0,1)$ has the same cardinality as the interval (a,b), where $a < b$; the interval $(0,1)$ has the same cardinality as the set of all positive reals; and the interval $(0,1)$ has the same cardinality as the set of all reals. Show by a similar construction that the unit square $(0,1) \times (0,1)$ has the same cardinality as the plane R^2.

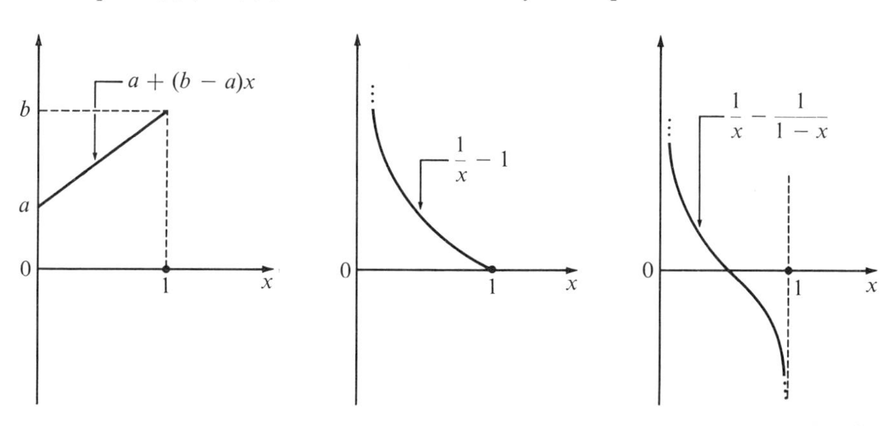

Fig. 10

4-3 AN UNCOUNTABLE SET

We wish to prove that some sets of real numbers are uncountable. In order to simplify the mechanics of a proof we select the **interval** $(0,1]$ of all reals x satisfying $0 < x \le 1$ and prove that it is uncountable. We first need a way to represent these numbers.

If x belongs to $(0,1]$ then x has an infinite decimal expansion $.b_1b_2b_3\cdots$; that is, the number x has an expansion in an infinite series

$$x = \frac{b_1}{10} + \frac{b_2}{10^2} + \frac{b_3}{10^3} + \cdots$$

where each b_k is one of $0, 1, 2, \ldots, 9$. Now unfortunately those numbers which are multiples of a power of .1 have two such expansions. For example, $.37000\cdots = .36999\cdots$. That is, using equation 3 of Appendix 1, we obtain

$$\begin{aligned}\frac{3}{10} + \frac{6}{10^2} + \frac{9}{10^3} + \frac{9}{10^4} + \cdots &= \frac{3}{10} + \frac{6}{10^2} + \frac{9}{10^3}\left(1 + \frac{1}{10} + \frac{1}{10^2} + \cdots\right) \\ &= \frac{3}{10} + \frac{6}{10^2} + \frac{9}{10^3}\frac{1}{1 - 1/10} \\ &= \frac{3}{10} + \frac{7}{10^2} + \frac{0}{10^3} + \frac{0}{10^4} + \cdots.\end{aligned}$$

For such a case we will use only the expansion $.36999\cdots$. It is a fact that every real in the interval $(0,1]$ has a *unique* decimal expansion $.b_1b_2b_3\cdots$ with the property that an infinite number of the b_i are nonzero. That is, if a decimal expansion is zero after some point we use the other decimal expansion with all 9's after a point. *We will consider the set* $(0,1]$ *of reals to be just the set of such decimal expansions.*

We wish to prove that the interval $(0,1]$ *is an uncountable set*, that is, that there cannot exist a one-to-one correspondence between $I = \{1, 2, \ldots\}$ and $(0,1]$. We do this by showing that if we are given *any* function $f\colon I \to (0,1]$ then by appropriate manipulations performed on this function we can construct a real number x_0 belonging to $(0,1]$ such that none of the real numbers $f(1), f(2), \ldots$ is equal to x_0. This shows that f is not onto $(0,1]$. Since this technique can be applied to any $f\colon I \to (0,1]$, we can conclude that $(0,1]$ is uncountable.

Given any $f\colon I \to (0,1]$, we consider the real numbers $f(1), f(2), \ldots,$ where each $f(n) = .b_{n1}b_{n2}b_{n3}\cdots$ is an infinite decimal expansion. We will show how to construct an infinite decimal expansion $.b_1b_2\cdots$ which is different from every one of $f(1), f(2), \ldots$. For each n select some b_n different from b_{nn}. For example, let

$$b_n = \begin{cases} 1 & \text{if } b_{nn} \neq 1 \\ 2 & \text{if } b_{nn} = 1. \end{cases}$$

That is, consider each $f(n)$ written out as a decimal expansion

$$\begin{aligned} f(1) &= .\mathbf{b}_{11}b_{12}b_{13}b_{14}\cdots \\ f(2) &= .b_{21}\mathbf{b}_{22}b_{23}b_{24}\cdots \\ f(3) &= .b_{31}b_{32}\mathbf{b}_{33}b_{34}\cdots \\ f(4) &= .b_{41}b_{42}b_{43}\mathbf{b}_{44}\cdots \\ &\cdots\cdots\cdots\cdots\cdots\cdots \end{aligned}$$

and concentrate attention on the main diagonal $b_{11}, b_{22}, b_{33} \ldots$. We have defined b_n in a fairly arbitrary way, but in such a way that $b_1 \neq b_{11}, b_2 \neq b_{22}, b_3 \neq b_{33}, \ldots$. Also $.b_1b_2 \cdots$ is not zero after some point. For example, if

$$f(1) = .35721 \cdots$$
$$f(2) = .52998 \cdots$$
$$f(3) = .951005 \cdots$$
$$f(4) = .0000015 \cdots$$
$$\cdots\cdots\cdots\cdots\cdots\cdots$$

then $.b_1b_2b_3 \cdots = .1121 \cdots$. Clearly for each integer n we have $.b_1b_2b_3 \cdots \neq f(n)$, since b_n differs from the nth digit b_{nn} in the expansion of $f(n)$. Thus *we have proved that* $(0,1]$ *is an uncountable set*. This method of proof is called the *Cantor diagonal method*, or the second diagonal method.

As indicated by Exercise 2, every R^n with $n \geq 1$ has the same cardinality. Thus the real line R has just as many points as does three-dimensional space R^3. From Example 3 (Sec. 2-2) we know that if $A = \{a_1, a_2, \ldots, a_n\}$ is a finite set of n distinct objects and B is the class of all subsets of A then B has 2^n members. Exercise 4 shows that if A is *any* nonempty set and B is the class of all subsets of A then A and B do not have the same cardinality. This, together with other facts in the general theory, justifies the statement that there is no largest infinite set.

EXERCISES

1. The cartesian product $\mathscr{X}_1 \times \mathscr{X}_2 \times \cdots$ of a sequence $\mathscr{X}_1, \mathscr{X}_2, \ldots$ of sets is just the set of sequences $x_1, x_2, \ldots$ such that each x_i belongs to $\mathscr{X}_i$. Thus if $\{0, 1, \ldots, 9\} = \mathscr{X}_1 = \mathscr{X}_2 = \cdots$ then a member of $\mathscr{X}_1 \times \mathscr{X}_2 \times \cdots$ looks like $(3, 5, 0, 9, 8, \ldots)$. Show by an example that Theorem 1*c* (Sec. 4-2) cannot be extended to an infinite sequence of finite sets.

2. The purpose of this exercise is to prove that *the interval* $(0,1]$ *has the same cardinality as the unit square* $(0,1] \times (0,1]$. Note that we want a function $f(x) = (y,z)$, and if we use decimal expansions we might define f by $f(.d_1d_2d_3d_4 \cdots) = (.d_1d_3d_5d_7 \cdots, .d_2d_4d_6d_8 \cdots)$. However $.d_1d_3d_5 \cdots$ might have only zeros after some point. If we divide $.d_1d_2d_3 \cdots$ into adjacent blocks such that each block contains only one nonzero digit we can define $f(x)$ as in Fig. 1. Is f as so defined a one-to-one correspondence between $(0,1]$ and $(0,1] \times (0,1]$?

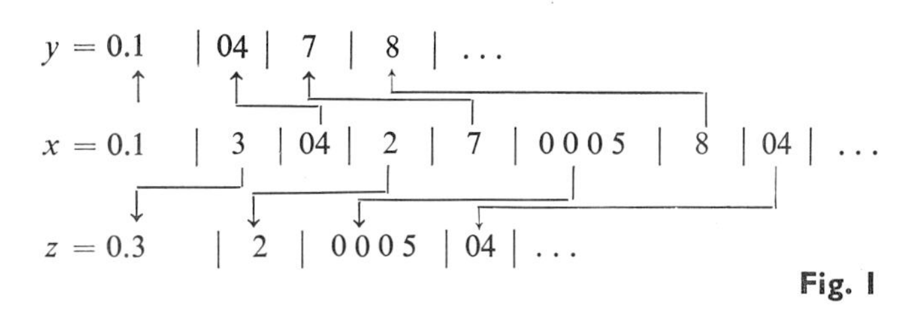

Fig. 1

3. The set R' of rational numbers was defined just prior to Example 3 (Sec. 4-2) and the set A of algebraic numbers was defined in Example 4 (Sec. 4-2). We clearly have $R' \subset A \subset R$. A real number is said to be *irrational* iff it does not belong to R' and *transcendental* iff it does not belong to A. Prove that there are transcendental real numbers.

COMMENTS: The numbers $\pi = 3.141 \cdots$ and $e = 2.718 \cdots$ are transcendental, but it is very difficult to prove that they are. The number $\sqrt{2}$ is clearly algebraic, and it is easy to prove that $\sqrt{2}$ is irrational, as follows. If a rational number $x = m/n$, where m and n are integers, satisfies the equation $x^2 = 2$, then $m^2 = 2n^2$. Using the factorizations of $m = p_1p_2 \cdots p_s$ and $n = q_1q_2 \cdots q_t$ into products of primes, we have

$$p_1p_1p_2p_2 \cdots p_sp_s = 2q_1q_1q_2q_2 \cdots q_tq_t.$$

But this equation says that the same integer can be factored into primes in two different ways, one containing an even number of factors equal to 2 and the other containing an odd number of factors equal to 2. Thus the assumption that a rational number x can satisfy the equation $x^2 = 2$ leads to a contradiction with the unique-prime-factorization theorem, and so $\sqrt{2}$ must be irrational.

4. The purpose of this exercise is to prove that *if A is any nonempty set*, finite or infinite, *and B is the class of all subsets of A then A and B do not have the same cardinality*. The proof consists in showing that if we are given *any* function $f: A \to B$, then by appropriate manipulations on this function we can construct a member b_0 of B such that $b_0 \neq f(x)$ holds for every x in A. For each x in A the function f assigns to x a subset $f(x)$ of A, and the point x may or may not happen to belong to the subset $f(x)$ assigned to it. Let b_0 denote the set of all those x such that x does not belong to $f(x)$.

(*a*) Exhibit an example of a finite set A and a function $f: A \to B$. Exhibit b_0 and note that $b_0 \neq f(x)$ holds for every x in A.

(*b*) Prove in general that $b_0 \neq f(x)$ holds for every x in A.

5. The purpose of this exercise is to show that *an interval of real numbers can be broken up into* (that is, is the union of) *an uncountable collection of disjoint subsets, with each subset being uncountable*. This exercise has relevance for Exercise 8 (Sec. 17-1). Let (0,1] denote the unit interval $0 < x \leq 1$ and represent any x in (0,1] by its binary expansion:

$$x = .x_1x_2x_3 \cdots = \sum_{n=1}^{\infty} \frac{x_n}{2^n} \qquad \text{each } x_i = 0 \text{ or } 1.$$

To make the expansion unique we may assume that an infinite number of the x_i are equal to 1, since $.01000 \cdots = .00111 \cdots$. The uncountably many points of the unit interval are thus placed into one-to-one correspondence with the collection of all infinite sequences of zeros and ones having infinitely many ones. For any real p satisfying $0 < p < 1$ let A_p consist of those members $x = .x_1x_2 \cdots$ in (0,1] for which

$$\lim_{n\to\infty} \frac{1}{n} \sum_{i=1}^{n} x_i = p.$$

That is, A_p consists of those x which have, asymptotically, fraction p of ones in their binary expansion. A sequence can have at most one limit, so that if $p \neq p'$ then A_p and $A_{p'}$ are disjoint. Clearly each A_p is nonempty. If, incidentally, we let B consist of those x for which the limit above is either 0 or 1 or does not exist, then every x in (0,1] belongs to B or to exactly one A_p. Prove that every A_p is uncountable.

HINT: Fix p with $0 < p < 1$. We wish to show how to assign to every x in (0,1] a $y = f_p(x)$ belonging to this A_p in such a way that f_p is a one-to-one function, hence A_p contains an uncountable subset. To define f_p we take any $z = .z_1z_2 \cdots$ belonging to this A_p. Now define $y = f_p(x)$ by utilizing z as follows:

$$y = .\underbrace{z_1}_{1}\, x_1\, \underbrace{z_2\, z_3}_{2}\, x_2\, \underbrace{z_4\, z_5\, z_6}_{3}\, x_3\, \underbrace{z_7\, z_8\, z_9\, z_{10}}_{4}\, x_4 \cdots$$

The idea is to insert the digits of $x = .x_1x_2 \cdots$ into fixed locations between the digits of $z = .z_1z_2 \cdots$ but to make these insertions at such a decreasingly slow rate that the asymptotic fraction of ones in y will be the same as the asymptotic fraction of ones in z regardless of which x is used. Clearly f_p as so defined is one-to-one, so we need only show that every such y belongs to A_p. For any $n = 1, 2, \ldots$ let k_n be the number of locations among the first n digits of y which are assigned to x. Thus $k_1 = 0$, $k_2 = k_3 = k_4 = 1$, $k_5 = k_6 = k_7 = k_8 = 2$, $k_9 = \cdots k_{13} = 3, \ldots$, and clearly k_n/n goes to zero as n goes to infinity. Complete the proof.

5

Absolutely Convergent Series

In this chapter we first review the standard definitions concerning the convergence of sequences and series. We then state and discuss two theorems which state in effect that for *absolutely convergent series* there is the same freedom in changing the order of terms and interchanging the order of summation as is available for finite sums. The material of this chapter is not really needed until Chap. 9.

5-1 REVIEW OF SERIES

This section is a brief review of some definitions concerning infinite series. If $A_1, A_2, \ldots$ is a sequence of real numbers and A is a real number we say that the sequence **converges to** A iff for every $\epsilon > 0$ there is an integer N such that for all integers $n \geq N$ we have $|A_n - A| \leq \epsilon$. In this case we also write $\lim_{n\to\infty} A_n = A$, or $A_n \to A$ as $n \to \infty$. That is, regardless of how small a tolerance band $A \pm \epsilon$ is specified, as in Fig. 1, there is a point N such that every A_n beyond this point is within the tolerance band. That is, regardless of how stringent our interpretation of the word "close," eventually A_n will become and remain close to A.

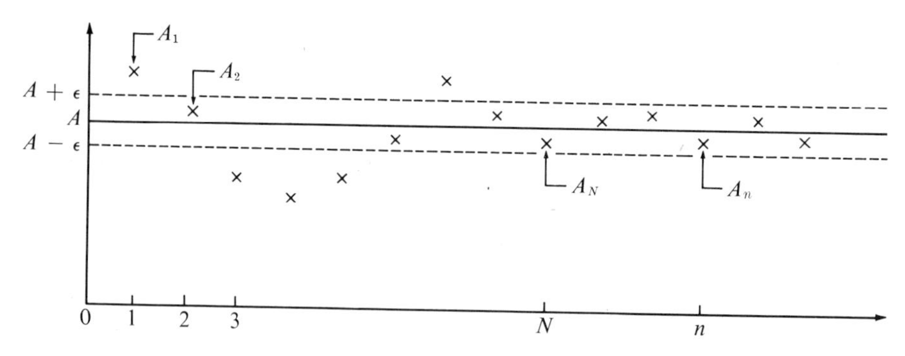

Fig. 1

If $A_1, A_2, \ldots$ is a sequence of real numbers we say that the sequence **converges to infinity** iff for every real number L there is an integer N such that for all $n \geq N$ we have $A_n \geq L$. In this case we also write $\lim_{n\to\infty} A_n = \infty$, or $A_n \to \infty$ as $n \to \infty$. We do *not* consider infinity or minus infinity to be real numbers. As shown in Fig. 2, regardless of how large a number L we select there is a point N such that every A_n beyond this point is greater than L. Thus regardless of how large we interpret the word "large" to be, eventually A_n will become and remain large. Naturally we say that a sequence $A_1, A_2, \ldots$ **converges to minus infinity** iff for every real number L there is an integer N such that for all $n \geq N$ we have $A_n \leq L$. In this case we also write $\lim_{n\to\infty} A_n = -\infty$, or $A_n \to -\infty$ as $n \to \infty$.

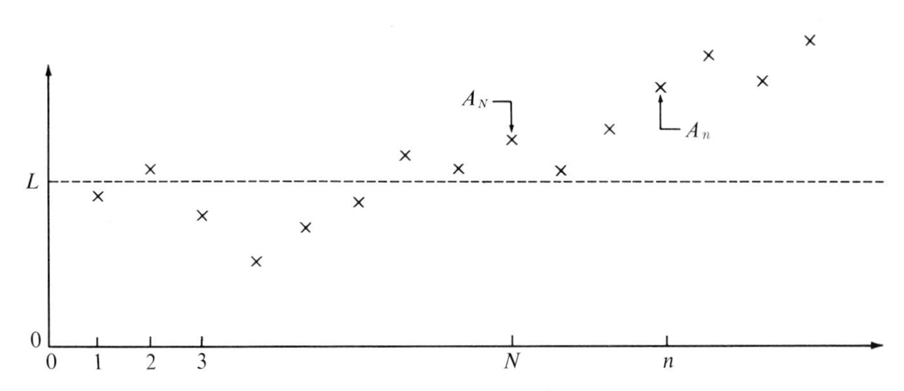

Fig. 2

Let $a_1 + a_2 + \cdots$ be an infinite series and let $A_1, A_2, \ldots$ be the corresponding sequence of partial sums, where $A_n = a_1 + a_2 + \cdots + a_n$. The series $a_1 + a_2 + \cdots$ is said to **converge to the real number** A (or to converge to infinity or to minus infinity) iff the sequence $A_1, A_2, \ldots$ converges to A (or to infinity or to minus infinity). Actually by an infinite series $a_1 + a_2 + \cdots$ all we usually mean is a sequence $a_1, a_2, \ldots$ whose terms we intend to add. We adhere to the poor tradition of using the symbol $\sum_{n=1}^{\infty} a_n$ both for the series and for its sum when it has one. Hopefully the context will make the proper interpretation of $\sum_{n=1}^{\infty} a_n$ clear.

Given a sequence $A_1, A_2, \ldots,$ it is either true or false that there exists a real number B such that $|A_n| \leq B$ for all $n = 1, 2, \ldots$. If the statement is true we say that the sequence is **bounded,** and if the statement is false we say that the sequence is **unbounded.** Thus a sequence $A_1, A_2, \ldots$ of reals is said to be bounded iff there is a number B such that $|A_n| \leq B$ for all $n = 1, 2, \ldots$. A sequence $A_1, A_2, \ldots$ of reals is said to be unbounded iff it is not bounded, that is, iff *for every* B there is an n such that $|A_n| > B$.

A sequence $A_1, A_2, \ldots$ is said to be **nondecreasing (nonincreasing)** iff $A_n \leq A_{n+1} (A_n \geq A_{n+1})$ for all n.

Some obvious relationships between classes of sequences are indicated in Fig. 3. For example, if a sequence converges to a real number it must be bounded, so the class of all sequences which converge to some real number must be a subset of the class of all bounded sequences. A basic property of the set of real numbers is that a *nondecreasing (nonincreasing) sequence of real numbers either converges to infinity (minus infinity) or is bounded and converges to a real number.*

EXERCISE

1. Figure 3 partitions the class of all infinite sequences of real numbers into 10 subclasses. Subclass 7, for example, consists of those sequences which converge to infinity but are not

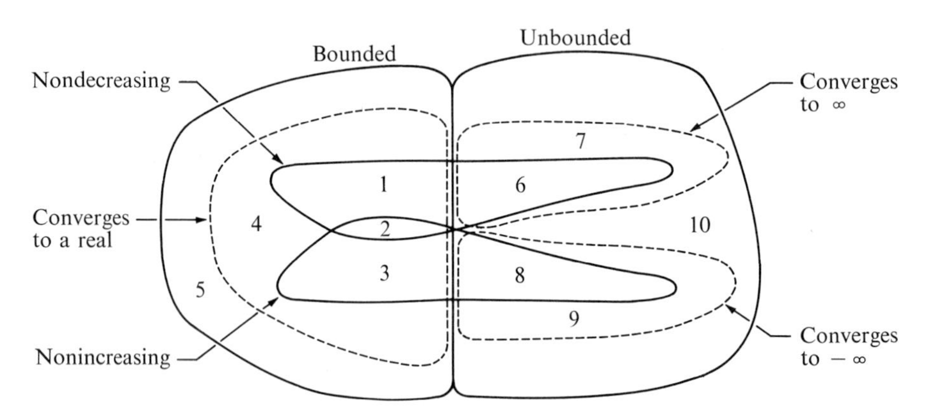

Fig. 3

nondecreasing. Thus the sequence 0, 3, 2, 5, 4, 7, . . . , $n + (-1)^n$, . . . belongs to subclass 7. Exhibit one member of each of the other nine subclasses.

5-2 ABSOLUTELY CONVERGENT SERIES

Two standard theorems about absolutely convergent series are recalled in this section and the next one.

We wish to define when one sequence $b_1, b_2, \ldots$ is considered to be a reordering of another sequence $a_1, a_2, \ldots$. We have in mind that each term in the sequence $a_1, a_2, \ldots$ appears exactly once in the sequence $b_1, b_2, \ldots$. Now if $a_1 = a_5 = -1.2$ and no other a_i equals -1.2 then we expect -1.2 to occur *twice* in $b_1, b_2, \ldots$. What concerns us here is not so much what the a_i numbers are, but the fact that if $b_1, b_2, \ldots$ is written out like $a_7, a_3, a_{12}, a_1, a_5, \ldots$ every positive integer should appear exactly once as a subscript. If $g: I \to I$ is a one-to-one correspondence between the set I of positive integers and itself then the sequence $a_{g(1)}, a_{g(2)}, \ldots$ is called a **reordering** of the sequence $a_1, a_2, \ldots$. Thus $g(1), g(2), \ldots$ contains each positive integer exactly once, and we form the sequence $b_1, b_2, \ldots$, where $b_n = a_{g(n)}$.

In Sec. 4-1 it was shown that the relationship of one-to-one correspondence is reflexive, symmetric, and transitive. Because of this the same properties hold for reordering. Thus a sequence $a_1, a_2, \ldots$ is a *reordering of itself* [since we can let $g(n) = n$ for all n]. *If* $b_1, b_2, \ldots$ is a reordering of $a_1, a_2, \ldots$ *then* $a_1, a_2, \ldots$ is a reordering of $b_1, b_2, \ldots$ (since we can use g^{-1}). Also *if* $b_1, b_2, \ldots$ is a reordering of

$a_1, a_2, \ldots$ *and* $c_1, c_2, \ldots$ is a reordering of $b_1, b_2, \ldots$ *then* $c_1, c_2, \ldots$ is a reordering of $a_1, a_2, \ldots$.

The series $b_1 + b_2 + \cdots$ is said to be a **reordering** of the series $a_1 + a_2 + \cdots$ iff the sequence $b_1, b_2, \ldots$ is a reordering of the sequence $a_1, a_2, \ldots$.

A series $a_1 + a_2 + \cdots$ is said to be **absolutely convergent** iff there is a number B such that $|a_1| + |a_2| + \cdots + |a_n| \leq B$ for all n. Theorem 1 relates reordering and absolute convergence. The theorem is stated and then discussed; its proof is left to Appendix 3.

Theorem 1 Every reordering of an absolutely convergent series is absolutely convergent and converges to the same real number. ■

Those series which always converge to the same real number regardless of reordering are precisely those series which are absolutely convergent. This property makes absolutely convergent series of special interest to us.

Example 1 The following series is not absolutely convergent, and its limiting behavior *is* affected by reordering. Let $a_1 + a_2 + \cdots$ denote the series which is completely defined by the array below in the sense that the number a_i is uniquely determined for each i (for example, $a_{130} = -1/100$):

$$\begin{aligned}
&1 - 1\\
&+ \underbrace{\frac{1}{10} + \frac{1}{10} + \cdots + \frac{1}{10}}_{10 \text{ terms}} \underbrace{- \frac{1}{10} - \frac{1}{10} - \cdots - \frac{1}{10}}_{10 \text{ terms}}\\
&+ \underbrace{\frac{1}{100} + \frac{1}{100} + \cdots + \frac{1}{100}}_{100 \text{ terms}} \underbrace{- \frac{1}{100} - \frac{1}{100} - \cdots - \frac{1}{100}}_{100 \text{ terms}} \qquad (1)\\
&\cdots\cdots\cdots\cdots\cdots\cdots\cdots\cdots\cdots\cdots\\
&+ \underbrace{\frac{1}{10^k} + \cdots + \frac{1}{10^k}}_{10^k \text{ terms}} \underbrace{- \frac{1}{10^k} - \cdots - \frac{1}{10^k}}_{10^k \text{ terms}}\\
&\cdots\cdots\cdots\cdots\cdots\cdots\cdots\cdots\cdots\cdots
\end{aligned}$$

Let $A_n = a_1 + a_2 + \cdots + a_n$, so that $A_1, A_2, \ldots$ is the sequence of partial sums. This sequence $A_1, A_2, \ldots$ is plotted in Fig. 1, and clearly it fluctuates forever between 0 and 1.

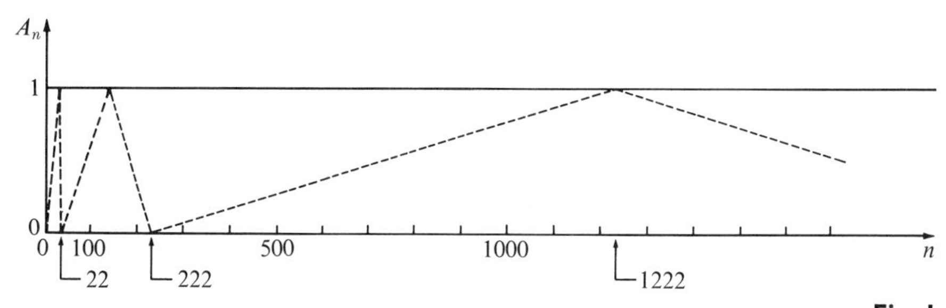

Fig. 1

In particular $A_n = 0$ iff $n = 2, 22, 222, 2222, \ldots$. Similarly $A_n = 1$ iff $n = 1, 12, 122, 1222, 12222, \ldots$. The series (2) below clearly converges to zero, and clearly series (2) is a reordering of the nonconvergent series (1).

$$
\begin{aligned}
&1-1\\
&+\underbrace{\frac{1}{10}-\frac{1}{10}+\frac{1}{10}-\cdots-\frac{1}{10}}_{20\text{ terms}}\\
&+\underbrace{\frac{1}{100}-\frac{1}{100}+\frac{1}{100}-\cdots-\frac{1}{100}}_{200\text{ terms}}\\
&\cdots\cdots\cdots\cdots\cdots\cdots\cdots\cdots\\
&+\underbrace{\frac{1}{10^k}-\frac{1}{10^k}+\frac{1}{10^k}-\frac{1}{10^k}+\cdots-\frac{1}{10^k}}_{2\times 10^k\text{ terms}}\\
&\cdots\cdots\cdots\cdots\cdots\cdots\cdots\cdots \quad \blacksquare
\end{aligned}
\tag{2}
$$

Separate consideration of the nonnegative and negative terms of a series is a technique for understanding the possible effects of reordering on the limiting behavior. If a series has only a finite number of nonnegative terms, it converges to a real or to minus infinity and its limit is unaffected by reordering. If a series has only a finite number of negative terms it converges to a real or to infinity and its limit is unaffected by reordering. If a series $a_1 + a_2 + \cdots$ has an infinite number of both nonnegative and negative terms it is possible to extract these as two series, $p_1 + p_2 + \cdots$ and $q_1 + q_2 + \cdots$. For example, if $a_1 + a_2 + \cdots$ is $1 - 1 + \frac{1}{2} - \frac{1}{2} + \frac{1}{3} - \frac{1}{3} + \cdots$ then $p_1 + p_2 + \cdots$ is $1 + \frac{1}{2} + \frac{1}{3} + \cdots$ and $q_1 + q_2 + \cdots$ is $(-1) + (-\frac{1}{2}) + (-\frac{1}{3}) + \cdots$. If $p_1 + p_2 + \cdots$ converges to the real number p and $q_1 + q_2 + \cdots$ converges to the real number q then all reorderings of $a_1 + a_2 + \cdots$ converge to $p + q$, and this is the case which exactly corresponds to absolute convergence. If $p_1 + p_2 + \cdots$ converges to the real number p and $q_1 + q_2 + \cdots = -\infty$ then all reorderings of $a_1 + a_2 + \cdots$ converge to minus infinity, since the nonnegative terms can contribute at most p to a partial sum, and the negative terms must eventually overwhelm the partial sums. Similarly if $p_1 + p_2 + \cdots = \infty$ and $q_1 + q_2 + \cdots$ converges to the real number q then all reorderings of $a_1 + a_2 + \cdots$ converge to infinity. Thus only in the case where $p_1 + p_2 + \cdots = \infty$ and $q_1 + q_2 + \cdots = -\infty$, as in Example 1, can reordering $a_1 + a_2 + \cdots$ affect the limit. In this case different limiting behaviors can always be obtained. For example, we can always construct a reordering $b_1 + b_2 + \cdots$ of $a_1 + a_2 + \cdots$ such that $b_1 + b_2 + \cdots = \infty$ by properly making the appearance of negative terms among $b_1 + b_2 + \cdots$ rare enough.

EXERCISE

1. Find a reordering converging to infinity for series (1).

5-3 ITERATED SUMMATION

A **double series** $\sum_{n=1}^{\infty}\sum_{m=1}^{\infty} a_{nm}$ is said to be **absolutely convergent** iff there is a number B such that $\sum_{n=1}^{N}\sum_{m=1}^{M} |a_{nm}| \leq B$ for all N and M; that is, if the terms a_{nm} are thought of as forming an infinite matrix, as in Fig. 1, then the sum of the absolute vaules of the terms in any upper-left rectangle is at most B. Our interpretation of the **sum** $\sum_{n=1}^{\infty}\sum_{m=1}^{\infty} a_{nm}$ **of an absolutely convergent double series** is that the terms are given an arbitrary order, as in Fig. 1 or in Fig. 2, and then added in that order, just as for a single-index series. More formally, we select any one-to-one correspondence $g: I \to (I \times I)$ so that every ordered pair of positive integers appears exactly once in the sequence $g(1)$, $g(2)$, $g(3)$, . . . , where $g(k) = (n,m)$ is the kth ordered pair. We then define the sum of the double series to be equal to

$$a_{g(1)} + a_{g(2)} + a_{g(3)} + \cdots.$$

Theorem 1 (Sec. 5-2) easily applies to show that *an absolutely convergent double series does converge to a real number, and this limit is independent of the ordering used.*

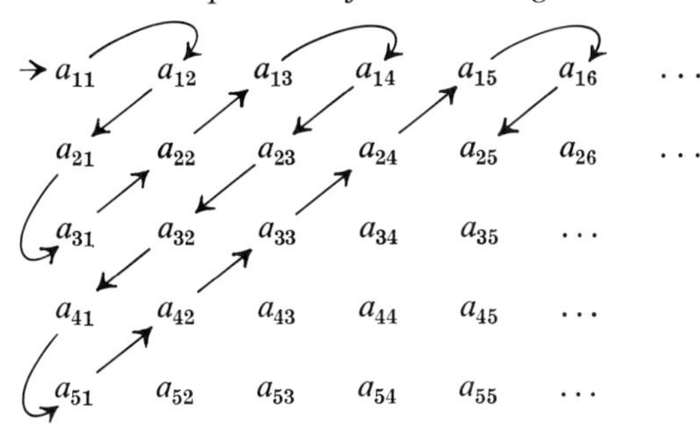

Fig. 1

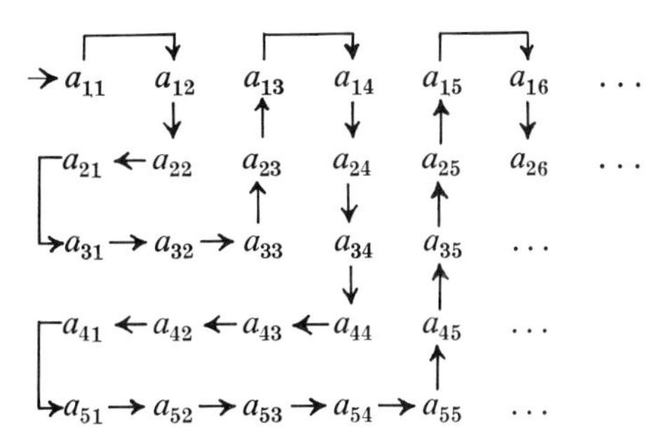

Fig. 2

Example 1 Let a_{nm} be the number in the nth row and mth column of the array of Fig. 3, where $a_{nm} = (.1)^m$ if $n \leq m$, that is, if a_{nm} is on or above the main diagonal, and $a_{nm} = -(.1)^n$ if $n > m$, that is, if a_{nm} is below the main diagonal.

$$\begin{array}{ccccc}
(.1) & (.1)^2 & (.1)^3 & (.1)^4 & \cdots \\
-(.1)^2 & (.1)^2 & (.1)^3 & (.1)^4 & \cdots \\
-(.1)^3 & -(.1)^3 & (.1)^3 & (.1)^4 & \cdots \\
-(.1)^4 & -(.1)^4 & -(.1)^4 & (.1)^4 & \cdots \\
\cdots & \cdots & \cdots & \cdots & \cdots
\end{array}$$

Fig. 3

We first show that this double series is absolutely convergent. Considering an upper-left square with N terms on a side, we see that

$$\begin{aligned}
\sum_{n=1}^{N}\sum_{m=1}^{N}|a_{nm}| &= .1 + (3)(.1)^2 + \cdots + (2N-1)(.1)^N \\
&\le (2)(.1) + (4)(.1)^2 + \cdots + (2N)(.1)^N \\
&= [.2][1 + (2)(.1) + \cdots + N(.1)^{N-1}] \\
&< [.2][1 + (2)(.1) + \cdots + N(.1)^{N-1} + \cdots] = \frac{.2}{(1-.1)^2}
\end{aligned}$$

where the last equality is equation (4) of Appendix 1 with $q = .1$ and $r = 2$. Thus the sum of the absolute values of the terms in any upper-left square is at most equal to $B = .2/(1-.1)^2$, but any upper-left rectangle is contained in a sufficiently large upper-left square so that the double series is absolutely convergent.

Since the double series is absolutely convergent, we know that we can add the terms in any arbitrary order, as in Figs. 1 and 2, and this summation will converge to a real number, and this limiting number will be the same regardless of which ordering we use. For this particular array the ordering of Fig. 2 conveniently reduces to the sum of the terms on the main diagonal, so

$$\sum_{n=1}^{\infty}\sum_{m=1}^{\infty} a_{nm} = (.1) + (.1)^2 + (.1)^3 + \cdots = \frac{.1}{1-.1} = \frac{1}{9}$$

from equation (3) of Appendix 1 for the sum of a geometric series. ∎

We often use Theorem 1 below (the proof is left to Appendix 3), which asserts that the sum of an absolutely convergent double series may also be obtained by iterated summation.

Theorem 1 The double series $\sum_{n=1}^{\infty}\sum_{m=1}^{\infty} a_{nm}$ is absolutely convergent iff for each n the nth row-sum series $\sum_{m=1}^{\infty}|a_{nm}|$ of absolute values is convergent *and* the series $\sum_{n=1}^{\infty}\left[\sum_{m=1}^{\infty}|a_{nm}|\right]$ consisting of the sum of these row sums is convergent. The same is true for the iterated sum obtained by summing column sums. If the double series is absolutely convergent then

$$\sum_{n=1}^{\infty}\sum_{m=1}^{\infty} a_{nm} = \sum_{n=1}^{\infty}\left[\sum_{m=1}^{\infty} a_{nm}\right] = \sum_{m=1}^{\infty}\left[\sum_{n=1}^{\infty} a_{nm}\right].$$

In particular if $a_{nm} = b_n c_m$, where $b_n c_m \neq 0$ for at least one pair of n and m, then the double series is absolutely convergent iff both $\sum_{n=1}^{\infty} b_n$ and $\sum_{m=1}^{\infty} c_m$ are, in which case

$$\sum_{n=1}^{\infty}\sum_{m=1}^{\infty} b_n c_m = \left[\sum_{n=1}^{\infty} b_n\right]\left[\sum_{m=1}^{\infty} c_m\right]. \quad ∎$$

For the special $b_n c_m$ result we had to rule out $0 = c_1 = c_2 = \cdots$, since absolute convergence of $\sum_{n=1}^{\infty} b_n c_m$ implies absolute convergence of $\sum_{n=1}^{\infty} b_n$ only when $c_m \neq 0$.

Example 2 In this example we apply Theorem 1 to the absolutely convergent double series of Example 1. For the nth row sum we find that

$$\begin{aligned}\sum_{m=1}^{\infty} a_{nm} &= -(n-2)(.1)^n + (.1)^{n+1} + (.1)^{n+2} + \cdots \\ &= -(n-2)(.1)^n + (.1)^{n+1}[1 + (.1) + (.1)^2 + \cdots] \\ &= -(n-2)(.1)^n + \frac{(.1)^n}{9} = \left(\frac{19}{9} - n\right)(.1)^n.\end{aligned}$$

where we used equation (3) of Appendix 1 for the next-to-last equality. The sum of these row sums is

$$\sum_{n=1}^{\infty} \left(\frac{19}{9} - n\right)(.1)^n$$

which we could now evaluate directly. However, from Theorem 1 we know that this sum equals the sum of the double series calculated as $1/9$ in Example 1. ■

EXERCISES

1. If a and b are real numbers for which $|a| < 1$ and $|b| < 1$, show that $\sum_{n=0}^{\infty} \sum_{m=0}^{\infty} a^n b^m$ is absolutely convergent and find its sum.

2. Prove that if $|\rho| < 1$ then $(1 - \rho)^{-2} = 1 + 2\rho + 3\rho^2 + \cdots$ by applying Theorem 1 to the array of Fig. 4.

$$\begin{matrix} 1 & \rho & \rho^2 & \rho^3 & \cdots \\ 0 & \rho & \rho^2 & \rho^3 & \cdots \\ 0 & 0 & \rho^2 & \rho^3 & \cdots \\ 0 & 0 & 0 & \rho^3 & \cdots \\ \cdots & \cdots & \cdots & \cdots & \cdots \end{matrix}$$

Fig. 4

3. As a check on an assertion made in Example 2, use equations (3) and (4) of Appendix 1 to prove that $\sum_{n=1}^{\infty} \left(\frac{19}{9} - n\right)(.1)^n = \frac{1}{9}$.

6
Discrete Probability Densities

This chapter initiates the study of probability theory in the discrete case. Section 6-1 introduces the notion of an idealized random-wheel experiment. Hopefully the reader will find this elementary notion useful for the development of probabilistic intuition. Many of our later definitions and results can be given simple interpretations in terms of such experiments. Section 6-2 formalizes the simple but basic concept of a discrete probability density. Relative frequencies, which are introduced in Sec. 6-3, assist in interpreting probability theory. Section 6-4 introduces discrete probability measures, and the important Theorem 2 (Sec. 6-4) collects their fundamental properties.

6-1 THE RANDOM-WHEEL EXPERIMENT

Consider the wheel in Fig. 1*a*, which has been divided into six sectors labeled $\omega_1, \omega_2, \ldots, \omega_6$. Let p_i be the fraction of the circumference which is cut off by the ith sector. Naturally each $p_i > 0$, and their sum is 1. Assume that the wheel is on a well-lubricated, almost frictionless axle which is fixed to a wall and that there is a pointer attached to the wall. Consider the experiment of spinning the wheel and observing which sector is opposite the pointer when the wheel finally comes to a stop. If the sector labeled ω_i is opposite the pointer when the wheel stops we say that the outcome of the experiment is ω_i. Assume that the wheel revolves many times before it stops, and that it is a very well balanced and fair wheel in gambling terminology.

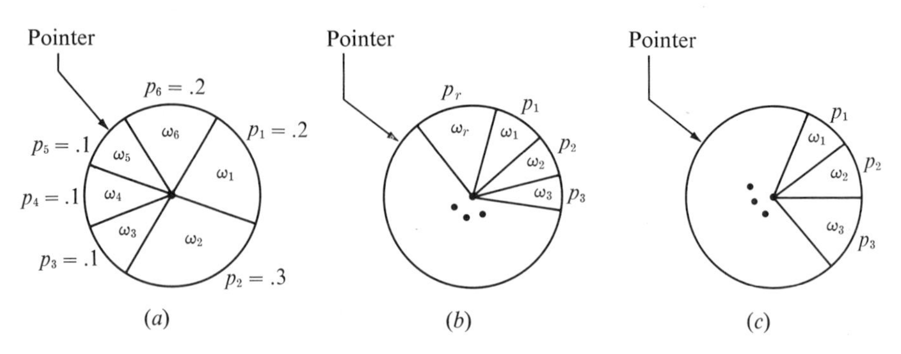

Fig. 1

If the experiment is performed a large number of times we anticipate that roughly 30 percent of the experiments should result in outcome ω_2. A special case of the law of large numbers provides, within our theory, considerable justification for such an anticipation. Naturally there may be instances when the experiment is performed, say, 1,000 times and ω_2 occurs in only five of the trials. This possibility of unlikely results is always present in probability theory and its applications and should never be forgotten.

If we are gambling we might bet that ω_2 or ω_5 will occur; that is, we win iff the outcome of the experiment is either ω_2 or ω_5. This *event* that the outcome is either ω_2 or ω_5 should occur in about 40 percent of the trials in a large number of experiments because $p_2 + p_5 = .4$.

Many of the results and examples in this book can be interpreted in terms of the repeated spinning of such a random wheel. It appears impossible to predict the outcome of one such experiment by solving some appropriate set of equations in the context of classical mechanics. We may say that the outcome of the experiment is random, or has some degree of randomness about it, or is not precisely predictable. Words such as "random" and "predictable" are used in their usual vague sense. There are a tremendous number of phenomena of interest which possess some randomness. We wish to set up a mathematical model which will help us to understand and analyze such phenomena.

Let us generalize the wheel with six sectors to the one in Fig. 1*b* with r sectors, where r is an integer greater than or equal to 1. Of course, $p_1 + p_2 + \cdots + p_r = 1$, each $p_i > 0$. Still more generally, the random wheel is permitted to have an infinite sequence of sectors, as in Fig. 1*c*, in which case $p_1 + p_2 + \cdots = 1$ and each $p_i > 0$. We ignore the exact position of the pointer within the arc of the sector in which it has stopped. Such factors as temperature and air currents are also ignored. That is, the model used to describe the random-wheel experiment does not go into fine details, and it considers "ω_i occurred" to be a complete enough description of the experimental outcome. Our basic mathematical model will essentially consist of just the ω_i and the p_i. In later chapters more structure will be added to such models, and in many applications the distinct ω_i labels for the sectors will have some internal structure; each ω_i might be a tuple, for example.

EXERCISES

1. Gamblers G_1, G_2, and G_3 each bet on 1,000 spins, or trials, of the wheel of Fig. 1*a*. On a particular trial G_1 wins iff the outcome belongs to $\{\omega_1,\omega_2,\omega_3\}$, G_2 wins iff the outcome belongs to $\{\omega_2,\omega_3,\omega_4,\omega_5\}$, and G_3 wins iff the outcome belongs to $\{\omega_1,\omega_2,\omega_5,\omega_6\}$. In about how many trials would you expect the following outcomes?
(*a*) G_1 wins.
(*b*) Both G_1 and G_2 win.
(*c*) All three win.
(*d*) At least one of G_1 or G_2 wins.
This exercise indicates that sets and such standard operations as union and intersection arise naturally in probability theory.

2. Wheel 1 has two sectors, labeled α_1 and α_2, with $p_1 = .7$ and $p_2 = .3$. Wheel 2 has three sectors, labeled β_1, β_2, and β_3 and assigned fractions $q_1 = .2$, $q_2 = .3$, and $q_3 = .5$ of the circumference. A *trial* consists of spinning both wheels, and its outcome is the ordered pair (α_i,β_j) iff wheel 1 stops at α_i and wheel 2 stops at β_j. Assume that the outcomes for all spins are independent of each other. This rules out, for example, giving wheel 2 only a half turn when wheel 1 has just yielded the outcome α_2. Sketch a wheel with six sectors labeled (α_i,β_j) such that one spin of this wheel is equivalent to one trial as described above. Exhibit the numerical fraction of the circumference assigned to each sector. It is standard in probability theory to create one wheel such that one spin of this wheel is equivalent to two or many spins of other wheels.

6-2 DISCRETE PROBABILITY DENSITIES

Let us now formalize and generalize slightly the ideas behind the random-wheel experiment.

A **discrete probability density** is a real-valued function $p: \Omega \to R$ defined on some set Ω and satisfying the following three conditions:

1. $p(\omega) \geq 0$ for all ω in Ω; that is, p has only nonnegative values.
2. $\Omega_0 = \{\omega: p(\omega) > 0\}$ is a countable nonempty set. Thus Ω_0 can be written as a finite or infinite sequence, where Ω_0 is the set of those points of Ω to which p assigns positive values. The set Ω may be an uncountable set such as R^n.
3. If $\Omega_0 = \{\omega_1, \omega_2, \ldots, \omega_r\}$ is a finite set then $p(\omega_1) + p(\omega_2) + \cdots + p(\omega_r) = 1$, and if $\Omega_0 = \{\omega_1, \omega_2, \ldots\}$ is denumerable then $p(\omega_1) + p(\omega_2) + \cdots = 1$; that is, if we add the values of p for all those points of Ω where p is positive, the sum is 1.

We often abbreviate to "discrete density." The word "discrete" refers to the fact that Ω_0 is countable. Thus a discrete density is a function $p: \Omega \to R$ which has only nonnegative values, takes positive values only on a countable subset Ω_0 of Ω, and sums to 1 over this subset.

A discrete probability density can always be interpreted as a random-wheel experiment. If $\Omega_0 = \{\omega_1, \omega_2, \ldots\}$ is denumerable we define p_i by $p_i = p(\omega_i)$ and just use Fig. 1*c* (Sec. 6-1), where the ith sector is labeled ω_i and has fraction p_i of the wheel assigned to it. Similarly if Ω_0 is finite we get a random wheel like that of Fig. 1*b* (Sec. 6-1). Thus in interpreting $p: \Omega \to R$ as a random-wheel experiment we ignore the points of ω for which $p(\omega) = 0$. In general the subset Ω_0 of Ω is the crucial subset, and at times we will alter a setup in order to make Ω_0 the same as Ω. However, it is often convenient to permit p to be defined on a larger set Ω.

Figure 1 depicts a discrete probability density. The points of Ω_0 are shown as crosses and other points of Ω are shown as dots. The fact that p does not take negative values is indicated by the lack of arrows terminating at negative numbers. The fact that p sums to 1 over Ω_0 cannot conveniently be indicated in such a picture. The function $p: \Omega \to R$ consists of the bundle of all arrows from Ω to R. We may imagine that from every point of Ω_0 there is an arrow to a point of the interval $0 < x \leq 1$ and from every point of Ω which does not belong to Ω_0 there is an arrow to the origin of the real line R.

If $p: \Omega \to R$ is a discrete probability density the set Ω is called the **sample space,** an element ω of Ω is called a **sample point,** and each subset of Ω is called an **event.**

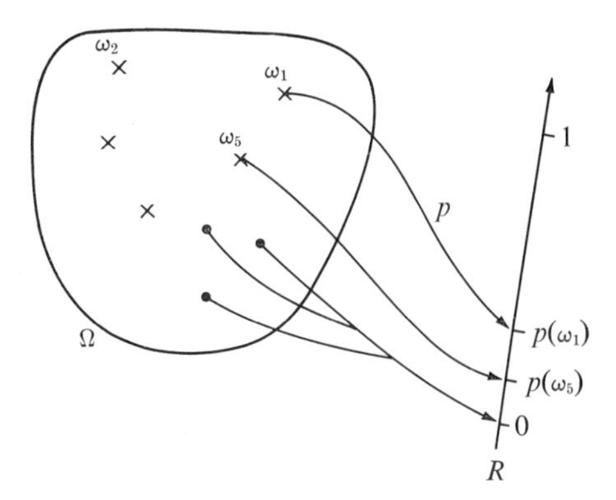

Fig. 1

If A is an event the **probability of** A is the real number denoted by $P(A)$ and defined to be the sum of all of the positive values $p(\omega)$ for all of the ω in A. In more detail, $P(A)$ is defined as follows:

1. If $p(\omega) = 0$ for all points of A then $P(A) = 0$.
2. If $p(\omega) > 0$ for only a finite number of points of A and $\omega_1, \omega_2, \ldots, \omega_r$ is an enumeration of these points then $P(A) = p(\omega_1) + p(\omega_2) + \cdots + p(\omega_r)$.
3. If $p(\omega) > 0$ for an infinite number of points of A and $\omega_1, \omega_2, \ldots$ is an enumeration of these points then $P(A) = p(\omega_1) + p(\omega_2) + \cdots$.

A discrete probability density $p: \Omega \to R$ can be thought of as a distribution of discrete mass points over Ω; that is, mass $p(\omega)$ is concentrated at point ω, so that $P(A)$ is the total mass of the set A. The mass $P(\Omega)$ of the whole space is 1.

Our usual interpretation of a discrete probability density $p: \Omega \to R$ is as follows: An experiment is to be performed and the outcome is in some sense random or unpredictable. A sample point ω is a complete description of a conceivable outcome of the experiment in the sense that the outcome of the experiment must correspond to precisely one sample point. The collection Ω of all these sample points is called the *sample space*. If A is an event we say that A occurred iff the sample point describing the outcome of the experiment belongs to A. The probability that A will occur is the real number $P(A)$.

Figure 2 depicts a discrete probability density which gives rise to the random-wheel experiment of Fig. 1*a* (Sec. 6-1). The event A contains only the two sample

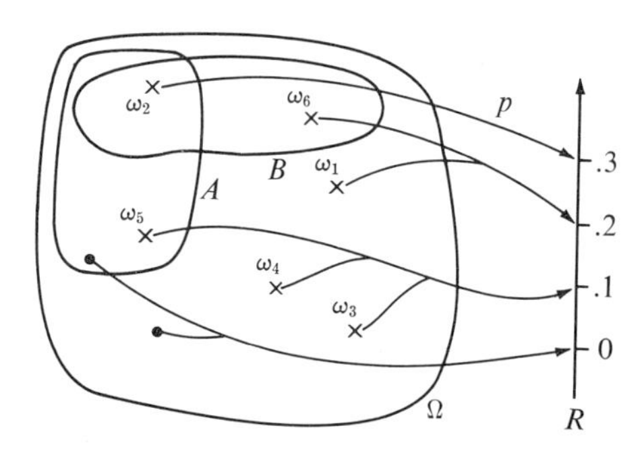

Fig. 2

points ω_2 and ω_5 of $\Omega_0 = \{\omega: p(\omega) > 0\}$. To find the probability of an event A we just add all the values of p for those points of Ω_0 which belong to A so $P(A) = .4$.

Several properties of probabilities of events are obvious. The empty set φ is an event, and since it contains no sample points its probability is 0, $P(\varphi) = 0$. The sample space Ω is an event, and clearly $P(\Omega) = 1$. The set $\Omega_0 = \{\omega: p(\omega) > 0\}$ is an event, and clearly $P(\Omega_0) = 1$. Also if A is an event then $0 \le P(A) \le 1$. Thus the probability of any event is between 0 and 1, and the empty set has probability 0, while the whole sample space has probability 1. Furthermore the probability is 1 that the outcome of the experiment will belong to the important set Ω_0. Also if $\{\omega\}$ is the event consisting of the single sample point ω then $P(\{\omega\}) = p(\omega)$. Also $P(A \cup B) = P(A) + P(B)$ *whenever A and B are disjoint.* This is also clear from the interpretation of $P(A)$ as the mass of A. Note that $.6 = P(A \cup B)$ does *not* equal $P(A) + P(B) = .9$ for Fig. 2, where A and B are *not* disjoint, because $p(\omega_2)$ contributes once to $P(A \cup B)$ but twice to $P(A) + P(B)$.

Example 1 An experiment consists of measuring the volume and temperature of a certain gas, and the resulting instrument readings of the volume and temperature are considered to be an adequate description of the outcome of the experiment. Thus the outcome of the experiment will be an ordered pair (v,t) of reals, so we can let the sample space be the plane, $\Omega = R^2$. The selection of an appropriate discrete probability density $p: R^2 \to R$ would depend on the particular physical experiment. It might be based on physical and probabilistic theoretical considerations, or on past data concerning similar experiments, or on both. Many densities which arise frequently in applications in almost all fields will be introduced later, and probabilistic reasons for their popularity will be considered in various appropriate chapters. In Sec. 6-3 some consideration will be given to the use of certain kinds of experimental data in arriving at a density. For the purpose of this example let us arbitrarily use the discrete probability density $p: R^2 \to R$ of Fig. 3, for which Ω_0 has only six points. Figure 3 happens to give rise to the random wheel of Fig. 1*a* (Sec. 6-1). The reasoning and the nature of the following interpretations would be the same if Ω_0 had 10^{100} closely spaced points, or consisted of an infinite sequence of points in the plane.

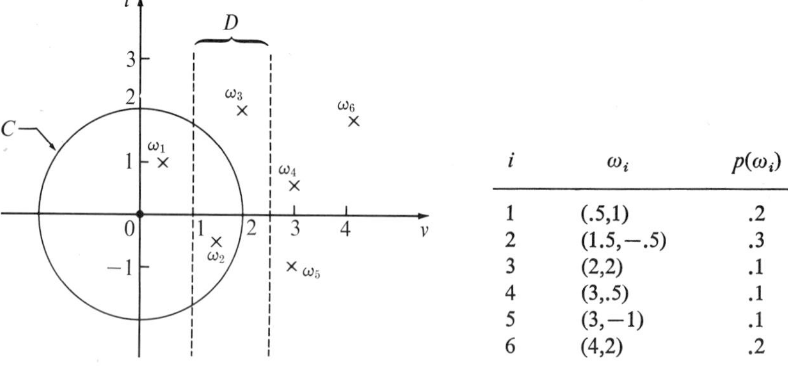

i	ω_i	$p(\omega_i)$
1	(.5,1)	.2
2	(1.5,−.5)	.3
3	(2,2)	.1
4	(3,.5)	.1
5	(3,−1)	.1
6	(4,2)	.2

Fig. 3 $C = \{(v,t): v^2 + t^2 \le 4\}$, $D = \{(v,t): 1 < v < 2.5\}$.

The event that both the volume and the temperature have the value 2 consists of the single sample point $\omega_3 = (2,2)$ and has probability .1. The event C that $[\text{volume}]^2 +$

$[\text{temperature}]^2 \le 4$ has probability $p(\omega_1) + p(\omega_2) = .2 + .3 = .5$. If D is the event that the volume is strictly between 1 and 2.5 then

$$P(D) = P(\{\omega_2,\omega_3\}) = p(\omega_2) + p(\omega_3) = .3 + .1 = .4.$$

The usual interpretation of $P(D) = .4$ is that the event D should happen in about 40 percent of a large number of independent repetitions or trials of the experiment. By independent repetitions we mean essentially that the experiments should be equivalent to repeatedly spinning the wheel of Fig. 1*a* (Sec. 6-1), in which case knowledge of the outcomes of some of the experiments does not assist us in predicting the outcomes of the other experiments. Some repeated physical experiments would satisfy this independence assumption and others would not. When there is a significant probabilistic dependence between repetitions it may be necessary to incorporate this dependence into the mathematical model. The event that either C or D occurs is just their union, and $P(C \cup D) = p(\omega_1) + p(\omega_2) + p(\omega_3) = .2 + .3 + .1 = .6$. The event that both C and D occur is their intersection, and $P(CD) = p(\omega_2) = .3$. Clearly the probability is 0 that the temperature is over 3 and the probability is 1 that the temperature is over -2. Also the event that D does not occur is the complement of D in R^2 and has probability $1 - P(D) = 1 - .4 = .6$. ■

This example illustrated a number of points. It is often natural to use R^2, or more generally R^n, as a sample space. Complicated mathematical expressions (such as $v^2 \cos e^t$) involving the variables of an experiment may arise naturally, depending on the applicational interest, and may lead to the definition of an involved event whose probability may be hard to calculate. Our initial description of the experiment was in terms of a discrete-probability-density function p, but natural answers use P. Figure 1*a* (Sec. 6-1) is the random-wheel experiment corresponding to Fig. 3, and the naturalness of the events considered is reduced in making the transition to a random-wheel experiment.

If $p: \Omega \to R$ is a discrete probability density and A is an event which contains $\Omega_0 = \{\omega : p(\omega) > 0\}$ we may say that p **is concentrated on** A. Thus p, or P, is said to be concentrated on A iff $P(A) = 1$. For example, we may assert that something is true for every P which is concentrated on the set of positive reals.

EXERCISES

Several of the following exercises use the discrete probability density $p: \Omega \to R$ of Fig. 4, for which $\Omega = \Omega_0 = \{0,1\}^4$ has $2^4 = 16$ sample points. This discrete probability density was selected quite arbitrarily.

i	ω_i	$p(\omega_i)$	i	ω_i	$p(\omega_i)$
1	(0,0,0,0)	.01	9	(1,0,0,0)	.08
2	(0,0,0,1)	.03	10	(1,0,0,1)	.01
3	(0,0,1,0)	.20	11	(1,0,1,0)	.05
4	(0,0,1,1)	.01	12	(1,0,1,1)	.01
5	(0,1,0,0)	.01	13	(1,1,0,0)	.03
6	(0,1,0,1)	.05	14	(1,1,0,1)	.03
7	(0,1,1,0)	.20	15	(1,1,1,0)	.03
8	(0,1,1,1)	.05	16	(1,1,1,1)	.20

Fig. 4

1. *Telephone Exchange* (*Fig. 4*). An experiment consists of observing a telephone exchange for four consecutive seconds. The ith coordinate of a sample point is 1 or 0, depending on whether the exchange did or did not put through at least one call during the ith second. In particular $p(\omega_7) = .2$ and $\omega_7 = (0,1,1,0)$ is interpreted as follows:

During second 1 the exchange put through no calls.
During second 2 the exchange put through at least one call.
During second 3 the exchange put through at least one call.
During second 4 the exchange put through no calls.

(*a*) We are interested in the chance of heavy loads on the exchange, so we let A be the event that the exchange is used for at least 3 of the 4 seconds. Exhibit A as a collection of sample points and find $P(A)$.
(*b*) Let $B = \{\omega_1,\omega_{16}\}$ and check that $P(A \cup B) = P(A) + P(B) - P(AB)$.
(*c*) Prove that $P(A \cup B) = P(A) + P(B) - P(AB)$ holds for any events A and B and any discrete probability density p. Is $P(A \cup B) \leq P(A) + P(B)$ always true?
(*d*) A factory normally receives its power from power line α but has power line β as an alternative. α is inoperative .3 percent of the time, β is inoperative .2 percent of the time, and at least one of the two is inoperative .4 percent of the time. What percentage of the time is the factory without any power?

2. *A Circuit with Random Relays* (*Fig. 4*). Let the sample point (0,1,1,0) be interpreted as follows: relay 1 is open, relay 2 is closed, relay 3 is closed, and relay 4 is open. The relay configuration is shown in Fig. 5. Find the probability of a closed connection (at least one path through closed relays) between points x and y.

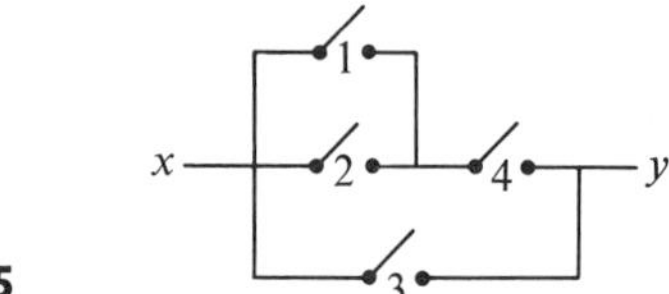

Fig. 5

3. *Firing Practice* (*Fig. 4*). The sample point (0,1,1,0) is interpreted as follows: in a barrage of shells fired at the target none landed in interval 1, at least one landed in interval 2, at least one landed in interval 3, and none landed in interval 4. As explained by Fig. 6, a barrage will be said to closely bracket the target iff at least one shell lands in interval 2 *and* at least one shell lands in interval 3. Find the probability of closely bracketing the target.

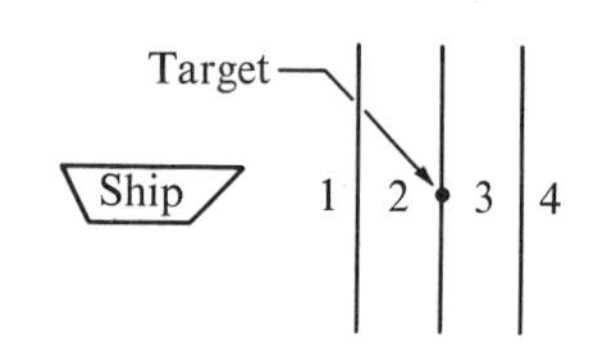

Fig. 6

4. *Coin Tossing with Elaborate Cheating* (*Fig. 4*). A coin is tossed four times, and the outcome is (x_1,x_2,x_3,x_4), where $x_i = 1$ or 0 is interpreted as head or tail, respectively, on the ith toss. Find the following probabilities:
(*a*) The first toss results in a head.
(*b*) The second toss results in a head.
(*c*) The number of heads equals the number of tails.

5. *Independent Tosses of a Fair Coin.* The experiment consists of four independent tosses of a fair coin. The sample space is the same as in Fig. 4, but the discrete probability density is $p(\omega) = 1/16$ for all ω. A is the event that the number of heads equals the number of tails.

(*a*) Find $P(A)$.
(*b*) Find the odds[1] for A.
(*c*) If the odds for B are x to y find $P(B)$.

6. Let $x_1, x_2, \ldots$ be the enumeration of the positive rationals described in Exercise 1 (Sec. 4-2). Define $p: R \to R$ as follows: if x is not a positive rational then $p(x) = 0$, while $p(x_n) = 2^{-n}$ for the nth positive rational in the enumeration. Is this p acceptable as a discrete probability density within our theory?

6-3 RELATIVE FREQUENCIES

Two of the coordinates of the 6-tuple $(\alpha,-4,0,\beta,-4,3,)$ are equal to -4, so we say that the relative frequency of -4 in this 6-tuple is $2/6$ or $1/3$, since this is the fraction of the coordinates of the 6-tuple which are equal to -4. More generally, if $e = (e_1, e_2, \ldots, e_n)$ is any n-tuple and ω is any object we define $m_e(\omega)$ by

$$m_e(\omega) = \frac{1}{n} \text{ [the number of coordinates of } e \text{ which equal } \omega].$$

We call $m_e(\omega)$ **the relative frequency of** ω **in** e. We use the letter e for the n-tuple because we often think of e as having been obtained experimentally. Thus the result of seven coin tosses might be $e = (H,H,T,H,T,T,H)$, in which case the relative frequency of heads is $m_e(H) = 4/7$ while for tails $m_e(T) = 3/7$. The term "relative frequency" is traditional in probability theory, where it is often applied to infinite rather than finite sequences. Thus it is sometimes said that the relative frequency of heads in an infinite sequence of fair coin tosses should be .5.

If a random wheel has $p(\omega_8) = .15$ we expect about 15 percent occurrence of ω_8 in a long sequence of spins. In our new terminology, we expect the relative frequency $m_e(\omega_8)$ to be near .15 if the number n of trials is large enough, where $e = (e_1, e_2, \ldots, e_n)$ and e_i is the outcome of the ith trial or spin. Thus we can use the idea of relative frequency in describing *guesses* or predictions about experimental outcomes.

If p is a discrete probability density it determines corresponding probabilities $P(A)$ for events A. We can construct a function M_e similar to P by setting

$$M_e(A) = \frac{1}{n} \text{ [the number of coordinates of } e \text{ which belong to } A]$$

where A is a set and $e = (e_1, e_2, \ldots, e_n)$ is an n-tuple. We call $M_e(A)$ the **relative frequency of** A **in** e. If $A = \{\omega_1,\omega_2\}$ and

$$e = (\omega_3,\boldsymbol{\omega}_2,\boldsymbol{\omega}_2,\omega_3,\boldsymbol{\omega}_1,\boldsymbol{\omega}_2,\boldsymbol{\omega}_1,\omega_3,\omega_3,\boldsymbol{\omega}_2,\boldsymbol{\omega}_2,\boldsymbol{\omega}_1)$$

then $M_e(A) = m_e(\omega_1) + m_e(\omega_2) = 8/12$, where the boldface $\boldsymbol{\omega}$ indicate those coordinates belonging to A, so that $M_e(A)$ is just the fraction of boldface coordinates. Clearly the relationship between M_e and m_e is similar to the relationship between P and p, so that the definition of $P(A)$ can be reinforced by considering relative frequencies.

[1] We will almost never use this common gambling terminology, but it is defined as follows: If B has probability $P(B)$ then the odds for B are $P(B)/P(B^c)$. Thus if $P(B) = 1/3$ the odds for B are $(1/3)/(2/3) = 1/2$. Usually this is expressed as "the odds for B are 1 to 2."

Relative frequencies can also be used to *construct a model* from experimental data. Suppose a large number n of trials were made on a random-wheel experiment and a record e of the outcomes were kept. Suppose the wheel were destroyed and that all that remained was the record e; that is, the discrete probability density p has been lost. If we are asked to construct a wheel which is experimentally equivalent to the one destroyed we need a reasonable guess for the unknown p. One reasonable guess, provided n is large enough, is given by the relative frequencies m_e. Clearly m_e is nonnegative and sums to 1 if it is summed over the things which occur as coordinates of e. Thus m_e is similar to a discrete probability density p. For example, if $e = (H,H,T,H,T,T,H)$, so that $n = 7$ in order to keep the computation simple, then we might construct a wheel with two sectors labeled H and T, with corresponding assigned fractions $4/7$ and $3/7$ of the circumference.

In discussing the possible use of relative frequencies in guessing a discrete probability density we have been essentially considering matters of statistical inference. It seems appropriate to say a few words about the subject of statistics. Theoretical statistics is concerned with theories, justifications, and techniques of reasonable guesses or statistical *inferences* from statistical, or random, data. In statistics the framework and results of probability theory are used in analyzing the properties of various inference procedures under various assumptions. This book is concerned with probability theory and does not go into statistical theory. We essentially assume that there is one and only one known discrete probability density, which may depend on parameters, under consideration at any one time, and we *deduce* conclusions involving it.

In this book we use the terms "statistics" and "probability theory" in the senses in which they are normally understood in course descriptions for present-day courses offered in a mathematics or statistics department of a university. The meaning of the word "statistics" depends very much on the field to which the word is applied. For example, statistical mechanics in physics does not normally concern itself with "statistics" as we use the word. The problems normally considered in statistical-mechanics courses are probability theory problems with very special structures. Statistical communication theory and statistical control theory are largely probabilistic but use more statistics now than they did some years ago.

Although we do not go into the study of statistics, it is worth keeping in mind that in some cases it is reasonable to use relative frequencies to guess a discrete probability density and then make deductions starting from there. Thus we can think of some of our discrete probability densities, particularly in the numerical examples, as having been arrived at by the use of relative frequencies from a long sequence of trials.

In probability theory we deduce general results from the assumption of a given model. Nothing in the theory requires any specific interpretation for the quantities considered. Our motivation for various definitions and our interest in various phenomena within the model usually derive from our interpretation of probability as "chance" in some vague sense. Our models can be thought of as idealized mathematical representations of various real-world situations. In some cases we may want to think of our model as having been arrived at by way of experimentally obtained relative frequencies. At times we may wish to interpret various statements about our model as reasonable predictions concerning experimental relative frequencies in particular real-world situations. The theory itself is a deductive mathematical theory.

Example 1 Suppose that we have a random wheel with 10 equal-sized sectors labeled 0, 1, 2, . . . , 9. We spin the wheel a total of $n = 20$ times and obtain the sequence

$$e = (7,3,4,7,0,0,1,7,9,5,7,2,1,8,5,0,0,1,1,6).$$

Clearly $m_e(7) = 4/20$ If we break e into consecutive nonoverlapping pairs

$$e' = ((7,3), (4,7), (0,0), (1,7), (9,5), (7,2), (1,8), (5,0), (0,1), (1,6))$$

we find that the relative frequency $m_{e'}((1,7))$ of (1,7) in e' is $1/10$. If we were to spin the wheel a very large number n of times and obtain $e = (e_1, e_2, \ldots, e_n)$ we would expect that each of $m_e(0), m_e(1), \ldots, m_e(9)$ would be close to $1/10$. Similarly we would expect to obtain an $e = (e_1, e_2, \ldots, e_n)$ for which the corresponding $e' = ((e_1,e_2), (e_3,e_4), \ldots, (e_{n-1},e_n))$ obtained by breaking e into consecutive pairs has each of $m_{e'}((0,0)), m_{e'}((0,1)), \ldots, m_{e'}((9,9))$ close to $(1/10)^2 = 10^{-2}$. For convenience we assume here that n is a multiple of 2. Furthermore we would expect, for example, that the relative frequency of (7,0,5) in $((e_1,e_2,e_3), (e_4,e_5,e_6), \ldots, (e_{n-2},e_{n-1},e_n))$ should be close to 10^{-3}. We will see that a special case of the law of large numbers provides considerable justification for such predictions. How large an n is required for such a prediction depends on the number r of consecutive digits which we collect together, $((e_1, e_2, \ldots, e_r), (e_{r+1}, \ldots, e_{2r}), \ldots)$; on how close we want the relative frequency to be to 10^{-r}; and on how confident we want to be of our prediction. ■

Example 2 Relative frequencies are also of interest for sequences which are determined in a completely deterministic nonrandom manner. For example, the napierian base e to 20 decimal places is

$$e = 2.71828182845904523536 \cdots.$$

[It is only a coincidence that the same symbol e is used both for $2.718 \cdots$ and for $(e_1, e_2, \ldots, e_n)$ as used in this section.] If we consider only the digits to the right of the decimal place then the relative frequency of 5 in these 20 digits is $3/20$. Including digits to the left of the decimal place would have no significant effect on any of the following considerations. If we break the 20 digits into consecutive pairs of digits

$$(7,1), (8,2), (8,1), (8,2), \ldots, (3,5), (3,6)$$

then the relative frequency of (8,2) in these 10 pairs is $2/10$.

Every decimal digit in the infinite decimal expansion of e is uniquely determined by the mathematical definition of e. Although there is nothing random about the decimal digits in the expansion of e, this sequence of decimal digits is remarkably similar to what might have been obtained from repeated spins of the random wheel of Example 1. The same is true for other irrational numbers such as π and $\sqrt{2}$. In particular the first 60,000 decimal places of e have passed a variety of statistical tests[1] meant to discriminate against finite sequences which are unlikely to be obtained by random selections as made in Example 1. For example, it was found that the relative frequency of 0 in the first $n = 60{,}000$ decimal places of e is .09903, while the relative frequency of (0,0) in the 30,000 corresponding pairs (7,1), (8,2), . . . is

[1] R. G. Stoneham, "A Study of 60,000 Digits of the Transcendental e," *The American Mathematical Monthly*, vol. 72, no. 5, pp. 483–500 (May, 1965).

.0098. We will see later from the central-limit theorem that deviations as large as $|.1 - .09903|$ are fairly common for $n = 60{,}000$ spins of the random wheel of Example 1. Similarly .0098 is "close enough" to .01.† ■

EXERCISES

1. *Fair-coin-tossing Frequencies.* Toss a coin 24 times and write down your results as an element $e = (e_1, e_2, \ldots, e_{24})$ of $\{H,T\}^{24}$, where H stands for heads and T stands for tails.

(*a*) Find $m_e(H)$. What number should the class average of this quantity be close to?

(*b*) Consider your data e as broken into pairs so that

$$e' = ((e_1,e_2), (e_3,e_4), \ldots, (e_{23},e_{24})).$$

Thus e' belongs to $\{(H,H), (H,T), (T,H), (T,T)\}^{12}$. Find $m_{e'}((H,T))$. What number should the class average of this quantity be close to?

(*c*) Sketch the random wheel which you think should correspond to the idealized experiment of tossing a coin twice.

(*d*) Sketch the random wheel corresponding to the relative frequencies of part (*b*).

6-4 DISCRETE PROBABILITY MEASURES

Some basic properties of P are derived in this section. It should already be fairly clear that although we started with a discrete density $p: \Omega \to R$, it is often more natural to deal with the corresponding P. It turns out that in nondiscrete cases there is always a P to work with, although a corresponding discrete density p may not exist. In Chap. 16 continuous densities will be available, but in Chap. 17 we will see that even when the sample space is the real line, mixtures of discrete and continuous densities are not sufficient for a general theory. Distribution functions on R^n are introduced in Chap. 17 as generalizations of densities. Even more dramatically, in Chap. 22 to 24 the sample spaces are essentially function spaces, and although there is always a P, it will not naturally correspond to any density or distribution function. Because of this naturalness and generality we will consistently tend to emphasize P rather than p.

† The following standard definition makes some of these notions more precise. A real number x in the unit interval $0 \le x \le 1$ is said to be a *normal number* (to the base 10) iff its decimal expansion $x = \sum_{i=1}^{\infty} b_i/10^i$ has the property that the relative frequency of $(d_1, d_2, \ldots, d_r)$ in

$$((b_1, \ldots, b_r), (b_{r+1}, \ldots, b_{2r}), \ldots, (b_{(j-1)r+1}, \ldots, b_{jr}))$$

converges to 10^{-r} as j goes to infinity, and this holds true for every r-tuple $(d_1, d_2, \ldots, d_r)$ of decimal digits and for every $r = 1, 2, \ldots$. The strong law of large numbers shows that "almost every" real number x in the unit interval is normal (to every base 2, 3, 4, ...); that is, the set of normal numbers in the unit interval has length (Lebesgue measure) 1. However, it is not known whether or not the particular numbers $e - 2$, $\pi - 3$, and $\sqrt{2} - 1$ are normal. If the number $x = \sum_{i=1}^{\infty} b_i/10^i$ is defined by letting $b_1, b_2, \ldots$ be what is left when parentheses are eliminated from the sequence

$$0, 1, 2, \ldots, 9, \underbrace{(0,0), (0,1), \ldots, (9,9)}_{10^2 \text{ pairs}}, \underbrace{(0,0,0), (0,0,1), \ldots, (9,9,9)}_{10^3 \text{ triples}}, (0,0,0,0), \ldots$$

then it is easily shown that x is normal. Thus examples of normal numbers are easily constructed.

Admittedly the concept of the P function is somewhat more abstract than that of densities. Therefore we do not carry the early emphasis on P as far as is possible. After many of the basic ideas of probability theory have been introduced in the discrete case we will review and generalize our results in Chap. 15 with greater emphasis on P, as a transition to nondiscrete cases. For this reason the discussions, definitions, statements of theorems, etc., for the discrete case are worded in such a way as to facilitate this later transition. The reader may wish to consider these comments merely a vague advance clue to the reason for a bias in favor of P, as opposed to p, in some contexts where either could as well be used.

If $p: \Omega \to R$ is a discrete probability density and A is an event, we have defined the number $P(A)$, which we call the *probability of A*. Thus we have assigned a unique real number $P(A)$ to each event A. Therefore we have defined a real-valued function P whose domain is the class of events. This function P is called the **discrete probability measure** corresponding to the discrete probability density $p: \Omega \to R$. The function P is sometimes said to be a *set function* because each member of its domain is a set. The ordered pair (Ω, P), where Ω is the sample space and P is the discrete probability measure, is called the **discrete probability space** corresponding to $p: \Omega \to R$. There is no basic difference between P and (Ω, P), since if we are given the function P, we can take the union of all of the members of its domain and so get Ω. However, the notation (Ω, P) is convenient because it simultaneously introduces notation for the sample space and for the discrete probability measure.

If the sample space $\Omega = \{\omega_1, \omega_2, \omega_3\}$ of a discrete probability density $p: \Omega \to R$ has only three sample points there are exactly $2^3 = 8$ events. The three nonnegative numbers $p(\omega_1)$, $p(\omega_2)$, and $p(\omega_3)$ sum to 1. The value of P at each of the eight events is $P(\varphi) = 0$, $P(\{\omega_1\}) = p(\omega_1)$, $P(\{\omega_2\}) = p(\omega_2)$, $P(\{\omega_3\}) = p(\omega_3)$, $P(\{\omega_1,\omega_2\}) = p(\omega_1) + p(\omega_2)$, $P(\{\omega_1,\omega_3\}) = p(\omega_1) + p(\omega_3)$, $P(\{\omega_2,\omega_3\}) = p(\omega_2) + p(\omega_3)$, $P(\Omega) = 1$. More generally, if $\Omega = \{\omega_1, \omega_2, \ldots, \omega_n\}$ has n sample points there are 2^n events, while if Ω is infinite there are an infinite number of events but $P(A)$ is still well defined for each event A.

Figure 1 depicts both p and P. An arrow for the P function originates on a subset of Ω and terminates on the interval $0 \le x \le 1$. The P function can be thought of as the whole bundle of such arrows, one for each subset of Ω.

Clearly $P(\{\omega\}) = p(\omega)$ for every ω in Ω, so that the same discrete probability measure P cannot arise from a different discrete probability density $p': \Omega \to R$.

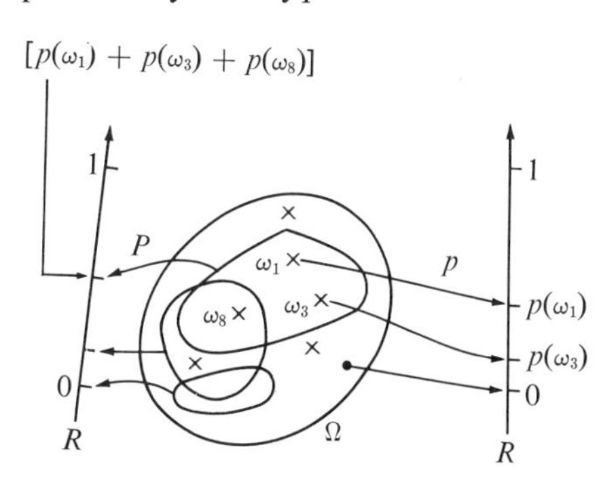

Fig. 1

Our next task is to take a first step toward characterizing discrete probability measures. Theorem 1 states four properties which every discrete probability measure must have. Theorem 3 shows that any function with these four properties must indeed be a discrete probability measure.

Theorem 1 If (Ω,P) is a discrete probability space the following properties hold:

(*a*) $P(\Omega) = 1$.

(*b*) $P(A) \geq 0$ for every event A.

(*c*) If $A_1, A_2, \ldots$ is any sequence of disjoint events then

$$P\left(\bigcup_{n=1}^{\infty} A_n\right) = \sum_{n=1}^{\infty} P(A_n).$$

(*d*) The set $\Omega_0 = \{\omega : P(\{\omega\}) > 0\}$ is countable, and $P(\Omega_0) = 1$. ■

REMARKS: Properties (*a*), (*b*), and (*d*) are obvious and were noted earlier. Property (*c*), which is called the **countable additivity** of P, is new and of fundamental importance. It is intuitively obvious from the interpretation of $P(A)$ as the mass of A. From the disjointness shown in Fig. 2 it is clear that summing p over all of $\bigcup_{n=1}^{\infty} A_n$ should yield

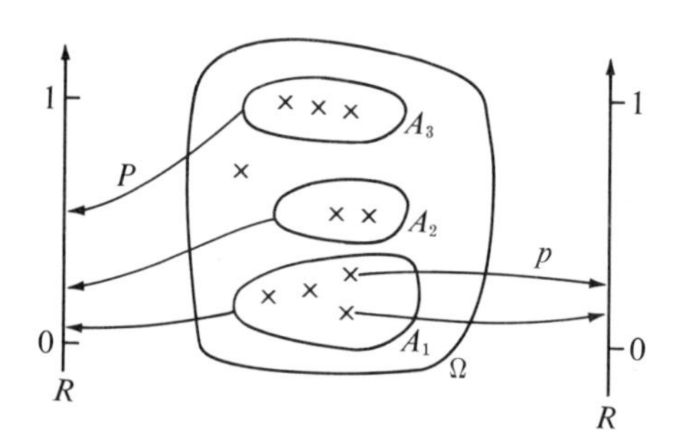

Fig. 2

the same result as summing p over A_n to get $P(A_n)$ and then summing these numbers. Now $P(\varphi) = 0$, so if we let $\varphi = A_{N+1} = A_{N+2} = \cdots$ in property (*c*) we get a finite version as a special case: if $A, A_2, \ldots, A_N$ are disjoint events then

$$P\left(\bigcup_{n=1}^{N} A_n\right) = \sum_{n=1}^{N} P(A_n).$$

Proof of Theorem 1 We have already observed that a discrete probability measure has properties (*a*), (*b*), and (*d*), so we need only prove that P is countably additive. Let $A_1, A_2, \ldots$ be a sequence of disjoint subsets of Ω and let $A = A_1 \cup A_2 \cup \cdots$ be their union. We wish to prove that $P(A) = P(A_1) + P(A_2) + \cdots$, where P is the discrete probability measure corresponding to the discrete probability density $p: \Omega \to R$. By Theorem 1*a* (Sec. 4-2) for each n the set of points ω of A_n for which $p(\omega) > 0$ is countable, and for a moment let us assume that this set is denumerable rather than finite. As in Fig. 3, let $\omega_{n1}, \omega_{n2}, \ldots$ be an enumeration of all those points

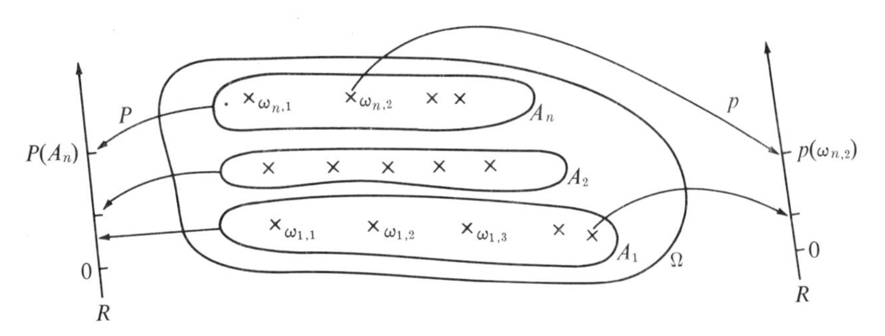

Fig. 3

ω of A_n for which $p(\omega) > 0$ so that $P(A_n) = p(\omega_{n1}) + p(\omega_{n2}) + \cdots$. The result now follows from Theorem 1 of Sec. 5-3 by letting $a_{nm} = p(\omega_{nm})$, since

$$P(A) = \sum_{n=1}^{\infty} \sum_{m=1}^{\infty} a_{nm} = \sum_{n=1}^{\infty} \left[\sum_{m=1}^{\infty} a_{nm} \right] = \sum_{n=1}^{\infty} P(A_n).$$

We needed the fact that the A_n are disjoint in order to state that $P(A) = \sum_{n=1}^{\infty} \sum_{m=1}^{\infty} a_{nm}$ for if $\omega_{3,2} = \omega_{5,8}$ this point occurs only once in an enumeration of those points of A for which $p(\omega) > 0$, so that the summation for $P(A)$ would contain only one of the terms $a_{3,2}$ or $a_{5,8}$; hence $P(A) < \sum_{n=1}^{\infty} \sum_{m=1}^{\infty} a_{nm}$.

Now if for some n the set of those points ω belonging to A_n for which $p(\omega) > 0$ is finite, and $\omega_{n1}, \omega_{n2}, \ldots, \omega_{nr}$ is an enumeration of these points, we simply let $a_{n1} = p(\omega_{n1}), \ldots, a_{nr} = p(\omega_{nr})$ and set $0 = a_{n,r+1} = a_{n,r+2} = \cdots$, and the same proof applies. For instance, if $A_n = \varphi$ then $0 = a_{n1} = a_{n2} = \cdots$. ■

In nondiscrete cases we will use various techniques to assign a probability $P(A)$ to each event A. In each case P will have properties (a), (b), and (c) of Theorem 1, and these properties are essentially all that is needed in order to develop probability theory in the standard way. The discrete case consists of those P which satisfy property (d) as well.

This paragraph and Exercises 6 to 8 show how to construct some nondiscrete P. Let $f\colon R^n \to R$ be a continuous real-valued function on R^n for which $f(x) \geq 0$ for all x in R^n and for which the integral of f over all of R^n equals 1. Now for each subset A of R^n we define $P_f(A)$ to equal the integral of f over the set A. It is intuitively reasonable that P_f satisfies properties (a), (b), and (c) of Theorem 1. However, $P_f(A) = 0$ if A is any countable subset of R^n. Thus Ω_0 is empty and $P_f(\Omega_0) = 0$, so property (d) does not hold for P_f.

In our treatment of the discrete case we will use properties (a), (b), and (c) freely, but we use (d) only where necessary, or when significant simplication results. In this way we achieve greater generality and emphasize a way of thinking which requires little change later.

Theorem 2 collects together a number of simple properties of P which are true in the most general cases. The only properties of P used in the proof are that P assigns a real number $P(A) \geq 0$ to each event A, $P(\Omega) = 1$, and P is countably additive.

The notation of Theorem 2 would be more complicated if p were used instead of P. The interpretation of $P(A)$ as the mass of A should make the conclusions appear to be reasonable.

Theorem 2 If (Ω,P) is a discrete probability space the following properties hold:

(*a*) $P(\varphi) = 0$.

(*b*) $P(A_1 \cup A_2 \cup \cdots \cup A_N) = P(A_1) + P(A_2) + \cdots + P(A_N)$ for any disjoint events $A_1, A_2, \ldots, A_N$.

(*c*) $0 \leq P(A) \leq 1$ and $P(A) + P(A^c) = 1$ for any event A.

(*d*) $P(A) \leq P(B)$ whenever event A is a subset of event B.

(*e*) $P(A \cup B) = P(A) + P(B) - P(AB)$ for any events A and B.

(*f*) $P(A_1 \cup A_2 \cup \cdots \cup A_N) \leq P(A_1) + P(A_2) + \cdots + P(A_N)$ for any events $A_1, A_2, \ldots, A_N$.

(*g*) $P(A_1 \cup A_2 \cup \cdots) \leq P(A_1) + P(A_2) + \cdots$ for any events $A_1, A_2, \ldots$.

(*h*) If $A_1 \subset A_2 \subset \cdots$ is an increasing sequence of events then $P(A_1 \cup A_2 \cup \cdots) = \lim_{N\to\infty} P(A_N)$.

(*i*) If $A_1 \supset A_2 \supset \cdots$ is a decreasing sequence of events then $P(A_1 \cap A_2 \cap \cdots) = \lim_{N\to\infty} P(A_N)$. ■

Proof

(*a*) Clearly $\varphi = \varphi \cup \varphi \cup \cdots$, so that $P(\varphi) = P(\varphi) + P(\varphi) + \cdots$ by the countable additivity of P, but this implies that $P(\varphi) = 0$.

(*b*) If $A_1, A_2, \ldots, A_N$ are disjoint events then so are $A_1, A_2, \ldots, A_N, \varphi, \varphi, \ldots$, and clearly $A_1 \cup A_2 \cup \cdots \cup A_N = A_1 \cup A_2 \cup \cdots \cup A_N \cup \varphi \cup \varphi \cup \cdots$, so that by the countable additivity of P we have

$$\begin{aligned} P(A_1 \cup A_2 \cup \cdots \cup A_N) &= P(A_1 \cup A_2 \cup \cdots \cup A_N \cup \varphi \cup \varphi \cup \cdots) \\ &= P(A_1) + P(A_2) + \cdots + P(A_N) + P(\varphi) + P(\varphi) + \cdots \\ &= P(A_1) + P(A_2) + \cdots + P(A_N). \end{aligned}$$

(*c*) If A^c is the complement of A with respect to Ω then A and A^c are disjoint and $\Omega = A \cup A^c$. Thus from the preceding result and the fact that $P(\Omega) = 1$ we get

$$1 = P(\Omega) = P(A \cup A^c) = P(A) + P(A^c).$$

But the probability of any event is nonnegative, so $P(A^c) \geq 0$, and hence $P(A) \leq 1$.

(*d*) If A is a subset of B let $C = BA^c$; that is, C consists of those points which belong to B but not to A. As indicated in Fig. 4, A and C are disjoint and $B = A \cup C$, so that $P(B) = P(A) + P(C)$; but $P(C) \geq 0$, so $P(B) \geq P(A)$.

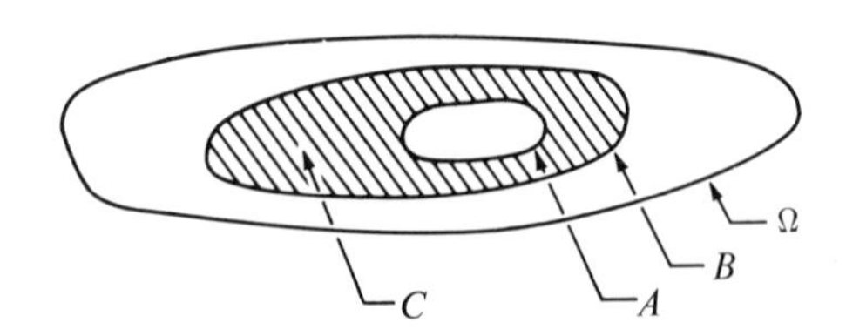

Fig. 4

(*e*) If A and B are two events let $C = AB^c$, $D = AB$, and $E = A^cB$. As indicated in Fig. 5, $A \cup B = C \cup D \cup E$, and C, D, and E are disjoint. Using part (*b*), we get

$$P(A \cup B) = P(C \cup D \cup E) = P(C) + P(D) + P(E)$$
$$= [P(C) + P(D)] + [P(E) + P(D)] - P(D).$$

But $A = C \cup D$ and C and D are disjoint, so $P(A) = P(C) + P(D)$. Also $B = E \cup D$ and E and D are disjoint, so $P(B) = P(E) + P(D)$. Thus $P(A \cup B) = P(A) + P(B) - P(D)$, where $D = AB$.

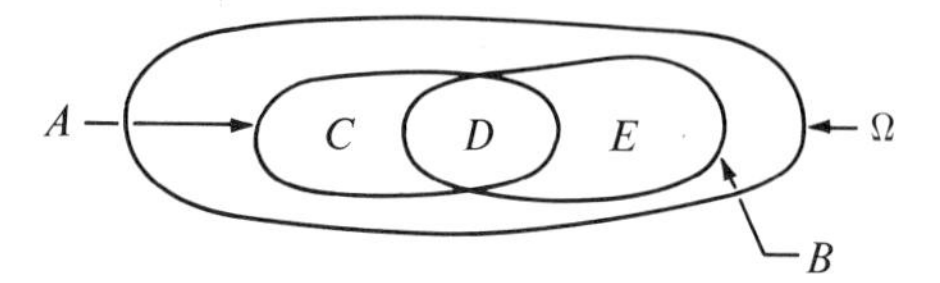

Fig. 5

(*f*) Let $A = A_1 \cup A_2 \cup \cdots \cup A_N$. As in Example 5 (Sec. 1-4), let $B_n = A_nA_{n-1}^cA_{n-2}^c \cdots A_1^c$, so that B_n consists of those points which belong to A_n but to none of $A_1, A_2, \ldots, A_{n-1}$. Clearly $B_1, B_2, \ldots, B_N$ are disjoint and $A = B_1 \cup B_2 \cup \cdots \cup B_N$, so that $P(A) = P(B_1) + P(B_2) + \cdots + P(B_N)$. But B_n is a subset of A_n, so $P(B_n) \leq P(A_n)$, and hence $P(A) \leq P(A_1) + P(A_2) + \cdots + P(A_N)$.

(*g*) As in the proof of property (*f*), let $B_n = A_nA_{n-1}^cA_{n-2}^c \cdots A_1^c$, so that $P(B_n) \leq P(A_n)$, while $B_1, B_2, \ldots$ are disjoint and $A_1 \cup A_2 \cup \cdots = B_1 \cup B_2 \cup \cdots$. Thus

$$P(A_1 \cup A_2 \cup \cdots) = P(B_1 \cup B_2 \cup \cdots)$$
$$= P(B_1) + P(B_2) + \cdots \leq P(A_1) + P(A_2) + \cdots.$$

(*h*) Starting as in the proof of property (*g*), we have

$$P(A_1 \cup A_2 \cup \cdots) = P(B_1 \cup B_2 \cup \cdots)$$
$$= \lim_{N\to\infty} \{P(B_1) + \cdots + P(B_N)\}$$
$$= \lim_{N\to\infty} P(A_1 \cup \cdots \cup A_N).$$

Thus for *any* sequence $A_1, A_2, \ldots$ of events we always have $P(A_1 \cup A_2 \cup \cdots) = \lim_{N\to\infty} P(A_1 \cup \cdots \cup A_N)$. Now if $A_1, A_2, \ldots$ is an increasing sequence of events then $A_1 \cup \cdots \cup A_N = A_N$, and we get property (*h*).

(*i*) Let $A_1, A_2, \ldots$ be any sequence of events and let $A = A_1 \cap A_2 \cap \cdots$. From Example 4 (Sec. 1-4) we have $A^c = A_1^c \cup A_2^c \cup \cdots$. From the proof of property (*h*) and the De Morgan laws

$$1 - P(A) = P(A^c) = \lim_{N\to\infty} P(A_1^c \cup \cdots \cup A_N^c) = \lim_{N\to\infty} P((A_1 \cap \cdots \cap A_N)^c)$$
$$= \lim_{N\to\infty} [1 - P(A_1 \cap \cdots \cap A_N)].$$

Thus for *any* sequence $A_1, A_2, \ldots$ of events we always have $P(A_1 \cap A_2 \cap \cdots) = \lim_{N \to \infty} P(A_1 \cap \cdots \cap A_N)$. Now if $A_1, A_2, \ldots$ is a decreasing sequence of events then $A_1 \cap \cdots \cap A_N = A_N$, and we get property ($i$). ■

Henceforth we will often use results from Theorem 2 without comment. Properties (h) and (i) will be used primarily in nondiscrete contexts later.

Verbalizing results like those in Theorem 2 can assist in remembering them and making them seem more natural. Thus $P(\varphi) = 0$ can be interpreted as saying that the probability is 0 that the experiment will not have an outcome. The countable additivity of P and Theorem 2b mean that if at most one of a countable collection of events can occur, the probability that at least one of them will occur equals the sum of their probabilities. The probability that A will not occur is $1 - P(A)$. If the occurrence of A implies the occurrence of B (that is, A is a subset of B) then A must have less probability than B. Theorem 2e is easily recalled by visualizing a picture like Fig. 5, which shows that $P(AB)$ contributes once to $P(A \cup B)$ but twice to $P(A) + P(B)$. The probability of the occurrence of at least one of a countable collection of events is at most the sum of their probabilities. If a sequence of events either increases or decreases to an event, then the same is true of the probabilities.

We can now easily show that every P function which satisfies Theorem 1 is obtainable from a discrete probability density. Thus we always deal with P which satisfy properties (a), (b), and (c) of Theorem 1, and the discrete case corresponds to those P which also satisfy property (d).

Theorem 3 Let every subset of a nonempty set Ω be called an event. Let P be *any* real-valued function whose domain is the class of all events, and for which all properties of Theorem 1 hold. Define the function $p: \Omega \to R$ by $p(\omega) = P(\{\omega\})$ for all ω in Ω. Then p is a discrete probability density, and the given function P is its corresponding discrete probability measure. ■

***Proof** Let P and p be as stated in the theorem. Now $P(A) \geq 0$, so $p(\omega) = P(\{\omega\}) \geq 0$ for all ω in Ω, and $\Omega_0 = \{\omega : p(\omega) > 0\}$ is countable. Thus we need only show that the sum of p over Ω_0 is 1 and that P is related to p by $P(A) =$ [the sum of all of the values $p(\omega)$ for all ω in A for which $p(\omega) > 0$]. The proof of Theorem 2 used only properties (a), (b), and (c) of Theorem 1, so we know that our P satisfies (b), (c) and (d) of Theorem 2. If A is any event let $A_0 = A\Omega_0$ consist of those points of A which belong to Ω_0. Let $A_1 = A\Omega_0{}^c$ be the remainder of A. Clearly A_0 and A_1 are disjoint and $A = A_0 \cup A_1$. Thus $P(A) = P(A_0) + P(A_1)$. Also $P(A_1) \leq P(\Omega_0{}^c)$, since A_1 is a subset of $\Omega_0{}^c$. But $1 = P(\Omega_0) + P(\Omega_0{}^c)$ and $P(\Omega_0) = 1$, so $P(A_1) \leq P(\Omega_0{}^c) = 0$. Thus $P(A_1) = 0$, and hence $P(A) = P(A_0)$.

If we let $\omega_1, \omega_2, \ldots$ be an enumeration of A_0 then

$$P(A) = P(A_0) = P(\{\omega_1\} \cup \{\omega_2\} \cup \cdots) = P(\{\omega_1\}) + P(\{\omega_2\}) + \cdots$$
$$= p(\omega_1) + p(\omega_2) + \cdots.$$

Thus P and p are properly related, and by letting $A = \Omega$ we see that the sum of p over Ω_0 is 1. ■

EXERCISES

1. Let $\Omega = \{\omega_1, \omega_2, \omega_3, \omega_4\}$ and define a discrete density $p: \Omega \to R$ by $p(\omega_i) = .3$ for $i = 1, 2, 3$ and $p(\omega_4) = .1$. Exhibit its corresponding discrete probability measure P; that is, tabulate $P(A)$ for all $2^4 = 16$ events.

2. In a particular geographical region the probability of a significant amount of rain during any particular day in May is at most $1/155$, and for June it is at most $1/300$ for each day (May has 31 days and June has 30 days). Show that we can be "fairly" certain that during the months of May and June no day will have a significant amount of rain.

3. (*a*) Find a formula analogous to Theorem 2*e* that expresses $P(A_1 \cup A_2 \cup A_3)$ as a sum, with coefficients of ± 1, of the $P(A_i)$ and probabilities of various intersections $P(A_i A_j)$ and $P(A_1 A_2 A_3)$.

(*b*) Interpret event A_i as meaning that power line i is inoperative, as was done in Exercise 1*d* (Sec. 6-2). What probabilities does part (*a*) of this exercise show to be sufficient for the calculation of the probability $P(A_1 \cup A_2 \cup A_3)$ that at least one of three lines is inoperative?

4. (*a*) Prove that $P(A_1 A_2 \cdots A_N) \geq 1 - \{P(A_1^c) + P(A_2^c) + \cdots + P(A_N^c)\}$ for any $p: \Omega \to R$ and any events $A_1, \ldots, A_N$. Thus if we are certain enough that *each* of N events will occur, $P(A_i) \doteq 1$, then we are fairly certain that they will all occur, $P(A_1 A_2 \cdots A_N) \doteq 1$. No assumptions concerning independence are needed for this *basic and frequently useful result.*

(*b*) An amplifier will operate properly for 10,000 hours if every one of its seven transistors and five vacuum tubes operates properly for the specified time period. For each transistor (vacuum tube) the probability is at least .99 (.98) that it will operate properly for the specified time. Show that the probability is fairly large that the amplifier will operate properly for 10,000 hours.

COMMENT: We are assuming, for example, that the probability is at least .99 that transistor 1 will function properly for 10,000 hours *in the amplifier and environment being considered.* Therefore the calculation does not apply to superb transistors and vacuum tubes used in an amplifier which is so wired that it "explodes" when all components are inserted. Similarly, corrections may be required for life-test data on individual components if the amplifier is such that acceptable deterioration of one of the components drastically increases the chance of failure of another component.

5. We show that the assumption in Theorem 3 of property *d* of Theorem 1 can be replaced by the weaker assumption that $P(\Omega_0) = 1$, where $\Omega_0 = \{\omega: P(\{\omega\}) > 0\}$. Let every subset of a nonempty set Ω be called an event. Let P be *any* real-valued function whose domain is the class of all events and for which properties (*a*), (*b*) and (*c*) of Theorem 1 hold. Let $\Omega_0 = \{\omega: P(\{\omega\}) > 0\}$. Prove that Ω_0 is countable. It then immediately follows that $P(\Omega_0) = P(\{\omega_1\}) + P(\{\omega_2\}) + \cdots$ if $\omega_1, \omega_2, \ldots$ is an enumeration of Ω_0.

6. This exercise illustrates the fact that many nondiscrete cases differ from the discrete case primarily in the replacement of sums by integrals. Let $f: R^2 \to R$ be defined by

$$f(x,y) = \begin{cases} e^{-x-y} & \text{if } x > 0 \text{ and } y > 0 \\ 0 & \text{otherwise.} \end{cases}$$

An event is a subset of the sample space R^2. The probability $P(A)$ of an event A is defined to equal the integral of f over A:

$$P(A) = \iint_A f(x,y)\, dx\, dy.$$

Thus $P(A)$ equals the volume which is under the surface f and above the subset A of the xy plane. We interpret the sample point (x,y) to mean that unit 1 breaks down x time units after it is put into operation and unit 2 breaks down y time units after it is put into operation. An $f: R^2 \to R$ of the form we have assumed applies to many life-testing problems.

(*a*) Find the probability of event A that unit 1 breaks down prior to time 2 if it is put into operation at time 0.

(*b*) Find the probability of event B that both units break down prior to time 2. Assume that they are both put into operation at time 0.

(*c*) Unit 1 is put into operation, and when it breaks down unit 2 is put into operation. Find the probability of event C that the total operating time $x + y$ is less than 2 time units.

7. This exercise illustrates the fact that *each* sample point can correspond to the outcome of an infinite sequence of spins of a random wheel. Let the sample space be the unit interval $0 \le x \le 1$. Our experiment consists of an infinite sequence of spins of a random wheel which has 10 sectors labeled $0, 1, 2, \ldots, 9$. Our interpretation of a sample point x is that if x is written as an infinite decimal expansion $x = .a_1a_2a_3 \cdots$ then spin n resulted in sector a_n. Thus for sample point $x = .239875805 \cdots$; sector 8 occurred on spins 4 and 7, etc. We define $P(A)$ to be the total length of event A. In particular if A is the union of countably many disjoint intervals then $P(A)$ is the sum of their lengths. We wish to provide some justification for the fact that this definition of P corresponds to the 10-sectored wheel having all sectors of equal size.

(*a*) Let B be the event that the first trial results in sector 7, so that B consists of all those $x = .a_1a_2a_3 \cdots$ for which $a_1 = 7$. Find $P(B)$.

(*b*) Let C be the event that the second trial results in sector 5, so that C consists of all those $x = .a_1a_2a_3 \cdots$ for which $a_2 = 5$. Find $P(C)$.

(*c*) Let $D = BC$ be the event that sector 7 occurs on trial 1 and sector 5 occurs on trial 2; find $P(D)$. Note that an experiment consisting of n spins of the wheel would have 10^n possible outcomes and each should have probability 10^{-n}.

(*d*) Our setup has the difficulty, as noted in Chap. 4, that the one real number $.36000 \cdots = 0.35999 \cdots$ has two decimal expansions, and so it corresponds to two different conceivable outcomes of our conceptual experiment. This difficulty is not serious, because our P assigns zero probability to the set of such ambiguities. To see this *prove that the set of such ambiguities is denumerable.* We can therefore let $\{x_1, x_2, \ldots\}$ be an enumeration of them, so that by our assumed countable additivity of P we have

$$P(\{x_1, x_2, \ldots\}) = P(\{x_1\}) + P(\{x_2\}) + \cdots = 0 + 0 + \cdots = 0$$

since the length of a point is 0.

8. If a plant has an initial height of A units at time $t = 0$ and receives sunshine at a constant intensity of α units per unit time its height at time t is $Ae^{\alpha t}$. We assume that $0 \le A \le 1$, $0 \le t \le 1$, $0 \le \alpha \le 1$, so that the units have been normalized; for example, the unit height might equal 3 inches, while a unit of time might equal 4 hours. Note that

$$\left(\frac{d}{dt}\right)(Ae^{\alpha t}) = (\alpha)(Ae^{\alpha t})$$

so that the rate of growth is proportional to the present height and to the intensity of sunshine. Let the sample space be $\Omega = \{(A,\alpha): 0 \le A \le 1, 0 \le \alpha \le 1\}$. To each sample point (A,α) in the unit square we associate the *function* of time, $Ae^{\alpha t}$ for $0 \le t \le 1$, which is the growth behavior of a plant with initial height A receiving sunshine of intensity α. Three sample points are singled out, and their associated functions are shown in Fig. 6.

6-4, 6

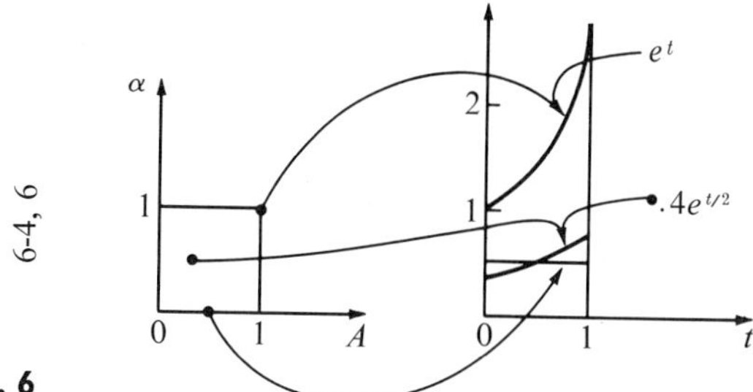

Fig. 6

We could use the collection of all such functions as our sample space instead of Ω, in which case we would have a function space for a sample space. Such an approach is taken in Chap. 22 where the collection of permitted functions is extremely varied, rather than quite restricted as we have assumed here. We are interested in the event F that the final height of a plant is over one unit, that is, that $Ae^{\alpha} > 1$.

(*a*) Assume that the discrete probability density $p: \Omega \to R$ of Fig. 7 applies, and find $P(F)$.

(A,α)	$p(A,\alpha)$
(.2,.9)	.1
(.8,0)	.3
(1,1)	.2
(.9,.8)	.4

Fig. 7

(*b*) Find the area of the event F, where F is the subset of those points (A,α) of the unit square satisfying $Ae^{\alpha} > 1$. Interpreting this number as the probability of F corresponds roughly to assuming that all pairs (A,α) are equally likely.

7
Independent Events

This chapter introduces the basic idea of independent events. The definition of independence is motivated by the simple and even more fundamental idea of conditional probability.

7-1 CONDITIONAL PROBABILITIES

If (Ω,P) is a discrete probability space and A and B are events with $P(A) > 0$ then $P(B \mid A)$ is defined by

$$P(B \mid A) = \frac{P(AB)}{P(A)}.$$

We call $P(B \mid A)$ the **conditional probability of B given A**, or the probability of B given A. Since

$$P(AB) = [P(B \mid A)][P(A)]$$

we can say that the probability of "A and B" is equal to the probability of B given A, times the probability of A.

Consider the random-wheel experiment of Fig. 1 with 23 sectors, and suppose that we are interested in the event $B = \{\omega_1,\omega_2,\omega_{22},\omega_{23}\}$. Let $A = \{\omega_1, \omega_2, \ldots, \omega_{12}\}$

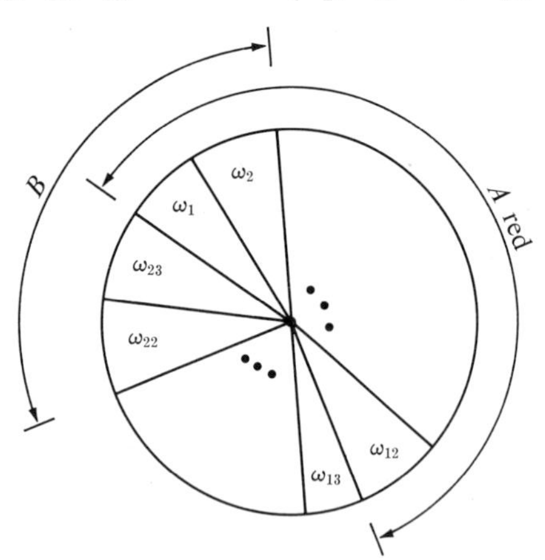

Fig. 1

and assume that these and only these sectors are colored red. Suppose the wheel is spun once and we learn *only* that the outcome was a red sector. Then we have a new experimental situation, and we can ask for the probability that under these new conditions the outcome will belong to B. A reasonable answer is that under these new conditions the probability of B is just the fraction of the red-colored circumference which is assigned to B. For Fig. 1 we would get $[p(\omega_1) + p(\omega_2)]/[p(\omega_1) + p(\omega_2) + \cdots + p(\omega_{12})]$. Thus a reasonable definition for the probability of B, given that A has occurred, is $[1/P(A)]$ times the probability of that part of B which is contained in A. This is how $P(B \mid A)$ was pictured in Fig. 2. We assumed that the wheel had actually

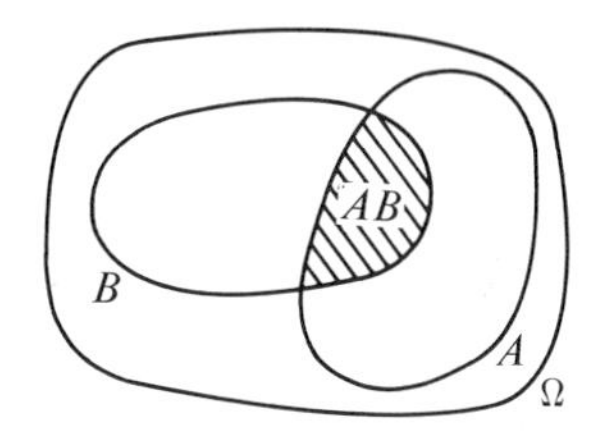

Fig. 2

been spun and red had been observed. There are many other situations where we might be interested in $P(B \mid A)$. We might, for instance, bet on B and agree to ignore all experiments which do not have a red outcome.

In probability theory considerable attention is devoted to independence. We motivate the definition of independence in terms of conditional probabilities. Conditional probabilities are also fundamental in their own right. For these reasons we motivate the interpretation of the number $P(B \mid A)$ as "the probability of B given A," in several ways.

We now take a slightly different point of view and arrive at the same interpretation for $P(B \mid A)$. We start with the density $p: \Omega \to R$ and with event A, $P(A) > 0$, of Fig. 3*a*. We wish to construct a new density to correspond to the experiment $p: \Omega \to R$ with the additional requirement that A must occur. It would be reasonable to use A for the new sample space, but there is no harm in letting the new sample space be Ω if we require that the new probability of A be 1. Thus we want a new density $p_A: \Omega \to R$ and we require that $p_A(\omega) = 0$ if ω does not belong to A. In Fig. 3*b* the density was made 0 off A but left unchanged on A. This "altered p" of Fig. 3*b* seems to be a reasonable candidate for the new density, except that it does not sum to 1. We can take care of this and still keep the relative chances (ratios) for points within A the same by multiplying each value by $1/P(A)$. We have thus arrived at the density $p_A: \Omega \to R$ of Fig. 3*c*, where

$$p_A(\omega) = \begin{cases} \dfrac{p(\omega)}{P(A)} & \text{if } \omega \in A \\ 0 & \text{if } \omega \notin A. \end{cases}$$

If, as indicated in Fig. 3*d*, we compute the probability of an event B from this new density $p_A: \Omega \to R$, which has a corresponding discrete probability space (Ω, P_A),

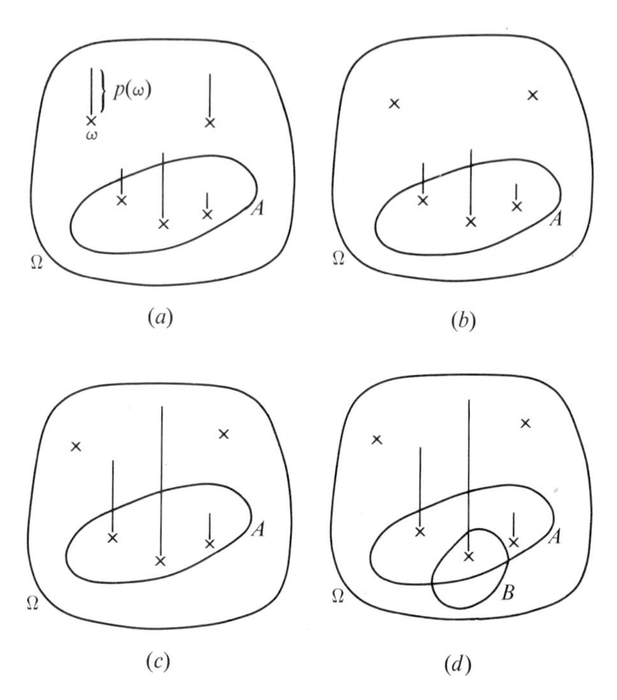

Fig. 3 (*a*) $p\colon \Omega \to R$; (*b*) altered p; (*c*) $p_A\colon \Omega \to R$; (*d*) $p_A\colon \Omega \to R$.

we see that $P_A(B)$ equals $[1/P(A)]$ times the sum of $p(\omega)$ over AB, so that $P_A(B) = P(B \mid A)$. Thus the density p_A gives us the same probability for every event B that we would get by using $P(B \mid A)$. Therefore $P(B \mid A)$ is what we get for the probability of B if we use the density obtained by setting $p = 0$ off A and normalizing what is left.

Let us now formalize these ideas. If $p\colon \Omega \to R$ is a discrete probability density and A is an event with $P(A) > 0$, the function $p_A\colon \Omega \to R$ defined by

$$p_A(\omega) = \begin{cases} \dfrac{p(\omega)}{P(A)} & \text{if } \omega \in A \\ 0 & \text{if } \omega \notin A \end{cases}$$

is called the **discrete probability density obtained by conditioning** p **by,** or on, **the event** A. If P_A is the discrete proability measure corresponding to p_A we call (Ω,P_A) the **discrete probability space obtained by conditioning** (Ω,P) **by the event** A. We have seen that P_A and P satisfy the basic relation

$$P_A(B) = P(B \mid A) = \frac{P(AB)}{P(A)} \qquad \text{for every event } B.$$

Applying Theorem 2*e* (Sec. 6-4) to (Ω,P_A) of the last paragraph, we see that $P_A(B \cup C) = P_A(B) + P_A(C) - P_A(BC)$ for any two events B and C. But $P_A(D) = P(D \mid A)$ for every event D. Therefore for any events A, B, and C with $P(A) > 0$ we have $P(B \cup C \mid A) = P(B \mid A) + P(C \mid A) - P(BC \mid A)$. This is one of the many identities that can be obtained by conditioning every probability in an identity by an event A with $P(A) > 0$. This identity can be obtained with a little effort from blind manipulations proceeding directly from the definition of $P(B \mid A)$, but its existence and truth are obvious, since it is nothing more than the original identity for the measure P_A. In essence $P_A(B) = P(B \mid A)$ is a discrete probability measure if A is kept fixed while B is allowed to vary.

We wish to indicate that our interpretation of $P(B \mid A)$ is reasonable from relative-frequency considerations. Let $\Omega = \{\omega_1, \omega_2, \ldots, \omega_{10}\}$ and let $A = \{\omega_2,\omega_3,\omega_4,\omega_5\}$ and $B = \{\omega_2,\omega_4,\omega_8\}$ be two subsets of interest. Let e be obtained from n spins of a random wheel with 10 sectors labeled $\omega_1, \omega_2, \ldots, \omega_{10}$.

$$e = \begin{pmatrix} & \surd & \surd & & & \surd & & & \surd & \surd & & & & \surd & & \\ \times & \times & \times & \times & & & \times & & \times & & & \times & & \times & & \times \\ \omega_3, & \omega_2, & \omega_2, & \omega_3, & \omega_1, & \omega_8, & \omega_3, & \omega_1, & \omega_4, & \omega_8, & \omega_9, & \omega_5, & \omega_1, & \omega_2, & \ldots, & \omega_5 \\ 1 & 2 & 3 & 4 \ldots & & & & & & & & & & & & n \end{pmatrix}.$$

We have placed a cross ($\times$) over elements of A and a check ($\surd$) over elements of B. Recall that the relative frequency $M_e(A)$ of A in e is just $1/n$ times the number of crosses. We define α as the number of coordinates with both a check and a cross, divided by the number of coordinates with crosses. Thus α is the fraction of coordinates checked *if* we look at *only* those coordinates which already have a cross. We may think of α as a reasonable candidate for the chance of B if we restrict our attention to those cases where A occurred and use only the information in e. Clearly $\alpha = M_e(AB)/M_e(A)$, since the $1/n$ factors cancel. Thus $\alpha = P(B \mid A)$ *if we define* p by the relative frequencies in e, that is, if we let $p = m_e$. However, *if we have* p then $\alpha \doteq P(B \mid A)$, *if* the relative frequencies behaved properly, that is, if $M_e(AB) \doteq P(AB)$ and $M_e(A) \doteq P(A)$ for the e obtained.

Example 1 Consider the density of Fig. 4 and Exercise 1 (Sec. 6-2) with the telephone-exchange interpretation. For $i = 1, 2, 3, 4$ let A_i be the event that the exchange put through at least one call during the ith second; for example, $P(A_1) = .44$ and $P(A_2) = .60$. The probability that the exchange will be used during second 2, given that it was used during second 1, is

$$P(A_2 \mid A_1) = \frac{P(A_2A_1)}{P(A_1)} = \frac{.29}{.44} \doteq .659.$$

Thus learning that A_1 has occurred increases the chance of A_2. Similarly

$$P(A_3 \mid A_2A_1) = \frac{P(A_3A_2A_1)}{P(A_2A_1)} = \frac{.23}{.29} \doteq .791$$

while $P(A_3) = .75$, so that there is a slight increase in the chance of A_3 if we learn that A_1 *and* A_2 have occurred. For any three events we have

$$\begin{aligned} P(A_3A_2A_1) &= P(A_3 \mid A_2A_1)P(A_2A_1) \\ &= P(A_3 \mid A_2A_1)\, P(A_2 \mid A_1)P(A_1) \end{aligned}$$

by using the definition of conditional probability twice. For our example this equation can be interpreted as follows:

[the probability that the exchange will be used for all three of the first 3 seconds] = [the probability that it will be used during the 1st second]

× [the probability that it will be used during the 2d second, given that it was used during the 1st second]

× [the probability that it will be used during the 3d second, given that it was used during both the 1st and 2d seconds].

The conditional density $p_{A_1}: \Omega \to R$ obtained by conditioning on the event A_1: for example, is defined by $p_{A_1}(\omega_i) = 0$ if $i = 1, 2, \ldots, 8$ and $p_{A_1}(\omega_j) = p(\omega_j)/P(A_1)$ if $j = 9, 10, \ldots, 16$. ■

EXERCISES

1. Jones has received a letter at his office from a friend. The probability is .5 that he will take the letter home with him at the end of the day. If he does take it home the probability is .3 that he will write an answer that evening. If he takes it home and writes an answer the probability is .8 that he will remember to take the answer with him in the morning, in which case the probability is .7 that he will mail it on the way to work. What is the probability that this whole chain of events will take place?

2. Three dice are cast. They are colored red, white, and blue. What is the conditional probability that the white die will come to rest with one dot on its upward face, given that the sum of the dots on the three faces is 5?

3. (*a*) What is the probability, to three decimal places, that the first toss of a fair coin was heads if the first 100 tosses yielded exactly 32 heads?

(*b*) What is the conditional density on $\{H,T\}^{100}$, given that 32 heads occurred?

4. *Conditional Probabilities for Random Relays* (*Fig. 5, Sec. 6-2*). Let B, A, and D be the events that relay 1 is closed, there is a closed connection between x and y, and relay 3 is closed, respectively.

(*a*) Use the density $p: \Omega \to R$ of Fig. 4 (Sec. 6-2) to find $P(B)$, $P(B \mid A)$, $P(B \mid AD)$, and $P(B \mid AD^c)$. Note the paradoxical fact that $P(B) > P(B \mid A)$.

(*b*) Assign a new probability density $q: \Omega \to R$, with associated measure Q, to the same sample space in such a way that $Q(B) = .9$ and $Q(B \mid A) = 0$.

5. *Conditional Probabilities of Rain.* For the density $p: \Omega \to R$ of Fig. 4 let $A_i = \{\omega$: the ith coordinate of ω equals $1\}$ and interpret A_i as the event that it is raining in town T_i.

ω	$p(\omega)$
(0,0,0)	.6
(0,0,1)	.02
(0,1,0)	.02
(0,1,1)	.05
(1,0,0)	.05
(1,0,1)	.08
(1,1,0)	.08
(1,1,1)	.1

Fig. 4

(*a*) Find the probability $P(A_1 \mid A_2)$ that it is raining in T_1, given that it is raining in T_2.
(*b*) Find the probability $P(A_1 \mid A_2 \cup A_3)$ that it is raining in T_1, given that it is raining in at least one of the other two towns.
(*c*) Find the probability that it is raining in T_1 and T_2, given that it is not raining in T_3.
(*d*) Evaluate the probabilities $P(A_2A_3{}^c \mid A_1{}^c)$, $P(A_1A_2A_3 \mid A_1{}^c)$, and $P(A_2 \cup A_3{}^c \mid A_1)$ and give a verbal interpretation for each.

6. Let (Ω,P) be a discrete probability space and let A be an event for which $P(A) > 0$. Let $P_A(B) = P(AB)/P(A)$ for every event B. We know that P_A is a discrete probability measure, so that if $B_1, B_2, \ldots$ are disjoint then $P_A(B_1 \cup B_2 \cup \cdots) = P_A(B_1) + P_A(B_2) + \cdots$. Prove this same result directly from the definition $P_A = P(AB)/P(A)$.

7. Let (Ω,P) be a discrete probability space and let $B_1, B_2, \ldots$ be a sequence of disjoint events whose union equals the whole sample space Ω. Assume that $P(B_n) > 0$ for all n.

(*a*) Prove that if A is any event then $P(A) = \sum_{n=1}^{\infty} P(A \mid B_n)P(B_n)$. COMMENT: This result is used often. It says that if precisely one of the events B_n must occur then the probability of A equals the sum of: the probability of A given that the nth event occurs, times the chance that the nth event will occur.
(*b*) Let B_n be the event that the first breakdown of a machine occurs during its nth hour of operation. Let A be the event that the first breakdown is repaired within 3 hours. Assume that $P(B_n) = \frac{1}{2}^n$ and $P(A \mid B_n) = .3 + (.6)^n$. Find $P(A)$.
(*c*) Deduce the equation of part (*a*) under the assumption that $A \subset \left(\bigcup_{n=1}^{\infty} B_n\right)$, where $B_1, B_2, \ldots$ are disjoint events, with each $P(B_n) > 0$.

7-2 BAYES' THEOREM

We now state, prove, discuss, and apply a famous formula which is controversial in some applicational contexts.

Theorem 1: Bayes' theorem Let $B_1, B_2, \ldots, B_n$ be disjoint events for which each $P(B_i) > 0$. Let A be an event such that $P(A) > 0$ and such that A is contained in the union $B_1 \cup B_2 \cup \cdots \cup B_n$. Then

$$P(B_i \mid A) = \frac{P(A \mid B_i)P(B_i)}{\sum_{k=1}^{n} P(A \mid B_k)P(B_k)} \qquad \text{if } i = 1, 2, \ldots, n. \quad \blacksquare$$

Proof We have

$$P(A) = P(AB_1) + P(AB_2) + \cdots + P(AB_n)$$

since, as indicated in Fig. 1, the events $AB_1, AB_2, \ldots, AB_n$ are disjoint and their union equals A. From the definition of conditional probability we have

$$P(B_i \mid A) = \frac{P(AB_i)}{P(A)} = \frac{P(AB_i)}{\sum_{k=1}^{n} P(AB_k)}$$

and by using the definition of conditional probability on each probability in the right-hand side we complete the proof of the theorem. ■

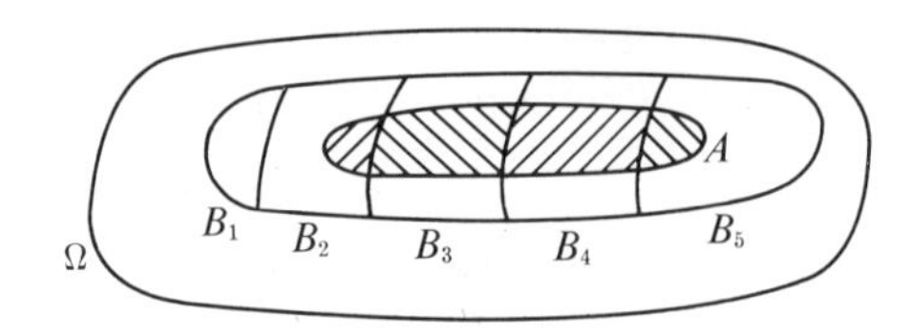

Fig. 1

In applications of Bayes' theorem the union of $B_1, B_2, \ldots, B_n$ usually equals the whole sample space. Sometimes the events $B_1, B_2, \ldots, B_n$ are thought of as n hypotheses, exactly one of which is true. Thus if we have n such hypotheses with probabilities $P(B_1), P(B_2), \ldots, P(B_n)$, and if an event A occurs, and we can calculate its probability $P(A \mid B_k)$ under each hypothesis, then Theorem 1 permits us to calculate the revised set $P(B_1 \mid A), P(B_2 \mid A), \ldots, P(B_n \mid A)$ of probabilities for the hypotheses, taking into account the fact that A has occurred.

Bayes' theorem is a straightforward deduction, within our mathematical theory, from definitions and assumptions. Its application in some contexts has been highly controversial. For example, most statisticians would not consider it appropriate to assign a numerical probability to the hypothesis that the sun will rise tomorrow.

Example 1 We spin wheel 0 of Fig. 2; if we obtain f, for 'fair,'' then we spin wheel 1, but if we obtain b, for "biased," then we spin wheel 2. Thus the outcome of an experiment is one of the four sample points (f,H), (f,T), (b,H), and (b,T). We have, essentially, a fair coin and a biased coin with probability .2 for heads. We select one of these two coins with probabilities

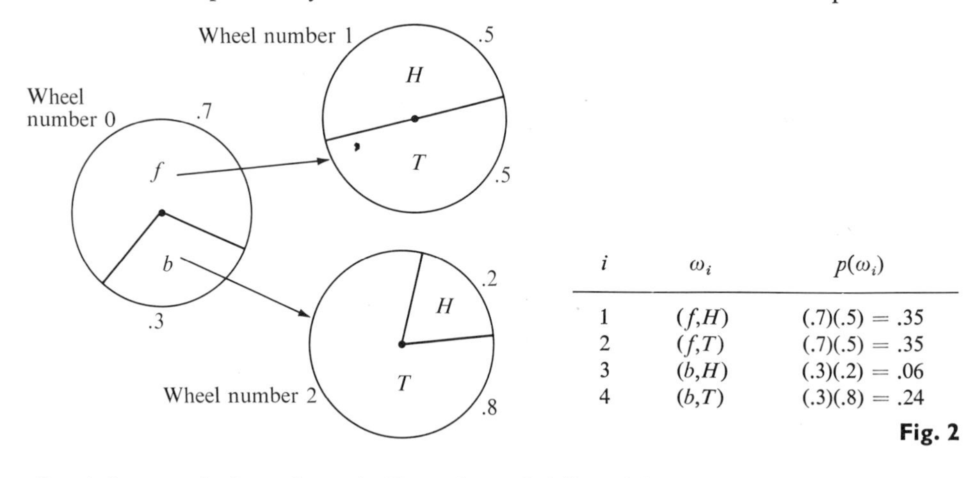

i	ω_i	$p(\omega_i)$
1	(f,H)	$(.7)(.5) = .35$
2	(f,T)	$(.7)(.5) = .35$
3	(b,H)	$(.3)(.2) = .06$
4	(b,T)	$(.3)(.8) = .24$

Fig. 2

.7 and .3, respectively, and toss it. Now, the probability of f is .7 and the probability of H, given coin f, is .5, so that from the definition of conditional probability we have $(.7)(.5) = .35$ for the probability of f and H. The previous sentence is really just motivation for the definition $p((f,H)) = .35$, since we have not yet defined a density on the sample space. Proceeding in this fashion, we define the density shown in Fig. 2. Let $B_1 = \{(f,H), (f,T)\}$ and $B_2 = \{(b,H), (b,T)\}$ be the events that wheels 1 and 2 are used, respectively. Let $A = \{(f,H), (b,H)\}$ be the event that H is obtained. *From the density of Fig. 2* we find that

$$P(B_1) = .35 + .35 = .7 \qquad P(B_2) = .06 + .24 = .3$$

$$P(A \mid B_1) = \frac{.35}{.7} = .5 \qquad P(A \mid B_2) = \frac{.06}{.30} = .2$$

so that the unique density which we defined does yield the appropriate numbers for the experiment we described. *From the density of Fig. 2* we get

$$P(A) = .35 + .06 = .41$$

$$P(B_1 \mid A) = \frac{.35}{.41} \doteq .854 \qquad P(B_2 \mid A) = \frac{.06}{.41} \doteq .146.$$

Thus initially the probability of selecting the fair coin is .7, but if we learn that the experiment yielded a head the probability is .854 that the fair coin was used. This is sometimes stated as follows: the a priori probability for the fair coin is .7, but after a head is obtained the a posteriori probability for the fair coin is .854. Thus, among those experiments which yield heads, about 85 percent of the heads should result from the fair coin.

Usually in a problem of this type, Bayes' theorem is used and the density of Fig. 2 is not even calculated. Using $n = 2$ in Bayes' theorem, we again find that

$$P(B_1 \mid A) = \frac{P(A \mid B_1)P(B_1)}{P(A \mid B_1)P(B_1) + P(A \mid B_2)P(B_2)} = \frac{(.5)(.7)}{(.5)(.7) + (.2)(.3)} \doteq .854. \quad \blacksquare$$

Many problems like this one are most naturally stated in terms of the probabilities $P(B_i)$ for the various hypotheses and in terms of the probability measures P_{B_i} given the various hypotheses, while what we wish to find is the probability $P(B_i \mid A)$ of hypothesis B_i given that a particular event occurs. Bayes' theorem, then, is particularly convenient, since it permits the calculation of $P(B_i \mid A)$ almost directly from the given specifications.

EXERCISES

1. Three coins have probabilities .9, .8, and .5 for heads, respectively. One of them is selected at random, that is, with equal chance for each, and tossed. If the outcome is a head what is the a posteriori probability that the coin with probability .9 for heads was selected?

2. *False Alarms for Rare Events*

(*a*) If there is a burglary on a particular night the probability is .99 that the burglar alarm will ring. If there is no burglary on a particular night the probability is .005 that the burglar alarm will ring, falsely. The probability is .001 that a burglary will occur on a particular night. What is the probability that there is a burglary, given that the alarm just rang?

(*b*) What is the answer to part (*a*) if the numbers .99, .005, and .001 are replaced by 1, 10^{-4}, and 10^{-7}? Explain this odd answer in one sentence.

3. *Inferences from Blown Fuses.* An amplifier blows one or both of its fuses if one or both of its vacuum tubes is defective. Three events are defined as follows:

C_1 = tube α is defective and tube β is not defective; $P(C_1) = .5$.
C_2 = tube α is not defective and tube β is defective; $P(C_2) = .4$.
C_3 = both tubes are defective; $P(C_3) = .1$.

Let F_i be the event that the ith fuse is blown. The conditional probabilities for blown fuses, given vacuum-tube conditions, are as below:

i	$P(F_1F_2^c \mid C_i)$	$P(F_1^cF_2 \mid C_i)$	$P(F_1F_2 \mid C_i)$
1	.6	.2	.2
2	.1	.6	.3
3	.1	.1	.8

(*a*) What is the probability that both tubes are defective if both fuses have blown?

(*b*) If you found both fuses blown and could replace only one tube, which one would you replace and why?

4. There are $n + 1$ urns labeled $0, 1, 2, \ldots, n$ and each urn contains n balls. The ith urn contains i white balls and $n - i$ black balls.

(*a*) An urn is selected at random and a ball is drawn from it. What is the probability of obtaining a white ball?

(*b*) If a white ball is drawn what is the a posteriori probability distribution for the urn selected? Sketch it; that is, plot P[urn k was selected | a white ball was drawn] as a function of k.

(*c*) An urn is selected at random and $N \geq 1$ balls are drawn from it with replacement. What is the probability that all N balls are white?

5. If the next student you meet is a coed, what is the probability that she is single? More precisely, find $P(M^c \mid F)$ if

F = the next student you meet is female.
M = the next student you meet is married.
G = the next student you meet is a graduate student.

$$P(F) = \frac{8}{25}, \; P(M) = \frac{9}{25}, \; P(FG \mid M) = P(FG^c \mid M) = \frac{1}{9}.$$

7-3 INDEPENDENCE OF TWO EVENTS

If $P(A) > 0$ and $P(B) > 0$, it easily follows from the definition of $P(B \mid A)$ that either all three of the following are true or all three are false:

$$P(B \mid A) = P(B) \qquad P(AB) = P(A)P(B) \qquad P(A \mid B) = P(A)$$

If $P(B \mid A) = P(B)$ then the chance that B will occur is unaltered by (that is, independent of) the fact that A occurred, so we say that B is independent of A. From these equations it is clear that B is independent of A iff A is independent of B, so that independence is a symmetric relationship. For this reason, plus the advantage of not worrying about division by 0, we adopt the middle of the equations above for our formal definition.

Let (Ω,P) be a discrete probability space and let A and B be events. The events A and B are said to be **independent** iff $P(AB) = P(A)P(B)$. Clearly A and B are independent iff B and A are independent. We may say that A and B are **dependent** instead of saying that A and B are not independent.

Events A and B will be intimately related in some examples but still independent as defined above. Thus, as indicated by Figs. 1 and 2 (Sec. 7-1), the knowledge that A has occurred can tell us quite a bit about B, namely, that no outcome belonging to B but not to A can have occurred. But A and B can still be independent if $P(B \mid A) = P(B)$, so that the chance that B will occur has been unaffected by the information that A has occurred. Independence of events is primarily a numerical fact about probabilities rather than a fact about the relationship between the sets A and B without regard to P. To emphasize this we may call them **stochastically independent** or **statistically independent** rather than just independent. The adjectives "stochastically" and "statistically" may also be used later with independence of $n \geq 2$ events and independence of random variables.

Example 1 We illustrate conditional probability and independence by the very simple density $p: \Omega \to R$ of Fig. 1 with corresponding discrete probability space (Ω, P)

Fig. 1

Ω	p	p'
(R,R)	.27	$.123 = (.3)(.41)$
$(R,\bar{R})$	.03	$.177 = (.3)(.59)$
$(\bar{R},R)$	.14	$.287 = (.7)(.41)$
$(\bar{R},\bar{R})$	.56	$.413 = (.7)(.59)$

We interpret the sample points as follows, where α and β are towns which are geographically close:

(R,R) = it is raining in α and it is raining in β
$(R,\bar{R})$ = it is raining in α and it is *not* raining in β
$(\bar{R},R)$ = it is *not* raining in α and it is raining in β
$(\bar{R},\bar{R})$ = it is *not* raining in α and it is *not* raining in β

Let $A = \{(R,R), (R,\bar{R})\}$ = [it is raining in α] and $B = \{(R,R), (\bar{R}, R)\}$ = [it is raining in β]. Any one of the three facts below shows that A and B are not independent (that is, A and B are dependent):

$.9 \quad = P(B \mid A) \neq P(B) = .41$
$.27 \quad = P(AB) \neq P(A)P(B) = .123$
$.659 = P(A \mid B) \neq P(A) = .3$

Since $P(B \mid A) = .9$, a person should certainly take his umbrella if he plans an immediate trip to β and has just learned from the radio that it is raining in α. It is quite likely that he planned to take his umbrella anyway, since generally β is even worse, $P(B) = .41$, than α, which is notorious for its rain, $P(A) = .3$.

Now we use the facts $P(A) = .3$ and $P(B) = .41$ to construct the density $p': \Omega \to R$ of Fig. 1 with the same sample space Ω. Density p' has corresponding discrete probability space (Ω, P'). We find that $P'(A) = .3$ and $P'(B) = .41$, and A and B are stochastically independent. Thus the p' model makes each of α and β just as wet as the p model did, but now a statistically sophisticated traveler to β is not depressed or delighted, $P'(B \mid A) = P'(B \mid A^c) = P'(B) = .41$, by the report of the weather in α even though there may be intimate meteorological relations between the weather of the two towns. ■

For an event A to be independent of itself we must have $P(A \cap A) = P(A)P(A)$, that is, $P(A) = [P(A)]^2$, and this can occur only if $P(A) = 0$ or 1. *Thus if $0 < P(A) < 1$ then A is not independent of A.* We certainly would not usually expect an event to be independent of itself.

If a pair of events are independent then the complement of one of them is independent of the other. This is proved as follows: We always have $P(B) = P(AB) + P(A^cB)$, since B equals the disjoint union of those points of B which belong to A and those points of B which do not belong to A. Thus if (A,B) is a pair of events which are independent then

$$P(A^cB) = P(B) - P(AB) = P(B) - P(A)P(B) = P(B)[1 - P(A)] = P(B)P(A^c)$$

so (A^c,B) is an independent pair of events. This result is easily generalized to the following: *if one of the four pairs* (A,B) or (A^c,B) or (A,B^c) or (A^c,B^c) *is an independent pair of events then all four are.* For example, if (A,B^c) is an independent pair we can repeatedly use the fact stated in the first sentence of this paragraph, so that (A,B) is an independent pair, and hence (A^c,B) is an independent pair. Thus, for example, we can say that if A is independent of the occurrence of B then A is also independent of the nonoccurrence of B.

Let (Ω,P) be a discrete probability space and let $A_1, A_2, \ldots, A_n$ be a finite sequence of $n \geq 2$ events. The events $A_1, A_2, \ldots, A_n$ are said to be **pairwise independent** iff A_i and A_j are two independent events whenever $i \neq j$. In this case $P(A_i \mid A_j) = P(A_i)$, so that the probability of any one of the n events is unaffected by the knowledge of the occurrence of *any one* of the other $n - 1$ events. In discussing pairwise independence of events or random variables we never drop the adjective "pairwise."

The notion of pairwise independence is used only rarely. In the next section we define mutual independence of $n \geq 2$ events. Mutual independence is widely used and normally abbreviated to "independence." Very roughly, mutual independence of events $A_1, A_2, \ldots, A_n$ means that the probability of any one of them is unaffected by the knowledge of the occurrence of *any one or more* of the other $n - 1$ events. Thus, for example, mutual independence of $A_1, A_2, \ldots, A_{10}$ implies that $P(A_7 \mid A_1A_3A_9) = P(A_7)$. If $n = 2$, pairwise independence and mutual independence are equivalent. If $n \geq 3$ events are mutually independent they are pairwise independent, but as shown in Example 2, pairwise independent events *may not* be mutually independent. Thus mutual independence is stronger than pairwise independence. Mutual independence is usually considered intuitively to be more natural than pairwise independence. For the sake of simplicity we will usually follow tradition and seldom mention pairwise independence, even though some later results which we state and prove under the assumption of mutual independence also hold, with the same proof, under the weaker assumption of pairwise independence of the events or random variables involved.

Example 2 Let $\Omega = \{H,T\}^2$ and let $p: \Omega \to R$ correspond to two independent tosses of a fair coin. As shown in Fig. 2, A_1 is heads on first toss, A_2 is heads on second toss, and A_3 is the event

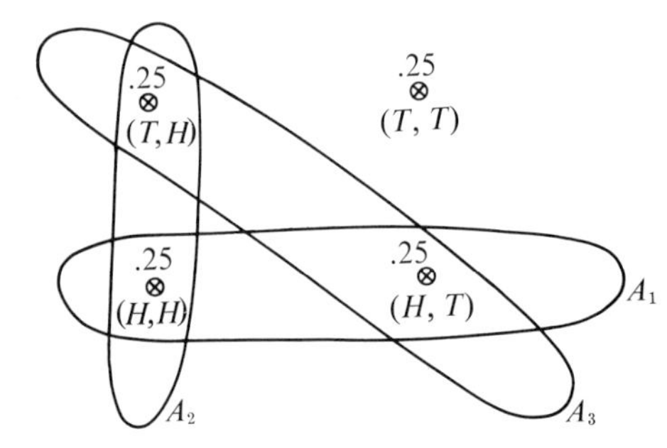

Fig. 2

that the outcomes of the two tosses are different. Clearly

$$P(A_1A_2) = P(A_1)P(A_2) \qquad P(A_1A_3) = P(A_1)P(A_3) \qquad P(A_2A_3) = P(A_2)P(A_3)$$

so that A_1, A_2, A_3 are pairwise independent. However $0 = P(A_3 \mid A_1A_2) \neq P(A_3) = .5$, so that the two events A_3 and A_1A_2 are dependent, as is obvious from the interpretation. Thus the three events A_1, A_2, and A_3 are not mutually independent, as defined in the next section. ∎

EXERCISES

1. Let $A_1, A_2, \ldots, A_7$ be a sequence of seven pairwise-independent events.

(*a*) Is each of the following a finite sequence of pairwise-independent events?

A_2, A_3, A_5, A_7

$A_3, A_1, A_2, A_4, A_5, A_6, A_7$

$A_1, A_2^c, A_3, A_4^c, A_5, A_6^c, A_7$

(*b*) How many numerical epuations are needed to guarantee the pairwise independence of $A_1, A_2, \ldots, A_7$?

(*c*) Are A_1A_2 and A_3 two independent events?

2. Prove that two independent events cannot be disjoint if each has positive probability.

7-4 INDEPENDENCE OF n EVENTS

For $n = 10$ we wish to define the mutual independence of $A_1, A_2, \ldots, A_{10}$ in such a way that A_3 is independent of the event $A_5A_7A_9$, for example. Thus we want $P(A_3 \mid A_5A_7A_9) = P(A_3)$, and we want $P(A_5 \mid A_7A_9) = P(A_5)$, and we want $P(A_7 \mid A_9) = P(A_7)$. Now we always have

$$\begin{aligned} P(A_3A_5A_7A_9) &= P(A_3 \mid A_5A_7A_9)P(A_5A_7A_9) \\ &= P(A_3 \mid A_5A_7A_9)P(A_5 \mid A_7A_9)P(A_7 \mid A_9)P(A_9) \end{aligned}$$

if the conditioning events all have positive probability, so that the conditional probabilities are defined. Thus the above requirements imply that

$$P(A_3A_5A_7A_9) = P(A_3)P(A_5)P(A_7)P(A_9).$$

We will say that $A_1, A_2, \ldots, A_{10}$ are mutually independent iff all equations like this last one are true.

Let (Ω,P) be a discrete probability space and let $A_1, A_2, \ldots, A_n$ be a finite sequence of $n \geq 2$ events. The events $A_1, A_2, \ldots, A_n$ are said to be **mutually independent** or **independent,** iff

$$P(A_{n_1}A_{n_2} \cdots A_{n_k}) = P(A_{n_1})P(A_{n_2}) \cdots P(A_{n_k})$$

for every subsequence of two or more of the events. We call $A_{n_1}, A_{n_2}, \ldots, A_{n_k}$ a **subsequence** of $A_1, A_2, \ldots, A_n$ iff the subscript integers satisfy $1 \leq n_1 < n_2 < \cdots < n_k \leq n$.

For $n = 2$ this definition agrees with our previous definition for the independence of two events. For $n = 3$ we see that A_1, A_2, A_3 are mutually independent iff the four equations below are true:

$$P(A_1A_2) = P(A_1)P(A_2) \qquad P(A_1A_3) = P(A_1)P(A_3)$$
$$P(A_2A_3) = P(A_2)P(A_3) \qquad P(A_1A_2A_3) = P(A_1)P(A_2)P(A_3)$$

Selecting a subsequence of k of the n events is clearly equivalent to selecting a subset $\{n_1, n_2, \ldots, n_k\}$ of k integers from $\{1, 2, \ldots, n\}$. We do not use the empty subset, $k = 0$, or any of the n subsets containing only one integer, $k = 1$. From Example 3 (Sec. 2-2) we know that $\{1, 2, \ldots, n\}$ has 2^n subsets, so that it has $2^n - n - 1$ subsets of size at least 2. Thus *the finite sequence* $A_1, A_2, \ldots, A_n$ *has exactly*

$2^n - n - 1$ *subsequences of two or more events.* Hence the mutual independence of $A_1, A_2, \ldots, A_n$ is equivalent to the validity of $2^n - n - 1$ numerical equations. For $n = 4$ we have 11 equations.

An example of $n \geq 2$ mutually independent events arises if an experiment consists of spinning "independently" each of n random wheels once, and letting the occurrence or nonoccurrence of A_k be determined solely by the outcome for the kth wheel. For such an experiment we certainly expect $A_2, A_3, \ldots, A_n$ to be $n - 1$ independent events, since ignoring the first wheel should not destroy the mutual independence of the rest. Also we expect that A_1A_2 and $A_3A_4 \cdots A_n$ should be two independent events. However, A_1A_2 and $A_2A_3 \cdots A_n$ may be dependent, since they both involve the second wheel. Thus our intuitive notion of independent random results has a number of properties. We now easily show that our definition leads to such properties.

If $A_1, A_2, \ldots, A_{10}$ are 10 independent events then certainly A_3, A_7, A_8 are three independent events, because the equations needed to guarantee their independence are already included among the equations guaranteeing the independence of $A_1, A_2, \ldots, A_{10}$. Thus *if $A_1, A_2, \ldots, A_n$ is a finite sequence of $n \geq 2$ independent events then so is any subsequence of two or more of them.*

If A_1, A_2, A_3, A_4 are independent then so are A_3, A_2, A_4, A_1, since reordering does not change the intersection of sets or the product of real numbers. Thus *reordering independent events does not destroy their independence.*

If $A_1, A_2, \ldots, A_m, A_{m+1}, \ldots, A_n$ is a sequence of independent events then the two events $A_1A_2 \cdots A_m$ and $A_{m+1}A_{m+2} \cdots A_n$ are independent. To prove this we first let $B = A_1A_2 \cdots A_m$ and $C = A_{m+1}A_{m+2} \cdots A_n$. Then

$$P(BC) = P(A_1)P(A_2) \cdots P(A_m)P(A_{m+1}) \cdots P(A_n).$$

But we also have

$$P(B) = P(A_1)P(A_2) \cdots P(A_m) \qquad P(C) = P(A_{m+1})P(A_{m+2}) \cdots P(A_n).$$

Thus $P(BC) = P(B)P(C)$, so that the two events B and C are independent.

We can easily generalize this last property and use it in conjunction with reordering to show that if $A_1, A_2, \ldots, A_{10}$ is an independent sequence of 10 events, the three events A_3A_8, A_5, and $A_1A_7A_{10}$ are independent. More generally, if $A_1, A_2, \ldots, A_n$ are independent and if $n_1, \ldots, n_k, r_1, \ldots, r_s$, and $t_1, \ldots, t_m$ are *distinct* positive integers less than or equal to n, the three events B_1, B_2, B_3 defined by

$$B_1 = A_{n_1}A_{n_2} \cdots A_{n_k} \qquad B_2 = A_{r_1}A_{r_2} \cdots A_{r_s} \qquad B_3 = A_{t_1}A_{t_2} \cdots A_{t_m}$$

are independent. Naturally this property extends to more than three events. Thus *if $A_1, A_2, \ldots, A_n$ are independent and if each of $B_1, B_2, \ldots, B_N$ equals the intersection of some of the A_k, and if no A_k is used more than once, then $B_1, B_2, \ldots, B_N$ are independent.*

If $A_1, A_2, \ldots, A_8$ are independent then the two events $A_2A_5A_6$ and A_8 are independent, so from the previous section $A_2A_5A_6$ and $A_8{}^c$ are two independent events, and hence $P(A_2A_5A_6A_8{}^c) = P(A_2A_5A_6)P(A_8{}^c) = P(A_2)P(A_5)P(A_6)P(A_8{}^c)$, where we used the independence of A_2, A_5, A_6 for the last equality. This computation easily generalizes to derive the remaining equations needed to show that $A_1, A_2, \ldots, A_7, A_8{}^c$ are independent. Thus *the independence of n events is not destroyed by complementing some or all of the events.*

Example 1 The events A_1, A_2, A_3, A_4, A_5 are independent, and $P(A_1) = P(A_3) = P(A_5) = .3$ and $P(A_2) = P(A_4) = .2$. The event A_k corresponds to the sales for a company exceeding \$14,000 on the kth day, where the first day is Monday. The outcomes for Monday, Wednesday, and Friday are independent and $P(A_1A_3A_5) = (.3)^3 = .027$ is the probability of high sales on all three days. Similarly $P(A_2{}^cA_4{}^c) = (.8)^2 = .64$ is the probability of lower sales on both Tuesday and Thursday. These two events are independent, so that

$$P(A_1A_3A_5A_2{}^cA_4{}^c) = (.027)(.64) = .01728.$$

The two events A_1A_2 and $A_2{}^cA_3$ both involve A_2 and are dependent. Analytically we have

$$0 = P(\varphi) = P[(A_1A_2) \cap (A_2{}^cA_3)] \neq P(A_1A_2)P(A_2{}^cA_3) = [(.3)(.2)][(.8)(.3)]. \quad \blacksquare$$

EXERCISES

1. Exhibit every subset of at least two members for the set $\{1,2,3,4\}$. For three of these subsets show the corresponding equations which must be satisfied if A_1, A_2, A_3, A_4 are to be independent events.

2. The events $A_1, A_2, \ldots, A_{10}$ are independent. Which of the finite sequences of events below *must* be independent?

(*a*) A_2, A_5, A_7, A_9

(*b*) A_5, A_7, A_2, A_9

(*c*) $A_5, A_7{}^c, A_2, A_9{}^c$

(*d*) $A_1A_3, A_5{}^cA_7A_2, A_8$

(*e*) $A_1A_3, A_3{}^cA_7, A_8$

(*f*) $A_1A_3 \cup A_1A_3{}^c, A_3{}^cA_7, A_8$

(*g*) $A_1A_3 \cup A_5, A_3{}^cA_7, A_8$

3. *At Least One of Many Probable Independent Events Should Occur.* In general if we have a large number $A_1, A_2, \ldots, A_n$ of events and each is fairly probable, say each $P(A_k) \geq .1$, then we *cannot* be quite certain that at least one of them will occur. They might be highly dependent events, as is the case if $A_1 = A_2 = \cdots = A_n$, whence $P(A_1 \cup A_2 \cup \cdots \cup A_n) = P(A_1)$. However, we can draw the desired conclusion if the events are independent.

(*a*) Prove that if $A_1, A_2, \ldots, A_n$ are independent events then

$$P(A_1 \cup A_2 \cup \cdots \cup A_n) \geq 1 - e^{-[P(A_1)+P(A_2)+\cdots+P(A_n)]}.$$

Hint: Use inequality (9) of Appendix 1.

(*b*) Events $A_1, A_2, \ldots, A_{100}$ are independent and each $P(A_k) \geq .1$. Event A_k occurs iff the kth of a fleet of new taxicabs has a major motor breakdown during its first year of operation. Show that we can be quite sure that at least one of these 100 events will occur.

4. (*a*) Give a short proof showing that A_1A_2 and A_3A_4 are independent if A_1, A_2, A_3, A_4 are.

(*b*) Show exactly how the proof of part (*a*) breaks down if we attempt to prove that A_1A_2 and A_2A_3 are independent, given that A_1, A_2, A_3 are independent.

(*c*) Exhibit a discrete probability space (Ω,P) and events A_1, A_2, A_3 such that A_1, A_2, A_3 are independent but A_1A_2 and A_2A_3 are dependent.

5. Three relays are connected as in Fig. 1. For $i = 1, 2, 3$ let A_i be the event that relay i is closed. Assume that A_1 and A_2 are independent and that $P(A_1) = 1/2$ and $P(A_2) = 9/20$. The conditional probabilities for A_3, given the condition of the other relays, are as follows:

$$P(A_3 \mid A_1A_2) = 5/9 \qquad P(A_3 \mid A_1{}^cA_2) = 4/9$$

$$P(A_3 \mid A_1A_2{}^c) = 5/11 \qquad P(A_3 \mid A_1{}^cA_2{}^c) = 6/11$$

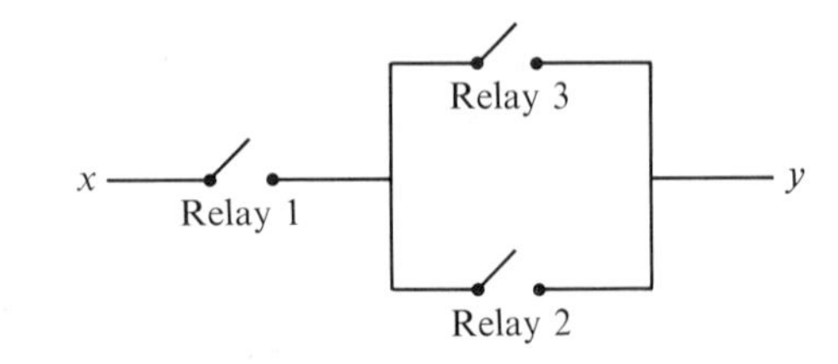

Fig. 1

(*a*) Are A_1, A_2, A_3 mutually independent events? Justify your answer.
(*b*) Are A_1, A_2, A_3 pairwise independent events? Justify your answer.
(*c*) Find the probability of a closed connection between x and y.
(*d*) Find the conditional probability of A_2, given that there is a closed connection between x and y.

6. Prove that the following "alternate definition of independence" is equivalent to the definition which we made.

ALTERNATE DEFINITION. Let (Ω,P) be a discrete probability space and let $A_1, A_2, \ldots, A_n$ be a finite sequence of $n \geq 2$ events. The events $A_1, A_2, \ldots, A_n$ are said to be independent iff *whenever* one event is selected from each of the n sets $\{A_1,A_1{}^c\}, \{A_2,A_2{}^c\}, \ldots, \{A_n,A_n{}^c\}$ then for the n events selected the probability of their intersection equals the product of their probabilities.

Clearly there are 2^n ways to select the n events, and for each such selection the corresponding equation must be true. Thus A_1, A_2, A_3 are independent, according to the alternate definition, iff the eight equations below are true.

$$\begin{aligned}
P(A_1A_2A_3) &= P(A_1)P(A_2)P(A_3) & P(A_1A_2{}^cA_3{}^c) &= P(A_1)P(A_2{}^c)P(A_3{}^c) \\
P(A_1A_2A_3{}^c) &= P(A_1)P(A_2)P(A_3{}^c) & P(A_1{}^cA_2A_3{}^c) &= P(A_1{}^c)P(A_2)P(A_3{}^c) \\
P(A_1A_2{}^cA_3) &= P(A_1)P(A_2{}^c)P(A_3) & P(A_1{}^cA_2{}^cA_3) &= P(A_1{}^c)P(A_2{}^c)P(A_3) \\
P(A_1{}^cA_2A_3) &= P(A_1{}^c)P(A_2)P(A_3) & P(A_1{}^cA_2{}^cA_3{}^c) &= P(A_1{}^c)P(A_2{}^c)P(A_3{}^c).
\end{aligned}$$

8
Independent Experiments

Section 8-1 explains the construction of models for independent experiments. This construction is then utilized in Secs. 8-2, 8-3, and 8-4, which introduce the binomial, Poisson, and Pascal densities. The binomial and Poisson densities and the normal density, which is introduced in Sec. 13-1, are often considered the three most important distributions in probability theory. The term **distribution** will often be used to refer to any function, such as a density or a measure, which determines the probability of every event.

8-1 INDEPENDENT EXPERIMENTS

Suppose we have $n \geq 2$ discrete densities $p_i: F_i \to R$ for $i = 1, 2, \ldots, n$. We may think of p_i as corresponding to the ith of n random wheels. We wish to construct a model for a large experiment which contains *independent* subexperiments corresponding to each wheel. It is natural to make our sample space $\Omega = F_1 \times F_2 \times \cdots \times F_n$, so that a sample point corresponds to a complete description of the outcomes of all wheels; that is, $\omega = (\alpha_1, \ldots, \alpha_n)$ occurs iff α_i is the outcome for the ith wheel, for all $i = 1, 2, \ldots, n$. We need a discrete density $p: \Omega \to R$ from which we can calculate probabilities of events involving more than one wheel. For example, if each α_i is real we might want the probability of the event that $\alpha_2 + \alpha_3 + \alpha_7 = 13$ occurs. We must show that an appropriate density $p: \Omega \to R$ *exists* and is *unique*.

The *existence* of an appropriate density is physically obvious and mathematically trivial. Nevertheless it is worthwhile to indicate in this simple case the logical necessity for showing existence, since this necessity also arises in more complicated situations. Suppose, in a somewhat different context, that we start an analysis by assuming that we have independent events $A_1, \ldots, A_{100}$, with each $P(A_i) \geq .1$ and with $P(A_1 \cup \cdots \cup A_{100}) \leq .99$. From these assumptions we can, by legitimate calculations, easily arrive at a contradiction, such as finding that some event has a negative probability. This can happen because we made assumptions which can never be satisfied in any discrete probability space, since Exercise 3*a* (Sec. 7-4) shows that $P(A_1 \cup \cdots \cup A_{100}) \geq 1 - e^{-10}$ for every (Ω, P) with independent events having each $P(A_i) \geq .1$. In our general theoretical development we always assume for each calculation that all events and their probabilities are obtained from *the same* (Ω,P). Thus if we wish to make certain assumptions about events and their probabilities and then apply our theoretical results, we must first assure ourselves that these assumptions are not inherently contradictory with our theory; that is, we must show that there exists at least one (Ω,P) which can satisfy our assumptions.

It is physically obvious that the assumption of independent spins of wheels corresponding to discrete densities $p_1, \ldots, p_n$ should determine an essentially *unique* large experiment. If this were not the case, different people might make this same assumption and legitimately arrive at different numbers for the probability of some event described in terms of the outcomes of these n wheels. Thus a proof of uniqueness assures us that we have incorporated enough assumptions in our model to make it correspond to our intuition in this instance and permits us to assert that *the* probability of such an event must be a certain number.

We return now to our goal for this section. We have discrete densities $p_i: F_i \to R$ and we define $\Omega = F_1 \times F_2 \times \cdots \times F_n$ to be the sample space for our large experiment. We want to define a discrete density $p: \Omega \to R$ in such a way that (Ω,P) can be interpreted as a model for the assumption of independent spins of wheels corresponding to $p_1, \ldots, p_n$. Now *in* Ω we can define for each α_1 in F_1 the event

$$A_1(\alpha_1) = \{(\alpha_1', \ldots, \alpha_n'): \alpha_1' = \alpha_1\}$$

that α_1 is the outcome for the first wheel; that is, $A_1(\alpha_1)$ consists of all n-tuples in Ω for which the first coordinate is α_1. One requirement on $p: \Omega \to R$ is that the probability assigned by p to this event must be consistent with the given data and desired interpretation; that is, when $P[A_1(\alpha_1)]$ is calculated from $p: \Omega \to R$ it must equal the given $p_1(\alpha_1)$. The same is true for the other wheels. Thus we require that when P is calculated from $p: \Omega \to R$ it must satisfy

$$P[A_i(\alpha_i)] = p_i(\alpha_i) \qquad \text{for all } \alpha_i \in F_i \text{ and all } i = 1, \ldots, n$$

where

$$A_i(\alpha_i) = \{(\alpha_1', \ldots, \alpha_n'): \alpha_i' = \alpha_i\}.$$

This condition on $p: \Omega \to R$ means that if $p: \Omega \to R$ is used to calculate the probability of the outcome of any *one* wheel then $p: \Omega \to R$ will yield the *same* answer as would be obtained from the given data $p_1, \ldots, p_n$.

We also require that the density $p: \Omega \to R$ always make the events $A_1(\alpha_1), \ldots, A_n(\alpha_n)$ independent. For given $\alpha_1, \ldots, \alpha_n$ this requires that $2^n - n - 1$ equations be satisfied by P. One of these equations is

$$P[A_1(\alpha_1) \cdots A_n(\alpha_n)] = P[A_1(\alpha_1)] \cdots P[A_n(\alpha_n)] = p_1(\alpha_1) \cdots p_n(\alpha_n).$$

But clearly $(\alpha_1, \ldots, \alpha_n)$ is the only sample point belonging to the intersection of $A_1(\alpha_1), \ldots, A_n(\alpha_n)$, so that we require $p: \Omega \to R$ to satisfy

$$p(\alpha_1, \alpha_2, \ldots, \alpha_n) = p_1(\alpha_1)p_2(\alpha_2) \cdots p_n(\alpha_n) \qquad \text{for all } \alpha_i \in F_i.$$

Thus our conditions imply a *unique* candidate function $p: \Omega \to R$, which we say is defined by the **product rule** applied to $p_1, \ldots, p_n$. Note that we are *not* multiplying functions with a common domain of definition, as in Sec. 3-3, since our product rule yields a function p on $F_1 \times \cdots \times F_n$.

Clearly this candidate function $p: \Omega \to R$ is nonnegative, positive on a countable subset by Theorem 1c (Sec. 4-2), and sums to 1 by Theorem 1 (Sec. 5-3). Thus this candidate function *is* a discrete density. Also it is consistent with the basic data, and it does make $A_1(\alpha_1), \ldots, A_n(\alpha_n)$ independent. For example, if $n = 3$ then

$$P[A_1(\alpha_1)] = \sum_{\alpha_2}\sum_{\alpha_3} p(\alpha_1,\alpha_2,\alpha_3) = \sum_{\alpha_2}\sum_{\alpha_3} p_1(\alpha_1)p_2(\alpha_2)p_3(\alpha_3)$$

$$= p_1(\alpha_1)\left[\sum_{\alpha_2} p_2(\alpha_2)\right]\left[\sum_{\alpha_3} p_3(\alpha_3)\right] = p_1(\alpha_1).$$

Also if α_1, α_2, α_3 are given, one of the $2^n - n - 1 = 4$ equations required for the independence of $A_1(\alpha_1)$, $A_2(\alpha_2)$, $A_3(\alpha_3)$ is verified as follows:

$$P[A_1(\alpha_1) \cap A_3(\alpha_3)] = \sum_{\alpha_2} p(\alpha_1,\alpha_2,\alpha_3) = \sum_{\alpha_2} p_1(\alpha_1)p_2(\alpha_2)p_3(\alpha_3)$$

$$= p_1(\alpha_1)p_3(\alpha_3) = P[A_1(\alpha_1)]P[A_3(\alpha_3)]$$

In summary, if we are given discrete densities $p_i: F_i \to R$ then there is a unique discrete density $p: \Omega \to R$, given by the product rule, on $\Omega = F_1 \times \cdots \times F_n$ such that p is consistent with the given probabilities for the individual wheels and makes their outcomes independent. Thus the assumption of independent spins corresponding to discrete densities $p_1, \ldots, p_n$ can always be satisfied within our theory and by a unique model with Ω as above. There are, of course, many "larger" models that contain this model essentially as a subexperiment. In particular, then, if any event, such as the occurrence of $\alpha_2 + \alpha_3 + \alpha_7 = 13$, is defined in terms of the outcomes of these wheels, the event has a well-defined unique probability. We summarize in the following theorem.

Theorem 1 Let p_i: $F_i \to R$ for $i = 1, 2, \ldots, n$ be $n \geq 2$ discrete densities. Let $\Omega = F_1 \times F_2 \times \cdots \times F_n$ and let $p: \Omega \to R$ be defined by the product rule as

$$p(\alpha_1, \ldots, \alpha_n) = p_1(\alpha_1)p_2(\alpha_2) \cdots p_n(\alpha_n) \qquad \text{for all } (\alpha_1, \ldots, \alpha_n) \text{ in } \Omega.$$

Then $p: \Omega \to R$ is a discrete density with corresponding discrete probability space (Ω,P). *Also*, if

$$A_i(\alpha_i) = \{(\alpha_1', \ldots, \alpha_n'): \alpha_i' = \alpha_i\}$$

is the event in Ω that the ith subexperiment yields α_i then $A_1(\alpha_1), \ldots, A_n(\alpha_n)$ are always independent, and $P[A_i(\alpha_i)] = p_i(\alpha_i)$ always holds. *Furthermore* p is the unique discrete density on Ω with these properties. ■

We have shown that an independent-subexperiments model is obtained by using the cartesian product space with probabilities assigned by the product rule. If the subexperiment densities are the same, $p_1 = p_2 = \cdots = p_n$, we may interpret the large experiment either as single independent spins of n essentially identical wheels or as n repeated independent spins of the same wheel. We refer to the latter case as *n independent trials*, or *repetitions*.

Example 1 We desire a model corresponding to the assumption of one independent spin of each of the n two-sectored random wheels of Fig. 1. Let $F_1 = F_2 = \cdots = F_n = \{0,1\}$. Define the n

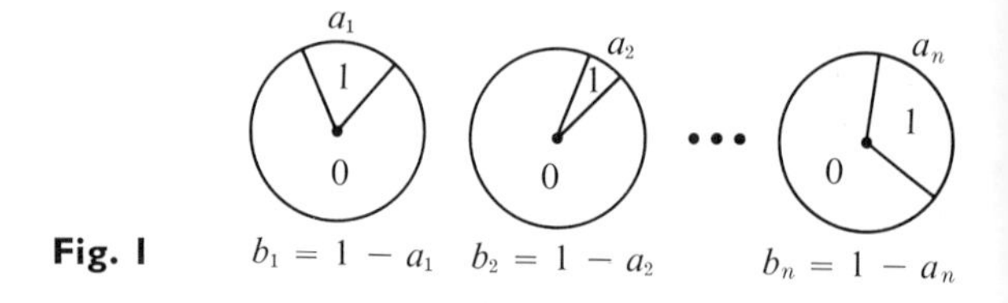

Fig. 1

densities $p_i\colon F_i \to R$ by $p_i(1) = a_i$ and $p_i(0) = b_i$, where each of the numbers $a_1, a_2, \ldots, a_n$ satisfies $0 < a_i < 1$ and b_i is defined by $b_i = 1 - a_i$. As in Theorem 1, we let the sample space be $\Omega = F_1 \times F_2 \times \cdots \times F_n = \{0,1\}^n$ and define $p\colon \Omega \to R$ by the product rule. Thus, for example,

$$p(1, 0, 1, \ldots, 0) = p_1(1)p_2(0)p_3(1) \cdots p_n(0) = a_1b_2a_3 \cdots b_n.$$

The density $p\colon \Omega \to R$ is exhibited in Fig. 2 for the case $n = 3$. Now $A_1(0) = \{\omega_1,\omega_2,\omega_3,\omega_4\}$ is

i	ω_i	$p(\omega_i)$
1	(0,0,0)	$b_1b_2b_3$
2	(0,0,1)	$b_1b_2a_3$
3	(0,1,0)	$b_1a_2b_3$
4	(0,1,1)	$b_1a_2a_3$
5	(1,0,0)	$a_1b_2b_3$
6	(1,0,1)	$a_1b_2a_3$
7	(1,1,0)	$a_1a_2b_3$
8	(1,1,1)	$a_1a_2a_3$

Fig. 2

the event that the first wheel yields 0, and we can check directly that

$$P(A_1(0)) = p(\omega_1) + p(\omega_2) + p(\omega_3) + p(\omega_4) = p_1(0) = b_1$$

as assured by Theorem 1. Also, from Theorem 1 and the properties of independence derived in Sec. 7-4 we know that $A_1(0)$ and $A_3(1)$ are independent events, so that b_1a_3 is the probability that wheels 1 and 3 will yield 0 and 1, respectively.

For $k = 0, 1, 2, \ldots, n$ let B_k be the event that exactly k of the n wheels yield 1. Then $P(B_k)$, when considered as a function of k, is called the **Poisson binomial density.** It depends on the parameters n and $a_1, a_2, \ldots, a_n$ and does not have a simple analytical form. For $n = 3$ we have, for example,

$$P(B_1) = a_1b_2b_3 + b_1a_2b_3 + b_1b_2a_3$$

for the probability of obtaining exactly one 1 on the three wheels. If $a_1 = a_2 = \cdots = a_n$ we have n independent spins of the same two-sectored wheel, and this case will be considered in detail in the next section. ■

Example 2 Let $p_1\colon F_1 \to R$ be uniform on $F_1 = \{\alpha_1, \ldots, \alpha_k\}$, so that each $p_1(\alpha_i) = 1/k$. Let $p_1 = \cdots = p_n$ and let p be defined as in Theorem 1, so that we have n independent spins of the same random wheel having k equal sectors. Then we always have

$$p(\alpha_1, \ldots, \alpha_n) = p_1(\alpha_1) \cdots p_n(\alpha_n) = \frac{1}{k} \cdots \frac{1}{k} = \frac{1}{k^n}$$

so that p is uniform on $(F_1)^n$. Thus p can also be interpreted as the density for n draws with replacement from an urn containing k balls having distinct labels $a_1, \ldots, a_k$. ■

Example 3 Given $p_1: F_1 \to R$, let $p_1 = p_2 = \cdots = p_n$ and let $\Omega = (F_1)^n$. We want to define a density $p': \Omega \to R$ which does *not* correspond to independent subexperiments, but rather to complete dependence of the form that all n wheels must yield the same outcome; for example, the n wheels might be aligned and welded to the same axle. We naturally define $p': \Omega \to R$ by

$$p'(\alpha_1, \ldots, \alpha_n) = \begin{cases} p_1(\alpha_1) & \text{if } \alpha_1 = \cdots = \alpha_n \\ 0 & \text{otherwise.} \end{cases}$$

As in Theorem 1, let $A_i(\alpha_i)$ be the event in Ω that the ith wheel has outcome α_i. Then it is intuitively obvious and easily shown that $P'[A_i(\alpha_i)] = p_i(\alpha_i)$ is always true, so that $p': \Omega \to R$, as well as $p: \Omega \to R$ of Theorem 1, determines probabilities *for individual wheels* consistent with the given data $p_1, \ldots, p_n$. In general there are many densities on Ω which are consistent with the basic densities and correspond to various "kinds" and "amounts" of dependence. ■

EXERCISES

1. Find $P(B_8)$ for the special case of Example 1, where $a_1 = a_2 = \cdots = a_n = 2/3$ and $n = 100$.
2. Find $P(B_0)$ and $P(B_1)$ for the special case of Example 1, where $n = 10$ and $a_i = .1$ for $i = 1, 2, 3, 4, 5$ and $a_i = .2$ for $i = 6, 7, 8, 9, 10$.
3. *Independent Trials, with Dependence within a Trial.* Let $F_1 = \{1,2,3,4,5,6\}^2$ and let the density $p_1: F_1 \to R$ be defined by $p_1(1,1) = 0$ and $p_1(x,y) = 1/35$ for all other pairs (x,y) in F_1. The pair (x,y) can be interpreted as the outcome that the red die has x dots on its upward face and the black die has y dots on its upward face, where the two dice are "fixed" so that $(1,1)$ never happens but all other results are equally likely. Let (x,y) and (x',y') denote the outcomes of two independent trials of this subexperiment. What is the probability that $x + y + x' + y' = 6$?

8-2 THE BINOMIAL DENSITY

In the remainder of this chapter we introduce several fundamental, related discrete probability densities. All of them arise in connection with experiments involving the random wheel of Fig. 1, which has two sectors labeled 1 and 0, with probabilities p and $q = 1 - p$, respectively. If we interpret 1 and 0 as heads and tails, respectively, then we have biased-coin tossing if $p \neq .5$ and fair-coin tossing if $p = .5$. Sometimes we will interpret 1 and 0 as success and failure, or as something happening or something not happening, respectively. We may say that a physical experiment is a "success" iff $[\text{volume}]^2 + \cos\,[\text{temperature}] \leq 12.2$. Thus if the possible outcomes of a complicated experiment are divided into two types, and if repeated experiments are independent and essentially the same (that is, the parameter p does not change with

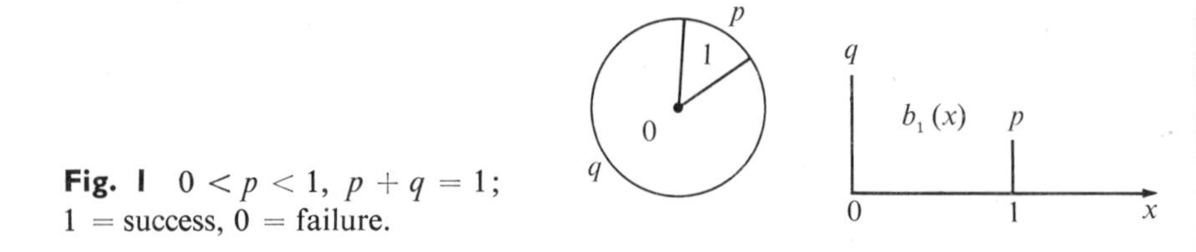

Fig. 1 $0 < p < 1$, $p + q = 1$;
1 = success, 0 = failure.

experiments), then we essentially have repeated spins of the wheel of Fig. 1. The importance and applicability of results concerning such experiments should be apparent.

It is traditional to use the letters p (and $q = 1 - p$) for the parameters of Fig. 1, so we use the notation b_1 for the corresponding density. The **Bernoulli discrete probability density** b_1**:** $R \to R$ **with parameter** p is defined by

$$b_1(0) = q$$
$$b_1(1) = p$$
$$b_1(x) = 0 \qquad \text{if } x \text{ is not 0 or 1}$$

where p is a real number satisfying $0 < p < 1$ and $q = 1 - p$. We will see shortly that the Bernoulli density $b_1: R \to R$ is the special case $n = 1$ of the binomial density $b_n: R \to R$, and this is the reason for the subscript 1 on the symbol for the density.

We will refer to n independent spins of the wheel of Fig. 1 as n **Bernoulli trials with parameter** p. As in Sec. 8-1, a model for Bernoulli trials is obtained by letting $\Omega = R^n$ and defining the density $p': \Omega \to R$ by

$$p'(x_1, x_2, \ldots, x_n) = b_1(x_1)b_1(x_2) \cdots b_1(x_n).$$

We use the symbol p' for the density in order to distinguish it from the parameter p, but we will use the symbol P, without a prime, for the measure corresponding to p'. Clearly the value of p' is 0 unless every x_i belongs to $\{0,1\}$, since otherwise at least one of the n factors equals 0. If every x_i belongs to $\{0,1\}$ then

$$p'(1, 1, 0, 1, \ldots, 1, 0) = ppqp \cdots pq = p^k q^{n-k}$$

where $k = x_1 + x_2 + \cdots + x_n$ is the number of coordinates of $(x_1, x_2, \ldots, x_n)$ which equal 1. The nonzero values of p' are shown in Fig. 2 for the case $n = 4$.

If B_k is the event that there are exactly k successes in n Bernoulli trials,

$$B_k = \{(x_1, x_2, \ldots, x_n): x_1 + x_2 + \cdots + x_n = k\}$$

then the density p' has the value $p^k q^{n-k}$ at any $(x_1, x_2, \ldots, x_n)$ belonging to B_k for which $p'(x_1, x_2, \ldots, x_n)$ is positive. From Theorem 1 (Sec. 2-3) there are exactly $\binom{n}{k}$ such $(x_1, x_2, \ldots, x_n)$, so that

$$P(B_k) = \binom{n}{k} p^k q^{n-k} \qquad \text{for } k = 0, 1, \ldots, n.$$

The events $B_0, B_1, \ldots, B_n$ are disjoint and the probability of their union is 1, since every $(x_1, x_2, \ldots, x_n)$ for which p' is positive belongs to exactly one of these events.

$$P(B_0) = q^4 = \binom{4}{0} q^4$$

$$P(B_1) = 4pq^3 = \binom{4}{1} pq^3$$

$$P(B_2) = 6p^2q^2 = \binom{4}{2} p^2q^2$$

$$P(B_3) = 4p^3q = \binom{4}{3} p^3q$$

$$P(B_4) = p^4 = \binom{4}{4} p^4$$

(x_1,x_2,x_3,x_4)	$p'(x_1,x_2,x_3,x_4)$	$x_1 + x_2 + x_3 + x_4$
(0,0,0,0)	q^4	0
(0,0,0,1)	pq^3	1
(0,0,1,0)	pq^3	1
(0,0,1,1)	p^2q^2	2
(0,1,0,0)	pq^3	1
(0,1,0,1)	p^2q^2	2
(0,1,1,0)	p^2q^2	2
(0,1,1,1)	p^3q	3
(1,0,0,0)	pq^3	1
(1,0,0,1)	p^2q^2	2
(1,0,1,0)	p^2q^2	2
(1,0,1,1)	p^3q	3
(1,1,0,0)	p^2q^2	2
(1,1,0,1)	p^3q	3
(1,1,1,0)	p^3q	3
(1,1,1,1)	p^4	4

Fig. 2

Thus we must have $P(B_0) + P(B_1) + \cdots + P(B_n) = 1$. This equation also follows from the expansion of $1 = (p + q)^n$ by the binomial theorem.

Let us now summarize this information in traditional terminology. The **binomial discrete probability density** b_n: $R \to R$ **with parameters** n **and** p is defined by

$$b_n(k) = \binom{n}{k} p^k q^{n-k} \qquad \text{if } k = 0, 1, \ldots, n$$

$$b_n(x) = 0 \qquad \text{otherwise}$$

where n is a positive integer, p is a real number satisfying $0 < p < 1$, and $q = 1 - p$. There is a different density for each pair n,p. We have shown that *the probability of obtaining exactly k successes in n Bernoulli trials with parameter p is given by the value $b_n(k)$ of the binomial density b_n with parameters n and p.*

The binomial density b_n for $n = 40$ and $p = .08$ is shown[1] in Fig. 3 and Table 1. We may at times drop the subscript n and use b instead of b_n. For $n = 40$ and $p = .08$

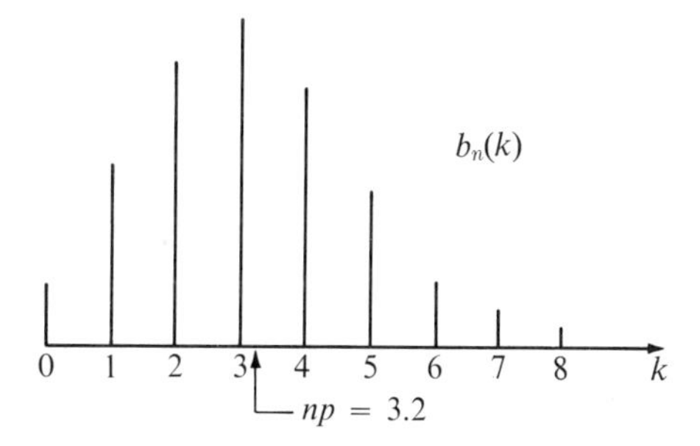

Fig. 3 Binomial density $n = 40$ and $p = .08$.

[1] One of the better binomial tables is Tables of the Cumulative Binomial Probability Distribution, by the staff of the Computation Laboratory, Harvard University Press, Cambridge, Mass., 1955.

we see that $\sum_{r=0}^{13} b(r) \doteq 1$ to within 5 decimal places even though $b(14), b(15), \ldots, b(40)$ are all positive.

The random wheel of Fig. 4 corresponds to the binomial density with $n = 40$ *and* $p = .08$. Observing the outcome of one spin of Fig. 4 is equivalent to observing the number of ones obtained on $n = 40$ independent spins of Fig. 1 with $p = .08$.

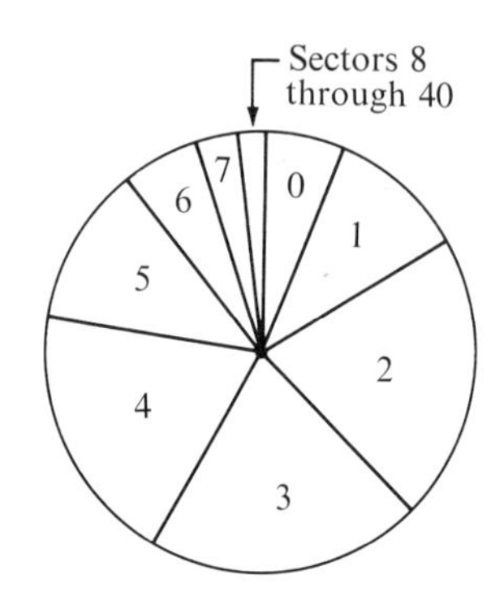

Fig. 4

The probability of a success on one Bernoulli trial is p, so if there is a large number n of such trials then we expect that the fraction of successes will usually be close to p, and hence the total number of successes should usually be close to np. We call np the **mean of the number of successes,** or the **expected number of successes,** on n Bernoulli trials with parameter p. This definition will be further justified in later chapters. For $n = 40$ and $p = .08$ we have $np = 3.2$, and as can be seen from Fig. 3 or 4, the density of the number of successes is concentrated near to np. In particular if

Table 1 The binomial density for $n = 40$ and $p = .08$

k	$b(k)$	$\sum_{r=0}^{k} b(r)$	k	$b(k)$	$\sum_{r=0}^{k} b(r)$
0	.03561	.03561	8	.00895	.99626
1	.12384	.15945	9	.00277	.99903
2	.21000	.36945	10	.00075	.99978
3	.23130	.60075	11	.00017	.99995
4	.18604	.78679	12	.00004	.99999
5	.11648	.90327	13	.00001	1.00000
6	.05909	.96236	14	.00000	1.00000
7	.02495	.98731			

$p = .08$ and we spin the wheel of Fig. 1 a total of $n = 40$ times, the probability that the total number k of successes will satisfy $1 \leq k \leq 5$ is given by $.90327 - .03561 = .86766$.

As indicated in Fig. 3 and Table 1, if $n = 40$ and $p = .08$ then the following chain of inequalities holds:

$$b_n(0) < b_n(1) < b_n(2) < b_n(3) > b_n(4) > b_n(5) > \cdots > b_n(40)$$

that is, as k increases from zero to a point near to np the values $b_n(k)$ increase, but after that point they decrease. Thus when only integer arguments are considered this density b_n is unimodal and its maximum value occurs at a k near to np. We now show that every binomial density has this shape. The result is almost immediate from the easily checked identity

$$\frac{b_n(k)}{b_n(k-1)} = 1 + \frac{(n+1)p - k}{kq} \qquad \text{if } k = 1, 2, \ldots, n. \tag{1}$$

We consider two cases:

1. Given n and p, if $(n+1)p$ is not an integer let m be the largest integer less than $(n+1)p$. In this case

$$\begin{aligned} &b_n(0) < b_n(1) < \cdots < b_n(m-1) \\ &\qquad < b_n(m) > b_n(m+1) > \cdots > b_n(n) \end{aligned} \tag{2}$$

so that $b_n(k)$ has a unique maximum at $k = m$.
Note that if $n = 40$ and $p = .08$ then $(n+1)p = 3.28$, so $m = 3$.

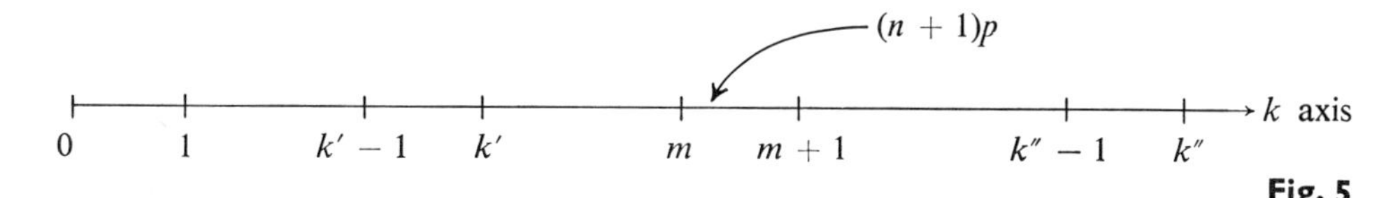

Fig. 5

If, as in Fig. 5, we select any integer k' satisfying $1 \le k' \le m$, so that $k' < (n+1)p$, we see from equation (1) that $b_n(k')/b_n(k'-1) > 1$. Thus the values $b_n(k')$ are increasing as k' increases, as long as k' is to the left of m. If we select any integer k'' satisfying $m < k'' \le n$ then $(n+1)p < k''$, so from equation (1) we have $b_n(k'')/b_n(k''-1) < 1$. Thus the values $b_n(k'')$ are decreasing as k'' increases, as long as k'' is to the right of m. The proof of this case is completed.

2. Given n and p, if $(n+1)p$ is an integer m then

$$b_n(0) < b_n(1) < \cdots < b_n(m-1) = b_n(m) > \cdots > b_n(n) \tag{3}$$

so that $b_n(k)$ attains its maximum value at the two points $k = (n+1)p$ and $k = (n+1)p - 1$.
The proof of this case is left as an exercise.

EXERCISES

1. *Crime Clusters.* Over a period of several years in a large city there is a major crime committed every 12.5 days, on the average. A simple analysis is being made to see if such crimes tend to come in clusters because one crime may tend to stimulate others. A model is to be constructed under the assumption that each crime is committed completely independently of all other crimes, and then the actual data are to be compared to the model to see if the actual clustering of crimes is indeed far greater than the probable clustering just from chance if the crimes were independent. The model assumes that every day the random wheel of Fig. 1, with $p = .08 = 1/12.5$, is spun, and a success is interpreted as the outcome that a major crime is committed on that day. Now the actual data are divided up into a large number of disjoint 40-day periods, and it is noted that in about 10 percent of these periods six or more crimes were actually committed. Does this large a percentage of actual "high crime periods" tend to indicate that the assumptions used in constructing the model are inappropriate for this situation?

2. Verify identity (1) for the binomial density.

3. Prove the inequality chain (3) for the case when $(n + 1)p$ is an integer.

4. *The Binary Symmetric Channel.* The binary symmetric channel in information theory is pictured in Fig. 6 which is interpreted as meaning that if the input to the channel is 0 then

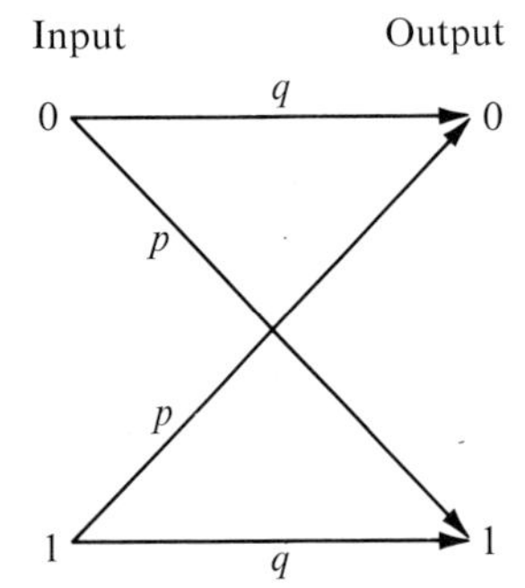

Fig. 6 Binary symmetric channel.

the probability of output 0 is q and the probability of output 1 is p. Thus the probability of an error in the transmission of a digit is p. When the probability of making an error is not the same for the two inputs the channel is called a *binary asymmetric channel.* Physically, 1 and 0 might correspond to pulse and no pulse. It is assumed that the errors are independent from digit to digit, so that if the input sequence is 01100010 then the output sequence 01101000 contains two errors and so has probability p^2q^6. Thus the errors are determined by Bernoulli trials.

(*a*) List the output sequences which correspond to at most two errors being made if the input sequence is 11001.

(*b*) If $n = 5$ digits are sent through the channel and $p = .1$, find the probability of making at most two errors.

5. *The Mean of the Binomial is np.* A total of nr Bernoulli trials with parameter p are divided into a large number r of disjoint subexperiments each consisting of n Bernoulli trials. Let m_k be the number of subexperiments for which exactly k successes are obtained. Thus $m_0 + m_1 + \cdots + m_n = r$, and the total number of successes in the nr trials is $(0)m_0 + (1)m_1 + (2)m_2 + \cdots + (n)m_n$. We expect that the fraction of the successes on all nr trials will typically be close to p,

$$\frac{1}{nr}\sum_{k=0}^{n} km_k \doteq p.$$

But m_k/r is the fraction of subexperiments yielding the outcome k, where there are a total of r independent subexperiments and each essentially consists of spinning once a random wheel corresponding to the binomial density b_n with parameters n and p. Thus we expect m_k/r

typically to be close to $b_n(k)$. Therefore we suspect that

$$\sum_{k=0}^{n} kb_n(k) = np$$

is an identity. Prove it.

6. *A Biased Galton Device.* If the device of Example 5 (Sec. 2-3) is modified so that a movement of the ball to the right at any junction has probability p, where $0 \le p \le 1$, and the various choices are independent, what is the probability distribution for the channel in which the ball comes to rest?

7. *A Generalization of Bertrand's Theorem* (*Example 6, Sec. 2-3*). Let the sample space $\Omega = \{1, -1\}^n$ correspond to n Bernoulli trials with probability p for a head, $0 < p < 1$, where 1 and -1 correspond to heads and tails, respectively. For $0 \le r \le n$ let B_r be the event that there are exactly r heads in the n trials. Thus $\epsilon = (\epsilon_1, \ldots, \epsilon_n)$ belongs to B_r iff r of its coordinates are 1, and the remaining $n - r$ are -1.

(*a*) Show that the conditional density on Ω, given B_r, is uniform over B_r. Thus if we learn that the outcome had exactly r heads then every outcome with exactly r heads is equally likely, regardless of the value of p.

(*b*) Let L be the event that heads led all the way; that is, ϵ belongs to L iff

$$\epsilon_1 + \epsilon_2 + \cdots + \epsilon_k \ge 1 \qquad \text{for all } k = 1, 2, \ldots, n.$$

Use the previous case ($p = \frac{1}{2}$) of Bertrand's theorem to show that

$$P(L \mid B_r) = \begin{cases} 0 & \text{if } r \le n - r \\ \dfrac{r - (n - r)}{n} & \text{if } r > n - r. \end{cases}$$

(*c*) A particle of dust in the air of a room without air currents is subjected to random impacts from air molecules. Its vertical motion can be described roughly as follows: each second it moves downward 1 centimeter with probability .8 or upward 1 centimeter with probability .2. Its movement from second to second is independent. The particle is found at time $t = 100$ seconds to be 12 centimeters above its initial position at time $t = 0$. What is the probability that the particle was always above its initial position after its first movement was made?

8-3 THE POISSON DENSITY

In this section we introduce the Poisson density. There is also some indication of the theoretical basis for the very frequent appearance of the Poisson distribution in almost every area in which probability theory is applied.

The power-series expansion for the exponential function $e^\lambda = \sum_{k=0}^{\infty} \lambda^k/k!$ converges for all real λ. Multiplying both sides by $e^{-\lambda}$, we obtain $1 = \sum_{k=0}^{\infty} e^{-\lambda}\lambda^k/k!$. If $\lambda > 0$ every term in the series is positive. Therefore we can define the **Poisson discrete probability density** p_λ**:** $R \to R$ **with parameter** $\lambda > 0$ by

$$p_\lambda(k) = \frac{e^{-\lambda}\lambda^k}{k!} \qquad \text{if } k = 0, 1, 2, \ldots$$

$$p_\lambda(x) = 0 \qquad \text{otherwise.}$$

There is a different density p_λ for each real $\lambda > 0$, but when there is little chance of confusion we drop the subscript λ on p_λ. Often the Poisson density is considered to be defined only on the nonnegative integers rather than on the whole real line. Normally it makes little difference whether we make the sample space R or $\{0, 1, 2, \ldots\}$.

The Poisson densities[1] for $\lambda = 1.8$ and $\lambda = 3.2$ are plotted in Fig. 1 and tabulated in Table 1.

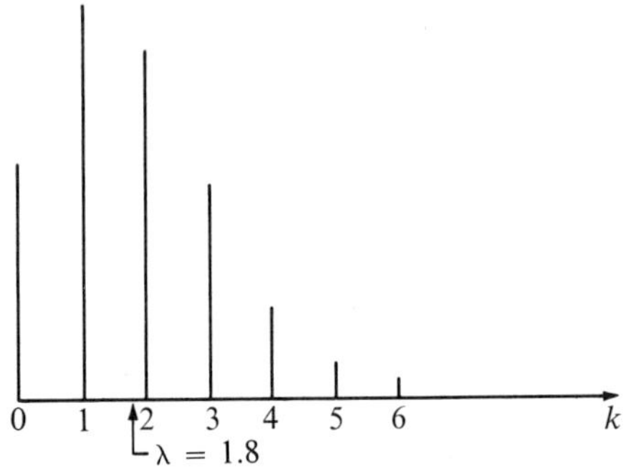

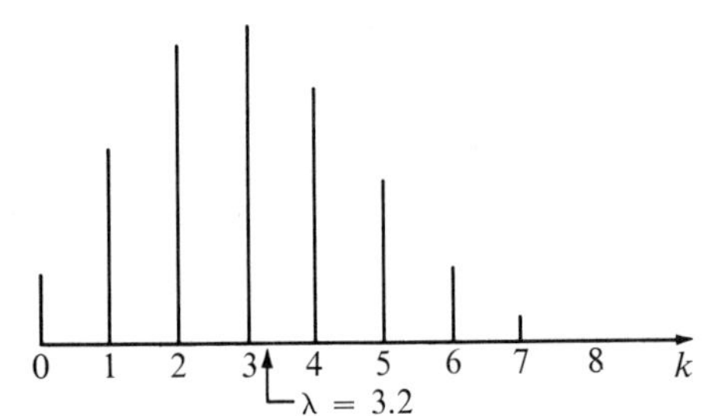

Fig. 1 Two Poisson densities.

For the $\lambda = 3.2$ density, for example,

$$\sum_{r=0}^{14} \frac{(3.2)^r e^{-3.2}}{r!} = 1.00000.$$

Thus the sum of the first 15 terms equals 1 to within five decimal places, even though $p_\lambda(k) > 0$ for every positive integer k. Thus the random wheel for $p_{3.2}$ has roughly 15 sectors of significant size, while the infinitely many remaining sectors have a total fraction of less than 10^{-5} of the circumference assigned to them. Even though $\Omega_0 = \{x: p(x) > 0\}$ may be infinite, we often find that most of the probability is located on a "small" finite subset of Ω_0. Such order-of-magnitude facts are frequently helpful in developing a realistic understanding of a particular problem.

Note that the expected number of successes for the binomial density of Fig. 3 (Sec. 8-2) is $np = (40)(.08) = 3.2$. This binomial density is approximately equal to the Poisson density of Fig. 1 with $\lambda = 3.2$. The binomial density with parameters $n = 400$ and $p = .008$ also has 3.2 for the expected number of successes and is even more nearly

[1] One of the better Poisson tables is "Tables of the Individual and Cumulative Terms of Poisson Distribution," Defense Systems Department, General Electric Company, D. Van Nostrand Company, Inc., New York, 1962.

Table 1 Two Poisson densities

k	$p_\lambda(k)$	$\sum_{r=0}^{k} p_\lambda(r)$	$p_\lambda(k)$	$\sum_{r=0}^{k} p_\lambda(r)$
0	.16530	.16530	.04076	.04076
1	.29754	.46284	.13044	.17120
2	.26778	.73062	.20870	.37990
3	.16067	.89129	.22262	.60252
4	.07230	.96359	.17809	.78061
5	.02603	.98962	.11398	.89459
6	.00781	.99743	.06079	.95538
7	.00201	.99944	.02779	.98317
8	.00045	.99989	.01112	.99429
9	.00009	.99998	.00395	.99824
10	.00002	1.00000	.00126	.99950
11	.00000	1.00000	.00037	.99987
12			.00010	.99997
13			.00002	.99999
14			.00001	1.00000
15			.00000	1.00000
	$\lambda = 1.8$		$\lambda = 3.2$	

equal to the Poisson density with $\lambda = 3.2$. We now show that *the Poisson density with parameter λ is close to every one of the binomial densities for which $np = \lambda$ and n is large enough.* More precisely, we show that *for every fixed $\lambda > 0$ and every nonnegative integer k we have*

$$\lim_{n\to\infty} \binom{n}{k}\left(\frac{\lambda}{n}\right)^k\left(1-\frac{\lambda}{n}\right)^{n-k} = \frac{e^{-\lambda}\lambda^k}{k!}.$$

We first obtain a more convenient form for the binomial density with parameters n and λ/n:

$$\binom{n}{k}\left(\frac{\lambda}{n}\right)^k\left(1-\frac{\lambda}{n}\right)^{n-k} = \frac{n(n-1)\cdots(n-k+1)}{n^k}\frac{\lambda^k}{k!}\left(1-\frac{\lambda}{n}\right)^{n-k}$$

$$= \left[1\left(1-\frac{1}{n}\right)\cdots\left(1-\frac{k-1}{n}\right)\left(1-\frac{\lambda}{n}\right)^{-k}\right]\frac{\lambda^k}{k!}\left(1-\frac{\lambda}{n}\right)^n.$$

Clearly [ ] $\to 1$ as $n \to \infty$, since k and λ are fixed, and $(1 - \lambda/n)^n \to e^{-\lambda}$, as asserted in equation (6) of Appendix 1. Thus if p is small and n is large enough, the binomial density with parameters n and p can be approximated by the Poisson density with $\lambda = np$.

The result of the last paragraph and the way that the binomial density was obtained from Bernoulli trials permit us to describe a simple random-wheel experiment which yields as close an approximation to a Poisson density as is desired. This experiment provides some intuitive indication of the nature of many of the situations where we might expect to find an approximately Poisson distribution. Suppose we are given a positive number λ. Select some integer n so large that λ/n is very much less than 1.

Consider the experiment of making n independent spins of the random wheel of Fig. 2, so that the expected number of successes is $n(\lambda/n) = \lambda$. If n is large enough the distribution of the number k of successes on the n spins is approximately Poisson with parameter λ. Thus if the probability of success on each trial is small enough, and if

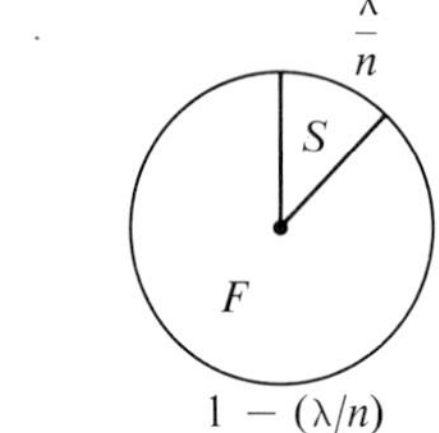

Fig. 2

there are enough independent trials, the distribution of the number of successes is approximately Poisson with parameter λ equal to the expected number of successes, which equals the number of trials times the probability of a success on one trial.

If some variable, such as the number of successes, has a Poisson density p_λ we may call λ the *expected value* or mean value of the variable. This interpretation will be justified in Chap. 9.

Note that the two Poisson densities of Fig. 1 increase as k increases, have a maximum value $p_\lambda(k)$ for k near to λ, and then decrease. Every Poisson density has this shape, and this is almost immediate from the identity

$$\frac{p_\lambda(k)}{p_\lambda(k-1)} = \frac{\lambda}{k} \qquad \text{for all } k = 1, 2, \ldots \tag{1}$$

which is true by inspection. More precisely, it is left to an exercise to prove that if $\lambda > 0$ is not an integer and if m_λ is defined to be the largest integer less than λ then

$$p_\lambda(0) < p_\lambda(1) < \cdots < p_\lambda(m_\lambda) > p_\lambda(m_\lambda + 1) > \cdots \tag{2}$$

and if λ is an integer then

$$p_\lambda(0) < p_\lambda(1) < \cdots < p_\lambda(\lambda - 1) = p_\lambda(\lambda) > p_\lambda(\lambda + 1) > \cdots. \tag{3}$$

Example 1 A remarkable number of phenomena are described quite accurately by the Poisson distribution. An example is radioactive disintegrations under quite general conditions. For example, if we have a gram of radioactive substance, and our experiment consists of counting with an instrument the number of alpha particles reaching the instrument during a particular 1-second time interval, and from past data we know that the average number of these emissions per second is 3.2, then the Poisson density with $\lambda = 3.2$ will usually provide a satisfactory description. If A is the event that the number of emissions counted during the second is between 2 and 5, inclusive, then

$$P(A) = p_{3.2}(2) + p_{3.2}(3) + p_{3.2}(4) + p_{3.2}(5) = .89459 - .17120 = .72339.$$

Thus A should occur in about 72 percent of a large number of experiments. We can imagine that there is some large unknown number n of atoms, and each one has probability $\lambda/n = 3.2/n$ of disintegrating and sending an alpha particle into our counter during the second being considered. ■

Example 2 An experiment consists of observing the number of emissions from each of two blocks of radioactive material during a 1-second interval. For verbal simplicity we assume that we have two instruments and each one counts all emissions from one of the blocks. The experiment is described by the density $p: R^2 \to R$ given by

$$p(i,j) = \frac{e^{-\lambda_1}(\lambda_1)^i}{i!} \frac{e^{-\lambda_2}(\lambda_2)^j}{j!} \quad \text{for } i \text{ and } j \text{ nonnegative integers}$$
$$p(x,y) = 0 \quad \text{otherwise} \tag{4}$$

where $\lambda_1 > 0$ and $\lambda_2 > 0$; that is, $p(i,j)$ is the probability that block 1 will emit i particles and block 2 will emit j particles. Thus, in accordance with Theorem 1 (Sec. 8-1), we defined our density by the product rule, so that the emissions from the two blocks are independent, and the number of emissions from block i is Poisson with parameter λ_i.

Figure 3 exhibits part of the joint density $p: R^2 \to R$ in the case $\lambda_1 = 3.2$ and $\lambda_2 = 1.8$. If A is the event that neither block emits more than one particle during the second then

$$P(A) = p(0,0) + p(0,1) + p(1,0) + p(1,1)$$
$$= .00674 + .01213 + .02156 + .03881 = .07924.$$

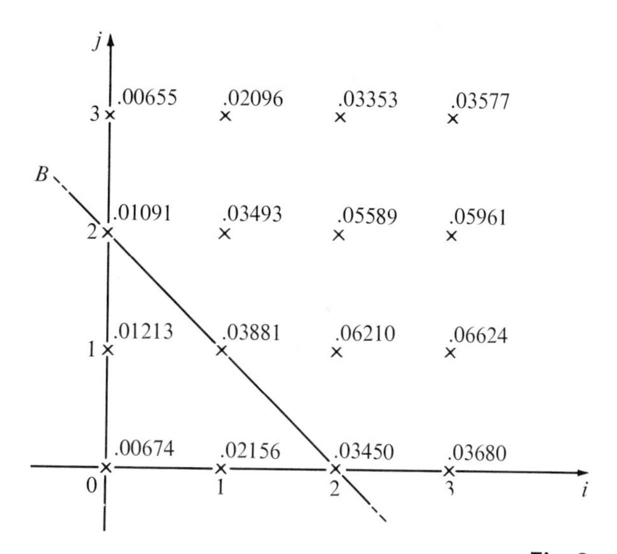

Fig. 3

If B is the event that the total number of emissions from both blocks together equals 2 then

$$p(B) = p(0,2) + p(1,1) + p(2,0) = .01091 + .03881 + .03450 = .08422.$$

We can find $P(B)$ analytically as follows:

$$P(B) = e^{-3.2}\frac{e^{-1.8}(1.8)^2}{2} + e^{-3.2}(3.2)e^{-1.8}(1.8) + \frac{e^{-3.2}(3.2)^2}{2}e^{-1.8}$$
$$= \frac{e^{-5}}{2}[(1.8)^2 + (2)(1.8)(3.2) + (3.2)^2] = \frac{e^{-5}(1.8 + 3.2)^2}{2!}$$

This calculation suggests that the total number of emissions in 1 second from the two blocks together has a Poisson distribution with parameter $\lambda_1 + \lambda_2 = 3.2 + 1.8 = 5$ emissions per second. We now verify that this is true in general.

Let k be a nonnegative integer and let C_k be the event $\{(x,y): x + y = k\}$ that the sum of the emissions from blocks 1 and 2 equals k. Thus C_k is the line through $(0,k)$ having a slope of -1. Then $P(C_k) = p(0,k) + p(1, k - 1) + p(2, k - 2) + \cdots + p(k,0)$, so that

$$P(C_k) = e^{-\lambda_1-\lambda_2}\sum_{r=0}^{k}\frac{(\lambda_1)^r}{r!}\frac{(\lambda_2)^{k-r}}{(k-r)!}$$

$$= \frac{e^{-(\lambda_1+\lambda_2)}}{k!}\sum_{r=0}^{k}\binom{k}{r}(\lambda_1)^r(\lambda_2)^{k-r} = \frac{e^{-(\lambda_1+\lambda_2)}}{k!}(\lambda_1+\lambda_2)^k$$

where the binomial theorem was used for the last equality. Thus *if* $p: R^2 \to R$ is given by equation (4) *then*

$$P\{(x,y): x + y = k\} = \frac{e^{-(\lambda_1+\lambda_2)}(\lambda_1+\lambda_2)^k}{k!} \qquad \text{if } k = 0, 1, 2, \ldots. \tag{5}$$

Therefore the distribution of the total number of emissions from the two blocks together is Poisson with mean $\lambda_1 + \lambda_2$. This result is intuitively not unreasonable. It extends to any number of blocks in an obvious fashion. In the terminology of Chap. 11, we have been considering sums of independent random variables. It is an unusual property of the Poisson distribution that such a sum has a distribution of the same form as one of the component variables. We will see that the binomial, Pascal, and normal distributions have similar properties. ■

EXERCISES

1. For example 2 prove that the distribution of the number of particles emitted by block 2 is Poisson with mean λ_2.

2. A random wheel corresponds to the Poisson density p_λ; that is, the sector labeled k has probability $p_\lambda(k)$, where $k = 0, 1, 2, \ldots$. Give an intuitive explanation for the fact that the probability is $p_{3\lambda}(m)$ that the sum of the three sector labels obtained on three independent spins will be m, where $p_{3\lambda}$ is the Poisson density with mean 3λ. HINT: Consider $3n$ spins of the two-sectored wheel of Fig. 2.

3. *Motor-vehicle Traffic.* Let X stand for the number of motor vehicles traveling north that pass a point on a certain highway during a particular 1-minute period. Assume that X has a Poisson distribution with $\lambda = 3.2$. Thus, on the average, 3.2 cars per minute pass the point. Such a Poisson assumption is common in theoretical studies of traffic flow.

(*a*) Find the probability that $X \geq 8$. For about how many minutes during a 1-hour period would you expect such "heavy" traffic?

(*b*) Find the probability that $1 \leq X \leq 6$. For about how many minutes during a 1-hour period would you expect such "normal" traffic?

(*c*) If the road has only one lane north, and all cars are 1 foot long and traveling bumper to bumper at 60 miles per hour, then 5,280 cars pass the point in 1 minute. Are you concerned by the fact that our assumptions assign positive probability to the event $X > 5{,}280$?

4. *Fluctuations in Bacteria Counts.* A certain type of bacteria is dangerous if there are eight or more of them per gallon of water. A reservoir has an average concentration of 1.8 of these bacteria per gallon. A gallon of water is selected at random from the reservoir and the number X of these bacteria in this gallon is determined.

(*a*) Find, approximately, the probability that $X \geq 8$. We may *assume* that if the gallon is divided into $n = 1{,}000$ equal disjoint subvolumes $V_1, \ldots, V_{1,000}$ then each V_i has none or one of these bacteria. Also, whether or not V_i has one is equivalent to whether or not we get a success on the ith of $n = 1{,}000$ spins of Fig. 2 with probability $p = 1.8/1{,}000$ for a success on one spin, so that the mean number of successes is $np = 1.8$.

(*b*) Do part (*a*) using analogous assumptions about a subdivision into 10^6 subvolumes. Is your estimate of the probability affected by the number n of subvolumes you contemplate?

5. *Many Unlikely Attempts.* One thousand infantrymen fire at a low-flying airplane and each has probability 10^{-4} of hitting the plane. What is the probability that at least one of them will hit the plane?

6. *Computer Failures.* A particular well-maintained digital computer makes mistakes independently from hour to hour. The distribution of the number of mistakes made in an hour is Poisson with a mean of .02 mistakes per hour. What is the probability that the computer will not make a mistake during the particular time intervals having the following lengths?

(*a*) 1 hour (*b*) 24 hours (*c*) (7)(24) = 168 hours

7. Prove the chains of inequalities exhibited in formulas (2) and (3).

8. Let p_λ be the Poisson density with parameter λ.

(*a*) Prove that $\sum_{k=0}^{\infty} k p_\lambda(k) = \lambda$.

(*b*) A random wheel corresponds to the density p_λ, and a gambler receives a reward of k units if the outcome of a spin is the kth sector, which has probability $p_\lambda(k)$. Explain why the arithmetic average of the gambler's n rewards on n independent spins of the wheel should typically be close to λ. HINT: Use the result of part (*a*).

9. This exercise shows that if n is large the Poisson density approximately describes the number of successes in n Bernoulli trials with probability $p_n = \lambda/n + 1/n^2$ for a success on one trial, for example; that is, instead of $p_n = \lambda/n$ for all n, we only need $np_n \to \lambda$ as $n \to \infty$.

(*a*) By direct multiplication we find that

$$\frac{1}{1+x}e^x = [1 - x + x^2 - \cdots]\left[1 + x + \frac{x^2}{2} + \cdots\right]$$

$$= 1 + \frac{x^2}{2} - \cdots \leq e^{x^2} = 1 + x^2 + \cdots$$

where the inequality must hold for all small x. Hence if

$$g(x) = e^{x-x^2} \qquad \text{for all real } x$$

then $g(x) \leq 1 + x$ for small x [a glance at Fig. 4 shows that $g(x) \leq 1 + x$ if $x \geq -.68\cdots$]. Prove that

$$e^{x-x^2} \leq 1 + x \leq e^x \qquad \text{for all real } x \geq -.2.$$

HINT: Observe that

$$g(0) = g'(0) = 1$$

$$g'(x) \leq 0 \qquad \text{if } x \geq 1/2$$

$$g''(x) < 0 \qquad \text{if } -.2 \leq x \leq 1.$$

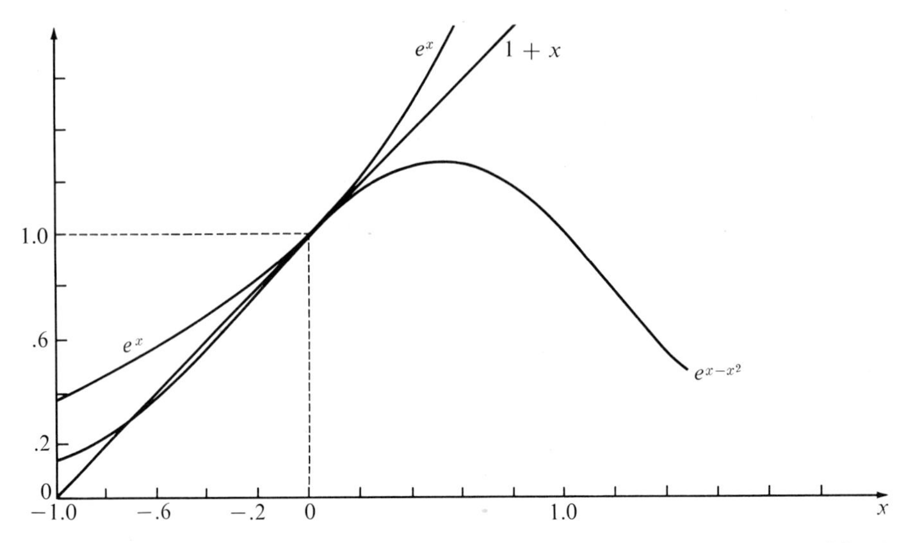

Fig. 4

(*b*) For real y and positive integer n prove that

$$e^{y-(y^2/n)} \leq \left[1 + \frac{y}{n}\right]^n \leq e^y \qquad \text{if } \frac{y}{n} \geq -.2.$$

Therefore if real b_n converges to b as n goes to infinity then

$$\lim_{n\to\infty} \left[1 + \frac{b_n}{n}\right]^n = e^b.$$

(*c*) Assume that $p_1, p_2, \ldots$ is a sequence of real numbers such that np_n converges to some $\lambda > 0$ as n goes to infinity. For any fixed integer $k \geq 0$ show that

$$\lim_{n\to\infty} \binom{n}{k} (p_n)^k (1 - p_n)^{n-k} = \frac{e^{-\lambda}\lambda^k}{k!}.$$

10. This exercise generalizes Exercise 9*b* for use in the proof of Lemma 1 (Sec. 18-3).

(*a*) Prove that if complex numbers δ_n go to zero as n goes to infinity then $[1 + (\delta_n/n)]^n \to 1$ as $n \to \infty$. HINT: Use the binomial theorem to show that

$$|(1 + z)^n - 1| \leq (1 + |z|)^n - 1 \qquad \text{for all complex } z.$$

(*b*) Prove that if complex numbers c_n converge to a real number b as n goes to infinity then

$$\left[1 + \frac{c_n}{n}\right]^n \to e^b \qquad \text{as } n \to \infty.$$

Actually the result is true for b complex. HINT: Define δ_n by $1 + c_n/n = [1 + (b/n)] \times [1 + (\delta_n/n)]$.

11. For integers $0 \leq k \leq n$ let

$$P(A_k \mid B_n) = \binom{n}{k} p^k q^{n-k} \qquad \text{and} \qquad P(B_n) = \frac{e^{-\lambda}\lambda^n}{n!}$$

while $P(A_k \mid B_n) = 0$ if $k > n$. Prove that

$$P(A_k) = \frac{e^{-\lambda p}(\lambda p)^k}{k!} \qquad \text{for } k = 0, 1, \ldots .$$

INTERPRETATION: The Poisson density with mean λ describes the number of eggs laid (particles emitted, babies born, . . .) in 1 day. Each has probability p, independent of the number and fates of the others, of hatching (being recorded, living for 1 year, . . .). Then the Poisson density with mean λp describes the number of eggs which hatch (recorded particles, babies living 1 year, . . .).

8-4 THE PASCAL DENSITY

Let r be a positive integer and let $q = 1 - p$, where p is a real number satisfying $0 < p < 1$. The **Pascal discrete probability density** p_r: $R \to R$ **with parameters** r **and** p is defined by

$$p_r(k) = \begin{cases} \dbinom{k + r - 1}{k} p^r q^k & \text{if } k = 0, 1, 2, \ldots \\ 0 & \text{otherwise.} \end{cases}$$

The **geometric density** p_1: $R \to R$ **with parameter** p is the special case $r = 1$ of the Pascal density, so

$${}_1(k) = \begin{cases} pq^k & \text{if } k = 0, 1, 2, \ldots \\ 0 & \text{otherwise.} \end{cases}$$

Equations (3) and (4) of Appendix 1 show that these obviously nonnegative functions do indeed sum to 1.

In this section we will see that $p_1(k)$, and more generally $p_r(k)$, equals the probability of a certain simple event defined in terms of a large enough number n of Bernoulli trials. Example 1 then indicates how p_1 and p_r arise in failure and life-testing problems, etc. It is then proved that p_r and p_1 have the same relationship as have the binomial b_n and Bernoulli b_1 densities. This relationship is shown to be intuitively apparent in the context of an infinite sequence of Bernoulli trials. The section closes with a proof that a model for an infinite sequence of Bernoulli trials cannot be constructed within our present discrete theory.

Naturally $p_1(k) = pq^k$ decreases as k increases, since $0 < q < 1$. Consider n Bernoulli trials with probability p for a success on one trial. If $0 \leq k \leq n$ then clearly $q^k p$ is the probability that the first k trials are failures *and* the $(k + 1)$st trial is a success. Therefore the value $p_1(k) = pq^k$ of the geometric density can be interpreted as the probability that exactly $k \geq 0$ failures occur prior to the first success in n Bernoulli trials, where n is any integer satisfying $n \geq k + 1$.

More generally we now relate the Pascal density for any $r \geq 1$ to Bernoulli trials. Let $A_{k,r}$ be the event that the $(k + r)$th Bernoulli trial is a success *and* the first $k + r$ trials contain k failures and r successes. The outcome shown below belongs to $A_{k,r}$:

$$\begin{matrix} & & 1 & & & & 2 & 3 & & & & & & & & r & & & & & \\ (0, & 0, & 1, & 0, & 0, & 0, & 1, & 1, & 0, & \ldots, & 1, & 0, & 0, & 0, & 1, & 0, & 1, & \ldots, & 0, & 1) \\ 1 & 2 & 3 & 4 & 5 & & & & & & & & & & k+r & & & & & n \end{matrix}$$

The probability of a success on the $(k + r)$th trial is p and the probability of exactly $r - 1$ successes in the preceding $k + r - 1$ trials is given by the binomial density as $\binom{k+r-1}{k} p^{r-1}q^k$ so

$$P(A_{k,r}) = \binom{k+r-1}{k} p^r q^k = p_r(k).$$

Therefore the value $p_r(k)$ of the Pascal density can be interpreted as the probability that exactly $k \geq 0$ failures occur prior to the rth (≥ 1) success in n Bernoulli trials, where n is any integer satisfying $n \geq k + r$.

If an experiment consists of a fixed number n of Bernoulli trials and we fix r then the values $p_r(k)$ have their natural interpretation as above *only* for $k \leq n - r$, even though $p_r(k)$ was defined for all $k = 0, 1, \ldots$ and is a discrete density and does not depend on n. We will return to this point and show that $p_r(k)$ has its natural interpretation for all $k = 0, 1, \ldots$ for an *infinite* sequence of Bernoulli trials. Alternatively we may, if we wish, assume that for fixed r and p the number n of Bernoulli trials considered is so large that

$$\sum_{k=n-r+1}^{\infty} p_r(k) \doteq 0.$$

For the large enough n, then, $p_r(k)$ has its natural interpretation except for $k > n - r$, but all these k have a total probability of almost 0.

Example 1 The geometric density, and the exponential density which is its continuous analog, has been found to provide a satisfactory approximation to an astonishingly large number of length-of-time and similar distributions. For example, $p_1(k)$ has been interpreted as the probability that exactly k time units of trouble-free service occur prior to first breakdown in an electrical component, a propeller, a complicated electrical or mechanical apparatus, etc. Also $p_1(k)$ has been interpreted as the probability of obtaining exactly k time units between the third and fourth radioactive emission, electron emission from the cathode of a vacuum tube, airplane arrivals at an airport, customer arrivals at a service counter, failures of a computer, etc.

For a particular parameter p suppose p_1 describes the number of minutes of illumination obtainable from a certain type of light bulb under specified conditions of service; that is, $p_1(k)$ is the probability that a bulb will burn out during its $(k + 1)$st minute of use. Then the density p_1 can be considered to be a simple mathematical model for the experiment of observing the number k of minutes of illumination prior to burnout. We could construct a random wheel corresponding to p_1, so that observing the outcome k from one spin of this wheel would then be "equivalent" to observing k from the bulb experiment. We could construct a more "atomic" or "refined" "equivalent" experiment by observing the number k of 0's prior to the first 1 in repeated independent spins of a two-sectored random wheel having probability p for a 1. By equivalent we mean only that the probability of obtaining any particular k is theoretically the same for each of the experiments. We might assume that whether or not a bulb burns out during its ith minute of use is equivalent to whether or not we obtain a 1 on the ith spin of our two-sectored wheel, given that there are no earlier 1's.

If bulb i yields x_i minutes of illumination, in accordance with p_1, prior to burnout, then $x_1 + \cdots + x_r$ is the number of minutes of illumination from r bulbs. Similarly $x_1 + \cdots + x_r$ can be interpreted as the waiting time to the arrival of the rth airplane, etc. We show below, under the assumption of independence, that the Pascal density p_r describes this variable. Roughly speaking, observing the sum of the outcomes from r repeated independent spins from p_1 is equivalent to observing the outcome from one spin of p_r. Thus the Pascal and geometric densities, having the same parameter p, are related by the basic fact that p_r is the density for the sum of r independent variables, each of which has density p_1. In other words, the Pascal and geometric densities are related in the same way as are the binomial and Bernoulli densities. ■

Let $r \geq 2$ and let p_r and p_1 be the Pascal and geometric densities *having the same parameter* p. We show that $p_r(k)$ for every $k = 0, 1, 2, \ldots$ *is the probability of the event that* $x_1 + x_2 + \cdots + x_r = k$ *occurs when* $x_1, x_2, \ldots, x_r$ *are obtained from* r *repeated independent spins of a random wheel corresponding to the geometric density* p_1. By independence the probability of any outcome $(x_1, \ldots, x_r)$ is

$$p_1(x_1)p_1(x_2) \cdots p_1(x_r) = pq^{x_1}pq^{x_2} \cdots pq^{x_r} = p^r q^{x_1+\cdots+x_r}.$$

We must show that $p_r(k)$ equals the sum of those probabilities which correspond to $(x_1, \ldots, x_r)$ satisfying $x_1 + \cdots + x_r = k$. But each of these $(x_1, \ldots, x_r)$ has the same probability $p^r q^k$. Thus we need only show that $\binom{k+r-1}{k}$ equals the number of different r-tuples $(x_1, \ldots, x_r)$ of nonnegative integers satisfying $x_1 + \cdots + x_r = k$. This is just formula (1) of Sec. 2-4, which can be proved as follows. In the expression $x_1 + \cdots + x_r$ we write each x_i as a sum of x_i ones and insert a 0 between each x_i and x_{i+1}; for example,

$$2 + 3 + 0 + 2 + \cdots + 1 = 1 + 1 + 0 + 1 + 1 + 1 + 0 + 0 + 0 + 1 + 1 + 0 + \cdots + 0 + 1.$$

By this technique we correspond to $(x_1, x_2, \ldots, x_r)$ a tuple having k 1's and $(r - 1)$ 0's. From Theorem 1 (Sec. 2-3) we know that there are $\binom{k+r-1}{k}$ such tuples of 0's and 1's, and since the correspondence is one-to-one *the proof is completed.*

This result and its proof are rigorous and completely within our present discrete theory. Let us next show that this result arises naturally and is intuitively obvious in the context of an infinite sequence of Bernoulli trials. We then show that an infinite sequence of Bernoulli trials cannot be fitted into our present discrete theory.

Consider an experiment consisting of an infinite sequence of Bernoulli trials, with probability p for a success. *One* performance of the experiment yields an infinite sequence $z_1, z_2, \ldots$ of 0's and 1's. If $k \geq 0$ let A_k be the event that the first k trials are failures and the $(k + 1)$st trial is a success, so that $P(A_k) = pq^k$. For $j \geq 0$ let D_j be the event that exactly j failures occur between the first and second successes. Clearly an outcome $(z_1, z_2, \ldots)$ belongs to $A_k D_j$ iff its first $k + j + 2$ coordinates are

$$(\overbrace{\underset{1}{0}, \underset{2}{0}, \ldots, \underset{k}{0}}^{k\ 0\text{'s}}, 1, \overbrace{0, 0, \ldots, 0}^{j\ 0\text{'s}}, 1, \ldots)$$

so that $P(A_k D_j) = q^k p q^j p$. Now D_j is the union of the disjoint events $A_0 D_j, A_1 D_j, A_2 D_j, \ldots$, so

$$P(D_j) = \sum_{k=0}^{\infty} P(A_k D_j) = \sum_{k=0}^{\infty} q^k p q^j p = pq^j.$$

Thus $P(A_k D_j) = P(A_k)P(D_j)$, so that these two types of events are independent, and each gives rise to the geometric density, as should be intuitively obvious after a little thought. For example, $P(D_8 \mid A_6) = P(D_8 \mid A_2)$, so that knowledge of the time of occurrence of the first success has no effect on the probabilities for various possible waiting times between the first and second successes.

Now fix $r \geq 2$, and for any outcome $(z_1, z_2, \ldots)$ define x_i for $i = 1, 2, \ldots, r$ to be the number of 0's between the $(i - 1)$st and ith 1's for this outcome:

$$(\underbrace{0, 0, 0, 0}_{x_1 = 4}, \overset{1}{1}, \underbrace{0, 0, 0, 0, 0}_{x_2 = 5}, \overset{2}{1}, \ldots, 1, \underbrace{0, 0, 0}_{x_r = 3}, \overset{r}{1}, \ldots)$$

Then the set of outcomes for which $x_1 + \cdots + x_r = k$ occurs is nothing more than the event that exactly k failures occur prior to the rth success, so this event has probability $p_r(k)$. Although the x_i above are defined in terms of Bernoulli trials, we know from the preceding paragraph that a probabilistically equivalent experiment can be constructed by obtaining $x_1, \ldots, x_r$ from r repeated independent spins of a random wheel corresponding to the geometric density, so this latter experiment must yield the same probability for obtaining $x_1 + \cdots + x_r = k$. This completes our intuitive explanation of the relationship between p_1 and p_r.

For later comparison we state the preceding result in the terminology of Chap. 11 and later chapters. Let $Z_1, Z_2, \ldots$ be an infinite sequence of Bernoulli trials and let X_i be the number of failures between the $(i - 1)$st and ith successes. Then $P[X_1 + \cdots + X_r = k] = p_r(k)$, and furthermore $X_1, X_2, \ldots$ are independent and each has density p_1.

We will show that an infinite sequence of Bernoulli trials cannot fit within our present discrete theory. Example 7 (Sec. 16-1) shows that a model for an infinite sequence of Bernoulli trials can be constructed if probabilities are assigned by integration. In this paragraph we consider the case $p = 1/2$ from an intuitive standpoint. It is reasonable that every possible outcome $(0, 1, 0, 0, 1, \ldots)$ for an infinite sequence of fair-coin tosses should be equally likely. But it is obviously impossible to assign the *same* probability (positive or zero) to *each* of an infinite collection of sample points and have these probabilities *add* to 1. Furthermore we know from Sec. 4-3 that this collection of equally likely outcomes is uncountable.

More generally, if we had an infinite sequence of Bernoulli trials we could let B_i be the event that the ith trial is a success, and these events would be independent and have the same probability. Therefore we assume that $0 < p < 1$ and that we have a discrete probability density $p': \Omega \to R$ and an *infinite* sequence of events $B_1, B_2, \ldots$ in Ω such that $P(B_i) = p$ for all i and such that $B_1, \ldots, B_n$ are independent events for every n. From these assumptions we will deduce that $p'(\omega) = 0$ for *all* ω in Ω; thus these assumptions lead to a contradiction, and hence no such model can exist. Select any sample point ω in Ω. We wish to show that $p'(\omega) = 0$. Let δ equal the larger of p and $1 - p$, so that $0 < \delta < 1$. For each i the sample point ω belongs to either B_i or B_i^c. If, for example, ω belongs to B_1 and B_2 and B_3^c and B_4 then ω belongs to their intersection, so

$$p'(\omega) \le P(B_1B_2B_3^cB_4) = P(B_1)P(B_2)P(B_3^c)P(B_4) = ppqp \le \delta^4.$$

More generally, $0 \le p'(\omega) \le \delta^n$ for all n, so $p'(\omega) = 0$. Therefore *a model for an infinite sequence of Bernoulli trials cannot be constructed within the discrete case.*

EXERCISES

1. Let a success on the ith of $n = 60$ Bernoulli trials with parameter $p = .7$ correspond to at least one vehicle passing a point on a highway during the ith minute.

(*a*) What is the probability of exactly 10 failures prior to the first success?

(*b*) What is the probability of exactly 10 failures prior to the fifth success?

(*c*) What is the probability of exactly 10 failures and 5 successes for the first 15 minutes?

(*d*) Exhibit an outcome which belongs to the event of part (*c*) but not to the event of part (*b*).

2. The ith of 10 light bulbs lasts x_i hours before it burns out and is replaced by the $(i + 1)$st bulb. What is the probability that the total hours $x_1 + x_2 + \cdots + x_{10}$ of illumination will exceed 10^4 if the $x_1, x_2, \ldots, x_{10}$ are obtained from independent trials from a geometric density with $p = .005$?

3. *Lack of Wearout under the Geometric Distribution*

(*a*) Let the discrete probability space (R,P) correspond to the geometric density $p_1: R \to R$ with parameter p. Let $A_r = \{x: x \ge r\}$ and prove that $P(A_{r+t} \mid A_r) = P(A_t)$ for all nonnegative integers r and t. COMMENT: If we interpret x as the number of trouble-free hours prior to the first breakdown, and if the geometric density applies, then the probability of at least t trouble-free hours equals the probability of at least t additional trouble-free hours if we have already had r trouble-free hours. This peculiar "lack of wearout" under the geometric density is reasonable from our derivation of this density from Bernoulli trials.

(*b*) For $t = 1$ the equation of part (*a*) reduces to $P(A_{r+1}A_r) = P(A_1)P(A_r)$. Clearly $A_{r+1}A_r = A_{r+1}$, so that the geometric density satisfies

$$P(A_{r+1}) = P(A_1)P(A_r) \qquad \text{if } A_r = \{x: x \ge r\}, \quad r = 0, 1, 2, \ldots.$$

Prove that if P satisfies this condition, where P corresponds to a discrete probability density $p': R \to R$ with $p'(0) + p'(1) + p'(2) + \cdots = 1$ and $0 < p'(0) < 1$, then p' is geometric with parameter $p'(0)$. Thus the geometric density is the unique discrete density with this lack-of-wearout property. In Sec. 16-1 we will see that the continuous exponential density has this same property.

4. *Guaranteed Wearout under the Poisson Distribution.* Let P correspond to the Poisson density with fixed parameter λ. Prove that $\lim_{r\to\infty} P(\{r\} \mid x \geq r) = 1$, where r has the values $0, 1, 2, \ldots$. Thus if r is large enough, and if we have enjoyed r trouble-free hours, then it is almost certain that there will be a breakdown during the next hour.

5. Let p_r be the Pascal density with parameters r and p.

(*a*) Prove that

$$\frac{p_r(k)}{p_r(k-1)} = 1 + \frac{p}{k}\left[(r-1)\frac{q}{p} - k\right] \qquad \text{if } k = 1, 2, \ldots$$

(*b*) State in detail and prove that the Pascal density, like the binomial and Poisson, increases to a maximum value and then decreases.

6. If (Ω,P) is a discrete probability space and $A_0, A_1, \ldots$ are disjoint events whose union has probability 1 then obviously $1 = \sum_{k=0}^{\infty} P(A_k)$. Therefore if $p'\colon R \to R$ is defined by

$$p'(k) = \begin{cases} P(A_k) & \text{if } k = 0, 1, \ldots \\ 0 & \text{otherwise} \end{cases}$$

then p' is a discrete density. Often our discrete densities arise most naturally in this fashion. The formula $1 = \sum_{k=0}^{\infty} p'(k)$, which is obtained almost trivially in such a circumstance, may be complicated and difficult to obtain by other methods. Let (Ω,P) correspond to r repeated independent spins of a random wheel for the geometric density p_1. Show that this section essentially contains a proof, by the above technique, of the equation $1 = \sum_{k=0}^{\infty} p_r(k)$. Thus we have a probabilistic proof that the Pascal density p_r is indeed a density; that is, we have a new proof of equation (4) of Appendix 1.

7. Each of three gamblers has a two-sectored random wheel. The wheel for gambler i has probabilities p_i and $q_i = 1 - p_i$ for the sectors labeled 1 and 0, respectively. Assume that at least one of p_1, p_2, p_3 is positive. The three gamblers spin their wheels independently, one after another, in the order $1, 2, 3, 1, 2, 3, 1, \ldots$. The first gambler to obtain a 1 wins, and the game stops. The sample space may be taken to be

$$\Omega = \{(1), (0,1), (0,0,1), (0,0,0,1), (0,0,0,0,1), \ldots\}$$

where, for example, the sample point (0,0,0,0,1) contributes probability $q_1q_2q_3q_1p_2$ to the number P[gambler 2 wins].

(*a*) For each gambler find the probability of a win. Note that these three numbers sum to 1, so that we have indeed defined a discrete probability density on Ω.

(*b*) Find necessary and sufficient conditions on p_1, p_2, p_3 guaranteeing that each gambler has the same chance of winning the game. Also exhibit numbers p_1, p_2, p_3 having p_1 as large as possible and satisfying these conditions.

(*c*) Find necessary and sufficient conditions on p_1, p_2, p_3 guaranteeing that *each* gambler has the same chance of winning on *his* first spin.

9
The Mean and Variance

The mean, variance, and moment-generating function of a density are introduced in this chapter. They are later related to random variables. *This chapter is largely restricted to densities* $p: \Omega \to R$ *for which* $\{\omega: p(\omega) > 0\}$ *is a set of real numbers.* Usually $\Omega = R$.

9-1 THE MEAN OF A DENSITY

If $p: R \to R$ is a discrete probability density and $\{x: p(x) > 0\} = \{x_1, x_2, \ldots, x_r\}$ is finite then the **mean of** p is the real number μ defined by

$$\mu = x_1p(x_1) + x_2p(x_2) + \cdots + x_rp(x_r).$$

Thus if $p(x)$ is positive we form the product $xp(x)$ and add all such products. Similarly if $\{x: p(x) > 0\} = \{x_1, x_2, \ldots\}$ is denumerable then the mean of p is defined by the series $\mu = x_1p(x_1) + x_2p(x_2) + \cdots$ if it is absolutely convergent, so that by Chap. 5 the sum does not depend on the order in which the terms are added. If the series is not absolutely convergent the mean is not defined.

Example 1 Figure 1a shows a random wheel *whose sample points are real numbers.* This discrete density is plotted in Fig. 1b and its mean is

$$\mu = (-3.5)(.2) + (-1)(.1) + (1)(.1) + (2)(.3) + (4.6)(.3) = 1.28.$$

Note that the definition of the mean of $p: R \to R$ is the same as the definition of the *center of gravity* of the discrete mass distribution p on the real axis.

If, as in Fig. 1c, the real axis is a weightless rigid horizontal rod, and we hang weight $p(x)$ at x, the rod should balance without rotation if we support it with a knife edge at the unique point μ. This physical interpretation of the mean of p sometimes enables us to make a fairly accurate guess at the number or to catch a computational mistake. The location of the *mean* of p on the real axis is often considered to be a reasonable location to interpret as the *center* of the distribution. That is, the mass or probability is considered to be distributed around, or centered on, the mean of p. Later in this section the *variance* of p will be introduced as a measure of the *spread* of p about its mean, and it is just the moment of inertia of the mass distribution p as defined in mechanics. This mechanical interpretation of the variance does not seem to be as helpful to intuition. ■

One interpretation for a random wheel whose sample points x_i are reals is that if the outcome of a spin is x_i a gambler receives the reward x_i. With this interpretation

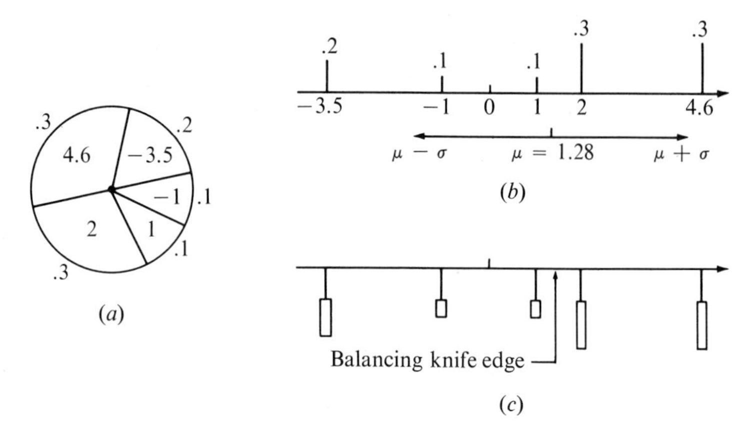

Fig. 1 Mean $\mu = 1.28$, second moment $= 10.198$, variance $= 8.5596$, and standard deviation $\doteq 2.92$.

of the wheel we now wish to give a preliminary justification for the following fundamental interpretation of the mean μ. *In a large number of independent spins of the wheel the arithmetic average of the rewards should be near to* μ. For simplicity we assume that $\Omega_0 = \{x_1, x_2, \ldots, x_r\}$ is finite and let $e = (e_1, e_2, \ldots, e_n)$ be the outcomes of n spins. Each e_i belongs to Ω_0 and the player receives reward e_i on trial i. The arithmetic average reward is

$$\frac{1}{n}(e_1 + e_2 + \cdots + e_n) = x_1 m_e(x_1) + x_2 m_e(x_2) + \cdots + x_r m_e(x_r)$$

since by definition the relative frequency is

$$m_e(x_i) = \frac{1}{n} \times \text{[the number of coordinates of } e \text{ which equal } x_i\text{]}.$$

Thus if n is large enough and *if* the relative frequencies behave properly, that is, if $m_e(x_i)$ is close enough to $p(x_i)$ for each i, *then* the arithmetic average reward should be close to μ. The average reward from a large number of spins of the random wheel of Fig. 1*a* should be near to the mean, which is 1.28. A rigorous version of this kind of result is called the *law of large numbers*, and it will be proved later within our mathematical model.

If $p: R \to R$ *is concentrated on some interval*, that is, if $|x| > B$ implies $p(x) = 0$, *then* p *has a mean and it is in this interval.* To prove this, note that for any n

$$\sum_{i=1}^{n} |x_i|\, p(x_i) \le \sum_{i=1}^{n} Bp(x_i) = B \sum_{i=1}^{n} p(x_i) \le B$$

so that the series $\sum_{i=1}^{\infty} x_i p(x_i)$ is absolutely convergent. Furthermore for every n

$$-B \le -B \sum_{i=1}^{n} p(x_i) \le \sum_{i=1}^{n} x_i p(x_i) \le B \sum_{i=1}^{n} p(x_i) \le B$$

so that $-B \leq \mu \leq B$. For example, if $\{x: p(x) > 0\}$ consists of all points $-1 + (1/k)$ $1 - (1/k)$, where k can be any positive integer, then $-1 \leq \mu \leq 1$. This result is certainly desirable for the interpretation of μ as the center of p.

Incidentally, as in the last paragraph, we will often exhibit a result or argument in one case, say when Ω_0 is denumerable, and neglect even to mention other cases, as when Ω_0 is finite, if these other cases are then fairly obvious.

If $p: R \to R$ is symmetric about some real number b,

$$p(b + x) = p(b - x) \qquad \text{for all real } x$$

and if p has a mean, then it is equal to b. This is easily proved, since the set $\Omega_0 = \{x: p(x) > 0\}$ can be enumerated in the form

$$b + x_1, b - x_1, b + x_2, b - x_2, b + x_3, b - x_3, \ldots .$$

Furthermore the series defining μ is assumed to be absolutely convergent, so that the terms can be added in this order, yielding

$$\begin{aligned}(b + x_1)p(b + x_1) + (b - x_1)p(b - x_1) + (b + x_2)p(b + x_2)& \\ + (b - x_2)p(b - x_2) + \cdots& \\ = b[p(b + x_1) + p(b - x_1)] + b[p(b + x_2) + p(b - x_2)] + \cdots = b.&\end{aligned}$$

Certainly if p is symmetric about a number b and has a "center" this center should equal b.

Example 2 If $p_\lambda: R \to R$ is the Poisson density with parameter λ its mean is

$$\mu = \sum_{k=0}^{\infty} kp_\lambda(k) = \sum_{k=1}^{\infty} k \frac{e^{-\lambda}\lambda^k}{k!} = \lambda \sum_{k=1}^{\infty} \frac{e^{-\lambda}\lambda^{k-1}}{(k-1)!} = \lambda$$

where for the last equality we used the fact that p_λ sums to 1. Thus *the Poisson density with parameter λ has λ for its mean*, as was hinted in Sec. 8-3. The density p_λ is often called the Poisson density with mean λ.

In order to find the mean of the geometric density we differentiate the geometric series $1/(1 - q) = 1 + q + q^2 + \cdots$ and get

$$\frac{1}{(1 - q)^2} = 1 + 2q + 3q^2 + 4q^3 + \cdots .$$

Multiplying both sides by $pq = (1 - q)q$, we get

$$\frac{q}{p} = pq + 2pq^2 + 3pq^3 + \cdots$$

for the mean. Thus *the mean is q/p for the geometric density with parameter p.* ■

Exercise 5 (Sec. 8-2) showed that *np is the mean of the binomial density with parameters n and p.* In Exercise 5 it is shown that *rq/p is the mean of the Pascal density with parameters r and p.* Figure 2 summarizes some of the results of this chapter. Section 9-4 presents a different, somewhat indirect technique for calculating means.

Name and parameters	Discrete probability density	Mean	Variance	Moment-generating function
Poisson $\lambda > 0$	$p_\lambda(k) = e^{-\lambda}\lambda^k/k!$ $k = 0, 1, \ldots$	λ	λ	$\exp\{\lambda(e^t - 1)\}$, $-\infty < t < \infty$
Bernoulli $0 < p < 1$	$b_1(0) = q,\ b_1(1) = p$	p	pq	$q + pe^t$, $-\infty < t < \infty$
Binomial $n = 1, 2, \ldots, 0 < p < 1$	$b_n(k) = \binom{n}{k} p^k q^{n-k}$ $k = 0, 1, \ldots, n$	np	npq	$(q + pe^t)^n$, $-\infty < t < \infty$
Geometric $0 < p < 1$	$p_1(k) = pq^k$ $k = 0, 1, \ldots$	q/p	q/p^2	$p/(1 - qe^t)$, $qe^t < 1$
Pascal $r = 1, 2, \ldots, 0 < p < 1$	$p_r(k) = \binom{r+k-1}{k} p^r q^k$ $k = 0, 1, \ldots$	$r(q/p)$	$r(q/p^2)$	$[p/(1 - qe^t)]^r$, $qe^t < 1$

Fig. 2 In each case $q = 1 - p$ and $p(x) = 0$ except when x equals one of the values of k shown. The lower expression in the last column states the constraint on t needed for convergence. The Bernoulli density is the special case $n = 1$ of the binomial density. The geometric density is the special case $r = 1$ of the Pascal density.

EXERCISES

1. Let $p(0) = .999$ and $p(10^6) = .001$, so that $\mu = 1{,}000$. Do you think the probability is close to 1 for the arithmetic average of the rewards on n trials to be near to μ, in each of the following cases?
(*a*) $n = 10$ (Justify your answer.)
(*b*) $n = 10^5$

2. Prove that if $p: R \to R$ is a discrete density for which $A \le x \le B$ whenever $p(x) > 0$ where $A < B$ are two real numbers, then p has a mean and it satisfies $A \le \mu \le B$.

3. Let $p: R \to R$ be defined by

$$p(n) = \frac{1 - \rho}{2\rho} \rho^{|n|} \qquad \text{if } n \text{ is one of } \pm 1, \pm 2, \pm 3, \ldots$$

$$p(x) = 0 \qquad \text{otherwise}$$

where ρ is a fixed parameter satisfying $0 < \rho < 1$. Find the mean of p.

4. A gambler receives reward x^2 if the outcome of the wheel of Fig. 1 is x. To what number should his arithmetic average reward converge as the number of trials grows?

5. Prove that rq/p is the mean of the Pascal density by deducing the identity

$$\sum_{k=0}^{\infty} k \binom{r+k-1}{k} p^r q^k = r\frac{q}{p} \sum_{k=1}^{\infty} \binom{r+k-1}{k-1} p^{r+1} q^{k-1}.$$

6. Give a brief intuitive argument explaining why the mean of the Pascal density should be r times the mean of the geometric density.

7. A marksman's shots constitute Bernoulli trials with probability $p = .1$ for hitting the bull's-eye. He practices each day until he gets his twentieth bull's-eye. About how many shots per day would you expect him to fire over a large number of days?

8. A crystal-growing process produces one crystal or none in each cycle of operation. The probability of a success on one cycle is .1 and the cycles constitute Bernoulli trials. Two different operating strategies are under consideration:

1. Make 10 cycles each day.
2. Each day make as many cycles as are necessary in order to produce one crystal, and then stop production for the day.

Compare the properties below for the two strategies:

(*a*) Mean number of crystals produced per day
(*b*) Variability of the number of crystals produced each day
(*c*) Mean number of cycles per day
(*d*) Variability of the number of cycles per day

9. Suppose a gambler receives reward $k(2^k)$ with probability $1/2^k$, where $k = 1, 2, \ldots$, so this density does not have a mean because $\mu = 1 + 2 + 3 + \cdots$ is infinite. Such models arise as manageable approximations to real-world problems and so cannot be ignored. If $e = (e_1, e_2, \ldots, e_n)$ is the sequence of n outcomes due to n trials, and $m_e(5) \doteq 1/2^5$, then $(1/n) \sum_{i=1}^{n} e_i \geq (5)(2^5)m_e(5) \doteq 5$. Generalize this argument to show that if r is *any* fixed positive integer and $m_e(r)$ converges to $1/2^r$ as the number n of trials increases then $(1/n) \sum_{i=1}^{n} e_i$ will be essentially larger than r if n is large enough. Thus as n goes to infinity the arithmetic average reward should grow arbitrarily large.

9-2 THE VARIANCE OF A DENSITY

If $p: R \to R$ is a discrete probability density for which $\Omega_0 = \{\omega: p(\omega) > 0\} = \{x_1, x_2, \ldots, x_r\}$ is finite then the **variance of** p is the real number defined by

$$\operatorname{Var} p = (x_1 - \mu)^2 p(x_1) + (x_2 - \mu)^2 p(x_2) + \cdots + (x_r - \mu)^2 p(x_r)$$

where μ is the mean of this p. We will shortly make the obvious generalization of this definition to infinite series. We multiply, or weight, $p(x_i)$ by the square of the distance $(x_i - \mu)^2$ of x_i from μ and then sum all such products; that is, as indicated in Fig. 1,

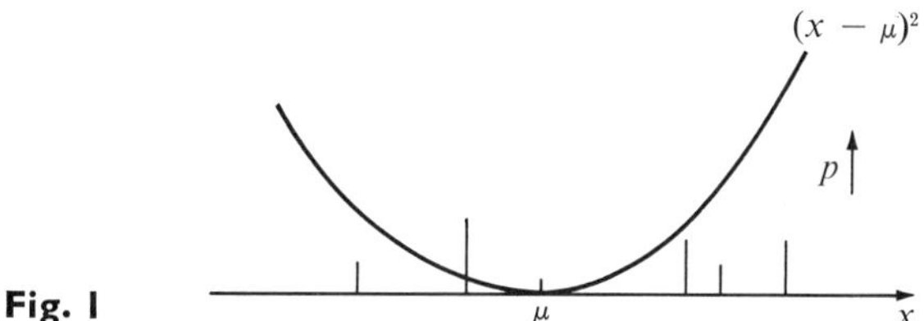

Fig. 1

we obtain the variance of p by multiplying p by the weighting function $(x - \mu)^2$ and then summing. The farther x_i is from μ, the greater the weight given to the probability $p(x_i)$. The variance of p is interpreted as a numerical measure of the *spread*, or *dispersion*, of p about its center μ.

If some one x_i happens to equal the mean μ then the corresponding term $(x_i - \mu)^2 p(x_i)$ equals 0 and all other terms are positive. Thus if $\{x: p(x) > 0\}$ contains just one point b then $p(b) = 1$, $\mu = b$, and $\operatorname{Var} p = 0$. If $\{x: p(x) > 0\}$ *contains two or more points then the variance of p is positive.*

By simple manipulations we find that

$$\begin{aligned}\operatorname{Var} p &= \sum_{i=1}^{r} (x_i - \mu)^2 p(x_i) = \sum_{i=1}^{r} [x_i^2 - (2\mu)x_i + \mu^2]p(x_i) \\ &= \sum_{i=1}^{r} x_i^2 p(x_i) - (2\mu)\sum_{i=1}^{r} x_i p(x_i) + \mu^2 \\ &= \sum_{i=1}^{r} x_i^2 p(x_i) - (2\mu)(\mu) + \mu^2 = \left[\sum_{i=1}^{r} x_i^2 p(x_i)\right] - \mu^2\end{aligned}$$

This last expression gives us an alternate way of calculating the variance. The mean μ of p is sometimes called the **first moment of** p, and the number $\sum_{i=1}^{r} x_i^2 p(x_i)$ is called the **second moment of** p. Thus the variance and first and second moments are related by

$$\left[\sum_{i=1}^{r} (x_i - \mu)^2 p(x_i)\right] = \left[\sum_{i=1}^{r} x_i^2 p(x_i)\right] - \left[\sum_{i=1}^{r} x_i p(x_i)\right]^2$$

$$[\text{variance of } p] = [\text{second moment of } p] - [\text{mean of } p]^2.$$

Since $0 \leq \operatorname{Var} p$, we see that

$$[\text{mean of } p]^2 \leq [\text{second moment of } p]$$

and this inequality is often useful.

Example 1 The two methods for computing the variance are shown below for the example of Fig. 1 (Sec. 9-1) with $\mu = 1.28$:

$$\begin{aligned}\operatorname{Var} p &= \sum_{i=1}^{r} (x_i - \mu)^2 p(x_i) \\ &= (-3.5 - 1.28)^2(.2) + (-1 - 1.28)^2(.1) + (1 - 1.28)^2(.1) \\ &\qquad + (2 - 1.28)^2(.3) + (4.6 - 1.28)^2(.3) \\ &= (22.8484)(.2) + (5.1984)(.1) + (.0784)(.1) + (.5184)(.3) + (11.0224)(.3) \\ &= 8.5596\end{aligned}$$

and

$$\begin{aligned}[\text{second moment of p}] &= \sum_{i=1}^{r} x_i^2 p(x_i) \\ &= (-3.5)^2(.2) + (-1)^2(.1) + (1)^2(.1) + (2)^2(.3) + (4.6)^2(.3) \\ &= (12.25)(.2) + .1 + .1 + (4)(.3) + (21.16)(.3) \\ &= 10.198\end{aligned}$$

so that

$$\operatorname{Var} p = 10.198 - (1.28)^2 = 10.198 - 1.6384 = 8.5596.$$

For this example the second computation is simpler. ■

If $p: R \to R$ is a discrete probability density and $\{x: p(x) > 0\} = \{x_1, x_2, \ldots\}$ is denumerable we naturally define the second moment of p to equal $\sum_{i=1}^{\infty} x_i^2 p(x_i)$ if this series is convergent; otherwise we say that p does not have a second moment. We now easily show that *if p has a second moment it also has a mean.* Note that if $|x| \leq 1$ then $|x| \leq 1 + x^2$ since $0 \leq x^2$, and if $|x| > 1$ then $|x| \leq x^2 \leq 1 + x^2$; thus $|x| \leq 1 + x^2$ for all real x. Therefore

$$\sum_{i=1}^{\infty} |x_i|\, p(x_i) \leq \sum_{i=1}^{\infty} (1 + x_i^2) p(x_i) = 1 + \sum_{i=1}^{\infty} x_i^2 p(x_i).$$

Thus if p has a second moment it has a mean, and we can define the variance of p by

$$\operatorname{Var} p = \sum_{i=1}^{\infty} (x_i - \mu)^2 p(x_i) = \left[\sum_{i=1}^{\infty} x_i^2 p(x_i)\right] - \mu^2$$

where μ is the mean of p, and these two expressions are easily shown to be equal, as before. A density p is said to have a variance iff it has a second moment.

Since the variance is nonnegative, its square roots are real. The **standard deviation of** p is defined to equal the *nonnegative* square root of the variance of p,

$$[\text{standard deviation of } p] = [\operatorname{Var} p]^{\frac{1}{2}} \geq 0.$$

Mean-square entities such as variance are common in physics, engineering, and many other fields. *Root-mean-square* (rms) entities such as standard deviation are also common. Usually we will denote the mean of p by μ, the variance of p by σ^2, and the standard deviation of p by σ.

Many other weighting functions, such as $|x - \mu|$ and $(x - \mu)^4$, could be used in place of $(x - \mu)^2$ to define a measure of spread about μ. However, among such measures of spread the variance has a unique usefulness in probability theory, since, as will be seen later, the variance of a sum of independent random variables is equal to the sum of their variances. This and similar formulas are most naturally handled in terms of variances rather than standard deviations. However, the standard deviation arises more naturally in other contexts, such as in normalizing for applications of the central-limit theorem.

Both σ^2 and σ are interpreted as measures of the spread of p about μ. If we have two discrete probability densities p_1 and p_2 with respective variances σ_1^2 and σ_2^2, then clearly $\sigma_1^2 \leq \sigma_2^2$ iff $\sigma_1 \leq \sigma_2$. Thus our interpretation of which density is more spread about its mean would be the same whether we use the variance or the standard deviation; that is, both σ^2 and σ determine the same ordering of densities with respect to spread. However, as indicated in Example 2 below and in the exercises, it is usually preferable to think of the numerical value of σ rather than the numerical value of σ^2 as measuring the spread.

It is often useful, as in Fig. 1 (Sec. 9-1), to indicate the two points which differ from the mean by one standard deviation. We will see in Sec. 10-1 that *every density concentrates most of the probability within a few standard deviations of its mean.* Still another point of view supporting the use of σ as a numerical measure of spread is that if x_i has the dimension of inches then so does $x_i p(x_i)$, since $p(x_i)$ is dimensionless, and so μ has the dimension inches, as it should. However, σ^2 has the dimension square inches, while σ measures spread about μ in inches.

Example 2 Let p be a uniform discrete density on the integers $0, 1, 2, \ldots, n$; that is, let $p(0) = p(1) = \cdots = p(n) = 1/(n+1)$. This corresponds to a random wheel with $n+1$ equal-sized sectors labeled $0, 1, 2, \ldots, n$. "Select an integer at random from among $0, 1, 2, \ldots, n$" can normally be translated as "spin the wheel just described and observe the outcome." Evaluating $\sum_{k=0}^{n} k$ and $\sum_{k=0}^{n} k^2$ by use of equation (1) of Appendix 1 and Exercise 4*b* (Sec. 2-2), we find that

$$\mu = \frac{1}{n+1}\sum_{k=0}^{n} k = \frac{1}{n+1}\,\frac{n(n+1)}{2} = \frac{n}{2}$$

The second moment of p is

$$\frac{1}{n+1}\sum_{k=0}^{n} k^2 = \frac{1}{n+1}\,\frac{n(n+1)(2n+1)}{6} = \frac{n(2n+1)}{6}$$

hence

$$\sigma^2 = \frac{n(2n+1)}{6} - \left[\frac{n}{2}\right]^2 = \frac{n^2}{12} + \frac{n}{6}$$

$$\sigma = \left[\frac{n^2}{12} + \frac{n}{6}\right]^{1/2} = \frac{n}{\sqrt{12}}\left[1 + \frac{2}{n}\right]^{1/2}$$

Thus $\mu = n/2$, which is intuitively reasonable, and if n is large then $\sigma \doteq n/\sqrt{12}$. As shown in Fig. 2, p *is* "roughly" spread out over the interval $[\mu - \sigma, \mu + \sigma]$, but not so for the interval $[\mu - \sigma^2, \mu + \sigma^2]$. ■

Fig. 2

EXERCISES

1. Let $p(-2) = .3$, $p(0) = .3$, and $p(1) = .4$, so that $\mu = -.2$.
(*a*) Find σ^2 by both computational methods.
(*b*) If we interpret the outcome x as the voltage across a unit resistor then we can interpret x^2 as the power. Thus $q(4) = .3$, $q(0) = .3$, $q(1) = .4$ is the discrete density for the power. Find its mean and variance. Note that

$$[\text{mean of } q] = [\text{second moment of } p] > [\text{mean of } p]^2$$

so that our new variable is the square of the old variable, but the mean of the new variable does *not* equal the square of the mean of the old variable.

2. Prove that if p has variance σ^2 then

$$\sum_{i=1}^{\infty} (x_i - b)^2 p(x_i) = \sigma^2 + (b - \mu)^2$$

for every real number b. Thus if we define the spread about b by the left-hand side of this equation then it is minimized iff $b = \mu$, indicating that μ is the natural center of p for such a measure of spread.

3. Prove that the Poisson density with parameter $\lambda > 0$ has variance λ. Thus, for example, the Poisson density with mean 10,000 has standard deviation 100. HINT: $k^2 = k(k - 1) + k$.

4. Assume that p has the property that $|x - \mu| \leq B$ whenever $p(x) > 0$, so that all the probability is within B units of the mean.

(*a*) Prove that $\sigma \leq B$. Clearly this is a desirable property for a measure of spread about μ.

(*b*) Exhibit an example for which $\sigma^2 > B$.

5. Assume that p has the property that $|x - \mu| \geq B$ whenever $p(x) > 0$, so that all the probability is at least B units away from the mean.

(*a*) Prove that $\sigma \geq B$. Clearly this is a desirable property for a measure of spread about μ.

(*b*) Exhibit an example for which $\sigma^2 < B$.

6. Let r and s be any two real numbers for which $0 < r < s$. Clearly

$$\frac{|x|^s}{|x|^r} = |x|^{s-r} \geq 1 \qquad \text{iff } |x| \geq 1.$$

Thus if $0 < r < s$ and if $|x| \geq 1$ then $|x|^r \leq |x|^s$. Use this fact to prove that if $p: R \to R$ is a density for which the series

$$\sum_{i=1}^{\infty} |x_i|^s p(x_i)$$

converges for a particular real s then replacing s by any smaller real r, where $0 < r < s$, results in a convergent series. Note that the special case $r = 1$ and $s = 2$ was proved in this section.

9-3 A LINEAR TRANSFORMATION OF A DENSITY

Linear transformations of densities arise often. They illuminate the nature of the mean and variance. In Chap. 11 we will see that they are special cases of the induced density of a function of a random variable.

If the outcome of an experiment is some real number x we may be interested in the real number $y = ax + b$ for $a \neq 0$, which is obtained by applying a linear transformation to x. For example, if x is the temperature as measured in degrees centigrade then $y = 1.8x + 32$ is the temperature in degrees Fahrenheit. Suppose x is obtained according to a discrete density $p_X: R \to R$. We use a subscript X for conformity with later notation for random variables. Thus $p_X(x)$ is the probability that x will be observed. If $p_X(x_0)$ is the probability of observing x_0 and if $y_0 = ax_0 + b$ then $p_X(x_0)$ must be the probability of obtaining this transformed y_0, since we obtain this y_0 iff we obtain x_0. If we are interested in the probabilities of various $y = ax + b$ values, where $a \neq 0$, then we can introduce the **transformed discrete probability density** $p_Y: R \to R$, for which

$$p_Y(y) = p_X(x) \qquad \text{if } x \text{ and } y \text{ are related by } y = ax + b$$

that is, $p_Y: R \to R$ is defined by

$$p_Y(ax + b) = p_X(x) \qquad \text{for all real } x$$

or equivalently by

$$p_Y(y) = p_X\left(\frac{y-b}{a}\right) \qquad \text{for all real } y.$$

As shown in Fig. 1, we obtain p_Y from p_X by transferring the probability $p_X(x)$ from the point x on the x axis to the point $y = ax + b$ on the y axis.

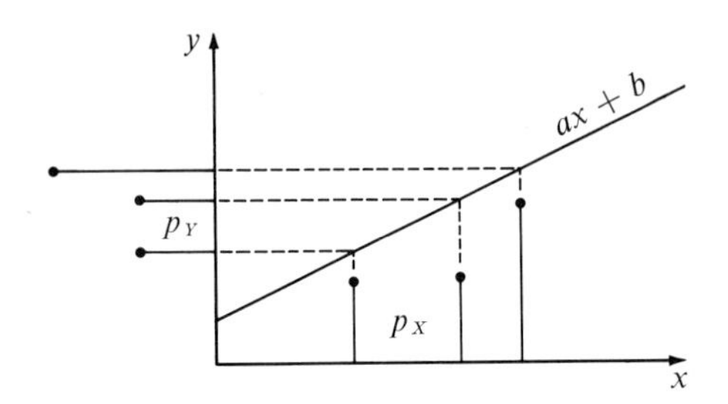

Fig. 1

To find the mean μ_Y of p_Y in terms of the mean μ_X of p_X we let $\{x: p_X(x) > 0\} = \{x_1, x_2, \ldots\}$, so that $\{y: p_Y(y) > 0\} = \{y_1, y_2, \ldots\}$, where $y_i = ax_i + b$. Then

$$\mu_Y = \sum_{i=1}^{\infty} y_i p_Y(y_i) = \sum_{i=1}^{\infty} (ax_i + b) p_Y(ax_i + b)$$
$$= \sum_{i=1}^{\infty} (ax_i + b) p_X(x_i) = a\mu_X + b.$$

Thus if we transform our observable x into y by *a linear transformation* $y = ax + b$, with $a \neq 0$, then we can use *this same* transformation $\mu_Y = a\mu_X + b$ on the mean μ_X to get the mean μ_Y of the transformed observable.

For example, if the mean temperature is $\mu_X = 10°\text{C}$ and we decide to use the Fahrenheit scale for the observable then the mean temperature will be $\mu_Y = (1.8)(10) + 32 = 50°\text{F}$. As another example, if $y = x - \mu_X$ then $\mu_Y = \mu_X - \mu_X = 0$, and this transformation corresponds to measuring the observable in terms of units to the right of its old mean. Equivalently it corresponds to considering the old mean μ_X as our new origin.

Similarly we can obtain the variance σ_Y^2 of p_Y in terms of the variance σ_X^2 of p_X as follows:

$$\sigma_Y^2 = \sum_{i=1}^{\infty} [y_i - \mu_Y]^2 p_Y(y_i) = \sum_{i=1}^{\infty} [ax_i + b - (a\mu_X + b)]^2 p_Y(ax_i + b)$$
$$= \sum_{i=1}^{\infty} a^2 [x_i - \mu_X]^2 p_X(x_i) = a^2 \sigma_X^2.$$

The standard deviation is the nonnegative square root of the variance, so

$$\sigma_Y = |a|\, \sigma_X.$$

If we are to measure the length x of girders in feet, and we consider the excess $y = x - 100$ of this length over the desired or nominal length of 100 feet, then

$\mu_Y = \mu_X - 100$ and $\sigma_Y = \sigma_X$, so that the spread remains the same, as it should. However, if $y = 12x$ then $\mu_Y = 12\mu_X$ and $\sigma_Y = 12\sigma_X$, so that a spread of 3 feet becomes a spread of 36 inches, as it should. Similarly if we are measuring distances from a reference point along a line, and we interchange our positive and negative directions $y = -x$, then $\mu_Y = -\mu_X$ and $\sigma_Y = \sigma_X$, so that the new spread about the new mean is the same as the old spread about the old mean, as it should.

Thus *if* $y = ax + b$ *then* $\mu_Y = a\mu_X + b$, *and* $\sigma_Y^2 = a^2\sigma_X^2$, *and* $\sigma_Y = |a|\,\sigma_X$, so that the mean is transformed in exactly the same way as the observables, while the standard deviation is unaffected by a change of direction or a change of origin but is proportional to the expansion of the scale.

If we wish to make a linear transformation $y = ax + b$ and we want to select a and b such that $\mu_Y = 0$ we must have $a\mu_X + b = 0$, so that $b = -a\mu_X$. Thus $\mu_Y = 0$ iff $y = a(x - \mu_X)$. If we now insist that $\sigma_Y^2 = 1$, or equivalently that $\sigma_Y = 1$, we must have $|a|\,\sigma_X = 1$, so that $a = \pm(1/\sigma_X)$. Therefore if σ_X exists and is positive then $y = (x - \mu_X)/\sigma_X$ is the unique linear transformation $y = ax + b$ for which $a > 0$, $\mu_Y = 0$, and $\sigma_Y = 1$. To conform with tradition, and for later convenience, we use the notation x^* instead of y. Therefore if $p_X: R \to R$ is a discrete probability density with a nonzero variance and if $x^* = (x - \mu_X)/\sigma_X$, then $\mu_{X^*} = 0$ and $\sigma_{X^*} = 1$, where $p_{X^*}: R \to R$ is defined by

$$p_{X^*}\left(\frac{x - \mu_X}{\sigma_X}\right) = p_X(x) \qquad \text{for all real } x$$

or equivalently by

$$p_{X^*}(x^*) = p_X((\sigma_X)x^* + \mu_X) \qquad \text{for all real } x^*.$$

We call p_{X^*} the **standardized** or **normalized discrete probability density corresponding** to p_X. As shown in Fig. 2, the transformation $x^* = (x - \mu_X)/\sigma_X$ sends the mean into

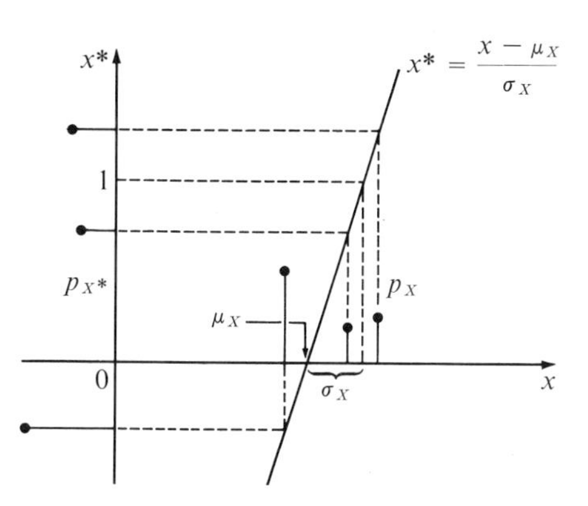

Fig. 2

0 and sends the point $\mu_X + \sigma_X$, which is one standard deviation to the right of the mean, into 1. Thus p_{X^*} is centered at the origin and has unit spread.

It is sometimes convenient, as in statements of the central-limit theorem, to consider standardized densities. Most properties of p_X of interest are easily converted into properties of p_Y if $y = ax + b$.

EXERCISES

1. It will be shown later that if X_n represents the total number of heads in n fair-coin tosses then $\mu_{X_n} = n/2$ and $\sigma^2_{X_n} = n/4$. Let Y_n represent the fraction of heads, so that $y_n = x_n/n$. Find μ_{Y_n} and σ_{Y_n}.

2. Assume that $\mu_X = 3$ hours and 42 minutes and $\sigma_X = .2$ hours, where X represents the time for a train to travel between Boston and New York. The schedule states that this time should be 3 hours and 20 minutes. Let Y represent the lateness of the train, in minutes, and find μ_Y and σ_Y.

3. For any real $r > 0$ and density $p: R \to R$ let

$$S^{(r)} = \left[\sum_{i=1}^{\infty} |x_i - \mu|^r p(x_i)\right]^{1/r}$$

assuming that it is finite. We can think of $S^{(r)}$ as the rth measure of the spread of p about μ. Clearly $S^{(2)} = \sigma$. If p_Y is obtained by applying $y = ax + b$ to p_X show that $S_Y^{(r)} = |a|S_X^{(r)}$ for every $r > 0$. Thus every one of these measures of spread has this desirable property.

4. Sketch a picture like Fig. 2 where p_X is given by Fig. 1 (Sec. 9-1). Exhibit all important numbers. Check that $\mu_{X^*} = 0$ and $\sigma_{X^*} = 1$.

5. Sketch p_{X^*} where p_X is the Poisson density for $\lambda = 3.2$ given in Fig. 1 (Sec. 8-3). Is p_{X^*} a Poisson density?

6. Sketch the standardized density corresponding to the uniform density on $0, 1, 2, \ldots, n$.

9-4 MOMENT-GENERATING FUNCTIONS

Moments other than the first and second are considered in this section. The moment-generating function is introduced as a useful entity for calculating moments. This function is a transform which is essentially like the Fourier and Laplace transforms. Its more general usefulness will not be apparent until sums of independent random variables are considered.

If $p: R \to R$ is a discrete probability density and if r is any positive real number we define the **rth absolute moment of p** to be the sum

$$\sum_{j=1}^{\infty} |x_j|^r p(x_j) \qquad \text{where } \{x: p(x) > 0\} = \{x_1, x_2, \ldots\}$$

if this series converges; otherwise we say that p does not have an rth absolute moment, or the rth absolute moment of p is infinite.

If $n = 1, 2, \ldots$ is a positive integer we define the **nth moment of p** to be the sum of the series

$$\sum_{j=1}^{\infty} (x_j)^n p(x_j)$$

if the series is absolutely convergent; otherwise we say that p does not have an nth moment.

The rth absolute moment is defined for *every* positive real r for which the series converges, since each term in the series is a positive real number. However, we define nth moments *only* for positive *integer* n. We do not, for example, attempt to define the $n = 2.7$th moment, since it involves terms like $(x_j)^{2.7}p(x_j)$, and if x_j is negative then

$$(x_j)^{2.7} = |x_j|^{2.7}(-1)^{2.7} = |x_j|^{2.7}(e^{\pi i})^{2.7} = |x_j|^{2.7}(\cos 2.7\pi + i \sin 2.7\pi)$$

is a complex number.

Clearly if $n = 2, 4, 6, \ldots$ is an even integer the nth moment and nth absolute moment are equal. Furthermore for any $n = 1, 2, 3, \ldots$ it is clear that p has an nth moment iff it has an nth absolute moment.

The first and second moment have already been used. The third absolute moment $r = 3$ will be useful when we consider the central-limit theorem. In this book we have only occasional need for moments other than these three. However, any and all moments are of use in various parts of probability theory. Most of the simple useful probability densities, such as the binomial, Poisson, Pascal, and normal densities, have an rth absolute moment for every real $r > 0$. Also there is clearly an rth absolute moment for every $r > 0$ if the density is concentrated on some interval, $p(x) = 0$ if $|x| > B$. However, some discrete densities of theoretical and practical importance do not have all moments.

Example 1 Recall that $\sum_{k=1}^{\infty} 1/k^{1+\alpha}$ converges iff $\alpha > 0$, so that for positive α we can denote its sum by $1/C_\alpha$. We are not interested in the numerical value of C_α. Thus if $\alpha > 0$ then $p_\alpha: R \to R$ is a discrete density, where p_α is defined by

$$p_\alpha(k) = \frac{C_\alpha}{k^{1+\alpha}} \qquad \text{if } k = 1, 2, \ldots$$

$$p_\alpha(x) = 0 \qquad \text{otherwise.}$$

Now the rth absolute moment of p_α is given by

$$\sum_{k=1}^{\infty} k^r \frac{C_\alpha}{k^{1+\alpha}} = C_\alpha \sum_{k=1}^{\infty} \frac{1}{k^{1+\alpha-r}}$$

and this is finite iff $r < \alpha$. Thus p_α has a finite rth absolute moment iff $r < \alpha$. In particular p_1 does not have a mean, while $p_{1.7}$ has a mean but not a second moment, and p_3 has a mean and a second moment but not a third moment. Theorem 1 below shows that some of the properties of this example are true for every density.

Moments can be interpreted probabilistically in the same way that the mean was interpreted. Let the kth sector of a random wheel be labeled k^r and have probability $p_\alpha(k)$. A gambler receives reward k^r on a trial if the trial results in the kth sector. We consider the arithmetic average of his rewards over n independent trials for large n. If $r < \alpha$, with high probability his average reward will be close to the rth absolute moment of p_α. If $r \geq \alpha$, with high probability his average reward will grow arbitrarily large with increasing n. Similarly if a special-purpose digital computer has to solve one of a special class of problems each hour, and if α and r are fixed and the probability is $p_\alpha(k)$ that it will be given a problem of type k which requires k^r computations, then we anticipate overloading if $r \geq \alpha$. ■

Exercise 6 (Sec. 9-2) showed that if p has a finite sth absolute moment for some real $s > 0$ then it has a finite rth absolute moment for every r with $0 < r < s$. This result is also contained in Theorem 1 below. For fixed $|x| \geq 1$ clearly $|x|^r$ is an increasing function of $r > 0$. Thus we might hope that for a fixed density p the rth absolute moment is an increasing function of r, but this is shown to be false in Exercise 1. However, Theorem 1 shows that the rth root of the rth absolute moment of p *is* an increasing function of r. We use this last part of Theorem 1 only rarely, but it is simple and basic enough to be explicitly stated and proved, so we do so here.

Theorem 1 If a discrete probability density $p: R \to R$ has a finite sth absolute moment for some real $s > 0$ then it has a finite rth absolute moment for any real r satisfying $0 < r < s$, and furthermore

$$\left[\sum_{j=1}^{\infty} |x_j|^r\, p(x_j)\right]^{1/r} \leq \left[\sum_{j=1}^{\infty} |x_j|^s\, p(x_j)\right]^{1/s}. \quad ■$$

Proof We first prove that if t and x are reals then

$$1 + t(x - 1) \leq x^t \qquad \text{if } t > 1,\ x \geq 0 \tag{1}$$

We fix $t > 1$ and define $h(x) = x^t$ for all $x \geq 0$. Then

$$\frac{d}{dx}\,h(x) = tx^{t-1} \qquad \text{and} \qquad \frac{d^2}{dx^2}\,h(x) = t(t-1)x^{t-2} \geq 0$$

so h is convex downward, as shown in Fig. 1. Now $h'(1) = t$, so that the line $1 + t(x - 1)$ of slope t is tangent to h at the point (1,1) of the plane and below h

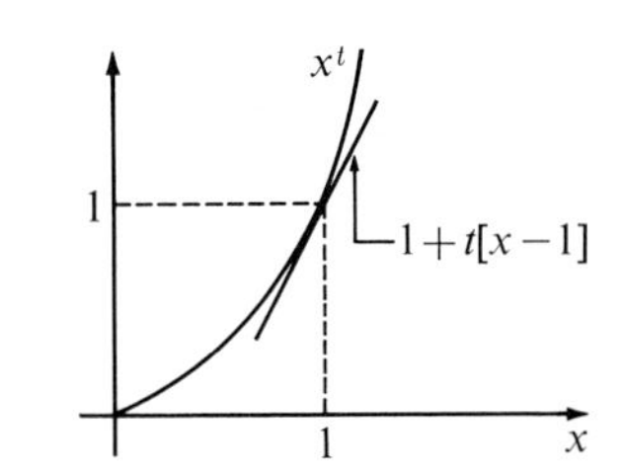

Fig. 1

elsewhere, proving inequality (1). In inequality (1) replace t by s/r and replace x by $|x|^r$ to obtain, for real r, s, and x,

$$1 + \frac{s}{r}\,(|x|^r - 1) \leq |x|^s \qquad \text{if } 0 < r < s \tag{2}$$

Assume that p has a finite sth absolute moment. We can replace x in inequality (2) by x_j, multiply the resulting inequality by $p(x_j)$, and sum over j to conclude that p has a finite rth absolute moment. Now we replace x in inequality (2) by

$$\frac{x_i}{\left[\sum_{j=1}^{\infty} |x_j|^r\, p(x_j)\right]^{1/r}},$$

multiply the resulting inequality by $p(x_i)$, and sum over i to get

$$1 \le \frac{\sum_{i=1}^{\infty} |x_i|^s \, p(x_i)}{\left[\sum_{j=1}^{\infty} |x_j|^r \, p(x_j)\right]^{s/r}}. \quad \blacksquare$$

Throughout probability theory various statements and proofs of results, such as the law of large numbers and the central-limit theorem, require that several moments exist. From Theorem 1 we see that it is only necessary to assume that the highest moment needed exists, since all lower moments must then also exist. Thus it is standard practice in the hypothesis of a theorem to state that a certain moment exists and then to use, without comment, lower moments in the remainder of the statement of the theorem and in its proof.

In this paragraph we motivate the introduction of the moment-generating function. If a density $p: R \to R$ is nonzero only on the integers $\Omega_0 = \{1, 2, \ldots\}$ then we can introduce the power series $f(s) = \sum_{k=1}^{\infty} p(k)s^k$ in the variable s. Clearly $f(1) = 1$. Also

$$\frac{d}{ds} f(s) = \sum_{k=1}^{\infty} kp(k)s^{k-1} \qquad \left[\frac{d}{ds} f(s)\right]_{s=1} = \sum_{k=1}^{\infty} kp(k) = \mu.$$

Thus the mean μ can be obtained from f by $f'(1) = \mu$. Many densities have a nice enough form that f and f' are simple, and $f'(1) = \mu$ provides a useful way to calculate μ. We can generalize this approach to any $\Omega_0 = \{x_1, x_2, \ldots\}$ as follows:

$$f(s) = \sum_{k=1}^{\infty} p(x_k)s^{x_k} \qquad f'(s) = \sum_{k=1}^{\infty} x_k p(x_k)s^{x_k - 1} \qquad f'(1) = \mu.$$

However, if we let $s = -1$ and we have $x_1 = \frac{1}{2}$ then $s^{x_1} = \sqrt{-1}$ and we are led to complex numbers; that is, the x_k need not be integers, so that the restriction of s to positive numbers is needed if f is to be real-valued. Therefore it is reasonable to replace s by e^t, since e^t is positive for all real t, and in addition $1 = e^{\circ}$, so that the important point $s = 1$ becomes the origin $t = 0$.

If $p: R \to R$ is a discrete probability density and $\{x: p(x) > 0\} = \{x_1, x_2, \ldots\}$ then the **moment-generating function** (referred to hereafter as the MGF) of p is defined by

$$m(t) = \sum_{k=1}^{\infty} p(x_k)e^{tx_k}.$$

The MGF is defined for all real t for which the series converges. It is positive valued, since every term is positive, and is always defined at $t = 0$, and $m(0) = 1$. For some densities the MGF is defined *only* at $t = 0$, and for such densities it is useless.

We now show that if an MGF is defined for some $t_0 > 0$ then it is defined in the interval $0 \le t \le t_0$. It is apparent from Fig. 2 that if $0 \le t \le t_0$ then

$$e^{tx} \le 1 + e^{t_0 x} \qquad \text{for all real } x. \tag{3}$$

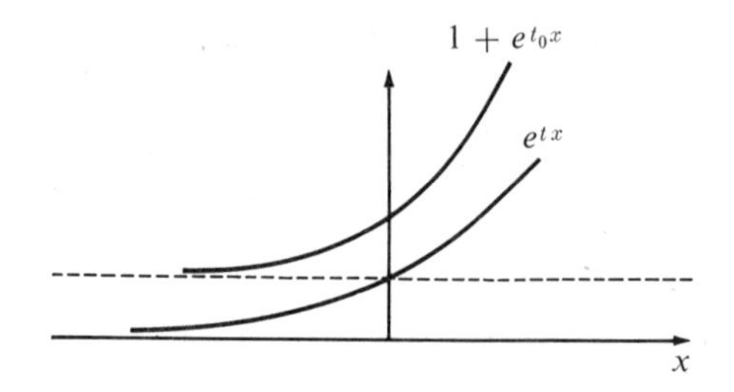

Fig. 2

That is, if $x \geq 0$ then $e^{tx} \leq e^{t_0 x}$, and if $x < 0$ then $e^{tx} \leq 1$, so inequality (3) holds. Therefore if $0 \leq t \leq t_0$ and the MGF is defined at t_0 then the right-hand series

$$\sum_{k=1}^{\infty} p(x_k)e^{tx_k} \leq \sum_{k=1}^{\infty} p(x_k)(1 + e^{t_0 x_k})$$

is convergent, so from inequality (3) the MGF is defined at t. A similar proof shows that if an MGF is defined for some particular $t_1 < 0$ then it is defined in the interval $t_1 \leq t \leq 0$. Thus if $t_0 > 0$ and an MGF is defined at both t_0 and $-t_0$ then it is defined in the interval $-t_0 \leq t \leq t_0$ containing the origin. This fact is stated as part of Theorem 2 below. We will utilize the MGF almost exclusively for densities for which it is defined at least in some interval containing the origin, $-t_0 \leq t \leq t_0$ for $t_0 > 0$.

In this paragraph we largely ignore questions of convergence and formally derive some properties of the MGF. Note that

$$\frac{d^n}{dt^n} m(t) = \sum_{k=1}^{\infty} (x_k)^n p(x_k)e^{tx_k} \qquad \left[\frac{d^n}{dt^n} m(t)\right]_{t=0} = \sum_{k=1}^{\infty} (x_k)^n p(x_k)$$

so that $m^{(n)}(0)$ equals the nth moment of p. Thus $\sigma^2 = m^{(2)}(0) - [m^{(1)}(0)]^2$, since the variance equals the second moment minus the square of the mean. Furthermore, the power-series expansion

$$m(t) = \sum_{n=0}^{\infty} \frac{m^{(n)}(0)}{n!} t^n$$

shows that the moments of p are obtainable from or *generated* by the power-series expansion of its MGF. This explains the name "moment-generating function." If the MGF converges in an interval which contains the origin then p has all moments, and the conclusions above can be justified. This is so stated in Theorem 2 below.

Sometimes we will define a density $p_1: R \to R$ in such an indirect fashion—say, as the uniquely determined density for a sum of independent random variables—that we may not know what p_1 really looks like. However, in such cases it is sometimes possible to obtain the MGF m_1 of p_1 explicitly. Suppose that we find m_1 and notice that $m_1 = m_2$, where m_2 is the MGF of a density $p_2: R \to R$ which is already familiar to us. We would like, then, to assert that $p_1 = p_2$. But perhaps different densities can have the same MGF. We motivated MGFs in terms of power series, and the sum of a power series determines its coefficients uniquely. Thus we suspect that different densities must have different MGFs. They do, according to Theorem 2, at least when there is convergence in some common interval containing the origin, in which case we can conclude that $p_1 = p_2$.

Theorem 2 collects together some facts about MGFs. *Theorem 2 is not proved in this book.* Actually only the proof of the last paragraph in the statement of Theorem 2 is difficult, and this is proved in Example 8 (Sec. 11-3) for a large class of densities.

Theorem 2 Let

$$m(t) = \sum_{k=1}^{\infty} e^{tx_k} p(x_k) \qquad \text{where } \{x: p(x) > 0\} = \{x_1, x_2, \ldots\}$$

be the MGF of a discrete probability density $p: R \to R$. Assume that $m(t_0)$ and $m(-t_0)$ are finite, where $t_0 > 0$. Then m is finite and has derivatives of all orders whenever $|t| < t_0$. Differentiation under the summation sign is permitted, so

$$m^{(n)}(t) = \frac{d^n}{dt^n} m(t) = \sum_{k=1}^{\infty} (x_k)^n e^{tx_k} p(x_k) \qquad \text{if } n = 1, 2, \ldots; \quad |t| < t_0.$$

In particular all moments of p exist and $m^{(n)}(0)$ equals the nth moment, so

$$\mu = m^{(1)}(0) \qquad \text{and} \qquad \sigma^2 = m^{(2)}(0) - [m^{(1)}(0)]^2.$$

Furthermore the power series

$$m(t) = \sum_{n=0}^{\infty} \frac{m^{(n)}(0)}{n!} t^n$$

converges absolutely for $|t| < t_0$.

If the MGFs m_1 and m_2 of discrete probability densities $p_1: R \to R$ and $p_2: R \to R$ are finite whenever $|t| \le t_0$, where $t_0 > 0$, then $p_1 = p_2$ iff $m_1(t) = m_2(t)$ for all $|t| \le t_0$. ■

Example 2 Let $p_\lambda: R \to R$ be the Poisson density with parameter λ. The MGF for p_λ is

$$\begin{aligned} m(t) &= \sum_{k=0}^{\infty} e^{tk} p_\lambda(k) = \sum_{k=0}^{\infty} e^{tk} \frac{e^{-\lambda}\lambda^k}{k!} \\ &= e^{-\lambda} \sum_{k=0}^{\infty} \frac{(\lambda e^t)^k}{k!} = e^{-\lambda} e^{\lambda e^t} \\ &= e^{\lambda(e^t - 1)} \end{aligned}$$

where we used the fact that the power series $e^y = \sum_{k=0}^{\infty} y^k/k!$ for the exponential function converges for all real y. Thus m is defined for all real t.

We now find μ and σ^2 from the MGF:

$$\frac{d}{dt} m(t) = \lambda e^t m(t)$$

so

$$\mu = \left[\frac{d}{dt} m(t)\right]_{t=0} = \lambda.$$

Also

$$\frac{d^2}{dt^2} m(t) = \lambda^2 e^{2t} m(t) + \lambda e^t m(t)$$

so the second moment of p is

$$\left[\frac{d^2}{dt^2} m(t)\right]_{t=0} = \lambda^2 + \lambda$$

hence

$$\sigma^2 = (\lambda^2 + \lambda) - \lambda^2 = \lambda.$$

Thus $\mu = \sigma^2 = \lambda$, as was also proved in Example 2 (Sec. 9-1) and Exercise 3 (Sec. 9-2). ■

EXERCISES

1. Let $p(-.1) = p(.1) = .5$ and show that the rth absolute moment of p is a decreasing function of r but that its rth root is a nondecreasing function of r, as assured by Theorem 1.

2. Prove that if $x_1, x_2, \ldots, x_n$ are any n fixed real numbers and f is a real-valued function defined for positive real r by

$$f(r) = \left[\frac{1}{n}(|x_1|^r + |x_2|^r + \cdots + |x_n|^r)\right]^{1/r}$$

then f is nondecreasing as r increases.

3. (*a*) Let m_α be the MGF for p_α as defined in Example 1. For what values of t is m_α defined?

(*b*) Let $q_\alpha: R \to R$ be the density obtained by reflecting p_α in the origin; that is, $q_\alpha(x) = p_\alpha(-x)$, so that

$$q_\alpha(k) = \frac{C_\alpha}{|k|^{1+\alpha}} \qquad \text{if } k = -1, -2, \ldots$$

$$q_\alpha(x) = 0 \qquad \text{otherwise.}$$

For what values of t is the MGF of q_α defined?

(*c*) Let the density $r_\alpha: R \to R$ be defined by $r_\alpha = (p_\alpha + q_\alpha)/2$, so that

$$r_\alpha(k) = \begin{cases} \dfrac{1}{2}\dfrac{C_\alpha}{|k|^{1+\alpha}} & \text{if } k = \pm 1, \pm 2, \ldots \\ 0 & \text{otherwise.} \end{cases}$$

For what values of t is the MGF of r_α defined?

4. Prove that the MGF of $p: R \to R$ is defined for all t if p is zero outside some interval; that is, if $p(x) = 0$ whenever $|x| > B$.

Exercises 5 to 7, together with the Poisson results of Example 2, derive the facts summarized in Fig. 2 (Sec. 9-1).

5. Derive the MGF, and from it the mean and variance, for the following:
(*a*) Bernoulli density (*b*) Geometric density

6. For $i = 1, 2$ let the discrete density $p_i: R \to R$ have MGF m_i, mean μ_i, and variance σ_i^2. Assume that m_2 happens to equal the nth power of m_1, where n is a positive integer; that is, for some $t_0 > 0$ we have

$$m_2(t) = [m_1(t)]^n \qquad \text{whenever } |t| \leq t_0.$$

Prove that $\mu_2 = n\mu_1$ and $\sigma_2^2 = n\sigma_1^2$. COMMENT: We will see later that this result is true because the assumption $m_2 = (m_1)^n$ implies that p_2 is the density for the sum of n independent random variables each having density p_1.

7. Derive the MGF and then use Exercises 5 and 6 to find the mean and variance of the following:

(*a*) Binomial density (*b*) Pascal density

HINT: If p_r is the Pascal density then $\sum_{k=0}^{\infty} p_r(k) = 1$, so that by multiplying this equation by p^{-r} we get

$$\sum_{k=0}^{\infty} \binom{r+k-1}{k} q^k = \frac{1}{(1-q)^r} \qquad \text{if } 0 < q < 1;\ r = 1, 2, \ldots.$$

8. Prove that an MGF is convex downward.

9. Sketch the three functions $.2e^{-t}$, $.3e^{1.5t}$, and $.5e^{4t}$ and then sketch their sum, which is the MGF of $p(-1) = .2$, $p(1.5) = .3$, and $p(4) = .5$.

10. The inequality

$$\frac{|y|^n}{n!} \leq e^y + e^{-y}$$

is true for any real y and any positive integer n. It can be proved as follows: If $y \geq 0$ then

$$\frac{|y|^n}{n!} = \frac{y^n}{n!} \leq \sum_{k=0}^{\infty} \frac{y^k}{k!} = e^y \leq e^y + e^{-y}$$

since each term in the series is positive, and one term is $y^n/n!$. If $y < 0$ then

$$\frac{|y|^n}{n!} = \frac{(-y)^n}{n!} \leq \sum_{k=0}^{\infty} \frac{(-y)^k}{k!} = e^{-y} \leq e^y + e^{-y}$$

since each term in the series is positive. Thus the inequality above is proved. Use it to prove that if MGF m of a density p is defined for some positive t_0 and also for $-t_0$ then p has an absolute rth moment for every real $r > 0$.

10
The Law of Large Numbers for Bernoulli Trials

In Sec. 10-1 we prove Chebychev's inequality, which is a simple, crude, but often useful tool in probability theory. This inequality is then utilized in Sec. 10-2 to deduce the law of large numbers for Bernoulli trials. This last result yields in Sec. 10-3 a probabilistic proof of a basic polynomial approximation theorem in mathematical analysis.

10-1 CHEBYCHEV'S INEQUALITY

We begin by describing a simple technique which is widely used in proving inequalities such as Theorems 1 and 2 below. By definition $P(A)$ equals the sum of all positive $p(x)$ for $x \in A$. Certainly $P(A)$ is less than or equal to the sum over A of $f(x)p(x)$ as long as $f(x) \geq 1$ when $x \in A$. If, in addition, $f(x) \geq 0$ for all x then summing $f(x)p(x)$ over all x can only increase the sum. Thus the sum of $f(x)p(x)$ over all x is an upper bound to $P(A)$ if $f \geq 0$ and $f(x) \geq 1$ for $x \in A$. With proper selection of the function f this result leads to many useful bounds which are often more easily calculated and manipulated than the probabilities which they bound. The lemma below states this result formally. The restriction to the real line for sample space is not essential.

Lemma 1 Let $p\colon R \to R$ be a discrete probability density and let A be any event. If a function $f\colon R \to R$ is nonnegative valued, $f \geq 0$, and satisfies $f(x) \geq 1$ whenever $x \in A$, then

$$P(A) \leq \sum_{i=1}^{\infty} f(x_i)p(x_i) \qquad \text{where } \{x\colon p(x) > 0\} = \{x_1, x_2, \ldots\}$$

if the series has a finite sum. ■

Proof As in Fig. 1, let $g\colon R \to R$ have the values 1 on A and 0 off A. Clearly $g \leq f$, since if $x \in A$ then $1 = g(x) \leq f(x)$ and if $x \notin A$ then $0 = g(x) \leq f(x)$. Therefore

$$P(A) = \sum_{i=1}^{\infty} g(x_i)p(x_i) \leq \sum_{i=1}^{\infty} f(x_i)p(x_i)$$

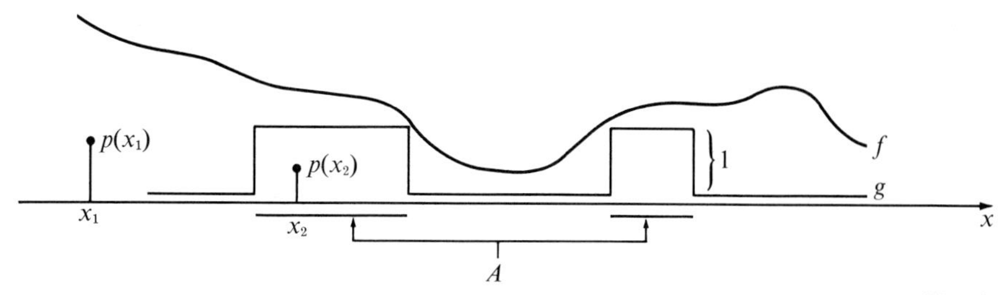

Fig. 1

since $g(x_i)p(x_i)$ equals either $p(x_i)$ or 0, depending on whether x_i does or does not belong to A. ■

Chebychev's inequality below is an immediate consequence of applying Lemma 1 to the function f of Fig. 2. The inequality states in effect that most of the probability must be within a few standard deviations of the mean.

Theorem 1: Chebychev's inequality Let $p: R \to R$ be a discrete probability density with mean μ and finite variance σ^2. If ϵ is any positive real then

$$1 - P\{x: |x - \mu| < \epsilon\} = P\{x: |x - \mu| \geq \epsilon\} \leq \frac{\sigma^2}{\epsilon^2}. \quad \blacksquare$$

Proof The event $A = \{x: |x - \mu| \geq \epsilon\}$ consisting of all points x differing from the mean μ by at least ϵ is shown in Fig. 2, as is its complement A^c. Certainly $P(A) + P(A^c) = 1$,

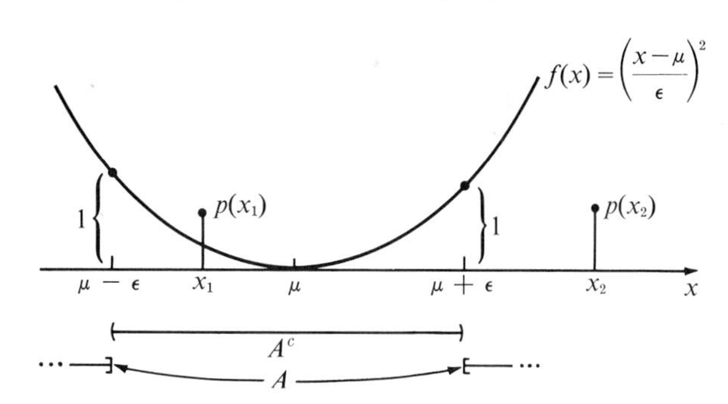

Fig. 2 $A = \{x: |x - \mu| \geq \epsilon\} = \{x: x \leq \mu - \epsilon\} \cup \{x: x \geq \mu + \epsilon\}$;
$A^c = \{x: |x - \mu| < \epsilon\} = \{x: \mu - \epsilon < x < \mu + \epsilon\}$.

so the equality in the statement of the theorem is true and trivial. The function $f(x) = [(x - \mu)/\epsilon]^2$ is nonnegative and satisfies $f(x) \geq 1$ if $x \in A$, so from Lemma 1

$$P(A) \leq \sum_{i=1}^{\infty} \left(\frac{x_i - \mu}{\epsilon}\right)^2 p(x_i) = \frac{\sigma^2}{\epsilon^2}. \quad \blacksquare$$

Example 1 A block of radioactive material emits an average of $\lambda = 100$ particles per minute, according to a Poisson density. Using $\epsilon = 30$ in Chebychev's inequality with $\lambda = \mu = \sigma^2 = 100$, we obtain

$$1 - P\{x: |x - 100| < 30\} = P\{x: |x - 100| \geq 30\} \leq 1/9$$

without recourse to tables of the Poisson density. The inequality $P\{x: |x - 100| \geq 30\} \leq 1/9$ states that the chance of obtaining an x far from the mean is small. The equivalent inequality

$$P\{x: |x - 100| < 30\} \geq 1 - 1/9 = 8/9$$

states that we can be quite certain of obtaining an x near to the mean. By definition $|x - 100| < 30$ is equivalent to $-30 < x - 100 < 30$, so that

$$P\{x: 70 < x < 130\} \geq 8/9.$$

The Poisson density is concentrated on the nonnegative integers, so

$$\sum_{k=71}^{129} \frac{e^{-100}(100)^k}{k!} \geq 8/9. \quad \blacksquare$$

Example 2 Applying Chebychev's inequality to the Poisson density with $\lambda = 3.2$ and letting $\epsilon = 2.2$, we obtain

$$P\{x: |x - 3.2| < 2.2\} \geq 1 - \frac{3.2}{(2.2)^2} \doteq .339.$$

From Table 1 (Sec. 8-3) we find that

$$\begin{aligned} P\{x: |x - 3.2| < 2.2\} &= P\{x: -2.2 < x - 3.2 < 2.2\} = P\{x: 1 < x < 5.4\} \\ &= p(2) + p(3) + p(4) + p(5) = .72339. \end{aligned}$$

Thus the actual value of the probability is about .723, while Chebychev's inequality only asserted that it was at least .339. Although the Chebychev inequality is a powerful theoretical and practical tool, in particular applications its conclusion may be deceptively weak. This "weakness" is to be expected, since the bound $P\{x: |x - 3.2| < 2.2\} \geq .339$ is valid for *every* discrete density with $\mu = \sigma^2 = 3.2$. It follows from Exercise 8*b* that no bound of the form $P\{x: |x - \mu| \geq \epsilon\} \leq F(\sigma^2, \epsilon)$, which is valid for *every* discrete density $p: R \to R$, can be smaller than the Chebychev bound when $\epsilon \geq \sigma$. $\blacksquare$

Example 3 For any positive real δ we can let $\epsilon = \delta\sigma$ in Chebychev's inequality to obtain

$$P\{x: |x - \mu| \geq \delta\sigma\} \leq \frac{1}{\delta^2} \qquad \text{and} \qquad P\{x: |x - \mu| < \delta\sigma\} \geq 1 - \frac{1}{\delta^2}.$$

If $\delta = 3$, for example, we obtain

$$\begin{aligned} P\{x: |x - \mu| < 3\sigma\} &= P\{x: -3\sigma < x - \mu < 3\sigma\} \\ &= P\{x: \mu - 3\sigma < x < \mu + 3\sigma\} \geq 8/9. \end{aligned}$$

Thus every density $p: R \to R$ must assign probability $8/9$ or more to the interval consisting of those points which are within three standard deviations of its mean. In this sense *every density must concentrate most of its probability within a few standard deviations of its mean.*

Letting $\delta = 1$ we see that $P\{x: |x - \mu| < \sigma\} \geq 0$, while $\delta = 1/2$ yields

$$P\{x: |x - \mu| \geq .5\sigma\} \leq 4 \qquad \text{and} \qquad P\{x: |x - \mu| < .5\sigma\} \geq -3.$$

These bounds are true enough, but uninformative and trivial, since we know that $0 \leq P(A) \leq 1$ for every event A. Thus Chebychev's inequality is true but uninformative if $0 < \delta \leq 1$, or if $0 < \epsilon \leq \sigma$ in the formal statement of the theorem. In particular Chebychev's inequality does not assure us that there is positive probability assigned to the interval $\mu \pm \sigma$. Indeed it could not, since the density $p(-1) = p(1) = .5$ has $\mu = 0$ and $\sigma = 1$, so the interval strictly within one standard deviation of the mean has zero probability. ■

Example 4 This example suggests the way in which Chebychev's inequality is useful in relation to the weak law of numbers as considered in Chap. 12. A gambler plays the wheel of Fig. 1 (Sec. 9-1), with $\mu_1 = 1.28$ and $\sigma_1^2 = 8.56$, a total of $n = 10{,}000$ times independently. It will be shown in Chap. 11 that if he wins X_i on trial i then his arithmetic average reward $Z = (1/n)(X_1 + X_2 + \cdots + X_n)$ has a unique, even though intractable, density whose mean and variance can be shown to be $\mu = \mu_1 = 1.28$ and $\sigma^2 = \sigma_1^2/n = .000856$. Using $\epsilon = .1$ in Chebychev's inequality, we obtain

$$P\{z: 1.18 < z < 1.38\} \geq 1 - \frac{\sigma^2}{\epsilon^2} = 1 - .0856 = .9144$$

so that there is more than a 90 percent chance that his arithmetic average reward will be within $\pm .1$ of the mean $\mu_1 = 1.28$ for the wheel. Naturally X_i can be interpreted as an error, or voltage, or pressure, etc.[1] ■

Given a density $p: R \to R$, we can calculate μ and σ^2 and then apply Chebychev's inequality. However, if we know the density we can calculate $P\{x: |x - \mu| \geq \epsilon\}$ exactly, so why bother to bound a number we know exactly? One reason is indicated by Example 1, where a nonobvious bound was obtained with very little calculation. Furthermore from the *two numbers* μ and σ^2 we can conclude that $P\{x: |x - \mu| \geq \epsilon\} \leq \sigma^2/\epsilon^2$ holds *for every real* $\epsilon > 0$ without having to perform a separate calculation for each ϵ. A different kind of answer is indicated by Example 4; we may "have" a density in the sense that it is uniquely defined, but it may be hard to learn much about it. In some such cases μ and σ^2 can be obtained fairly easily, so that Chebychev's inequality can be applied to learn more about the density.

A ONE-SIDED CHEBYCHEV INEQUALITY

As an introduction to the next theorem, note that the event $\{x: x - \mu \geq \epsilon\}$ is a subset of $\{x: |x - \mu| \geq \epsilon\}$, so from Chebychev's inequality

$$P\{x: x - \mu \geq \epsilon\} \leq P\{x: |x - \mu| \geq \epsilon\} \leq \frac{\sigma^2}{\epsilon^2}.$$

By this technique Chebychev's inequality can be used to bound probabilities of events which are not symmetric with respect to the mean. A sharper bound than is obtainable by this technique is exhibited in the theorem below for one-sided events of the form "at least ϵ units to the right of the mean," as depicted in Fig. 3. Similarly for events of the form "at least ϵ units to the left of the mean." Note that the bounds of Theorem 2 are always nontrivial, since $\sigma^2/(\epsilon^2 + \sigma^2) < 1$ whenever $\epsilon > 0$.

[1] In this chapter we occasionally use X, Y, Z, etc., for random variables. Although random variables are not formally defined until Sec. 11-1, their intuitive meaning should be apparent enough for now. Note the Comment on Memorization at the beginning of Chap. 12.

Theorem 2 Let $p: R \to R$ be a discrete probability density with mean μ and finite variance σ^2. If ϵ is any positive real then

$$1 - P\{x: x < \mu + \epsilon\} = P\{x: x \geq \mu + \epsilon\} \leq \frac{\sigma^2}{\epsilon^2 + \sigma^2}$$

and

$$1 - P\{x: x > \mu - \epsilon\} = P\{x: x \leq \mu - \epsilon\} \leq \frac{\sigma^2}{\epsilon^2 + \sigma^2} \quad \blacksquare$$

Proof Let $A = \{x: x \geq \mu + \epsilon\}$, so that the upper left-hand equality asserts the obvious fact that $1 - P(A^c) = P(A)$. We wish to prove that $P(A) \leq \sigma^2/(\epsilon^2 + \sigma^2)$, so we seek a function $f: R \to R$ in order to apply Lemma 1. As in Fig. 3, we let $f_b(x) = [(x - b)/$

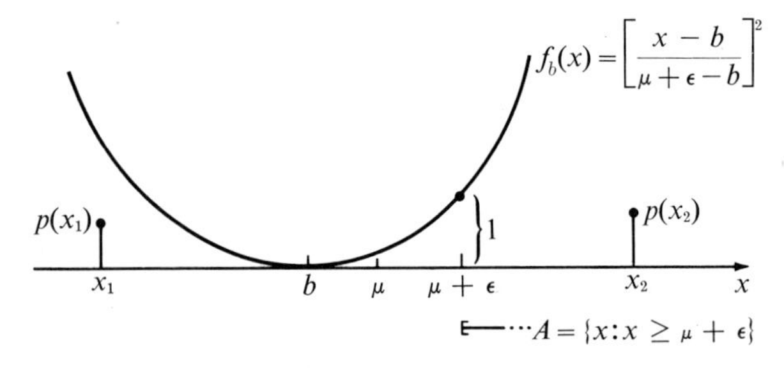

Fig. 3

$(\mu + \epsilon - b)]^2$, where b is some real number satisfying $b < \mu + \epsilon$, so that the slope of the line [ ] is positive. That is, we take the square function $(x - b)^2$ centered at b and divide by its value at $x = \mu + \epsilon$, so that $f_b(\mu + \epsilon) = 1$ and $f_b(x) \geq 1$ if $x \in A$, and $f_b \geq 0$. Applying Lemma 1, we obtain

$$\begin{aligned} P\{x: x \geq \mu + \epsilon\} &\leq \sum_{i=1}^{\infty} f_b(x_i)p(x_i) \\ &= \frac{1}{[\epsilon - (b - \mu)]^2} \sum_{i=1}^{\infty} [(x_i - \mu) - (b - \mu)]^2 p(x_i) \\ &= \frac{\sigma^2 + (b - \mu)^2}{[\epsilon - (b - \mu)]^2} = h(b) \end{aligned}$$

so $h(b)$ is the bound obtained by using f_b. We want the b which minimizes $h(b)$. Differentiating, we find that $h'(b_0) = 0$ leads to $b_0 = \mu - (\sigma^2/\epsilon)$. Then, denoting f_{b_0} by f, we have

$$f(x) = \left[\frac{\epsilon(x - \mu) + \sigma^2}{\epsilon^2 + \sigma^2}\right]^2$$

for the best function to use, and the bound becomes $h(b_0) = \sigma^2/(\epsilon^2 + \sigma^2)$, proving the first half of the theorem. The second half is proved similarly. ■

Example 5 Between 7:30 and 8:30 A.M. X_i motor vehicles enter a city by way of road i, where $X_1, X_2, \ldots, X_n$ are independent and X_i has a Poisson density with mean λ_i. From Sec. 8-2 we know that the total entering traffic $S = X_1 + \cdots + X_n$ has a Poisson distribution with mean $\lambda = \lambda_1 + \cdots + \lambda_n$. Applying Theorem 2 to this distribution, we obtain

$$P\{s: s \geq \lambda + \epsilon\} \leq \frac{\lambda}{\epsilon^2 + \lambda}$$

for a bound to the probability that the total traffic is at least ϵ greater than the mean λ. ■

EXERCISES

1. *Traffic Flow.* In Example 5 assume that the mean total traffic entering the city is $\lambda = 10^4$. Show that the probability
(*a*) $P\{s: s \geq 10{,}300\}$ of "excessive" traffic is small.
(*b*) $P\{s: s \leq 9{,}700\}$ of "low" traffic is small.
(*c*) $P\{s: 9{,}700 < s < 10{,}300\}$ of "normal" traffic is large.
(*d*) $P\{s: 9{,}670 < s < 10{,}300\}$ is larger than your answer to part (*c*).

2. *Total Operating Time.* The number X_i of operating hours prior to breakdown for the ith of $r = 100$ machines has a geometric distribution with a mean of 40 hours. The hour during which the ith machine breaks down is used in replacing it with the $(i + 1)$st machine. Thus $S_r = X_1 + \cdots + X_r$ is the total number of operating hours obtained from all r machines. Assume that independence prevails among machines.
(*a*) Show that the probability of the event $|S_r - 4{,}000| < 1{,}000$ is large.
(*b*) Show that the probability of $S_r \leq 3{,}200$ is small.

3. *Is the Coin Biased?* Let S be the number of heads obtained from $n = 100$ independent coin tosses with probability p for a head on one toss, so S is binomial $n = 100, p$. Suppose we do not know the value of p but we are certain that either $p = .3$ or $p = .5$. We decide to use the following arbitrary *decision rule* to be applied to the outcome of a planned experiment:

If $S \geq 40$ we will decide that $p = .5$.
If $S \leq 39$ we will decide that $p = .3$.

There are *two kinds of mistakes* we might make.
(*a*) Show that the probability of the event $S \leq 39$ is small, if we *assume* that $p = .5$.
(*b*) Show that the probability of $S \geq 40$ is small, if we *assume* that $p = .3$.

COMMENT: Head and tail might correspond to defective and nondefective, respectively. We might be testing whether a production facility was operating properly, $p = .3$, as opposed to whether a common breakdown, $p = .5$, occurred prior to the run of the batch of $n = 100$ items. Similarly in a communication context $p = .3$ might correspond to both noise and signal in the reception, and $p = .5$ to noise but no signal.

4. *Distribution of the Gambler's Reward* (*Example 4*). A gambler plays the wheel of Fig. 1 (Sec. 9-1). His total reward on n plays is $S_n = X_1 + X_2 + \cdots + X_n$.
(*a*) Find the distribution of S_2; that is, for each real s find the probability of the event that $S_2 = s$ and plot the resulting function of s.
(*b*) Find the probability of the event $S_4 = 0$, that the sum of the rewards on four plays is 0.

5. *A Generalization of Chebychev's Inequality.* Prove the following theorem, which reduces to Theorem 1 when $r = 2$: Let p be a discrete density with mean μ and finite rth absolute moment, where r is a positive real number. If ϵ is any positive real number then

$$P\{x: |x - \mu| \geq \epsilon\} \leq \frac{1}{\epsilon^r} \sum_{i=1}^{\infty} |x_i - \mu|^r p(x_i).$$

COMMENT: Letting $\epsilon = \delta S^{(r)}$, we obtain

$$P\{x: |x - \mu| \geq \delta S^{(r)}\} \leq \frac{1}{\delta^r} \qquad \text{where } S^{(r)} = \left[\sum_{i=1}^{\infty} |x_i - \mu|^r p(x_i)\right]^{1/r}$$

so the probability is large that we will obtain an x within a few $S^{(r)}$ units of the mean. This result, together with that of Exercise 3 (Sec. 9-3), encourages the use of every $S^{(r)}$ as a measure of spread, as well as just $S^{(2)} = \sigma$. The value $r = 2$ is unique, however, as will be seen in Chap. 11, because variances add under independence. ∎

For the remaining exercises, which are devoted to the graphical interpretation and sharpness of Theorems 1 and 2, consider the following: For any discrete density $p: R \to R$ having a finite variance define the two functions

$$h_p(\delta) = P\{x: |x - \mu| \geq \delta\sigma\} \qquad \text{and} \qquad g_p(\delta) = P\{x: x \geq \mu + \delta\sigma\}$$

for all $\delta \geq 0$. In particular $g_p(\delta)$ is the probability of obtaining an x at least δ standard deviations greater than the mean. Letting $\epsilon = \delta\sigma$ in Theorems 1 and 2, we obtain $h_p(\delta) \leq 1/\delta^2$ and $g_p(\delta) \leq 1/(1 + \delta^2)$ for all $\delta > 0$. Furthermore we know that $0 \leq P(A) \leq 1$ for every event A. Therefore for every density p we know that the graphs of h_p and g_p must lie in the shaded regions of Figs. 4a and b, respectively.

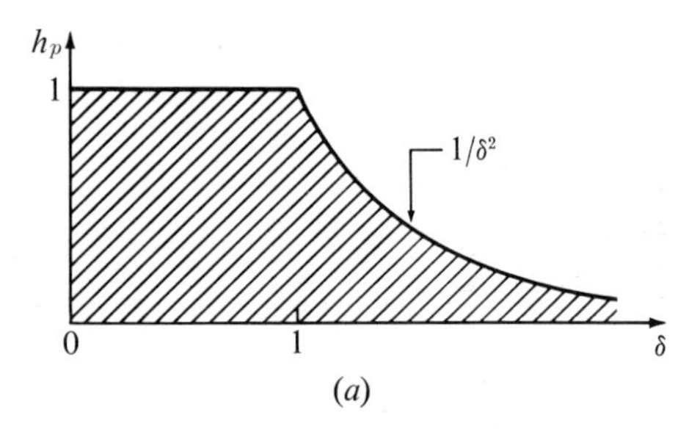

(a)

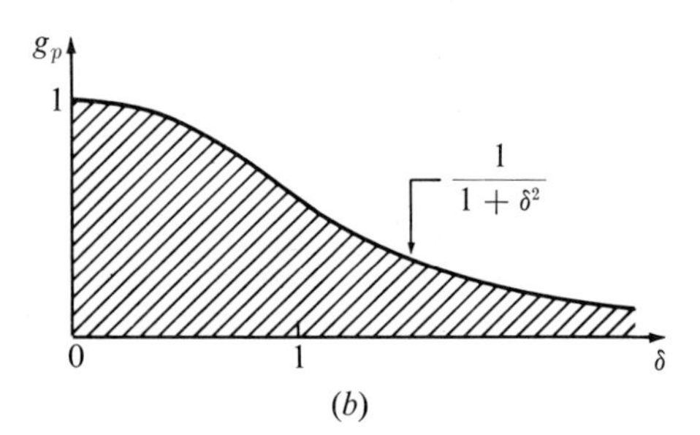

(b)

Fig. 4

6. For the density p of Fig. 1 (Sec. 9-1):
(a) Plot h_p on a reproduction of Fig. 4a.
(b) Plot g_p on a reproduction of Fig. 4b.

7. Prove that the inequality of Lemma 1 becomes an equality iff p, A, and f are related at every real x by

$$f(x) = \begin{cases} 1 & \text{if } p(x) > 0,\ x \in A \\ 0 & \text{if } p(x) > 0,\ x \notin A. \end{cases}$$

In particular if we have f and A then any p which is to achieve equality must assign positive probability only to some or all of those x for which $f(x) = 0$ or 1.

8. (*a*) Use Exercise 7 to prove that if there is equality $h_p(\delta) = 1/\delta^2$, essentially equality in Theorem 1, for a particular density p with mean μ and variance σ^2 and for a particular $\delta \geq 1$, then p can assign positive probability to at most the three points μ and $\mu \pm \delta\sigma$.

(*b*) For each $\delta \geq 1$ define a density p_δ: $R \to R$ for which $h_{p_\delta}(\delta) = 1/\delta^2$. For $\delta = 1, 2, 3$ plot the corresponding three functions h_{p_δ} on a reproduction of Fig. 4*a*.

9. (*a*) Use Exercise 7 to prove that if there is equality $g_p(\delta) = 1/(1 + \delta^2)$, essentially equality in Theorem 2, for a particular density p with mean μ and variance σ^2 and for a particular $\delta > 0$, then p can assign positive probability to at most the two points $\mu - (\sigma/\delta)$ and $\mu + \delta\sigma$.

(*b*) For each $\delta > 0$ define a density p_δ: $R \to R$ for which $g_{p_\delta}(\delta) = 1/(1 + \delta^2)$. For $\delta = .5, 1, 2$ plot the corresponding three functions g_{p_δ} on a reproduction of Fig. 4*b*.

10-2 THE LAW OF LARGE NUMBERS FOR BERNOULLI TRIALS

If we consider the binomial density b_n: $R \to R$ with fixed p, and let n grow then we have the relationships indicated in Fig. 1. The mean np grows as n, but the standard deviation $\sqrt{npq}$ grows only as $\sqrt{n}$, so that the interval $np \pm \sqrt{npq}$, consisting of those points within 1 standard deviation of the mean, is only a small fraction of the interval

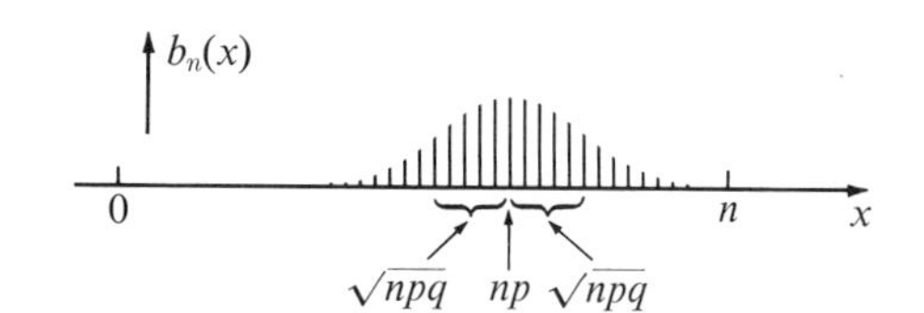

Fig. 1

$0 \leq k \leq n$ over which the density is spread. But from Chebychev's inequality we know that most of the probability must be within a few standard deviations of the mean. Thus for large n the probability is large that the number x of successes will be relatively near to the mean number np of successes. It is traditional, as in Theorem 1 below, to state this result in terms of the fraction x/n of successes.

Theorem 1: The law of large numbers for Bernoulli trials Let P correspond to the binomial density b_n: $R \to R$ with parameters n and p. If ϵ is any positive real then

$$P\left\{x: \left|\frac{x}{n} - p\right| \geq \epsilon\right\} \leq \frac{pq}{n\epsilon^2} \leq \frac{1}{4\epsilon^2 n}. \quad \blacksquare$$

Proof Application of Chebychev's inequality, Theorem 1 (Sec. 10-1), to b_n with $\mu = np$, $\sigma^2 = npq$, and ϵ replaced by $n\epsilon$ proves the first inequality of the theorem. Then, as shown in Fig. 2, $pq = p(1 - p) \leq \frac{1}{4}$ for all p with $0 \leq p \leq 1$. $\blacksquare$

For any fixed p and ϵ the bound $pq/n\epsilon^2$ converges to zero as n goes to infinity. Thus for any tolerance $\epsilon > 0$ we can, by making n large, be certain that the probability

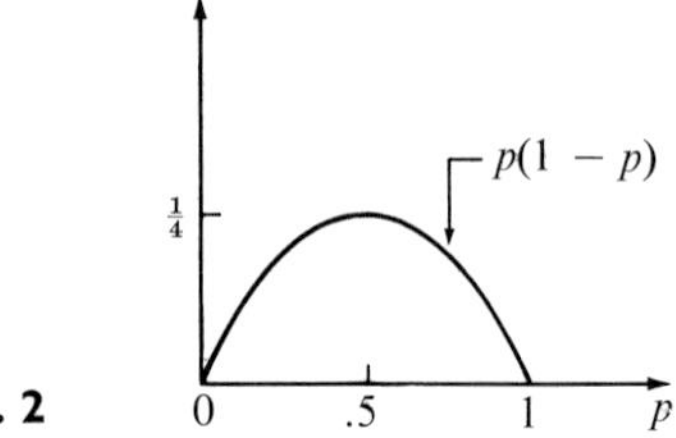

Fig. 2

is as small as we like for obtaining a fraction x/n of successes outside the tolerance limits $p \pm \epsilon$. For large enough n we can be "almost certain" that the fraction x/n of successes will be near to the probability p of a success on one trial. This is the qualitative intuitive meaning of Theorem 1.

Example 1 The random wheel of Fig. 1*a* (Sec. 6-1) assigns probability .3 to sector ω_2 of its six sectors. Let us call outcome ω_2 a success, and all other outcomes failure, so that we have essentially a two-sectored wheel with probability $p = .3$ for a success. Letting $n = 1{,}000$ and $\epsilon = .1$ in Theorem 1 yields

$$1 - P\left\{x: \left|\frac{x}{n} - .3\right| < .1\right\} = P\left\{x: \left|\frac{x}{n} - .3\right| \geq .1\right\} \leq .021.$$

Thus the probability is at least .979 that in $n = 1{,}000$ independent spins of the wheel, the relative frequency x/n of occurrence of sector ω_2 will satisfy $.2 < x/n < .4$. ■

Example 2 The second bound $1/4\epsilon^2 n$ of Theorem 1 is weaker than the first bound, but it has the advantage of not depending on the parameter p. If we let $\epsilon = .1$ and $n = 1{,}000$ then $1/4\epsilon^2 n = .025$, so that

$$P\left\{x: p - .1 < \frac{x}{1{,}000} < p + .1\right\} \geq .975$$

for *every* p. In some problems the true p is not known. Nevertheless the probability is at least .975 that the fraction $x/1{,}000$ of successes in 1,000 independent trials will differ by less than .1 from the true unknown probability p of a success on one trial. This kind of result helps to justify the use of an "experimentally obtained fraction of successes" as an "estimate" for the unknown parameter p in statistical problems. ■

EXERCISES

1. *Frequency of Telephone-exchange Overloads.* A trial consists of observing a telephone exchange for four consecutive seconds. A success corresponds to the exchange being used for at most 3 of the 4 seconds. The trials constitute Bernoulli trials, although there may be dependence within one trial. Assume that the density of Fig. 4 (Sec. 6-2) applies for one trial and show that the probability of the event $S_n \geq 7{,}000$ is large, where S_n is the total number of successes on $n = 10^4$ trials.

2. *Number of Successes.* Each trial of an experiment yields a volume-temperature pair (v,t) according to the discrete density of Fig. 3. A trial is a success iff $v^2 + t^2 \leq 9$. Show that the probability is large that $n = 100$ repeated independent trials yield at least 21 successes.

(v,t)	$p(v,t)$
$(-3,1)$	.3
$(-1,1)$	.2
$(1,1.5)$	.1
$(4,2)$	.4

Fig. 3

3. *Products of Independent Random Variables.* Let α be a real number satisfying $0 < \alpha < 1$ and let the density $p_1: R \to R$ be defined by $p_1(1/3) = 1 - \alpha$ and $p_1(9) = \alpha$. A gambler always bets everything he has on each of n independent spins of the wheel corresponding to p_1. If the outcome of a trial is X then his capital after the trial is X times his capital before the trial. Thus either he loses $2/3$ of his capital or he increases it by a factor of 9. If he starts with one unit, his capital after n trials is the *product* $X_1X_2 \cdots X_n$, where the outcome of the ith trial is X_i. Show that if n is large enough, then the probability is almost 1 that the gambler's capital is large if $\alpha = .4$, and small if $\alpha = .3$.

HINT: The mean of p_1 is at least 1 iff $\alpha \geq 1/13$, but do not let this mislead you.

*10-3 THE WEIERSTRASS APPROXIMATION THEOREM

This theorem asserts that any continuous real-valued function f defined on the real interval $0 \leq p \leq 1$ can be approximated arbitrarily closely over the whole interval by a polynomial. We give S. Bernstein's proof, which exhibits the required polynomials and is essentially based on the law of large numbers for Bernoulli trials. The theorem is included here principally as an interesting mathematical digression, although Corollary 1 is used in Example 8 (Sec. 11-3) to prove the uniqueness of certain MGFs. In this section Corollary 1 yields the uniqueness of Laplace transforms.

Given such a function f and a positive integer n, define

$$B_n(p) = \sum_{k=0}^{n} \left[f\left(\frac{k}{n}\right)\right]\binom{n}{k}p^k(1-p)^{n-k} \qquad \text{for } 0 \leq p \leq 1$$

in terms of the values $f(0/n), f(1/n), \ldots, f(n/n)$ of this function at the $n + 1$ equispaced points of the unit interval. The dependence of $B_n(p)$ on the fixed function f is not exhibited in the notation. If $(1-p)^{n-k}$ is expanded by the binomial theorem then $B_n(p)$ can be expressed as an nth-degree polynomial $B_n(p) = \sum_{k=0}^{n} a_k p^k$ in the real variable p with coefficient a_k depending on k, n, and the values $f(0/n), f(1/n), \ldots, f(n/n)$. We call B_n the **nth Bernstein polynomial for f.**

Suppose that a gambler has a reward function f in the sense that he receives reward $f(k/n)$ if k/n is the fraction of successes on n Bernoulli trials with parameter p. He receives reward $f(k/n)$ with probability $\binom{n}{k}p^k(1-p)^{n-k}$, so $B_n(p)$ is his mean, or expected, reward. If n is large then with high probability k/n is near to p, in which case $f(k/n)$ is near to $f(p)$, since f is continuous. Therefore we anticipate that his *expected reward* $B_n(p)$ for n trials will converge to $f(p)$ as the number n of trials considered goes to infinity. The theorem below shows that the Bernstein polynomials do indeed converge to f and the convergence is uniform in p.

Theorem 1: The Weierstrass approximation theorem Let f be a continuous real-valued function defined on the real interval $0 \le p \le 1$ and let

$$B_n(p) = \sum_{k=0}^{n} f\left(\frac{k}{n}\right)\binom{n}{k} p^k(1-p)^{n-k} \qquad \text{for } 0 \le p \le 1$$

be the nth Bernstein polynomial for f. Then $B_n(p) \to f(p)$ as $n \to \infty$, and the convergence is uniform in p. That is, there is a sequence of constants $\delta_1, \delta_2, \ldots,$ determined solely by f, which converge to zero and satisfy

$$|B_n(p) - f(p)| \le \delta_n \qquad \text{for all } p,\ 0 \le p \le 1;\ \text{all } n = 1, 2, \ldots . \quad \blacksquare$$

Thus, as shown in Fig. 1, the polynomial B_n is everywhere inside the band $f \pm \delta_n$, and the width $2\delta_n$ of the band can be made as small as we like by selecting a polynomial of high enough degree n.

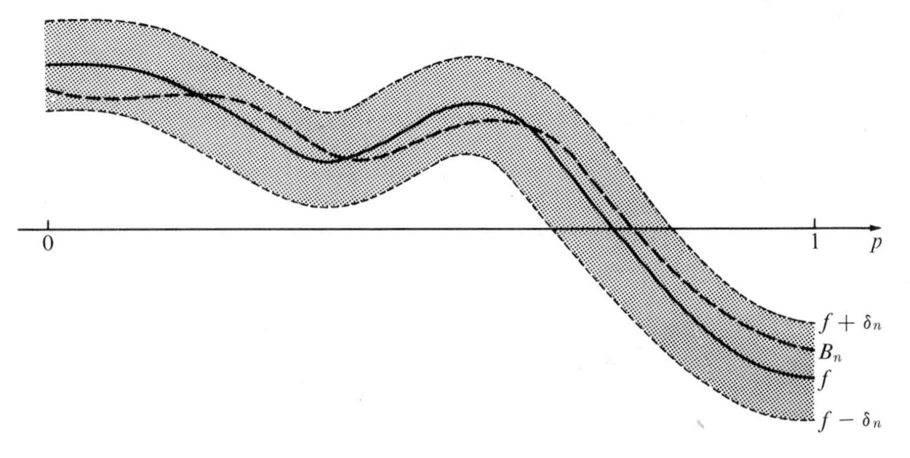

Fig. 1 The Bernstein polynomial B_n approximation to f.

Proof For this proof, but not for the proof of the following corollary, we need the following two facts from real analysis: If f is as in the statement of the theorem then

1. f is bounded; that is, there is a real number B such that $|f(p)| \le B$ whenever $0 \le p \le 1$.
2. f is uniformly continuous; that is, there is a function g such that

$$|f(p') - f(p)| \le g(\epsilon) \qquad \text{whenever } |p' - p| \le \epsilon$$

where g is defined for all real ϵ satisfying $0 < \epsilon \le 1$ and $g(\epsilon) \to 0$ as $\epsilon \to 0$. Thus for *any* two points p' and p which are within distance ϵ of each other the corresponding two values of f can differ by at most $g(\epsilon)$, and this bound goes to zero with ϵ.

Let n and p be fixed and let $b_n(k) = \binom{n}{k} p^k(1-p)^{n-k}$. In terms of a given ϵ, to be specified shortly, divide the integers $0, 1, 2, \ldots, n$ into the two disjoint sets

$$C = \left\{k: \left|\frac{k}{n} - p\right| < \epsilon\right\} \qquad D = \left\{k: \left|\frac{k}{n} - p\right| \ge \epsilon\right\}.$$

Then we have the following inequality chain, whose justification appears below:

$$|B_n(p) - f(p)| = \left|\sum_{k=0}^{n}\left[f\left(\frac{k}{n}\right) - f(p)\right]b_n(k)\right|$$

$$\leq \sum_{k\in C}\left|f\left(\frac{k}{n}\right) - f(p)\right| b_n(k) + \sum_{k\in D}\left|f\left(\frac{k}{n}\right) - f(p)\right| b_n(k)$$

$$\overset{1}{\leq} g(\epsilon)\sum_{k\in C} b_n(k) + 2B\sum_{k\in D} b_n(k)$$

$$\overset{2}{\leq} g(\epsilon) + \frac{2B}{4\epsilon^2 n} \overset{3}{=} g(n^{-1/3}) + \frac{B}{2}n^{-1/3} \overset{4}{=} \delta_n.$$

The first term of $\overset{1}{\leq}$ used the uniform continuity of f, since k is restricted to C, and the second term used the boundedness of f,

$$\left|f\left(\frac{k}{n}\right) - f(p)\right| \leq \left|f\left(\frac{k}{n}\right)\right| + |f(p)| \leq B + B.$$

The first term of $\overset{2}{\leq}$ used

$$\sum_{k\in C} b_n(k) \leq \sum_{k=0}^{n} b_n(k) = 1$$

and the second term follows from Theorem 1 (Sec. 10-2). Then arbitrarily setting $\epsilon = n^{-1/3}$ yields $\overset{3}{=}$, and $\overset{4}{=}$ defines δ_n. Clearly δ_n depends on n and on f through g and B but does *not* depend on p. Also δ_n converges to zero as n goes to infinity. ■

We specialize the proof of Theorem 1 in order to obtain the following more explicit version for the function of Fig. 2.

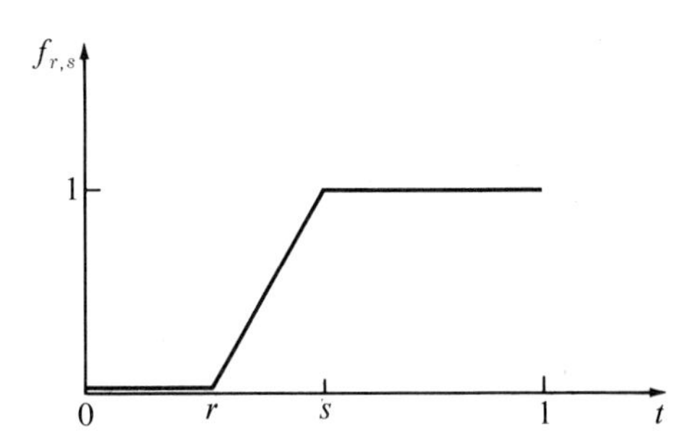

Fig. 2 The function $f_{r,s}$ of Corollary 1.

Corollary 1 For $0 \leq r < s \leq 1$ let the function $f_{r,s}$ be defined by

$$f_{r,s}(t) = \begin{cases} 0 & \text{if } 0 \leq t \leq r \\ \dfrac{t-r}{s-r} & \text{if } r \leq t \leq s \\ 1 & \text{if } s \leq t \leq 1. \end{cases}$$

For any such r and s and for any positive integer n there is an nth-degree polynomial $F_n(t) = \sum_{k=0}^{n} a_{n,k}t^k$ with coefficients $a_{n,k}$ depending on r, s, and n such that

$$|f_{r,s}(t) - F_n(t)| \le n^{-1/3}\left(1 + \frac{1}{s-r}\right) \qquad \text{for } 0 \le t \le 1. \quad \blacksquare$$

Proof In the proof of Theorem 1 we can let $B = 1$ and $g(\epsilon) = \epsilon/(s - r)$, since $1/(s - r)$ is the slope of the rising portion of $f_{r,s}$. Then the bound becomes

$$\delta_n = g(n^{-1/3}) + \frac{1}{2}n^{-1/3} = n^{-1/3}\left(\frac{1}{2} + \frac{1}{s-r}\right)$$

and for simplicity we may replace $\frac{1}{2}$ by 1. $\blacksquare$

Thus for each r and s we have a sequence of polynomials which converge uniformly to $f_{r,s}$. Except for the trivial case $r = 0$ and $s = 1$, the coefficient $a_{n,k}$ of t^k in F_n must depend on n, since otherwise we would have a convergent power series, which is impossible, since $f_{r,s}$ does not have a derivative at r or s. Either Theorem 1 or Corollary 1 yields the uniqueness for Laplace transforms, where the transforms are only required to be equal on the integers.

Uniqueness of the Laplace transform Let f and g be real-valued functions defined on the positive real axis $x \ge 0$. Assume that $|f|$ and $|g|$ have finite integrals over $0 \le x < \infty$. If

$$\int_0^\infty e^{-nx}f(x)\,dx = \int_0^\infty e^{-nx}g(x)\,dx \qquad \text{for all } n = 1, 2, \ldots$$

then

$$\int_0^y f(x)\,dx = \int_0^y g(x)\,dx \qquad \text{for all real } y \ge 0.$$

In particular by taking d/dy we see that $f(y) = g(y)$ at any $y \ge 0$ at which both are continuous. $\blacksquare$

Proof Let $h = f - g$. We must show that if

$$\int_0^\infty |h(x)|\,dx < \infty \qquad \text{and} \qquad \int_0^\infty e^{-nx}h(x)\,dx = 0 \qquad \text{for } n = 1, 2, \ldots$$

then

$$\int_0^y h(x)\,dx = 0 \qquad \text{for all real } y \ge 0.$$

As in Fig. 3, for fixed r and s define the function W and the two numbers y and z by

$$W(x) = f_{r,s}(e^{-x}) \qquad \text{for all } x \ge 0$$

so that $0 \le W \le 1$ and

$$W(x) = \begin{cases} 1 & \text{if } 0 \le x \le y,\ e^{-y} = s \\ 0 & \text{if } z \le x, \qquad e^{-z} = r. \end{cases}$$

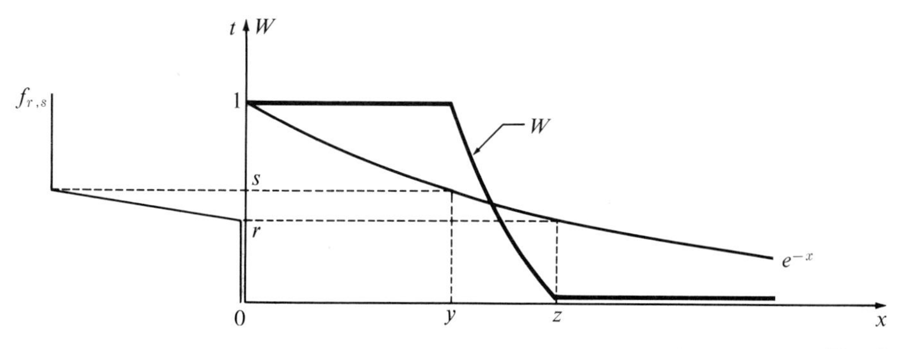

Fig. 3

Corollary 1 yields

$$|W(x) - F_n(e^{-x})| = \left|W(x) - \sum_{k=0}^{n} a_{n,k}e^{-kx}\right|$$

$$\leq n^{-1/3}\left(1 + \frac{1}{s-r}\right) \qquad \text{for all } x \geq 0$$

so we have approximated W uniformly *over all* $x \geq 0$ by a polynomial in e^{-x}. In particular, the exhibited bound applies to $|F_n(e^{-x})|$ whenever $x \geq z$. By assumption

$$\int_0^\infty F_n(e^{-x})h(x)\,dx = 0$$

hence

$$\left|\int_0^\infty W(x)h(x)\,dx\right| = \left|\int_0^\infty [W(x) - F_n(e^{-x})]h(x)\,dx\right|$$

$$\leq n^{-1/3}\left(1 + \frac{1}{s-r}\right)\int_0^\infty |h(x)|\,dx \to 0 \qquad \text{as } n \to \infty.$$

Thus for any fixed r and s we have

$$0 = \int_0^\infty W(x)h(x)\,dx = \int_0^y h(x)\,dx + \int_y^z W(x)h(x)\,dx.$$

But if s and y are fixed and we let r go to s and z go to y then the last term goes to zero, and hence the uniqueness is proved. ■

EXERCISES

1. One play in a game consists of a fixed large number n of Bernoulli trials with parameter p. A gambler receives reward $e^{k/n}\cos(k/n)$ on the play if the fraction of successes is k/n. What would you expect his arithmetic average reward to be close to on a large number N of plays?

2. Exhibit a continuous real-valued function f defined on $0 \leq p < 1$ such that no sequence of polynomials converges to f uniformly on $0 \leq p < 1$.

3. Prove that if a real-valued f is defined and bounded on $0 \leq p \leq 1$ then $B_n(p) \to f(p)$ as $n \to \infty$ for every p at which f is continuous.

4. Let $f: R \to R$ and $g: R \to R$ be two functions such that there are two real numbers T and r for which $\int_T^\infty e^{-rt} |f(t)| \, dt$ and $\int_T^\infty e^{-rt} |g(t)| \, dt$ are finite; that is, there is some large enough damping constant r which guarantees integrability. In applications it is usually assumed that f and g are zero to the left of some time T. Assume that

$$\int_T^\infty e^{-st} f(t) \, dt = \int_T^\infty e^{-st} g(t) \, dt \qquad \text{for all real } s \geq r$$

and prove that $f(y) = g(y)$ at any $y \geq T$ at which both are continuous.

11 Random Variables

This chapter introduces the idea of a random point, which can be thought of as a variable associated with an experiment. If the values of the variable are real numbers we call the random point a *random variable.* Section 11-1 is devoted to the probability distribution induced by a random point. Section 11-2 extends the notion of independence to random points. Section 11-3 concerns the expectation of a random variable, and this is naturally and simply defined to be the mean of its induced density. Section 11-4 considers correlations and linear prediction. This chapter extends some of our earlier concepts and essentially completes the basic framework of probability theory in the discrete case.

11-1 RANDOM POINTS AND RANDOM VARIABLES

The next few paragraphs are devoted to an intuitive introduction to the main concepts of this section. The more systematic development starts with the formal definition of a random point. Let us introduce the concepts in connection with a fairly concrete example.

A MOTIVATING EXAMPLE: METAL TABS

Suppose a sheet of metal has been cut up into small rectangular pieces which we will call tabs. Each tab has a length l, width w, and one-sided area lw. It may be helpful to imagine that we have orders of magnitude such as those stated below, since these would cause significant "data reductions" in the ensuing discussion. We have a total of 10^{10} tabs and the number of different tab lengths, widths, length-width pairs, and areas are 10^5, 10^4, 10^7, and 10^2, respectively. Suppose that each tab has a distinct label ω, and let the sample space Ω consist of the set of these labels. For example, the label ω_i on the ith tab might just be the integer i. Assume that we have a discrete density $p: \Omega \to R$ and that a tab is to be selected according to it. For example, p might be uniform on Ω, or p might be proportional to tab area.

Let L and W denote the length and width, respectively, of the tab which is to be selected according to p. We call L a *random variable*, since it is a variable which can have one of many values and the determination of its value involves some randomness. The same is true of W. If the experiment is performed we will obtain some real number l for the value of the random variable L. In our mathematics we let L be the real-valued function on Ω whose value $L(\omega)$ is just the length of the tab labeled ω. Similarly $W: \Omega \to R$ is also a real-valued function on Ω. We call (L,W) a *random vector* and at times abbreviate it to $f = (L,W)$. Thus $f: \Omega \to R^2$ and the value $f(\omega) = (L(\omega), W(\omega))$ is the length-width pair for tab ω.

The event that f has the value (2.3,4.7) consists of all sample points ω having this length and width. If we add $p(\omega)$ over these ω we obtain the probability

$$P[f = (2.3,4.7)] = P\{\omega : f(\omega) = (2.3,4.7)\}$$

of this event. Let $p_f(2.3,4.7)$ equal this number. More generally, define $p_f: R^2 \to R$ by

$$p_f(l,w) = P[f = (l,w)] = P\{\omega : L(\omega) = l \text{ and } W(\omega) = w\}$$

so that $p_f(l,w)$ is the probability that the tab selected will have length l and width w. Clearly p_f is a discrete density *on the plane*, since only finitely many (l,w) can occur, and these events $\{\omega : L(\omega) = l,\ W(\omega) = w\}$ are disjoint and have Ω for their union. We call p_f the *density induced by f with associated* $p: \Omega \to R$. When the notation $p_{L,W}$ is used for p_f we may call $p_{L,W}$ the *joint density of L and W*, or *induced by L and W*.

Imagine a picture of p_f as the plane, together with the assignment of numbers $p_f(l,w)$ to points (l,w). This picture (the function p_f) summarizes some of the relevant aspects of the model. From $p: \Omega \to R$ and $f: \Omega \to R^2$ we can calculate $p_f: R^2 \to R$, but of course, if we know only p_f we cannot find p and f. At times we may think of p_f as a "reduction of the theoretical data" consisting of p and f.

Let A be some subset of R^2, such as $\{(l,w) : l < 4,\ l + w < 5\}$. The expression "$f$ has a value in A" specifies an event *describable in terms of f*. The probability of such an event can be calculated in two ways: this event can be thought of as the subset A of R^2 and then its probability can be calculated by summing p_f over A, or it can be thought of as the subset

$$A_\Omega = \{\omega : (L(\omega), W(\omega)) \in A\}$$

of Ω and its probability can be calculated by summing p over A_Ω. Naturally, as guaranteed by Theorem 1 below, both calculations yield the same number. Thus whenever the probability of an event can be calculated from various induced densities, besides always being calculable from the basic $p: \Omega \to R$, all calculations must agree. In the future we will always assume that we have a basic (Ω,P); we will introduce many induced distributions but will always require that an induced distribution must yield the same numbers as obtained by going back to (Ω,P). (Section 15-2 shows that this consistency requirement essentially uniquely defines induced distributions.)

Let the random variable $X: \Omega \to R$ be the one-sided area of the tab selected according to $p: \Omega \to R$. Naturally $X(\omega) = L(\omega)W(\omega)$, and X has an induced density obtained from $p: \Omega \to R$ by

$$p_X(x) = P[X = x] = P\{\omega : L(\omega)W(\omega) = x\} \qquad \text{for all real } x.$$

Thus $p_X(7.2)$ is the probability that the area of the selected tab will be 7.2. If we have already calculated $p_{L,W} = p_f$ we can also find $p_X(7.2)$ by summing $p_{L,W}(l,w)$ over those (l,w) for which $lw = 7.2$. Theorem 2 guarantees that this second calculation of $p_X(7.2)$ from the *intermediate density* $p_{L,W}$ will yield the same number as obtained from $p\colon \Omega \to R$. Thus once again all different but sensible calculations of the same entity must yield the same result.

In the same vein $p_L(2.3)$ can be obtained either by summing $p\colon \Omega \to R$ over $\{\omega\colon L(\omega) = 2.3\}$ or by summing $p_{L,W}(2.3,w)$ over all real w, and Theorem 3 guarantees that the two calculations yield the same number.

GENERAL DEFINITIONS

Several definitions will be made and then used in examples and discussed. Any function $f\colon \Omega \to F$ defined on the sample space of a discrete probability space (Ω,P) is said to be a **random point f with associated discrete probability space** (Ω,P). The set F is quite arbitrary. Usually the probability space (Ω,P) is not mentioned, and we say "the random point f in F," or "the random point f." However, the probability space (Ω,P) *is* part of the definition of a random point. If $F = R^n$ then f may be called a random vector. If $F = R$ then f may be called a random variable, in which case we use symbols such as X, Y, T, R in place of symbols such as f, g, and t. Thus a **random variable X with associated discrete probability space** (Ω,P) is any real-valued function $X\colon \Omega \to R$ defined on the sample space of the discrete probability space (Ω,P). Its value at the sample point ω is the real number $X(\omega)$. Random vectors and random variables are of greatest interest to us, but at times F may be a set of letters or a set of functions. A random point may be depicted as in Fig. 1.

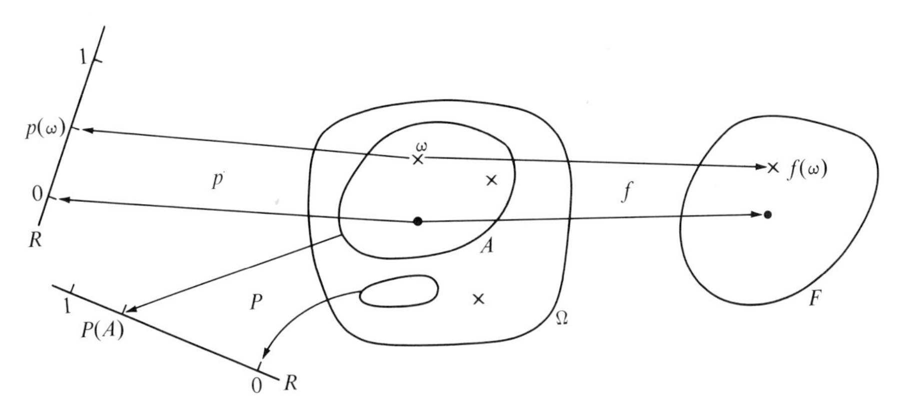

Fig. 1 A random point f.

Example 1 *This example will be amplified and used frequently in this section and in Exercise 1* to illustrate many of the ideas of this section in a simple numerical context. Figure 2 represents a simplified model of a production process which is observed at three consecutive instants in time. The significant variables are to be the temperature, in degrees Fahrenheit, and the pressure, in pounds per square inch. We assume that these are read from meters which monitor the process. Six random variables X_1, X_2, X_3 and Y_1, Y_2, Y_3 are defined on Ω as follows: if $\omega = [(x_1,y_1), (x_2,y_2), (x_3,y_3)]$ then the values of the random variables at this ω are

$$X_1(\omega) = x_1 \qquad X_2(\omega) = x_2 \qquad X_3(\omega) = x_3$$

$$Y_1(\omega) = y_1 \qquad Y_2(\omega) = y_2 \qquad Y_3(\omega) = y_3.$$

Our interpretation is that if ω is the outcome then $X_i(\omega)$ was the temperature and $Y_i(\omega)$ was the pressure at time i for $i = 1, 2, 3$. Thus if the outcome is ω_4 the sequence of temperatures was 3, 5, 4 and the sequence of pressures was 3, 1, 0. The extra parentheses used in labeling sample points are meant to assist in interpreting the points.

The values of the random vector (X_1,Y_1) are tabulated in Fig. 2. We might be interested

i	ω_i	$p(\omega_i)$	(X_1,Y_1)	Z	M
1	[(3,1), (5,2), (4,2)]	.05	(3,1)	18	2
2	[(3,1), (5,2), (4,3)]	.1	(3,1)	18	3
3	[(3,1), (5,1), (4,2)]	.05	(3,1)	18	2
4	[(3,3), (5,1), (4,0)]	.1	(3,3)	0	3
5	[(4,1), (4,2), (4,2)]	.05	(4,1)	2	2
6	[(4,1), (4,2), (4,3)]	.05	(4,1)	2	3
7	[(4,1), (4,1), (4,2)]	.1	(4,1)	2	2
8	[(4,2), (4,2), (4,2)]	.1	(4,2)	1	2
9	[(5,1), (4,2), (3,2)]	.05	(5,1)	2	2
10	[(5,1), (4,3), (3,3)]	.1	(5,1)	2	3
11	[(5,1), (4,1), (3,2)]	.2	(5,1)	2	2
12	[(5,3), (4,1), (3,1)]	.05	(5,3)	0	3

Fig. 2 $X_i(\omega) = x_i$ and $Y_i(\omega) = y_i$ for $i = 1, 2, 3$ where $\omega = [(x_1,y_1), (x_2,y_2), (x_3,y_3)]$; $Z = (3 - Y_1)(2X_1 - 9)^2$ and $M = \max\{Y_1, Y_2, Y_3\}$.

in the random variable Z defined by $Z = (3 - Y_1)(2X_1 - 9)^2$; that is, $Z(\omega) = [3 - Y_1(\omega)] \times [2X_1(\omega) - 9]^2$, so that in particular

$$Z(\omega_6) = [3 - Y_1(\omega_6)][2X_1(\omega_6) - 9]^2 = [3 - 1][2(4) - 9]^2 = 2.$$

Repeating this computation, we obtain all the values of Z, as shown in Fig. 2. Another random variable of interest might be the maximum pressure $M = \max\{Y_1,Y_2,Y_3\}$; that is, $M(\omega) = \max\{Y_1(\omega),Y_2(\omega),Y_3(\omega)\}$ for all ω in Ω. Its values are shown in Fig. 2.

Clearly the event that the temperature at time 1 equals 3 is

$$[X_1 = 3] = \{\omega: X_1(\omega) = 3\} = \{\omega_1,\omega_2,\omega_3,\omega_4\}$$

so that

$$P[X_1 = 3] = .05 + .1 + .05 + .1 = .3.$$

Similarly we find that $P[X_1 = 4] = .3$ and $P[X_1 = 5] = .4$. If we define the function $p_{X_1}: R \to R$ by

$$p_{X_1}(x_1) = P\{\omega: X_1(\omega) = x_1\} \qquad \text{for all real } x_1$$

then p_{X_1} gives us the probabilities of X_1 having various values. Similarly we can let

$$p_Z(z) = P\{\omega: Z(\omega) = z\} \qquad \text{for all real } z$$

so that, for example, $p_Z(18) = P\{\omega_1,\omega_2,\omega_3\} = .05 + .1 + .05 = .2$. These two functions p_{X_1} and p_Z are shown in Fig. 3 and will be called the *densities induced by X_1 and Z.* ■

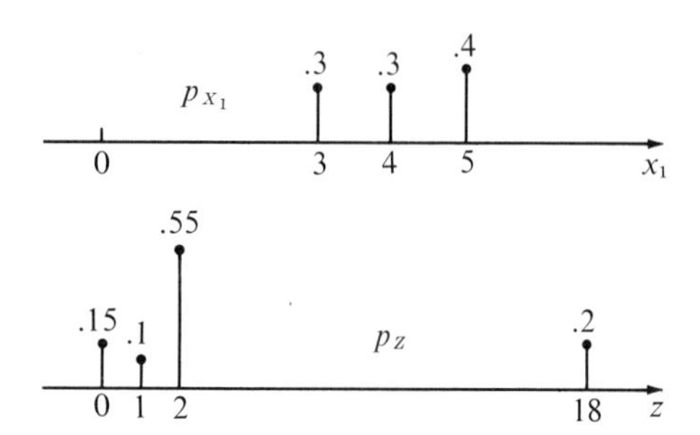

Fig. 3 The densities induced by X_1 and Z of Fig. 2.

Our use of the word "variable" in "random variable" is very similar to its use in science. In mechanics, for instance, we might have a dissipative mechanical system involving many particles in motion, and we may think of total kinetic energy as a variable. If we are given a complete enough description of the system for some particular instant of time we can then evaluate the total kinetic energy as a number. Thus total kinetic energy can be thought of as a real-valued function, and an element of the domain of this function is a sufficiently complete description of the system. In the same way a sample point ω can be thought of as a complete description of the outcome of an experiment, so that $X(\omega)$ is the value of the variable X if the outcome is ω. Contrary to a deterministic situation, however, we think of our outcome ω as being determined by probabilities, so that we have a *random* variable. A random variable can be thought of as a random point on the real line, while a general random point $f: \Omega \to F$ can be thought of as a random point in the set F. Thus if the experiment is performed and the outcome is ω then the point $f(\omega)$ belonging to F is obtained. The discrete probability measure on the sample space determines the chances of various sample points and hence the chances of various points of F.

INDUCED DENSITIES

If $f: \Omega \to F$ is a random point with associated discrete probability space (Ω,P) then the function $p_f: F \to R$ defined by

$$p_f(\alpha) = P\{\omega: f(\omega) = \alpha\} \qquad \text{for each } \alpha \in F$$

is called the **discrete probability density of** f, or for f, or induced by f, or induced by f with associated (Ω,P). If (F,P_f) is the discrete probability space corresponding to $p_f: F \to R$ we also say that (F,P_f) is induced by f. As shown in Fig. 4, the event $\{\omega: f(\omega) = \alpha\}$ is just the event that the value of f is α, so that $P\{\omega: f(\omega) = \alpha\}$ is the probability that f has the value α, and p_f just assigns this number $p_f(\alpha)$ to the point α. Thus $p_f(\alpha)$ is the probability that the random point f will have the value α. Figure 3 shows the densities induced by X_1 and Z of Fig. 2.

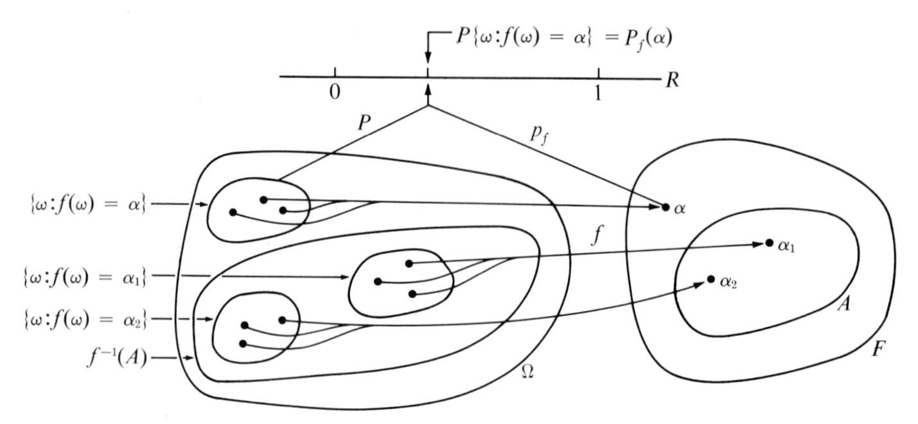

Fig. 4

As shown in Exercise 4 (Sec. 4-2), $\{\alpha : p_f(\alpha) > 0\}$ is countable because $\{\omega : p(\omega) > 0\}$ is countable.

If we think of $p: \Omega \to R$ as a distribution of unit mass on Ω then we can think of $f: \Omega \to F$ as transferring this mass to F and yielding the transferred distribution $p_f: F \to R$ of unit mass on F.

We will often be interested in the probability that the random point $f: \Omega \to F$ has some value in a particular subset A of F; that is, we want $P[f^{-1}(A)]$, where $f^{-1}(A) = \{\omega : f(\omega) \in A\}$ is the inverse image of A as defined in Sec. 3-4. As indicated in Fig. 4, let $\alpha_1, \alpha_2, \ldots$ be an enumeration of those points of A for which $p_f(\alpha)$ is positive. By adding the probabilities of f having these various values and using the definition of p_f for $\overset{1}{=}$ we get

$$\begin{aligned} p_f(\alpha_1) + p_f(\alpha_2) + \cdots &\overset{1}{=} P\{\omega : f(\omega) = \alpha_1\} + P\{\omega : f(\omega) = \alpha_2\} + \cdots \\ &\overset{2}{=} P[\{\omega : f(\omega) = \alpha_1\} \cup \{\omega : f(\omega) = \alpha_2\} \cup \cdots] \\ &\overset{3}{=} P[f^{-1}(A)]. \end{aligned}$$

For $\overset{2}{=}$ we used the fact, as indicated in Fig. 4, that the events $\{\omega : f(\omega) = \alpha_1\}$, $\{\omega : f(\omega) = \alpha_2\}, \ldots$ are disjoint, while $\overset{3}{=}$ follows from the fact that their union contains every point ω of $f^{-1}(A)$ for which $p(\omega)$ is positive. If $A = F$ in this result we obtain

$$p_f(\alpha_1) + p_f(\alpha_2) + \cdots = P[f^{-1}(F)] = P(\Omega) = 1$$

showing that p_f *is indeed a discrete probability density*. For any subset A of F the number $P_f(A)$ is defined as usual to be the sum of all of the positive values $p_f(\alpha) > 0$ for α in A, so that *we have proved that* $P_f(A) = P[f^{-1}(A)]$. We state this as a theorem.

Theorem I Let the discrete probability space (F, P_f) be induced by the random point $f: \Omega \to F$ with associated (Ω, P). Then

$$P_f(A) = P\{\omega : f(\omega) \in A\}$$

for every subset A of F. ■

Thus if we want the probability that f will have a value in A it is natural, as indicated in Fig. 5, to compute this number *from* P, or *from* P_f. Theorem 1 assures us that both computations yield the same number.

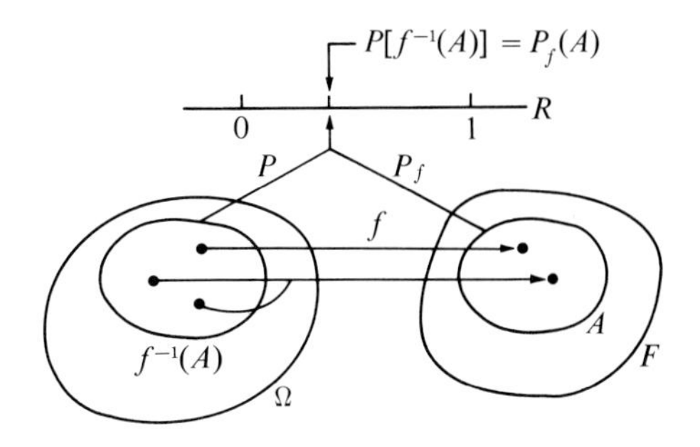

Fig. 5

Example 2 The densities induced by X_1 and Z of Fig. 2 are shown in Fig. 3. The density induced by the random vector $f = (X_1, Y_1)$ of Fig. 2 is shown in Fig. 6. One of the calculations needed to obtain Fig. 6 is

$$p_f(4,1) = P\{\omega: (X_1(\omega), Y_1(\omega)) = (4,1)\} = P\{\omega_5, \omega_6, \omega_7\} = .2.$$

We can think of p_f as being obtained by "data reduction" in that the set $\{\omega_5, \omega_6, \omega_7\}$ is "reduced" or "collapsed" into the point (4,1).

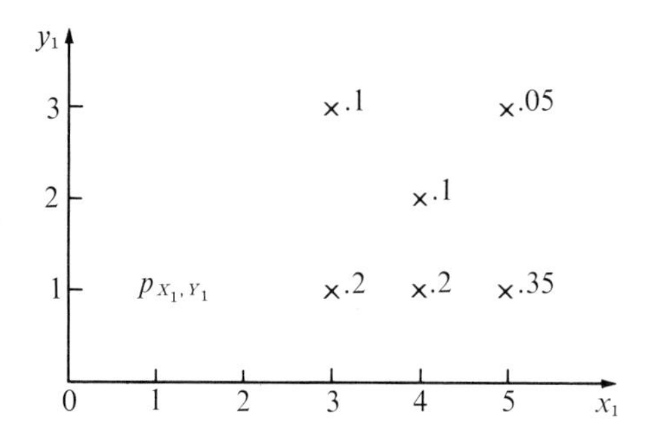

x_1 \ y_1	1	2	3	p_{X_1}
3	.2	0	.1	.3
4	.2	.1	0	.3
5	.35	0	.05	.4
p_{Y_1}	.75	.1	.15	1

Fig. 6 The density induced by (X_1, Y_1) of Fig. 2.

Let A be the set of reals z satisfying $.5 < z < 3$. Then *from Fig. 2* we get

$$P\{\omega: .5 < Z(\omega) < 3\} = \sum_{i=5}^{11} p(\omega_i) = .65.$$

Also *from Fig. 3* we find that

$$P_Z(A) = p_Z(1) + p_Z(2) = .1 + .55 = .65.$$

Thus, as assured by Theorem 1, we get the same number if we use the induced density p_Z as we do by going back to the basic density p. ■

Several induced densities were obtained in earlier chapters, although this terminology was not used. Figure 2 (Sec. 8-2) tabulated the density p' for $n = 4$ Bernoulli trials with probability p for success. If we let S_4 be the random variable which is the total number of successes then the values of S_4 at each of the 16 sample points are as listed in the column labeled $x_1 + x_2 + x_3 + x_4$. The *induced density* p_{S_4} was shown to be binomial with parameters $n = 4$ and p.

Example 2 (Sec. 8-3) showed that the random variable $S: R^2 \to R$ defined by $S(x,y) = x + y$ for all (x,y) in R^2, with associated density given as the product of two Poisson densities, has the Poisson density with parameter $\lambda_1 + \lambda_2$ as its induced density p_S.

In Sec. 8-4 it was shown that the induced density of the random variable $S_r: R^r \to R$ is Pascal with parameters r and p if S_r is defined on R^r by

$$S_r(x_1, \ldots, x_r) = x_1 + x_2 + \cdots + x_r$$

where the associated density was the product of r geometric densities with parameter p.

INTERMEDIATE DISTRIBUTIONS

New random points are often obtained from given random points by composition, as defined in Sec. 3-3. The next theorem applies when *a random point* $g(f)$ *is a function of a random point* f.

Theorem 2: Intermediate distributions If $f: \Omega \to F$ is a random point with associated density $p: \Omega \to R$ and if

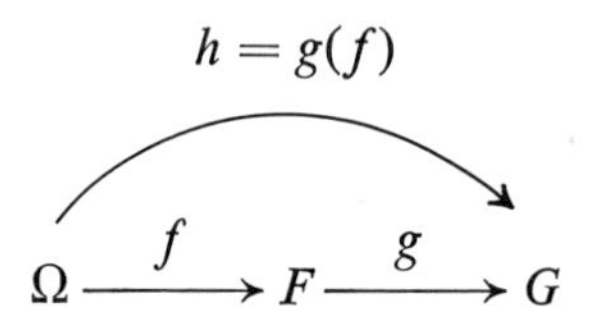

then $p_h = p_g$, where p_g is induced by g with associated p_f. ■

Thus f with associated p induces density p_f on F. Similarly h with associated p induces density p_h on G. But g can be regarded as a random point with associated density p_f, and so g induces a density p_g on G. Theorem 2 tells us that $p_h = p_g$; that is, the density p_h can be thought of as having been induced by g, with the *intermediate density* p_f used as associated density. It will often be the case that we know g and p_f but ignore or do not know p, and we find p_h by finding p_g.

Proof of Theorem 2 Given any point β in G, let A_F be defined, as indicated in Fig. 7, by

$$A_F = \{\alpha : \alpha \in F, g(\alpha) = \beta\}$$

and let A_Ω be defined by

$$A_\Omega = f^{-1}(A_F) = \{\omega : \omega \in \Omega, f(\omega) \in A_F\}.$$

Then

$$p_g(\beta) \overset{1}{=} P_f(A_F) \overset{2}{=} P(A_\Omega) \overset{3}{=} p_h(\beta)$$

where $\overset{1}{=}$ follows from the definition of p_g and $\overset{2}{=}$ follows from Theorem 1.

From Sec. 3-4 we know that $A_\Omega = \{\omega : h(\omega) = \beta\}$, as is obvious from Fig. 7, so $\overset{3}{=}$ follows from the definition of p_h. ■

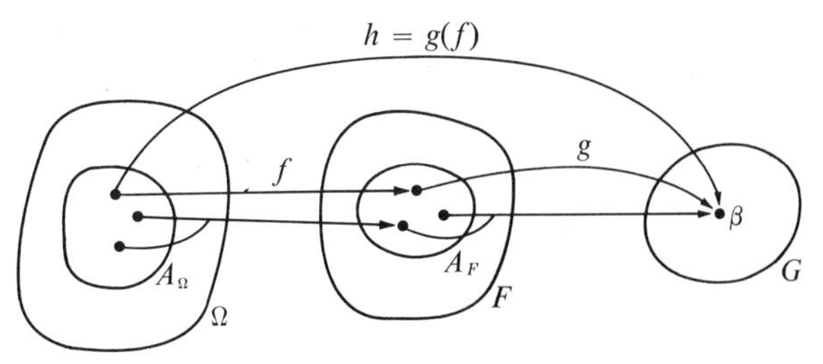

Fig. 7

Example 3 If we specialize Theorem 2 to the case $F = G = R$ we have

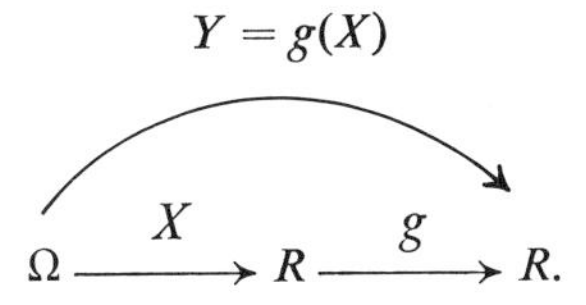

Thus if X is a random variable and the random variable Y is defined by $Y = g(X)$, so that $Y(\omega) = g(X(\omega))$ for every sample point ω, then $p_Y(y)$ equals the sum of p_X over the subset $\{x: g(x) = y\}$ of the reals. Suppose g is given as in Fig. 8 and p_X has already been found to be as shown in Fig. 8. To find $p_Y(1.8)$ we first find $\{x: g(x) = 1.8\} = \{2.5, 5, 7.6\}$, so that

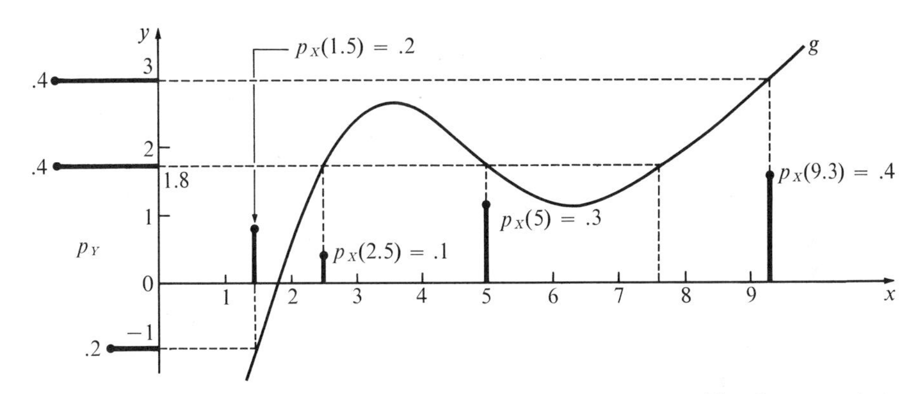

Fig. 8 $Y = g(X)$.

$p_Y(1.8) = p_X(2.5) + p_X(5) + p_X(7.6) = .1 + .3 + 0 = .4$. Similarly $p_Y(3) = p_X(9.3) = .4$ and $p_Y(-1) = p_X(1.5) = .2$. If y is not one of 1.8, 3, or -1 then the set $\{x: g(x) = y\}$ contains only points for which $p_X(x) = 0$, so $p_Y(y) = 0$. For convenience we plot p_Y along the vertical axis. Theorem 2 assures us that the same p_Y would have been obtained if we went back to the basic density $p: \Omega \to R$, which we do not happen to be given in this example.

Among the most important functions g for which we find the density $p_{g(X)}$ of $g(X)$ are the straight lines $g(x) = ax + b$. This case was considered in Sec. 9-3, which might also have been titled "A Linear Transformation of a Random Variable." ■

JOINT PROBABILITY DENSITIES

If $f_i: \Omega \to F_i$ for $i = 1, 2, \ldots, n$ are n random points with the same associated density $p: \Omega \to R$ we can form the random point $f = (f_1, f_2, \ldots, f_n)$, where $f: \Omega \to F$

is into the cartesian product $F = F_1 \times F_2 \times \cdots \times F_n$ and f is defined by

$$f(\omega) = (f_1(\omega), f_2(\omega), \ldots, f_n(\omega)) \qquad \text{for all } \omega \in \Omega.$$

Naturally the density induced by f is defined as usual by

$$p_f(\alpha) = P\{\omega : f(\omega) = \alpha\} \qquad \text{for all } \alpha \in F$$

or equivalently by

$$p_f(\alpha_1, \alpha_2, \ldots, \alpha_n) = P\{\omega : f_1(\omega) = \alpha_1, f_2(\omega) = \alpha_2, \ldots, f_n(\omega) = \alpha_n\}$$
$$\text{for all } \alpha_1 \in F_1,\ \alpha_2 \in F_2, \ldots, \alpha_n \in F_n.$$

Often the notation $p_{f_1, f_2, \ldots, f_n}$ is used instead of p_f, and this density is called the **joint density of** $f_1, f_2, \ldots, f_n$. If f_i is thought of as the outcome of the ith subexperiment then $p_{f_1, \ldots, f_n}$ is the density induced by the joint outcomes of all n subexperiments.

If we have the joint density p_{f_1, f_2} of two random points then, as shown in the next theorem, we can obtain the density p_{f_1} of one of them by summing out the extra variable in p_{f_1, f_2}.

Theorem 3 If $f_1 : \Omega \to F_1$ and $f_2 : \Omega \to F_2$ are two random points with the same associated discrete probability density $p : \Omega \to R$ then

$$p_{f_1}(\alpha_1) = \sum_{\alpha_2} p_{f_1, f_2}(\alpha_1, \alpha_2) \qquad \text{for each } \alpha_1 \in F_1. \quad \blacksquare$$

In the equation above α_1 appears on both sides, so that a separate computation is needed for each α_1. For fixed α_1 the number $p_{f_1}(\alpha_1)$ is obtained by adding all the positive values $p_{f_1, f_2}(\alpha_1, \alpha_2)$ as α_2 varies over F_2. This is certainly a countable summation. Naturally the analogous formula

$$p_{f_2}(\alpha_2) = \sum_{\alpha_1} p_{f_1, f_2}(\alpha_1, \alpha_2) \qquad \text{for each } \alpha_2 \in F_2$$

is also true.

Proof of Theorem 3 The proof is indicated in Fig. 9 for the case $F_1 = F_2 = R$ for which $F_1 \times F_2$ is the plane. Given α_1 in F_1, we define the subset A_{α_1} of $F_1 \times F_2$ by $A_{\alpha_1} = \{(\alpha_1', \alpha_2') : \alpha_1' = \alpha_1\}$, so that A_{α_1} is the set of all pairs (α_1, α_2') whose first coordinate is α_1. In Fig. 9 the set A_{α_1} is the vertical line through the point $(\alpha_1, 0)$. *The theorem states*

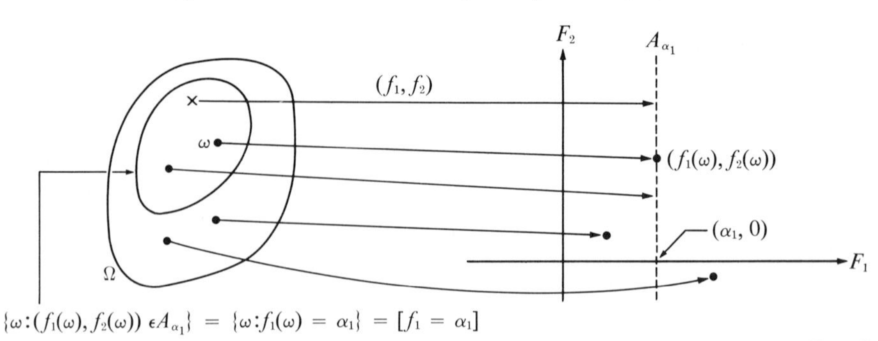

Fig. 9

the fairly obvious fact that $p_{f_1}(\alpha_1)$ *is the sum of* p_{f_1,f_2} *over* A_{α_1}. Clearly

$$\{\omega: (f_1(\omega), f_2(\omega)) \in A_{\alpha_1}\} = \{\omega: f_1(\omega) = \alpha_1\}$$

so that

$$\sum_{\alpha_2} p_{f_1,f_2}(\alpha_1,\alpha_2) \overset{1}{=} P_{f_1,f_2}(A_{\alpha_1}) \overset{2}{=} P\{\omega: f_1(\omega) = \alpha_1\} \overset{3}{=} p_{f_1}(\alpha_1)$$

where $\overset{1}{=}$ follows from the usual relation of a measure to its density, $\overset{2}{=}$ follows from Theorem 1 and the set identity above, and $\overset{3}{=}$ is just the definition of p_{f_1}. ■

A common case of Theorem 3 occurs when $f_1 = (X_1, X_2, \ldots, X_m)$ and $f_2 = (X_{m+1}, X_{m+2}, \ldots, X_n)$ are random vectors, in which case the theorem reduces to the following: If $X_1, \ldots, X_m$ and $X_{m+1}, \ldots, X_n$ are all random variables with the same associated discrete probability density then

$$p_{X_1,\ldots,X_m}(x_1, \ldots, x_m) = \sum_{x_{m+1},\ldots,x_n} p_{X_1,\ldots,X_n}(x_1, \ldots, x_m, \ldots, x_n)$$
$$\text{for all } (x_1, \ldots, x_m) \in R^m.$$

As before, we fix $(x_1, \ldots, x_m)$ and sum all the positive values of $p_{X_1,\ldots,X_n}(x_1, \ldots, x_m, x_{m+1}, \ldots, x_n)$ as $(x_{m+1}, \ldots, x_n)$ varies over R^{n-m}. *Thus a joint density is obtained from a larger joint density just by summing out the extra coordinates.*

If $f_1: \Omega \to F_1$ and $f_2: \Omega \to F_2$ are two random points with the same associated density, and if the subset $\{\alpha: p_{f_1}(\alpha) > 0\} = \{\alpha_1, \alpha_2, \ldots, \alpha_k\}$ of F_1 and the subset $\{\beta: p_{f_2}(\beta) > 0\} = \{\beta_1, \beta_2, \ldots, \beta_m\}$ of F_2 are both finite, then the joint density p_{f_1,f_2} can be exhibited in the standard tabular form of Table 1.

Table 1 The joint probability table for f_1 and f_2

$p(\alpha_i,\beta_j)$ is an abbreviation for $p_{f_1,f_2}(\alpha_i,\beta_j)$

α \ β	β_1	β_2	...	β_m	p_{f_1}
α_1	$p(\alpha_1,\beta_1)$	$p(\alpha_1,\beta_2)$	...	$p(\alpha_1,\beta_m)$	$p_{f_1}(\alpha_1)$
α_2	$p(\alpha_2,\beta_1)$	$p(\alpha_2,\beta_2)$	...	$p(\alpha_2,\beta_m)$	$p_{f_1}(\alpha_2)$
...	......	......	...	......	.....
α_i	$p(\alpha_i,\beta_1)$	$p(\alpha_i,\beta_2)$	...	$p(\alpha_i,\beta_m)$	$p_{f_1}(\alpha_i)$
...	......	......	...	......	.....
α_k	$p(\alpha_k,\beta_1)$	$p(\alpha_k,\beta_2)$	...	$p(\alpha_k,\beta_m)$	$p_{f_1}(\alpha_k)$
p_{f_2}	$p_{f_2}(\beta_1)$	$p_{f_2}(\beta_2)$	...	$p_{f_2}(\beta_m)$	1

If $p_{f_1,f_2}(\alpha,\beta) > 0$ for a particular point (α,β) then from Theorem 3 we have both $p_{f_1}(\alpha) > 0$ and $p_{f_2}(\beta) > 0$. This shows that all points (α,β) at which $p_{f_1,f_2}(\alpha,\beta)$ is positive occur in the table, so that

$$\sum_{i=1}^{k} \sum_{j=1}^{m} p_{f_1,f_2}(\alpha_i,\beta_j) = 1.$$

From Theorem 3 and the construction of the table it is clear that the sum of the entries in the ith row is $p_{f_1}(\alpha_i)$ and the sum of the entries in the jth column is $p_{f_2}(\beta_j)$. For this reason the table has a marginal column on the right for p_{f_1} and a marginal row on the bottom for p_{f_2}. Sometimes, especially when it is obtained in this way, p_{f_i} is called the **marginal density of** f_i.

Example 4 Figure 3 exhibits the density p_Z of Z of Fig. 2 as found directly from the sample space in Example 1. Now $Z = (3 - Y_1)(2X_1 - 9)^2$, and if we are curious about why p_Z turned out as it did it is natural to look at p_{X_1,Y_1}, which is shown in Fig. 6. Thus we define $g\colon R^2 \to R$ by

$$g(x_1,y_1) = (3 - y_1)(2x_1 - 9)^2 \qquad \text{for all } (x_1,y_1) \in R^2$$

so that

$$Z = g(f)$$

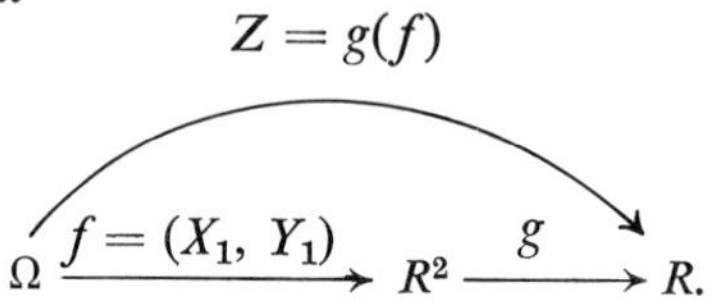

Then from Theorem 2 we have $p_Z(2) = p_g(2) = p_{X_1,Y_1}(4,1) + p_{X_1,Y_1}(5,1) = .2 + .35 = .55$, where we obtained $p_Z(2) = p_g(2)$ from the intermediate density p_{X_1,Y_1} rather than from the basic density p of Fig. 2. The relationships involved are exhibited clearly in Fig. 10, which

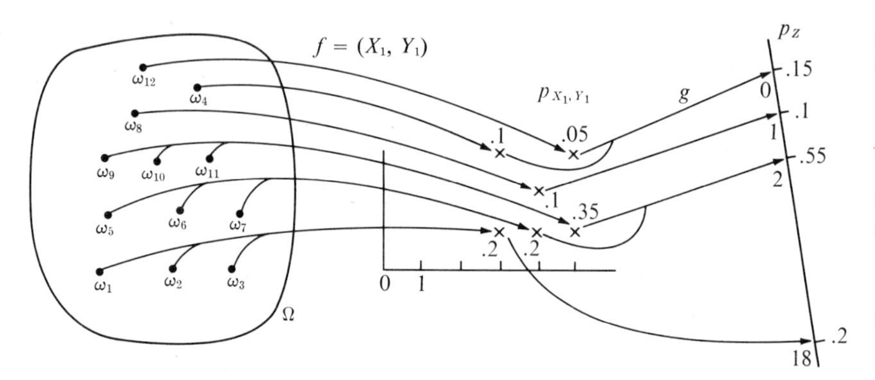

Fig. 10

shows that $\{\omega_5,\omega_6,\omega_7\}$ is the event $\{\omega\colon (X_1(\omega), Y_1(\omega)) = (4,1)\}$ and $\{\omega_9,\omega_{10},\omega_{11}\} = \{\omega\colon (X_1(\omega), Y_1(\omega)) = (5,1)\}$, and these two points in the plane, (4,1) and (5,1), are essentially collapsed together to yield $p_Z(2) = .55$. ■

Random points $f_1, f_2, \ldots, f_n$ into the same set F are said to be **identically distributed** iff $p_{f_1} = p_{f_2} = \cdots = p_{f_n}$. The f_i need not all have the same associated discrete density; however, they must all be into the same set F, which is the domain of every p_{f_i}. Thus several random points are identically distributed if the probability for taking on various values is the same for each of them.

If $\Omega = F_1 \times F_2 \times \cdots \times F_n$ is the cartesian product of $n \geq 2$ sets and $1 \leq i \leq n$ then the **ith coordinate function on** Ω is the function $f_i\colon \Omega \to F_i$ defined by

$$f_i(\alpha_1, \alpha_2, \ldots, \alpha_n) = \alpha_i \qquad \text{for all } (\alpha_1, \alpha_2, \ldots, \alpha_n) \in \Omega.$$

Thus the value of f_i at any point in Ω is just the ith coordinate of that point. For example, if $\Omega = \{0,1\}^4$ then

$$f_1(0,1,1,0) = 0 \qquad f_2(0,1,1,0) = 1 \qquad f_3(0,1,1,0) = 1 \qquad f_4(0,1,1,0) = 0.$$

Similarly the third coordinate function f_3 on R^5 assigns the value 3.2 to the point $(-1,8,3.2,-12,8)$ of R^5. Our random variables will sometimes be coordinate functions.

If $f\colon \Omega \to F$ is a random point with associated discrete probability space (Ω,P) and A is an event with $P(A) > 0$ then the function $p_{f|A}\colon F \to R$ defined by

$$p_{f|A}(\alpha) = P_A\{\omega : f(\omega) = \alpha\} = P(\{\omega : f(\omega) = \alpha\} \mid A)$$

$$= \frac{P\{\omega : f(\omega) = \alpha, \omega \in A\}}{P(A)} \qquad \text{for each } \alpha \in F$$

is called the **conditional discrete probability density of f given A**. Thus $p_{f|A}(\alpha)$ is the probability that f will have the value α if we are given that A occurred.

Example 5 In Fig. 2 we let $A = [M = 3]$ be the event that the random variable M has the value 3 so $P(A) = .4$. The conditional density of Z, given $[M = 3]$, is

$$p_{Z|A}(0) = \frac{.15}{.4} \qquad p_{Z|A}(2) = \frac{.15}{.4}$$

$$p_{Z|A}(18) = \frac{.1}{.4} \qquad p_{Z|A}(z) = 0 \qquad \text{otherwise.} \quad ■$$

Henceforth, unless stated otherwise, it will always be assumed that all the random points in a discussion have the same associated discrete probability density.

This is a standard natural convention which we have already used.

Henceforth sample points will often be suppressed in the notation.

This is a standard convention in probability theory. Often the notation [statement] denotes the set of sample points which make the statement true. We use such a convention when we refer to "the event that the third toss is heads." Thus we can define the induced density p_f by

$$p_f(\alpha) = P[f = \alpha] \qquad \text{for each } \alpha \in F$$

where we replaced $\{\omega : f(\omega) = \alpha\}$ by $[f = \alpha]$. Clearly $[f = \alpha]$ must be an event, since P is defined only for events. Similarly for $\{\omega : f(\omega) \in A\} = [f \in A]$ and $\{\omega : X(\omega) > -.5, Y(\omega) > -.5\} = [X > -.5, Y > -.5]$. We have previously refrained from using this convention too often, since it can lead to confusion if the proper interpretation is not constantly recalled.

EXERCISES

1. *The Production Process (Fig. 2).*

(*a*) Define the random variable m by $m = \min\{Y_1, Y_2, Y_3\}$ and define the random variable R by $R = M - m$, so that R is the pressure range in the experiment; that is,

$R =$ [the largest pressure] $-$ [the smallest pressure].

Tabulate m and R for Fig. 2 and plot the density induced by each. Exhibit the event $[R \geq 1.5]$ as a set of sample points and find $P[R \geq 1.5]$ from Fig. 2. Find $P[R \geq 1.5]$ from p_R. Find the joint-probability table for m and M and exhibit p_m as a marginal density.

(*b*) Exhibit the joint-probability table for "(X_1, Y_1) and M" so that p_{X_1,Y_1} appears as a marginal density.

(*c*) Find the range of the random variable $X_1^2 - 9X_1 - 2X_2$.

(*d*) Find the density induced by

$$U = X_1 + (X_2 - 4)\left[\sin\left(\frac{\pi}{2}M\right)\right]\exp\left(-M + \frac{X_1}{2} + \frac{Y_2^2}{4}\right)$$

(*e*) Find the density induced by the random vector $f = (X_1, Y_1)$ if we are given the event $A = [Z \neq 2] \cup [M \neq 2]$ that at least one of Z and M has a value different from 2.

2. Theorem 1 (Sec. 8-1) shows that the coordinate random variables X_1, X_2, X_3, X_4 of Fig. 2 (Sec. 8-2) are identically distributed with distribution $p_{X_1}(1) = p$ and $p_{X_1}(0) = q = 1 - p$. The event $[X_2 = 1]$ corresponds to heads on the second trial, and contains eight sample points. Adding their probabilities yields

$$p_{X_2}(1) = P[X_2 = 1] = pq^3 + p^2q^2 + p^2q^2 + p^3q + p^2q^2 + p^3q + p^3q + p^4.$$

Reduce this expression for $p_{X_2}(1)$ to p.

11-2 INDEPENDENT RANDOM VARIABLES

In this section we generalize the material of Sec. 7-4 on independent events to independent random points. We will define, discuss, and deduce intuitively natural properties of independent random points.

Random points $f_1, f_2, \ldots, f_n$, with $n \geq 2$, having the same associated discrete probability space (Ω, P) are said to be **mutually independent** or **independent** iff

$$P\{\omega : f_1(\omega) \in A_1, f_2(\omega) \in A_2, \ldots, f_n(\omega) \in A_n\}$$
$$= P\{\omega : f_1(\omega) \in A_1\} P\{\omega : f_2(\omega) \in A_2\} \cdots P\{\omega : f_n(\omega) \in A_n\}$$

for all subsets A_i of F_i where $f_i : \Omega \to F_i$. At times we may call such f_i **stochastically independent** or **statistically independent** rather than just independent; "dependent" means "not independent."

It is clear from the definition that *reordering does not destroy independence.* Thus if f_1, f_2, f_3 are independent then so are f_3, f_1, f_2.

If we let $A_2 = F_2$ in the defining equations then $P\{\omega : f_2(\omega) \in F_2\} = P(\Omega) = 1$ and the resulting equations are just those which show that $f_1, f_3, f_4, \ldots, f_n$ are independent. Thus *if $f_1, f_2, \ldots, f_n$ is a finite sequence of $n \geq 2$ independent random points then any subsequence of two or more of them is independent.*

The trivial identity

$$\{\omega : f_1(\omega) \in A_1, \ldots, f_n(\omega) \in A_n\} = \{\omega : f_1(\omega) \in A_1\} \cap \cdots \cap \{\omega : f_n(\omega) \in A_n\}$$

shows that we can write the defining equation for independence as

$$P(\{\omega : f_1(\omega) \in A_1\} \cap \cdots \cap \{\omega : f_n(\omega) \in A_n\}) = P\{\omega : f_1(\omega) \in A_1\} \cdots P\{\omega : f_n(\omega) \in A_n\}.$$

Using the fact that a subsequence of an independent sequence $f_1, \ldots, f_n$ is independent, and recalling the definition of independence of events, we see that *random points $f_1, f_2, \ldots, f_n$ are independent iff the events*

$$\{\omega : f_1(\omega) \in A_1\}, \{\omega : f_2(\omega) \in A_2\}, \ldots, \{\omega : f_n(\omega) \in A_n\}$$

are independent for all subsets A_i of F_i, where $f_i : \Omega \to F_i$. This fact adequately motivates our definition of independence for random points. We may think of the value of f_i as describing the outcome of the ith subexperiment, so that $\{\omega : f_i(\omega) \in A_i\}$ is an event defined solely in terms of the outcome of the ith subexperiment. Then we can say that the subexperiments are independent iff the events $D_1, D_2, \ldots, D_n$ are independent *whenever* each $D_i = \{\omega : f_i(\omega) \in A_i\}$ is defined in terms of the outcome of the ith subexperiment; that is, $f_1, f_2, \ldots, f_n$ are independent iff *any* events defined in terms of different individual f_i are always independent.

It is intuitively clear, as stated in the next theorem, that functions of different independent random points or subexperiments must be independent.

Theorem 1 If $f_1, f_2, \ldots, f_n$ are $n \geq 2$ independent random points and

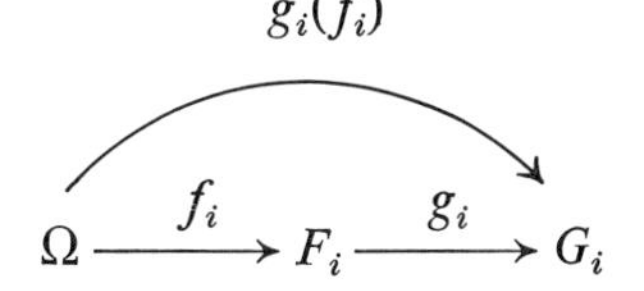

then $g_1(f_1), g_2(f_2), \ldots, g_n(f_n)$ are n independent random points. ∎

Proof Let B_i be an arbitrary subset of G_i. We wish to prove that

$$P\{\omega : g_1(f_1(\omega)) \in B_1, \ldots, g_n(f_n(\omega)) \in B_n\} = P\{\omega : g_1(f_1(\omega)) \in B_1\} \cdots P\{\omega : g_n(f_n(\omega)) \in B_n\}.$$

As in Fig. 1, let

$$A_i = g_i^{-1}(B_i) = \{\alpha_i : \alpha_i \in F_i \text{ and } g_i(\alpha_i) \in B_i\}$$

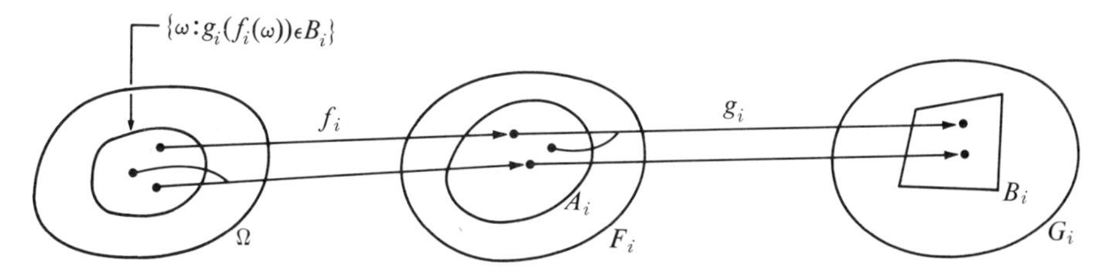

Fig. 1

so that $g_i(f_i(\omega))$ belongs to B_i iff $f_i(\omega)$ belongs to A_i. The desired equation becomes

$$P\{\omega: f_1(\omega) \in A_1, \ldots, f_n(\omega) \in A_n\} = P\{\omega: f_1(\omega) \in A_1\} \cdots P\{\omega: f_n(\omega) \in A_n\}$$

which is true by the assumption that $f_1, f_2, \ldots, f_n$ are independent. ■

The definition of independence can be restated in terms of induced distribution as follows: random points $f_1, f_2, \ldots, f_n$ are said to be independent iff

$$P_{f_1,\ldots,f_n}(A_1 \times A_2 \times \cdots \times A_n) = P_{f_1}(A_1)P_{f_2}(A_2) \cdots P_{f_n}(A_n)$$

for all subsets A_i of F_i where $f_i: \Omega \to F_i$. If we now let $A_i = \{\alpha_i\}$ consist of just one point of F_i then $P_{f_i}(\{\alpha_i\}) = p_{f_i}(\alpha_i)$. Thus if $f_1, f_2, \ldots, f_n$ are independent then

$$p_{f_1,\ldots,f_n}(\alpha_1, \alpha_2, \ldots, \alpha_n) = p_{f_1}(\alpha_1)p_{f_2}(\alpha_2) \cdots p_{f_n}(\alpha_n) \qquad \text{for all } \alpha_i \in F_i.$$

We now show that this property $p_{f_1,\ldots,f_n} = p_{f_1}p_{f_2} \cdots p_{f_n}$ implies independence. For notational simplicity we let $n = 2$. If $f: \Omega \to F$ and $g: \Omega \to G$ satisfy $p_{f,g} = p_f p_g$ and if A_F and A_G are subsets of F and G respectively, let

$$\{\alpha: \alpha \in A_F \text{ and } p_f(\alpha) > 0\} = \{\alpha_1, \alpha_2, \ldots\}$$
$$\{\beta: \beta \in A_G \text{ and } p_g(\beta) > 0\} = \{\beta_1, \beta_2, \ldots\}.$$

Then, using $p_{g,f} = p_f p_g$ for the second equality, we have

$$P_{f,g}(A_F \times A_G) = \sum_{i=1}^{\infty} \sum_{j=1}^{\infty} p_{f,g}(\alpha_i, \beta_j) = \sum_{i=1}^{\infty} \sum_{j=1}^{\infty} p_f(\alpha_i)p_g(\beta_j)$$
$$= \left[\sum_{i=1}^{\infty} p_f(\alpha_i)\right]\left[\sum_{j=1}^{\infty} p_g(\beta_j)\right] = P_f(A_F)P_g(A_G)$$

showing that f and g are independent. Thus *random points $f_1, f_2, \ldots, f_n$ are independent iff*

$$p_{f_1,\ldots,f_n}(\alpha_1, \alpha_2, \ldots, \alpha_n) = p_{f_1}(\alpha_1)p_{f_2}(\alpha_2) \cdots p_{f_n}(\alpha_n) \text{ for all } \alpha_1 \in F_1, \ldots, \alpha_n \in F_n$$

where $f_i: \Omega \to F_i$; that is, *independence is equivalent to the joint density equaling the product rule applied to the marginal densities.*

As shown in Fig. 2, the joint-probability table for two independent random points has any two rows, or columns, proportional to each other. Furthermore every entry in the table must be positive, since α_i and β_j are included in the table iff $p_f(\alpha_i) > 0$ and $p_g(\beta_j) > 0$.

$\alpha \backslash \beta$	β_1	$\cdots$	β_j	$\cdots$	β_m	p_f
α_1	$p_f(\alpha_1)p_g(\beta_1)$	$\cdots$	$p_f(\alpha_1)p_g(\beta_j)$	$\cdots$	$p_f(\alpha_1)p_g(\beta_m)$	$p_f(\alpha_1)$
$\cdots$	$\cdots$	$\cdots$	$\cdots$	$\cdots$	$\cdots$	$\cdots$
α_i	$p_f(\alpha_i)p_g(\beta_1)$	$\cdots$	$p_f(\alpha_i)p_g(\beta_j)$	$\cdots$	$p_f(\alpha_i)p_g(\beta_m)$	$p_f(\alpha_i)$
$\cdots$	$\cdots$	$\cdots$	$\cdots$	$\cdots$	$\cdots$	$\cdots$
α_k	$p_f(\alpha_k)p_g(\beta_1)$	$\cdots$	$p_f(\alpha_k)p_g(\beta_j)$	$\cdots$	$p_f(\alpha_k)p_g(\beta_m)$	$p_f(\alpha_k)$
p_g	$p_g(\beta_1)$	$\cdots$	$p_g(\beta_j)$	$\cdots$	$p_g(\beta_m)$	1

Fig. 2 Joint-probability table for independent random points f and g. $\{\alpha: p_f(\alpha) > 0\} = \{\alpha_1, \alpha_2, \ldots, \alpha_k\}$ and $\{\beta: p_g(\beta) > 0\} = \{\beta_1, \beta_2, \ldots, \beta_m\}$.

In general there can be many different joint densities $p_{f_1,\ldots,f_n}$ having the same marginals $p_{f_1}, \ldots, p_{f_n}$. We think of p_{f_i} as a complete description of the probabilistic behavior of f_i when f_i is considered by itself. Also $p_{f_1,\ldots,f_n}$ can be thought of as a complete description of the probabilistic relationship, or dependence, between the random points $f_1, \ldots, f_n$. Therefore knowledge of the probabilistic behavior of each of several *independent* random points completely determines their unique probabilistic relationship by the product rule $p_{f_1,\ldots,f_n} = p_{f_1}p_{f_2}\cdots p_{f_n}$; that is, under independence the marginals determine the joint table.

If $f_1, f_2, \ldots, f_m, f_{m+1}, \ldots, f_n$ are independent random points then the two random points $(f_1, \ldots, f_m)$ and $(f_{m+1}, \ldots, f_n)$ are independent. To prove this we note that $f_1, f_2, \ldots, f_m$ are independent, so that $p_{f_1,\ldots,f_m} = p_{f_1}p_{f_2}\cdots p_{f_m}$. Similarly $p_{f_{m+1},\ldots,f_n} = p_{f_{m+1}}p_{f_{m+2}}\cdots p_{f_n}$, so

$$p_{f_1,\ldots,f_n} = p_{f_1}\cdots p_{f_m}p_{f_{m+1}}\cdots p_{f_n} = p_{f_1,\ldots,f_m}p_{f_{m+1},\ldots,f_n}.$$

Clearly we could break $f_1, \ldots, f_n$ into more than two *disjoint* parts and still have independence.

For example, if $X_1, X_2, \ldots, X_{10}$ are 10 independent random variables then $(X_3,X_8,X_5,)$, X_4, and (X_2,X_9) are three independent random points, but (X_3,X_8,X_5) and (X_4,X_8) may well be dependent. The strong nature of independence as we defined it should be fairly apparent. Two events such as $[(f_1, f_2, \ldots, f_m) \in A]$ and $[(f_{m+1}, \ldots, f_n) \in B]$ must always be independent, where A and B are subsets of $F_1 \times \cdots \times F_m$ and $F_{m+1} \times \cdots \times F_n$, respectively; that is, any event defined in terms of some of the f_i must be independent of any event defined in terms of the other f_i. Thus probabilities about some of them are never affected by knowledge concerning the others; that is,

$$P[(f_1, \ldots, f_m) \in A \mid (f_{m+1}, \ldots, f_n) \in B] = P[(f_1, \ldots, f_m) \in A]$$

whenever $P[(f_{m+1}, \ldots, f_n) \in B] > 0$.

Random points $f_1, f_2, \ldots, f_n$ with $n \geq 2$ are said to be **pairwise independent** iff f_i and f_j are two independent random points whenever $i \neq j$. In this case knowledge about any *one* f_j does not alter the probabilities about any *one* other f_i; that is,

$$P[f_i \in A_i \mid f_j \in A_j] = P[f_i \in A_i] \qquad \text{if } P[f_j \in A_j] > 0,\ i \neq j.$$

In discussing pairwise independence we never drop the adjective "pairwise." As shown in Example 2 (Sec. 7-3) for events, it is possible for f_1, f_2, f_3 to be pairwise independent and still be dependent. Exercise 6 shows how to construct such an example. We seldom use the notion of pairwise independence.

Theorem 1 (Sec. 8-1) showed how to construct a model for independent subexperiments. It is restated below in the terminology of this section.

Theorem 2 Let $p_i: F_i \to R$ be a discrete probability density for each $i = 1, 2, \ldots, n$, where $n \geq 2$. Let $\Omega = F_1 \times F_2 \times \cdots \times F_n$ and let p be defined by

$$p(\alpha_1, \alpha_2, \ldots, \alpha_n) = p_1(\alpha_1)p_2(\alpha_2)\cdots p_n(\alpha_n) \qquad \text{for all } (\alpha_1, \ldots, \alpha_n) \in \Omega.$$

Then $p: \Omega \to R$ is a discrete probability density, and the coordinate functions $f_1, f_2, \ldots, f_n$ on Ω are independent, and $p_{f_i} = p_i$ for $i = 1, 2, \ldots, n$. ■

Example 1 Suppose we have constructed a simple model with independent random variables $X_1, \ldots, X_{100}$, where we interpret X_k as the sales of a store on the kth operating day. Assume that the store is open 5 days a week, so that $(X_1, X_2, \ldots, X_5)$ is its sales as a 5-tuple for the first week. From what we now know we can, for instance, assert that its sales in the first, second, and fourth weeks are independent, that is, that $(X_1, \ldots, X_5)$, $(X_6, \ldots, X_{10})$, and $(X_{16}, \ldots, X_{20})$ are three independent random points. We can also say that its sales on Mondays, $(X_1, X_6, X_{11}, \ldots, X_{96})$, are independent of its sales on Tuesdays, $(X_2, X_7, X_{12}, \ldots, X_{97})$. Theorem 1 shows that its total sales on Mondays $X_1 + X_6 + X_{11} + \cdots + X_{96}$ are independent of its total sales $X_2 + X_7 + X_{12} + \cdots + X_{97}$ on Tuesdays. Naturally $X_1 + X_2 + X_3$ and $X_2 + X_5$ may be dependent, since (X_1,X_2,X_3) and (X_2,X_5) may be dependent. ■

Example 2 Data sources are located at three different locations, as shown in Fig. 3. Each has a different data-transmission link to a central location. The first two sources use shortwave radio links, so that the corresponding received data R_1 and R_2 are influenced by the random

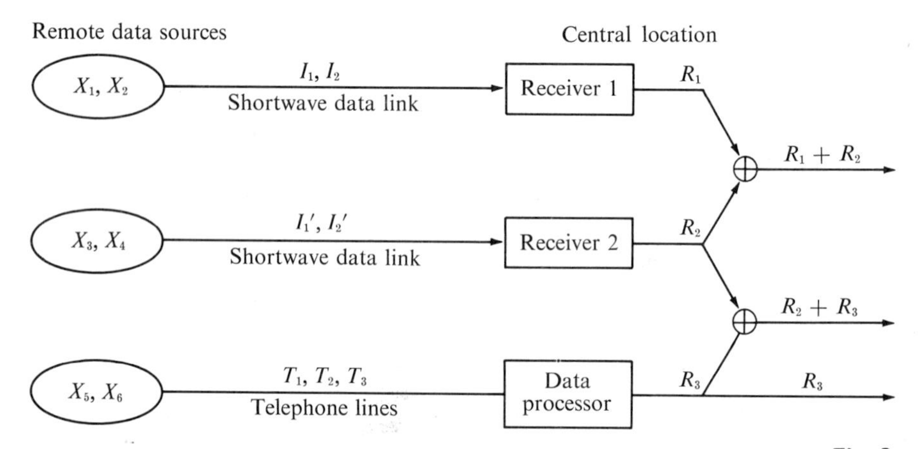

Fig. 3

variables I_1, I_2, I_1', and I_2', which are related to the ionosphere. The third source uses telephone lines and its corresponding received data are influenced by the random variables T_1, T_2, T_3, which are related to telephone-line noise. The R_i are defined as follows in terms of the data and noise variables

$$R_1 = (X_1e^{I_1} + X_2e^{I_2}) \cos (I_1 + I_2)$$

$$R_2 = (X_3e^{.9I_1'} + X_4e^{.8I_2'}) \sin (I_1' + 3I_2')$$

$$R_3 = (X_5e^{T_1+T_2} + X_6e^{T_3}) \sin (T_1^2 + T_2 + e^{T_3})$$

Let us assume that the eight random vectors

$$X_1, X_2, X_3, X_4, X_5, X_6, (I_1,I_2,I_1',I_2'), (T_1,T_2,T_3)$$

are independent. We then know, for example, that the two random vectors

$$(X_1,X_2,X_3,X_4,I_1,I_2,I_1',I_2') \qquad \text{and} \qquad (X_5,X_6,T_1,T_2,T_3)$$

are independent; hence from Theorem 1 we know that $R_1 + R_2$ and e^{R_3} are two independent random variables. However, R_1 and R_2 may well be dependent, since they both involve (I_1,I_2,I_1',I_2'). ■

Example 3: A powerful technique We seek the density induced by $S = Y_1 + Y_2 + Y_3$ when Y_1, Y_2, Y_3 are independent and p_{Y_i} is binomial with parameters n_i and p. Since the binomial density originally arose from Bernoulli trials, it is reasonable to introduce $n_1 + n_2 + n_3$ Bernoulli trials

$$X_1, X_2, \ldots, X_{n_1+n_2+n_3}$$

with parameter p, so that $P[X_i = 1] = p$ and $P[X_i = 0] = q = 1 - p$. From our general properties of independence we know that the random variables Y_1, Y_2, Y_3 defined by

$$Y_1 = X_1 + X_2 + \cdots + X_{n_1}$$

$$Y_2 = X_{n_1+1} + X_{n_1+2} + \cdots + X_{n_1+n_2}$$

$$Y_3 = X_{n_1+n_2+1} + X_{n_1+n_2+2} + \cdots + X_{n_1+n_2+n_3}$$

are independent. We also know that the distribution of the sum of n_i Bernoulli trials with parameter p is binomial with parameters n_i and p, so that p_{Y_i} must be binomial with parameters n_i and p, as required. But then $S = Y_1 + Y_2 + Y_3$ is the sum of $n_1 + n_2 + n_3$ Bernoulli trials with parameter p, so that p_S must be binomial with parameters $n_1 + n_2 + n_3$ and p. Clearly this approach generalizes to prove that *the density* p_S of $S = Y_1 + Y_2 + \cdots + Y_d$ *is binomial with parameters* $n_1 + n_2 + \cdots + n_d$ *and* p *if* $Y_1, Y_2, \ldots, Y_d$ *are independent and* p_{Y_i} *is binomial* n_i and p.

In Exercise 4 this same technique is used to derive an analogous result for Pascal densities.

The technique used in this example is very simple and powerful, as indicated by the ease with which the result was obtained. We now discuss a special case of this example in a little more detail. Suppose we want the density of $S = Y_1 + Y_2$, where Y_1 and Y_2 are independent and Y_1 is binomial $n_1 = 4$ and p while Y_2 is binomial $n_2 = 5$ and p. Thus Y_1 has the same distribution as the sum of four Bernoulli trials with parameter p and Y_2 has the same distribution as the sum of five Bernoulli trials with the same parameter p, *and* Y_1 and Y_2 are independent. It seems reasonable to consider nine Bernoulli trials, with Y_1 equal to the sum of the outcomes of the first four and Y_2 equal to the sum of the outcomes of the last five, so that $S = Y_1 + Y_2$ is the sum of the outcomes of all nine trials; hence S must be binomial $n = 9$ and p. This approach makes the result intuitively obvious and almost trivial. However, as indicated in Exercise 3, a fair amount of effort is required for an unimaginative brute-force proof of the same result. ■

Often in probability theory, as in the preceding example, problems are posed in such a way that the answer depends only on some specified induced distributions. In such a case any convenient experiment generating these induced distributions can be used to solve the problem, *and* the answer obtained will not depend on the experiment selected. *Although the ideas involved are quite simple, they are fundamental enough to deserve the detailed discussion which follows.*

If two different random points f and f', with possibly different associated densities or experiments, happen to have the same induced densities $p_f = p_{f'}$, then so do functions of them $p_{g(f)} = p_{g(f')}$. That is, if we have the two setups

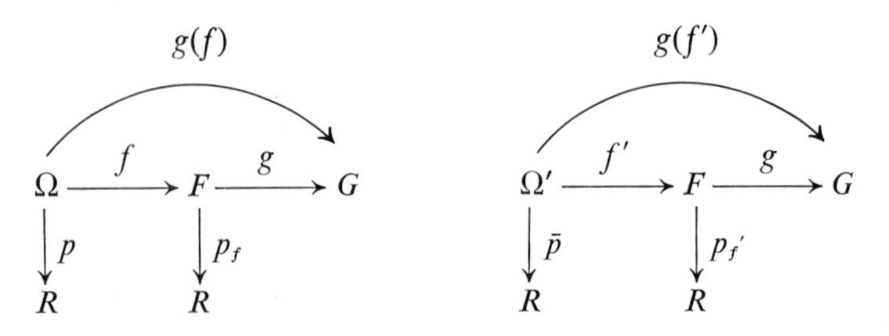

and $p_f = p_{f'}$, then $p_{g(f)} = p_{g(f')}$. This result is immediate from Theorem 2 (Sec. 11-1) because the *intermediate densities* are equal $p_f = p_{f'}$, so p_g must be the same in both setups. Thus if we are given some function $g\colon F \to G$ and some density denoted by $p'\colon F \to R$ and we wish to find the density p_g induced by g with associated p', we can introduce *any* convenient experiment $p\colon \Omega \to R$ and random point $f\colon \Omega \to F$ having the specified density $p_f = p'$ and be sure that the induced density $p_{g(f)}$ will be the same as for any other such experiment; that is, the density of some function $g(f)$ of a random point f is uniquely determined by the density p_f induced by the random point f. Now if $Y_1, Y_2, \ldots, Y_d$ are independent random variables their joint density is determined by their marginals

$$p_f = p_{Y_1,\ldots,Y_d} = p_{Y_1} p_{Y_2} \cdots p_{Y_d}.$$

Therefore the density induced by *any* function of them, such as

$$g(f) = g(Y_1, Y_2, \ldots, Y_d) = Y_1 + Y_2 + \cdots + Y_d$$

is uniquely determined by the marginals $p_{Y_1}, p_{Y_2}, \ldots, p_{Y_d}$, so in calculating $p_{g(f)}$ we can utilize *any* experiment having independent random variables with these marginals.

The preceding general technique is analogous to the following technique, which is sometimes employed in engineering analysis. We will use the symbols p, p', p_g, f, and g to suggest the analogy, although probabilities, etc., may not be involved. An input p' is given for a given system g, and the uniquely determined but unknown output p_g of g is desired:

$$\xrightarrow{p'} \boxed{g} \xrightarrow{p_g}$$

Some system f and input p to f are created in such a way that the output p_f of (f with input p) equals the given p':

$$\xrightarrow{p} \boxed{f} \xrightarrow{p'} \boxed{g} \xrightarrow{p_g}$$

Then in order to find p_g we need only find the output of the system [f followed by g] when the input is p. Naturally the new problem should be more manageable, in some sense, than the original problem. For example, p might be much simpler than p'.

INDICATORS AND THEIR INDEPENDENCE

We now introduce a device which permits us to consider the independence of events as a special case of the independence of random variables. If Ω is any set and A is a subset of Ω then let I_A be the real-valued function defined on Ω by

$$I_A(\omega) = \begin{cases} 1 & \text{if } \omega \in A \\ 0 & \text{if } \omega \notin A. \end{cases}$$

The function $I_A: \Omega \to R$ is called the **indicator of A on Ω**. As shown in Fig. 4*a*, I_A has the value 1 on A and 0 off A. If A is the set of positive real numbers then the indicator of A on R is as shown in Fig. 4*b* and is called the *unit step function* in many fields. If it is clear what set Ω is being considered we may abbreviate to the *indicator of A*.

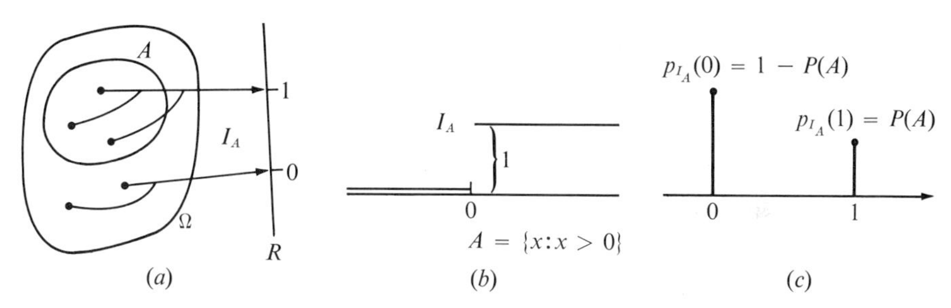

Fig. 4

In many branches of mathematics, and by some authors in probability theory, I_A is called the *characteristic function of A*. In probability theory the term "characteristic function" traditionally has another meaning, the Fourier transform of a probability distribution.

The trivial idea of I_A turns out to be quite useful and at times powerful, especially when used with the expectation operation introduced in the next section. If (Ω,P) is a discrete probability space and A is an event then I_A is a random variable. Clearly A is the event that I_A has the value 1, and the complement of A is the event that I_A has the value 0; that is, $[I_A = 1] = \{\omega: I_A(\omega) = 1\} = A$, $[I_A = 0] = \{\omega: I_A(\omega) = 0\} = A^c$, and $[I_A = x] = \varphi$ if x is not 0 or 1. Thus knowledge of the value of I_A is equivalent to knowledge of whether or not A has occurred. The density p_{I_A} induced by I_A is shown in Fig. 4*c*.

It is not surprising that *events $A_1, A_2, \ldots, A_n$ are independent iff the random variables $I_{A_1}, I_{A_2}, \ldots, I_{A_n}$ are independent*. In particular if $I_{A_1}, \ldots, I_{A_4}$ are independent then we know that $I_{A_1}, I_{A_3}, I_{A_4}$ are independent, so that

$$P[I_{A_1} = x_1, I_{A_3} = x_3, I_{A_4} = x_4] = P[I_{A_1} = x_1]P[I_{A_3} = x_3]P[I_{A_4} = x_4]$$

and using $x_1 = x_3 = x_4 = 1$, we get

$$P[A_1A_3A_4] = P(A_1)P(A_3)P(A_4)$$

which is one of the equations needed to prove that A_1, A_2, A_3, A_4 are independent. Thus we have shown that if $I_{A_1}, \ldots, I_{A_n}$ are independent then $A_1, \ldots, A_n$ are independent. The proof of the converse is left as Exercise 5.

EXERCISES

1. *Communication Links* (*Example 2*).

(*a*) Prove that the four random vectors (X_1,X_3), X_4, I_2', and $(T_1,e^{R_3},(X_2)^4)$ are independent, or indicate why they may be dependent.

(*b*) Prove that the two random variables $(X_1 - R_1)^2$ and I_2' are independent, or indicate why they may be dependent.

2. *Bernoulli Trials Generalized.* We say that random variables $X_1, X_2, \ldots, X_n$ constitute n **Bernoulli trials with parameter** p iff they are independent and identically distributed and $P[X_1 = 1] = p$ and $P[X_1 = 0] = q = 1 - p$. We have generalized the notion of Bernoulli trials in that we no longer specify the experiment involved. In Sec. 8-2 Bernoulli trials were supposed to arise from spins of a two-sectored wheel.

(*a*) A six-sectored random wheel has probability $p = .1$ assigned to sector 3. The wheel is spun $n = 100$ times independently. Define the random variable X_i by

$$X_i = \begin{cases} 1 & \text{if the } i\text{th spin resulted in sector 3} \\ 0 & \text{otherwise.} \end{cases}$$

Are $X_1, X_2, \ldots, X_n$ Bernoulli trials and is $X_1 + X_2 + \cdots + X_n$ binomial n and p?

(*b*) For the production process of Fig. 2 (Sec. 11-1) define the random variable V by

$$V = \begin{cases} 1 & \text{if } Z \le 1.5 \\ 0 & \text{otherwise.} \end{cases}$$

Consider $n = 30$ independent repetitions of this experiment resulting in $T = V_1 + V_2 + \cdots + V_{30}$, where V_i is from the ith repetition. Find p_T.

(*c*) Can X_i of part (*a*) and V_i of part (*b*) be defined as indicator functions of events?

3. *Sum of Two Independent Binomials* (*the Hard Way*). Let $S = Y_1 + Y_2$, where Y_1 and Y_2 are independent and p_{Y_i} is binomial n_i and p. Now $P[S = k]$ is computed from p_{Y_1,Y_2}, as in the Poisson case of Example 2 (Sec. 8-3), by

$$\begin{aligned} P[S = k] &= p_{Y_1}(0)p_{Y_2}(k) + p_{Y_1}(1)p_{Y_2}(k-1) + p_{Y_1}(2)p_{Y_2}(k-2) + \cdots + p_{Y_1}(k)p_{Y_2}(0) \\ &= \sum_{i=0}^{k} p_{Y_1}(i)p_{Y_2}(k-i) = \sum_{i=0}^{k} \binom{n_1}{i} p^i q^{n_1-i} \binom{n_2}{k-i} p^{k-i} q^{n_2-k+i} \\ &= p^k q^{n_1+n_2-k} \sum_{i=0}^{k} \binom{n_1}{i}\binom{n_2}{k-i}. \end{aligned}$$

From Example 3 we know that S is binomial $n_1 + n_2$ and p, so we have proved that

$$\sum_{i=0}^{k} \binom{n_1}{i}\binom{n_2}{k-i} = \binom{n_1+n_2}{k}.$$

Prove this last formula directly by expanding each of the three powers in

$$(1 + t)^{n_1+n_2} = (1 + t)^{n_1}(1 + t)^{n_2}$$

by the binomial theorem.

4. *Sums of Independent Pascal Variables.* Let $S = Y_1 + Y_2 + \cdots + Y_d$, where $Y_1, Y_2, \ldots, Y_d$ are independent and p_{Y_i} is Pascal with parameters r_i and p.

(*a*) Find the density induced by S.

(*b*) Describe the following three different random-wheel experiments, each of which has a random variable S having this density p_S; assume that each random wheel of an experiment is spun only once:

Experiment 1 has only one random wheel.

Experiment 2 has d random wheels.

Experiment 3 has $r_1 + r_2 + \cdots + r_d$ identical random wheels.

(*c*) What conditions must the parameters $r_1, r_2, \ldots, r_d$ and p satisfy in order to make $Y_1, Y_2, \ldots, Y_d$ identically distributed?

(*d*) Let $Z_1, Z_2, \ldots$ be an infinite sequence of Bernoulli trials with parameter $p = P[Z_i = 1]$, let $q = 1 - p = P[Z_i = 0]$, and define random variables $Y_1, Y_2, \ldots, Y_d$ having the properties specified.

(*e*) Use the result of part (*a*) with $d = 2$ to derive a binomial coefficient identity, as was done in Exercise 3.

5. Use Exercise 6 (Sec. 7-4) to prove that if $A_1, \ldots, A_n$ are independent then $I_{A_1}, \ldots, I_{A_n}$ are independent.

6. Prove that the three random variables $I_{A_1}, I_{A_2}, I_{A_3}$ are pairwise independent, but dependent, where A_1, A_2, A_3 are the events of Example 2 (Sec. 7-3).

7. This exercise exhibits one method for constructing many discrete densities having the same marginals. Let $p: R^2 \to R$ be a discrete density. Select two points (x_1,y_1) and (x_2,y_2) with positive probability and not on the same vertical or horizontal line, $x_1 \neq x_2$ and $y_1 \neq y_2$. Select any real A satisfying $0 \leq A \leq p(x_1,y_1)$ and $0 \leq A \leq p(x_2,y_2)$. Define the function $p': R^2 \to R$ as in Fig. 5, at the four points shown, with p and p' agreeing everywhere else.

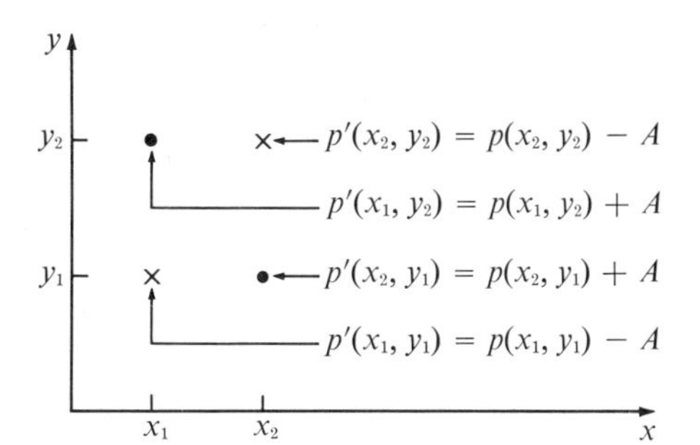

Fig. 5

Clearly p' is a discrete density and has the same marginal distributions as p. Assume that we know only that two random variables X and Y have specified discrete densities p_X and p_Y, neither of which is concentrated on one point. Show that it is possible for X and Y to be independent, and it is possible for X and Y to be dependent.

11-3 EXPECTATION

If X is a random variable with induced density p_X then the mean of p_X is also denoted by EX and may be called the **expectation of** X, or the expected value of X, or the mean value of X; that is, we define EX by

$$EX = \sum_{i=1}^{\infty} x_i p_X(x_i) \qquad \text{where } \{x: p_X(x) > 0\} = \{x_1, x_2, \ldots\}$$

if the series is absolutely convergent. If the series is not absolutely convergent we say that X does not have an expectation, or EX does not exist. Clearly *identically distributed random variables have the same expectation*; that is, if $p_X = p_Y$ then $EX = EY$.

Our intuitive interpretation of EX is the same as described in Sec. 9-1 and as justified in Chap. 12; that is, if X_i is obtained from the ith of a large number n of independent repetitions of an experiment then $(1/n)(X_1 + X_2 + \cdots + X_n)$ should, with high probability, be close to the number EX.

If A is an event and I_A is the indicator of A then $EI_A = P(A)$ because

$$EI_A = \sum_x xp_{I_A}(x) = (0)p_{I_A}(0) + (1)p_{I_A}(1) = p_{I_A}(1) = P(A).$$

This simple fact can be quite useful, as illustrated in Exercise 11.

If c is a real number and $p: \Omega \to R$ is a density we can define a random variable $X: \Omega \to R$ by $X(\omega) = c$ for all ω in Ω; that is, X has the constant value c on all of Ω. Clearly the induced density of X is given by

$$p_X(c) = P[X = c] = P(\Omega) = 1$$

$$p_X(x) = P[X = x] = P(\varphi) = 0 \qquad \text{if } x \neq c$$

so $EX = cp_X(c) = c$; that is, if a random variable has the constant value c on the sample space then its expected value is c, which is reasonable. If the number 5.1 appears in an expression such as $Z^2 + 5.1$ we will often interpret 5.1 as a random variable and write $E(5.1) = 5.1$. Except for this dual way of thinking about numbers, we never take the expectation of anything except random variables. Thus if we have E (something) the "something" must be a random variable. In particular $E(Z^2 + Y + e^W \cos T)$ is the expectation of the random variable $(Z^2 + Y + e^W \cos T)$, that is, the mean of its induced density.

Note that

$$E|X| = 0 \qquad \text{iff} \qquad p_X(0) = 1 \qquad \text{iff} \qquad P\{\omega: X(\omega) = 0\} = 1$$

so the expectation of the absolute value of X is 0 iff X is essentially the constant zero.

EXPECTATIONS FROM THE SAMPLE SPACE, OR FROM AN INDUCED DENSITY

Section 11-1 showed that probabilities of events defined in terms of a random variable $X: \Omega \to R$ could be calculated from the induced density p_X, or from the basic associated density $p: \Omega \to R$, or from an appropriate intermediate density if there happened to be one. These properties are so fundamental in probability theory that after making them explicit in theorems we used them almost unconsciously. Similarly, as shown in the next two theorems, the expectation of $X: \Omega \to R$ can be calculated from p_X as defined, or from its associated $p: \Omega \to R$, or from an intermediate density if there is one. In the case of probabilities and induced densities we are dealing with convergent series of *positive* numbers. As explained in Chap. 5, there are no serious convergence difficulties in this case when the terms are reordered or regrouped, as is done in constructing an induced density. In the case of expectations the series can have both positive and negative terms, so some care is needed. For example, a random variable always has an induced density p_X, but it may not have an expectation. Thus for induced densities we stated that certain fairly obvious equalities held, and now for expectations we will state that certain fairly obvious equalities hold *if* some appropriate series converges absolutely.

Suppose a sample space $\Omega = \{\omega_1, \omega_2, \ldots, \omega_{300}\}$ has 300 sample points and a random variable $X: \Omega \to R$ is defined by

$$X(\omega_i) = -1.3 \qquad \text{if } 1 \le i \le 100$$
$$X(\omega_j) = 5 \qquad \text{if } 101 \le j \le 200$$
$$X(\omega_k) = 13 \qquad \text{if } 201 \le k \le 300.$$

Consider the following manipulation:

$$\begin{aligned} X(\omega_1)p(\omega_1) + X(\omega_2)p(\omega_2) + \cdots + X(\omega_{300})p(\omega_{300}) \\ = -1.3[p(\omega_1) + \cdots + p(\omega_{100})] + 5[p(\omega_{101}) + \cdots + p(\omega_{200})] \\ + 13[p(\omega_{201}) + \cdots + p(\omega_{300})] \\ = -1.3P[X = -1.3] + 5P[X = 5] + 13P[X = 13] \\ = -1.3p_X(-1.3) + 5p_X(5) + 13p_X(13) = EX \end{aligned}$$

We formed the product $X(\omega)p(\omega)$ for each sample point ω and added these numbers, then by collecting terms appropriately and factoring out common factors, we got EX. Clearly this kind of computation can be performed in general. Thus we can calculate EX on the sample space Ω, and the computation is similar to that when p_X is used. On Ω we form the product $X(\omega)p(\omega)$, ignore the sample points where this product is 0, and add all the nonzero numbers so obtained. On R we form the product $xp_X(x)$, ignore any x where this product is 0, and add all the nonzero numbers so obtained. All that we did was rearrange and collect together terms of a sum, so that the result should certainly hold for absolutely convergent series, as asserted in the next theorem. Example 1 (Sec. 5-2) and Exercise 14 show that serious difficulties can arise in rearranging non–absolutely convergent series.

Theorem 1 Let X be a random variable with associated density $p: \Omega \to R$ and let $\{\omega: p(\omega) > 0\} = \{\omega_1, \omega_2, \ldots\}$ and $\{x: p_X(x) > 0\} = \{x_1, x_2, \ldots\}$. Then

$$\sum_{j=1}^{\infty} X(\omega_j)p(\omega_j) = \sum_{n=1}^{\infty} x_n p_X(x_n)$$

in the sense that if either series is absolutely convergent then so is the other, and equality holds. ■

Proof As in Fig. 1, let $\omega_{n1}, \omega_{n2}, \ldots$ be an enumeration of the event $\{\omega: p(\omega) > 0, X(\omega) = x_n\}$. Thus $X(\omega_{nm}) = x_n$ for all $m = 1, 2, \ldots$. Therefore

$$x_n p_X(x_n) = x_n \sum_{m=1}^{\infty} p(\omega_{nm}) = \sum_{m=1}^{\infty} X(\omega_{nm})p(\omega_{nm}).$$

Using this fact, we see that we wish to prove that

$$\sum_{j=1}^{\infty} X(\omega_j)p(\omega_j) = \sum_{n=1}^{\infty}\left[\sum_{m=1}^{\infty} X(\omega_{nm})p(\omega_{nm})\right]$$

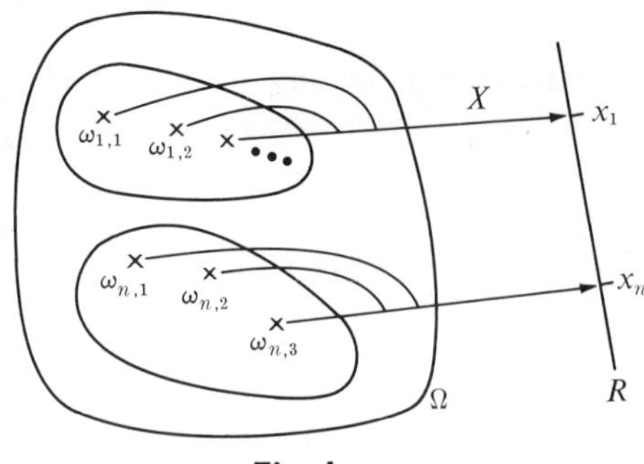

Fig. 1

where $\omega_1, \omega_2, \ldots$ is some enumeration, such as that exhibited in Fig. 1 or 2 (Sec. 5-3), of all the ω_{nm}. The result now follows from Theorem 1 (Sec. 5-3). ■

The next theorem asserts that if p_X can be obtained from an intermediate density p_f then EX can be calculated from p_f essentially by just using p_f as though it were the basic density p. In the statement of the theorem $\sum_{j=1}^{\infty} g(f(\omega_j))p(\omega_j)$ could be written as $\sum_{j=1}^{\infty} X(\omega_j)p(\omega_j)$.

Theorem 2 If

$$X = g(f)$$

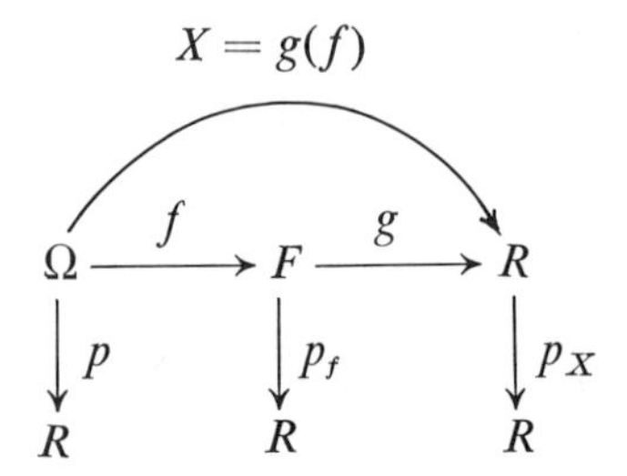

and

$$\begin{aligned}
&\omega_1, \omega_2, \ldots \text{ is an enumeration of } \{\omega : p(\omega) > 0\}\\
&\alpha_1, \alpha_2, \ldots \text{ is an enumeration of } \{\alpha : p_f(\alpha) > 0\}\\
&x_1, x_2, \ldots \text{ is an enumeration of } \{x : p_X(x) > 0\}
\end{aligned}$$

then

$$\sum_{j=1}^{\infty} g(f(\omega_j))p(\omega_j) = \sum_{i=1}^{\infty} g(\alpha_i)p_f(\alpha_i) = \sum_{n=1}^{\infty} x_n p_X(x_n)$$

in the sense that if any one of the three series is absolutely convergent then all three are, and the equalities hold. ■

Proof Application of Theorem 1 to X with associated density p shows that the left- and right-hand series are absolutely convergent together, and equal. If p_g is induced by g with associated p_f then $p_g = p_X$ by the intermediate distribution Theorem 2 (Sec. 11-1). Theorem 1 applied to g with associated p_f shows that the middle and right-hand series are absolutely convergent together, and equal. ■

If we introduce the temporary notation

$$E_pX = \sum_{j=1}^{\infty} X(\omega_j)p(\omega_j)$$

to indicate that the expectation of the random variable $X: \Omega \to R$ is calculated from its associated density $p: \Omega \to R$, we can write the conclusion of Theorem 2 as

$$E_pX = E_{p_f}g = E_{p_X}x$$

where x stands for the identity function on R. In each case the calculation of the expectation is by the same rule: take the product of the random variable and its associated density and sum all nonzero values.

We exhibit the conclusion of Theorem 2 in the important case when $F = R^n$, so that $f = (X_1, \ldots, X_n)$ is a random vector. If $X_1, \ldots, X_n$ are random variables with associated density $p: \Omega \to R$ and $g: R^n \to R$ then

$$\begin{aligned} Eg(X_1, \ldots, X_n) &= \sum_{\omega} g(X_1(\omega), \ldots, X_n(\omega))p(\omega) \\ &= \sum_{x_1, \ldots, x_n} g(x_1, \ldots, x_n)p_{X_1, \ldots, X_n}(x_1, \ldots, x_n) \\ &= \sum_{x} xp_{g(X_1, \ldots, X_n)}(x). \end{aligned}$$

For example, if $X = X_1 + e^{3X_2} \cos \pi X_3$ then

$$\begin{aligned} EX &= E[X_1 + e^{3X_2} \cos \pi X_3] \\ &= \sum_{\omega} [X_1(\omega) + e^{3X_2(\omega)} \cos \pi X_3(\omega)]p(\omega) \\ &= \sum_{x_1, x_2, x_3,} [x_1 + e^{3x_2} \cos \pi x_3]p_{X_1, X_2, X_3}(x_1, x_2, x_3) = \sum_{x} xp_X(x). \end{aligned}$$

We specialize Theorem 2 still further to the case $F = R$. If Y is a random variable with associated density p and $g: R \to R$ then

$$Eg(Y) = \sum_{\omega} g(Y(\omega))p(\omega) = \sum_{y} g(y)p_Y(y) = \sum_{x} xp_{g(Y)}(x).$$

For example,

$$EY^6 = \sum_{\omega} [Y(\omega)]^6 p(\omega) = \sum_{y} y^6 p_Y(y) = \sum_{x} xp_{Y^6}(x).$$

We often calculate $Eg(Y)$ from $Eg(Y) = \sum_{y} g(y)p_Y(y)$, and in this case if

$$\{y: p_Y(y) > 0\} = \{y_1, y_2, \ldots, y_k\}$$

then

$$Eg(Y) = g(y_1)p_Y(y_1) + g(y_2)p_Y(y_2) + \cdots + g(y_k)p_Y(y_k)$$

so $Eg(Y)$ may be described as a probability weighted average of the values $g(y_1), \ldots, g(y_k)$.

Example 1 For the graphical example of Fig. 8 (Sec. 11-1) we may calculate $Eg(X)$ from p_X or from $p_{g(X)}$

$$\sum_{k=1}^{4} g(x_k)p_X(x_k) = g(1.5)(.2) + g(2.5)(.1) + g(5)(.3) + g(9.3)(.4)$$

$$= (-1)(.2) + (1.8)(.1) + (1.8)(.3) + (3)(.4) = 1.72$$

$$\sum_{i=1}^{3} y_i p_Y(y_i) = (-1)(.2) + (1.8)(.4) + (3)(.4) = 1.72 \quad \blacksquare$$

Example 2 As an example of Theorem 2 with $F = R^2$ we take the tabulated p, $f = (X_1, Y_1)$, and $Z = g(f) = g(X_1, Y_1)$ of the production process of Figs. 2, 3, 6, and 10 of Sec. 11-1. The three methods

$$E_p Z = E_{p_f} g = E_{p_Z} z$$

for calculating EZ are as follows:

$$EZ = \sum_{i=1}^{12} [3 - Y_1(\omega_i)][2X_1(\omega_i) - 9]^2 p(\omega_i)$$

$$= (18)(.05) + (18)(.1) + (18)(.05) + (0)(.1) + (2)(.05) + (2)(.05)$$

$$+ (2)(.1) + (1)(.1) + (2)(.05) + (2)(.1) + (2)(.2) + (0)(.05) = 4.8$$

$$EZ = \sum_{x_1, y_1} [3 - y_1][2x_1 - 9]^2 p_{X_1,Y_1}(x_1, y_1)$$

$$= (0)(.05) + (0)(.1) + (1)(.1) + (2)(.35) + (2)(.2) + (18)(.2) = 4.8$$

$$EZ = \sum_z z p_Z(z) = (0)(.15) + (1)(.1) + (2)(.55) + (18)(.2) = 4.8 \quad \blacksquare$$

From Theorem 1 we know that X has an expectation iff the series $\sum_{j=1}^{\infty} X(\omega_j)p(\omega_j)$ is absolutely convergent, that is, iff the series

$$E|X| = \sum_{j=1}^{\infty} |X(\omega_j)|\, p(\omega_j)$$

of *positive* numbers converges. Thus X has an expectation iff the absolute value of X has an expectation. We can abbreviate this to *EX exists iff $E|X| < \infty$*. It is often more convenient to write $E|X| < \infty$ than to state that X has an expectation.

If r is any positive real number then from Theorem 2 we can define **the rth absolute moment of the random variable** X by any of the following:

$$E|X|^r = \sum_{j=1}^{\infty} |X(\omega_j)|^r\, p(\omega_j) = \sum_{n=1}^{\infty} |x_n|^r\, p_X(x_n) = \sum_{i=1}^{\infty} y_i p_{|X|^r}(y_i)$$

if these series are absolutely convergent; otherwise we say that X does not have an rth absolute moment. Thus the rth absolute moment of the random variable X equals the rth absolute moment of its induced density p_X as defined in Sec. 9-4.

For any $n = 1, 2, \ldots$ we define the **nth moment of a random variable** X by

$$EX^n = \sum_{j=1}^{\infty} [X(\omega_j)]^n p(\omega_j) = \sum_{k=1}^{\infty} (x_k)^n p_X(x_k) = \sum_{i=1}^{\infty} y_i p_{X^n}(y_i)$$

if these series are absolutely convergent. Clearly X has an nth moment iff X has an nth absolute moment. Also the nth moment of X equals the nth moment of p_X as defined in Sec. 9-4.

Theorem 1 (Sec. 9-4) shows that if X has an sth absolute moment then X has an rth absolute moment for any r satisfying $0 < r < s$. Thus if X has a second moment $s = 2$ then X has a mean $r = 1$, so we can define the **variance of** X by

$$\operatorname{Var} X = E[X - EX]^2 = \sum_{j=1}^{\infty} [X(\omega_j) - EX]^2 p(\omega_j) = \sum_{n=1}^{\infty} [x_n - EX]^2 p_X(x_n).$$

Clearly the variance of X equals the variance of p_X as defined in Sec. 9-2. As before,

$$\operatorname{Var} X = E(X^2) - (EX)^2 \qquad [\text{standard deviation of } X] = \sqrt{\operatorname{Var} X} \geq 0.$$

The assumption that $\sigma^2 = \operatorname{Var} X < \infty$ is equivalent to the assumption that X has a finite second moment.

Example 3 If $Y = aX + b$ then

$$EY = E[aX + b] = \sum_{j=1}^{\infty} [aX(\omega_j) + b]p(\omega_j) = a\left[\sum_{j=1}^{\infty} X(\omega_j)p(\omega_j)\right] + b = a[EX] + b.$$

Thus calculation of expectations on the sample space, or similarly from p_X, yields a transparent proof of

$$E[aX + b] = a[EX] + b$$

which was essentially obtained in Sec. 9-3 from the induced density p_{aX+b}.

Similarly, as in Sec. 9-3, we get

$$\operatorname{Var} [aX + b] = \sum_{j=1}^{\infty} \{[aX(\omega_j) + b] - E[aX + b]\}^2 p(\omega_j)$$

$$= a^2 \sum_{j=1}^{\infty} \{X(\omega_j) - EX\}^2 p(\omega_j) = a^2 \operatorname{Var} X. \quad \blacksquare$$

If X has mean μ_X and variance $\sigma_X^2 > 0$ then $X^* = (X - \mu_X)/\sigma_X$ is called the **standardized** or **normalized random variable corresponding to** X. Naturally $EX^* = 0$ and $\operatorname{Var} X^* = 1$.

Exercises 1, 2, 3, 12, 13, and 14 may be done before reading the remainder of this section.

BASIC PROPERTIES OF EXPECTATIONS

The more basic properties of expectation are collected together in Theorem 3 below. *These properties all hold true in the most general cases in probability theory.* There is, of course, arbitrariness in the selection of the facts to appear in such a theorem.

The interpretation of Theorem 3 is that if the hypotheses of a property are true then any expectations or variances which appear in the conclusion do indeed exist and do satisfy the conclusion. For example, property (c) means that if each $E|X_i| < \infty$ then $E|X_1 + X_2 + \cdots + X_n| < \infty$ and $E(X_1 + X_2 + \cdots + X_n) = (EX_1) + (EX_2) + \cdots + (EX_n)$. Since the reader may be seeing these expressions for the first time, parentheses are used more liberally than usual in the theorem to reduce the uncertainty of meaning. Thus the conclusion of property (c) is often written as $E\sum_{i=1}^{n} X_i = \sum_{i=1}^{n} EX_i$. The various parts of Theorem 3 are discussed and utilized after the theorem is proved.

Theorem 3: Properties of expectation Let X, Y, and $X_1, \ldots, X_n$ be random variables.

(a) If $E|Y| < \infty$ and $|X| \leq |Y|$ then $E|X| \leq E|Y|$.
(b) If $E|X| < \infty$ and a is a real then $E(aX) = a(EX)$ and $|EX| \leq E|X|$.
(c) If each $E|X_i| < \infty$ then $E(X_1 + X_2 \cdots + X_n) = (EX_1) + (EX_2) + \cdots + (EX_n)$.
(d) If each $E|X_i| < \infty$ and $X_1, X_2, \ldots, X_n$ are independent then $E(X_1X_2 \cdots X_n) = (EX_1)(EX_2) \cdots (EX_n)$.
(e) If $E|X| < \infty$, $E|Y| < \infty$, and $X \leq Y$ then $EX \leq EY$.
(f) If $E|X|^s < \infty$ and $0 < r < s$ then $[E(|X|^r)]^{1/r} \leq [E(|X|^s)]^{1/s}$.
(g) If $0 < r$ and each $E|X_i|^r < \infty$ then $E|X_1 + X_2 + \cdots + X_n|^r < \infty$.
(h) If $EX^2 < \infty$ and $EY^2 < \infty$ then $(EXY)^2 \leq (EX^2)(EY^2)$.
(i) If each $EX_i^2 < \infty$ and $X_1, X_2, \ldots, X_n$ are independent then $\text{Var}(X_1 + X_2 + \cdots + X_n) = (\text{Var } X_1) + (\text{Var } X_2) + \cdots + (\text{Var } X_n)$. ■

Proof

(a) If $E|Y| < \infty$ and $|X| \leq |Y|$ then

$$\sum_{j=1}^{N} |X(\omega_j)|\, p(\omega_j) \leq \sum_{j=1}^{N} |Y(\omega_j)| p(\omega_j) \leq E|Y| < \infty$$

so $E|X| < \infty$, and letting N go to infinity, we have $E|X| \leq E|Y|$.

(b) We get $E(aX) = a(EX)$ by letting N go to infinity in

$$\sum_{j=1}^{N} [aX(\omega_j)]p(\omega_j) = a\sum_{j=1}^{N} X(\omega_j)p(\omega_j).$$

We get $|EX| \leq E|X|$ by letting N go to infinity in

$$\left|\sum_{j=1}^{N} X(\omega_j)p(\omega_j)\right| \leq \sum_{j=1}^{N} |X(\omega_j)|\, p(\omega_j).$$

(c) If each $E|X_i| < \infty$ then

$$\begin{aligned}\sum_{j=1}^{N} |X_1(\omega_j) + \cdots + X_n(\omega_j)|\, p(\omega_j) &\leq \sum_{j=1}^{N} [|X_1(\omega_j)| + \cdots + |X_n(\omega_j)|]p(\omega_j) \\ &= \left[\sum_{j=1}^{N} |X_1(\omega_j)|\, p(\omega_j)\right] + \cdots + \left[\sum_{j=1}^{N} |X_n(\omega_j)|\, p(\omega_j)\right] \\ &\leq [E|X_1|] + [E|X_2|] + \cdots + [E|X_n|] < \infty\end{aligned}$$

so $E(|X_1 + X_2 + \cdots + X_n|) < \infty$. We get $E(X_1 + \cdots + X_n) = EX_1 + \cdots + EX_n$ by letting N go to infinity in

$$\begin{aligned}&\sum_{j=1}^{N} [X_1(\omega_j) + X_2(\omega_j) + \cdots + X_n(\omega_j)]p(\omega_j) \\ &\quad = \left[\sum_{j=1}^{N} X_1(\omega_j)p(\omega_j)\right] + \left[\sum_{j=1}^{N} X_2(\omega_j)p(\omega_j)\right] + \cdots + \left[\sum_{j=1}^{N} X_n(\omega_j)p(\omega_j)\right].\end{aligned}$$

(*d*) For the previous proofs it seemed easiest to calculate expectations on the sample space, but here it is easiest to introduce the assumption of independence by using the fact that the joint density $p_{X_1,\ldots,X_n}$ equals the product of the marginals $p_{X_1}p_{X_2}\cdots p_{X_n}$. Assume that each $E|X_i| < \infty$. Theorem 1 (Sec. 5-3) asserts that

$$\left[\sum_{n=1}^{\infty} b_n\right]\left[\sum_{m=1}^{\infty} c_m\right] = \sum_{n=1}^{\infty}\sum_{m=1}^{\infty} b_n c_m$$

in the sense that if the *two* series on the left-hand side are absolutely convergent then the series on the right-hand side is absolutely convergent, and equality holds. This fact is easily extended to more than two factors, and with this extension we obtain the result as follows:

$$\begin{aligned}(EX_1)(EX_2)\cdots(EX_n) &= \left[\sum_{x_1} x_1 p_{X_1}(x_1)\right]\left[\sum_{x_2} x_2 p_{X_2}(x_2)\right]\cdots\left[\sum_{x_n} x_n p_{X_n}(x_n)\right]\\ &= \sum_{x_1}\sum_{x_2}\cdots\sum_{x_n}\{[x_1x_2\cdots x_n]p_{X_1}(x_1)p_{X_2}(x_2)\cdots p_{X_n}(x_n)\}\\ &= \sum_{x_1}\sum_{x_2}\cdots\sum_{x_n}\{[x_1x_2\cdots x_n]p_{X_1,\ldots,X_n}(x_1, x_2, \ldots, x_n)\}\\ &= E(X_1X_2\cdots X_n).\end{aligned}$$

(*e*) If $X \le Y$ then $EX \le EY$ if both series converge absolutely, since we can let N go to infinity in

$$\sum_{i=1}^{N} X(\omega_i)p(\omega_i) \le \sum_{i=1}^{N} Y(\omega_i)p(\omega_i).$$

(*f*) This property follows from Theorem 1 (Sec. 9-4).

(*g*) We first show that if r is a positive real and $x_1, x_2, \ldots, x_n$ are reals then

$$(|x_1 + x_2 + \cdots + x_n|)^r \le n^r(|x_1|^r + |x_2|^r + \cdots + |x_n|^r).$$

Assume for notational convenience that $|x_1|$ is the largest of the numbers $|x_1|, |x_2|, \ldots, |x_n|$. Then

$$\begin{aligned}(|x_1 + x_2 + \cdots + x_n|)^r &\le (|x_1| + |x_2| + \cdots + |x_n|)^r\\ &\le (n\,|x_1|)^r = n^r(|x_1|^r)\\ &\le n^r(|x_1|^r + |x_2|^r + \cdots + |x_n|^r).\end{aligned}$$

Replacing x_i by X_i, we obtain

$$(|X_1 + X_2 + \cdots + X_n|)^r \le n^r(|X_1|^r + |X_2|^r + \cdots + |X_n|^r)$$

and since we assume that each $E(|X_i|^r) < \infty$, we know from property (*c*) that $E(|X_1|^r + |X_2|^r + \cdots + |X_n|^r) < \infty$, so from property (*a*) we have the conclusion of property (*g*).

(*h*) If x and y are reals then $0 \le (x - y)^2 = x^2 - 2xy + y^2$ so $2xy \le x^2 + y^2$, and $0 \le (x + y)^2 = x^2 + 2xy + y^2$ so $-(x^2 + y^2) \le 2xy$. Thus $2\,|XY| \le X^2 + Y^2$, so that if $EX^2 < \infty$ and $EY^2 < \infty$ then $E\,|XY| < \infty$, and from properties (*b*) and (g) we have $E(aX \pm bY)^2 < \infty$. These convergence facts justify the following calculation:

$$0 \le E\left\{\frac{X}{(EX^2)^{1/2}} \pm \frac{Y}{(EY^2)^{1/2}}\right\}^2$$

$$= E\left\{\frac{X^2}{EX^2} \pm \frac{2XY}{[(EX^2)(EY^2)]^{1/2}} + \frac{Y^2}{EY^2}\right\}$$

$$= 1 \pm \frac{2E(XY)}{[(EX^2)(EY^2)]^{1/2}} + 1 = 2\left\{1 \pm \frac{E(XY)}{[(EX^2)(EY^2)]^{1/2}}\right\}$$

Now if A is any real number for which $0 \le \{1 + A\}$ and $0 \le \{1 - A\}$ then $-1 \le A \le 1$, so $|A| \le 1$, hence $A^2 \le 1$, and property (h) is proved, at least when both EX^2 and EY^2 are positive. If one of them equals 0, it quickly follows that $E(XY) = 0$.

(i) If $X_1, \ldots, X_n$ are independent and each $EX_i^2 < \infty$ we let $\mu_i = EX_i$. From property (c) the expectation of $X_1 + \cdots + X_n$ is $\mu_1 + \cdots + \mu_n$. Therefore

$$\begin{aligned}
\operatorname{Var}(X_1 + \cdots + X_n) &= E[(X_1 + \cdots + X_n) - (\mu_1 + \cdots + \mu_n)]^2 \\
&= E[(X_1 - \mu_1) + \cdots + (X_n - \mu_n)]^2 \\
&= E\left\{\left[\sum_{i=1}^{n} (X_i - \mu_i)\right]\left[\sum_{j=1}^{n} (X_j - \mu_j)\right]\right\} \\
&= E\left[\sum_{i=1}^{n}\sum_{j=1}^{n} (X_i - \mu_i)(X_j - \mu_j)\right] \\
&= \sum_{i=1}^{n}\sum_{j=1}^{n} E(X_i - \mu_i)(X_j - \mu_j) \\
&= \operatorname{Var} X_1 + \operatorname{Var} X_2 + \cdots + \operatorname{Var} X_n
\end{aligned}$$

where property (c) was used for the next-to-last equality. For the last equality note that if $i \ne j$ then $X_i - \mu_i$ and $X_j - \mu_j$ are independent, so from property (d) we have

$$E(X_i - \mu_i)(X_j - \mu_j) = [E(X_i - \mu_i)][E(X_j - \mu_j)] = [0][0] = 0.$$

Therefore for each $i = 1, 2, \ldots, n$ we have

$$\sum_{j=1}^{n} E(X_i - \mu_i)(X_j - \mu_j) = E(X_i - \mu_i)(X_i - \mu_i) = \operatorname{Var} X_i$$

so property (i) is proved, hence the proof of Theorem 3 is complete. ■

APPLICATION OF THE BASIC PROPERTIES OF EXPECTATION

In the remainder of this section we discuss Theorem 3 and illustrate its use. Our principal way of proving that a random variable X has an expectation $E|X| < \infty$ is to find a dominating $|Y| \ge |X|$ and show that $E|Y| < \infty$, so that property (a) applies.

Although it is not profound, this method is used frequently. In terms of series this amounts to proving that $\sum_{j=1}^{\infty} |a_j| < \infty$ by finding $|b_j| \geq |a_j|$ for which $\sum_{j=1}^{\infty} |b_j| < \infty$.

Combining properties (*b*) and (*c*), we see that if each $E|X_i| < \infty$ then

$$E(a_1X_1 + a_2X_2 + \cdots + a_nX_n) = a_1EX_1 + a_2EX_2 + \cdots + a_nEX_n.$$

This result is referred to as the *linearity of expectation* and it is usually considered to be the most important property of expectation. It is always true if each $E|X_i| < \infty$. It does *not* require independence, as does property (*d*). Clearly if we can find EX_1, $EX_2, \ldots, EX_n$ then by linearity we can find $E\left(\sum_{i=1}^{n} a_iX_i\right)$ without ever finding the density of $\sum_{i=1}^{n} a_iX_i$, which might be unmanageable, as indicated in Example 5.

Example 4 The linearity of expectation yields *the following frequently used fact*: if

$$S_n = X_1 + \cdots + X_n$$

where $E|X_1| < \infty$, and *if* $X_1, X_2, \ldots, X_n$ *are identically distributed, then* $ES_n = nEX_1$. Independence need not be assumed.

If X_1 has a Bernoulli density with parameter p then $EX_1 = p$, so $ES_n = np$. If X_1 has a geometric density with parameter p then $EX_1 = q/p$, so $ES_n = n(q/p)$. If X_1 has a Poisson density with $EX_1 = \lambda$ then $ES_n = n\lambda$.

Considering the Bernoulli case in more detail, we assume *only* that $p_{X_i}(1) = p$ and $p_{X_i}(0) = q = 1 - p$ for $i = 1, 2, \ldots, n$. The variables $X_1, X_2, \ldots, X_n$ may have any one of many different joint densities, depending on the nature of their dependence. The induced density p_{S_n} of S_n is determined by $p_{X_1,\ldots,X_n}$ and in general will be different for different joint densities, as shown in Exercise 3. However, the mean of p_{S_n} is np in *all* cases, regardless of the nature of the dependence. We obtained this result without calculating even one induced density p_{S_n}. In Sec. 8-2 we found that for Bernoulli trials, where by definition $X_1, X_2, \ldots, X_n$ are independent, S_n has a binomial density; we *then* found the mean by direct calculation from p_{S_n} in Exercise 5 (Sec. 8-2) and from the MGF in Exercise 7 (Sec. 9-4). Thus in earlier chapters we made the common simplifying assumption of independence and we were able to find p_{S_n} and then find ES_n from p_{S_n}. The linearity of expectation does not yield p_{S_n}, but it does show that the mean of p_{S_n} must be np even when independence is not assumed. A similar discussion applies to the case where the X_i are geometric, in which case S_n is Pascal *if* $X_1, X_2, \ldots, X_n$ are independent. ■

Example 5 Theorem 3*i* asserts that independence implies that the variance of a sum equals the sum of the variances. This property is often used, especially in the special case contained in the following frequently used fact: If $S_n = X_1 + \cdots + X_n$, where $EX_1^2 < \infty$, and *if* $X_1, X_2, \ldots, X_n$ *are independent and identically distributed then* $ES_n = nEX_1$ *and* $\text{Var } S_n = n \text{ Var } X_1$.

If X_1 is Bernoulli then $EX_1 = p$ and $\text{Var } X_1 = pq$, so $ES_n = np$ and $\text{Var } S_n = npq$. Thus simple properties of expectation immediately yield the mean and variance of p_{S_n} without calculation of p_{S_n}. We found earlier that in this case p_{S_n} is binomial n and p, so we have a new, almost trivial derivation of the mean and variance of binomial densities.

If X_1 is geometric with parameter p then $EX_1 = q/p$ and $\text{Var } X_1 = q/p^2$, so $ES_n = nq/p$ and $\text{Var } S_n = nq/p^2$. Thus simple properties of expectation immediately yield the mean and variance of p_{S_n} without calculation of p_{S_n}. We found earlier that in this case p_{S_n} is Pascal n and p, so we have a new, almost trivial derivation of the mean and variance of Pascal densities.

If X_1 has the density of Fig. 1 (Sec. 9-1) then $EX_1 = 1.28$ and $\text{Var } X_1 = 8.5596$, so $ES_n = n(1.28)$ and $\text{Var } S_n = n(8.5596)$. Thus simple properties of expectation immediately yield the mean and variance of p_{S_n} without calculation of p_{S_n}. In this case we will never exhibit p_{S_n} for large n in any standard simple analytic form, although p_{S_n} is uniquely defined for each n. ■

Simple manipulations with expectations, independence, etc., are sometimes startlingly powerful. This power appears to be intimately related to the equivalence of various operations which can be performed on different related distributions, as shown, for example, in Theorems 1 and 2 in this section and in Sec. 11-1. In addition, assumptions concerning various induced distributions are natural to the interpretation and application of probabilistic statements.

Theorem 3d states that *for independent random variables* the expectation of a product equals the product of the expectations. This is the principal way in which we will utilize available independence. Note that the result can be applied to functions of different independent random points, since these are independent by Theorem 1 (Sec. 11-2). Thus if $f_1, f_2, \ldots, f_n$ are independent random points and

$$X_i = g_i(f_i)$$

$$\Omega \xrightarrow{f_i} F_i \xrightarrow{g_i} R$$

then $X_1, X_2, \ldots, X_n$ are independent, so that $E\prod_{i=1}^{n} X_i = \prod_{i=1}^{n} EX_i$, and of course, $\text{Var} \sum_{i=1}^{n} X_i = \sum_{i=1}^{n} \text{Var } X_i$. For example, if (Y_1, Y_2, Y_3), (Z_1, Z_2), W are three independent random vectors then

$$E(e^{Y_2})(\cos 3Y_3)(Z_1 + e^{Z_2})(W^5) = [E(e^{Y_2}\cos 3Y_3)][(EZ_1) + (Ee^{Z_2})][EW^5].$$

If $X_1, X_2, \ldots, X_n$ are independent then so are $e^{tX_1}, e^{tX_2}, \ldots, e^{tX_n}$ for any real t, so from Theorem 3d we have

$$Ee^{t(X_1+X_2+\cdots+X_n)} = Ee^{tX_1}e^{tX_2}\cdots e^{tX_n} = (Ee^{tX_1})(Ee^{tX_2})\cdots(Ee^{tX_n}).$$

The **moment-generating function** (MGF) **of a random variable** X is naturally defined to equal the MGF of its density p_X,

$$m(t) = Ee^{tX} = \sum_x e^{tx}p_X(x).$$

Thus *the MGF of a sum of independent random variables is equal to the product of their MGFs*. In more detail, if $X_1, \ldots, X_n$ are independent and have finite MGFs $m_{X_i}(t)$ at t, and if $S_n = X_1 + X_2 + \cdots + X_n$ then the MGF of S_n is finite at this t and

$$m_{S_n}(t) = m_{X_1}(t)m_{X_2}(t)\cdots m_{X_n}(t).$$

Example 6 Let $S = Y_1 + Y_2 + Y_3$ where Y_1, Y_2, Y_3 are independent and p_{Y_i} is binomial n_i and p. We will use MGFs to deduce the same density for S as was found in Example 3 (Sec. 11-2).

The MGF of Y_i is $(q + pe^t)^{n_i}$, so the MGF of S is their product,

$$m_S(t) = (q + pe^t)^{n_1}(q + pe^t)^{n_2}(q + pe^t)^{n_3}$$
$$= (q + pe^t)^n \qquad \text{where } n = n_1 + n_2 + n_3.$$

We observe that m_S happens to equal the MGF of the binomial density with parameters n and p. From the last sentence of Theorem 2 (Sec. 9-4) we conclude that p_S is binomial n and p.

More generally the identity

$$(q + pe^t)^{n_1+n_2+\cdots+n_d} = \prod_{i=1}^{d} (q + pe^t)^{n_i}$$

shows that $S = Y_1 + Y_2 + \cdots + Y_d$ is binomial $n_1 + n_2 + \cdots + n_d$ and p if the Y_i are independent and binomial n_i and p.

Similarly the identity

$$\left(\frac{p}{1 - qe^t}\right)^{r_1+r_2+\cdots+r_d} = \prod_{i=1}^{d}\left(\frac{p}{1 - qe^t}\right)^{r_i}$$

shows that $S = Y_1 + Y_2 + \cdots + Y_d$ is Pascal $r_1 + r_2 + \cdots + r_d$ and p if the Y_i are independent and Pascal r_i and p. ■

Theorem 3*e* states that the expectation of a smaller random variable must be smaller. Thus if one gambler never receives more, $X(\omega) \le Y(\omega)$, than another gambler, regardless of the outcome ω of the experiment, then his expected reward certainly satisfies $EX \le EY$.

Example 7 If $g\colon R \to R$ and $h\colon R \to R$ satisfy $g \le h$, that is,

$$g(x) \le h(x) \qquad \text{for all real } x$$

then $g(X) \le h(X)$ for any random variable $X\colon \Omega \to R$; that is,

$$g(X(\omega)) \le h(X(\omega)) \qquad \text{for all } \omega \in \Omega.$$

Then from Theorem 3*e* we have $Eg(X) \le Eh(X)$. This is obvious, since we can sum the inequality $g(x)p_X(x) \le h(x)p_X(x)$ over real x. Therefore if $g \le h$ then $Eg(X) \le Eh(X)$, and this result is used often. For example, $1 - x \le e^{-x}$ for all real x from equation (7) of Appendix 1, so $1 - EX \le Ee^{-X}$. Similarly $-1 \le \cos x \le 1$ for all real x, so $-1 \le E(\cos X) \le 1$ for every random variable X. ■

According to Theorem 3*g*, the property of having an rth absolute moment is preserved in taking sums. Thus $X_1 + X_2 + \cdots + X_n$ has an expectation, or second moment, if each X_i does.

By Theorem 3*h* we know that $E|XY| < \infty$ if $EX^2 < \infty$ and $EY^2 < \infty$; that is, we know that the *product* XY has an expectation if each random variable has a *second* moment. We certainly cannot conclude that $E|XY| < \infty$ if we know only that $E|X| < \infty$ and $E|Y| < \infty$, since if $X = Y$ this specializes to $EX^2 < \infty$ if $E|X| < \infty$, which is false. Thus if $EX^2 < \infty$ and $EY^2 < \infty$ then each of the four [ ] random variables in the equation

$$[(X + Y)^2] = [X^2] + [2XY] + [Y^2]$$

has an expectation. Therefore in considering second moments or variances of sums it is natural to assume that each $EX_i^2 < \infty$.

***Example 8** We show that if $X \geq 0$ then its MGF determines its distribution. We first show that *for bounded random variables* the moments determine the distribution.

Assume that $P[0 \leq X \leq 1] = P[0 \leq Y \leq 1] = 1$ and that

$$EX^n = EY^n \qquad \text{for all } n = 1, 2, \ldots.$$

We show that $P[X \geq s] = P[Y \geq s]$ for all real s. Obviously we need only consider $0 < s \leq 1$, so we fix such an s. Using the polynomial F_n and function $f_{r,s}$ of Corollary 1 (Sec. 10-3), we have $EF_n(X) = EF_n(Y)$; hence we may add and subtract this quantity within the following expression

$$|Ef_{r,s}(X) - Ef_{r,s}(Y)| \leq E\,|f_{r,s}(X) - F_n(X)| + E\,|f_{r,s}(Y) - F_n(Y)|$$
$$\leq 2n^{-1/3}\left(1 + \frac{1}{s - r}\right) \to 0 \qquad \text{as } n \to \infty.$$

Thus $Ef_{r,s}(X) = Ef_{r,s}(Y)$ for all $0 \leq r < s \leq 1$; and this easily yields the desired result as follows. Applying Theorem 3*e* to indicator functions and $f_{r,s}(X)$ yields

$$P[X \geq s] = EI_{[X \geq s]} \leq Ef_{r,s}(X) \leq EI_{[X \geq r]} = P[X \geq r].$$

Here the right-hand side is $P[X \geq r] = P[X \geq s] + P[r \leq X < s]$, and clearly $P[r \leq X < s]$ goes to zero as r approaches s. Therefore replacing r by $r_k = s - (1/k)$ yields

$$P[X \geq s] = \lim_{k \to \infty} Ef_{r_k,s}(X) = \lim_{k \to \infty} Ef_{r_k,s}(Y) = P[Y \geq s]$$

as was asserted. This proof is valid in all nondiscrete cases as well. In the discrete case it now follows from Exercise 15 that X and Y must be identically distributed; the same conclusion follows in general from Theorem 2 (Sec. 15-1).

If $P[X \geq 0] = P[Y \geq 0] = 1$ and

$$Ee^{tX} = Ee^{tY} \qquad \text{for all real } t \leq 0$$

then we show that X and Y are identically distributed. Note that these MGFs are finite for all $t \leq 0$, since if $x \geq 0$ then $tx \leq 0$, so $Ee^{tX} \leq 1$. Also $P[0 \leq e^{-X} \leq 1] = 1$. Applying the preceding paragraph to e^{-X} and e^{-Y} yields the center equality below

$$P[X \leq \beta] = P[e^{-X} \geq e^{-\beta}] = P[e^{-Y} \geq e^{-\beta}] = P[Y \leq \beta] \qquad \text{for any real } \beta$$

hence X and Y are identically distributed.

This proof easily extends to random variables for which there is a constant c for which $P[X \geq c] = 0$ or 1. However, it does not extend, for example, to a discrete density which is positive on all integers $n = 0, \pm 1, \pm 2, \ldots$, even if it is assumed that the MGF is finite for $|t| \leq t_0$ with $t_0 > 0$. ■

EXERCISES

1. *Expectations in the Production Process* (*Fig. 2, Sec. 11-1*). Find the density and from it the expectation for each of the three random variables X_1, Y_1, and $X_1 + Y_1$. Check that $E(X_1 + Y_1) = EX_1 + EY_1$.

2. *Eg(X) Graphically* (*Fig. 2*). Make a rough estimate of $Eg(X)$ where g and p_X are as graphed in Fig. 2.

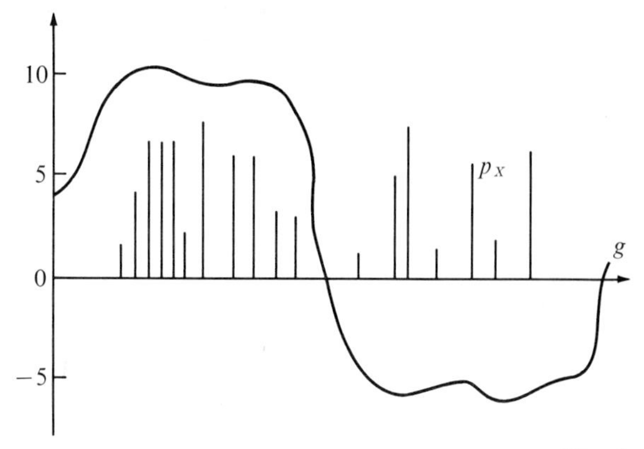

Fig. 2

3. *Dependent Zero-One Trials.* The joint distribution for X_1 and X_2 is as in Fig. 3, so that

x_1 \ x_2	0	1	p_{X_1}
0	$.3q$	$.3p$	.3
1	$.3p$	$.7-.3p$	.7
p_{X_2}	.3	.7	1

Fig. 3 The joint-probability table for X_1 and X_2. $0 \le p \le 1$ and $p + q = 1$.

X_1 and X_2 are identically distributed and each has a Bernoulli density with parameter .7. However X_1 and X_2 do not in general constitute "Bernoulli trials," which are by definition independent. The parameter p can be any number satisfying $0 \le p \le 1$. We can interpret the outcome (x_1,x_2) as the outcomes of two connected, or related, or dependent random wheels.

(*a*) The three independent random wheels of Fig. 4 can be used to provide an alternate experiment which generates the same joint density p_{X_1,X_2}. Wheel A is spun and yields x_1. If $x_1 = 0$ wheel B_0 is spun and yields x_2. If $x_1 = 1$ wheel B_1 is spun and yields x_2. Thus X_i is the outcome of the ith wheel spun. Show that this experiment does indeed yield the joint density of Fig. 3.

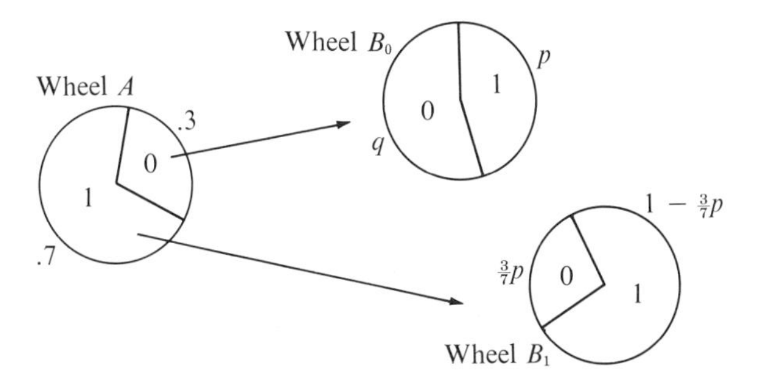

Fig. 4

(*b*) Find the density induced by $X_1 + X_2$ and note that it is different for different values of the dependence parameter p.
(*c*) Find $E(X_1 + X_2)$ from $p_{X_1+X_2}$ and note that $E(X_1 + X_2) = EX_1 + EX_2$ for every value of p.

4. *The Expectation of a Mess.* Let $Y = X_1 + X_2 + X_3$, where X_1, X_2, X_3 are highly dependent and p_{X_1} is binomial n and p, p_{X_2} is Poisson λ, p_{X_3} is Pascal r and p'. Can you find EY and/or Var Y?

5. *Expectations in Data Transmission* (*Example 1, Sec. 11-2*). Assume that all random variables which are considered have expectations. Which of the following *must* be true?
(*a*) $E(R_1 + R_2)e^{R_3} = (ER_1 + ER_2)Ee^{R_3}$
(*b*) $ER_1R_2 = (ER_1)(ER_2)$
(*c*) $EX_1e^{X_2}T_1 \cos e^{R_3} = (EX_1)(Ee^{X_2})(ET_1)(E \cos e^{R_3})$
(*d*) $EX_1e^{X_2}T_1 \cos e^{R_3} = (EX_1)(Ee^{X_2})E(T_1 \cos e^{R_3})$

6. Use MGFs to prove that $S_n = X_1 + X_2 + \cdots + X_n$ is Poisson with parameter $\lambda = \lambda_1 + \lambda_2 + \cdots + \lambda_n$ if each X_i is Poisson with parameter λ_i and $X_1, X_2, \ldots, X_n$ are independent.

7. *Expectations for Bernoulli Trials.* Let X_1, X_2, X_3 be $n = 3$ Bernoulli trials with parameter p.
(*a*) Find $E\{(X_2 - 3)^2[\cos(\pi X_3/4)]e^{2X_1}\}$.
(*b*) Find Var $\{e^{2X_1} + \cos(\pi X_3/4)\}$.

8. *An Application of the Weak Law of Large Numbers.* In Example 4 (Sec. 10-1) it was assumed that $\mu = \mu_1$ and $\sigma^2 = \sigma_1^2/n$. Prove it.

9. *Double or Nothing.* If a gambler bets D dollars on the first play of a game then he receives DX_1 dollars at the end of the play, where X_1 is a random variable for which $P[X_1 \geq 0] = 1$. Thus he receives an amount proportional to his bet. If $E(DX_1) = D$ then his expected return equals his bet, so in a sense the game is fair; that is, we will call the game fair iff $EX_1 = 1$. If the gambler starts with \$1 and always bets everything that he has then

$$T_n = X_1X_2 \cdots X_{n-1}X_n$$

is his total capital at the end of n plays. If the plays are independent then

$$ET_n = (EX_1)(EX_2) \cdots (EX_n).$$

Thus if the plays are independent and each play is fair, $EX_i = 1$, then the game as he played it is fair, $ET_n = 1$. If $p_{X_1}(0) + p_{X_1}(2) = 1$ we can interpret X_1 as a play of "double or nothing," since his return at the end of the play is either 0 or twice his bet. In this case $EX_1 = 2p_{X_1}(2)$, so that the play is fair iff $p_{X_1}(0) = p_{X_1}(2) = .5$, as seems reasonable.

Let $X_1, X_2, \ldots, X_n$ be independent and identically distributed by $p_{X_1}(0) = p_{X_1}(2) = .5$. Let $T_n = X_1X_2 \cdots X_n$, so that $ET_n = 1$ and the game is, in a sense, fair. Find the density induced by T_n and comment on the strategy of always betting everything on double or nothing for large n.

10. *Don't Bet Three Times, or, Pairwise Independence Isn't Enough.* Define the random variable $X_i = 2I_{A_i}$ to equal twice the indicator function of the event A_i, where A_1, A_2, A_3 are the events of Example 2 (Sec. 7-3). Abbreviating the joint density p_{X_1,X_2,X_3} to p, we find that

$$p(0,0,0) = p(0,2,2) = p(2,0,2) = p(2,2,0) = .25$$

so

$$p_{X_i}(0) = p_{X_i}(2) = .5 \qquad \text{for } i = 1, 2, 3.$$

Thus X_1, X_2, X_3 are identically distributed. It is easily shown directly that X_1, X_2, X_3 are pairwise independent, but dependent. Find the density and expected value of each of the two products X_1X_2 and $X_1X_2X_3$. Interpret the results in terms of the double-or-nothing game discussed in Exercise 9.

11. *Indicators and Expectations.* If A and B are subsets of Ω observe that $I_\varphi I_A = I_A I_\varphi = I_\varphi$, so I_φ acts like the number 0; $I_\Omega I_A = I_A I_\Omega = I_A$ so I_Ω acts like the number 1; $I_A + I_{A^c} = I_\Omega$; and $(I_A)(I_B) = I_{AB}$ so the indicator of an intersection of sets equals the product of their indicators. We can manipulate indicator functions algebraically, as with any random variables. With indicators we are essentially manipulating sets. By using $EI_A = P(A)$ we obtain results about probabilities of events. Let $A_1, \ldots, A_n$ be any events and let $A = A_1 \cup A_2 \cup \cdots \cup A_n$. Prove that the equation below is true and derive a formula for $P(A_1 \cup A_2 \cup A_3)$ similar to the formula $P(A_1 \cup A_2) = P(A_1) + P(A_2) - P(A_1A_2)$. [Section 17–4 an expansion of this exercise.]

$$I_\Omega - I_A = (I_\Omega - I_{A_1})(I_\Omega - I_{A_2}) \cdots (I_\Omega - I_{A_n})$$

12. *$X \le Y$ Expressed in Terms of Their Joint Density.* Let $X: \Omega \to R$ and $Y: \Omega \to R$ be random variables with the same associated (Ω,P).

(*a*) Sketch the subset $D = \{(x,y): x \le y\}$ of the plane and prove that if $X \le Y$ then $p_{X,Y}$ must be concentrated on D.

(*b*) Prove that if $p_{X,Y}$ is concentrated on D then $P[X \le Y] = 1$.

13. *The Properties of Expectation Are Obvious.* This exercise demonstrates that basic expectation properties are *easily* proved (and insight gained) if convergence difficulties are ignored. Let X and Y be random variables with joint density $p_{X,Y}$ and let $g: R \to R$ and $h: R \to R$. Assume that all needed series are absolutely convergent. Assuming Theorems 1 and 2, exhibit the manipulations whose omission is indicated by ellipses in the following derivations. These derivations are self-contained in the sense that they might well have appeared prior to Theorem 3. Assume that X and Y are independent, except for parts (*a*) and (*e*), where they may be dependent.

(*a*) $$E(aX + bY) = \sum_x\sum_y (ax + by)p_{X,Y}(x,y) = \cdots$$
$$= a\sum_x xp_X(x) + b\sum_y yp_Y(y) = aEX + bEY.$$

(*b*) $$EXY = \sum_x\sum_y xyp_{X,Y}(x,y) = \cdots = \left[\sum_x xp_X(x)\right]\left[\sum_y yp_Y(y)\right] = [EX][EY].$$

(*c*) $$\text{Var}\,(X + Y) = \sum_x\sum_y[(x + y) - (\mu_X + \mu_Y)]^2 p_{X,Y}(x,y) = \cdots$$
$$= \sum_x [x - \mu_X]^2 p_X(x) + \sum_y [y - \mu_Y]^2 p_Y(y) = \text{Var}\,X + \text{Var}\,Y.$$

(*d*) $$Ee^{t(X+Y)} = \sum_x\sum_y e^{t(x+y)}p_{X,Y}(x,y) = \cdots$$
$$= \left[\sum_x e^{tx}p_X(x)\right]\left[\sum_y e^{ty}p_Y(y)\right] = [Ee^{tX}][Ee^{tY}].$$

(*e*) $$E[ag(X) + bh(Y)] = \sum_x\sum_y [ag(x) + bh(y)]p_{X,Y}(x,y) = \cdots$$
$$= a\sum_x g(x)p_X(x) + b\sum_y h(y)p_Y(y) = aEg(X) + bEh(Y).$$

(*f*) $$Eg(X)h(Y) = \sum_x\sum_y g(x)h(y)p_{X,Y}(x,y) = \cdots$$
$$= \left[\sum_x g(x)p_X(x)\right]\left[\sum_y h(y)p_Y(y)\right] = [Eg(X)][Eh(Y)].$$

$$(g)\ \mathrm{Var}\,[g(X)+h(Y)] = \sum_x\sum_y \{[g(x)+h(y)] - [Eg(X)+Eh(Y)]\}^2 p_{X,Y}(x,y) = \cdots$$
$$= \sum_x [g(x) - Eg(X)]^2 p_X(x) + \sum_y [h(y) - Eh(Y)]^2 p_Y(y)$$
$$= \mathrm{Var}\, g(X) + \mathrm{Var}\, h(Y).$$

$$(h)\ Ee^{t[g(X)+h(Y)]} = \sum_x\sum_y e^{t[g(x)+h(y)]} p_{X,Y}(x,y) = \cdots$$
$$= \left[\sum_x e^{tg(x)} p_X(x)\right]\left[\sum_y e^{th(y)} p_Y(y)\right] = [Ee^{tg(X)}][Ee^{th(Y)}].$$

14. *Expectations with Non-Absolutely Convergent Series May Lead to Disaster.* If $h: R \to R$ and $h(x) = 0$ whenever x is not an integer then for this exercise define $\sum_x h(x)$ by

$$\sum_x h(x) = \lim_{N\to\infty} \sum_{n=-N}^{N} h(n)$$

if this limit exists. This is the widely used *Cauchy sum*, which corresponds to performing the summation by using the ordering

$$h(0) + h(-1) + h(1) + h(-2) + h(2) + h(-3) + \cdots.$$

If this series is absolutely convergent we know that every ordering will give us the same sum. However, if the series is not absolutely convergent the particular Cauchy ordering may still assign it a sum. For example, if h is an odd function, $h(-x) = -h(x)$ for all x, then $\sum_x h(x)$ is defined and equals 0, even if the series is not absolutely convergent.

If $p: R \to R$ is a discrete density which is concentrated on the integers then for this exercise we define

$$[\text{the mean of } p] = \sum_x x p_X(x)$$

when this Cauchy sum exists. Naturally we also define EX to equal the mean of p_X, when p_X has a mean. This new definition agrees with our previous definition when the series is absolutely convergent, and it assigns a mean to many densities which did not have one previously. For example, if p_X is an even function, $p_X(-x) = p(x)$ for all x, then p_X has a mean and it equals 0. This is certainly a very nice property. However, part (a) of this exercise shows that this extended definition of EX *may violate linearity of expectation*, $E(X + Y) \neq EX + EY$. Furthermore part (b) shows that we may have

$$\sum_x g(w)p_W(w) \neq Eg(W)$$

so that *we cannot calculate $Eg(W)$ in various natural fashions*, as in Theorem 2.

(a) Define the constant c by

$$c = \frac{1}{4\sum_{n=1}^{\infty} 1/n^2}$$

Define the discrete density $p_{X,Y}: R^2 \to R$, as in Fig. 5, by

$$p_{X,Y}(0,-n) = \frac{2c}{n^2} \quad\text{and}\quad p_{X,Y}(-n,n) = p_{X,Y}(n,n) = \frac{c}{n^2} \quad \text{if } n = 1, 2, \ldots$$
$$p_{X,Y}(x,y) = 0 \qquad \text{otherwise.}$$

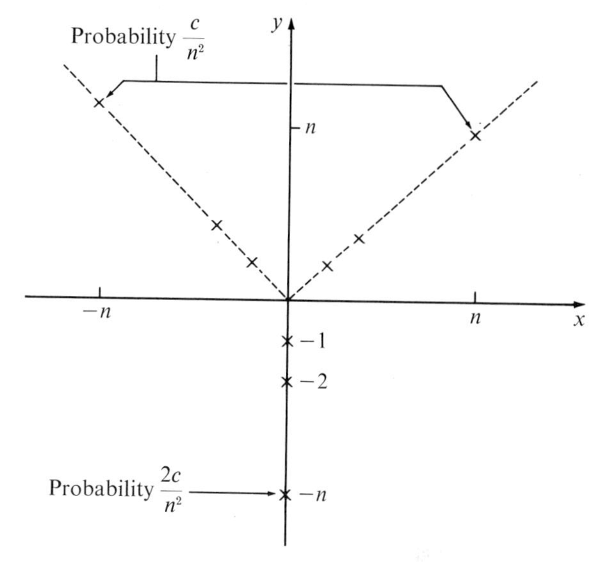

Fig. 5

Observe that p_X and p_Y are even functions, so that $EX = EY = 0$.

Let $Z = X + Y$ and observe that

$$p_Z(0) = \tfrac{1}{4}$$

$$p_Z(-n) = \frac{2c}{n^2} \qquad \text{and} \qquad p_Z(2n) = \frac{c}{n^2} \qquad \text{if } n = 1, 2, \ldots$$

$$p_Z(z) = 0 \qquad \text{otherwise.}$$

Sketch p_Z and observe that

$$\sum_{|z| \le 2N} z p_Z(z) = \sum_{n=1}^{2N} (-n) \frac{2c}{n^2} + \sum_{n=1}^{N} 2n \frac{c}{n^2} = - \sum_{n=N+1}^{2N} \frac{2c}{n} .$$

For each of the N terms in the last sum we have $-2c/n \le -2c/2N$, so that

$$\sum_{|z| \le 2N} z p_Z(z) \le N\left(\frac{-2c}{2N}\right) = -c.$$

Therefore we cannot have $EZ = 0$. Show that $EZ = -2c \log 2$, as a Cauchy sum.

(*b*) Define the discrete density $p_W \colon R \to R$ by

$$p_W(0) = \tfrac{1}{4}$$

$$p_W(-n) = \frac{2c}{n^2} \qquad \text{and} \qquad p_W(n) = \frac{c}{n^2} \qquad \text{if } n = 1, 2, \ldots$$

$$p_W(w) = 0 \qquad \text{otherwise.}$$

Define the function $g \colon R \to R$ by $g(w) = w$ if $w < 0$, and $g(w) = 2w$ if $w \ge 0$. Sketch p_W

and g. Observe that $g(w)p_W(w)$ is an odd function of w, so that $\sum_w g(w)p_W(w) = 0$. Observe that the density induced by $g(W)$ is just the density p_Z of part (*a*), which does not have zero mean.

15. Prove that if $P[X < \beta] = P[Y < \beta]$ for all real β then both random variables induce the same discrete density. Naturally we permit $p_X(x)$ to be positive for *all* rational numbers x, for example.

11-4 CORRELATION AND LINEAR PREDICTION

This section introduces the covariance between two random variables as a quantity which arises naturally in calculating the variance of a sum of dependent random variables. The correlation coefficient is a standardized covariance. It is shown to be interpretable as a standardized measure of the proportionality of the statistical dependence between the two random variables.

If X and Y have finite variances then

$$\begin{aligned}\operatorname{Var}(X+Y) &= E[(X+Y)-(\mu_X+\mu_Y)]^2 = E[(X-\mu_X)+(Y-\mu_Y)]^2 \\ &= E[(X-\mu_X)^2 + 2(X-\mu_X)(Y-\mu_Y) + (Y-\mu_Y)^2] \\ &= \operatorname{Var} X + 2E(X-\mu_X)(Y-\mu_Y) + \operatorname{Var} Y\end{aligned}$$

so we name the center expectation. If X and Y have finite variances then

$$\operatorname{Cov}(X,Y) = E[(X-EX)(Y-EY)]$$

is called the **covariance of,** or between, X **and** Y. It exists by Theorem 3*h* (Sec. 11-3). Clearly $\operatorname{Cov}(X,Y) = \operatorname{Cov}(Y,X)$ and $\operatorname{Cov}(X,X) = \operatorname{Var} X$. Using this notation, we have

$$\operatorname{Var}(X+Y) = \operatorname{Var} X + \operatorname{Var} Y + 2\operatorname{Cov}(X,Y).$$

We say that $X_1, \ldots, X_n$ are **uncorrelated** iff $\operatorname{Cov}(X_i,X_j) = 0$ whenever $i \neq j$.

Another form for the covariance is obtained as follows:

$$\begin{aligned}\operatorname{Cov}(X,Y) &= E[(X-\mu_X)(Y-\mu_Y)] = E[XY - \mu_X Y - \mu_Y X + \mu_X\mu_Y] \\ &= E(XY) - \mu_X\mu_Y - \mu_Y\mu_X + \mu_X\mu_Y = E(XY) - (EX)(EY).\end{aligned}$$

Therefore *if X and Y are independent they are uncorrelated*, since the expectation of their product equals the product of their expectations. As shown in Exercise 3, we *cannot* conclude that X and Y are independent just because the number $\operatorname{Cov}(X,Y)$ happens to equal 0. Independence is equivalent to an equality between two functions, $p_{X,Y} = p_X p_Y$. If $EX = EY = 0$ then

$$\operatorname{Cov}(-X,Y) = E(-X)(Y) = -EXY = -\operatorname{Cov}(X,Y)$$

showing that these two covariances are of opposite sign. Thus $\operatorname{Cov}(X,Y)$ may be positive, negative, or zero.

***Example 1** An urn contains n balls labeled $1, 2, \ldots, n$ and X_i is the label on the ith ball drawn *without* replacement. A real-valued reward function f is given on $\{1, 2, \ldots, n\}$ and we receive reward $f(k)$ if the ball labeled k is drawn. Thus $Y_i = f(X_i)$ is the reward from the ith draw. We seek Cov (Y_i, Y_j), and we will calculate it from p_{X_i,X_j}.

Essentially by definition, $p_{X_1,\ldots,X_n}(x_1, \ldots, x_n) = 1/n!$ for each of the $n!$ permutation $(x_1, \ldots, x_n)$ of the integers $1, 2, \ldots, n$. For the density of X_2 we have $p_{X_2}(k)$ equal to $1/n!$ times the number of permutations $(x_1, k, x_3, \ldots, x_n)$, with the second coordinate equal to k. Thus we remove k from $\{1, \ldots, n\}$ and permute the remaining $n - 1$ integers, so that $p_{X_2}(k) = (n-1)!/n! = 1/n$, as anticipated. Clearly *each* induced density p_{X_i} is uniform on $1, 2, \ldots, n$. Similarly for fixed $s \neq t$ we have $p_{X_1,X_3}(s,t)$ equal to $1/n!$ times the number of permutations $(s, x_2, t, x_4, \ldots, x_n)$, of which there are $(n-2)!$, so $p_{X_1\,X_3}(s,t) = 1/n(n-1)$. Thus in general if $i \neq j$ then $p_{X_i,X_j}(s,s) = 0$ while

$$p_{X_i,X_j}(s,t) = \frac{1}{n(n-1)} \qquad \text{if } s \neq t.$$

From the induced density of X_i we obtain

$$\mu = EY_i = \sum_{k=1}^{n} f(k) p_{X_i}(k) = \frac{1}{n} \sum_{k=1}^{n} f(k)$$

and

$$\sigma^2 = \text{Var } Y_j = \sum_{k=1}^{n} [f(k) - \mu]^2 p_{X_i}(k) = \frac{1}{n} \sum_{k=1}^{n} [f(k) - \mu]^2.$$

Similarly for $i \neq j$ we find that

$$\begin{aligned}
\text{Cov } (Y_i, Y_j) &= E\{[f(X_i) - \mu][f(X_j) - \mu]\} \\
&= \sum_{s=1}^{n} \sum_{t=1}^{n} [f(s) - \mu][f(t) - \mu] p_{X_i,X_j}(s,t) \\
&= \sum_{s=1}^{n} \left\{ [f(s) - \mu] \left(\sum_{t=1}^{n} [f(t) - \mu] p_{X_i,X_j}(s,t) \right) \right\}.
\end{aligned}$$

Let $a_t = [f(t) - \mu]/n(n-1)$, so that

$$\begin{aligned}
\left(\sum_{t=1}^{n} [f(t) - \mu] p_{X_i,X_j}(s,t) \right) &= a_1 + \cdots + a_{s-1} + 0 + a_{s+1} + \cdots + a_n \\
&= \left(\sum_{t=1}^{n} a_t \right) - a_s = \frac{1}{n-1} (\mu - \mu) - a_s = -a_s.
\end{aligned}$$

Hence Cov $(Y_i, Y_j) = -\sigma^2/(n-1)$. Note that if $\sigma^2 > 0$ then the correlation coefficient, as defined below, is independent of the reward function f,

$$\rho_{Y_i,Y_j} = -\frac{1}{n-1} \qquad \text{if } i \neq j. \quad \blacksquare$$

For a sum of $n \geq 2$ random variables we have

$$\begin{aligned}
\left[\left(\sum_{i=1}^{n} X_i \right) - \left(\sum_{i=1}^{n} \mu_i \right) \right]^2 = \left[\sum_{i=1}^{n} (X_i - \mu_i) \right]^2 &= \left[\sum_{i=1}^{n} (X_i - \mu_i) \right] \left[\sum_{j=1}^{n} (X_j - \mu_j) \right] \\
&= \sum_{i=1}^{n} \sum_{j=1}^{n} (X_i - \mu_i)(X_j - \mu_j).
\end{aligned}$$

Taking expectations and using the linearity of expectation yields

$$\operatorname{Var}\left(\sum_{i=1}^{n} X_i\right) = \sum_{i=1}^{n}\sum_{j=1}^{n} \operatorname{Cov}(X_i,X_j) \qquad \text{if each } EX_i^2 < \infty.$$

In other words, the variance of $\sum_{i=1}^{n} X_i$ equals the sum of the n^2 numbers in the square array of Fig. 1. If $X_1, \ldots, X_n$ are independent the terms off the main diagonal are 0,

$\operatorname{Var} X_1$	$\operatorname{Cov}(X_1,X_2)$	$\operatorname{Cov}(X_1,X_3)$	$\ldots$	$\operatorname{Cov}(X_1,X_n)$
$\operatorname{Cov}(X_2,X_1)$	$\operatorname{Var} X_2$	$\operatorname{Cov}(X_2,X_3)$	$\ldots$	$\operatorname{Cov}(X_2,X_n)$
$\operatorname{Cov}(X_3,X_1)$	$\operatorname{Cov}(X_3,X_2)$	$\operatorname{Var} X_3$	$\ldots$	$\operatorname{Cov}(X_3,X_n)$
$\ldots$	$\ldots$	$\ldots$	$\ldots$	$\ldots$
$\operatorname{Cov}(X_n,X_1)$	$\operatorname{Cov}(X_n,X_2)$	$\operatorname{Cov}(X_n,X_3)$	$\ldots$	$\operatorname{Var} X_n$

Fig. 1

so we have again proved Theorem 3*i* (Sec. 11-3). In general by separating the diagonal nonnegative terms and using $\operatorname{Cov}(X_i,X_j) = \operatorname{Cov}(X_j,X_i)$ we obtain *the frequently used form*

$$\operatorname{Var}\left(\sum_{i=1}^{n} X_i\right) = \sum_{k=1}^{n} \operatorname{Var} X_k + 2 \sum_{1\le i<j\le n} \operatorname{Cov}(X_i,X_j)$$

which is valid whenever each $EX_i^2 < \infty$. In other words, $\operatorname{Var}\left(\sum_{i=1}^{n} X_i\right)$ equals the sum of the diagonal terms plus twice the sum of the terms above the diagonal.

There are n terms on the diagonal, so there are $n^2 - n$ terms off the diagonal, of which half, $(n^2 - n)/2 = \binom{n}{2}$, are above the diagonal. An alternate notation for the summation is

$$\sum_{1\le i<j\le n} = \sum_{i=1}^{n-1}\sum_{j=i+1}^{n}$$

***Example 2** Continuing from Example 1, we see that the total reward $S_t = Y_1 + \cdots + Y_t$ from the first t draws $1 \le t \le n$ has $ES_t = t\mu$ and

$$\operatorname{Var} S_t = t\sigma^2 + 2\binom{t}{2}\left(-\frac{\sigma^2}{n-1}\right) = \frac{n-t}{n-1}\,t\sigma^2.$$

In particular $\operatorname{Var} S_1 = \sigma^2$ and $\operatorname{Var} S_n = 0$, as is obvious. For large n the variance of the total reward S_t on a small number t of draws is close to the variance $t\sigma^2$ for draws *with* replacement. ■

If X and Y have finite *nonzero* variances then the covariance between their standardizations is called their **correlation coefficient** and is denoted by

$$\rho_{X,Y} = \operatorname{Cov}(X^*, Y^*) = EX^*Y^* = E\left(\frac{X-\mu_X}{\sigma_X}\right)\left(\frac{Y-\mu_Y}{\sigma_Y}\right) = \frac{\operatorname{Cov}(X,Y)}{\sigma_X\sigma_Y}.$$

Clearly $\rho_{X,Y} = \rho_{Y,X}$ and $\rho_{X,X} = 1$. Often $\rho_{X,Y}$ is abbreviated to ρ. Certainly

$\rho = 0$ iff Cov $(X, Y) = 0$, so X and Y are uncorrelated iff their correlation coefficient equals zero. Theorem 3h (Sec. 11-3) yields $-1 \leq \rho \leq 1$, since

$$\rho^2 = [E(X^* Y^*)]^2 \leqq [E(X^*)^2][E(Y^*)^2] = 1.$$

Exercises 4 to 6 and 10 to 13 may be done before reading the remainder of this section.

LINEAR PREDICTION

The remainder of this section is devoted to linear least-squares prediction of the value of one random variable when we are permitted to observe the value of *one* other random variable [Theorem 1 (Sec. 18-1) permits the observation of more than one other random variable]. In particular we show that ρ can be interpreted as a standardized measure of the degree of proportionality of the statistical dependence between X and Y, and that Cov $(X, Y) = \sigma_X \sigma_Y \rho$ is therefore an unstandardized measure of this.

Let X and Y be random variables having finite nonzero variances. We seek the two real numbers a and b which minimize $E[Y - (aX + b)]^2$. This problem turns out to have a unique solution a_0, b_0. We call the line $f(x) = a_0 x + b_0$ the **least-squares regression line** of, or for, Y on X. The problem can be interpreted as a prediction problem as follows: Suppose, as in Fig. 2, that we decide on some line $y_p = ax + b$ to be used

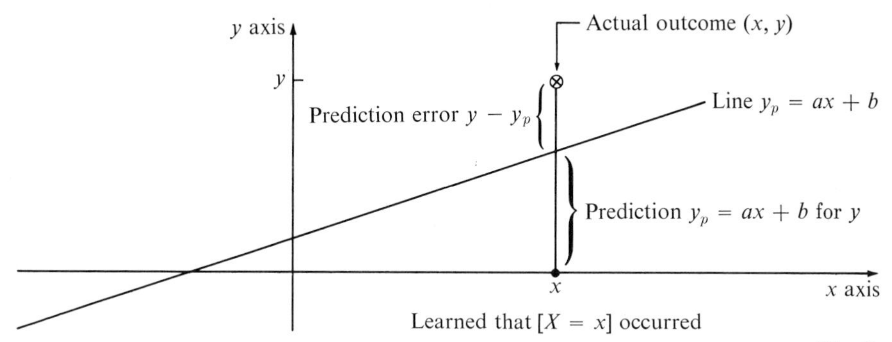

Fig. 2

for predicting Y whenever we learn X; that is, if the outcome of an experiment is the event $[(X, Y) = (x, y)]$ and we learn only that $[X = x]$ occurred, then we will use $y_p = ax + b$ as our prediction of the unknown value of Y. Then our prediction error is $y - y_p = y - (ax + b)$. It is not unreasonable to seek that linear function $aX + b$ of X which yields the minimum expected squared error. General *nonlinear* predictors $f(X)$ for Y, given X, are considered in Sec. 14-1.

We first solve this problem for the standardized random variables X^* and Y^*. Thus we seek the α and β which minimize

$$\begin{aligned} E[Y^* - (\alpha X^* + \beta)]^2 &= E[(Y^* - \alpha X^*) - \beta]^2 \\ &= E(Y^* - \alpha X^*)^2 - 2\beta E(Y^* - \alpha X^*) + \beta^2 \\ &= E[(Y^*)^2 - 2\alpha X^* Y^* + \alpha^2 (X^*)^2] + \beta^2 \\ &= 1 - 2\alpha\rho + \alpha^2 + \beta^2 \end{aligned}$$

where the correlation coefficient ρ is assumed to be known. Clearly we want $\beta = 0$. For the resulting quadratic function in α we set

$$\frac{d}{d\alpha}(1 - 2\alpha\rho + \alpha^2) = -2\rho + 2\alpha = 0$$

so we want $\alpha = \rho$. Therefore $E[Y^* - (\alpha X^* + \beta)]^2$ is minimized by the unique values $\beta = 0$ and $\alpha = \rho$, and this minimum is

$$E(Y^* - \rho X^*)^2 = 1 - \rho^2.$$

Since the left-hand side is nonnegative, we have $\rho^2 \leq 1$, as was shown earlier. Thus the line $y_p^* = \rho x^*$ must lie in the shaded region of Fig. 3. Clearly the observed x^*

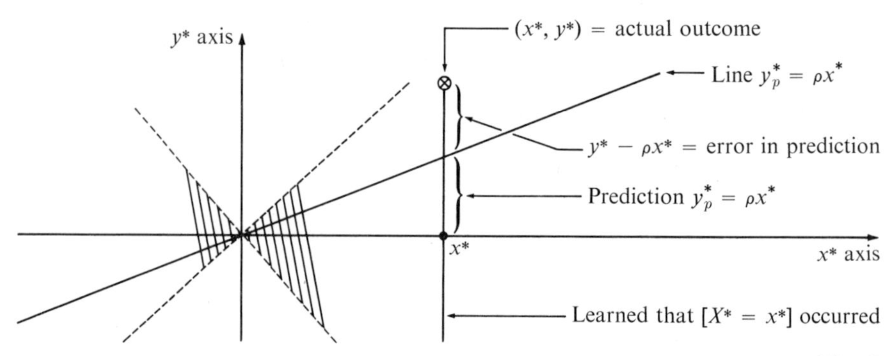

Fig. 3

and predicted $y_p^* = \rho x^*$ have the same sign if $\rho > 0$ and opposite signs if $\rho < 0$. Furthermore the prediction is always closer to the origin than is the observed, $|y_p^*| = (|\rho|)(|x^*|) \leq |x^*|$.

For any random variable W we have $EW^2 = 0$ iff $P[W = 0] = 1$, so that $\rho^2 = 1$ iff $P[Y^* - \rho X^* = 0] = 1$. Thus $\rho = 1$ is equivalent to the two random variables being equal, $Y^* = X^*$, with probability 1; that is, their joint distribution is concentrated on the line of slope 1 through the origin. Similarly $\rho = -1$ is equivalent to one random variable being the negative of the other, $Y^* = -X^*$.

If we wish to predict X^* from Y^* then the α and β which minimize $E[X^* - (\alpha Y^* + \beta)]^2$ are $\beta = 0$ and $\alpha = \rho$, and the resulting minimum expected squared error is $1 - \rho^2$. Thus we have two ways to justify interpreting ρ as a normalized measure of the degree of proportionality or linearity of the statistical dependence between X and Y. In the first place, the least-squares regression line $y_p^* = \rho x^*$ for Y^* on X^* has slope ρ (for y_p^* with respect to x^*); similarly the line $x_p^* = \rho y^*$ for X^* on Y^* has slope ρ (for x_p^* with respect to y^*). In the second place, both of the predictors have expected squared error $1 - \rho^2$ satisfying $0 \leq 1 - \rho^2 \leq 1$. Thus $\rho = 0$ is interpreted as zero linear statistical dependence and $\rho^2 = 1$ corresponds to complete linear statistical dependence. Naturally ρ measures both degree, $|\rho|$, and sign, $\pm$. In Fig. 4 both predictor lines are shown in the same plane. Clearly they coincide iff $\rho = \pm 1$.

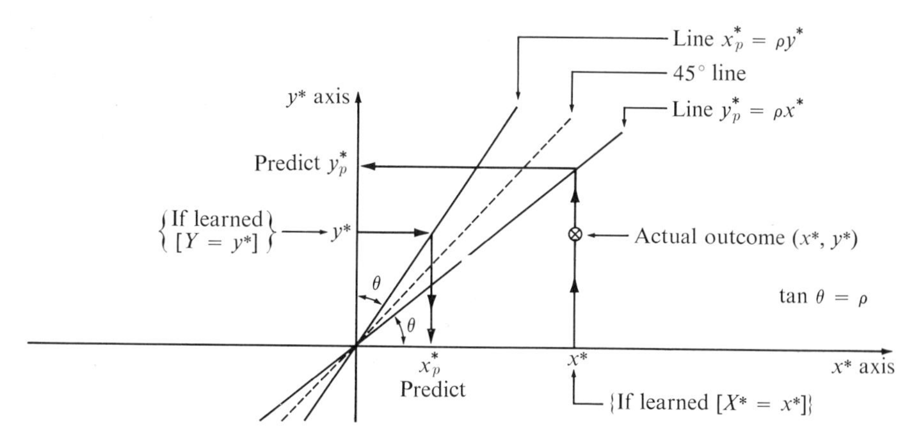

Fig. 4

Now consider the unnormalized random variables X and Y. Clearly

$$Y^* - (\alpha X^* + \beta) = \frac{Y - \mu_Y}{\sigma_Y} - \left(\alpha\frac{X - \mu_X}{\sigma_X} + \beta\right)$$

$$= \frac{1}{\sigma_Y}\left[Y - \left(\mu_Y + \sigma_Y\left\{\alpha\frac{X - \mu_X}{\sigma_X} + \beta\right\}\right)\right]$$

which has the form $(1/\sigma_Y)[Y - (aX + b)]$. By varying α and β we get every possible a and b, so that the minimum of $1 - \rho^2$ for $E[Y^* - (\alpha X^* + \beta)]^2$ over all α and β is the same as the minimum for $E\{[1/\sigma_Y][Y - (aX + b)]\}^2$ over all a and b. Thus the minimum of $E[Y - (aX + b)]^2$ over all a and b is $\sigma_Y^2(1 - \rho^2)$. This is achieved when $\beta = 0$ and $\alpha = \rho$ in the expression above, so that

$$Y_p = \mu_Y + \sigma_Y\rho\frac{X - \mu_X}{\sigma_X}$$

is the unique linear function of X for which we obtain the minimum value

$$E(Y - Y_p)^2 = \sigma_Y^2(1 - \rho^2).$$

Applying this result to the problem of predicting X from Y, we see that

$$X_p = \mu_X + \sigma_X\rho\frac{Y - \mu_Y}{\sigma_Y}$$

is the unique linear function of Y for which we obtain the minimum value

$$E(X - X_p)^2 = \sigma_X^2(1 - \rho^2).$$

Thus the two predictor lines are given by

$$\frac{y_p - \mu_Y}{\sigma_Y} = \rho\frac{x - \mu_X}{\sigma_X} \quad \text{and} \quad \frac{x_p - \mu_X}{\sigma_X} = \rho\frac{y - \mu_Y}{\sigma_Y}.$$

If in the same xy plane we plot the locus (x,y_p) and the locus (x_p,y) then, as in Fig. 5,

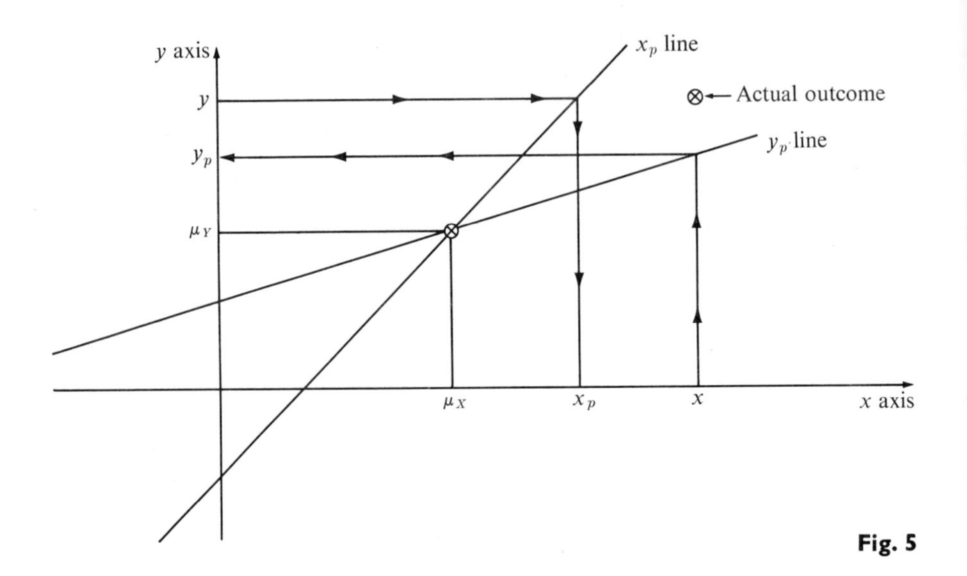

Fig. 5

both prediction lines go through the center of mass (μ_X, μ_Y) of the mass density $p_{X,Y}$. Also, the lines coincide iff $\rho = \pm 1$.

We next note that the random error of the best predictor must have zero expectation and be uncorrelated with the variable used for prediction. This is quite reasonable, since otherwise such correlation could be used to further improve the predictor. In more detail, if

$$Y_p = \mu_Y + \sigma_Y \rho \frac{X - \mu_X}{\sigma_X}$$

then $E(Y - Y_p) = 0$, and also

$$\begin{aligned} \text{Cov}\,[(Y - Y_p), X] &= E(Y - Y_p)(X - \mu_X) \\ &= \sigma_X \sigma_Y E(Y^* - \rho X^*) X^* = \sigma_X \sigma_Y (\rho - \rho) = 0. \end{aligned}$$

Theorem 1 (Sec. 18-1) shows that this property generalizes to predictors which use more than one variable.

EXERCISES

1. For Example 1 sketch in the same plane the least-squares regression line of Y_3 on Y_1, and of Y_1 on Y_3, for the following cases

(*a*) $n = 3$ (*b*) $n = 100$

2. Find and sketch ρ_{X_1,X_2} as a function of the dependence parameter p for the density p_{X_1,X_2} of Fig. 3 (Exercise 3, Sec. 11-3). For $p = .1$ sketch the two least-squares regression lines and the density p_{X_1,X_2}.

3. The joint density of two random variables is

$$p_{X,Y}(\sqrt{2},1) = p_{X,Y}(-\sqrt{2},1) = \tfrac{1}{4} \qquad p_{X,Y}(0,-1) = \tfrac{1}{2}.$$

(*a*) Find the least-squares regression line of Y on X. Are X and Y uncorrelated, or independent? Can you find a better nonlinear predictor $Y_p = f(X)$ of Y in terms of X?

(*b*) Find the least-squares regression line of X on Y. Can you find a better nonlinear predictor $X_p = g(Y)$ of X in terms of Y?

COMMENT: The least-squares regression line may or may not be a reasonably good predictor.

4. Prove that $\rho_{aX+b,cY+d} = \rho_{X,Y}$ if $ac > 0$. Thus the correlation coefficient is invariant to changes of origin and scale, $a > 0$ and $c > 0$.

5. In Fig. 6 "True" means true for all joint distributions [and functions for part (*b*)] and "False" means not true.

If $X_1, \ldots, X_n$ are:	*then $X_1, \ldots, X_n$ must be:* Independent	Pairwise independent	Uncorrelated
Independent	True	True	True
Pairwise independent	False	True	True
Uncorrelated	False	False	True

Fig. 6 $EX_i^2 < \infty$ and $n \geq 3$.

(*a*) Which, if any, of the nine entries must be changed to make the assertion valid?

(*b*) Assume that "then $X_1, \ldots, X_n$" has been replaced by "then $f_1(X_1), \ldots, f_n(X_n)$," where $f_i: R \to R$ and each $f_i(X_i)$ has a finite variance. Which of the nine entries in the resulting figure must be changed in order to make the assertion valid?

(*c*) Prove that if X_i for $i = 1, \ldots, n$ are uncorrelated then so are $Y_i = a_iX_i + b_i$ for $i = 1, \ldots, n$.

(*d*) Prove that if $X_1, \ldots, X_n$ are uncorrelated then

$$\operatorname{Var}\left[\sum_{i=1}^{n} (a_iX_i + b_i)\right] = \sum_{i=1}^{n} a_i^2 \operatorname{Var} X_i.$$

6. Assume that X has a symmetric density $p(x) = p(-x)$ for all real x. To eliminate trivial cases assume that $p_X(x) > 0$ for at least three different x. Also assume that all needed covariances exist.

(*a*) Show that X and $Y = X^2$ are dependent but uncorrelated even though Y is a function of X.

(*b*) Sketch the joint density p_{X,X^2} when $p_X(x) = \tfrac{1}{7}$ for $x = 0, \pm 1, \pm 2, \pm 3$.

(*c*) Show that X and $f(X)$ are uncorrelated if $f: R \to R$ is an even function $f(x) = f(-x)$ for all real x.

7. You learn that $[X = x]$ occurred and use the least-squares regression line of Y on X to predict

$$y_p = \mu_Y + \sigma_Y \rho \frac{x - \mu_X}{\sigma_X}$$

for Y. A friend hears of your prediction y_p and believes that you actually observed the outcome $[Y = y_p]$. He uses the least-squares regression line of X on Y in order to guess that

X had the value

$$x' = \mu_X + \sigma_X \rho \frac{y_p - \mu_Y}{\sigma_Y}.$$

(a) Show that $x' \geq \mu_X$ iff $x \geq \mu_X$, so that his guess x' is on the same side of the mean as your actual observation x.
(b) Assume that $x \neq \mu_X$ and $\rho^2 < 1$. Show that $|x' - \mu_X| < |x - \mu_X|$, so that his guess x' is closer to the mean than your observation x.

8. Let $f(b_0, \ldots, b_n) = E(Y - Y_p)^2 \geq 0$ be the expected squared error for the predictor $Y_p = b_0 + \sum_{i=1}^{n} b_i X_i$ of Y. Expanding $(Y - Y_p)^2$ and using the linearity of expectation leads to a quadratic expression for f in the variables $b_0, \ldots, b_n$ with coefficients such as EYX_i and EX_iX_j. If $b_0^*, \ldots, b_n^*$ is a set of numbers minimizing f then $y_p^* = b_0^* + \sum_{i=1}^{n} b_i^* x_i$ is called a *least-squares* ***linear*** *regression function for* Y *on* $X_1, \ldots, X_n$. Let $Y_p^* = b_0^* + \sum_{i=1}^{n} b_i^* X_i$ and prove that the prediction error is uncorrelated with every one of the random variables on which the prediction is based, that is, that $\text{Cov}\,[(Y - Y_p^*), X_i] = 0$ for $i = 1, 2, \ldots, n$. HINT: For every Z we know that

$$EZ^2 = \text{Var}\, Z + (EZ)^2 \qquad \text{and} \qquad \min_{\alpha,\beta} E[Z - (\alpha X_i + \beta)]^2 = (\text{Var}\, Z)(1 - \rho_{Z,X_i}^2).$$

9. Let $\sigma_{Y|X} = \sigma_Y [1 - \rho^2]^{1/2} \geq 0$ be the square root of the expected squared error of the regression line for Y on X.
(a) Let $\bar{X} = aX + b$ and $\bar{Y} = cY + d$ and find $\sigma_{\bar{Y}|\bar{X}}$ in terms of $\sigma_{Y|X}$.
(b) Interpret the impact of changes of dimension (such as degrees centigrade to degrees Fahrenheit) on this numerical measure of "goodness of prediction."

10. Let X and Y be random variables with finite variances. Prove that

$$|\sigma_X - \sigma_Y| \leq \sigma_{X+Y} \leq \sigma_X + \sigma_Y.$$

Thus the standard deviation of a sum is always between the sum and difference of the standard deviations.

11. *Correlation When There Is a Common Component*
(a) Let U, V, and W be independent and each have a finite positive variance. Find the correlation coefficient between $U + V$ and $W + V$.
(b) Let $X_1, X_2, \ldots$ be independent and each have the same distribution satisfying $0 < \text{Var}\, X_1 < \infty$. Let $n \geq 1$ be fixed and define the "running sums" $T_1, T_2, \ldots$ by $T_k = X_k + X_{k+1} + \cdots + X_{k+n-1}$. For example, X_i might be the yield from the ith trial, so that T_k is the total yield from the n trials starting with the kth; or X_i might be the noise output from a communication channel which is being monitored by observing $T_1, T_2, \ldots$. Find the correlation coefficient between T_k and T_{k+r} for all $k \geq 1$ and $r \geq 0$. Note that if $0 \leq r \leq n$ then T_k and T_{k+r} have an "overlap" of $n - r$ variables in common.

12. *Correlation between Two Events.* Let A and B be two events such that $0 < P(A) < 1$ and $0 < P(B) < 1$. Let $\rho(A,B)$ be the correlation coefficient between their indicator functions I_A and I_B.
(a) Express $\rho(A,B)$ in terms of $P(A)$, $P(B)$, and $P(AB)$. Observe that $\rho(A,B) = 0$ iff A and B are independent. What if $A = B$, or if $A = B^c$? Relate $\rho(A,B) \lesseqgtr 0$ to whether there is greater or less chance for one event to occur if we learn that the other has occurred.

(*b*) A fair die is rolled, and $A = \{1,2,3\}$ and $B =$ [an even face up] $= \{2,4,6\}$. Find $\rho(A,B)$ $\rho(A,B^c)$, $\rho(A^c,B)$, and $\rho(A^c,B^c)$.

(*c*) Show that

$$\rho(A,B) = \frac{P(AB)P(A^cB^c) - P(AB^c)P(A^cB)}{[P(A)P(A^c)P(B)P(B^c)]^{1/2}}.$$

Relate each of $\rho(A,B^c)$, $\rho(A^c,B)$, and $\rho(A^c,B^c)$ to $\rho(A,B)$. Is the correlation between the occurrence of two events the same as the correlation between the nonoccurrence?

13. *Correlation is Unchanged by Independent Additions*

(*a*) The random vector (X_1, Y_1) has a distribution in the plane such that both X_1 and Y_1 have finite positive variances. Let (X_1, Y_1), (X_2, Y_2), . . . , (X_n, Y_n) be independent and identically distributed random vectors. For example, X_m and Y_m might be the pressure and volume from the mth independent repetition of an experiment. Let $S_n = X_1 + \cdots + X_n$ and $T_n = Y_1 + \cdots + Y_n$. Show that

$$\text{Cov}\,(S_n, T_n) = n\,\text{Cov}\,(X_1, Y_1)$$

$$\text{Cov}\left(\frac{S_n - nEX_1}{\sqrt{n}}, \frac{T_n - nEY_1}{\sqrt{n}}\right) = \text{Cov}\,(X_1, Y_1)$$

$$\rho_{aS_n+b,cT_n+d} = \rho_{X_1,Y_1} \qquad \text{if } a > 0,\ c > 0.$$

COMMENT: Section 13-2 shows that each of $(S_n - nEX_1)/\sqrt{n}$ and $(T_n - nEY_1)/\sqrt{n}$ (which have zero means and variances Var X_1 and Var Y_1, respectively) has a normal distribution in the limit as n goes to infinity. Section 18-4 shows that in the limit their joint distribution in the plane approaches a bivariate normal distribution whose covariance equals Cov (X_1, Y_1).

(*b*) In n independent rolls of a fair die let S_n be the number of rolls which yield a 1 or a 2 or a 3 and let T_n be the number of rolls which yield a 2 or a 4 or a 6. Find the correlation between S_n and T_n.

12
The Law of Large Numbers

The term "law of large numbers" refers to theorems like those in this chapter, along with many generalizations. They are most easily understood for sequences of independent random variables, and we will consider them primarily in this context, although there are many extensions to sequences of dependent random variables. The basic theorems are stated as inequalities, since they seem more concrete in this form, which also permits the calculation of numerical bounds. The more typical, weaker asymptotic forms are easily obtained from the inequalities. All the theorems of this chapter, together with their formal proofs, are valid in the most general cases in probability theory.

Sampling experiments are introduced in Sec. 12-2. They are useful in many areas of probability and statistics. In particular they should be helpful in gaining an appreciation for the law of large numbers.

COMMENT ON MEMORIZATION

Theorem 1 (Chebychev's inequality) and Theorem 2 of Sec. 10-1 state that if $\epsilon > 0$ then

$$P[|X - EX| \geq \epsilon] \leq \frac{\text{Var } X}{\epsilon^2}$$

$$P[X \geq EX + \epsilon] \leq \frac{\text{Var } X}{\epsilon^2 + \text{Var } X} \qquad P[X \leq EX - \epsilon] \leq \frac{\text{Var } X}{\epsilon^2 + \text{Var } X}.$$

Each inequality states that the probability is small that X will have a value in a certain set of reals not containing EX. The sets are "*at least distance* ϵ away from EX, or to the right of EX, or to the left of EX." The weak law of large numbers below follows immediately from application of Chebychev's inequality to $(1/n)S_n = (1/n)\sum_{i=1}^{n} X_i$ and from the two facts

$$E\left(\frac{1}{n} S_n\right) = \frac{1}{n}\sum_{i=1}^{n} EX_i,$$

$$\text{Var}\left(\frac{1}{n} S_n\right) = \frac{1}{n^2}\sum_{i=1}^{n} \text{Var } X_i \qquad \text{if } X_1, \ldots, X_n \text{ are independent.}$$

The theorems are explicitly exhibited for reasons of tradition and for convenience of discussion and interpretation. Many readers will probably find it profitable to memorize the three simple inequalities above, which can be used as the basis for the solution of quite a few exercises in this and some later chapters. When needed, the traditional form of the weak law of large numbers can then be quickly rederived from Chebychev's inequality.

12-1 THE WEAK LAW OF LARGE NUMBERS

If $X_1, \ldots, X_n$ are independent and identically distributed and $S_n = X_1 + X_2 + \cdots + X_n$ we know that $ES_n = nEX_1$ and $\operatorname{Var} S_n = n \operatorname{Var} X_1$. Thus the mean of S_n grows as n, while the standard deviation $\sigma_{S_n} = \sqrt{n}\sigma_{X_1}$ grows only as $\sqrt{n}$. But from Chebychev's inequality we know that most of the probability for S_n must be within a few standard deviations of its mean. Thus for large n the probability is large that S_n will take a value relatively near to its mean.

It is traditional to state this kind of result in terms of the arithmetic average $S_n/n = (X_1 + X_2 + \cdots + X_n)/n$ of the random variables. Working with S_n/n, we have

$$E\left(\frac{1}{n} S_n\right) = \frac{1}{n} ES_n = \frac{1}{n} nEX_1 = EX_1$$

$$\operatorname{Var}\left(\frac{1}{n} S_n\right) = \frac{1}{n^2} n \operatorname{Var} X_1 = \frac{\operatorname{Var} X_1}{n}.$$

Thus the expectation of S_n/n is equal to the expectation of one of the X_i and the standard deviation of S_n/n is $\sigma_{X_1}/\sqrt{n}$, which goes to zero as n goes to infinity, so that the distribution of S_n must become more and more concentrated near to the number EX_1. Chebychev's inequality states that

$$1 - P[|X - EX| < \epsilon] = P[|X - EX| \geq \epsilon] \leq \frac{\operatorname{Var} X}{\epsilon^2}$$

for *any* random variable X. Applying this result to the random variable S_n/n, we prove the following **weak law of large numbers** (WLLN), which reduces to Theorem 1 (Sec. 10-2) for Bernoulli trials.

Theorem 1: A WLLN for independent identically distributed random variables Let $X_1, X_2, \ldots, X_n$ be independent identically distributed random variables, with $EX_1^2 < \infty$, and let $S_n = X_1 + X_2 + \cdots + X_n$. If $\epsilon > 0$ then

$$1 - P\left[EX_1 - \epsilon < \frac{1}{n} S_n < EX_1 + \epsilon\right] = P\left[\left|\frac{1}{n} S_n - EX_1\right| \geq \epsilon\right] \leq \frac{\operatorname{Var} X_1}{\epsilon^2 n}. \quad \blacksquare$$

Theorem 1 implies that

$$P\left[EX_1 - \epsilon < \frac{1}{n} S_n < EX_1 + \epsilon\right] \geq 1 - \frac{\operatorname{Var} X_1}{\epsilon^2 n} \to 1 \qquad \text{as } n \to \infty$$

that is, regardless of how small a tolerance interval $EX_1 \pm \epsilon$ we insist on, if n is large enough we can be quite certain that the value of S_n/n will be inside the tolerance interval. Thus if X_i is the reward on the ith spin of a random wheel then we can be quite certain that the arithmetic average reward will be close to the expected reward on one trial if n is large enough.

If we apply Theorem 1 to the case $P[X_1 = -1] = P[X_1 = 1] = .5$ then $EX_1 = 0$ and $\text{Var } X_1 = 1$, so $P[|(1/n)S_n| < \epsilon] \geq 1 - (1/\epsilon^2 n)$. Letting $\epsilon = 2/\sqrt{n}$, we get $P[|S_n| < 2\sqrt{n}] \geq .75$ for all n. Clearly S_n can be as large as $\pm n$, but most of the probability is concentrated in $-2\sqrt{n} < s < 2\sqrt{n}$. Thus when we add independent random variables there is *smoothing* caused by the fact that values on either side of the mean tend to cancel each other, so that although values of S_n differing from ES_n by the order of n may be possible, in actuality the value of S_n should not differ from ES_n by more than the order of $\sqrt{n}$.

Example 1 We found the mean and variance of the density corresponding to the random wheel of Fig. 1 (Sec. 9-2) to be 1.28 and 8.5596, respectively. If X_i is the reward obtained on the ith independent spin of that wheel then from Theorem 1, with $n = 100$ and $\epsilon = 1$, we get

$$P[.28 < (1/100)S_{100} < 2.28] \geq .914.$$

Thus the probability is at least .914 that our total reward on 100 trials will be between 28 and 228. This calculation was based only on the assumptions that $EX_1 = 1.28$, $\text{Var } X_1 = 8.5596$, and $X_1, X_2, \ldots, X_n$ are independent and identically distributed. Therefore the calculation applies to *any* discrete density for X_1 as long as it has this mean and variance. We will see that Theorem 1 and this calculation remain valid in all nondiscrete cases as well. Just as in this example, it will typically be the case that a particular calculation utilizes only some of the properties of the explicit example considered, and so the same calculation applies to any example having the properties utilized. ■

Example 2 If $X_1, X_2, \ldots, X_n$ are independent and identically distributed then so are $f(X_1)$, $f(X_2), \ldots, f(X_n)$, where $f\colon R \to R$. Thus Theorem 1 can be applied to this latter sequence. Let p_{X_1} be as in the preceding example and interpret X_i as the voltage across a resistor at time i, so that $Y_i = (X_i)^2$ is of interest. Neglecting slight numerical approximations, we have $EY_1 = EX_1^2 = 10.2$ and

$$\text{Var } Y_1 = E(Y_1^2) - (EY_1)^2 = E(X_1^4) - (EX_1^2)^2 = 214 - (10.2)^2 = 110.$$

Applying Theorem 1 to $X_1^2, X_2^2, \ldots, X_n^2$ we obtain

$$P\left[\left|\frac{1}{n}(X_1^2 + X_2^2 + \cdots + X_n^2) - 10.2\right| \geq \epsilon\right] \leq \frac{110}{\epsilon^2 n}$$

which in a quantitative form shows that the arithmetic average of the squared voltages should be near to $EX_1^2 = 10.2$ if n is large enough. ■

Example 3 For the production process of Fig. 2 (Sec. 11-1) we let $Z = (3 - Y_1)(2X_1 - 9)^2$ and exhibited p_Z in Fig. 3 (Sec. 11-1). In this case we can easily compute

$$EZ = 4.8 \qquad EZ^2 = 67.1 \qquad \text{Var } Z = 67.1 - 23.04 = 44.06.$$

Thus if Z_i is this random variable for the ith independent trial of the whole production process then

$$P\left[\left|\frac{1}{n}(Z_1 + Z_2 + \cdots + Z_n) - 4.8\right| \geq \epsilon\right] \leq \frac{44.06}{\epsilon^2 n}. \quad \blacksquare$$

The proof of Theorem 1 extends to independent random variables which may not have the same distribution. Theorem 2 below reduces to Theorem 1 in the identically distributed case.

Theorem 2: A WLLN for independent random variables Let $X_1, X_2, \ldots, X_n$ be independent random variables, with all $EX_i^2 < \infty$, and let $S_n = X_1 + X_2 + \cdots + X_n$ and $m_n = E[(1/n)S_n] = (1/n)(EX_1 + EX_2 + \cdots + EX_n)$. If $\epsilon > 0$ then

$$P\left[\left|\frac{1}{n}S_n - m_n\right| \geq \epsilon\right] \leq \frac{1}{\epsilon^2 n^2}(\text{Var } X_1 + \text{Var } X_2 + \cdots + \text{Var } X_n). \quad \blacksquare$$

Proof Apply Chebychev's inequality to $(1/n)S_n$ and use the fact that

$$\text{Var}\left(\frac{1}{n}S_n\right) = \frac{1}{n^2}\text{Var } S_n = \frac{1}{n^2}\sum_{i=1}^{n}\text{Var } X_i. \quad \blacksquare$$

The most important case of Theorem 2 is when the variances are all bounded, $\text{Var } X_i \leq A$, by a constant which does not depend on n, so that

$$P\left[\left|\frac{1}{n}S_n - m_n\right| \geq \epsilon\right] \leq \frac{A}{\epsilon^2 n} \to 0 \qquad \text{as } n \to \infty.$$

Example 4 Let X_k be the sales on the kth operating day of a store. Assume that $X_1, X_2, \ldots, X_n$ are independent, but because of weekly and seasonal variations, they have different distributions. Assume that for large n, $m_n = \$1{,}000$ is a satisfactory approximation for the arithmetic average of the expected daily sales. Past experience indicates that every standard deviation is bounded by \$600, so that $\text{Var } X_i \leq (600)^2 = A$. Then

$$P\left[\left|\frac{1}{n}S_n - 1{,}000\right| \geq \epsilon\right] \leq \frac{(600)^2}{\epsilon^2 n}.$$

For $n = 2{,}000$ and $\epsilon = 60$ we find that

$$P\left[940 < \frac{1}{n}S_n < 1{,}060\right] \geq .95. \quad \blacksquare$$

PRODUCTS OF INDEPENDENT RANDOM VARIABLES

We now consider *products of random variables*, as in Exercise 3 (Sec. 10-2). Let $Y_1, Y_2, \ldots, Y_n$ be independent and identically distributed with

$$\{y_1 : p_{Y_1}(y_1) > 0\} = \{\alpha_1, \alpha_2, \ldots, \alpha_k\} \qquad \text{where each } \alpha_i > 0$$

and let $T_n = Y_1 Y_2 \cdots Y_n$ equal their product. We can, as in Exercise 9 (Sec. 11-3), interpret T_n as the capital at time n of a gambler who starts with \$1 at time 0 and always bets his total capital in a game where his capital at time $n - 1$ is multiplied by the *random factor* Y_n to yield his capital $T_n = Y_n T_{n-1}$ at time n. Similar problems arise when plant growth, or animal population growth, is proportional to present size.

Because of independence

$$ET_n = (EY_1)(EY_2) \cdots (EY_n) = (EY_1)^n.$$

Now the WLLN shows that for a *sum* $S_n = Y_1 + Y_2 + \cdots + Y_n$ the dominant behavior of S_n is that it should grow with n like $ES_n = nEY_1$. A first guess for a product is that T_n should grow with n like $ET_n = (EY_1)^n$. However, the following simple argument shows this guess to be false. If the experiment yielded the value y_i for Y_i then the value of the product T_n would be $y_1y_2 \cdots y_n$. For example, $y_1y_2 \cdots y_n$ might equal $\alpha_3\alpha_1\alpha_1\alpha_2 \cdots \alpha_1$. Now if exactly n_i of these factors are equal to α_i then

$$\begin{aligned} y_1y_2 \cdots y_n &= (\alpha_1)^{n_1}(\alpha_2)^{n_2} \cdots (\alpha_k)^{n_k} = [(\alpha_1)^{n_1/n}(\alpha_2)^{n_2/n} \cdots (\alpha_k)^{n_k/n}]^n \\ &\doteq [(\alpha_1)^{p_{Y_1}(\alpha_1)}(\alpha_2)^{p_{Y_1}(\alpha_2)} \cdots (\alpha_k)^{p_{Y_1}(\alpha_k)}]^n \\ &= \{\exp [p_{Y_1}(\alpha_1) \log \alpha_1 + p_{Y_1}(\alpha_2) \log \alpha_2 + \cdots + p_{Y_1}(\alpha_k) \log \alpha_k]\}^n \\ &= (e^{E \log Y_1})^n \end{aligned}$$

where for the approximate equality, $\doteq$, we assumed that the relative frequency n_i/n of occurrence of α_i was close to $p_{Y_1}(\alpha_i)$, and the next equality used the identity $\alpha_i = e^{\log \alpha_i}$. Therefore T_n should grow exponentially with n like $(e^{E \log Y_1})^n$ rather than like $(EY_1)^n$. Thus T_n should grow (or decrease) exponentially if $E \log Y_1$ is positive (or negative).

Note that we assumed that each α_i is positive so that the product would be relatively stable; if we could obtain a factor $\alpha_1 = -1$ on the nth trial it would change a large positive product into a large negative product. Also the assumption $P[Y_1 > 0] = 1$ permitted us to define the random variable $\log Y_1$, since the real-valued log function is defined only on the positive axis. Naturally we could have used any base $b > 1$ for the logarithm, $\alpha_i = b^{\log_b \alpha_i}$, yielding the number $b^{E \log_b Y_1}$, which does not vary with b.

We easily make our preceding rough-relative-frequency argument precise in the following theorem.

Theorem 3 Let $Y_1, Y_2, \ldots, Y_n$ be independent and identically distributed random variables with $P[Y_1 > 0] = 1$ and $E(\log_b Y_1)^2 < \infty$, where $b > 1$ is any base. If $\epsilon > 0$ then

$$P[b^{n(E \log_b Y_1 - \epsilon)} < Y_1Y_2 \cdots Y_n < b^{n(E \log_b Y_1 + \epsilon)}] \geq 1 - \frac{\text{Var}(\log_b Y_1)}{n\epsilon^2}. \quad \blacksquare$$

Proof If we take logarithms we can write this event as

$$[n(E \log_b Y_1 - \epsilon) < \sum_{i=1}^{n} \log_b Y_i < n(E \log_b Y_1 + \epsilon)]$$

and then apply Theorem 1 to the random variables $X_1, X_2, \ldots, X_n$ where $X_i = \log_b Y_i$. $\blacksquare$

Example 5 We apply Theorem 3 to the case

$$p_{Y_1}(.5) = .3 \qquad p_{Y_1}(1) = .2 \qquad p_{Y_1}(2) = .5.$$

Letting $X_i = \log_{10} Y_i$ we get

$$EX_1 = .0602 \qquad EX_1^2 = .0725 \qquad \text{Var } X_1 = .0689$$

so that $T_n = Y_1 Y_2 \cdots Y_n$ should be near to $10^{.0602n}$. That is, for large enough n we can be quite certain that T_n will take a value in the interval $10^{(.0602 \pm \epsilon)n}$. For $n = 100$ we get $10^{.0602n} = 10^{6.02}$. As our argument preceding Theorem 3 showed, this just amounts to guessing that about 30, 20, and 50 of the $n = 100$ factors will be .5, 1, and 2, respectively, so that the product will be $(\frac{1}{2})^{30}(1)^{20}(2)^{50} = 2^{20} = 10^{6.02}$.

To show that the probability is large that the capital at the end of $n = 100$ trials should have increased by at least 85 percent, define ϵ by

$$10^{(100)(.0602-\epsilon)} = 1.85.$$

We find that $\epsilon = .0575$, so that from Theorem 3

$$P[1.85 < Y_1 Y_2 \cdots Y_{100}] \geq 1 - \frac{.0689}{100\epsilon^2} = .792.$$

The bound .792 is probably much smaller than the number $P[1.85 < T_{100}]$. Theorem 3 shows for this example that T_n should grow like

$$(e^{E \log Y_1})^n = (10^{.0602})^n = (1.15)^n$$

rather than like $(EY_1)^n = (1.35)^n$. ■

We now show that if $E\,|Y| < \infty$ and $P[Y > 0] = 1$, then $e^{E \log Y} \leq EY$, so that by Theorem 3 the true growth $e^{nE \log Y_1}$ of the product $T_n = Y_1 Y_2 \cdots Y_n$ is always less than the wrong guess $(EY_1)^n$. From inequality (8) of Appendix 1 we have $\log x \leq x - 1$ for $x > 0$. Replacing x by Y/EY and taking expectations, we get

$$E\left[\log\left(\frac{Y}{EY}\right)\right] \leq E\left(\frac{Y}{EY}\right) - 1 = 0$$

so that $E[\log Y - \log EY] \leq 0$. Thus $E \log Y \leq \log EY$, which is the desired inequality. Note that if f is the log function we have $Ef(X) \leq f(EX)$. *A common mistake to be avoided* is the assumption that the expectation of every function of a random variable is equal to the value of the function evaluated at the expectation, $Ef(X) = f(EX)$.

EXERCISES

1. *100 Days of Production* (*Fig. 2, Sec. 11-1*). The random vector $f = (X_1, Y_1, X_2, Y_2, X_3, Y_3)$ of Fig. 2 (Sec. 11-1) describes the process for 1 day. Assume that $f_1, f_2, \ldots, f_n$ are independent and each has the distribution of f. Introducing superscript (n), we interpret

$$f_n = (X_1^{(n)}, Y_1^{(n)}, X_2^{(n)}, Y_2^{(n)}, X_3^{(n)}, Y_3^{(n)})$$

as describing the process on the nth day of operation.

(*a*) Let $S_n = \sum_{k=1}^{n} V_k$, where $V_k = (X_1^{(k)} - 3)^2$, and show that $P[.9 < (\frac{1}{100})S_{100} < 2.5]$ is large.[1]

[1] We often say "show that $P(A)$ is large" rather than "show that $P(A)$ is near 1" or "find a nontrivial lower bound for $P(A)$."

(*b*) A certain large unit in the plant corrodes at a rate proportional to the pressure so that the total corrosion is proportional to

$$T_n = Y_1^{(1)} + Y_2^{(1)} + Y_3^{(1)} + Y_1^{(2)} + Y_2^{(2)} + Y_3^{(2)} + \cdots + Y_1^{(n)} + Y_2^{(n)} + Y_3^{(n)}.$$

This large unit must be replaced at the end of $n = 100$ days iff $[T_{100} \geq 600]$ occurs. Show that $P[T_{100} \geq 600]$ is small. CAUTION: $Y_1^{(1)}$ and $Y_2^{(1)}$ are dependent.

(*c*) A certain small unit must be replaced at the end of the kth day iff $[Y_1^{(k)} + Y_2^{(k)} + Y_3^{(k)} \geq 6.2]$ occurs. Show that the probability is small that more than 25 such units will be replaced during $n = 100$ days.

2. *Roundoff Error.* Let $X_1, \ldots, X_n$ be independent, with each $P[-.5 \leq X_k \leq .5] = 1$. We interpret X_k as the roundoff error of the kth of n numbers which are added. Thus $(1/n)S_n = (1/n)(X_1 + X_2 + \cdots + X_n)$ is the arithmetic average roundoff error in the sum. Clearly $P[-.5 \leq (1/n)S_n \leq .5] = 1$. Let $m_n = E(1/n)S_n$ and let $n = 10^5$.

(*a*) Assume that $m_n = 0$ and show that $P[|(1/n)S_n| < .025]$ is large. Note that $P[-c \leq X \leq c] = 1$ always implies that $EX^2 \leq c^2$, and hence that $\text{Var } X \leq c^2$. Thus, although S_n might be as large as $\pm 50{,}000$, the positive and negative roundoff errors tend to cancel each other, so S_n is quite certain to take a value in the interval $-2{,}500 < x < 2{,}500$.

(*b*) Assume that $m_n = .05$ and show that $P[|(1/n)S_n| < .025]$ is small.

Thus if there is independence, then we can apply the WLLN to the average roundoff error. However, if the errors have a bias, say each $EX_k = .05$, then the average roundoff error will typically have this bias.

3. *The Smoothing Effect of Averaging Can Overcome Gradually Increasing Fluctuations.* Let $Y_1, Y_2, \ldots, Y_n$ be independent and identically distributed, with $EY_1 = 0$ and $\text{Var } Y_1 = 1$. Interpret $X_k = \mu + Y_k(k)^{1/4}$ as the yield of a process on the kth day. Thus the expected yield $EX_k = \mu$ is the same each day, and the distribution of the yield about its mean $X_k - \mu = Y_k(k)^{1/4}$ is always the same, except that it is spreading out as the fourth root of the number of days. Let $S_n = \sum_{k=1}^{n} X_k$ be the total yield and show that

$$P\left[\left|\frac{1}{n} S_n - \mu\right| \geq \epsilon\right] \to 0 \qquad \text{as } n \to \infty \qquad \text{for any } \epsilon > 0.$$

HINT: $\sum_{k=1}^{n} a_k \leq nb$ if b is the largest of $a_1, a_2, \ldots, a_n$.

4. *A Ruinously Deceptive Investment.* Let $T_n = Y_1 Y_2 \cdots Y_n$, where $Y_1, Y_2, \ldots, Y_n$ are independent and identically distributed, with $p_{Y_1}(.1) = .6$ and $p_{Y_1}(1) = p_{Y_1}(10) = .2$. Thus T_n can be interpreted as the capital at time n, given that all presently available capital is always immediately invested. As in Theorem 3 we have a product of random variables.

(*a*) Find ET_n and $\text{Var } T_n$. Note that both go to infinity as n goes to infinity.

(*b*) Show that $P[T_n < 10^{-.2n}]$ converges to 1 as n goes to infinity. In particular show that $P[T_{100} < 10^{-20}]$ is large.

(*c*) Now assume that we have unlimited capital, but we only invest one unit of capital at the kth opportunity, and Y_k is the total return from this investment, so that $Y_k - 1$ is the net gain. Thus $S_n - n$ is the net gain after n investments, where $S_n = Y_1 + Y_2 + \cdots + Y_n$. Show that $P[(S_n - n) > n]$ converges to 1 as n goes to infinity. Thus by proper strategy we can take advantage of $EY_1 > 1$.

5. *A Generalization of Theorems 1 and 2.* For the three conditions

$$X_1, X_2, \ldots, X_n \text{ are independent} \tag{1}$$

$$X_1, X_2, \ldots, X_n \text{ are pairwise independent} \tag{2}$$

$$X_1, X_2, \ldots, X_n \text{ are uncorrelated} \tag{3}$$

condition (1) implies condition (2) which implies condition (3), so that the third is the weakest condition. That is, all entries in Fig. 6 (Exercise 5, Sec. 11-4) are correct.

(*a*) Prove that Theorems 1 and 2 apply to uncorrelated random variables.

(*b*) If $X_1, X_2, \ldots, X_n$ are pairwise independent then so are $f(X_1), f(X_2), \ldots, f(X_n)$, where $f: R \to R$, so that Theorems 1 and 2 apply to functions of pairwise-independent random variables. Construct an example of a joint density $p_{X,Y}$ and a function $f: R \to R$ for which $\text{Cov}(X, Y) = 0$ but $\text{Cov}(f(X), f(Y)) \neq 0$. Thus condition (3) can be destroyed by taking functions.

6. *Electron Impacts.* A trial of an experiment consists of shooting an electron at a target located at the origin of the xy plane. The discrete density for the impact point (X, Y) is shown in Fig. 1. Let D_i be the distance of the impact point from the target on the ith of a large

(x,y)	$p_{X,Y}(x,y)$
(0,0)	.1
(0,1)	.4
(1,2)	.3
(−2,1)	.2

Fig. 1

number of independent trials and let $\bar{D}_n = (1/n)(D_1 + \cdots + D_n)$ be the arithmetic average of the first n of these distances. Show that $P[\bar{D}_{100} < 1.7]$ is large.

7. *A Sum of Products.* The kth cube of metal has sides X_k, Y_k, Z_k and volume $V_k = X_k Y_k Z_k$ where $X_1, Y_1, Z_1, X_2, Y_2, Z_2, \ldots$ are all independent and identically distributed, with

$$p_{X_1}(.9) = .1 \qquad p_{X_1}(1) = .8 \qquad p_{X_1}(1.1) = .1.$$

Let $T_n = V_1 + \cdots + V_n$ be the total volume of the first n cubes. Show that $P[992 < T_{1,000} < 1{,}008]$ is large.

12-2 SAMPLING EXPERIMENTS

In this section the WLLN for Bernoulli trials is illustrated by a sampling experiment. Sampling experiments (the Monte Carlo method) are valuable in many fields. The ideas of the present section should prove particularly helpful in understanding the next section. "Sampling experiment" is a vague phrase meaning a "random" experiment which is used to simulate or gain knowledge about some fact or other experiment. Simulating radioactive emissions by coin tosses is a sampling experiment. Tossing a coin can become tedious, however. Also, the coin and method of tossing may not be "fair." An alternative is to try just to write down numbers, say the decimal digits 0, 1, . . . , 9 in a random manner. It has been found that people are not very good at this. For example, one person may include much too many 7's in his list, while another may have 5 followed by 3 much more often than theory predicts. It has been found convenient to have tables of numbers which are supposed to be random. These have been checked for many obvious forms of bias. A table of random sampling numbers, say the digits 0 to 9, is supposed to be a table of what might typically occur if we repeatedly selected these digits independently and each with probability $1/10$. A useful precise definition of a random sequence of digits of finite length does not exist.

Table 1 shows some random numbers from "Random Sampling Numbers."[1] The foreword to this work states in part:

> . . . Practical experiment has demonstrated that it is impossible to mix the balls or shuffle the tickets between each draw adequately. Even if marbles be replaced by the more manageable beads of commerce their very differences in size and weight are found to tell on the sampling. The dice of commerce are always loaded, however imperceptibly. The records of whist, even those of long experienced players, show how the shuffling is far from perfect, and to get theoretically correct whist returns we must deal the cards without playing them. In short, tickets and cards, balls and beads fail in large scale random sampling tests; it is as difficult to get artificially true random samples as it is to sample effectively a cargo of coal or of barley.
>
> In order to get over the difficulty of random sampling for experimental purposes in the Biometric Laboratory its Director suggested to Mr. L. H. C. Tippett, when he was struggling with "ticket" sampling, that he should replace the whole system of tickets by a single random system of numbers ranging from 0000 to 9999. These numbers, if truly random, could be used in a very great variety of ways for artificial sampling . . . .
>
> Of course the value of the present series of numbers depends on their being truly a random selection. Any such series may be a random sample, and yet a very improbable sample. We have no reason whatever to believe that Tippett's numbers are such, they have conformed to the mathematical expectation in a variety of cases, and we would suggest to the user who finds a discordance between the results provided by the table and the theory of sampling he has adopted, first to investigate whether his theory is really sound, and if he be certain that it is, only then to question the randomness of the numbers. ■

We wish to simulate, by using a random-number table, $n = 400$ Bernoulli trials with $p = .3$. Let us take a first block of 20 random digits from the table

2 9 5 2
4 1 6 7
2 7 3 0
0 5 6 0
2 7 5 4

and arrange them in a row:

2 9 5 2 4 1 6 7 2 7 3 0 0 5 6 0 2 7 5 4

Now replace any 0, 1, or 2 by a 1 and any 3, 4, 5, 6, 7, 8, 9 by a 0. We get

1 0 0 1 0 1 0 0 1 0 0 1 1 0 0 1 1 0 0 0.

The chance that a digit would be less than or equal to 2 in the original sequence was supposed to be .3, so that our new sequence is a simulation of 20 Bernoulli trials with

[1] L. H. C. Tippett, "Random Sampling Numbers," p. 1 (with a foreword by E. S. Pearson), Cambridge University Press, New York, 1950.

Table 1 1,600 random sampling numbers

1 2 3 4	5 6 7 8	9 10 11 12	13 14 15 16	17 18 19 20	21 22 23 24	25 26 27 28	29 30 31 32
2 9 5 2	6 6 4 1	3 9 9 2	9 7 9 2	7 9 7 9	5 9 1 1	3 1 7 0	5 6 2 4
4 1 6 7	9 5 2 4	1 5 4 5	1 3 9 6	7 2 0 3	5 3 5 6	1 3 0 0	2 6 9 3
2 7 3 0	7 4 8 3	3 4 0 8	2 7 6 2	3 5 6 3	1 0 8 9	6 9 1 3	7 6 9 1
0 5 6 0	5 2 4 6	1 1 1 2	6 1 0 7	6 0 0 8	8 1 2 6	4 2 3 3	8 7 7 6
2 7 5 4	9 1 4 3	1 4 0 5	9 0 2 5	7 0 0 2	6 1 1 1	8 8 1 6	6 4 4 6
5 8 7 0	2 8 5 9	4 9 8 8	1 6 5 8	2 9 2 2	6 1 6 6	6 0 6 9	2 7 6 3
9 2 6 3	2 4 6 6	3 3 9 8	5 4 4 0	8 7 3 8	6 0 2 8	5 0 4 8	2 6 8 3
2 0 0 2	7 8 4 0	1 6 9 0	7 5 0 5	0 4 2 3	8 4 3 0	8 7 5 9	7 1 0 8
9 5 6 8	2 8 3 5	9 4 2 7	3 6 6 8	2 5 9 6	8 8 2 0	1 9 5 5	6 5 1 5
8 2 4 3	1 5 7 9	1 9 3 0	5 0 2 6	3 4 2 6	7 0 8 8	3 9 9 1	7 1 5 1
5 6 6 7	3 5 1 3	9 2 7 0	6 2 9 8	6 3 9 6	7 3 0 6	7 8 9 8	7 8 4 2
1 0 1 8	6 8 9 1	1 2 1 2	6 5 6 3	2 2 0 1	5 0 1 3	0 7 3 0	2 4 0 5
6 8 4 1	5 1 1 1	5 6 8 8	3 7 7 7	7 3 5 4	3 4 3 4	8 3 3 6	6 4 2 4
2 0 4 1	2 2 0 7	4 8 8 9	7 3 4 6	2 8 6 5	1 5 5 0	5 9 6 0	5 4 7 9
5 5 6 5	4 7 6 4	2 6 1 7	5 2 8 1	1 8 7 0	6 4 9 7	5 7 4 4	9 5 7 6
4 5 0 8	1 8 0 8	3 2 8 9	3 9 9 3	9 4 8 5	4 2 4 0	2 8 3 5	9 9 5 5
2 1 5 2	6 4 7 3	5 6 9 2	9 3 0 9	7 6 6 1	1 6 6 8	5 4 3 1	7 6 5 8
6 9 1 7	4 1 1 3	7 3 4 0	6 8 5 3	1 1 7 2	7 2 2 9	1 2 7 9	5 0 8 5
8 2 4 1	4 1 2 4	4 1 3 1	9 5 0 0	5 6 5 7	3 9 3 2	5 9 4 2	3 3 1 7
7 9 1 3	3 7 0 9	5 9 4 4	9 7 6 3	2 7 5 5	4 2 1 1	4 9 9 5	8 6 5 7
9 3 8 5	7 1 2 5	3 2 3 0	0 7 3 7	2 9 5 7	1 0 1 3	6 3 6 9	4 4 9 4
3 4 3 6	6 2 9 3	6 0 2 5	9 3 8 4	3 3 4 3	1 0 7 1	1 4 6 8	4 8 0 1
9 0 9 4	1 6 3 4	5 0 7 0	0 6 6 4	6 5 1 0	0 9 1 8	4 6 0 1	4 2 9 4
9 2 2 6	9 2 9 6	2 7 9 6	7 0 9 7	4 0 5 7	2 0 7 4	6 2 9 7	2 5 8 7
7 7 8 1	3 7 6 0	2 8 9 5	7 6 5 3	0 0 9 1	7 0 1 2	1 3 0 8	1 9 4 6
9 7 4 2	9 6 9 4	7 3 4 7	0 0 1 7	9 5 7 2	1 8 5 0	0 1 1 6	1 8 9 9
9 4 2 0	9 2 1 0	8 7 8 7	9 3 7 5	4 6 6 3	0 3 9 6	6 7 1 7	5 5 6 2
1 1 7 9	3 5 7 1	5 9 9 2	3 0 5 9	9 0 1 5	5 6 0 8	2 3 4 8	8 1 4 4
0 7 0 8	4 0 1 1	4 0 5 7	1 5 5 0	1 6 7 4	1 3 7 6	5 2 4 3	4 4 2 7
6 3 5 0	3 9 9 6	3 7 9 5	2 1 7 6	8 1 8 2	4 5 1 4	6 3 4 9	3 4 8 3
1 4 1 4	7 1 5 2	3 6 5 8	1 6 3 6	0 6 3 8	3 4 4 3	4 4 4 0	3 0 8 6
7 0 4 1	8 9 8 5	7 0 1 1	5 6 7 6	7 5 7 0	6 6 8 5	1 7 7 6	3 1 5 4
3 2 4 3	2 7 8 3	0 8 4 0	9 0 5 4	8 8 6 2	5 1 7 3	8 4 3 3	9 1 1 7
7 9 2 2	4 9 3 1	5 7 5 3	6 1 6 0	6 5 6 6	8 6 0 2	3 4 2 3	9 0 7 4
8 7 6 9	3 5 1 3	8 9 7 6	0 7 8 0	6 3 8 2	0 0 2 9	2 6 1 9	5 9 8 2
2 5 1 0	7 2 7 4	8 7 4 3	0 0 0 0	1 8 5 0	2 4 0 8	3 6 0 2	5 1 7 9
0 2 2 4	2 4 0 4	9 8 1 1	6 6 4 1	9 7 3 2	1 6 6 2	9 1 5 8	1 4 0 4
3 0 0 9	8 5 1 6	7 2 4 5	9 4 0 9	2 8 4 4	0 7 1 7	1 0 7 2	3 1 3 7
7 4 8 9	0 2 2 1	7 9 2 1	2 3 5 1	2 6 9 6	4 9 0 6	2 4 8 4	3 8 6 8
5 1 8 8	1 8 2 5	2 2 2 0	9 3 8 2	0 5 3 2	1 9 1 5	1 7 9 0	2 0 8 1
1 1 9 8	2 5 4 5	2 4 8 2	9 6 0 7	0 0 6 7	3 7 4 4	9 8 6 6	5 0 9 6
3 9 0 8	4 6 7 6	7 8 1 6	6 5 1 7	9 1 2 1	3 1 7 1	4 1 1 9	3 6 1 5
1 0 9 4	2 2 2 3	1 6 7 5	2 2 8 2	3 7 1 2	8 1 9 1	1 3 3 0	1 4 5 4
1 8 1 7	7 7 2 3	5 5 8 2	7 1 5 3	9 5 1 8	0 2 3 1	7 7 8 2	5 7 4 2
6 2 0 8	9 5 9 8	9 6 2 3	2 1 1 4	7 7 4 7	2 0 9 6	5 0 2 7	0 5 6 1
4 7 5 2	4 5 1 9	2 7 4 9	8 0 2 0	4 6 4 2	1 1 9 0	7 3 0 2	8 3 5 0
0 4 8 6	6 9 9 3	3 1 1 5	5 0 2 5	4 8 8 7	1 5 7 1	9 8 1 9	6 8 0 4
4 9 4 2	3 0 0 4	1 4 4 2	2 8 1 0	1 4 7 9	0 9 7 0	7 3 0 2	3 7 7 5
4 9 3 0	9 7 8 5	7 4 6 0	3 9 9 6	2 8 6 4	0 5 5 9	3 9 8 5	8 0 9 2
2 3 4 9	1 5 9 4	7 1 5 2	0 2 5 7	4 0 4 1	4 1 0 5	3 1 8 0	9 8 0 6

probability .3 for a 1. Eight 1's occurred on the first 20 trials. Looking at the next block of 20 random digits,

6 6 4 1
9 5 2 4
7 4 8 3
5 2 4 6
9 1 4 3

we see that four 1's occur in these 20 trials. Figures 1 and 2 summarize some aspects

Sampling experiment 1			*Sampling experiment 2*		
k	$\sum_{i=1}^{k} \alpha_i$	$\frac{1}{k}\sum_{i=1}^{k} \alpha_i$	k	$\sum_{i=1}^{k} \beta_i$	$\frac{1}{k}\sum_{i=1}^{k} \beta_i$
20	8	.400	20	7	.350
40	12	.300	40	12	.300
60	21	.350	60	15	.250
80	29	.363	80	19	.238
100	36	.360	100	27	.270
120	45	.375	120	34	.283
140	53	.378	140	39	.278
160	56	.350	160	42	.262
180	63	.350	180	47	.261
200	68	.340	200	56	.280
220	73	.331	220	61	.277
240	78	.325	240	63	.262
260	85	.327	260	67	.258
280	92	.328	280	73	.261
300	96	.320	300	81	.270
320	103	.322	320	84	.263
340	110	.324	340	91	.268
360	118	.328	360	104	.289
380	126	.332	380	110	.290
400	129	.322	400	115	.288

Fig. 1

of this sampling experiment. We let $\alpha_1, \alpha_2, \ldots, \alpha_{400}$ denote the sequence of four hundred 0's and 1's which result from the simulation. In Fig. 1 the line

k	$\sum_{i=1}^{k} \alpha_i$	$\frac{1}{k}\sum_{i=1}^{k} \alpha_i$
200	68	.340

means that sixty-eight 1's occurred in the first 200 trials. This sampling experiment was performed a second time with a different set of 400 random digits, and the result of the simulation was a sequence $\beta_1, \beta_2, \ldots, \beta_{400}$ of four hundred 0's and 1's. $(1/k) \sum_{i=1}^{k} \alpha_i$ for

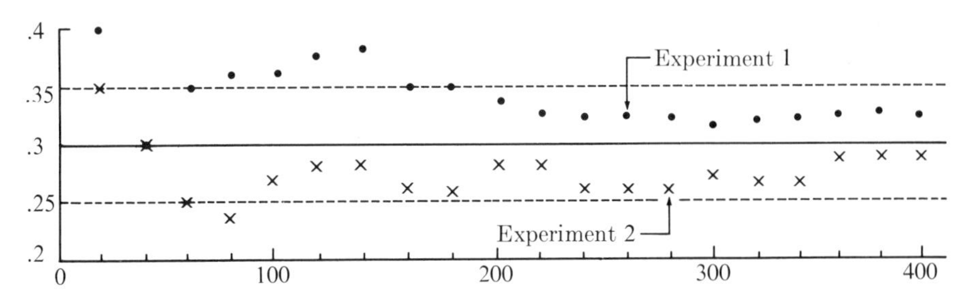

Fig. 2

$k = 20, 40, 60, \ldots, 400$ is plotted against k for experiment 1. Similarly $(1/k) \sum_{i=1}^{k} \beta_i$ for $k = 20, 40, 60, \ldots, 400$ is plotted against k for experiment 2.

From binomial tables for $n = 400$ and $p = .3$ we find, for example, that $P[101 < S_{400} < 138] = .95024$. Thus $P[|(1/400)S_{400} - .3| < .0475] = .95024$, so that in about 95 percent of such sampling experiments we should find the terminal fraction of 1's in the interval $.3 \pm .0475$. In our two sampling experiments we found $.3 + .022$ and $.3 - .012$ for the terminal fraction of 1's.

EXERCISES

1. *A Sampling Experiment.* $Y_1, Y_2, \ldots, Y_{40}$ are independent random variables with $Y_1, Y_2, \ldots, Y_{20}$ identically distributed and $Y_{21}, Y_{22}, \ldots, Y_{40}$ identically distributed, and $T_k = Y_1 + Y_2 + \cdots + Y_k$. Y_k is the net profit, in some appropriate units, of a business enterprise for the kth day of operation. The enterprise is in phase 1 of its operation for the first 20 days and in the more profitable phase 2 during the next 20 days. Let $D = [T_{20} > -4]$ and $B = [T_{40} > 8]$. The manager of the enterprise is particularly anxious for DB to occur. If D occurs he will, on the basis of the potential indicated by phase 1, be able to borrow money for phase 2 at a favorable rate of interest. He feels that the whole enterprise is a success iff the final profit T_{40}, disregarding interest on the loan, exceeds eight units *and* he was able to borrow at a favorable rate at the start of phase 2. Thus the enterprise is a success iff DB occurs. Naturally D and B are dependent events. Assume the distributions of Fig. 3.

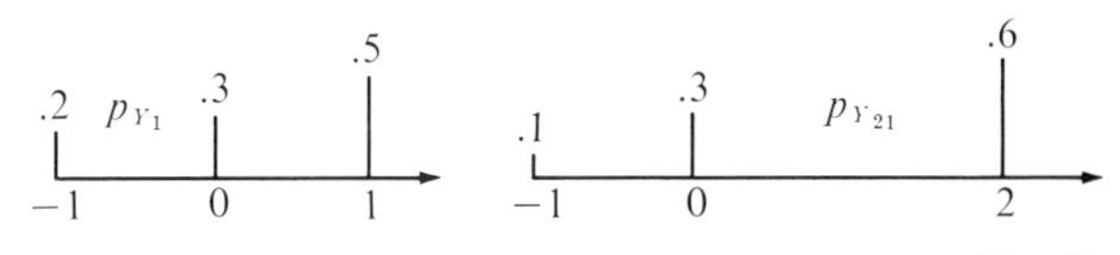

Fig. 3

(*a*) We wish to show that $P(DB)$ is large. Let A equal the complement of DB so that $A = (D^c) \cup (B^c)$; that is, the enterprise is unsuccessful iff either D or B fails to occur. Thus $P(A) \leq P(D^c) + P(B^c)$. Show that $P(A)$ is small.

(*b*) Use random numbers to simulate this setup and explain briefly how you did it. Give your result, which is essentially a point of R^{40}, in some tabular form such as the one in Fig. 4 for the result of such a simulation.

n	y_n	$\sum_{k=1}^{n} y_k$	n	y_n	$\sum_{k=1}^{n} y_k$
1	−1	−1	21	2	6
2	−1	−2	22	2	8
3	0	−2	23	0	8
4	0	−2	24	2	10
5	1	−1	25	2	12
6	1	0	26	2	14
7	1	1	27	2	16
8	1	2	28	2	18
9	1	3	29	0	18
10	−1	2	30	2	20
11	1	3	31	−1	19
12	−1	2	32	2	21
13	1	3	33	0	21
14	−1	2	34	0	21
15	0	2	35	2	23
16	1	3	36	0	23
17	−1	2	37	2	25
18	1	3	38	0	25
19	0	3	39	2	27
20	1	4	40	−1	26

Fig. 4

(*c*) Plot $\sum_{k=1}^{n} y_k$ against n for $n = 1, 2, \ldots, 40$ in a plot like Fig. 5 and state whether or not the manager would consider the whole enterprise a success if it resulted in what you obtained by simulation. For example, for the Fig. 4 simulation, $\sum_{k=1}^{20} y_k = 4 > -4$ and $\sum_{k=1}^{40} y_k = 26 > 8$, so this simulation resulted in a "success."

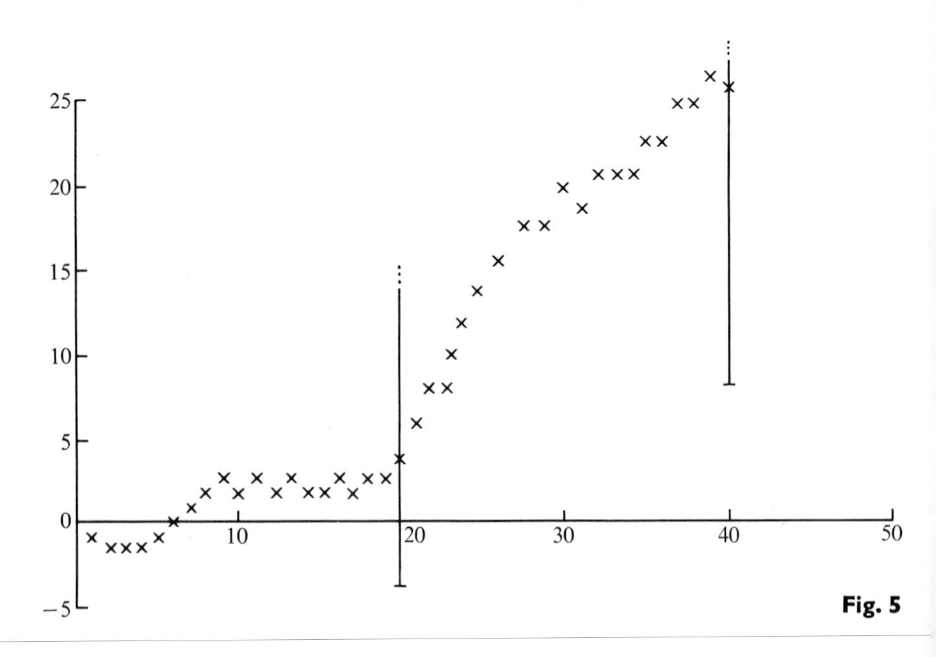

Fig. 5

(*d*) Assume that 1,000 students are going to do this exercise correctly and independently, each one using completely different random numbers. Show that the probability is large that over 600 students report success on part (*c*).

12-3 THE STRONG LAW OF LARGE NUMBERS

The WLLN asserts that under certain conditions if n is large enough we can be quite sure that $(1/n)S_n$ will be close to EX_1. The **strong law of large numbers** (SLLN) asserts that under certain conditions if n is large enough we can be quite sure that every one of

$$\frac{1}{n}S_n, \frac{1}{n+1}S_{n+1}, \frac{1}{n+2}S_{n+2}, \ldots$$

will be close to EX_1. The lengthy proof is left to the next section. In this section we state, interpret, and apply one version of the SLLN.

Theorem 1: A SLLN for independent identically distributed random variables Let $X_1, X_2, \ldots, X_n, \ldots, X_m$ be independent and identically distributed random variables, with $EX_1^2 < \infty$, and let $S_k = X_1 + X_2 + \cdots + X_k$. If $\epsilon > 0$ then

$$1 - P\left[\left|\frac{1}{n}S_n - EX_1\right| < \epsilon, \left|\frac{1}{n+1}S_{n+1} - EX_1\right| < \epsilon, \ldots, \left|\frac{1}{m}S_m - EX_1\right| < \epsilon\right]$$

$$= P\left[\left|\frac{1}{k}S_k - EX_1\right| \geq \epsilon \text{ for at least one } k \text{ satisfying } n \leq k \leq m\right]$$

$$\leq \frac{2\text{Var } X_1}{\epsilon^2 n}. \quad \blacksquare$$

The bound of our SLLN is just twice the bound of our WLLN, but the conclusion is much stronger. The bound is independent of m, so we have the same bound for all $m \geq n$. We will usually think of m as an astronomically large number, such as $m = n(10^{1,000})$. If we perform an experiment a sample point ω occurs, and we observe the numbers $X_1(\omega), X_2(\omega), \ldots, X_m(\omega)$. We can calculate the successive arithmetic averages

$$\frac{1}{1}S_1(\omega) = X_1(\omega), \frac{1}{2}S_2(\omega) = \frac{1}{2}[X_1(\omega) + X_2(\omega)], \ldots,$$

$$\frac{1}{m}S_m(\omega) = \frac{1}{m}[X_1(\omega) + \cdots + X_m(\omega)]$$

and so plot $(1/k)S_k(\omega)$ as a function of k. Two such plots for two different experimental outcomes ω_1 and ω_2 are shown in Fig. 1.

Let A be the event that the plot of $(1/k)S_k$ remains strictly inside the rectangle shown in Fig. 1 for all k satisfying $n \leq k \leq m$. Clearly $\omega_1 \in A$ and $\omega_2 \notin A$. Let B be the event that the plot of $(1/k)S_k$ is outside or on the boundary of the rectangle for at least one k satisfying $n \leq k \leq m$. Clearly $\omega_1 \notin B$ and $\omega_2 \in B$. Also A is the complement of B, so that $1 - P(A) = P(B)$. The conclusion of Theorem 1 is just $1 - P(A) = P(B) \leq 2(\text{Var } X_1)/\epsilon^2 n$. By selecting a small $\epsilon > 0$ we specify a tolerance

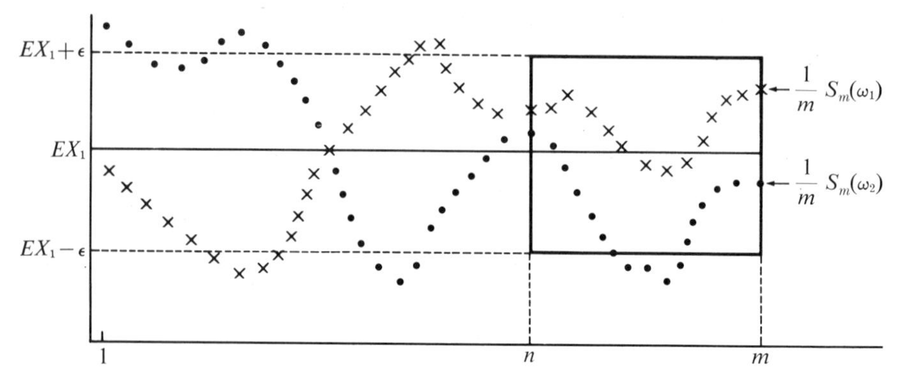

Fig. 1

band $EX_1 \pm \epsilon$. If we select an n so large that $2(\text{Var } X_1)/\epsilon^2 n$ is $\leq .01$, for example, we have $P(B) \leq .01$ and $P(A) \geq .99$, and this is true regardless of how large m is. Thus the SLLN states that we can select an $\epsilon > 0$ as small as we like and then select an n large enough so that regardless of how large we then make m, we have probability almost 1 for the event

$$EX_1 - \epsilon < \frac{1}{k} S_k < EX_1 + \epsilon \qquad \text{for all } k \text{ satisfying } n \leq k \leq m.$$

Thus by making n large enough we can be as certain as we like that every one of

$$\frac{1}{n} S_n, \frac{1}{n+1} S_{n+1}, \ldots, \frac{1}{m} S_m$$

will be as close to EX_1 as we like. Speaking intuitively, we can be quite certain that the sequence $(1/1)S_1, (1/2)S_2, (1/3)S_3, \ldots$ will become *and remain* arbitrarily close to EX_1. In dealing with an infinite-sequence space this last statement can be made precise by showing that the probability is 1 that the sequence $(1/k)S_k$ converges to EX_1.

Results such as Theorem 1 are particularly relevant when there are tolerance limits which should constantly be satisfied, such as "total capital should never be negative," "total sales should never exceed stock in inventory," "the angular error in a radar antenna should not exceed 2° or the target will be lost," or "a controlled system should never get too far out of control."

Note that if $m = n$ Theorem 1 reduces to Theorem 1 (Sec. 12-1), except that the bound it gives is not as good.

Example 1 As in Example 1 (Sec. 12-1), let X_1 have the density of Fig. 1 (Sec. 9-2) and let $\epsilon = 1$ and $n = 100$, so that if $m \geq 100$, then

$$P\left[.28 < \frac{1}{100} S_{100} < 2.28, \ldots, .28 < \frac{1}{m} S_m < 2.28\right] \geq .828.$$

Thus the probability is at least .828 that after the ninety-ninth trial the average reward will always be within one unit of the expected reward per trial of 1.28. ■

Theorem 1 specialized to Bernoulli trials immediately yields the following result.

Theorem 2: An SLLN for Bernoulli Trials If $X_1, X_2, \ldots, X_n, \ldots, X_m$ are m Bernoulli trials with parameter p, and if $S_k = X_1 + X_2 + \cdots + X_k$ and $\epsilon > 0$ then

$$P\left[p - \epsilon < \frac{1}{k} S_k < p + \epsilon \text{ whenever } n \le k \le m\right] \ge 1 - \frac{2pq}{\epsilon^2 n} \ge 1 - \frac{1}{2\epsilon^2 n}. \quad ■$$

Thus the fraction of successes must become *and remain* close to p.

Example 2 A fair coin is tossed $m = 10^{100}$ times, and a gambler wins the game iff from the $n = 10^6$ toss on, the fraction of heads is always between $.5 \pm .01$. The probability that he will win is at least .995. ■

We now briefly discuss the traditional and often referred to asymptotic form of the SLLN, but we will leave its deduction from Theorem 1 to Sec. 20-3. Suppose that *one* trial of an experiment consists of an *infinite* sequence of independent spins of the same random wheel and hence results in an infinite sequence $x_1, x_2, \ldots$ of rewards. We say that the trial is successful iff

$$\lim_{n\to\infty} \frac{1}{n} \sum_{k=1}^{n} x_k = EX_1$$

that is, iff the arithmetic average reward converges to the expectation EX_1 of the reward for one spin. The asymptotic form of the SLLN states that the probability of obtaining a successful trial is 1. For example, the probability is 1 that the fraction of heads will converge to ½ in an infinite sequence of fair-coin tosses.

More precisely, Theorem 1 (Sec. 20-3) shows that if $X_1, X_2, \ldots$ is an infinite sequence of independent and identically distributed random variables, with $EX_1^2 < \infty$, then the probability is 1 for the event consisting of those ω for which

$$\lim_{n\to\infty} \frac{1}{n} [X_1(\omega) + \cdots + X_n(\omega)] = EX_1.$$

In Sec. 20-3 it is also shown that the assumption $EX_1^2 < \infty$ may be relaxed to $E|X_1| < \infty$.

Results of this kind are sometimes interpreted, as in Exercise 4, as meaning that with probability 1 the time average is equal to the ensemble average EX_1 at one time; that is, x_n is thought of as obtained at time n, so the limit of $(x_1 + \cdots + x_n)/n$ as n goes to infinity, if it exists, is the time average for this one outcome. Now we consider the ensemble, or collection, of all possible outcomes and think of $P[X_1 = x_1']$ as the fraction of these outcomes $x_1, x_2, \ldots$ having $x_1 = x_1'$, so that EX_1 can be interpreted roughly as the average, over the ensemble, of the first coordinates x_1.

EXERCISES

1. State the result obtained by applying the SLLN, Theorem 1, rather than the WLLN to each of the following in Sec. 12-1:

(*a*) Example 2.
(*b*) Example 3.
(*c*) Exercise 2*a*, with the roundoff errors assumed to be identically distributed.
(*d*) Exercise 6; show that $P[\bar{D}_k < 1.7$ for all k satisfying $100 \le k \le 1{,}000]$ is large.
(*e*) Exercise 7; show that $P[|(1/k)T_k - 1| < .008$ for all k satisfying $10^3 \le k \le 10^4]$ is large.

2. (*a*) Use Theorem 1 to state and prove an SLLN for products similar to Theorem 3 (Sec. 12-1).
(*b*) Apply part (*a*) to Example 5 (Sec. 12-1).
(*c*) Apply part (*a*) to Exercise 4*b* (Sec. 12-1).

3. Let D_i and R_i be the number of Democrats and Republicans, respectively, who register to vote on the ith day. Therefore $X_i = D_i - R_i$ is the excess of registering Democrats over Republicans, so $S_k = X_1 + \cdots + X_k$ is the total registered excess from the first k days. Assume that $EX_1 = 2$ and $\operatorname{Var} X_1 = 4$ and that $X_1, X_2, \ldots$ are independent and identically distributed. Show that the probability is large that starting with day $n = 20$ the excesses $S_n, S_{n+1}, \ldots$ will *all* be positive.

4. This exercise illustrates by a trivial example the ideas of time average and ensemble average. Let the joint discrete probability density for X_1, X_2, X_3 be uniform on the five points (1,5,0), (1,2,3), (2,1,0), (3,1,2), and (3,−5,8). Let $Z = (1/3)(X_1 + X_2 + X_3)$ and find $P[Z = EX_1]$.

COMMENTS: Assume that an experiment results in observing the value of (X_1,X_2,X_3), where X_i is observed at time i. Then we will observe some outcome from the *ensemble* of five possible outcomes. Also, EX_1 is the arithmetic average, over the ensemble, of our possible observations at time 1. Naturally the value of Z for an experiment is just the time average of our separate time observations for that experiment. Thus the event $[Z = EX_1]$ may be described as the event that the time average will equal the "ensemble average at time 1."

Let EX_i be called the ensemble average at time i. Assume that the random variables are identically distributed and so have the same ensemble average for every time. This may then be called *the* ensemble average. Assume that $EX_1^2 < \infty$ and that the random variables are independent. The WLLN states that if the observation time n is large enough, then with high probability the time average $(1/n)S_n$ at time n will be close to the ensemble average EX_1. The SLLN states that from some large enough time n onward, with high probability, all later time averages $(1/k)S_k$ will be close to the ensemble average EX_1. Theorem 1 (Sec. 20-3) yields the following alternate interpretation of the SLLN: with probability 1, the time average will converge to the ensemble average as the "averaging time" goes to infinity.

*12-4 THE PROOF OF THE STRONG LAW OF LARGE NUMBERS

In this section we prove Theorem 1 below, which is an SLLN for independent random variables which need not have the same distribution. The tolerance limits ϵ_k in Theorem 1 can vary with k. We then deduce Theorem 1 (Sec. 12-3) from Theorem 1 below.

Theorem 1: The Kolmogorov-Hájek-Rényi (KHR) inequality[1] Let $X_1, X_2, \ldots, X_n, \ldots, X_m$ be independent random variables, with all $EX_j^2 < \infty$, and let $S_k = X_1 + X_2 + \cdots + X_k$.

[1] Theorem 1 is from J. Hájek and A. Rényi, Generalization of an Inequality of Kolmogorov, *Acta Math. Acad. Sci. Hungar.*, vol. 6, 1955. The special case of Theorem 1 obtained by setting $\epsilon_n = \epsilon_{n+1} = \cdots = \epsilon_m$ was known earlier; A. N. Kolmogorov, Sur la loi fort des grands nombres, *C. R. Acad. Sci. Paris*, vol. 191, 1930.

If $0 < \epsilon_n \leq \epsilon_{n+1} \leq \cdots \leq \epsilon_m$ then

$$P[|S_k - ES_k| \geq \epsilon_k \text{ for at least one } k \text{ satisfying } n \leq k \leq m]$$
$$\leq \frac{1}{(\epsilon_n)^2}\sum_{i=1}^{n} \operatorname{Var} X_i + \sum_{k=n+1}^{m} \frac{\operatorname{Var} X_k}{(\epsilon_k)^2}. \quad \blacksquare$$

We specialize Theorem 1 to the case $n = 1$, which we call the KHR lemma.

KHR lemma Let $X_1, X_2, \ldots, X_m$ be independent random variables, with all $EX_j^2 < \infty$, and let $S_k = X_1 + X_2 + \cdots + X_k$ and $T_k = S_k - ES_k$. If $0 < \epsilon_1 \leq \epsilon_2 \leq \cdots \leq \epsilon_m$ then

$$P\left\{\bigcup_{k=1}^{m} [|T_k| \geq \epsilon_k]\right\} \leq \sum_{k=1}^{m} \frac{\operatorname{Var} X_k}{(\epsilon_k)^2}. \quad \blacksquare$$

If $X_1, X_2, \ldots, X_n, \ldots, X_m$ are independent and we apply the KHR lemma to the sequence $(X_1 + X_2 + \cdots + X_n), X_{n+1}, X_{n+2}, \ldots, X_m$ of $m - n + 1$ random variables we obtain Theorem 1 in a different notation. Thus the *KHR lemma implies Theorem 1.*

In the proof of the KHR lemma we will need the formula for summation by parts. We make the convention that $\sum_{r=i}^{j} b_r = 0$ if $j < i$; that is, a summation without any terms equals 0.

Summation by parts Given reals $W_1, \ldots, W_n$ and $a_1, \ldots, a_n$ for $n \geq 1$ and partial sums $A_r = \sum_{k=1}^{r} a_k$ then

$$\sum_{k=1}^{n} W_k a_k = W_n A_n + \sum_{r=1}^{n-1} (W_r - W_{r+1}) A_r. \quad \blacksquare$$

Proof of summation by parts Writing out the right side and rearranging yields

$$W_n A_n + (W_{n-1} - W_n)A_{n-1} + (W_{n-2} - W_{n-1})A_{n-2}$$
$$+ \cdots + (W_2 - W_3)A_2 + (W_1 - W_2)A_1$$
$$= W_n(A_n - A_{n-1}) + W_{n-1}(A_{n-1} - A_{n-2}) + \cdots + W_2(A_2 - A_1) + W_1 A_1$$

which equals the left-hand side, so the formula for summation by parts is proved. $\blacksquare$

The above formula is often used to prove the formula for *integration by parts*, $\int Fg = FG - \int fG$, by letting F and G correspond to W_k and A_k, respectively.

Proof of the KHR lemma Let $C = \bigcup_{k=1}^{m} [|T_k| \geq \epsilon_k]$; that is, C is the event that at least one of $|T_1| \geq \epsilon_1, |T_2| \geq \epsilon_2, \ldots, |T_m| \geq \epsilon_m$ is true. For $1 \leq k \leq m$ define the event C_k by

$$C_k = [|T_1| < \epsilon_1, |T_2| < \epsilon_2, \ldots, |T_{k-1}| < \epsilon_{k-1}, |T_k| \geq \epsilon_k].$$

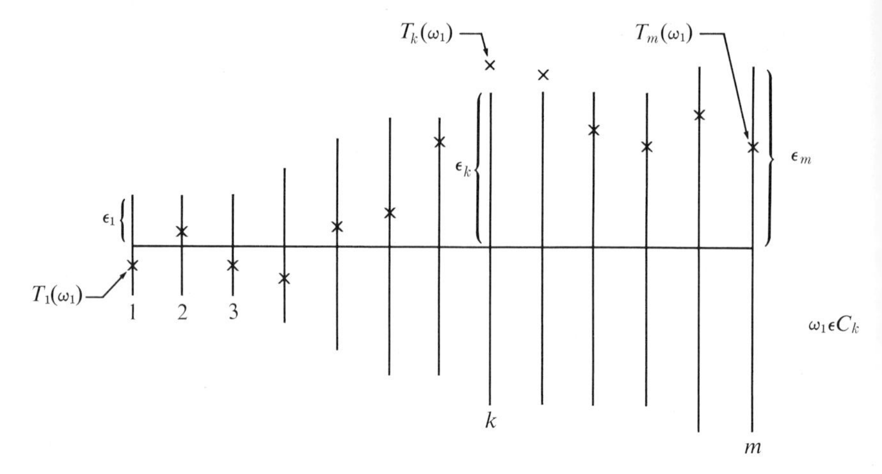

Fig. 1

Thus $C_k \subset [|T_k| \geq \epsilon_k] \subset C$, and as indicated in Fig. 1, C_k is the event that the first of the inequalities $|T_1| < \epsilon_1, |T_2| < \epsilon_2, \ldots, |T_m| < \epsilon_m$ to be violated is the kth one.

Clearly $C_1, \ldots, C_m$ are disjoint and $\bigcup_{k=1}^{m} C_k = C$. That is, $\omega \in C$ iff there is a unique k satisfying $1 \leq k \leq m$ such that $|T_1(\omega)| < \epsilon_1, |T_2(\omega)| < \epsilon_2, \ldots, |T_{k-1}(\omega)| < \epsilon_{k-1}, |T_k(\omega)| \geq \epsilon_k$.

We first prove that

$$ET_r^2 \leq \sum_{k=1}^{r} \epsilon_k^2 P(C_k) \qquad \text{for every } r = 1, 2, \ldots, m. \tag{1}$$

Let I_k be the indicator function of the event C_k. For integers k and r satisfying $1 \leq k \leq r \leq m$ we have

$$\begin{aligned} E(T_r^2 I_k) &= E[T_k + (T_r - T_k)]^2 I_k \\ &= E(T_k^2 I_k) + 2E(T_k I_k)(T_r - T_k) + E(T_r - T_k)^2 I_k \\ &\overset{1}{=} E(T_k^2 I_k) + E(T_r - T_k)^2 I_k \overset{2}{\geq} E(T_k^2 I_k) \overset{3}{\geq} \epsilon_k^2 E I_k = \epsilon_k^2 P(C_k). \end{aligned}$$

For $\overset{1}{=}$ we used the fact that both T_k and I_k are functions of $X_1, X_2, \ldots, X_k$, so that the random variable $T_k I_k$ is independent of

$$T_r - T_k = (X_{k+1} + X_{k+2} + \cdots + X_r) - E(X_{k+1} + \cdots + X_r)$$

so

$$E[(T_k I_k)(T_r - T_k)] = [E(T_k I_k)][E(T_r - T_k)] = [E(T_k I_k)][0] = 0.$$

Clearly $(T_r - T_k)^2 I_k \geq 0$, so we get $\overset{2}{\geq}$. For $\overset{3}{\geq}$ we note that $(T_k{}^2 I_k) \geq \epsilon_k{}^2 I_k$, since $|T_k(\omega)| \geq \epsilon_k$ if $I_k(\omega) = 1$.

We have shown that if $1 \leq k \leq r \leq m$ then $E(T_r{}^2 I_k) \geq \epsilon_k{}^2 P(C_k)$. Now for any r satisfying $1 \leq r \leq m$ we know that $C_1, C_2, \ldots, C_r$ are disjoint, so that $I_1 + I_2 + \cdots + I_r \leq 1$, and hence $T_r{}^2 \sum_{k=1}^{r} I_k \leq T_r{}^2$. Thus

$$ET_r{}^2 \geq E\left(T_r{}^2 \sum_{k=1}^{r} I_k\right) = \sum_{k=1}^{r} E(T_r{}^2 I_k) \geq \sum_{k=1}^{r} \epsilon_k{}^2 P(C_k)$$

so that *inequality* (1) *is proved.*

Letting $W_k = 1/\epsilon_k{}^2$ in the formula for summation by parts, we get

$$\sum_{k=1}^{m} \frac{a_k}{\epsilon_k{}^2} = \frac{A_m}{\epsilon_m{}^2} + \sum_{r=1}^{m-1}\left(\frac{1}{\epsilon_r{}^2} - \frac{1}{\epsilon_{r+1}^2}\right)A_r \qquad \text{where } A_r = \sum_{j=1}^{r} a_j. \tag{2}$$

We will use equation (2) twice in the remainder of the proof.

Let $b_k = \epsilon_k{}^2 P(C_k)$ and $B_r = \sum_{k=1}^{r} b_k$, so that inequality (1) becomes $B_r \leq ET_r{}^2$ for every $r = 1, 2, \ldots, m$. Consider the following inequality chain:

$$P(C) \overset{1}{=} \sum_{k=1}^{m} P(C_k) \overset{2}{=} \sum_{k=1}^{m} \frac{b_k}{\epsilon_k{}^2} \overset{3}{=} \frac{B_m}{\epsilon_m{}^2} + \sum_{r=1}^{m-1}\left(\frac{1}{e_r{}^2} - \frac{1}{\epsilon_{r+1}^2}\right)B_r$$

$$\overset{4}{\leq} \frac{ET_m{}^2}{\epsilon_m{}^2} + \sum_{r=1}^{m-1}\left(\frac{1}{\epsilon_r{}^2} - \frac{1}{\epsilon_{r+1}^2}\right)ET_r{}^2 \overset{5}{=} \sum_{k=1}^{m} \frac{\text{Var } X_k}{\epsilon_k{}^2}.$$

$\overset{1}{=}$ and $\overset{2}{=}$ are obvious from what has already been said about the C_k and the definition of b_k. For $\overset{3}{=}$ we used equation (2) and for $\overset{4}{\leq}$ we used inequality (1) and the fact that all the coefficients are nonnegative, since $0 < \epsilon_r \leq \epsilon_{r+1}$. Now $ET_r{}^2 = \text{Var } T_r = \sum_{k=1}^{r} \text{Var } X_k$, so we can apply equation (2) to get $\overset{5}{=}$. The *KHR lemma is proved: therefore Theorem 1 is proved.* ∎

In Theorem 1 assume that the random variables are identically distributed and let $\epsilon_k = k\epsilon$, so that the bound of Theorem 1 becomes

$$\frac{\text{Var } X_1}{\epsilon^2}\left(\frac{1}{n} + \sum_{k=n+1}^{m} \frac{1}{k^2}\right).$$

For the summation note that

$$\sum_{k=n+1}^{m} \frac{1}{k^2} \leq \int_n^\infty \frac{dx}{x^2} = \left[-\frac{1}{x}\right]_n^\infty = \frac{1}{n}$$

where the inequality is clear from Fig. 2, since the summation equals the shaded area. Therefore Theorem 1 (Sec. 12-3) is proved. ∎

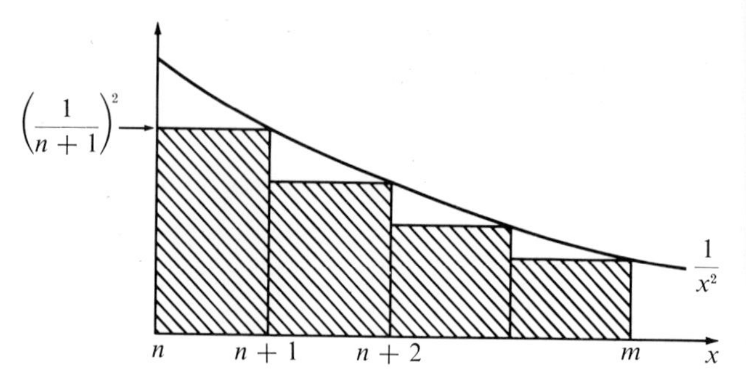

Fig. 2

13
The Central-limit Theorem

The central-limit theorem is a fundamental result in the theory and applications of probability theory. It asserts that under quite general conditions the standardized sum of independent random variables has the normal distribution in the limit. In this chapter we state, apply, and motivate this theorem; the proof is left for Chap. 19. Essentially the same proof could be given here, but it seems preferable to postpone the proof until after continuous densities have been considered.

The normal density is studied in Sec. 13-1, and Sec. 13-2 is devoted to the statement, interpretation, and application of the central-limit theorem. In Sec. 13-3 some analytic justification is given for the special case of this theorem for Bernoulli trials. In Sec. 13-4 a particular, fairly arbitrary numerical example is calculated in detail. This example demonstrates quite convincingly that convergence toward the normal distribution does indeed take place, and in some cases quite rapidly.

13-1 THE NORMAL DENSITY

In this section we introduce the normal continuous probability density and establish some of its properties. Continuous probability densities will be studied in a later chapter. As far as we are concerned in this chapter, the normal density is just a useful function for obtaining approximations to, and limiting values of, certain probabilities. The function $\varphi: R \to R$ defined in Fig. 1 is called the **normal density**, or when more explicitness is needed, the normal continuous probability density with zero mean and unit variance. The normal density is also called the gaussian density or the Laplace density. We note some of the more obvious properties of φ. The exponential function has only positive values, so that $\varphi(x) > 0$ for all real x. Also $x^2/2 \geq 0$, so that $e^{-x^2/2} \leq e^0 = 1$. Thus $0 < \varphi(x) \leq 1/\sqrt{2\pi}$ for all x. Clearly $\varphi(-x) = \varphi(x)$, so that φ is an even function; that is, it is symmetric about the origin and hence we need only examine it for $x \geq 0$. As $x \geq 0$ increases we have $x^2/2$ increasing, so $\varphi(x)$ decreases.

Let us compare φ to the simpler exponential function $f: R \to R$ defined by $f(x) = (1/\sqrt{2\pi})e^{-x/2}$ and shown in Fig. 1a. Clearly $\varphi(0) = f(0)$ and $\varphi(1) = f(1)$. If $x > 1$ then $x^2/2 > x/2$, so that $\varphi(x) < f(x)$. Clearly $\varphi(x)$ is very much less than $f(x)$ if x is large. For example, for $x = 4$ we have $\varphi(4) = (1/\sqrt{2\pi})e^{-8} \doteq .0001$ and

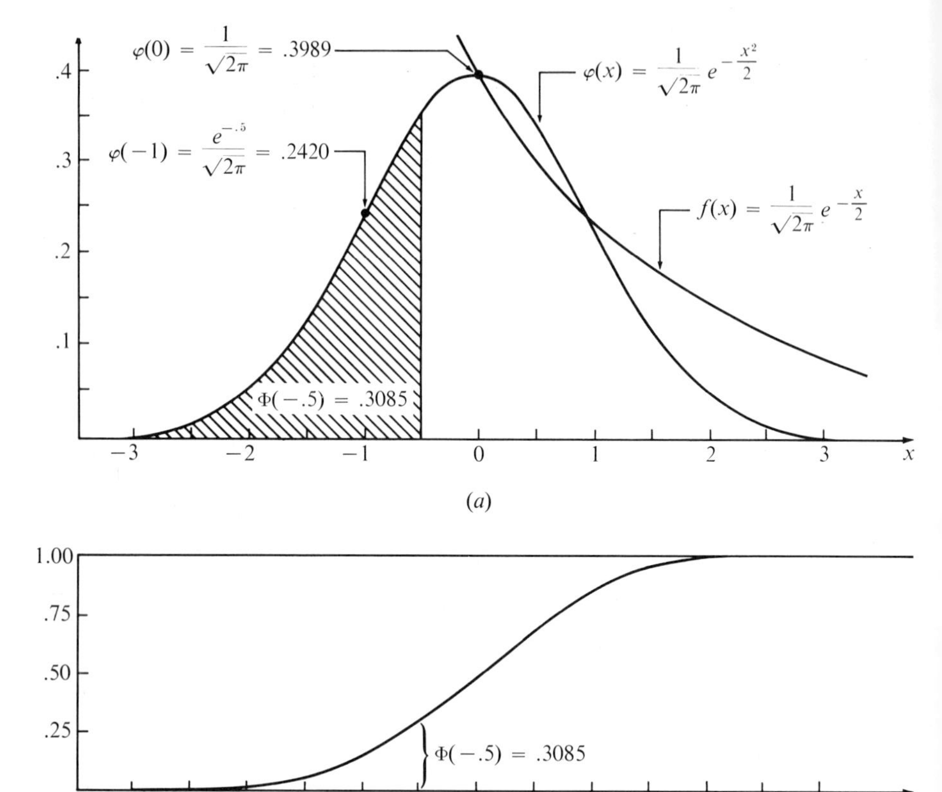

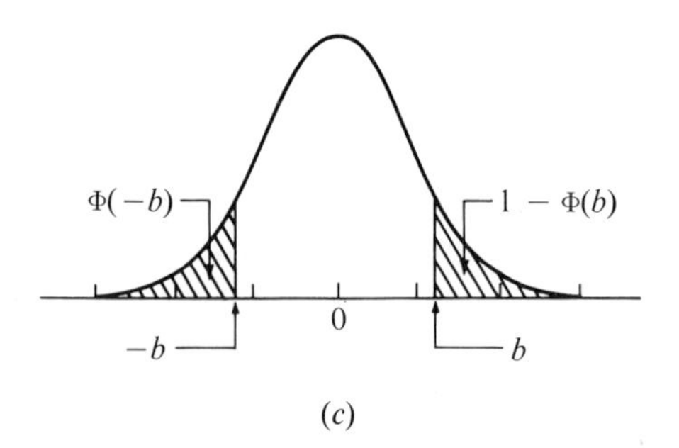

(c)

Fig. 1 (*a*) The normal density φ with zero mean and unit variance; (*b*) the cumulative normal $\Phi(b) = \int_{-\infty}^{b} \varphi(x)\,dx$; (*c*) $\Phi(-b) = 1 - \Phi(b)$.

$f(4) = (1/\sqrt{2\pi})e^{-2} = .0540$. If $0 \le x \le 1$ then $x^2/2 \le x/2$, so that $\varphi(x) \ge f(x)$ in this range.

Taking derivatives of φ, we find that $\varphi'(x) = -x\varphi(x)$ and $\varphi''(x) = (x^2 - 1)\varphi(x)$. Thus $\varphi''(x) < 0$ if $-1 < x < 1$, so φ is convex upward in this range. Also $\varphi''(x) > 0$ if $|x| > 1$, so that φ is convex downward on the rest of the real axis.

We now consider the area $\int_{-\infty}^{\infty} \varphi(x)\, dx$ above the x axis and below φ. Clearly $\int_0^{\infty} f(x)\, dx$ is finite, so from Fig. 1 we see that $\int_0^{\infty} \varphi(x)\, dx$ must be finite. But φ is even, so

$$\int_{-\infty}^{\infty} \varphi(x)\; dx = 2\int_0^{\infty} \varphi(x)\, dx.$$

Thus the area $\int_{-\infty}^{\infty} \varphi(x)\, dx$ is finite.

We now show that the value of the area $\int_{-\infty}^{\infty} \varphi(x)\, dx$ is 1, and this is why the multiplicative constant $1/\sqrt{2\pi}$ is included in the definition of φ. Consider the surface $\varphi(x)\varphi(y) = (1/2\pi)e^{-(x^2+y^2)/2} = (1/2\pi)e^{-r^2/2}$ over the xy plane, as shown in Fig. 2, where $r^2 = x^2 + y^2$. The volume below the surface and above the xy plane is given by

$$\int_{-\infty}^{\infty}\int_{-\infty}^{\infty} \varphi(x)\varphi(y)\; dx\; dy = \left[\int_{-\infty}^{\infty} \varphi(x)\;\; dx\right]\left[\int_{-\infty}^{\infty} \varphi(y)\;\; dy\right] = \left[\int_{-\infty}^{\infty} \varphi(x)\;\; dx\right]^2$$

so that we need only show that this volume is 1 in order to conclude that $\int_{-\infty}^{\infty} \varphi(x)\; dx = 1$. The annulus between r and $r + \Delta r$ in the xy plane is shaded in Fig. 2 and has area $2\pi r\; \Delta r$, so the volume above this annulus and below the surface is $2\pi r\; \Delta r(1/2\pi)e^{-r^2/2}$.

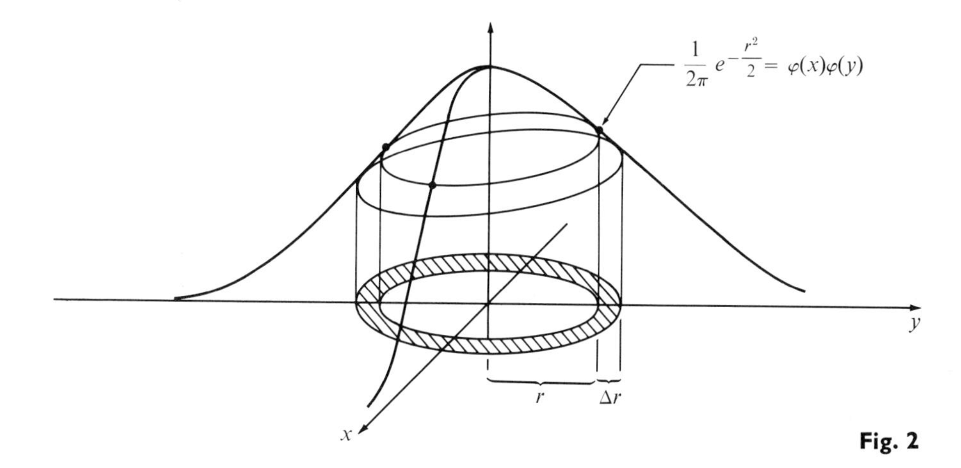

Fig. 2

Adding such disjoint volumes, we get

$$\int_0^\infty re^{-r^2/2}\,dr = -e^{-r^2/2}\Big|_0^\infty = 1$$

so we have shown that $\int_{-\infty}^{\infty} \varphi(x)\,dx = 1.$

Thus we have proved the first of the following three equations:

$$\int_{-\infty}^{\infty} \varphi(x)\,dx = 1 \qquad \int_{-\infty}^{\infty} x\varphi(x)\,dx = 0 \qquad \int_{-\infty}^{\infty} x^2\varphi(x)\,dx = 1. \tag{1}$$

The center equation is obvious, since the integrand is an odd function. As will be seen in a later chapter, the first equation, together with the fact that $\varphi \geq 0$, shows φ to be a continuous probability density. The other two equations show φ to have zero mean and unit variance. From one point of view functions of the form $Ce^{-B(x-A)^2}$ arise in studying limiting distributions, and by requiring that the standardizing equations (1) hold we find that the resulting values of the constants A, B, and C yield the function φ.

In order to derive the last of equations (1) we take any positive real t and make the transformation $x = y\sqrt{t}$ and $dx = dy\sqrt{t}$ in the integral

$$\frac{1}{\sqrt{2\pi}}\int_{-\infty}^{\infty} e^{-x^2/2}\,dx = 1$$

and so obtain the useful identity

$$\frac{1}{\sqrt{2\pi}}\int_{-\infty}^{\infty} e^{-ty^2/2}\,dy = \frac{1}{\sqrt{t}} \qquad \text{for all } t > 0. \tag{2}$$

Differentiating this identity with respect to t, we get

$$-\frac{1}{\sqrt{2\pi}}\int_{-\infty}^{\infty} \frac{y^2}{2}\,e^{-ty^2/2}\,dy = -\frac{1}{2}\,t^{-3/2}$$

and setting $t = 1$ yields the last of equations (1), with y rather than x for the dummy variable of integration.

Let the function $\Phi\colon R \to R$ be the indefinite integral of φ, as shown in Fig. 1b; that is, $\Phi(b) = \int_{-\infty}^{b} \varphi(x)\,dx$, so that $\Phi(b)$ is the area under φ to the left of b. The function Φ is called the **cumulative normal,** or the normal distribution function, or the normal cumulative distribution function for zero mean and unit variance. Clearly Φ is always strictly increasing, and $\Phi(b) \to 0$ as $b \to -\infty$, $\Phi(0) = .5$, and $\Phi(b) \to 1$ as $b \to \infty$. The function Φ cannot be expressed in closed form in terms of elementary functions, but it is tabulated in Table 1 for various *nonnegative* arguments b. We can use Table 1 to evaluate Φ for negative arguments by using the following identity, which is obvious from Fig. 1c, since φ is an even function:

$$\Phi(-b) = 1 - \Phi(b) \qquad \text{for any real } b. \tag{3}$$

For example, $\Phi(-2.03) = 1 - \Phi(2.03) = 1 - .9788 = .0212.$

Table 1 The cumulative normal $\Phi(x)$

x	.00	.01	.02	.03	.04	.05	.06	.07	.08	.09
.0	.5000	.5040	.5080	.5120	.5160	.5199	.5239	.5279	.5319	.5359
.1	.5398	.5438	.5478	.5517	.5557	.5596	.5636	.5675	.5714	.5753
.2	.5793	.5832	.5871	.5910	.5948	.5987	.6026	.6064	.6103	.6141
.3	.6179	.6217	.6255	.6293	.6331	.6368	.6406	.6443	.6480	.6517
.4	.6554	.6591	.6628	.6664	.6700	.6736	.6772	.6808	.6844	.6879
.5	.6915	.6950	.6985	.7019	.7054	.7088	.7123	.7157	.7190	.7224
.6	.7257	.7291	.7324	.7357	.7389	.7422	.7454	.7486	.7517	.7549
.7	.7580	.7611	.7642	.7673	.7704	.7734	.7764	.7794	.7823	.7852
.8	.7881	.7910	.7939	.7967	.7995	.8023	.8051	.8078	.8106	.8133
.9	.8159	.8186	.8212	.8238	.8264	.8289	.8315	.8340	.8365	.8389
1.0	.8413	.8438	.8461	.8485	.8508	.8531	.8554	.8577	.8599	.8621
1.1	.8643	.8665	.8686	.8708	.8729	.8749	.8770	.8790	.8810	.8830
1.2	.8849	.8869	.8888	.8907	.8925	.8944	.8962	.8980	.8997	.9015
1.3	.9032	.9049	.9066	.9082	.9099	.9115	.9131	.9147	.9162	.9177
1.4	.9192	.9207	.9222	.9236	.9251	.9265	.9279	.9292	.9306	.9319
1.5	.9332	.9345	.9357	.9370	.9382	.9394	.9406	.9418	.9429	.9441
1.6	.9452	.9463	.9474	.9484	.9495	.9505	.9515	.9525	.9535	.9545
1.7	.9554	.9564	.9573	.9582	.9591	.9599	.9608	.9616	.9625	.9633
1.8	.9641	.9649	.9656	.9664	.9671	.9678	.9686	.9693	.9699	.9706
1.9	.9713	.9719	.9726	.9732	.9738	.9744	.9750	.9756	.9761	.9767
2.0	.9772	.9778	.9783	.9788	.9793	.9798	.9803	.9808	.9812	.9817
2.1	.9821	.9826	.9830	.9834	.9838	.9842	.9846	.9850	.9854	.9857
2.2	.9861	.9864	.9868	.9871	.9875	.9878	.9881	.9884	.9887	.9890
2.3	.9893	.9896	.9898	.9901	.9904	.9906	.9909	.9911	.9913	.9916
2.4	.9918	.9920	.9922	.9925	.9927	.9929	.9931	.9932	.9934	.9936
2.5	.9938	.9940	.9941	.9943	.9945	.9946	.9948	.9949	.9951	.9952
2.6	.9953	.9955	.9956	.9957	.9959	.9960	.9961	.9962	.9963	.9964
2.7	.9965	.9966	.9967	.9968	.9969	.9970	.9971	.9972	.9973	.9974
2.8	.9974	.9975	.9976	.9977	.9977	.9978	.9979	.9979	.9980	.9981
2.9	.9981	.9982	.9982	.9983	.9984	.9984	.9985	.9985	.9986	.9986
3.0	.9987	.9987	.9987	.9988	.9988	.9989	.9989	.9989	.9990	.9990
3.1	.9990	.9991	.9991	.9991	.9992	.9992	.9992	.9992	.9993	.9993
3.2	.9993	.9993	.9994	.9994	.9994	.9994	.9994	.9995	.9995	.9995
3.3	.9995	.9995	.9995	.9996	.9996	.9996	.9996	.9996	.9996	.9997
3.4	.9997	.9997	.9997	.9997	.9997	.9997	.9997	.9997	.9997	.9998

For any real μ and any positive real σ we define the **normal continuous probability density $\varphi_{\mu,\sigma}: R \to R$ with mean μ and standard deviation** σ, or variance σ^2, by

$$\varphi_{\mu,\sigma}(x) = \frac{1}{\sigma\sqrt{2\pi}} e^{-(x-\mu)^2/2\sigma^2} \qquad \text{for all real } x.$$

By making the transformation $y = (x - \mu)/\sigma$, so $dy = dx/\sigma$, in the equations below, we see that these equations reduce to equations (1),

$$\int_{-\infty}^{\infty} \varphi_{\mu,\sigma}(x)\,dx = 1 \qquad \int_{-\infty}^{\infty} x\varphi_{\mu,\sigma}(x)\,dx = \mu \qquad \int_{-\infty}^{\infty} (x-\mu)^2\varphi_{\mu,\sigma}(x)\,dx = \sigma^2$$

and so they are true. By analogy with the discrete case these equations justify the names of the parameters. Thus φ is an abbreviation for $\varphi_{0,1}$. By making the transformation $y = (x-\mu)/\sigma$ we see that

$$\int_a^b \varphi_{\mu,\sigma}(x)\,dx = \int_{(a-\mu)/\sigma}^{(b-\mu)/\sigma} \varphi(y)\,dy = \Phi\left(\frac{b-\mu}{\sigma}\right) - \Phi\left(\frac{a-\mu}{\sigma}\right) \qquad \text{if } a < b$$

so that areas under any $\varphi_{\mu,\sigma}$ can be obtained from Table 1.

The function $\varphi_{\mu,\sigma}$ is plotted in general form in Fig. 3, which also shows three of these functions for three different parameter pairs. Thus $\varphi_{\mu,\sigma}$ is centered at μ and has a unique maximum $\varphi_{\mu,\sigma}(\mu) = 1/(\sigma\sqrt{2\pi})$ at $x = \mu$. It drops off at $x = \mu \pm \sigma$ to $e^{-.5}$, or 61 percent of its maximum value. Thus varying μ just shifts $\varphi_{\mu,\sigma}$ back and

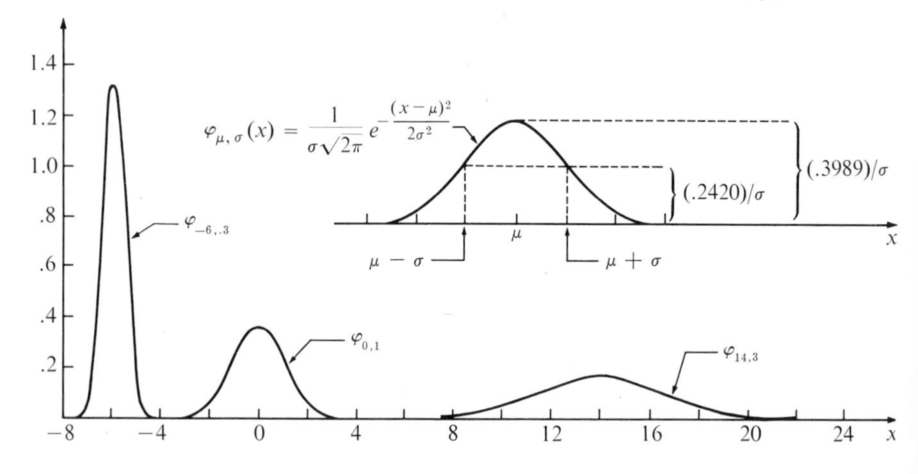

Fig. 3

forth, while increasing $\sigma > 0$ increases its spread. We see that increasing σ increases its spread *and* decreases its height in order to keep the area under $\varphi_{\mu,\sigma}$ always equal to 1. Thus we have here a different phenomenon from the discrete case. By making a linear transformation $y = ax + b$ in the discrete case we found that the new density p_Y was essentially obtained by shifting the density p_X and altering its spread. However, in the discrete case the heights, or values, of p_Y were the same as for p_X; they just occurred at different locations. In the continuous case we see that changing the spread also requires a change in the heights in order to keep the area equal to 1. In the continuous case areas are more fundamental than heights.

EXERCISES

1. Let $\mu = -75$ and $\sigma = 10$ and find $\int_{-90}^{-55} \varphi_{\mu,\sigma}(x)\,dx$:
(*a*) From Table 1
(*b*) By estimating the appropriate area under φ as graphed in Fig. 1.

2. *The Moments of the Normal Distribution.* Prove that if k is a positive integer then

$$\int_{-\infty}^{\infty} x^k \varphi(x)\,dx = \begin{cases} 0 & \text{if } k \text{ is odd} \\ (1)(3)(5)\cdots(k-1) & \text{if } k \text{ is even.} \end{cases}$$

3. *Bounding the Tail of the Normal Distribution.* It is easily verified that the derivative of $\varphi(x)/x$ equals $-\varphi(x)\{1 + (1/x^2)\}$, so that by the fundamental theorem of the calculus if $x > 0$ then

$$\frac{\varphi(x)}{x} = \int_x^{\infty} \varphi(t)\left\{1 + \frac{1}{t^2}\right\} dt$$

since the constant of integration must equal 0 because both sides go to zero as x goes to infinity. Now $1 + (1/t^2) > 1$, so we get $\varphi(x)/x > 1 - \Phi(x)$. Similarly the derivative of $\varphi(x)\{(1/x) - (1/x^3)\}$ equals $-\varphi(x)\{1 - (3/t^4)\}$, so that if $x > 0$ then

$$\varphi(x)\left\{\frac{1}{x} - \frac{1}{x^3}\right\} = \int_x^{\infty} \varphi(t)\left\{1 - \frac{3}{t^4}\right\} dt$$

Now $1 - (3/t^4) < 1$, so $\varphi(x)\{(1/x) - (1/x^3)\} < 1 - \Phi(x)$. Collecting these two facts together, we see that *if* $x > 0$ *then*

$$\frac{\varphi(x)}{x}\left\{1 - \frac{1}{x^2}\right\} < 1 - \Phi(x) < \frac{\varphi(x)}{x}.$$

If we evaluate these three expressions for $x = 3$ we get $.00098 < .000135 < .00148$, so that if x is large the inequalities permit us to find $1 - \Phi(x)$ quite accurately by evaluating the two extremes. *Use these inequalities to find bounds on* $1 - \Phi(5)$ and compare the bounds to the value $1 - \Phi(5) = 2.87 \times 10^{-7}$. FACT: $e^{-12.5} \doteq 3.73 \times 10^{-6}$. COMMENT: If $x \geq 1$ then $1 - \Phi(x) < \varphi(x)/x \leq \varphi(x)$, so φ is an upper bound for the tail probability in this range.

13-2 THE CENTRAL-LIMIT THEOREM

A **central-limit theorem** (CLT) asserts that under quite general conditions the standardized sum of independent random variables has an approximately normal distribution. One interpretation of a CLT is that the total error is approximately normally distributed if it is the sum of a "fairly large" number of "relatively small" independent errors. Theorem 1 makes the conditions "fairly large" and "relatively small" precise in terms of third absolute moments. In this section we apply Theorem 1; its proof is left until Chap. 19. In Corollary 1 we specialize Theorem 1 to identical distributions. After applying Corollary 1 in Examples 1 to 6, we derive and apply Corollary 2 for random variables which need not be identically distributed.

Theorem 1 Let $S_n = X_1 + \cdots + X_n$ where $X_1, \ldots, X_n$ are independent random variables with each $E|X_k|^3 < \infty$. Assume that $s_n^{\,2} = \operatorname{Var} S_n > 0$ and let $S_n^* = (S_n - ES_n)/s_n$. Then for every real b

$$|P[S_n^* < b] - \Phi(b)| \leq 3\left\{\sum_{k=1}^{n} E\left|\frac{X_k - EX_k}{s_n}\right|^3\right\}^{1/4}. \quad ■$$

The condition $s_n^2 > 0$ is equivalent to requiring that at least one Var X_k be positive. Thus we assume that not all the X_k are constants. The notation s_n^2 is used, rather than σ_n^2, since it should be apparent that s_n^2 is the variance of S_n, and not of X_n.

The right-hand side is a bound which does not depend on b, and it is calculable from the individual distributions of the X_k, without the distributions of sums. The theorem is of interest only when the bound is much less than 1, in which case $|P[S_n^* < b] - \Phi(b)|$ is small for all b. Thus Theorem 1 bounds the error in the approximation $P[S_n^* < b] \doteq \Phi(b)$. This bound is much too weak to be useful for most numerical examples. In particular, Exercise 8 shows the bound to be uninformative if $n \leq$ 6,561. However, as explained after Examples 6 and 9, a bound of this form is far more useful than typical asymptotic versions which it implies. Naturally

$$P\left[\frac{S_n - ES_n}{s_n} < b\right] = P[S_n^* < b]$$

so that we need only the two numbers

$$ES_n = EX_1 + \cdots + EX_n \qquad \text{and} \qquad \text{Var } S_n = \text{Var } X_1 + \cdots + \text{Var } X_n$$

each calculable from the individual X_k distributions, to obtain approximations to the distribution of S_n.

Theorem 1 immediately yields the asymptotic conclusion that $P[S_n^* < b]$ converges to $\Phi(b)$ if { } goes to zero as n goes to infinity. For example, $P[S_n^* < 2] \to \Phi(2) = .9772$. *The exact distribution of* S_n^* *depends intimately on the distributions of* $X_1, \ldots, X_n$ *so it is remarkable that the* CLT *holds under such general conditions.* Under our assumption of independence, the distribution of every one of $S_1^*, S_2^*, \ldots$ is uniquely determined once the distribution of every one of $X_1, X_2, \ldots$ is specified, and *all* of these must be specified for the asymptotic version. In particular for identically distributed $X_1, X_2, \ldots$ the distribution of X_1 determines the distribution of every S_n^* and S_n.

It is shown in Exercise 3 that if

$$\lim_{n\to\infty} P[S_n^* < b] = \Phi(b) \qquad \text{for all real } b$$

then

$$\lim_{n\to\infty} P[S_n^* = b] = 0 \qquad \text{for all real } b.$$

Thus *in a limiting sense* each real b has zero probability. (In the discrete case if n is large enough then the density of S_n is spread out over many points, each of which has small probability.) Therefore

$$P[S_n^* \leq b] = P[S_n^* < b] + P[S_n^* = b] \to \Phi(b) + 0 \qquad \text{as } n \to \infty$$

so that the asymptotic version could use $P[S_n^* \leq b]$ in place of $P[S_n^* < b]$. We also have

$$P[S_n^* > b] = 1 - P[S_n^* \leq b] \to 1 - \Phi(b),$$

$$P[a < S_n^* < b] = P[S_n^* < b] - P[S_n^* \leq a] \to \Phi(b) - \Phi(a) \qquad \text{as } n \to \infty$$

where the inclusion or exclusion of end points does not affect the limiting value. Thus if the CLT applies then the limiting value of the probability that S_n^* belongs in interval is just the area under φ and above the interval.

THE IDENTICALLY DISTRIBUTED CASE

If $X_1, \ldots, X_n$ are identically distributed then { } in Theorem 1 becomes

$$nE\left|\frac{X_1 - EX_1}{\sigma_{X_1}\sqrt{n}}\right|^3 = \frac{1}{\sqrt{n}}E|X_1^*|^3 \to 0 \qquad \text{as } n \to \infty$$

Thus the CLT always applies to independent and identically distributed random variables if $E|X_1|^3 < \infty$. This traditional asymptotic form of the CLT is stated as part (*a*) of the following corollary; part (*b*) is the inequality just obtained.

Corollary 1 Let $S_n = X_1 + \cdots + X_n$ where $X_1, \ldots, X_n$ are independent and identically distributed random variables satisfying $E|X_1|^3 < \infty$ and $\sigma_{X_1}^2 = \text{Var } X_1 > 0$. If $S_n^* = (S_n - ES_n)/\sigma_{X_1}\sqrt{n}$ then for every real b

(*a*) $\lim_{n\to\infty} P[S_n^* < b] = \Phi(b)$

(*b*) $|P[S_n^* < b] - \Phi(b)| \leq 3\left\{\frac{E|X_1^*|^3}{\sqrt{n}}\right\}^{1/4}$

where $X_1^* = (X_1 - EX_1)/\sigma_{X_1}$. ∎

Actually the asymptotic CLT applies to any distribution for X_1 for which $0 < \text{Var } X_1 < \infty$. That is, Corollary 1*a* holds if the assumption $E|X_1|^3 < \infty$ is relaxed to $EX_1^2 < \infty$. This is proved by Corollary 2 (Sec. 20-2), or by Theorem 1 (Sec. 18-3), which uses a different technique of proof.

We now consider examples of the application of the CLT of Corollary 1. Example 1 compares the CLT to Chebychev's inequality. Example 2 makes an important distinction between *bounds* obtained from Chebychev's inequality and *estimates* obtained from the CLT. Examples 3 and 4 involve the binomial density in a practical context. Example 5 applies the CLT to products of random variables, and Example 6 obtains new information about Poisson densities.

Example 1 Chebychev's inequality tells us that $P[|S_n^*| \geq \delta] \leq 1/\delta^2$ for all $\delta > 0$. For $0 < \delta \leq 1$ this inequality is true but uninformative, since we know that $P(A) \leq 1$ for every event A. Thus Chebychev's inequality tells us nothing about $P[-1 < S_n^* < 1]$, but from Corollary 1 we know that as n goes to infinity $P[-1 < S_n^* < 1]$ converges to $\Phi(1) - \Phi(-1) = \Phi(1) - [1 - \Phi(1)] = 2\Phi(1) - 1 = (2)(.8413) - 1 = .6826$. For $\delta = 2$ Chebychev's inequality tells us that $P[|S_n^*| \geq 2] \leq .25$ so that $P[-2 < S_n^* < 2] \geq .75$, while from Corollary 1 we find that $P[-2 < S_n^* < 2]$ approaches .9544. Making similar calculations for $\delta = 3, 4, 5$ and using the values $\Phi(4) = 1 - (3.17 \times 10^{-5})$ and $\Phi(5) = 1 - (2.87 \times 10^{-7})$, which are not available in Table 1 (Sec. 13-1), we find that

$$\begin{aligned}
0 &\leq P[-1 < S_n^* < 1] \to .6826 \\
.7500 &\leq P[-2 < S_n^* < 2] \to .9544 \\
.8888 &\leq P[-3 < S_n^* < 3] \to .9974 \\
.9375 &\leq P[-4 < S_n^* < 4] \to 1 - (6.34 \times 10^{-5}) \\
.9600 &\leq P[-5 < S_n^* < 5] \to 1 - (5.74 \times 10^{-7})
\end{aligned}$$

Thus for large δ, say $\delta > 3$, the limiting value of $P[-\delta < S_n^* < \delta]$ is much closer to 1 than the guaranteed Chebychev bound. ■

Example 2 Let $X_1, X_2, \ldots, X_n$ be independent and identically distributed with the density of Fig. 1 (Sec. 9-1), so that $EX_1 = 1.28$ and $\sigma_{X_1} \doteq 2.92$. From Corollary 1*a* we know that if $a < b$ then

$$P\left[a < \frac{S_n - n(1.28)}{2.92\sqrt{n}} < b\right] \to \Phi(b) - \Phi(a) \qquad \text{as } n \to \infty.$$

For $n = 100$, $a = -3.42$, and $b = 3.42$ the left-hand side reduces to $P[28 < S_{100} < 228]$, which we estimate by $\Phi(3.42) - \Phi(-3.42) = .9994$. In Example 1 (Sec. 12-1) we applied Chebychev's inequality and obtained the bound $P[28 < S_{100} < 228] \geq .914$. This calculation indicates that $P[28 < S_{100} < 228]$ is much closer to 1 than is indicated by the bound .914. This is undoubtedly the case, and we will utilize calculations of this kind. However, it is essential to be clear about the nature of such calculations, and such clarity encourages a needed cautiousness. In particular from Corollary 1*a* we *know* that

$$P\left[-3.42 < \frac{S_n - n(1.28)}{2.92\sqrt{n}} < 3.42\right] \to .9994 \qquad \text{as } n \to \infty.$$

We do *not* know how close the left-hand side is to the limit of .9994 when $n = 100$. If we assume that the left side is approximately equal to .9994 when $n = 100$ then substituting 100 for n gives us $P[28 < S_{100} < 228] \doteq .9994$; we really do not know how good this approximation is, although it probably is "reasonably good." Thus Corollary 1*a* gives us a precise limit statement, but it does not tell us what the error is if we use a particular n, a, b, and distribution for X_1. There are many known bounds on such errors. Such bounds make various assumptions about the distribution of X_1, and some of the bounds are quite complicated. For the "nice" distributions for X_1 which are often of interest, the normal approximations are often better than is shown by the error bounds. However, *it should be kept in mind that we do not have a rigorous justification for making an approximation like the one made in this example.* Particular caution is needed in making such approximations if n is small, or if a or b is too far from zero, or if $E|X_1^*|^3$ is large. Thus we have the *estimate* .9994 for $P[28 < S_{100} < 228]$, and we *know* that $.914 \leq P[28 < S_{100} < 228] \leq 1$. ■

Example 3 It has been found that 10 percent of a certain kind of vacuum tube perform unsatisfactorily. A "lot" of 1,000 such tubes is to be used, and we want to be fairly certain that at most 115 of them will perform unsatisfactorily. We assume that we have Bernoulli trials with $n = 1{,}000$ and $p = .1$, so $np = 100$, and $npq = 90$, and $\sqrt{npq} = 9.49$. We want an estimate for $P[S_n \leq 115]$. From Corollary 1*a* we find that

$$P[S_n \leq 115] = P[S_n - np \leq 15] = P\left[\frac{S_n - np}{\sqrt{npq}} \leq 1.58\right] \doteq \Phi(1.58) = .9429.$$

From the Harvard tables we find the exact value to be $P[S_n \le 115] = .94655$. Thus the estimate is quite close in this case, which is one of the few where the accuracy of the estimate can easily be checked. ■

Example 4 Continuing from the preceding example, suppose the 5 percent chance of getting $[S_n > 115]$ concerns us, so we ask for that number β such that $P[S_n \le \beta] \doteq .99$; that is, the number β should have the property that the probability is about .01 that more than β tubes are unsatisfactory. As before, we have

$$P[S_n \le \beta] = P\left[\frac{S_n - np}{\sqrt{npq}} \le \frac{\beta - 100}{9.49}\right] \doteq \Phi\left(\frac{\beta - 100}{9.49}\right).$$

From Table 1 (Sec. 13-1) we find that $\Phi(2.326) \doteq .99$, so we define β by $(\beta - 100)/9.49 = 2.326$ and find that $\beta = 122.1 \doteq 122$. Thus we conclude that $P[S_n \le 122] \doteq .99$, so that in only about 1 percent of such experiments should the number of unsatisfactory tubes exceed 122. Thus in this example we specified the value of the probability, and by using the CLT we found an appropriate event. As a check on the accuracy we find from the Harvard tables that $P[S_n \le 122] = .98964$ and $P[S_n \le 123] = .99207$. ■

Example 5 We apply the CLT to Example 5 (Sec. 12-1), where $T_{100} = Y_1 Y_2 \cdots Y_{100}$ is the product of 100 random variables. T_{100} can be interpreted as the total capital at the end of 100 months. We found that $P[T_{100} > 1.85] \ge .792$. In the course of this calculation we found $EX_1 = .0602$ and $\sigma_{X_1} = .262$, where $X_i = \log_{10} Y_i$. Thus if $S_{100} = X_1 + X_2 + \cdots + X_{100}$ we have

$$P[T_{100} > 1.85] = P[\log_{10} T_{100} > .267] = P[S_{100} > .267]$$

$$= P\left[\frac{S_{100} - nEX_1}{\sigma_{X_1}\sqrt{n}} > \frac{.267 - 6.02}{2.62}\right] = P[S_{100}^* > -2.2]$$

$$\doteq 1 - \Phi(-2.2) = \Phi(2.2) = .9861.$$

Thus from these two computations we have the definite bound $P[T_{100} > 1.85] \ge .792$ and the estimate $P[T_{100} > 1.85] \doteq .9861$. ■

Example 6 In this example we show that if X has a Poisson distribution with parameter λ, and λ is much larger than 1 then X^* is approximately normally distributed. We can apply the CLT to X, since we can represent X as the sum of a large number of independent random variables, each having a Poisson density. The result is easily stated in terms of λ converging to infinity, so we exhibit λ as a subscript of X. *If X_λ has a Poisson distribution with parameter λ we show that $P[X_\lambda^* < b]$ converges to $\Phi(b)$ as λ goes to infinity.*

In this paragraph we apply the result; the proof is left to the next paragraph. Note that $X_\lambda^* = (X_\lambda - \lambda)/\sqrt{\lambda}$, so if $a < b$ then

$$P[\lambda + a\sqrt{\lambda} < X_\lambda < \lambda + b\sqrt{\lambda}] = P[a < X_\lambda^* < b] \to \Phi(b) - \Phi(a) \qquad \text{as } \lambda \to \infty.$$

In particular if $a = -3$ and $b = 3$ then, as in Example 1, $P[\lambda - 3\sqrt{\lambda} < X_\lambda < \lambda + 3\sqrt{\lambda}]$ approaches .9974 as λ goes to infinity. Thus for large λ more than 99 percent of the probability is in the interval $\lambda \pm 3\sqrt{\lambda}$. For example, if $\lambda = 25$ we may use the estimate .9974 for $P[10 < X_\lambda < 40]$. From tables for $\lambda = 25$ we find that $P[10 < X_\lambda < 40] = .9963$, so that the approximation is fairly good in this case.

To prove the result we fix $\lambda > 0$ and let X_1 be Poisson with mean λ/n. Then *the distribution of S_n* is Poisson with mean λ and so *does not depend on n*. The bound of Corollary 1*b* becomes

$$3\left\{\frac{1}{\sqrt{\lambda}}\frac{E|X_1 - EX_1|^3}{\lambda/n}\right\}^{1/4}$$

which *does* depend on n and is valid for all n. It is left to Exercise 7 to show that $(n/\lambda)E|X_1 - EX_1|^3$ converges to 1 as n goes to infinity. Therefore if X_λ is Poisson with mean λ then

$$|P[X_\lambda^* < b] - \Phi(b)| \le \frac{3}{\lambda^{1/8}}$$

and the asymptotic result follows by letting λ go to infinity. ■

The proof in the preceding example illustrates the advantage of the bound of Corollary 1*b* over the asymptotic version of Corollary 1*a*. For the asymptotic version we must *fix* the distribution of X_1 and then let n go to infinity. However, from Corollary 1*b* we know, for example, that

$$|P[S_n^* < b] - \Phi(b)| \le .03 \qquad \textit{whenever}\ \frac{E|X_1^*|^3}{\sqrt{n}} \le 10^{-8}.$$

Thus Corollary 1*b* yields simultaneous results for large classes of distributions for X_1. In particular, as in Example 6, we may let the distribution for X_1 depend on n. Thus if $f(0) = 0$ and f is continuous for $x > 0$ then a bound of the form $f(E|X_1^*|^3/\sqrt{n})$ is vastly more powerful than the asymptotic conclusion of Corollary 1*a*. In particular such a bound proves that convergence toward normality is uniform with respect to a large class of initial distributions for X_1, say for $E|X_1^*|^3 \le 10$. As shown after Example 9, such bounds are particularly useful when the collection of random variables being added is permitted to change drastically as n increases.

All the exercises at the end of this section may be done before reading the remainder of the section.

*THE NONIDENTICALLY DISTRIBUTED CASE

The remainder of this section is devoted to the CLT for the case where $X_1, \ldots, X_n$ need not be identically distributed. We will derive Corollary 2 from Theorem 1 and apply it to three examples, but first let us consider an intuitively obvious context in which normal approximations may *not* be valid. Taking the variance of

$$S_n^* = \frac{S_n - ES_n}{s_n} = \frac{X_1 - EX_1}{s_n} + \cdots + \frac{X_n - EX_n}{s_n}$$

we obtain

$$1 = \text{Var}\ S_n^* = \frac{\text{Var}\ X_1}{\text{Var}\ S_n} + \cdots + \frac{\text{Var}\ X_n}{\text{Var}\ S_n}.$$

If $(\text{Var } X_1)/(\text{Var } S_n)$ is approximately 1, so that all the other ratios are small, then we *may* have $P[S_n^* < b] \doteq P[X_1 < b]$. Thus, as in Exercise 5, if $X_2, X_3, \ldots, X_n$ have very little variability then $S_n \doteq X_1 + EX_2 + \cdots + EX_n$, so $S_n - ES_n \doteq X_1 - EX_1$. That is, the dominance of one, or a very few, of the random variables in a sum can easily prevent $P[S_n^* < b]$ from being close to $\Phi(b)$. In terms of errors, if one error is typically drastically greater than all others then the distribution of the sum of errors might well be about the same as the distribution of the one large error.

One of the most common assumptions in asymptotic CLTs is that δ_n goes to zero as n goes to infinity, where δ_n is the largest of the n numbers $(\text{Var } X_1)/(\text{Var } S_n), \ldots, (\text{Var } X_n)/\text{Var } S_n$. Thus $(\text{Var } X_k)/(\text{Var } S_n)$ is a rough measure of the relative contribution of the kth term to S_n^*, and we require that no one of these be too large. In particular $(\text{Var } X_1)/(\text{Var } S_n)$ must go to zero, so $\text{Var } S_n$ must go to infinity. We now show that such an assumption is implicit in Theorem 1. Applying Theorem 3*f* (Sec. 11-3) to $Y_k = (X_k - EX_k)/s_n$, we have $(EY_k^2)^{1/2} \le (E\,|Y_k|^3)^{1/3}$, so that

$$\frac{\text{Var } X_k}{\text{Var } S_n} = EY_k^2 \le (E\,|Y_k|^3)^{2/3} \le \left\{\sum_{k=1}^{n} E\,|Y_k|^3\right\}^{2/3}$$

Thus for Theorem 1 to yield a useful conclusion, { } must be much less than 1, and this implies that each of $(\text{Var } X_k)/(\text{Var } S_n)$ is much less than 1.

Corollary 2 Let $S_n = X_1 + \cdots + X_n$ where $X_1, \ldots, X_n$ are independent random variables with each $E\,|X_k|^3 < \infty$. Assume that $s_n^2 = \text{Var } S_n > 0$ and let $S_n^* = (S_n - ES_n)/s_n$. Assume that M is a constant such that

$$E\,|X_k - EX_k|^3 \le M \text{ Var } X_k \qquad \text{whenever } 1 \le k \le n.$$

Then for every real b

$$|P[S_n^* < b] - \Phi(b)| \le 3\left\{\frac{M}{s_n}\right\}^{1/4}. \quad ■$$

Proof The hypothesized inequality shows that { } in Theorem 1 is dominated by

$$\frac{1}{s_n^3}\sum_{k=1}^{n} M \text{ Var } X_k = \frac{Ms_n^2}{s_n^3} = \frac{M}{s_n}. \quad ■$$

Example 7 If a gambler bets \$$A$ on the kth play of a game his gain is AZ_k; that is, AZ_k = [how much he is paid] − [the \$$A$ he bet]. Assume that $Z_1, Z_2, \ldots, Z_n$ are independent and identically distributed with $EZ_1 = 0$ and $EZ_1^2 = 1$ and $E\,|Z_1|^3 < \infty$. The gambler adopts the strategy of betting \$$k$ on the kth play, so his sequence of gains is $X_1, X_2, X_3, \ldots$, where $X_k = kZ_k$. Let $S_n = X_1 + X_2 + \cdots + X_n$ be his total gain on the first n plays so $ES_n = 0$ while $EX_k^2 = k^2$ and $E\,|X_k|^3 = k^3E\,|Z_k|^3$.

Clearly the smallest constant M that can be used in Corollary 2 is just the largest of the n numbers

$$\frac{E\,|X_k - EX_k|^3}{\text{Var } X_k} \qquad \text{for } 1 \le k \le n$$

and M could have been defined in this way.

For our example we may use $M = nE|Z_1|^3$. From Exercise 4 (Sec. 2-4)

$$s_n^2 = \text{Var } S_n = 1 + 2^2 + 3^2 + \cdots + n^2 = (1/6)n(n+1)(2n+1) \geq \frac{(n)(n)(2n)}{6} = \frac{n^3}{3}$$

so that M/s_n goes to zero as n goes to infinity. Thus $P[S_n < bs_n]$ converges to $\Phi(b)$ as n goes to infinity.

If $b = -1$ then $P[S_n < -s_n]$ approaches .1587 as n goes to infinity. For $n = 100$ we find that there is *about* a 16 percent chance that the gambler will have lost over \$600 on the first 100 trials, at which time the total of his bets would be $1 + 2 + \cdots + n = n(n+1)/2 =$ \$5,050. The validity of this *estimate* depends, of course, on the distribution of Z_1. ■

Example 8 If c is a constant such that $P[|X_k| \leq c] = 1$ then $E|X_k| \leq c$. Furthermore on the event $[|X_k| \leq c]$ we have

$$|X_k - EX_k|^3 = |X_k - EX_k|\,(X_k - EX_k)^2 \leq (c + c)(X_k - EX_k)^2.$$

Therefore we can let $M = 2c$ in Corollary 2 and use $(3)(2^{1/4}) \leq 4$ to obtain the following result:

Let $S_n = X_1 + \cdots + X_n$ where $X_1, \ldots, X_n$ are independent, with each $P[|X_k| \leq c] = 1$, where c is a constant. If $s_n^2 = \text{Var } S_n > 0$ then for every real b

$$|P[S_n^* < b] - \Phi(b)| \leq 4\left\{\frac{c}{s_n}\right\}^{1/4}.$$

Thus the CLT always applies to the sum of independent uniformly bounded random variables, whenever the variance of the sum goes to infinity. ■

Example 9 We call $X_1, \ldots, X_n$ **generalized Bernoulli trials** iff they are independent and $P[X_k = 0] + P[X_k = 1] = 1$ whenever $1 \leq k \leq n$. The probability of success may vary from trial to trial. The notation

$$p_k = P[X_k = 1] \qquad q_k = P[X_k = 0]$$

seems natural. Using $c = 1$ in the preceding example, we obtain the following result:

If $S_n = X_1 + \cdots + X_n$ is the sum of n generalized Bernoulli trials and if

$$\text{Var } S_n = s_n^2 = \sum_{k=1}^{n} p_k q_k > 0$$

then for every real b

$$|P[S_n^* < b] - \Phi(b)| \leq \frac{4}{(s_n)^{1/4}}.$$

Thus $s_n \to \infty$ guarantees normal convergence for generalized Bernoulli trials. For example, if $p_k = 1/k$ then

$$s_n^2 = \sum_{k=2}^{n} \frac{1}{k}\left(1 - \frac{1}{k}\right) \geq \sum_{k=2}^{n} \frac{1}{k}\frac{1}{2} \to \infty \qquad \text{as } n \to \infty.$$

Clearly $p_k = 1/k^2$ decreases too rapidly to make s_n go to infinity. We recall from Example 1 (Sec. 8-1) that the exact probability density for S_n was called the *Poisson binomial*, and it is quite complicated.

The conclusions of Theorems 1 and Corollary 2 depend on n essentially only through the bounding quantities. In particular we may restate our last result as follows:

If S is the sum of some number of generalized Bernoulli trials and if Var $S > 0$ then for every real b

$$|P[S^* < b] - \Phi(b)| \leq \frac{4}{(\text{Var } S)^{1/8}}.$$

For example, S might be the number of motor vehicles which will have flat tires on a particular day in a particular city. We need not know the number n of vehicles involved or the conditions p_k of their tires, but we may be willing to assume independence and have large enough Var S to indicate normal approximations to the distribution of S.

If $p_k \leq 1/2$ then $p_k q_k \geq p_k/2$, and summing over k, we see that Var $S \geq ES/2$. If we use this assumption and note that $(4)(2^{1/8}) \leq 5$ we obtain the following pleasing result:

If S is the sum of some number of generalized Bernoulli trials for which each trial has at most probability $1/2$ for a success, and $ES > 0$, then for every real b

$$|P[S^* < b] - \Phi(b)| \leq \frac{5}{(ES)^{1/8}}.$$

Thus a large mean suggests the normal approximation to the distribution of the number of successes in many rare independent events. Section 8-3 showed that a Poisson approximation is valid for some such cases. There is no contradiction, since Example 6 showed the normal approximation to be applicable to a Poisson distribution with a large mean. ■

Let $k_1, k_2, \ldots$ be a sequence of positive integers. We normally think of them as increasing, but they need not. For each n let $X_{n,1}, X_{n,2}, \ldots, X_{n,k_n}$ be k_n independent random variables and let S_n be their sum. Thus S_n is the sum of the random variables in the nth row of Fig. 1.

$$\begin{array}{llllll} X_{1,1} & X_{1,2} & \cdots & X_{1,k_1} & & \\ X_{2,1} & X_{2,2} & \cdots & & X_{2,k_2} & \\ \cdots & \cdots & \cdots & \cdots & \cdots & \\ X_{n,1} & X_{n,2} & \cdots & & & X_{n,k_n} \\ \cdots & \cdots & \cdots & \cdots & \cdots & \cdots \end{array}$$

Fig. 1

Assume that each $E|X_{n,k}|^3 < \infty$ and that each $s_n^2 = \text{Var } S_n > 0$. Theorem 1 applied to the nth row yields

$$|P[S_n^* < b] - \Phi(b)| \leq 3(B_n)^{1/4}$$

where

$$B_n = \sum_{j=1}^{k_n} E\left|\frac{X_{n,j} - EX_{n,j}}{s_n}\right|^3.$$

Therefore, if B_n goes to zero as n goes to infinity then $P[S_n^* < b]$ converges to $\Phi(b)$ as n goes to infinity, for each real b. This is a CLT for a *general triangular array*. Such formulations are traditional. They deal only with the sequence of distributions of $S_1, S_2, \ldots$, so only the joint distribution of variables in a row is of relevance. Theorem 3 (Sec. 20-2) is such a CLT, and it requires the finiteness of only second moments.

If $k_n = n$ and $X_{n,j} = X_j$ then Fig. 1 reduces to Fig. 2, which just yields our previous

$$\begin{array}{lll} X_1 & & \\ X_1 & X_2 & \\ X_1 & X_2 & X_3 \\ \cdots & \cdots & \cdots \end{array}$$

Fig. 2

consecutive sum formulation, $S_n = X_1 + \cdots + X_n$. As was just demonstrated, an inequality like that of Theorem 1 immediately applies to Fig. 1. A limit theorem for Fig. 2 does not normally yield a corresponding theorem for Fig. 1. One simple way in which Fig. 1 arises is in applying varying weightings to a sequence $X_1, X_2, \ldots$;

$$S_n = \sum_{j=1}^{k_n} a_{n,j} X_j \qquad \text{so } X_{n,j} = a_{n,j} X_j.$$

We have already essentially used arrays like Fig. 1. In Sec. 8-3 we let $k_n = n$ and made each random variable in the nth row Bernoulli with parameter λ/n. We found for large n that S_n was approximately Poisson with mean λ. In Example 6 we let $k_n = n$ and made each random variable in the nth row Poisson with mean λ/n. In both cases the distribution of $X_{n,1}$ depends on n, so that we cannot reduce to Fig. 2. In Example 9 we observed that the number k_n of random variables added is sometimes best left unspecified.

EXERCISES

1. *Estimates from the CLT Instead of Bounds from Chebychev's Inequality*

(*a*) Use the CLT to estimate the number $P[.9 < (1/100)S_{100} < 2.5]$ of Exercise 1*a* (Sec. 12-1). Chebychev's inequality showed it to be $\geq .914$.

(*b*) Use the CLT to estimate the number $P[T_{100} \geq 600]$ of Exercise 1*b* (Sec. 12-1). Chebychev's inequality showed it to be $\leq .011$.

(*c*) Use the CLT to estimate the number P[more than 25 units will be replaced during 100 days] of Exercise 1*c* (Sec. 12-1). Chebychev's inequality showed it to be $\leq .04$.

2. *Sums of Random Decimal Digits.* Let $X_1, X_2, \ldots, X_{50}$ be independent and identically distributed, with $P[X_1 = 0] = P[X_1 = 1] = \cdots = P[X_1 = 9] = .1$, and let $S_{50} = X_1 + X_2 + \cdots + X_{50}$. Clearly $EX_1 = 4.5$, so $ES_{50} = 225$. We easily find that $EX_1^2 = 28.5$, so $\text{Var } X_1 = 8.25$ and $\sigma_{X_1} = 2.872$.

(*a*) Find a number k such that $P[225 - k \leq S_{50} \leq 225 + k] \doteq .8$

(*b*) Select a column of 50 random decimal digits from Table 1 (Sec. 12-2) and add them. Is your sum in the interval $225 \pm k$ for the k obtained from part (*a*)?

3. *Demonstrating that $P[S_n^* = b]$ Goes to Zero as n Goes to Infinity.* Let Z_n be any sequence of random variables such that for every real b we have $P[Z_n < b]$ converging to $\Phi(b)$ as n approaches infinity. Show that for every real b, $P[Z_n = b]$ goes to zero as n goes to infinity. HINT: If $\epsilon > 0$ then clearly $P[Z_n = b] \leq P[b - \epsilon \leq Z_n < b + \epsilon]$.

4. *Nonuniformity of CLT Convergence over Too Large a Class of Distributions.* Under the assumptions of Corollary 1, for any n there are distributions for X_1 such that $P[S_n^* < b]$ is not yet close to $\Phi(b)$. For a certain p_{X_1} we may find that $n = 8$ is large enough for Φ to yield good approximations, while for another p_{X_1} we may need $n = 200$, etc. To prove that this phenomenon can occur, for each n let

$$S_n = X_{n,1} + X_{n,2} + \cdots + X_{n,n}$$

where $X_{n,1}, X_{n,2}, \ldots, X_{n,n}$ are independent and each is Poisson with mean $1/(2n)$. Then each S_n has the same Poisson distribution with mean $\frac{1}{2}$. In particular $P[S_n = 0] = e^{-1/2}$, so that the distribution of S_n^* is not close to Φ in any reasonable sense. If n is large then

$$P[X_{n,1} = 0] = e^{-1/(2n)} \doteq 1$$

so that each $X_{n,k}$ is "almost" the constant zero; hence n of them must be added before there is appreciable variability.

(*a*) On the same plot sketch both $P[S_n^* < b]$ and $\Phi(b)$ as functions of b.

(*b*) Show that this same phenomenon arises if $X_{n,1}, \ldots, X_{n,n}$ are n Bernoulli trials with probability $1/(2n)$ for a success on each trial.

(*c*) Prove or disprove the statement that: there is a sequence $A_1, A_2, \ldots$ of real numbers converging to zero and satisfying $|P[S_n^* < 0] - .5| \leq A_n$ *whenever* $S_n = X_1 + \cdots + X_n$ where $X_1, \ldots, X_n$ are independent and identically distributed, with $E|X_1|^3 < \infty$ and $\operatorname{Var} X_1 > 0$.

5. *The Nonapplicability of the CLT When X_1 Dominates.* Let $Y_1, Y_2, \ldots, Y_n$ be independent and identically distributed with $p_{Y_1}(-1) = p_{Y_1}(1) = .5$ so that $EY_1 = 0$ and $\operatorname{Var} Y_1 = 1$. Let $X_k = \sqrt{15}\, Y_k/4^k$ and let $S_n = X_1 + X_2 + \cdots + X_n$, so that

$$S_n = \frac{\sqrt{15}}{4} Y_1 + \frac{\sqrt{15}}{4^2} Y_2 + \frac{\sqrt{15}}{4^3} Y_3 + \cdots + \frac{\sqrt{15}}{4^n} Y_n$$

and

$$\operatorname{Var} S_n = 15[\tfrac{1}{16} + (\tfrac{1}{16})^2 + (\tfrac{1}{16})^3 + \cdots + (\tfrac{1}{16})^n]$$

$$= \frac{15}{16}\frac{1 - (\frac{1}{16})^n}{1 - (\frac{1}{16})} = 1 - \left(\frac{1}{16}\right)^n$$

from equation (2) of Appendix 1. Thus $ES_n = 0$, and $\operatorname{Var} S_n \doteq 1$ if n is large, which is why the factor of $\sqrt{15}$ was introduced. Prove that $P[|S_n| \leq .5] = 0$ for every $n \geq 1$, so that convergence to Φ is impossible.

6. *Normal Convergence for Weighted Sums.* Let $Y_1, Y_2, \ldots, Y_n$ be independent and identically distributed, with $EY_1 = 0$ and $\operatorname{Var} Y_1 = 1$ and $E|Y_1|^3 < \infty$. Exercise 5 contains an example where $S_n = a_1 Y_1 + a_2 Y_2 + \cdots + a_n Y_n$ does not have the normal distribution in the limit because $a_1 Y_1$ dominates, since the a_k go to zero very rapidly. In this exercise we let the a_k go to zero, but so slowly that we are near enough to the case $a_1 = a_2 = a_3 = \cdots$ for normal convergence to be obtained from Theorem 1. Find an interval of values of $\alpha > 0$ such that if $S_n = Y_1 + Y_2/2^\alpha + Y_3/3^\alpha + \cdots + Y_n/n^\alpha$ then $P[S_n^* < b]$ converges to $\Phi(b)$ as n approaches infinity.

7. *Completing the Proof of Example 6.* Let X_λ be Poisson with mean λ and prove that $(1/\lambda)\, E|X_\lambda - \lambda|^3$ converges to 1 as λ goes to zero. HINT:

$$\frac{1}{\lambda} E|X_\lambda - \lambda|^3 = \frac{1}{\lambda}|0 - \lambda|^3 e^{-\lambda} + \frac{1}{\lambda}|1 - \lambda|^3 e^{-\lambda}\lambda + \sum_{k=2}^{\infty} \frac{1}{\lambda}|k - \lambda|^3 \frac{e^{-\lambda}\lambda^k}{k!}$$

and

$$\frac{1}{\lambda}|k - \lambda|^3 \frac{e^{-\lambda}\lambda^k}{k!} \leq k^3 \lambda^{k-1} \qquad \text{if } 0 < \lambda \leq 1 \text{ and } k \geq 1.$$

8. *The Bound of Theorem 1 is Worthless if $1 \leq n \leq 6{,}561$.* The two numbers $P[S_n^* < b]$ and $\Phi(b)$ belong to the unit interval, so $|P[S_n^* < b] - \Phi(b)| \leq 1$ is always true. Thus the bound of Theorem 1 is informative only when it is less than 1. Prove that if the bound of Theorem 1 is less than 1 then $n > 6{,}561$.

A USEFUL FACT: First note that if $0 < r < s$ and $0 \le a_k \le b_k$ for each k then

$$\left[\frac{1}{n}(a_1^r + \cdots + a_n^r)\right]^{1/r} \le \left[\frac{1}{n}(a_1^s + \cdots + a_n^s)\right]^{1/s} \le \left[\frac{1}{n}(b_1^s + \cdots + b_n^s)\right]^{1/s}.$$

The first inequality is obtained by applying Theorem 3*f* (Sec. 11-3) to the random variable $X(k) = a_k$ with associated uniform density on $\Omega = \{1, 2, \ldots, n\}$. The second inequality follows from the fact that x^s is an increasing function of $x \ge 0$ for fixed $s > 0$. If $a_k = (E\,|\,Y_k|^r)^{1/r}$ and $b_k = (E\,|\,Y_k|^s)^{1/s}$ then $a_k \le b_k$ from Theorem 3*f* (Sec. 11-3). *Therefore* if $0 < r < s$ then

$$\left[\frac{1}{n}(E\,|\,Y_1|^r + \cdots + E\,|\,Y_n|^r)\right]^{1/r} \le \left[\frac{1}{n}(E\,|\,Y_1|^s + \cdots + E\,|\,Y_n|^s)\right]^{1/s}.$$

Theorem 3*f* (Sec. 11-3) is the special case $n = 1$ of this result. Now let $r = 2$, $s = 3$, and $Y_k = (X_k - EX_k)/[\mathrm{Var}\,(X_1 + \cdots + X_n)]^{1/2}$. The left-hand side becomes $1/\sqrt{n}$, and raising both sides to the third power yields

$$E\,|\,Y_1|^3 + \cdots + E\,|\,Y_n|^3 \ge \frac{1}{\sqrt{n}}.$$

Exercises 9 and 10 permit the application of some of the techniques of Chaps. 6 to 13.

9. The kth of $n = 100$ blocks of metal has linear dimensions X_k, Y_k, and Z_k in feet, and density ρ_k in pounds per cubic foot. The 100 random vectors $f_1, f_2, \ldots, f_{100}$ are independent and identically distributed, with $f_k = (X_k, Y_k, Z_k, \rho_k)$. The joint density of X_1, Y_1, Z_1, ρ_1 is given in Fig. 3. Let $T = W_1 + W_2 + \cdots + W_{100}$ be the total weight of the 100 blocks,

x_1	y_1	z_1	ρ_1	$p_{f_1}(x_1, y_1, z_1, \rho_1)$
1	2	2	1	.1
2	1	2	.75	.4
1	1	2	1	.3
1	2	1	2	.2

Fig. 3

where $W_k = X_k Y_k Z_k \rho_k$ is the weight of the kth block.

(*a*) Prove that $P[T \ge 315]$ must be small.

(*b*) Use the CLT to estimate $P[T \ge 315]$.

(*c*) The kth block is given a thin protective plastic coating δ_k feet thick. Thus $V_k = A_k \delta_k$ cubic feet of the liquid plastic is needed for the kth block, where

$$A_k = 2[X_k Y_k + X_k Z_k + Y_k Z_k]$$

is the surface area of the kth block. Assume that $f_1, f_2, \ldots, f_{100}, \delta_1, \delta_2, \ldots, \delta_{100}$ are 200 independent random vectors, and $\delta_1, \delta_2, \ldots, \delta_{100}$ are identically distributed by

$$p_{\delta_1}(.001) = .2 \qquad p_{\delta_1}(.002) = .7 \qquad p_{\delta_1}(.004) = .1$$

Let $S = V_1 + V_2 + \cdots + V_{100}$ be the total volume of plastic needed and show that $P[S \ge 3]$ must be small.

10. A sample space Ω has eight sample points with probabilities as shown in Fig. 4. Four

i	$p(\omega_i)$	$X(\omega_i)$	$Y(\omega_i)$	$Z(\omega_i)$	$W(\omega_i)$
1	.1	0	−1	0	0
2	.05	0	−1	−1	1
3	.05	−1	−1	0	1
4	.3	−1	0	1	0
5	.1	1	1	0	−1
6	.1	−1	0	1	1
7	.1	2	0	−1	0
8	.2	2	1	0	−1

Fig. 4

random variables X, Y, Z, and W are defined on Ω as shown. The two random variables S and T are defined on Ω by $S = X + Y + Z + W$ and $T = X^2 + Y^2 + Z^2 + W^2$. For interpretation we might assume that an experiment consists of presenting a stimulus to a subject and then observing the four voltages X, Y, Z, and W at different electrodes on his scalp.

(*a*) Find and plot the discrete densities $p_{S,T}$ and p_S and p_T. Find the least-squares regression line of T on S and plot it on the same plot with $p_{S,T}$. Find the expected squared error if the least-squares regression line of T on S is used as a prediction of T, given S.

A total of $m = 10^3$ independent repetitions of the experiment are to be performed. Therefore $(S_1,T_1), (S_2,T_2), \ldots, (S_m,T_m)$ are m independent random vectors, each having the same density $p_{S,T}$ as obtained above. If $1 \le k \le m$ let $U_k = S_1 + S_2 + \cdots + S_k$ and $V_k = T_1 + T_2 + \cdots + T_k$.

(*b*) Let $n = 100$ and show that $P[U_n \ge 80]$ is small.

(*c*) Let $n = 100$ and use the CLT to find a number b such that $P[U_n < b] \doteq .67$.

(*d*) Use the SLLN to show that the probability is large that every one of $(1/100)U_{100}$, $(1/101)U_{101}, \ldots, (1/1{,}000)U_{1{,}000}$ will be within the limits .2 to .8.

(*e*) Let $n = 100$ and show that $P[V_n e^{(1/n)U_n} < 900]$ is large. HINT: Recall Exercise 4*a* (Sec. 6-4).

13-3 THE NORMAL APPROXIMATION TO THE BINOMIAL

In this section we first consider approximations for unstandardized densities. After Example 2 we exhibit a self-contained demonstration (without appealing to the CLT) that the normal density arises when the binomial is approximated.

Let X be a random variable, $EX = \mu$, $\operatorname{Var} X = \sigma^2$, for which the normal approximation for the standardized X^*, $P[X^* < b] \doteq \Phi(b)$ for all b, is satisfactory. If $\alpha < \beta$ and we want an approximation to the probability $P[\alpha < X < \beta]$ that the unstandardized X belongs to an interval, we can describe this probability in terms of X^* and obtain

$$P[\alpha < X < \beta] = P\left[\frac{\alpha-\mu}{\sigma} < \frac{X-\mu}{\sigma} < \frac{\beta-\mu}{\sigma}\right]$$

$$= P\left[\frac{\alpha-\mu}{\sigma} < X^* < \frac{\beta-\mu}{\sigma}\right] \doteq \Phi\left(\frac{\beta-\mu}{\sigma}\right) - \Phi\left(\frac{\alpha-\mu}{\sigma}\right)$$

$$= \int_{(\alpha-\mu)/\sigma}^{(\beta-\mu)/\sigma} \varphi(y)\,dy = \int_\alpha^\beta \frac{1}{\sigma}\varphi\left(\frac{x-\mu}{\sigma}\right) dx = \int_\alpha^\beta \varphi_{\mu,\sigma}(x)\,dx$$

where for the next-to-last equality we made the transformation $y = (x - \mu)/\sigma$. Therefore the usual normal approximations for X^* are equivalent to

$$P[\alpha < X < \beta] \doteq \Phi\left(\frac{\beta - \mu}{\sigma}\right) - \Phi\left(\frac{\alpha - \mu}{\sigma}\right) = \int_\alpha^\beta \varphi_{\mu,\sigma}(x)\,dx \qquad \alpha < \beta.$$

Thus when we want numerical approximations concerning the distribution of X we use the function Φ, which is easily tabulated since it involves no parameters, but this is equivalent to approximating $P[\alpha < X < \beta]$ by the area under the function $\varphi_{\mu,\sigma}$ and above the interval $\alpha < x < \beta$, where we use μ and σ given by $\mu = EX$ and $\sigma = \sqrt{\operatorname{Var} X}$.

If we know the points $\{x: p_X(x) > 0\} = \{x_1, x_2, \ldots, x_n\}$ where the discrete density p_X is positive, at times it is reasonable, as in Fig. 1, to introduce the midpoints

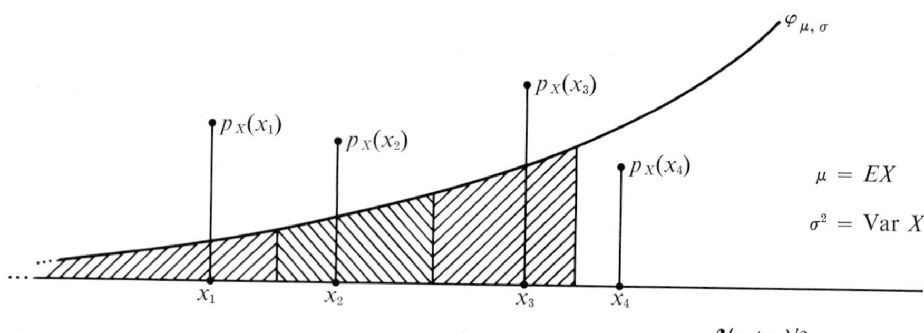

Fig. 1 $p_X(x_2) \doteq \int_{(x_1+x_2)/2}^{(x_2+x_3)/2} \varphi_{\mu,\sigma}(x)\,dx.$

$(x_k + x_{k+1})/2$ and use the approximation

$$p_X(x_k) \doteq \int_{(x_{k-1}+x_k)/2}^{(x_k+x_{k+1})/2} \varphi_{\mu,\sigma}(x)\,dx$$

for $p_X(x_k)$. We can then approximate the probability of an interval by adding the approximations to the probabilities of those x_k in the interval. For example, if S_n is binomial n and p then we are led to the approximations

$$\sum_{k=i}^{j} \binom{n}{k} p^k q^{n-k} = P[i \le S_n \le j] = P[i - .5 \le S_n \le j + .5]$$
$$\doteq \int_{i-.5}^{j+.5} \varphi_{\mu,\sigma}(x)\,dx = \Phi\left(\frac{j + .5 - np}{\sqrt{npq}}\right) - \Phi\left(\frac{i - .5 - np}{\sqrt{npq}}\right)$$
$$\text{if } 0 \le i \le j \le n.$$

Thus the present approach leads to the incorporation of the .5 terms, and usually such an approximation is better than one which does not incorporate them.

Example 1 Let S_n be binomial with $n = 1{,}000$ and $p = .1$, so that $np = 100$, $npq = 90$, and $\sqrt{npq} = 9.49$. In Example 3 (Sec. 13-2) we found that

$$P[S_n \le 115] \doteq \Phi\left(\frac{115 - np}{\sqrt{npq}}\right) = \Phi(1.58) = .9429.$$

From tables, $P[S_n \leq 115] = .94655$. From the approximation technique which has just been described we get

$$P[S_n \leq 115] = P[S_n \leq 115.5] \doteq \Phi\left(\frac{115.5 - np}{\sqrt{npq}}\right) = \Phi(1.63) = .9484$$

which is somewhat closer than the first approximation. ■

If a discrete density p_X has positive values at only the points $k - 1$, k, and $k + 1$ in the interval $k - 1 \leq x \leq k + 1$ our approximation technique yields

$$p_X(k) = P[k - \tfrac{1}{2} < X < k + \tfrac{1}{2}] \doteq \int_{k-\frac{1}{2}}^{k+\frac{1}{2}} \varphi_{\mu,\sigma}(x)\, dx \doteq \varphi_{\mu,\sigma}(k)$$

if $\varphi_{\mu,\sigma}$ changes very little over the interval, so that the area above the interval is approximately equal to the length of the interval times the height of $\varphi_{\mu,\sigma}$ at the center of it. That is, *in the special case* where the points where p_X has positive values happen to be just *one unit apart* then we may hope that these values lie on or close to the continuous curve $\varphi_{\mu,\sigma}$. As is evident from Fig. 3 (Sec. 13-1), where $\varphi_{-6,.3}$ have values larger than 1, this kind of fit cannot be hoped for in general. The next example and the numerical example of the next section involve discrete densities of this special kind. Figure 2 of this section and Fig. 5 of the next section show that the positive values for these densities do indeed fall close to the continuous curve $\varphi_{\mu,\sigma}$.

Example 2 The binomial density for $n = 40$ and $p = .08$ of Fig. 3 (Sec. 8-2) is fairly asymmetrical because n must be made still larger for this small $p = .08$ before the limiting normal shape becomes clearer. Figure 2 shows that the normal approximation is quite good for $n = 50$ and $p = .3$. ■

As illustrated by Fig. 2, the approximation

$$b(k) = \binom{n}{k} p^k q^{n-k} \doteq \varphi_{np,\sqrt{npq}}(k) = \frac{1}{\sqrt{2\pi npq}} e^{-(k-np)^2/2npq} \tag{1}$$

of the individual terms of the binomial density by the values of the normal density is often quite good. *The remainder of this section is devoted to justifying this normal approximation to the binomial*, in certain cases by direct fairly simple manipulations, without invoking the CLT.

For convenience we consider only very large n, and we will make the mean np an integer. From Sec. 8-2 we know that $b(k)$ is maximized for $k = np$, in which case approximation (1) becomes

$$b(np) = \binom{n}{np} p^{np} q^{n-np} \doteq \frac{1}{\sqrt{2\pi npq}} \tag{2}$$

which can also be justified directly, as in Exercise 1. Thus if np is an integer then approximation (1) is similar to

$$b(k) = \binom{n}{k} p^k q^{n-k} \doteq b(np) e^{-(k-np)^2/2npq} \tag{3}$$

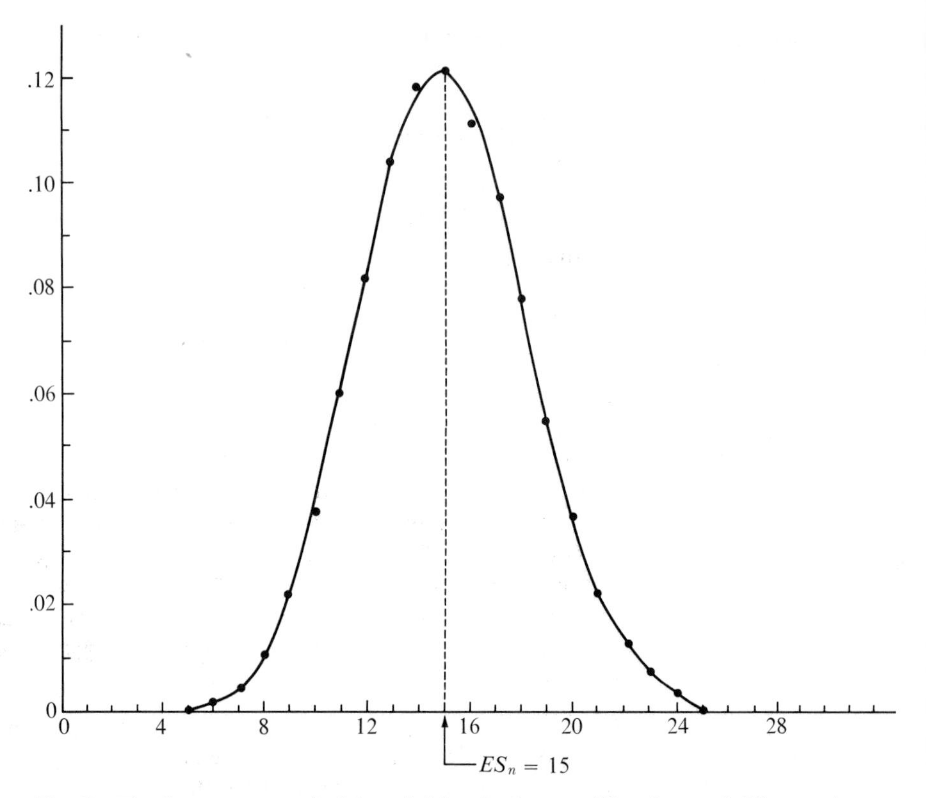

Fig. 2 The dots represent the binomial density for $n = 50$ and $p = .3$. The continuous curve represents $(1/\sqrt{2\pi npq})e^{-(k-np)^2/2npq}$ where $np = 15$, $npq = 10.5$, and k is treated as a continuous variable.

and *it is approximation (3) which we will justify*. We are not interested in precise approximations or multiplicative constants; our principal interest is to show that over a range of values of k that $b(k)$ has a reasonable approximation by some function of the form $Ce^{-B(k-A)^2}$. As we saw in Sec. 13-1, the constants A, B, and C can then be determined by standardizing the function, with k treated as a continuous variable.

We first note that if $k = np \pm 2\sqrt{npq}$ is two standard deviations from the mean then approximation (3) becomes $b(k) \doteq b(np)e^{-2} = .135b(np)$, so that b has already dropped to about 14 percent of its peak value of $b(np)$. We will attempt to approximate $b(k)$ only in the range $np \pm 2\sqrt{npq}$. Actually we will consider k only in the range $np \leq k \leq np + 2\sqrt{npq}$, since an analogous argument works on the other side of the mean. In order to keep our deduction simple we select particular values for n and p, but it will be clear that the approximations apply somewhat more generally.

We arbitrarily select $p = .3$ and $n = 210{,}000$, so that the mean is 63,000 and the standard deviation is $\sqrt{npq} = 210$. We want to approximate this binomial density $b(k)$ for k satisfying $63{,}000 \leq k \leq 63{,}420$. We select such a k and keep it fixed. Then, as in Fig. 3, we define r by $k = np + r$, so r is fixed and

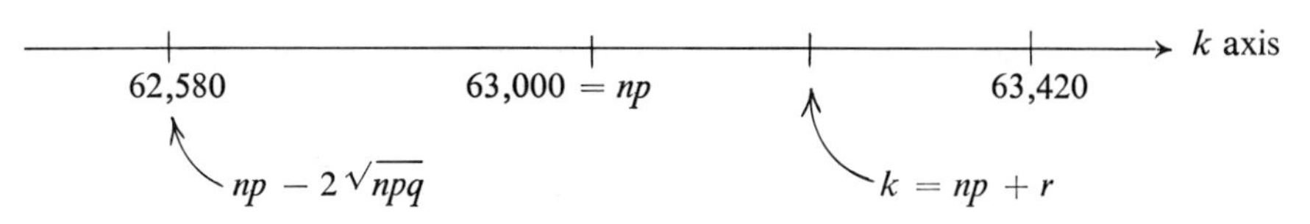

Fig. 3

satisfies $0 \le r \le 420$. Now the fact that $b(k)/b(np)$ can be written as

$$\frac{b(k)}{b(np)} = \frac{b(np+r)}{b(np)} = \frac{b(np+1)}{b(np)}\frac{b(np+2)}{b(np+1)}\frac{b(np+3)}{b(np+2)}\cdots\frac{b(np+r)}{b(np+r-1)} \tag{4}$$

is easily seen by cancelling each numerator with the next denominator. As noted in Sec. 8-2, an easy calculation shows that

$$\frac{b(t)}{b(t-1)} = 1 - \frac{t-(n+1)p}{tq} \qquad \text{where } t = 1, 2, \ldots, n.$$

If we let $t = np + j$ we get for each factor in equation (4) the simple form

$$\frac{b(np+j)}{b(np+j-1)} = 1 - \frac{j-p}{npq+jq} \doteq 1 - \frac{j-p}{npq}$$

where the approximation is obtained by observing that for every factor in equation (4) we have $npq = 44{,}100$ and $jq \le (420)(.7) = 294$. Using this approximation in equation (4), we get

$$\frac{b(k)}{b(np)} \doteq \left(1 - \frac{1-p}{npq}\right)\left(1 - \frac{2-p}{npq}\right)\left(1 - \frac{3-p}{npq}\right)\cdots\left(1 - \frac{r-p}{npq}\right). \tag{5}$$

Now we want to approximate each of these factors by an exponential so that we can add exponents. From inequality (7) of Appendix 1, $1 - x \le e^{-x}$ and we have approximate equality if x is close to zero. For example, $1 - .01 = e^{-.010050}$, where the exponent is shown to six decimal places, so that $1 - .01 \doteq e^{-.01}$. Using this result in approximation (5) and recalling that $1 + 2 + 3 + \cdots + r = r(r+1)/2$ from equation (1) of Appendix 1, we obtain

$$\begin{aligned}\frac{b(k)}{b(np)} &\doteq \exp\left\{-\frac{1-p}{npq} - \frac{2-p}{npq} - \frac{3-p}{npq} - \cdots - \frac{r-p}{npq}\right\}\\ &= \exp\left\{-\frac{1}{npq}[1 + 2 + \cdots + r - rp]\right\} = \exp\left\{-\frac{1}{npq}\left[\frac{r(r+1)}{2} - rp\right]\right\}\\ &= \exp\left\{-\frac{r}{2npq}[r + 1 - 2p]\right\} \doteq \exp\left\{-\frac{r^2}{2npq}\right\}\end{aligned}$$

where the last $\doteq$ is reasonable, at least for $r \ge 10$. For $0 \le r \le 10$ both sides of the last $\doteq$ are approximately $e^\circ = 1$, so the last $\doteq$ is still reasonable. *Thus approximation (3) is reasonable* for these and similar conditions on n, p, and k.

EXERCISES

1. *The Maximum Value of the Binomial.* Recalling Sirling's formula $n! \sim \sqrt{2\pi n}e^{-n}n^n$, where $\sim$ means that the ratio of the two sides converges to 1, prove that

$$\binom{n}{np} p^{np} q^{nq} \sim \frac{1}{\sqrt{2\pi npq}}$$

in the sense that the ratio of the two sides converges to 1 as n, np, and nq all go to infinity, always requiring np to be an integer. This exercise justifies approximation (2) of this section.

2. *The Normal Density from the Poisson.* Let X have a Poisson density p with parameter λ, where λ is a large integer. The CLT was used in Example 6 (Sec. 13-2) to show that $P[X^* < b] \doteq \Phi(b)$, so from the approximation technique of this section we expect that

$$p(k) = \frac{e^{-\lambda}\lambda^k}{k!} \doteq \frac{1}{\sqrt{2\pi\lambda}} e^{-(k-\lambda)^2/2\lambda}$$

From this we see that $p(\lambda) \doteq 1/\sqrt{2\pi\lambda}$, so that

$$p(k) \doteq p(\lambda)e^{-(k-\lambda)^2/2\lambda}$$

Without using the CLT, justify this last approximation if $\lambda = 10{,}000 \le k \le \lambda + 2\sqrt{\lambda} = 10{,}200$.

13-4 A NUMERICAL EXAMPLE

In this section we consider a numerical example and find that a few simple calculations provide startling evidence for the phenomena described by the CLT. The calculations themselves may help to make more concrete the idea of the distribution of a sum of independent random variables.

Let $X_1, X_2, \ldots, X_8$ be eight independent identically distributed random variables, where p_{X_1} is given by Fig. 1, and let $S_k = X_1 + X_2 + \cdots + X_k$. The purpose of this section is to calculate p_{S_8} and to show that S_8 is approximately normally distributed. We now comment on the results of the calculations to be described

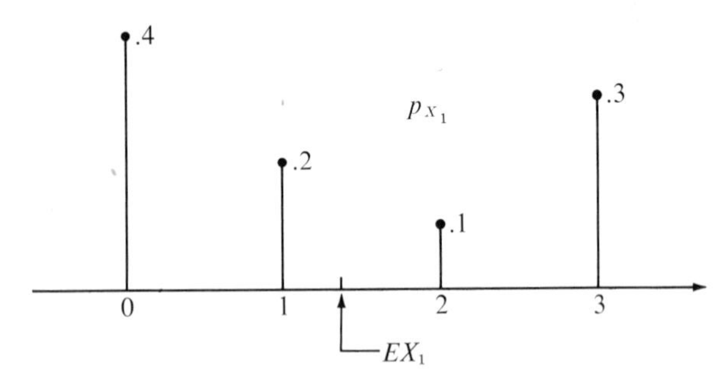

Fig. 1 $EX_1 = 1.3$, $\sigma^2_{X_1} = 1.61$, and $\sigma_{X_1} \doteq 1.269$.

shortly. The discrete densities $p_{X_1}, p_{S_2}, p_{S_4}$, and p_{S_8} are tabulated in Fig. 4 and plotted in Fig. 5. As explained in the previous section, we should have $p_{S_8}(k) \doteq \varphi_{\mu,\sigma}(k)$ for $k = 0, 1, \ldots, 24$, where $\mu = ES_8 = 10.4$ and $\sigma = \sigma_{S_8} = 3.589$. A glance at Fig. 5 or the $p_{S_8}, \varphi_{\mu,\sigma}$ columns of Fig. 4 shows that this approximation is quite good. Figure 5 also contains a table which makes more concrete the fact from Example 1 (Sec. 13-2) that $P[|S_n - ES_n| \le \sigma_{S_n}]$ converges to .6826 as n approaches infinity. Figure 5 shows plots of Φ and $P[S_8^* < b]$ as a function of b. Clearly $P[S_8^* < b] \doteq \Phi(b)$ is a fairly good approximation. According to Sec. 13-3, a reasonable method for approximating $P[S_8 < k]$, where $k = 0, 1, \ldots, 24$, is to use $\Phi([k + .5 - ES_8]/\sigma_{S_8})$; that is,

$$P[S_8 \le k] = P[S_8 \le k + .5] = P\left[\frac{S_8 - ES_8}{\sigma_{S_8}} < \frac{k + .5 - 10.4}{3.589}\right]$$

$$= P[S_8^* < b_k] \doteq \Phi(b_k)$$

where $b_k = (k - 9.9)/3.589$. Figure 4 tabulates this approximation, whose error is never as large as .01. *The remainder of this section is devoted to the calculation of* p_{S_8}.

The density p_{X_1} was kept simple enough that any one calculation may be checked quite easily. For the same reason only eight such random variables were added. Within the constraint that the calculations be kept simple, the density was selected to be one with a fairly arbitrary shape.

We will perform three separate computations to get p_{S_8}. We know the density for X_1, and we know that it equals the density for X_2, so by the product rule we know the joint density of X_1 and X_2. Our first computation consists of finding p_{S_2} from the joint density of X_1 and X_2. We then know the density for $S_2 = (X_1 + X_2)$, and we know that it equals the density for $(X_3 + X_4)$, so by the product rule we know the joint density of $(X_1 + X_2)$ and $(X_3 + X_4)$. Our second computation consists of finding p_{S_4} from the joint density of $(X_1 + X_2)$ and $(X_3 + X_4)$. We then know the density for $S_4 = (X_1 + X_2 + X_3 + X_4)$, and we know that it equals the density for $(X_5 + X_6 + X_7 + X_8)$, so by the product rule we know the joint density of $(X_1 + X_2 + X_3 + X_4)$ and $(X_5 + X_6 + X_7 + X_8)$, and our third computation gets p_{S_8} from this joint density. In general this method requires k such computations in order to obtain the density of S_n, where $n = 2^k$.

We now turn to the first of these computations. The joint probability for p_{X_1,X_2}, as shown in Fig. 2, was obtained by first filling in the two known marginal densities p_{X_1} and p_{X_2} and then using the product rule to get the remaining entries.

x_1 \ x_2	0	1	2	3	p_{X_1}
0	.16	.08	.04	.12	.4
1	.08	.04	.02	.06	.2
2	.04	.02	.01	.03	.1
3	.12	.06	.03	.09	.3
p_{X_2}	.4	.2	.1	.3	1

Fig. 2 Joint-probability table for X_1 and X_2.

y \ z	0	1	2	3	4	5	6	p_Y
0	.0256	.0256	.0192	.0448	.0208	.0096	.0144	.16
1	.0256	.0256	.0192	.0448	.0208	.0096	.0144	.16
2	.0192	.0192	.0144	.0336	.0156	.0072	.0108	.12
3	.0448	.0448	.0336	.0784	.0364	.0168	.0252	.28
4	.0208	.0208	.0156	.0364	.0169	.0078	.0117	.13
5	.0096	.0096	.0072	.0168	.0078	.0036	.0054	.06
6	.0144	.0144	.0108	.0252	.0117	.0054	.0081	.09
p_Z	.16	.16	.12	.28	.13	.06	.09	1

Fig. 3 Joint-probability table for Y and Z. $Y = X_1 + X_2$ and $Z = X_3 + X_4$.

k	$p_{X_1}(k)$	$p_{S_2}(k)$	$p_{S_4}(k)$	$p_{S_8}(k)$	$\varphi_{\mu,\sigma}(k)$	b_k	$P[S_8^* < b_k]$	$\Phi(b_k)$
0	.4	.16	.0256	.0007	.0016	−2.76	.0007	.0029
1	.2	.16	.0512	.0026	.0036	−2.48	.0033	.0066
2	.1	.12	.0640	.0059	.0072	−2.20	.0092	.0139
3	.3	.28	.1280	.0131	.0133	−1.92	.0223	.0274
4		.13	.1456	.0247	.0227	−1.65	.0470	.0495
5		.06	.1280	.0378	.0359	−1.37	.0848	.0853
6		.09	.1576	.0562	.0524	−1.09	.1410	.1379
7			.1160	.0757	.0710	−.81	.2167	.2090
8			.0721	.0897	.0888	−.53	.3064	.2981
9			.0660	.1032	.1030	−.25	.4096	.4013
10			.0270	.1093	.1105	.03	.5189	.5120
11			.0108	.1043	.1096	.31	.6232	.6217
12			.0081	.0974	.1006	.58	.7206	.7190
13				.0834	.0855	.86	.8040	.8051
14				.0647	.0672	1.14	.8687	.8729
15				.0497	.0506	1.42	.9184	.9222
16				.0341	.0329	1.70	.9525	.9554
17				.0213	.0205	1.98	.9738	.9761
18				.0133	.0118	2.25	.9871	.9878
19				.0070	.0063	2.53	.9941	.9943
20				.0033	.0031	2.81	.9974	.9975
21				.0017	.0014	3.09	.9991	.9990
22				.0006	.0006	3.37	.9997	.9996
23				.0002	.0002	3.65	.9999	.9999
24				.0001	.0001	3.92	1.0000	1.0000
Mean	1.3	2.6	5.2	10.4	$\mu = 10.4$			
Variance	1.61	3.22	6.44	12.88	$\sigma = 12.88$			
Standard deviation	1.269	1.794	2.538	3.589	$\sigma = 3.589$			

Fig. 4

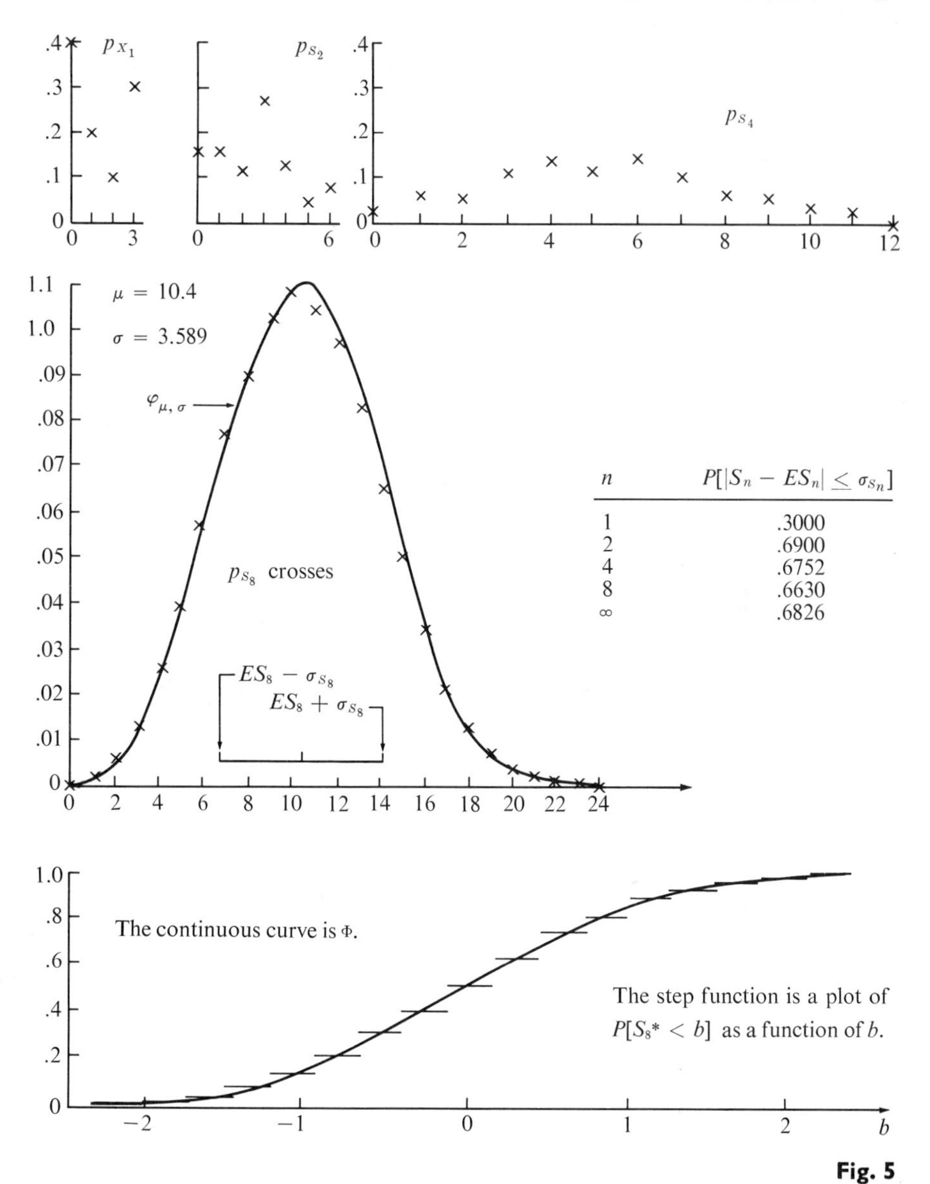

n	$P[\lvert S_n - ES_n\rvert \leq \sigma_{S_n}]$
1	.3000
2	.6900
4	.6752
8	.6630
∞	.6826

Fig. 5

We easily find p_{S_2} from Fig. 2 as follows:

$$
\begin{aligned}
p_{S_2}(0) &= p_{X_1,X_2}(0,0) = .16\\
p_{S_2}(1) &= p_{X_1,X_2}(1,0) + p_{X_1,X_2}(0,1) = .08 + .08 = .16\\
p_{S_2}(2) &= p_{X_1,X_2}(2,0) + p_{X_1,X_2}(1,1) + p_{X_1,X_2}(0,2) = .04 + .04 + .04 = .12\\
p_{S_2}(3) &= p_{X_1,X_2}(3,0) + p_{X_1,X_2}(2,1) + p_{X_1,X_2}(1,2) + p_{X_1,X_2}(0,3)\\
&= .12 + .02 + .02 + .12 = .28\\
p_{S_2}(4) &= p_{X_1,X_2}(3,1) + p_{X_1,X_2}(2,2) + p_{X_1,X_2}(1,3) = .06 + .01 + .06 = .13\\
p_{S_2}(5) &= p_{X_1,X_2}(3,2) + p_{X_1,X_2}(2,3) = .03 + .03 = .06\\
p_{S_2}(6) &= p_{X_1,X_2}(3,3) = .09
\end{aligned}
$$

Thus, for example, we found $p_{S_2}(3) = P[S_2 = 3] = P[X_1 + X_2 = 3]$ by summing $p_{X_1,X_2}(x_1,x_2)$ over all pairs (x_1,x_2) such that $x_1 + x_2 = 3$. In each case this amounts to adding entries in the table along a 45° line. We check for errors by checking that $p_{S_2}(0) + p_{S_2}(1) + \cdots + p_{S_2}(6) = 1$. The density p_{S_2} obtained by this computation is shown in Fig. 4, and *this completes our first computation.*

Turning to our second computation, for notational convenience we let $Y = X_1 + X_2$ and $Z = X_3 + X_4$. Clearly Y and Z are independent and each has the density p_{S_2}. The joint probabilities of Fig. 3 were obtained by filling in the two known marginals p_Y and p_Z and then obtaining the other entries by the product rule. Since Y and Z are independent and identically distributed, we see that

$$p_{Y,Z}(y,z) = p_Y(y)p_Z(z) = p_Y(z)p_Z(y) = p_{Y,Z}(z,y)$$

so that the table is symmetric about the main diagonal. Thus, for example, we get

$$
\begin{aligned}
p_{S_4}(8) &= p_{Y,Z}(6,2) + p_{Y,Z}(5,3) + p_{Y,Z}(4,4) + p_{Y,Z}(3,5) + p_{Y,Z}(2,6)\\
&= (2)(.0108 + .0168) + .0169 = .0721.
\end{aligned}
$$

The density p_{S_4} obtained by these calculations is shown in Fig. 4, and *this completes our second computation.*

Turning to the third computation, we let $U = S_4$ and $V = X_5 + X_6 + X_7 + X_8$, so that U and V are independent and $p_U = p_V = p_{S_4}$, which is known. The same method is again used to find the density for $S_8 = U + V$. We do not exhibit or actually need the joint-probability table for U and V. The computation of $p_{S_8}(5)$ for instance is as follows:

$$
\begin{aligned}
p_{S_8}(5) &= p_{U,V}(5,0) + p_{U,V}(4,1) + p_{U,V}(3,2) + p_{U,V}(2,3) + p_{U,V}(1,4)\\
&\qquad + p_{U,V}(0,5)\\
&= 2\{p_{U,V}(5,0) + p_{U,V}(4,1) + p_{U,V}(3,2)\}\\
&= 2\{p_U(5)p_V(0) + p_U(4)p_V(1) + p_U(3)p_V(2)\}\\
&= 2\{(.1280)(.0256) + (.1456)(.0512) + (.1280)(.0640)\} \doteq .0378.
\end{aligned}
$$

The density p_{S_8} obtained by these calculations is shown in Fig. 4, and *this completes our third and final computation.*

14
Conditional Expectations and Branching Processes

The most important and widely used concepts of probability theory discussed in this chapter are the conditional distribution and expectation of a random variable, given a random point, as defined in Sec. 14-1. Very roughly, the conditional expectation of X, given a random point g, is the expectation for X after all probabilities have been conditioned on knowledge of the value of g. The other two sections of this chapter are devoted to less fundamental but still quite widely used ideas. In Sec. 14-2 we generalize a sum $S_N = X_1 + \cdots + X_N$ of random variables to the case where the number N of random variables summed is also random. This generalization is often useful and sometimes essential in practical problems. Section 14-3 specializes the preceding section to a special class of population-growth problems for which some interesting nontrivial results can be obtained without too much effort.

*14-1 CONDITIONAL EXPECTATION

In this section we study the density, expectation, and variance of a random variable when all probabilities are conditioned on a random point. The formulas which are derived are few and simple. However, the concepts involved are subtle and arise frequently in more advanced theoretical and applicational work in probability, so considerable space is devoted to their interpretation. Conditional expectation can be defined and interpreted as a function of the given random point or as a random variable on the basic sample space. Both definitions and interpretations are used widely. This section is based on the first approach; the second approach is considered in Example 2.

Let $f: \Omega \to F$ and $g: \Omega \to G$ be two random points with associated discrete probability space (Ω,P). For any α in F and for any β in G for which $P[g = \beta] = p_g(\beta) > 0$ we can consider the number

$$P[f = \alpha \mid g = \beta] = \frac{P[f = \alpha, g = \beta]}{P[g = \beta]} = \frac{p_{f,g}(\alpha,\beta)}{p_g(\beta)}\,.$$

Clearly for fixed β this defines a function of α which is a discrete probability density on

F, and this density is naturally called the **conditional density of f given the event** $[g = \beta]$. We also use the notations

$$p_{f|[g=\beta]}(\alpha) = p_{f|g=\beta}(\alpha) = p_{f|g}(\alpha \mid \beta) = \frac{p_{f,g}(\alpha,\beta)}{p_g(\beta)} \qquad \text{where } p_g(\beta) > 0.$$

Clearly, then,

$$p_f(\alpha) = \sum_\beta p_{f,g}(\alpha,\beta) = \sum_\beta p_{f|g=\beta}(\alpha)p_g(\beta) \qquad \text{for all } \alpha$$

which shows that the density p_f is a probabilistically weighted average or **mixture** of the densities $p_{f|g=\beta}$, where the density $p_{f|g=\beta}$ is given weight $p_g(\beta)$.

In practice many problems are naturally described initially in terms of $p_{f|g}$ and p_g rather than in terms of some basic (Ω,P). Thus we specify some discrete density $p_g\colon G \to R$ and some function $p_{f|g}(\alpha \mid \beta)$ which is a discrete density on F for each fixed β having $p_g(\beta) > 0$. Then we assert that random points f and g induce these densities. Such a problem description is meaningful because we can let $\Omega = F \times G$ and define the discrete density $p\colon \Omega \to R$ by $p(\alpha,\beta) = p_{f|g}(\alpha \mid \beta)p_g(\beta)$, so that the coordinate functions f and g on Ω do indeed induce the specified $p_{f|g}$ and p_g.

For example, we might say that experiment β is selected with probability $p_g(\beta)$, and f has density $p_{f|g=\beta}$ if experiment β is selected. In particular we may say that the density $p_X(1) = p_X(2) = .2$ and $p_X(3) = .6$ determines which block X of radioactive material is selected, and the conditional density $p_{Y|X=i}$ for the number of emissions Y, given that block i is selected, is Poisson with mean λ_i. This same problem could be introduced in a less natural fashion by the joint density

$$p_{Y,X}(k,i) = \frac{e^{-\lambda_i}(\lambda_i)^k}{k!}p_i \qquad \text{for } k = 0, 1, \ldots; i = 1, 2, 3$$

where $p_1 = p_2 = .2$ and $p_3 = .6$.

Sometimes we may use $p_{f|g}$ and p_g to define p_f, and *only* p_f is of interest. We may then say that p_f is obtained by **randomizing the parameter** β in $p_{f|g}(\alpha \mid \beta)$ according to p_g. For example, $p_\lambda(k) = e^{-\lambda}\lambda^k/k!$ is a discrete Poisson density in $k = 0, 1, \ldots$ for each real $\lambda > 0$, so that if p is some discrete density on $1, 2, \ldots$ we may define the density p' by randomizing the parameter λ in p_λ according to p, hence

$$p'(k) = \sum_\lambda p_\lambda(k)p(\lambda) = \sum_{i=1}^{\infty} \frac{e^{-i}(i)^k}{k!}p(i) \qquad \text{for } k = 0, 1, \ldots .$$

These ideas are most frequently employed, as in the remainder of this section, in the case where $f = X$ is a random variable, so that we can take expectations and variances of X with or without conditioning on g. Suppose we start with a random variable X and a random point g having the same associated discrete probability space (Ω,P), so that

$$p_{X|g}(x \mid \beta) = \frac{p_{X,g}(x,\beta)}{p_g(\beta)} \qquad \text{for } p_g(\beta) > 0$$

$$p_X(x) = \sum_\beta p_{X|g}(x \mid \beta)p_g(\beta) \qquad \text{for all real } x.$$

Multiplying the last equation by x and then summing over x yields

$$EX = \sum_x xp_X(x) = \sum_\beta \left[\sum_x xp_{X|g}(x \mid \beta)\right] p_g(\beta).$$

This simple equation is so basic that we introduce some relevant notation. If $E|X| < \infty$ and $p_g(\beta) > 0$ we can let

$$\mu_{X|g}(\beta) = [\text{the mean of } p_{X|g=\beta}] = \sum_x xp_{X|g}(x \mid \beta),$$

and the series converges absolutely. We call $\mu_{X|g}(\beta)$ the **conditional expectation of X given $g = \beta$**, and we also use the notation $\mu_{X|g=\beta}$ and $\mu_{X|[g=\beta]}$. We call the *function* $\mu_{X|g}$ the **conditional expectation of X given g**, or the **regression function of X on g**. We have

$$EX = \sum_\beta \mu_{X|g}(\beta) p_g(\beta)$$

with the series converging absolutely.

If we think of $\mu_{X|g}$ as a random variable with associated density p_g, this equation says that the unconditional expectation of X equals the expectation of the conditional expectation of X given g; that is, the mean of a mixture of densities is the same mixture of the means of the densities. We will also use the alternate standard notation

$$E[X \mid g = \beta] = \mu_{X|g}(\beta) \qquad \text{and} \qquad E[X \mid g] = \mu_{X|g}$$

so

$$EX = \sum_\beta E[X \mid g = \beta] p_g(\beta).$$

We next obtain an analogous formula for Var X. If $EX^2 < \infty$ then let $\sigma^2_{X|g=\beta} = \sigma^2_{X|g}(\beta) = \text{Var}\, p_{X|g=\beta}$ where $p_g(\beta) > 0$, and call the *function* $\sigma^2_{X|g}$ the **conditional variance of X given g**. We will also use the standard notation

$$\text{Var}\,[X \mid g = \beta] = \sigma^2_{X|g}(\beta) \qquad \text{and} \qquad \text{Var}\,[X \mid g] = \sigma^2_{X|g}.$$

Now we find Var X starting from the formula

$$p_X(x) = \sum_\beta p_{X|g=\beta}(x) p_g(\beta)$$

expressing p_X as a mixture of the conditional densities. Multiplying by $(x - \mu_X)^2$ and summing over x, we get

$$\text{Var}\, X = \sum_\beta \left[\sum_x (x - \mu_X)^2 p_{X|g=\beta}(x)\right] p_g(\beta)$$

In the expression [ ] we can replace $(x - \mu_X)^2$ by

$$\begin{aligned}(x - \mu_X)^2 &= \{[x - \mu_{X|g}(\beta)] + [\mu_{X|g}(\beta) - \mu_X]\}^2 \\ &= [x - \mu_{X|g}(\beta)]^2 + 2[\mu_{X|g}(\beta) - \mu_X][x - \mu_{X|g}(\beta)] \\ &\qquad + [\mu_{X|g}(\beta) - \mu_X]^2\end{aligned}$$

and the middle term will drop out because $\mu_{X|g}(\beta)$ is the mean of $p_{X|g=\beta}$. *Therefore*

$$\operatorname{Var} X = \sum_{\beta} \sigma^2_{X|g}(\beta)p_g(\beta) + \sum_{\beta} [\mu_{X|g}(\beta) - \mu_X]^2 p_g(\beta).$$

Thus the variance of a mixture of densities exceeds the same mixture of the variances by a nonnegative number which can be interpreted as the variance of the regression function. Said differently, the variance of X equals the mean of the conditional variance, plus the variance of the conditional mean.

For interpretation suppose that fraction $p_g(\beta)$ of a large population have incomes described by the density $p_{X|g=\beta}$. Thus $g(\omega)$ might equal the sex and type of employment of individual ω. If we interpret the terms "average" and "variability" properly then the variability of the income in such a mixture exceeds the average of the subpopulation variabilities by the variability of the mean incomes among the subpopulations.

If X and g are independent then $p_{X|g=\beta} = p_X$, so that

$$E[X \mid g = \beta] = EX \qquad \text{and} \qquad \operatorname{Var}[X \mid g = \beta] = \operatorname{Var} X \qquad \text{if } p_g(\beta) > 0$$

that is, if X is independent of g then its conditional expectation and variance just equal the constant unconditional expectation and variance. In particular if $X = c$ is a constant then $E[c \mid g] = c$. The following theorem summarizes our results. The convergence conditions follow easily from Theorem 1 (Sec. 5-3).

Theorem 1 Let X and g be a random variable and a random point with the same associated discrete probability space, so that p_X can be expressed as a mixture

$$p_X(x) = \sum_{\beta} p_{X|g=\beta}(x)p_g(\beta).$$

Then p_X has a mean iff "each $p_{X|g=\beta}$ with $p_g(\beta) > 0$ has a mean and the series below converges absolutely," in which case

$$EX = \sum_{\beta} \mu_{X|g}(\beta)p_g(\beta) = \sum_{\beta} E[X \mid g = \beta]p_g(\beta).$$

Also p_X has a variance iff "each $p_{X|g=\beta}$ with $p_g(\beta) > 0$ has a variance and both series below converge," in which case

$$\operatorname{Var} X = \sum_{\beta} \sigma^2_{X|g}(\beta)p_g(\beta) + \sum_{\beta} [\mu_{X|g}(\beta) - \mu_X]^2 p_g(\beta). \quad \blacksquare$$

Let us now represent graphically the quantities we have introduced. For this purpose we make $g = V$ a random variable and start with the joint density $p_{X,V}$ which is not shown in Fig. 1, where the x axis is vertical so that the regression function can be plotted in the usual fashion. The two marginal densities are plotted on their respective axes. For convenience these and other functions are plotted as though they were continuous functions of a continuous variable.

Three of the conditional densities of X, given the event $[V = v]$,

$$p_{X|V=v}(x) = \frac{p_{X,V}(x,v)}{p_V(v)} \qquad \text{for } p_V(v) > 0$$

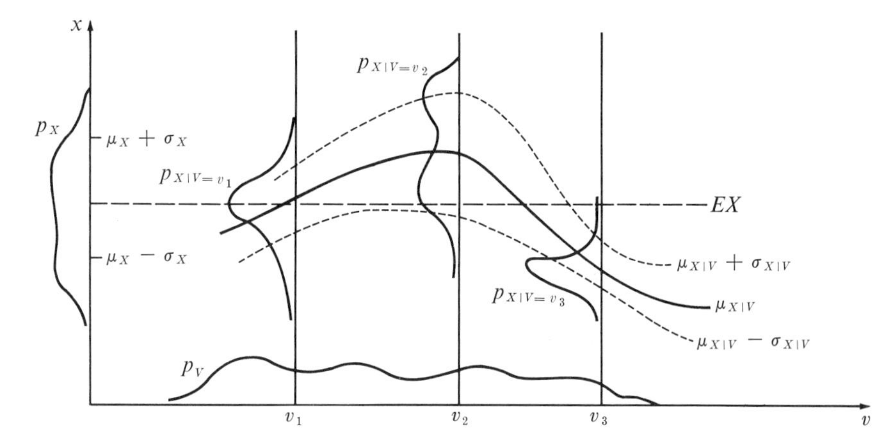

Fig. 1 The function $\mu_{X|V}$ is the conditional expectation of X given V, while the function $\sigma^2_{X|V}$ is the conditional variance of X given V.

are shown. If we consider this last equation as a function of x, with v fixed, it is clear that $p_{X|V=v}$ just equals $p_{X,V}$ restricted to the vertical line through v, except for the normalizing constant $1/p_V(v)$. For $p_{X|V=v_1}$ we have the mean $\mu_{X|V}(v_1)$ and the two points $\mu_{X|V}(v_1) \pm \sigma_{X|V}(v_1)$, one standard deviation on each side, and so we obtain the *functions* shown as a solid line and two dashed lines.

Let us now interpret Fig. 1 from the point of view of predicting the value of X. If the experiment corresponding to $p_{X,V}$ is to be performed, an obvious prediction for the value of X is its expectation. From Chebychev's inequality the probability is at least $8/9$ that X will have some value inside the interval $\mu_X \pm 3\sigma_X$, so we expect that this will usually happen in many independent repetitions of the experiment. If the experiment is performed and we learn that $[V = v_1]$ has occurred then we can use the mean $\mu_{X|V}(v_1)$ of $p_{X|V=v_1}$ as a prediction for the unknown value of X. The probability, conditioned on the known event $[V = v_1]$, is high that X will have a value within a few standard deviations, $\mu_{X|V}(v_1) \pm 3\sigma_{X|V}(v_1)$, of this prediction. Thus the regression function $\mu_{X|V}$ of X on V can be interpreted as a predictor of X when we know V, in the sense that we predict that X will have the value $\mu_{X|V}(v)$ when we learn that $[V = v]$ has occurred. If there are many independent repetitions of the experiment and for each repetition we learn V and predict X, we may be interested in the expected value of our predictor. Now $[V = v]$ occurs with probability $p_V(v)$, in which case we predict $\mu_{X|V}(v)$, so our prediction has expectation $\sum_v \mu_{X|V}(v)p_V(v)$, which we saw equaled EX. Thus the expectation of "our prediction of X, given V," is just the expectation of X; that is, the expectation, with respect to p_V, of the regression of X on V is EX.

Returning to the more general case when g is not necessarily a random variable, we still consider

$$\mu_{X|g}(\beta) = [\text{mean of } p_{X|g=\beta}]$$

a reasonable prediction for the unknown value of X if we learn that $[g = \beta]$ occurred.

Often g is a random vector. For example, $g = (X_1, \ldots, X_n)$, and we learn the values $\beta = (\beta_1, \ldots, \beta_n)$ of all random variables corresponding to times $t = 1, 2, \ldots, n$. Then we may use $\mu_{X|g}(\beta) = \mu_{X|X_1, \ldots, X_n}(\beta_1, \ldots, \beta_n)$ to predict the value of X. In particular X might correspond to some future time $X = X_{n+3}$, or to some past time $X = X_{-7}$, or to some other quantity $X = \sum_{i=1}^{n+3} (X_i)^2$ of interest. Similarly we might observe the value of the random vector $\left(X_2, \sum_{i=1}^{n} X_i, \sum_{i=1}^{n} X_i^2\right)$ and try to predict, or guess, what the largest value of any of the random variables $X_1, \ldots, X_n$ was. Thus we are using the word "predict" in the loose sense of "guess," and no sense of time, past or future, need be involved.

We next derive the linearity of conditional expectation. After that it is shown that $\mu_{X|g}$ is the best predictor of X, given g, in the sense that it minimizes the expected squared error. The remainder of this section is then devoted to Example 1, which illustrates these ideas in detail, and to Example 2, which describes the alternate definition of conditional expectation.

Let $X_1, \ldots, X_n$ be random variables and $g: \Omega \to G$ be a random point with the same associated discrete probability space (Ω, P). Let $X = \sum_{i=1}^{n} a_i X_i$ where $a_1, \ldots, a_n$ are reals. If $p_g(\beta) > 0$ then $E[X \mid g = \beta]$ is the mean of the conditional density $p_{X|g=\beta}$. Equivalently $E[X \mid g = \beta]$ is the expectation of X with associated $(\Omega, P_{[g=\beta]})$, where $P_{[g=\beta]}$ is obtained by conditioning P on the event $[g = \beta]$. The linearity of expectation yields

$$E\left(\sum_{i=1}^{n} a_i X_i \mid g\right) = \sum_{i=1}^{n} a_i E(X_i \mid g)$$

in the sense that whenever $P[g = \beta] > 0$ then the functions on the two sides of this equation are defined and equal at β. This fundamental fact is referred to as the *linearity of conditional expectation.* In particular if we always use conditional expectations in making predictions then $E(8X \mid g) = 8E(X \mid g)$, so that our prediction for $8X$ is always eight times our prediction for X. Furthermore if we learn that $[g = \beta]$ has occurred and we predict that X and Y will have values -3 and 7.2, respectively, then 4.2 is our prediction for the value of $X + Y$, *regardless* of any dependencies involved.

We now show that $\mu_{X|g}$ is the unique best predictor of X, given g, in the sense of minimizing the expected squared error. The expected squared error is an *arbitrary but often used* measure of merit for a predictor. Suppose we have a model for an experiment to be performed a year from now. The model incorporates a random variable $X: \Omega \to R$, a random point $g: \Omega \to G$, and their joint density $p_{X,g}$ on $R \times G$. In the following discussion we assume that $R \times G$ is the sample space and X and g are the coordinate functions. At the present time we use EX as our prediction for the value of X. As justification for this prediction note that if the number c is a prediction for X then $[X - c]^2$ is the random squared error which has expectation

$$\begin{aligned} E[X - c]^2 &= E[(X - \mu_X) + (\mu_X - c)]^2 \\ &= E[(X - \mu_X)^2 + 2(\mu_X - c)(X - \mu_X) + (\mu_X - c)^2] \\ &= \operatorname{Var} X + (\mu_X - c)^2. \end{aligned}$$

Thus we see that $E[X - c]^2$ equals the variance of X, which does not depend on c, plus the square of the distance of the prediction from the mean μ_X. Therefore $c = EX$ is the unique real-number predictor which minimizes the expected squared error $E[X - c]^2$, and this minimum is Var X.

Assume that 1 month after the experiment is performed we will learn which event $[g = \beta]$ occurred, and then use the number $h(\beta)$ for a revised prediction of the value of X. We can select any function $h: G \to R$ to use as a revised predictor. If the outcome of the experiment is (x,β) then $x - h(\beta)$ is the error of the revised predictor, as shown in Fig. 2 for the case where $g = V$ is a random variable. Thus the expectation of the square of the error for the revised predictor is

$$\sum_\beta \sum_x [x - h(\beta)]^2 p_{X,g}(x,\beta) = \sum_\beta \left\{\sum_x [x - h(\beta)]^2 p_{X|g=\beta}(x)\right\} p_g(\beta)$$

which we wish to minimize. For each β having $p_g(\beta) > 0$ we must define the number $h(\beta)$ so as to minimize the nonnegative coefficient { } of $p_g(\beta)$. But in the last paragraph

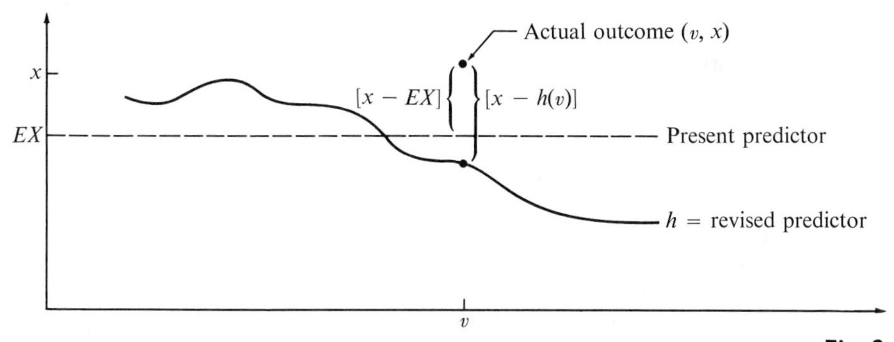

Fig. 2

we saw that { } is minimized by letting $h(\beta)$ be the mean of $p_{X|g=\beta}$. Therefore the regression function $\mu_{X|g}$ is the unique [at least for β with $p_g(\beta) > 0$] function of β achieving the minimum value of

$$\sum_\beta \sigma^2_{X|g}(\beta) p_g(\beta)$$

for the expectation of the squared error. Recall that

$$p_X(x) = \sum_x p_{X|g=\beta}(x) p_g(\beta)$$

is a mixture of the conditional densities, so we now see that the same mixture of their variances can be interpreted as the minimum (achieved by $\mu_{X|g}$) expected squared error for a predictor which can use g. Furthermore the expectation of this revised predictor $\mu_{X|g}$ is just EX, while the equation

$$\sum_\beta [\mu_{X|g}(\beta) - \mu_X]^2 p_g(\beta) = \text{Var } X - \sum_\beta \sigma^2_{X|g}(\beta) p_g(\beta)$$

shows that its variance equals the amount by which the expected squared error can be reduced because of knowing g. In a sense this measures the benefit obtained by observing the value of g before guessing X. If X and g are independent then $\mu_{X|g} = \mu_X$, and observing the value of g does not help.

If $g = Y$ is a random variable it may happen that there are two real numbers a and b such that $\mu_{X|Y}(y) = ay + b$ whenever $p_Y(y) > 0$. In this case $\mu_{X|Y}$ is just the least-squares *linear* regression of X on Y, as defined in Sec. 11-4. Otherwise the nonlinear predictor $\mu_{X|Y}$ achieves a strictly smaller expected squared error than the linear regression, since $\mu_{X|Y}$ is the unique predictor achieving a minimum.

***Example I** A game is based on the three random wheels of Fig. 3. One play of the game can be interpreted roughly as follows: You leave home, and on your way to the casino to receive reward Y from one spin of wheel 3 you encounter N obstacles, where N is Poisson, and you

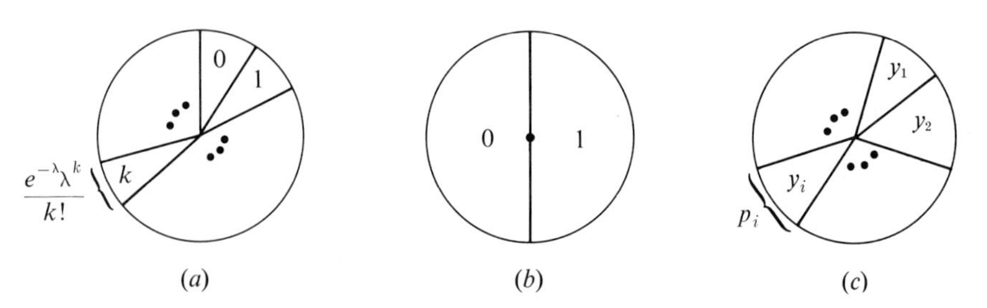

Fig. 3 (*a*) wheel 1: spin once, outcome N; (*b*) wheel 2: spin N times, success iff all N trials result in 1's; (*c*) wheel 3: spin once, outcome Y.

overcome each with probability $\frac{1}{2}$. The reward X for your effort is Y if you are successful in reaching the casino, but if you are not successful then $X = 0$. Thus wheel 1 is spun once and yields an integer N having a Poisson distribution with mean λ. Then wheel 2 is spun N times, where N is determined by wheel 1, and the final result Z from wheel 2 is a Success $[Z = S]$ iff all $N \geq 1$ trials yielded 1's, or if $N = 0$. Wheel 3 is spun once and yields a nonzero real number Y having distribution $P[Y = y_j] = p_j$. All spins are independent, but the final result Z from wheel 2 depends, of course, on the outcome from wheel 1. The joint density of N, Z, and Y is given by

$$p_{N,Z,Y}(n,z,y_j) = P[Y = y_j \mid N = n, Z = z]P[Z = z \mid N = n]P[N = n]$$

$$= p_j P[Z = z \mid N = n] \frac{e^{-\lambda}\lambda^n}{n!}$$

where Z can have the values S for Success, or F for Failure, so

$$P[Z = S \mid N = n] = (\tfrac{1}{2})^n \qquad \text{and} \qquad P[Z = F \mid N = n] = 1 - (\tfrac{1}{2})^n.$$

Thus the game is essentially defined in terms of conditional probabilities which we can use to define a joint density which has these conditional probabilities. On one play of the game you receive reward $X = f(Z, Y)$, where

$$f(S,y_j) = y_j \qquad \text{and} \qquad f(F,y_j) = 0$$

so your reward X equals Y iff you obtain a success on wheel 2; otherwise you get nothing.

As another interpretation of this game, the yield X of a process is Y iff all N breakdowns during the setup time are rapidly corrected. As still another interpretation, the output

X of a communication channel will equal the input Y iff all N noise impulses encountered in the channel are detected and eliminated.

In this example we will assume that predictions of the reward X may sometimes be based on the number N of obstacles, so we find the conditional density, expectation, and variance of X, given N. Clearly (N,Z) and Y are independent, but $X = f(Z,Y)$, so X and N may be dependent, and it is intuitively obvious that they are.

For convenience let $x_j = y_j$, so that $0, x_1, x_2, \ldots$ are possible values for the reward X. For $p_{X,N}$ we obtain

$$p_{X,N}(x_j,n) = P[Y = x_j,\, Z = S,\, N = n] = p_j \frac{1}{2^n} \frac{e^{-\lambda}\lambda^n}{n!}$$

$$p_{X,N}(0,n) = P[Z = F,\, N = n] = P[Z = F \mid N = n]P[N = n] = \left(1 - \frac{1}{2^n}\right)\frac{e^{-\lambda}\lambda^n}{n!}$$

since for simplicity we assumed that no $y_j = 0$, so that $[X = 0] = [Z = F]$.

At a casual glance $p_{X,N}$ appears to factor into a function of j times a function of n, but the same factorization does not hold for all argument values, as is necessary for independence. The power series (5) of Appendix 1, for the exponential function, yields

$$\sum_{n=0}^{\infty} \frac{1}{2^n} p_N(n) = e^{-\lambda} \sum_{n=0}^{\infty} \frac{(\lambda/2)^n}{n!} = e^{-\lambda/2}$$

which will be useful at several points. For p_X we get

$$p_X(x_j) = \sum_{n=0}^{\infty} p_X(x_j,n) = p_j \sum_{n=0}^{\infty} \frac{1}{2^n} p_N(n) = p_j e^{-\lambda/2}$$

$$p_X(0) = \sum_{n=0}^{\infty} p_X(0,n) = \sum_{n=0}^{\infty} \left(1 - \frac{1}{2^n}\right) p_N(n) = 1 - e^{-\lambda/2}.$$

Therefore, in summary, p_N is Poisson λ and

$$p_{X|N}(x_j \mid n) = \frac{p_{X,N}(x_j,n)}{p_N(n)} = \frac{p_j}{2^n} \qquad p_{X|N}(0 \mid n) = 1 - \frac{1}{2^n}$$

while

$$p_X(x_j) = p_j e^{-\lambda/2} \qquad p_X(0) = 1 - e^{-\lambda/2}$$

for $j = 1, 2, \ldots$ and $n = 0, 1, \ldots$. These results could have been obtained in many different ways. As indicated in Fig. 4, the conditional density $p_{X|N=0}$ of the reward, given that no obstacles arise, is just p_Y. However, for fixed $n \geq 1$ the conditional density $p_{X|N=n}$ is essentially p_Y scaled down by the factor $1/2^n$ corresponding to all n obstacles being overcome, and of course, $p_{X|N}(0 \mid n) = 1 - (1/2^n)$ is the probability that not all obstacles will be overcome. The unconditional density p_X is p_Y scaled down by the factor $e^{-\lambda/2}$, with the probability so lost appearing at $p_X(0)$.

We now find the conditional expectation and variance of the reward X, given the number N of obstacles. Assume that $EY = 1$ and $EY^2 = 2$, so that $\text{Var } Y = 1$. The computation for the moments of $p_{X|N=n}$ is simplified by the fact that $p_{X|N}(0 \mid n)$ does not make a

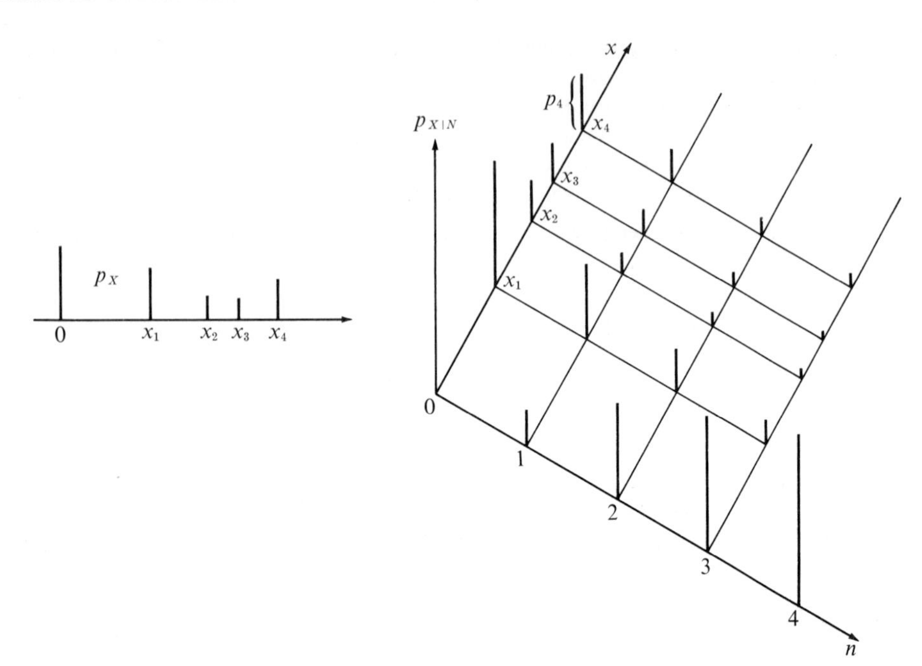

Fig. 4

contribution, so

$$[\text{the } r\text{th moment of } p_{X|N=n}] = \sum_x x^r p_{X|N}(x \mid n) = \sum_{j=1}^{\infty} (x_j)^r \frac{p_j}{2^n} = \frac{EY^r}{2^n}$$

$$[\text{the mean of } p_{X|N=n}] = \mu_{X|N}(n) = \frac{1}{2^n}$$

$$\text{Var}\, p_{X|N=n} = \sigma^2_{X|N}(n) = [\text{the second moment of } p_{X|N=n}] - [\text{the mean of } p_{X|N=n}]^2$$

$$= \frac{1}{2^{n-1}} - \frac{1}{2^{2n}}\,.$$

The regression function $\mu_{X|N}$ is sketched in Fig. 5, as are the functions $\mu_{X|N} \pm \sigma_{X|N}$ one standard deviation above and below it.

The derivations of the four formulas below are left as an exercise:

$$EX = e^{-\lambda/2} \qquad \text{Var}\, X = 2e^{-\lambda/2} - e^{-\lambda}$$

$$\sum_{n=0}^{\infty} \sigma^2_{X|N}(n) p_N(n) = EX^2 - e^{-3\lambda/4} \qquad \sum_{n=0}^{\infty} [\mu_{X|N}(n) - EX]^2 = e^{-3\lambda/4} - (EX^2).$$

Note that the sum of the last two expressions equals Var X, as it should.

As a numerical illustration suppose a friend plays the game and the expected number of obstacles is $EN = \lambda = 2$. Then the expectation and variance of his reward are

$$EX = e^{-1} \doteq .368 \qquad \text{and} \qquad \text{Var}\, X = 2e^{-1} - e^{-2} \doteq .600.$$

If we learn that he actually encountered two obstacles then the expectation and variance of his reward, conditioned on this event $[N = 2]$, are

$$\mu_{X|N}(2) = \frac{1}{2^2} = .250 \qquad \text{and} \qquad \sigma^2_{X|N}(2) = \frac{1}{2} - \frac{1}{2^4} \doteq .437.$$

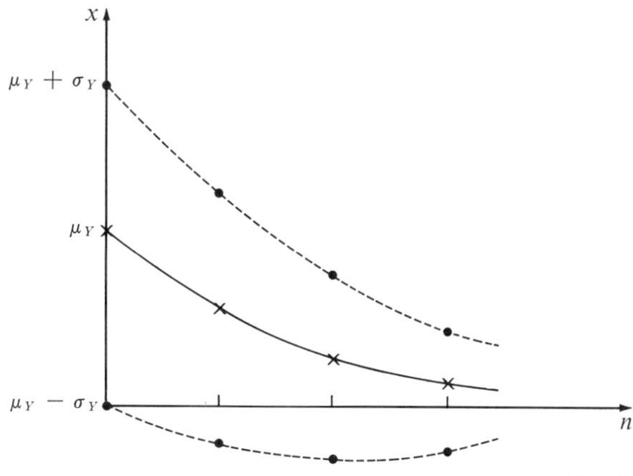

Fig. 5

If he is going to play the game again, and we will learn the value n of N and predict $\mu_{X|N}(n)$, then the expectation, variance, and expected squared error for our prediction are, respectively,

$$EX = e^{-1} \doteq .368 \qquad e^{-3/2} - e^{-2} \doteq .088 \qquad 2e^{-1} - e^{-3/2} \doteq .513.$$

In particular the predictor $\mu_{X|N}$ has an expected squared error of about .513, compared to .600 for the constant predictor EX. ■

***Example 2** We have defined the conditional expectation of $X: \Omega \to R$, given $g: \Omega \to G$, as a function of $\beta \in G$. Conditional expectation is often defined and interpreted in this way, particularly in literature primarily concerned with applications. In some of the literature, and particularly in most advanced theoretical work in probability theory, conditional expectation is defined as a random variable on the basic sample space Ω. There is a considerable theoretical advantage in having all random entities defined on the same sample space. In particular it is then easier to examine the limiting behavior of a sequence of predictors $E(X \mid Y_1)$, $E(X \mid Y_1, Y_2)$, $E(X \mid Y_1, Y_2, Y_3), \ldots$, which have more and more information available. In this example we show how conditional expectation may be defined in an essentially equivalent fashion on the sample space.

We will continue to call the previously defined function $\mu_{X|g}$ the regression function of X on g, but in the remainder of this example we reserve the notation $E(X \mid g)$ and the name "conditional expectation of X given g," for a related function to be defined now. Let $X: \Omega \to R$ be a random variable and let $g: \Omega \to G$ be a random point having the same associated discrete probability space (Ω, P). The real-valued function $E(X \mid g)$ defined at ω in Ω by

$$E(X \mid g)(\omega) = \mu_{X|g}(g(\omega))$$

is called *the conditional expectation of X given g*. That is, we define $E(X \mid g)$ as the composition of the two functions g and $\mu_{X|g}$ as in the following diagram:

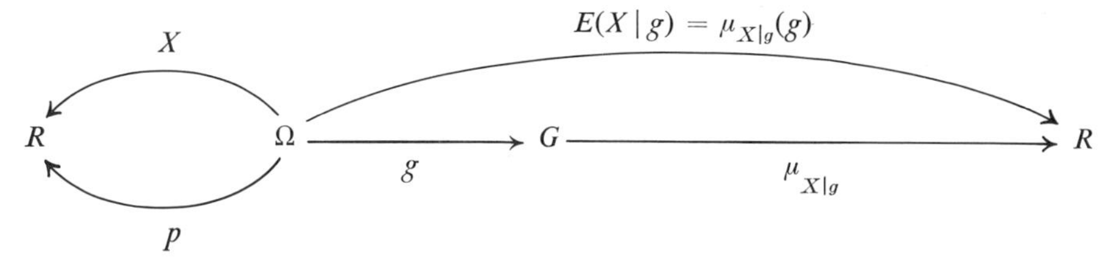

Thus the value of $E(X \mid g)$ at ω is the prediction we would make for the unknown value $X(\omega)$ if the actual outcome was ω but all that we learned was the value $g(\omega)$ of g at this ω.

Now $\mu_{X|g}(\beta)$ is defined only when $p_g(\beta) > 0$, so $E(X|g)$ may not be defined for some ω in Ω. However, if $p(\omega) > 0$ and $g(\omega) = \beta$ then $p_g(\beta) > 0$, so $\mu_{X|g}(\beta)$ is defined, and hence $E(X|g)$ is defined at this ω. Thus $E(X|g)$ is certainly defined at every sample point having positive probability.

We can describe $E(X \mid g)$ more directly. Let $A_\omega = \{\omega' \colon g(\omega') = g(\omega)\}$ consist of those sample points ω' where g has the same value that it has at ω. If ω is a sample point for which $P(A_\omega) > 0$ then

$$E(X \mid g)(\omega) = [\text{the expectation of } X \text{ with associates } (\Omega, P_{A_\omega})]$$

if this expectation exists. That is, given an ω, we first condition all probabilities on the event A_ω and then take the expectation of X, and the number so obtained is the value of $E(X \mid g)$ at ω.

The relationship between $E(X \mid g)$ and $\mu_{X|g}$ is depicted in Fig. 6. We enumerate the points $\beta_1, \beta_2, \ldots$, for which $p_g(\beta_k) = P[g = \beta_k] > 0$. Then the events $[g = \beta_1], [g = \beta_2], \ldots$ are disjoint and their union has probability 1. The function $E(X \mid g)$ is constant on each

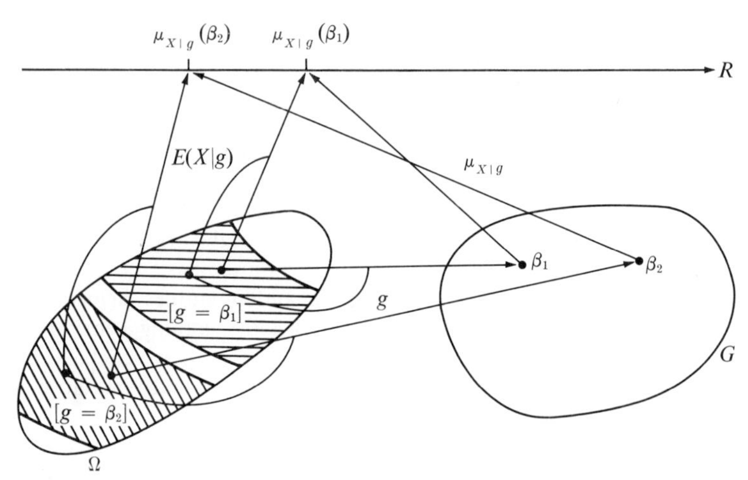

Fig. 6 The regression function $\mu_{X|g}$ is defined on G while the conditional expectation $E(X \mid g)$ is defined on Ω.

such event $[g = \beta_k]$, and its constant value there is just the expectation of X taken with respect to the $P_{[g=\beta_k]}$ obtained by conditioning P onto that event.

Note that $E(X \mid g)$ and X are both random variables with the same associated (Ω, P), and

$$E[E(X \mid g)] = \sum_{\omega} \mu_{X|g}(g(\omega))p(\omega) = \sum_{\beta} \mu_{X|g}(\beta)p_g(\beta) = EX.$$

We can similarly define the random variable $\text{Var}\,(X \mid g)$ on Ω by setting its value at ω equal to $\sigma^2_{X|g}(g(\omega))$, and our previous equation for $\text{Var}\,X$ becomes

$$\text{Var}\,X = E[\text{Var}\,(X \mid g)] + \text{Var}\,[E(X \mid g)]. \quad \blacksquare$$

EXERCISES

1. For Example 1:
(*a*) Derive the four formulas which were left underived.
(*b*) Check that EX equals the expectation of $\mu_{X|N}$.
2. For the density of Exercise 3 (Sec. 11-4) compare the regression function and the least-squares regression line for the following:
(*a*) Y on X (*b*) X on Y
3. Show that if $\{y: p_Y(y) > 0\}$ consists of only one or two points then $\mu_{X|Y}$ agrees with the least-squares regression line for X on Y.
4. Discuss the use of $\mu_{Y|X}$ as a predictor for $Y = f(X)$, given X. Comment on Exercise 6 (Sec. 11-4).
5. For Example 1 find the least-squares regression *line* for X on N, as described in Sec. 11-4. Also find its expected squared error in predicting X, given N.
6. Let $X: \Omega \to R$ and $g: \Omega \to G$ be a random variable and a random point with the same associated discrete probability space. If a predictor for X can be based on g then any function $h: G \to R$ can be used as follows: if we learn that $[g = \beta]$ has occurred we use $h(\beta)$ to predict X; that is, the composite function $h(g)$ can be interpreted as a random variable which uses only the value of g in predicting the value of X.
(*a*) Show that

$$\sum_{\beta}\sum_{x} [x - h(\beta)]^2 p_{X,g}(x,\beta) = \sum_{\beta} \sigma^2_{X|g}(\beta)p_g(\beta) + \sum_{\beta} [\mu_{X|g}(\beta) - h(\beta)]^2 p_g(\beta).$$

Thus the expected squared error for the predictor h of X exceeds the expected squared error of the predictor $\mu_{X|g}$ by an amount equal to the last term. The last term can be interpreted as the expected squared error when h is used to predict $\mu_{X|g}$.
(*b*) Let $h(\beta) = EX$, for all β, and relate the formula in part (*a*) to Theorem 1.
(*c*) Let $g = Y$ be a random variable and let h be the least-squares *linear* regression of X on Y, as described in Sec. 11-4. Deduce from the formula in part (*a*) the conditions under which a nonlinear predictor achieves an improvement.
7. There is uncertainty as to which of the two different densities p_g and p'_g to use for the distribution of the random point g. Both densities assign positive probability to the same points

$$\{\beta: p_g(\beta) > 0\} = \{\beta: p'_g(\beta) > 0\} = \{\beta_1, \beta_2, \ldots\}$$

and the conditional densities $p_{X|g=\beta_i}$ have all been specified and fixed.
(*a*) Consider all the quantities appearing in Theorem 1 and, assuming a prediction context, interpret the possible impact or lack of impact of using p'_g rather than p_g.
(*b*) Let $g = Y$ be a random variable and exhibit an example with

$$\{y: p_Y(y) > 0\} = \{y: p'_Y(y) > 0\} = \{y_1, y_2, y_3\}$$

for which the least-squares regression *line* of X on Y, as defined in Sec. 11-4, is different if p'_Y rather than p_Y is assumed.
8. Let $X: \Omega \to R$ and $g: \Omega \to G$ be a random variable and a random point. Let $h(g)$ be the composition of h applied after g, where $h: G \to R$. Assuming that all expectations below exist, prove that

$$E[h(g)X \mid g] = [h]E[X \mid g].$$

More explicitly, prove that

$$E[h(g)X \mid g = \beta] = h(\beta)E[X \mid g = \beta]$$

for every β for which $p_g(\beta) > 0$. COMMENT: This result is a natural generalization of $E[aX \mid g] = aE[X \mid g]$. It shows that any function of g factors out of the conditional expectation, given g, just as a constant factors out. For example, if $g = (Z,W)$ then our prediction of the value of $Z^2e^{WZ}X$, given $[(Z,W) = (3,-4)]$, is just 3^2e^{-12} times our prediction of the value of X.

9. Let $X: \Omega \to R$ and $g: \Omega \to G$ be a random variable and a random point. Let $f(g)$ be the composition of f applied after g, where $f: G \to F$. Use the definition of conditional expectation given in Example 2, with the assumption that all expectations below exist, and prove that

$$E\{E(X \mid g) \mid f(g)\} = E\{X \mid f(g)\} = E(E\{X \mid f(g)\} \mid g)$$

in the sense that all three random variables are equal at every sample point which has positive probability.

INTERPRETATION: A friend can base his prediction of X on $g = (Y,Z,W)$, but you can base your prediction of X only on a function, $f(g) = (Y, Z^2 + W^2)$, of what he can use. Then your predictor $E\{X \mid f(g)\}$ for X is no different from your predictor of his predictor $E(X \mid g)$. Furthermore his predictor of your predictor is precisely your predictor.

*14-2 THE SUM OF A RANDOM NUMBER OF RANDOM VARIABLES

We first introduce the random selection of a random variable and then specialize to the sum of a random number of random variables. Let $N, S_0, S_1, \ldots, S_m$ be $m + 2$ random variables such that $P[N = 0 \text{ or } N = 1 \text{ or } \cdots \text{ or } N = m] = 1$. Assume that N and S_n are two independent random variables, for every $n = 0, 1, \ldots, m$. In applications we usually assume even more—namely, that N and $(S_0, \ldots, S_m)$ are two independent random vectors. Now define a random variable S_N by $S_N(\omega) = S_{N(\omega)}(\omega)$; that is,

$$S_N(\omega) = S_n(\omega) \qquad \text{if } N(\omega) = n.$$

In other words, on the event $[N = 0]$ we make S_N agree with S_0, and on the event $[N = 1]$ we make S_N agree with S_1, etc. Thus we use N to select one of the random variables $S_0, S_1, \ldots, S_m$, and the value of S_N is then the value of that random variable selected. Therefore S_N can be described as *a randomly selected random variable.* The context makes it clear that N is here a random variable rather than a fixed integer. In general S_N and N are dependent, since S_N involves N in its definition.

If $P[N = n] > 0$ then

$$P[S_N = s \mid N = n] = P[S_n = s \mid N = n] = P[S_n = s]$$

where for the first equality we used the fact that S_N and S_n agree on the event $[N = n]$ and for the second equality we used the independence of S_n and N. For p_{S_N} we find

$$p_{S_N}(s) = \sum_{n=0}^{m} p_{S_N,N}(s,n) = \sum_{n=0}^{m} p_{S_N|N}(s \mid n)p_N(n) = \sum_{n=0}^{m} p_{S_n}(s)p_N(n)$$

so that p_{S_N} is the mixture of $p_{S_0}, p_{S_1}, \ldots, p_{S_m}$ determined by p_N, and this is intuitively reasonable. Note that p_{S_N} is determined by the marginal densities $p_{S_0}, \ldots, p_{S_m}$

regardless of whether or not the $S_0, \ldots, S_m$ are independent, as long as N is independent of every random variable which might be selected. Multiplying by s and summing over s, or using our earlier result for the mean of a mixture, we get

$$ES_N = \sum_{n=0}^{m} (ES_n)p_N(n)$$

and Theorem 1 (Sec. 14-1) yields

$$\text{Var } S_N = \sum_{n=0}^{m} (\text{Var } S_n)p_N(n) + \sum_{n=0}^{m} (ES_n - ES_N)^2 p_N(n).$$

Example 1 Let the joint density p_{S_0,S_1} on the plane describe the joint distribution of the lengths S_0 and S_1 of two adjacent sides, labeled 0 and 1, of rectangular sheets of metal. We select only one side and measure its length. The selection of the side N is by an independent biased coin with $P[N = 0] = .3$ and $P[N = 1] = .7$. Hence S_N is the length of the selected side and

$$p_{S_N}(s) = .3p_{S_0}(s) + .7p_{S_1}(s)$$

$$ES_N = .3ES_0 + .7ES_1$$

$$\text{Var } S_N = [(\text{Var } S_0)(.3) + (\text{Var } S_1)(.7)] + [(ES_0 - ES_N)^2(.3) + (ES_1 - ES_N)^2(.7)].$$

Note that p_{S_N} depends only on p_N and the marginals p_{S_0} and p_{S_1}, regardless of whether S_0 and S_1 are dependent. ■

We now introduce the kind of S_N which interests us most. Let $N, X_1, X_2, \ldots, X_m$ be $m + 1$ independent random variables such that $P[N = 0 \text{ or } N = 1 \text{ or} \cdots \text{or } N = m] = 1$ and for which $X_1, X_2, \ldots, X_m$ have the same distribution. Let S_0 equal the constant zero and let $S_n = X_1 + \cdots + X_n$ for $1 \le n \le m$. As before, let S_N be the random selection of one of $S_0, S_1, \ldots, S_m$ according to N; that is,

$$S_N(\omega) = S_n(\omega) = X_1(\omega) + \cdots + X_n(\omega) \qquad \text{if } N(\omega) = n$$

so

$$S_N(\omega) = S_{N(\omega)}(\omega) = X_1(\omega) + \cdots + X_{N(\omega)}(\omega).$$

In this case we can write $S_N = X_1 + \cdots + X_N$ and interpret S_N *as the sum of a random number N of random variables.*

We can describe S_N in terms of the two random wheels of Fig. 1. We spin the first wheel once and obtain an integer n satisfying $0 \le n \le m$. Then we spin the second

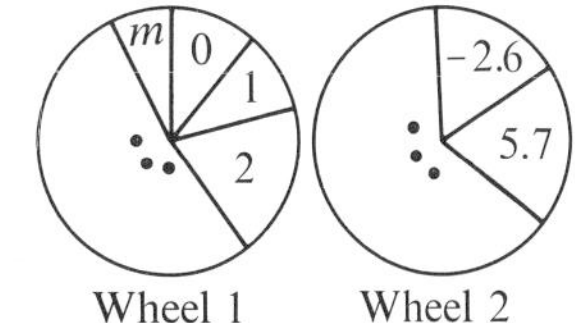

Fig. 1

wheel m times, and the sum of the values obtained on the first n spins is the value of $S_N = X_1 + \cdots + X_N$ for this performance of the experiment. If $n = 0$ the value of S_N is 0, which is a natural convention.

Such an $S_N = X_1 + \cdots + X_N$ can arise in many ways. For example, X_i might be the size of the ith of a total of N orders received by a factory during a day; or X_i might be the time required to repair the ith of N airplanes, or the ith of N breakdowns of a computer, during a month. Thus in many examples the number N of random variables which are summed is also quite naturally a random variable.

Now $S_n = X_1 + \cdots + X_n$, where the random variables are independent, so $ES_n = nEX_1$ and $\text{Var } S_n = n \text{ Var } X_1$. Using these in our formulas for ES_N and $\text{Var } S_N$, we get

$$ES_N = (EN)(EX_1)$$

$$\text{Var } S_N = (EN)(\text{Var } X_1) + \sum_{n=0}^{m} [n(EX_1) - (EN)(EX_1)]^2 p_N(n)$$

$$= (EN)(\text{Var } X_1) + (EX_1)^2(\text{Var } N).$$

Therefore the expectation of $S_N = X_1 + \cdots + X_N$ is just the expectation of one of the random variables EX_1 times the expected number EN of them which are summed. However, $\text{Var } S_N$ exceeds "the variance of one of them times the expected number summed" by $(EX_1)^2 \text{ Var } N$, which is due to the variability of the number of terms.

Let $N, X_1, X_2, \ldots$ be an infinite sequence of independent random variables such that $P[N = 0 \text{ or } N = 1 \text{ or } \cdots] = 1$ and let $S_N = X_1 + \cdots + X_N$ so that

$$p_{S_N}(s) = \sum_{n=0}^{\infty} p_{S_n}(s) p_N(n).$$

In general we cannot define N and $X_1, X_2, \ldots$ all on the same discrete probability space. However, if we have densities p_N and $p_{X_1}, p_{X_2}, \ldots$ then for *each* n we can define p_{S_n} within the discrete case. Thus p_N and the p_{S_n} are well defined, and we use the expression above to define p_{S_N} which is clearly a discrete probability density. Since the formulas for ES_N and $\text{Var } S_N$ were derived from p_{S_N} expressed as such a mixture, they still apply. We summarize this in the theorem below, which also includes a fact about probability-generating functions which will be proved later in this section.

Theorem 1: The sum $S_N = X_1 + \cdots + X_N$ of a random number N of random variables.

Let p_N and p_{X_1} be discrete probability densities on R such that $\sum_{n=0}^{\infty} p_N(n) = 1$. Define the discrete density p_{S_0} by $p_{S_0}(0) = 1$, and for $n = 1, 2, \ldots$ let p_{S_n} be the density of $S_n = X_1 + \cdots + X_n$ where $X_1, \ldots, X_n$ are independent and each has density p_{X_1}. Define the discrete density $p_{S_N}: R \to R$ by

$$p_{S_N}(s) = \sum_{n=0}^{\infty} p_{S_n}(s) p_N(n).$$

Then p_{S_N} has a mean iff both p_N and p_{X_1} do, and in this case

$$ES_N = (EX_1)(EN).$$

Also p_{S_N} has a variance iff both p_N and p_{X_1} do, and in this case

$$\text{Var } S_N = (EN)(\text{Var } X_1) + (EX_1)^2(\text{Var } N).$$

Furthermore if p_{X_1} is concentrated on the nonnegative reals and G_N, G_{X_1}, and G_{S_N} are the probability-generating functions for p_N, p_{X_1}, and p_{S_N}, respectively, then

$$G_{S_N}(z) = G_N(G_{X_1}(z)) \qquad \text{for } 0 \leq z \leq 1. \quad \blacksquare$$

If $\sum_{n=0}^{m} p_N(n) = 1$ then p_{S_N} as defined above is the density induced by the sum $S_N = X_1 + \cdots + X_N$ of a random number N of random variables, where N, $X_1, \ldots, X_m$ are independent and $X_1, \ldots, X_m$ have the same density p_{X_1}. By convention we interpret p_{S_N} in this same way when $p_N(n) > 0$ for infinitely many n.

Example 2 Consider a factory with $m \geq 1$ separate numbered units producing the same chemical. At 5 P.M. the factory shuts down with unit number k having produced X_k tons, and the assistant manager starts to inspect the units in their numbered order. By 7 P.M. he has inspected a random number N of the units, where $0 \leq N \leq m$. He then telephones the manager and tells him the number N of units inspected and the total yield for them, $S_N = X_1 + X_2 \cdots + X_N$. Thus S_N equals the sum of a random number N of random variables.

In general we would expect the yields $X_1, \ldots, X_m$ to be dependent; the units might, for example, receive their power from the same fluctuating source. If the assistant manager, upon finding that the first few units had low yields, typically hurried to inspect many additional units quickly, then N and $(X_1, \ldots, X_m)$ would also be dependent. However, if N is determined, for example, solely by the condition of the vehicle used to travel from one unit to another then N and $(X_1, \ldots, X_m)$ may be independent, so at least

$$ES_N = \sum_{n=0}^{m} (ES_n)p_N(n)$$

would apply. If $\mu = EX_1 = \cdots = EX_m$ then

$$ES_n = EX_1 + \cdots + EX_n = n\mu$$

so $ES_N = (\mu)(EN)$. Thus the mean of the total inspected yield equals the common mean yield per unit, times the mean number of units inspected, even though the yields may be dependent, as long as N and $(X_1, \ldots, X_m)$ are independent.

Assume that N, $X_1, \ldots, X_m$ are independent and $X_1, \ldots, X_m$ have the same distribution. If $EX_1 = 2$ tons and $EN = 10$ units then $ES_N = 20$ tons. In the extreme case that N is a constant, $P[N = 10] = 1$, then $\text{Var } N = 0$, so $\text{Var } S_N = (EN)(\text{Var } X_1) = 10 \text{ Var } X_1$. Thus the case of a sum of a nonrandom number of random variables is subsumed within our present framework. In the extreme case that there is no variability in production, so that $P[X_1 = 2] = 1$, then $\text{Var } X_1 = 0$ and $\text{Var } S_N = 4 \text{ Var } N$, as it should, since $S_N = 2N$ in this case. In a nonextreme case we might have $\text{Var } X_1 = 1.7$ and $\text{Var } N = 4.5$, so that

$$\text{Var } S_N = (EN)(\text{Var } X_1) + (EX_1)^2(\text{Var } N) = (10)(1.7) + (2^2)(4.5) = 17 + 18 = 35$$

and $\text{Var } S_N$ has roughly equal contributions from the variability of the yields and the variability of the number of units inspected. $\blacksquare$

Example 3 Let p_N be Poisson with mean λ and let p_{X_1} be Bernoulli so that $p_{X_1}(1) = p$ and $p_{X_1}(0) = q = 1 - p$. Then p_{S_N} as defined in Theorem 1 has many interpretations. Obviously if we make N independent tosses of a coin having probability p for a head then $S_N = X_1 + \cdots + X_N$ is the observed number of heads; if there are N radioactive disintegrations and each is recorded with probability p then S_N is the number recorded; if each of N insects has probability p of surviving an experiment then S_N is the number surviving; etc.

We now show that p_{S_N} is Poisson with mean λp. To evaluate $p_{S_N}(k)$ at a fixed nonnegative integer k note that p_{S_n} is binomial n and p, so that $p_{S_n}(k) = 0$ if $n < k$. Thus

$$p_{S_N}(k) = \sum_{n=k}^{\infty} \binom{n}{k} p^k q^{n-k} \frac{e^{-\lambda}\lambda^n}{n!}$$

$$= \frac{(\lambda p)^k e^{-\lambda p}}{k!} \sum_{n=k}^{\infty} \frac{e^{-\lambda q}(\lambda q)^{n-k}}{(n-k)!} = \frac{(\lambda p)^k e^{-\lambda p}}{k!} \sum_{i=0}^{\infty} \frac{e^{-\lambda q}(\lambda q)^i}{i!}$$

which completes the proof, since the sum over all $i = 0, 1, 2, \ldots$ of the Poisson density, with mean λq, is 1. ■

Exercises 1 to 5a may be done before reading the remainder of this section.

The special but important discrete compound Poisson densities and infinitely divisible densities are considered in Examples 4 and 6, which may be skipped. Compound Poisson densities arise naturally in Sec. 23-4. Example 4 provides a brief exposure to some of the ideas in Sec. 23-4. After Example 4 probability-generating functions are introduced as slight modifications of MGFs, and the proof of Theorem 1 is completed. Then Examples 5 and 6 apply probability-generating functions to Examples 3 and 4, respectively. In the next section probability-generating functions are employed in the proof of a basic theorem.

***Example 4** We call p_{S_N} a **compound Poisson density** iff N is Poisson; that is, a discrete density $p: R \to R$ is a compound Poisson density iff there is some positive real λ and some discrete density $p_{X_1}: R \to R$ such that

$$p(s) = \sum_{n=0}^{\infty} p_{S_n}(s) \frac{e^{-\lambda}\lambda^n}{n!} \quad \text{for all real } s$$

where p_{S_n} is defined in terms of p_{X_1} as in Theorem 1. To construct such a density we can select an arbitrary p_{X_1} and $\lambda > 0$, so there are many compound Poisson densities. If $p_{X_1}(1) = 1$ then $p_{S_n}(n) = 1$, so $p_{S_N} = p_N$; hence a Poisson density is a compound Poisson density, as was also shown by the preceding example. An exercise shows that a Pascal density is a compound Poisson density.

A discrete density $p: R \to R$ is said to be **infinitely divisible** iff for every integer $n = 2, 3, \ldots$ there is some density p_n such that p is induced by the sum of n independent random variables each having density p_n. We might have a density p and wonder if it could have been induced by the sum of some large number n of independent and identically distributed random variables. If p is infinitely divisible the answer is yes, for every n. If p is infinitely divisible and not concentrated on just one real $p(c) = 1$, then it is shown in Exercise 8 that $p(x) > 0$ for infinitely many x. Therefore a binomial density is not infinitely divisible. If p is Poisson with mean λ then p is infinitely divisible, since we can let p_n be Poisson with mean λ/n. We will show that every compound Poisson density is infinitely divisible. More precisely, we show that if p is a compound Poisson density obtained from p_{X_1} and λ then we can let p_n be the compound Poisson density obtained from p_{X_1} and λ/n. We first indicate the important context considered in Sec. 23-4, where infinite divisibility and compound Poisson densities arise naturally.

Suppose that each sample point ω corresponds to a real-valued function f_ω on the unit interval $0 \le t \le 1$. Three such are shown in Fig. 2*a*, and each performance of the experiment results in such a function or waveform. For each t in the interval $0 \le t \le 1$ define a random variable X_t by $X_t(\omega) = f_\omega(t)$ as in Fig. 2*b*; that is, X_t is the value at time t of the function obtained. Therefore, as indicated in Fig. 2*c*, we can express the total increase of the function over the interval $0 \le t \le 1$ as a sum of increments

$$X_1 - X_0 = \sum_{k=1}^{n} (X_{k/n} - X_{(k-1)/n}).$$

Sometimes these n increment random variables can reasonably be assumed to be independent and identically distributed, and this is true for every $n = 2, 3, \ldots$. Such an assumption is often made when X_t equals the number of radioactive disintegrations occurring during the interval from 0 to t. For such an example, as in Fig. 2*d*, the jumps of the function f_ω would correspond to disintegrations. A jump at time t can also be interpreted as meaning that at time t a motor vehicle drove by, or a baby was born, or an electron was emitted, or a nerve impulse was initiated, or an order was received, or a telephone conversation was initiated. Independent identically distributed increments clearly imply that the distribution of the total increase $X_1 - X_0$ is infinitely divisible. In Sec. 23-4 we will see that every compound Poisson density arises in this context.

We wish to show that *every compound Poisson density is infinitely divisible*. This result follows easily from an intuitively obvious property which we now introduce, state, and prove. Consider $p_{X_1}: R \to R$ to be a fixed density. Suppose that $N_1, N_2; X_1, X_2, \ldots ; Y_1, Y_2, \ldots$ are independent, each X_i and Y_j has the same density p_{X_1}, and $\sum_{n=0}^{\infty} p_{N_1}(n) = \sum_{n=0}^{\infty} p_{N_2}(n) = 1$. Let $S_{N_1} = X_1 + \cdots + X_{N_1}$, $S_{N_2} = Y_1 + \cdots + Y_{N_2}$, and $S_{N_1+N_2} = X_1 + \cdots + X_{N_1+N_2}$. Clearly $S_{N_1} + S_{N_2}$ and $S_{N_1+N_2}$ have the same distribution, since both random variables equal the sum of the random (and independent) number $N_1 + N_2$ of random variables each having density p_{X_1}. If $\sum_{n=0}^{m} p_{N_1}(n) = \sum_{n=0}^{m} p_{N_2}(n) = 1$ we can fit everything into the discrete case, and the argument above is rigorous. Otherwise, as we will shortly show, the same property can be derived when densities are defined as in Theorem 1.

Thus under widespread independence we can show more generally that

$$(X_1 + \cdots + X_{N_1}) + (Y_1 + \cdots + Y_{N_2}) + (Z_1 + \cdots + Z_{N_3})$$

and $X_1 + \cdots + X_N$ have the same density if $N = N_1 + N_2 + N_3$, where the X_i, Y_j, and Z_k have the same density p_{X_1}. But if N_i is Poisson λ_i then N is Poisson $\lambda = \lambda_1 + \lambda_2 + \lambda_3$. Let p_λ denote the density induced by $X_1 + \cdots + X_N$ when p_N is Poisson λ, while p_{X_1} always remains the same. We see, then, that p_λ is the density induced by $W_1 + W_2 + W_3$ when W_1, W_2, W_3 are independent and each has density $p_{\lambda/3}$. Since this argument is not restricted to $n = 3$, we see that p_λ is infinitely divisible. We now formalize this proof.

Property Let p_{N_1} and p_{N_2} be discrete densities such that $\sum_{n=0}^{\infty} p_{N_1}(n) = \sum_{n=0}^{\infty} p_{N_2}(n) = 1$. Let p_N be the density induced by $N = N_1 + N_2$, where N_1 and N_2 are independent and have densities p_{N_1} and p_{N_2}. Let $p_{S_{N_1}}$, $p_{S_{N_2}}$, and p_{S_N} be defined as in Theorem 1 by using the same p_{X_1} in all three cases. Then p_{S_N} is induced by $S_{N_1} + S_{N_2}$ when S_{N_1} and S_{N_2} are independent and have densities $p_{S_{N_1}}$ and $p_{S_{N_2}}$. ■

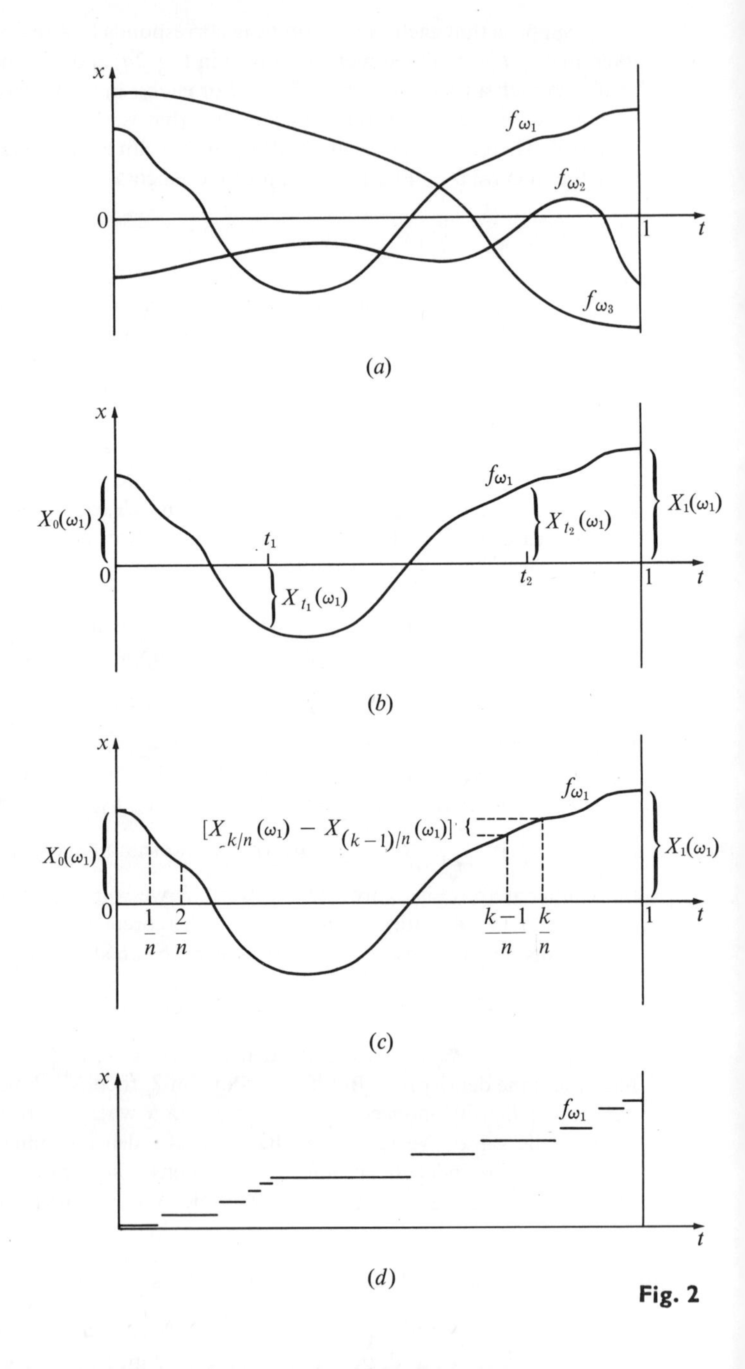

x
t
0
1
$f\omega_1$
$f\omega_2$
$f\omega_3$
(a)
$X_0(\omega_1)$
$X_1(\omega_1)$
t_1
t_2
$X_{t_1}(\omega_1)$
$X_{t_2}(\omega_1)$
(b)
$[X_{k/n}(\omega_1) - X_{(k-1)/n}(\omega_1)]$
$\frac{1}{n}$
$\frac{2}{n}$
$\frac{k-1}{n}$
$\frac{k}{n}$
(c)
(d)

Fig. 2

Proof For each fixed real s we have

$$\begin{aligned} p_{S_{N_1}+S_{N_2}}(s) &= \sum_{s'} p_{S_{N_1}}(s')p_{S_{N_2}}(s-s') \\ &= \sum_{s'} \sum_{n=0}^{\infty} p_{S_n}(s')p_{N_1}(n) \sum_{m=0}^{\infty} p_{S_m}(s-s')p_{N_2}(m) \\ &= \sum_{n=0}^{\infty} \sum_{m=0}^{\infty} p_{N_1}(n)p_{N_2}(m) \sum_{s'} p_{S_n}(s')p_{S_m}(s-s') \\ &= \sum_{n=0}^{\infty} \sum_{m=0}^{\infty} p_{N_1}(n)p_{N_2}(m)p_{S_{n+m}}(s). \end{aligned}$$

In this double series collect together the terms for which $n + m$ equals a fixed integer k; that is, sum the series in the order of Fig. 1 (Sec. 5-3) to get

$$\begin{aligned} p_{S_{N_1}+S_{N_2}}(s) &= \sum_{k=0}^{\infty} p_{S_k}(s)\left[\sum_{\substack{n,m \\ n+m=k}} p_{N_1}(n)p_{N_2}(m)\right] \\ &= \sum_{k=0}^{\infty} p_{S_k}(s)p_{N_1+N_2}(k) = p_{S_N}(s). \quad \blacksquare \end{aligned}$$

Fix p_{X_1} and let p_λ be the compound Poisson density p_{S_N} defined as in Theorem 1 with p_N Poisson $\lambda > 0$. The preceding property shows that $W_1 + W_2$ has density $p_{\lambda_1+\lambda_2}$ if W_1 and W_2 are independent and $p_{W_i} = p_{\lambda_i}$. Applying this result repeatedly, we see that $W_1 + \cdots + W_n$ has density p_λ if $W_1, \ldots, W_n$ are independent and each has density $p_{\lambda/n}$. *We have proved that every compound density is infinitely divisible.* The relationship between compound Poisson and infinitely divisible densities is even more intimate. For example, it can be shown[1] that any infinitely divisible discrete density concentrated on $n = 0, 1, \ldots$ is a compound Poisson density. This completes Example 4. ■

It turns out that MGFs are particularly useful for the study of densities p_{S_N}. In $\sum_i e^{tx_i}p(x_i)$ replace e^t by z to get $\sum_i z^{x_i}p(x_i)$. To avoid problems of convergence, and complex numbers, we require $0 \le z \le 1$ and $x_i \ge 0$, so that $0 \le z^{x_i} \le 1$ and the series consists of positive terms and converges. If $p: R \to R$ is a discrete density concentrated on the nonnegative reals $x \ge 0$ then the real-valued function G defined on the unit interval $0 \le z \le 1$ by

$$G(z) = \sum_i z^{x_i}p(x_i)$$

is called the **probability-generating function for** p, or for any random variable inducing p. Various definitions of generating functions are widely used as a matter of convenience. Among the more popular are the MGF and the probability-generating function $\sum_{n=0}^{\infty} z^n p(n)$ defined for $-1 \le z \le 1$ when $\sum_{n=0}^{\infty} p(n) = 1$. We permit noninteger x_i (for example, $x_i = \frac{1}{2}$) and so restrict z to $0 \le z \le 1$. The convenient form of the formula in Theorem 1 is our principal reason for using probability-generating functions instead of MGFs in this section and the next one.

[1] W. Feller, "An Introduction to Probability Theory and Its Applications," 2d ed., vol. I, p. 271, John Wiley & Sons, Inc., New York, 1957.

If $p(0) > 0$ let $x_1 = 0$ in the enumeration $x_1, x_2, \ldots$, so that

$$G(z) = z^0 p(0) + \sum_{i=2}^{\infty} z^{x_i} p(x_i).$$

If $0 < z \leq 1$ then $G(z)$ is well defined, and the first term is $p(0)$. However, if $z = 0$ then 0^0 is undefined. We naturally interpret it here as equal to 1, so $G(0) = p(0)$ and G is *continuous from the right at* $z = 0$. Thus

$$0 \leq p(0) \leq G(z) \leq G(1) = 1 \qquad \text{for } 0 \leq z \leq 1.$$

For the next section it is crucial that $p(0) > 0$ be permitted.

If X is a random variable for which $P[X \geq 0] = 1$ then for each real z in $0 < z \leq 1$ the expectation of the random variable z^X is just the value at z of the probability-generating function of p_X or of X,

$$G_X(z) = Ez^X = \sum_x z^x p_X(x) \qquad \text{for } 0 < z \leq 1.$$

If $e^t = z \leq 1$ then $t \leq 0$, so the last part of Theorem 2 (Sec. 9-4) does not apply. Nevertheless *a probability-generating function determines its density*. That is, as shown in Example 8 (Sec. 11-3), if two discrete densities are concentrated on $x \geq 0$ and their probability-generating functions are equal for all z, with $0 < z \leq 1$, then the densities are equal. For densities concentrated on $0, 1, 2, \ldots$ this fact is an obvious property of power series. Note that in Theorem 1 if p_{X_1} is concentrated on $0, 1, 2, \ldots$ then so are p_{S_n} and p_{S_N}.

Probability-generating functions behave as do MGFs for sums of independent random variables. If $X_1, \ldots, X_n$ are independent and $P[X_i \geq 0] = 1$ for each i then

$$\begin{aligned} G_{X_1 + \cdots + X_n}(z) &= Ez^{X_1 + \cdots + X_n} = Ez^{X_1} z^{X_2} \cdots z^{X_n} \\ &= Ez^{X_1} Ez^{X_2} \cdots Ez^{X_n} = G_{X_1}(z) G_{X_2}(z) \cdots G_{X_n}(z) \end{aligned}$$

since $z^{X_1}, z^{X_2}, \ldots, z^{X_n}$ are independent. In particular if the random variables are also identically distributed then

$$G_{S_n}(z) = G_{X_1 + \cdots + X_n}(z) = [G_{X_1}(z)]^n.$$

Let p_{S_N} be defined as in Theorem 1 with p_{X_1} concentrated on $x \geq 0$, so that p_{S_n} and p_{S_N} are also concentrated on $x \geq 0$. For p_{S_N} we find

$$\begin{aligned} G_{S_N}(z) &= \sum_s z^s p_{S_N}(s) = \sum_s z^s \sum_{n=0}^{\infty} p_{S_n}(s) p_N(n) \\ &= \sum_{n=0}^{\infty} \left[\sum_s z^s p_{S_n}(s) \right] p_N(n) = \sum_{n=0}^{\infty} G_{S_n}(z) p_N(n) \\ &= \sum_{n=0}^{\infty} [G_{X_1}(z)]^n p_N(n) = G_N(G_{X_1}(z)). \end{aligned}$$

Thus $G_{S_N}(z)$ is obtained by replacing z in $G_N(z)$ by $G_{X_1}(z)$. *We have proved that* G_{S_N} *is the composition of* G_N *and* G_{X_1}, *as stated in Theorem 1.* ∎

Example 5 As in Example 3, let N be Poisson λ and let p_{X_1} be Bernoulli, so that

$$G_N(z) = \sum_{n=0}^{\infty} z^n \frac{e^{-\lambda}\lambda^n}{n!} = e^{-\lambda} \sum_{n=0}^{\infty} \frac{(\lambda z)^n}{n!} = e^{-\lambda+\lambda z}$$

$$G_{X_1}(z) = z^0 p_{X_1}(0) + z^1 p_{X_1}(1) = q + pz$$

as could also be obtained from their MGFs by replacing e^t by z. Define p_{S_N} as in Theorem 1, so that from the last part of that theorem we have

$$G_{S_N}(z) = G_N(G_{X_1}(z)) = e^{-\lambda+\lambda(q+pz)} = e^{-(\lambda p)+(\lambda p)z}.$$

Thus the probability-generating function for p_{S_N} is the same as for a Poisson density with mean λp. Hence p_{S_N} must be Poisson λp, as was also shown in Example 3. ■

***Example 6** In this example we apply probability-generating functions to Example 4, where we dealt directly with densities. To avoid convergence difficulties with probability-generating functions we restrict ourselves in this example to densities concentrated on $x \geq 0$. If p with corresponding G is infinitely divisible, as defined in Example 4, then for each n there is a probability-generating function G_n such that $G = (G_n)^n$, but then $G_n = G^{1/n}$. Thus discrete density p concentrated on $x \geq 0$ with probability-generating function G is *infinitely divisible* iff for each $n = 1, 2, \ldots$ the function $G^{1/n}$ is the probability-generating function for a discrete density concentrated on $x \geq 0$.

In conformance with Example 4 and Theorem 1, a density p concentrated on $x \geq 0$ is a compound Poisson density iff there exists a positive real λ and a discrete density p_{X_1} concentrated on $x \geq 0$ and such that

$$G(z) = e^{-\lambda+\lambda G_{X_1}(z)} \qquad \text{for all } 0 < z \leq 1$$

where G and G_{X_1} are the probability-generating functions of p and p_{X_1}.

Now if p is compound Poisson corresponding to $\lambda > 0$ and p_{X_1} then

$$G^{1/n}(z) = e^{-(\lambda/n)+(\lambda/n)G_{X_1}(z)}.$$

Therefore $G^{1/n}$ is the probability-generating function for the compound Poisson density corresponding to λ/n and p_{X_1}. Hence every compound Poisson density concentrated on $x \geq 0$ is infinitely divisible. ■

EXERCISES

1. Let $S_N = X_1 + \cdots + X_N$ where N, X_1, X_2 are independent and

$$p_N(0) = .1 \qquad p_N(1) = .2 \qquad p_N(2) = .7$$

$$p_{X_1}(0) = .2 \qquad p_{X_1}(1) = .8 \qquad p_{X_2}(0) = .4 \qquad p_{X_2}(1) = .6$$

(*a*) Sketch the compositions of three urns which can be used to generate observations of (N,X_1,X_2). How is the value of S_N defined in terms of the outcome (n,x_1,x_2) of such a triple drawing?

(*b*) For each of the 12 points (n,x_1,x_2) tabulate the values of S_0, S_1, S_2, S_N and the joint density p_{N,X_1,X_2}. Also find the density induced by S_N.

2. Under the assumptions of Theorem 1 let $EN = 10^3$, Var $N = 400$, $EX_1 = 100$, and Var $X_1 = 900$ and show that $P[S_N \leq 5{,}000]$ is small. We might interpret X_i as the length of the ith of the N telephone calls made within an hour.

3. Under the assumptions of Theorem 1 let $EN = \text{Var } N = 100$, $EX_1 = 10$, and $\text{Var } X_1 = 100$, so that $ES_N = 10^3$. Interpret S_N as the number of tons ordered on the *first* day, where X_i is the size of the ith order. Assume that day-to-day operations are independent and identically distributed. Let Y be the total quantity of orders received in 100 days, so that $EY = 10^5$.
(*a*) Show that $P[|Y - 10^5| < 5{,}000]$ is large.
(*b*) Find a number A such that $P[|Y - 10^5| < A] \doteq .80$.
4. Let $p_{f,Z}$ be the joint density of a random point $f: \Omega \to F$ and a random variable $Z: \Omega \to R$ having the same associated discrete probability space (Ω,P). Assume that $P[Z = 0 \text{ or } Z = 1 \text{ or } \cdots \text{ or } Z = m] = 1$. This exercise shows that any joint density $p_{f,Z}$ can be obtained as the joint density of a randomly selected random point, and the selection random variable. Given $p_{f,Z}$, define a new experiment (Ω',P') with a random variable N and random points $f_0, f_1, \ldots, f_m$ such that $N, f_0, f_1, \ldots, f_m$ are independent, and $p_N = p_Z$, and $p_{f_i} = p_{f|Z=i}$ for $i = 0, 1, \ldots, m$. Thus we use the conditional distribution of f, given $[Z = 4]$, for the distribution of f_4 in our new experiment. For the new experiment let the random point f_N be obtained by a random selection according to N of one of $f_0, \ldots, f_m$; that is,

$$f_N(\omega) = f_n(\omega) \qquad \text{if } \omega \in [N = n].$$

(*a*) Prove that $p_{f_N,N} = p_{f,Z}$. In particular, then, $p_{f_N} = p_f$.
(*b*) Sketch the density p defined by

$$p(j,i) = \frac{1/6}{i+1} \qquad \text{for } i = 0, 1, \ldots, 5; j = 0, 1, \ldots, i.$$

Describe an experiment using seven urns and having independent random variables N, $Y_0, Y_1, \ldots, Y_5$ such that if Y_N is obtained by randomly selecting one of $Y_0, Y_1, \ldots, Y_5$ according to N then $p_{Y_N,N}$ equals the given density p.
5. In Theorem 1 let N be Poisson $\lambda = 1$ and let $p_{X_1}(0) = .2$, $p_{X_1}(.2) = .1$, $p_{X_1}(.3) = .2$, $p_{X_1}(.6) = .1$, and $P[X_1 > .6] = .4$. Find $P[S_N = .6]$.
6. (*a*) In Theorem 1 let N be geometric with mean q_0/p_0 and let X_1 be Bernoulli with mean p. Show that S_N is geometric with mean $(q_0/p_0)p$.
(*b*) In the equation

$$p_{S_N}(k) = \sum_{n=0}^{\infty} p_{S_n}(k) p_N(n) \qquad \text{for } k = 0, 1, \ldots$$

replace p_{S_N}, p_{S_n}, and p_N by the expressions corresponding to part (*a*). Naturally the truth of part (*a*) is equivalent to the truth of these equations.

Exercises 7 to 9 refer to Examples 4 and 6.

7. Use probability-generating functions to prove the property in Example 4, under the additional assumption that p_{X_1} is concentrated on $x \geq 0$.
8. Let p_{S_n} be induced by $S_n = X_1 + \cdots + X_n$ where $X_1, \ldots, X_n$ are independent and identically distributed. If p_{X_1} is concentrated on one point $p_{X_1}(c) = 1$ then so is p_{S_n}, since $p_{S_n}(nc) = 1$. Prove the following:
(*a*) If $p_{X_1}(x) > 0$ for at least two different x then $p_{S_n}(s) > 0$ for at least $n + 1$ different s.
(*b*) If $p: R \to R$ is an infinitely divisible density which is not concentrated on one point, then the set $\{x: p(x) > 0\}$ contains an infinite number of members.

9. Discrete density p_X is concentrated on $x \geq 0$ and satisfies $p_X(0) < 1$. Let p_Y be obtained by conditioning p_X onto $x > 0$, so that

$$p_Y(0) = 0 \qquad \text{and} \qquad p_Y(y) = \frac{1}{1 - p_X(0)} p_X(y) \qquad \text{if } y > 0.$$

If $\lambda > 0$ show that there is a $\lambda' > 0$ such that

$$e^{-\lambda + \lambda G_X(z)} = e^{-\lambda' + \lambda' G_Y(z)} \qquad \text{for } 0 \leq z \leq 1.$$

COMMENT: It follows immediately that density p concentrated on $x \geq 0$ is a compound Poisson density iff there is a real $\lambda > 0$ and a density p_{X_1} *concentrated on* $x > 0$ such that

$$G(z) = e^{-\lambda + \lambda G_{X_1}(z)} \qquad \text{for } 0 \leq z \leq 1$$

where G and G_{X_1} are the probability-generating functions of p and p_{X_1}, respectively. Evaluating this equation at $z = 0$ leads to $p(0) = G(0) = e^{-\lambda}$, so λ and hence G_{X_1} and p_{X_1} are unique in this representation of G.

10. Integrating the geometric series $1/(1 - x) = \sum_{k=0}^{\infty} x^k$ from 0 to q yields

$$\log \frac{1}{1 - q} = \sum_{n=1}^{\infty} \frac{q^n}{n} \qquad \text{whenever } 0 \leq q < 1. \qquad (*)$$

Therefore if $0 < p < 1$ and $p + q = 1$ we can define the **logarithmic discrete probability density** with parameter $0 < p < 1$ by

$$l(n) = \frac{q^n}{n \log (1/p)} \qquad \text{for } n = 1, 2, \ldots.$$

(*a*) Show that the logarithmic density has

$$\frac{\log [1/(1 - qz)]}{\log (1/p)}$$

for its probability-generating function.

(*b*) In Theorem 1 let p_{X_1} be the logarithmic density with parameter p and let N be Poisson with mean $\lambda = r \log (1/p)$, where r is *any positive real number*. Show that the resulting density p_{S_N} has $[p/(1 - qz)]^r$ for its probability-generating function. The [compound Poisson] discrete probability density corresponding to this probability-generating function $[p/(1 - qz)]^r$ is called the **negative binomial density** with parameters $r > 0$ and $0 < p < 1$. Of course, positive *integer* r correspond to Pascal densities. For general r and p we have mean rq/p and variance rq/p^2 and closure under independent additions, just as shown for the Pascal densities in Example 6 (Sec. 11-3).

(*c*) Show that the negative binomial density p_r: $R \to R$ with parameters $r > 0$ and $0 < p < 1$ is given by

$$p_r(k) = \frac{(r + k - 1)(r + k - 2) \cdots (r + 1)r}{k!} p^r q^k \qquad \text{for } k = 0, 1, \ldots.$$

Note also the alternate forms

$$p_r(k) = \binom{r + k - 1}{k} p^r q^k = \binom{-r}{k} p^r (-q)^k \qquad \text{for } k = 0, 1, \ldots$$

where the appearance of $\binom{-r}{k}$ explains the name given to these densities. For these alternate forms we must generalize the definition of the binomial coefficient $\binom{b}{k}$ to arbitrary real b, where Sec. 2-3 required integer $b \geq 0$. For *any real number b* let $\binom{b}{0} = 1$ and let

$$\binom{b}{k} = \frac{b(b-1)\cdots(b-k+1)}{k!} \qquad \text{for any positive integer } k.$$

Thus the numerator of $\binom{b}{k}$ has k factors, starting at b and decreasing in integer steps. Note that one of these factors is zero iff b is a nonnegative integer with $b < k$.

HINT: Recall the binomial series

$$(1+x)^b = \sum_{k=0}^{\infty} \binom{b}{k} x^k \qquad \text{if } -1 < x < 1$$

which is valid for all real b. In particular if b is a positive integer then every $\binom{b}{k}$ with $k > b$ must be zero; hence the series then agrees with the binomial theorem, which is valid for all real x. Letting $b = -r$ and $x = -q$ yields

$$(1-q)^{-r} = \sum_{k=0}^{\infty} \binom{-r}{k} (-q)^k \qquad \text{if } -1 < q < 1$$

valid for all real r. Also

$$\binom{-r}{k} = (-1)^k \frac{r(r+1)\cdots(r+k-1)}{k!} = (-1)^k \binom{r+k-1}{k}$$

so the binomial series implies formula (4) of Appendix 1.

*14-3 BRANCHING PROCESSES

Consider the following experiment. In the spring of year n you plant W_n seeds. A seed experiencing favorable conditions may grow into a plant which produces seeds prior to its death within 1 year. You collect the seeds from all the plants that lived and bore seeds, so that in the spring of year $n + 1$ you have W_{n+1} seeds, which you then plant; and so on. Call W_n the size of the nth generation and assume that $W_0 = 1$, so that you start with 1 seed in year 0. The number W_n of seeds planted is random, as is the number of seeds produced by a planted seed, so W_{n+1} is the sum of a random number W_n of random variables. We assume that the number of seeds produced by a planted seed is independent of the history of the generations that preceded it, and independent of the number and fates of other seeds in its own generation, and the distribution of the number of seeds produced by any one seed is the same as that for any other seed planted in any year. The event $[W_n = 0]$ can be interpreted as extinction within n generations. We concentrate attention on $P[W_n = 0]$ for large n.

Our assumptions are, of course, idealized. They ignore climatic conditions which might cause most plants in a particular year to have low yields, thus making the yields dependent. Also if the total number of seeds became astronomical there would be competition among plants, and the distribution of offspring per parent would depend on the size of the generation. Nevertheless the model is a useful first approximation in many applied problems where, for example, W_n might be the number of particles at time n in a nuclear chain reaction. Our basic assumptions were introduced by Francis Galton in connection with human male offspring and the survival of family names. The first results were obtained by Watson and Galton[1] in 1874, and the basic results of this section were proved by Steffensen[2] in 1930. Historical comments, an extensive bibliography, and many generalizations and recent results can be found in the book by Harris.[3]

Example 1 Processes of this kind can be depicted as in Fig. 1. They are often called *branching processes*, and each line between a pair of nodes is called a *branch*. Figure 1 shows the outcome out to generation 3 for such an experiment, where W_n is the number of nodes on the vertical line through n. Suppose that the number of seeds produced by a plant has a uniform distribution over the 10 integers 0, 1, . . . , 9. The yield from the ith seed is equivalent to the number

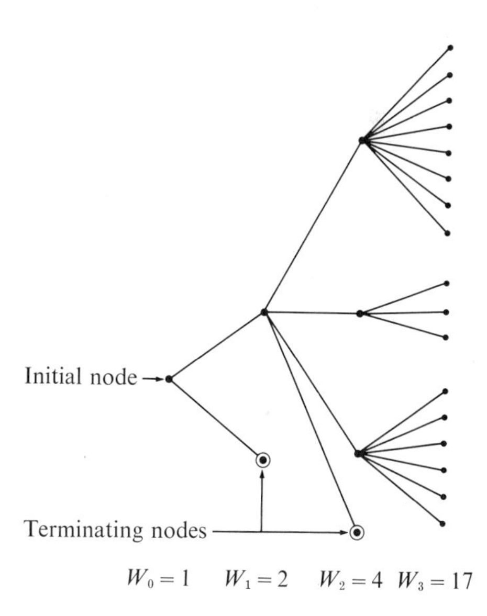

Fig. 1 The simulation of a branching process out to the third generation; random digits 2, 4, 0, 8, 3, 6, 0 were used.

on the ith ball drawn with replacement from an urn containing 10 balls labeled 0, 1, . . . , 9. Drawing balls from such an urn (actually we have used the table of random numbers in Chap. 12) we obtain the numbers 2, 4, 0, 8, 3, 6, 0, 2, 5, 1, 7, 9 on the first 12 draws. Thus the initial seed produced 2 seeds, so Fig. 1 has 2 branches leaving the initial node. These 2 seeds produced 4 and 0 seeds, respectively, so $[W_2 = 4]$ occurred. These 4 seeds produced 8, 3, 6, and 0 seeds, respectively, so $[W_3 = 17]$ occurred.

[1] H. W. Watson and Francis Galton, On the Probability of the Extinction of Families, *J. Anthropol. Inst. Great Britain and Ireland*, 4: pp. 138–144, 1874.

[2] J. F. Steffensen, Om sandsynligheden for at afkommed uddor, *Matematisk Tidsskrift*, 1 pp. 19–23, 1930.

[3] Theodore E. Harris, "The Theory of Branching Processes," Springer-Verlag, Prentice-Hall, Inc., Englewood Cliffs, N.J., 1963.

There are many different ways in which such a fluctuating population, starting at zero, might become extinct, that is, have only a finite number of branches in its complete graph. Extinction occurs earliest if the initial seed produces no seeds, and this has probability $P[W_1 = 0] = .1$. The next simplest way in which extinction can occur is for the initial seed to produce 1 seed which produces no seeds, and this has probability (.1)(.1). Thus the probability of eventual extinction exceeds $.1 + .01 = .11$, and we will see that it actually equals .113. Typically this population either dies out almost immediately or it soon becomes so large that its immortality is almost a certainty. Note that the expected number of offspring per parent is $\mu = 4.5$. In general if β is the probability of eventual extinction we show that $\beta = 1$ iff $\mu \leq 1$. When $\mu > 1$ we describe a method for obtaining the number $\beta < 1$. ∎

A sample point for such an experiment could be a complete description of a possibly infinite graph, like the initial portion shown in Fig. 1. Such a setup can become fairly involved, and if $\mu > 1$ it would in general require an uncountable number of nonextinction outcomes. Our principal interest is in the numbers $p_{W_1}(0), p_{W_2}(0), p_{W_3}(0), \ldots$, so we define only the sequence of densities $p_{W_1}, p_{W_2}, \ldots$. We use the notation p_{W_n} even though we do not define $W_1, W_2, \ldots$ as random variables on the same discrete-probability space.

The basic datum is the density p, concentrated on $0, 1, 2, \ldots$, where $p(k)$ is the probability that a parent will have k offspring. Assume that $0 < p(0) < 1$, so that extinction may occur, but not necessarily immediately. We start with one seed $W_0 = 1$, so the density p_{W_1} of the number of seeds in the first generation is the given density p. If p_{W_n} has been defined then $p_{W_{n+1}}$ is defined to be the density induced by

$$W_{n+1} = X_1 + X_2 + \cdots + X_{W_n}$$

where W_n and the X_i are independent, W_n has density p_{W_n}, and each X_i has density p. Thus X_i is the number of seeds produced by the ith seed from the nth generation. In other words, let $p_{S_0}(0) = 1$ and let p_{S_n} be the distribution of the size $S_n = X_1 + \cdots + X_n$ of the next generation if the present generation has size n. Then the densities $p = p_{W_1}, p_{W_2}, \ldots$ are defined recursively by

$$p_{W_{n+1}}(w) = \sum_{r=0}^{\infty} p_{S_r}(w) p_{W_n}(r)$$

and we call this sequence $p_{W_1}, p_{W_2}, \ldots$ the **branching process determined by** $p = p_{W_1}$. Clearly

$$p_{W_{n+1}}(0) \geq p_{S_0}(0) p_{W_n}(0) = p_{W_n}(0).$$

Thus the **probability** $\beta_n = p_{W_n}(0)$ **of extinction within** n **generations** is an increasing function of n,

$$0 < p(0) = \beta_1 \leq \beta_2 \leq \beta_3 \leq \cdots \leq 1.$$

We naturally interpret $\beta = \lim_{n\to\infty} \beta_n$ as the **probability of eventual extinction** for the branching process $p_{W_1}, p_{W_2}, \ldots$.

We now easily show that $\beta = 1$ if the expected number $\mu = \sum_{k=0}^{\infty} kp(k)$ of offspring per parent is less than 1. From Theorem 1 (Sec. 14-2) we have

$$EW_n = \mu[EW_{n-1}] = \mu[\mu(EW_{n-2})] = \cdots = \mu^n.$$

For any random variable X concentrated on $x \geq 0$ we have $P[X \geq 1] \leq EX$, which can be proved by using Lemma 1 (Sec. 10-1) with $A = \{x: x \geq 1\}$ and $f(x) = |x|$. Hence

$$\beta_n = p_{W_n}(0) = 1 - P[W_n \geq 1] \geq 1 - EW_n \doteq 1 - \mu^n.$$

Therefore if $\mu < 1$ then β_n converges to 1 as n goes to infinity. If $\mu > 1$ then $EW_n = \mu^n$ goes to infinity as n does, suggesting, as stated below, that $\beta < 1$.

Theorem 1: Fundamental theorem for branching processes Let the discrete density $p: R \to R$ satisfy $0 < p(0) < 1$ and $\sum_{k=0}^{\infty} p(k) = 1$, and have probability-generating function G. Let $p = p_{W_1}, p_{W_2}, \ldots$ be the branching process determined by p, so that $p_{W_n}(0) \leq p_{W_{n+1}}(0)$ and the limit $\beta = \lim_{n \to \infty} p_{W_n}(0)$ is interpretable as the probability of eventual extinction. Let $\mu = \sum_{k=0}^{\infty} kp(k)$ be the expected number of offspring per parent. If $\mu \leq 1$ then $\beta = 1$, and if $\mu > 1$ then β is the unique real number satisfying $0 < \beta < 1$ and $\beta = G(\beta)$. ■

We now interpret the analytical conclusion of the theorem, apply it to Example 1, exhibit an *intuitive* derivation of the conclusion, and finally prove the theorem.

Differentiating $G(z) = \sum_{k=0}^{\infty} z^k p(k)$, we get

$$G'(z) = p(1) + 2zp(2) + 3z^2p(3) + \cdots + kz^{k-1}p(k) + \cdots$$
$$G''(z) = 2p(2) + (3)(2)zp(3) + \cdots + k(k-1)z^{k-2}p(k) + \cdots.$$

Clearly $G' > 0$ and $G'' \geq 0$, so G is strictly increasing and convex. Furthermore $G(1) = 1$ and the slope $G'(1) = \mu$ of G at $z = 1$ is equal to the mean number of offspring per parent. If $\mu < 1$ then, as in Fig. 2*a*, G is always on or above the dashed line tangent to it at $z = 1$. However, if $\mu > 1$ then, as in Fig. 2*b*, the slope of G near $z = 1$ is greater than that of the exhibited diagonal, which has slope 1; hence $G(z) < z$ for z near 1. But $G(0) = p(0) > 0$, so G must intersect the diagonal at least at one intermediate point, while the convexity of G guarantees that there is at most one intersection. The theorem asserts that G intersects the diagonal at $z = \beta$, so β may be found by plotting G.

Example 2 We now apply Theorem 1 to Example 1, where $p(0) = p(1) = \cdots = p(9) = .1$. We always have $p(0) \leq \beta$, so $.1 \leq \beta$ for this example. The expected number $\mu = 4.5$ of offspring per parent exceeds 1, and hence $\beta < 1$. For G we have

$$G(z) = (.1)(1 + z + \cdots + z^9) = (.1)\left(\frac{1 - z^{10}}{1 - z}\right).$$

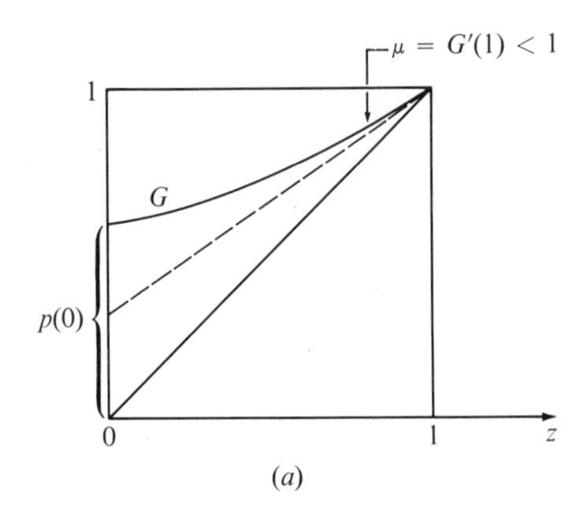

(a)

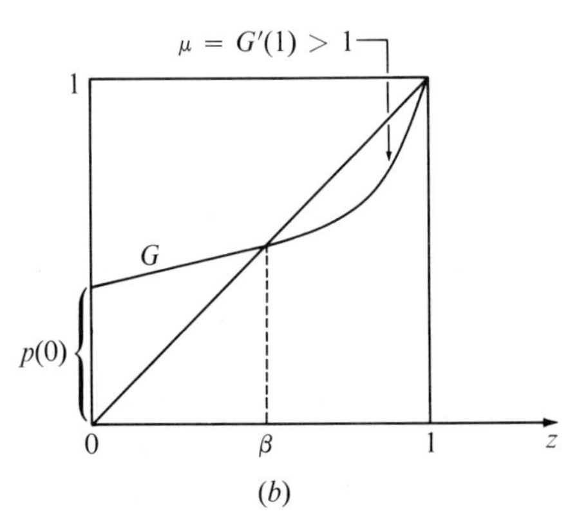

Fig. 2 (b)

Assume for a moment that the solution $\beta = G(\beta)$ will yield a small enough β so that β^{10} is negligible. Therefore we solve the quadratic equation $\beta' = .1/(1 - \beta')$ and find $\beta' = .5 - .5\sqrt{.6} \doteq .1127$, so our approximation is justifiable, and $\beta \doteq \beta' \doteq .1127$. As a check we evaluate $G(z) - z$ at one point on either side of β' and find that $G(.112) - .112 > 0$ and $G(.114) - .114 < 0$, so that β *must* satisfy $.112 < \beta < .114$.

It should be clear intuitively that if β is the probability of extinction for a branching process with one initiator then β^r is the probability of extinction for a branching process with $r \geq 1$ initiators, $W_0 = r$, since each of the subpopulations generated by each initiator must become extinct independently. More generally the probability of eventual extinction, conditioned on the event $[W_n = k]$, with $P[W_n = k] > 0$, is β^k for any $n \geq 0$ and $k \geq 0$ and any number of initiators. Thus $(.113)^{17}$ is the probability of eventual extinction for the indefinite continuation of the sampling experiment of Fig. 1. ■

The basic equation used to prove Theorem 1 is $\beta_{n+1} = G(\beta_n)$, so that

$$p(0) = \beta_1 \quad \text{and} \quad \beta_2 = G(\beta_1), \quad \beta_3 = G(\beta_2), \ldots.$$

Let us now describe an intuitive probabilistic derivation of this recursion equation $\beta_{n+1} = G(\beta_n)$. We decompose the event $[W_{n+1} = 0]$ into disjoint events depending on the size W_1 of the first generation:

$$P[W_{n+1} = 0] = \sum_{k=0}^{\infty} P[W_{n+1} = 0, W_1 = k]$$
$$= \sum_{k=0}^{\infty} P[W_{n+1} = 0 \mid W_1 = k]p(k) = \sum_{k=0}^{\infty} \{P[W_n = 0]\}^k p(k).$$

This is the desired equation $\beta_{n+1} = G(\beta_n)$ so we need only justify

$$P[W_{n+1} = 0 \mid W_1 = k] = \{P[W_n = 0]\}^k.$$

This last equation can be interpreted as saying that if an initial seed is planted at time $t = 0$ and produces k seeds, which are planted at time $t = 1$, then the total population will be extinct at time $t = n + 1$ iff each of the k subpopulations starting at time $t = 1$ becomes extinct, independently and each with probability $P[W_n = 0]$, by time $t = n + 1$. Having explained the equation $\beta_{n+1} = G(\beta_n)$, we note that G is a convergent power series on $0 \le z \le 1$ and so is continuous there. Hence β must satisfy

$$\beta = \lim_{n\to\infty} \beta_n = \lim_{n\to\infty} G(\beta_{n-1}) = G(\beta).$$

In the proof of the theorem it is shown that if $\mu > 1$ then $\beta \ne 1$ even though $1 = G(1)$.

Proof of Theorem 1 Let G_n be the probability-generating function for p_{W_n}, so that $G_1 = G$. Theorem 1 (Sec. 14-2) implies that $G_{n+1}(z) = G_n(G(z))$, since G_{n+1} corresponds to the sum of W_n random variables each having density p. Therefore

$$G_2(z) = G(G(z))$$
$$G_3(z) = G_2(G(z)) = G(G(G(z)))$$
$$G_{n+1}(z) = G_n(G(z)) = G(G(\cdots(G(z))\cdots)).$$

In particular $G_{n+1}(z) = G(G_n(z))$. Evaluating this equation at $z = 0$ and using $\beta_n = p_{W_n}(0) = G_n(0)$, we obtain $\beta_{n+1} = G(\beta_n)$.

Thus β satisfies $\beta = G(\beta)$. We must show that β is the smallest number satisfying this equation; that is, we must show that if β^* is any number such that $0 < \beta^* \le 1$ and $G(\beta^*) = \beta^*$ then $\beta_n \le \beta^*$ for all n. We do this by induction on n, using the equation $\beta_{n+1} = G(\beta_n)$ and the fact that G is an increasing function. Now $0 < \beta^*$ so

$$\beta_1 = G(0) \le G(\beta^*) = \beta^*$$

and hence $\beta_1 \le \beta^*$. Also if $\beta_n \le \beta^*$ then

$$\beta_{n+1} = G(\beta_n) \le G(\beta^*) = \beta^*.$$

Therefore $\beta_n \le \beta^*$ for all n. Thus we know that $0 < \beta \le 1$ and $\beta = G(\beta)$ and that β is the smallest real number satisfying these two conditions.

The remainder of the proof is devoted to showing that if $\mu \le 1$ then the conditions $0 < z \le 1$ and $z = G(z)$ are satisfied only by $z = 1$, and if $\mu > 1$ then there is exactly one solution in addition to $z = 1$. This result is graphically obvious from Fig. 2 and the discussion immediately following the statement of Theorem 1.

Define $F(z) = G(z) - z$ for $0 \leq z \leq 1$. From the standard theorems on power series, which we use repeatedly, we know that F is continuous on $0 \leq z \leq 1$ and infinitely differentiable on $0 \leq z < 1$. Clearly $F(0) = p(0) > 0$ and $F(1) = 0$. If $F'(1)$ is the limit from the left of F' then $F'(1) = \mu - 1$. If $p(0) + p(1) = 1$ then G is a straight line and $\mu = p(1) < 1$, so the desired conclusion is true for this case. In all other cases at least one of $p(2), p(3), \ldots$ is positive, so that $F''(z) = G''(z) > 0$ if $0 < z < 1$. Hence F is *strictly* convex and F' is *strictly* increasing.

As in Fig. 3*a*, if $\mu \leq 1$ then $F'(z) < F'(1) \leq 0$ for $0 < z < 1$, so we cannot have a z satisfying $0 < z < 1$ and $F(z) = 0$. For $\mu = 1$ we used the strictness of the increasing of F'.

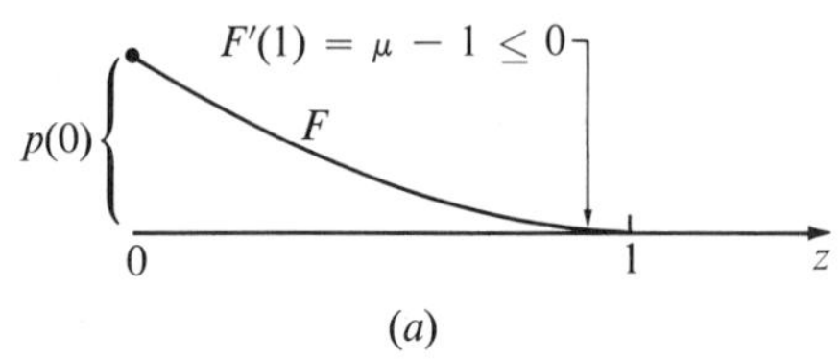

(*a*)

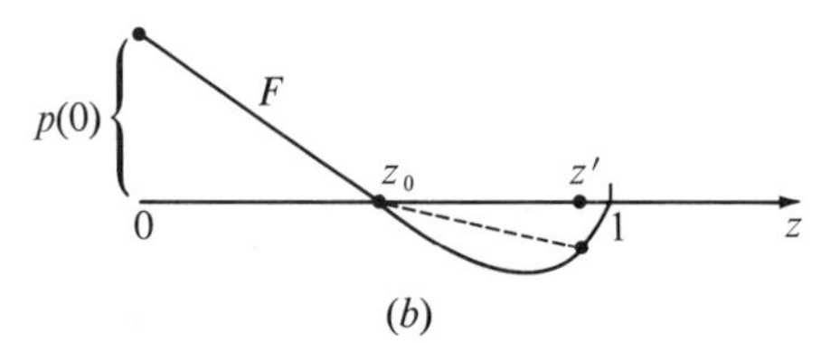

(*b*)

Fig. 3

If $\mu > 1$ then $F'(1) > 0$, so, as in Fig. 3*b*, we must have $F'(z) > 0$ for z near 1, and hence F must be negative near $z = 1$; that is, there is a z' such that $0 < z' < 1$ and $F(z) < 0$ whenever $z' \leq z < 1$. Now, $F(0) > 0$ and $F(z') < 0$, so there is at least one z_0 such that $0 < z_0 < z'$ and $F(z_0) = 0$. There cannot be two such numbers z_0 and z_0', since if $z_0 < z_0'$ then by convexity F must be below the dashed-line segment shown in Fig. 3*b*, so $F(z_0') < 0$. ■

EXERCISES

1. A rare plant produces seeds according to $p(0) = p(1) = .3$ and $p(2) = .4$. Find the probability of extinction if the initial number of seeds is

(*a*) $W_0 = 1$ (*b*) $W_0 = 100$

2. For a branching process with $\mu < 1$ we have the formal manipulation

$$E(W_0 + W_1 + W_2 + \cdots) = 1 + \mu + \mu^2 + \cdots = \frac{1}{1 - \mu}$$

suggesting (as can be proved) that $1/(1 - \mu)$ is the expectation of the total number of individuals (including the given initiator) who ever exist prior to extinction. Let the distribution of the number of offspring per parent be given by

$$p(0) = .45 \qquad p(1) = .4 \qquad p(2) = .1 \qquad p(3) = .05$$

so that $\mu = \frac{3}{4}$ and $1/(1 - \mu) = 4$.

(*a*) Find the probability β_4 of extinction within four generations.
(*b*) Use the random-number table of Chap. 12 to simulate this branching process to extinction, 10 times. Let v_i be the total number of individuals who ever exist in your ith simulation. Note whether or not

$$\frac{1}{10}(v_1 + v_2 + \cdots + v_{10}) \doteq 4 = \frac{1}{1 - \mu}.$$

3. Let $p(k) = bc^{k-1}$ for $k = 1, 2, \ldots$, where $0 < b$ and $0 < c < 1$. Then

$$\sum_{k=0}^{\infty} p(k) = p(0) + \frac{b}{1 - c} = 1 \qquad \text{so} \qquad p(0) = 1 - \frac{b}{1 - c}$$

and we add the constraint $b < 1 - c$ in order to make $0 < p(0)$. Clearly the geometric density corresponds to $c = q$ and $b = pq$. Show the following:

(*a*) $G(z) = p(0) + [bz/(1 - cz)]$ for $0 \le z \le 1$.
(*b*) $\mu = G'(1) = b/(1 - c)^2$.
(*c*) If $\mu \ge 1$ then $\beta = p(0)/c$ and *in particular* $\beta = p/q = 1/\mu$ *for a geometric density with* $\mu \ge 1$.

COMMENT: Lotka[1] used the density of this exercise with $b = .2126$ and $c = .5893$ to fit 1920 census data so $p(0) = .4825$, and $\beta = .819$ becomes the probability of extinction for American male lines of descent. The family of two parameter densities of this exercise is convenient in that[2] if the initial density p belongs to this family then so does the density p_{W_n} for any generation.

4. If W_n is the size of the nth generation for a branching process and $EW_1 = \mu$ and $\operatorname{Var} W_1 = \sigma^2$, show that $\operatorname{Var} W_n = \sigma^2\mu^{n-1}(1 + \mu + \cdots + \mu^{n-1})$. Thus

$$\operatorname{Var} W_n = \begin{cases} n\sigma^2 & \text{if } \mu = 1 \\ \dfrac{\sigma^2\mu^n(\mu^n - 1)}{\mu^2 - \mu} & \text{if } \mu \ne 1. \end{cases}$$

5. Show that the construction of Fig. 4 does actually yield the numbers $\beta_1, \beta_2, \ldots$.

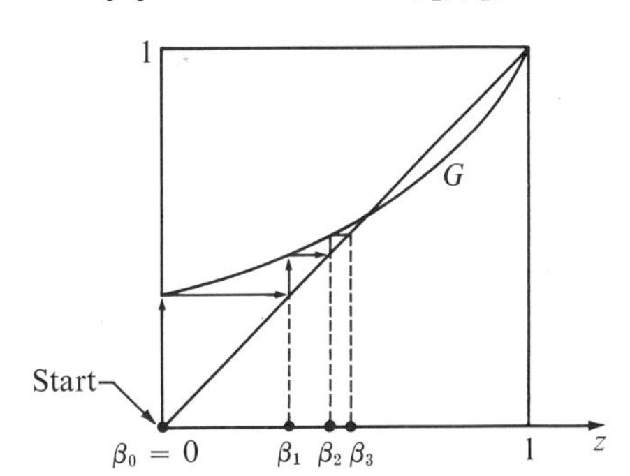

Fig. 4

[1] A. J. Lotka, "Théorie analytique des associations biologiques, deuxième partie," *Actualités Scientif. Industrielles*, vol. 780, Herman & Cie, Paris, 1939.
[2] Harris, *op. cit.*, pp. 9, 10.

6. Let W_n be the size of the nth generation of a branching process and prove that $\lim_{n\to\infty} P[1 \le W_n \le r] = 0$ for all $r = 1, 2, \ldots$. Thus, regardless of the value of β, if n is large enough then we can be quite certain that the population is either extinct or enormous.

7. (*a*) Consider the branching process $p(0) = q$, $p(2) = p$, and $p + q = 1$ in which each organism either dies with probability q, or with probability p it subdivides into two offspring. Show that $\beta = 1$ if $p \le \frac{1}{2}$ and $\beta = q/p$ if $p > \frac{1}{2}$.

(*b*) You start with \$1, and as long as you have any money you bet \$1 on each toss of a biased coin. Let $Y_1, Y_2, \ldots$ be independent and identically distributed by

$$P[Y_1 = -1] = q \qquad P[Y_1 = 1] = p \qquad p + q = 1$$

so that Y_n is the gain from the nth toss for anyone who bets \$1 on this toss. Let $B = \bigcup_{n=1}^{\infty} [Y_1 + \cdots + Y_n = -1]$ be the event that you eventually go broke. Show that $P(B) = 1$ if $p \le \frac{1}{2}$ and $P(B) = q/p$ if $p > \frac{1}{2}$. This result is used in Example 1 (Sec. 23-7). HINT: A friend inspects your capital after the first toss and finds it to be $W_1 = 1 + Y_1$. If on his nth inspection he finds that you have \$18 then he next inspects you after 18 additional tosses, since you could not go broke before then. Thus your capital W_{n+1} on his $(n + 1)$st inspection has the distribution induced by

$$W_n + Y_1' + \cdots + Y_{W_n}'$$

where Y_i are distributed like Y_1 while W_n and the Y_i' are independent.

15
Probability Spaces

In this chapter the structure of the general theory is developed in a fashion applicable to discrete and continuous densities and all other cases. As was explained in the introduction to Sec. 6-4, this presentation differs from that for the discrete case primarily in its added emphasis on probability measures, which are available in the most general case, and its lack of emphasis on densities, which are available only in special cases. Except for this necessary increase in abstractness, the presentation is similar to, and sometimes simpler than, the previous one.[1]

Much of the remainder of the book is not rigorous, primarily because a rigorous development valid for nondiscrete cases requires that $P(A)$ be defined for only some but not all subsets A of the sample space, and that some but not all real-valued functions $X: \Omega \to R$ defined on the sample space be called random variables. Except in Chap. 21, we normally ignore these two complications. However, as explained early in Sec. 16-1, these restrictions are more a technical nuisance than a practical limitation.

15-1 PROBABILITY SPACES

This section includes obvious generalizations of our earlier (Ω,P) and conditional distributions. A **probability space** (Ω,P) consists of a nonempty set Ω and a function P. The set Ω is called the **sample space**, every point ω in Ω is called a **sample point**, and *every* subset A of Ω is called an **event.** P is a real-valued function defined on the collection of all events, and if A is an event then $P(A)$ is the **probability of** A. The function P is called a **probability measure** and is required to satisfy the following three conditions:

1. $P(\Omega) = 1$.
2. $P(A) \geq 0$ for every event A.
3. If $A_1, A_2, \ldots$ is any sequence of disjoint events then $P\left(\bigcup_{n=1}^{\infty} A_n\right) = \sum_{n=1}^{\infty} P(A_n)$.

Thus a probability measure is a normed, nonnegative, countably additive set function defined on the class of all events.

Theorem 3 (Sec. 6-4) showed that the discrete case corresponds to a P which is concentrated on a countable subset of the sample space. The proof of Theorem 2

[1] More immediate utilization of this chapter may be provided by studying Chaps. 15 and 16 together in the order Secs. 15-1, 16-1, 15-2, 16-2, 16-3, 15-3, 16-4, 15-4, 16-5, 15-5, 16-6.

(Sec. 6-4) did not assume that P was concentrated on a countable subset. Therefore the properties of P described in that theorem are valid in general and will be used freely.

The sample space Ω can be thought of as containing exactly one sample point ω for each possible experimental outcome. For technical advantages we insist on countable additivity rather than just finite additivity. Sometimes P can be calculated from a more manipulable entity, such as a discrete density, or a continuous density, or a distribution function. As in earlier chapters, it is often more natural to work with P directly, and this is essentially unavoidable when Ω is a function space, as in Chaps. 22 to 24.

The definition of conditional probability remains the same,

$$P(B \mid A) = \frac{P(AB)}{P(A)} \qquad \text{if } P(A) > 0$$

as do the statement and proof of Bayes' theorem 1 (Sec. 7-2) and the definition, proofs, and properties of independent events given in Chap. 7.

Section 7-1 showed how to define a conditional density p_A such that if P_A is calculated from p_A we always have $P_A(B) = P(B \mid A)$. A more basic approach for nondensity cases is to fix A and *define* P_A by $P_A(B) = P(B \mid A)$; that is, we fix A and consider the probability $P(AB)$ of that part of B which is contained in A, and except for normalization, this function of B is a probability measure, as shown in the next theorem.

Theorem 1 Let (Ω,P) be a probability space and let A be an event with $P(A) > 0$. If the function P_A is defined for every event B by

$$P_A(B) = P(B \mid A) = \frac{P(AB)}{P(A)}$$

then (Ω,P_A) is a probability space, which is said to be obtained by **conditioning** (Ω,P) **by** A. ■

Proof Clearly $P_A(B) \geq 0$ and $P_A(\Omega) = P(A\Omega)/P(A) = 1$. If $B_1, B_2, \ldots$ are disjoint events then so are $AB_1, AB_2, \ldots$; hence

$$P_A\left(\bigcup_{n=1}^{\infty} B_n\right) = \frac{1}{P(A)} P\left[(A) \cap \left(\bigcup_{n=1}^{\infty} B_n\right)\right] = \frac{1}{P(A)} P\left[\bigcup_{n=1}^{\infty} (AB_n)\right]$$

$$= \frac{1}{P(A)} \sum_{n=1}^{\infty} P(AB_n) = \sum_{n=1}^{\infty} P_A(B_n)$$

So p_A is countably additive. ■

If $a = (a_1, \ldots, a_n)$ and $b = (b_1, \ldots, b_n)$ are two points of R^n then the set

$$R_{a,b} = \{(x_1, \ldots, x_n)\colon a_i \leq x_i \leq b_i \text{ for all } i = 1, 2, \ldots, n\}$$

is called a **closed rectangle in** R^n. The word "closed" refers to the fact that we used only $\leq$. If we replace every $\leq$ by $<$ the set is called an **open rectangle** in R^n. For every mixture of the inequality signs $\leq$ and $<$ the set is always called a **rectangle** in R^n. Clearly a closed rectangle is empty iff $a_i > b_i$ for at least one i. The intersection of two closed (or open) rectangles is also a closed (or open) rectangle. If $a_i = b_i$ for $i = 1, \ldots n$, then the closed rectangle $R_{a,a}$ contains only the one point $a = (a_1, \ldots, a_n)$. When $n = 1$ we may call the set an *interval* instead of a rectangle. Thus for real $a < b$ we have the four intervals $\{x: a \leq x \leq b\}$, $\{x: a \leq x < b\}$, $\{x: a < x \leq b\}$, and $\{x: a < x < b\}$, where the first one is closed and the last one is open.

We next show that if two probability measures assign the same probabilities to closed rectangles they must assign the same probabilities to all events.

Theorem 2 If (R^n,P) and (R^n,P') are probability spaces for which $P(R_{a,b}) = P'(R_{a,b})$ for every closed rectangle $R_{a,b}$ in R^n then $P = P'$. ■

Proof Since $R_{a,a}$ contains only the one point a, we know that agreement on all closed rectangles implies agreement on all one-point events, and in the discrete case this is enough to prove the theorem.

In general if A is any event for which $P(A) = P'(A)$ then

$$P(A^c) = 1 - P(A) = 1 - P'(A) = P'(A^c).$$

Furthermore if $A_1, A_2, \ldots$ are disjoint events and $P(A_n) = P'(A_n)$ for all n then

$$P\left(\bigcup_{n=1}^{\infty} A_n\right) = \sum_{n=1}^{\infty} P(A_n) = \sum_{n=1}^{\infty} P'(A_n) = P'\left(\bigcup_{n=1}^{\infty} A_n\right).$$

Thus if we have some collection of events for which P and P' agree we can enlarge this collection by adding to it every event which is the complement or countable disjoint union of events in the collection. By applying this result to the collection of all closed rectangles we get a much larger collection of events, and then we can apply the same result to this larger collection, and so on. By this procedure we can show that P and P' agree for virtually any event which may be of interest.

From a less constructive point of view, we can directly consider the whole collection of events A for which $P(A) = P'(A)$. This collection must contain all closed rectangles, together with the complement and countable disjoint union of any events in the collection. Hopefully it is intuitively acceptable that every "reasonable" set must belong to such a collection. On this basis *we will consider Theorem 2 to be proved.* ■

In the future we restrict random points $f: \Omega \to F$ to be random vectors, with $F = R^n$. This restriction is fairly common and will not limit us in applications. Usually, however, we let the sample space Ω be quite general. In Sec. 22-3 the sample space is a function space. Any function $f: \Omega \to R^n$ defined on the sample space of a probability space (Ω,P) is said to be a **random vector** f **with associated probability space** (Ω,P). If $n = 1$ then f may be called a **random variable.** At times we permit f to be defined only on an event having probability 1.

EXERCISES

1. Let A and B events for which $P(AB) > 0$. The experiment (Ω,P) is performed and we learn that A occurred, so we focus attention on (Ω,P_A) obtained by conditioning (Ω,P) by A.

We then learn that B also occurred, so we form the probability space $(\Omega,(P_A)_B)$ obtained by conditioning (Ω,P_A) by B.

(a) Prove that $(P_A)_B = P_{AB}$. Thus our final probability space does not depend on the order in which we applied the information, and it is the same as would be obtained if we learned and applied all our information simultaneously.

(b) Five Bernoulli trials are performed, with probability $P[X_i = 1] = 1 - P[X_i = 0] = .4$ for a success on each trial. We learn that $A = [X_1 + \cdots + X_5 = 2]$ occurred. Tabulate the discrete density p_A obtained by conditioning the joint discrete density $p_{X_1,\ldots,X_5}$ by A.

(c) We then learn that $B = [X_1 = 0]$ also occurred. Tabulate the density $(p_A)_B$ obtained by conditioning p_A by B. Note that $(p_A)_B = p_{AB}$.

2. The use of "closed" rectangles in the hypothesis of Theorem 2 is not essential. For example, if $n = 2$ and P and P' agree on all open rectangles $\{(x_1,x_2): a_1 < x_1 < b_1, a_2 < x_2 < b_2\}$ we can use a decreasing sequence of these

$$A_k = \left\{(x_1,x_2): a_1 - \frac{1}{k} < x_1 < b_1 + \frac{1}{k}, a_2 - \frac{1}{k} < x_2 < b_2 + \frac{1}{k}\right\}$$

to obtain the hypothesis of Theorem 2, as follows:

$$\begin{aligned} P\{(x_1,x_2): a_1 \le x_1 \le b_1, a_2 \le x_2 \le b_2\} &= P\left(\bigcap_{k=1}^{\infty} A_k\right) \\ &= \lim_{k\to\infty} P(A_k) = \lim_{k\to\infty} P'(A_k) = P'\left(\bigcap_{k=1}^{\infty} A_k\right) \\ &= P'\{(x_1,x_2): a_1 \le x_1 \le b_1, a_2 \le x_2 \le b_2\}. \end{aligned}$$

Justify the steps in this proof.

3. For each real $t > 0$ let A_t be an event and assume that these events are increasing in t; that is, if $0 < t < t'$ then $A_t \subset A_{t'}$. Let $A = \bigcup_{t>0} A_t$ be their uncountable union.

(a) Prove that $\lim_{t\to\infty} P(A_t) = P(A)$. Thus we can extend Theorem 2h (Sec. 6-4) to events indexed by a continuous parameter; Theorem 2i can be similarly extended.

(b) An experiment is started at time $t = 0$. It is inspected at $t = 1$, and under certain conditions it may be terminated then. Let T be the termination time for the experiment. Assume that $P[T < 1] = 1 - e^{-t}$ if $0 < t < 1$, and that $P[T = 1] = .2$. Find $P[T < 1]$ and $P[T \le 1]$.

15-2 INDUCED DISTRIBUTIONS

This section introduces the induced P_f of a random vector. It includes the basic theorem on intermediate distributions and the fact that P_f can be obtained from $P_{f,g}$.

Section 11-1 contained the definition of the discrete density p_f induced by a random vector $f: \Omega \to R^n$ with associated discrete probability space (Ω,P). It was shown that if D is a subset of R^n and $P_f(D)$ is calculated from p_f then $P_f(D) = P[f \in D]$. Therefore the probability of every event $[f \in D]$ describable in terms of f can be calculated from p_f or P_f as well as from P. Thus P_f can be thought of as summarizing all the information concerning probabilities of such events. Often in more general cases there is nothing like an induced density p_f. However, we can use anything equivalent to knowledge of the probabilities of all events $[f \in D]$. Rather than seek something equivalent to this knowledge, a more fundamental point of view is to consider the distribution of f to be the assignment of the number $P[f \in D]$ to each subset D of R^n. We take this point of view, and the next theorem shows that this assignment is a probability measure. Thus *we define* P_f by the basic property $P_f(D) = P[f \in D]$.

In later chapters we are able in special cases to find more manipulable entities, such as a density or a distribution function, equivalent to P_f. Our search for such an entity is guided by, and ultimately must be justified by, its being equivalent to P_f. These ideas and the statement and proof of the next theorem are extremely simple yet fundamental.

Theorem 1 If $f: \Omega \to R^n$ is a random vector with associated probability space (Ω,P) and if P_f is defined for every subset D of R^n by

$$P_f(D) = P[f \in D] = P\{\omega : f(\omega) \in D\} = P[f^{-1}(D)]$$

then (R^n,P_f) is a probability space, which is said to be **induced** by $f: \Omega \to R^n$ with associated (Ω,P). ■

Proof The definition of $P_f(D)$ is depicted in Fig. 1*a*. Clearly $P_f(D) \geq 0$ and $P_f(R^n) = P[f^{-1}(R^n)] = P(\Omega) = 1$.

As shown in Fig. 1*b*, if $D_1, D_2, \ldots$ are disjoint subsets of R^n then $f^{-1}(D_1 \cup D_2 \cup \cdots) = f^{-1}(D_1) \cup f^{-1}(D_2) \cup \cdots$, and these events are disjoint, so that

$$P_f\left[\bigcup_{n=1}^{\infty} D_n\right] = P[f^{-1}(D_1 \cup D_2 \cup \cdots)] = P[f^{-1}(D_1) \cup f^{-1}(D_2) \cup \cdots]$$

$$= \sum_{n=1}^{\infty} P[f^{-1}(D_n)] = \sum_{n=1}^{\infty} P_f(D_n)$$

hence P_f is countably additive. ■

Naturally random vectors $f_1, f_2, \ldots, f_k$ into the same R^n are said to be **identically distributed** iff $P_{f_1} = P_{f_2} = \cdots = P_{f_k}$; that is, they must have the same induced distribution. The f_i need not all have the same associated probability space, and in particular this occurs for $P_h = P_g$ in Theorem 2 below.

If $f = (f_1, \ldots, f_n)$, where $f_1, \ldots, f_n$ are random vectors with the same associated probability space (Ω,P), then, as before, we may write $P_f = P_{f_1,\ldots,f_n}$ and call it the **joint distribution** of $f_1, \ldots, f_n$.

Let (Ω,P_A) be obtained by conditioning the probability space (Ω,P) by the event A with $P(A) > 0$. Then the *random vector* $f: \Omega \to R^n$ *with associated* (Ω,P_A) induces a probability space $(R^n,P_{f|A})$, called the **conditional distribution of f given A**. Therefore $P_{f|A}$ is defined by

$$P_{f|A}(D) = P_A[f \in D] = P[f \in D \mid A] = \frac{P([f \in D] \cap A)}{P(A)}$$

for every subset D of R^n. That is, $P_{f|A}(D)$ is the conditional probability that f will belong to D if we are given that A has occurred.

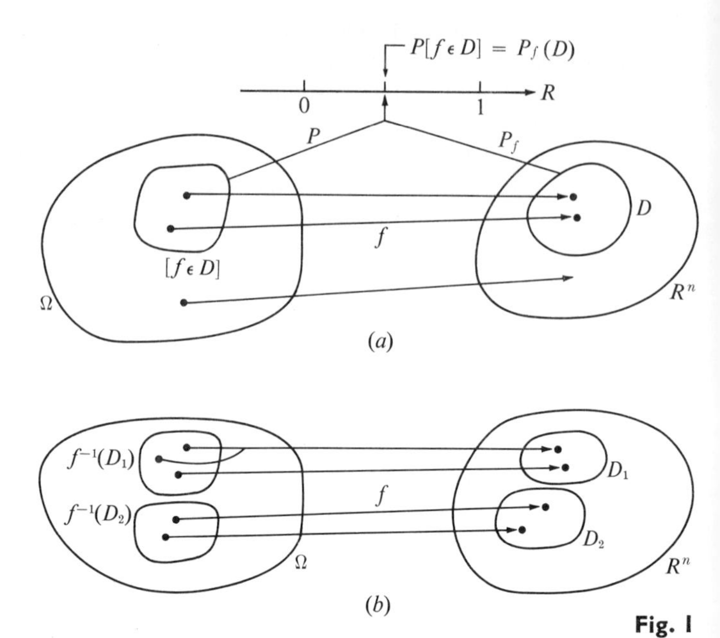

Fig. 1

Theorem 2: Intermediate distributions If (Ω,P) is a probability space and if

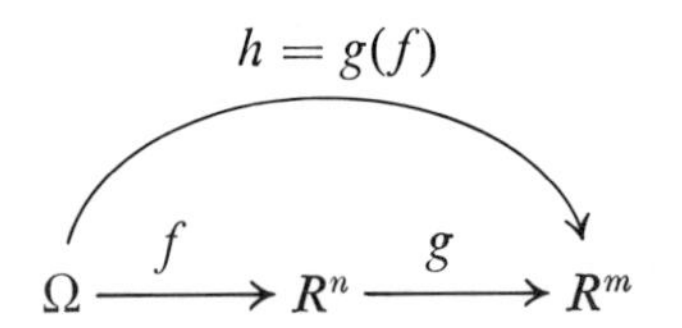

then $P_h = P_g$, where P_g is induced by g with associated (R^n,P_f). ■

Proof If D is any subset of R^m then, as indicated in Fig. 2,

$$P_g(D) = P_f[g^{-1}(D)] = P[f^{-1}(g^{-1}(D))] = P[h^{-1}(D)] = P_h(D). \quad ■$$

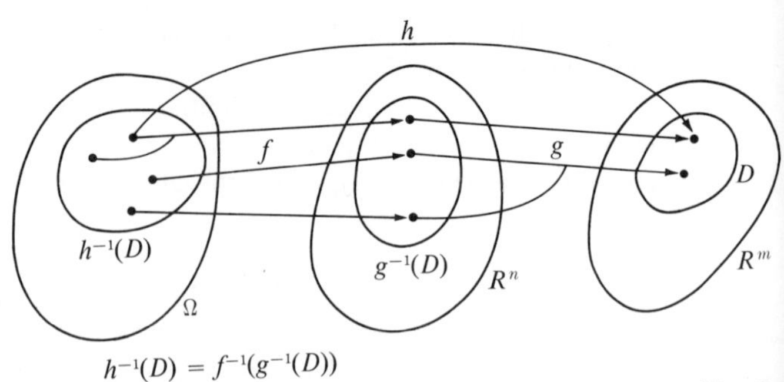

Fig. 2

The usefulness of intermediate distributions was particularly apparent in Example 3 (Sec. 11-2). We indicate the most simple application of Theorem 2. Assume that we are given (R^n, P_f) and $g: R^n \to R^m$. *These data completely determine* P_g. Suppose that we have some experiment (Ω, P) with a random vector $f: \Omega \to R^n$ having this assumed induced P_f. Theorem 2 assures us that the function $h = g(f)$ of this random vector induces this same P_g, and this is true for *every* such experiment. Thus the problem of finding the distribution of $h = g(f)$ when only P_f and g are given makes sense, since it has the same answer $P_h = P_g$ for every experiment from which the problem might arise. We sometimes find P_g directly from the basic data P_f and g, while at other times it is simpler to introduce some convenient experiment and find P_h, which Theorem 2 assures us will equal P_g.

We know that a discrete density p_f can be obtained from a larger joint density $p_{f,g}$. The next theorem shows, without considering densities, that an induced P_f can be obtained from a larger joint $P_{f,g}$ since any event described in terms of f can also be described in terms of (f,g).

Theorem 3 If $f: \Omega \to R^n$ and $g: \Omega \to R^m$ are random vectors with the same associated probability space (Ω, P) then

$$P_f(D) = P_{f,g}(D \times R^m) \qquad \text{for every subset } D \text{ of } R^n. \quad \blacksquare$$

Proof The proof is shown in Fig. 3 for the case $n = m = 1$. Now, $D \times R^m$ is the cartesian product of D with all of R^m, so that a point (x,y) belongs to $D \times R^m$ iff $x \in D$ and

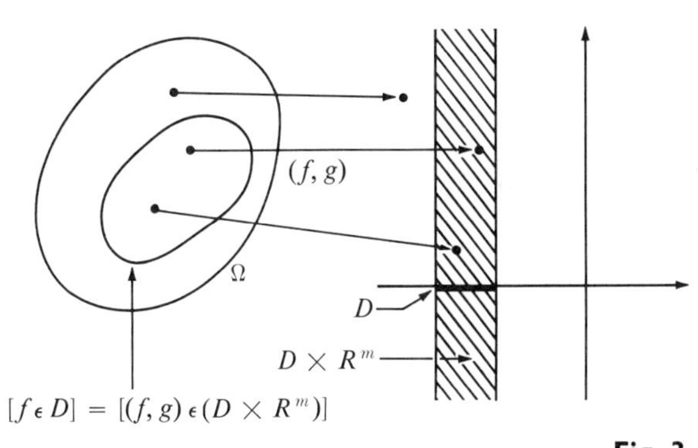

Fig. 3

$y \in R^m$. Therefore the two events $[f \in D]$ and $[(f,g) \in (D \times R^m)]$ are equal, hence $P_f(D) = P[f \in D] = P_{f,g}(D \times R^m)$. $\blacksquare$

EXERCISES

1. Two quite different models $P_{X,Y}$ and $Q_{X,Y}$ are considered for an experiment, and both yield the same distribution for the observable random variable $X - Y$. Will both models necessarily assign the same probability to the following events?

(*a*) $[1 \leq X^2 + Y^2 \leq 3]$

(*b*) $[e^X \geq e^Y]$

2. The distribution P_X is known, where X is in yards. The distribution P_Y of $Y = 3X$ feet is found from P_X. The distribution P_Z of $Z = 12Y$ inches is found from P_Y. Show that the same P_Z would be obtained if it were calculated from $Z = 36X$ and P_X.

3. The distribution P_X of a random variable X is given and P_Y is found for $Y = e^X$. It is then discovered that log Y is relevant, and $P_{\log Y}$ is found from P_Y. Show that $P_{\log Y} = P_X$.

4. The joint distribution $P_{X,Y}$ of two random variables is known and the distribution P_W of $W = e^{X+Y}$ is desired. Show that P_W can be obtained by either of the following methods:

(*a*) Let $Z = X + Y$ and find P_Z from $P_{X,Y}$. Then find the distribution of $W = e^Z$ from P_Z.

(*b*) Let $U = e^X$ and $V = e^Y$ and find $P_{U,V}$ from $P_{X,Y}$. Then find the distribution of $W = UV$ from $P_{U,V}$.

5. A search is to be made for a missing boat whose unknown location (X, Y) is considered to be uniformly distributed over the lake L of area A shown in Fig. 4. Thus for any subset B of the plane, $P[(X, Y) \in B]$ equals the area of $B \cap L$, divided by A. Estimate $P[X \in D]$, where $D = \{x: 1 \leq x \leq 3\}$. Relate your calculation to Theorem 3.

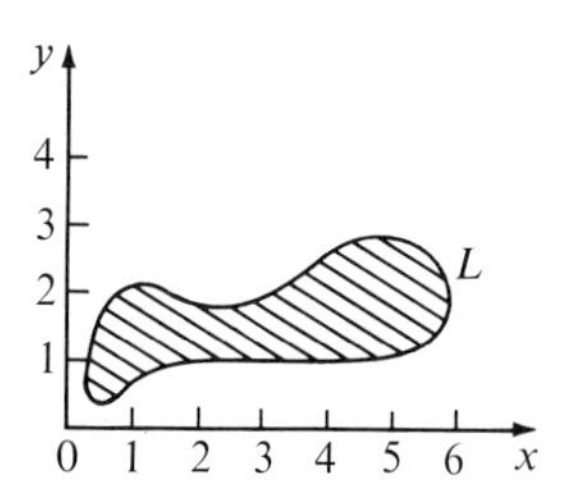

Fig. 4

15-3 INDEPENDENCE

As stated in Sec. 15-1, the definition, proofs, and basic results for independent events remain as in Chap. 7. In this section it is shown that for independent random vectors the definition, results, and all the proofs except for Theorem 2 below remain as in Sec. 11-2.

As before, $n \geq 2$ random vectors $f_1, \ldots, f_n$ having the same associated probability space (Ω, P) are said to be **independent** iff

$$P[f_1 \in D_1, \ldots, f_n \in D_n] = P[f_1 \in D_1]P[f_2 \in D_2] \cdots P[f_n \in D_n]$$

for all subsets D_i of R^{k_i} where $f_i: \Omega \to R^{k_i}$. This defining equation can be written in any of the forms below:

$$P\{\omega: f_1(\omega) \in D_1, \ldots, f_n(\omega) \in D_n\}$$
$$= P\{\omega: f_1(\omega) \in D_1\}P\{\omega: f_2(\omega) \in D_2\} \cdots P\{\omega: f_n(\omega) \in D_n\},$$
$$P[f_1^{-1}(D_1) \cap \cdots \cap f_n^{-1}(D_n)] = P[f_1^{-1}(D_1)]P[f_2^{-1}(D_2)] \cdots P[f_n^{-1}(D_n)],$$
$$P_{f_1, \ldots, f_n}(D_1 \times D_2 \times \cdots \times D_n) = P_{f_1}(D_1)P_{f_2}(D_2) \cdots P_{f_n}(D_n).$$

As before, taking a subsequence of two or more, or reordering, does not destroy independence, so that $f_1, \ldots, f_n$ are independent iff the events $[f_1 \in D_1], \ldots, [f_n \in D_n]$ are always independent.

As before, events $A_1, \ldots, A_n$ are independent iff the random variables $I_{A_1}, \ldots, I_{A_n}$ are independent, where I_{A_i} is the indicator of A_i on Ω.

As before, $n \geq 2$ random vectors $f_1, \ldots, f_n$ are **pairwise independent** iff f_i and f_j are two independent random vectors whenever $i \neq j$.

In Chap. 17 it is shown that for given probability spaces (R^{n_i},P_i) it is possible, as for the discrete case in Theorem 2 (Sec. 11-2), to construct a probability space (R^n,P), with $n = n_1 + \cdots + n_d$, and random vectors $f_i: R^n \to R^{n_i}$ such that $f_1, \ldots, f_d$ are independent and $P_{f_i} = P_i$. Thus it is always possible to construct an (Ω,P) with independent random vectors $f_1, \ldots, f_d$ which have specified marginals $P_{f_1}, \ldots, P_{f_d}$.

If we know only the marginals $P_{f_1}, \ldots, P_{f_d}$ of d independent random vectors $f_i: \Omega \to R^{n_i}$ with associated (Ω,P) then the definition of independence determines the value of their joint probability measure $P_{f_1,\ldots,f_d}$ for every closed rectangle in R^n, where $n = n_1 + \cdots + n_d$. Hence by Theorem 2 (Sec. 15-1) the value of $P_{f_1,\ldots,f_d}$ is determined for every event in R^n. Thus if $f = (f_1, \ldots, f_d)$ and $g: R^n \to R^m$ then the induced distribution $P_{g(f)}$ is uniquely determined by g and the marginals $P_{f_1}, \ldots, P_{f_d}$ of the independent $f_1, \ldots, f_d$. Therefore we can find $P_{g(f)}$ directly from P_f and g, or we can introduce a convenient experiment, as was discussed immediately after Theorem 2 (Sec. 15-2).

The next theorem shows that, as before, functions of independent random vectors are independent.

Theorem 1 If $f_1, f_2, \ldots, f_d$ are $d \geq 2$ independent random vectors with the same associated probability space (Ω,P), and if

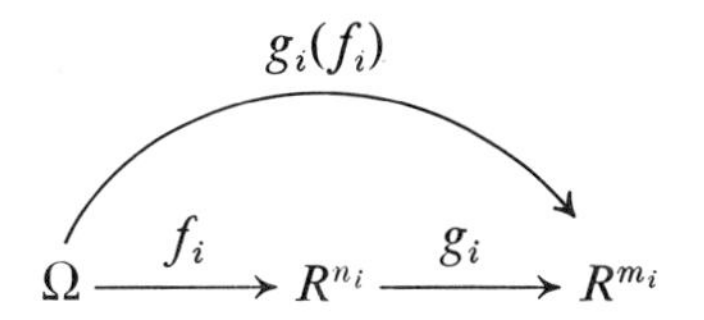

then $g_1(f_1), g_2(f_2), \ldots, g_d(f_d)$ are d independent random vectors. ∎

Proof Let C_i be any subset of R^{m_i} and, as in Fig. 1, let $D_i = g_i^{-1}(C_i)$ and $h_i = g_i(f_i)$. Then $h_i(\omega) \in C_i$ iff $f_i(\omega) \in D_i$, so that the events $[h_1 \in C_1], \ldots, [h_d \in C_d]$ are the same as the events $[f_1 \in D_1], \ldots, [f_d \in D_d]$, which are assumed to be independent. ∎

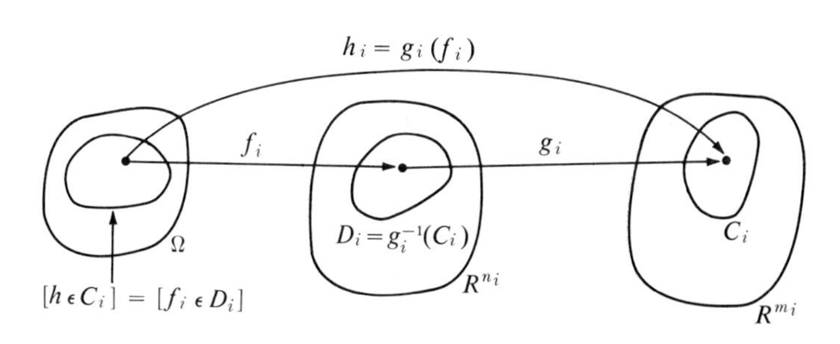

Fig. 1

We saw in Sec. 11-2 that if X, Y, Z are three independent random variables then X^2 and $Y^4 + Z^7$ are two independent random variables. This is because of the basic fact that if X, Y, Z are three independent random variables then X and (Y,Z) are two independent random vectors, so Theorem 1 can be applied to them. This basic fact is true in general, as stated in the next theorem.

Theorem 2 If $f_1, \ldots, f_m, \ldots, f_n$ are $n \geq 3$ independent random vectors with associated probability space (Ω,P) then $(f_1, \ldots, f_m)$ and $(f_{m+1}, \ldots, f_n)$ are two independent random vectors. ■

If X, Y, Z are independent then the independence of X and (Y,Z) in the discrete case was proved as follows. We have $p_{X,Y,Z} = p_X p_Y p_Z$ and $p_{Y,Z} = p_Y p_Z$, so $p_{X,Y,Z} = p_X p_{Y,Z}$, and hence X and (Y,Z) are independent. Essentially this same simple proof applies to continuous densities, and also in the general case in Sec. 17-3 to distribution functions, once their properties are established. In this section more effort is required to prove Theorem 2.

*THE PROOF OF THEOREM 2

We first prove a special case which indicates the ideas involved. We then state and prove Lemmas 1 and 2, which lead to a proof of Theorem 2. *The proof of Theorem 2 and its associated Lemmas* 1 *and* 2 *may well be postponed to a later reading.*

Let $W = (Y,Z)$, where X, Y, Z are three independent random variables. We wish to prove that X and W are independent. If D_1, D_2, D_3 are sets of reals then

$$P[X \in D_1, Y \in D_2, Z \in D_3] = P[X \in D_1]P[Y \in D_2]P[Z \in D_3]$$

and if $D_1 = R$ we get

$$P[Y \in D_2, Z \in D_3] = P[Y \in D_2]P[Z \in D_3].$$

If we let $D' = D_2 \times D_3$ these equations imply that

$$P[X \in D_1, W \in D'] = P[X \in D_1]P[W \in D'].$$

For X and W to be independent this last equation must hold for all subsets D_1 of R and D' of R^2. We have only proved it for those D' of the form $D' = D_2 \times D_3$. However, as shown in Exercise 3 (Sec. 2-1), many subsets D' of the plane, such as a circular disk, cannot be represented as $D_2 \times D_3$. Thus independence of X, Y, Z immediately implies that any $[X_1 \in D_1]$ is independent of any event $[W \in D']$ describable in terms of separate conditions on Y and Z. In the next paragraph we show that $[X \in D_1]$ is independent of any event describable in terms of Y and Z jointly.

For any fixed set D_1 of reals with $P[X \in D_1] > 0$ define the probability space (R^2,P') by

$$P'(D') = \frac{P[X \in D_1, W \in D']}{P[X \in D_1]} \qquad \text{for every subset } D' \text{ of } R^2.$$

As in Sec. 15-2, $P'(D')$ is the conditional probability of $[W \in D']$, given the fixed event $[X \in D_1]$. From the preceding discussion we know that $P'(D') = P_W(D')$

whenever D' is a closed rectangle in R^2, so from Theorem 2 (Sec. 15-1) we have this same equality,

$$\frac{P[X \in D_1, W \in D']}{P[X \in D_1]} = P_W(D') = P[W \in D']$$

for any event D' in R^2, which proves that X and W are independent. This proof is generalized in Lemmas 1 and 2 below.

Lemma 1 Let $f\colon \Omega \to R^n$ be a random vector and A an event such that A and $[f \in D]$ are two independent events whenever D is a closed rectangle in R^n. Then A and $[f \in D]$ are two independent events whenever D is any subset of R^n. ■

Proof If $P(A) = 0$ then $P(AB) = 0 = P(A)P(B)$ for every event B, so the lemma is true in this case. If $P(A) > 0$ then the conditional probability

$$P[f \in D \mid A] = \frac{P([f \in D] \cap A)}{P(A)}$$

considered as a function of the subset D of R^n, is a probability measure. By assumption it agrees with P_f when D is a closed rectangle, so by Theorem 2 (Sec. 15-1) it equals P_f. ■

Lemma 2 Random vectors $f_1, \ldots, f_d$ are independent if

$$P[f_1 \in R_1, \ldots, f_d \in R_d] = P[f_1 \in R_1]P[f_2 \in R_2] \cdots P[f_d \in R_d]$$

for all closed rectangles R_i of R^{n_i} where $f_i\colon \Omega \to R^{n_i}$. ■

Proof If $R_1^{(1)}, R_1^{(2)}, \ldots$ is a sequence of increasing closed rectangles in R^{n_1} and their union is R^{n_1} then from the hypothesis of Lemma 2 and from Theorem 2h (Sec. 6-4) we have

$$\begin{aligned} P[f_2 \in R_2, \ldots, f_d \in R_d] &= \lim_{i\to\infty} P[f_1 \in R_1^{(i)}, f_2 \in R_2, \ldots, f_d \in R_d] \\ &= \lim_{i\to\infty} P[f_1 \in R_1^{(i)}]P[f_2 \in R_2] \cdots P[f_d \in R_d] \\ &= P[f_2 \in R_2] \cdots P[f_d \in R_d]. \end{aligned}$$

Thus if $A = [f_2 \in R_2, \ldots, f_d \in R_d]$ then A and $[f_1 \in D_1]$ are independent if D_1 is a closed rectangle, so Lemma 1 applies. Therefore

$$P[f_1 \in D_1, f_2 \in R_2, \ldots, f_d \in R_d] = P[f_1 \in D_1]P[f_2 \in R_2] \cdots P[f_d \in R_d]$$

for every subset D_1 of R^{n_1} and closed rectangles $R_2, \ldots, R_d$. Similarly this last equation is valid if we replace R_2 by any subset D_2 of R^{n_2}, since we can apply Lemma 1 with $A = [f_1 \in D_1, f_3 \in R_3, \ldots, f_d \in R_d]$. Repeating this procedure, we complete the proof of Lemma 2. ■

Proof of Theorem 2 Let $f_1, \ldots, f_m, \ldots, f_n$ be independent random vectors with $f_i\colon \Omega \to R^{n_i}$. If R_i is a closed rectangle in R^{n_i} then $P[f_1 \in R_1, \ldots, f_n \in R_n]$ equals both of the following expressions:

$$P[(f_1, \ldots, f_m) \in D, (f_{m+1}, \ldots, f_n) \in D']$$
$$= P[(f_1, \ldots, f_m) \in D]P[f_{m+1}, \ldots, f_n) \in D']$$

where $D = R_1 \times \cdots \times R_m$ and $D' = R_{m+1} \times \cdots \times R_n$ are closed rectangles. Applying Lemma 2 to this last equation completes the proof of Theorem 2. ■

EXERCISES

1. A probability space (R,P_1) is given. For each $n \geq 1$ let P_n be the distribution of $S_n = X_1 + \cdots + X_n$ where $X_1, \ldots, X_n$ are independent and each has distribution P_1.

(*a*) Let Y_1, Y_2, Y_3 be independent, where Y_i has distribution P_{n_i}. Prove that the sum $Y_1 + Y_2 + Y_3$ has distribution $P_{n_1+n_2+n_3}$.

(*b*) Let Q_n be the distribution of e^{S_n}, where S_n has distribution P_n. Let W_1, W_2, W_3 be independent, where W_i has distribution Q_{n_i}. Prove that the product $W_1W_2W_3$ has distribution $Q_{n_1+n_2+n_3}$.

COMMENT: Part (*a*) is the basic fact used in Example 3 (Sec. 11-2).

2. Let the random vector $h = g(f)$ be a function of the random vector f. Exclude the trivial case where P_h is concentrated on one point by assuming that for at least one set D_h that $0 < P[h \in D_h] < 1$. Prove that h and f are dependent. Thus in all nontrivial cases a function of a random vector cannot be statistically independent of it. HINT: Let $D_f = g^{-1}(D_h)$.

15-4 EXPECTATION

This section completes the structure of the general theory. The mean of a probability measure on the real line is defined in a rigorous and general fashion, and then EX is defined as the mean of the induced P_X. Some basic properties of expectation are obtained, and indications are given for the validity of others. General proofs are not attempted, although the results collected together in Theorem 3 (Sec. 11-3) hold in general and are thereafter assumed. The laws of large numbers stated and proved in Chap. 12 hold in general. The central limit theorems stated in Chap. 13 hold in general.

Let (R,P) be any probability space with the real line for its sample space. Theorem 1 below defines the mean of P. Theorem 1 (Sec. 16-5) shows that in the discrete case the present definition is equivalent to our earlier one. In this paragraph we motivate the definition. If we alter by P concentrating all of the probability $P\{x: a \leq x < b\}$ on some point x_0 in this interval then we would naturally use $x_0P\{x: a \leq x < b\}$ as the contribution from this interval to the mean of the altered P. Therefore the contribution to the mean of P from the interval $a \leq x < b$ should be some number between $aP\{x: a \leq x < b\}$ and $bP\{x: a \leq x < b\}$. This approach leads to natural approximations to the mean of P. We handle the contributions to the mean of P from the positive and negative half-axes separately. Let A_n^+ denote the nth approximation to that part of the mean of P which is contributed by the positive half-axis. The same is true for A_n^-. To define A_n^+ we use the subdivision of the interval $0 \leq x < n$ into adjacent disjoint subintervals each of length $1/2^n$, and we let any one, $a \leq x < b$, of these $n2^n$ subintervals contribute the quantity $aP\{x{:}a \leq x < b\}$ to

A_n^+. Thus, as shown in Fig. 1, to go from A_n^+ to A_{n+1}^+ we increase the interval over which we are approximating from $0 \le x < n$ to $0 \le x < n + 1$ and simultaneously halve the length of the subintervals. It is then an easy matter to show that the sequence A_n^+ increases to some limit $A^+ \ge 0$. The sequence A_n^- decreases to some limit $A^- \le 0$, so we define the mean of P to be $A^+ + A^-$, as in Theorem 1.

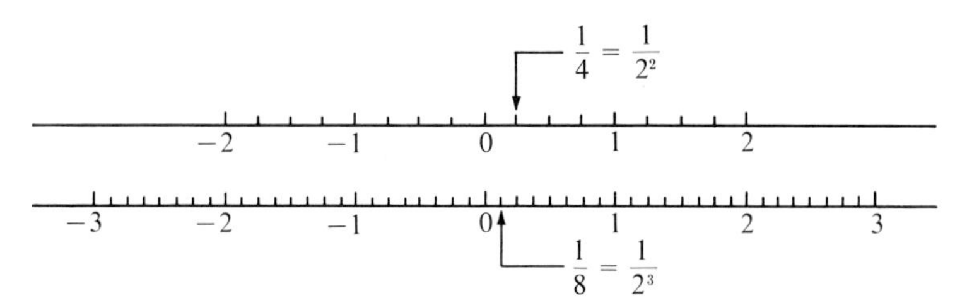

Fig. 1 Subdivisions for $n = 2$ and $n = 3$.

It was shown in Exercise 14 (Sec. 11-3) that linearity of expectation may be violated if non–absolutely convergent series are permitted. In order to prevent such violations we must insist on some condition on P before we attempt to define its mean. The condition that both A^+ and A^- be finite is a natural one. Theorem 1 shows this condition to be equivalent to requiring that $\sum_{k=1}^{\infty} kP\{x: k - 1 \le |x| < k\}$ have a finite sum. This later condition depends only on probabilities of intervals between integers, and not on the fine detail within P.

Theorem 1 Let (R,P) be a probability space with the real line for sample space. For each $n = 1, 2 \ldots$ define A_n^+ and A_n^- by

$$A_n^+ = \sum_{j=1}^{M_n} \frac{j}{2^n} P\left\{x: \frac{j}{2^n} \le x < \frac{j+1}{2^n}\right\}$$

$$A_n^- = \sum_{j=1}^{M_n} \frac{-j}{2^n} P\left\{x: -\frac{j+1}{2^n} < x \le -\frac{j}{2^n}\right\}$$

where $M_n = n2^n - 1$. Then $0 \le A_1^+ \le A_2^+ \le \cdots$ and $0 \ge A_1^- \ge A_2^- \ge \cdots$. Also the two sequences A_n^+ and A_n^- both converge to finite limits A^+ and A^- iff the series

$$\sum_{k=1}^{\infty} kP\{x: k - 1 \le |x| < k\}$$

has a finite sum B. We say that P has a mean iff this is the case, and we define

the mean of $P = A^+ + A^-$

and it satisfies

$$|\text{the mean of } P| \le A^+ - A^- \le B. \quad \blacksquare$$

***Proof** We first show that $A_n^+ \le A_{n+1}^+$. The contribution to A_n^+ from an interval can be broken into the contributions from the left and right halves of this interval, so that

$$\frac{j}{2^n} P\left\{x: \frac{j}{2^n} \le x < \frac{j+1}{2^n}\right\}$$

$$= \frac{j}{2^n} P\left\{x: \frac{j}{2^n} \le x < \frac{j}{2^n} + \frac{1}{2^{n+1}}\right\} + \frac{j}{2^n} P\left\{x: \frac{j}{2^n} + \frac{1}{2^{n+1}} \le x < \frac{j+1}{2^n}\right\}$$

$$\le \frac{j}{2^n} P\left\{x: \frac{j}{2^n} \le x < \frac{j}{2^n} + \frac{1}{2^{n+1}}\right\} + \left(\frac{j}{2^n} + \frac{1}{2^{n+1}}\right) P\left\{x: \frac{j}{2^n} + \frac{1}{2^{n+1}} \le x < \frac{j+1}{2^n}\right\}$$

where this last sum is the contribution to A_{n+1}^+ from this same interval $j/2^n \le x < (j+1)/2^n$. Also A_{n+1}^+ has additional nonnegative contributions from the interval $n \le x < n+1$. Clearly, then, $A_n^+ \le A_{n+1}^+$, while $A_n^- \ge A_{n+1}^-$ is proved similarly.

Now A_n^+ and $-A_n^-$ are both nondecreasing sequences, so that both have finite limits A^+ and $-A^-$ iff their sum $A_n^+ + (-A_n^-)$ does. If $1 \le k \le n$ then the contribution to A_n^+ from terms corresponding to subintervals of the interval $k-1 \le x < k$ all have coefficients between $k-1$ and k, so $kP\{x: k-1 \le x < k\}$ exceeds this contribution. [If $k = 1$ the bound $P\{x: 0 < x < 1\}$ may be used in order to preserve disjointness shortly.] Similarly if we reflect the interval $k-1 \le x < k$ we easily see that the contribution to $-A_n^-$ from those terms corresponding to subintervals of the interval $-k < x \le -k+1$ is less than $kP\{x: -k < x \le -k+1\}$. Adding these two bounds, and then summing over k and defining B_n as below, yields

$$A_n^+ + (-A_n^-) \le B_n = \sum_{k=1}^{n} kP\{x: k-1 \le |x| < k\}.$$

Furthermore if we use the *lower* bound $k-1$ for subinterval contributions to A_n^+ then

$$A_n^+ \ge \sum_{k=1}^{n} (k-1)P\{x: k-1 \le x < k\} \ge \sum_{k=1}^{n} kP\{x: k-1 \le x < k\} - 1.$$

Similarly for $-A_n^-$ so that $B_n - 2 \le A_n^+ - A_n^- \le B_n$. Clearly, then, $A_n^+ - A_n^-$ has a finite limit $A^+ - A^-$ iff B_n has a finite limit B. In this case $B - 2 \le A^+ - A^- \le B$. Also

$$|A^+ + A^-| \le |A^+| + |A^-| = A^+ + (-A^-) \le B. \quad \blacksquare$$

The **expectation** EX **of the random variable** $X: \Omega \to R$ **with associated probability space** (Ω, P) is defined to be the mean of P_X, if P_X has a mean; otherwise we say that X does not have an expectation.

Observe that the definition and proof of Theorem 1 only require that P be evaluated for very simple sets, such as unions of finitely many disjoint intervals. Because of this, as observed in Sec. 21-4, our definition and proof are quite valid within the rigorous general theory of probability. Thus EX has been defined in general, and in agreement with the general theory. In an addendum to this section, an alternate interpretation of our definition of EX is described in terms of approximating random variables. This interpretation is not needed in this book. However, some readers will encounter it in later reading.

The following properties of expectation are almost immediate. Identically distributed random variables $P_X = P_Y$ have the same expectation $EX = EY$, since both expectations are defined to equal the mean of the same $P_X = P_Y$. If a random variable is essentially a constant, $P[X = c] = 1$, then $EX = c$. If I_A is the indicator function of A on Ω then $EI_A = P(A)$. If there is a number B such that $P[|X| > B] = 0$ then X has an expectation and $-B \leq EX \leq B$. Also $E\,|X| = 0$ iff $P[|X| = 0] = 1$.

Furthermore X has an expectation iff $|X|$ does, and we use $E\,|X| < \infty$ to indicate this case. In this case $|EX| \leq E\,|X|$. This result is immediate, since $|A_n^+ + A_n^-| \leq A_n^+ - A_n^-$, while $A_n^+ - A_n^-$ equals the nth approximation to the mean of $P_{|X|}$.

Theorem 2 If (Ω,P) is a probability space and if

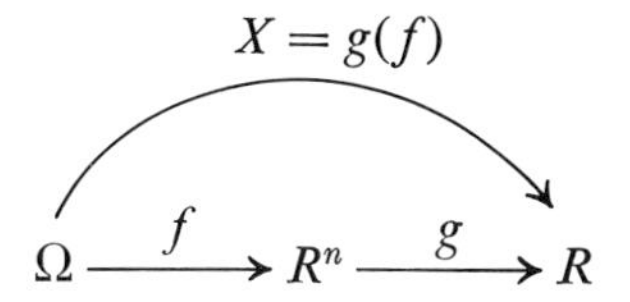

then $EX = Eg$. That is, the expected value of $X\colon \Omega \to R$ with associated (Ω,P) equals the expected value of g with associated (R^n,P_f), in the sense that if either expectation exists then both exist and are equal. ■

Proof From Theorem 2 (Sec. 15-2) we know that X and g have the same induced distribution $P_X = P_g$, and both expectations are just defined to equal the mean of this $P_X = P_g$. ■

Under the assumptions of the preceding theorem, and with an obvious interpretation of notation, we may write

$$E_P g(f) = E_{P_f} g = E_{P_{g(f)}} x$$

where x denotes the identity function on R.

General properties of expectation were collected in Theorem 3 (Sec. 11-3). These hold in general and will henceforth be assumed.

The proofs of parts (f) to (i) of Theorem 3 (Sec. 11-3) depend primarily only on parts (a) to (e). We do *not* attempt general proofs of parts (a) to (e) of this theorem. The interested reader may consult Sec. 21-4 and the addendum to the present section for comments. The theorem can be extended to continuous densities in Sec. 16-5 by proofs like those for the discrete case.

If r is a positive real, and $E\,|X|^r$ is finite then we call it the **rth absolute moment of** X. The **moment-generating function** (MGF) **of a random variable** X is defined by $m(t) = Ee^{tX}$ for any real t for which the positive-valued random variable e^{tX} has a finite expectation. The availability of Theorem 3 (Sec. 11-3) justifies the same definitions, interpretations, and facts as stated in the summary of Chap. 11 for EX^n, σ_X^2, σ_X, X^*, MGFs, Cov (X, Y), ρ, and the least-squares regression lines.

Theorem 3 below is Chebychev's inequality with a proof that is valid in general. Theorem 4 is a natural version of the MGF Theorem 2 (Sec. 9-4) but uses a notation which applies generally. Theorem 4 is not proved in this book. Example 8 (Sec. 11-3) does prove the last part of Theorem 4 for nonnegative random variables.

Theorem 3: Chebychev's inequality If $EX^2 < \infty$ and ϵ is any positive real then

$$P[|X - EX| \geq \epsilon] \leq \frac{\text{Var } X}{\epsilon^2}. \quad \blacksquare$$

Proof Let $\epsilon > 0$ and let I_A be the indicator function of the event $A = [|X| \geq \epsilon]$. If $r > 0$ then $I_A \leq (|X|/\epsilon)^r$, since if $\omega \notin A$ then $I_A(\omega) = 0$, and if $\omega \in A$ then $1 \leq (|X(\omega)|/\epsilon)^r$. Taking expectations yields

$$P[|X| \geq \epsilon] \leq \frac{E\,|X|^r}{\epsilon^r}$$

for any $\epsilon > 0$, $r > 0$, and X with $E\,|X|^r < \infty$. Let $r = 2$ and replace X by $X - EX$ to complete the proof. $\blacksquare$

Theorem 4 Let $m(t) = Ee^{tX}$ be the MGF of a random variable X. Assume that $m(t_0)$ and $m(-t_0)$ are finite, where $t_0 > 0$. Then m is finite and has derivatives of all orders whenever $|t| < t_0$. Differentiation under the expectation sign is permitted, so

$$m^{(n)}(t) = \frac{d^n}{dt^n} m(t) = EX^n e^{tX} \qquad \text{if } n = 1, 2, \ldots; \ |t| < t_0.$$

In particular all moments of X exist and $m^{(n)}(0)$ equals the nth moment, so

$$EX = m^{(1)}(0) \qquad \text{and} \qquad \text{Var } X = m^{(2)}(0) - [m^{(1)}(0)]^2.$$

Furthermore the power series

$$m(t) = \sum_{n=0}^{\infty} \frac{m^{(n)}(0)}{n!} t^n$$

converges absolutely for $|t| < t_0$.

If Ee^{tX} and Ee^{tY} are finite whenever $|t| \leq t_0$ with $t_0 > 0$ then

$$P_X = P_Y \text{ iff } Ee^{tX} = Ee^{tY} \text{ for all } |t| \leq t_0. \quad \blacksquare$$

*ADDENDUM

This addendum is devoted to an alternate widely used interpretation of our definition of EX. We defined EX as the mean of P_X. Such a definition has the advantage that identically distributed random variables obviously have the same expectation, and hence the important Theorem 2 becomes almost trivial. The mean of P_X was defined essentially as the limit of the means of special discrete approximations to P_X. Clearly we could equivalently use special discrete approximations to X and define EX as the limit of the expectations of these approximating random variables. We describe this approach in detail.

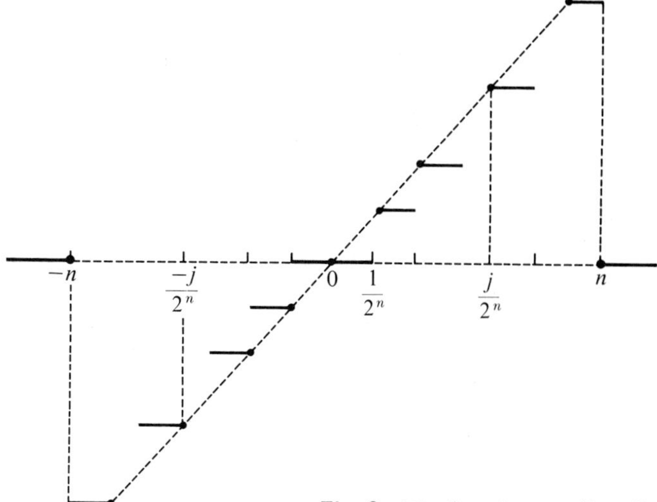

Fig. 2 The function $g_n: R \to R$.

For each positive integer n we define $g_n: R \to R$, as shown in Fig. 2, as follows: If $|x| \geq n$ let $g_n(x) = 0$. For some $j = 0, 1, 2, \ldots, M_n = n2^n - 1$; if $j/2^n \leq x < (j+1)/2^n$ let $g_n(x) = j/2^n$, and if $-(j+1)/2^n < x \leq -j/2^n$ let $g_n(x) = -j/2^n$. Thus g_n takes only finitely many values on the interval $|x| < n$ and is zero off that interval. The function g_n can be thought of as a finitely many valued approximation to the identity function on R. In particular

$$|g_n(x) - x| \leq \frac{1}{2^n} \qquad \text{if } |x| < n.$$

For fixed x we will have $|x| < n$ for all large n, so

$$\lim_{n\to\infty} g_n(x) = x \qquad \text{for all real } x.$$

Furthermore

$$|g_n(x)| \leq |x| \qquad \text{for all } x \text{ and } n.$$

Also g_n is an odd function, $g_n(-x) = -g_n(x)$, and $g_n(x) \geq 0$ for $x \geq 0$, so that

$$|g_n(x)| = g_n(|x|) \qquad \text{for all real } x.$$

In addition $g_n(x)$ is nondecreasing (or nonincreasing) as n increases if $x \geq 0$ (or $x \leq 0$).

Let X be a random variable with associated probability space (Ω,P). For each n introduce the random variable $X_n = g_n(X)$ with the same associated probability space. Each X_n has only finitely many values. The following facts are immediate:

$$|X_n(\omega) - X(\omega)| = |g_n(X(\omega)) - X(\omega)| \leq \frac{1}{2^n} \qquad \text{if } |X(\omega)| < n,$$

$$\lim_{n\to\infty} X_n(\omega) = \lim_{n\to\infty} g_n(X(\omega)) = X(\omega) \qquad \text{for all } \omega \in \Omega,$$

$$|X_n(\omega)| = |g_n(X(\omega))| = g_n(|X(\omega)|) \leq |X(\omega)| \qquad \text{for all } \omega \in \Omega.$$

Thus each $X_n = g_n(X)$ has only finitely many values, and X_n converges to X as n approaches infinity. Also $|X_n| = g_n(|X|) \leq |X|$.

Now, X_n induces a discrete density concentrated on a finite number of points. Therefore EX_n, as defined in the discrete case, is just the sum of $xp_{X_n}(x)$. A contribution $xp_{X_n}(x)$ can be expressed in terms of P_X or P as follows, where $0 \leq j \leq M_n$:

$$\frac{j}{2^n} p_{X_n}\left(\frac{j}{2^n}\right) = \frac{j}{2^n} P_X\left\{x: \frac{j}{2^n} \leq x < \frac{j+1}{2^n}\right\} = \frac{j}{2^n} P\left[\frac{j}{2^n} \leq X < \frac{j+1}{2^n}\right]$$

Thus we see that EX_n and $E|X_n|$, as defined in the discrete case, are given by $EX_n = A_n^+ + A_n^-$ and $E|X_n| = A_n^+ - A_n^-$, where A_n^+ and A_n^- are obtained by applying the definitions of Theorem 1 to P_X. Also, A_n^+ and A_n^- can be expressed in terms of P. Thus Theorem 1 essentially takes this special sequence $X_n = g_n(X)$ of random variables, for which X_n converges to X as n approaches infinity, and Theorem 1 defines EX as the limit of EX_n, where EX_n is calculated as in the discrete case. Applying this definition of expectation to $|X|$, we see that

$$|EX_n| \leq E|X_n| = Eg_n(|X|) \to E|X| \qquad \text{as } n \to \infty.$$

We collect these ideas, together with the conclusion of Theorem 1, into the following *alternate definition of expectation, essentially equivalent to Theorem 1.*

Alternate definition of expectation Let X be a random variable with associated probability space (Ω, P). For random variables which take only finitely many values let expectation be defined as the mean of the induced discrete probability density. Then the sequence $E|g_n(X)| = Eg_n(|X|)$ is nondecreasing as n increases, and it converges to a finite limit iff

$$\sum_{k=1}^{\infty} kP[k-1 \leq |X| < k]$$

has a finite sum B. We say that X has an expectation iff this is the case. Clearly X has an expectation iff $|X|$ does. If this is the case then $Eg_n(X)$ has a limit, and we define **expectation** by

$$EX = \lim_{n\to\infty} Eg_n(X)$$

and we have $|EX| \leq E|X| \leq B$. ■

Note that $g_n(X) \to X$; hence if X has an expectation then by definition

$$E\left[\lim_{n\to\infty} g_n(X)\right] = \lim_{n\to\infty} Eg_n(X)$$

so that we are interchanging expectation and limit.

The general theory of expectation essentially defines EX precisely as in Theorem 1 and shows that under quite general conditions if $Z_n \to X$ then $EZ_n \to EX$. Such results can be used to justify the following *proof* of linearity of expectation in the general case:

$$\begin{aligned} E(X+Y) &\overset{1}{=} \lim_{n\to\infty} Eg_n(X+Y) \ominus \lim_{n\to\infty} \{E[g_n(X)+g_n(Y)]\} \\ &\overset{2}{=} \lim_{n\to\infty} \{Eg_n(X)+Eg_n(Y)\} = \lim_{n\to\infty} Eg_n(X) + \lim_{n\to\infty} Eg_n(Y) \\ &\overset{3}{=} EX+EY. \end{aligned}$$

For $\overset{1}{=}$ and $\overset{3}{=}$ we used our definition of expectation, while $\overset{2}{=}$ is the known linearity in the discrete case. For $\ominus$ we *assumed* that the sequence $g_n(X) + g_n(Y)$, which clearly converges to $X + Y$, will have its expectations converge to $E(X + Y)$. Thus for this proof we must use an approximating random variable $g_n(X) + g_n(Y)$ which cannot be written as a function of $X + Y$, let alone as our special function $g_n(X + Y)$. In a sense, linearity involves $P_{X,Y}$ and not just the P_X, P_Y, and P_{X+Y} which determine EX, EY, and $E(X + Y)$.

Similarly if X and Y are independent and we use the discrete-case result then we can prove the general result

$$\begin{aligned} E(XY) &\overset{1}{=} \lim_{n\to\infty} Eg_n(XY) \ominus \lim_{n\to\infty} [Eg_n(X)g_n(Y)] \\ &\overset{2}{=} \lim_{n\to\infty} [Eg_n(X)][Eg_n(Y)] \overset{3}{=} [EX][EY]. \end{aligned}$$

Thus the general theory reduces proofs of general facts to the same fact in the simplest discrete case by first proving that under general conditions, if two sequences of random variables have the same limit then so do their expectations.

EXERCISES

1. Let (R,P) be a probability space and define, in terms of P, a discrete density p' concentrated on the integers according to $p'(k) = P\{x: k - 1 \le |x| < k\}$ for $k = 1, 2, \ldots$. Show that P has a mean in the following cases:
(*a*) Iff p' does
(*b*) If $p'(k) \le 10^{100}/k^{2.1}$ whenever $k \ge 10^{10}$
(*c*) If $P\{x: |x| < 10^{100}\} = 1$
COMMENT: Thus P has a mean iff it does not assign too much probability too far out, and this criterion does not depend on how intricately P distributes its probability over finite intervals.

2. Sketch proofs of the following properties of expectation:
(*a*) If $P[X = c] = 1$ then $EX = c$.
(*b*) If I_A is the indicator of an event A then $EI_A = P(A)$.

(*c*) If $P[-B \le X \le B] = 1$ then $-B \le EX \le B$.
(*d*) X has an expectation iff $|X|$ does.
(*e*) $E|X| = 0$ iff $P[|X| = 0] = 1$.

3. Let $X: R \to R$ be the identity function on R; that is, $X(x) = x$ for all x, so X is the line of slope 1 through the origin. If (R,P) is a probability space show that the mean of P equals the expectation of X with associated (R,P). Thus the mean of P is just the expectation of a particular random variable.

4. State whether or not each of the following entities is uniquely defined by the assumption that both P_X and P_Y are binomial $n = 100$ and $p = .1$. Note that X and Y may be dependent.

(*a*) P_{X+Y}
(*b*) $E(X + Y)$
(*c*) $E(X + Y)^2$
(*d*) $E[(X + Y)^2 + (X - Y)^2]$
(*e*) $Ee^{t(X+Y)}$ for all real t
(*f*) $(Ee^{tX})(Ee^{tY})$ for all real t
(*g*) $E(e^X + \cos Y)$

5. Are any of your answers for the preceding exercise changed if the question asked is changed to read: "Which of the entities below is uniquely defined by the specification of P_X and P_Y, which are both concentrated on $-100 \le x \le 100$"?

6. Let R_n be the range of the function $g_n: R \to R$ defined in the addendum to this section. Thus

$$R_n = \left\{\frac{j}{2^n} : j = 0, \pm 1, \ldots, \pm M_n\right\}.$$

The R_n are increasing sets and $g_n(x) = x$ if $x \in R_n$. Show that

$$g_m(g_n(x)) = \begin{cases} g_m(x) & \text{if } m < n \\ g_n(x) & \text{if } m \ge n. \end{cases}$$

Using discrete-case expectations, conclude that $Eg_m(g_n(X)) \to Eg_n(X)$ as $m \to \infty$. Thus the new definition of expectation, when applied to $g_n(X)$, agrees with the discrete definition of $Eg_n(X)$.

*15-5 CONDITIONAL EXPECTATION

This section extends Sec. 14-1 to the case where X is a random variable with an arbitrary distribution and g is a random vector which induces a discrete density. Theorem 1 shows that in constructing a model for dependent X and g we may use any p_g for g and any family of conditional distributions $P_{X|g=\beta}$ for X. Theorem 2 justifies the use of our previous formulas for EX and Var X in terms of $E[X \mid g = \beta]$ and Var $[X \mid g = \beta]$, which are easily defined. Theorem 3 extends the results of Sec. 14-2 for sums of a random number of random variables.

Let $f: \Omega \to R^n$ and $g: \Omega \to R^m$ be random vectors with the same associated probability space (Ω,P). Assume that g induces a discrete probability density, although f need not. For a point β for which $P[g = \beta] = p_g(\beta) > 0$ the **conditional distribution of** f **given** $[g = \beta]$ was defined in Sec. 15-2 to be the distribution of f with associated $(\Omega,P_{[g=\beta]})$; that is,

$$P_{f|g=\beta}(D) = P[f \in D \mid g = \beta] = \frac{P[f \in D, g = \beta]}{p_g(\beta)}$$

for every subset D of R^n. Therefore

$$P_f(D) = \sum_\beta P[f \in D, g = \beta] = \sum_\beta p_g(\beta) P_{f|g=\beta}(D)$$

so that the distribution P_f of f is a **mixture,** according to discrete density p_g, of the conditional distributions $P_{f|g=\beta}$, and we write

$$P_f = \sum_\beta p_g(\beta) P_{f|g=\beta}.$$

At times we may say that subexperiment β is selected with probability $p_g(\beta)$, and if this occurs then f has conditional distribution $P_{f|g=\beta}$.

Similarly any mixture $P = \sum_\beta p(\beta) P_\beta$ is easily shown to define a probability measure P, which we may say is obtained by **randomizing the parameter** β. The next theorem implies that such a mixture can always be interpreted as in the preceding paragraph. More important, the next theorem shows that in constructing a model with random vectors f and g it is permissible to use *any* discrete density for p_g and *any* collection of conditional distributions $P_{f|g=\beta}$ for f, given g. Often these are the natural objects to use in specifying a model. Independence, of course, corresponds to $P_{f|g=\beta} = P_f$ for all β with $p_g(\beta) > 0$.

Theorem 1 Let $p: R^m \to R$ be a discrete density, and for each β for which $p(\beta) > 0$ let (R^n, P_β) be a probability space. Then there is a probability space (Ω, P) and two random vectors $g: \Omega \to R^m$ and $f: \Omega \to R^n$ such that $p_g = p$ and $P_{f|g=\beta} = P_\beta$ whenever $p_g(\beta) > 0$. Therefore

$$P_f = \sum_\beta p_g(\beta) P_{f|g=\beta} = \sum_\beta p(\beta) P_\beta \quad \blacksquare$$

***Proof** To simplify the notation let $F = R^n$ and $G = R^m$. We have discrete density $p: G \to R$ while (F, P_β) is a probability space whenever $p(\beta) > 0$. Let $\Omega = F \times G$ and define f and g on Ω by

$$f(\alpha,\beta) = \alpha \qquad \text{and} \qquad g(\alpha,\beta) = \beta \quad \text{if } (\alpha,\beta) \in \Omega.$$

Let A be some subset of Ω, as depicted in Fig. 1*a* for the case $n = m = 1$. For each β in G define the subset A_β of F by $A_\beta = \{\alpha : (\alpha,\beta) \in A\}$, then define P by

$$P(A) = \sum_\beta P_\beta(A_\beta) p(\beta).$$

Clearly $P(A) \geq 0$. Also $P(\Omega) = 1$ since $\Omega_\beta = F$, and hence $P_\beta(\Omega_\beta) = 1$ for all β. If $A_1, A_2, \ldots$ are disjoint subsets of Ω and $A = \bigcup_{n=1}^{\infty} A_n$ then for each β, as indicated in Fig. 1*b*, the sets $(A_1)_\beta, (A_2)_\beta, \ldots$ are disjoint and their union equals A_β so

$$\begin{aligned} P(A) &= \sum_\beta P_\beta\left[\bigcup_{n=1}^{\infty} (A_n)_\beta\right] p_g(\beta) \\ &= \sum_\beta \left\{ \sum_{n=1}^{\infty} P_\beta[(A_n)_\beta] \right\} p_g(\beta) = \sum_{n=1}^{\infty} \left\{ \sum_\beta P_\beta[(A_n)_\beta] p_g(\beta) \right\} = \sum_{n=1}^{\infty} P(A_n). \end{aligned}$$

Therefore (Ω, P) is a probability space.

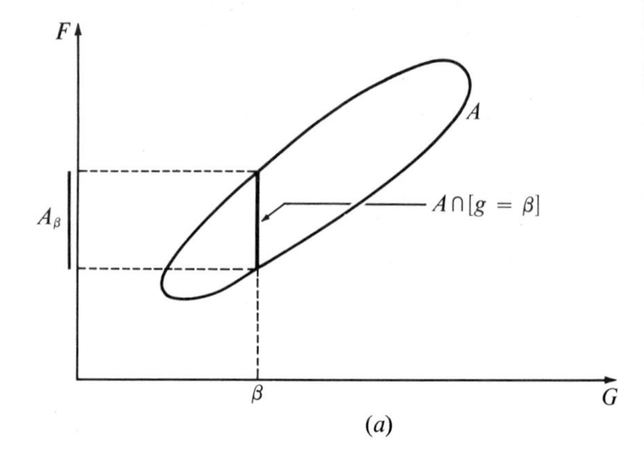
F
A
A_β
$A \cap [g = \beta]$
β
G
(a)

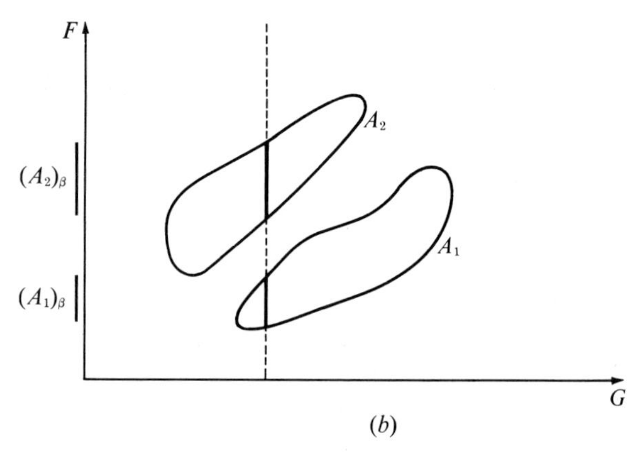
F
A_2
$(A_2)_\beta$
A_1
$(A_1)_\beta$
G
(b)

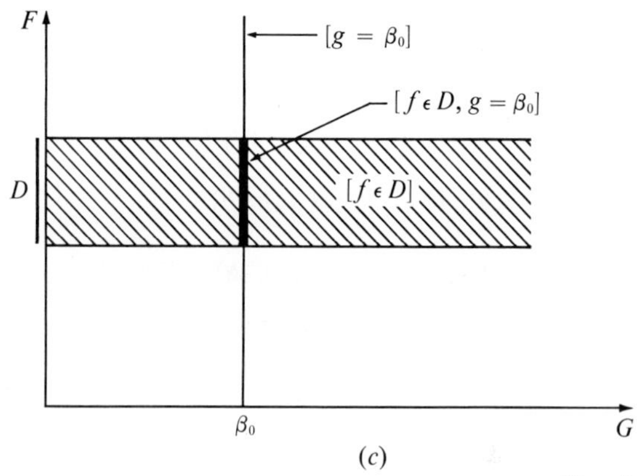
F
$[g = \beta_0]$
$[f \in D, g = \beta_0]$
D
$[f \in D]$
β_0
G
(c)

Fig. 1

Fix β_0 and let $A = [g = \beta_0]$, as shown in Fig. 1*c*. Every A_β is empty except for $A_{\beta_0} = F$, so $P[g = \beta_0] = p(\beta_0)$ and g has the desired distribution. Fix a subset D of F and a β_0 in G. Let $A = [f \in D, g = \beta_0]$, as in Fig. 1*c*. Then every A_β is empty except for $A_{\beta_0} = D$, so $P[f \in D, g = \beta_0] = P_{\beta_0}(D)p(\beta_0)$, hence f has the desired conditional distributions. ■

If $f = X$ is a random variable we can take expectations and variances with or without conditioning. Thus let X be a random variable and g a random vector with the same associated probability space (Ω,P). Assume that g induces a discrete density. If $E|X| < \infty$ and $p_g(\beta) > 0$ we introduce the notation

$$\mu_{X|g=\beta} = \mu_{X|g}(\beta) = E[X \mid g = \beta] = [\text{the mean of } P_{X|g=\beta}]$$
$$= [\text{the expectation of } X \text{ with associated } (\Omega, P_{[g=\beta]})].$$

We call $\mu_{X|g}(\beta)$ the **conditional expectation of** X **given** $g = \beta$. The function $\mu_{X|g}$ is called the **conditional expectation of** X **given** g, or the **regression function of** X **on** g. Clearly $E(c \mid g) = c$, and if X and g are independent then $E(X \mid g) = EX$. Similarly if $EX^2 < \infty$ we introduce, for $p_g(\beta) > 0$, the notation

$$\sigma^2_{X|g=\beta} = \sigma^2_{X|g}(\beta) = \text{Var}\,[X \mid g = \beta] = [\text{the variance of } P_{X|g=\beta}]$$
$$= [\text{the variance of } X \text{ with associated } (\Omega, P_{[g=\beta]})].$$

The function $\sigma^2_{X|g}$ is called the **conditional variance of** X **given** g. Clearly if X and g are independent then $\text{Var}\,(X \mid g) = \text{Var}\, X$.

If P_X has a mean we substitute the expression for

$$P_X = \sum_\beta p_g(\beta) P_{X|g=\beta}$$

as a mixture into Theorem 1 (Sec. 15-4). We then see that

$$[\text{the mean of } P_X] = \sum_\beta p_g(\beta)\mu_{X|g}(\beta)$$

so that the mean of a mixture is the same mixture of the means. We easily obtain the *linearity of conditional expectation*,

$$E\left(\sum_{i=1}^n a_i X_i \mid g\right) = \sum_{i=1}^n a_i E(X_i \mid g)$$

which is valid whenever each $E|X_i| < \infty$.

If $h: R \to R$ then

$$Eh(X) = \sum_\beta p_g(\beta) E[h(X) \mid g = \beta].$$

Letting $h(x) = (x - \mu_X)^2$ we see that

$$\text{Var}\, X = \sum_\beta p_g(\beta) E[(X - \mu_X)^2 \mid g = \beta].$$

Using the identity

$$(X-\mu_X)^2 = [(X-\mu_{X|g}(\beta)) + (\mu_{X|g}(\beta)-\mu_X)]^2$$
$$= [X-\mu_{X|g}(\beta)]^2 + 2[\mu_{X|g}(\beta)-\mu_X][X-\mu_{X|g}(\beta)] + [\mu_{X|g}(\beta)-\mu_X]^2$$

we obtain the same formula for Var X as in Sec. 14-1. The interpretations in Sec. 14-1 of predictions of X, given g, still apply. We summarize in the following theorem, which extends Theorem 1 (Sec. 14-1).

Theorem 2 Let X be a random variable and g a random vector with the same associated probability space. Assume that g induces a discrete density, so that P_X can be expressed as a mixture,

$$P_X = \sum_\beta p_g(\beta) P_{X|g=\beta}.$$

Then P_X has a mean iff "each $P_{X|g=\beta}$ with $p_g(\beta) > 0$ has a mean and the series below converges absolutely," in which case

$$EX = \sum_\beta \mu_{X|g}(\beta) p_g(\beta) = \sum_\beta E[X \mid g = \beta] p_g(\beta).$$

Also P_X has a variance iff "each $P_{X|g=\beta}$ with $p_g(\beta) > 0$ has a variance and both series below converge," in which case

$$\text{Var } X = \sum_\beta \sigma^2_{X|g}(\beta) p_g(\beta) + \sum_\beta [\mu_{X|g}(\beta) - \mu_X]^2 p_g(\beta). \quad \blacksquare$$

If $g = Y$ is a random variable we can use either the regression $\mu_{X|Y}$ of X on Y, or the linear regression $\mu_X + \sigma_X \rho(y - \mu_Y)/\sigma_Y$ of X on Y, as predictors of X, given Y. Either they agree whenever $p_Y(y) > 0$, or the former has a strictly smaller expected squared error.

As in Example 2 (Sec. 14-1) the function defined for all ω in Ω by $\mu_{X|g}(g(\omega))$ is referred to in many books on probability theory as the *conditional expectation of X given g*.

Section14-2 can easily be extended to our present context, as stated in the theorem below. If sufficiently general probability spaces are permitted it can be shown that the defined P_{S_N} can always be interpreted as induced by $S_N = X_1 + \cdots + X_N$ where N, $X_1, X_2, \ldots$ are independent and $X_1, X_2, \ldots$ have the same distribution P_{X_1}.

If P_X is concentrated on the nonnegative reals we naturally define the **probability-generating function** for P_X by

$$G_X(z) = Ez^X \qquad \text{for } 0 \le z \le 1$$

where X has distribution P_X. Then $G_X(0) = G_X(0+) = P[X = 0]$, $0 \le G_X(z) \le 1$, and $G_X(1) = 1$.

Theorem 3: The sum $S_N = X_1 + \cdots + X_N$ of a random number N of random variables Let (R, P_{X_1}) be a probability space and $p_N\colon R \to R$ be a discrete density for which

$\sum_{n=0}^{\infty} p_N(n) = 1$. Let $P_{S_0}(\{0\}) = 1$, and for each $n = 1, 2, \ldots$ let P_{S_n} be induced by $S_n = X_1 + \cdots + X_n$ where $X_1, \ldots, X_n$ are independent and each has distribution P_{X_1}. Define the probability space (R, P_{S_N}) by the mixture

$$P_{S_N} = \sum_{n=0}^{\infty} p_N(n) P_{S_n}$$

Let random variables X_1, N, and S_N have distributions P_{X_1}, P_N, and P_{S_N}. Then S_N has an expectation iff both N and X_1 do, and in this case

$$ES_N = (EX_1)(EN).$$

Also S_N has a variance iff both N and X_1 do, and in this case

$$\text{Var } S_N = (EN)(\text{Var } X_1) + (EX_1)^2(\text{Var } N).$$

Furthermore if P_{X_1} is concentrated on the nonnegative reals then the probability-generating functions satisfy

$$G_{S_N}(z) = G_N(G_{X_1}(z)) \qquad \text{for } 0 \leq z \leq 1. \quad \blacksquare$$

Proof From Theorem 1 and the definition of P_{S_N} as a mixture, we may assume that we have random variables S_N and N with the same associated probability space (R^2, P) such that N has density p_N and $P_{S_N|N=n} = P_{S_n}$. From Theorem 2 we get

$$ES_N = \sum_{n=0}^{\infty} (nEX_1) p_N(n) = (EX_1)(EN),$$

$$\begin{aligned} \text{Var } S_N &= \sum_{n=0}^{\infty} (n \text{ Var } X_1) p_N(n) + \sum_{n=0}^{\infty} [(nEX_1) - (EX_1)(EN)]^2 p_N(n) \\ &= (EN)(\text{Var } X_1) + (EX_1)^2(\text{Var } N). \end{aligned}$$

Also

$$\begin{aligned} G_{S_N}(z) &= Ez^{S_N} = \sum_{n=0}^{\infty} E[z^{S_N} \mid N = n] p_N(n) \\ &= \sum_{n=0}^{\infty} G_{S_n}(z) p_N(n) = \sum_{n=0}^{\infty} [G_{X_1}(z)]^n p_N(n). \quad \blacksquare \end{aligned}$$

16
Continuous Probability Densities

In this chapter we essentially replace the summations in the discrete-density case by integrations for the continuous-density case.

16-1 CONTINUOUS PROBABILITY DENSITIES

Let us first consider some phenomena and technicalities which arise in nondiscrete cases. Then we will define continuous densities and their means, variances, MGFs, etc., essentially by analogy with discrete densities. Most of these entities are studied more systematically later in the chapter, but they are introduced in this section so that they can be calculated for the basic uniform, exponential, and normal densities used as illustrations here.

Consider the random wheel of Fig. 1, whose circumference is continuously and uniformly calibrated from 0 to 1. In essence the outcome of the experiment of spinning the wheel once is that unique real number x satisfying $0 \leq x < 1$ and such that after the wheel has come to rest, a clockwise rotation of the wheel by $2\pi x$ radians brings the zero reference mark opposite the pointer.

The physical performance of such an experiment could not result in the observation of $x = \frac{1}{2} = .500 \cdots$, for example, since the outcome can be observed only to some limited accuracy, say 10^{100} significant figures. Idealizations, such as consideration of length, angle, mass, intensity, or income, to be "continuously variable over an interval "are, of course, common. They make available the techniques of calculus, and lead to simpler, more comprehensible conclusions than seem obtainable from awkward, more realistic models. Almost by definition, such an idealized model has some peculiar properties, such as the above-mentioned unlimited precision. The next paragraph is concerned with such a property of Fig. 1.

The random-wheel experiment of Fig. 1 yields a real number x. If $a < b$ are reals in the unit interval we naturally interpret the length $b - a$ of the interval $a \leq x \leq b$ as the probability of obtaining some x in that interval. What is the probability p that the outcome is $\frac{1}{2}$? This probability is less than $2/n$, which is the probability of obtaining an x in the interval $.5 - (1/n) \leq x \leq .5 + (1/n)$. Thus p must satisfy $0 \leq p \leq 2/n$ for all $n = 1, 2, \ldots$, so $p = 0$. For every particular number x in the unit interval the probability is 0 that the experiment will result in x. Clearly this idealized experiment cannot be described by a discrete probability density. In particular the probability of an interval $a \leq x \leq b$ is unchanged if we remove one or both of its end points, or any countable number of points. Thus the probability of each permissible outcome is 0 even though the probability of obtaining some permissible outcome is 1; that is, the unit interval has length 1 even though each point has length 0.

Sometimes the labels "certain" or "impossible" are applied to events of probability 1 or 0, respectively. With such terminology this experiment is certain to yield an impossible outcome. We usually avoid such terminology. The normal English usage of the word "impossible" seems to imply that if each of a collection of events is impossible then the event that at least one of them will occur is also impossible. However, as just

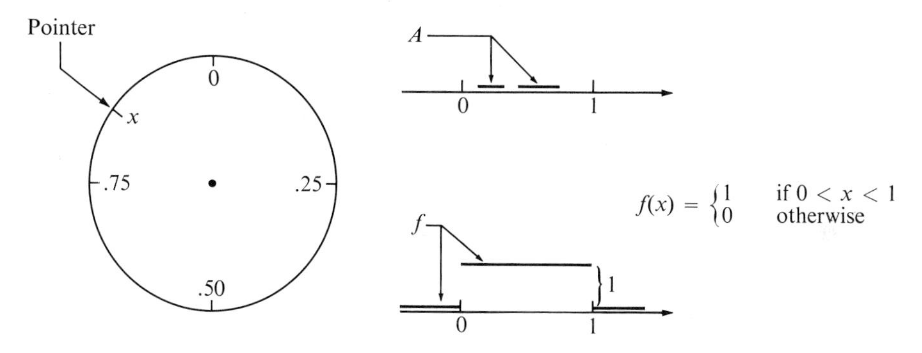

Fig. 1 $P(A) = \int_A f(x)\,dx$ is the length of A.

indicated, each of an *uncountable* collection of events can have probability 0, and yet their union may have probability 1. This property of Fig. 1 is actually no more troublesome than its corresponding interpretation in terms of idealized length.

Almost every technical complication in probability theory appears in connection with Fig. 1, which we now consider in more detail. The unit interval is the sample space Ω. Any point x in Ω is a sample point, while an event A is a subset of Ω. The probability $P(A)$ of an event A is the total length of A. The apparent simplicity of the experiment of Fig. 1 is deceptive in that, as in Exercise 7 (Sec. 6-4), this same Ω and P can be used for the experiment of making an infinite number of independent spins with a 10-sectored wheel. Some subsets of the unit interval are unbelievably weird. In the discrete case no difficulty arises in assigning a probability to every subset and obtaining a P which has all the desired properties. However, this is unfortunately not the case for our present example, as is suggested by the following nonobvious fact. It is impossible to assign a nonnegative "length" to *every* subset of the real line (some subsets naturally have infinite length) in such a way that the length of every finite interval $a \leq x \leq b$ is $b - a$, and the length assignment is countably additive on disjoint subsets and is translation invariant; that is, the length of any subset A of reals is the same as the length of every subset obtained by shifting A to the right or left. Because of such unfortunate facts the usual rigorous attitude in probability theory, as in measure theory, is that probability assignments P should be countably additive in order to have a powerful theory of expectation, but then the penalty must be accepted that P be defined only for some subsets. This is irksome but can be lived with. Thus $P(A)$ is defined for some subsets A which are then called events, and $P(A)$ is not defined for other subsets A which are not called events. For examples like Fig. 1 and others considered in this book, it is possible to make all subsets of any interest events. Thus these considerations are more a technical nuisance than a practical limitation.

A concomitant technical nuisance is the fact that if a real-valued function g is defined on the unit interval then the number $P\{x:g(x) < d\}$ may not be defined for all real d, since the subset $\{x: g(x) < d\}$ may not be an event. Therefore not every g is called a random variable, but only those g concerning which obvious probabilities of interest are defined. A random vector is then an n-tuple of random variables. However, once again, this is not a practical limitation. For example, in the case of Fig. 1 every continuous function, every possibly discontinuous limit of continuous functions, etc., is a random variable. Thus in a rigorous presentation of probability theory some subsets of the sample space are *not* called events, and some real-valued functions defined on the sample space are *not* called random variables. We normally ignore these complications, except in Chap. 21.

The preceding discussion is not meant to imply that with a wave of the hand all technical difficulties may be ignored. On the contrary, once examples like Fig. 1 have been admitted, some complications of a kind not arising in the discrete case cannot be avoided by even the most casual approach. For example, Theorem 1 (Sec. 15-2) shows that every random variable X has an induced probability measure P_X, however Example 4 (Sec. 17-1) shows that even with the simple (Ω,P) of Fig. 1 it is possible to obtain an induced P_X which does not correspond to any mixture of discrete and continuous probability densities. For such examples we must deal directly with the set function P_X, or some equivalent entity such as the distribution function as defined in Sec. 17-1. The discrete case is "closed" in the sense that if (Ω,P) is any discrete probability space and f: $\Omega \to R^n$ is a random vector then the induced (R^n,P_f) is again a discrete probability space. The continuous-density case is *not* closed in this sense. This is one reason for considering probability measures to be more basic than densities.

In this chapter we are concerned primarily with situations where the basic (R^n,P) and all induced distributions which we consider can be described by continuous probability densities. We place only as much emphasis on probability measures as is needed to maintain a conceptually general framework that can be made rigorous.

GENERAL DEFINITIONS

A **continuous probability density** is a function p: $R^n \to R$, where $n \geq 1$, satisfying

$$p(x_1, \ldots, x_n) \geq 0 \qquad \text{for all } (x_1, \ldots, x_n) \text{ in } R^n$$

and

$$\int_{-\infty}^{\infty} \cdots \int_{-\infty}^{\infty} p(x_1, \ldots, x_n)\, dx_1\, dx_2 \cdots dx_n = 1.$$

Thus p is nonnegative and its integral over the whole of R^n is 1. The **probability space**

(R^n,P) **corresponding to a continuous probability density** $p\colon R^n \to R$ is defined by setting

$$P(A) = \int \cdots \int_A p(x_1, \ldots, x_n)\, dx_1\, dx_2 \cdots dx_n$$

for each event A in the sample space R^n. Thus the probability $P(A)$ of a subset A of R^n is just the integral of p over A. Clearly $P(\Omega) = 1$ and $P(A) \geq 0$, and the countable additivity of P should be intuitively acceptable from the most basic ideas of integration. The adjective "continuous" in "continuous probability density" is only meant to distinguish the present case based on integration from the discrete case based on summation. A continuous probability density may be highly discontinuous. We require only that it be continuous enough for the integrals to be defined. Note the reasonableness of all the conclusions of Theorem 2 (Sec. 6-4) for a P obtained from a continuous probability density.

If (R^n,P) corresponds to a continuous density and A contains only one or a countable number of sample points then $P(A) = 0$, since the integral over a point is zero. Therefore such a P cannot fall within the discrete case.

If we change a continuous density at one or a countable number of points then the integral over any event A remains unchanged, so that the original and altered densities determine the same probability measure P. The probabilities of events are most basic, so we consider two continuous probability densities, $p\colon R^n \to R$ and $p'\colon R^n \to R$, to be **equivalent** iff $P = P'$; that is, $P(A) = P'(A)$ for every event A. We let $p \equiv p'$ denote the fact that p is equivalent to p'. Clearly this relation is reflexive ($p \equiv p$), symmetric (if $p \equiv p'$ then $p' \equiv p$), and transitive (if $p = p'$ and $p' \equiv p''$ then $p \equiv p''$). Theorem 2 (Sec. 15-1) shows that $p \equiv p'$ iff P and P' agree on all closed rectangles.

This phenomenon is different from the discrete case, where different densities determined different measures. However, equivalent densities cannot be very different. For example, they must be equal at every point x_0 in R^n where both are continuous, since if $p(x_0) < p'(x_0)$ then $P(R_{a,b}) < P'(R_{a,b})$ for a small enough closed rectangle centered on x_0. In particular if $p \equiv p'$ and both are continuous functions at every point of R^n then they must be equal $p = p'$. Hence there is at most one everywhere continuous density corresponding to a probability space (R^n,P).

The normal density of Sec. 13-1 is a continuous probability density on R, and it is continuous on all of R. As shown in Fig. 2, when $n = 1$ the probability $P(A)$ from continuous density $p\colon R \to R$ is just the area above A and below p. The example of Fig. 1 is subsumed under the continuous-density case by using the continuous density f exhibited in Fig. 1. The values $f(0)$ and $f(1)$ of f at $x = 0$ and $x = 1$ are arbitrary and unimportant. This density is called the *uniform density over the interval* $0 \leq x \leq 1$. It is continuous at every real x except $x = 0, 1$. No density equivalent to f can be

Fig. 2 $P(A) = \int_A p(x)\, dx$ is the area of shaded set.

continuous at either of these two points, since if a density $p\colon R \to R$ is continuous at x_0 then for its corresponding P both $P\{x\colon x_0 - \Delta \le x \le x_0\}/\Delta$ and $P\{x\colon x_0 \le x \le x_0 + \Delta\}/\Delta$ must converge to the *same* limit $p(x_0)$ as Δ goes to zero; while for Fig. 1 these two limits are 0 and 1 if x_0 is 0, and similarly if $x_0 = 1$. Thus, as is natural and traditional, we have selected a density f which is as continuous as possible in that every equivalent density must be discontinuous where f is. Where a density cannot be continuous we typically define it to be zero.

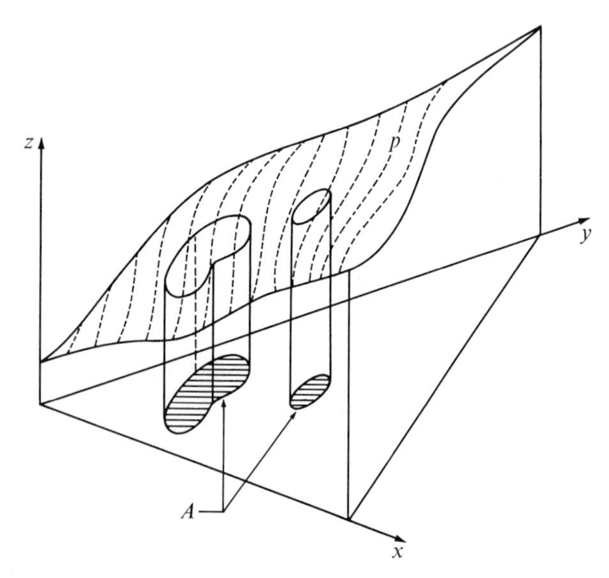

Fig. 3 $P(A) = \iint_A p(x,y)\,dx\,dy$ is the volume above A and below the surface p where $p\colon R^2 \to R$ and $p(x,y) = z$.

As shown in Fig. 3, if $n = 2$ the probability $P(A)$ assigned to a set A of points in the plane by a continuous density $p\colon R^2 \to R$ is just the volume above A and below the surface p. Such a density was utilized in Exercise 6 (Sec. 6-4).

If $p\colon R^n \to R$ is a continuous density and A is an event with $P(A) > 0$ then the function $p_A\colon R^n \to R$ defined by

$$p_A(x_1, \ldots, x_n) = \begin{cases} \dfrac{1}{P(A)}\, p(x_1, \ldots, x_n) & \text{if } (x_1, \ldots, x_n) \in A \\ 0 & \text{if } (x_1, \ldots, x\epsilon) \notin A \end{cases}$$

is called the **continuous probability density obtained by conditioning** p **by,** or on, **the event** A. If (R^n, P_A) is the probability space corresponding to p_A we call (R^n, P_A) the **probability space obtained by conditioning** (R^n, P) **by the event** A. We have

$$P_A(B) = \frac{P(AB)}{P(A)} = P(B \mid A) \qquad \text{for every event } B.$$

This last equation is true because by definition $P_A(B)$ is equal to the integral of p_A over the event B. But p_A is zero off A so $P_A(B)$ is equal to the integral of p_A over AB. However, on AB we have $p_A = p/P(A)$, so $P_A(B)$ equals $1/P(A)$ times the integral of p over AB. Thus $P_A(B) = P(AB)/P(A)$. This equation shows that our present definition of P_A, by way of the density p_A, is equivalent to the more general approach of Theorem 1 (Sec. 15-1), where P_A was defined by $P_A(B) = P(AB)/P(A)$.

Exercises 2 to 6, 10, 11, 15, and 17 may be done before reading the remainder of this section.

We will shortly introduce several continuous densities which arise often in theory and applications. In this section most of our densities are defined on the real line R. When we introduce such densities we wish to calculate their means, variances, and MGFs. Therefore we now introduce and define these quantities in advance of the more general discussion of expectation in Sec. 16-5. If $p: R \to R$ is a continuous density and Δ is a small positive number then

$$P\{x' : x \le x' \le x + \Delta\} = \int_x^{x+\Delta} p(x')\, dx' \doteq p(x)\Delta$$

if p is continuous at x. In the discrete case, if we had probability $p(x)\Delta$ concentrated at x, this would give rise to a contribution of $x(p(x)\Delta)$ to the mean of the distribution. If we partition the large interval $-n \le x \le n$ into many adjacent subintervals of length Δ and add such contributions we have an approximation to the mean of the distribution, and also to the integral $\int_{-\infty}^{\infty} xp(x)\, dx$. Therefore if $p: R \to R$ is a continuous probability density we say that p **has a mean** iff $xp(x)$ is absolutely integrable, that is,

$$\int_{-\infty}^{\infty} |x| p(x)\, dx$$

is finite; and in this case we define the **mean of** p by

$$[\text{the mean of } p] = \int_{-\infty}^{\infty} xp(x)\, dx.$$

Therefore we do not assign a mean to p if either of the contributions from the negative or positive half-axes is infinite:

$$\int_{-\infty}^{0} xp(x)\, dx = -\infty \qquad \int_{0}^{\infty} xp(x)\, dx = \infty.$$

This definition will be shown in Theorem 1 (Sec. 16-5) to be consistent with the more general approach of Sec. 15-4 to expectations.

Many of the formulas for discrete densities, such as those for $P(A)$ or the mean of $p: R \to R$, apply to continuous densities if sums are replaced by integrals. We proceed now in a fashion analogous to Chap. 9.

If p is zero outside some interval $-B \le x \le B$ then

$$\left|\int_{-\infty}^{\infty} xp(x)\, dx\right| = \left|\int_{-B}^{B} xp(x)\, dx\right| \le \int_{-B}^{B} |x|\, p(x)\, dx \le B \int_{-B}^{B} p(x)\, dx = B.$$

Thus if $p(x) = 0$ whenever $|x| > B$ then p has a mean and $-B \le [\text{the mean of } p] \le B$.

Just as in Chap. 9, if $p: R \to R$ is symmetric about some real number b,

$$p(b + x) = p(b - x) \qquad \text{for all real } x$$

and if p has a mean, then it is equal to b.

If $\int_{-\infty}^{\infty} x^2 p(x)\,dx$ is finite we call it the **second moment of** p. It was shown in Sec. 9-2 that $|x| \leq 1 + x^2$ for all real x. Thus if p has a second moment it has a mean μ, and we can define the **variance of** p by

$$\operatorname{Var} p = \int_{-\infty}^{\infty} (x - \mu)^2 p(x)\,dx = \int_{-\infty}^{\infty} x^2 p(x)\,dx - \mu^2$$

where the equality is immediate from the identity $(x - \mu)^2 = x^2 - 2\mu x + \mu^2$. Since $\operatorname{Var} p \geq 0$, we see that $\mu^2 \leq$ [the second moment of p]. Actually $\operatorname{Var} p > 0$ for every continuous probability density $p: R \to R$. The **standard deviation** of p is the *nonnegative* square root of the variance of p.

For every positive real r we define the **rth absolute moment of** p to be

$$\int_{-\infty}^{\infty} |x|^r p(x)\,dx$$

if it is finite. If $n = 1, 2, \ldots$ is a positive integer we define the **nth moment of** p to be

$$\int_{-\infty}^{\infty} x^n p(x)\,dx$$

if p has a finite nth absolute moment.

The following theorem is proved by replacing sums with integrals in the proof of Theorem 1 (Sec. 9-4).

Theorem 1 If a continuous probability density $p: R \to R$ has a finite sth absolute moment for some real $s > 0$ then it has a finite rth absolute moment for any real r satisfying $0 < r < s$, and furthermore

$$\left[\int_{-\infty}^{\infty} |x|^r p(x)\,dx\right]^{1/r} \leq \left[\int_{-\infty}^{\infty} |x|^s p(x)\,dx\right]^{1/s}. \quad \blacksquare$$

If $p: R \to R$ is a continuous density then the **moment-generating function** (MGF) **of** p is defined by

$$m(t) = \int_{-\infty}^{\infty} e^{tx} p(x)\,dx$$

for all those real t for which the integral is finite. The integrand is nonnegative. Naturally if $p(x) = 0$ outside some interval $-B \leq x \leq B$ then in this interval $e^{tx} \leq e^{|t|B}$, so $m(t)$ is defined for all real t and $m(t) \leq e^{|t|B}$. Theorem 2 (Sec. 9-4) contains various properties of MGFs in the discrete case. The obvious analog of that theorem holds for continuous densities where

$$m^{(n)}(t) = \frac{d^n}{dt^n} m(t) = \int_{-\infty}^{\infty} x^n e^{tx} p(x)\, dx \qquad \text{if } n = 1, 2, \ldots ;\ |t| < t_0.$$

In the last conclusion of that theorem $p_1 = p_2$ becomes $p_1 \equiv p_2$ for continuous densities.

SPECIAL DENSITIES

We now introduce the uniform, exponential, and normal continuous probability densities on R and calculate their means, variances, and MGFs, as exhibited in Table 1 in the summary of Chap. 16.

The **uniform continuous probability density** over an interval is defined in Fig. 4.

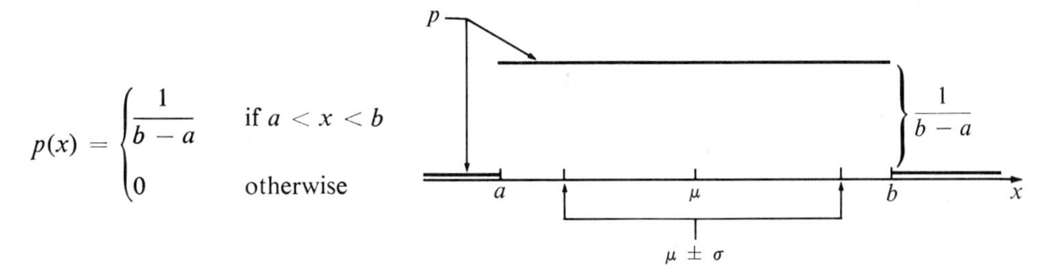

Fig. 4 The uniform continuous probability density $p\colon R \to R$ over the interval $a < x < b$, with parameters a and b, where $a < b$.

Its mean must be the center $\mu = (a + b)/2$ of the interval, since the density is symmetric about the center. For the variance and MGF we obtain

$$[\text{the second moment of } p] = \int_{-\infty}^{\infty} x^2 p(x)\, dx = \frac{1}{b-a}\int_a^b x^2\, dx = \frac{1}{b-a}\left[\frac{x^3}{3}\right]_{x=a}^{x=b}$$

$$= \frac{b^3 - a^3}{3(b-a)} = (1/3)(b^2 + ab + a^2).$$

$$\operatorname{Var} p = [\text{the second moment of } p] - [\text{the mean of } p]^2$$

$$= \frac{1}{3}(b^2 + ab + a^2) - \frac{(a+b)^2}{4} = \frac{(b-a)^2}{12}.$$

$$m(t) = \int_{-\infty}^{\infty} e^{tx} p(x)\, dx = \frac{1}{b-a}\int_a^b e^{tx}\, dx = \frac{1}{b-a}\left[\frac{e^{tx}}{t}\right]_{x=a}^{x=b} = \frac{e^{tb} - e^{ta}}{t(b-a)}.$$

The **exponential continuous probability density** with parameter $\lambda > 0$ is defined in Fig. 5.

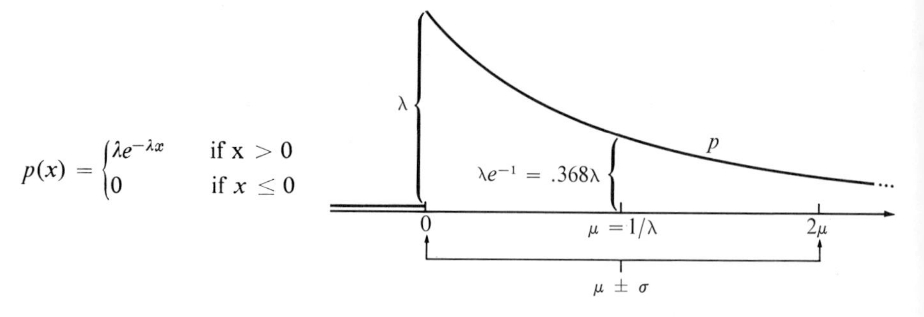

Fig. 5 The exponential continuous probability density $p: R \to R$ with parameter $\lambda > 0$ and mean $1/\lambda$.

Making the change of variables $y = \lambda x$ we see that

$$\int_{-\infty}^{\infty} p(x)\, dx = \int_0^{\infty} \lambda e^{-\lambda x}\, dx = \int_0^{\infty} e^{-y}\, dy = 1.$$

$$[\text{the mean of } p] = \int_{-\infty}^{\infty} xp(x)\, dx = \int_0^{\infty} x\lambda e^{-\lambda x}\, dx = \frac{1}{\lambda}\int_0^{\infty} ye^{-y}\, dy$$

$$= \frac{1}{\lambda}\,[-(1+y)e^{-y}]_{y=0}^{y=\infty} = \frac{1}{\lambda}\,.$$

We often refer to this density as the *exponential density with mean* $1/\lambda$. We use λ instead of the mean $\mu = 1/\lambda$ as the parameter because it is then easier to relate the exponential density to the Poisson process in Chap. 23. For the MGF we obtain

$$m(t) = \int_{-\infty}^{\infty} e^{tx} p(x)\, dx = \int_0^{\infty} e^{tx}\lambda e^{-\lambda x}\, dx$$

$$= \lambda \int_0^{\infty} e^{-(\lambda - t)x}\, dx = \frac{\lambda}{(\lambda - t)} = \frac{1}{1 - t/\lambda} \qquad \text{if } t < \lambda.$$

Clearly $m(t)$ is finite iff $t < \lambda$, since if $t \geq \lambda$ the integrand is greater than or equal to 1 for all $x \geq 0$. We expand $m(t)$ in a geometric series

$$m(t) = \sum_{n=0}^{\infty} \left(\frac{t}{\lambda}\right)^n = \sum_{n=0}^{\infty} \frac{t^n}{\lambda^n}$$

so that from Theorem 2 (Sec. 9-4) the nth moment of p is $m^{(n)}(0) = n!/\lambda^n$ for $n = 1, 2, \ldots$. Thus the second moment of p is $2/\lambda^2$, so the variance of p is $(2/\lambda^2) - (1/\lambda)^2 = 1/\lambda^2$.

The **normal continuous probability density** $\varphi_{\mu,\sigma}: R \to R$ with parameters μ and $\sigma > 0$ was defined and graphed in Fig. 3 (Sec. 13-1). Its mean is μ and its variance is σ^2. Its MGF is

$$m(t) = \int_{-\infty}^{\infty} e^{tx}\varphi_{\mu,\sigma}(x)\, dx = \frac{1}{\sigma\sqrt{2\pi}}\int_{-\infty}^{\infty} \exp\left\{tx - \frac{(x-\mu)^2}{2\sigma^2}\right\} dx.$$

Completing the square of the quadratic in x, we have

$$tx - \frac{(x-\mu)^2}{2\sigma^2} = \mu t + \frac{\sigma^2 t^2}{2} - \frac{[x - (\mu + t\sigma^2)]^2}{2\sigma^2}$$

so that

$$m(t) = \left[\exp\left\{\mu t + \frac{\sigma^2 t^2}{2}\right\}\right] \frac{1}{\sigma\sqrt{2\pi}} \int_{-\infty}^{\infty} \exp\left\{-\frac{[x-(\mu+t\sigma^2)]^2}{2\sigma^2}\right\} dx$$

$$= \exp\left\{\mu t + \frac{\sigma^2 t^2}{2}\right\} \qquad \text{for } -\infty < t < \infty$$

where we used the fact that the integral of the normal density, with mean $\mu + t\sigma^2$ and standard deviation σ, integrates to 1 over R.

All exercises except 12, 13, and 14 (which presume Examples 4, 6, and 7, respectively) may be done before reading the remainder of this section.

Example 1 The uniform density over the interval $a \le x \le b$ corresponds intuitively to selecting a real number at random from this interval, with each point of the interval being equally likely. It can also be thought of as the continuous idealization of a discrete uniform distribution on a large finite number of closely equispaced points on an interval of the real axis. In Example 2 (Sec. 9-2) we essentially defined the discrete density $p_{X_n}: R \to R$ by

$$p_{X_n}(0) = p_{X_n}(1) = \cdots = p_{X_n}(n) = \frac{1}{n+1}$$

and $p_{X_n}(x) = 0$ otherwise; we found that

$$EX_n = \frac{n}{2} \qquad \text{and} \qquad \operatorname{Var} X_n = \frac{n^2}{12} + \frac{n}{6}.$$

Let $Y_n = X_n/n$ so that

$$p_{Y_n}\left(\frac{k}{n}\right) = \begin{cases} \dfrac{1}{n+1} & \text{if } k = 0, 1, \ldots, n \\ 0 \text{ otherwise.} \end{cases}$$

Thus p_{Y_n} is uniform on the $n + 1$ equispaced points

$$\frac{0}{n}, \frac{1}{n}, \ldots, \frac{n-1}{n}, \frac{n}{n}$$

of the unit interval. To *some* extent, if n is large then p_{Y_n} and the uniform continuous density p on $0 \le x \le 1$ can be thought of as approximations to each other. In particular

$$EY_n = E\frac{X_n}{n} = \frac{1}{2} = [\text{the mean of } p]$$

$$\operatorname{Var} Y_n = \operatorname{Var}\left(\frac{X_n}{n}\right) = \frac{1}{12} + \frac{1}{6n} \xrightarrow[n\to\infty]{} \frac{1}{12} = \operatorname{Var} p.$$

Caution must be exercised in thinking in terms of such approximations since, for example, p assigns probability 0 to every countable set and hence to the set of all rational numbers, while every p_{Y_n} assigns probability 1 to this set. ■

Example 2 A real number x is selected at random from the unit interval by the random-wheel experiment of Fig. 1. Its decimal expansion $x = .x_1x_2 \cdots$ is calculated and we are told only that its second digit x_2 equals 8. What, then, is the probability that its first digit x_1 is 5? That is, what is $P(B \mid A)$ if

$$A = \{.x_1x_2 \cdots : x_2 = 8\} \qquad \text{and} \qquad B = \{.x_1x_2 \cdots : x_1 = 5\}?$$

Clearly

$$\begin{aligned} P(B) &= P\{x\colon .5 \le x \le .6\} = .1 \\ P(A) &= P\{x\colon .08 \le x < .09,\ .18 \le x < .19, \ldots, .98 \le x < .99\} \\ &= P\{x\colon .08 \le x < .09\} + P\{x\colon .18 \le x < .19\} + \cdots + P\{x\colon .98 \le x < .99\} \\ &= (10)(.01) = .1 \\ P(AB) &= P\{x\colon .58 \le x < .59\} = .01. \end{aligned}$$

Thus A and B are independent, and

$$P(B \mid A) = \frac{P(AB)}{P(A)} = .1 = P(B). \quad ■$$

Example 3 Let $p\colon R \to R$ be the uniform density on the interval $a \le x \le b$ and let A be the subinterval $A = \{x\colon a' \le x \le b'\}$, where $a \le a' < b' \le b$. Then the density p_A obtained by conditioning p on A is given by

$$p_A(x) = \begin{cases} \dfrac{p(x)}{P(A)} = \dfrac{1/(b-a)}{(b'-a')/(b-a)} = \dfrac{1}{b'-a'} & \text{if } x \in A \\ 0 \text{ otherwise} \end{cases}$$

so that p_A is the uniform density on the subinterval $a' \le x \le b'$.

In radio communications it is common to consider a message of the form $\cos(\omega t + \theta)$ over the interval $-\infty < t < \infty$, where ω is a known frequency and the random phase θ is assumed to have a uniform density over the interval from 0 to 2π. If we learn that $.1 \le \theta \le .3$ then our conditional density for θ is uniform over this subinterval. ■

Example 4 The exponential density can be thought of as the continuous analog of the geometric density pq^k for $k = 0, 1, 2, \ldots$. The analogy may be emphasized by writing

$$pq^k = pe^{-\alpha k} \qquad \text{where } \alpha = -\log q > 0.$$

As mentioned earlier, the geometric and exponential distributions are used extensively in failure and life-testing problems. We might use the exponential density with mean $1/\lambda = 500$ hours to describe the in-service lifetime for a light bulb. The outcome of the experiment is x hours iff the bulb malfunctioned for the first time after precisely x hours of use. Then, for example, the probability that the lifetime will exceed the mean lifetime is

$$\int_{1/\lambda}^{\infty} \lambda e^{-\lambda x}\, dx = \int_1^{\infty} e^{-y}\, dy = [-e^{-y}]_{y=1}^{\infty} = e^{-1}.$$

We now show that the continuous exponential density has the same unique "lack of wearout" property noted in the discrete case for the geometric density in Exercise 3 (Sec. 8-4).

Let (R,P) correspond to the exponential density with mean $1/\lambda$. If r is a nonnegative real number let $A_r = \{x: x \geq r\}$, so that

$$P(A_r) = \int_r^\infty \lambda e^{-\lambda x}\, dx = e^{-\lambda r}.$$

If t is another nonnegative real then

$$P(A_{r+t} \mid A_r) = \frac{P(A_{r+t} \cap A_r)}{P(A_r)} = \frac{P(A_{r+t})}{P(A_r)} = \frac{e^{-\lambda(r+t)}}{e^{-\lambda r}} = e^{-\lambda t}.$$

Therefore if we learn that the bulb is burning at time $r > 0$ then $e^{-\lambda t}$ is the probability that it will burn for at least t more hours, *and* this probability does not depend on r. We next show that the exponential density is the unique distribution having this property for all real $r \geq 0$ and $t \geq 0$.

We have shown that an exponential density has the property that $P(A_{r+t} \mid A_r) = P(A_t)$; that is,

$$P(A_{r+t}) = P(A_r)P(A_t) \qquad \text{for all real } r \geq 0,\ t \geq 0.$$

Note that the geometric density satisfied this equation for nonnegative *integers*. If $P(A_r) = 0$ for all $r \geq 0$ the equation is satisfied trivially, so to rule this out we assume $P(A_1) > 0$. We now show that any P with sample space R which satisfies both the equation above and $P(A_1) > 0$ must correspond to an exponential density. For all real $x > 0$ we define $g(x) = P(A_x)$, so that

$$g(r + t) = g(r)g(t) \qquad \text{if } r > 0,\ t > 0.$$

Thus g satisfies the functional equation which is shown in the addendum to this section to be satisfied only by the exponential function. Hence for some real λ

$$P(A_x) = g(x) = e^{-\lambda x} \qquad \text{for all } x > 0,$$

and clearly $\lambda > 0$ since $P(A_x) \leq 1$. Therefore P corresponds to the exponential density with mean $1/\lambda$, where λ is determined by $P(A_1) = e^{-\lambda}$. ■

Example 5 A manufacturer produces rods whose lengths have a normal distribution with mean $\mu = 100$ feet and standard deviation $\sigma = 2$ feet. The normal density is symmetric about its mean, so the probability is ½ that a rod will have a length less than the mean length. If we learn that a rod has a length less than 101 feet then the conditional probability that its length is less than 100 feet is

$$P(\{x: x < 100)\} \mid \{x: x < 101\}) = \frac{P\{x: x < 100\}}{P\{x: x < 101\}} = \frac{.5}{P\{x: x < 101\}}$$

$$= \frac{.5}{P\{x: (x - 100)/2 < (101 - 100)/2\}} = \frac{.5}{\Phi(.5)}$$

$$= \frac{.5}{.6915} \doteq .723$$

where we used Table 1 (Sec. 13-1) to evaluate the cumulative normal Φ. ■

Example 6 It is always possible to interpret "the selection of a real number x according to a continuous density $p: R \to R$" as the observation of the outcome x of one spin of a properly *continuously* calibrated random wheel. For example, the wheel of Fig. 6*a* will yield an x corresponding to the continuous density of Fig. 6*b*. We now describe a general calibration technique which yields a wheel equivalent to a given continuous density $p: R \to R$. We must assign a real number x to each point on the circumference. The outcome of a spin is the real number x if, when the wheel stops, the pointer is opposite the point assigned the number x. Assume that the wheel always has an initial uniform labeling near the outside, as in Fig. 7*a*. A calibration consists of assigning an inner labeling as shown. Thus a calibration is a real-valued function h defined on the unit interval $0 < y < 1$, and a point on the circumference having outer label y is assigned inner label $x = h(y)$. The wheel of Fig. 7*a* is calibrated by the function h of Fig. 7*b*.

Given a continuous density $p: R \to R$ as in Fig. 8*a*, we define $F: R \to R$ by

$$F(x) = \int_{-\infty}^{x} p(x')\,dx' \qquad \text{for all real } x$$

so that the value of F at x is the area under p to the left of x; that is, $F(x)$ is the probability of obtaining some x' less than x. For convenience assume that F is strictly increasing over an interval, as shown. We will show that the wheel calibrated by $h = F^{-1}$ is equivalent to p; that is, for a point with outer label y we will find from Fig. 8*a* the unique x satisfying $y = F(x)$ and use this x for the inner label. The calibration function of Fig. 7*b* was obtained in this way from Fig. 8*a*. More naturally, the graph of F in Fig. 8*a* can be taken as the calibration function, where the vertical axis is used for arguments and the horizontal axis is used for values. For this calibration we note, as in Fig. 8*b*, that for fixed $F(x_1) = y_1 < y_2 = F(x_2)$ we will obtain with probability $y_2 - y_1$ a point with outer label y satisfying $y_1 < y < y_2$, and this is equivalent to obtaining an inner label x satisfying $x_1 < x < x_2$. Thus this calibration yields an x satisfying $x_1 < x < x_2$ with probability

$$y_2 - y_1 = F(x_2) - F(x_1) = \int_{x_1}^{x_2} p(x')\,dx'$$

which is the probability assigned to $x_1 < x < x_2$ by p. Therefore the wheel calibrated in this way is equivalent to p.

Example 7 (Sec. 17-2) shows that for *every* (R,P) there is a corresponding random wheel-calibration function h. Thus every distribution on R arises once we include a uniform density and arbitrary functions.

From the point of view of performing sampling experiments as in Sec. 12-2, suppose that we want independent observations $x_1, \ldots, x_n$ according to (R,P). All we need are independent observations $y_1, \ldots, y_n$ from a uniform density on $0 < y < 1$, and we can then let $x_k = h(y_k)$, with the proper calibration function h. In particular this technique has been used, as in Exercise 13*e*, to construct tables of random normal deviates.

It is physically impossible to obtain an observation from a uniform density on $0 < y < 1$; however, using the top row of the random-number table (Sec. 12-2), we may for example, consider

.295266413992 .979279795911

to be two "independent" observations from an "approximately" uniform density on $0 < y < 1$. Somewhat more realistically, we may think of a uniform density on 10^{12} equispaced points in $0 \le y < 1$. ■

Example 7 Let Ω be the unit interval $0 \le x \le 1$ and let P correspond to the uniform density on Ω, as in Fig. 1; that is, if A is a subset of the unit interval then $P(A)$ equals the total length of A.

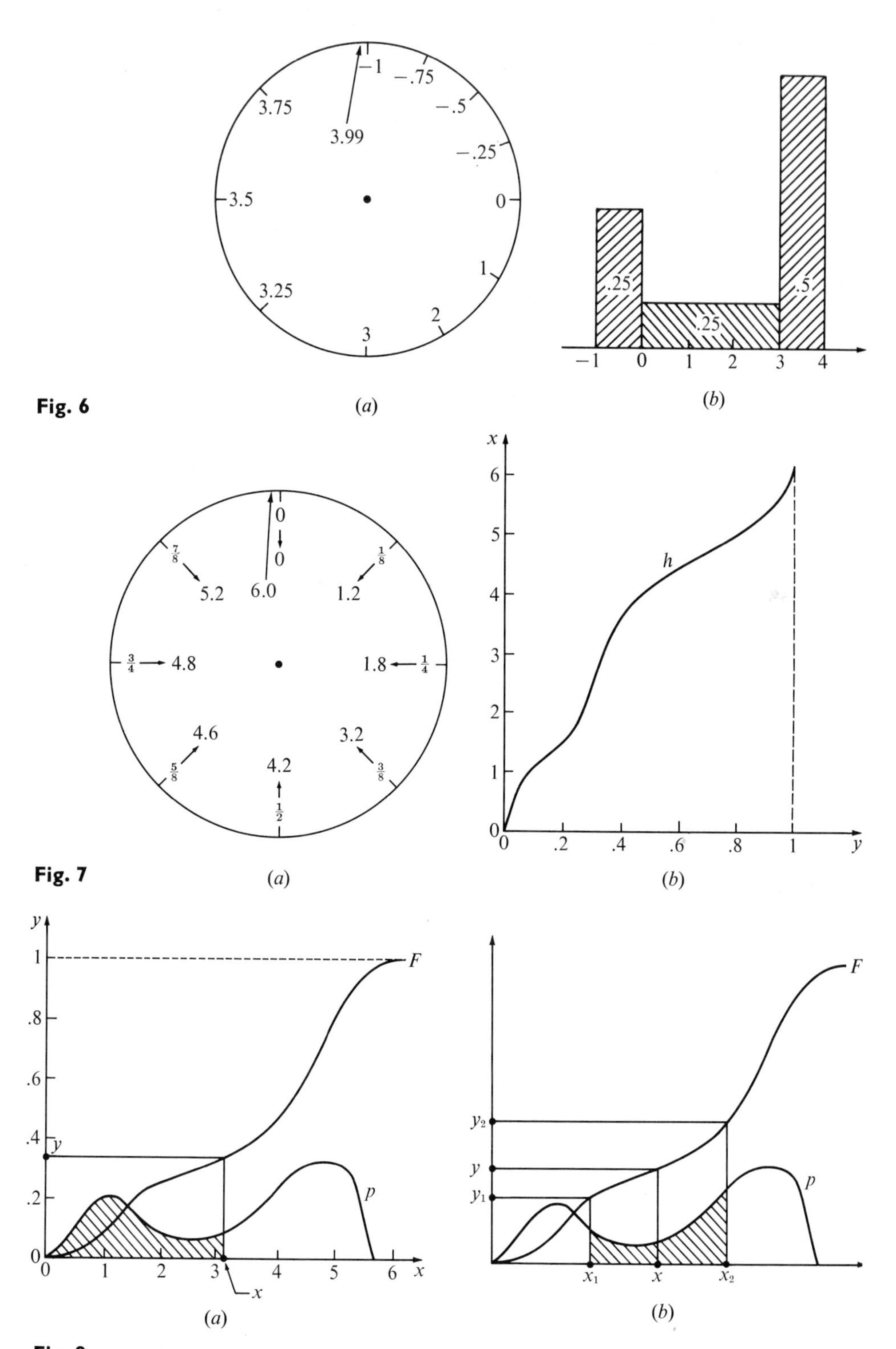

Fig. 6 (a) (b)

Fig. 7 (a) (b)

Fig. 8 (a) (b)

For each real p satisfying $0 < p < 1$ we will show how to construct a corresponding *infinite* sequence $X_1, X_2, \ldots$ of random variables $X_n: \Omega \to R$, all with this same associated (Ω, P), such that $X_1, X_2, \ldots$ are Bernoulli trials with probability p for a success on each trial; that is, $P[X_n = 1] = p$ and $P[X_n = 0] = q = 1 - p$ for all n, and $X_1, \ldots, X_n$ are independent for every n. Each X_i will have only the values 0 and 1.

Define X_1 and X_2 as in Figs. 9*a* and *b*. Suppose that $X_1, \ldots, X_n$ have been defined and that the 2^n events $[X_1 = x_1, \ldots, X_n = x_n]$ are disjoint intervals, where the x_i can each be 0 or 1. If one of these events $[X_1 = x_1, \ldots, X_n = x_n]$ is the interval $\alpha < x \leq \beta$ then on this interval define X_{n+1}, as in Fig. 9*c*, by

$$X_{n+1}(x) = \begin{cases} 0 & \text{if } \alpha < x \leq \alpha + q(\beta - \alpha) \\ 1 & \text{if } \alpha + q(\beta - \alpha) \leq x < \beta. \end{cases}$$

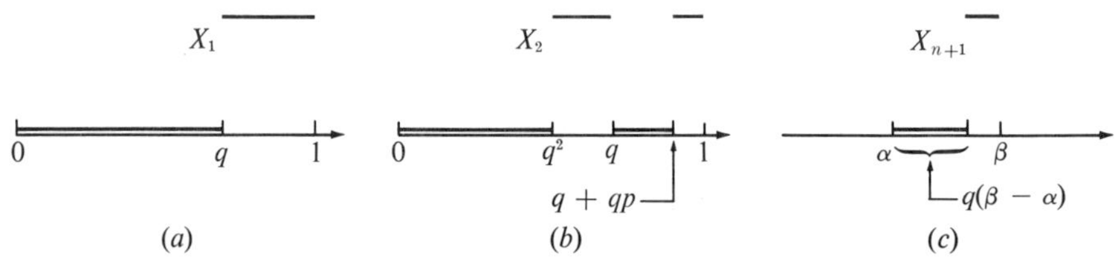

Fig. 9

Thus X_{n+1} has the value 0 on fraction q of this interval and the value 1 on the remainder. Define X_{n+1} in this way on each of the 2^n intervals. Then $P[X_{n+1} = 0] = q$, since X_{n+1} has the value 0 on fraction q of each of the 2^n intervals. Also we certainly have

$$P[X_{n+1} = x_{n+1} \mid X_n = x_n, X_{n-1} = x_{n-1}, \ldots, X_1 = x_1] = P[X_{n+1} = x_n]$$

whenever $x_1, \ldots, x_{n+1}$ are 0's and 1's. As in Exercise 17, this is enough to make $X_1, \ldots, X_{n+1}$ a total of $n + 1$ Bernoulli trials with probability p for a success. Thus for each p we have shown inductively how to define $X_1, X_2, \ldots$ having the desired properties.

If the outcome of the experiment is the real number x, where $0 \leq x \leq 1$, then $X_1(x)$, $X_2(x), \ldots$ is the particular *infinite* sequence of 1's and 0's obtained from this one performance of the experiment of spinning the wheel of Fig. 1 once. Define the event B by

$$B = \left\{x: \lim_{n\to\infty} \frac{1}{n} [X_1(x) + \cdots + X_n(x)] = p\right\}.$$

Then $P(B) = 1$ from the asymptotic form of the SLLN, as stated in Sec. 12-3; that is, with probability 1 we will obtain an x for which the relative frequency of successes converges to p.

The case $p = \frac{1}{2}$ is particularly interesting, and X_1, X_2, X_3 are shown in Fig. 10 when $p = \frac{1}{2}$. In this case if the real number x, with $0 < x \leq 1$, is expanded in binary form as

$$x = .x_1x_2x_3 \cdots = \sum_{n=1}^{\infty} \frac{x_n}{2^n} \qquad \text{where each } x_n \text{ is 0 or 1}$$

then a little thought shows that $X_1(x) = x_1, X_2(x) = x_2, \ldots$. That is, when $p = \frac{1}{2}$ the random variable X_n assigns to each x the nth digit x_n in its binary expansion $x = .x_1x_2x_3 \ldots$. The SLLN shows that if $A_{1/2}$ is defined to consist of those reals in the unit interval which have a binary expansion $x = .x_1x_2 \cdots$ having exactly one-half 1's in the limit, $\lim_{n\to\infty} (x_1 + \cdots + x_n)/n = \frac{1}{2}$, then $P(A_{1/2}) = 1$. That is, if we select a number "at random" from the unit interval

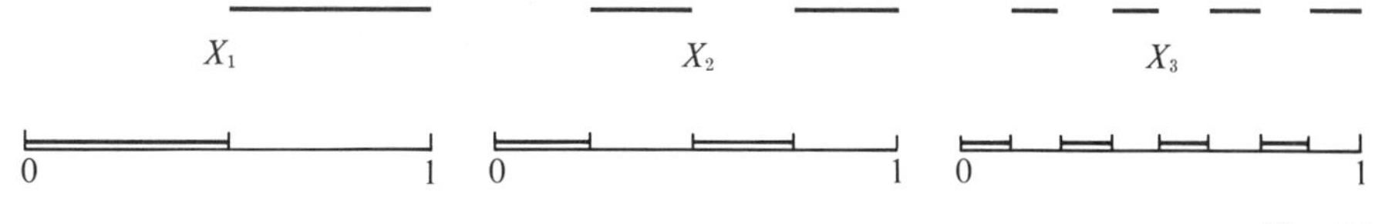

Fig. 10

then we are "certain" to get one whose binary expansion has one-half 1's in the limit. In other words, the length (Lebesgue measure) of this set $A_{1/2}$ is 1. ■

***ADDENDUM**

The functional equation $g(r + t) = g(r)g(t)$ arose in Example 4 in connection with the exponential density. This addendum shows, as stated in Lemma 2, that the only function g which can satisfy this equation is either identically 0 or

$$g(x) = [g(1)]^x = e^{[\log g(1)]x}.$$

We do not assume that g has a derivative, or even that it is continuous, but only that it is bounded on an interval. If we let $f(x) = \log g(x)$ then $f(r + t) = f(r) + f(t)$, so f satisfies the equation for a linear function $f(x) = cx$. Lemma 1 shows that only a linear function can satisfy this equation, then Lemma 1 is applied to $f(x) = \log g(x)$ to obtain Lemma 2.

Lemma 1 Let f be a real-valued function defined for all positive reals and bounded on the unit interval

$$|f(x)| \le M \qquad \text{whenever } 0 < x \le 1.$$

If f satisfies the equation

$$f(x + y) = f(x) + f(y) \qquad \text{whenever } x > 0, y > 0$$

then $f(x) = f(1)x$ for all $x > 0$. ■

Proof Clearly

$$[f(x + y) - f(1)(x + y)] = [f(x) - f(1)x] + [f(y) - f(1)y] \qquad \text{if } x > 0, y > 0.$$

Define F by $F(x) = f(x) - f(1)x$ for all $x > 0$, so that $F(1) = 0$ and F is bounded on the unit interval and satisfies the same functional equation

$$F(x + y) = F(x) + F(y) \qquad \text{if } x > 0, y > 0.$$

We must show that F is identically 0. If each $x_i > 0$ then repeated application of the functional equation gives us

$$F\left(\sum_{i=1}^{n} x_i\right) = F\left(\sum_{i=1}^{n-1} x_i\right) + F(x_n)$$
$$= F\left(\sum_{i=1}^{n-2} x_i\right) + F(x_{n-1}) + F(x_n) = \cdots = \sum_{i=1}^{n} F(x_i).$$

Letting $x_2 = x_3 = \cdots = x_n = 1$, we see that $F(x_1 + (n - 1)) = F(x_1)$ for all $x_1 > 0$ and all $n - 1 = 1, 2, \ldots$. Thus F is periodic, with period 1, as suggested by Fig. 11. Hence F is bounded on the whole positive axis,

$$|F(x)| \leq M' \qquad \text{if } x > 0.$$

Letting $x_1 = x_2 = \cdots = x_n = x$ gives us $F(nx) = nF(x)$. Therefore if $F(x_0) \neq 0$ for some $x_0 > 0$ then $|F(nx_0)| = n\,|F(x_0)| > M'$ if n is large enough, contradicting the boundedness of F. ■

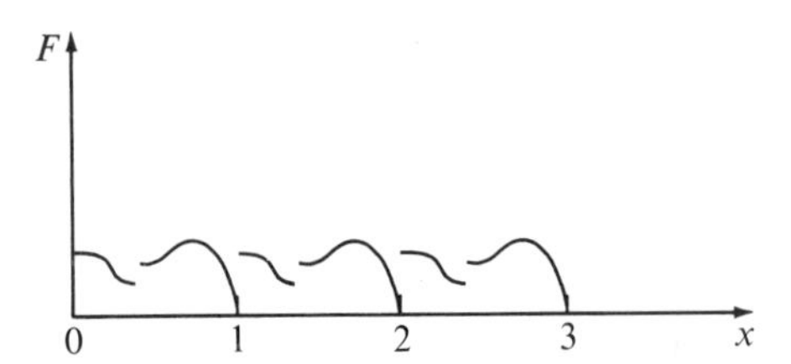

Fig. 11

We note that the boundedness assumption is satisfied if f is assumed to be defined and continuous on the closed interval $0 \leq x \leq 1$; or if $f(x) \geq 0$ whenever $0 < x < 1$, since then $f(1) = f(x) + f(1 - x) \geq f(x)$.

Lemma 2 Let g be a real-valued function defined for all positive reals and bounded on the unit interval

$$|g(x)| \leq M \qquad \text{whenever } 0 < x \leq 1.$$

If g satisfies the equation

$$g(x + y) = g(x)g(y) \qquad \text{whenever } x > 0, y > 0$$

then $g(1) \geq 0$ and $g(x) = [g(1)]^x$ for all $x > 0$. ■

Proof If we let $x = y$ we have $g(2x) = [g(x)]^2 \geq 0$ for all $x > 0$, so $g \geq 0$. If g is identically 0 the theorem is true. Otherwise there is an $x_0 > 0$ for which $g(x_0) > 0$. In this case

$$g(nx_0) = g(\overbrace{x_0 + \cdots + x_0}^{n \text{ times}}) = \overbrace{g(x_0) \cdots g(x_0)}^{n \text{ times}} = [g(x_0)]^n > 0.$$

Therefore

$$0 < g(nx_0) = g(x)g(nx_0 - x) \qquad \text{if } 0 < x < nx_0$$

so $g(x) > 0$; but n is arbitrary, so $g(x) > 0$ for all $x > 0$. Therefore we can define $f(x) = \log g(x)$ for all $x > 0$. Furthermore

$$0 < g(1) = g(x)g(1 - x) \leq g(x)M \qquad \text{if } 0 < x < 1$$

so that g is bounded away from zero on $0 < x \leq 1$. Thus

$$\log \frac{g(1)}{M} \leq \log g(x) \leq \log M \qquad \text{if } 0 < x < 1$$

so that Lemma 1 can be applied to f to obtain $\log g(x) = [\log g(1)]x$ for all $x > 0$. ■

Historical references for this addendum may be found in G. S. Young, The Linear Functional Equation, *American Mathematical Monthly*, vol. 65, pp. 37–38, 1958.

EXERCISES

1. A company guarantees machines to be trouble free for at least the first 5 weeks of service, even though the distribution of operating time to first failure is exponential with mean $1/\lambda = 10$ weeks.

(*a*) What is the probability that a machine will violate the guarantee?

(*b*) A machine is placed in service in an isolated location, and upon inspection exactly 8 weeks later it is found to have failed at some earlier time. What is the probability that it actually failed in the first 5 weeks?

2. The input to a device is a voltage X having a normal density with mean $\mu = 3$ volts and standard deviation $\sigma = 2$ volts. The device passes positive voltages without distortion. Find and sketch the conditional density of the output voltage, given that $X > 0$.

3. Rods whose length have a normal density with mean 100 and variance 4 pass inspection iff their length is between 96 and 106. Find and sketch the density of the length for those passing inspection.

4. For the density $p_{X,Y}: R^2 \to R$ of Example 1 (Sec. 16-2) find the probability of the event $\{(x,y): x < .5, y < .5\}$.

5. Find $P(C \mid B)$ for Exercise 6 (Sec. 6-4).

6. Let $p: R \to R$ be as in Fig. 12 and estimate the numbers $P(A)$, $P(B)$, $P(AB)$, $P(A \mid B)$, and $P(B \mid A)$.

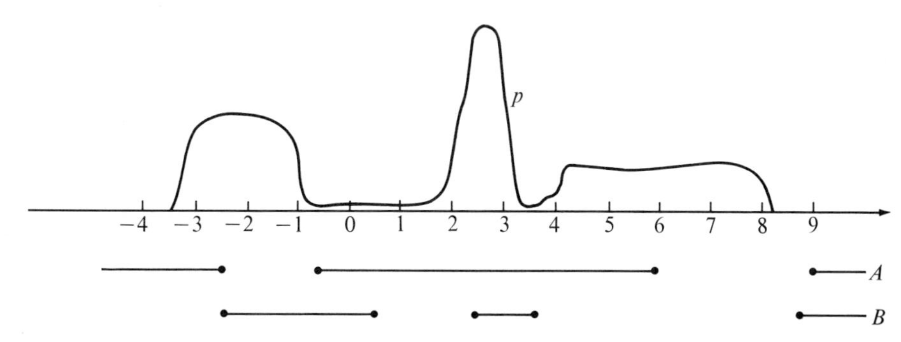

Fig. 12

7. An airplane is flying at altitude h, as in Fig. 13, and its pilot looks out a side window in a direction making angle θ with the vertical, where θ is uniformly distributed over $0 < \theta < \pi/2$. For each real $d > 0$ find and plot as a function of d the probability $P[D < d]$, where D is as shown in the figure. What is $P[D < h]$?

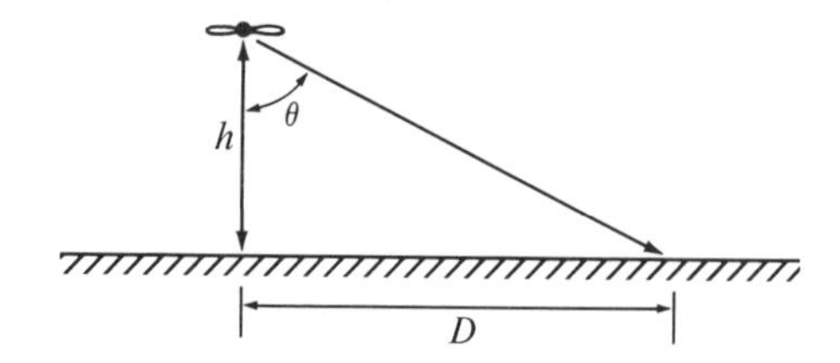

Fig. 13

8. Let $\varphi_{0,\sigma}: R \to R$ be normal with mean 0 and variance σ^2. Any odd moment is zero, since the density is an even function and x^n is an odd function for n odd. Expand the MGF in a Taylor series and deduce that if n is even then the nth moment is given by $(1)(3)(5)(7)\cdots(n-1)\sigma^n$. Thus for $n = 2, 4, 6$ we get σ^2, $3\sigma^4$, $15\sigma^6$.

9. Find the mean and variance of the lognormal density, which is defined in Example 6 (Sec. 16-2). Check your answers with Example 3 (Sec. 16-5). HINT: Make the transformation $x = \log y$ and recall the MGF of the normal density.

10. The coefficients b and c of the quadratic equation $x^2 + bx + c$ are each selected independently and uniformly over the unit interval; that is,

$$p(b,c) = \begin{cases} 1 & \text{if } 0 < b < 1, 0 < c < 1 \\ 0 & \text{otherwise.} \end{cases}$$

Find the probability that the two roots of $x^2 + bx + c$ are
(*a*) Real (*b*) Equal (*c*) Complex.

11. Solve Exercise 10 when the coefficients are independent and exponential with mean 1; that is,

$$p(b,c) = \begin{cases} e^{-b-c} & \text{if } b > 0, c > 0 \\ 0 & \text{otherwise.} \end{cases}$$

12. This exercise exhibits the lack-of-wearout property of the exponential density in a form somewhat different from that shown in Example 4. Prove that

$$p_{A_r}(t) = \begin{cases} \lambda e^{-\lambda(t-r)} & \text{if } 0 \le r \le t \\ 0 & \text{otherwise.} \end{cases}$$

is the density obtained by conditioning the exponential density on the event $A_r = \{x: x \ge r\}$. Thus p_{A_r} is essentially the same exponential density, starting at time r rather than at time 0.

13. Sketch a random wheel which has been calibrated as in Example 6 to correspond to the following:

(*a*) The normal density with mean 0 and variance 1
(*b*) The exponential density with mean 1
(*c*) The density of Fig. 12
(*d*) Use part (*b*) and the random-number table (Sec. 12-2) to obtain (in approximation) three independent observations from an exponential density with mean 1.
(*e*) The first four 5-tuples y_1, y_2, y_3, y_4 of random decimal digits from the Rand table, referred to in Table 1 (Sec. 24-1), are

.10097 .37542 .08422 .99019

where we have inserted decimal points so that we may interpret them as four independent observations from the uniform density over the interval $0 < y < 1$. Thus if x_i is defined by $y_i = \Phi(x_i)$ then x_1, x_2, x_3, x_4 may be interpreted as four independent observations from the standard normal density. Use Table 1 (Sec. 13-1) to find the x_i. Check your x_i against Table 1 (Sec. 24-1), since this is how the Rand normal deviates were actually constructed.

14. For Example 7 with $p = 1/3$ exhibit the random variables X_1, X_2, X_3 and check that $P[X_3 = 1 \mid X_2 = 0, X_1 = 1] = p$.

15. Show that probability theory cannot contain models corresponding to the following descriptions:

(*a*) Select a positive integer from among all positive integers, with each integer being equally likely. That is, show that no (R,P) can satisfy

$$P\{1\} = P\{2\} = \cdots \qquad \text{and} \qquad 1 = P\{1\} + P\{2\} + \cdots.$$

(*b*) Select a real from among all positive reals, with every two equal-length intervals being equally probable. That is, show that no (R,P) can satisfy

$$P\{x: 0 \le x < 1\} = P\{x: 1 \le x < 2\} = \cdots$$

and

$$1 = P\{x: 0 \le x < 1\} + P\{x: 1 \le x < 2\} + \cdots.$$

16. (*a*) Prove Chebychev's inequality for a continuous probability density $p: R \to R$:

$$\int_{\mu-\epsilon}^{\mu+\epsilon} p(x)\, dx \ge 1 - \frac{\sigma^2}{\epsilon^2} \qquad \text{for all } \epsilon > 0.$$

When $\epsilon = 2\sigma$ or 3σ the right-hand side is .75 or $8/9 = .888 \cdots$. Evaluate the left-hand side for these two values of ϵ when p is as follows:

(*b*) Normal μ and σ^2

(*c*) Uniform over the unit interval

(*d*) Exponential with mean $\mu = 1/\lambda$

17. Prove that $n \ge 2$ random variables $X_1, X_2, \ldots, X_n$ having a joint *discrete* probability density are independent iff for each k satisfying $2 \le k \le n$ we have

$$P[X_k = x_k \mid X_{k-1} = x_{k-1}, X_{k-2} = x_{k-2}, \ldots, X_1 = x_1] = P[X_k = x_k]$$

for all $x_1, x_2, \ldots, x_k$ for which

$$P[X_{k-1} = x_{k-1}, X_{k-2} = x_{k-2}, \ldots, X_1 = x_1] > 0.$$

Do not ignore points $x_1, x_2, \ldots, x_n$ for which $P[X_n = x_n, \ldots, X_1 = x_1] = 0$. Thus $X_1, X_2, \ldots, X_n$ are jointly independent if each is independent of all earlier ones. This method of proving that random variables are independent is sometimes useful, as in Example 7.

16-2 THE DENSITY OF $Y = g(X)$

This section is devoted primarily to finding the continuous density induced by a random variable $Y = g(X)$ which is a function of a random variable X having a continuous density p_X. We develop and apply a simple technique which, for many examples of practical interest, yields p_Y in terms of p_X and dg/dx. Before turning to this topic we first obtain in Theorem 1 the obvious analog of the discrete-case formula $p_X(x) = \sum_y p_{X,Y}(x,y)$ and apply it in Example 1.

Theorem 1 (Sec. 15-2) showed that every random vector $f: \Omega \to R^n$ with associated probability space (Ω,P) induces a probability space (R^n,P_f), where P_f is defined by

$$P_f(D) = P[f \in D] \qquad \text{for all subsets } D \text{ of } R^n.$$

It *sometimes* happens that P_f corresponds to a continuous probability density $p_f\colon R^n \to R$,

$$P_f(D) = \int \cdots \int_D p_f(x_1, \ldots, x_n)\, dx_1 \cdots dx_n \qquad \text{for every subset } D \text{ of } R^n.$$

In this case we call p_f the **continuous probability density induced by** $f\colon \Omega \to R^n$ with associated (Ω,P). This will be the case in all examples in this and the next section, except for Example 2 below.

Naturally if random vectors $f_1, \ldots, f_k$ into the same R^n all induce continuous densities they are *identically distributed* iff $p_{f_1} \equiv p_{f_2} \equiv \cdots \equiv p_{f_k}$.

If $f = (f_1, \ldots, f_k)$ induces a continuous density p_f, where each f_i is a random vector, we may write p_f as $p_{f_1,\ldots,f_k}$ and call it the **joint density of** $f_1, \ldots, f_k$.

In the discrete case p_f could be obtained from a larger joint density by summing out the extra coordinates. The next theorem shows that if f and g have a joint continuous density $p_{f,g}$ then f has a continuous density obtainable by integrating out the coordinates corresponding to g in $p_{f,g}$. For notational convenience in the statement of the theorem, the earlier coordinates correspond to f and the later coordinates correspond to g.

Theorem 1 Let $f\colon \Omega \to R^n$ and $g\colon \Omega \to R^m$ be random vectors with the same associated probability space (Ω,P). If the random vector $(f,g)\colon \Omega \to R^{n+m}$ induces a continuous density $p_{f,g}\colon R^{n+m} \to R$ then f induces a continuous density obtainable by integrating out the extra coordinates in $p_{f,g}$:

$$p_f(x_1, \ldots, x_n) = \int_{-\infty}^{\infty} \cdots \int_{-\infty}^{\infty} p_{f,g}(x_1, \ldots, x_n, x_{n+1}, \ldots, x_{n+m})\, dx_{n+1}\, dx_{n+2} \cdots dx_{n+m}. \qquad \blacksquare$$

Proof For notational simplicity assume that $n = m = 1$. From Theorem 3 (Sec. 15-2) we have

$$P_f(D) = P_{f,g}(D \times R) = \iint_{D\times R} p_{f,g}(x_1,x_2)\, dx_1\, dx_2 = \int_D \left[\int_{-\infty}^{\infty} p_{f,g}(x_1,x_2)\, dx_2\right] dx_1$$

for every set D of reals. This equation shows that P_f corresponds to the density [ ]. $\blacksquare$

Example 1 A randomly selected sheet of metal has length X' inches and width Y', where x_0 and $x_0 + 1$ are the minimum and maximum possible lengths. Thus $X = X' - x_0$ and $Y = Y' - y_0$ are the excesses over minimum. Assume that they have joint continuous density

$$p_{X,Y}(x,y) = \begin{cases} 2[(1-x)(1-y) + xy] & \text{if } 0 < x < 1, 0 < y < 1 \\ 0 & \text{otherwise.} \end{cases}$$

As can be seen from Fig. 1, X and Y tend to be small (or large) together.

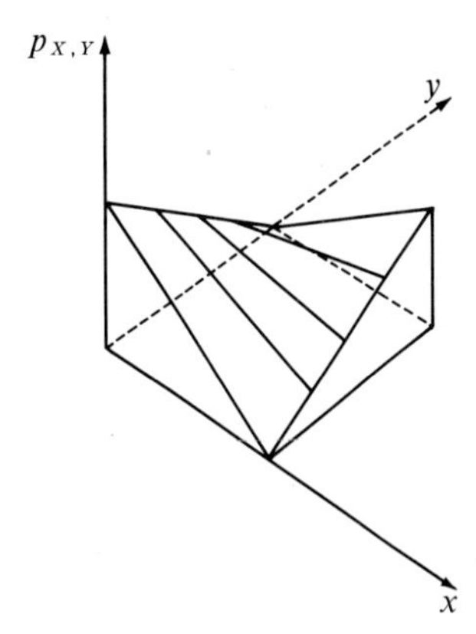

Fig. 1

From Theorem 1 if $0 < x < 1$ then

$$p_X(x) = \int_{-\infty}^{\infty} p_{X,Y}(x,y)\,dy = 2\int_0^1 (1 - x - y + 2xy)\,dy$$

$$= 2\left[y - xy - \frac{y^2}{2} + xy^2\right]_{y=0}^{y=1} = 1$$

so that X is uniformly distributed over the unit interval. From symmetry, or direct calculation, so is Y. ■

Exercise 1 may be done before reading the remainder of this section.

Example 2 As in Fig. 2, let X have an exponential density with mean $1/\lambda$, where we interpret X as the length of service, in days, of a machine prior to its first breakdown. Let Y be the whole number of days prior to breakdown. Thus $Y = g(X)$, where $g: R \to R$ is the step function shown in Fig. 2. The only values that g can have are 0, 1, 2, . . . , so that Y must induce a discrete density. We have here an example where *$g: R \to R$ with associated continuous density $p_X: R \to R$ induces a discrete density $p_g: R \to R$.*

If $k = 0, 1, 2, \ldots$ then

$$P[Y = k] = P[k \le X < k + 1] = \int_k^{k+1} \lambda e^{-\lambda x}\,dx = [-e^{-\lambda x}]_{x=k}^{x=k+1}$$

$$= e^{-\lambda k} - e^{-\lambda(k+1)} = (1 - e^{-\lambda})(e^{-\lambda})^k.$$

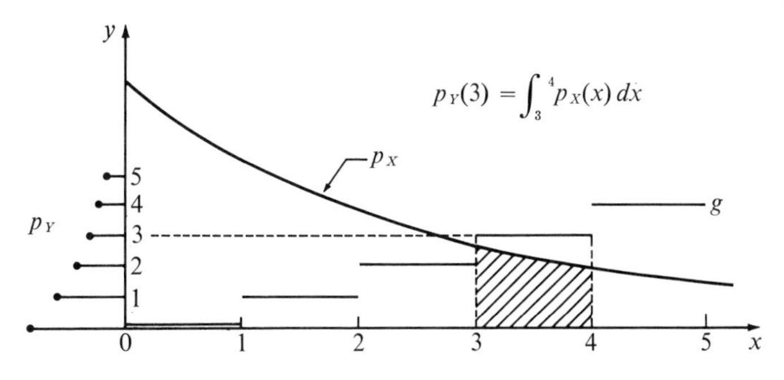

Fig. 2

Let $p = 1 - e^{-\lambda}$ so $P[Y = k] = p(1 - p)^k$, and hence Y has a geometric density with parameter p.

In this example X might be the voltage into a quantizer defined by g, so that $Y = g(X)$ is the output voltage. Or Y might be obtained from X by "rounding off." ■

The remainder of this section is devoted to finding the distribution of a real-valued function of a random variable. Assume that we have a random variable $X: \Omega \to R$ with associated probability space (Ω, P), and assume that X has an induced continuous density p_X. We also have a function $g: R \to R$ and define the random variable $Y = g(X)$ by

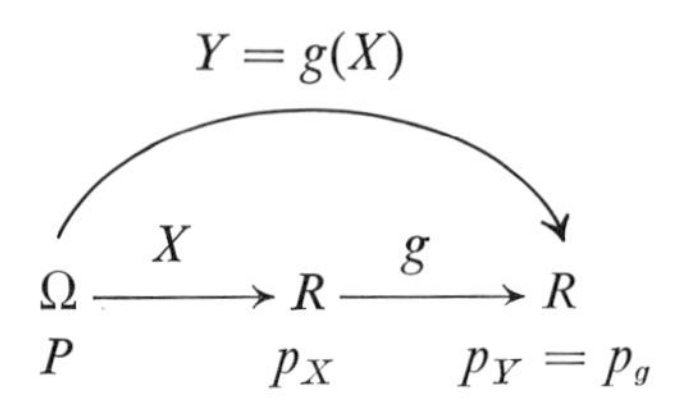

From Theorems 1 and 2 (Sec. 15-2) there are always induced probability measures P_Y and P_g, which are equal, $P_Y = P_g$. In this section we are interested in the case where these correspond to a continuous density $p_Y \equiv p_g$. Example 2 was intended to help prevent the false impression that $Y = g(X)$ has a continuous density whenever X does. We do not give general conditions under which P_Y will correspond to a continuous density p_Y. Instead we develop a technique for finding the continuous density p_Y in "reasonable cases" where there is such a p_Y, and then we apply this technique to problems where it quickly becomes obvious that the function p_Y so obtained is actually a continuous density which corresponds to P_Y. Assume that our basic data consist of g and p_X and we wish to find $p_Y = p_g$ from g and p_X.

THE DENSITY OF $Y = g(X)$ IN SOME CASES WHEN g IS ONE-TO-ONE

We first consider the case where g has a continuous positive derivative everywhere $g' > 0$ and so is strictly increasing, or g has a continuous negative derivative everywhere $g' < 0$ and so is strictly decreasing. The technique for finding p_Y is shown in Fig. 3. We obtain a y' in the interval $y \leq y' \leq y + \Delta y$ iff we obtain an x' in the interval

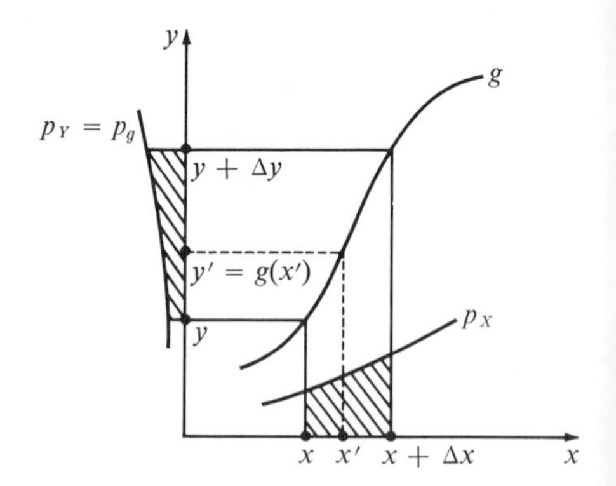

Fig. 3

$x \le x' \le x + \Delta x$. Thus p_Y must satisfy

$$\int_y^{y+\Delta y} p_Y(y')\, dy' = \int_x^{x+\Delta x} p_X(x')\, dx'$$

since the probability of this event must be the same whether it is calculated from p_X or from p_Y. The graphic construction of Fig. 3 is intuitively obvious and can be used for rough approximations and analytically intractable cases. If Δx is small enough and p_X is continuous at the points of interest then

$$p_Y(y)\, \Delta y \doteq p_X(x)\, \Delta x$$

where y is related to x by $y = g(x)$. As Δx goes to zero the ratio $\Delta y/\Delta x$ converges to $g'(x)$ so

$$p_Y(y) = \frac{1}{dg(x)/dx} p_X(x) \qquad \text{where } y = g(x).$$

Thus to find the value $p_Y(y)$ of p_Y at a particular point y we first find the unique x for which $y = g(x)$ and then evaluate $[1/g'(x)]p_X(x)$.

If g is decreasing, as in Fig. 4, and $\Delta x > 0$ then $\Delta y < 0$ and we want p_Y to satisfy

$$p_Y(y)\,(-\Delta y) \doteq \int_{y+\Delta y}^{y} p_Y(y')\, dy' = \int_x^{x+\Delta x} p_X(x')\, dx' \doteq p_X(x)\, \Delta x$$

so that

$$p_Y(y) = \frac{1}{-dg(x)/dx} p_X(x) \qquad \text{where } y = g(x).$$

In this case $g' < 0$ so $-g' = |g'|$.

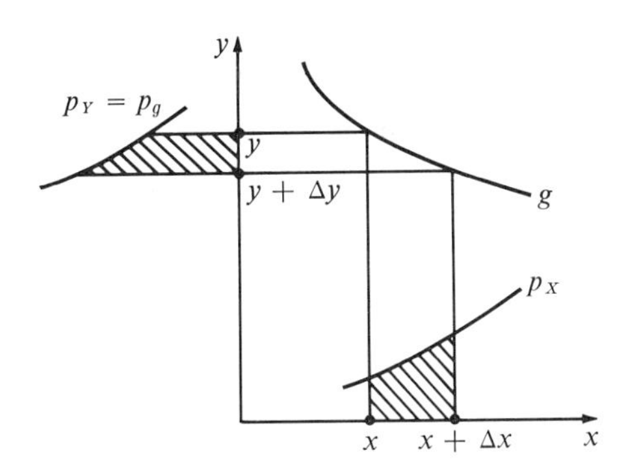

Fig. 4

Thus if a random variable X has a continuous density p_X and if $Y = g(X)$ where $g: R \to R$ has a continuous positive, or negative, derivative everywhere, then

$$p_Y(y) = \frac{1}{\left|\dfrac{dg(x)}{dx}\right|} p_X(x) \qquad \text{where } y = g(x),$$

holds for those y belonging to the range of g. *This is our basic formula under these conditions.* This formula is also derived in Example 5 (Sec. 17-2).

The range $\mathscr{R}$ of g consists of all the values which g can take on. Thus $P[g(X) \in \mathscr{R}] = 1$, and hence $p_Y(y) = 0$ if y does not belong to $\mathscr{R}$. As shown in Fig. 5, the range of g can have one of only four forms.

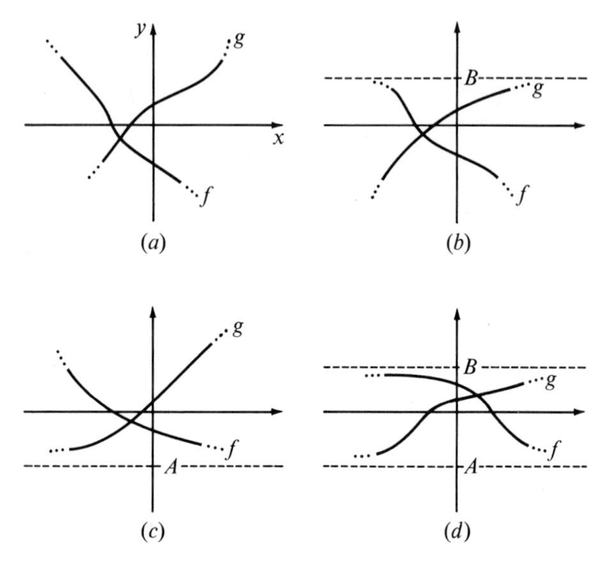

Fig. 5 $\mathscr{R}$ = range of f = range of g, where $g' > 0$ and $f' < 0$; (a) $\mathscr{R} = R$; (b) $\mathscr{R} = \{y: y < B\}$; (c) $\mathscr{R} = \{y: y > A\}$; (d) $\mathscr{R} = \{y: A < y < B\}$.

In Examples 3 to 7 we apply the basic formula just derived. Then we extend it to somewhat more general functions g and apply this extension in Examples 8 and 9.

Example 3 Let $g(x) = ax + b$ for all real x, where a and b are reals and $a \neq 0$, so that $g' = a$. If $Y = g(X) = aX + b$ then

$$p_Y(y) = \frac{1}{|a|} p_X(x) \qquad \text{where } y = ax + b.$$

We can replace y in $p_Y(y)$ by $ax + b$, or we can replace x in $p_X(x)$ by $(y - b)/a$. *Thus* if X has a continuous density and if $a \neq 0$ then $Y = aX + b$ has continuous density

$$p_Y(ax + b) = \frac{1}{|a|} p_X(x) \qquad \text{for all real } x$$

or equivalently

$$p_Y(y) = \frac{1}{|a|} p_X\left(\frac{y - b}{a}\right) \qquad \text{for all real } y.$$

Several examples are depicted in Fig. 6. The factor $1/|a|$ arises because the continuous case is based on areas under p_Y. In contrast to the discrete case, the value $p_Y(y)$ of the density p_Y at the point y does *not* equal the value $p_X(x) > 0$ of p_X at that $x = (y - b)/a$ which is carried into y, unless $|a| = 1$.

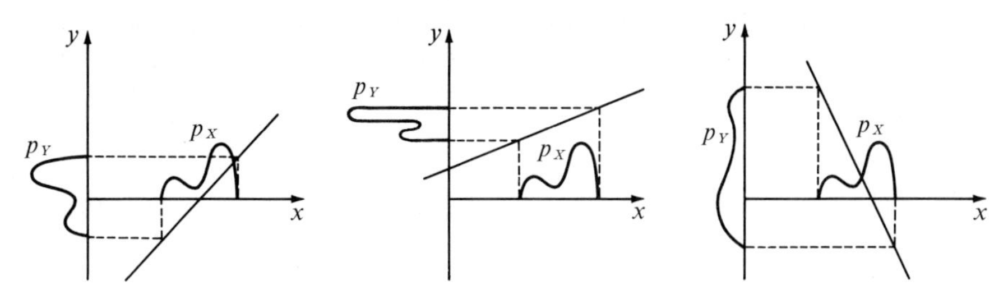

Fig. 6 $Y = aX + b.$

We now find the mean and variance of p_Y when $Y = aX + b$, $a > 0$, in terms of those of p_X. Making the transformation of variables $y = ax + b$ and $dy = a\,dx$ in the integrals, we obtain

$$\mu_Y = \int_{-\infty}^{\infty} y p_Y(y)\,dy = \int_{-\infty}^{\infty} y p_X\left(\frac{y-b}{a}\right)\frac{dy}{a} = \int_{-\infty}^{\infty} (ax+b)p_X(x)\,dx = a\mu_X + b,$$

$$\sigma_Y^2 = \int_{-\infty}^{\infty} (y-\mu_Y)^2 p_Y(y)\,dy = \int_{-\infty}^{\infty} (y-\mu_Y)^2 p_X\left(\frac{y-b}{a}\right)\frac{dy}{a}$$

$$= \int_{-\infty}^{\infty} [(ax+b) - (a\mu_X + b)]^2 p_X(x)\,dx = a^2\sigma_X^2.$$

In an exercise these same formulas are naturally obtained when $a < 0$. Thus if X has discrete or continuous density p_X and $Y = aX + b$ with $a \neq 0$ then

$$\mu_Y = a\mu_X + b \qquad \sigma_Y^2 = a^2\sigma_X^2 \qquad \sigma_Y = |a|\,\sigma_X.$$

These formulas actually hold in all cases in probability theory. The random variable $X^* = (X - \mu_X)/\sigma_X$ is **normalized** in the sense that $\mu_{X^*} = 0$ and $\sigma_{X^*} = 1$.

If $y = aX$ with $a > 0$ then

$$p_Y(y) = \frac{1}{a}\,p_X\left(\frac{y}{a}\right) \qquad \text{for all real } y.$$

In particular if X is exponential with mean $\mu_X = 1/\lambda > 0$ and $Y = aX$ with $a > 0$ then

$$p_Y(y) = \begin{cases} \dfrac{\lambda}{a}\,e^{-(\lambda/a)y} & \text{if } y > 0 \\ 0 & \text{otherwise} \end{cases}$$

so p_Y is exponential with mean $a\mu_X$.

We now show that if X is normal with mean μ and variance σ^2 then $Y = aX + b$, $a \neq 0$, is normal with mean $a\mu + b$ and variance $a^2\sigma^2$:

$$p_Y(y) = \frac{1}{|a|}\,p_X\left(\frac{y-b}{a}\right) = \frac{1}{|a|}\,\frac{1}{\sigma\sqrt{2\pi}}\exp\left\{\frac{-1}{2\sigma^2}\left[\frac{y-b}{a} - \mu\right]^2\right\}$$

$$= \frac{1}{|a|\,\sigma\sqrt{2\pi}}\exp\left\{-\frac{1}{2a^2\sigma^2}\,[y - (a\mu + b)]^2\right\}$$

for all real y. Therefore the class of normal distributions is closed under nonsingular, $a \neq 0$, linear transformations. The linear transformations are, of course, applied to the random variables. Exercise 9 shows that *any* symmetric density naturally generates such a class of densities. ■

Example 4 If $n = 3, 5, \ldots$ is an odd integer and $g(x) = x^n$ then g is strictly increasing everywhere and $g'(x) = nx^{n-1} > 0$ if $x \neq 0$. [Here $g'(0) = 0$, but this single point causes no difficulty and $p_Y(0)$ may be left undefined. It is often possible to extend our basic formula to examples where the initial hypotheses are not quite satisfied.] Therefore if $y \neq 0$ then

$$p_Y(y) = \frac{1}{nx^{n-1}} p_X(x) \qquad \text{where } y = x^n.$$

We may solve for the unique $x = y^{1/n}$ as a function of y. *Thus* if X has a continuous density and n is some odd integer $3, 5, \ldots$ then $Y = X^n$ has continuous density

$$p_Y(y) = \frac{1}{n(y^{1/n})^{n-1}} p_X(y^{1/n}) \qquad \text{for all real } y \neq 0. \quad ■$$

Exercises 1 to 4a, 9, and 10 may be done before reading the remainder of this section.

Example 5 If $g(x) = e^{ax+b}$ with $a \neq 0$ then the range of g is the positive axis. If $a > 0$ then $g'(x) = ae^{ax+b} > 0$, and if $a < 0$ then $g'(x) < 0$. Therefore

$$p_Y(y) = \frac{1}{|a|e^{ax+b}} p_X(x) \qquad \text{where } y = e^{ax+b}.$$

Thus if X has a continuous density and $a \neq 0$ then $Y = e^{aX+b}$ has continuous density

$$p_Y(y) = \frac{1}{|a|\, y} p_X\left(\frac{(\log y) - b}{a}\right) \qquad \text{for all } y > 0,$$

and $p_Y(y) = 0$ if $y \leq 0$. In particular if $Y = e^X$ then

$$p_Y(y) = \frac{1}{y} p_X(\log y) \qquad \text{for all } y > 0.$$

If p_X is uniform over the unit interval $0 < x < 1$ then $p_X(\log y) = 1$ if $0 < \log y < 1$, that is, if $1 < y < e$. Therefore if an experiment is started when a plant is at unit height, and the plant grows to height $Y = e^X$ if it is exposed to X units of sunshine, and X is uniformly distributed over the unit interval $0 < x < 1$, then

$$p_Y(y) = \begin{cases} \dfrac{1}{y} & \text{if } 1 < y < e \\ 0 & \text{otherwise.} \end{cases} \quad ■$$

Example 6 From the preceding example, if X is normal with mean μ and variance σ^2 then $Y = e^X$ has continuous density

$$p_Y(y) = \begin{cases} \dfrac{1}{y\sigma\sqrt{2\pi}} \exp\left\{-\dfrac{1}{2\sigma^2}[(\log y) - \mu]^2\right\} & \text{if } y > 0 \\ 0 & \text{if } y \leq 0. \end{cases}$$

This density is called the **lognormal density** with parameters μ and σ^2. Several of this rich class of densities are shown in Fig. 7. Their means, etc., are found in Example 3 (Sec. 16-5). It is traditional and convenient to parameterize p_Y by the mean and variance of X rather

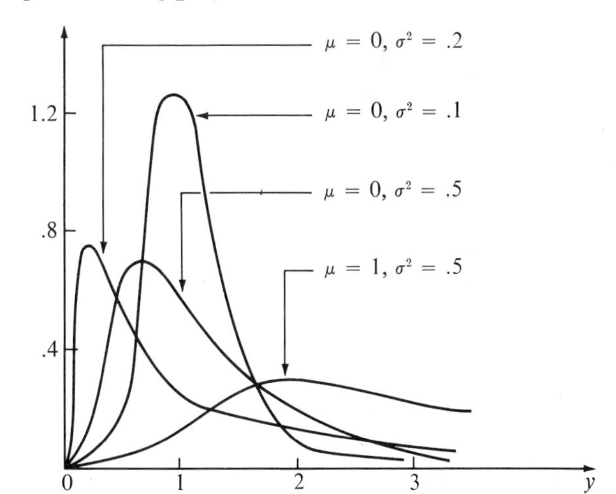

Fig. 7 Lognormal densities p_Y with parameters μ and σ^2. Thus $Y = e^X$, where X is normal μ and σ^2.

than by moments of p_Y. Thus Y has a lognormal density with parameters μ and σ^2 iff it has the density induced by e^X, where X is normal μ and σ^2. That is, "Y has a lognormal density" means that "log Y has a normal density."

If Y is lognormal with parameters μ and σ^2, and $0 < a < b$ then

$$P[a \le Y \le b] = P[a \le e^X \le b] = P[\log a \le X \le \log b]$$

$$= P\left[\frac{(\log a) - \mu}{\sigma} \le \frac{X - \mu}{\sigma} \le \frac{(\log b) - \mu}{\sigma}\right]$$

$$= \Phi\left(\frac{(\log b) - \mu}{\sigma}\right) - \Phi\left(\frac{(\log a) - \mu}{\sigma}\right).$$

Thus the probability that Y belongs to an interval can be obtained from Table 1 (Sec. 13-1) for the cumulative normal Φ.

We note prematurely that if $Y = X_1 X_2 \cdots X_n$ is the product of n independent random variables satisfying $P[X_i > 0] = 1$, then $Y = e^X$, where $X = \log X_1 + \log X_2 + \cdots + \log X_n$. From the CLT we know that X will often be approximately normally distributed, in which case Y will be approximately lognormal. Therefore the lognormal distribution is related to products in a fashion analogous to that in which the normal distribution is related to sums. Thus it is not surprising that the lognormal distribution is used often, as, for example[1] in the distribution of small-particle size, income size, or human bodyweights. ∎

Example 7 We seek the distribution of $Z = cY^a$, where $c > 0$, $a \ne 0$, and p_Y is lognormal with parameters μ and σ^2. Now $Y = e^X$, where X is normal μ and σ^2, so $Z = c(e^X)^a = e^{aX + \log c}$. But $aX + \log c$ is normal with parameters $a\mu + \log c$ and $a^2\sigma^2$. *Thus* Z *is lognormal* $a\mu + \log c$ and $a^2\sigma^2$.

[1] J. Aitchison and J. A. C. Brown, "The Lognormal Distribution," Cambridge University Press, London, 1963.

We have shown that the distribution and parameters of Z are essentially immediate from inspection of the equation

$$Z = cY^a = c(e^X)^a = e^{aX+\log c}$$

and hence can be rederived mentally. The simple relationship between the parameters of cY^a and the parameters μ and σ^2 of Y indicates the utility of these parameters rather than others, such as moments of p_Y.

The distribution of the diameters of an enormous collection of spherical particles is approximately lognormal; that is, p_Y is lognormal μ and σ^2, where $p_Y(y)\,dy$ is the fraction of the number of particles with diameters y' satisfying $y \le y' \le y + dy$. Thus if we select a particle at random (equal chance for each, regardless of size) then the distribution of its diameter Y can be "satisfactorily" approximated by a lognormal density. If c_2 and c_3 are appropriate positive constants then $Z_2 = c_2Y^2$ is the surface area and $Z_3 = c_3Y^3$ is the weight of the particle. The distribution p_{Z_2} of the surface area is then also lognormal $2\mu + \log c_2$ and $(2\sigma)^2$, and the distribution p_{Z_3} of numbers of particles according to weight is also lognormal $3\mu + \log c_3$ and $(3\sigma)^3$.

In Exercise 7 the density of $Z = cY^a$ is derived "directly" from the function $g(y) = cy^a$ with associated density p_Y. The direct derivation is more involved than the derivation above, and it provides little aid in comprehending, or predicting, the basic fact that *the class of lognormal densities is closed under multiplication by positive numbers and under non-zero powers*, including reciprocals $a = -1$. More precisely, if Y has a lognormal density then so does cY^a whenever $c > 0$ and $a \ne 0$. We now show that the theorem on intermediate distributions justifies the breezy derivation in the first paragraph of this example.

Let $c > 0$ and $a \ne 0$ and consider the diagram,

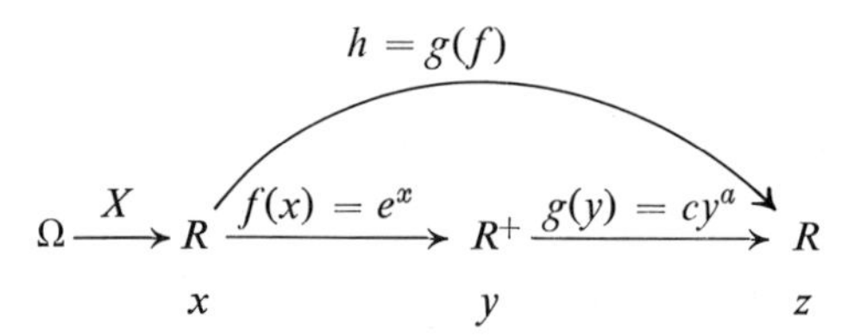

where R^+ is the positive half-axis $y > 0$, since g is defined only there. (We could equivalently replace R^+ by R and define g arbitrarily on $y \le 0$.) Clearly

$$h(x) = g(f(x)) = c(f(x))^a = c(e^x)^a = e^{ax+\log c}.$$

Assume that p_X is normal μ and σ^2 and let $Y = f(X) = e^X$. By definition p_Y is lognormal μ and σ^2. Also, if $Z = h(X) = e^{aX+\log c}$ then $aX + \log c$ is normal $a\mu + \log c$ and $a^2\sigma^2$, so that, again by the definition of the lognormal, we know that p_Z is lognormal $a\mu + \log c$ and $a^2\sigma^2$. From Theorem 2 (Sec. 15-2) we know that $p_g = p_Z$, where p_g is induced by g with associated p_Y, and this is the desired conclusion. Thus we were given g and p_Y, and in essence we introduced an experiment with a normal random variable X, since p_Y is most easily handled as the induced density of $Y = e^X$. ∎

All exercises except 4b and 5c may be done before reading the remainder of this section.

THE DENSITY OF $Y = g(X)$ IN SOME CASES WHEN g IS NOT ONE-TO-ONE

Assume that we have p_X: $R \to R$ and g: $R \to R$, as in Fig. 8, and let $Y = g(X)$. We now show how $p_Y(y)$ can be found at a y for which finitely many x satisfy $y = g(x)$.

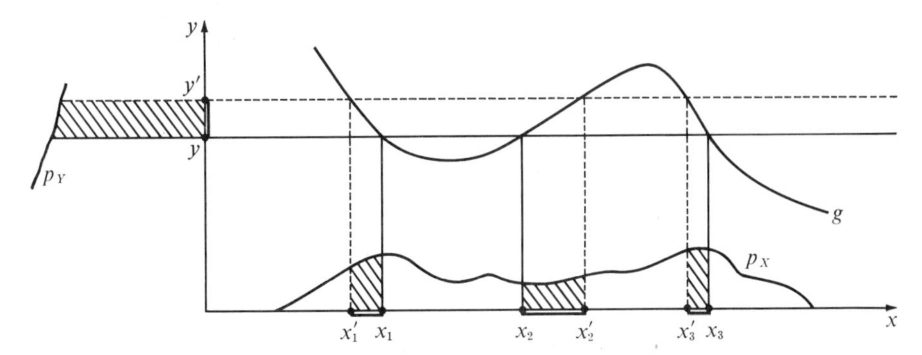

Fig. 8

Note that for Fig. 8 a y might yield one or no solutions to $y = g(x)$. For the y shown $y = g(x_1) = g(x_2) = g(x_3)$. Select a y' greater than but close to y, and find the three points satisfying $y' = g(x_1') = g(x_2') = g(x_3')$. We obtain a y'' in the interval $y \le y'' \le y'$ iff we obtain an x'' in one of the three x intervals shown, so p_Y must satisfy

$$\int_y^{y'} p_Y(y)\,dy = \int_{x_1'}^{x_1} p_X(x)\,dx + \int_{x_2}^{x_2'} p_X(x)\,dx + \int_{x_3'}^{x_3} p_X(x)\,dx.$$

In other words, the shaded area shown under p_Y must equal the sum of the three shaded areas shown under p_X. The shaded area under p_Y is approximately $p_Y(y)(y' - y)$ and the ith shaded area under p_X is approximately $p_X(x_i)\,|x_i' - x_i|$. We need the absolute value $|x_i' - x_i|$, since this equals the length of the interval, regardless of whether $x_i < x_i'$ or $x_i' < x_i$. Thus

$$p_Y(y) \doteq \frac{1}{\left|\dfrac{y' - y}{x_1' - x_1}\right|} p_X(x_1) + \frac{1}{\left|\dfrac{y' - y}{x_2' - x_2}\right|} p_X(x_2) + \frac{1}{\left|\dfrac{y' - y}{x_3' - x_3}\right|} p_X(x_3)$$

and letting $y' \to y$, we obtain

$$p_Y(y) = \frac{1}{\left|\left[\dfrac{dg}{dx}\right]_{x_1}\right|} p_X(x_1) + \frac{1}{\left|\left[\dfrac{dg}{dx}\right]_{x_2}\right|} p_X(x_2) + \frac{1}{\left|\left[\dfrac{dg}{dx}\right]_{x_3}\right|} p_X(x_3)$$

Particular points where g' equals zero, or does not exist, can often be handled separately.

Thus if a random variable X has a continuous density p_X, and $Y = g(X)$ has continuous density p_Y, where $g\colon R \to R$ and dg/dx are "reasonably continuous," and if for a particular y there are a finite number $N_y \ge 1$ of solutions $y = g(x_1) = g(x_2) = \cdots = g(x_{N_y})$ to the equation $y = g(x)$, and if each $g'(x_i) \ne 0$, then

$$p_Y(y) = \sum_{i=1}^{N_y} \frac{1}{\left|\left[\dfrac{dg}{dx}\right]_{x=x_i}\right|} p_X(x_i).$$

Example 8 If $g(x) = x^n$, where $n = 2, 4, 6, \ldots$ is an even integer, then $g'(x) = nx^{n-1}$ and the range of g is the set of $y \geq 0$. As shown in Fig. 9, for any $y > 0$ there are two solutions to

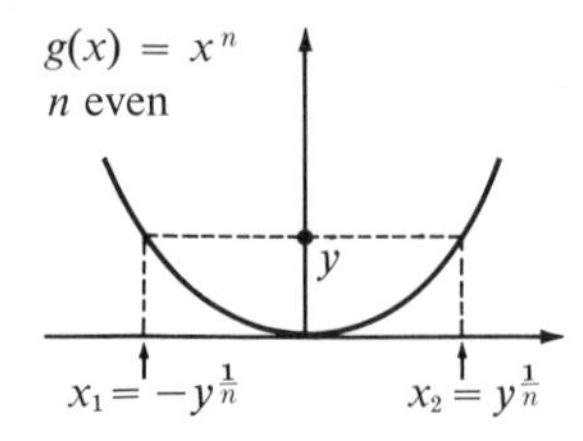

Fig. 9

$y = g(x) = x^n$, and they are $x_1 = -y^{1/n}$ and $x_2 = y^{1/n}$ so

$$p_Y(y) = \frac{1}{|nx_1^{n-1}|} p_X(x_1) + \frac{1}{|nx_2^{n-1}|} p_X(x_2).$$

Thus if X has a continuous density and n is some even integer $2, 4, \ldots$ then $Y = X^n$ has continuous density

$$p_Y(y) = \begin{cases} \dfrac{1}{ny^{1-1/n}} [p_X(-y^{1/n}) + p_X(y^{1/n})] & \text{if } y > 0 \\ 0 & \text{if } y \leq 0. \end{cases}$$

The most common application of this formula is when $n = 2$. Thus if X has a continuous density and $Y = X^2$ then

$$p_Y(y) = \begin{cases} \dfrac{1}{2\sqrt{y}} [p_X(-\sqrt{y}) + p_X(\sqrt{y})] & \text{if } y > 0 \\ 0 & \text{if } y \leq 0. \end{cases}$$

In particular if X is normal with mean 0 and variance σ^2 and $Y = X^2$ then

$$p_Y(y) = \begin{cases} \dfrac{1}{\sigma\sqrt{2\pi y}} e^{-y/2\sigma^2} & \text{if } y > 0 \\ 0 & \text{if } y \leq 0. \end{cases}$$

Naturally Y can be interpreted as squared error. Also, if at a particular time a random noise voltage X is observed across a 1-ohm resistor then X^2 is the instantaneous power flowing into the resistor at that time. ■

Example 9 If $g(x) = |x|$ then, as in Fig. 10, for every $y > 0$ there are two solutions to $y = |x|$, and they are $x_1 = -y$ and $x_2 = y$. Also, $|g'(x)| = 1$ if $x \neq 0$. *Thus* if X has a continuous

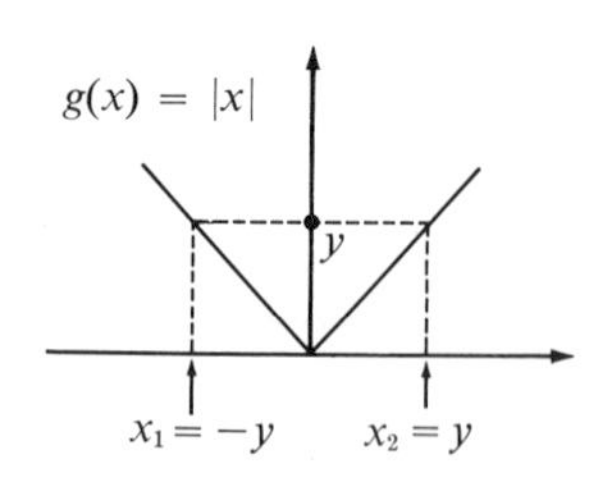

Fig. 10

density then $Y = |X|$ has continuous density

$$p_Y(y) = \begin{cases} p_X(-y) + p_X(y) & \text{if } y > 0 \\ 0 & \text{if } y \leq 0. \end{cases}$$

In particular if X is normal with mean 0 and variance σ^2 and $Y = |X|$ then

$$p_Y(y) = \begin{cases} \dfrac{2}{\sigma\sqrt{2\pi}} e^{-y^2/2\sigma^2} & \text{if } y > 0 \\ 0 & \text{if } y \leq 0. \end{cases}$$

For example, if X is the landing point on a road of parachutist whose goal was $[X = 0]$ then $Y = |X|$ is the distance from his landing point to his goal. ■

EXERCISES

1. The bivariate normal density $p_{Y_1,Y_2}: R^2 \to R$ is exhibited in Example 1 (Sec. 18-2), where it is shown that Y_1 and Y_2 are independent iff $\rho = 0$. Use Theorem 1 to show that Y_1 is normal with mean μ_1 and variance σ_1^2. From symmetry it is then clear that Y_2 is normal μ_2 and σ_2^2, so that the two marginal densities are unaffected by the extent ρ of dependence. HINT: Check that

$$p_{Y_1,Y_2}(y_1,y_2) = \frac{1}{\sigma_1\sqrt{2\pi}} \exp\left\{-\frac{1}{2}\left[\frac{y_1 - \mu_1}{\sigma_1}\right]^2\right\}$$
$$\times \frac{1}{\sigma_2\sqrt{2\pi(1-\rho^2)}} \exp\left\{-\frac{1}{2}\left[\frac{y_2 - h(y_1)}{\sigma_2\sqrt{1-\rho^2}}\right]^2\right\}$$

where $h(y_1) = \mu_2 + \rho\sigma_2(y_1 - \mu_1)/\sigma_1$.

2. Find and sketch the density of the standardization X^* of X:
(*a*) When X is uniform over $a < x < b$ (*b*) When X is exponential with mean $1/\lambda$

3. Each cube has all its sides of equal length. A randomly selected cube has sides of length X, where X is normal with $\mu > 0$ and σ^2. (Naturally μ/σ is very much larger than 1, so that $P[X < 0]$ is negligible.) Find the distribution of the volume $V = X^3$.

4. Let $p_X: R \to R$ be as in Fig. 11 and sketch the density induced by the following:
(*a*) $Y = g(X)$ (*b*) $Z = f(X)$

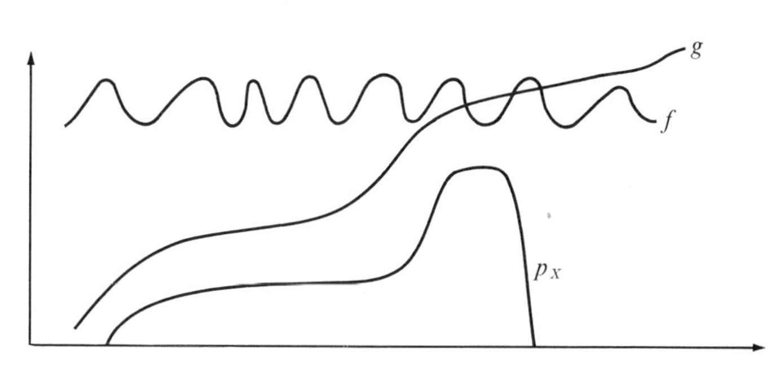

Fig. 11

5. Find the probability density of the magnitude $F = 1/X^2$ of the force exerted on an electron located at the origin by an electron located at X if X is as follows:
(*a*) Uniform over $0 \leq x \leq 1$
(*b*) Exponential with mean 1
(*c*) Normal $\mu = 0$ and σ^2
6. Let Y be lognormal with parameters μ and σ^2.
(*a*) Prove that

$$P[Y < e^{\mu-\sigma}] = \Phi(-1) = .1586 \qquad P[Y < e^{\mu}] = .5 \qquad P[Y < e^{\mu+\sigma}] = \Phi(1) = .8413.$$

Also sketch a picture of the functions e^x and $\varphi_{\mu,\sigma}$ and note that the facts above are graphically obvious.
(*b*) The diameter Y in centimeters of a randomly selected small spherical particle is lognormal with parameters $\mu = -6$ and $\sigma^2 = 4$, so that

$$P[Y < e^{-8}] = \Phi(-1) \qquad P[Y < e^{-6}] = .5 \qquad P[Y < e^{-4}] = \Phi(1).$$

Note the drastic asymmetry in that both the intervals $e^{-8} \leq y \leq e^{-6}$ and $e^{-6} \leq y \leq e^{-4}$ on either side of the median e^{-6} have the same probability but quite different lengths—approximately e^{-6} and e^{-4}.

Let $Z = \pi Y^2$ be the surface area of the particle and find the numbers a, b, and c such that

$$P[Z < e^{a}] = \Phi(-1) \qquad P[Z < e^{b}] = .5 \qquad P[Z < e^{c}] = \Phi(1).$$

7. Let Y have continuous density p_Y satisfying $p_Y(y) = 0$ if $y \leq 0$. Let $c > 0$ and $a \neq 0$, so that $cy^a = ce^{a \log y}$ is uniquely defined and positive valued if $y > 0$.
(*a*) Show that $Z = cY^a$ has continuous density

$$p_Z(z) = \frac{(z/c)^{1/a}}{|a|\, z}\, p_Y\left(\left(\frac{z}{c}\right)^{1/a}\right) \qquad \text{for all } z > 0.$$

(*b*) Use part (*a*) to find p_Z when Y is lognormal. Check that this density is the same as was derived in Example 7.
(*c*) The lifetime Y of a \$1 light bulb is exponential with mean $1/\lambda$ hours. Find the probability density for the cost per hour of illumination, $Z = 1/Y$.
8. Let $X = g(Y) = \log Y$, where Y has the lognormal density with parameters μ and σ^2, as in Example 6. Show that p_X is normal μ and σ^2 by finding p_X directly from g and p_Y. Compare to Exercise 3 (Sec. 15-2).
9. Let $p_{0,1}: R \to R$ be any continuous density with zero mean and unit variance. For any reals $\sigma > 0$ and μ define $p_{\mu,\sigma}$ to be the density induced by $Y = \sigma X + \mu$ when X has density $p_{0,1}$. Similarly $p_{\mu,-\sigma}$ is induced by $Y = -\sigma X + \mu$. Thus

$$p_{\mu,\sigma}(y) = \frac{1}{\sigma} p_{0,1}\left(\frac{y-\mu}{\sigma}\right) \qquad \text{and} \qquad p_{\mu,-\sigma}(y) = \frac{1}{\sigma} p_{0,1}\left(-\frac{y-\mu}{\sigma}\right)$$

for all real y. Both densities have mean μ and variance σ^2.
(*a*) Prove that if $Z = aY + b$ with $a \neq 0$, then Z induces one of these densities if Y does.
COMMENT: Thus every normalized continuous density generates a smallest class of densities closed under nonsingular linear transformations. Actually the linear transformations are applied to the random variables. Note that if $p_{0,1}$ is symmetric, as for the normal density, then $p_{\mu,\sigma} = p_{\mu,-\sigma}$ so the second parameter can be taken to be positive.

(*b*) Let $p_{0,1}$ be defined by

$$p_{0,1}(x) = \begin{cases} e^{-(x+1)} & \text{if } x > -1 \\ 0 & \text{otherwise.} \end{cases}$$

Sketch $p_{5,1}$ and $p_{0,-1}$ and $p_{5,-1}$ and $p_{\lambda,\lambda}$ for $\lambda > 0$.

10. Let $Y = g(X)$, where X has continuous density p_X and $g' > 0$ or $g' < 0$. In this section it was shown that Y has continuous density

$$p_Y(y) = \begin{cases} \dfrac{1}{|dg(x)/dx|}\, p_X(x) & \text{if } y = g(x) \text{ is in the range of } g \\ 0 & \text{otherwise.} \end{cases}$$

Use the relationship between g and its inverse g^{-1} to obtain the following four equivalent forms for p_Y, valid for all x, and for all y in the range of g:

$$p_Y(g(x)) = \frac{1}{|[dg/dx]_x|}\, p_X(x) = \left| \left[\frac{dg^{-1}}{dy}\right]_{y=g(x)} \right| p_X(x),$$

$$p_Y(y) = \frac{1}{|[dg/dx]_{x=g^{-1}(y)}|}\, p_X(g^{-1}(y)) = \left| \left[\frac{dg^{-1}}{dy}\right]_y \right| p_X(g^{-1}(y)).$$

HINT: We recall here several facts which are useful for this exercise. If $g: R \to R$ is strictly increasing, $g(x) < g(x')$ whenever $x < x'$, or strictly decreasing, $g(x) > g(x')$ whenever $x < x'$, then clearly g is a one-to-one correspondence between R and the range $\mathscr{R}$ of g. Therefore, as in Sec. 3-5, there is a unique inverse function $g^{-1}: \mathscr{R} \to R$, as shown in Fig. 12*a*, satisfying

$$g^{-1}(g(x)) = x \qquad \text{for all real } x$$

$$g(g^{-1}(y)) = y \qquad \text{for all } y \in \mathscr{R}.$$

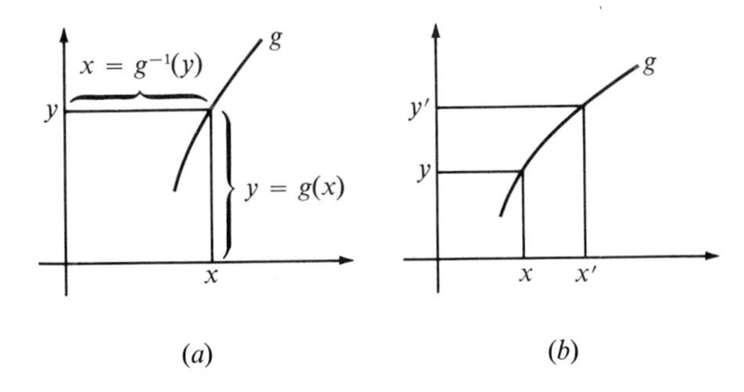

Fig. 12

If $x < x'$ and $y = g(x)$ and $y' = g(x')$ then we can write the ratio $(y' - y)/(x' - x)$ in several ways:

$$\frac{g(x') - g(x)}{x' - x} = \frac{y' - y}{x' - x} = \frac{y' - y}{g^{-1}(y') - g^{-1}(y)} = \frac{1}{[g^{-1}(y') - g^{-1}(y)]/[y' - y]}$$

If g is everywhere differentiable and we let x' converge to x then we see that g^{-1} is everywhere differentiable and that

$$\left[\frac{dg}{dx}\right]_x = \frac{1}{[dg^{-1}/dy]_{y=g(x)}}$$

holds for all x; that is, the derivative of g at x equals the reciprocal of the derivative of the inverse g^{-1} evaluated at $y = g(x)$. This formula can also be obtained by differentiating the identity $g^{-1}(g(x)) = x$ with respect to x.

16-3 INDUCED DENSITIES, $n \geq 2$

In Sec. 16-2 we found the density of a random variable which was a function of another random variable, $Y = g(X)$. In the next three examples we find the density of a random variable $Y = g(X_1,\ldots,X_n)$ which is a function of a random vector, where we use $n =$ 3, 2, 2. The result of Example 3 for a sum $Y = X_1 + X_2$ is basic. We then seek the density of a random vector $(Y_1, \ldots, Y_m) = g(X_1, \ldots, X_n)$ which is a function of a random vector, with emphasis on the case $m = n \geq 2$. At that point we investigate in more detail the basic ideas involved in our past and future examples. We find in particular that the jacobian of the transformation g appears.

***Example 1** A particle has velocity vector $V = (V_x, V_y, V_z)$ having continuous density $p_V: R^3 \to R$ given by

$$p_V(v_x,v_y,v_z) = \varphi_{0,\sigma}(v_x)\varphi_{0,\sigma}(v_y)\varphi_{0,\sigma}(v_z) = \left(\frac{1}{\sigma\sqrt{2\pi}}\right)^3 e^{-(v_x^2+v_y^2+v_z^2)/2\sigma^2}$$
$$= \left(\frac{1}{\sigma\sqrt{2\pi}}\right)^3 e^{-s^2/2\sigma^2} \qquad \text{where } s = (v_x^2 + v_y^2 + v_z^2)^{1/2}.$$

This density corresponds to the velocity components being independent and each normally distributed with mean 0 and variance σ^2. We want the density of the magnitude

$$S = (V_x^2 + V_y^2 + V_z^2)^{1/2} \geq 0$$

of the velocity. We obtain an s' satisfying $s \leq s' \leq s + ds$ iff we obtain a (v_x,v_y,v_z) inside the spherical shell of inner radius s and outer radius $s + ds$. The probability of this event is the integral of p_V over the shell. But this p_V is constant on the shell, so this probability equals this constant value of p_V times the volume of the shell. The inner sphere has area $4\pi s^2$ and the shell is ds thick and so has volume $4\pi s^2\, ds$. Thus

$$p_S(s)\, ds = \left(\frac{1}{\sigma\sqrt{2\pi}}\right)^3 e^{-s^2/2\sigma^2} 4\pi s^2\, ds.$$

Therefore

$$p_S(s) = \begin{cases} \sqrt{\dfrac{2}{\pi}}\, \dfrac{s^2}{\sigma^3}\, e^{-s^2/2\sigma^2} & \text{if } s > 0 \\ 0 & \text{if } s \leq 0 \end{cases}$$

and this is the *Maxwell density* for velocities in statistical mechanics. ■

Example 2 We desire the density induced by $Z = X + Y$, where $\lambda > 0$ and

$$p_{X,Y}(x,y) = \begin{cases} \lambda^2 e^{-\lambda(x+y)} & \text{if } x > 0,\ y > 0 \\ 0 & \text{otherwise.} \end{cases}$$

This density corresponds to X and Y being independent and each exponential with mean $1/\lambda$. Z might be the total number of hours of illumination obtained from two light bulbs. If $z > 0$ then $p_Z(z)\,dz$ must equal the volume under $p_{X,Y}$ and over the shaded strip shown in Fig. 1. This shaded strip has length $z\sqrt{2}$, thickness $dz/\sqrt{2}$, and area $z\,dz$. Over this strip

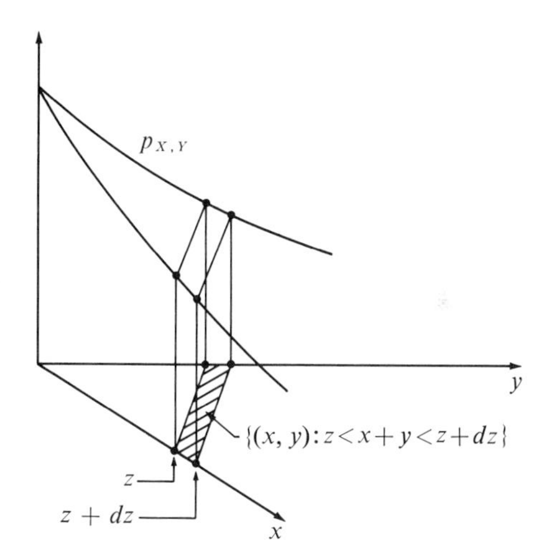

Fig. 1

$p_{X,Y}$ has constant height $\lambda^2 e^{-\lambda z}$ so

$$p_Z(z) = \begin{cases} \lambda^2 z e^{-\lambda z} & \text{if } z > 0 \\ 0 & \text{otherwise.} \end{cases}$$

Figure 2 shows this induced density p_Z. ■

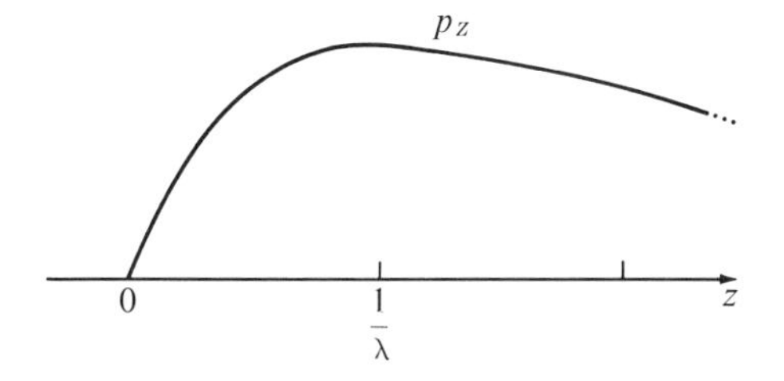

Fig. 2

Example 3 We want the continuous density p_Z of $Z = X + Y$ in terms of the joint continuous density $p_{X,Y}: R^2 \to R$. We fix a z and wish to find $p_Z(z)$. For fixed $dz > 0$ we will obtain a z' satisfying $z < z' \leq z + dz$ iff we obtain an (x,y) in the shaded strip of the xy plane shown in Fig. 3a. The probability of this is the integral of $p_{X,Y}$ over the strip, that is, the volume indicated in Fig. 3b above the strip and below the $p_{X,Y}$ surface. Select two close points x and

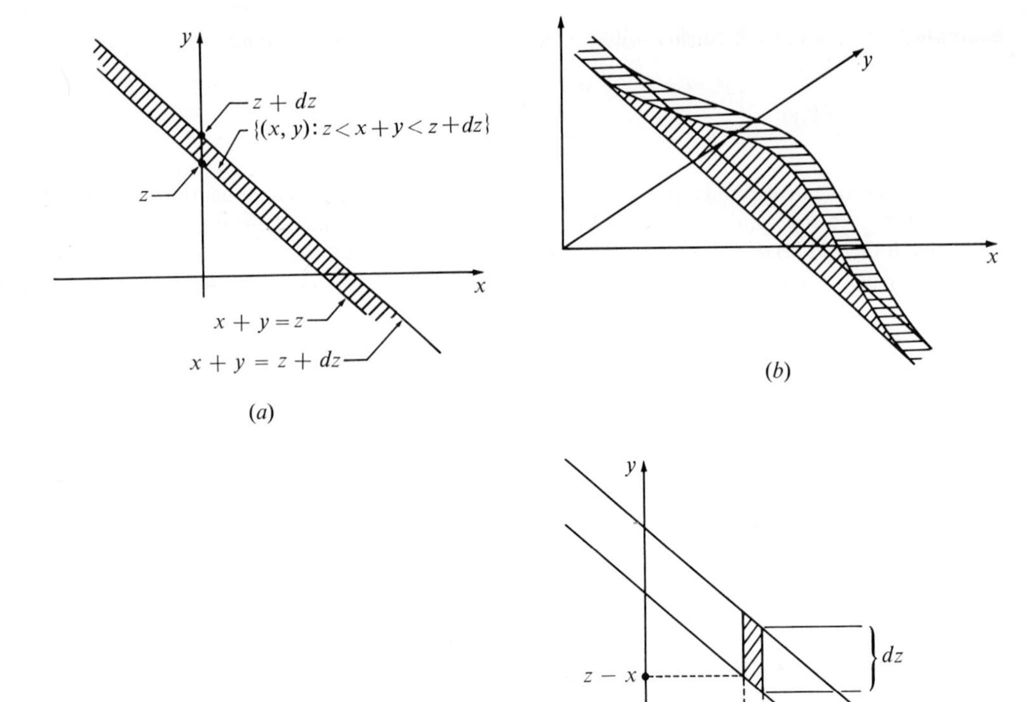

Fig. 3

$x + dx$. Consider the shaded parallelogram of Fig. 3c, whose area is the base dz times the altitude dx. The parallelogram contributes probability $p_{X,Y}(x, z - x)\,dx\,dz$. Adding such contributions over x, we obtain

$$dz \int_{-\infty}^{\infty} p_{X,Y}(x, z - x)\,dx$$

for the probability of the strip of Fig. 3a, and this must equal $p_Z(z)\,dz$. *Thus if* X and Y have joint continuous density $p_{X,Y}$ then $Z = X + Y$ has continuous density

$$p_Z(z) = \int_{-\infty}^{\infty} p_{X,Y}(x, z - x)\,dx = \int_{-\infty}^{\infty} p_{X,Y}(z - y, y)\,dy \qquad \text{for each real } z.$$

The last expression was obtained by making, for each fixed z, the change of variables $y = z - x$ and $dy = -dx$.

As will be seen in the next section, a joint continuous density of the product form $p_{X,Y}(x,y) = p_X(x)p_Y(y)$ corresponds to independence of X and Y. In this case $Z = X + Y$ has continuous density

$$p_Z(z) = \int_{-\infty}^{\infty} p_X(x)p_Y(z - x)\,dx = \int_{-\infty}^{\infty} p_X(z - y)p_Y(y)\,dy \qquad \text{for all real } z.$$

When a function $p_Z: R \to R$ is obtained from two functions $p_X: R \to R$ and $p_Y: R \to R$ by the formula above we say that p_Z is the **convolution of p_X and p_Y**, or p_Z is obtained by convolving p_X and p_Y. This operation occurs in many branches of mathematics.

We apply the convolution formula

$$p_Z(z) = \int_{-\infty}^{\infty} p_X(x)p_Y(z-x)\,dx \qquad \text{for all real } z$$

to Example 2, where p_X and p_Y are both exponential with mean $1/\lambda$. Now $p_X(x) = 0$ if $x \leq 0$ so

$$p_Z(z) = \int_0^{\infty} p_X(x)p_Y(z-x)\,dx \qquad \text{for all real } z.$$

If we select a $z < 0$ then $z - x < 0$ as x varies over $0 < x < \infty$, so that $p_Y(z - x)$ is always zero, and hence $p_Z(z) = 0$. If we select a $z > 0$ then $p_Y(z - x) = 0$ when $x > z$. Thus if $z > 0$ then

$$p_Z(z) = \int_0^{z} p_X(x)p_Y(z-x)\,dx = \int_0^z \lambda e^{-\lambda x}\lambda e^{-\lambda(z-x)}\,dx = \lambda^2 e^{-\lambda z}\int_0^z dx = \lambda^2 z e^{-\lambda z}$$

as was found in Example 2. ■

Exercises 1 to 6, and 12 may be done before reading the remainder of this section, but Exercises 7 to 11 fall into the general case involving jacobians.

THE BASIC IDEAS BEHIND CALCULATIONS OF INDUCED DENSITIES

We now consider the basic ideas involved in past and future calculations of induced densities, after which we find densities induced by some random vectors. Assume that we have a continuous density $p_f: R^n \to R$ and a function $g: R^n \to R^m$,

$$\begin{array}{ccc} R^n & \xrightarrow{\quad g \quad} & R^m \\ p_f & & P_g \end{array}$$

Then g, with associated p_f, induces P_g.

Suppose P_g happens to correspond to a continuous density p_g. Then both p_f and p_g must assign the same probability to the event that g has a value in the closed rectangle $R_{a,b}$ in R^m; that is, the integral of p_g over $R_{a,b}$ must equal the integral of p_f over $g^{-1}(R_{a,b})$.

Suppose we do not know whether or not P_g corresponds to a continuous density. We may still somehow be able to find a continuous density $p': R^m \to R$ such that the integral of p' over every $R_{a,b}$ equals the integral of p_f over $g^{-1}(R_{a,b})$. In this case if P' corresponds to p' then we have

$$P'(R_{a,b}) = P_f(g^{-1}(R_{a,b})) = P_g(R_{a,b})$$

where the last equality is the definition of P_g. But then $P' = P_g$, from Theorem 2 (Sec. 15-1), so that P_g does correspond to a continuous density, and p' is it. Often problems of this kind arise, as in Theorem 2 (Sec. 15-2), from an intermediate density, so the following theorem states our conclusion in this context.

Theorem 1 Assume that the random vector $f: \Omega \to R^n$ with associated probability space (Ω,P) induces a continuous density $p_f: R^n \to R$, and that

$$\Omega \xrightarrow{f} R^n \xrightarrow{g} R^m, \qquad h = g(f).$$

If a continuous density $p': R^m \to R$ can be found for which

$$\int \cdots \int_{R_{a,b}} p'(y_1, \ldots, y_m)\, dy_1 \cdots dy_m = \int \cdots \int_{g^{-1}(R_{a,b})} p_f(x_1, \ldots, x_n)\, dx_1 \cdots dx_n$$

holds for every closed rectangle $R_{a,b}$ in R^m, then h and g induce continuous densities and $p_h \equiv p_g \equiv p'$. ■

If we find a p' satisfying the equation of Theorem 1 for all "small" rectangles we can essentially add such equations to obtain the equation for large rectangles. Therefore we let $R_{a,b}$ be a so-called "infinitesimal rectangle," where $a = y$ and $b = y + dy$. We henceforth assume that g and p_f are "continuous enough" to justify our approach.

Thus under "appropriate conditions of continuity" we have the following *basic formula:* Assume that the random vector $f: \Omega \to R^n$ with associated probability space (Ω,P) induces a continuous density $p_f: R^n \to R$ and that

$$\Omega \xrightarrow{f} R^n \xrightarrow{g} R^m, \qquad h = g(f).$$

If h and g induce a continuous density $p_h \equiv p_g$ then this density is given by

$$p_g(y)\, dv' = \int \cdots \int_{g^{-1}(R_{y,y+dy})} p_f(x_1, \ldots, x_n)\, dx_1 \cdots dx_n$$

where $y = (y_1, \ldots, y_m)$ and $dy = (dy_1, \ldots, dy_m)$ and $dv' = dy_1\, dy_2 \cdots dy_m$ is the volume of the closed rectangle $R_{y,y+dy}$. That is, as shown in Fig. 4, $P_f[g \in R_{y,y+dy}]$ must be the same whether it is calculated from p_f, or as $p_g(y)\, dv'$ from p_g. Naturally if $g^{-1}(R_{y,y+dy})$ is empty then $p_g(y) = 0$.

We may select a y and find that g sends infinitely many x into this one point y. This occurred in the preceding three examples, where we essentially applied the basic formula in the form above. This is not the case, however, for many examples when $n = m$, and we next derive a more explicit formula for such cases.

We select a y and wish to evaluate $p_g(y)$. Assume, as in Fig. 5*a*, that there are a finite number $N_y \geq 1$ of points carried into this y by g. That is, $y = g(x^{(1)}) = g(x^{(2)}) = \cdots = g(x^{(N_y)})$ and $y \neq g(x)$ for all other x. Assume, as in Fig. 5*b*, that $g^{-1}(R_{y,y+dy})$ consists of N_y small disjoint sets $A_1, \ldots, A_{N_y}$, where A_k is centered on $x^{(k)}$. If A_k has volume dv_k and p_f is continuous at $x^{(k)}$ then the integral of p_f over

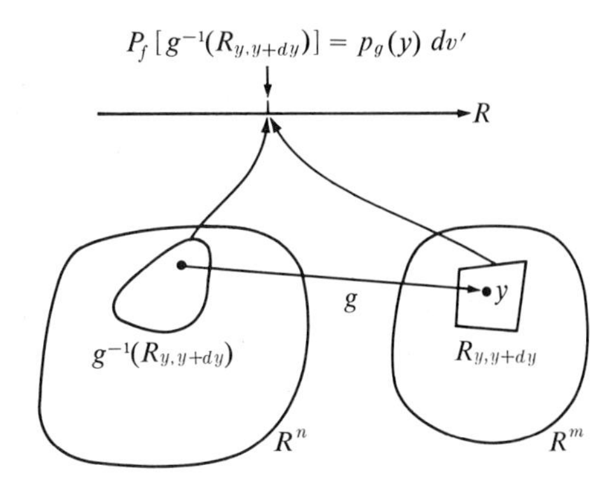

Fig. 4

A_k equals $p_f(x^{(k)})\,dv_k$. For such a case our basic formula reduces to

$$p_g(y)\,dv' = \sum_{k=1}^{N_y} p_f(x^{(k)})\,dv_k.$$

We rewrite this *derived formula* as follows:

$$p_g(y) = \sum_{k=1}^{N_y} \frac{dv_k}{dv'} p_f(x^{(k)}) = \sum_{k=1}^{N_y} \frac{1}{dv'/dv_k} p_f(x^{(k)}).$$

For most of our examples $N_y = 1$. However $N_y > 1$ in Example 8 below, and in Examples 8 and 9 (Sec. 16-2).

Thus to find $p_g(y)$ we must calculate the local scale factor dv'/dv_k by which small volumes near $x^{(k)}$ are multiplied when transformed by g. When $n = m = 1$ we found in the previous section that $dv'/dv_k = |g'(x^{(k)})|$. More generally if $n = m \geq 1$ then it is well known that $dv'/dv_k = |J_g(x^{(k)})|$, where J_g is the *jacobian* of g. If $y = g(x) = Tx + b$ consists of a translation and a nonsingular linear transformation T then $N_y = 1$ and J_g reduces to the constant determinant of T. We first examine the case $Tx + b$, which is used in connection with the multivariate normal densities in Sec. 18-2. After applying the result for $Tx + b$ in the case $n = 2$ in Examples 4 to 6, we then state and use the formula for the general case involving jacobians. This last result is not used in an essential fashion later in the book.

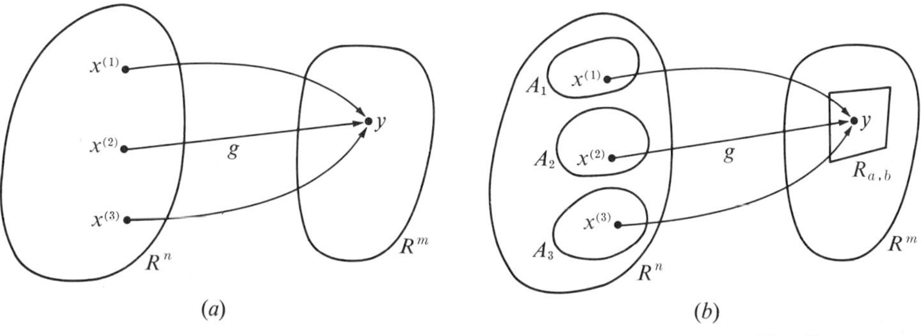

Fig. 5 $N_y = 3$.

THE CASE $Y = TX + b$

For notational convenience in the remainder of this section we often let $X = (X_1, \ldots, X_n)$ and $Y = (Y_1, \ldots, Y_n)$ denote random *vectors*.

Let $g: R^n \to R^n$ be defined by $y = g(x) = Tx + b$, where b is a constant vector and T is a nonsingular linear transformation:

$$\begin{bmatrix} y_1 \\ y_2 \\ \cdot \\ \cdot \\ \cdot \\ y_n \end{bmatrix} = \begin{bmatrix} t_{11} & t_{12} & \cdots & t_{1n} \\ t_{21} & t_{22} & \cdots & t_{2n} \\ \cdots & \cdots & \cdots & \cdots \\ t_{n1} & t_{n2} & \cdots & t_{nn} \end{bmatrix} \begin{bmatrix} x_1 \\ x_1 \\ \cdot \\ \cdot \\ \cdot \\ x_n \end{bmatrix} + \begin{bmatrix} b_1 \\ b_2 \\ \cdot \\ \cdot \\ \cdot \\ b_n \end{bmatrix}$$

We write $y = (y_1, \ldots, y_n)$ but show y as a column vector in a displayed formula, as above. It is well known that g is a one-to-one correspondence between R^n and R^n, so $N_y = 1$ for all y. Also the scale factor dv'/dv, for volumes, is the absolute value of the determinant of T, so

$$p_Y(y) = \frac{1}{|\det T|} p_X(x) \qquad \text{where } y = Tx + b.$$

Now T has an inverse T^{-1} and $x = T^{-1}(y - b)$, so in the formula above, x can be replaced by $T^{-1}(y - b)$. Also, we have $TT^{-1} = T^{-1}T = I$, where I is the $n \times n$ identity matrix with 1's on the main diagonal and 0's elsewhere. The product rule for determinants yields $(\det T)(\det T^{-1}) = 1$. *Thus* if X has continuous density $p_X: R^n \to R$ and $Y = TX + b$, where b is a constant vector and T is a nonsingular linear transformation of R^n into R^n, then Y has continuous density

$$p_Y(y) = \frac{1}{|\det T|} p_X(T^{-1}(y - b)) = |\det T^{-1}|\, p_X(T^{-1}(y - b))$$

for all y in R^n. Note that if $n = 1$ then $y = ax + b$ and we obtain our previous result $p_Y(y) = (1/|a|)p_X((y - b)/a)$.

We have indicated a proof of the transformation-of-variables formula[1]

$$\int_{A'} \frac{1}{|\det T|} p_X(T^{-1}(y - b))\, dy = \int_A p_X(x)\, dx$$

where $A = g^{-1}(A') = \{x: (Tx + b) \text{ belongs to } A'\}$. Naturally the formula holds for many functions p_X which are not continuous probability densities. If we let p_X be

[1] Proofs and precise statements of transformation formulas like those of this section can be found in standard references such as W. Rudin, "Principles of Mathematical Analysis," 2d ed., p. 206, McGraw-Hill Book Company, New York, 1964, and, at a more advanced level, in A. C. Zaanen, "An Introduction to the Theory of Integration," p. 166, Interscience Publishers, Inc., New York, 1958.

identically 1 then the formula becomes

$$\frac{1}{|\det T|}\int_{A'} dy = \int_A dx$$

which reduces to

$$[\text{the volume of } A'] = |\det T|\ [\text{the volume } A]$$

and this is how we interpreted $|\det T|$ initially.

If $n = 2$ the general result for $Y = TX + b$ specializes as follows:

$$\begin{aligned} y_1 &= t_{11}x_1 + t_{12}x_2 + b_1 \\ y_2 &= t_{21}x_1 + t_{22}x_2 + b_2 \end{aligned} \qquad T = \begin{bmatrix} t_{11} & t_{12} \\ t_{21} & t_{22} \end{bmatrix}, \ \det T = t_{11}t_{22} - t_{21}t_{12} \neq 0$$

$$T^{-1} = \frac{1}{\det T}\begin{bmatrix} t_{22} & -t_{12} \\ -t_{21} & t_{11} \end{bmatrix} \qquad \det T^{-1} = \frac{1}{\det T}$$

$$\begin{bmatrix} x_1 \\ x_2 \end{bmatrix} = x = T^{-1}(y - b) = \frac{1}{\det T}\begin{bmatrix} t_{22} & -t_{12} \\ -t_{21} & t_{11} \end{bmatrix}\begin{bmatrix} y_1 - b_1 \\ y_2 - b_2 \end{bmatrix}$$

$$p_{Y_1,Y_2}(y_1,y_2) = \frac{1}{|\det T|} \times p_{X_1,X_2}\left(\frac{t_{22}(y_1 - b_1) - t_{12}(y_2 - b_2)}{\det T}, \frac{-t_{21}(y_1 - b_1) + t_{11}(y_2 - b_2)}{\det T}\right).$$

We apply this result for $n = 2$ in Examples 4 to 6.

Example 4 If

$$T = \begin{bmatrix} 1 & -1 \\ 1 & 1 \end{bmatrix} \quad \text{and} \quad b = \begin{bmatrix} 0 \\ 0 \end{bmatrix}$$

then we have $\det T = 2$ and $Y_1 = X_1 - X_2$ and $Y_2 = X_1 + X_2$, while

$$p_{Y_1,Y_2}(y_1,y_2) = (1/2)p_{X_1,X_2}\left(\frac{y_1 + y_2}{2}, \frac{y_2 - y_1}{2}\right)$$

is the joint density of the difference and sum. If X_i is the length of the ith experiment then $Y_2 = X_1 + X_2$ is the sum of the lengths of the first two experiments, and $Y_1 = X_1 - X_2$ is the amount by which the first is longer than the second. ■

Example 5 If $Y_1 = X_1$ and $Y_2 = X_1 + X_2$ then

$$p_{Y_1,Y_2}(y_1,y_2) = p_{X_1,X_2}(y_1, y_2 - y_1)$$

is the joint density of Y_1 and Y_2. Integrating over y_1, we find the density of $Y_2 = X_1 + X_2$ to be

$$p_{Y_2}(y_2) = \int_{-\infty}^{\infty} p_{X_1,X_2}(y_1, y_2 - y_1)\, dy_1$$

as was obtained in Example 3. A common procedure for finding the density of some random variable $Z = h(X_1, \ldots, X_n)$ is to proceed as above and first find the joint density of $X_1, X_2, \ldots, X_{n-1}, Z$ and then integrate over $x_1, x_2, \ldots, x_{n-1}$. ■

***Example 6** We apply the result for $Y = TX$ with $n = 2$ to a clockwise rotation, by angle θ where $0 \leq \theta < 2\pi$, of the point (X_1, X_2) into the new point (Y_1, Y_2), as shown in Fig. 6.

$$y_1 = x_1 \cos\theta + x_2 \sin\theta$$

$$y_2 = -x_1 \sin\theta + x_2 \cos\theta$$

$$\det T = \cos^2\theta + \sin^2\theta = 1$$

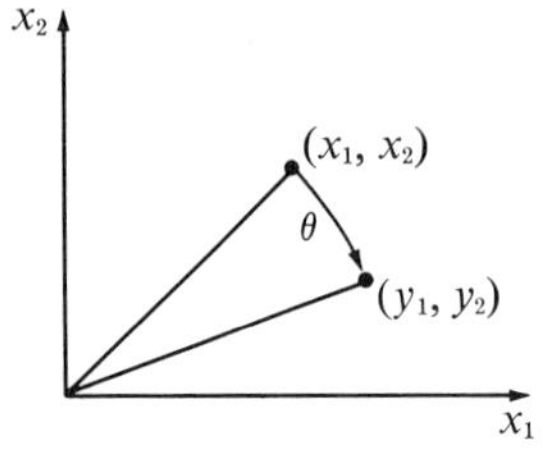

Fig. 6

Multiplying the first equation by $\cos\theta$, and the second by $\sin\theta$, then subtracting, yields x_1. Similarly for x_2. Thus

$$p_{Y_1,Y_2}(y_1,y_2) = p_{X_1,X_2}(y_1 \cos\theta - y_2 \sin\theta, y_1 \sin\theta + y_2 \cos\theta).$$

A function $p: R^n \to R$ is said to be **radially symmetric** iff there exists a function $f: R \to R$ such that

$$p(x_1, \ldots, x_n) = f(x_1^2 + \cdots + x_n^2) \qquad \text{for all } (x_1, \ldots, x_n) \text{ in } R^n.$$

That is, p is essentially only a function of the distance from the origin. If p_{X_1,X_2} is radially symmetric and (y_1,y_2), (x_1,x_2) are as above, then $y_1^2 + y_2^2 = x_1^2 + x_2^2$ so that

$$p_{Y_1,Y_2}(y_1,y_2) = p_{X_1,X_2}(x_1,x_2) = p_{X_1,X_2}(y_1,y_2).$$

Thus if (X_1,X_2) has a radially symmetric continuous density then (Y_1, Y_2) has this same density if (Y_1, Y_2) is obtained by applying an orthogonal transformation to (X_1,X_2). Note that the product rule applied to the same mean zero normal density, yields a radially symmetric density $\prod_{i=1}^{n} \varphi_{0,\sigma}(x_i)$ on R^n. ■

*THE GENERAL CASE INVOLVING JACOBIANS

We next exhibit the formula for the continuous density p_Y in terms of the continuous density p_X when $Y = g(X)$, where $g: R^n \to R^n$ and $N_Y \equiv 1$ on the range of g, so g is one-to-one. If $g: R^n \to R^n$ then $y = g(x)$ when written out is

$$\begin{bmatrix} y_1 \\ y_2 \\ \cdot \\ \cdot \\ \cdot \\ y_n \end{bmatrix} = \begin{bmatrix} g_1(x) \\ g_2(x) \\ \cdot \\ \cdot \\ \cdot \\ g_n(x) \end{bmatrix} = \begin{bmatrix} g_1(x_1, x_2, \ldots, x_n) \\ g_2(x_1, x_2, \ldots, x_n) \\ \cdot \\ \cdot \\ \cdot \\ g_n(x_1, x_2, \ldots, x_n) \end{bmatrix}.$$

If at a particular x all the partial derivatives $\partial g_i/\partial x_j$ exist and are continuous then we define the **jacobian** of g at $x = (x_1, \ldots, x_n)$ to be

$$J_g(x) = \det \begin{bmatrix} \dfrac{\partial g_1}{\partial x_1} & \dfrac{\partial g_1}{\partial x_2} & \cdots & \dfrac{\partial g_1}{\partial x_n} \\ \dfrac{\partial g_2}{\partial x_1} & \dfrac{\partial g_2}{\partial x_2} & \cdots & \dfrac{\partial g_2}{\partial x_n} \\ \cdots & \cdots & \cdots & \cdots \\ \dfrac{\partial g_n}{\partial x_1} & \dfrac{\partial g_n}{\partial x_2} & \cdots & \dfrac{\partial g_n}{\partial x_n} \end{bmatrix}$$

where all the entries of the matrix are evaluated at this x. If the absolute value $|J_g(x)|$ of the jacobian is nonzero at this x and $y = g(x)$ then $|J_g(x)|$ is the local (depending on x) scale factor by which the volume dv of a small set near x must be multiplied in order to find the volume dv' of the small set near $y = g(x)$ that it is transformed into by g, so $dv' = |J_g(x)|\, dv$. *Thus* we have

$$p_Y(y) = \frac{1}{|J_g(x)|} p_X(x) \qquad \text{where } y = g(x).$$

Note that if $g(x) = Tx + b$ then $J_g(x) = \det T$ and we have the previous result.

Now, g is one-to-one and so has an inverse function g^{-1} where $y = g(x)$ iff $x = g^{-1}(y)$. We want a version of the formula above in terms of g^{-1}. If for each i and j we take the partial derivative of the ith component of the vector equation $g(g^{-1}(y)) = y$ with respect to y_j and use the chain rule for differentiation, we obtain a product of two $n \times n$ matrices equal to the identity matrix, and taking determinants we get

$$\{[J_g(x)]_{x=g^{-1}(y)}\}\{J_{g^{-1}}(y)\} = 1.$$

Thus for each y in the range of g we have

$$p_Y(y) = \frac{1}{|[J_g(x)]_{x=g^{-1}(y)}|} p_X(g^{-1}(y)) = |J_{g^{-1}}(y)|\, p_X(g^{-1}(y)).$$

We apply this formula in Example 7, while Example 8 uses the obvious extension of this formula to the case $N_y = 2$.

***Example 7** Let $(r,\theta) = g(x_1,x_2)$ transform a point (x_1,x_2) in the plane into polar coordinates so that

$$r = g_1(x_1,x_2) = \sqrt{x_1^2 + x_2^2}$$

$$\theta = g_2(x_1,x_2) = [\text{the angle between the vector } (x_1,x_2) \text{ and the abscissa}].$$

Thus $0 \leq \theta < 2\pi$, and sometimes g_2 is written as $\theta = \tan^{-1}(x_2/x_1)$, although strictly speaking, $\tan^{-1}(1/1) = \tan^{-1}(-1/-1)$, so $\tan^{-1}$ does not distinguish between the first and third quadrants. The inverse function g^{-1} is given by

$$x_1 = g_1^{-1}(r,\theta) = r\cos\theta \qquad x_2 = g_2^{-1}(r,\theta) = r\sin\theta$$

where $r > 0$ and $0 \leq \theta < 2\pi$. Therefore

$$J_{g^{-1}}(r,\theta) = \det \begin{bmatrix} \dfrac{\partial g_1^{-1}}{\partial r} & \dfrac{\partial g_1^{-1}}{\partial \theta} \\ \dfrac{\partial g_2^{-1}}{\partial r} & \dfrac{\partial g_2^{-1}}{\partial \theta} \end{bmatrix} = \det \begin{bmatrix} \cos\theta & -r\sin\theta \\ \sin\theta & r\cos\theta \end{bmatrix} = r$$

hence

$$p_{R,\theta}(r,\theta) = r p_{X_1,X_2}(r\cos\theta, r\sin\theta) \qquad \text{for } r > 0, 0 \leq \theta < 2\pi.$$

If p_{X_1,X_2} describes the location of a missing airplane then $p_{R,\theta}$ may be more useful for determining the strategy of search to be employed by a radar station located at the origin. ■

***Example 8** An experiment is supposed to terminate at time 0 but actually terminates at some random time X_1 and yields amount $X_2 e^{X_1}$ of bacteria. Thus the growth is exponential in time, while the random factor X_2 might, for example, be due to the initial number of bacteria being random. We are given the joint continuous density $p_{X_1,X_2}(x_1,x_2)$, which is 0 when $x_2 < 0$. We want the joint density of the absolute value $Y_1 = |X_1|$ of the time error and the terminal amount $Y_2 = X_2 e^{X_1}$ of bacteria.

Clearly $p_{Y_1,Y_2}(y_1,y_2) = 0$ unless both $y_1 > 0$ and $y_2 > 0$. Now $y_1 = g_1(x_1,x_2) = |x_1|$ and $y_2 = g_2(x_1,x_2) = x_2 e^{x_1}$ so

$$|J_g(x_1,x_2)| = \left| \det \begin{bmatrix} \pm 1 & 0 \\ x_2 e^{x_1} & e^{x_1} \end{bmatrix} \right| = e^{x_1}.$$

For a particular $y = (y_1,y_2)$ with $y_1 > 0$ and $y_2 > 0$ there are, as indicated in Fig. 7, two points $x = (x_1,x_2)$ and $x' = (x_1',x_2')$ satisfying $y = g(x) = g(x')$, and they are found by solving

$$y_1 = x_1 \qquad\qquad y_1 = -x_1'$$

and

$$y_2 = x_2 e^{x_1} \qquad\qquad y_2 = x_2' e^{x_1'}$$

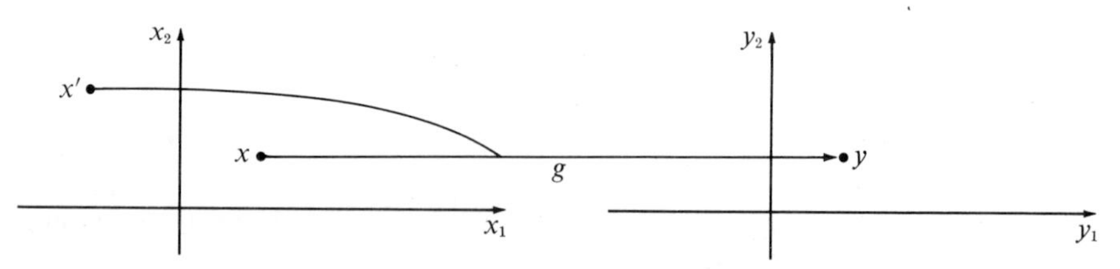

Fig. 7

so that

$$x_1 = y_1 \qquad\qquad x_1' = -y_1$$

$$x_2 = y_2 e^{-y_1} \qquad\qquad x_2' = y_2 e^{y_1}.$$

Thus if $y_1 > 0$ and $y_2 > 0$ then

$$p_{Y_1,Y_2}(y_1,y_2) = \frac{1}{|J_g(x_1,x_2)|} p_{X_1,X_2}(x_1,x_2) + \frac{1}{|J_g(x_1',x_2')|} p_{X_1,X_2}(x_1',x_2')$$

$$= \frac{p_{X_1,X_2}(y_1,y_2 e^{-y_1})}{e^{y_1}} + \frac{p_{X_1,X_2}(-y_1,y_2 e^{y_1})}{e^{-y_1}}. \quad ■$$

EXERCISES

1. A particle of mass m has velocity V distributed as in Example 1. Find the distribution of its kinetic energy

$$E = \frac{m}{2}(V_x^2 + V_y^2 + V_z^2).$$

2. Let $S_n = X_1 + \cdots + X_n$, where $X_1, \ldots, X_n$ are independent and each is exponential with mean $1/\lambda$, so that

$$p_{X_1,\ldots,X_n}(x_1, \ldots, x_n) = \begin{cases} \lambda^n \exp\left\{-\lambda \sum_{i=1}^{n} x_i\right\} & \text{if each } x_i > 0 \\ 0 & \text{otherwise.} \end{cases}$$

Clearly $p_{S_n}(s) = 0$ if $s \leq 0$. In Examples 2 and 3 we derived p_{S_2} in two different ways.
(*a*) Show by a derivation similar to that used in Example 2 that

$$p_{S_3}(s) = \frac{\lambda^3}{2} s^2 e^{-\lambda s} \qquad \text{if } s > 0.$$

(*b*) Find p_{S_3} by convolving p_{S_2} with p_{X_3}.
(*c*) Find p_{S_4} by convolving p_{S_3} with p_{X_4}.
(*d*) Find p_{S_4} by convolving $p_{X_1+X_2}$ with $p_{X_3+X_4}$.

3. Show that if X and Y have a continuous joint density then the continuous density of $Z = XY$ is given by

$$p_Z(z) = \int_{-\infty}^{\infty} p_{X,Y}\left(x, \frac{z}{x}\right) \frac{dx}{|x|} = \int_{-\infty}^{\infty} p_{X,Y}\left(\frac{z}{y}, y\right) \frac{dy}{|y|} \qquad \text{for all real } z.$$

HINT. If we think of z and $dz > 0$ as fixed then $p_Z(z)\, dz$ equals the sum of the probabilities of many small rectangles like the shaded one in Fig. 8, where $dx > 0$, so that

$$p_Z(z)\, dz = \int_{-\infty}^{\infty} [p_{X,Y}(x,y)\, |dy|]\, dx.$$

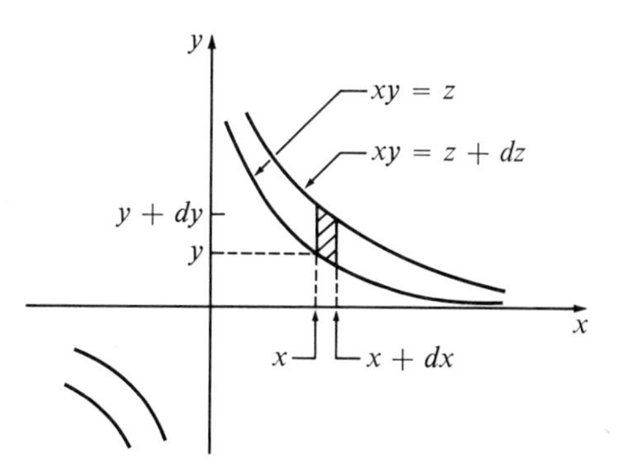

Fig. 8

4. Let X and Y be independent and both lognormal with parameters 0, 1, so that

$$p_{X,Y}(x,y) = \frac{1}{x\sqrt{2\pi}} e^{(-\frac{1}{2})(\log x)^2} \frac{1}{y\sqrt{2\pi}} e^{(-\frac{1}{2})(\log y)^2} \qquad \text{if } x > 0,\ y > 0.$$

Use Exercise 3 to show that $Z = XY$ is lognormal with parameters 0 and 2.

5. Show that if X and Y have a continuous joint density then the continuous density of $Z = X/Y$ is given by

$$p_Z(z) = \frac{1}{z^2}\int_{-\infty}^{\infty} p_{X,Y}\left(x, \frac{x}{z}\right) |x|\, dx = \int_{-\infty}^{\infty} p_{X,Y}(yz,y)\, |y|\, dy$$

6. Let X and Y be independent errors, each normal 0, 1. Use Exercise 5 to show that $Z = X/Y$ has the *Cauchy density*

$$\frac{1}{\pi}\frac{1}{1+z^2} \qquad \text{for all real } z.$$

7. Find the continuous joint density of X and $Z = XY$ in terms of the continuous joint density of X and Y. Then integrate x out to get the density of Z, as in Exercise 3.

8. Find the continuous joint density of $Z = X/Y$ and Y in terms of the continuous joint density of X and Y. Then integrate y out to get the density of Z, as in Exercise 5.

9. Find the continuous joint density of $Y_1 = X_1 + X_2$ and $Y_2 = X_2/X_1$ when

$$p_{X_1,X_2}(x_1,x_2) = \begin{cases} e^{-(x_1+x_2)} & \text{if } x_1 > 0,\ x_2 > 0 \\ 0 & \text{otherwise.} \end{cases}$$

Note that Y_1 and Y_2 turn out to be independent even though they are both functions of (X_1,X_2). State why this is reasonable for this particular p_{X_1,X_2}.

10. Find the joint density of $Y_1 = \sqrt{X_1^2 + X_2^2}$ and $Y_2 = X_1/X_2$ when X_1 and X_2 are independent and each is normal 0 and σ^2. Note that Y_1 and Y_2 turn out to be independent. State why this is reasonable for this particular p_{X_1,X_2}.

11. Find the joint density of $U = X^2$ and $V = Y^3$ when $p_{X,Y}$ is as in Example 1 (Sec. 16-2).

12. Let $Z = XY$, where X and Y are independent and have continuous densities. Assume that $P[X > 0] = P[Y > 0] = 1$, so $P[Z > 0] = 1$. From Exercise 3

$$p_Z(z) = \int_0^{\infty} p_X(x)p_Y\left(\frac{z}{x}\right)\frac{dx}{x} \qquad \text{if } z > 0.$$

The purpose of this exercise is to derive this formula by converting to a sum

$$Z = e^{\log Z} = e^{\log X + \log Y}$$

and convolving. Find the densities of $\log X$ and $\log Y$, then convolve them to find the density of $\log Z$, and then find the density of $Z = e^{\log Z}$ from that for $\log Z$.

16-4 INDEPENDENCE

In this section we consider independent random vectors when continuous densities are available. We first show that in this case independence is equivalent to the product rule for densities. The general results of Sec. 15-3 for reordering, taking subsequences, taking functions, and collecting together random variables always hold. As in the discrete case, we can easily construct models of independence for the continuous-density case. With these general properties established, we then concentrate on the convolution formula, which was derived in Example 3 (Sec. 16-3) and is here applied to the normal and gamma densities. The gamma densities are introduced in this section. They include the exponential densities, and as indicated in the exercises, they are intimately related to many other densities of importance in probability and statistics.

Theorem 1*a* below states that if there is a *joint* continuous density then independence is equivalent to the joint density being given by the product rule applied to the marginals. Theorem 1*b* asserts that if *independent* random vectors *each* have continuous densities then they have a joint continuous density. As shown in Exercise 1, two *dependent* random variables can each have continuous marginal densities even though their joint distribution does not correspond to a continuous density.

Theorem 1 Let $f_1, \ldots, f_k$ where $f_i: \Omega \to R^{n_i}$ be random vectors with the same associated probability space (Ω, P) and let $n = n_1 + \cdots + n_k$.

(*a*) If $f_1, \ldots, f_k$ induce a joint continuous density on R^n then from Theorem 1 (Sec. 16-2) each f_i induces a continuous density. In this case $f_1, \ldots, f_k$ are independent iff the joint density is equivalent to the density $p_{f_1}(\alpha_1)p_{f_2}(\alpha_2) \cdots p_{f_k}(\alpha_k)$, where each $\alpha_i \in R^{n_i}$, obtained from the product rule applied to the marginals.

(*b*) If $f_1, \ldots, f_k$ are independent and each induces a continuous density $p_{f_i}: R^{n_i} \to R$ then they have a joint continuous density, given as in part (*a*). ■

Proof Let A_i be a closed rectangle in R^{n_i} so that $A = A_1 \times A_2 \times \cdots \times A_k$ is a closed rectangle in R^n. Clearly every closed rectangle in R^n can be expressed in this form.

If there is a joint continuous density satisfying the product rule $p_{f_1, \ldots, f_k} = p_{f_1}p_{f_2} \cdots p_{f_k}$ then $P_{f_1, \ldots, f_k}(A)$ equals the integral of it over A. But this equals the product, over i, of the integral of p_{f_i} over A_i so

$$P_{f_1, \ldots, f_k}(A_1 \times \cdots \times A_k) = P_{f_1}(A_1)P_{f_2}(A_2) \cdots P_{f_k}(A_k)$$

and hence $f_1, \ldots, f_k$ are independent. This proves half of part (*a*).

If $f_1, \ldots, f_k$ are independent and each induces a continuous density $p_{f_i}: R^{n_i} \to R$ then, using a casual but obvious notation for integrals, we have

$$\begin{aligned} P_{f_1, \ldots, f_k}(A) &= P_{f_1}(A_1)P_{f_2}(A_2) \cdots P_{f_k}(A_k) \\ &= \left(\int_{A_1} p_{f_1}\right)\left(\int_{A_2} p_{f_2}\right) \cdots \left(\int_{A_k} p_{f_k}\right) = \int_A p_{f_1}p_{f_2} \cdots p_{f_k}. \end{aligned}$$

Therefore $P_{f_1, \ldots, f_k}$ and the density $p_{f_1}p_{f_2} \cdots p_{f_k}$ assign the same probability to each A, so Theorem 2 (Sec. 15-1) implies that $p_{f_1}p_{f_2} \cdots p_{f_k}$ is a continuous density for $P_{f_1, \ldots, f_k}$. This proves part (*b*), which implies the remaining half of part (*a*). ■

Thus if each f_i has a continuous density p_{f_i} then saying that $f_1, \ldots, f_k$ are independent is equivalent to saying that $p_{f_1} \cdots p_{f_k}$ is a joint density for $f_1, \ldots, f_k$. In this case

$$p_{f_1, \ldots, f_k} = (p_{f_1} \cdots p_{f_m})(p_{f_{m+1}} \cdots p_{f_k}) = p_{f_1, \ldots, f_m} p_{f_{m+1}, \ldots, f_k}$$

so $(f_1, \ldots, f_m)$ and $(f_{m+1}, \ldots, f_k)$ are independent. This was proved in Theorem 2 (Sec. 15-3), which required a lengthy proof since it did not assume the existence of densities. Obviously reordering and taking a subsequence do not destroy independence.

As shown in Theorem 1 (Sec. 15-3), $g_1(f_1), \ldots, g_k(f_k)$ are independent if $f_1, \ldots, f_k$ are.

Thus the general properties of independence as derived in Sec. 15-3 are certainly available in the continuous-density case. We next note that in the continuous-density case we can construct models of independent experiments in the same way we did for the discrete case in Theorem 2 (Sec. 11-2).

For example, if we have two continuous densities $p'\colon R^2 \to R$ and $p''\colon R \to R$ then we can define a density $p\colon R^3 \to R$ by the product rule,

$$p(x_1,x_2,x_3) = p'(x_1,x_2)p''(x_3) \qquad \text{for all real } x_1, x_2, x_3.$$

Let X_i be the ith coordinate function on R^3, so that $X_i(x_1,x_2,x_3) = x_i$ and X_1, X_2, X_3 are random variables with the same associated density $p\colon R^3 \to R$. Clearly X_1, X_2, X_3 have joint density p, X_1 and X_2 have joint density p', X_3 has density p'', and (X_1,X_2) and X_3 are independent. Thus if we start with continuous densities $p_1, \ldots, p_k$ and we want to construct random vectors $f_1, \ldots, f_k$ with the same associated continuous density p such that f_i induces p_i, and $f_1, \ldots, f_k$ are independent, we can do so by defining p by the product rule and letting f_i be the appropriate coordinate vectors. *Therefore* it is always possible to construct an experiment with independent random vectors having arbitrarily specified continuous marginal densities. In particular if $p'\colon R \to R$ is a continuous density on R and $p\colon R^k \to R$ is defined by

$$p(x_1, \ldots, x_k) = p'(x_1)p'(x_2) \cdots p'(x_k)$$

and $X_1, \ldots, X_k$ are the coordinate functions on R^k, then they are independent and each has p' for induced density.

Exercises 1 to 3 may be done before reading the remainder of this section.

CONVOLUTIONS

We now turn our attention to the convolution formula, apply it to normal densities, and then introduce and convolve gamma densities.

Example 3 (Sec. 16-3) derived the basic fact that if X and Y are independent random variables having continuous densities then $Z = X + Y$ has a continuous density obtainable by convolving p_X and p_Y,

$$p_Z(z) = \int_{-\infty}^{\infty} p_X(x)p_Y(z - x)\,dx = \int_{-\infty}^{\infty} p_X(z - y)p_Y(y)\,dy \qquad \text{for each real } z.$$

We sometimes denote convolution by $*$ so that the convolution formula becomes

$$p_Z = p_X * p_Y = p_Y * p_X$$

and the operation is commutative. This operation is also associative, so that we can drop the parentheses used to indicate which of two convolutions is performed first,

$$(p_{X_1} * p_{X_2}) * p_{X_3} = p_{X_1} * (p_{X_2} * p_{X_3}) = p_{X_1} * p_{X_2} * p_{X_3}.$$

For some examples, as for the gamma densities below, the range of integration in a convolution may be reduced. Clearly if $p_X(x) = 0$ whenever $x < 0$ then

$$p_Z(z) = \int_0^\infty p_X(x) p_Y(z - x)\, dx.$$

If, in addition, $p_Y(y) = 0$ whenever $y < 0$ then for $z < 0$ the integrand is always zero, and for $z > 0$ the integrand is zero whenever $x > z$. A similar argument may be applied to the other form of convolution. *Thus* if $Z = X + Y$, where X and Y are independent with continuous densities p_X and p_Y for which $p_X(x) = p_Y(y) = 0$ whenever $x < 0$ and $y < 0$, then

$$p_Z(z) = \begin{cases} \displaystyle\int_0^z p_X(x) p_Y(z - x)\, dx = \int_0^z p_X(z - y) p_Y(y)\, dy & \text{for } z > 0. \\ 0 & \text{for } z \le 0 \end{cases}$$

Example 1 We now derive the fundamental fact that *the collection of normal densities is closed under independent additions.* More precisely, the collection of normal densities is closed under the operation of convolution. Equivalently any sum of a finite number of independent normally distributed random variables has a normal distribution.

Let $Z = X_1 + X_2$ where X_1 and X_2 are independent and X_i has normal density $\varphi_{\mu_i,\sigma}$ with mean μ_i and variance σ_i^2. Let $\mu = \mu_1 + \mu_2$ and $\sigma^2 = \sigma_1^2 + \sigma_2^2$. We show that Z has normal density $\varphi_{\mu,\sigma}$. By the convolution formula, for each fixed z, we have

$$\begin{aligned} p_Z(z) &= \int_{-\infty}^{\infty} \varphi_{\mu_1,\sigma_1}(x_1) \varphi_{\mu_2,\sigma_2}(z - x_1)\, dx_1 \\ &= \frac{1}{\sigma_1 \sigma_2 2\pi} \int_{-\infty}^{\infty} \exp\left\{ -\frac{1}{2\sigma_1^2}(x_1 - \mu_1)^2 - \frac{1}{2\sigma_2^2}(z - x_1 - \mu_2)^2 \right\} dx_1 \\ &= \frac{1}{\sigma_1 \sigma_2 2\pi} \int_{-\infty}^{\infty} \exp\left\{ -\frac{1}{2}\left[\frac{y^2}{\sigma_1^2} + \frac{[y - (z - \mu)]^2}{\sigma_2^2} \right] \right\} dy \end{aligned}$$

where for the last equality we made the transformation $y = x_1 - \mu_1$ and $dy = dx_1$. The algebraic identity below is easily verified:

$$\frac{y^2}{\sigma_1^2} + \frac{[y - (z - \mu)]^2}{\sigma_2^2} = \frac{(z - \mu)^2}{\sigma^2} + \frac{(y - a)^2}{b^2}$$

where $a = (\sigma_1^2/\sigma^2)(z - \mu)$ and $b = \sigma_1 \sigma_2 / \sigma$. Substituting this identity into the last expression for p_Z, we get

$$p_Z(z) = \int_{-\infty}^{\infty} \varphi_{\mu,\sigma}(z) \varphi_{a,b}(y)\, dy = \varphi_{\mu,\sigma}(z)$$

since z, a, and b are constants in so far as integration with respect to y is concerned. Thus $p_Z = \varphi_{\mu,\sigma}$ as was to be shown.

This paragraph provides an alternate[1] geometrically flavored proof of the same result. As a preliminary reduction note that

$$Z = X_1 + X_2 = \sigma Y + (\mu_1 + \mu_2) \qquad \text{where} \qquad Y = \frac{\sigma_1}{\sigma} X_1^* + \frac{\sigma_2}{\sigma} X_2^*$$

[1] Suggested by David A. Freedman.

and $X_i^* = (X_i - \mu_i)/\sigma_i$. Clearly Z is normal if Y is normal. Define $0 < \theta < \pi/2$ by $\cos\theta = \sigma_1/\sigma$. Thus we need only show that

$$Y = X_1^* \cos\theta + X_2^* \sin\theta$$

is normal. But this was proved in Example 6 (Sec. 16-3). As shown there the normality of Y, and hence of Z, follows from the radial symmetry of the joint density of two independent standardized normal random variables. This radial symmetry is clearly related to the uniqueness property stated after Example 5 (Sec. 16-5), and to the functional equation of Lemma 2 in the Addendum to Sec. 16-1, essentially $p(x^2)p(y^2) = p(x^2 + y^2)$.

If X_1, X_2, X_3 are independent and X_i has density φ_{μ_i,σ_i} we now know that $Z = X_1 + X_2$ has normal density with mean $\mu_1 + \mu_2$ and variance $\sigma_1^2 + \sigma_2^2$. But Z and X_3 are independent, so the same result shows that $X_1 + X_2 + X_3 = Z + X_3$ has normal density with mean $(\mu_1 + \mu_2) + \mu_3$ and variance $(\sigma_1^2 + \sigma_2^2) + \sigma_3^2$.

More generally, if $X_1, \ldots, X_n$ are independent and X_i has normal density with mean μ_i and variance σ_i^2 then $\sum_{i=1}^{n} X_i$ has normal density with mean $\sum_{i=1}^{n} \mu_i$ and variance $\sum_{i=1}^{n} \sigma_i^2$. In terms of convolutions we have shown that

$$\varphi_{\mu_1,\sigma_1} * \varphi_{\mu_2,\sigma_2} * \cdots * \varphi_{\mu_n,\sigma_n} = \varphi_{\mu,\sigma}$$

where $\mu = \sum_{i=1}^{n} \mu_i$ and $\sigma^2 = \sum_{i=1}^{n} \sigma_i^2$.

We now easily show that *the collection of lognormal densities is closed under independent products*. If X_i is normal μ_i and σ_i^2 then by definition $Y_i = e^{X_i}$ is lognormal μ_i and σ_i^2. If $X_1, \ldots, X_n$ are independent then so are $e^{X_1}, e^{X_2}, \ldots, e^{X_n}$, and their product equals

$$(e^{X_1})(e^{X_2}) \cdots (e^{X_n}) = \exp\left\{\sum_{i=1}^{n} X_i\right\}$$

where $\sum_{i=1}^{n} X_i$ is normal. Thus if $Y_1, Y_2, \ldots, Y_n$ are independent and Y_i is lognormal μ_i and σ_i^2 then their product $Y_1 Y_2 \cdots Y_n$ is lognormal $\sum_{i=1}^{n} \mu_i$ and $\sum_{i=1}^{n} \sigma_i^2$. ■

Exercises 1 to 7 may be done before reading the remainder of this section.

THE GAMMA DENSITIES

The gamma densities form a two-parameter family of continuous probability densities on the positive real axis. As indicated in Example 3 and in the exercises, this family contains many densities of importance in statistical theory. Densities having the gamma-density form arise perhaps most naturally as induced by the sum of independent random variables all having the same exponential density. In particular it is in this fashion that gamma densities appear in Theorem 1 (Sec. 23-3), where they are related to Poisson densities and Poisson processes.

Neglecting constants, gamma densities are zero for $x < 0$ and of the form $x^b e^{-x}$ for $x > 0$. The cases $b = -\frac{1}{2}$ and $b = 1$ have already arisen in Example 8 (Sec. 16-2) and Example 2 (Sec. 16-3), respectively, while $b = 0$ is an exponential density. We first find the values of b which make such nonnegative functions integrable.

For x positive but near zero the function $x^b e^{-x}$ is almost equal to x^b. Therefore we must have $b > -1$, since $\int_0^1 (1/x)\,dx$ is infinite and $\int_0^1 x^b\,dx$ is finite if $b > -1$. We replace b by $\nu - 1$, so that we will be constrained to positive real ν. Now $x^{\nu-1}e^{-x}$ is positive and $\nu > 0$ guarantees a finite integral over $0 \le x \le 1$. Furthermore the negative exponential dominates for large x and guarantees a finite integral over $1 \le x < \infty$. Therefore we can divide this function by its integral over $0 < x < \infty$ and so obtain for each positive ν a continuous density which we denote by

$$\gamma_{1,\nu}(x) = \frac{x^{\nu-1}e^{-x}}{\displaystyle\int_0^\infty t^{\nu-1}e^{-t}\,dt} \qquad \text{for } x > 0.$$

Clearly $\gamma_{1,1}$ is the exponential density with mean 1. Several of these densities are shown in Fig. 2.

It will prove convenient to enlarge this class by including those densities obtainable by a positive scale change. If X has density $\gamma_{1,\nu}$ and $Y = X/\lambda$, where $\lambda > 0$, then Y has density $p_Y(y) = \lambda\gamma_{1,\nu}(\lambda y)$, which we will denote by $\gamma_{\lambda,\nu}$. Thus if $\lambda > 0$ and $\nu > 0$ then

$$\gamma_{\lambda,\nu}(x) = \frac{\lambda^\nu x^{\nu-1}e^{-\lambda x}}{\displaystyle\int_0^\infty t^{\nu-1}e^{-t}\,dt} \qquad \text{if } x > 0.$$

Clearly $\gamma_{\lambda,1}$ is the exponential density with mean $1/\lambda$. The normalizing function of ν in the denominator is the classical **gamma function**, which is defined by the integral

$$\Gamma(\nu) = \int_0^\infty t^{\nu-1}e^{-t}\,dt \qquad \text{for all positive real } \nu.$$

We will shortly derive the only properties of Γ which we need.

Thus we have introduced the **gamma continuous probability density** $\gamma_{\lambda,\nu}\colon R \to R$ with parameters λ and ν by

$$\gamma_{\lambda,\nu}(x) = \begin{cases} \dfrac{\lambda^\nu}{\Gamma(\nu)}\, x^{\nu-1}e^{-\lambda x} & \text{if } x > 0 \\ 0 & \text{if } x \le 0 \end{cases}$$

where λ and ν can be any positive real numbers and $\Gamma(\nu)$ is the gamma function. If X has density $\gamma_{\lambda,\nu}$ and if $a > 0$ then aX has the gamma density with parameters λ/a and ν.

It is immediate from integration by parts that the gamma function satisfies the *recursion relation*

$$\Gamma(\nu) = (\nu - 1)\Gamma(\nu - 1) \qquad \text{for all real } \nu > 1.$$

Clearly $\Gamma(1) = 1$, so $\Gamma(2) = (2 - 1)\Gamma(1) = 1$, and by using the relation above repeatedly, we see that if $\nu = 1, 2, \ldots$ is a positive integer then

$$\Gamma(\nu) = (\nu - 1)(\nu - 2)\Gamma(\nu - 2) = \cdots = (\nu - 1)(\nu - 2)\cdots(3)(2)(1) = (\nu - 1)!$$

where, as usual, $0! = 1$. The Γ function is graphed in Fig. 1. The only remaining property of Γ which we need is $\Gamma(\frac{1}{2}) = \sqrt{\pi}$. To show this, note that the normal

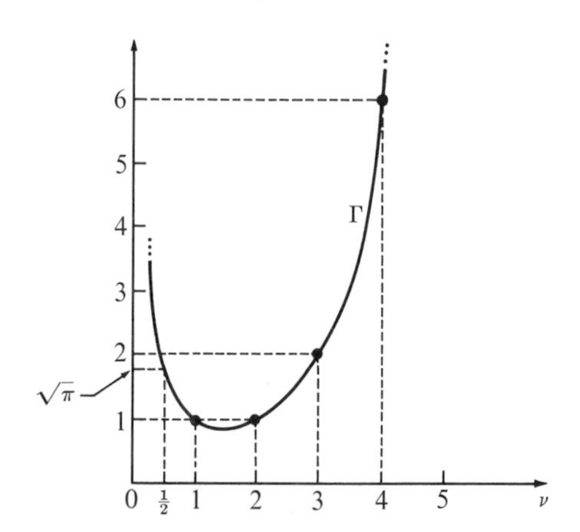

Fig. 1 The gamma function. $\Gamma(\nu) = (\nu - 1)!$ for $\nu = 1, 2, 3, \ldots$.

density $\varphi_{0,1}$ is symmetric, and make the change of variables $y = x^2/2$ and $dy = x\,dx = \sqrt{2y}\,dx$ to get

$$1 = \frac{2}{\sqrt{2\pi}} \int_0^\infty e^{-x^2/2}\,dx = \frac{1}{\sqrt{\pi}} \int_0^\infty \frac{1}{\sqrt{y}} e^{-y}\,dy = \frac{1}{\sqrt{\pi}}\,\Gamma(\tfrac{1}{2}).$$

Thus if $\nu = 1, 2, \ldots$ is a positive integer and $\lambda > 0$ then

$$\gamma_{\lambda,\nu}(x) = \frac{\lambda^\nu}{(\nu - 1)!}\, x^{\nu-1} e^{-\lambda x} \qquad \text{if } x > 0.$$

This is the case of greatest interest to us, but the case where ν equals an integer plus $\frac{1}{2}$ is also of interest.

Gamma densities with various parameter values arise in a variety of statistical contexts. Let us proceed to find their moments, verify their shape as shown in Fig. 2, and convolve them. We first show that if r is any, possibly negative, real number satisfying $r > -\nu$ then

$$[\text{the } r\text{th moment of } \gamma_{\lambda,\nu}] = \int_0^\infty x^r \gamma_{\lambda,\nu}(x)\,dx = \frac{\Gamma(r + \nu)}{\lambda^r \Gamma(\nu)}.$$

For x near zero $\gamma_{\lambda,\nu}$ acts like x^ν/x, so for a finite integral we must have $r + \nu > 0$. If $r + \nu > 0$ then the identity

$$x^r \gamma_{\lambda,\nu}(x) = \frac{\Gamma(r + \nu)}{\lambda^r \Gamma(\nu)}\, \gamma_{\lambda,r+\nu}$$

is easily verified. Integrating over all x yields the stated result.

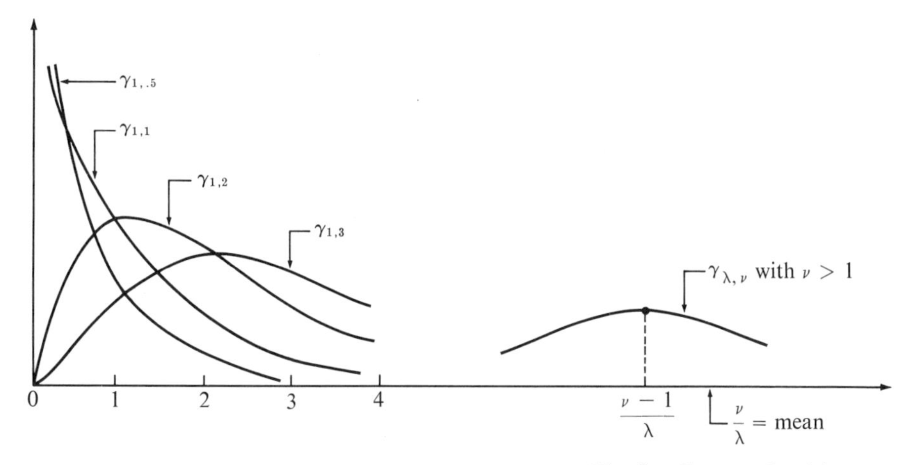

Fig. 2 Gamma densities $\gamma_{\lambda,\nu}$.

If $r = 1, 2, \ldots$ is a positive integer then the recursion relation for Γ yields

$$[\text{the } r\text{th moment of } \gamma_{\lambda,\nu}] = (r + \nu - 1)(r + \nu - 2) \cdots (\nu + 1)\nu \frac{1}{\lambda^r}.$$

In particular the mean is ν/λ and the second moment is $(\nu + 1)\nu/\lambda^2$ so the variance is

$$\frac{(\nu + 1)\nu}{\lambda^2} - \left(\frac{\nu}{\lambda}\right)^2 = \frac{\nu}{\lambda^2}.$$

Thus the gamma density $\gamma_{\lambda,\nu}$ has mean ν/λ and variance ν/λ^2.

We now show that gamma densities have the qualitative shapes shown in Fig. 2. If $x > 0$ the identity

$$\gamma'_{\lambda,\nu}(x) = \frac{d}{dx}\gamma_{\lambda,\nu}(x) = \left\{\frac{\lambda^{\nu+1}}{\Gamma(\nu)} x^{\nu-2}e^{-\lambda x}\right\}\left[\frac{\nu - 1}{\lambda} - x\right]$$

is easily verified. Now $\{\ \} > 0$ so $\gamma'_{\lambda,\nu}$ has the same sign as $[\ ]$. Therefore if $0 < \nu \le 1$ then $\gamma'_{\lambda,\nu}(x) < 0$ for all $x > 0$, so $\gamma_{\lambda,\nu}$ is strictly decreasing. If $\nu > 1$ then $\gamma'_{\lambda,\nu}(x) > 0$ iff $(\nu - 1)/\lambda > x$, so $\gamma_{\lambda,\nu}$ increases to a maximum occurring at $x = (\nu - 1)/\lambda$ and then decreases. The location $x = (\nu - 1)/\lambda$ of its maximum is to left of the mean ν/λ.

We now show that the family of gamma densities having the same parameter λ is closed under convolution. More particularly,

$$\gamma_{\lambda,\nu} * \gamma_{\lambda,\mu} = \gamma_{\lambda,\nu+\mu} \qquad \text{if } \nu > 0,\ \mu > 0$$

so that the second parameter adds under convolution. If we let $f = \gamma_{\lambda,\nu} * \gamma_{\lambda,\mu}$ then for each fixed $z > 0$ we have

$$\begin{aligned} f(z) &= \int_0^z \gamma_{\lambda,\nu}(x)\gamma_{\lambda,\mu}(z - x)\,dx = \frac{\lambda^{\nu+\mu}}{\Gamma(\nu)\Gamma(\mu)} \int_0^z x^{\nu-1}e^{-\lambda x}(z - x)^{\mu-1}e^{-\lambda(z-x)}\,dx \\ &= \frac{\lambda^{\nu+\mu}e^{-\lambda z}}{\Gamma(\nu)\Gamma(\mu)} \int_0^z x^{\nu-1}(z - x)^{\mu-1}\,dx. \end{aligned}$$

We make the change of variables $t = x/z$ and $dt = dx/z$ and get $f(z) = A\lambda^{\nu+\mu} \times z^{\nu+\mu-1} e^{-\lambda z}$ where

$$A = \frac{1}{\Gamma(\nu)\Gamma(\mu)} \int_0^1 t^{\nu-1}(1-t)^{\mu-1}\, dt$$

so A depends on ν and μ but not z. Thus $f = A\Gamma(\nu+\mu)\gamma_{\lambda,\nu+\mu}$ where both f and $\gamma_{\lambda,\nu+\mu}$ are probability densities. *Note* that if two continuous densities $p: R^n \to R$ and p': $R^n \to R$ are proportional, $p = Bp'$, then by integrating over R^n we get $B = 1$. Therefore $A\Gamma(\nu+\mu) = 1$ and the proof is complete. Incidentally we have also evaluated A to be equal to $1/\Gamma(\nu+\mu)$. Using the result just obtained, twice, yields

$$(\gamma_{\lambda,\nu_1} * \gamma_{\lambda,\nu_2}) * \gamma_{\lambda,\nu_3} = \gamma_{\lambda,\nu_1+\nu_2} * \gamma_{\lambda,\nu_3} = \gamma_{\lambda,\nu_1+\nu_2+\nu_3}.$$

More generally, if $X_1, X_2, \ldots, X_n$ are independent random variables, where each X_i has gamma density γ_{λ,ν_i} with the same $\lambda > 0$ parameter, then $X_1 + X_2 + \cdots + X_n$ has the gamma density $\gamma_{\lambda,\nu}$ where $\nu = \nu_1 + \nu_2 + \cdots + \nu_n$. Naturally the means and variances add, as they should,

$$\frac{\nu}{\lambda} = \sum_{i=1}^{n} \frac{\nu_i}{\lambda} \qquad \frac{\nu}{\lambda^2} = \sum_{i=1}^{n} \frac{\nu_i}{\lambda^2}.$$

Example 2 If $X_1, \ldots, X_n$ are independent and each X_i has the same exponential density $\gamma_{\lambda,1}$ with mean $1/\lambda$ and variance $1/\lambda^2$, then $X_1 + \cdots + X_n$ has gamma density $\gamma_{\lambda,n}$ with mean n/λ and variance n/λ^2. ■

Example 3 If $Y = X^2$ where X is normal with mean 0 and variance 1, then from Example 8 (Sec. 16-2) we have

$$p_Y(y) = \frac{1}{\sqrt{2\pi y}} e^{-y/2} \qquad \text{if } y > 0$$

so $p_Y = \gamma_{1/2,1/2}$. Therefore if we add n independent random variables each having this density the sum will have density $\gamma_{1/2,n/2}$, which has a traditional name. For each $n = 1, 2, \ldots$ define the continuous **chi-square density** with n degrees of freedom by

$$f_n(x) = \begin{cases} \gamma_{1/2,n/2}(x) = \dfrac{1}{\Gamma(n/2)2^{n/2}} x^{(n/2)-1} e^{-x/2} & \text{if } x > 0 \\ 0 & \text{if } x \le 0. \end{cases}$$

From the above we know that if $X_1, \ldots, X_n$ are independent and each is normal with mean 0 and variance 1 then $\chi_n{}^2 = X_1{}^2 + X_2{}^2 + \cdots + X_n{}^2$ has a chi-square density with n degrees of freedom.

The quantity $\Gamma(n/2)$ appears in f_n. If n is an even integer then $n/2$ is an integer, so $\Gamma(n/2) = [(n/2) - 1]!$ If n is an odd integer then $n/2 = [(n-1)/2] + \frac{1}{2} = [\text{an integer}] + \frac{1}{2}$, in which case the recursion formula for Γ yields

$$\Gamma\left(\frac{n}{2}\right) = \left(\frac{n}{2} - 1\right)\left(\frac{n}{2} - 2\right) \cdots \left(\frac{3}{2}\right)\left(\frac{1}{2}\right)\Gamma\left(\frac{1}{2}\right)$$

where $\Gamma\left(\frac{1}{2}\right) = \sqrt{\pi}$. Exercise 9 justifies the alternate form

$$\Gamma\left(\frac{n}{2}\right) = \frac{(n-1)!\,\sqrt{\pi}}{[(n-1)/2]!\,4^{(n-1)/2}} \qquad \text{if } n = 1, 3, 5, \ldots. \quad \blacksquare$$

EXERCISES

1. *Nondensity Joint Distributions Can Have Marginals That Are Discrete or Continuous Densities.* A point (X, Y) is selected according to a uniform distribution on the straight road between two towns located at (0,0) and (1,1). Thus for every subset A of the plane we have

$$P_{X,Y}(A) = \frac{1}{\sqrt{2}}\,[\text{length of } A \cap D]$$

where D is the diagonal segment shown in Fig. 3. Clearly $P_{X,Y}$ assigns zero probability to every point in the plane and so cannot correspond to a discrete density. If $P_{X,Y}$ corresponded

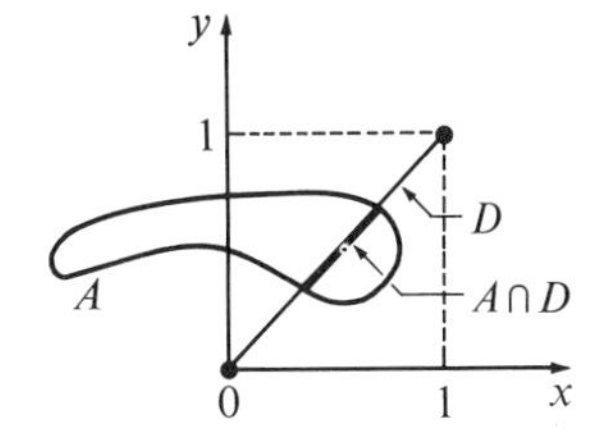

Fig. 3

to a continuous density on the plane then the integral of $p_{X,Y}$ over D would have to equal 1. But D has zero area, and the integral over a set of zero area of any function $f: R^2 \to R$ equals 0. Thus $P_{X,Y}$ does not correspond to a discrete or continuous density on R^2.

(*a*) Find the continuous densities induced by X and by Y. COMMENT: Each of X and Y has a continuous density even though they do not have a joint continuous density.

(*b*) Suppose that the road is a nice continuous meandering curve of finite length between the two towns, and that (X, Y) is distributed uniformly according to length on the road. Once again, $P_{X,Y}$ cannot correspond to any density. Sketch roads such that

neither, exactly one, both

of X and Y induce continuous densities.

(*c*) Let X be normal 0 and 1 and define $Y = aX + b$ for $a \neq 0$, so that Y is also normal. Describe the joint distribution of X and Y and note that it cannot correspond to a density on the plane. What is the distribution of $X + Y$? Note the case $a = -1$. COMMENT: Such distributions $P_{X,Y}$ are sometimes called "singular" bivariate normal distributions. Each is concentrated on a straight line. The nonsingular bivariate normal distributions all have continuous densities and are exhibited in Example 1 (Sec. 18-2).

2. *Different Joint Densities with the Same Continuous Density Marginals.* Exercise 7 (Sec. 11-2) showed how to construct many discrete densities having the same marginals. This exercise applies that technique to continuous densities. The continuous densities p_X and p_Y are specified, as are the numbers a, b, $\delta > 0$, and $d > 0$ satisfying

$$p_X(x) \geq d \qquad \text{if } |x - a| \leq \delta$$

$$p_Y(y) \geq d \qquad \text{if } |y - b| \leq \delta.$$

For each real λ satisfying $|\lambda| \leq 1$ define the function $p_\lambda: R^2 \to R$ by

$$p_\lambda(x,y) = \begin{cases} p_X(x)p_Y(y) + \lambda d^2 \left(\dfrac{x-a}{\delta}\right)\left(\dfrac{y-b}{\delta}\right) & \text{if } |x-a| \leq \delta, |y-b| \leq \delta \\ p_X(x)p_Y(y) & \text{otherwise.} \end{cases}$$

(*a*) Show that every p_λ is a continuous density having p_X and p_Y for marginals.
(*b*) Assume that p_λ is the joint density of X and Y. Exhibit two sets A and B of reals such that the events $[X \in A]$ and $[Y \in B]$ are dependent whenever $\lambda \neq 0$.

3. *A Special Case of Theorem 1.* State and prove Theorem 1 for the special case $k = 2$, $n_1 = 2$, and $n_2 = 1$, so that the statement starts as follows: "let (X_1,X_2) and X_3 be two random vectors with the same associated probability space (Ω,P). . . ." Exhibit the probabilities of events as integrals of densities.

4. *The Possible Impact of "Poor" Instruments.* Three instruments are to be used in estimating a length l. The ith instrument yields $l + X_i$ where the errors X_1, X_2, X_3 are independent and normal. The variable

$$\bar{X} = \frac{1}{3}\sum_{i=1}^{3}(l + X_i) = l + \tfrac{1}{3}(X_1 + X_2 + X_3)$$

is to be used to estimate l. Assume that

$$\mu_1 = 0 \qquad \mu_2 = -1 \qquad \mu_3 = 1$$

$$\sigma_1 = 1 \qquad \sigma_2 = 2 \qquad \sigma_3 = 5.$$

(*a*) Find $P[|\bar{X} - l| < 2]$.
(*b*) For this example would it be preferable to just use the first instrument?

5. *The Counteracting Effect of Small and Large Random Factors.* Let $V = XYZ$ be the volume of a block of metal having sides X, Y, and Z, where X is lognormal with parameters $\mu = 4$, $\sigma^2 = .09$. A process is supposed to produce cubes with each side having length e^4, but there are random errors. Find numbers a, b, and c such that

$$P[V < e^a] = \Phi(-1) \qquad P[V < e^b] = .5 \qquad P[V < e^c] = \Phi(1)$$

under the following assumptions:
(*a*) The block is a cube, so $X = Y = Z$.
(*b*) X, Y, and Z are independent and identically distributed.
(*c*) State briefly why it is intuitively reasonable that V should have less spread under assumption (*b*) than under (*a*).

6. *Products of Lognormal Voltage Gains.* Three amplifiers are to be purchased and connected in series. The resulting amplifier will have voltage gain Z, so that the output in volts will be Z times the input in volts, where the gain is measured at a specified center frequency. Assume that $Z = Y_1Y_2Y_3$ where Y_1, Y_2, Y_3 are independent and each is lognormal with parameters $\mu = 2$ and $\sigma^2 = .09$ (the assumption that Y_1 is lognormal is equivalent to the gain $10 \log_{10} Y_1$, as measured in decibels, being normal).

(*a*) Prove the inequality

$$P[Y_1 > e^2,\ Y_2 > e^2,\ Y_3 > e^2] \le P[Z > e^6]$$

and evaluate both sides.

(*b*) Find $P[Z > e^5]$.

7. *Sums of Independent Uniformly Distributed Random Variables.* Let $S_n = X_1 + \cdots + X_n$ where $X_1, \ldots, X_n$ are independent and each is uniformly distributed over $0 < x < 1$.

(*a*) Find p_{S_2} from $p_{S_2} = p_{X_1} * p_{X_2}$.

(*b*) Find p_{S_3} from $p_{S_3} = p_{S_2} * p_{X_3}$.

8. *The Distribution of the Average Interarrival Time in a Poisson Process.* $X_1, \ldots, X_n$ are independent, and each is exponential with mean $1/\lambda = 3$ minutes. Interpret X_1 as the time, measured in minutes after 2 P.M., of arrival of the first airplane after 2 P.M., and X_k for $k \ge 2$ as the time between the arrivals of the kth and $(k-1)$st airplanes. These assumptions are common for airplanes, electron emissions, automobile traffic, etc. What is the distribution of the arithmetic average $(1/n)\sum_{k=1}^{n} X_k$ of the interarrival times?

9. *The Constant $\Gamma(n/2)$ in the Chi-square Density of Example 3.* Prove that

$$\Gamma\left(\nu + \frac{1}{2}\right) = \frac{(2\nu)!\,\sqrt{\pi}}{\nu!\,4^\nu} \qquad \text{if } \nu = 0, 1, 2, \ldots.$$

10. *An Equivalent Simulation Experiment.* Voltages $X_1, \ldots, X_6$ are independent and each is normal 0 and 1. A simulation experiment requires random observations having the distribution of $Z = \sum_{i=1}^{6} X_i^2$. Independent random variables $Y_1, Y_2, \ldots$ each uniformly distributed over $0 < y < 1$ are available in an approximate form, as from a random-number table. Someone asserts that $-2 \log Y_1 Y_2 Y_3$ has the same distribution as Z.

(*a*) Prove it.

(*b*) Use the random-number table (Sec. 12-2) to obtain two independent observations from the distribution of Z.

11. *Total Kinetic Energy.* The magnitude of the velocity of the ith molecule of a gas is S_i where $S_1, \ldots, S_n$ are independent and each has a Maxwell density with parameter σ as in Example 1 (Sec. 16-3). Each has the same mass m, so that their total kinetic energy is $K = (m/2)\sum_{i=1}^{n} S_i^2$. Find the density induced by K.

12. *A "Geometric" Derivation of the Chi-square Density, with the Volume of a Sphere in R^n as a By-product.* Let $\chi_n^2 = X_1^2 + \cdots + X_n^2$ where $X_1, \ldots, X_n$ are independent and each is normal 0 and 1.

(*a*) Ignore Example 3 and give a simple derivation, similar to that of Example 1 (Sec. 16-3), of

$$p_{\chi_n^2}(x) = \left(\frac{1}{\sqrt{2\pi}}\right)^n \frac{n}{2} V_n(1) x^{(n/2)-1} e^{-x/2} \qquad \text{if } x > 0$$

where $V_n(1)$ is the volume of a unit sphere in R^n. HINT: The volume $V_n(r)$ of a unit sphere in R^n of radius $r > 0$ can be defined as

$$V_n(r) = \int \cdots \int_{\sum_{i=1}^{n} x_i^2 \le r^2} dx_1\, dx_2 \cdots dx_n.$$

The transformation of variables $y_1 = x_1/r, \ldots, y_n = x_n/r$ shows that $V_n(r) = V_n(1)r^n$. The surface area $S_n(r)$ of a sphere of radius r in R^n can be defined by

$$S_n(r) = \frac{dV_n(r)}{dr} = nV_n(1)r^{n-1}$$

Hence $S_n(1) = nV_n(1)$. Thus the volume $V_n(r + \Delta) - V_n(r) \doteq S_n(r)\Delta$ of a thin spherical shell approximately equals the surface area of the inner sphere times the thickness of the shell.

(*b*) Since $p_{\chi_n^2}$ of part (*a*) is a probability density, we must have $\int_0^\infty p_{\chi_n^2}(x)\,dx = 1$, which yields

$$V_n(1) = \frac{2}{n}(\sqrt{2\pi})^n \frac{1}{\displaystyle\int_0^\infty x^{(n/2)-1}e^{-x/2}\,dx} = \frac{\pi^{n/2}}{(n/2)\Gamma(n/2)} = \frac{\pi^{n/2}}{\Gamma((n/2)+1)}$$

where we made the change of variables $y = x/2$. Thus we have shown that $p_{\chi_n^2} = f_n$ of Example 3. The derivation yielded $V_n(1)$ as a side-product. Define C_n by $V_n(1) = C_n\pi^{n/2}$ if n is even and by $V_n(1) = C_n\pi^{(n-1)/2}$ if n is odd. Tabulate C_n for $n = 1, 2, \ldots, 7$.

13. *The Derivation of Student's t, a Basic Density in Statistical Theory.*

(*a*) Let X and W be independent, where X is normal 0 and 1 and W is gamma $\lambda = 1$ and $\nu > 0$. Let $Z = X/\sqrt{W}$ and use Exercise 5 (Sec. 16-3) to show that

$$p_Z(z) = \frac{\Gamma(\nu + \frac{1}{2})}{\sqrt{2\pi}\,\Gamma(\nu)}\left(1 + \frac{z^2}{2}\right)^{-[\nu+(1/2)]} \qquad \text{for all real } z.$$

COMMENT: p_Z is an even function. For large z, $p_Z(z)$ is like $c/z^{2\nu+1}$ and so has a finite rth moment for $r > 0$ iff $r < 2\nu$. In particular the MGF of p_Z is finite only for $t = 0$.

(*b*) If X and Y are independent, where X is normal 0 and 1, and Y is chi-square with n degrees of freedom then $T = X/\sqrt{(1/n)Y}$ is said to have a **Student's *t* distribution** with n degrees of freedom. Show that

$$p_T(t) = \frac{\Gamma((n+1)/2)}{\sqrt{n\pi}\,\Gamma(n/2)}\left(1 + \frac{t^2}{n}\right)^{-(n+1)/2} \qquad \text{for all real } t.$$

14. *The Derivation of the Snedecor F.* Let X and Y be independent random variables with respective gamma densities $\gamma_{1,\nu}$ and $\gamma_{1,\mu}$. Let $W = X/Y$ and $T = X/(X + Y)$.

(*a*) Use Exercise 5 (Sec. 16-3) to show that

$$p_W(w) = \frac{\Gamma(\nu + \mu)}{\Gamma(\nu)\Gamma(\mu)}\frac{w^{\nu-1}}{(1 + w)^{\mu+\nu}} \qquad \text{if } w > 0.$$

COMMENT: If we denote this density by $p_{\nu,\mu}$ then $1/W$ has density $p_{\mu,\nu}$.

(*b*) If $\nu = k_1/2$ and $\mu = k_2/2$ then the density of $F = (X/k_1)/(Y/k_2) = (k_2/k_1)W$ is called the **Snedecor *F*-distribution** with k_1 and k_2 degrees of freedom. We adhere to tradition by denoting this random variable by F. Show that

$$p_F(f) = \frac{\Gamma((k_1 + k_2)/2)}{\Gamma(k_1/2)\Gamma(k_2/2)}\left(\frac{k_1}{k_2}\right)^{k_1/2} f^{(k_1/2)-1}\left(1 + \frac{k_1}{k_2}f\right)^{-(k_1+k_2)/2} \qquad \text{if } f > 0.$$

(*c*) Let X' and Y' be independent and each chi-square with k_1 and k_2 degrees of freedom. Show that $(X'/k_1)/(Y'/k_2)$ is Snedecor k_1 and k_2. CAUTION: The chi-square density was defined with parameter $\lambda = .5$. Also state how the Snedecor distribution can be obtained from $k_1 + k_2$ independent random variables each normal 0 and 1.

(*d*) If F is Snedecor k_1 and k_2 the density of $Z = (\frac{1}{2}) \log F$ is called the **Fisher's Z distribution** with k_1 and k_2 degrees of freedom. Show that

$$p_Z(z) = 2\frac{\Gamma(k_1 + k_2)}{\Gamma(k_1/2)\Gamma(k_2/2)} k_1{}^{k_1/2} k_2{}^{k_2/2} e^{k_1 z} (k_2 + k_1 e^{2z})^{-(k_1+k_2)/2} \qquad \text{for all real } z.$$

(*e*) Show that $T = X/(X + Y) = 1/[1 + (1/W)]$ has the **beta density** with parameters ν and μ:

$$p_T(t) = \frac{\Gamma(\nu + \mu)}{\Gamma(\nu)\Gamma(\mu)} t^{\nu-1}(1 - t)^{\mu-1} \qquad \text{if } 0 < t < 1.$$

COMMENT: The beta function is defined by

$$B(\nu,\mu) = \int_0^1 t^{\nu-1}(1 - t)^{\mu-1}\, dt \qquad \text{if } \nu > 0,\ \mu > 0$$

and since p_T is a probability density and must integrate to 1, we have proved that $B(\nu,\mu) = \Gamma(\nu)\Gamma(\mu)/\Gamma(\nu + \mu)$. In particular $B(\nu,\mu) = B(\mu,\nu)$.

(*f*) If F has the Snedecor distribution with k_1 and k_2 degrees of freedom show that $(k_1F)/(k_2 + k_1F)$ has the beta density with parameters $\nu = k_1/2$ and $\mu = k_2/2$.

(*g*) Find the joint density of $U = X + Y$ and $W = X/Y$ and conclude that U and W are independent. Are $U = X + Y$ and $T = X/(X + Y)$ independent?

16-5 EXPECTATIONS

This section is devoted primarily to expectations when continuous densities are available. Theorem 1 shows that if P_X corresponds to a continuous density p_X then the mean of P_X, as defined in Theorem 1 (Sec. 15-4), equals $\int_{-\infty}^{\infty} xp_X(x)\, dx$, as introduced in Sec. 16-1. Theorem 2 shows that, just as in the discrete case, we can calculate $Ef(Y)$ from p_Y rather than just from $P_{f(Y)}$. These two basic theorems permit us to demonstrate the consistency of the general approach of Sec. 15-4 with the temporary definitions of Sec. 16-1. They also make the general properties of expectation obvious in the continuous-density case. These properties are applied in Examples 1 to 5, which include the MGF approach to sums of independent normals and gammas. After Example 5 a unique property of the normal density is related to the CLT. The remainder of this section is devoted to Cauchy densities which dramatize the odd behavior that may occur if relevant moments do not exist.

If (R,P) is a probability space whose sample space is the real line then the mean of P is defined in Theorem 1 (Sec. 15-4). We now show that if P happens to correspond to a continuous density $p\colon R \to R$ then the mean of P, as defined in that theorem, equals $\int_{-\infty}^{\infty} xp(x)\, dx$.

We define $f(x)$ for $x \geq 0$ as in Fig. 1, so that

$$x - (1/2^n) \leq f(x) \leq x \qquad \text{for all } x \geq 0.$$

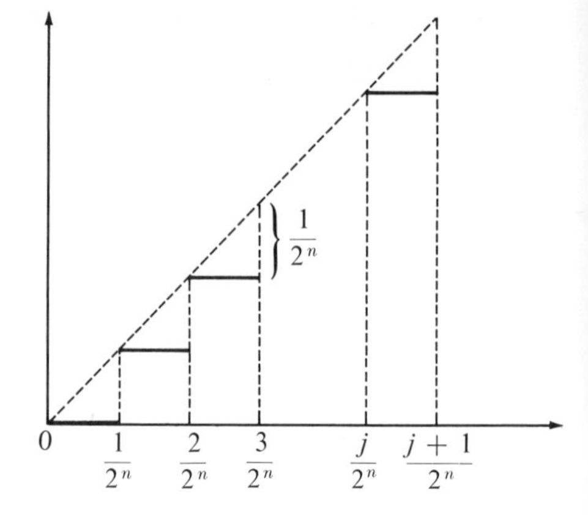

Fig. 1 $f(x) = j/2^n$ if $j/2^n \leq x < (j+1)/2^n$ for $j = 0, 1, \ldots$.

We multiply this inequality by $p(x)$ and integrate to get

$$\int_0^n xp(x)\,dx - \frac{1}{2^n}\int_0^n p(x)\,dx \leq \int_0^n f(x)p(x)\,dx \leq \int_0^n xp(x)\,dx.$$

Now $\int_0^n p(x)\,dx \leq 1$ and the center expression above equals A_n^+ of Theorem 1 (Sec. 15-4), so

$$\int_0^n xp(x)\,dx - \frac{1}{2^n} \leq A_n^+ \leq \int_0^n xp(x)\,dx$$

and hence

$$\lim_{n\to\infty} A_n^+ = \lim_{n\to\infty}\int_0^n xp(x)\,dx$$

in the sense that both limits are infinite or both limits are finite and equal. Similarly

$$\lim_{n\to\infty} A_n^- = \lim_{n\to\infty}\int_{-n}^0 xp(x)\,dx.$$

Clearly both limits are finite iff $xp(x)$ is absolutely integrable over R. We state this result in the theorem below, which includes the analogous result for the discrete case, where we essentially need only replace the above integrals by summations. [In the discrete case we might let $x_1, x_2, \ldots$ be an enumeration of those real x satisfying $k - 1 \leq x < k$ and $p(x) > 0$, and then let c_k equal the summation of these $x_i p(x_i)$ and replace $\int_0^n xp(x)\,dx$ by $c_1 + c_2 + \cdots + c_n$.]

Theorem 1 Let (R,P) be a probability space and let the mean of P be defined as in Theorem 1 (Sec. 15-4). If P corresponds to a discrete density $p: R \to R$ which is positive only at $x_1, x_2, \ldots$ then the mean of P exists iff

$$\sum_{i=1}^{\infty} |x_i|\, p(x_i) < \infty$$

and if this is the case then [the mean of P] $= \sum_{i=1}^{\infty} x_i p(x_i)$.

If P corresponds to a continuous density $p: R \to R$ then the mean of P exists iff

$$\int_{-\infty}^{\infty} |x|\, p(x)\, dx < \infty$$

and if this is the case then [the mean of P] $= \displaystyle\int_{-\infty}^{\infty} xp(x)\, dx$. ■

The next theorem asserts that if a random vector $(Y_1, \ldots, Y_k)$ has a continuous density then $Ef(Y_1, \ldots, Y_k)$ can be calculated from it by a formula analogous to the discrete one.

Theorem 2: The expectation, as calculated from an intermediate density If the random variable $X = f(Y_1, \ldots, Y_k)$ is a function of the random vector $(Y_1, \ldots, Y_k)$ which has a continuous density,

$$X = f(Y_1, \ldots, Y_k)$$

$$\Omega \xrightarrow{\;Y = (Y_1, \ldots, Y_k)\;} R^k \xrightarrow{\;f\;} R$$

$$P \qquad\qquad p_{Y_1, \ldots, Y_k}$$

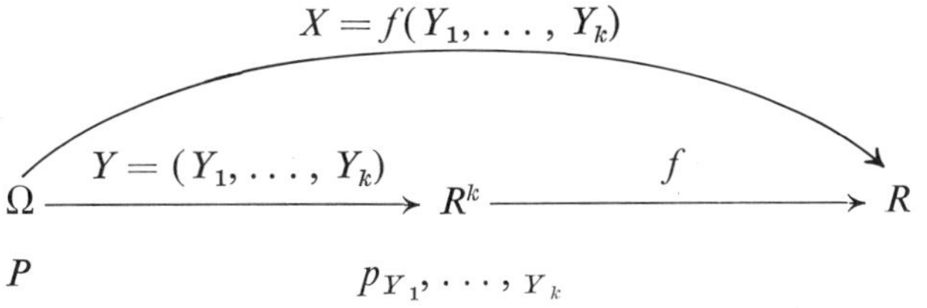

then X has an expectation iff

$$\int_{-\infty}^{\infty} \cdots \int_{-\infty}^{\infty} |f(y_1, \ldots, y_k)|\, p_{Y_1,\ldots,Y_k}(y_1, \ldots, y_k)\, dy_1 \cdots dy_k < \infty$$

and if this is the case then

$$\begin{aligned} EX &= Ef(Y_1, \ldots, Y_k) \\ &= \int_{-\infty}^{\infty} \cdots \int_{-\infty}^{\infty} f(y_1, \ldots, y_k) p_{Y_1,\ldots,Y_k}(y_1, \ldots, y_k)\, dy_1 \cdots dy_k. \end{aligned}$$ ■

Before we prove this theorem we will briefly discuss its implications and the simple idea behind its proof. Theorem 2 is stated in a form convenient for applications. Its basic conclusion is that if f is a random variable with associated continuous density $p_{Y_1,\ldots,Y_k}$ then the expectation of f, which is defined to equal the mean of P_f, can actually be calculated from f and $p_{Y_1,\ldots,Y_k}$ by the exhibited formula.

From Theorem 1 we can calculate EX from the continuous density p_X if X has a continuous density. From Theorem 2 if $X = f(Y)$ we can calculate $Ef(Y)$ from the continuous density p_Y if Y has a continuous density. Thus an expectation can always be calculated from the standard formula in terms of *any* discrete or continuous density which exists. In particular examples, some of the distributions may correspond to densities and others may not. Example 4 (Sec. 17-1) shows that $X: R \to R$ *with associated continuous density* $p: R \to R$ may have a distribution P_X which does *not* correspond to *any* mixture of densities. Exercise 1 (Sec. 16-4) shows that $X: R^2 \to R$ with associated (R^2, P) *may have a continuous density* p_X even though P does *not* correspond to *any* density. Usually in our future examples *both* $X = f(Y)$ *and* Y have continuous densities, and in that case

$$EX = \int_{-\infty}^{\infty} xp_X(x)\,dx$$
$$= \int_{-\infty}^{\infty} \cdots \int_{-\infty}^{\infty} f(y_1, \ldots, y_k) p_{Y_1,\ldots,Y_k}(y_1, \ldots, y_k)\,dy_1 \cdots dy_k$$

which is analogous to the discrete case.

The idea behind the proof of Theorem 2 is indicated in Fig. 2. If D is that subset of R^k where f takes values between a and b then P_f is defined so that

$$\int_D p_Y(y)\,dy = P_f\{x\colon a \le x < b\}$$

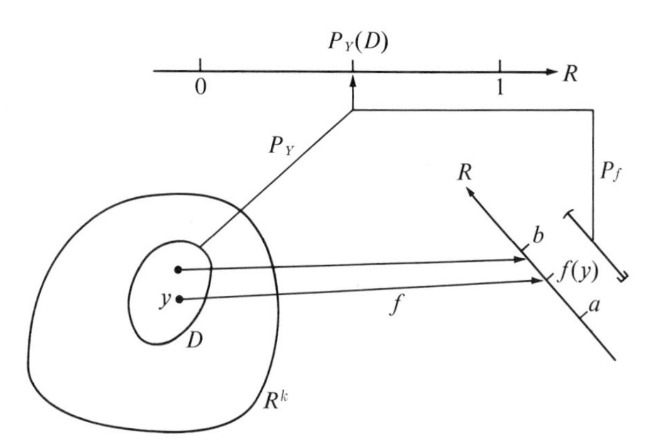

Fig. 2

where we use the abbreviations $y = (y_1, \ldots, y_k)$ and $dy = dy_1 \cdots dy_k$.

If $a \doteq b$ then $f(y) \doteq a$ when $y \in D$, so

$$\int_D f(y)p_Y(y)\,dy \doteq a\int_D p_Y(y)\,dy = aP_f\{x\colon a \le x < b\}.$$

Thus if we obtain an approximation to $\int_{R^k} f(y)p_Y(y)\,dy$ by adding up the contributions from a large collection of disjoint subsets like D then their sum equals an approximation to the mean of P_f as described in Sec. 15-4.

***Proof of Theorem 2** If $a < b$ let

$$D = \{(y_1, \ldots, y_k)\colon a \le f(y_1, \ldots, y_k) < b\}.$$

On D the function f takes values only between a and b, so, as in Fig. 2, we have

$$aP_f\{x\colon a \le x < b\} \le \int_D f(y)p_Y(y)\,dy \le bP_f\{x\colon a \le x < b\}$$

which can be written as

$$0 \leq \int_D f(y)p_Y(y)\,dy - [aP_f\{x\colon a \leq x < b\}] \leq (b-a)P_f\{x\colon a \leq x < b\}.$$

Let $a = j/2^n$ and $b = (j+1)/2^n$, so that [ ] is a term in the definition of A_n^+ in Theorem 1 (Sec. 15-4). Adding these inequalities over j and using the fact that the sum of the integrals over disjoint sets equals the integral over their union, we get

$$0 \leq \int_{B_n} f(y)p_Y(y)\,dy - A_n^+ \leq \frac{1}{2^n}$$

where $B_n = \{(y_1, \ldots, y_k)\colon 0 \leq f(y_1, \ldots, y_k) < n\}$. As n goes to infinity the sets B_n increase upward to the set $B^+ = \{(y_1, \ldots, y_k)\colon 0 \leq f(y_1, \ldots, y_k)\}$ on which f is nonnegative. Thus A_n^+ converges to a finite limit A^+ iff the exhibited integral over B_n does, and in this case

$$A^+ = \int_{B^+} f(y)p_Y(y)\,dy.$$

The same is true for A_n^- so Theorem 2 is proved. ■

Theorem 2 is a powerful result. The only additional assumption needed to make the statement of Theorem 2 rigorous, in the context of modern integration theory, is that both f and $p_{Y_1,\ldots,Y_n}$ be Borel measurable. This is *always* true in applications. [A non–Borel measurable function has never been explicitly exhibited, although they can be proved to exist.] Note that if f is a very complicated function then integration over an extremely irregular subset D of R^n must be defined.

Let Y with associated (Ω,P) be a random variable with any kind of distribution. The moments and MGF are defined in general by

$$[\text{the } n\text{th moment of } Y] = E(Y^n) = [\text{mean of } P_W] \qquad \text{where } W = Y^n,$$

$$m(t) = E(e^{tY}) = [\text{mean of } P_Z] \qquad \text{where } Z = e^{tY}.$$

If Y has a continuous density, then from Theorem 2 we have

$$Ef(Y) = \int_{-\infty}^{\infty} f(y)p_Y(y)\,dy.$$

Let $f(y) = y^n$ and $f(y) = e^{ty}$ to obtain

$$EY^n = \int_{-\infty}^{\infty} y^n p_Y(y)\,dy \qquad \text{and} \qquad Ee^{tY} = \int_{-\infty}^{\infty} e^{ty} p_Y(y)\,dy.$$

Thus our general definitions specialize in the continuous-density case to the temporary definitions of Scc. 16-1.

The general properties of expectation in the summary of Chap. 11 are mostly obvious, as indicated by Exercise 9 below, once continuous densities and Theorem 2 are available.

Chebychev's inequality is easily proved when X has a continuous density p_X:

$$P[|X-\mu| \geq \epsilon] = \int_{\{x:\,|x-\mu|\geq\epsilon\}} p_Y(x)\,dx \leq \int_{\{x:\,|x-\mu|\geq\epsilon\}} \left(\frac{x-\mu}{\epsilon}\right)^2 p_X(x)\,dx$$

$$\leq \int_{-\infty}^{\infty} \left(\frac{x-\mu}{\epsilon}\right)^2 p_X(x)\,dx = \frac{\sigma^2}{\epsilon^2}.$$

The first inequality uses the fact that $(x-\mu)^2/\epsilon^2 \geq 1$ over the range of integration, while the nonnegative integrand yielded the second inequality.

Example 1 Let $W = X + Y + Z$ where X is normal μ and σ^2, Y is exponential with mean $1/\lambda$, and Z is uniform over $0 \leq z \leq 1$. The linearity of expectation yields $EW = \mu + (1/\lambda) + .5$ even though their unspecified joint distribution may be any one of many. Different joint distributions having these specified marginals would induce different P_W, but every such P_W must have this mean.

If we assume that X, Y, and Z are independent then they have a joint density determined by the product rule, so p_W is uniquely determined and formally obtainable by performing two convolutions. We already know at least the mean, variance, and MGF of this p_W:

$$\text{Var } W = \sigma^2 + 1/\lambda^2 + 1/12$$

$$m_W(t) = e^{\mu t + (\sigma^2 t^2/2)} \frac{1}{1-(t/\lambda)} \frac{e^t - 1}{t} \qquad \text{for } \frac{t}{\lambda} \leq 1. \quad ■$$

Example 2 We want the moments of the density p_Y found in Example 8 (Sec. 16-2) for $Y = X^2$ where X is normal with mean 0 and variance σ^2. Thus

$$EY^n = \int_{-\infty}^{\infty} y^n p_Y(y)\,dy = \int_0^{\infty} y^n \frac{1}{\sigma\sqrt{2\pi y}} e^{-y/2\sigma^2}\,dy.$$

Instead of evaluating this integral, note that

$$EY^n = EX^{2n} = (1)(3)(5)\cdots(2n-1)\sigma^{2n}$$

where we used the result of Exercise 8 (Sec. 16-1) for the last equality. That is, the nth moment of Y equals the $(2n)$th moment of X, and we already have these. ■

Example 3 We want the moments of the lognormal density. As defined in Example 6 (Sec. 16-2), the lognormal density p_Y with parameters μ and σ^2 is the density induced by $Y = e^X$ when X is normal μ and σ^2. Since $p_Y(y) = 0$ if $y \leq 0$, the tth moment of p_Y can be considered for every real t. Clearly

$$EY^t = E(e^X)^t = Ee^{tX} = e^{t\mu + (t^2\sigma^2/2)}$$

so the tth moment of Y equals *the value* of the MGF of X at t. In particular

$$EY = e^{\mu + (\sigma^2/2)} \qquad \text{and} \qquad EY^2 = e^{2\mu + 2\sigma^2}$$

$$\text{Var } Y = (EY)^2\left[\frac{EY^2}{(EY)^2} - 1\right] = (EY)^2[e^{\sigma^2} - 1].$$

We do not know of a manageable expression for the MGF for Y,

$$m_Y(t) = Ee^{tY} = Ee^{te^X} = \frac{1}{\sigma\sqrt{2\pi}}\int_{-\infty}^{\infty} \exp\left\{te^x - \frac{(x-\mu)^2}{2\sigma^2}\right\} dx$$

so that we might *attempt* to differentiate it [it is easily seen that $m_Y(t) < \infty$ iff $t \le 0$] in order to find the moments for Y.

It should be apparent that simplicity and insight are aided by the fact that the same expectation can be calculated in many ways. In particular we could, but did not, calculate EY^t from p_Y by

$$EY^t = \int_0^{\infty} y^t p_Y(y)\,dy = \int_0^{\infty} y^t \frac{1}{y\sigma\sqrt{2\pi}} \exp\left\{-\frac{1}{2\sigma^2}[(\log y) - \mu]^2\right\} dy.$$

Nor did we calculate EY^t from p_{Y^t} by

$$EY^t = \int_0^{\infty} z p_{Y^t}(z)\,dz$$

where p_{Y^t} is lognormal $t\mu$ and $(t\sigma)^2$. ■

Example 4 If $X_1, \ldots, X_n$ are independent and X_i is normal the MGF of $S_n = X_1 + X_2 + \cdots + X_n$ is

$$m_{S_n}(t) = m_{X_1}(t)m_{X_2}(t)\cdots m_{X_n}(t) = \exp\left\{\sum_{i=1}^{n}\left(t\mu_i + \frac{t^2\sigma_i^2}{2}\right)\right\}$$

$$= \exp\left\{t\mu + \frac{t^2\sigma^2}{2}\right\} \qquad \text{where} \qquad \mu = \sum_{i=1}^{n}\mu_i \text{ and } \sigma^2 = \sum_{i=1}^{n}\sigma_i^2.$$

But m_{S_n} is the MGF of a normal μ and σ^2, so from Theorem 4 (Sec. 15-4) we conclude that p_{S_n} is normal μ and σ^2. Thus MGFs provide an alternate proof of the fact that a sum of independent normals is normal. This same fact was proved in Example 1 (Sec. 16-4) by performing a convolution.

There is a simple method for recalling and rederiving the normal MGF. We use the normal density with mean $\mu = \sigma^2$ to get

$$1 = \frac{1}{\sigma\sqrt{2\pi}}\int_{-\infty}^{\infty} e^{-(x-\sigma^2)^2/2\sigma^2}\,dx = \frac{1}{\sigma\sqrt{2\pi}}\int_{-\infty}^{\infty} e^{-x^2/2\sigma^2}e^{x-\sigma^2/2}\,dx$$

then multiply by $e^{\sigma^2/2}$. Thus we have the *basic fact that if X is normal with $EX = 0$ then* $Ee^X = e^{EX^2/2}$. Once this fact has been recalled then we can apply it to *any* normally distributed X as follows

$$Ee^X = e^{EX}\,Ee^{X-EX} = e^{EX+[(\text{Var }X)/2]}$$

and replacing X by tX yields the MGF

$$Ee^{tX} = e^{tEX+[(t^2\,\text{Var }X)/2]}. \quad ■$$

Example 5 The MGF of the gamma density $\gamma_{\lambda,\nu}$ is

$$\int_0^\infty e^{tx}\frac{\lambda^\nu}{\Gamma(\nu)}x^{\nu-1}e^{-\lambda x}\,dx = \frac{\lambda^\nu}{(\lambda-t)^\nu}\int_0^\infty \frac{(\lambda-t)^\nu}{\Gamma(\nu)}x^{\nu-1}e^{-(\lambda-t)x}\,dx$$

$$= \frac{\lambda^\nu}{(\lambda-t)^\nu}\int_0^\infty \gamma_{\lambda-t,\nu}(x)\,dx = \frac{1}{[1-(t/\lambda)]^\nu} \qquad \text{for } t<\lambda$$

where $\lambda - t$ must be positive in order to have a decreasing exponential in the integrand. The equation

$$\frac{1}{[1-(t/\lambda)]^{\nu_1}}\,\frac{1}{[1-(t/\lambda)]^{\nu_2}}\cdots\frac{1}{[1-(t/\lambda)]^{\nu_n}} = \frac{1}{[1-(t/\lambda)]^\nu}$$

where $\nu = \nu_1 + \nu_2 + \cdots + \nu_n$ proves that if independent random variables each have a gamma density with the same λ parameter then so does their sum. This result was obtained in Sec. 16-4 by performing convolution. ∎

All exercises except 12 *and* 13 *may be done before reading the remainder of this section.*

We next relate a property of normal distributions to the CLT. Let X and Y be independent and each induce the standardized normal density φ. From Example 4, if $a^2 + b^2 = 1$ then $aX + bY$ induces this same density φ. As shown below (or in Exercise 2), the normal distribution is *unique* in having this fundamental property.

The Uniqueness of a Property of Normal Distributions Let random variables X and Y be independent and each have the same distribution P_X such that $EX = 0$ and $EX^2 = 1$. Assume that $aX + bY$ induces this same distribution P_X whenever $a^2 + b^2 = 1$. Then P_X corresponds to the normal density φ. ∎

***Proof** Let X, Y, and Z be independent and each have this distribution P_X. If $a^2 + b^2 + c^2 = 1$ then

$$(aX + bY + cZ) = \sqrt{a^2+b^2}\left[\frac{aX}{\sqrt{a^2+b^2}} + \frac{bY}{\sqrt{a^2+b^2}}\right] + cZ$$

where [ ] must induce P_X. Applying our assumption to [ ] and Z shows that ( ) also induces P_X. If $c = 1$ the same result holds. More generally, by induction it follows that if $X_1, \ldots, X_n$ are independent and each has this distribution P_X then $\sum_{i=1}^n a_iX_i$ induces distribution P_X whenever $\sum_{i=1}^n a_i^2 = 1$. Letting $a_i = 1/\sqrt{n}$ we see that $S_n^* = (1/\sqrt{n})\sum_{i=1}^n X_i$ induces the same P_X for every n. Thus our assumption implies that no "smoothing" or change of any kind is caused by adding independent random variables each having this distribution, and then normalizing. But then CLT of Theorem 1 (Sec. 18-3) or Corollary 2 (Sec. 20-2) shows that $P[S_n^* < s] \to \Phi(s)$ as $n \to \infty$. [Corollary 1*a* (Sec. 13-2) yields the same conclusion if we assume that $E\,|X_1|^3 < \infty$.] But $P[S_n^* < s] = P[X_1 < s]$ for all n, so that we must have $P[X_1 < s] = \Phi(s)$ for all real s, as was asserted. ∎

We show that this same property arises naturally from a different but related point of view. Let $S_n^* = (1/\sqrt{n}) \sum_{i=1}^{n} X_i$ where $X_1, \ldots, X_n$ are independent and each has the same distribution P_X satisfying $EX = 0$ and $EX^2 = 1$. Assume that we do not know the CLT. We might wonder if the distribution of S_n^* converged, in some reasonable sense, to some distribution (R,P_1). Of course, for some P_X it might happen that $P_{S_n^*}$ does not converge to anything, while for other P_X there might be a limiting distribution, but different ones depending on P_X. We have the obvious identity

$$S_{n+m}^* = \sqrt{\frac{n}{n+m}}\, S_n^* + \sqrt{\frac{m}{n+m}} \left[\frac{1}{\sqrt{m}} (X_{n+1} + \cdots + X_{n+m}) \right]$$

where S_n^* and [ ] are independent. We can let m and n go to infinity in such a way that $n/(n+m)$ converges to a^2 and $m/(n+m)$ converges to b^2, where $a^2 + b^2 = 1$. Thus if for a particular P_X it were true that $P_{S_n^*}$ converged to some P_1, then the distributions of S_{n+m}^* and [ ] would also converge to P_1, so we would expect all such limit laws P_1 to have the property above. Of course, the CLT shows that there is only one such limit law, and furthermore it applies to every such P_X.

*CAUCHY DENSITIES

We have seen that translations and scale changes applied to φ generate a family of densities closed under independent additions. More precisely, let X and Y be independent and both have density φ. Then for arbitrary real a, b, c, and d there are real e and f such that $(aX + c) + (bY + d)$ and $eX + f$ have the same distribution. Namely, let $e = \sqrt{a^2 + b^2}$ and $f = c + d$. Furthermore if we require that $0 < \text{Var } X < \infty$, and $ab \neq 0$, then only the normal family has this property, which is intimately related to limiting distribution phenomena. The assumption of finite second moments is crucial to these considerations. There are many such families which do not have finite second moments. In particular the remainder of this section is devoted to Cauchy densities, which comprise such a closed family, to which the CLT and laws of large numbers do not apply. These densities are introduced in such a way as to show that they arise naturally in practical problems.

Example 6 As in Fig. 3*a*, let a light source be located at unit distance from a wall and let 0 be the closest point on the wall. The source emits a narrow beam of light in the plane of the figure and making an angle U with the perpendicular to the wall. Assume, as in Fig. 3*b*, that U has a uniform density over the interval from $-\pi/2$ to $\pi/2$. Then $V = \tan U$ is the distance, with sign, of the illuminated point on the wall, as measured from 0. The density induced by $V = g(U) = \tan U$ is given by

$$p_V(v) = \frac{1}{|dg(u)/du|} p_U(u) \qquad \text{where } v = \tan u.$$

But $g'(u) = 1/(\cos u)^2$ and $\cos u = (1 + v^2)^{-1/2}$ so that

$$p_V(v) = \frac{1}{\pi} \frac{1}{1 + v^2} \qquad \text{for all real } v,$$

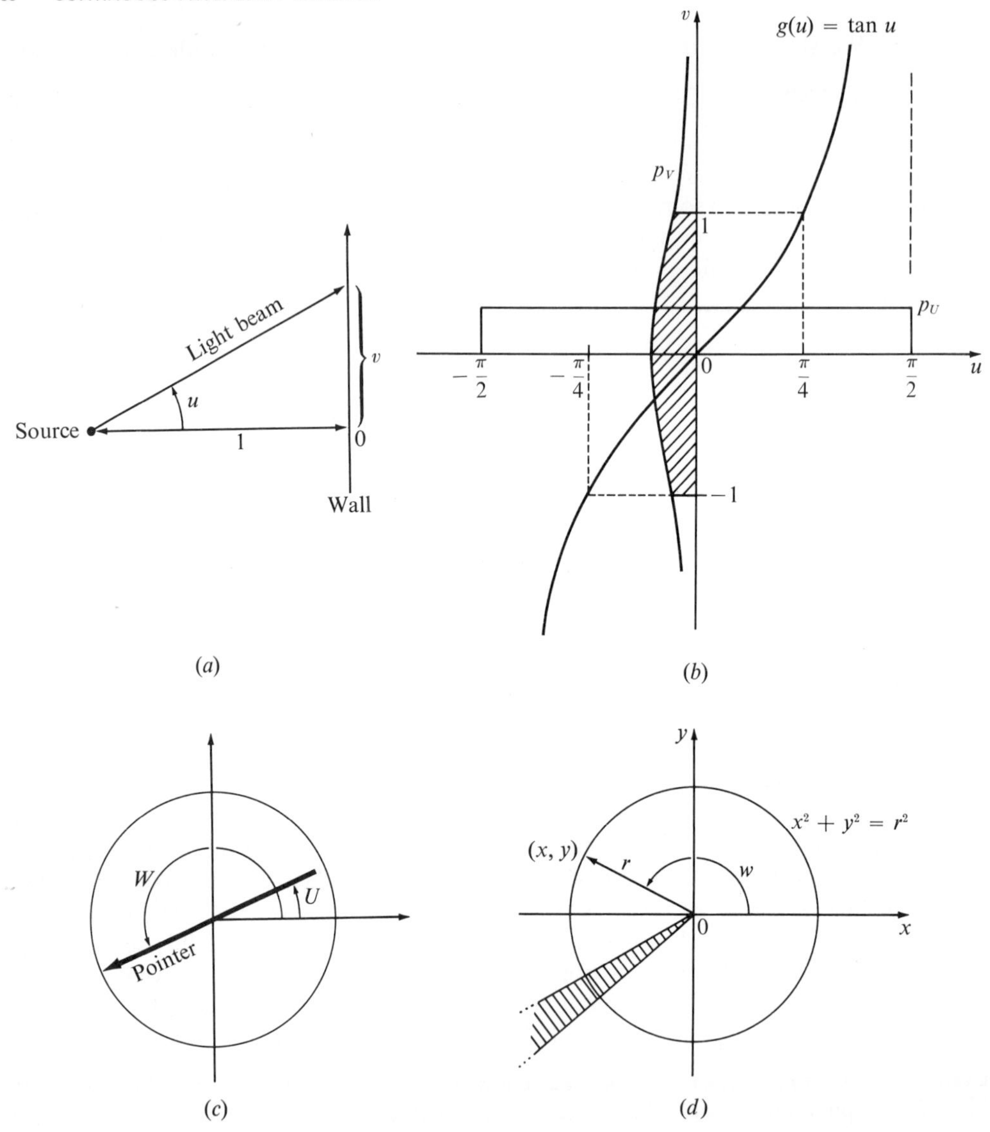

Fig. 3

which is called the **standardized Cauchy density.** [This is also Student's t distribution with $n = 1$, as defined in Exercise 13 (Sec. 16-4).]

As shown in Fig. 3*b*, this density is symmetric about the origin and looks somewhat like the normal density. However, it decreases so slowly as v goes to infinity, or minus infinity, that it does not have a mean; that is,

$$vp_V(v) \sim \frac{1}{\pi v} \qquad \text{as } v \to \infty$$

so that

$$\int_{-a}^{a} |v|\, p_V(v)\, dv = 2\int_0^a v p_V(v)\, dv \to \infty \qquad \text{as } a \to \infty.$$

Thus p_V is a continuous analog of the discrete density p_1 of Example 1 (Sec. 9-4). In particular the MGF is useless, since Ee^{tV} is finite iff $t = 0$. This can be seen directly by using inequality 7 of Appendix 1, so that

$$Ee^{tV} \geq \int_0^\infty e^{tv} p_V(v)\, dv \geq \int_0^\infty (1 + tv) p_V(v)\, dv = \infty \qquad \text{if } t > 0.$$

The same is true if $t < 0$.

Since U has a uniform density we find, as is obvious from Fig. 3*b*, that

$$P[V < v] = P[\tan U < v] = P[U < \arctan v]$$
$$= \frac{1}{\pi}\left[(\arctan v) - \left(-\frac{\pi}{2}\right)\right] = \frac{1}{2} + \frac{1}{\pi} \arctan v.$$

In particular for $v = -1$ or 1 we obtain $\frac{1}{4}$ or $\frac{3}{4}$. Thus p_V is symmetric about the origin so $P[V < 0] = \frac{1}{2}$, and furthermore p_V assigns probability $\frac{1}{2}$ to the interval $-1 < v < 1$. Since p_V does not have a mean or variance, we use the last sentence to justify considering p_V to be centered at the origin and to have roughly unit spread. ■

Example 7 As in Fig. 3*c*, a pointer is spun and comes to rest at angle W, assumed to have a uniform density over $0 < w < 2\pi$. We show that the slope, $\tan W$, of the pointer line has a standardized Cauchy density. Let U be the acute angle determined by that end of the pointer which is in the right half-plane. Clearly U is uniformly distributed over $-\pi/2 < u < \pi/2$. The slope of the pointer line is given by $\tan W = \tan U$, which has a standardized Cauchy density by Example 6. ■

Example 8 An electron is fixed at a target located at the origin of the xy plane, as in Fig. 3*d*. Then $V = Y/X$ is the slope of the line from the origin to the impact point (X, Y). If W is the angle to the impact point then $V = \tan W$, so that if W has a uniform density over $0 < w < 2\pi$ then V has a standard Cauchy density. In particular this must be the case if the random vector (X, Y) has a radially symmetric continuous density, that is, if there is some function $f: R \to R$ such that

$$p_{X,Y}(x,y) = f(x^2 + y^2) \qquad \text{for all real } x \text{ and } y.$$

This is geometrically obvious because the volume below such a $p_{X,Y}$ and above the narrow shaded sector in Fig. 3*d* must be the same as for any rotated version of the sector. This occurs, for example, if

$$p_{X,Y}(x,y) = \frac{1}{2\pi\sigma^2} e^{-(x^2+y^2)/2\sigma^2}.$$

Thus if X and Y are independent and each has the same normal density with mean 0 then $V = Y/X$ has a standardized Cauchy density, as was also shown in Exercise 6 (Sec. 16-3). Naturally then $1/V = X/Y$ also has a standard Cauchy density. Thus if V has a standardized Cauchy density then so does $1/V$. ■

Let $Z = \lambda V + \mu$ where $\lambda > 0$ and μ are real numbers and V has a standardized Cauchy density. Then

$$p_Z(z) = \frac{1}{\lambda} p_V\left(\frac{z-\mu}{\lambda}\right) = \frac{1}{\pi} \frac{\lambda}{\lambda^2 + (z-\mu)^2} \qquad \text{for all real } z$$

which is called the **Cauchy density with location parameter μ and spread parameter** $\lambda > 0$. Note that p_Z is symmetric about μ,

$$p_Z(\mu + t) = p_Z(\mu - t) \qquad \text{for all real } t.$$

Also p_Z assigns probability $\frac{1}{2}$ to the interval $\mu \pm \lambda$ since

$$P[\mu - \lambda < Z < \mu + \lambda] = P[-1 < V < 1] = \tfrac{1}{2}.$$

Clearly V and $-V$ have the same density, and hence if $\lambda < 0$ then $\lambda V + \mu = (-\lambda)(-V) + \mu$ has parameters μ and $-\lambda$. Thus if $\lambda \neq 0$ then $\lambda V + \mu$ has a Cauchy density with parameters μ and $|\lambda|$.

If Z is Cauchy μ and λ while $a > 0$ then

$$aZ + b = a(\lambda V + \mu) + b = (a\lambda)V + (a\mu + b)$$

so $aZ + b$ is Cauchy $a\mu + b$ and $a\lambda$. If $a < 0$ we can write $aV = (-a)(-V)$, where $-V$ is Cauchy 0 and 1. *Thus* if Z is Cauchy μ and λ while $a \neq 0$ then $aZ + b$ is Cauchy $a\mu + b$ and $|a|\,\lambda$.

By performing a difficult convolution, or by using characteristic functions as in Sec. 18-3, it can be shown that if Z_1 and Z_2 are independent and Z_i is Cauchy with parameters 0 and λ_i then $Z_1 + Z_2$ is Cauchy 0 and $\lambda_1 + \lambda_2$. By induction we see that if $Z_1, \ldots, Z_n$ are independent and Z_i is Cauchy μ_i and λ_i then

$$\sum_{i=1}^{n} Z_i = \sum_{i=1}^{n} (\lambda_i V_i + \mu_i) = \sum_{i=1}^{n} \lambda_i V_i + \sum_{i=1}^{n} \mu_i$$

has a Cauchy density with parameters $\sum_{i=1}^{n} \mu_i$ and $\sum_{i=1}^{n} \lambda_i$. Thus independent additions do not take us outside the family of Cauchy densities. However, the spread parameters add, while for normals it was the variances, and not the standard deviations, that added. This distinction is fundamental, as we now see.

Let $S_n = V_1 + \cdots + V_n$ where $V_1, \ldots, V_n$ are independent and each is Cauchy 0 and 1. Then S_n is Cauchy 0 and n, so $(1/n)S_n$ is Cauchy 0 and 1. Thus the arithmetic average has exactly the same distribution as one of the components. The arithmetic average of Cauchy errors is not small, and *the law of large numbers does not apply*. Similarly if we attempt to normalize in the usual way for the CLT then $(1/\sqrt{n})S_n$ is Cauchy 0 and $\sqrt{n}$ so its distribution continues to spread out as n grows. Therefore convergence to the normal distribution does not occur and the CLT does not apply.

Note that if X_1 and X_2 are independent and each is normal 0 and 1 then $X_1 + X_2$ has the same distribution as $(\sqrt{2})X_1$. Thus adding these two variables yields a distribution which is less spread than twice one of the variables. We may say that these independent errors tended to cancel. However, if V_1 and V_2 are independent and each is Cauchy 0 and 1 then $V_1 + V_2$ has the same distribution as $2V_1 = V_1 + V_1$.

EXERCISES

1. Find the mean and variance of $Y = X_1 + X_1X_2 + X_1X_2X_3$ when X_1, X_2, X_3 are independent and each is normal 0 and 1.

2. *A Stronger Version of the Unique Property of Normal Distributions.* Prove that we need only consider $a = b = 1/\sqrt{2}$ in the property stated after Example 5. That is, if X and Y are independent random variables such that X, Y, and $(X + Y)/\sqrt{2}$ have the same mean 0 and variance 1 distribution, then this distribution corresponds to the normal density φ.

3. Find the mean and variance of the volume V of the block of metal in Exercise 5 (Sec. 16-4) under the assumption of part (a) and under the assumption of part (b).

4. Find the MGF of the sum $S_2 = X_1 + X_2$ of two independent random variables, each uniformly distributed over $0 < x < 1$, from the density p_{S_2}. Check that this MGF equals the square of the MGF for X_1.

5. Show that *Student's t* distribution with n degrees of freedom, as defined in Exercise 13 (Sec. 16-4), has mean 0 if $n \geq 2$ and variance $n/(n - 2)$ if $n \geq 3$. What is the reason for the restrictions $n \geq 2$ and $n \geq 3$?

6. As in Exercise 14 (Sec. 16-4), let $W = X/Y$ where X and Y are independent with respective gamma densities $\gamma_{1,\nu}$ and $\gamma_{1,\mu}$.

(a) Show that

$$EW^r = \frac{\Gamma(\nu + r)\Gamma(\mu - r)}{\Gamma(\nu)\Gamma(\mu)} \qquad \text{if } -\nu < r < \mu.$$

What is the reason for constraining r by $-\nu < r < \mu$?

(b) Let $\nu = k_1/2$ and $\mu = k_2/2$ so that $F = (k_2/k_1)W$ has the *Snedecor F* distribution with k_1 and k_2 degrees of freedom. Show that

$$E(F^r) = \left(\frac{k_2}{k_1}\right)^r \frac{\Gamma((k_1/2) + r)\Gamma((k_2/2) - r)}{\Gamma(k_1/2)\Gamma(k_2/2)} \qquad \text{if } -k_1 < 2r < k_2.$$

(c) Show that

$$E(F) = \frac{k_2}{k_2 - 2} \qquad \text{if } k_2 > 2,$$

$$\text{Var } F = \frac{2(k_2)^2(k_1 + k_2 - 2)}{k_1(k_2 - 2)^2(k_2 - 4)} \qquad \text{if } k_2 > 4.$$

7. You bet \$1 on a game which terminates after a fixed large number ν of plays, at which time your return is the product $W_\nu = Y_1Y_2 \cdots Y_\nu$ where $Y_1, \ldots, Y_\nu$ are independent and identically distributed. Thus in effect on each play you must bet all your current playing capital, and your playing capital at the end of the play is a random factor times your bet. Let $M > 0$ and $Y_i = MX_i$ where X_i is uniform over $0 < x < 1$. Thus the ith random factor is uniformly distributed over the interval from zero to a maximum M. Among other things we are interested in discovering the values of M which make the game favorable.

(a) Show that

$$EW_\nu = \left(\frac{M}{2}\right)^\nu \qquad \text{and} \qquad \text{Var } W_\nu = \left(\frac{M^2}{3}\right)^\nu \left[1 - \left(\frac{3}{4}\right)^\nu\right].$$

COMMENT: If $M < \sqrt{3} \doteq 1.73$ then EW_ν and Var W_ν both go to zero as ν goes to infinity, so from Chebychev's inequality the game is clearly unfavorable. For $M > \sqrt{3}$ we have that Var W_ν goes to infinity as ν does, so we cannot learn much by applying Chebychev's inequality.

(*b*) Justify the following derivation, valid for any $\epsilon > 0$:

$$EW_\nu = \int_0^\infty w p_{W_\nu}(w)\,dw \geq \int_\epsilon^\infty w p_{W_\nu}(w)\,dw \geq \int_\epsilon^\infty \epsilon p_{W_\nu}(w)\,dw = \epsilon P[W_\nu > \epsilon].$$

COMMENT: If $M < 2$ then

$$P[W_\nu > \epsilon] \leq \frac{1}{\epsilon} EW_\nu \to 0 \qquad \text{as } \nu \to \infty$$

so the game is clearly unfavorable. For $M > 2$ we have that EW_ν and Var W_ν both go to infinity.

(*c*) Show that

$$P[W_\nu > 1] = P\left[\frac{1}{\nu} S_\nu < \log M\right]$$

where $S_\nu = \sum_{i=1}^{\nu} Z_i$ and $Z_i = -\log X_i$. Find the density, mean, and variance of Z_1. Conclude that the game is favorable if $M > e \doteq 2.73$ and unfavorable if $M < e$.

(*d*) Show that

$$P\left[W_\nu > \left(\frac{M}{e}\right)^{\nu/2}\right] = P\left[\frac{1}{\nu} S_\nu < \frac{1 + \log M}{2}\right]$$

and draw conclusions concerning the exponential growth of W_ν with respect to ν if $M > e$, and the exponential decrease of W_ν if $M < e$.

(*e*) Find the density of W_ν.

8. Assume that $X_i = f_i(Y_1, \ldots, Y_k)$ where $f_i: R^k \to R$ and $Y_1, \ldots, Y_k$ have a joint continuous density. In this case we can derive the *linearity of expectation* as follows:

$$E\left[\sum_{i=1}^n a_i X_i\right] \overset{1}{=} \int_{-\infty}^\infty \cdots \int_{-\infty}^\infty \left[\sum_{i=1}^n a_i f_i(y_1, \ldots, y_k)\right] p_{Y_1,\ldots,Y_k}(y_1, \ldots, y_k)\,dy_1 \cdots dy_k$$

$$\overset{2}{=} \sum_{i=1}^n a_i \int_{-\infty}^\infty \cdots \int_{-\infty}^\infty f_i(y_1, \ldots, y_k) p_{Y_1,\ldots,Y_k}(y_1, \ldots, y_k)\,dy_1 \cdots dy_k$$

$$\overset{3}{=} \sum_{i=1}^n a_i EX_i.$$

Justify these three equalities, assuming that all integrands are absolutely integrable. Which marginal or joint distributions of $X_1, \ldots, X_n$ must be assumed to correspond to continuous densities?

9. *The Properties of Expectation Are Obvious if There Is a Joint Continuous Density* $p_{X,Y}$. Carry Exercise 13 (Sec. 11-3) over to the continuous-density case by replacing sums by integrals and exhibiting omitted manipulations. Use the symbol $\overset{2}{=}$ for any equality which uses Theorem 2. Show incidentally that the continuous case of Theorem 1 is a special case of Theorem 2.

10. *Theorem.* Let X be a random variable with associated probability space (Ω,P) and let

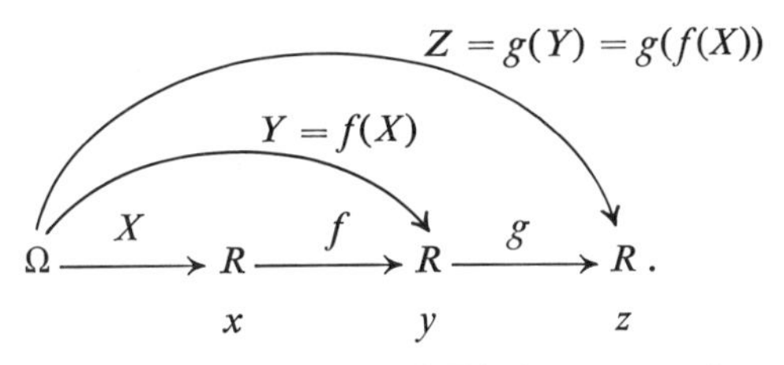

Assume that each of X, Y, and Z induces a continuous density. Then

$$EZ = \int_{-\infty}^{\infty} z p_Z(z)\,dz = \int_{-\infty}^{\infty} g(y) p_Y(y)\,dy = \int_{-\infty}^{\infty} g(f(x)) p_X(x)\,dx$$

in the sense that if any one of the integrands is absolutely integrable then all three are, and the equalities hold. ■

(*a*) Prove this theorem.
(*b*) Relate this theorem to Example 2.
(*c*) Relate this theorem to Example 3.

11. Calculate ES from both p_X and p_S where $S = \pi X^2$ is the surface area of a randomly selected sphere whose diameter X has continuous density $p_X(x) = 5x^4$ for $0 < x < 1$.

12. In Example 8 it was shown that if V has a standard Cauchy density then so does $1/V$. Use Exercise 7 (Sec. 16-2) to deduce this same fact.

13. The length μ of an object is unknown. Repeated measurements will yield $\mu + V_1$, $\mu + V_2, \ldots, \mu + V_n$ where the errors $V_1, V_2, \ldots, V_n$ are independent and each is Cauchy 0 and 1. One possible estimate for μ is the arithmetic average of the observations. How much better is this estimate when n is large than when $n = 1$?

Assume that n is odd. Another possible estimate for μ is the *sample median*, namely, that observation which has as many observations less than it, as greater than it. Would this seem to be a better estimate than the arithmetic average?

14. Measurement errors $X_1, \ldots, X_n$ are independent and each has the continuous density

$$p_X(x) = \begin{cases} \frac{1}{3} & \text{if} \quad -1 < x < 0 \\ \frac{2}{3} & \text{if} \quad 0 \le x < 1 \\ 0 & \text{otherwise.} \end{cases}$$

Let $S_n = X_1 + \cdots + X_n$ and use the CLT to estimate $P[S_n < 13]$ when $n = 60$.

15. Let $X_1, X_2, \ldots$ be independent and identically distributed. We may interpret $S_n = X_1 + \cdots + X_n$ as the point on the real axis where a man (or particle) will be located after n steps if he starts at the origin and if X_i is the size and direction of his ith step.

(*a*) Assume that X_1 has continuous density

$$p(x) = \begin{cases} (\frac{3}{2})x^2 & \text{if} \quad |x| < 1 \\ 0 & \text{otherwise.} \end{cases}$$

Find the limit of $P[S_n < \sqrt{n}]$ as $n \to \infty$. This number gives some indication of the tendency of the *random walk* to rapidly wander off to the right.

(*b*) An obvious approximation to the distribution of part (*a*) is to assume that he always takes unit steps, but with direction (right or left) determined by a fair coin. Assume that $P[X_1 = -1] = P[X_1 = 1] = .5$ and for this case find the limit of $P[S_n < \sqrt{n}]$ as $n \to \infty$.

16. Let $D_1, D_2, \ldots, D_n$ be independent and each have the continuous density

$$p_{D_1}(d_1) = \begin{cases} 4(d_1)^3 & \text{if } 0 < d_1 < 1 \\ 0 & \text{otherwise.} \end{cases}$$

Interpret D_i as the diameter of a small spherical droplet of liquid, slowly falling in a gaseous medium. The rate of a chemical reaction, between the droplet and the gas, is proportional to the surface area $S_i = \pi(D_i)^2$.

(*a*) Let $T_n = S_1 + \cdots + S_n$ be the total surface area of n droplets, and show that $P[T_n \leq 1180\pi]$ is small when $n = 1800$.

(*b*) Use the CLT to estimate $P[T_n \leq 1180\pi]$ when $n = 1800$.

(*c*) Find the density induced by $T_2 = S_1 + S_2$.

(*d*) Find $P[S_1 + S_2 < .5 \mid S_1 + S_2 < 1]$.

17. Two adjacent counties receive rainfalls X_i and Y_i, respectively, on the ith day of the rainy season in a particular year. Assume that the n random vectors $(X_1, Y_1), (X_2, Y_2), \ldots, (X_n, Y_n)$ are independent and identically distributed. Let

$$Z_n = \sum_{i=1}^{n} X_i \quad \text{and} \quad W_n = \sum_{i=1}^{n} Y_i$$

be the n-day rainfalls for each county, and let $T_n = Z_n + W_n$ be their sum.

(*a*) Find the correlation coefficient ρ_{Z_n, W_n} in terms of the correlation coefficient ρ_{X_1, Y_1} for one day.

(*b*) The water from these two counties supplies a reservoir which is supposed to provide water during the long dry season. The rainy season is considered a failure iff $[T_n \leq 80]$ occurs, where $n = 100$ days. Show that $P[T_n \leq 80]$ is small if

$$EX_1 = .5,\ EY_1 = .4,\ \text{Var } X_1 = \text{Var } Y_1 = .05,\ \rho_{X_1, Y_1} = .6.$$

(*c*) Assume that year-to-year rainfalls are independent and identically distributed. Show that the probability is small that there will be 20 or more failures in 100 years.

(*d*) Use the CLT to estimate $P[T_n \leq 80]$ in part (*b*).

COMMENTS. The joint distribution of (X_1, Y_1) in this exercise can be of any type whatsoever: discrete, continuous, or otherwise. If it were bivariate normal, as defined in Sec. 18-2, then $X_1 + Y_1$ would be normal, hence T_n would be normal so that the estimate of part (*d*) would be the exact answer in this case. More generally, if (X_1, Y_1) has a bivariate normal distribution then, as shown in Exercise 14 (Sec. 18-2), so will (Z_n, W_n) so that exact probabilities can easily be found for events involving both Z_n and W_n; for example, $[Z_n \leq z, W_n \leq w]$.

18. Let $(L_1, W_1), \ldots, (L_n, W_n)$ be $n = 1000$ independent and identically distributed random vectors. If L_i and W_i are the length and width, respectively, of the ith rectangular plot of ground, then this plot requires $2(L_i + W_i)$ miles of fence for complete enclosure. Thus a total of $S_n = \sum_{i=1}^{n} 2(L_i + W_i)$ miles of fence are required. Assume that $EL_1 = 2$, $\text{Var } L_1 = .09$, $EW_1 = 1$, $\text{Var } W_1 = .01$, Covariance $(L_1, W_1) = -.024$.

(*a*) Find ES_n and $\text{Var } S_n$.

(*b*) Prove that $P[S_n \geq 6020]$ is small.

(*c*) Use the CLT to estimate $P[S_n \geq 6020]$.

19. There is 1 quart of liquid oxygen at the beginning of an experiment. However X hours later when the liquid oxygen will actually be needed, there will only be $Y = e^{-X}$ quarts available because of evaporation. Assume that X has an exponential density with mean $1/\lambda$.

(*a*) Find EY from p_X.

(*b*) Find p_Y then find EY from p_Y.

(*c*) What is the density induced by $-\log Y$?

*16-6 CONDITIONAL EXPECTATIONS

Conditional distributions, expectations, and variances were introduced and studied in Sec. 14-1 in the discrete case. The ideas and results were then extended in Sec. 15-5 to the case where the random vector g induced a discrete density, while the conditional distributions of X, given g, were permitted to be arbitrary. In this section we assume that the distribution of g, as well as every conditional distribution of X given g, corresponds to a density. Some of these densities may be continuous and others may be discrete, so various cases arise. We first assume, as in Sec. 15-5, that p_g is discrete, and we find that P_X may correspond to a mixture of a discrete and a continuous density. After Example 2 we assume that p_g is continuous. Example 3 is devoted to a case where all conditional densities are discrete and p_g is continuous. The remainder of the section is devoted to the common case where all densities are continuous.

Let $f: \Omega \to R^n$ and $g: \Omega \to R^m$ be random vectors and assume, until after Example 2, that g induces a discrete density p_g. Then, as in Sec. 15-5, P_f is a mixture,

$$P_f(D) = \sum_\beta p_g(\beta) P_{f|g=\beta}(D) \qquad \text{for all subsets } D \text{ of } R^n.$$

If each $P_{f|g=\beta}$ for $p_g(\beta) > 0$ corresponds to a discrete density then so does P_f and we are back in the discrete case.

If each $P_{f|g=\beta}$ for $p_g(\beta) > 0$ corresponds to a continuous density then

$$P_f(D) = \sum_\beta p_g(\beta) \int_D p_{f|g=\beta}(\alpha)\, d\alpha = \int_D \left[\sum_\beta p_g(\beta) p_{f|g=\beta}(\alpha) \right] d\alpha$$

where $\alpha = (\alpha_1, \ldots, \alpha_n)$ and $d\alpha = d\alpha_1 \cdots d\alpha_n$. Thus in this case P_f corresponds to the continuous density

$$p_f(\alpha) = \sum_\beta p_g(\beta) p_{f|g=\beta}(\alpha)$$

which is a mixture, according to the discrete density p_g, of the continuous densities $p_{f|g=\beta}$.

Suppose that $p_g(\beta)$ is positive for only two values of β, so that

$$P_f(D) = p_g(\beta_1) P_{f|g=\beta_1}(D) + p_g(\beta_2) P_{f|g=\beta_2}(D).$$

If the two conditional distributions correspond, respectively, to discrete and continuous densities then

$$P_f(D) = p_g(\beta_1) \sum_{\alpha \in D} p_{f|g=\beta_1}(\alpha) + p_g(\beta_2) \int_D p_{f|g=\beta_2}(\alpha)\, d\alpha$$

so P_f corresponds to a mixture of a discrete and a continuous density. We say that (R^n, P) corresponds to a **mixture of a discrete and a continuous density** iff there is a discrete density $p_d: R^n \to R$ and a continuous density $p_c: R^n \to R$ and a real number

$0 < d < 1$ such that P is obtained from $dp_d + (1-d)p_c$ by

$$P(D) = d \sum_{\alpha \in D} p_d(\alpha) + (1-d)\int_D p_c(\alpha)\, d\alpha \qquad \text{for each subset } D \text{ of } R^n.$$

Clearly if each conditional distribution $P_{f|g=\beta}$ for $p_g(\beta) > 0$ corresponds to a discrete or continuous density then exactly one of the following three cases occurs. If every conditional density is discrete (continuous) then P_f corresponds to the discrete (continuous) density $p_f = \sum_{\beta} p_g(\beta) p_{f|g=\beta}$. For the third case at least one conditional density is discrete and at least one conditional density is continuous. In this case we can collect the discrete densities together and we can collect the continuous densities together to obtain a mixture, as in the preceding paragraph, with $d = P_g(A)$, where A consists of those β for which $P_{f|g=\beta}$ corresponds to a discrete density. In other words,

$$P_f(D) = P_g(A) \sum_{\beta \in A} \frac{p_g(\beta)}{P_g(A)} P_{f|g=\beta}(D) + P_g(A^c) \sum_{\beta \in A_c} \frac{p_g(\beta)}{P_g(A^c)} P_{f|g=\beta}(D)$$

where for $p_g(\beta) > 0$

$$P_{f|g=\beta}(D) = \begin{cases} \sum_{\alpha \in D} p_{f|g=\beta}(\alpha) & \text{if } \beta \in A \\ \int_D p_{f|g=\beta}(\alpha)\, d\alpha & \text{if } \beta \in A^c \end{cases}$$

so that p_d and p_c are given by

$$p_d = \sum_{\beta \in A} \frac{p_g(\beta)}{P_g(A)} p_{f|g=\beta} \qquad \text{and} \qquad p_c = \sum_{\beta \in A^c} \frac{p_g(\beta)}{P_g(A^c)} p_{f|g=\beta}.$$

If $f = X$ happens to be a random variable we can calculate unconditional and conditional means and variances, and the conclusions of Theorem 2 (Sec. 15-5) apply, of course, to all cases.

Example 1 Let $p_i\colon R \to R$ be a continuous density with mean μ_i and variance σ_i^2. Then the mixture $p = .2p_1 + .3p_2 + .5p_3$ is a continuous density with mean $\mu = .2\mu_1 + .3\mu_2 + .5\mu_3$ and variance $\sigma^2 = .2\sigma_1^2 + .3\sigma_2^2 + .5\sigma_3^3 + (\mu_1 - \mu)^2(.2) + (\mu_2 - \mu)^2(.3) + (\mu_3 - \mu)^2(.5)$. In particular if each p_i is normal and all means are different then p can have three local maxima, so p is not normal. Such a mixture may arise as the distribution of incomes, or of lifespans, in a population which has three subpopulations, each having normally distributed incomes.

■

Example 2 A facility produces rods with length exponentially distributed with mean $1/\lambda = 2$ feet. These rods go into a truncator which is supposed to reduce the length of each rod to the largest number of whole feet. From Example 2 (Sec. 16-2) we know that the output of the truncator should be geometric with parameter $p = 1 - e^{-.5} \doteq .393$. Because of a malfunction, the truncator either truncates properly with probability .7 or leaves the rod untouched with probability .3. Therefore the distribution of the length X of a rod leaving the truncator is a mixture $.7p_d + .3p_c$ where p_d is geometric and p_c is exponential. Hence the mean and variance of X can be found as follows:

$$[\text{the mean of } p_d] = \frac{q}{p} \doteq 1.54 \qquad [\text{the mean of } p_c] = 2$$

$$\text{Var}\, p_d = \frac{q}{p^2} = 3.93 \qquad \text{Var}\, p_c = 4$$

$$EX \doteq (.7)(1.54) + (.3)(2) \doteq 1.86$$

$$\begin{aligned}\text{Var}\, X &\doteq \{(.7)(3.93) + (.3)(4)\} + (.7)(1.54 - 1.86)^2 + (.3)(2 - 1.86)^2 \\ &\doteq 3.95 + .13 = 4.08.\end{aligned}$$

Naturally the law of large numbers and the CLT can be applied to $S_n = X_1 + \cdots + X_n$ where the X_i are independent and each has this same distribution. ■

In the remainder of this section we consider mixtures according to a continuous density.

If $\bar{p}$: $R^m \to R$ is any continuous density and (R^n,P_β) is a probability space for each β for which $\bar{p}(\beta) > 0$ then we then can define the probability space (R^n,P) by

$$P(D) = \int_{R^m} P_\beta(D)\bar{p}(\beta)\, d\beta \qquad \text{for all subsets } D \text{ of } R^n.$$

Clearly $P(D) \geq 0$ and $P(R^n) = 1$, and the countable additivity of P follows by interchanging a summation and integration.

If for each β for which $\bar{p}(\beta) > 0$ we assume that every P_β corresponds to a discrete density p_β concentrated on the *same* countable subset C of R^n then

$$P(D) = \int_{R^m} \left[\sum_{\alpha \in DC} p_\beta(\alpha)\right] \bar{p}(\beta)\, d\beta = \sum_{\alpha \in DC} \int_{R^m} p_\beta(\alpha)\bar{p}(\beta)\, d\beta$$

so P corresponds to the discrete density p: $R^n \to R$ concentrated on C and obtained by taking the same mixture

$$p(\alpha) = \int_{R^m} p_\beta(\alpha)\bar{p}(\beta)\, d\beta.$$

If each P_β corresponds to a continuous density p_β then

$$P(D) = \int_{R^m} \left[\int_D p_\beta(\alpha)\, d\alpha\right] \bar{p}(\beta)\, d\beta = \int_D \left[\int_{R^m} p_\beta(\alpha)\bar{p}(\beta)\, d\beta\right] d\alpha$$

showing that P corresponds to the continuous density p: $R^n \to R$ obtained by taking the same mixture

$$p(\alpha) = \int_{R^m} p_\beta(\alpha)\bar{p}(\beta)\, d\beta.$$

As before, it is possible for P to correspond to a mixture of discrete and continuous densities.

If $n = 1$ and the mixture P has a mean then, just as in Sec. 15-5, we can substitute

$$P(D) = \int_{R^m} P_\beta(D)\bar{p}(\beta)\, d\beta$$

into the definition of the mean of P in Theorem 1 (Sec. 15-4) in order to obtain expressions for the mean and variance of P analogous to those of Theorem 2 (Sec. 15-5).

In Example 3 each conditional density is discrete. We then consider the important case of a mixture of continuous densities according to a continuous density. In this case we interpret $p_\beta(\alpha)\bar{p}(\beta)$ as a joint continuous density.

Example 3 Fix n and consider the binomial densities with parameters n and p. Let the discrete density p_d be obtained by randomizing the parameter p according to the uniform density on the interval $0 < p < 1$. Thus

$$p_d(k) = \int_0^1 \binom{n}{k} p^k q^{n-k}\, dp = \binom{n}{k} \int_0^1 p^k(1-p)^{n-k}\, dp.$$

While showing in Sec. 16-4 that the gamma densities are closed under convolution, we observed that

$$\int_0^1 t^{\nu-1}(1-t)^{\mu-1}\, dt = \frac{\Gamma(\nu)\Gamma(\mu)}{\Gamma(\nu+\mu)} \qquad \text{if } \nu > 0,\ \mu > 0.$$

Using this formula with $\nu = k + 1$ and $\mu = n - k + 1$, we get

$$p_d(k) = \frac{n!}{k!\,(n-k)!} \frac{k!\,(n-k)!}{(n+1)!} = \frac{1}{n+1}$$

so p_d is uniform over the integers $k = 0, 1, \ldots, n$. We can interpret this result in several ways.

Suppose an initial point is marked on a random wheel, as in Fig. 1, and then the wheel is spun once and a new mark is made opposite the pointer. Let p be the fraction of the circumference which passes the pointer when the wheel is rotated clockwise until the initial mark is opposite the pointer, and label this sector "success." Clearly p is uniformly distributed over the unit interval. Now make n independent spins of this same two-sectored wheel and let k be the number of successes. From the previous paragraph we see that for the total experiment of double-marking the wheel and then spinning it n more times, every possible outcome for $k = 0, 1, \ldots, n$ is equally likely.

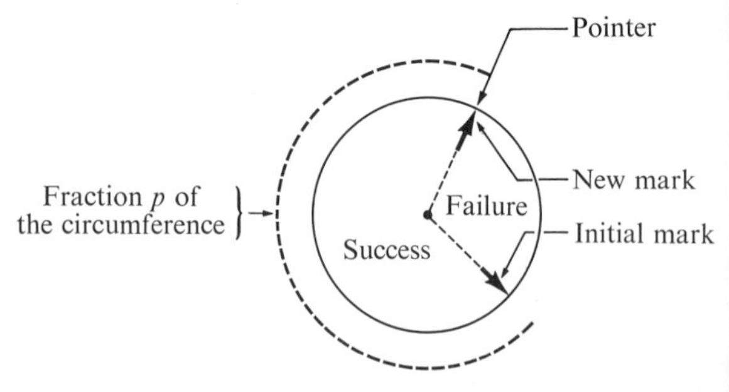

Fig. 1

In other words, if a single biased coin is selected in such a way that every bias $0 < p < 1$ is equally likely, and *the coin so obtained* is tossed n times, then the density p_d of the number of heads is uniform over $k = 0, 1, \ldots, n$. In particular if $n = 1$ then $p_d(0) = p_d(1) = .5$, as would be the case if we just selected a fair coin and tossed it once. However, if $n = 2$ then $p_d(0) = p_d(1) = p_d(2) = \frac{1}{3}$, which is quite different from two independent tosses of a fair coin. This example is generalized in Example 9 (Sec. 17-2). ■

All exercises may be done before reading the remainder of this section.

CONDITIONAL EXPECTATIONS WHEN ALL DENSITIES ARE CONTINUOUS

Suppose that random variables X and Y have a joint continuous density and we are interested in X having a value in some set D of reals, given that we have observed Y to have a value near to a number y. Then

$$P[X \in D \mid y \le Y \le y + \Delta] = \frac{\int_D \left[\int_y^{y+\Delta} p_{X,Y}(x,y')\,dy'\right] dx}{\int_y^{y+\Delta} p_Y(y')\,dy'}$$

$$\doteq \frac{\int_D [p_{X,Y}(x,y)\Delta]\,dx}{p_Y(y)\Delta} = \int_D \frac{p_{X,Y}(x,y)}{p_Y(y)}\,dx$$

where the approximate equality $\doteq$ indicates that $\Delta > 0$ is "small enough" and the integrands are "continuous enough." Thus $p_{X,Y}(x,y)/p_Y(y)$, considered as a function of x with y fixed, can be interpreted as the conditional density of X, given $[Y \doteq y]$. However, we are essentially conditioning on an event having zero probability; this is tricky and requires caution. Note that if we start with some joint density $p'_{X,Y}$ which is equivalent to $p_{X,Y}$ then many of the conditional densities may be different. However, a calculation like

$$P[X \in D,\ Y \in D'] = \int_{D'} \left[\int_D p_{X,Y}(x,y)\,dx\right] dy$$

$$= \int_{D'} p_Y(y) \left[\int_D \frac{p_{X,Y}(x,y)}{p_Y(y)}\,dx\right] dy$$

$$= \int_{D'} p_Y(y) \left[\int_D p_{X|Y=y}(x)\,dx\right] dy$$

would be valid for any one of the equivalent joint densities. Thus in the continuous-density case a collection of conditional densities is unique only in the sense that any equivalent collection will yield the same results for all such integrations as above. A similar lack of uniqueness naturally applies to conditional means and variances.

We proceed with the obvious definitions, assuming that there is enough continuity available to prevent any serious problems. If $f: \Omega \to R^n$ and $g: \Omega \to R^m$ are random vectors having a joint continuous density $p_{f,g}: R^{n+m} \to R$ then, as usual,

$$p_g(\beta) = \int_{R^n} p_{f,g}(\alpha,\beta)\,d\alpha \qquad \text{for each } \beta \text{ in } R^m.$$

If $p_g(\beta) > 0$ then the continuous density defined on R^n by

$$p_{f|g}(\alpha \mid \beta) = p_{f|g=\beta}(\alpha) = \frac{p_{f,g}(\alpha,\beta)}{p_g(\beta)} \qquad \text{for all } \alpha \text{ in } R^n$$

is called the **conditional density of f given** $g = \beta$. Clearly

$$p_f(\alpha) = \int_{R^m} p_{f,g}(\alpha,\beta)\, d\beta = \int_{R^m} p_{f|g=\beta}(\alpha)\, p_g(\beta)\, d\beta$$

so that p_f is the mixture, determined by p_g, of these conditional densities.

If a random variable $X\colon \Omega \to R$ and a random vector $g\colon \Omega \to R^m$ have a joint continuous density $p_{X,g}\colon R^{n+1} \to R$ and if $E\,|X| < \infty$ and $p_g(\beta) > 0$ then

$$\begin{aligned}\mu_{X|g=\beta} = \mu_{X|g}(\beta) &= E[X \mid g = \beta] = [\text{the mean of } p_{X|g=\beta}] \\ &= \int_{-\infty}^{\infty} x p_{X|g=\beta}(x)\, dx = \frac{1}{p_g(\beta)} \int_{-\infty}^{\infty} x p_{X,g}(x,\beta)\, dx\end{aligned}$$

is called the **conditional expectation of X given** $g = \beta$. The function $\mu_{X|g}$ is called the **conditional expectation of X given** g, or the **regression function of X on** g. We have

$$EX = \int_{R^m} \mu_{X|g}(\beta) p_g(\beta)\, d\beta$$

so once again EX equals the expectation of its conditional expectation.

Similarly if $EX^2 < \infty$ and $p_g(\beta) > 0$ then let

$$\sigma^2_{X|g=\beta} = \sigma^2_{X|g}(\beta) = \text{Var}\,[X \mid g = \beta] = [\text{the variance of } p_{X|g=\beta}]$$

and call the function $\sigma^2_{X|g}$ the **conditional variance of X given** g. As in Theorem 2 (Sec. 15-5), we easily see that if X has a variance then

$$\text{Var}\, X = \int_{R^m} \sigma^2_{X|g}(\beta) p_g(\beta)\, d\beta + \int_{R^m} [\mu_{X|g}(\beta) - EX]^2 p_g(\beta)\, d\beta.$$

The interpretations in Sec. 14-1 in terms of predictions of X, given g, still apply. As in Example 2 (Sec. 14-1), the function defined by

$$\mu_{X|g}(g(\omega)) \qquad \text{for all } \omega \text{ in } \Omega$$

is often called the *conditional expectation of X given g*.

Example 4 In Example 1 (Sec. 16-2) it was shown that if X and Y have joint continuous density

$$p_{X,Y}(x,y) = \begin{cases} 2[(1-x)(1-y) + xy] & \text{if } 0 < x < 1,\ 0 < y < 1 \\ 0 & \text{otherwise} \end{cases}$$

then X and Y are each uniformly distributed over the unit interval. Therefore if $0 < y < 1$

then

$$p_{X|Y=y}(x) = \frac{p_{X,Y}(x,y)}{p_Y(y)} = p_{X,Y}(x,y),$$

$$\mu_{X|Y}(y) = \int_0^1 x p_{X|Y=y}(x)\,dx = 2\int_0^1 x[1 - x - y + 2xy]\,dx$$

$$= 2\left[\frac{x^2}{2} - \frac{x^3}{3} - \frac{x^2y}{2} + \frac{2x^3y}{3}\right]_{x=0}^{x=1} = \frac{1+y}{3}.$$

Now in general if there are two real numbers a and b such that

$$\int_{-\infty}^{\infty} |\mu_{X|Y}(y) - (ay + b)|\, p_Y(y)\,dy = 0$$

then $\mu_{X|Y}$ is essentially equal to $ay + b$, which must be the linear regression of X on Y. Therefore the regression function $\mu_{X|Y}(y) = (1 + y)/3$ equals the linear regression

$$\mu_X + \rho\frac{\sigma_X}{\sigma_Y}(y - \mu_Y) = \tfrac{1}{2} + \rho(y - \tfrac{1}{2})$$

so $\rho = \frac{1}{3}$, as can be checked directly from $p_{X,Y}$. ■

Example 5 Let Y_1 and Y_2 have a bivariate normal density

$$p_{Y_1,Y_2}(y_1,y_2) = \frac{1}{2\pi\sigma_1\sigma_2\sqrt{1-\rho^2}}e^{-(\frac{1}{2})Q(y_1,y_2)}$$

as in Example 1 (Sec. 18-2). Then Y_2 is known to be normal μ_2 and $\sigma_2{}^2$ so

$$p_{Y_1|Y_2}(y_1 \mid y_2) = \frac{p_{Y_1,Y_2}(y_1,y_2)}{p_{Y_2}(y_2)} = \frac{1}{\sqrt{2\pi\sigma_1{}^2(1-\rho^2)}}e^{-(\frac{1}{2})[Q(y_1,y_2)-((y_2-\mu_2)/\sigma_2)^2]}.$$

After a little simplification we find that [ ] equals

$$\frac{1}{\sigma_1{}^2(1-\rho^2)}\left\{y_1 - \left[\mu_1 + \rho\frac{\sigma_1}{\sigma_2}(y_2 - \mu_2)\right]\right\}^2.$$

Thus $p_{Y_1|Y_2=y_2}$ is normal with

$$\mu_{Y_1|Y_2=y_2} = \mu_1 + \rho\frac{\sigma_1}{\sigma_2}(y_2 - \mu_2) \qquad \text{and} \qquad \sigma^2_{Y_1|Y_2=y_2} = \sigma_1{}^2(1-\rho^2).$$

In particular the conditional mean, given $Y_2 = y_2$, is linear in y_2. Furthermore, as indicated in Fig. 2, each conditional density is normal with the same variance.

We may interpret Y_1 as the height of an oldest son whose father's height is Y_2. Then the regression function is linear and so equals the linear regression function. Therefore the expected squared error cannot be reduced by using a nonlinear predictor. If we use the regression function for predictions,

$$Y_p = \mu_1 + \rho\frac{\sigma_1}{\sigma_2}(Y_2 - \mu_2)$$

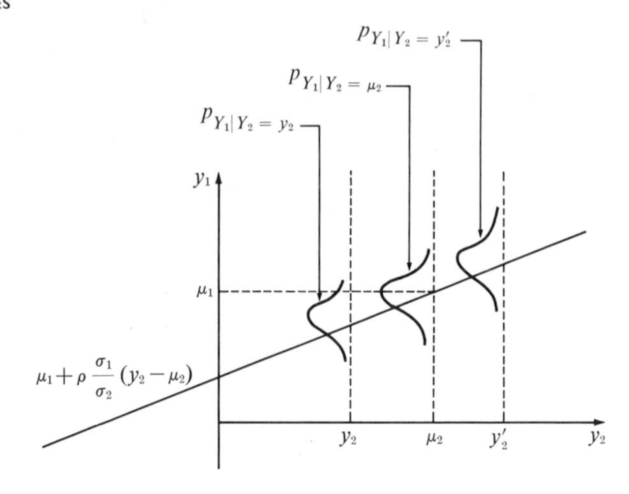

Fig. 2 Conditional densities from a bivariate normal.

then the conditional density of the error $Y_1 - Y_p$, given $Y_2 = y_2$, is normal 0 and $\sigma_1^2(1 - \rho^2)$ for every y_2. In general the conditional distribution of a prediction error $Y_1 - Y_p$, given $Y_2 = y_2$, would be different if $y_2 = 4$ feet from that for $y_2 = 6$ feet, *but* in the bivariate normal case each of the distributions is the same normal distribution. Naturally then $E(Y_1 - Y_p)^2 = \sigma_1^2(1 - \rho^2)$. ∎

EXERCISES

1. Let $N = 1 + X$ where X has a geometric density with parameter p. Thus $p_N(n) = pq^{n-1}$ for $n = 1, 2, \ldots$.

(*a*) Take the continuous gamma densities $\gamma_{\lambda,\nu}$ and randomize the parameter ν by the discrete density p_N. Show that the density p' so obtained is the exponential density with mean $1/(\lambda p)$. COMMENT: With the assumption of widespread independence this exercise can be given the following interpretation. The lifetime X of a nondefective light bulb has an exponential density with mean $1/\lambda$. Therefore the total number of hours of illumination from n nondefective bulbs is $\gamma_{\lambda,n}$. Each bulb has probability p of being defective. We start counting with a nondefective bulb, and N is the number of successive nondefective bulbs prior to the first defective bulb after the start of our count. Let $S_N = X_1 + \cdots + X_N$ be the total hours of illumination from these N nondefective bulbs. Then S_N is exponential with mean $1/(\lambda p)$.

(*b*) Find p' of part (*a*) by using probability-generating functions and Theorem 3 (Sec. 15-5).

2. Let $S_n = X_1 + \cdots + X_n$ where $X_1, \ldots, X_n$ are independent and each has the distribution of X of Example 2.

(*a*) Show that $P[S_{100} \leq 140]$ is small.

(*b*) Find $\lim_{n\to\infty} P[S_n < n(EX_1) + 20\sqrt{n}]$.

3. A professional coffee taster is to participate in a 10-day experiment which is supposed to decide which of 100 marginal coffee blends are to be considered acceptable, or a success. On the first morning he passes judgment on each of the first 10 blends, one after another; and on the second morning on the next 10 blends; etc. Actually the 100 blends are as identical as is physically possible, and the psychologist conducting the experiment is interested in the persistence during the morning of the taster's initial good mood, or inclination to judge coffee as acceptable. Let X_i equal 1 or 0, depending on whether the taster judges the ith blend to be acceptable or unacceptable. Let $S_n = X_1 + \cdots + X_n$. One possible model makes $X_1, \ldots, X_n$ a total of $n = 100$ Bernoulli trials with probability $\frac{1}{2}$ for a success, in which case $ES_n = n(\frac{1}{2}) = 50$ and $\text{Var } S_n = n(\frac{1}{2})(\frac{1}{2}) = 25$.

Find ES_n and $\text{Var } S_n$ for the following model which incorporates perfect persistence of the taster's initial mood during each morning's subexperiment. The 10 subexperiments, corresponding to the 10 mornings, are independent repetitions of the same subexperiment. For the first morning p is uniformly distributed over the unit interval $0 < p < 1$. For a given p the conditional distribution of $X_1, \ldots, X_{10}$ is that of 10 Bernoulli trials with probability p for a success on each trial.

4. Let Y be a random variable with a continuous density. Let $X = Y$ so that X has this same continuous density $p_X = p_Y$. The conditional *discrete* density $p_{X|Y=y}$ is concentrated on the one point y.

(*a*) Verify that

$$P_X(D) = \int_{-\infty}^{\infty} P_{X|Y=y}(D) p_Y(y)\, dy \qquad \text{where} \qquad P_{X|Y=y}(D) = \sum_{x \in D} p_{X|Y}(x).$$

(*b*) In this exercise we have a mixture of discrete densities according to a continuous density. Yet the mixture P_X does *not* correspond to a discrete density, as occurred in Example 3. Why?

(*c*) Will the same phenomenon emphasized in part (*b*) occur if $X = Y + Z$, where Y and Z are independent and $P[Z = 1] = P[Z = -1] = .5$?

5. Let p_λ be the Poisson density with mean λ. Show that a geometric density is obtained if we randomize the parameter λ by the exponential density with mean $1/\lambda'$.

17
Distribution Functions

A distribution function $F\colon R^n \to R$ is a *point function*, as opposed to the *set function* P. There is a one-to-one correspondence between such $F\colon R^n \to R$ and the family of probability spaces (R^n,P). Thus every (R^n,P) can be described by its corresponding $F\colon R^n \to R$. A consequent advantage of distribution functions is that they permit a terminology which applies to all cases which have R^n as sample space. However, manipulations with distribution functions are convenient only in special situations. Densities are often more convenient when available; otherwise P is often as convenient as is F.

Section 17-1 discusses distribution functions on R and introduces the *continuous-singular case*, which is covered by our theory. It turns out that this new third case, together with the discrete- and continuous-density cases, exhausts all possible cases. Example 4 (Sec. 17-1) shows that manipulations which are common in fields such as communication theory and operations research can easily lead (whether realized or not) to continuous singular distributions.

Section 17-2 shows that distribution functions yield generality and simplicity in finding the distribution of $Y = g(X)$ for certain g. In Sec. 17-3 we study the properties of distribution functions on R^n. Section 17-4, on the method of indicators, is included in this chapter primarily because the method is illustrated by deriving a formula needed in Sec. 17-3.

17-1 DISTRIBUTION FUNCTIONS ON R

This section is concerned primarily with probability spaces (R,P) whose sample space is the real line R. The distribution function $F\colon R \to R$ corresponding to (R,P) is introduced. It is shown that knowledge of the more concrete point function F is equivalent to knowledge of the more elusive set function P. The class of such functions F is characterized, and we see that any such F, or corresponding P, can be decomposed into a unique mixture of three components, two of which correspond to discrete and continuous densities, respectively. The third component is called *continuous singular*. Means and variances are found for such mixtures, in terms of the components. Naturally the laws of large numbers and central-limit theorems apply to all such cases.

Thus in a sense a general (R,P) is a mixture of three cases, two of which are the discrete- and continuous-density cases already studied. The continuous-singular case does not normally arise in problems of the kind typically considered at this level. However, knowledge of its existence is desirable for a proper initial perspective. Example 4 shows that a continuous singular P can arise naturally as the induced distribution of a random variable which is the weighted sum of an *infinite* sequence of Bernoulli trials.

If (R,P) is any probability space whose sample space is the real line then the function $F: R \to R$ defined by

$$F(x) = P\{x' : x' < x\} \qquad \text{for all real } x$$

is called the **distribution function,** or cumulative distribution function, **corresponding to** (R,P). Thus the value of F at x is equal to the probability of obtaining some real x' less than x. If $X: \Omega \to R$ is a random variable then the **distribution function** $F_X: R \to R$ **induced by** X with associated (Ω,P) is just the one corresponding to P_X. Thus

$$F_X(x) = P_X\{x' : x' < x\} = P\{\omega: X(\omega) < x\} = P[X < x]$$

so $F_X(x)$ is the probability that X will have a value less than x.

If F_i corresponds to (R,P_i) and there is a point x_0 such that $F_1(x_0) \neq F_2(x_0)$ then

$$P_1\{x' : x' < x_0\} = F_1(x_0) \neq F_2(x_0) = P_2\{x' : x' < x_0\}$$

so that $P_1 \neq P_2$. Thus, in contrast to the phenomenon associated with continuous densities, if two distribution functions differ at one point then their corresponding probability measures are different.

We say that F is **discrete** iff (R,P) corresponds to a discrete density $p: R \to R$ so that

$$F(x) = \sum_{x' < x} p(x') \qquad \text{for all real } x.$$

We say that F is **absolutely continuous** iff (R,P) corresponds to a continuous density $p: R \to R$ so that

$$F(x) = \int_{-\infty}^{x} p(x')\, dx' \qquad \text{for all real } x.$$

In this case

$$\frac{dF(x)}{dx} = p(x),$$

at least at every x where p is continuous, so we can obtain p by differentiating F. We say that F is **continuous** iff it is continuous at every point. An absolutely continuous F is necessarily continuous. As shown in Example 4, there are continuous distribution functions which are not absolutely continuous.

Example 1 Let p be a Poisson discrete density with mean $\lambda = 3.2$. Clearly $F(x) = 0$ if $x \leq 0$. From Table 1 (Sec. 8-3) we obtain

$$F(x) = \begin{cases} p(0) = .04076 & \text{if } 0 < x \leq 1 \\ p(0) + p(1) = .17120 & \text{if } 1 < x \leq 2 \\ \cdots\cdots\cdots\cdots\cdots\cdots & \end{cases}$$

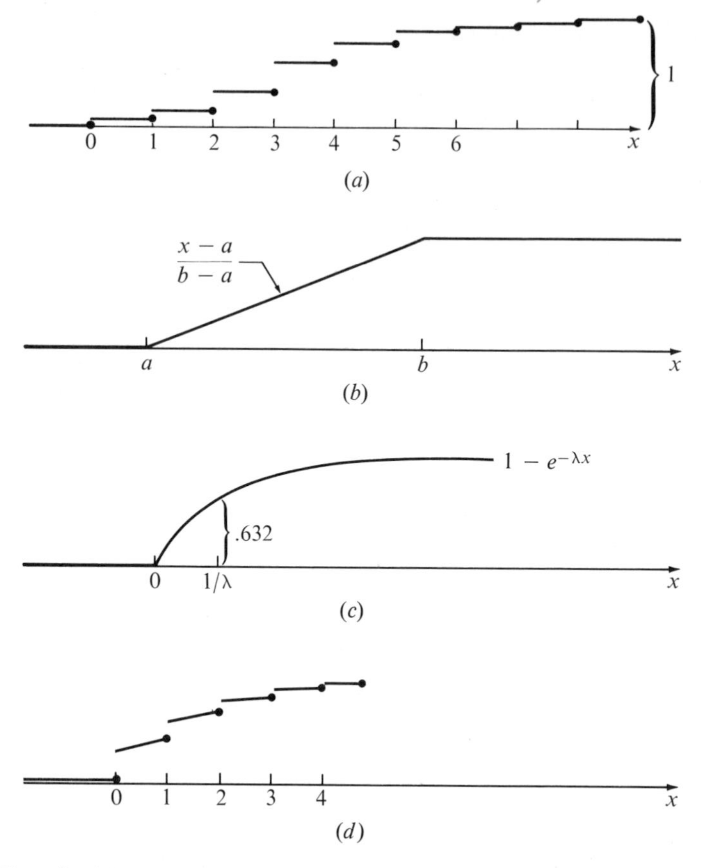

Fig. 1 (*a*) Poisson $\lambda = 3.2$; (*b*) uniform distribution over $a < x < b$; (*c*) exponential with mean $1/\lambda$; (*d*) a mixture of a geometric and an exponential.

so that F is as shown in Fig. 1*a*. If n is a positive integer then $F(n) = p(0) + p(1) + \cdots + p(n-1)$, so that F has a jump of magnitude $p(n)$ at $x = n$, but the value $F(n)$ of F at $x = n$ does not include this jump. This will always be the case because of our definition of F. Thus the value $F(x)$ equals the limit of F from the left, $F(x) = \lim_{\epsilon \to 0} F(x - \epsilon)$ where ϵ is constrained to be positive. Some writers define F by $F(x) = P\{x: x' \leq x\}$, and with such a definition the value $F(x)$ equals the limit of F from the right. The reader should be aware of this lack of convention.

The step function of Fig. 5 (Sec. 13-4) is the distribution function corresponding to the standardized discrete density $p_{S_8^*}$.

The CLT was stated essentially in terms of the distribution function $F_{S_n^*}$ induced by $S_n^* = (S_n - ES_n)/\sigma_{S_n}$; that is,

$$F_{S_n^*}(b) = P[S_n^* < b] \to \Phi(b) \qquad \text{as } n \to \infty, \text{ for all real } b,$$

is a form of the conclusion of Corollary 1*a* (Sec. 13-2).

Figure 1*b* shows the distribution function corresponding to a uniform continuous density over the interval $a < x < b$. Clearly F has a derivative equal to the density, except, of course, at the two end points of the interval.

If $p: R \to R$ is exponential with parameter λ then

$$F(x) = \int_0^x \lambda e^{-\lambda x'}\, dx' = 1 - e^{-\lambda x}$$

for $x > 0$, as shown in Fig. 1*c*.

Figure 1 (Sec. 13-1) shows the distribution function Φ corresponding to the standardized normal density.

Figure 1*d* is the distribution function corresponding to a mixture of a geometric and an exponential density, as described in Example 3. ■

Example 2 Let $X_1, X_2, \ldots, X_n$ be independent and identically distributed by $F: R \to R$. Let $Y = \max\{X_1, \ldots, X_n\}$ so that Y is the largest of the n observations; for example, Y might be the maximum length of n rods, or the maximum lifetime of n bulbs or men. The maximum observation is less than y iff every observation is less than y so

$$\begin{aligned} F_Y(y) &= P[Y < y] = P[X_1 < y, X_2 < y, \ldots, X_n < y] \\ &= P[X_1 < y]P[X_2 < y] \cdots P[X_n < y] = F^n(y). \end{aligned}$$

Thus F_Y is just the nth power of F. If F corresponds to continuous density $p: R \to R$ then

$$p_Y(y) = \frac{d}{dy} F_Y(y) = nF^{n-1}(y)p(y).$$

In particular if F is exponential with mean $1/\lambda$ then

$$p_Y(y) = \begin{cases} 0 & \text{for } y \le 0 \\ n(1 - e^{-\lambda y})^{n-1}\lambda e^{-\lambda y} & \text{for } y > 0. \end{cases}$$ ■

A function $F: R \to R$ is called a **distribution function on** R iff the following four conditions hold:

1. $\lim\limits_{x \to \infty} F(x) = 1.$
2. $\lim\limits_{x \to -\infty} F(x) = 0.$
3. $\lim\limits_{\substack{\epsilon \to 0 \\ \epsilon > 0}} F(x - \epsilon) = F(x)$ for each real x.
4. $F(x_1) \le F(x_2)$ whenever $x_1 < x_2$.

Thus $F: R \to R$ is a distribution function on R iff it converges to 1 to the right and to zero to the left, is continuous from the left at every point, and is nondecreasing.

As already suggested by this terminology, we now prove that if (R,P) is any probability space then the distribution function F corresponding to P *is indeed* a distribution function on R; that is, F has the four properties just stated. We use parts (*d*), (*h*), and (*i*) of Theorem 2 (Sec. 6-4). If $x_1 < x_2$ then $\{x' : x < x_1\}$ is a subset of $\{x' : x' < x_2\}$ so

$$F(x_1) = P\{x' : x' < x_1\} \le P\{x' : x' < x_2\} = F(x_2)$$

and hence F is nondecreasing. The events

$$\{x' : x' < -1\}, \{x' : x' < -2\}, \{x' : x' < -3\}, \ldots$$

are decreasing and their intersection is the empty set, so $F(-n) = P\{x' : x' < -n\}$ converges to $P(\varphi) = 0$ as n goes to infinity. This, together with the fact that F is nondecreasing, shows that F converges to zero to the left. The events $\{x' : x' < 1\}$, $\{x' : x' < 2\}, \ldots$ are increasing to R, so $F(n) = P\{x' : x' < n\}$ converges to $P(R) = 1$ as n goes to infinity. Thus F converges to 1 to the right. Similarly

$$F\left(x - \frac{1}{n}\right) = P\left\{x' : x' < x - \frac{1}{n}\right\} \to P\{x' : x' < x\} = F(x) \qquad \text{as } n \to \infty$$

so F is continuous from the left. *We have proved that "the distribution function $F: R \to R$ corresponding to a probability space (R,P)" is a distribution function on R.* We will see shortly that *every distribution function on R is the distribution function corresponding to a probability space (R,P).*

If F is the distribution function corresponding to a probability space (R,P) we often wish to calculate $P(A)$ from F. Let us now investigate how this can be done for some simple sets A of reals. We first consider the case where A is an interval of the form $a \leq x < b$, then we let A consist of a single real number, and then we consider other intervals and sets. The results are summarized in Fig. 2.

If $a < b$ then $\{x : x < a\}$ and $\{x : a \leq x < b\}$ are disjoint with union $\{x : x < b\}$ so that $P\{x : x < a\} + P\{x : a \leq x < b\} = P\{x : x < b\}$. Therefore

$$P\{x : a \leq x < b\} = P\{x : x < b\} - P\{x : x < a\} = F(b) - F(a) \qquad \text{if } a < b.$$

The fact that F is nondecreasing guarantees that F has both right- and left-hand limits at each point. Therefore the only kind of discontinuity that F can have is a jump. The magnitude of the jump of F at $x = a$ is defined to equal the difference between the right- and left-hand limits of F at $x = a$; that is,

$$[\text{the } \mathbf{jump} \text{ of } F \text{ at } x = a] = \lim_{\substack{\epsilon \to 0 \\ \epsilon > 0}} F(a + \epsilon) - \lim_{\substack{\epsilon \to 0 \\ \epsilon > 0}} F(a - \epsilon)$$
$$= F(a + 0) - F(a).$$

Clearly [the jump of F at $x = a$] $= 0$ iff F is continuous at $x = a$, in which case a is called a **point of continuity** of F. If [the jump of F at $x = a$] > 0 then a is called a **point of discontinuity** of F. Observe that

$$[\text{the jump of } F \text{ at } x = a] = \lim_{n \to \infty} \left\{F\left(a + \frac{1}{n}\right) - F(a)\right\}$$
$$= \lim_{n \to \infty} P\left\{x : a \leq x < a + \frac{1}{n}\right\} = P(\{a\})$$

where $\{a\}$ is the event consisting of the single point a. Thus the jump of F at $x = a$ is equal to the probability that the outcome will be a. In particular, F corresponding to (R,P) is continuous iff every real a has zero probability, $P(\{a\}) = 0$.

We now find the probability of a closed interval from F,

$$P\{x: a \le x \le b\} = P\{x: a \le x < b\} + P(\{b\})$$
$$= F(b) - F(a) + [\text{the jump of } F \text{ at } x = b]$$
$$= F(b + 0) - F(a).$$

Thus we can handle end points of intervals by adding or subtracting appropriate jumps of F. The first three formulas of Fig. 2 yield the remaining ones by inspection.

$P\{x: x < a\} = F(a)$
$P\{x: a \le x < b\} = F(b) - F(a)$
$P(\{a\}) = [\text{the jump of } F \text{ at } x = a]$
$P\{x: x \ge a\} = 1 - F(a)$
$P\{x: x \le a\} = F(a) + [\text{the jump of } F \text{ at } x = a] = F(a + 0)$
$P\{x: x > a\} = 1 - F(a) - [\text{the jump of } F \text{ at } x = a] = 1 - F(a + 0)$
$P\{x: a \le x \le b\} = F(b) - F(a) + [\text{the jump of } F \text{ at } x = b] = F(b + 0) - F(a)$
$P\{x: a < x < b\} = F(b) - F(a) - [\text{the jump of } F \text{ at } x = a] = F(b) - F(a + 0)$
$P\{x: a < x \le b\} = F(b) - F(a) + [\text{the jump of } F \text{ at } x = b] - [\text{the jump of } F \text{ at } x = a]$
$= F(b + 0) - F(a + 0)$

Fig. 2 F is the distribution function corresponding to (R,P).

If (R,P_1) and (R,P_2) are two probability spaces with corresponding distribution functions F_1 and F_2 then $P_1 = P_2$ *iff* $F_1 = F_2$. This is because the probability of any interval can be calculated from $F_1 = F_2$, so P_1 and P_2 must assign the same probability to every closed interval; hence Theorem 2 (Sec. 15-1) implies that $P_1 = P_2$. Therefore X and Y are identically distributed iff $F_X = F_Y$.

Thus the correspondence which "assigns a distribution function F to each probability space (R,P)" is clearly one-to-one, since different P generate different F. This same correspondence is also *onto*; that is, for every distribution function F on R there is a probability space (R,P) such that F is the distribution function corresponding to (R,P). [Thus probability spaces (R,P) are no more (or less) general than distribution functions on R.] This can be made intuitively reasonable as follows. Given a distribution function $F: R \to R$, we can define $P(A)$ for every interval A by using Fig. 2. If A is an interval we can define $P(A^c)$ by $P(A^c) = 1 - P(A)$, and if $A_1, A_2, \ldots$ are disjoint intervals we can define $P\left(\bigcup_{n=1}^{\infty} A_n\right) = \sum_{n=1}^{\infty} P(A_n)$. This technique can be used repeatedly to extend the definition of P to the complement or countable disjoint union of any events for which P has already been defined. It should appear reasonable that we have enough conditions in the definition of a distribution function F on R to guarantee that F determines a unique number $P(A)$ for every reasonable set A of reals, and that the resulting P is a probability measure, so that F is the distribution function corresponding to this (R,P).

Exercises 1, 2, 5, and 9 may be done before reading the remainder of this section.

We now define the continuous-singular case and state the decomposition, or representation, of a general distribution function $F: R \to R$. Only part of this decomposition, Lemma 1, is proved. An exercise suggests the nature of the remainder of the proof. Then, just as in Sec. 15-5, means and variances are obtained for mixtures and the results are applied in Example 3. After that Example 4 shows how the continuous-singular case can arise.

A set A of reals is said to have **zero length** (Lebesgue measure) iff for every $\epsilon > 0$ there is a finite or infinite sequence $I_n = \{x: a_n \leq x \leq b_n\}$ of intervals with $a_n \leq b_n$, such that A is a subset of $\bigcup_{n=1}^{\infty} I_n$ and $\sum_{n=1}^{\infty} (b_n - a_n) < \epsilon$. That is, A has zero length iff we can cover it by a sequence of intervals the sum of whose lengths is arbitrarily small. If $A = \{x_1, x_2, \ldots\}$ is countable we can let $a_n = b_n = x_n$, so that A has zero length. Example 4, and Exercise 5 (Sec. 4-3), exhibit uncountable sets which have zero length. A distribution function $F: R \to R$ is said to be **continuous singular** iff it is continuous and there is a set A of reals of zero length such that $P(A) = 1$, where P is the probability measure corresponding to F. That is, continuous singular F are those corresponding to probability measures which assign probability 0 to every point but are nevertheless concentrated on a set of zero length. It turns out that such an F has a derivative equal to zero at every point, except those belonging to some set of zero length.

If A is a set of zero length then for any (measurable) function $f: R \to R$ we have $\int_A f(x)\,dx = 0$, according to any of the standard theories of integration. That is, the integral, with respect to length dx, of any real-valued function is zero when the integral is over a set of zero length. Therefore if (R,P) corresponds to a continuous singular $F: R \to R$ then P cannot correspond to any continuous density $p: R \to R$, since there is a set A of zero length and

$$P(A) = 1 \neq 0 = \int_A p(x)\,dx.$$

(We note that δ functions are of no assistance for such a P, since every point has zero probability. However, δ functions are used by some writers to assign positive probability to a point, which we have easily done without their use.)

Basic decomposition of a distribution function on R Every distribution function $F: R \to R$ has a representation $F = \beta_d F_d + \beta_{ac} F_{ac} + \beta_{cs} F_{cs}$ where $\beta_d + \beta_{ac} + \beta_{cs} = 1$ and each $\beta \geq 0$. The F_d, F_{ac}, and F_{cs} are, respectively, discrete, absolutely continuous, and continuous singular distribution functions on R. Furthermore the representation is unique in the sense that the numbers β_d, β_{ac}, and β_{cs} are unique, and each distribution function with a positive coefficient is unique. ■

Thus F_d in the representation corresponds to a discrete density $p_d: R \to R$ so

$$F_d(x) = \sum_{x' < x} p_d(x').$$

Also F_{ac} corresponds to a continuous density $p_{ac}: R \to R$ so

$$F_{ac}(x) = \int_{-\infty}^{x} p_{ac}(x')\,dx'.$$

Furthermore there is a set A of reals having zero length and satisfying $P_{cs}(A) = 1$, where P_{cs} corresponds to F_{cs} which is continuous.

Naturally F corresponds to a discrete (continuous) density iff $\beta_d = 1$ ($\beta_{ac} = 1$). Also F is a **mixture of a discrete and a continuous density** iff $\beta_d > 0$, $\beta_{ac} > 0$, and $\beta_d + \beta_{ac} = 1$. In this case if P corresponds to F then

$$P(D) = \beta_d \sum_{x \in D} p_d(x) + (1 - \beta_d) \int_D p_{ac}(x)\, dx$$

for every set D of reals.

Often an F is said to be *singular* iff $\beta_{ac} = 0$. Its corresponding P is also said to be singular (with respect to Lebesgue measure). Thus P is singular iff there exists an event having zero length and probability 1. We will not use this terminology.

***Lemma 1: Part of the decomposition** Every distribution function $F: R \to R$ has a representation $F = \beta_d F_d + (1 - \beta_d) F_c$ where $0 \le \beta_d \le 1$ and F_d and F_c are, respectively, discrete and continuous distribution functions on R. Furthermore the representation is unique in the sense that β_d is unique, and each distribution function with a positive coefficient is unique. ■

It can be shown, as indicated in Exercise 6, that every continuous $F_c: R \to R$ has a representation

$$F_c = bF_{ac} + (1 - b)F_{cs} \qquad \text{with } 0 \le b \le 1$$

where F_{ac} is absolutely continuous, F_{cs} is continuous singular, and the representation is unique as usual. Lemma 1, together with this unproved fact, immediately yields the basic decomposition stated above.

***Proof of Lemma 1** We first prove the uniqueness of the representation. If there are two such then

$$F = \beta_d F_d + (1 - \beta_d) F_c = \beta'_d F'_d + (1 - \beta'_d) F'_c$$

so

$$\beta_d F_d - \beta'_d F'_d = (1 - \beta'_d) F'_c - (1 - \beta_d) F_c$$

where the right-hand side is a continuous function. Therefore the left-hand side is continuous and so cannot have any positive jumps. But F_d and F'_d are discrete, so $\beta_d F_d - \beta'_d F'_d$ must be a constant, and if we let x go to minus infinity we see that this constant is zero. But then if we let x go to infinity we get $\beta_d - \beta'_d = 0$ so $\beta_d = \beta'_d$; hence

$$\beta_d(F_d - F'_d) = 0 = (1 - \beta_d)(F_c - F'_c)$$

yielding the uniqueness of the representation.

As a first step toward proving the existence of the representation we show that if J is the set of reals at which a distribution function $F: R \to R$ has a positive jump then J is countable. Let J_n consist of those points at which F jumps by at least $1/n$. Recall that $F(\infty) = 1$, $F(-\infty) = 0$, and F is nondecreasing. Therefore if $a_1 < a_2 < \cdots < a_k$

all belong to J_n then

$$1 \geq F(a_k + 0) - F(a_1 - 0) \geq \sum_{i=1}^{k} [\text{the jump of } F \text{ at } x = a_i] \geq k \frac{1}{n}$$

and hence $k \leq n$. Thus J_n contains at most n points. But the union $J = J_1 \cup J_2 \cup \cdots$ of a sequence of finite sets is a countable set.

We now *sketch* the proof of the existence of the representation. Continual reference to Fig. 3 should make the construction transparent. First define $h: R \to R$ by

$$h(a) = [\text{the jump of } F \text{ at } x = a]$$

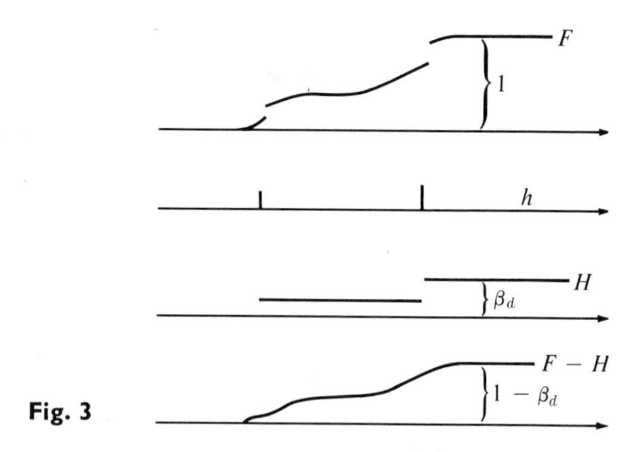

Fig. 3

and let $a_1, a_2, \ldots$ be an enumeration of the set J of points at which h has a positive value. Let

$$\beta_d = \sum_{i=1} h(a_i)$$

so that $0 \leq \beta_d \leq 1$. Clearly $\beta_d = 0$ iff F is continuous. Also $\beta_d = 1$ iff h is a discrete density and F corresponds to it. If $0 < \beta_d < 1$ we define $H: R \to R$ by

$$H(x) = \sum_{x' < x} h(x')$$

so that $H(x)$ equals the sum of the jumps of F to the left of x. The function $F - H$ is continuous since H has the same positive jumps as F. Also

$$F = H + F - H = \beta_d\left[\frac{1}{\beta_d} H\right] + (1 - \beta_d)\left[\frac{1}{1 - \beta_d}(F - H)\right]$$

is the desired representation. ■

THE MEAN AND VARIANCE OF A MIXTURE

We now relate distribution functions, and mixtures of them, to expectations. **The mean (variance) of a distribution function** $F: R \to R$ is just the mean (variance) of its corresponding (R,P), if this mean (variance) exists. The mean of P was defined in

Theorem 1 (Sec. 15-4) in terms of sequences A_n^+ and A_n^-, which used expressions such as

$$P\left\{x: \frac{j}{2^n} \leq x < \frac{j+1}{2^n}\right\} = F\left(\frac{j+1}{2^n}\right) - F\left(\frac{j}{2^n}\right)$$

so that A_n^+ and A_n^- can easily be expressed in terms of F rather than P. At times it will be convenient to think of A_n^+ and A_n^- as so expressed in terms of F. Also EX was defined to equal the mean of P_X so EX may be thought of as defined directly in terms of F_X.

If $F_1, \ldots, F_n$ are distribution functions on R and $b_1, \ldots, b_n$ are nonnegative real numbers whose sum is 1 then the **mixture** $F = \sum_{i=1}^{n} b_i F_i$ is easily seen to be a distribution function on R. That is, F goes to 1 to the right and to zero to the left and is left-continuous and nondecreasing, since each F_i has these properties. Let F and F_i correspond to (R,P) and (R,P_i), respectively, so that $P = \sum_{i=1}^{n} b_i P_i$ equals the mixture of the P_i. If P_i has mean μ_i and variance σ_i^2 then Theorems 1 and 2 (Sec. 15-5) show that the mean and variance of P, or F, are given by

$$\mu = \sum_{i=1}^{n} b_i\mu_i \qquad \sigma^2 = \sum_{i=1}^{n} \sigma_i^2 b_i + \sum_{i=1}^{n} (\mu_i - \mu)^2 b_i.$$

(Exercise 3a sketches an explicit fairly direct proof of these formulas for the important case considered in the next paragraph.) Thus the mean of the mixture $F = \sum_{i=1}^{n} b_i F_i$ equals the same mixture of the means, while variance of F exceeds the same mixture of the variances by an amount interpretable as the variance of the means.

In particular if $F = \beta_d F_d + (1 - \beta_d) F_{ac}$ for $0 < \beta_d < 1$ is a mixture of a discrete F_d and an absolutely continuous F_{ac} then the mean and variance of F are given by

$$\mu = \beta_d \mu_d + (1 - \beta_d)\mu_{ac}$$
$$\sigma^2 = \beta_d \sigma_d^2 + (1 - \beta_d)\sigma_{ac}^2 + \beta_d(\mu_d - \mu)^2 + (1 - \beta_d)(\mu_{ac} - \mu)^2$$

where

$$\mu_d = \sum_x x p_d(x) \qquad \sigma_d^2 = \sum_x (x - \mu_d)^2 p_d(x)$$

$$\mu_{ac} = \int_{-\infty}^{\infty} x p_{ac}(x)\, dx \qquad \sigma_{ac}^2 = \int_{-\infty}^{\infty} (x - \mu_{ac})^2 p_{ac}(x)\, dx.$$

Example 3 Example 2 (Sec. 16-6), which may be read independently of the remainder of Sec. 16-6, yields a random variable X with distribution function $F = .7F_d + .3F_{ac}$ where F_d and F_{ac} are geometric and exponential, respectively. EX and Var X are calculated, then, by the formulas described above. Exercise 4, as well as Exercise 2 (Sec. 16-6), apply the laws of large numbers and the CLT to this F. Fig. 1d shows F. ■

*__Example 4__ This example shows that a continuous singular distribution function can be induced by a random variable which is a weighted sum of an infinite sequence of Bernoulli trials. As shown in Example 7 (Sec. 16-1), the unit interval can be used for the sample space for an infinite sequence of Bernoulli trials. Thus such distributions can arise when using the unit interval, with length for probability, as the probability space.

Let $X_1, X_2, \ldots$ be an infinite sequence of Bernoulli trials with

$$P[X_i = 1] = p \qquad P[X_i = 0] = q = 1 - p \qquad 0 < p < 1.$$

Let ρ be any real number satisfying $0 < \rho < \frac{1}{2}$. For any infinite sequence $x_1, x_2, \ldots$ of 0's and 1's the series $\sum_{n=1}^{\infty} x_n \rho^n$ of nonnegative numbers is clearly convergent, so that we can define the random variable Y by

$$Y = \sum_{n=1}^{\infty} X_n \rho^n.$$

We might describe Y as the sum of a geometric series whose terms have been eliminated, or not, according to biased-coin tosses. We show that the distribution function $F_Y : R \to R$ induced by Y is continuous singular.

The continuous function F_Y is partially plotted in Fig. 4. Roughly speaking, it is constant over a sequence of disjoint intervals having total length 1, and it makes its increase on an uncountable collection of points having total length 0.

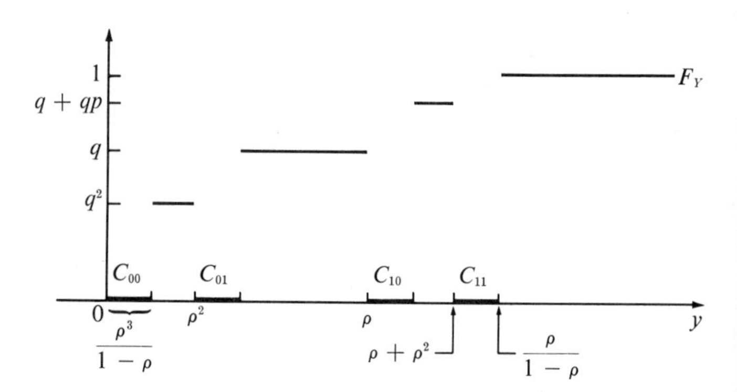

Fig. 4

Before turning to the proof we comment that the usual expectation manipulations can be justified for this infinite series, so that

$$EY = \sum_{n=1}^{\infty} (EX_n)\rho^n = \sum_{n=1}^{\infty} p\rho^n = p\rho[1 + \rho + \rho^2 + \cdots] = \frac{p\rho}{1 - \rho}.$$

Thus EY is easily found in this way. The same result could be obtained, with more effort, by finding the mean of F_Y directly from the definition of Theorem 1 (Sec. 15-4) in terms of A_n^+. Similarly

$$\text{Var } Y = \sum_{n=1}^{\infty} \text{Var } (X_n \rho^n) = \sum_{n=1}^{\infty} qp\rho^{2n} = pq\rho^2[1 + \rho^2 + (\rho^2)^2 + \cdots] = \frac{pq\rho^2}{1 - \rho^2}$$

so Y is quite manageable. In particular, the laws of large numbers and CLT can be applied to sums of independent random variables each distributed like Y. Such random variables are often dealt with, without suspecting that they induce a distribution for which density manipulations do not apply. We now turn to the proof that F_Y is continuous singular.

Rather than restricting ourselves to Bernoulli trials, we will prove more generally that F_Y is continuous singular if

$$Y = \sum_{n=1}^{\infty} X_n \rho^n \qquad \text{with } 0 < \rho < \tfrac{1}{2}$$

and if the sequence $X_1, X_2, \ldots$ of random variables satisfies $P[X_i = 0 \text{ or } 1] = 1$ for all i, and $P[X_1 = x_1, X_2 = x_2, \ldots] = 0$ for every infinite sequence $x_1, x_2, \ldots$ of reals. The X_i may be dependent and have different distributions.

We note the obvious implication that

$$P\{\omega: X_i(\omega) = 0 \text{ or } 1 \text{ for } every\ i = 1, 2, \ldots\} = 1,$$

since the complement has probability at most

$$\sum_{i=1}^{\infty} P\{\omega: X_i(\omega) \neq 0 \text{ and } X_i(\omega) \neq 1\} = \sum_{i=1}^{\infty} 0 = 0.$$

Thus we assume, in effect, that $X_1, X_2, \ldots$ *must* yield *some* infinite sequence $x_1, x_2, \ldots$ of 0's and 1's, but that *each* particular sequence $x_1, x_2, \ldots$ has 0 probability.

Bernoulli trials satisfy these assumptions because if δ is the larger of p and $1 - p$ then for any $x_1, x_2, \ldots$ we have

$$\begin{aligned} P[X_1 = x_1, X_2 = x_2, \ldots] &\leq P[X_1 = x_1, \ldots, X_n = x_n] \\ &= P[X_1 = x_1] \cdots P[X_n = x_n] \leq \delta^n, \end{aligned}$$

for each n, but δ^n converges to 0 as n goes to infinity.

Let C consist of those reals y for which there exists at least one sequence $x_1, x_2, \ldots$ of zeros and ones satisfying $y = \sum_{n=1}^{\infty} x_n \rho^n$. Thus C is defined in terms of the parameter ρ, but C does not depend on the joint distribution of the X_n. Clearly C contains the set of possible values of Y, in that $P[Y \in C] = 1$. Naturally $P[Y \in C^c] = 0$. We show that if $y \in C$ then there is *only one* sequence $x_1, x_2, \ldots$ of 0's and 1's such that $y = \sum_{n=1}^{\infty} x_n \rho^n$. Therefore

$$P[Y = y] = P[X_1 = x_1, X_2 = x_2, \ldots] = 0$$

so F_Y is continuous. We then show that C has zero length, so F_Y is continuous singular.

Let $x_1, x_2, \ldots$ and $x_1', x_2', \ldots$ be two *different* sequences of 0's and 1's, and let

$$y = \sum_{n=1}^{\infty} x_n \rho^n \qquad y' = \sum_{n=1}^{\infty} x_n' \rho^n.$$

We must show that $y \neq y'$. Let k be the first coordinate in which the sequences differ, and assume that $x_k = 0$, $x_k' = 1$, while $x_1 = x_1', \ldots, x_{k-1} = x_{k-1}'$. We will show that $y < y'$, so that the *relative sizes of the sums y and y′ are determined solely by the first coordinate in which there is a difference.* Now $0 < \rho < \frac{1}{2}$ implies that $\rho/(1 - \rho) < 1$, which yields the strict inequality sign $<$ in the chain

$$\begin{aligned} y &\leq \sum_{n=1}^{k-1} x_n \rho^n + \rho^{k+1} + \rho^{k+2} + \cdots = \sum_{n=1}^{k-1} x_n \rho^n + \rho^{k+1}(1 + \rho + \rho^2 + \cdots) \\ &= \sum_{n=1}^{k-1} x_n \rho^n + \frac{\rho^{k+1}}{1 - \rho} < \sum_{n=1}^{k-1} x_n \rho^n + \rho^k \leq y'. \end{aligned}$$

Thus $y < y'$, and hence $P[Y = y] = 0$ for all y.

Let $x_1, \ldots, x_k$ be any finite sequence of 0's and 1's. Let $C_{x_1,\ldots,x_k}$ consist of those real y such that there exists a sequence $x'_{k+1}, x'_{k+2}, \ldots$ of 0's and 1's such that

$$y = \sum_{n=1}^{k} x_n \rho^n + \left[\sum_{n=k+1}^{\infty} x'_n \rho^n \right].$$

Thus $C_{x_1,\ldots,x_k}$ consist of those y in C which arise from sequences starting with $x_1, \ldots, x_k$. As before, $[\quad] \le \rho^{k+1}/(1-\rho)$ so that

$$\sum_{n=1}^{k} x_n \rho^n \qquad \text{and} \qquad \sum_{n=1}^{k} x_n \rho^n + \frac{\rho^{k+1}}{1-\rho}$$

are the smallest and largest numbers in $C_{x_1,\ldots,x_k}$. Therefore $C_{x_1,\ldots,x_k}$ is contained in an interval of length $\rho^{k+1}/(1-\rho)$. But C equals the union of the 2^k sets like $C_{x_1,\ldots,x_k}$, so C is subset of the union of 2^k intervals having total length

$$\frac{2^k \rho^{k+1}}{1-\rho} = (2\rho)^k \frac{\rho}{1-\rho}.$$

But this is true for each k, and $(2\rho)^k$ goes to zero as k goes to infinity, so C has length 0 by definition. *Therefore F_Y is continuous singular.*

Exercise 7 shows that simple finite computations yield arbitrarily close approximations to F_Y.

Exercise 8 shows that if

$$Y = \sum_{n=1}^{\infty} \frac{X_n}{2^n}$$

where $X_1, X_2, \ldots$ are Bernoulli trials with $0 < p < 1$ and $p \ne \frac{1}{2}$, then F_Y is continuous singular and *strictly* increasing over the unit interval $0 \le y \le 1$. Thus a continuous singular F_Y need not be constant over "many" intervals. ■

EXERCISES

1. At 3 P.M. daily a scheduled airplane is supposed to request permission from the control tower to take off. If its request is more than 1 minute late then its departure may be held up for quite a while. Let X be its time of request, in minutes after 3 P.M., and assume that F_X is as in Fig. 5. Find $P[X = 0]$, $P[-1 \le X \le 1]$, and $P[X > 0]$.

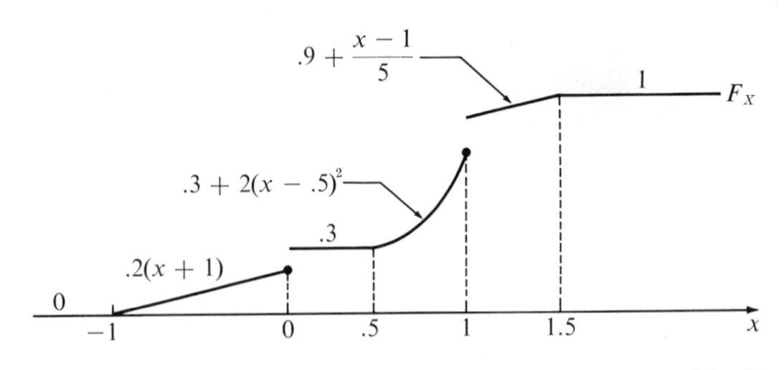

Fig. 5

2. Let $Z = \min \{X_1, \ldots, X_n\}$ be the smallest of $n \geq 2$ random variables having common distribution function F.
(*a*) Show that

$$F_Z(z) = 1 - [1 - F(z)]^n \qquad \text{for all real } z.$$

(*b*) If F has corresponding continuous density p show that Z has continuous density

$$p_Z(z) = n[1 - F(z)]^{n-1}p(z) \qquad \text{for all real } z.$$

Interpret this formula in terms of the ith observation being the smallest and satisfying $z \leq X_i \leq z + dz$, etc.
(*c*) Find the density of the maximum and the density of the minimum of n independent random variables, each uniformly distributed over the unit interval $0 \leq x \leq 1$. Find their means and sketch the densities for a large n.

3. Assume that a random variable X has distribution $F: R \to R$, which is a mixture $F = \beta F_d + (1 - \beta)F_{ac}$ with $0 < \beta < 1$, where distribution functions F_d and F_{ac} correspond, respectively, to discrete and continuous densities $p_d: R \to R$ and $p_{ac}: R \to R$. If $h: R \to R$ then, as should be expected, it can be shown that $Eh(X)$ exists and is given by

$$Eh(X) = \beta_d \sum_x h(x)p_d(x) + (1 - \beta_d)\int_{-\infty}^{\infty} h(x)p_{ac}(x)\,dx,$$

assuming that the sum is absolutely convergent and the integrand absolutely integrable. *Assume this formula*, but do *not* assume any other results from Secs. 14-1, 15-5, and 16-6. In particular, we can let $h(x) = x^n$ to get EX^n for $n = 1, 2, \ldots$.
(*a*) Let $h(x) = [x - \mu]^2$ where $\mu = EX$, and derive

$$\text{Var } X = \beta_d\sigma_d^2 + (1 - \beta_d)\sigma_{ac}^2 + \beta_d(\mu_d - \mu)^2 + (1 - \beta_d)(\mu_{ac} - \mu)^2.$$

HINT: $x - \mu = (x - \mu_d) + (\mu_d - \mu) = (x - \mu_{ac}) + (\mu_{ac} - \mu)$.

COMMENT: The assumed formula may be derived as follows: Introduce the discrete density $p: R \to R$ by $p(0) = \beta_d$ and $p(1) = 1 - \beta_d$. Let (R,P), (R,P_0), and (R,P_1) correspond to F, F_d, and F_{ac}, respectively. Theorem 1 (Sec. 15-5) justifies assuming that we have random variables X and g with the same associated probability space, and satisfying

$$P_{X|g=0} = P_0 \qquad P_{X|g=1} = P_1 \qquad P_g = p \qquad P_X = P.$$

Applying Theorem 2 (Sec. 15-5) to $Y = h(X)$ yields

$$EY = \beta_d E[Y \mid g = 0] + (1 - \beta_d)E[Y \mid g = 1]$$

where $E[Y \mid g = 1]$ is the expectation of $h(X)$ when X induces density p_{ac}.
(*b*) Let F_d be Poisson with mean $\lambda = 3.2$ and let F_{ac} be exponential with mean 1. Sketch $F = .4F_d + .6F_{ac}$ and find its mean and variance.

4. In Example 3 let $S_n = X_1 + \cdots + X_n$ so that $(1/n)S_n$ is the average length of the rods.
(*a*) Apply the WLLN of Theorem 1 (Sec. 12-1) to $(1/n)S_n$ with $n = 100$ and $\epsilon = \frac{1}{2}$.
(*b*) Apply the SLLN of Theorem 1 (Sec. 12-3) to $(1/n)S_n$ with $n = 100$ and $\epsilon = \frac{1}{2}$.
(*c*) Apply the CLT of Corollary 1*a* (Sec. 13-2) to S_n, using $b = 2$.

5. This exercise indicates the extent of arbitrariness in our definition of a distribution function on R. If X is a random variable define, for the purposes of this exercise, a *generalized* *distribution function* induced by X to be *any* function $G_X: R \to R$ such that

$$P[X < x] \le G_X(x) \le P[X \le x] \qquad \text{for all real } x.$$

The function $G_X(x) = P[X < x]$, which in this book is called the *distribution function induced by* X, may be called the *smallest generalized distribution function*. Similarly $G_X(x) = P[X \le x]$ may be called the largest, and

$$G_X(x) = \tfrac{1}{2}\{P[X < x] + P[X \le x]\} \qquad \text{for all real } x$$

may be called the middle one.

(*a*) Prove that every generalized distribution function induced by X goes to zero to the left, and to 1 to the right, and is nondecreasing, and that all have the same right- and left-hand limits at every point. COMMENT: Actually G_X could be left undefined at each x for which $P[X = x] > 0$.

(*b*) Express the probability of each interval of Fig. 2 in terms of the one-sided limits of any generalized distribution function. Thus the probabilities of intervals, and hence P_X, are determined with almost equal ease from any G_X.

(*c*) Let $H_X = G_X + c$, where G_X is a generalized distribution function induced by X, and c is a real number. Show that such a vertical translation of a G_X can be used to find the probabilities of intervals, and that for finite intervals the formulas are the same as for a generalized distribution function. Note that a function $H: R \to R$ can be a vertical translation of some generalized distribution function iff H is nondecreasing and has a total increase of 1, $H(\infty) - H(-\infty) = 1$.

6. Let $F_c: R \to R$ be the continuous distribution function corresponding to (R,P). This exercise indicates how the decomposition $F_c = bF_{ac} + (1 - b)F_{cs}$ can be obtained. This decomposition was mentioned after the statement of Lemma 1. The idea is to find, by some unspecified method, a continuous density which accounts for as much of P as possible, and let F_{cs} account for the rest. If $f: R \to R$ is any nonnegative function for which

$$\int_A f(x)\,dx \le P(A) \qquad \text{for } \textit{every} \text{ event } A$$

then let $b_f = \int_{-\infty}^{\infty} f(x)\,dx$ so that $0 \le b_f \le P(R) = 1$. *Assume* that among such functions there is one, f_0, for which b_{f_0} is as large as for any other such function (that is, assume that the supremum of b_f over such f is attained). Let $b = b_{f_0}$ and to eliminate the two extreme cases assume that $0 < b < 1$. The equation

$$P(A) = b\int_A \frac{1}{b} f_0(x)\,dx + (1 - b)Q(A) \qquad \text{for every event } A$$

then defines a probability measure Q; and F_{ac} and F_{cs} correspond to $(1/b)f_0$ and Q, respectively.

(*a*) Prove that Q, as defined above, is a probability measure.

(*b*) Prove that if $h: R \to R$ is a nonnegative function for which

$$\int_A h(x)\,dx \le Q(A) \qquad \text{for every event } A$$

then

$$\int_{-\infty}^{\infty} h(x)\,dx = 0.$$

7. This exercise continues the study of Example 4. Define the ordering $(x_1, x_2, \ldots) < (x'_1, x'_2, \ldots)$ between infinite sequences to mean that for some k we have $x_1 = x'_1, \ldots, x_{k-1} = x'_{k-1}$ and $x_k < x'_k$. This is the usual decimal ordering $.8739 \cdots < .8750 \cdots$ (if we neglect the cases $.7999 \cdots = .8000 \cdots$), or the usual dictionary lexicographic ordering, *alarm* < *also*. In Example 4 it was shown that for sequences of 0's and 1's the above ordering is equivalent to the usual real-number ordering between their corresponding sums; that is, $(x_1, x_2, \ldots) < (x'_1, x'_2, \ldots)$ iff

$$\sum_{n=1}^{\infty} x_n \rho^n = y < y' = \sum_{n=1}^{\infty} x'_n \rho^n \qquad \text{where } 0 < \rho < \tfrac{1}{2}.$$

Interpret the finite sequence $(x_1, \ldots, x_n)$ as the beginning of $(x_1, \ldots, x_n, 0, 0, \ldots)$. Then, for example,

$$(0,0,0) < (0,0,1) < (0,1,0) < (0,1,1) < (1,0,0) < (1,0,1) < (1,1,0) < (1,1,1)$$

and the same inequality chain holds between the eight corresponding sums $x_1\rho + x_2\rho^2 + x_3\rho^3$.

(*a*) Prove that if $(x_1, \ldots, x_k) < (x'_1, \ldots, x'_k)$ then

$$\sum_{n=1}^{k} x_n \rho^n + \frac{\rho^{k+1}}{1-\rho} < \sum_{n=1}^{k} x'_n \rho^n.$$

Note that the left end points of the 2^k sets $C_{x_1,\ldots,x_k}$ have the same ordering as do their subscripts $(x_1, \ldots, x_k)$, *and* from the above we see that the right end point of $C_{x_1,\ldots,x_k}$ is less than the next left end point.

(*b*) Show that if $y = \sum_{n=1}^{k} x_n \rho^n$ then

$$F_Y(y) = P[Y < y] = \sum P[X_1 = x'_1, \ldots, X_k = x'_k]$$

where the summation is over all $(x'_1, \ldots, x'_k)$ satisfying

$$(x'_1, \ldots, x'_k) < (x_1, \ldots, x_k).$$

In particular if $y = 0\rho + 1\rho^2 + 0\rho^3$ then

$$F_Y(y) = P[X_1 = 0, X_2 = 0, X_3 = 0] + P[X_1 = 0, X_2 = 0, X_3 = 1].$$

(*c*) Show that if $y = \sum_{n=1}^{\infty} x_n \rho^n$ then

$$F_Y(y) = P[X_1 < x_1] + P[X_1 = x_1, X_2 < x_2] + P[X_1 = x_1, X_2 = x_2, X_3 < x_3] + \cdots.$$

In particular if $y = \sum_{n=1}^{k} x_n \rho^n$ note that all terms after $P[X_1 = x_1, \ldots, X_{k-1} = x_{k-1}, X_k < x_k]$ are zero. For example, if $y = \rho + \rho^2 + \rho^5$ then

$$F_Y(y) = P[X_1 = 0] + P[X_1 = 1, X_2 = 0] + 0 + 0 + P[X_1 = 1, X_2 = 1, X_3 = 0, X_4 = 0, X_5 = 0].$$

(*d*) Assume Bernoulli trials. Make a sketch showing the 16 end points, in proper order, of the sets C_{x_1,x_2,x_3}. Indicate the resulting subintervals over which F_Y is constant. Exhibit the corresponding values of F_Y, in terms of p and q, over each of these subintervals, as calculated in part (*b*) and in part (*c*).

(*e*) Specialize the sketch of part (*d*) to $\rho = \frac{1}{3}$ and $p = \frac{1}{2}$.

(*f*) Show that for Bernoulli trials if $p = \frac{1}{2}$ and $y = \sum_{n=1}^{\infty} x_n \rho^n$ then $F_Y(y) = \sum_{n=1}^{\infty} x_n/2^n$.

(*g*) If $\rho = \frac{1}{3}$ and we multiply each number in C by 2 we get the set C_0 of all real y such that there exists a sequence $x_1, x_2, \ldots$ of 0's and 2's such that $y = \sum_{n=1}^{\infty} x_n/3^n$. Thus y belongs to C_0 iff y has only 0's and 2's in its ternary expansion. The set C_0 is called the *Cantor set*. Explain the following common description of C_0: Take the unit interval $0 \leq y \leq 1$ and remove its middle third; take the remaining two intervals and remove their middle thirds; etc.; C_0 is what is finally left. Note that the sum of the lengths of the intervals so removed is 1.

8. In this exercise we use the SLLN to exhibit continuous singular distribution functions different from those in Example 4. Let $X_1, X_2, \ldots$ be Bernoulli trials with $0 < p < 1$ and let

$$Y_p = \sum_{n=1}^{\infty} \frac{X_n}{2^n}.$$

The distribution of the sum, as well as that of each X_n, depends on p. For convenience we exhibit p as a subscript, Y_p. Thus Y_p can be thought of as a number in the unit interval $0 \leq y \leq 1$ obtained by selecting the digits in its binary expansion by independent coin tosses, with probability p for a 1. Now for most y, where $0 \leq y \leq 1$, there is exactly one sequence $x_1, x_2, \ldots$ of 0's and 1's such that $y = \sum_{n=1}^{\infty} x_n/2^n$. For such a y we have $P[Y = y] = 0$. Two sequences, such as $0, 1, 1, 0, 0, 0, \ldots$ and $0, 1, 0, 1, 1, 1, \ldots$, yield the same sum y, but both have zero probability, so $P[Y = y] = 0$ in these other cases as well. Thus F_{Y_p} is continuous whenever $0 < p < 1$. For convenience we may restrict ourselves to the event that $X_1, X_2, \ldots$ are not all zero after some term. Let A_p consist of those $y = \sum_{n=1}^{\infty} x_n/2^n$ in $0 < y < 1$ for which

$$\lim_{k \to \infty} \frac{1}{k} \sum_{n=1}^{k} x_n = p.$$

That is, y has asymptotic relative frequency p of 1's in its binary expansion. Clearly $p \neq p'$ implies that A_p and $A_{p'}$ are disjoint. The SLLN of Theorem 3 (Sec. 12-3) or Theorem 1 (Sec. 20-3) implies that $P[Y_p \in A_p] = 1$ for each p in $0 < p < 1$. Thus each P_{Y_p} is concentrated on a different disjoint set A_p.

(*a*) Show that

$$F_{Y_{.5}}(y) = y \qquad \text{if } 0 \leq y \leq 1.$$

Therefore $Y_{.5}$ has a uniform continuous density and $P_{Y_{.5}}$ corresponds to length. Hence $A_{.5}$ has length 1, so all other A_p with $p \neq .5$ have zero length. Thus F_{Y_p} is continuous singular whenever $0 < p < 1$ with $p \neq .5$.

(*b*) Show that every F_{Y_p} is strictly increasing, that is, $0 \leq y < y' \leq 1$ implies $F_{Y_p}(y) < F_{Y_p}(y')$. Therefore each P_{Y_p} assigns positive probability to each subinterval of $0 < y < 1$.

(*c*) Find EY_p and Var Y_p.

9. For each real x define a function $f_x \colon R \to R$ as in Fig. 6*a* by

$$f_x(x') = \begin{cases} 1 & \text{if } x' < x \\ 0 & \text{if } x' \geq x. \end{cases}$$

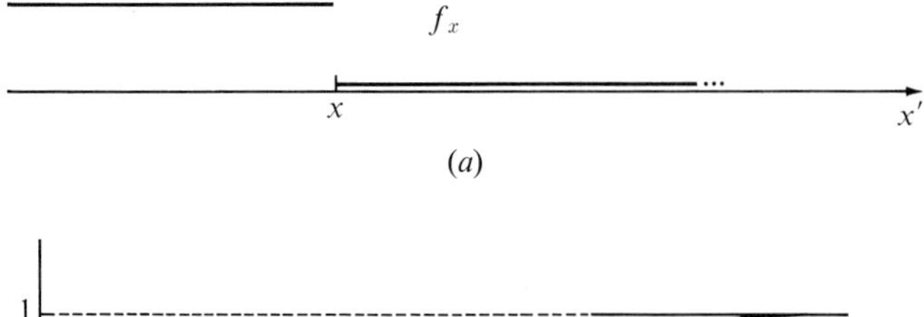

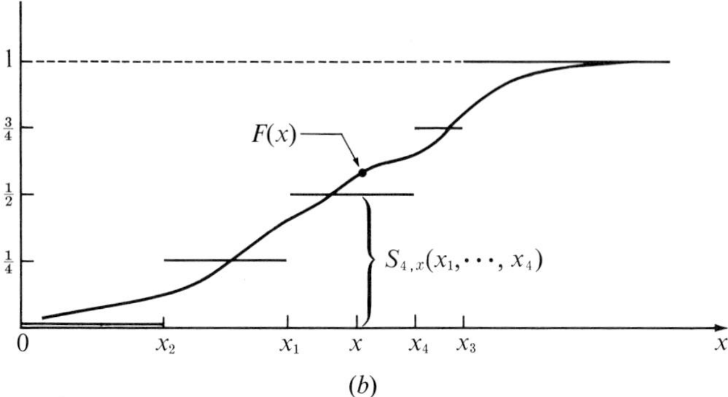

Fig. 6

For real $x, x_1, \ldots, x_n$ define

$$S_{n,x}(x_1, \ldots, x_n) = \frac{1}{n} \sum_{k=1}^{n} f_x(x_k) = \frac{1}{n} \text{ [the number of } x_i < x].$$

As indicated in Fig. 6*b*, suppose that we obtain observations $x_1, \ldots, x_n$ from some unknown distribution function $F: R \to R$. Then a reasonable guess for an approximation to F is *the empirical distribution function* corresponding to $x_1, \ldots, x_n$, and this is essentially the distribution function corresponding to a uniform discrete density on the observed points $x_1, \ldots, x_n$. Clearly $S_{n,x}(x_1, \ldots, x_n)$, when considered as a function of x, is this function, by definition. If $x_1, \ldots, x_n$ are distinct this function has n jumps, each of height $1/n$, and is otherwise constant.

Let $X_1, X_2, \ldots$ be independent random variables each having the same distribution function $F: R \to R$. For each $n \geq 1$ and real x we can define the random variable $S_{n,x}(X_1, \ldots, X_n)$, and this is clearly the random height at x of the empirical distribution function from the first n observations. Also

$$P[f_x(X_k) = 1] = F(x) \qquad P[f_x(X_k) = 0] = 1 - F(x)$$

so for fixed x we have Bernoulli trials, with outcomes to the left of x being successes. Thus

$$P\left[S_{n,x}(X_1, \ldots, X_n) = \frac{k}{n}\right] = \binom{n}{k}[F(x)]^k[1 - F(x)]^{n-k}.$$

The WLLN of Theorem 1 (Sec. 10-2) shows that

$$P[|S_{n,x}(X_1, \ldots, X_n) - F(x)| \geq \epsilon] \leq \frac{1}{4\epsilon^2 n} \qquad \text{for every } F, n, \epsilon > 0.$$

Thus for every fixed x if n is large then we can be quite certain that the empirical distribution function will be close to the true distribution function at this x. Also n depends on how close we want to be (say, $\epsilon = 10^{-3}$), and on how certain we want to be (.9975 for $n = 10^8$ and $\epsilon = 10^{-3}$), but *not* on the unknown F. Exhibit and interpret the analogous conclusion obtainable from the SLLN of Theorem 2 (Sec. 12-3).

17-2 THE DISTRIBUTION FUNCTION OF $Y = g(X)$

This section is mainly devoted to finding F_Y in terms of F_X when $Y = g(X)$. The result is particularly simple and useful when $g: R \to R$ is continuous and strictly increasing, or decreasing. The decreasing case is left as an exercise. Examples 8 to 11 are all related to the probability-integral transformation and order statistics.

Assume that $g: R \to R$ is continuous and strictly increasing. We make no restrictions on F_X. As in Fig. 5 (Sec. 16-2), the range of g is either the whole real line or a set of one of the forms

$$\{y: y < B\} \qquad \{y: y > A\} \qquad \{y: A < y < B\}.$$

If $g(x) < B$ for all x then $P[Y < B] = 1$ so $F_Y(B) = 1$, and naturally $F_Y(y) = 1$ whenever $y \geq B$. Similarly if $g(x) > A$ for all x then $F_Y(y) = 0$ if $y \leq A$. Thus without knowing F_X we easily find F_Y off the range of g, for such a $g: R \to R$.

As in Fig. 1, select a y belonging to the range of g, then there is a unique x satisfying $y = g(x)$. If X has the value x' then Y will have the value $y' = g(x')$, and obviously $y' < y$ iff $x' < x$. Therefore $P[Y < y] = P[X < x]$. Hence if $Y = g(X)$, where $g: R \to R$ is continuous and strictly increasing, and if y belongs to the range of g then

$$F_Y(y) = F_X(x) \qquad \text{where } y = g(x)$$

that is,

$$F_Y(y) = F_X(g^{-1}(y)).$$

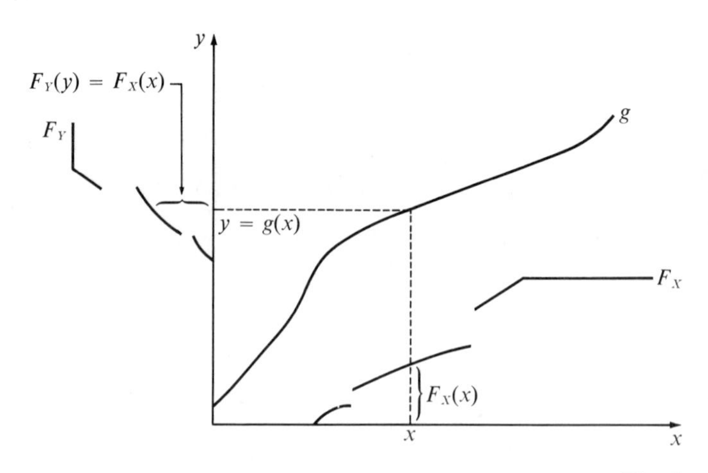

Fig. 1

The graphical construction of F_Y is shown in Fig. 1. To find $F_Y(y)$ first find the x for which $y = g(x)$ and then evaluate $F_X(x)$ and set $F_Y(y)$ equal to this number. The simplicity and transparency of this result recommends the use of distribution functions for such functions g even if X happens to have a continuous density.

The same formula can be derived as follows:

$$F_Y(y) = P[Y < y] = P[g(X) < y] = P[X < g^{-1}(y)] = F_X(g^{-1}(y)).$$

As will be seen in the following examples, it is often simpler to perform such a calculation than to recall a formula.

Examples 1 to 3 are devoted to $Y = X^n$ with n odd, $Y = e^X$, and $Y = aX + b$. In Example 4 we consider $Y = g(X) = X^n$ with n even, so that g is not strictly increasing or decreasing. Examples 5 and 6 use distribution functions in order to derive general formulas in the continuous-density case. Example 7 shows that *every* distribution can arise from the unit interval with length for probability. Example 8 introduces the basic probability-integral transformation. This transformation can be used, as in Examples 9 to 11, to reduce certain kinds of problems to simpler problems involving only uniformly distributed random variables.

Example 1 If $n = 3, 5, \ldots$ is an odd integer, so that x^n is strictly increasing, and $Y = X^n$ then

$$F_Y(y) = P[X^n < y] = P[X < y^{1/n}] = F_X(y^{1/n}) \qquad \text{for all real } y.$$

The formula above is valid for every F_X.

If F_X happens to be absolutely continuous we can differentiate F_Y to find p_Y, at least at those points where p_Y is continuous. Then

$$p_Y(y) = \frac{dF_Y(y)}{dy} = \left[\frac{dF_X(x)}{dx}\right]_{x=y^{1/n}} \frac{1}{n} y^{(1/n)-1} = \frac{1}{ny^{1-(1/n)}} p_X(y^{1/n}) \qquad \text{for all real } y. \quad \blacksquare$$

Example 2 If $Y = e^X$ then $F_Y(y) = 0$ if $y \le 0$, and if $y > 0$ then

$$F_Y(y) = P[e^X < y] = P[X < \log y] = F_X(\log y).$$

If F_X is absolutely continuous then $p_Y(y) = 0$ if $y \le 0$, and if $y > 0$ then

$$p_Y(y) = \frac{dF_Y(y)}{dy} = \frac{dF_X(\log y)}{dx}\frac{1}{y} = \frac{1}{y} p_X(\log y). \quad \blacksquare$$

Example 3 Let $Y = aX + b$ with $a \neq 0$. If $a > 0$ then

$$F_Y(y) = P[aX + b < y] = P\left[X < \frac{y-b}{a}\right] = F_X\left(\frac{y-b}{a}\right) \qquad \text{for all real } y.$$

However, if $a < 0$ then, as in Fig. 2, we see that $y' = ax' + b < y$ iff $x' > (y - b)/a$ so

$$F_Y(y) = P[aX + b < y] = P[aX < y - b] = P\left[X > \frac{y-b}{a}\right]$$

$$= 1 - F_X\left(\frac{y-b}{a}\right) - \left[\text{the jump of } F_X \text{ at } x = \frac{y-b}{a}\right] \qquad \text{for all real } y.$$

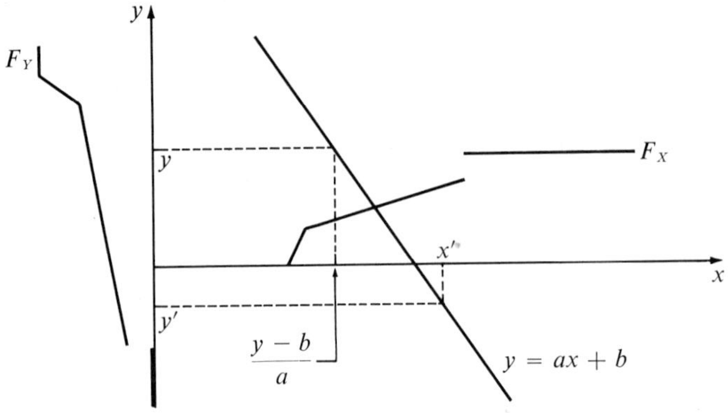

Fig. 2

If F_X is continuous then

$$F_Y(y) = \begin{cases} F_X\left(\dfrac{y-b}{a}\right) & \text{if } a > 0 \\ 1 - F_X\left(\dfrac{y-b}{a}\right) & \text{if } a < 0. \end{cases}$$

If F_X is absolutely continuous and $a \neq 0$ then

$$p_Y(y) = \frac{dF_Y(y)}{dy} = \frac{dF_X((y-b)/a)}{dx}\frac{1}{|a|} = \frac{1}{|a|}p_X\left(\frac{y-b}{a}\right). \quad \blacksquare$$

Example 4 Let $Y = X^n$ where $n = 2, 4, 6, \ldots$ is an even integer. If $y > 0$ then, as in Fig. 3, we have $x^n < y$ iff $-y^{1/n} < x < y^{1/n}$. Thus $F_Y(y) = 0$ if $y \leq 0$, while if $y > 0$ then

$$\begin{aligned} F_Y(y) &= P[X^n < y] = P[-y^{1/n} < X < y^{1/n}] \\ &= F_X(y^{1/n}) - F_X(-y^{1/n}) - [\text{the jump of } F_X \text{ at } x = -y^{1/n}]. \end{aligned}$$

If F_X is continuous then

$$F_Y(y) = F_X(y^{1/n}) - F_X(-y^{1/n}) \qquad \text{if } y > 0.$$

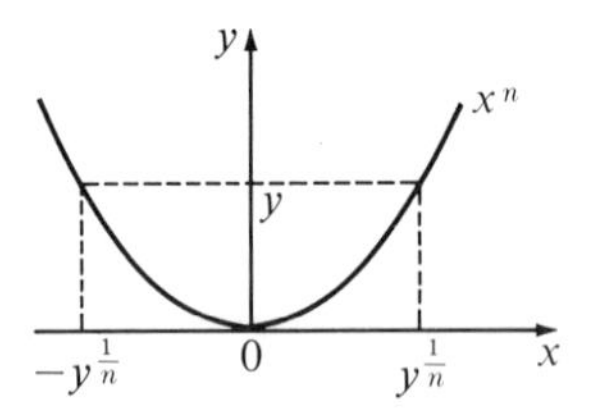

Fig. 3 x^n for n even.

If F_X is absolutely continuous then $p_Y(y) = 0$ if $y \le 0$, while

$$p_Y(y) = F_Y'(y) = \frac{1}{n} y^{(1/n)-1}[F_X'(y^{1/n}) + F_X'(-y^{1/n})]$$

$$= \frac{1}{ny^{1-(1/n)}} [p_X(y^{1/n}) + p_X(-y^{1/n})] \qquad \text{if } y > 0. \quad ■$$

Exercises 1 to 8 may be done before reading the remainder of this section.

***Example 5** In this example a transformation formula for continuous densities is stated. This formula was used extensively in Sec. 16-2. Here distribution functions are used to derive the formula in the case of a positive derivative.

Let $Y = g(X)$ where $g: R \to R$ has a continuous positive derivative everywhere, or a continuous negative derivative everywhere. If X has continuous density $p_X: R \to R$ then Y has continuous density $p_Y: R \to R$ given by $p_Y(y) = 0$ if y is not in the range of g, and

$$p_Y(y) = \left| \frac{dg^{-1}(y)}{dy} \right| p_X(g^{-1}(y)) \qquad \text{if } y \text{ is in the range of } g.$$

To prove this in the case where $g' > 0$ note that g is strictly increasing so

$$F_Y(y) = F_X(g^{-1}(y)) = \int_{-\infty}^{g^{-1}(y)} p_X(x)\, dx$$

for all y in the range of g. Since $g' > 0$, we can make the transformation of variables $t = g(x)$ so

$$x = g^{-1}(t) \qquad \text{and} \qquad dx = \frac{dg^{-1}(t)}{dt}\, dt$$

and hence

$$F_Y(y) = \int_{g(-\infty)}^{y} \frac{dg^{-1}(t)}{dt}\, p_X(g^{-1}(t))\, dt$$

for y in the range of g. Therefore the integrand is a continuous density for Y. ■

***Example 6** If $Z = X + Y$ where X and Y have a joint continuous density then

$$F_Z(z) = P[X + Y < z] = \iint_{x+y<z} p_{X,Y}(x,y)\, dx\, dy.$$

The integration is over the lower left half-plane determined by the line $x + y = z$, as shown in Fig. 4. For fixed x we integrate with respect to dy over the vertical strip of Fig. 4 and then integrate over all x, so that

$$F_Z(z) = \int_{-\infty}^{\infty} \left[\int_{-\infty}^{z-x} p_{X,Y}(x,y)\, dy \right] dx.$$

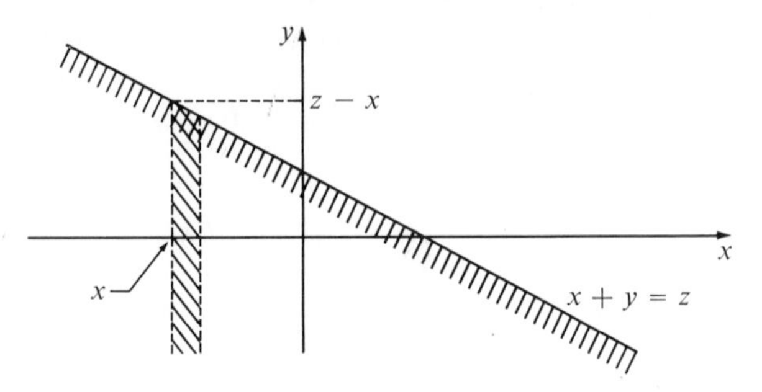

Fig. 4

Differentiating under the first integral sign yields

$$p_Z(z) = F'_Z(z) = \int_{-\infty}^{\infty} p_{X,Y}(x, z - x)\, dx$$

as was obtained in Example 3 (Sec. 16-3). ■

Example 7 In this example we show that every possible distribution for a random variable already appears in the class of distributions induced by functions of a uniformly distributed random variable. Let $F: R \to R$ be a distribution function. For simplicity assume for the moment, as in Fig. 5, that F is continuous and strictly increasing. Then F has a continuous strictly increasing inverse F^{-1} defined on $0 < y < 1$. Let Y be uniformly distributed on $0 < y < 1$, so that

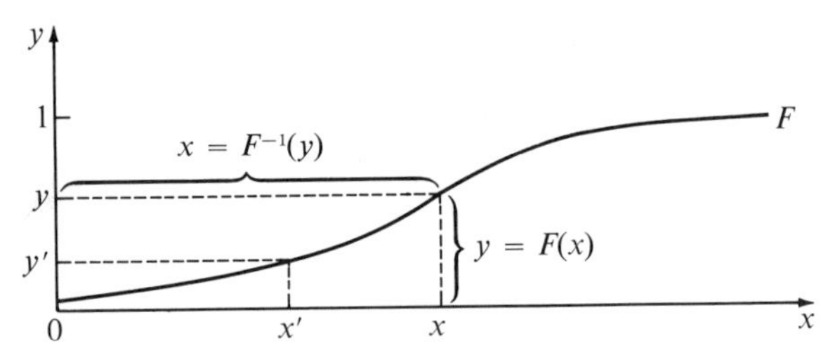

Fig. 5

$P[Y < y] = y$ whenever $0 < y < 1$. Define the random variable X by $X = F^{-1}(Y)$. Then for every real x we have

$$P[X < x] = P[F^{-1}(Y) < x] = P[F(F^{-1}(Y)) < F(x)] = P[Y < F(x)] = F(x).$$

Thus $F^{-1}(Y)$ has distribution function F. The result is obvious from Fig. 5, since for each fixed real x, if $y = F(x)$ then Y has some value $y' < y$ iff X has some value $x' < x$, and this event has probability y.

In the terminology of Example 6 (Sec. 16-1), we have shown that for any continuous strictly increasing F it is possible to define a random variable X in terms of a continuously calibrated random wheel in such a way that X has distribution function F, the calibration function being $h = F^{-1}$.

We indicate now how this can be done for *any* distribution function F, as in Fig. 6. We wish to define on $0 < y < 1$ a function h which reduces to F^{-1} when F^{-1} exists. If $0 < y < 1$ let A_y be the set of x such that $F(x) = y$. If for some y_1 the set A_{y_1} contains exactly

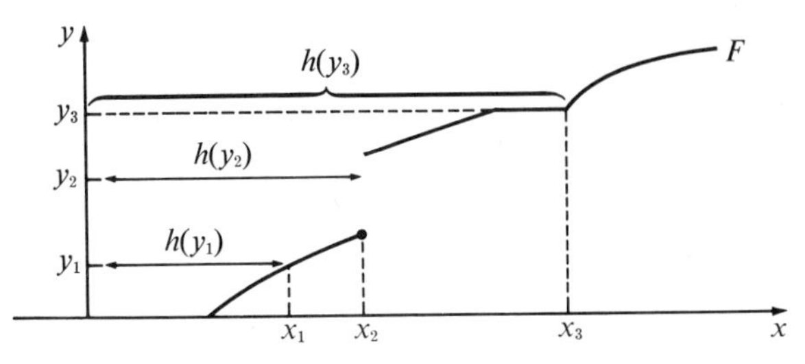

Fig. 6

one member x_1 we define $h(y_1) = x_1$. If for some y_2 the set A_{y_2} is empty then there is a unique largest x_2 such that $F(x_2) < y_2$, and in this case we set $h(y_2) = x_2$. If for some y_3 the set A_{y_3} has more than one member we let $h(y_3)$ equal any one of them, for example, the largest. Thus h has been uniquely defined on $0 < y < 1$. As is graphically obvious, for every real x

$$\{y' : h(y') < x\} = \{y' : 0 < y' < F(x)\}.$$

Therefore if Y is uniformly distributed over $0 < y < 1$ then $h(Y)$ has distribution function F.

In order to perform a sampling experiment requiring independent observations $x_1, \ldots,$ x_n according to F, we may obtain the x_k by letting $x_k = h(y_k)$ where $y_1, \ldots, y_n$ are independent observations from a uniform density on $0 < y < 1$. This technique was discussed in Example 6 (Sec. 16-1). ∎

THE PROBABILITY INTEGRAL TRANSFORMATION

Example 8 Let X have distribution function $F_X: R \to R$. Define the random variable Y by $Y(\omega) = F_X(X(\omega))$. Thus $Y = g(X)$, where g is the distribution function of X. We show that if $F_X: R \to R$ is continuous then $Y = F_X(X)$ is uniformly distributed over $0 < y < 1$. This transformation of X is called the *probability-integral transformation* because if X has a continuous density then F_X is its indefinite integral. The continuity assumption for F_X is essential, since if $P[X = a] > 0$ then Y takes the value $F_X(a)$ with positive probability. Given y satisfying $0 < y < 1$, then, as in Fig. 7, there is a unique smallest x_y such that $F_X(x_y) = y$.

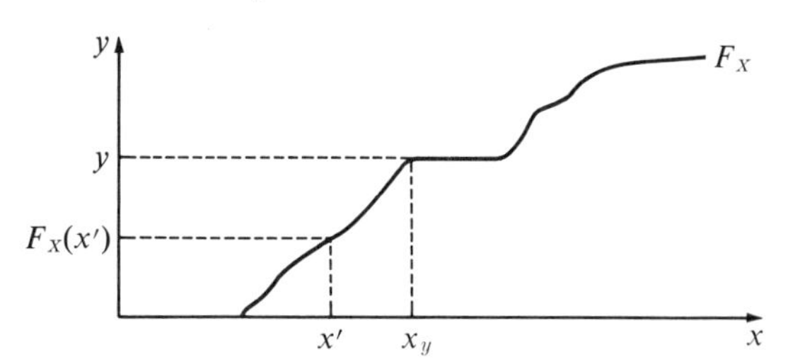

Fig. 7

Clearly $F_X(x') < y$ iff $x' < x_y$ so

$$P[F_X(X) < y] = P[X < x_y] = F_X(x_y) = y$$

as was to be shown.

For interpretation let F_X describe the distribution of heights in a population. Select a person and measure his height X; that is, select X according to F_X. Then $Y = F_X(X)$

is the fraction of the population having heights less than the observed height X. Thus "the random fraction of the population shorter than the observed person" is as likely to be any fraction as any other fraction. ■

***Example 9** Let $X_0, X_1, \ldots, X_n$ be independent and identically distributed according to a continuous distribution function F. For simplicity assume that F is also strictly increasing on $a < x < b$, where $F(a) = 0$ and $F(b) = 1$. Interpret $X_0, X_1, \ldots, X_n$ as the heights of the next $n + 1$ people you meet. For $0 \le k \le n$ let A_k be the event that exactly k of these people are shorter than the first person. Thus A_k is the event that exactly k of the events $[X_1 < X_0]$, $\ldots, [X_n < X_0]$ occur. Let $Y_i = F(X_i)$, so that $Y_0, Y_1, \ldots, Y_n$ are independent and uniformly distributed on $0 < y < 1$. Now if x and x' are in the interval $a < x < b$ then $x < x'$ iff $F(x) < F(x')$. Therefore A_k is the event that exactly k of the events $[Y_1 < Y_0], \ldots,$ $[Y_n < Y_0]$ will occur. Thus the probability-integral transformation has reduced the problem to uniformly distributed random variables. In particular we see that $P(A_k)$ does not depend on F. Naturally X_i might be the ith voltage observed, or the rainfall in the ith year.

If $0 < p < 1$ then the event $[p < Y_0 < p + dp]$ has probability dp. Given this event, the conditional probability of $[Y_i < Y_0]$ is p, so the conditional probability of A_k is $\binom{n}{k} p^k(1 - p)^{n-k}$. Therefore

$$P(A_k) = \int_0^1 \binom{n}{k} p^k(1 - p)^{n-k}\, dp = \frac{1}{n + 1} \qquad \text{for } 0 \le k \le n$$

where this integral was evaluated in Example 3 (Sec. 16-6). (The evaluation was a direct application of a result from Sec. 16-4.) Thus every value for k is equally likely.

If F corresponds to continuous density p_c the result can be made intuitively obvious as follows. Select $n + 1$ distinct reals $h_0 < h_1 < \cdots < h_n$ for the heights to be encountered *in some order*. Let $(x_0, \ldots, x_n)$ be *any one* of the $(n + 1)!$ permutations of $h_0, \ldots, h_n$. Then

$$p_{X_0,\ldots,X_n}(x_0, \ldots, x_n) = p_c(x_0)p_c(x_1) \cdots p_c(x_n) = p_c(h_0)p_c(h_1) \cdots p_c(h_n)$$

so each of the $(n + 1)!$ orders in which these heights might be observed is equally likely. Therefore, as in Exercise 13 (Sec. 2-2), the probability is $1/(n + 1)$ that we will observe h_k first. In essence if $(x_0, \ldots, x_n)$ is any permutation of $h_0, \ldots, h_n$ and if we *unjustifiably* condition on an event of zero probability, then

$$P\,[X_0 = x_0, \ldots, X_n = x_n \mid (X_0, \ldots, X_n) \text{ is some permutation of } (h_0, \ldots, h_n)] = \frac{1}{(n + 1)}$$

for any $h_0, h_1, \ldots, h_n$. ■

***Example 10** This example applies the probability-integral transformation to the fundamental problem of discovering an interval such that (R,P) is largely concentrated on the interval, when we do not know (R,P) but we can obtain independent observations from (R,P). Assume that $X_1, \ldots, X_n$ are independent and identically distributed according to a continuous distribution function F: $R \to R$. Define the random variable Z_n to be the probability assigned by this F to the interval between the smallest and largest of $X_1, \ldots, X_n$. That is, if $n = 5$ and the X_i have the values x_i then Z_5 has the value $F(\bar{x}) - F(\underline{x})$, where, as in Fig. 8, $\underline{x}$ and $\bar{x}$ are the smallest and largest numbers among $x_1, \ldots, x_5$, respectively. More precisely

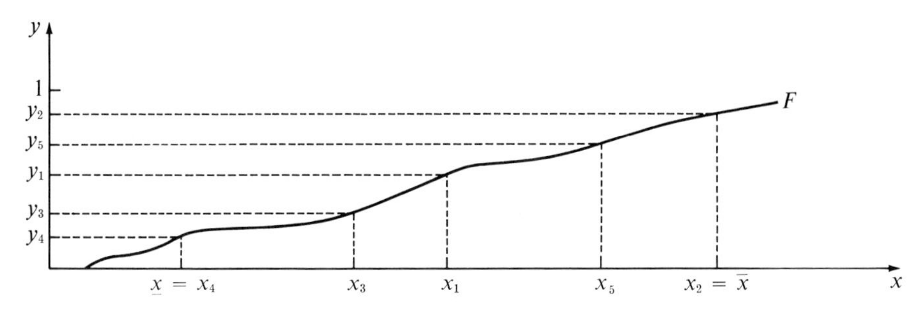

Fig. 8

we define a function g_n: $R^n \to R$ by

$$g_n(x_1, \ldots, x_n) = F(\max\{x_1, \ldots, x_n\}) - F(\min\{x_1, \ldots, x_n\})$$

and let $Z_n = g_n(X_1, \ldots, X_n)$. Equivalently we have $g_n(x_1, \ldots, x_n) = \max\{F(x_1), \ldots, F(x_n)\} - \min\{F(x_1), \ldots, F(x_n)\}$. That is, the value of F at the largest of $x_1, \ldots, x_n$ is just the largest of $F(x_1), \ldots, F(x_n)$. Therefore if we let $Y_i = F(X_i)$ then $Z_n = \max\{Y_1, \ldots, Y_n\} - \min\{Y_1, \ldots, Y_n\}$, where $Y_1, \ldots, Y_n$ are independent and each is uniformly distributed over $0 < y < 1$. The probability-integral transformation has shown that the distribution of Z_n does not depend on F.

We derive the distribution function for Z_n. Fix z satisfying $0 < z < 1$. We seek $P[Z_n < z]$. The event $[Z_n < z]$ can occur with all of $y_1, \ldots, y_n$ being small, or with at least one of them large. More precisely, as in Fig. 9*a* where the crosses represent $y_1, \ldots, y_n$, if $[Y_1 < z, \ldots, Y_n < z]$ occurs then so does $[Z_n < z]$, and we have $P[Y_1 < z, \ldots, Y_n < z] = P[Y_1 < z] \cdots P[Y_n < z] = z^n$. Another way that $[Z_n < z]$ can occur is shown in Fig. 9*b*, where y_1 is to the right of z and each of $y_2, \ldots, y_n$ is in the interval between $y_1 - z$ and y_1. Clearly

$$P[y_1 - z < Y_2 < y_1, \ldots, y_1 - z < Y_n < y_1, y_1 < Y_1 < y_1 + dy_1] = z^{n-1}\,dy_1$$

and integrating over dy_1 from z to 1, we obtain $(1 - z)z^{n-1}$ for the probability of the event

$$[Z_n < z, z < Y_1, Y_1 \text{ is the largest of } Y_1, \ldots, Y_n].$$

But there are n such events, depending on which of $Y_1, \ldots, Y_n$ is largest. Clearly this takes care, without duplication, of all ways in which $[Z_n < z]$ can occur. Thus

$$P[Z_n < z] = z^n + n(1 - z)z^{n-1} \qquad \text{if } 0 < z < 1.$$

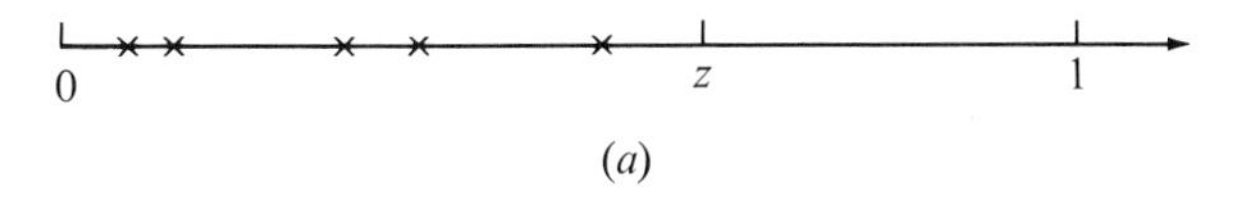

(*a*)

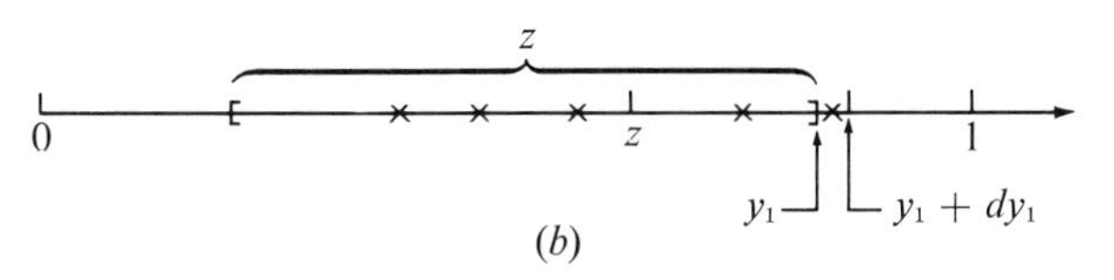

(*b*)

Fig. 9

Differentiating, we find the continuous density

$$p_{Z_n}(z) = \begin{cases} n(n-1)(1-z)z^{n-2} & \text{if } 0 < z < 1 \\ 0 & \text{otherwise.} \end{cases}$$

Obviously if n is large then we expect Z_n usually to have a value near to 1. In particular for $z = .9$ and $n = 50, 100$ we calculate

$$P[Z_{50} < .9] = .034 \qquad P[Z_{100} < .9] = .00032.$$

These numbers can be given the following interpretation. We intend to perform an experiment and observe $X_1, \ldots, X_n$, which are independent and identically distributed. We may or may not know their common distribution function $F: R \to R$, which *is* assumed to be continuous. That is, we are at least willing to assume that $P[X_1 = x_1] = 0$ for all real x_1 and that $X_1, \ldots, X_n$ are independent and identically distributed. After performing the experiment, we will consider the interval between the smallest and largest observations obtained. If $n = 100$ then the probability is .99968 that we will obtain some interval to which F assigns probability .9 or more. Thus if $n = 100$ then we can be very certain that the unknown F will assign at least 90 percent of its probability to the interval between our (as yet to be obtained) extreme observations. It is a crucial point that we are considering a random, as yet undetermined interval. If we have a particular F, say normal 0 and 1, and perform the experiment and obtain $\underline{x}$ and $\bar{x}$, *then* the probability is either 0 or 1 (depending on $\underline{x}$ and $\bar{x}$) that F assigns a probability of at least .9 to this interval. ■

Example 11 As seen in the preceding example, random variables such as the maximum and minimum arise naturally. The study of their distributions is often assisted by making the probability integral transformation or by using an equivalent type of reasoning. In this example we discuss natural generalizations of these random variables and find their joint distribution in certain simple cases. These so-called *order statistics* are used extensively in certain branches of statistics. We use them briefly in connection with the Poisson process in Theorem 2 (Sec. 23-3).

For any $n \geq 1$ define a function $f_n: R^n \to R^n$ as follows. For any reals $v_1 \leq v_2 \leq \cdots \leq v_n$ if $(x_1, \ldots, x_n)$ is any permutation of $(v_1, \ldots, v_n)$ then let $f_n(x_1, \ldots, x_n) = (v_1, \ldots, v_n)$. Thus $f_n(x_1, \ldots, x_n)$ is just the vector of arguments placed in nondecreasing order. Thus $f_3(2.1, -7, 1.3) = (-7, 1.3, 2.1)$. Similarly $f_4(2.1, 2.1, -7, 1.3) = (-7, 1.3, 2.1, 2.1)$, where the value is unique, although this value is equal to either of the two permutations (x_3, x_4, x_1, x_2) and (x_3, x_4, x_2, x_1). If $(X_1, \ldots, X_n)$ is any random vector then we call the random vector

$$(V_1, \ldots, V_n) = f_n(X_1, \ldots, X_n)$$

the **ordered vector determined by** or **corresponding to** $(X_1, \ldots, X_n)$. Usually V_k is called the **order statistic** of rank k from $X_1, \ldots, X_n$. Naturally V_k may be described as the kth smallest of the random numbers $X_1, \ldots, X_n$. Each V_k depends on every $X_1, \ldots, X_n$ so sometimes the notation $V_k^{(n)}$ is used. Naturally $V_1^{(n)}$ and $V_n^{(n)}$ are the minimum and maximum of $X_1, \ldots, X_n$, respectively.

Assume that $(X_1, \ldots, X_n)$ has a joint continuous density. If $i \neq j$ then (X_i, X_j) has a continuous density, so $P[X_i = X_j] = 0$, since this is the P_{X_i, X_j} probability of the 45° line in the plane. Therefore the probability is 1 that $X_1, \ldots, X_n$, and hence $V_1, \ldots, V_n$, will have distinct values. For any *distinct* $v_1 < v_2 < \cdots < v_n$ every one of its $n!$ permutations is distinct, and f_n takes these and only these points into $(v_1, \ldots, v_n)$. Therefore

$$P[v_1 < V_1 < v_1 + dv_1, \ldots, v_n < V_n < v_n + dv_n]$$
$$= \sum p_{X_1,\ldots,X_n}(x_1, \ldots, x_n)\, dv_1 \cdots dv_n$$

where the summation contains $n!$ terms, one for each permutation $(x_1, \ldots, x_n)$ of $(v_1, \ldots, v_n)$. Clearly f_n transforms $n!$ distinct infinitesimal sets into the volume element $dv_1 \cdots dv_n$, and all have this same volume. *Therefore* if $(X_1, \ldots, X_n)$ has a continuous density on R^n then its ordered vector $(V_1, \ldots, V_n)$ has a continuous density on R^n given by

$$p_{V_1,\ldots,V_n}(v_1, \ldots, v_n) = \begin{cases} \sum p_{X_1,\ldots,X_n}(x_1, \ldots, x_n) & \text{if } v_1 < v_2 < \cdots < v_n \\ 0 & \text{otherwise} \end{cases}$$

where the summation is over the $n!$ permutations $(x_1, \ldots, x_n)$ of $(v_1, \ldots, v_n)$.

If $X_1, \ldots, X_n$ are independent and identically distributed then the value of their joint density is unchanged by a permutation of its arguments. Therefore if $X_1, \ldots, X_n$ are independent and each has the same continuous density $p: R \to R$ then the corresponding ordered vector $(V_1, \ldots, V_n)$ has a continuous density on R^n given by

$$p_{V_1,\ldots,V_n}(v_1, \ldots, v_n) = \begin{cases} n!\, p(v_1)p(v_2) \cdots p(v_n) & \text{if } v_1 < v_2 < \cdots < v_n \\ 0 & \text{otherwise.} \end{cases}$$

Roughly speaking, we take the probability that $(X_1, \ldots, X_n) = (v_1, \ldots, v_n)$ and multiply by $n!$ because the same $(v_1, \ldots, v_n)$ is obtained, with the same probability, if $(X_1, \ldots, X_n)$ should equal any one of the $n!$ permutations of $(v_1, \ldots, v_n)$. As indicated by Fig. 10, if we

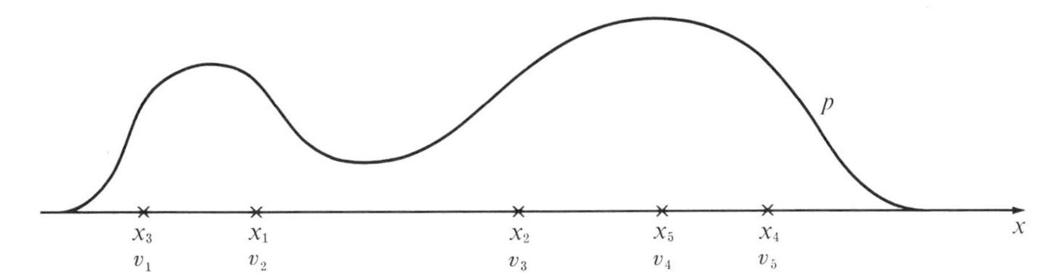

Fig. 10

think of selecting x_i at time i and then placing a cross at x_i then $(v_1, \ldots, v_n)$ may be interpreted as the final resulting configuration of crosses, *without* regard to the time order of selection.

In particular if $X_1, \ldots, X_n$ are independent and each is uniformly distributed over the same interval,

$$p_{X_1}(x_1) = \frac{1}{b - a} \qquad \text{if } a < x_1 < b \text{ where } a < b$$

then the corresponding ordered vector $(V_1, \ldots, V_n)$ has continuous density

$$p_{V_1,\ldots,V_n}(v_1, \ldots, v_n) = \begin{cases} \dfrac{n!}{(b - a)^n} & \text{if } a < v_1 < v_2 < \cdots < v_n < b \\ 0 & \text{otherwise.} \end{cases}$$

Therefore if $Y_1, \ldots, Y_n$ are independent and identically distributed according to a continuous distribution function $F: R \to R$, and if $X_i = F(Y_i)$, then the ordered vector $(V_1, \ldots, V_n)$ corresponding to $(X_1, \ldots, X_n)$ has joint density

$$p_{V_1,\ldots,V_n}(v_1, \ldots, v_n) = \begin{cases} n! & \text{if } 0 < v_1 < v_2 < \cdots < v_n < 1 \\ 0 & \text{otherwise.} \end{cases}$$ ■

EXERCISES

1. (*a*) Sketch the distribution function of $Y = g(X)$ when g and F_X are as in Fig. 11.
(*b*) Find and sketch F_Y and p_Y for $Y = X^2$ when X has continuous density

$$p_X(x) = \begin{cases} \frac{1}{2} & \text{if } -1 < x < 0 \\ \frac{1}{4} & \text{if } \quad 0 \le x < 2 \\ 0 & \text{otherwise.} \end{cases}$$

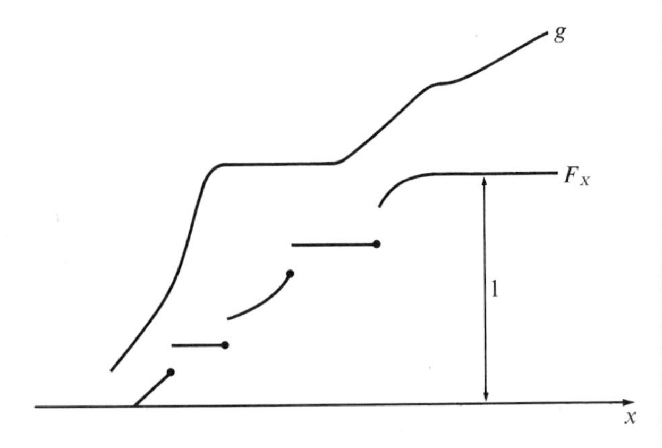

Fig. 11

2. Let $Y = cX^a$, where $c > 0$, $a > 0$, and $P[X > 0] = 1$.
(*a*) Find F_Y in terms of F_X.
(*b*) Assume that F_X has continuous density p_X and find the continuous density of Y by differentiating F_Y. Check your answer with Exercise 7 (Sec. 16-2).
(*c*) Find the probability density for the diameter of a randomly selected spherical raindrop whose volume is exponentially distributed with mean $1/\lambda$.

3. Let $Y = g(X)$, where $g: R \to R$ is strictly increasing.
(*a*) Show that

$$F_Y(y) = F_X(g^{-1}(y))$$

for all y in the range of g. Thus this same formula holds even if g has positive jumps.
(*b*) Sketch an F_X and a strictly increasing g both of which have positive jumps. Sketch the F_Y so determined.

4. Let $Y = g(X)$. Which, if any, of the four cases below is impossible?

Case	F_X is continuous	F_Y is continuous
(*a*)	Yes	Yes
(*b*)	Yes	No
(*c*)	No	Yes
(*d*)	No	No

5. Let $Y = g(X)$, where $g: R \to R$ is strictly decreasing.
(*a*) Show that

$$F_Y(y) = 1 - F_X(g^{-1}(y)) - [\text{the jump of } F_X \text{ at } x = g^{-1}(y)]$$

for all y in the range of g.
(*b*) Find the distribution function and density of $V = e^{-T}$ where T is exponential with mean $1/\lambda$.
6. Show that if $Y = |X|$ then

$$F_Y(y) = \begin{cases} 0 & \text{if } y \leq 0 \\ F_X(y) - F_X(-y) - [\text{the jump of } F_X \text{ at } x = -y] & \text{if } y > 0. \end{cases}$$

7. For Fig. 5 (Sec. 17-1) find and sketch the distribution function of the following:
(*a*) The time $Y = 60X$, in seconds
(*b*) The magnitude $Z = |X|$ of the time off schedule
(*c*) $W = e^X$
8. Let $Z = X + Y$ where X and Y are independent and X has a continuous density. Show that

$$F_Z(z) = \int_{-\infty}^{\infty} F_Y(z - x)p_X(x)\,dx \qquad \text{for all real } z.$$

Such a formula applies if X is normal and Y is Poisson, or continuous singular.
9. The density for Z_n was found in Example 10. Show that

$$EZ_n = \frac{n-1}{n+1} = 1 - \frac{2}{n+1}.$$

10. As in Example 11, let $Y_1, \ldots, Y_n$ be independent and identically distributed according to a continuous distribution function $F: R \to R$. Let $X_i = F(Y_i)$ so that X_i has a continuous uniform density over $0 < x < 1$. Let $(V_1, \ldots, V_n)$ be the ordered vector corresponding to $(X_1, \ldots, X_n)$. Show that V_k has continuous density

$$p_{V_k}(v) = \begin{cases} k\binom{n}{k} v^{k-1}(1-v)^{n-k} & \text{if } 0 < v < 1 \\ 0 & \text{otherwise.} \end{cases}$$

Then Example 9 yields $EV_k = k/(n+1)$. Also $EZ_n = E(V_n - V_1) = EV_n - EV_1$, which agrees with Exercise 9. HINT: Any of the n different events $[v < X_i < v + dv]$ may occur with any $k - 1$ of the remaining $n - 1$ different X_j to the left.

COMMENT: Let $(U_1, \ldots, U_n)$ be the ordered vector corresponding to the $(Y_1, \ldots, Y_n)$ introduced above. Clearly $V_k = F(U_k)$, so that

$$E[F(U_k) - F(U_{k-1})] = \frac{1}{n+1} \qquad \text{for } 2 \leq k \leq n.$$

Thus $1/(n+1)$ is the expected probability between any two adjacent observations from n independent identically distributed observations from any continuous $F\colon R \to R$.

17-3 DISTRIBUTION FUNCTIONS ON R^n

It is fairly common to refer to distribution functions in connection with general properties, such as the property that random variables are independent iff their joint distribution function satisfies the product rule. However, it is fairly uncommon to perform explicit calculations with distribution functions when $n \geq 2$. Densities are easier to manipulate, and if densities are not available it is often simpler just to use the probability measure. Part of the reason is that, as suggested by Theorem 1, calculations with distribution functions are fairly messy when $n \geq 2$.

In this section we first define distribution functions of random vectors and then see how to get the distribution function for a subsequence of $X_1, \ldots, X_n$ from that for $X_1, \ldots, X_n$. Theorem 1 shows how to calculate the probability of a rectangle from a distribution function. We then describe precisely which functions can be distribution functions of random vectors. Theorem 1 is proved in the next section by the method of indicator functions; this method of proof is probably of more interest than is Theorem 1.

The **distribution function** $F_X\colon R^n \to R$ **for** or induced by **a random vector** $X = (X_1, \ldots, X_n)$ with associated probability space (Ω,P) is defined by

$$F_{X_1,\ldots,X_n}(x_1, \ldots, x_n) = P[X_1 < x_1, \ldots, X_n < x_n] \qquad \text{for all reals } x_1, \ldots, x_n.$$

Thus $F_X(x_1, \ldots, x_n)$ equals the probability that every X_i takes a value less than the corresponding x_i. Naturally if (R^n,P) is any probability space, with sample space R^n, then its corresponding distribution function $F\colon R^n \to R$ is defined by

$$F(x_1, \ldots, x_n) = P\{(x'_1, \ldots, x'_n)\colon x'_1 < x_1, \ldots, x'_n < x_n\}.$$

Therefore $F_{X_1,\ldots,X_n}$ is also describable as the distribution function corresponding to the induced $P_{X_1,\ldots,X_n}$. As usual, if $F\colon R^n \to R$ corresponds to (R^n,P) then F can be thought of as induced by the random vector which is the identity function on R^n. If $F\colon R^2 \to R$ corresponds to (R^2,P) then $F(x_1,x_2)$ equals the probability of the shaded open set of Fig. 1.

If random variables $X_1, \ldots, X_n$ induce a continuous joint density then their joint distribution function can be obtained from it by integration,

$$F_{X_1,\ldots,X_n}(x_1, \ldots, x_n) = \int_{-\infty}^{x_1} \cdots \int_{-\infty}^{x_n} p_{X_1,\ldots,X_n}(x'_1, \ldots, x'_n)\, dx'_1 \cdots dx'_n.$$

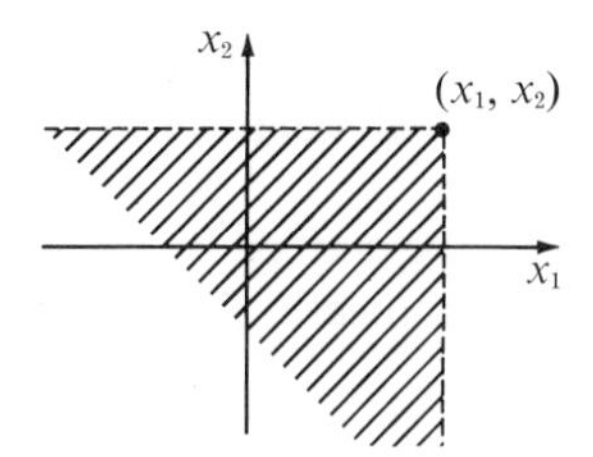

Fig. 1

Clearly F_X is continuous everywhere, and we can obtain the continuous density from it by differentiation,

$$p_{X_1,\ldots,X_n}(x_1, \ldots, x_n) = \frac{\partial^n F_{X_1,\ldots,X_n}(x_1, \ldots, x_n)}{\partial x_1\, \partial x_2 \cdots \partial x_n},$$

at least at points where the density is continuous.

A continuous density can be obtained from a larger continuous density by integrating out the extra coordinates. The corresponding result for distribution functions is obtained by letting the extra coordinates go to infinity. That is, since F is clearly nondecreasing in each coordinate, we get increasing events by letting x_i go to infinity to obtain

$$\begin{aligned} F_{X_1,\ldots,X_n}&(x_1, \ldots, x_{i-1}, \infty, x_{i+1}, \ldots, x_n) \\ &= \lim_{x_i\to\infty} F_{X_1,\ldots,X_n}(x_1, \ldots, x_n) \\ &= \lim_{x_i\to\infty} P[X_1 < x_1, \ldots, X_n < x_n] \\ &= P[X_1 < x_1, \ldots, X_{i-1} < x_{i-1}, X_{i+1} < x_{i+1}, \ldots, X_n < x_n] \\ &= F_{X_1,\ldots,X_{i-1},X_{i+1},\ldots,X_n}(x_1, \ldots, x_{i-1}, x_{i+1}, \ldots, x_n). \end{aligned}$$

More generally the distribution function for a subsequence of $X_1, \ldots, X_n$ can be obtained from that for $X_1, \ldots, X_n$ by letting the extra coordinates go to infinity.

Let (R^n,P) have distribution function $F: R^n \to R$. We now show for $n = 2$ how the probability of a rectangle can be calculated from F. Let $x_1^{(1)} < x_1^{(0)}$ and $x_2^{(1)} < x_2^{(0)}$. We use superscripts 1 and 0 for the smaller and larger values of a coordinate, respectively, in order to simplify the form of the general formula exhibited in Theorem 1 below. Now

$$\begin{aligned} P[x_1^{(1)} \le X_1 < x_1^{(0)}, x_2^{(1)} \le X_2 < x_2^{(0)}] \\ = P[x_1^{(1)} \le X_1 < x_1^{(0)}, X_2 < x_2^{(0)}] - P[x_1^{(1)} \le X_1 < x_1^{(0)}, X_2 < x_2^{(1)}] \end{aligned}$$

since this equation is of the form $P(A) = P(B) - P(C)$, where B equals the union of disjoint sets A and C. Applying essentially this same manipulation to each term in the right-hand side yields

$$\begin{aligned} P[x_1^{(1)} \le X_1 < x_1^{(0)}, x_2^{(1)} \le X_2 < x_2^{(0)}] \\ = \{F(x_1^{(0)},x_2^{(0)}) - F(x_1^{(1)},x_2^{(0)})\} - \{F(x_1^{(0)},x_2^{(1)}) - F(x_1^{(1)},x_2^{(1)})\} \end{aligned}$$

In terms of Fig. 2, this formula says that $P(A)$ is obtainable by starting with $F(x_1^{(0)},x_2^{(0)})$, then subtracting $P(D \cup E)$ and $P(C \cup E)$, and then adding $P(E)$.

The general formula for any $n \geq 1$ is stated in Theorem 1 below and could be derived by the method above. Theorem 1 is proved in the next section by means of indicator functions.

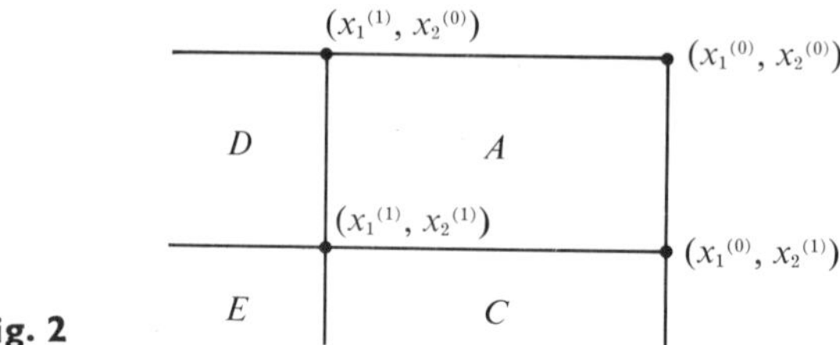

Fig. 2

Theorem 1 If a random vector $X = (X_1, X_2, \ldots, X_n)$ induces distribution function $F: R^n \to R$ then

$$P[x_1^{(1)} \leq X_1 < x_1^{(0)}, x_2^{(1)} \leq X_2 < x_2^{(0)}, \ldots, x_n^{(1)} \leq X_n < x_n^{(0)}] = \Delta_F(x^{(1)},x^{(0)})$$

for any $x^{(1)} = (x_1^{(1)}, \ldots, x_n^{(1)})$ and $x^{(0)} = (x_1^{(0)}, \ldots, x_n^{(0)})$ if $x^{(1)}$ is coordinatewise less than $x^{(0)}$, that is, if $x_i^{(1)} < x_i^{(0)}$ for all i. The function $\Delta_F: R^{2n} \to R$ is defined by

$$\Delta_F(x^{(1)},x^{(0)}) = \sum_{\epsilon_1=0}^{1} \sum_{\epsilon_2=0}^{1} \cdots \sum_{\epsilon_n=0}^{1} (-1)^{\epsilon_1+\cdots+\epsilon_n} F(x_1^{(\epsilon_1)},x_2^{(\epsilon_2)}, \ldots, x_n^{(\epsilon_n)}). \quad \blacksquare$$

According to Theorem 1, $\Delta_F(x^{(1)},x^{(0)})$ equals the sum of the 2^n terms

$$\pm F(x_1^{(\epsilon_1)}, \ldots, x_n^{(\epsilon_n)})$$

where the superscripts can be 0's and 1's. The sign is positive or negative, depending on whether an even or odd number of superscripts are 1's, where 1's correspond to smaller coordinates.

We now prove that *random vectors are identically distributed iff their distribution functions are equal;* in other words, the correspondence which assigns $F: R^n \to R$ to (R^n,P) is one-to-one. Note that as k goes to infinity the events

$$\left[x_1^{(1)} \leq X_1 < x_1^{(0)} + \frac{1}{k}, \ldots, x_n^{(1)} \leq X_n < x_n^{(0)} + \frac{1}{k}\right],$$

whose probabilities are calculable from F_X by Theorem 1, decrease to the event

$$[x_1^{(1)} \leq X_1 \leq x_1^{(0)}, \ldots, x_n^{(1)} \leq X_n \leq x_n^{(0)}]$$

so probabilities of closed rectangles can be found from F_X. Therefore if random vectors X and Y induce the same distribution function $F_X = F_Y$ then P_X and P_Y must agree on closed rectangles, and hence $P_X = P_Y$ by Theorem 2 (Sec. 15-1), *as was asserted.*

Random vectors are independent iff their joint distribution function satisfies the product rule. For example, the three random vectors $X = (X_1,X_2)$, $Y = (Y_1,Y_2,Y_3)$, Z are independent iff

$$F_{X,Y,Z}(x_1,x_2,y_1,y_2,y_3,z) = F_X(x_1,x_2)F_Y(y_1,y_2,y_3)F_Z(z)$$

for all real arguments. Half the proof of this fact is obvious, since, for simplicity, if (X_1,X_2) and Z are independent then

$$F_{X_1,X_2,Z}(x_1,x_2,z) = P[X_1 < x_1,\ X_2 < x_2,\ Z < z]$$
$$= P[X_1 < x_1,\ X_2 < x_2]P[Z < z] = F_{X_1,X_2}(x_1,x_2)F_Z(z)$$

so their distribution functions satisfy the product rule. The proof of the other half is left to Exercise 4.

Example 1 The distribution function corresponding to the continuous density of Example 1 (Sec. 16-2) is found, after a trivial integration, to be

$$F_{X,Y}(x,y) = \int_0^x \int_0^y 2[(1 - x')(1 - y') + x'y']\,dx'\,dy'$$
$$= xy[(1 - x)(1 - y) + 1] \qquad \text{if } 0 < x < 1,\ 0 < y < 1.$$

The obvious values of $F_{X,Y}$ on the rest of the plane are found in Exercise 1. We find F_X from $F_{X,Y}$ as follows:

$$F_X(x) = \lim_{y\to\infty} F_{X,Y}(x,y) = F_{X,Y}(x,1) = x \qquad \text{if } 0 < x < 1,$$

which agrees with the fact that X, as well as Y, was found to be uniformly distributed over the unit interval. Note that $p_{X,Y} \neq p_X p_Y$ and $F_{X,Y} \neq F_X F_Y$, and X and Y are obviously dependent.

From Theorem 1 we find that

$$P[0 \le X \le \tfrac{1}{2}, \tfrac{1}{4} \le Y < \tfrac{1}{2}] = F(\tfrac{1}{2},\tfrac{1}{2}) - F(0,\tfrac{1}{2}) - F(\tfrac{1}{2},\tfrac{1}{4}) + F(0,\tfrac{1}{4})$$
$$= \tfrac{5}{16} - 0 - \tfrac{11}{64} + 0 = \tfrac{9}{64}.$$

This differs from $P[0 \le X < \tfrac{1}{2}]P[\tfrac{1}{4} \le Y < \tfrac{1}{2}] = (\tfrac{1}{2})(\tfrac{1}{4})$, which would be the result if independence prevailed. ■

A distribution function corresponding to a discrete density must obviously have a discontinuity at every point having positive probability. More generally, even in mixed cases, if (x_1',x_2') has positive probability in Fig. 3 then

$$F(x_1,x_2) - F(x_1',x_2') = P(A) \ge P[X_1 = x_1',\ X_2 = x_2'] > 0$$

for all $x_1 > x_1'$ and $x_2 > x_2'$, so F has a discontinuity at (x_1',x_2').

However, if $n \ge 2$ then F corresponding to (R^n,P) *may* have discontinuities even though every point of R^n has zero probability. This phenomenon did not arise

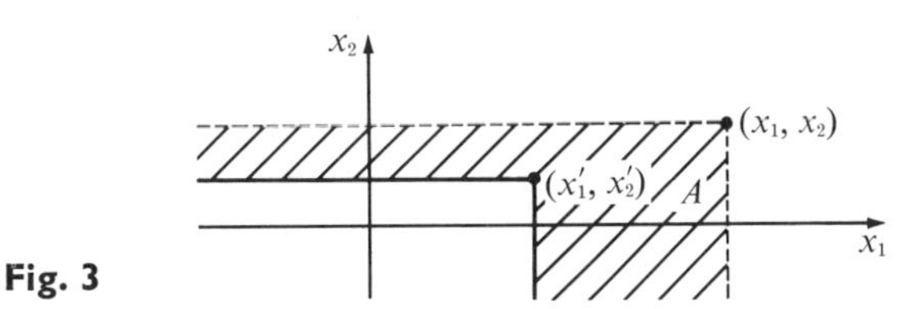

Fig. 3

for $n = 1$. For example, as indicated in Fig. 4*a*, let $P(A)$ be defined, for every event A in the plane, to equal the length of the intersection of A with the unit interval of the x axis. The corresponding $F: R^2 \to R$, whose values are indicated in Fig. 4*b*, is discontinuous at every point $(x,0)$ with $x > 0$. Thus discontinuities of F can be caused by lines, or planes, parallel to the coordinate axes and having positive probability.

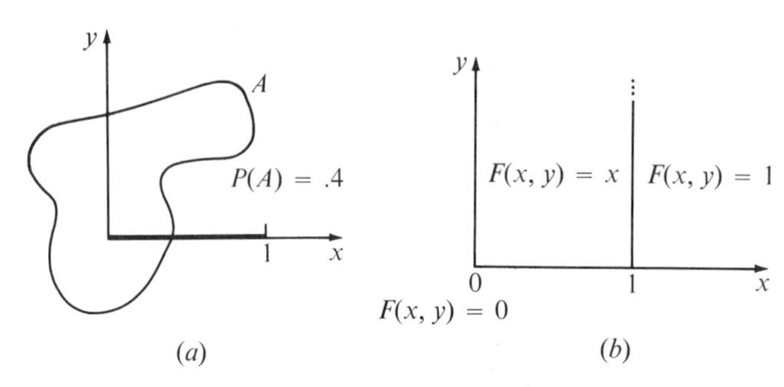

Fig. 4

We now characterize the class of functions $F: R^n \to R$ which are distribution functions corresponding to probability spaces (R^n,P); that is, we exhibit enough properties of such functions to guarantee that no other function could have all these properties.

A function $F: R^n \to R$ is called a **distribution function on R^n** iff the following four conditions hold:

1. $\lim_{x\to\infty} F(x, \ldots, x) = 1$.
2. $\lim_{x_i\to-\infty} F(x_1, \ldots, x_n) = 0$ for all i and $x_1, \ldots, x_{i-1}, x_{i+1}, \ldots, x_n$.
3. $F(x_1', \ldots, x_n') \to F(x_1, \ldots, x_n)$ whenever each $x_i' \to x_i$, where $x_i' < x_i$ is always satisfied.
4. $\Delta_F(x^{(1)},x^{(0)}) \geq 0$ whenever $x^{(1)} = (x_1^{(1)}, \ldots, x_n^{(1)})$ is coordinatewise less than $x^{(0)} = (x_1^{(0)}, \ldots, x_n^{(0)})$.

We now show that if $F: R^n \to R$ is the distribution function corresponding to a probability space (R^n,P) then F is a distribution function on R^n. Condition 1 obviously holds, since the corresponding events $[X_1 < x, \ldots, X_n < x]$ increase to the whole sample space. Thus F goes to 1 as all coordinates go to infinity simultaneously. Similarly the events corresponding to condition 2 decrease to the empty set, so their probabilities converge to zero. Thus F goes to zero when any one coordinate goes to minus infinity. Similarly condition 3 holds because the corresponding events are increasing to $[X_1 < x_1, \ldots, X_n < x_n]$. Thus F is lower-left continuous. Condition 4 follows from Theorem 1, so the assertion is proved.

Conditions 1 to 3 are obvious generalizations from the case $n = 1$. When $n = 1$, condition 4 is equivalent to requiring F to be nondecreasing. This was enough to ensure that in constructing an (R,P) from F we assign only *nonnegative* probabilities to intervals. However, for $n \geq 2$ it is not enough to require F to be nondecreasing in each coordinate, since not every such F is obtainable from an (R^n,P). We need the more complicated, stronger condition 4, $\Delta_F \geq 0$. For example, define F: $R^2 \to R$ to equal 1 on the open shaded set of Fig. 5 and let F equal 0 otherwise. Then F is nondecreasing and satisfies conditions 1 to 3. However, $\Delta_F(x',x) = 1 - 1 - 1 + 0 = -1$ for the x' and x of Fig. 5, so F cannot correspond to any (R^2,P).

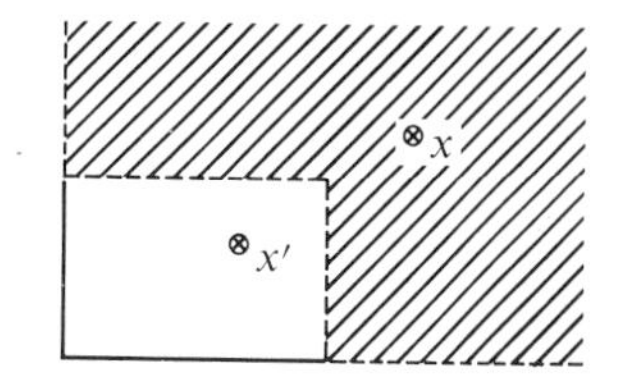

Fig. 5

It is a fact that for any distribution function F: $R^n \to R$ on R^n it is possible to use Δ_F to construct a probability space (R^n,P) such that F is the distribution function for (R^n,P). Exercises 5 to 8 attempt to make this fact intuitively plausible. Thus *the correspondence assigning* F: $R^n \to R$ *to* (R^n,P) *is a one-to-one correspondence between the class of all probability spaces* (R^n,P) *and the class of all distribution functions on* R^n.

EXERCISES

1. This exercise is concerned with the distribution function F of Example 1.

(*a*) Make a sketch indicating the value of F at every point of the plane.

(*b*) Find $P[x_1' \leq X_1 < x_1, x_2' \leq X_2 < x_2]$ by evaluating $\Delta_F(x',x)$ for the following arguments:

(1) $x' = (\frac{1}{4}, \frac{1}{2}), x = (\frac{3}{4}, \frac{3}{4})$

(2) $x' = (\frac{1}{2}, \frac{1}{2}), x = (1,1)$

(3) $x' = (\frac{1}{2}, \frac{1}{2}), x = (10,10)$

(*c*) Find an x' and x for which $\Delta_F(x',x) < 0$. Naturally x' cannot be coordinatewise less than x.

2. We know from this section that random vectors are independent iff their joint distribution function satisfies the product rule. Use this to prove that if X, Y, Z are three independent random variables then X and $W = (Y,Z)$ are two independent random vectors. Thus distribution functions provide a simple proof of Theorem 2 (Sec. 15-3).

3. Give a simple direct proof of the fact that if each random variable X_i induces a continuous density p_{X_i}, and if their joint distribution function satisfies the product rule then $p_{X_1}p_{X_2} \cdots p_{X_n}$, defined by the product rule, is a joint density for $X_1, \ldots, X_n$. In particular, then, $X_1, \ldots, X_n$ are independent.

4. In this exercise use Theorem 1 and Lemma 2 (Sec. 15-3) to deduce the independence of the following:

(*a*) Random variables $X_1, \ldots, X_n$ if their joint distribution function satisfies the product rule

(*b*) Random vectors $X = (X_1, \ldots, X_s)$ and $Y = (X_{s+1}, \ldots, X_n)$ if $F_{X,Y} = F_X F_Y$ in the usual sense.

For Exercises 5 to 8 assume that F: $R^n \to R$ is a distribution function on R^n, but do *not* assume that F corresponds to an (R^n,P). The purpose of these exercises is to make it seem more plausible that from F it is possible to construct a probability space (R^n,P) such that F corresponds to it. Let $x^{(1)}$, $y^{(1)}$, $y^{(0)}$, and $x^{(0)}$ be points in R^n, where $x^{(1)} = (x_1^{(1)}, \ldots, x_n^{(1)})$, and similarly for the others. Assume that

$$x_k^{(1)} \le y_k^{(1)} \le y_k^{(0)} \le x_k^{(0)} \qquad \text{for all } k = 1, 2, \ldots, n.$$

5. For any real α let

$$w^{(1)} = (\alpha, x_2^{(1)}, \ldots, x_n^{(1)}) \qquad w^{(0)} = (\alpha, x_2^{(0)}, \ldots, x_n^{(0)}).$$

Prove that

$$\Delta_F(x^{(1)},x^{(0)}) = \Delta_F(w^{(1)},x^{(0)}) + \Delta_F(x^{(1)},w^{(0)}).$$

IMPLICATIONS: If $x_1^{(1)} \le \alpha \le x_1^{(0)}$ we see, as indicated in Fig. 6 for the case $n = 3$, that if the probability of each left-closed and right-open rectangle is defined in terms of Δ_F, as

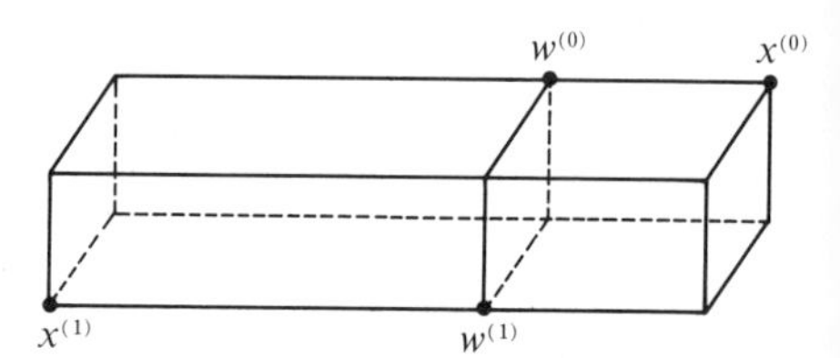

Fig. 6

suggested by Theorem 1, then we at least have additivity for two disjoint adjacent rectangles. Furthermore observe that if $x_1^{(1)} \le \alpha \le x_1^{(0)}$ then

$$\Delta_F(x^{(1)},x^{(0)}) \ge \Delta_F(w^{(1)},x^{(0)}) \qquad \text{and} \qquad \Delta_F(x^{(1)},x^{(0)}) \ge \Delta_F(x^{(1)},w^{(0)}).$$

Similar results hold for coordinates other than the first. Thus increasing a coordinate of the lower point $x^{(1)}$ or decreasing a coordinate of the higher point $x^{(0)}$ both result in decreasing $\Delta_F(x^{(1)},x^{(0)})$.

HINT: The proof depends only on the definition of Δ_F and applies to any function $F\colon R^n \to R$. From the definition of Δ_F we have

$$\Delta_F(x^{(1)},x^{(0)}) = \sum_{\epsilon_2=0}^{1} \cdots \sum_{\epsilon_n=0}^{1} (-1)^{\epsilon_2+\cdots+\epsilon_n} \times [F(x_1^{(0)}, x_2^{(\epsilon_2)}, \ldots, x_n^{(\epsilon_n)}) - F(x_1^{(1)}, x_2^{(\epsilon_2)}, \ldots, x_n^{(\epsilon_n)})]$$

and $F(\alpha, x_2^{(\epsilon_2)}, \ldots, x_n^{(\epsilon_n)})$ can be subtracted and added within the square brackets.

6. Prove that $\Delta_F(y^{(1)},y^{(0)}) \le \Delta_F(x^{(1)},x^{(0)})$. HINT: Use the observation made in Exercise 5.

7. Prove that $0 \le \Delta_F(x^{(1)},x^{(0)}) \le F(x_1^{(0)}, \ldots, x_n^{(0)})$. HINT: Let each $x_i^{(1)}$ go to minus infinity.

8. Prove that $0 \le F(y_1^{(0)}, \ldots, y_n^{(0)}) \le F(x_1^{(0)}, \ldots, x_n^{(0)}) \le 1$. HINT: The last two exercises yield $0 \le \Delta_F(y^{(1)},y^{(0)}) \le F(x_1^{(0)}, \ldots, x_n^{(0)}) \le 1$.

*17-4 THE METHOD OF INDICATORS

In this section we first describe the simple, but sometimes powerful, method of indicators. The method involves the use of Theorem 1 below, for which we exhibit an elementary proof. Theorem 1, together with simple indicator manipulations, yields Theorem 2, and this immediately implies Theorem 1 (Sec. 17-3), which exhibited the formula for Δ_F. This proof of the Δ_F formula, together with the standard combinatorial formula of Theorem 3, are our principal illustrations of the method of indicators. Theorem 3 is a generalization of Exercise 11 (Sec.11-3).

If $X = \sum_{i=1}^{n} b_i X_i$ then the linearity of expectation yields $EX = \sum_{i=1}^{n} b_i EX_i$. If the random variable I_B is the indicator function of an event B,

$$I_B(\omega) = \begin{cases} 1 & \text{if } \omega \in B \\ 0 & \text{if } \omega \notin B \end{cases}$$

then $EI_B = P(B)$. Therefore if $I_B = \sum_{i=1}^{n} b_i I_{B_i}$ then $P(B) = \sum_{i=1}^{n} b_i P(B_i)$, regardless of whether or not the events $B_1, \ldots, B_n$ are disjoint. That is, if the indicator of an event B can be expressed as a linear combination of indicators of other events,

$$I_B(\omega) = \sum_{i=1}^{n} b_i I_{B_i}(\omega) \qquad \text{for all } \omega \text{ in } \Omega$$

then $P(B)$ equals the same linear combination of the probabilities of the other events. Simple algebraic manipulations with indicators, such as those of Exercise 11 (Sec. 11-3), facilitate the construction of such linear combinations. *The techniques just described essentially comprise the method of indicators.*

We are in a sense addressing ourselves to problems dealing primarily with the probabilities of events rather than with general random variables. By introducing expectations and indicators, as above, we are better able to utilize negative numbers and handle nondisjoint sets. As a consequence a result like Theorem 3 should appear to be more comprehensible and generalizable than would be the case if it were presented, as it often is, prior to expectations.

Theorem 1 If events B and $B_1, \ldots, B_n$ in a probability space (Ω, P) satisfy

$$I_B = \sum_{i=1}^{n} b_i I_{B_i}$$

where $b_1, \ldots, b_n$ are reals, then

$$P(B) = \sum_{i=1}^{n} b_i P(B_i). \quad \blacksquare$$

Theorem 1 was proved above by using the linearity of expectations. Each of the random variables takes only two values, but (Ω, P) is unrestricted. Our treatment of expectation for a general (Ω, P) was sketchy, so we give below an elementary self-contained proof of Theorem 1. The useful technique of this proof introduces a finite collection of *disjoint* events sufficient for the problem at hand. Theorem 1 and a proof similar to the one below are often initial steps in the general introduction of expectation as described in Sec. 21-4.

Elementary proof of Theorem 1 We wish to prove that if $0 = \sum_{i=1}^{n} b_i I_{B_i}$ then $0 = \sum_{i=1}^{n} b_i P(B_i)$. Applying this to $0 = -I_B + \sum_{i=1}^{n} b_i I_{B_i}$ then yields Theorem 1.

As shown in Fig. 1*a*, two subsets B_1 and B_2 of a set Ω partition Ω in a natural fashion into the four disjoint sets B_1B_2, $B_1B_2{}^c$, $B_1{}^cB_2$, and $B_1{}^cB_2{}^c$. Similarly, as in Fig. 1*b*, three subsets, B_1, B_2, and B_3, of Ω partition Ω into $2^3 = 8$ disjoint sets. The

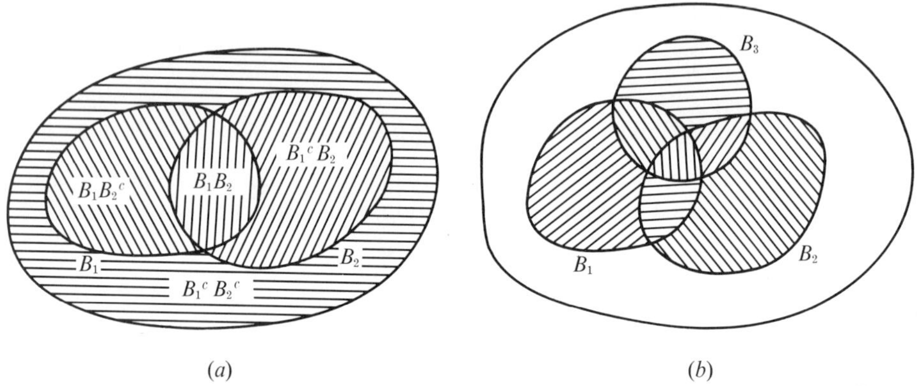

(*a*) (*b*)

Fig. 1

disjointness of $B_1B_2B_3{}^c$ and $B_1B_2{}^cB_3{}^c$, for example, follows from the fact that the first is included in B_2 and the second is included in $B_2{}^c$. If we select any point ω in Ω then for each i either ω belongs to B_i or $B_i{}^c$. Thus if ω belongs B_1, B_2, and $B_3{}^c$, for example, then ω belongs to $B_1B_2B_3{}^c$, and hence every point of Ω belongs to one of these eight sets.

In general if $B_1, \ldots, B_n$ are $n \geq 2$ subsets of a set Ω then we can define exactly 2^n sets like

$$B_1B_2{}^cB_3{}^cB_4 \cdots B_{n-1}^c B_n$$

where in this intersection of n sets exactly one of B_i and $B_i{}^c$ appears for $1 \leq i \leq n$. These 2^n sets are disjoint (some may be empty) and their union is Ω. We denote these 2^n disjoint sets by $A_1, A_2, \ldots, A_{2^n}$. Then B_i equals the union of half of them, namely, those which contain B_i rather than $B_i{}^c$ in their form as an intersection of n sets. Clearly a set equals the union of disjoint sets iff its indicator equals the sum of their indicators. Thus there are numbers c_{ij}, each either 0 or 1, such that

$$I_{B_i} = \sum_{j=1}^{2^n} c_{ij} I_{A_j} \qquad \text{and} \qquad P(B_i) = \sum_{j=1}^{2^n} c_{ij} P(A_j)$$

where we used the finite additivity of P for disjoint events.

Using our hypothesis for the first equality, then the first of the two equations above for the second equality, yields

$$0 = \sum_{i=1}^{n} b_i I_{B_i} = \sum_{i=1}^{n} b_i \sum_{j=1}^{2^n} c_{ij} I_{A_j} = \sum_{j=1}^{2^n} \left(\sum_{i=1}^{n} b_i c_{ij} \right) I_{A_j}.$$

For any j either A_j is empty or there is an ω belonging to A_j. In the latter case if we evaluate the right side at this ω we obtain

$$0 = \left(\sum_{i=1}^{n} b_i c_{ij}\right) 1$$

since the A_i sets are disjoint. Thus for each j, either I_{A_j} is zero or its coefficient ( ) is zero; hence *every term* ( ) I_{A_j} is zero, and so each I_{A_j} may be replaced by $P(A_j)$ since then every term will still be zero. We can then work backward to get

$$0 = \sum_{j=1}^{2^n} \left(\sum_{i=1}^{n} b_i c_{ij}\right) P(A_j) = \sum_{i=1}^{n} b_i \sum_{j=1}^{2^n} c_{ij} P(A_j) = \sum_{i=1}^{n} b_i P(B_i). \quad \blacksquare$$

Theorem 2 Let $A_1^{(1)}, A_2^{(1)}, \ldots, A_n^{(1)}$ and $A_1^{(0)}, A_2^{(0)}, \ldots, A_n^{(0)}$ be $2n$ events such that $A_i^{(1)}$ is a subset of $A_i^{(0)}$ for each i. If

$$A = \bigcap_{i=1}^{n} A_i^{(0)} (A_i^{(1)})^c$$

then

$$P(A) = \sum_{\epsilon_1=0}^{1} \sum_{\epsilon_2=0}^{1} \cdots \sum_{\epsilon_n=0}^{1} (-1)^{\epsilon_1 + \cdots + \epsilon_n} P(A_1{}^{(\epsilon_1)} A_2{}^{(\epsilon_2)} \cdots A_n{}^{(\epsilon_n)}). \quad \blacksquare$$

The event A of Theorem 2 is the outcome that every one of the larger events occurs but none of the smaller ones does. Then $P(A)$ equals the sum of 2^n numbers, one for each intersection,

$$A_1^{(\epsilon_1)} A_2^{(\epsilon_2)} \cdots A_n^{(\epsilon_n)}$$

where the superscripts can be 0's or 1's. The probability of such an intersection is added or subtracted, depending on whether an even or an odd number of the superscripts are 1's, where 1's correspond to smaller events.

Proof of Theorem 2 If C contains B then clearly $I_C - I_B$ is the indicator of CB^c. The first and last equalities below follow from the fact that the indicator of an intersection equals the product of the indicators,

$$\begin{aligned} I_A &= (I_{A_1{}^{(0)}} - I_{A_1{}^{(1)}})(I_{A_2{}^{(0)}} - I_{A_2{}^{(1)}}) \cdots (I_{A_n{}^{(0)}} - I_{A_n{}^{(1)}}) \\ &= \sum (-1)^{\epsilon_1 + \cdots + \epsilon_n} I_{A_1{}^{(\epsilon_1)}} I_{A_2{}^{(\epsilon_2)}} \cdots I_{A_n{}^{(\epsilon_n)}} \\ &= \sum (-1)^{\epsilon_1 + \cdots + \epsilon_n} I_{A_1{}^{(\epsilon_1)} A_2{}^{(\epsilon_2)} \cdots A_n{}^{(\epsilon_n)}}. \end{aligned}$$

The middle equality is just the fact, proved in Exercise 5 (Sec. 2-2), that a product of n factors, such as

$$\begin{aligned} &(a_0 + a_1)(b_0 + b_1)(c_0 + c_1) \\ &\quad = a_0 b_0 c_0 + a_0 b_0 c_1 + a_0 b_1 c_0 + a_0 b_1 c_1 + a_1 b_0 c_0 + a_1 b_0 c_1 + a_1 b_1 c_0 + a_1 b_1 c_1 \end{aligned}$$

equals the sum of 2^n terms, with one term for each selection of a number from each of the factors. [For our application we essentially write $\alpha - \beta$ as $\alpha + (-\beta)$.] Theorem 1, applied to the identity above, completes the proof of Theorem 2. ■

Proof of Theorem 1 (Sec. 17-3) In Theorem 2 let $A_i^{(0)} = [X_i < x_i^{(0)}]$ and $A_i^{(1)} = [X_i < x_i^{(1)}]$ so that

$$A_i^{(0)}(A_i^{(1)})^c = [x_i^{(1)} \le X_i < x_i^{(0)}]. \quad \blacksquare$$

Theorem 3 For any events $B_1, \ldots, B_n$ we have

$$P\left(\bigcup_{j=1}^{n} B_j\right) = \sum_{j=1}^{n} (-1)^{j-1} S_j = S_1 - S_2 + S_3 - \cdots + (-1)^{n-1} S_n$$

where

$$S_1 = \sum_{i=1}^{n} P(B_i) \qquad S_2 = \sum_{\substack{i=1 \\ i<k}}^{n} \sum_{k=1}^{n} P(B_i B_k) \qquad \cdots \qquad S_n = P(B_1 \cdots B_n). \quad \blacksquare$$

The number S_j can be described as equal to the sum of the probabilities of the $\binom{n}{j}$ events which can be written as an intersection

$$B_{i_1} B_{i_2} \cdots B_{i_j} \qquad \text{where } 1 \le i_1 < i_2 < \cdots < i_j \le n$$

of j events, using each of $B_1, \ldots, B_n$ at most once.

Proof of Theorem 3 In Theorem 2 let every $A_i^{(0)}$ equal the whole sample space Ω, so that

$$A = \bigcap_{i=1}^{n} (A_i^{(1)})^c \qquad \text{and} \qquad A^c = \bigcup_{i=1}^{n} A_i^{(1)}.$$

Furthermore the intersection $A_1^{(\epsilon_1)} A_2^{(\epsilon_2)} \cdots A_n^{(\epsilon_n)}$ equals Ω if every $\epsilon_i = 0$; otherwise the events with a zero superscript can be dropped. If we let S_j be the sum of the probabilities of the $\binom{n}{j}$ such events which have exactly j superscripts equal to 1 then

$$1 - P(A^c) = P(A) = \sum_{j=0}^{n} (-1)^j S_j = 1 + \sum_{j=1}^{n} (-1)^j S_j$$

and hence $P(A^c) = \sum_{j=1}^{n} (-1)^{j-1} S_j$. Now replace $A_i^{(1)}$ by B_i. $\blacksquare$

In the application of Theorem 3 it sometimes happens that for each j all the $\binom{n}{j}$ terms in S_j are equal. Then

$$P\left(\bigcup_{j=1}^{n} B_j\right) = \sum_{j=1}^{n} (-1)^{j-1} \binom{n}{j} P(B_1 B_2 \cdots B_j).$$

Example 1 A subject is given n cards labeled $1, 2, \ldots, n$. He cannot see the labels. On a table are written the integers $1, 2, \ldots, n$. The subject places one card on top of each integer. A *match* occurs at integer i iff the card labeled i is placed on top of integer i. The subject is presumably trying to make the maximum number of matches. We want the probability of at least one match, under the *assumption* that all placements are equally likely.

Let $x = (x_1, x_2, \ldots, x_n)$ be the outcome that x_i is the label on the card which is placed on top of integer i. Let Ω consist of all $n!$ permutations of $(1, 2, \ldots, n)$ and let P be uniform on Ω. Let B_i be the event that $x_i = i$; that is, there is a match at integer i. A member of $B_2B_3B_7$ looks like $(x_1, 2, 3, x_4, x_5, x_6, 7, x_8, \ldots, x_n)$, where the nonspecified coordinates can be any of $(n-3)!$ permutations. Thus $P(B_2B_3B_7) = (n-3)!/n!$ and every term in S_3 has this same probability. Approximating e^{-1} by using the first few terms in the power series of e^x, as exhibited in formula (5) of Appendix 1, yields

$$P\left(\bigcup_{j=1}^{n} B_j\right) = \sum_{j=1}^{n} (-1)^{j-1}\binom{n}{j}\frac{(n-j)!}{n!} = \sum_{j=1}^{n}\frac{(-1)^{j-1}}{j!} = -\sum_{j=1}^{n}\frac{(-1)^j}{j!}$$

$$= 1 - \sum_{j=0}^{n}\frac{(-1)^j}{j!} \doteq 1 - e^{-1} \doteq .6321.$$

For $n = 4$ the exact value of .625 is already close to the limit $1 - e^{-1}$. Therefore for *any* large n, about 63 percent of such experiments should yield at least one match *by mere chance*.

A variety of combinatorial problems[1] can be solved by the technique used in Theorems 1 and 2. ■

EXERCISES

1. Prove that if $\sum_{i=1}^{n'} b_i' I_{B_i'} = \sum_{i=1}^{n} b_i I_{B_i}$ then $\sum_{i=1}^{n'} b_i' P(B_i') = \sum_{i=1}^{n} b_i P(B_i)$ by using expectations and also by direct application of Theorem 1.

2. Assume that $0 = \sum_{i=1}^{r} a_i I_{A_i}$ where $A_1, \ldots, A_r$ are disjoint events. Show that for each i either A_i is empty or $a_i = 0$. Conclude that $0 = \sum_{i=1}^{r} a_i P(A_i)$. COMMENT: The elementary proof of Theorem 1 essentially reduced the general case to this case.

3. There are exactly n different white balls in an urn (chickens in a coop), and a sample of N of them, $N \geq n$, is drawn with replacement and colored red (inoculated). Show that the probability is

$$\sum_{j=1}^{n} (-1)^{j-1}\binom{n}{j}\left(1 - \frac{j}{n}\right)^N$$

that at the end of the experiment at least one ball is white (chicken is not inoculated). HINT: Let P be uniform on Ω which consists of all $(x_1, \ldots, x_N)$ where each coordinate is one of the integers $\{1, 2, \ldots, n\}$. Let B_i be the event that no coordinate equals i.

4. (*a*) Prove that $P\left(\bigcup_{j=1}^{n} B_j\right) = 1 - [1 - P(B_1)][1 - P(B_2)] \cdots [1 - P(B_n)]$ if $B_1, \ldots, B_n$ are independent events. This formula is simpler than Theorem 3, but Theorem 3 holds even if the events are dependent.

(*b*) From part (*a*) deduce the formula of Theorem 3 in the case of independence. This proof is *not* valid if there is dependence.

(*c*) Show that the events $B_1, \ldots, B_n$ in Example 1 are dependent if $n \geq 2$.

[1] See E. Parzen, "Modern Probability Theory and Its Applications," chap. 2, sec. 6, John Wiley & Sons, Inc., New York, 1960, and W. Feller, "An Introduction to Probability Theory and Its Applications," 2d ed., vol. 1, chap. 4, John Wiley & Sons, Inc., New York, 1957.

5. Show that the following chain of identities is essentially obvious by inspection, if T_j is correctly defined as a sum of $\binom{n}{j}$ indicators.

$$1 - I_{B_1 \cup \cdots \cup B_n} \overset{1}{=} I_{B_1^c \cdots B_n^c} \overset{2}{=} (1 - I_{B_1}) \cdots (1 - I_{B_n})$$

$$\overset{3}{=} \sum_{j=0}^{n} (-1)^j T_j.$$

Taking expectations yields Theorem 3.

18
Multivariate Normal Densities and Characteristic Functions

Section 18-1 is devoted to the study of covariance matrices of random vectors. For random vectors the covariance matrix plays a role similar to that played by the variance for random variables. In particular, dependent normal densities on R^k are specified in Sec. 18-2 by means and covariance matrices, which reduce to μ and σ^2 when $k = 1$. Normal densities on R^k are among the most widely used dependent distributions, in both theory and practice. Sections 18-1 and 18-2 make very little use of Chap. 17. Section 18-3 introduces the characteristic function of a random variable by replacing real t in the MGF, Ee^{tX}, by imaginary it to get Ee^{itX}. This introduction of complex numbers is so rewarding that characteristic functions are used extensively in probability theory. In particular they can be used as in Sec. 18-3 to deduce central-limit theorems and to obtain many other limiting distributions. The ideas and results from the first three sections are illustrated in Sec. 18-4 by a vector CLT.

18-1 COVARIANCE MATRICES AND LINEAR PREDICTION

In this section we first review a few simple facts concerning the transpose, inverse, and determinant of a matrix. Then the basic covariance matrix C of a random vector X is introduced. Lemma 1 shows that C is positive definite iff X is nondegenerate. Theorem 1 shows that C arises naturally in determining the minimum expected-squared-error linear predictor of a random variable. This theorem extends Sec. 11-4 to the case where $k \geq 1$ random variables may be used in the predictor. Lemmas 2 and 3 then show that every symmetric positive-definite matrix arises as a covariance matrix. These two lemmas are fundamental to the next section.

REVIEW OF MATRICES

Let us review a few facts concerning matrices. Let T be a $k \times n$ matrix whose entries are real numbers. Thus T has k rows and n columns, and the real number t_{ij} is the entry in the ith row and jth column (we usually omit the comma between i and j in $t_{i,j}$). The **transpose** T' of T is the $n \times k$ matrix with entry t_{ji} in its ith row and jth column. Clearly $(T')' = T$. In particular if t is $k \times 1$, so that t is a column vector, we denote the entry in its ith row by t_i so that $t' = (t_1, \ldots, t_k)$ is a $1 \times k$ row vector. If S is an $n \times r$ matrix it is easily seen that $(TS)' = (S')(T')$, so that the transpose of a product equals the product of the transposes in the reverse order. A matrix T is said to be **symmetric** iff it equals its transpose $T = T'$; that is, $t_{ij} = t_{ji}$ always holds. Clearly a symmetric matrix must be square, that is, $k \times k$. A square matrix is said to be **diagonal** iff all elements off its diagonal are 0's, $t_{ij} = 0$ whenever $i \neq j$. In particular the $k \times k$ **identity** matrix I is the diagonal matrix having all 1's on its diagonal, $\delta_{ii} = 1$ for $i = 1, \ldots, k$. Every diagonal matrix is symmetric.

A $k \times k$ matrix T is said to be **nonsingular** iff there exists a unique $k \times k$ matrix T^{-1}, called the **inverse** of T, such that $TT^{-1} = T^{-1}T = I$, where I is the identity matrix. Taking the transpose of this equation shows that T' is nonsingular and has inverse $(T^{-1})'$, so that $(T')^{-1} = (T^{-1})'$. In particular if T is symmetric then $(T^{-1})' = T^{-1}$, so T^{-1} is symmetric. As is well known, T is nonsingular iff $Tt \neq 0$ whenever $t \neq 0$. Here t is a column vector with k rows, and 0 is the zero column vector having 0's for all k of its entries. That is, T is singular iff there exists some nonzero vector t which is transformed into the zero vector, $Tt = 0$. We use this condition often. If S and T are nonsingular $k \times k$ matrices then so is their product ST, since $(T^{-1})(S^{-1})$ is obviously an inverse. Thus $(ST)^{-1} = T^{-1}S^{-1}$.

A square matrix T is nonsingular iff its determinant is nonzero, $\det T \neq 0$. For $k \times k$ matrices we have the product rule for determinants,

$$\det ST = (\det S)(\det T).$$

In particular if T is nonsingular then $TT^{-1} = I$, and hence $(\det T)(\det T^{-1}) = 1$. Also the determinant of any square matrix is equal to the determinant of its transpose, $\det T = \det T'$.

In this chapter we attempt to adhere to the usual conventions for row and column vectors.

Thus if X is a column random vector we usually exhibit its transpose $X' = (X_1, \ldots, X_k)$ for notational convenience. This distinction between X and X' is often probabilistically meaningless and somewhat confusing. In particular X and X' have the same distribution on R^k, as do $g(X)$ and $g(X')$ for every g on R^k. However, the distinction is useful when we are considering multiplication by a matrix.

COVARIANCE MATRICES AND QUADRATIC FORMS

We often multiply a matrix of real numbers by a random vector and then take the expectation of a component of the resulting random vector. Such calculations suggest the following natural definitions for sums and expectations of random vectors. We use only the simpler versions of the definitions extensively, but occasionally abbreviated alternate calculations are also indicated.

A $k \times n$ **random matrix** X has random variable X_{ij} for its (i,j) entry, and the transpose X' has X_{ji} for its (i,j) entry. If each $E|X_{ij}| < \infty$ we define EX to be the matrix of real numbers whose (i,j) entry is EX_{ij}. Most often we have random vectors, so $n = 1$ or $k = 1$:

$$X = \begin{bmatrix} X_1 \\ \cdot \\ \cdot \\ \cdot \\ X_k \end{bmatrix} \qquad EX = \begin{bmatrix} EX_1 \\ \cdot \\ \cdot \\ \cdot \\ EX_k \end{bmatrix} \qquad \begin{matrix} X' = (X_1, \ldots, X_k) \\ \\ \\ EX' = (EX_1, \ldots, EX_k). \end{matrix}$$

If X and Y are random $k \times n$ matrices then so is $aX + bY$ for real a and b. It has $aX_{ij} + bY_{ij}$ for its (i,j) entry. We obviously have linearity of expectation $E(aX + bY) = aEX + bEY$ if each of X and Y has an expectation. If A is an $r \times k$ matrix of real numbers and X is a random $k \times n$ matrix then $E(AX) = A(EX)$, where AX is a random $r \times n$ matrix whose (i,j) entry is, as usual, $\sum_{s=1}^{k} a_{is}X_{sj}$. Obviously we have linearity,

$$E(AXB + CYD) = A(EX)B + C(EY)D$$

if A, B, C, and D are matrices of proper dimensions with real entries. Random matrices are of course, multiplied by the same rules as constant matrices.

If $X' = (X_1, \ldots, X_k)$ is a random vector for which $EX_i^2 < \infty$ for $i = 1, \ldots, k$ then the **covariance matrix** for X', or for X, is the $k \times k$ matrix C whose (i,j) entry is the real number Cov (X_i, X_j). Clearly C is symmetric with diagonal entries

$$c_{ii} = \text{Cov}\,(X_i, X_i) = \text{Var}\,(X_i) \geq 0.$$

If $X_1, \ldots, X_k$ are independent then C is diagonal. If $X_1, \ldots, X_k$ are independent and each has unit variance then the covariance matrix is just the identity matrix. For linear combinations of components we have

$$\text{Var}\left[\sum_{i=1}^{k} t_i X_i\right] = E\left[\sum_{i=1}^{k} t_i(X_i - \mu_i)\right]^2$$

$$= E\sum_{i=1}^{k} t_i(X_i - \mu)\sum_{j=1}^{k} t_j(X_j - \mu_j) = \sum_{i=1}^{k}\sum_{j=1}^{k} t_i\,\text{Cov}\,(X_i, X_j)t_j$$

for all real $t_1, \ldots, t_k$. This is just the sum of the k^2 elements in the matrix whose (i,j) entry is t_i Cov $(X_i, X_j)t_j$. Thus we have the frequently used formula

$$0 \leq \text{Var}\left[\sum_{i=1}^{k} t_i X_i\right] = t'Ct \qquad \text{where } t = \begin{bmatrix} t_1 \\ \cdot \\ \cdot \\ \cdot \\ t_k \end{bmatrix}.$$

Therefore the vector $EX' = (EX_1, \ldots, EX_k)$ of means and the covariance matrix are precisely what are needed in order to find the mean and variance of every linear combination of components.

Note that if $Y' = (Y_1, \ldots, Y_k)$ and $EY = 0$ then the covariance matrix for Y is just the expectation of YY' whose (i,j) entry is Y_iY_j. The preceding formula can be derived in more abbreviated form as follows. The random variable $t'(X - EX)$ equals its transpose, so that

$$\begin{aligned}\text{Var}\,[t'X] &= E[t'(X - EX)]^2 = E[t'(X - EX)(X - EX)'t] \\ &= t'\{E[(X - EX)(X - EX)']\}t = t'Ct.\end{aligned}$$

If C is *any* $k \times k$ matrix of real numbers then its associated **quadratic form** is the function $Q\colon R^k \to R$ defined by

$$Q(t_1, \ldots, t_k) = t'Ct = \sum_{i=1}^{k} \sum_{j=1}^{k} c_{ij} t_i t_j.$$

Since

$$c_{12}t_1t_2 + c_{21}t_2t_1 = (c_{12} + c_{21})t_2t_1$$

it is clear, for example, that Q is unaffected if in C we replace c_{12} and c_{21} by 0 and $c_{12} + c_{21}$, respectively. However, if two *symmetric* $k \times k$ matrices have the same quadratic form then the two matrices are equal, because if $t'Ct = t'Dt$ for all t in R^k then letting only t_i and t_j be nonzero yields

$$c_{ii}t_i^2 + (c_{ij} + c_{ji})t_it_j + c_{jj}t_j^2 = d_{ii}t_i^2 + (d_{ij} + d_{ji})t_it_j + d_{jj}t_j^2.$$

We let (t_i, t_j) equal (0,1), then (1,0), and then (1,1) to show that the coefficients are equal. Thus we have a one-to-one correspondence between symmetric $k \times k$ real matrices and their corresponding quadratic forms on R^k.

Let C be a $k \times k$ matrix of real numbers and let Q be its quadratic form. We say that C and Q are **nonnegative** iff $Q \geq 0$ on R^k. Clearly $Q(0, \ldots, 0) = 0$. We say that C and Q are **positive definite** iff $Q(t_1, \ldots, t_k) > 0$ whenever at least one $t_i \neq 0$; that is, $t'Ct > 0$ whenever $t \neq 0$. A singular C cannot be positive definite because there is a $t \neq 0$ with $Ct = 0$, so $t'Ct = 0$. Thus a positive-definite matrix must be nonsingular. Furthermore if C is symmetric positive definite then so is C^{-1}, since if $t \neq 0$ then $(C^{-1}t) \neq 0$ and C^{-1} is symmetric, and hence

$$t'C^{-1}t = (C^{-1}t)'C(C^{-1}t) > 0.$$

We have seen that the covariance matrix for any random vector is symmetric and nonnegative. We next show that it is positive definite unless the random vector is degenerate. A *random variable* W, or the distribution induced by W, is said to be **degenerate** iff its distribution is concentrated on one point so that $P[W = EW] = 1$. Equivalently Var $W = 0$. A *random vector* $X' = (X_1, \ldots, X_k)$ is said to be **degenerate** iff there exist real numbers $t_1, \ldots, t_k$ not all zero, such that $t_1X_1 + \cdots + t_kX_k$ is a degenerate random variable, that is, has zero variance. This is the case iff one of $X_1, \ldots, X_k$ can be expressed as a constant plus a linear combination of the others, so that the distribution induced by X is concentrated on a space of dimension $k - 1$ or smaller. For example, $(X_1, X_2, X_1 + 6X_2)$ is degenerate. Thus X is degenerate iff

$$\text{Var}\left(\sum_{i=1}^{k} t_iX_i\right) = t'Ct = 0 \qquad \text{for some } t \neq 0.$$

Hence X is nondegenerate iff C is positive definite. The following lemma summarizes these facts and adds a further one.

Lemma 1 Let $X' = (X_1, \ldots, X_k)$ be a random vector for which $EX_j^2 < \infty$ for $j = 1, \ldots, k$, and let C be its covariance matrix. Then

$$0 \leq \text{Var}\left(\sum_{i=1}^{k} t_i X_i\right) = t'Ct \qquad \text{where } t = \begin{bmatrix} t_1 \\ \cdot \\ \cdot \\ \cdot \\ t_k \end{bmatrix}$$

so that C is nonnegative. Furthermore the following three conditions are equivalent:
(*a*) X is nondegenerate.
(*b*) C is positive definite.
(*c*) C is nonsingular. ∎

Proof We need only show that condition (*c*) implies condition (*b*). Assume that C is a *nonsingular covariance matrix*. We must show that $t'Ct = 0$ implies $t = 0$. If for some t we have $t'Ct = 0$ then $0 = \text{Var}\,(t'X)$, so that with probability 1, $t'X = E(t'X)$, hence

$$\sum_{i=1}^{k} t_i(X_i - EX_i) = 0.$$

Multiplying by $(X_j - EX_j)$ and taking expectations yields

$$\sum_{i=1}^{k} t_i \,\text{Cov}\,(X_j,X_i) = 0 \qquad \text{for } j = 1, \ldots, k$$

which is equivalent to $Ct = 0$. But C is nonsingular, so $Ct = 0$ implies $t = 0$. ∎

Example 1 We wish to find all covariance matrices when $k = 2$. Assume that $\sigma_1^2 = \text{Var}\, X_1$ and $\sigma_2^2 = \text{Var}\, X_2$ are finite. If both are positive then, as in Sec. 11-4, their correlation coefficient ρ is defined by

$$\rho = \frac{\text{Cov}\,(X_1,X_2)}{\sigma_1\sigma_2} \qquad \text{where } \rho^2 \leq 1.$$

Thus if $\sigma_1\sigma_2 > 0$ then there is a unique ρ such that

$$\text{Cov}\,(X_1,X_2) = \rho\sigma_1\sigma_2.$$

If $\sigma_1 = 0$ then $X_1 - EX_1 = 0$, so $\text{Cov}\,(X_1,X_2) = 0$, and hence the same equation holds for every number ρ. Therefore every covariance matrix can be written in the form

$$C = \begin{bmatrix} \sigma_1^2 & \rho\sigma_1\sigma_2 \\ \rho\sigma_1\sigma_2 & \sigma_2^2 \end{bmatrix} \qquad \text{where } \sigma_1 \geq 0,\ \sigma_2 \geq 0,\ \rho^2 \leq 1.$$

Furthermore every matrix of this form can arise as a covariance matrix, since we can define (X_1,X_2) by

$$X_1 = \sigma_1 Z_1 \qquad X_2 = \sigma_2[\rho Z_1 + \sqrt{1 - \rho^2}Z_2]$$

where Z_1 and Z_2 are assumed to be independent and both have zero mean and unit variance.

We have $\det C = \sigma_1^2\sigma_2^2(1 - \rho^2)$ so that C is nonsingular iff $\sigma_1 > 0, \sigma_2 > 0$, and $\rho^2 < 1$. Of course C is nonsingular iff C is positive definite, and iff (X_1, X_2) is nondegenerate. Furthermore every symmetric positive-definite 2×2 matrix is of this form since, as we see shortly in Lemma 3, all such matrices arise as covariance matrices. If C is nonsingular then

$$C^{-1} = \frac{1}{1 - \rho^2}\begin{bmatrix} 1/\sigma_1^2 & -\rho/\sigma_1\sigma_2 \\ -\rho/\sigma_1\sigma_2 & 1/\sigma_2^2 \end{bmatrix}$$

where the equation $CC^{-1} = I$ is easily checked. ∎

LINEAR PREDICTORS

Assume that we know the joint distribution of $k + 1$ random variables Y and $X_1, \ldots, X_k$. We desire some linear combination $Y_p = \beta_0 + \beta_1 X_1 + \cdots + \beta_k X_k$ of the X_i which can be used to predict Y with minimum expected squared error. Section 11-4 was devoted to $k = 1$. The following theorem shows that such Y_p are characterized by the property that the resulting prediction error $[Y - Y_p]$ has zero expectation and is uncorrelated with each of the variables $X_1, \ldots, X_k$ used for prediction. These conditions are easily expressed as linear equations with constants which are covariances. Thus for a solution to this problem only the means and covariances of $Y, X_1, \ldots, X_k$ need be known. For example, if $k = 2$ then $Y_p = \beta_0 + \beta_1 X_1 + \beta_2 X_2$ is such a best linear predictor iff β_1 and β_2 satisfy

$$(\operatorname{Var} X_1)\beta_1 + \operatorname{Cov}(X_1, X_2)\beta_2 = \operatorname{Cov}(X_1, Y)$$
$$\operatorname{Cov}(X_2, X_1)\beta_1 + (\operatorname{Var} X_2)\beta_2 = \operatorname{Cov}(X_2, Y)$$

while β_0 is determined by $EY_p = EY$. It turns out that if $(X_1, \ldots, X_k)$ is degenerate there is still a best linear predictor, but it is not unique. For example, if $\beta_0 + \beta_1 X_1$ is the best linear predictor using X_1, and if $X_2 = X_1$, then for every λ the predictor $\beta_0 + (\beta_1 - \lambda)X_1 + \lambda X_2$ is best among those using X_1 and X_2.

Theorem 1 Assume that the random variables Y and $X_1, \ldots, X_k$ with $k \geq 1$ have finite second moments. For *fixed* real $\beta_0, \ldots, \beta_k$ let

$$Y_p = \beta_0 + \sum_{i=1}^{k} \beta_i X_i.$$

Then the following three conditions are equivalent:
(*a*) Y_p is a best linear predictor for Y; that is,

$$E[Y - Y_p]^2 \leq E\left[Y - \left(b_0 + \sum_{i=1}^{k} b_i X_i\right)\right]^2 \qquad \text{for all real } b_0, \ldots, b_k.$$

(*b*) $E[Y - Y_p] = 0$ and $E[Y - Y_p]X_j = 0$ for $j = 1, \ldots, k$.
(*c*) $E[Y - Y_p] = 0$ and

$$C\begin{bmatrix} \beta_1 \\ \cdot \\ \cdot \\ \cdot \\ \beta_k \end{bmatrix} = \begin{bmatrix} \operatorname{Cov}(X_1, Y) \\ \cdot \\ \cdot \\ \cdot \\ \operatorname{Cov}(X_k, Y) \end{bmatrix}$$

where C is the covariance matrix for $(X_1, \ldots, X_k)$.
Furthermore there is a Y_p satisfying these conditions. Also this Y_p is unique iff C is nonsingular. ■

***Proof** Assume that condition (*a*) is satisfied. We deduce condition (*b*). If $\mu = E[Y - Y_p]$ then

$$E\{[Y - Y_p] - \mu\}^2 = \text{Var}\,[Y - Y_p] = E[Y - Y_p]^2 - \mu^2$$

so μ must equal zero, for otherwise $Y_p + \mu$ would be a better predictor than Y_p. Condition (*a*) implies that for all real λ we must have

$$E[Y - Y_p]^2 \le E\{[Y - Y_p] - \lambda X_j\}^2$$
$$= E[Y - Y_p]^2 - \lambda\{2E[Y - Y_p]X_j - \lambda EX_j^2\}$$

so that $\lambda\{\ \} \le 0$. If $E[Y - Y_p]X_j > 0$ then for small enough $\lambda > 0$ we have $\{\ \} > 0$, and hence $\lambda\{\ \} > 0$, which cannot occur. Similarly if $E[Y - Y_p]X_j < 0$ we can use small $\lambda < 0$. Thus condition (*a*) implies condition (*b*).

If condition (*b*) is true then for any real $b_0, \ldots, b_k$ let $V = \beta_0 - b_0 + \sum_{i=1}^{k} (\beta_i - b_i)X_i$ so that

$$E\left[Y - \left(b_0 + \sum_{i=1}^{k} b_i X_i\right)\right]^2 = E[Y - Y_p + V]^2$$
$$= E[Y - Y_p]^2 + 2E[Y - Y_p]V + EV^2$$
$$= E[Y - Y_p]^2 + EV^2 \ge E[Y - Y_p]^2$$

where by condition (*b*) the cross term equaled zero. Thus condition (*b*) implies condition (*a*).

Assume that $E[Y - Y_p] = 0$. If $Y' = Y - EY$ and $X_i' = X_i - EX_i$ then

$$E[Y - Y_p]X_j = E[Y - Y_p]X_j' = E[Y' - (Y_p - EY)]X_j'$$

$$= E\left[Y' - \sum_{i=1}^{k} \beta_i X_i'\right]X_j' = EY'X_j' - \sum_{i=1}^{k} (EX_j'X_i')\beta_i.$$

Thus $E[Y - Y_p]X_j = 0$ iff

$$\sum_{i=1}^{k} (EX_j'X_i')\beta_i = EX_j'Y'$$

showing that conditions (*b*) and (*c*) are equivalent. Thus we have shown that all three conditions are equivalent.

If C is nonsingular then obviously condition (*c*) has one and only one solution. If C is singular there is a $t \ne 0$ with $Ct = 0$. Thus if β satisfies condition (*c*) then so does $\beta + t$. Therefore if C is singular and if there is a Y_p satisfying the three conditions then there is more than one such Y_p.

It remains to prove that there is at least one best Y_p even if C is singular. [If C is singular there is, as is well known, a vector δ such that for no vector β does $C\beta = \delta$; we show that such δ cannot arise in the form of condition (*c*).] Note that if C is singular then $(X_1, \ldots, X_k)$ is degenerate. Therefore one of these random variables, say X_k, equals a constant plus some linear combination of $X_1, \ldots, X_{k-1}$, and making this substitution in any predictor using $X_1, \ldots, X_k$ yields a predictor using only $X_1, \ldots, X_{k-1}$ but having the same expected squared error. Now if $(X_1, \ldots, X_{k-1})$ is nondegenerate its covariance matrix is nonsingular, so there is a unique best predictor among those using $X_1, \ldots, X_{k-1}$, and this predictor will satisfy condition (*a*) for $X_1, \ldots, X_k$. If $(X_1, \ldots, X_{k-1})$ is degenerate we can similarly discard another variable, say X_{k-1}. We continue this process until we arrive at nondegenerate $(X_1, \ldots, X_r)$ with $1 \le r < k$, unless all X_i are degenerate, in which case $Y_p = EY$ is a best predictor. ■

Exercises 1 to 6 may be done before reading the remainder of this section.

EVERY SYMMETRIC POSITIVE-DEFINITE MATRIX IS A COVARIANCE MATRIX

Lemma 1 showed that every nondegenerate X has a covariance matrix which is positive definite. Suppose that we have some symmetric positive-definite matrix C. May we then construct a random vector having C for its covariance matrix? That is, does every such C arise as a covariance matrix? The answer is yes, but for the proof we need the next lemma, which is also useful in the next section.

Assume that S is some nonsingular $k \times k$ matrix and let $t \ne 0$ be some k vector. If $v = St$ then $v \ne 0$, so the square of the length of v is positive, hence

$$0 < \sum_{i=1}^{k} (v_i)^2 = v'v = (St)'(St) = t'S'St \qquad \text{for all } t \ne 0.$$

Thus for any nonsingular S the matrix $S'S$ is symmetric, since it equals its transpose, *and positive definite*. The following lemma shows that every symmetric positive-definite matrix arises in this way. Furthermore S may be restricted to be triangular.

Lemma 2 A real symmetric $k \times k$ matrix C is positive definite iff there exists a nonsingular real upper triangular $k \times k$ matrix

$$S = \begin{bmatrix} s_{11} & s_{12} & s_{13} & \cdots & s_{1k} \\ 0 & s_{22} & s_{23} & \cdots & s_{2k} \\ 0 & 0 & s_{33} & \cdots & s_{3k} \\ \cdots & \cdots & \cdots & \cdots & \cdots \\ 0 & 0 & \cdots & \cdots & s_{kk} \end{bmatrix} \qquad s_{ii} > 0;\ s_{ij} = 0 \text{ if } i > j$$

such that $C = S'S$. ■

***Proof** Because of earlier comments we need only prove that if C is symmetric positive definite then there exists an S of the exhibited form such that $C = S'S$; that is, there exists an S such that C and $S'S$ have the same quadratic form. In other words, there exist constants s_{ij} such that for all t we have

$$t'Ct = t'S'St = (St)'(St) = \cdots + (s_{k-1,k-1}t_{k-1} + s_{k-1,k}t_k)^2 + (s_{kk}t_k)^2.$$

Thus we must show that there are linear combinations such that

$$t'Ct = \sum_{i=1}^{k} [\text{linear combination of } t_i, t_{i+1}, \ldots, t_k]^2$$

for all $t_1, \ldots, t_k$. The coefficient of t_i in the ith linear combination must be positive. That is, $t'Ct$ can be expressed as the sum of the squares of k different linear combinations of $t_1, \ldots, t_k$, where the last involves only t_k, the next to last only t_{k-1} and t_k, etc.

The desired result is true if C is 1×1. We proceed by induction. Assume that for some integer $k - 1 \geq 1$ the result is true whenever C is $k - 1 \times k - 1$. Let C be a symmetric positive-definite $k \times k$ matrix with quadratic form

$$Q(t_1, \ldots, t_k) = t'Ct = \sum_{i=1}^{k} \sum_{j=1}^{k} c_{ij} t_i t_j.$$

Then $Q(1, 0, \ldots, 0) = c_{11}$ so c_{11} must be positive. The terms involving t_1 all come from the first row and the first column of C, and these terms are

$$c_{11} t_1^2 + 2t_1(c_{12}t_2 + c_{13}t_3 + \cdots + c_{1k}t_k) = c_{11}\left(t_1^2 + 2t_1 \sum_{r=2}^{k} \frac{c_{1r}}{c_{11}} t_r\right).$$

Completing the square of this quadratic in t_1 yields the *basic identity*

$$Q(t_1, \ldots, t_k) = c_{11}\left\{t_1 + \sum_{r=2}^{k} \frac{c_{1r}}{c_{11}} t_r\right\}^2 + Q_1(t_2, \ldots, t_k)$$

where Q_1 follows, although we do not need its explicit expression,

$$Q_1(t_2, \ldots, t_k) = \sum_{i=2}^{k} \sum_{j=2}^{k} c_{ij} t_i t_j - c_{11}\left(\sum_{r=2}^{k} \frac{c_{1r}}{c_{11}} t_r\right)^2 = \sum_{i=2}^{k} \sum_{j=2}^{k} \left(c_{ij} - \frac{c_{1i}c_{1j}}{c_{11}}\right) t_i t_j.$$

For any $t_2, \ldots, t_k$ not all zero we can define t_1 by setting $\{\ \}$, in the basic identity, equal to zero, making Q have the same value as Q_1, but this value must be positive because Q is positive definite. Therefore Q_1 is positive definite, so by the induction hypothesis

$$Q_1(t_2, \ldots, t_k) = \sum_{i=2}^{k} [\text{linear combination of } t_i, t_{i+1}, \ldots, t_k]^2$$

which, when substituted into the basic identity, yields the desired expression for Q. Thus there exists such an S with $C = S'S$. But C is nonsingular, and hence so is S. ∎

If $X' = (X_1, \ldots, X_k)$ is a random vector and T is a constant $k \times k$ matrix then $Y = TX$ is a random vector,

$$Y = \begin{bmatrix} Y_1 \\ \cdot \\ \cdot \\ \cdot \\ Y_k \end{bmatrix} = T \begin{bmatrix} X_1 \\ \cdot \\ \cdot \\ \cdot \\ X_k \end{bmatrix} = \begin{bmatrix} \sum_{j=1}^{k} t_{1j} X_j \\ \cdot \\ \cdot \\ \cdot \\ \sum_{j=1}^{k} t_{kj} X_j \end{bmatrix}$$

We want the covariance matrix for Y in terms of that for X. Clearly

$$\operatorname{Cov}(Y_i, Y_j) = E\left[\sum_{s=1}^{k} t_{is}(X_s - EX_s) \sum_{r=1}^{k} t_{jr}(X_r - EX_r)\right]$$

$$= \sum_{s=1}^{k} \sum_{r=1}^{k} t_{is} \operatorname{Cov}(X_s, X_r) t_{jr}.$$

For $k \times k$ matrices the (i,j) entry of the product TCA is easily seen to be the sum over s and r of $t_{is}c_{sr}a_{rj}$, which is what we have above if we let $A = T'$. *Therefore* if a random k vector X has covariance matrix C and if T is a $k \times k$ matrix with real entries then the random vector TX has covariance matrix TCT'. If T is nonsingular then TCT' is singular iff C is singular, and hence TX is degenerate iff X is degenerate. In particular if $C = I$ then TX has covariance matrix TT'.

Note that the same formula TCT' can be derived as follows, assuming for simplicity that $EX = 0$, and letting $Y = TX$:

$$EYY' = E(TX)(TX)' = ETXX'T' = T[E(XX')]T'.$$

Let C be any symmetric positive-definite $k \times k$ matrix of real numbers. From Lemma 2, with $T = S'$, there exists a *lower* triangular matrix T such that $C = TT'$. If X is any k-dimensional random vector having the identity matrix for its covariance matrix then TX has covariance matrix $TT' = C$. Therefore every symmetric positive-definite matrix arises as the covariance matrix of some random vector. We make this even more particularized as follows.

Lemma 3 If C is any real symmetric positive-definite $k \times k$ matrix then there exists a nonsingular real lower triangular matrix T such that $C = TT'$. Therefore C is the covariance matrix of TX where the components of $X' = (X_1, \ldots, X_k)$ are independent and each is normal 0 and 1. ■

EXERCISES

Many of the exercises in Sec. 11-4 are relevant to this section.

1. Specialize Theorem 1 to $k = 1$ and show that the best predictor is the same as in Sec. 11-4.

2. If Y_p in Theorem 1 is a best predictor for Y show that $aY_p + b$ is a best predictor for $aY + b$.

3. Show that the constant $Y_p = EY$ is a best predictor in Theorem 1 iff Y is uncorrelated with each of $X_1, \ldots, X_k$.

4. Without using determinants, prove that a $k \times k$ lower triangular matrix T is singular iff at least one of its diagonal entries is zero.

5. In Theorem 1 assume that C is nonsingular and let Y_p be the best predictor. Show that

$$E[Y - Y_p]^2 = \operatorname{Var} Y - \alpha' C^{-1} \alpha \qquad \text{where } \alpha = \begin{bmatrix} \operatorname{Cov}(X_1, Y) \\ \cdot \\ \cdot \\ \cdot \\ \operatorname{Cov}(X_k, Y) \end{bmatrix}.$$

6. In Theorem 1 assume that $\operatorname{Var} X_i > 0$ for $i = 1, \ldots, k$ so that we may introduce standardized X_i^*.

(*a*) Assume that

$$Y_p = \beta_0 + \sum_{i=1}^{k} \beta_i X_i^*$$

is a best predictor for Y in terms of $X_1^*, \ldots, X_k^*$. Show that if in this expression we replace each X_i^* by $(X_i - EX_i)/\sigma_{X_i}$ we have a best predictor for Y in terms of $X_1, \ldots, X_k$.

(*b*) Let $k = 2$ and assume that $\rho^2 = (EX_1^* X_2^*)^2 < 1$ so that (X_1, X_2) is nondegenerate. Let $\alpha_i = EX_i^* Y$ for $i = 1, 2$. Show that

$$Y_p = EY + \frac{1}{1 - \rho^2} [(\alpha_1 - \rho\alpha_2)X_1^* + (\alpha_2 - \rho\alpha_1)X_2^*]$$

is the best predictor for Y in terms of X_1^* and X_2^*.

(*c*) "Explain" the statement that the best predictor in terms of X_1^* and X_2^* may actually use X_2^*, with a nonzero coefficient, even though the best predictor in terms of X_2^* alone does not use X_2^*.

(*d*) Let $Y = X_1 + W + Z$ where X_1, W, Z are independent and each is standardized. Find the best predictor of Y in terms of X_1 in the sense of Theorem 1. Let $X_2 = X_1 + W$ and find the best predictor of Y in terms of X_1 and X_2. Make a true statement similar to the one in part (*c*).

7. A matrix P is orthogonal iff $PP' = I$; that is, P must be nonsingular and its inverse must equal its transpose. Let C be any $k \times k$ symmetric positive-definite matrix. Prove that there is an orthogonal matrix P and a diagonal matrix D such that C is the covariance matrix of

$$Y = PD \begin{bmatrix} X_1 \\ \cdot \\ \cdot \\ \cdot \\ X_k \end{bmatrix} = P \begin{bmatrix} d_1 X_1 \\ \cdot \\ \cdot \\ \cdot \\ d_k X_k \end{bmatrix}$$

where $X_1, \ldots, X_k$ are independent and each is normal 0 and 1. Therefore $P'Y$ has independent components. HINT: Recall the well-known fact that for any nonsingular T there are orthogonal P and Q, and diagonal D such that $T = PDQ$.

18-2 MULTIVARIATE NORMAL DENSITIES

This section is devoted to the definition and basic properties of the continuous normal densities on R^k. These densities form a very popular family of joint distributions incorporating dependence between coordinates. As we will see, they possess many unique properties which simplify their use. The first few paragraphs of Sec. 18-4 might well be read at this point, since they provide some theoretical explanation for the widespread occurrence in practical problems of distributions on R^k which are approximately normal.

Let $X' = (X_1, \ldots, X_k)$ be a random vector whose components $X_1, \ldots, X_k$ are independent and each distributed by the standardized normal 0 and 1 density φ, so that X has density

$$p_X(x) = p_X(x_1, \ldots, x_k) = \varphi(x_1) \cdots \varphi(x_k)$$

$$= \left(\frac{1}{\sqrt{2\pi}}\right)^k \exp\left\{-\frac{1}{2}\sum_{i=1}^{k} x_i^2\right\} = (2\pi)^{-k/2}e^{-x'x/2}$$

for all x in R^k. Let $Y = TX + \mu$, where T is a nonsingular $k \times k$ matrix with real entries and μ is a vector of k real numbers. Thus

$$\begin{bmatrix} Y_1 \\ \cdot \\ \cdot \\ \cdot \\ Y_k \end{bmatrix} = T \begin{bmatrix} X_1 \\ \cdot \\ \cdot \\ \cdot \\ X_k \end{bmatrix} + \begin{bmatrix} \mu_1 \\ \cdot \\ \cdot \\ \cdot \\ \mu_k \end{bmatrix}$$

so that $Y_i = \sum_{j=1}^{k} t_{ij}X_j + \mu_i$ for $i = 1, \ldots, k$. Clearly Y is nondegenerate and $EY = \mu$, and TT' is the positive-definite covariance matrix for Y. From Sec. 16-3 we know that $Y = TX + \mu$ had continuous density $p_Y\colon R^k \to R$ given by

$$p_Y(y) = \frac{1}{|\det T|} p_X(T^{-1}(y - \mu)) \qquad \text{where} \qquad (\det T)(\det T^{-1}) = 1.$$

We call these, and only these, densities the **normal densities** on R^k. That is, a random vector Y has a normal density iff there exist a T and μ as above such that Y has the same density as $TX + \mu$, where X is as above. Often we use the labels *multivariate* or *k-variate normal density* when emphasizing that $k > 1$ is permitted. Letting T be diagonal with $t_{ii} = \sigma_i \neq 0$ makes $Y_1, \ldots, Y_k$ independent with Y_i normal μ_i and σ_i^2.

We first note some immediate properties of normal densities.

Clearly Y has a multivariate normal density with mean μ iff there is a nonsingular T such that $T^{-1}(Y - \mu)$ has independent normal 0 and 1 components.

If Y is multivariate normal then so is any random vector obtained by permuting the coordinates of Y. That is, if (Y_1, Y_2, Y_3, Y_4) is normal then so is (Y_3, Y_1, Y_2, Y_4), since $Y = TX + \mu$ and we can permute the appropriate rows of T and μ to get this "permuted" random vector. If T is nonsingular then clearly so is any T_1 obtained by permuting rows of T, because if $T_1t = 0$ then $Tt = 0$ so $t = 0$.

If Y has a multivariate normal density then each of its component random variables has a normal marginal density, because from Example 4 (Sec. 16-5) we know that a sum of independent normal variables is also normal. (Exercise 12, or 13, shows the converse to be false.) More generally if $Y' = (Y_1, \ldots, Y_k)$ has a normal density then for real t_0, and real $t_1, \ldots, t_k$ not all zero, the random variable

$$Z = \sum_{i=1}^{k} t_iY_i + t_0 = \sum_{j=1}^{k}\left(\sum_{i=1}^{k} t_it_{ij}\right)X_j + \left(t_0 + \sum_{i=1}^{k} t_i\mu_i\right)$$

has a normal density. Thus for multivariate normal Y we can find the *density* of any linear combination Z of the coordinates of Y by merely finding the mean and variance of Z. For this we need only the mean and covariance matrix for Y, but we see shortly that these determine p_Y.

If Y has a k-variate normal density and if U is a nonsingular $k \times k$ matrix of real numbers and b is a vector of k reals then

$$UY + b = U(TX + \mu) + b = (UT)X + (U\mu + b)$$

has a normal density on R^k. Thus if U is nonsingular then $UY + b$ is normal iff Y is. In effect we have defined the collection of normal densities on R^k to be the smallest collection closed under such transformations, and containing the density $\varphi(x_1) \cdots \varphi(x_k)$.

In this book we have no real need to consider *degenerate* or *singular normal distributions*. However, they can be defined to be those distributions induced by $Y = TX + \mu$ just as above, but with T singular. For example, if $k = 1$ then $Y_1 = 0$ is degenerate normal. If $k = 2$ and $Y_1 = Y_2 = X_1$ then (Y_1, Y_2) is degenerate normal, and its distribution can be "described" by a normal density on the line $y_1 = y_2$ in the y_1y_2 plane. Naturally a degenerate distribution P on R^k cannot have a continuous density on R^k, since P is concentrated on a subspace of dimension $k - 1$ or smaller. If the class of normal distributions on R^k is enlarged to include all degenerate normal distributions on R^k then we obviously have closure under *all* linear transformations $T: R^k \to R^k$.

Let us examine the density of $Y = TX + \mu$ in greater detail. Now Y has covariance matrix $C = TT'$, so

$$\det C = (\det T)(\det T') = (\det T)^2 > 0$$

and hence $\det C$ is positive. Also $(\det C)(\det C^{-1}) = 1$, and hence

$$|\det T| = \sqrt{\det C} = \frac{1}{\sqrt{\det C^{-1}}}.$$

Furthermore $C = TT'$ is positive definite, and hence so is C^{-1} which can be expressed as

$$C^{-1} = (TT')^{-1} = (T')^{-1}T^{-1} = (T^{-1})'T^{-1}.$$

The exponent for p_Y is

$$-(\tfrac{1}{2})[T^{-1}(y - \mu)]'[T^{-1}(y - \mu)] = -(\tfrac{1}{2})(y - \mu)'(T^{-1})'T^{-1}(y - \mu).$$

Therefore if Y has a multivariate normal density with $EY = \mu$ and covariance matrix C then

$$p_Y(y) = \frac{\sqrt{\det C^{-1}}}{(2\pi)^{k/2}} e^{-(1/2)(y-\mu)'C^{-1}(y-\mu)} \qquad \text{where } (\det C)(\det C^{-1}) = 1.$$

Example 1 From the formula above and Example 1 (Sec. 18-1) we see that all bivariate normal densities have the form

$$p_{Y_1,Y_2}(y_1,y_2) = \frac{1}{2\pi\sigma_1\sigma_2\sqrt{1 - \rho^2}} e^{-(1/2)Q(y_1,y_2)} \qquad \text{for all real } y_1, y_2$$

where

$$Q(y_1,y_2) = \frac{1}{1-\rho^2}\left[\left(\frac{y_1-\mu_1}{\sigma_1}\right)^2 - 2\rho\frac{y_1-\mu_1}{\sigma_1}\frac{y_2-\mu_2}{\sigma_2} + \left(\frac{y_2-\mu_2}{\sigma_2}\right)^2\right].$$

The five real parameters have the interpretations

$$\mu_1 = EY_1 \qquad \mu_2 = EY_2 \qquad \sigma_1^2 = \text{Var } Y_1 \qquad \sigma_2^2 = \text{Var } Y_2 \qquad \rho = \frac{\text{Cov }(Y_1, Y_2)}{\sigma_1\sigma_2}$$

and must satisfy the restrictions $\sigma_1 > 0$, $\sigma_2 > 0$, and $\rho^2 < 1$, and only these restrictions.

Since C^{-1} is positive definite, Q achieves its unique minimum of zero at $y_1 = \mu_1$ and $y_2 = \mu_2$. The equation $Q(y_1,y_2) = \delta$ for fixed $\delta > 0$ determines an ellipse in the y_1y_2 plane, and a larger δ yields a larger ellipse with the same major and minor axes. Thus p_{Y_1,Y_2} is a bell-shaped surface whose "equiprobability curves" $p_{Y_1,Y_2}(y_1,y_2) = c$ are ellipses, one of which is shown in Fig. 1 where $\rho > 0$. The major axis of the ellipse is shown, along with the two regression lines which intersect the ellipse at points whose tangents are parallel to the axes.

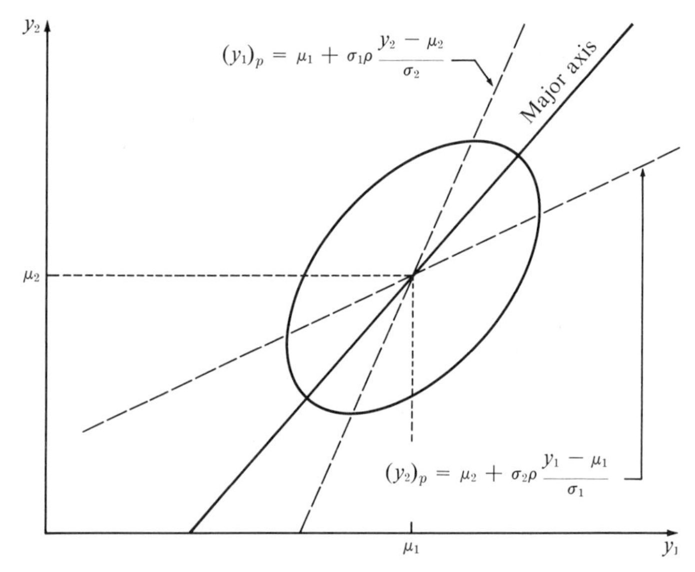

Fig. 1 Equiprobability ellipse for a bivariate normal density with $\rho > 0$.

For interpretation, Y_1 might be the height of a randomly selected man, while Y_2 is the height of his oldest son; or Y_1 and Y_2 might be voltages measured at the same time but at different locations in a communication or control system. ■

Thus p_Y has been exhibited in terms of parameters natural to Y, namely, $\mu = EY$ and the covariance matrix C of Y, or more precisely μ and C^{-1}. In particular we have the *basic fact* that two normal densities on R^k are equal if both densities have the same vector of means and the same covariance matrix. This fact and Lemma 3 (Sec. 18-1) show that we have a one-to-one correspondence between the family of normal densities on R^k and the set of pairs (μ,C) where μ is any k vector and C is any symmetric $k \times k$ positive-definite matrix.

Note that C and C^{-1} are symmetric and so are specified by only $k(k+1)/2$ real numbers. Thus p_Y has $k + k(k+1)/2$ natural parameters. However, T in $TX + \mu$ has k^2 entries constrained only to make T nonsingular. Thus many different T can give rise to the same p_Y, but the "extra parameters" in T do not determine different densities. For example, both of the following T matrices determine the same covariance matrix, and hence the same density for TX:

$$\begin{bmatrix} 1 & 0 \\ 0 & 1 \end{bmatrix} \qquad \frac{1}{\sqrt{2}}\begin{bmatrix} 1 & 1 \\ 1 & -1 \end{bmatrix}$$

If $Y = (Y_1, Y_2, Y_3)$ has some normal density on R^3 but we know only p_{Y_1,Y_2}, p_{Y_1,Y_3}, p_{Y_2,Y_3} then we can find μ and C for Y from these, and hence find p_Y. In other words, every multivariate normal density is uniquely determined by the collection of all joint densities of pairs of its coordinates. Thus *if* we are justified in assuming that we have a multivariate normal density then this joint density is determined by the bivariate marginals, and actually only the means and covariances are needed.

Independent random variables are uncorrelated. For joint normal densities the converse is also true. In other words, if Y is k-variate normal, and each of its pairs of coordinates are uncorrelated, then

$$C = \begin{bmatrix} \sigma_1^2 & & 0 \\ & \ddots & \\ 0 & & \sigma_k^2 \end{bmatrix} \qquad C^{-1} = \begin{bmatrix} 1/\sigma_1^2 & & 0 \\ & \ddots & \\ 0 & & 1/\sigma_k^2 \end{bmatrix}$$

$$\sqrt{\det C^{-1}} = \frac{1}{\sigma_1\sigma_2 \cdots \sigma_k} \qquad p_Y(y_1, \ldots, y_k) = \prod_{i=1}^{k} \varphi_{\mu_i,\sigma_i}(y_i).$$

We used the fact that the determinant of a triangular matrix equals the product of the diagonal entries. Thus if random variables $Y_1, \ldots, Y_k$ have a joint multivariate normal density on R^k then $Y_1, \ldots, Y_k$ are mutually independent random variables iff they are pairwise uncorrelated,

$$\text{Cov}(Y_i, Y_j) = 0 \qquad \text{whenever } 1 \le i < j \le k.$$

For a simpler proof just observe that if the coordinates of Y are uncorrelated then Y has the same covariance matrix as the normal density having the same coordinate marginals independent. In Exercise 5 this property is generalized to independence of jointly normal random *vectors*.

Every marginal density from a multivariate normal density is also a normal density. Since permutation of coordinates does not destroy normality, this reduces to the following. If $Y' = (Y_1, \ldots, Y_k)$ has a normal density then so does $Z' = (Y_1, \ldots, Y_r)$ whenever $1 \le r \le k$. This fundamental fact has already been observed when $r = 1$. For proof we note that if Y has $EY = \mu$ and covariance matrix C then from Lemma 3 (Sec. 18-1) there is a lower triangular T such that $C = TT'$. Hence Y has the same density as $TX + \mu$ *with this* T. Therefore we may assume that $Y = TX + \mu$ with this

T. But then the first r equations from

$$\begin{bmatrix} Y_1 \\ Y \\ \cdot \\ \cdot \\ \cdot \\ Y_k \end{bmatrix} = \begin{bmatrix} t_{11}X_1 \\ t_{21}X_1 + t_{22}X_2 \\ \cdot \\ \cdot \\ \cdot \\ t_{k1}X_1 + \cdots + t_{kk}X_k \end{bmatrix} + \begin{bmatrix} \mu_1 \\ \mu_2 \\ \cdot \\ \cdot \\ \cdot \\ \mu_k \end{bmatrix}$$

can be written as $Z = T_r(X_1, \ldots, X_r)' + (\mu_1, \ldots, \mu_r)'$. But T_r is triangular with positive diagonal entries. Hence T_r is nonsingular, so Z must be normal from the very definition of normal densities.

It may be helpful to indicate the crucial role played in our construction by the unique normal density $\varphi: R \to R$. Suppose that random variable X_1 has some distribution P_1 satisfying $EX_1 = 0$ and $EX_1^2 = 1$ but P_1 does *not* correspond to the normal density φ. For each positive integer k construct a family of distributions on R^k, just as above by $TX + \mu$ except that now each of the independent X_i has distribution P_1. Clearly the resulting family of distributions on R^k is closed under permutations of coordinates and nonsingular linear transformations on R^k. However, Exercise 11 shows that now some one-dimensional marginals do *not* belong to the $k = 1$ family, and the mean and covariance matrix do *not* determine the distribution. The proofs follow easily from the property that normal densities are unique in being closed under scale changes and independent additions, as precisely stated after Example 5 (Sec. 16-5). Of course if we do not require P_1 to have a finite variance we can let P_1 correspond to a Cauchy density, and then every one-dimensional marginal is also Cauchy, but means and covariances do not exist.

All exercises except 9 and 10 may be done before reading the remainder of this section.

If $Y' = (Y_1, \ldots, Y_k)$ is normal then the conditional density of any coordinate random variable, given the others, is also normal. More explicitly let $Y' = (Y_1, \ldots, Y_k)$ with $k \geq 2$ be normal with mean μ and covariance matrix C. If $D = C^{-1}$ then the conditional density of Y_1 given $Y_2, \ldots, Y_k$ is normal with

$$[\text{conditional mean}] = f(y_2, \ldots, y_k) = \mu_1 - \sum_{r=2}^{k} \frac{d_{1r}}{d_{11}} (y_r - \mu_r)$$

$$[\text{conditional variance}] = \frac{1}{d_{11}}.$$

For proof, as in the proof of Lemma 2 (Sec. 18-1), we have

$$p_Y(y_1, \ldots, y_k) = Ke^{-(d_{11}/2)\{y_1 - f\}^2 - Q_1(y_2, \ldots, y_k)}$$

with $f = f(y_2, \ldots, y_k)$ as above, while K is as before. The integral over R of any normal density on R is 1, so that

$$\int_{-\infty}^{\infty} \sqrt{\frac{d_{11}}{2\pi}}\, e^{-(d_{11}/2)\{y_1 - f\}^2}\, dy_1 = 1$$

for any fixed f, and hence

$$p_{Y_2,\ldots,Y_k}(y_2, \ldots, y_k) = \int_{-\infty}^{\infty} p_Y(y_1, \ldots, y_k)\, dy_1 = K\sqrt{\frac{2\pi}{d_{11}}}\, e^{-Q_1(y_2,\ldots,y_k)}.$$

Therefore the conditional density, which is defined [as in Section 16-6] by the first equality below, reduces to

$$p_{Y_1|Y_2,\ldots,Y_k}(y_1 \mid y_2, \ldots, y_k) = \frac{p_{Y_1,\ldots,Y_k}(y_1, \ldots, y_k)}{p_{Y_2,\ldots,Y_k}(y_2, \ldots, y_k)} = \sqrt{\frac{d_{11}}{2\pi}}\, e^{-(d_{11}/2)\{y_1-f\}^2}$$

as was asserted. Thus the conditional variance is a constant and the conditional mean is a constant plus a linear combination of $y_2, \ldots, y_k$.

As shown in Sec. 14-1, the function $g: R^{k-1} \to R$ which minimizes $E[Y_1 - g(Y_2, \ldots, Y_k)]^2$ is just the conditional expectation of Y_1 given $Y_2, \ldots, Y_k$. For "most" joint distributions the function g is complicated and nonlinear. However, if $Y' = (Y_1, \ldots, Y_k)$ is multivariate normal then the minimizing g is the f above, and the conditional distribution of the error $Y_1 - f(Y_2, \ldots, Y_k)$, given $Y_2 = y_2, \ldots, Y_k = y_k$, is normal with mean 0 and constant variance $1/d_{11}$. Therefore if $Y' = (Y_1, \ldots, Y_k)$ has a multivariate normal density then the minimum expected-squared-error predictor of Y_1 in terms of $Y_2, \ldots, Y_k$ is just the simple $f(Y_2, \ldots, Y_k)$ above, and the resulting expected squared error is

$$E[Y_1 - f(Y_2, \ldots, Y_k)]^2 = \frac{1}{d_{11}}.$$

In particular the best mean-squared predictor is the same as the best mean-squared *linear* predictor, of the kind considered in Theorem 1 (Sec. 18-1).

Example 2 By specializing the preceding results to $k = 2$, or by simple direct manipulation, we find that

$$p_{Y_1|Y_2}(y_1 \mid y_2) = \frac{p_{Y_1,Y_2}(y_1,y_2)}{p_{Y_2}(y_2)}$$

$$= \frac{1}{\sqrt{2\pi\sigma_1^2(1-\rho^2)}} \exp\left(\frac{-\left\{y_1 - \left[\mu_1 + \rho\dfrac{\sigma_1}{\sigma_2}(y_2 - \mu_2)\right]\right\}^2}{2\sigma_1^2(1-\rho)^2}\right)$$

Thus the conditional density of Y_1 given $Y_2 = y_2$ is normal with $\mu_1 + \rho(\sigma_1/\sigma_2)(y_2 - \mu_2)$ for conditional mean and $\sigma_1^2(1 - \rho^2)$ for conditional variance. Therefore the function $g: R \to R$ which minimizes $E[Y_1 - g(Y_2)]^2$ is just the linear function

$$g(y_2) = \mu_1 + \rho\frac{\sigma_1}{\sigma_2}(y_2 - \mu_2) \qquad \text{for all } y_2,$$

and for this function $E[Y_1 - g(Y_2)]^2 = \sigma_1^2(1 - \rho^2)$. These conditional densities are interpreted in Example 5 and Fig. 2 of Sec. 16-6. ∎

The **moment generating function of a random vector** $Y' = (Y_1, \ldots, Y_k)$ is the function

$$m_{Y_1,\ldots,Y_k}(t_1, \ldots, t_k) = Ee^{t_1Y_1+\cdots+t_kY_k}$$

which is defined for all real $t_1, \ldots, t_k$ for which this expectation is finite. Clearly $m_Y(0, \ldots, 0) = 1$ and $m_Y(1, 0, \ldots, 0) = Ee^{t_1 Y_1}$ so that the MGFs for marginals are obtainable from the joint MGF m_Y. If $k = 2$, for example, then Theorem 3h (Sec. 11-3) yields

$$Ee^{t_1 Y_1 + t_2 Y_2} = Ee^{t_1 Y_1} e^{t_2 Y_2} \leq (Ee^{2t_1 Y_1} Ee^{2t_2 Y_2})^{1/2}$$

so that finiteness of a joint MGF can be inferred from the finiteness of the MFGs of marginals for appropriate argument values. Once again the MFG determines the distribution, in the same sense as before. More precisely two random vectors $X' = (X_1, \ldots, X_k)$ and $Y' = (Y_1, \ldots, Y_k)$ are identically distributed if there exists a real $t_0 > 0$ such that

$$Ee^{t_1 X_1 + \cdots + t_k X_k} \qquad \text{and} \qquad Ee^{t_1 Y_1 + \cdots + t_k Y_k}$$

are finite and equal whenever $|t_i| \leq t_0$ for $i = 1, \ldots, k$.

As observed in Example 4 (Sec. 16-5), if a random variable Z has a normal density then

$$Ee^Z = e^{EZ + (1/2) \operatorname{Var} Z}$$

Therefore if $Y' = (Y_1, \ldots, Y_k)$ has a normal density with mean $\mu' = (\mu_1, \ldots, \mu_k)$ and covariance matrix C then Y has MFG

$$m_{Y_1, \ldots, Y_k}(t_1, \ldots, t_k) = e^{t'\mu + (1/2) t' C t} \qquad \text{where } t = \begin{bmatrix} t_1 \\ \cdot \\ \cdot \\ \cdot \\ t_k \end{bmatrix}$$

for all real $t_1, \ldots, t_k$; since if $Z = t_1 Y_1 + \cdots + t_k Y_k$ then Z is normal. For $k = 2$ we obtain

$$m_{Y_1, Y_2}(t_1, t_2) = \exp \{t_1 \mu_1 + t_2 \mu_2 + (1/2)(\sigma_1{}^2 t_1 + 2\rho\sigma_1\sigma_2 t_1 t_2 + \sigma_2{}^2 t_2{}^2)\}.$$

Naturally many properties of normal densities are easily obtained by use of MGFs. For example, two k-variate normal densities must be equal if they both have the same means and covariances, since then their MGFs are equal. Furthermore if $Y' = (Y_1, \ldots, Y_k)$ is normal and has pairwise uncorrelated components then C is diagonal, so

$$m_Y(t_1, \ldots, t_k) = \exp \left\{ \sum_{i=1}^{k} (t_i \mu_i + (1/2)\sigma_i{}^2 t_i{}^2) \right\} = \prod_{i=1}^{k} m_{Y_i}(t_1)$$

hence $Y_1, \ldots, Y_k$ must be independent, since m_Y equals the joint MGF for independent normals.

EXERCISES

1. Let (X, Y) have a normal density with $EX = EY$ and $EX^2 = EY^2$. Show that $X + Y$ and $X - Y$ are independent.

2. Let Y_1 and Y_2 have a bivariate normal density with $EY_1 = EY_2 = 0$ and correlation coefficient ρ.

(*a*) Prove that

$$P[Y_1 \geq 0,\ Y_2 \geq 0] = \tfrac{1}{4} + \frac{\arcsin \rho}{2\pi}.$$

HINT: Introduce (Y_1, Y_2) by $(Y_1/\sigma_1) = X_1$ and $(Y_2/\sigma_2) = \rho X_1 + \sqrt{1-\rho^2}X_2$ where X_1 and X_2 are independent and each normal 0 and 1, so that p_{X_1,X_2} is radially symmetric. Consider Fig. 2.

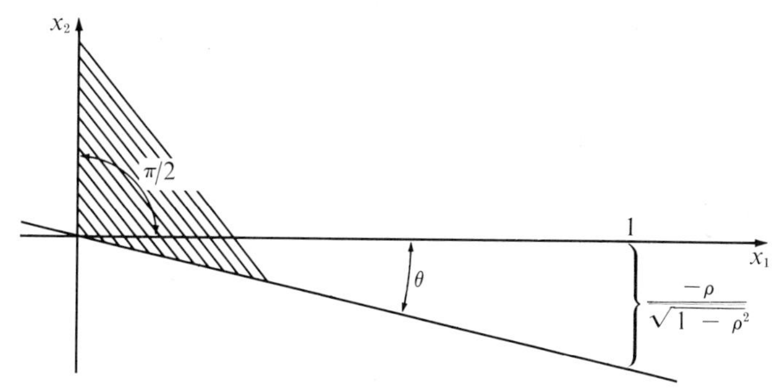

Fig. 2

(*b*) Plot this function of ρ on the interval $-1 < \rho < 1$ and relate its values for $\rho = 0$ and $\rho \doteq \pm 1$ to the surface p_{Y_1,Y_2}.

(*c*) Find $P[Y_1 Y_2 \geq 0]$ and $P[Y_1 Y_2 \leq 0]$ in terms of ρ. Evaluate when $\rho = 1/\sqrt{2}$.

3. Let μ be any k vector, let D be any symmetric positive-definite $k \times k$ matrix, and define $f\colon R^k \to R$ by

$$f(y) = e^{-(\frac{1}{2})(y-\mu)'D(y-\mu)}.$$

Prove that the integral of this positive-valued function f over R^k is finite. Hence there is a unique $K > 0$ such that Kf is a probability density. More precisely, prove the stronger result that there is a unique normal density with covariance matrix C such that $C^{-1} = D$ so that Kf is this normal density if

$$K = \frac{\sqrt{\det D}}{(2\pi)^{k/2}}$$

Thus not only is every normal density of the form Kf with f as above, but also every function of the form f is integrable and has a unique K which makes Kf a normal density.

4. Let $Y' = (Y_1, \ldots, Y_k)$ have a normal density with mean $\mu = EY$ and covariance matrix C. Then the quadratic form $(y - \mu)'C^{-1}(y - \mu)$ appears in the exponent of p_Y. Show that the random variable $(Y - \mu)'C^{-1}(Y - \mu)$, which is used in statistical theory, has the same distribution as the random variable $X_1^2 + \cdots + X_k^2$ where $X_1, \ldots, X_k$ are independent and each is normal 0 and 1. Example 3 (Sec. 16-4) called this the χ-square density with k degrees of freedom. HINT: $Y = TX + \mu$ and $C = TT'$.

5. Let $Y' = (Y_1, \ldots, Y_k)$ have a normal density and let $1 \leq r < k$. Prove that $(Y_1, \ldots, Y_r)$ and $(Y_{r+1}, \ldots, Y_k)$ are two independent random vectors iff they are componentwise uncorrelated,

$$\operatorname{Cov}(Y_i, Y_j) = 0 \qquad \text{whenever } 1 \leq i \leq r < j \leq k.$$

HINT: Each of the two random vectors has a normal density and hence a corresponding T matrix. Construct a $k \times k$ matrix from these two matrices.

6. This exercise proves a unique property of normal densities which is of importance in statistical theory. Let $Y = (X_1 + \cdots + X_k)/k$ where $X_1, \ldots, X_k$ are independent and each is normal 0 and 1. Prove that Y is independent of

$$Z = \sum_{i=1}^{k} (X_i - Y)^2.$$

HINT: The two transformations

$$(X_1, \ldots, X_k) \to (Y, X_2, \ldots, X_k) \to (Y, X_2 - Y, \ldots, X_k - Y)$$

are triangular and nonsingular. Furthermore

$$X_1 - Y = (kY - \sum_{i=2}^{k} X_i) - Y = \sum_{i=2}^{k} (Y - X_i)$$

so that Z can be expressed as a function of the random vector $(X_2 - Y, \ldots, X_k - Y)$.

7. Use the formula

$$p_{Y_1+Y_2}(z) = \int_{-\infty}^{\infty} p_{Y_1,Y_2}(y_1, z - y_1)\, dy_1$$

which was derived in Example 3 (Sec. 16-3), to show that if (Y_1, Y_2) has the bivariate normal density with parameters $\mu_1 = \mu_2 = 0$ and $\sigma_1 = \sigma_2 = 1$ and $-1 < \rho < 1$, then $Y_1 + Y_2$ is normal 0 and $2(1 + \rho)$. Note that it immediately follows that $EY_1Y_2 = \rho$.

8. Let X_1 and X_2 be independent and each normal 0 and 1. Then

$$\begin{bmatrix} Y_1 \\ Y_2 \end{bmatrix} = \begin{bmatrix} 1 & 0 \\ b & 0 \end{bmatrix} \begin{bmatrix} X_1 \\ X_2 \end{bmatrix} = \begin{bmatrix} X_1 \\ bX_1 \end{bmatrix}$$

has a degenerate normal distribution in the plane. Let $F_b: R^2 \to R$ be the distribution function for (X_1, bX_1).

(*a*) Sketch F_1 and briefly explain why F_b is continuous everywhere on R^2 if $b \neq 0$.

(*b*) Sketch F_0 and describe the set of points in the plane at which F_0 is continuous.

9. Let $Y' = (Y_1, \ldots, Y_k)$ have a normal density with MGF $m(t_1, \ldots, t_k)$. Use MGFs to prove the following:

(*a*) $Y_1 + \cdots + Y_k$ has a normal density.

(*b*) $(Y_1, \ldots, Y_r)$ has a normal density whenever $1 \leq r \leq k$.

HINT: Consider $m(t, \ldots, t)$ and $m(t_1, \ldots, t_r, 0, \ldots, 0)$.

10. Assume that $(Y_1, \ldots, Y_4)$ has a normal density with $EY_i = 0$ for $i = 1, \ldots, 4$. The density itself is expressible in terms of just covariances EY_iY_j, and hence so is every possible moment. In particular show that

$$EY_1Y_2Y_3Y_4 = (EY_1Y_2)(EY_3Y_4) + (EY_1Y_3)(EY_2Y_4) + (EY_1Y_4)(EY_2Y_3)$$

HINT: Utilizing the joint MGF, we have

$$EY_1Y_2Y_3Y_4 = \left[\frac{\partial^4}{\partial t_1\, \partial t_2\, \partial t_3\, \partial t_4} Ee^{t_1Y_1+t_2Y_2+t_3Y_3+t_4Y_4}\right]_{t_1=t_2=t_3=t_4=0}$$

Abbreviate $\sum_{i=1}^{4} t_i Y_i$ to $\sum$ and then expand the MGF in a power series,

$$Ee^{\Sigma} = e^{(\frac{1}{2})E(\Sigma)^2} = 1 + (\tfrac{1}{2})E(\textstyle\sum)^2 + \frac{[(\frac{1}{2})E(\sum)^2]^2}{2} + \cdots$$

and note that only the last term exhibited will make a final contribution.

11. *The Unique Role of the Normal Density* φ. Let the random variable X_1 have some distribution P_1 satisfying $EX_1 = 0$ and $EX_1^2 = 1$. Assume that P_1 does *not* correspond to the normal density φ. Let $X' = (X_1, \ldots, X_k)$ where $X_1, \ldots, X_k$ are independent and each has distribution P_1. Assume that $k \geq 2$. Exhibit a nonsingular $k \times k$ matrix T such that $Y = TX$ has the identity matrix for its covariance matrix but $P_{Y_1} \neq P_1$, where $Y' = (Y_1, \ldots, Y_k)$. HINT: The unique property of the normal density proved in Exercise 2 (Sec. 16-5) implies that if X_1 and X_2 are independent and each has distribution P_1 as above then $(X_1 + X_2)/\sqrt{2}$ does not have distribution P_1.

COMMENT: Clearly $EY_1 = 0$ and $EY_1^2 = 1$, so that the marginal distribution P_{Y_1} does not belong to the family of distributions induced by $\lambda X_1 + \mu$ for various λ and μ. Also X and Y have the same covariance matrix but different distributions. Thus if we use any such P_1 in the construction of this section, we lose closure under the taking of marginals and we lose the one-to-one correspondence between distributions and pairs (μ, C).

Exercises 12 and 13 exhibit cases where (X, Y) has a nonnormal distribution in R^2 even though each of X and Y is normal 0 and 1. For such distributions X and Y may be uncorrelated but dependent, and $X + Y$ may not be normal.

12. Use Exercise 2 (Sec. 16-4) to construct a continuous density $p_{X,Y}$ which is *not* normal but is such that each of X and Y is normal 0 and 1.

13. Let $Y = ZX$ where Z and X are independent, and $P[Z = -1] = P[Z = 1] = .5$ while X is normal 0 and 1. Show that Y is normal 0 and 1 and that X and Y are uncorrelated but dependent. Describe the joint distribution of (X, Y) and find the distribution of $X + Y$.

14. Let $(X_1, Y_1), \ldots, (X_n, Y_n)$ be n independent random vectors, each having the same bivariate normal density. Let $Z_n = \sum_{i=1}^{n} X_i$ and $W_n = \sum_{i=1}^{n} Y_i$. Show that (Z_n, W_n) has a bivariate normal density, as was mentioned in connection with Exercises 17 (Sec. 16-5).

18-3 CHARACTERISTIC FUNCTIONS

This section does not presume Secs. 18-1 and 18-2. In this section complex-valued random variables and their expectations are introduced. Their primary usefulness is in connection with the characteristic function of a real-valued random variable X, which is the complex-valued function of real t defined by

$$Ee^{itX} = E[\cos(tX)] + iE[\sin(tX)] \qquad \text{for all real } t.$$

Thus the characteristic function is defined formally in the same way as the MGF, Ee^{tX}, except that real t is replaced by imaginary it. However, *every* random variable has a characteristic function defined for *all real t, and* random variables having the same characteristic function are identically distributed. Characteristic functions are used extensively in probability theory. The characteristic function for a sum of n independent random variables is easily obtained as the product of the individual characteristic functions. Often such an easily obtained characteristic function can be shown to converge to a known characteristic function as n converges to infinity. The basic continuity theorem for characteristic functions then permits us to conclude that the corresponding distribution functions also converge. This technique yields central-limit theorems and limiting distributions in many other problems. In order to find the limiting characteristic function it is often necessary to use a polynomial approximation to a characteristic function. The facts indicated above are stated precisely below, and *some* of them are proved. Their usefulness is then indicated by a CLT, and in the next section by a vector CLT.

COMPLEX-VALUED RANDOM VARIABLES

If the expectation of a random vector is defined componentwise, $E(X_1, \ldots, X_k) = (EX_1, \ldots, EX_k)$, then we still have linearity of expectation. That is, if each component has an expectation and a and b are real then

$$\begin{aligned} E[a(X_1, \ldots, X_k) + b(Y_1, \ldots, Y_k)] &= E(aX_1 + bY_1, \ldots, aX_k + bY_k) \\ &= (aEX_1 + bEY_1, \ldots, aEX_k + bEY_k) \\ &= a(EX_1, \ldots, EX_k) + b(EY_1, \ldots, EY_k). \end{aligned}$$

If $k = 2$ we can interpret values of random vectors as complex numbers, with all their desirable properties, and so introduce *multiplication of these random vectors.* For complex numbers we use both notations (x,y) and $x + iy$, where x and y are real. Multiplication is defined by using $i^2 = -1$ as usual,

$$\begin{aligned} (x,y)(x',y') = (x + iy)(x' + iy') &= xx' - yy' + i(xy' + yx') \\ &= (xx' - yy', xy' + yx'). \end{aligned}$$

Thus when $k = 2$ we can multiply two random vectors, one of which may be a constant $a + ib$. Of course multiplication by a constant is componentwise, $a(x,y) = (ax,ay)$ when a is real, although this is not necessarily so in general.

We next introduce the obvious ideas for this case. We have no need for the common generalization to random vectors whose components are complex valued.

Let C denote the **set of all complex numbers** $z = (x,y) = x + iy$. A function $Z: \Omega \to C$ with associated probability space (Ω,P) is called a **complex-valued random variable.** *We continue to reserve the name "random variable" for real-valued functions.* Thus $Z = X + iY$, where X and Y are random variables. By the induced distribution of Z we mean the induced distribution of the random vector (X, Y) in the plane. We say that complex-valued random variables $Z_1, \ldots, Z_n$ are **independent** iff $(X_1, Y_1), \ldots, (X_n, Y_n)$ are n independent random vectors. We define **expectation** by

$$E(X + iY) = E(X) + iE(Y) \qquad \text{if } E|X| < \infty,\ E|Y| < \infty.$$

In other words,

$$\begin{aligned} &E[\text{the complex-valued random variable } Z] \\ &\qquad = E[\text{the real part of } Z] + iE[\text{the imaginary part of } Z] \end{aligned}$$

so we obtain the complex number EZ by taking the expectation of two random variables. Note that if $P[Z = i] = P[Z = -i] = .5$ then $P[Z^2 = -1] = 1$, and hence $EZ^2 = -1$, so that second moments and variances can be negative. If each Z_i has an expectation then the earlier proof yields $E\sum_{k=1}^{n} Z_k = \sum_{k=1}^{n} EZ_k$.

Furthermore if $Z_1 = X_1 + iY_1$ and $Z_2 = X_2 + iY_2$ are independent then each of X_1 and Y_1 is independent from each of X_2 and Y_2, yielding the circled equality in the following chain:

$$\begin{aligned}
E(Z_1Z_2) &= E(X_1 + iY_1)(X_2 + iY_2) \\
&= E[(X_1X_2 - Y_1Y_2) + i(X_1Y_2 + Y_1X_2)] \\
&= E(X_1X_2 - Y_1Y_2) + iE(X_1Y_2 + Y_1X_2) \\
&= EX_1X_2 - EY_1Y_2 + i(EX_1Y_2 + EY_1X_2) \\
&\ominus EX_1EX_2 - EY_1EY_2 + i(EX_1EY_2 + EY_1EX_2) \\
&= (EX_1 + iEY_1)(EX_2 + iEY_2) = (EZ_1)(EZ_2).
\end{aligned}$$

Thus if Z_1 and Z_2 are independent and have expectations then Z_1Z_2 has an expectation and $E(Z_1Z_2) = (EZ_1)(EZ_2)$. The proof can be specialized to let $Z_1 = c = a + ib$ be a complex constant, showing that $EcZ = cEZ$.

If $Z = X + iY$ then $|Z| = (X^2 + Y^2)^{1/2}$ is a random variable. But for any complex number $z = x + iy$ we have $|x| \le |z|$, $|y| \le |z|$, and $|z| \le |x| + |y|$. Therefore Z has an expectation iff $|Z|$ does. Furthermore if $E|Z| < \infty$ we prove $|EZ| < E|Z|$ as follows. Certainly there is a real number b such that $EZ = |EZ|\, e^{ib}$ and there is a random variable θ such that $Z = |Z|\, e^{i\theta}$. But $|EZ|$ has zero imaginary part, yielding the last equality in the chain

$$\begin{aligned}
|EZ| &= e^{-ib}EZ = E\,|Z|\, e^{i(\theta-b)} \\
&= E[|Z| \cos(\theta - b) + i\,|Z| \sin(\theta - b)] \\
&= E[|Z| \cos(\theta - b)] \le E\,|Z|
\end{aligned}$$

as was asserted.

We briefly summarize.

Summary of complex-valued random variables If X and Y are random variables then $Z = (X, Y) = X + iY$ is a **complex-valued random variable,** and $Z\colon \Omega \to C$ has associated probability space (Ω, P), and $|Z| = (X^2 + Y^2)^{1/2}$ is a random variable. By the distribution induced by Z we mean the distribution induced by the random vector (X, Y). We say that Z has an **expectation** iff $E|Z| < \infty$, or equivalently iff $E|X| < \infty$ and $E|Y| < \infty$. In this case we let $EZ = EX + iEY$ and we have $|EZ| \le E|Z|$.

If $c_1, \ldots, c_n$ are complex numbers, and $Z_1, \ldots, Z_n$ are complex-valued random variables, and each $E|Z_k| < \infty$ then the following sum has an expectation and we have linearity.

$$E\left(\sum_{k=1}^{n} c_k Z_k\right) = \sum_{k=1}^{n} c_k EZ_k$$

If $Z_1, \ldots, Z_n$ with $Z_k = X_k + iY_k$ are n *independent* complex-valued random variables, by which we mean that $(X_1, Y_1), \ldots, (X_n, Y_n)$ are n independent random vectors, and if each Z_k has an expectation, then so does their product, and we have

$$E\left(\prod_{i=1}^{n} Z_k\right) = \prod_{i=1}^{n} EZ_k \quad \blacksquare$$

CHARACTERISTIC FUNCTIONS: DEFINITION AND GENERAL PROPERTIES

If X is a random variable and t is a real number then

$$e^{itX} = \cos tX + i \sin tX = (\cos tX, \sin tX)$$

is a complex-valued random variable whose distribution happens to be concentrated on the unit circle in the plane, $|e^{itX}| = 1$. If X is a random variable then the function $\varphi: R \to C$ defined for all real t by

$$\varphi(t) = Ee^{itX} = E(\cos tX) + iE(\sin tX)$$

is called the **characteristic function of** X, or of its induced distribution or distribution function. Clearly $\varphi(0) = 1$, and $|e^{itX}| = 1$ so that $|\varphi(t)| \le 1$ for all t. Thus *every* random variable has a characteristic function defined for *all real t*. *These, and only these, functions are called characteristic functions*. For real a and b we have

$$\varphi_{aX+b}(t) = Ee^{it(aX+b)} = e^{itb}Ee^{itaX} = e^{itb}\varphi_X(ta).$$

It is shown in Exercise 5 that the obvious inequality

$$|\varphi(t + h) - \varphi(t)| \le E\,|e^{i(t+h)X} - e^{itX}| = E\,|e^{ihX} - 1|$$

implies that every characteristic function is uniformly continuous, since the bound must be small for h small. If $S_n = X_1 + \cdots + X_n$ where $X_1, \ldots, X_n$ are any $n \ge 2$ independent random variables, then $e^{itX_1}, \ldots, e^{itX_n}$ are obviously n independent complex-valued random variables, so that

$$\varphi_{S_n}(t) = Ee^{tS_n} = E\left(\prod_{k=1}^{n} e^{itX_k}\right) = \prod_{k=1}^{n} Ee^{itX_k} = \prod_{k=1}^{n} \varphi_{X_k}(t) \qquad \text{for all real } t.$$

Thus the characteristic function for the sum of *independent* random variables is just the product of their characteristic functions. For discrete and continuous densities $p: R \to R$ the characteristic function becomes

$$\varphi(t) = \sum_x e^{itx}p(x) = \sum_x (\cos tx)p(x) + i \sum_x (\sin tx)p(x)$$

$$\varphi(t) = \int_{-\infty}^{\infty} e^{itx}p(x)\,dx = \int_{-\infty}^{\infty} (\cos tx)p(x)\,dx + i\int_{-\infty}^{\infty} (\sin tx)p(x)\,dx.$$

Two random variables are identically distributed iff they have the same characteristic function. Thus if $\varphi_X(t) = \varphi_Y(t)$ for all real t then $P_X = P_Y$. This important fact is used often. We only *indicate* the method of proof. If φ is the characteristic function of a continuous density p then under quite general conditions p can be obtained from φ by using the inverse Fourier transform,

$$p(x) = \frac{1}{2\pi}\int_{-\infty}^{\infty} e^{-itx}\varphi(t)\,dt.$$

Interchanging orders of integration, we find, in terms of the distribution function F, that if $a < b$ then

$$F(b) - F(a) = \int_a^b p(x)\,dx = \frac{1}{2\pi}\int_{-\infty}^{\infty}\left(\int_a^b e^{-itx}\,dx\right)\varphi(t)\,dt$$

$$= \frac{1}{2\pi}\int_{-\infty}^{\infty}\frac{e^{-itb} - e^{-ita}}{-it}\,\varphi(t)\,dt.$$

This last formula can be proved for *any* F which is continuous at a and b. If F is not continuous at b then $F(b)$ should be replaced by one-half the left, $F(b)$, and right, $F(b + 0)$, limits of F at b, with a similar replacement for a. More precisely, *P. Lévy's inversion formula* states that if F is any distribution function on R, and φ is its characteristic function, and $a < b$ then

$$\frac{F(b) + F(b+0)}{2} - \frac{F(a) + F(a+0)}{2} = \lim_{c\to\infty}\int_{-c}^{c}\frac{e^{-itb} - e^{-ita}}{-2\pi it}\,\varphi(t)\,dt.$$

Thus we have an explicit formula for recovering F solely from φ. But this would be impossible if two different F's had the same φ. Therefore if two distribution functions F_1 and F_2 have the same characteristic function $\varphi_1 = \varphi_2$ then $F_1 = F_2$.

The characteristic function for $-X$ is the complex conjugate $\bar{\varphi}_X$ of the characteristic function for X, since

$$\varphi_{-X}(t) = Ee^{it(-X)} = E\cos(-tX) + iE\sin(-tX)$$

$$= E\cos tX - iE\sin tX = \overline{\varphi_X(t)}.$$

We say that X has a **symmetric distribution** iff X and $-X$ are identically distributed. A symmetric distribution must have a real-valued characteristic function because

$$\varphi_X(t) = \varphi_{-X}(t) = \bar{\varphi}_X(t) \qquad \text{for all real } t.$$

Conversely if $\varphi_X = \bar{\varphi}_X$ then X and $-X$ have the same characteristic function and so are identically distributed. Therefore a characteristic function is real valued iff it is the characteristic function of a symmetric distribution, and in this case $\varphi(t) = E(\cos tX)$.

All exercises except 1 to 3 and 7c may be done before reading the remainder of this section.

CHARACTERISTIC FUNCTIONS FOR SPECIAL DENSITIES

Suppose that we have a random variable X which has an MGF $m(t) = Ee^{tX}$ which is finite in some interval containing the origin; that is, $m(t_0)$ and $m(-t_0)$ are finite for some $t_0 > 0$. Then its characteristic function can be obtained by replacing t by it

in the MGF power-series expansion, at least for t for which the series converges (because the function Ee^{tX} of *complex* t is analytic in $|t| < t_0$). For all our previously obtained MGFs [such as those in Fig. 2 (Sec. 9-1) for the discrete case] we can obtain φ either directly; or more simply by formal substitution, $\varphi(t) = m(it)$, although we do not prove this. We next obtain φ for normal, gamma, and Cauchy densities.

For the normal density,

$$\varphi(t) = m(it) = e^{\mu it + [\sigma^2(it)^2/2]} = e^{it\mu - (\sigma^2 t^2)/2} \qquad \text{for all real } t.$$

If $\mu = 0$ then φ is real-valued, since the distribution is symmetric. This important characteristic function is also derived in Exercise 4. It is only necessary to remember that if X is normal and $EX = 0$ then $Ee^{iX} = e^{-(1/2)EX^2}$. Clearly, then, if Z has any normal distribution then

$$Ee^{iZ} = e^{iEZ}Ee^{i(Z-EZ)} = e^{iEZ-(1/2)\operatorname{Var} Z}$$

and replacing Z by tZ yields the characteristic function.

Just as for MGFs, if $S_n = X_1 + \cdots + X_n$ where $X_1, \ldots, X_n$ are independent and X_k is normal μ_k and σ_k^2, then

$$Ee^{itS_n} = Ee^{itX_1}e^{itX_2} \cdots e^{itX_n} = Ee^{itX_1}Ee^{itX_2} \cdots Ee^{itX_n}$$

$$= \prod_{k=1}^{n} e^{it\mu_k - (t^2/2)\sigma_k^2} = e^{it\mu - (t^2/2)\sigma^2}$$

where $\mu = \sum_{k=1}^{n} \mu_k$ and $\sigma^2 = \sum_{k=1}^{n} \sigma_k^2$. Thus S_n has the same characteristic function as a normal μ and σ^2 density, and hence S_n induces this density.

For the gamma density $\gamma_{\lambda,\nu}$ substituting it for t in its MGF yields the characteristic function

$$\varphi(t) = \frac{1}{[1 - (it/\lambda)]^\nu} \qquad \text{for all real } t.$$

The exponential density is obtained when $\nu = 1$.

The **Laplace continuous probability density** is defined by

$$p(x) = \frac{e^{-|x|}}{2} \qquad \text{for all real } x.$$

Our principal reason for introducing this density is that it leads to a fairly easy derivation of the characteristic function of the Cauchy density. The Laplace density is symmetric, so its characteristic function is

$$\frac{1}{2}\int_{-\infty}^{\infty} e^{itx-|x|}\,dx = \int_0^\infty (\cos tx)e^{-x}\,dx = \left[\frac{t(\sin tx) - \cos tx}{1+t^2}e^{-x}\right]_0^\infty = \frac{1}{1+t^2}$$

where it is easily checked that the derivative of the term in square brackets equals the integrand. Taking the inverse Fourier transform yields

$$\frac{e^{-|x|}}{2} = \frac{1}{2\pi}\int_{-\infty}^{\infty} \frac{e^{-itx}}{1+t^2}\,dt.$$

But the left side is real and so equals its conjugate, hence replacing $-t$ by t and then interchanging x and t shows that the characteristic function of the symmetric standardized Cauchy density is

$$\int_{-\infty}^{\infty} e^{itx} \frac{dx}{\pi(1+x^2)} = e^{-|t|} \qquad \text{for all real } t.$$

If V has a standard Cauchy density and $\lambda \neq 0$ then by definition $Z = \lambda V + \mu$ is Cauchy μ and $|\lambda|$. Clearly Z has characteristic function

$$Ee^{itZ} = e^{it\mu} Ee^{i(t\lambda)V} = e^{it\mu - |\lambda t|} \qquad \text{for all real } t.$$

The identity

$$\prod_{k=1}^{n} \exp\{it\mu_k - |t|\lambda_k\} = \exp\left\{it \sum_{k=1}^{n} \mu_k - |t| \sum_{k=1}^{n} \lambda_k\right\}$$

shows that if $Z_1, \ldots, Z_n$ are independent and Z_k is Cauchy μ_k and $\lambda_k > 0$ then $\sum_{k=1}^{n} Z_k$ is Cauchy with parameters $\sum_{k=1}^{n} \mu_k$ and $\sum_{k=1}^{n} \lambda_k$. This fact was used and interpreted at the end of Sec. 16-5.

DERIVATIVES OF CHARACTERISTIC FUNCTIONS

For discrete or continuous densities if we formally differentiate the characteristic function

$$\varphi(t) = \sum_{x} e^{itx} p(x) \qquad \text{or} \qquad \varphi(t) = \int_{-\infty}^{\infty} e^{itx} p(x)\, dx$$

with respect to t and interchange differentiation and expectation then $\varphi'(t) = EiXe^{itX}$. In particular $\varphi'(0) = iEX$. More generally we have

$$\varphi^{(k)}(t) = E(iX)^k e^{itX} \qquad \text{and} \qquad \varphi^{(k)}(0) = E(iX)^k$$

and this calculation can be justified if $E\,|X|^k < \infty$. We recall that a Cauchy density does *not* have a mean, and its characteristic function $e^{-|t|}$ is clearly *not* differentiable at $t = 0$.

Note that for a complex-valued function $\varphi = u + iv$ of a *real* variable t we have

$$\begin{aligned} \frac{d}{dt}\varphi(t) &= \lim_{\Delta \to 0} \frac{\varphi(t+\Delta) - \varphi(t)}{\Delta} \\ &= \lim_{\Delta \to 0} \left\{ \left[\frac{u(t+\Delta) - u(t)}{\Delta}\right] + i \left[\frac{v(t+\Delta) - v(t)}{\Delta}\right] \right\} \\ &= \frac{d}{dt} u(t) + i \frac{d}{dt} v(t) \end{aligned}$$

where u' and v' are real, since Δ and each of the terms in square brackets is real. Thus we easily obtain φ' from

$$\text{Re}\ \varphi' = (\text{Re}\ \varphi)' \qquad \text{and} \qquad \text{Im}\ \varphi' = (\text{Im}\ \varphi)'.$$

We collect these facts below; and also state that for small t the error is small if we approximate φ near $t = 0$ by a fitted polynomial in t, that is, by a polynomial of some degree m in the real variable t with complex coefficients selected so that the polynomial and φ have the same first m derivatives at $t = 0$.

Summary of differentiability of characteristic functions Let X be a random variable with characteristic function φ, let integers k and m satisfy $1 \leq k \leq m$, and assume that $E|X|^m < \infty$. Then φ has a continuous kth derivative obtainable by differentiating under the expectation sign,

$$\varphi^{(k)}(t) = E(iX)^k e^{itX} \qquad \text{for all real } t$$

hence

$$\varphi^{(k)}(0) = i^k EX^k.$$

Furthermore $r_m(t)$ goes to zero as t goes to zero if $r_m\colon R \to C$ is defined by the equation

$$\begin{aligned}\varphi(t) &= \sum_{k=0}^{m-1} \frac{\varphi^{(k)}(0)}{k!} t^k + \frac{t^m}{m!}[\varphi^{(m)}(0) + i^m r_m(t)] \\ &= \sum_{k=0}^{m-1} \frac{EX^k}{k!}(it)^k + \frac{(it)^m}{m!}[EX^m + r_m(t)].\end{aligned}$$

For notational convenience we have used $\varphi^{(0)}(0) = EX^0 = 1$. In particular for $m = 2$ we see that if $EX^2 < \infty$ and $EX = 0$ and r is defined by

$$Ee^{itX} = 1 - \frac{t^2}{2}[EX^2 + r(t)]$$

then $r(t)$ goes to zero as t goes to zero. ■

We shall *not* prove the facts just stated about $\varphi^{(k)}$. In this paragraph we *assume* that φ has a continuous mth derivative and then deduce the approximation result that $r_m(t)$ goes to zero as t does. If $\varphi = u + iv$ we apply Taylor's theorem with remainder, as stated in the proof of Lemma 1 (Chap. 19), to u. Then for every real t there is a number θ_t satisfying $0 \leq \theta_t \leq 1$ and such that

$$u(t) = \sum_{k=0}^{m-1} \frac{u^{(k)}(0)}{k!} t^k + \frac{t^m}{m!} u^{(m)}(t\theta_t).$$

We add this equation to i times the corresponding equation for v and subtract the resulting equation from the defining equation for r_m to get

$$\varphi^{(m)}(0) + ir_m(t) = u^{(m)}(t\theta_t) + iv^{(m)}(t\theta_t').$$

But $\varphi^{(m)}$ is continuous so the right-hand side converges to $\varphi^{(m)}(0)$ as t goes to zero, and hence $r_m(t)$ goes to zero, as was asserted.

It can be proved that the transformation of distribution functions into characteristic functions is a continuous transformation, in the sense that convergence of distribution functions to a distribution function is equivalent to convergence of the corresponding characteristic functions. The important following special case permits the deduction of convergence to the normal distribution.

Continuity theorem for convergence to the standardized normal distribution function Φ If $F_1, F_2, \ldots$ is a sequence of distribution functions on R, with corresponding characteristic functions $\varphi_1, \varphi_2, \ldots$ then

$$\lim_{n\to\infty} F_n(x) = \Phi(x) \qquad \text{for all real } x$$

iff

$$\lim_{n\to\infty} \varphi_n(t) = e^{-t^2/2} \qquad \text{for all real } t. \quad \blacksquare$$

For the sake of completeness we next state (again without proof) the more general result which implies the preceding result. The statement is made for general R^k, where we are now considering $k = 1$; the next section permits $k > 1$.

P. Lévy's continuity theorem Let $F_1, F_2, \ldots$ be a sequence of distribution functions on R^k and let $\varphi_1, \varphi_2, \ldots$ be the corresponding characteristic functions. If there is a distribution function F on R^k such that

$$\lim_{n\to\infty} F_n(x) = F(x) \qquad \text{for every } x \text{ at which } F \text{ is continuous,} \tag{†}$$

and if φ is the characteristic function for F then

$$\lim_{n\to\infty} \varphi_n(t) = \varphi(t) \qquad \text{for all } t \text{ in } R^k.$$

Conversely if there is some function $g: R^k \to C$ which is continuous at the origin of R^k and such that

$$\lim_{n\to\infty} \varphi_n(t) = g(t) \qquad \text{for all } t \text{ in } R^k$$

then there is a distribution function F on R^k such that equation (†) above holds, and hence g must be the characteristic function for F. $\blacksquare$

In attempting to apply the more useful converse portion of the theorem we may find characteristic functions converging to some function g which we may not recognize to be a characteristic function. Indeed the limit need not be a characteristic function. For example, if F_n is normal with mean 0 and variance n then $F_n(x)$ converges to $\frac{1}{2}$ for all x while

$$\lim_{n\to\infty} \varphi_n(t) = \lim_{n\to\infty} e^{-nt^2/2} = \begin{cases} 1 & \text{if } t = 0 \\ 0 & \text{if } t \neq 0 \end{cases}$$

so the limit is discontinuous and hence is *not* a characteristic function. However the theorem asserts that when the limit is continuous at $t = 0$ then it is a characteristic function and the desired conclusion follows.

A CHARACTERISTIC FUNCTION PROOF OF A CLT

The same CLT as Corollary 2 (Sec. 20-2) is easily deduced as Theorem 1 below, from the following lemma together with the continuity theorem.

Lemma 1 If the random variables $W_1, W_2, \ldots$ are independent and have the same distribution satisfying $EW_1^2 < \infty$ and $EW_1 = 0$ then

$$E \exp \left\{ \frac{i(W_1 + \cdots + W)_n}{\sqrt{n}} \right\} \to \exp \left\{ -\frac{1}{2} EW_1^2 \right\} \qquad \text{as } n \to \infty. \quad ■$$

Proof We use the independence, identical distributedness, and the quadratic approximation to a characteristic function,

$$E \exp \left\{ \frac{i}{\sqrt{n}} \sum_{m=1}^{n} W_m \right\} = \prod_{m=1}^{n} E \exp \left\{ \frac{i}{\sqrt{n}} W_m \right\} = \left[E \exp \left\{ \frac{i}{\sqrt{n}} W_1 \right\} \right]^n$$

$$= \left\{ 1 - \frac{1}{2n} \left[EW_1^2 + r\left(\frac{1}{\sqrt{n}} \right) \right] \right\}^n \to e^{-EW_1^2/2} \qquad \text{as } n \to \infty.$$

Exercise 10*b* (Sec. 8-3) justifies the limiting value, since we know that $r(1/\sqrt{n})$ converges to zero. ■

Theorem 1 If $S_n = X_1 + \cdots + X_n$ where $X_1 \, X_2, \ldots$ are independent and identically distributed random variables with $0 < \sigma_1^2 = \operatorname{Var} X_1 < \infty$, then for every real b

$$P\left[\frac{S_n - nEX_1}{\sigma_1 \sqrt{n}} < b \right] \to \Phi(b) \qquad \text{as } n \to \infty. \quad ■$$

Proof For any fixed real t we have

$$tS_n^* = t \frac{S_n - nEX_1}{\sigma_1 \sqrt{n}} = \frac{1}{\sqrt{n}} \sum_{m=1}^{n} W_m \qquad \text{where } W_m = t \frac{X_m - EX_1}{\sigma_1}$$

so that Lemma 1 yields

$$Ee^{itS_n^*} \to e^{-t^2/2} \qquad \text{as } n \to \infty$$

and hence the continuity theorem completes the proof. ■

EXERCISES

1. *WLLN Convergence.* Let random variable Y_n have distribution function F_n and characteristic function φ_n for $n = 1, 2, \ldots$. Prove that the following three conditions are equivalent:

(*a*) $\displaystyle\lim_{n \to \infty} P[|Y_n| \geq \epsilon] = 0 \qquad \text{for every } \epsilon > 0.$

(*b*) $\displaystyle\lim_{n \to \infty} F_n(y) = \begin{cases} 0 & \text{if } y < 0 \\ 1 & \text{if } y > 0. \end{cases}$

(*c*) $\displaystyle\lim_{n \to \infty} \varphi_n(t) = 1 \qquad \text{for all real } t.$

HINT: Find the distribution function and characteristic function for the identically zero random variable.

2. *A WLLN Permitting Infinite Second Moments.* Use the preceding exercise to prove the following theorem: If $S_n = X_1 + \cdots + X_n$ where the random variables $X_1, X_2, \ldots$ are independent and identically distributed according to a fixed distribution satisfying $E|X_1| < \infty$, then for any $\epsilon > 0$

$$P\left[\left|\frac{1}{n}S_n - EX_1\right| \geq \epsilon\right] \to 0 \qquad \text{as } n \to \infty.$$

3. Use characteristic functions to prove binomial convergence to the Poisson as stated in Exercise 9*c* (Sec. 8-3).

4. Let X have the symmetric normal 0 and 1 density φ. In this exercise the characteristic function $E(\cos tX)$ is derived without using complex numbers. We substitute the power series for $\cos tx$, noting that $e^{|tx|}\varphi(x)$ dominates the integrand, and interchange summation and integration to obtain

$$\int_{-\infty}^{\infty} (\cos tx)\varphi(x)\,dx = \int_{-\infty}^{\infty}\left(1 - \frac{t^2x^2}{2!} + \frac{t^4x^4}{4!} - \frac{t^6x^6}{6!} + \cdots\right)\varphi(x)\,dx$$
$$= 1 - \frac{t^2}{2!}EX^2 + \frac{t^4}{4!}EX^4 - \frac{t^6}{6!}EX^6 + \cdots = e^{-t^2/2}.$$

Use the moments of X from Exercise 8 (Sec. 16-1) to justify the last equality.

5. Prove that every characteristic function is uniformly continuous. In other words, let X be any random variable and let $\varphi(t) = Ee^{itX}$. Then for every $\epsilon > 0$ there is a $\delta > 0$ such that for all real t and h we have

$$|\varphi(t+h) - \varphi(t)| < \epsilon \qquad \text{whenever } |h| < \delta.$$

HINT: $|e^{ir} - 1| \leq 2$ for all real r. Also for every $\epsilon > 0$ there is a real $r_\epsilon > 0$ such that

$$|e^{ir} - 1| \leq \frac{\epsilon}{2} \qquad \text{for all real } r \text{ satisfying } |r| < r_\epsilon.$$

Furthermore for all $\epsilon > 0$ there is an integer n_ϵ large enough so that $P[|X| \geq n_\epsilon] \leq \epsilon/4$.

6. Let $Y = X_1 + X_2$ where X_1 and X_2 are independent and each has a symmetric uniform density over the interval $-\frac{1}{2} < x < \frac{1}{2}$.

(*a*) Consider the joint density of X_1 and X_2 in the plane. From a simple geometric argument show that Y has the triangular density

$$p_Y(y) = \begin{cases} 1 - |y| & \text{if } -1 < y < 1 \\ 0 & \text{otherwise.} \end{cases}$$

Plot this density.

(*b*) The characteristic function for X_1 is

$$\varphi_{X_1}(t) = \int_{-1/2}^{1/2} (\cos tx)\,dx = \frac{\sin(t/2)}{t/2} \qquad \text{for all real } t.$$

Find the characteristic function φ_Y from p_Y and observe that $\varphi_Y = (\varphi_{X_1})^2$ as it should.

COMMENT: Taking the inverse Fourier transform of φ_Y yields

$$p_Y(y) = \int_{-\infty}^{\infty} e^{-ity}\left\{\frac{1}{2\pi}\left[\frac{\sin(t/2)}{t/2}\right]^2\right\}dt$$

and letting $y = 0$ shows that the expression $\{\ \}$ is a continuous probability density.

Sometimes the distribution of a random variable U is such that it can be induced by the sum of independent nondegenerate random variables. This cannot be done if $P[U = 1] = P[U = -1] = .5$, as is easily proved. Of course if we permit the degenerate "constant random variables" then $U = (U - b) + b$ is such a representation. **Exercises 7 to 9** are concerned with the uniqueness of such representations, assuming there is at least one.

7. Using the comment in Exercise 6, we let V have the continuous symmetric density

$$p_1(v) = \frac{2\sin^2(v/2)}{\pi v^2} = \frac{1-\cos v}{\pi v^2} \qquad \text{for all real } v.$$

For real $\lambda > 0$ let the density p_λ with characteristic function φ_λ be the density induced by V/λ.

(*a*) Show that

$$p_\lambda(w) = \frac{1 - \cos \lambda w}{\lambda \pi w^2} \qquad \text{for all real } \lambda$$

$$\varphi_\lambda(t) = \begin{cases} 1 - \dfrac{|t|}{\lambda} & \text{if } |t| \le \lambda \\ 0 & \text{if } |t| > \lambda \end{cases}$$

and sketch both functions. Thus $\varphi_\lambda(0) = 1$ and φ_λ is triangular with interval $-\lambda \le t \le \lambda$ for base.

(*b*) Let X, Y, and Z be independent and have densities $(p_1 + p_2)/2$, $p_{4/3}$, and p_1, respectively. Show that $\varphi_X(t) = \varphi_Y(t)$ if $|t| \le 1$ and sketch both functions. Thus different distributions can have characteristic functions which agree on the interval $|t| \le 1$. This cannot happen for MGFs. That is, if two distributions have the same *finite-valued* MGF on $|t| \le 1$ then the distributions are equal.

(*c*) Show that the continuity theorem cannot be improved to imply convergence of the distributions if the characteristic functions converge only on the interval $|t| \le 1$.

(*d*) Show that $X + Z$ and $Y + Z$ are identically distributed even though X and Y are not.

8. Let $U = S + T$ where S and T are independent. Assume that we know the distribution of the noise, or error, T and the distribution of the sum U. Assume that the characteristic function of the error T is never zero; for example, T might have a normal or Cauchy density. Show that in this case, as opposed to Exercise 7*d* above, the distribution of S is uniquely determined by our assumptions.

9. Let $U = S + T$ where S and T are independent *and* identically distributed. Assume that we know the characteristic function for U and that it is real valued and never zero. Show that there is at most one distribution for S which can satisfy these assumptions. In particular the sum of independent and identically distributed random variables is normal *if and only if* each component is normal.

COMMENT: The uniqueness of the decomposition (if it exists) into equal independent components is false unless some restriction, such as the one above, is made. In particular there is[1] a real-valued characteristic function f which takes on some negative values but is such that $|f|$ is also a characteristic function. Therefore if S and T are independent and each has characteristic function f then $S + T$ has the same distribution as $S' + T'$, where S' and T' are independent and each has characteristic function $|f|$. Every characteristic function is continuous and equals 1 at $t = 0$, so that this f must, of course, take the value 0 at least once.

[1] W. Feller, "An Introduction to Probability Theory and Its Applications," vol. II, p. 479, John Wiley & Sons, Inc., New York, 1966.

10. Let U and Z be independent with U normal 0 and 1. Assume that Z has a symmetric distribution, so that $.5 + Z$ and $.5 - Z$ are identically distributed. Then

$$U = (.5 + Z)U + (.5 - Z)U$$

shows that the sum of two *dependent* identically nonnormally distributed random variables may have a normal distribution. Let $S = (.5 + Z)U$ and $T = (.5 - Z)U$.

(*a*) Let $P[Z = -1] = P[Z = 1] = .5$ and find the continuous density induced by S, and the correlation between S and T.

(*b*) Let $P[Z = -.5] = P[Z = .5] = .5$ and describe the joint distribution of S and T. Find the correlation between S and T.

18-4 A VECTOR CENTRAL-LIMIT THEOREM

In this section we discuss a vector CLT, introduce characteristic functions of random vectors, and then prove the vector CLT in much the same way as Theorem 1 (Sec. 18-3).

Let $(X_1, Y_1), (X_2, Y_2), \ldots$ be independent and identically distributed random vectors satisfying $EX_1 = EY_1 = 0$. From the CLT we know that the distribution function for $(X_1 + \cdots + X_n)/\sigma_{X_1}\sqrt{n}$ converges to Φ, with a similar convergence for $(Y_1 + \cdots + Y_n)/\sigma_{Y_1}\sqrt{n}$. More generally Theorem 1 below shows that the distribution function on R^2 of the random vector

$$\left(\frac{X_1 + \cdots + X_n}{\sigma_{X_1}\sqrt{n}}, \frac{Y_1 + \cdots + Y_n}{\sigma_{Y_1}\sqrt{n}}\right)$$

converges to the distribution function for the bivariate normal density with parameters $\mu_1 = \mu_2 = 0$, $\sigma_1 = \sigma_2 = 1$, and ρ equal to the correlation coefficient between X_1 and Y_1. With a slightly different normalization we can assert more simply that the limiting distribution of

$$\frac{1}{\sqrt{n}}(X_1 + \cdots + X_n, Y_1 + \cdots + Y_n) = \frac{1}{\sqrt{n}} \sum_{m=1}^{n} (X_m, Y_m)$$

is normal with the same mean (0,0) and covariance matrix as any one summand (X_1, Y_1).

For interpretation X_m and Y_m might be the pressure and volume from the mth independent repetition of an experiment. If we take sums $S_n = X_1 + \cdots + X_n$ and $T_n = Y_1 + \cdots + Y_n$, or arithmetic averages, we are often interested in the joint distribution of S_n and T_n rather than just their marginals. For large enough n the joint distribution is approximately bivariate normal. Naturally X_m and Y_m can be interpreted as noise voltages, or errors, measured at different points at time m, or as rainfalls in two counties, or as crop yields, etc. The vector CLT yields some theoretical insight, or "explanation," for the frequent appearance, in practical problems, of approximately multivariate normal distributions.

The **characteristic function of a random vector** $(Y_1, \ldots, Y_k)$ is the function $\varphi: R^k \to C$ defined by

$$\varphi(t_1, \ldots, t_k) = Ee^{i(t_1Y_1+\cdots+t_kY_k)} \qquad \text{for all real } t_1, \ldots, t_k.$$

As before, each Y_m is assumed to be real valued. Every random vector $(Y_1, \ldots, Y_k)$ has a characteristic function defined on all of R^k. Just as before, $\varphi(0, \ldots, 0) = 1$ and $|\varphi| \leq 1$ and φ is uniformly continuous. Clearly $\varphi(t_1, 0, \ldots, 0)$ is the characteristic function for Y_1. Similarly letting $t_1 = t_3 = 0$ yields the characteristic function for $(Y_2, Y_4, Y_5, \ldots, Y_k)$. Thus the characteristic function of every marginal is obtainable from the joint characteristic function. It is a fact that two random vectors into R^k are identically distributed iff they have the same characteristic function. The inversion formula generalizes naturally, but we ignore it. The continuity theorem has already been stated for R^k. Differentiability and other properties extend naturally, but we do not need them.

We know that if a random variable Z has a normal density then

$$Ee^{iZ} = e^{iEZ-(1/2)\operatorname{Var} Z}.$$

Therefore if $Y' = (Y_1, \ldots, Y_k)$ has a normal density with mean $\mu' = (\mu_1, \ldots, \mu_k)$ and covariance matrix C then Y has characteristic function

$$\varphi_{Y_1,\ldots,Y_k}(t_1, \ldots, t_k) = E\exp\left\{i\sum_{m=1}^{k} t_mY_m\right\} = \exp\{it'\mu - (1/2)t'Ct\}$$

because $Z = \sum_{m=1}^{k} t_mY_m$ has a normal density with $EZ = t'\mu$ and $\operatorname{Var} Z = t'Ct$. This proof is valid if at least one $t_m \neq 0$, and the formula is trivially true if all $t_m = 0$. We recall that every symmetric positive-definite matrix (and only these) is the covariance matrix for a normal density.

Theorem 1: A vector CLT Let $Y' = (Y_1, \ldots, Y_k)$ be a random vector which induces a nondegenerate distribution on R^k. Assume that $EY_m{}^2 < \infty$ for $m = 1, \ldots, k$ and let $\mu = EY$. Let $S_n = X_1 + \cdots + X_n$ where $X_1, X_2, \ldots$ are independent random vectors each having the same distribution as Y. Then, for each point of R^k, the distribution function for $(S_n - n\mu)/\sqrt{n}$ converges, as n goes to infinity, to the distribution function for the normal density with zero means and the same covariance matrix as Y. ■

Proof Let $X'_m = (X_{m1}, \ldots, X_{mk})$ and $T_n = S_n - n\mu$ so that

$$T_n = \begin{bmatrix} T_{n1} \\ \cdot \\ \cdot \\ \cdot \\ T_{nk} \end{bmatrix} = \left\{\sum_{m=1}^{n} \begin{bmatrix} X_{m1} \\ \cdot \\ \cdot \\ \cdot \\ X_{mk} \end{bmatrix}\right\} - n\begin{bmatrix} \mu_1 \\ \cdot \\ \cdot \\ \cdot \\ \mu_k \end{bmatrix} = \begin{bmatrix} \sum_{m=1}^{n} (X_{m1} - \mu_1) \\ \cdot \\ \cdot \\ \cdot \\ \sum_{m=1}^{n} (X_{mk} - \mu_k) \end{bmatrix}$$

For fixed real $t_1, \ldots, t_k$ let

$$W_m = t_1(X_{m1} - \mu_1) + \cdots + t_k(X_{mk} - \mu_k).$$

Then the characteristic function of $T_n/\sqrt{n}$ is

$$E \exp\left\{\frac{i}{\sqrt{n}}(t_1T_{n1} + \cdots + t_kT_{nk})\right\} = E \exp\left\{\frac{i}{\sqrt{n}} \sum_{m=1}^{n} W_m\right\}$$

which, by Lemma 1 (Sec. 18-3), converges to $e^{-(\frac{1}{2})EW_1^2}$ as n goes to infinity. But $EW_1^2 = t'Ct$ where C is the symmetric positive-definite covariance matrix for nondegenerate X_1, or Y, so the continuity theorem completes the proof. ■

We comment that the same proof applies when Y is degenerate, so that its covariance matrix is not positive definite. In this case the distribution function for $(S_n - n\mu)/\sqrt{n}$ converges to the distribution function (where continuous) of the degenerate normal distribution with zero means and the same covariance matrix as Y.

Characteristic functions are sometimes used to prove independence. Random variables $Y_1, \ldots, Y_k$ are independent iff the characteristic function φ_Y of the random vector $Y' = (Y_1, \ldots, Y_k)$ satisfies the equation

$$\varphi_Y(t_1, \ldots, t_k) = \prod_{m=1}^{k} \varphi_{Y_m}(t_m) \qquad \text{for all real } t_1, \ldots, t_k$$

where $\varphi_{Y_1}(t_1) = \varphi_Y(t_1, 0, \ldots, 0)$, with similar identities for the other Y_m. For proof we need only observe that if the equation holds then $(Y_1, \ldots, Y_k)$ has the same characteristic function as does that joint distribution which has the same marginals independent. It is not sufficient to check this equation only with all t_m equal. That is, if we have

$$Ee^{i(tY_1+tY_2)} = Ee^{itY_1}Ee^{itY_2} \qquad \text{for all real } t$$

then we know only that the joint distribution for (Y_1, Y_2) is such that $Y_1 + Y_2$ has the same distribution as it would if Y_1 and Y_2 were independent. But this can occur when $Y_1 = Y_2$ is Cauchy 0 and 1, so that Y_1 and Y_2 are dependent, but

$$Ee^{i(tY_1+tY_2)} = Ee^{i(2t)Y_1} = e^{-2|t|} = Ee^{itY_1}Ee^{itY_2}.$$

EXERCISES

1. Ceramic blocks which are roughly cubical are to be stacked on top of one another. The crucial vertical length of a block is called its length L and is required to satisfy $L \geq 10$. The horizontal cross section is a square of side S with $ES = 10$. Thus the volume of a block is $V = LS^2$. Assume that $L = 10 + U$ and $S = 10 + Z$, where U and Z are independent, U is exponential with $EU = 1$, and Z is normal 0 and 1.

(*a*) Find the mean, covariance matrix, and correlation coefficient $\rho_{L,V}$ for the random vector (L,V).

(*b*) A large number n of these blocks are to be generated independently and stacked on top of one another. The relationship between the resulting arithmetic average length and volume is of interest. Use Exercise 2 (Sec. 18-2) to show that

$$\lim_{n\to\infty} P\left[\frac{1}{n}\sum_{j=1}^{n} L_j \geq EL, \frac{1}{n}\sum_{j=1}^{n} V_j \geq EV\right] \doteq .317.$$

(*c*) Naturally this limit would be ¼ if the length and volume were actually uncorrelated. If L and V actually had a different joint distribution from that assumed above, but the true correlation coefficient between L and V happened to be the same as in part (*a*), would the limiting value of .317 still be valid?

2. A complex unit can partially fail, during an experiment, in two different ways, represented by events A = [type 1 failure] and B = [type 2 failure], where

$$P(AB) = .1 \qquad P(AB^c) = .1 \qquad P(A^cB) = .2 \qquad P(A^cB^c) = .6.$$

In n independent repetitions let S_n be the number of failures of type 1, and T_n the number of failures of type 2. Find the asymptotic distribution of the random vector (S_n, T_n), when properly normalized, as n goes to infinity.

3. Show that the mean and covariance matrix of $(S_n - n\mu)/\sqrt{n}$ of Theorem 1 is independent of n.

4. Let U, V, W be independent and each have finite positive variance. Apply Theorem 1 with $k = 2$ to $Y_1 = U + V$ and $Y_2 = W + V$ so that there is a common component.

5. Let $Z_1, Z_2, \ldots$ be independent and identically distributed random variables. It is sometimes of interest to find the asymptotic joint distribution between sums of different powers. For integers $1 \leq r < s$ assume that $EZ_1^{2s} < \infty$. Then Theorem 1 with $k = 2$ shows that asymptotically as n goes to infinity the distribution of the random vector

$$\frac{1}{\sqrt{n}}\left(\left[\sum_{m=1}^{n} Z_m^r\right] - nEZ_1^r, \left[\sum_{m=1}^{n} Z_m^s\right] - nEZ_1^s\right)$$

is bivariate normal with zero means and covariance matrix C. For simplicity we restrict ourselves to $k = 2$ so that C is 2×2. Clearly C is determined by P_{Z_1} and the two powers r and s.

(*a*) Exhibit C in terms of the moments of Z_1.

(*b*) Show how C simplifies if Z_1 has a symmetric distribution, and one of r and s is even while the other is odd. For example, Z_1 may be normal with $EZ_1 = 0$, and $r = 1$ and $s = 2$, or $r = 2$ and $s = 3$.

(*c*) Show how C simplifies if Z_1 has a symmetric distribution and both r and s are odd. Find the correlation coefficient.

(*d*) Let Z_1 be normal with $EZ_1 = 0$ and use Exercise 2 (Sec. 18-2) to find

$$\lim_{n\to\infty} P[Z_1 + \cdots + Z_n \geq 0, Z_1^3 + \cdots + Z_n^3 \geq 0].$$

19
Proof of the Central-limit Theorem

This chapter is devoted to the proof of Theorem 4, which is essentially a restatement of the central-limit theorem in the form of Theorem 1 (Sec. 13-2). The proof is essentially self contained and does *not* employ characteristic functions. This chapter consists of only one section.

We introduce a distance ρ between distribution functions on R and examine its properties. The conclusion of the CLT is essentially stated in terms of ρ. However, ρ lacks certain desirable properties, so we introduce another distance ρ_ϵ where the parameter ϵ is at our disposal. Theorem 1 then shows that small ρ_ϵ can imply a CLT. Theorem 2 shows that small third absolute moments can guarantee a small ρ_ϵ. Between the statement and proof of Theorem 2 is a discussion of the general strategy of this proof of the CLT, which goes back to Lindeberg.[1] Theorem 3 shows that properly related sums of independent random variables must have close distributions. The previous results then easily yield Theorem 4.

The definitions of ρ and ρ_ϵ utilize the supremum concept, as described in Appendix 2, but this use of the supremum is very simple, as explained below, so that Appendix 2 is not presumed. A simple form of Taylor's theorem with remainder is stated and discussed just prior to its use in the proof of Lemma 1.

TWO DISTANCES, ρ AND ρ_ϵ

For any two random variables U and V define the **distance** $\rho(U,V)$ **between them** by

$$\rho(U,V) = \sup_\alpha |P[U < \alpha] - P[V < \alpha]| = \sup_\alpha |F_U(\alpha) - F_V(\alpha)|.$$

Thus for each real α we have the number $|F_U(\alpha) - F_V(\alpha)|$ which is between 0 and 1. We form the set A of all such numbers and let $\rho(U,V)$ be the supremum of this set. As shown in Fig. 1, $\rho(U,V)$ is "essentially" the largest vertical distance between the two distribution functions F_U and F_V. More precisely, either the set A

[1] J. W. Lindeberg, Uber das Exponentialgesetz in der Wahrscheinlichkeitsrechnung, *Mathemat. Z.*, **15:** 211–225 (1922). For a "modernized" version see H. F. Trotter, An Elementary Proof of the Central Limit Theorem, *Archiv Mathemat.*, **10:** 226–234 (1959) or W. Feller, "An Introduction to Probability Theory and Its Applications," vol. II, chap. 8, sec. 4, John Wiley & Sons, Inc., New York, 1966.

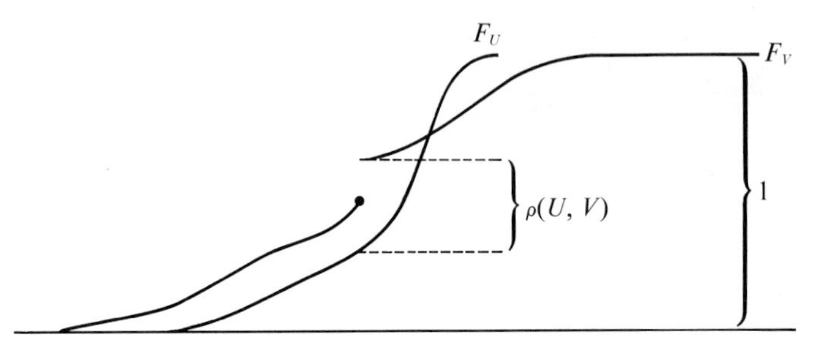

Fig. 1

has a largest number, in which case we let $\rho(U,V)$ equal this number; or, as in Fig. 1, the set A does not have a largest number, in which case we let $\rho(U,V)$ equal the smallest number that is larger than every number in A. For example, if $A = \{y: 0 \le y \le .2\}$ or if $A = \{y: 0 \le y < .2\}$ then $\rho(U,V) = .2$. In either case we can describe $\rho(U,V)$ as the smallest number such that

$$|F_U(\alpha) - F_V(\alpha)| \le \rho(U,V) \qquad \text{for all real } \alpha.$$

The CLT asserts that under certain conditions $\rho(U,V)$ must be small if V is normal 0 and 1, while U is the standardization of a sum of independent random variables.

Clearly if $F_U = F_{U'}$ and $F_V = F_{V'}$ then $\rho(U,V) = \rho(U',V')$; that is, $\rho(U,V)$ depends on the joint distribution $F_{U,V}$ only through the two marginals F_U and F_V. We often think of $\rho(U,V)$ as being just a measure of the distance between the two functions F_U and F_V on R, while the random variables with probability space may or may not be introduced and found useful in manipulations.

The function ρ, when considered on the space of distribution functions on R, is a metric. That is, $\rho(U,V) \ge 0$, and $\rho(U,V) = 0$ iff $F_U = F_V$, also $\rho(U,V) = \rho(V,U)$, and the triangle inequality holds. For the proof of the triangle inequality we perform a standard manipulation,

$$\begin{aligned} |F_U(\alpha) - F_V(\alpha)| &= |F_U(\alpha) - F_W(\alpha) + F_W(\alpha) - F_V(\alpha)| \\ &\le |F_U(\alpha) - F_W(\alpha)| + |F_W(\alpha) - F_V(\alpha)| \\ &\le \rho(U,W) + \rho(W,V) \qquad \text{for every } \alpha, \end{aligned}$$

and hence we have the triangle inequality,

$$\rho(U,V) \le \rho(U,W) + \rho(W,V) \qquad \text{for all } F_U, F_V, \text{ and } F_W.$$

If the function ρ is considered on the space of random variables on a fixed probability space, then it is not a metric, because it violates the condition that $\rho(U,V) = 0$ implies $U = V$.

The distance ρ is unchanged by simultaneous translation of the two distribution functions; that is,

$$\rho(U + \beta, V + \beta) = \rho(U,V) \qquad \text{for all real } \beta.$$

This fact, which is obvious from Fig. 1, is proved by observing that the sets

$$S_\beta = \{z: z = |F_U(\alpha - \beta) - F_V(\alpha - \beta)| \qquad \text{for some real } \alpha\}$$

are the same for every β, and hence have the same supremum. More generally, although we do not need this result, $\rho(U,V) = \rho(g(U),g(V))$ if $g: R \to R$ is continuous and strictly increasing (or strictly decreasing). In particular $\rho(aU + b, aV + b) = \rho(U,V)$ if $a \neq 0$. The distance ρ_ϵ, which will be introduced shortly, is invariant to translation but not to scale change.

The conclusion of the CLT can be naturally stated in terms of ρ, which has many desirable properties. However, ρ has characteristics which make it difficult to use in *proving* the CLT. For example, if $P[U = 0] = 1$ and $P[V = \delta] = 1$ then $\rho(U,V) = 1$ for any $\delta > 0$, since $F_U(\delta/2) = 1$ and $F_V(\delta/2) = 0$. Thus even though F_U may be close to F_V in some senses (when $\delta = 10^{-10}$, for example), the ρ distance says that F_U and F_V are as far apart as any two distribution functions can be. We wish to introduce a measure of distance which takes closeness of this kind more into account.

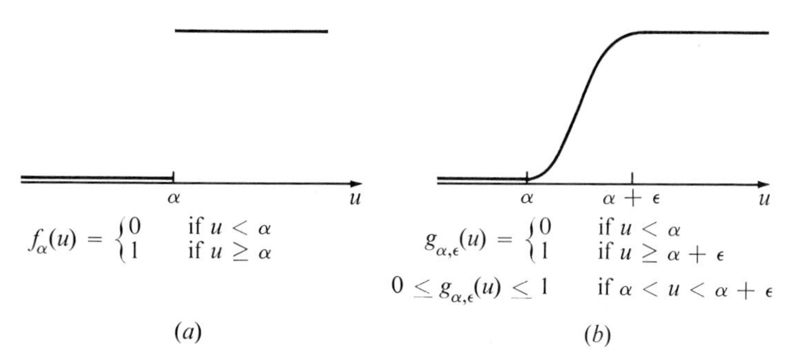

Fig. 2

Let $f_\alpha: R \to R$ be defined as in Fig. 2*a*, so that $f_\alpha(U)$ has only 0 and 1 as values. Also $Ef_\alpha(U) = P[U \geq \alpha]$ since $f_\alpha(U)$ has the value 1 with probability $P[U \geq \alpha]$. Now

$$|P[U < \alpha] - P[V < \alpha]| = |(1 - P[U \geq \alpha]) - (1 - P[V \geq \alpha])|$$
$$= |P[U \geq \alpha] - P[V \geq \alpha]|$$

so that

$$\rho(U,V) = \sup_\alpha |Ef_\alpha(U) - Ef_\alpha(V)|.$$

We changed from $[U < \alpha]$ to $[U \geq \alpha]$ for the slight convenience of permitting f_α to be a distribution function. In order to obtain a distance which is more sensitive to closeness as indicated in the last paragraph, we replace f_α in the above expression by a smoother function $g_{\alpha,\epsilon}$ as indicated in Fig. 2*b*. We then obtain a family of distances, one for each $\epsilon > 0$. Clearly $g_{\alpha,\epsilon}$ approximates f_α when ϵ is small, so these new distances can be related to ρ. We define a function G, corresponding to $\alpha = 0$ and $\epsilon = 1$ in Fig. 2*b*, and then let $g_{\alpha,\epsilon}(u) = G((u - \alpha)/\epsilon)$ to get Fig. 2*b*.

We wish to define a function $G: R \to R$ like the one shown in Fig. 3. Because we work with third absolute moments, it is desirable that G have a continuous third derivative *everywhere*. Thus we want to define a continuous probability density $p: R \to R$ such that p has a continuous second derivative everywhere and $p(x) > 0$ iff $0 < x < 1$.

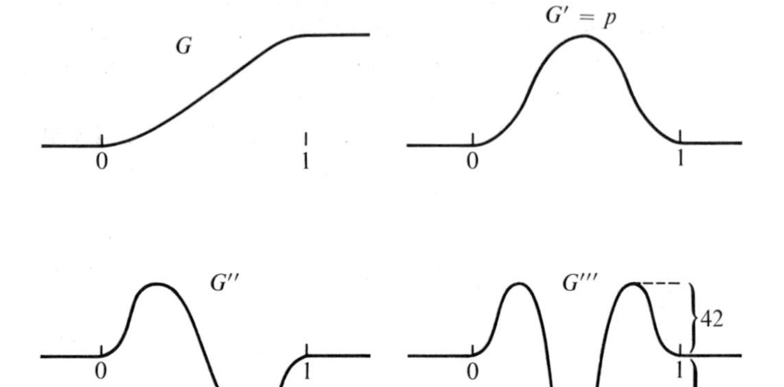

Fig. 3

We then let G be the distribution function for p. For simplicity we make p a polynomial in the interval $0 \leq x \leq 1$. Then p has continuous derivatives of all orders in $0 < x < 1$. But p and all its derivatives are zero outside this interval. Therefore we must make p and its first two derivatives equal to zero at the end points of the interval. This can be done by letting

$$p(x) = \begin{cases} Cx^3(1-x)^3 & \text{if } 0 \leq x \leq 1 \\ 0 & \text{otherwise} \end{cases}$$

where C is defined by $\int_0^1 p(x)\,dx = 1$. Then $G: R \to R$ is defined by $G(x) = \int_{-\infty}^{x} p(t)\,dt$ so

$$G(x) = \begin{cases} 0 & \text{if } x \leq 0 \\ 1 & \text{if } x \geq 1. \end{cases}$$

If $0 \leq x \leq 1$ then

$$\begin{aligned} G(x) &= C\int_0^x t^3(1-t)^3\,dt \\ &= C\int_0^x (t^3 - 3t^4 + 3t^5 - t^6)\,dt = C\left(\frac{x^4}{4} - 3\frac{x^5}{5} + 3\frac{x^6}{6} - \frac{x^7}{7}\right). \end{aligned}$$

Evaluating $G(1) = 1$, we find that $C = 140$, so G has been completely and explicitly defined.

The function G, together with its first three derivatives, is shown in Fig. 3. It is left to Exercise 1 to show that the maximum value of $|G'''(x)|$ is 52.5. For simplicity we weaken this to

$$|G'''(x)| \leq 60 \qquad \text{for all real } x,$$

a fact which will be used in the proof of Lemma 1.

For every $\epsilon > 0$ and random variables U and V define the **distance** $\rho_\epsilon(U,V)$ by

$$\rho_\epsilon(U,V) = \sup_\alpha \left| EG\left(\frac{U-\alpha}{\epsilon}\right) - EG\left(\frac{V-\alpha}{\epsilon}\right) \right|$$

where $G: R \to R$ is defined as above. Clearly $\rho_\epsilon(U,V)$ is determined by F_U and F_V, as was the case for ρ. Thus $\rho_\epsilon(U,V)$ can be thought of as a measure of distance between F_U and F_V. Clearly $\rho_\epsilon(U,V) \geq 0$, and $\rho_\epsilon(U,V) = 0$ if $F_U = F_V$. Also $\rho_\epsilon(U,V) = \rho_\epsilon(V,U)$. Furthermore the same proof as for ρ yields the triangle inequality

$$\rho_\epsilon(U,V) \leq \rho_\epsilon(U,W) + \rho_\epsilon(W,V)$$

for any three random variables U, V, and W. Also, just as before, each ρ_ϵ is invariant to simultaneous translation of the two distribution functions; that is,

$$\rho_\epsilon(U+\beta,\, V+\beta) = \rho_\epsilon(U,V) \qquad \text{for all real } \beta.$$

In proving the CLT we will show that under certain conditions $\rho_\epsilon(U,V)$ is small even when ϵ is small. We must then deduce that $\rho(U,V)$ is small, at least when V is normal. This last deduction follows from the following theorem.

Theorem 1 Let V be normally distributed with variance σ^2. If U is any random variable and if $\epsilon > 0$ then

$$\rho(U,V) \leq \frac{\epsilon}{2\sigma} + \rho_\epsilon(U,V). \quad \blacksquare$$

Proof We must prove that $(P[\alpha \leq V] - P[\alpha \leq U])$ and its negative are less than $\epsilon/2\sigma + \rho_\epsilon(U,V)$ for every real α. If we let $g_{\alpha,\epsilon}(u) = G((u-\alpha)/\epsilon)$ then $g_{\alpha,\epsilon}$ is as shown in Fig. 2b, so $f_{\alpha+\epsilon} \leq g_{\alpha,\epsilon} \leq f_\alpha$. Taking expectations yields

$$P[\alpha+\epsilon \leq U] \leq EG\left(\frac{U-\alpha}{\epsilon}\right) \leq P[\alpha \leq U]$$

for any U. Using these two inequalities, we see that for *any* U and *any* V

$$P[\alpha+\epsilon \leq V] \leq EG\left(\frac{V-\alpha}{\epsilon}\right) = EG\left(\frac{U-\alpha}{\epsilon}\right) + EG\left(\frac{V-\alpha}{\epsilon}\right) - EG\left(\frac{U-\alpha}{\epsilon}\right)$$

$$\leq EG\left(\frac{U-\alpha}{\epsilon}\right) + \rho_\epsilon(U,V) \leq P[\alpha \leq U] + \rho_\epsilon(U,V).$$

Thus for every U, V, and α we have

$$P[\alpha+\epsilon \leq V] - P[\alpha \leq U] \leq \rho_\epsilon(U,V). \tag{1}$$

Interchanging U and V and replacing α by $\alpha - \epsilon$ yields

$$P[\alpha \leq U] - P[\alpha - \epsilon \leq V] \leq \rho_\epsilon(U,V). \tag{2}$$

In equation (1) use $P[\alpha \leq V] = P[\alpha \leq V < \alpha + \epsilon] + P[\alpha + \epsilon \leq V]$ to obtain

$$P[\alpha \leq V] - P[\alpha \leq U] \leq P[\alpha \leq V < \alpha + \epsilon] + \rho_\epsilon(U,V). \tag{3}$$

In equation (2) use $P[\alpha - \epsilon \leq V] = P[\alpha - \epsilon \leq V < \alpha] + P[\alpha \leq V]$ to obtain

$$P[\alpha \leq U] - P[\alpha \leq V] \leq P[\alpha - \epsilon \leq V < \alpha] + \rho_\epsilon(U,V). \tag{4}$$

If V is normal then the maximum value of its density occurs at its mean and is $1/\sigma\sqrt{2\pi} < 1/(2\sigma)$. Therefore the P_V probability of any interval of length ϵ is at most $\epsilon/2\sigma$. Applying this fact to equations (3) and (4) completes the proof of Theorem 1. ■

The significance of the following theorem is discussed prior to its proof.

Theorem 2 Let U, V, Z be independent random variables such that U and V have finite third absolute moments and satisfy $EU = EV$ and $EU^2 = EV^2$. If $\epsilon > 0$ then

$$\rho_\epsilon(U + Z, V + Z) \leq \frac{10}{\epsilon^3}(E\,|U|^3 + E\,|V|^3). \quad ■$$

Observe that Theorem 2 does not require that the expectations of U and V be near to zero. However this will typically be the case in order to make $E\,|U|^3 + E\,|V|^3$ small, so that the inequality becomes useful.

THE STRATEGY FOR THIS PROOF OF THE CLT

For convenience in interpreting Theorem 2 we introduce the following two properties. Property 2 is the special case $Z = 0$ of Theorem 2.

Property 1 If U, V, Z are independent random variables and if $\epsilon > 0$ then

$$\rho_\epsilon(U + Z, V + Z) \leq \rho_\epsilon(U,V). \quad ■$$

Property 2 If $EU = EV$ and $EU^2 = EV^2$ and if $\epsilon > 0$ then

$$\rho_\epsilon(U,V) \leq \frac{10}{\epsilon^3}(E\,|U|^3 + E\,|V|^3). \quad ■$$

Property 1 asserts that if two random variables, U and V, have distribution functions which are close to each other then independent addition of the same random variable Z to each of them can only increase this closeness. We say that two random variables are **matched** iff they have the same mean and variance. In our application of Property 1 the two random variables U and V are matched, and their common variance is much smaller than the variance of Z, and $\rho_\epsilon(U,V)$ is small. In such a case Property 1 shows that if a random variable $U + Z$ is the sum of two independent components, one of which is small, then the small component U can be replaced by a close-matching V without drastically altering the distribution of the sum.

In essence our proof of the CLT amounts to taking a sum $U_1 + U_2 + \cdots + U_n = U_1 + Z$, replacing one small component U_1 by a matching normally distributed V_1, and then replacing U_2 in $V_1 + U_2 + \cdots + U_n$ by a matching normal V_2, etc., until we have a sum of independent normals, which we already know to be normal. We must show that each replacement effects the distribution of the sum only "slightly," and the *total* effect of all of these replacements is small. Property 1 shows that a replacement can be made if $\rho_\epsilon(U,V)$ is small, and Property 2 says that $\rho_\epsilon(U,V)$ is small for matching variables with small third absolute moments. Clearly Properties 1 and 2 together imply Theorem 2, which combines these two properties by showing that matched variables with small third absolute moments can replace one another in such sums.

Exercises 1 to 3, 6 and 8 may be done before reading the remainder of this section.

Theorem 2 is easily deduced from the following lemma.

Lemma 1 Let U, V, Z be independent random variables such that U and V have finite third absolute moments and satisfy $EU = EV$ and $EU^2 = EV^2$. Then

$$|EG(U + Z) - EG(V + Z)| \leq 10[E|U|^3 + E|V|^3]. \quad \blacksquare$$

For the proof of Lemma 1 we require the following special case of a standard result from advanced calculus.

Taylor's theorem with remainder Let $G: R \to R$ and let $n \geq 1$ be an integer. Assume that G has an nth derivative $G^{(n)}(z)$ at every real z. For any real z and u if $h_z(u)$ is defined by

$$h_z(u) = G(z + u) - \sum_{k=0}^{n-1} \frac{G^{(k)}(z)}{k!} u^k$$

then there exists a real number α between z and $z + u$ such that

$$h_z(u) = \frac{G^{(n)}(\alpha)}{n!} u^n. \quad \blacksquare$$

For interpretation of Taylor's theorem select some real z and consider it fixed. Suppose that we want to approximate $G(z + u)$ by a polynomial in u of degree $n - 1$. If we select that polynomial which at the point z has the same first $n - 1$ derivatives as does $G(z + u)$ then $h_z(u)$ is the error in the approximation. The theorem says that the error is "almost" the next higher term $G^{(n)}(z)u^n/n!$ which would be included if an approximating polynomial of degree n were being used. By "almost" we mean that the coefficient of $u^n/n!$ must be the value of $G^{(n)}$ at some intermediate point α, which depends on G, z, and u. No method is given for finding α. If $n = 1$ the theorem reduces to the mean-value theorem $G(z + u) - G(z) = G'(\alpha)u$, which says that the derivative of G at *some* intermediate point must equal the slope of that line which agrees with G at both z and $z + u$. The theorem for general $n \geq 1$ is usually deduced from the special case $n = 1$.

Proof of Lemma 1 We apply Taylor's theorem with remainder to our special function G with $n = 3$. Exercise 1 shows that $|G^{(3)}(\alpha)| \leq 60$ for all real α, hence

$$|h_z(u)| \leq \frac{60}{3!} |u|^3 = 10\, |u|^3 \qquad \text{for all real } z \text{ and } u.$$

A similar result holds for real z and v, so that

$$|h_z(u) - h_z(v)| \leq 10[|u|^3 + |v|^3] \qquad \text{for all real } z,\ u,\ v.$$

Now $|EX| \leq E\,|X|$ is always true, so replacing z, u, v by Z, U, V and taking expectations yields

$$|E[h_Z(U) - h_Z(V)]| \leq 10[E\,|U|^3 + E\,|V|^3]$$

where $h_Z(U) - h_Z(V)$ is given by

$$G(Z + U) - G(Z + V) - (U - V)G'(Z) - \frac{U^2 - V^2}{2} G''(Z).$$

Then $U - V$ and $G'(Z)$ are independent, so

$$E(U - V)G'(Z) = E(U - V)EG'(Z) = (0 - 0)EG'(Z) = 0.$$

Similarly $E(U^2 - V^2)G''(Z) = 0$, so Lemma 1 is proved. ■

Proof of Theorem 2 We apply Lemma 1 to U/ϵ, V/ϵ, $(Z - \alpha)/\epsilon$ to obtain

$$\left| EG\left(\frac{U + Z - \alpha}{\epsilon}\right) - EG\left(\frac{V + Z - \alpha}{\epsilon}\right) \right| \leq 10\left[E\left|\frac{U}{\epsilon}\right|^3 + E\left|\frac{V}{\epsilon}\right|^3\right]$$

for all real α, hence Theorem 2 is proved. ■

The following theorem shows that the two distributions of *any* two sums of independent matching random variables are close to each other if sums of third absolute moments are small.

Theorem 3 Let $U_1, \ldots, U_n, V_1, \ldots, V_n$ be $2n$ independent random variables having finite third absolute moments and satisfying

$$EU_k = EV_k \qquad EU_k^2 = EV_k^2$$

whenever $1 \leq k \leq n$. If $\epsilon > 0$ then

$$\rho_\epsilon\left(\sum_{k=1}^{n} U_k, \sum_{k=1}^{n} V_k\right) \leq \frac{10}{\epsilon^3} \sum_{k=1}^{n} [E\,|U_k|^3 + E\,|V_k|^3]. \quad ■$$

Proof The triangle inequality for ρ_ϵ yields

$$\rho_\epsilon(W_0, W_n) \leq \rho_\epsilon(W_0, W_1) + \rho_\epsilon(W_1, W_2) + \cdots + \rho_\epsilon(W_{n-1}, W_n).$$

Let $W_0 = U_1 + \cdots + U_n$ and $W_n = V_1 + \cdots + V_n$ while if $1 \leq k \leq n$ let

$W_k = V_1 + \cdots + V_k + U_{k+1} + \cdots + U_n$. In other words, the first k random variables in the summation for W_k come from $V_1, \ldots, V_n$. But if $1 \le k \le n$ then

$$\rho_\epsilon(W_k, W_{k-1}) = \rho_\epsilon(V_k + Z_k, U_k + Z_k) \le \frac{10}{\epsilon^3}[E|U_k|^3 + E|V_k|^3]$$

by Theorem 2, with $Z_k = V_1 + \cdots + V_{k-1} + U_{k+1} + \cdots + U_n$. ∎

Consider Theorem 3 with $V_1, \ldots, V_n$ fixed. Regardless of which independent matching random variables $U_1, \ldots, U_n$ we use, as long as the bound on the right-hand side is small, we see that the distribution of any resulting sum $\sum_{k=1}^{n} U_k$ will be close to the fixed distribution of $\sum_{k=1}^{n} V_k$. Thus all such sums $\sum_{k=1}^{n} U_k$ have distributions which are close to the distribution of any one such sum $\sum_{k=1}^{n} V_k$. But if $V_1, \ldots, V_n$ are independent and each is normal then so is $\sum_{k=1}^{n} V_k$. These observations are made precise in the following proof. The CLT in the form of Theorem 1 (Sec. 13-2) is obtained by applying Theorem 4 to

$$U_k = \frac{X_k - EX_k}{\sqrt{\text{Var } S_n}}.$$

Theorem 4: The central limit theorem Let $U = U_1 + \cdots + U_n$ where $U_1, \ldots, U_n$ are independent and each $EU_k = 0$ and $E|U_k|^3 < \infty$. If $EU^2 = 1$ and V is normal 0 and 1 then

$$\rho(U,V) \le 3B^{1/4} \qquad \text{where } B = \sum_{k=1}^{n} E|U_k|^3. \quad ∎$$

Proof Let $U_1, \ldots, U_n, V_1, \ldots, V_n$ be $2n$ independent random variables where $U_1, \ldots, U_n$ are as in the hypothesis of the theorem and $V_1, \ldots, V_n$ are each normally distributed with $EV_k = 0$ and $EV_k^2 = EU_k^2$. In other words, we introduce independent matching normal variables. Let $V = V_1 + \cdots + V_n$ so V is normal 0 and 1.

We first wish to show that $E|V_k|^3 \le 2E|U_k|^3$. We evaluate $E|V|^3$ by making the change of variables $x = v^2/2$ and $dx = v\,dv$,

$$\begin{aligned} E|V|^3 &= \frac{1}{\sqrt{2\pi}}\int_{-\infty}^{\infty} |v|^3 e^{-v^2/2}\,dv = \frac{2}{\sqrt{2\pi}}\int_0^{\infty} v^3 e^{-v^2/2}\,dv \\ &= \frac{4}{\sqrt{2\pi}}\int_0^{\infty} xe^{-x}\,dx = \frac{4}{\sqrt{2\pi}}[-(1+x)e^{-x}]_0^{\infty} = \frac{4}{\sqrt{2\pi}} \le 2. \end{aligned}$$

In calculating $E|V_k|^3$ we express V_k in the form

$$V_k = [EV_k^2]^{1/2}\left\{\frac{V_k}{[EV_k^2]^{1/2}}\right\}$$

where the term { } is normal 0 and 1, so that

$$E|V_k|^3 \le [EV_k^2]^{3/2}\,2 = 2\{[EU_k^2]^{1/2}\}^3.$$

But from Theorem 3f (Sec. 11-3) we know that $[E|U_k|^r]^{1/r}$ is an increasing function of r, so that

$$E|V_k|^3 \le 2\{[E|U_k|^3]^{1/3}\}^3 = 2E|U_k|^3$$

as was to be shown.

Applying Theorem 1, then Theorem 3, and then the result of the last paragraph, we obtain

$$\rho(U,V) \le \frac{\epsilon}{2} + \rho_\epsilon(U,V) \le \frac{\epsilon}{2} + \frac{10}{\epsilon^3}\sum_{k=1}^{n}[E|U_k|^3 + E|V_k|^3] \le \frac{\epsilon}{2} + \frac{30B}{\epsilon^3} = h(\epsilon).$$

Now ϵ, in the bound $h(\epsilon)$, is at our disposal, so we set $h'(\epsilon) = \frac{1}{2} - 90B/\epsilon^4$ equal to zero and find that $h(\epsilon)$ is minimized at $\epsilon_0 = (180B)^{1/4}$. The bound then becomes

$$h(\epsilon_0) = \frac{\epsilon_0}{2}\left(1 + \frac{60B}{\epsilon_0^4}\right) = \frac{\epsilon_0}{2}\frac{4}{3} = \frac{2}{3}(180B)^{1/4} \le 3B^{1/4}$$

since $180 \le 4^4$. ■

EXERCISES

1. Find G'' and G''' for $0 \le x \le 1$ and verify that both are zero at both ends of the interval. Show that for $0 \le x \le 1$

$$G'''(x) = 840(x - 6x^2 + 10x^3 - 5x^4) = 840y(1 - 5y)$$

where $y = x(1 - x)$. Sketch $y(1 - 5y)$ for $0 \le y \le \frac{1}{4}$ and note that $|y(1 - 5y)| \le \frac{1}{16}$ in this interval. Conclude that $|G'''| \le 52.5$.

2. Prove Property 1 in the case where each of U, V, Z has a discrete (continuous) density. Note that the proof merely requires that the summations (integrations) may be performed in the proper order, and it depends primarily on the invariance of ρ_ϵ to simultaneous translation, and hence applies to ρ as well. COMMENT: A proof valid in general can be given by taking the expectation of the conditional expectation given Z, once the properties of conditional expectation are established in general.

3. Let $P[U = -\delta] = .9$ and $P[U = 9\delta] = .1$ and let $V = -U$ so that U and V are matching random variables. Sketch F_U and F_V and find $\rho(U,V)$ and $[E|U|^3 + E|V|^3]$. Since $\rho(U,V) > 0$ is independent of $\delta \ne 0$, while $[E|U|^3 + E|V|^3]$ goes to zero as δ does, we see that ρ cannot satisfy the analog of Property 2 for ρ_ϵ.

4. Prove the following fact by direct substitution into Theorem 3. Let $X_1, \ldots, X_n$ be independent and identically distributed with $EX_1 = 0$, $EX_1^2 = 1$, and $E|X_1|^3 < \infty$. Let $S_n^* = (X_1/\sqrt{n}) + \cdots + (X_n/\sqrt{n})$ and let V be normal 0 and 1. If $\epsilon > 0$ then

$$\rho_\epsilon(S_n^*, V) \le \frac{10}{\epsilon^3\sqrt{n}}[E|X_1|^3 + E|V|^3]$$

COMMENT: Replace ϵ by ϵ_n where ϵ_n goes to zero slowly enough to make $(\epsilon_n)^3\sqrt{n}$ go to infinity as n goes to infinity. For example, let $\epsilon_n = n^{-1/8}$. Then $\rho_{\epsilon_n}(S_n^*, V)$ goes to zero, and hence $\rho(S_n^*, V)$ goes to zero by Theorem 1, yielding an asymptotic CLT.

5. In the proof of Lemma 1 if G were approximated by a linear rather than a quadratic expression, then under the assumptions of Theorem 3 we would have the modified result

$$\rho_\epsilon\left(\sum_{k=1}^{n} U_k, \sum_{k=1}^{n} V_k\right) \leq \frac{A}{\epsilon^2} \sum_{k=1}^{n} [EU_k^2 + EV_k^2].$$

Exercise 4 contained a proof of a CLT. A modified proof of a CLT can be attempted by starting with the result above rather than with Theorem 3. Indicate where such a modified proof breaks down.

6. Show that Theorem 2 immediately implies the following generalization of itself. Let U and Z_1 be two independent random variables, and let V and Z_2 be two independent random variables. Let U and V have finite third absolute moments and satisfy $EU = EV$ and $EU^2 = EV^2$. Let Z_1 and Z_2 be identically distributed. If $\epsilon > 0$ then

$$\rho_\epsilon(U + Z_1, V + Z_2) \leq \frac{10}{\epsilon^3} [E|U|^3 + E|V|^3].$$

7. Show that the independence assumption in Theorem 3 can immediately be weakened to $U_1, \ldots, U_n$ are independent and $V_1, \ldots, V_n$ are independent.

8. Let X and $X_1, X_2, \ldots$ be random variables and assume that X has a *continuous* distribution function F_X. Prove that if $F_{X_n}(x)$ converges to $F_X(x)$ as n converges to infinity, for each real x, then $\rho(X_n, X)$ converges to zero. Also show that the conclusion may be false if the assumption that F_X is continuous is dropped. Thus pointwise convergence of distribution functions to a continuous distribution function implies uniform convergence. In particular the conclusion of Corollary 1*a* (Sec. 13-2) can be strengthened to $\rho(S_n^*, V)$ goes to zero as n goes to infinity, where V is normal 0 and 1. HINT: For any integer M there are numbers x_k such that $F(x_k) = k/M$ for $k = 1, 2, \ldots, (M - 1)$.

20
The Method of Truncation

This chapter is devoted to three applications of the method of truncation. The method permits the generalization of some of our earlier results to random variables which have fewer finite moments than previously assumed. The method and several facts useful for its application are introduced in Sec. 20-1. The weak law of large numbers is then proved for random variables which need only have expectations. The WLLN was proved in Sec. 12-1 under the assumption of finite second moments. In Sec. 20-2 general central-limit theorems are proved for random variables having only finite second moments. The proof of the CLT in Chap. 19 required finite third absolute moments. Section 20-3 formulates the strong law of large numbers in a more detailed fashion, and then shows that it holds even when only $E|X| < \infty$, rather than requiring $EX^2 < \infty$, as we did in Chap. 12. The most natural statements of the conclusions of the WLLN and the SLLN are in terms of EX, while the CLT involves variances. Thus this chapter shows that we need only assume the finiteness of those moments which are naturally involved in the conclusions of these three results.

20-1 THE METHOD OF TRUNCATION AND THE WLLN

If $X_1, \ldots, X_n$ are random variables, with n fixed, let

$$B_m = [|X_1| \le m, \ldots, |X_n| \le m]$$

be the event that none of them has a value exceeding m. As m increases B_m increases to the whole sample space, so $P(B_m)$ converges to 1 as m goes to infinity. Therefore for m large enough B_m is almost certain to occur. If we condition our probabilities by B_m then on B_m every $|X_k|$ is bounded by m, that is,

$$P[|X_k| \le m \mid B_m] = 1, \qquad \text{and} \qquad P(B_m) \doteq 1 \qquad \text{for } m \text{ large.}$$

Thus if we use P_{B_m} then we may assume that the random variables are bounded and hence have finite moments of all orders, $E|X_k|^r < \infty$ for all $r > 0$.

Instead of using conditional probabilities, we might introduce a random variable X_k' which agrees with X_k on B_m but has the value 0 if B_m does not occur,

$$X_k'(\omega) = \begin{cases} X_k(\omega) & \text{if } \omega \in B_m \\ 0 & \text{if } \omega \notin B_m. \end{cases}$$

Clearly this "chopped-off" version X_k' of X_k satisfies $|X_k'| \leq m$ and so has all moments finite. This boundedness of X_k' often makes it a more manageable random variable than X_k.

Suppose, for example, that we want to show that $P(A)$ is small, where $A = [|S_n| \geq 3]$ and $S_n = X_1 + \cdots + X_n$. We may define $A' = [|S_n'| \geq 3]$ where $S_n' = X_1' + \cdots + X_n'$. Then S_n and S_n' are equal on B_m so $A \cap B_m = A' \cap B_m$. Hence $P(A) = P(A \cap B_m) + P(A \cap B_m{}^c) \leq P(A') + P(B_m{}^c)$. Therefore this technique of proof has two parts: we show that $P(B_m{}^c) \doteq 0$ and we show that $P(A') \doteq 0$. Thus under some conditions we may deduce a property of $X_1, \ldots, X_n$ by deducing a corresponding property for "chopped-off" versions $X_1', \ldots, X_n'$ and showing that the "chopped-off" variables are almost certain to equal the original variables.

Traditionally each random variable is "chopped off," or *truncated*, independently of the other variables. This technique will be seen to have the desirable property of preserving independence. The proof of results such as the WLLN, CLT, or SLLN by the method of truncation then has essentially two parts: one part consists of showing that the truncations have been carried out in such a way that only the truncated variables need be considered; the other part consists of showing that the truncated variables have the required property. For this latter purpose certain relations between the truncated and untruncated variables are needed, and most of these are gathered together in Lemma 2.

The truncation technique is formally introduced after Lemma 1, which relates certain infinite series to moments. Lemma 1 does not involve truncation but is useful in applying the method of truncation. As an introduction to Lemma 1, let X be a random variable, and in terms of X define a discrete density $p: R \to R$ by

$$p(0) = P[X = 0]$$

$$p(k) = P[k - 1 < |X| \leq k] \qquad \text{for } k = 1, 2, \ldots$$

$$p(x) = 0 \qquad \text{otherwise.}$$

Thus in effect, we assign all of the $P_{|X|}$ probability of the interval $k - 1 < x \leq k$ to the right end point of the interval. If $r > 0$ and $k - 1 < x \leq k$ then $x^r \leq k^r$ so that p may have a *somewhat* larger rth moment than $|X|$ does. Lemma 1 shows that $E|X|^r < \infty$ iff p has a finite rth moment.

Lemma 1 Let X be a random variable and r a positive real number. Then $E|X|^r < \infty$ iff $B_r < \infty$ where

$$B_r = \sum_{k=1}^{\infty} k^r P[k - 1 < |X| \leq k].$$

Furthermore if $E|X|^r < \infty$ then $E|X|^r \leq B_r$. In particular $E|X| < \infty$ iff $B_1 < \infty$ where

$$B_1 = \sum_{k=1}^{\infty} kP[k - 1 < |X| \leq k] = \sum_{n=0}^{\infty} P[|X| > n].$$

Furthermore if $E|X| < \infty$ then $B_1 - 1 \leq E|X| \leq B_1$. ∎

Proof Define a random variable $Z \geq 0$ as follows. For any ω if $X(\omega) = 0$ let $Z(\omega) = 0$. If $X(\omega) \neq 0$ find the unique integer $k \geq 1$ such that $k - 1 < |X(\omega)| \leq k$, and let $Z(\omega) = k$. Then $EZ^r = B_r$. Clearly $|X| \leq Z$ so $|X|^r \leq Z^r$. Thus if $B_r < \infty$ then $E|X|^r \leq B_r < \infty$. Clearly $Z \leq |X| + 1$ so $Z^r \leq (|X| + 1)^r$. Thus if $E|X|^r < \infty$ then $E(|X| + 1)^r < \infty$ from Theorem 3g (Sec. 11-3), so $B_r < \infty$. In particular taking the expectation of $Z \leq |X| + 1$ yields $B_1 = EZ \leq E|X| + 1$.

In addition B_1 equals the sum of the numbers in the triangular array

$$\begin{array}{lll} P[0 < |X| \leq 1] & & \\ P[1 < |X| \leq 2] & P[1 < |X| \leq 2] & \\ P[2 < |X| \leq 3] & P[2 < |X| \leq 3] & P[2 < |X| \leq 3] \\ \cdots & \cdots & \cdots \end{array}$$

since $kP[k - 1 < |X| \leq k]$ equals the sum of the numbers in the kth row. But $P[|X| > n - 1]$ equals the sum of the numbers in the nth column, and adding these "column sums" shows that $B_1 = \sum_{n=0}^{\infty} P[|X| > n]$. ∎

For any real number $c > 0$ define the two functions $t_c: R \to R$ and $d_c: R \to R$ as in Fig. 1. Clearly $|t_c| \leq c$. If c increases then $|t_c|$ increases and $|d_c|$ decreases. We

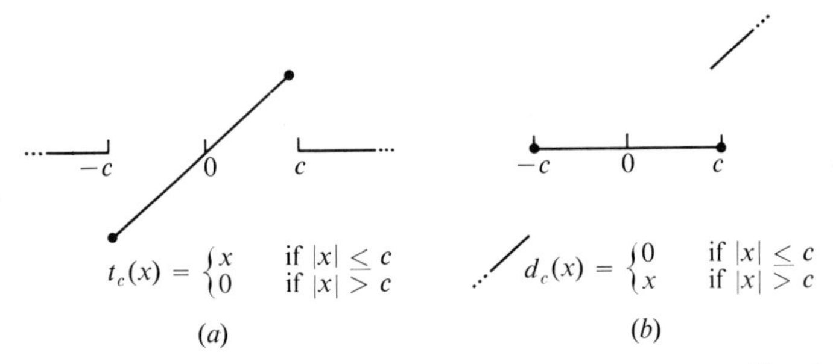

Fig. 1

may think of the function t_c as the *truncation* at level c of the identity function, $f(x) = x$ for all x. For each real x we have

$$x = t_c(x) + d_c(x).$$

Thus d_c is the *difference* between the identity function and its truncation at c. For each real x we have

$$t_c(-x) = -t_c(x) \qquad d_c(-x) = -d_c(x)$$

$$t_c(|x|) = |t_c(x)| \leq |x| \qquad d_c(|x|) = |d_c(x)| \leq |x|$$

$$t_c\left(\frac{x}{a}\right) = \frac{1}{a}\, t_{ac}(x) \quad \text{if } a > 0 \qquad d_c\left(\frac{x}{a}\right) = \frac{1}{a}\, d_{ac}(x) \quad \text{if } a > 0.$$

For example, in proving the bottom left equation note that if $|x/a| \leq c$ then $t_c(x/a) = x/a$, but in this case $|x| \leq ac$ so that $(1/a)t_{ac}(x) = (1/a)x$, as was asserted. Thus a scale factor can be taken outside the argument of t_c by altering the truncation level.

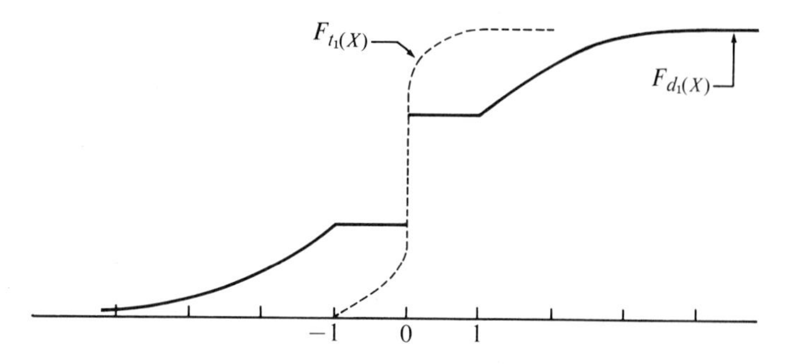

Fig. 2 X is normal $\mu = 0$ and $\sigma = 2$.

If $X: \Omega \to R$ is any random variable and $c > 0$ is a real number then by composition we can define the two random variables $t_c(X)$ and $d_c(X)$. We call $t_c(X)$ the **truncation of X at c.** Clearly

$$\text{if } |X(\omega)| \le c \text{ then } t_c(X(\omega)) = X(\omega) \quad \text{and} \quad d_c(X(\omega)) = 0$$

and

$$\text{if } |X(\omega)| > c \text{ then } t_c(X(\omega)) = 0 \quad \text{and} \quad d_c(X(\omega)) = X(\omega).$$

Thus $t_c(X)$ agrees with X on $[|X| \le c]$ and equals zero off $[|X| \le c]$. Also $d_c(X)$ equals zero on $[|X| \le c]$ and agrees with X off $[|X| \le c]$. Clearly $t_c(X)$ has finite moments of all orders, since $|t_c(X)| \le c$. Furthermore $X = t_c(X) + d_c(X)$, so that X has been decomposed into the sum of two random variables, one of which is bounded and so has finite moments of all orders. Therefore if $r > 0$ then $E\,|X|^r < \infty$ iff $E\,|d_c(X)|^r < \infty$.

We now consider the distributions of $t_c(X)$ and $d_c(X)$. Clearly

$$P[t_c(X) = 0] = P[X = 0] + P[|X| > c] \quad \text{and} \quad P[d_c(X) = 0] = P[|X| \le c]$$

so that both $t_c(X)$ and $d_c(X)$ can take the value 0 with positive probability even if X induces a continuous density. For example, if X has a normal distribution with $\mu = 0$ and $\sigma = 2$, and we truncate at $c = 1$, then we have the distribution functions of Fig. 2. If we let $Y = t_c(X)$ and $Z = d_c(X)$ then the joint distribution of (Y,Z) is concentrated on the three thickened lines of Fig. 3. Clearly $t_c(X)$ and $d_c(X)$ are highly dependent in general.

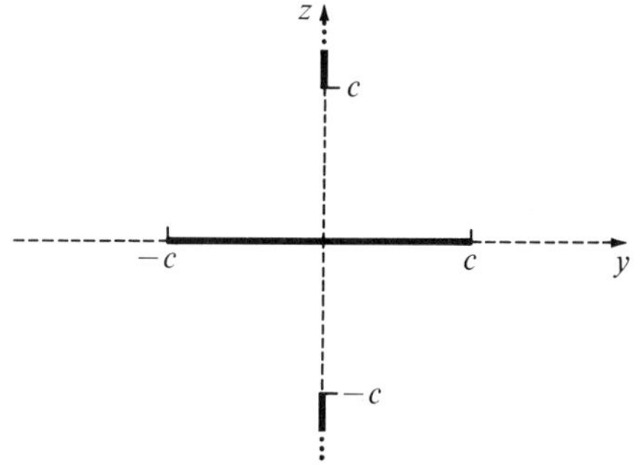

Fig. 3

If $X_1, \ldots, X_n$ are independent (or identically distributed) then so are $t_c(X_1), \ldots, t_c(X_n)$. However, if $EX_1 = 0$ we may have $Et_c(X_1) \neq 0$, so zero means may not be preserved under truncation.

Lemma 2 is briefly discussed immediately after its statement.

Lemma 2 If $E|X|^r < \infty$, where $r > 0$, then the following are true:

(*a*) $E\,|d_c(X)|^r \to 0$ as $c \to \infty$.

(*b*) For any $s > 0$, $(1/c^s)E\,|t_c(X)|^{r+s} \to 0$ as $c \to \infty$.

(*c*) For any $c > 0$, $P[|X| > c] \le (1/c^r)E\,|d_c(X)|^r$.

(*d*) If $0 < s < r$ and $c > 0$ then $E\,|d_c(X)|^s \le (1/c^{r-s})E\,|d_c(X)|^r$. ∎

In Lemma 2 we assume that $E\,|X|^r < \infty$ so that $E\,|d_c(X)|^r < \infty$, although higher moments of X and $d_c(X)$ may be infinite. Part (*d*) bounds lower moments of $d_c(X)$ in terms of $E\,|d_c(X)|^r$. Also by part (*a*) if $E\,|X|^r < \infty$ then the rth moment of the difference variable must go to zero as the truncation point increases. Parts (*a*) and (*c*) show that finiteness of $E\,|X|^r$ implies that if c goes to infinity then $P[|X| > c]$ goes to zero faster than $1/c^r$. All moments of $t_c(X)$ are finite, so larger moments than the rth may be utilized. Part (*b*) shows that if $E\,|X|^r < \infty$ and c approaches infinity then $E\,|t_c(X)|^{r+s}$ cannot grow as fast as c^s.

Proof of Lemma 2 To prove part (*a*), clearly $|d_c(X)| \le |d_n(X)|$ if $c \ge n$, so we need only prove that $E\,|d_n(X)|^r$ goes to zero as n goes to infinity. But $d_n(X)$ has no values between 0 and n, so applying Lemma 1 to $d_n(X)$ yields

$$E\,|d_n(X)|^r \le \sum_{k=n+1}^{\infty} k^r P[k-1 < |d_n(X)| \le k] = \sum_{k=n+1}^{\infty} k^r P[k-1 < |X| \le k]$$

which, by Lemma 1, goes to zero.

To prove part (*b*), if n is determined by c to be the unique integer larger than c and closest to c then

$$\frac{1}{c^s} E\,|t_c(X)|^{r+s} \le \left(\frac{n}{c}\right)^s \left\{\frac{1}{n^s} E\,|t_n(X)|^{r+s}\right\}$$

since $|t_c(X)|$ increases with c. Therefore to prove part (*b*) we need only prove that the term in braces goes to zero as n goes to infinity, since $(n/c)^s$ converges to 1. Now $|t_n(X)| \le n$, so applying Lemma 1 to $t_n(X)$ yields

$$\frac{1}{n^s} E\,|t_n(X)|^{r+s} \le \frac{1}{n^s} \sum_{k=1}^{n} k^s a_k \qquad \text{where } a_k = k^r P[k-1 < |X| \le k].$$

But from Lemma 1, $\sum_{k=1}^{\infty} a_k < \infty$ so applying the Kronecker lemma of Exercise 2 yields part (*b*).

To prove part (*c*) note that the inequality

$$I_{[|X|>c]} \le \left[\frac{|d_c(X)|}{c}\right]^r$$

is true, since it reduces to $0 \le 0$ when evaluated at an ω with $|X(\omega)| \le c$, and if

$|X(\omega)| > c$ the left-hand side equals 1 and the right-hand side exceeds 1. Taking expectations completes the proof of part (*c*).

Similarly for part (*d*), we take the expectation of

$$|d_c(X)|^s \leq \left[\frac{|d_c(X)|}{c}\right]^{r-s} |d_c(X)|^s. \quad \blacksquare$$

The following theorem, which is not needed later, is our first application of truncation.

Theorem 1: The WLLN for independent identically distributed random variables having expectations Let $X_1, X_2, \ldots$ be independent and identically distributed random variables satisfying $E|X_1| < \infty$. If $S_n = X_1 + \cdots + X_n$ and $\epsilon > 0$ then

$$P\left[\left|\frac{1}{n} S_n - EX_1\right| \geq \epsilon\right] \to 0 \qquad \text{as } n \to \infty. \quad \blacksquare$$

Proof We assume that $EX_1 = 0$, since this special result applied to the $X_k - EX_k$ yields the general result. Let

$$S'_n = \sum_{k=1}^{n} t_n(X_k)$$

be the sum of the n first random variables, each truncated at n. Let

$$A_n = \bigcup_{k=1}^{n} [|X_k| > n]$$

be the event that at least one of these n random variables differs from its truncated version. Clearly $S_n(\omega) = S'_n(\omega)$ if ω belongs to $A_n{}^c$. Thus

$$\begin{aligned} P[|S_n| \geq n\epsilon] &= P\{[|S_n| \geq n\epsilon] \cap A_n{}^c\} + P\{[|S_n| \geq n\epsilon] \cap A_n\} \\ &\leq P\{[|S'_n| \geq n\epsilon] \cap A_n{}^c\} + P(A_n) \\ &\leq P[|S'_n| \geq n\epsilon] + \sum_{k=1}^{n} P[|X_k| > n] \\ &= P[|S'_n| \geq n\epsilon] + nP[|X_1| > n] \end{aligned}$$

where the last equality uses the identical-distribution assumption. But by parts (*c*) and (*a*) of Lemma 2, $nP[|X_1| > n] \leq E|d_n(X_1)|$ converges to zero. *Thus we need only prove that* $P[|S'_n| \geq n\epsilon]$ goes to zero. Typically $ES'_n \neq 0$, but from the proof of Chebychev's inequality, Theorem 3 (Sec. 15-4), we have the first inequality in

$$P[|S'_n| \geq n\epsilon] \leq \frac{E(S'_n)^2}{n^2\epsilon^2} = \frac{\text{Var } S'_n + [ES'_n]^2}{n^2\epsilon^2} = \frac{1}{n\epsilon^2} \text{Var } t_n(X_1) + \frac{1}{n^2\epsilon^2} [nEt_n(X_1)]^2$$

But $\text{Var } t_n(X_1) \leq E[t_n(X_1)]^2$ so by Lemma 2*b* with $r = s = 1$ the first term goes to zero. Furthermore

$$0 = EX_1 = Et_n(X_1) + Ed_n(X_1)$$

and by Lemma 2*a*, $Ed_n(X_1)$ goes to zero, therefore $Et_n(X_1)$ converges to zero. $\blacksquare$

EXERCISES

1. Prove the *Toeplitz lemma:* If $A_1, A_2, \ldots$ are reals which converge to a real A as n goes to infinity, and if $w_1, w_2, \ldots$ are nonnegative reals with partial sums $W_n = \sum_{k=1}^{n} w_k$ converging to infinity as n does, then

$$\lim_{n\to\infty} \frac{1}{W_n} \sum_{k=1}^{n} w_k A_k = A.$$

HINT: $$\frac{1}{W_n} \sum_{k=1}^{n} w_k A_k = \frac{w_1}{W_n} A_1 + \frac{w_2}{W_n} A_2 + \cdots + \frac{w_n}{W_n} A_n$$

so we have a weighted average of $A_1, \ldots, A_n$, where the earlier terms are negligible since W_n goes to infinity, and for the later terms every $A_k \doteq A$. Clearly

$$\left| \frac{1}{W_n} \sum_{k=1}^{n} w_k A_k - A \right| \leq \frac{1}{W_n} \sum_{k=1}^{n} w_k \, |A_k - A|$$

and it seems natural to use $\sum_{k=1}^{n} = \sum_{k=1}^{r} + \sum_{k=r+1}^{n}$ on the right-hand side and let n go to infinity.

COMMENT: The most important specialization of the Toeplitz lemma is obtained by letting all $w_n = 1$, in which case we have the fact that if $\lim_{n\to\infty} A_n = A$ then $\lim_{n\to\infty} (1/n) \sum_{k=1}^{n} A_k = A$. For example, if $S_n = X_1 + \cdots + X_n$ and if EX_n converges to some limit μ as n goes to infinity then $E(S_n/n)$ converges to μ as n approaches infinity.

COMMENT: Given any sequence $A_1, A_2, \ldots$; if the limit

$$\lim_{n\to\infty} \frac{1}{W_n} \sum_{k=1}^{n} w_k A_k$$

happens to exist then we can call this the *limit of* $A_1, A_2, \ldots$ *as calculated by use of* $w_1, w_2, \ldots$. The Toeplitz lemma assures us that if $A_1, A_2, \ldots$ has a limit A in the usual sense then the limit of $A_1, A_2, \ldots$ as calculated by use of $w_1, w_2, \ldots$ exists and equals A. The sequence $A_n = (-1)^n$ does not have a limit in the usual sense, but the "natural limit" of zero is obviously assigned to it by the limit of $A_1, A_2, \ldots$ as calculated by use of all $w_n = 1$.

2. Prove the *Kronecker lemma:* If an infinite series of reals $\sum_{k=1}^{\infty} a_k$ has a finite sum, and $W_1 \leq W_2 \leq \cdots$ with W_n approaching infinity as n does, then

$$\lim_{n\to\infty} \frac{1}{W_n} \sum_{k=1}^{n} W_k a_k = 0.$$

HINT: We have assumed that the partial sums $A_n = \sum_{k=1}^{n} a_k$ converge to some real A as n converges to infinity. Divide the Sec. 12-4 formula for "summation by parts" by W_n, then apply the Toeplitz lemma.

20-2 THE CENTRAL-LIMIT THEOREM AND TRUNCATION

In this section we apply truncation to obtain quite general CLTs. In Chap. 19 we proved the CLT of Theorem 1 (Sec. 13-2), which was specialized to the case

$P[|X_k| \le c] = 1$ in Example 8 (Sec. 13-2). This "special case" is the elemental CLT from which we derive generalizations by truncation. Truncation alters means and variances. Therefore, in preparation for truncation, Theorem 1 extends this special case to permit ES to differ slightly from zero, and Var S to differ slightly from 1.

Theorem 1, together with truncation, yields the key Theorem 2, which assumes only finite variances, and bounds the error in the normal approximation by $9c^{1/4}$, where c is the truncation level used in the proof. This bound is valid whenever a certain condition holds for the second moments of the difference variables $d_c(X_k)$. Although such conditions appear awkward, they are the standard CLT conditions having great generality. They satisfy the criterion of being conditions on the individual X_k distributions rather than conditions on distributions of sums.

Theorem 2 yields general asymptotic CLTs, such as Theorem 3 and its corollaries, almost by inspection. Theorem 3 is the CLT, under the Lindeberg condition, for general triangular arrays as introduced at the end of Sec. 13-2.

Theorem 1 Let $S = X_1 + \cdots + X_n$ where $X_1, \ldots, X_n$ are independent random variables with each $P[|X_k| \le c] = 1$, where c is a constant. If $\mu = ES$ and $s^2 = \text{Var } S$ then for every real b

$$|P[S < b] - \Phi(b)| \le 5c^{1/4} + |\mu| + 2\,|s - 1|. \quad \blacksquare$$

We used S and s^2 instead of S_n and s_n^2 for the same reasons we dropped the subscript n in Example 9 (Sec. 13-2): because the conclusion need not exhibit n, and it is valid for every n and $X_1, \ldots, X_n$ satisfying the hypotheses. However, as was the case for Theorem 1 (Sec. 13-2), variables which satisfy the hypotheses and yield a *small* error bound cannot be found unless n is *large* enough. To prove this note that $P[|X_k| \le c] = 1$ implies that $EX_k^2 \le c^2$, so that $s^2 \le \sum_{k=1}^{n} EX_k^2 \le nc^2$, and hence $n > (s/c)^2$. Small error bounds require $c \doteq 0$ and $s \doteq 1$, and hence n must be large before we can find $X_1, \ldots, X_n$ with such c and s.

Proof of Theorem 1 If $0 \le s < \frac{1}{2}$ then $2\,|s - 1| > 1$, so the theorem is true, since no two distribution functions can differ by more than 1. Therefore we henceforth assume that $s \ge \frac{1}{2}$. Adding and subtracting the same expression to $P[S < b] - \Phi(b)$ and then taking absolute values yields

$$|P[S < b] - \Phi(b)| \le \left| P\left[S^* < \frac{b-\mu}{s}\right] - \Phi\left(\frac{b-\mu}{s}\right) \right| + \left| \Phi\left(\frac{b-\mu}{s}\right) - \Phi\left(\frac{b}{s}\right) \right| + \left| \Phi\left(\frac{b}{s}\right) - \Phi(b) \right|.$$

From Example 8 (Sec. 13-2) the first term is bounded by

$$4\left(\frac{c}{s}\right)^{1/4} \le 4(2c)^{1/4} \le 5c^{1/4}.$$

The normal density satisfies $|\varphi| \le \frac{1}{2}$, so that its integral over an interval of length $|\mu|/s$ is at most $(|\mu|/s)\frac{1}{2} \le |\mu|$, which bounds the second term.

It only remains to show that

$$\left|\Phi\left(\frac{b}{s}\right) - \Phi(b)\right| \le 2\,|s-1|.$$

As is obvious from Fig. 1, we have $|x\varphi(x)| \le 1$ for all real x. Now $|\Phi(b/s) - \Phi(b)|$

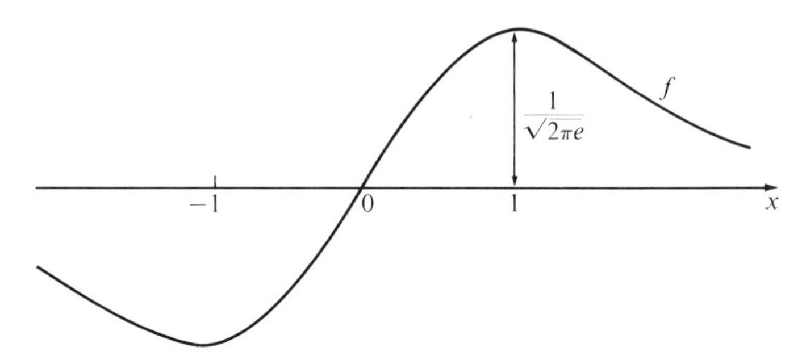

Fig. 1 $f(x) = x\varphi(x), f'(x) = (1 - x^2)\varphi(x)$.

equals the integral of φ over an interval of length $|(b/s) - b|$. If $s > 1$ then $\varphi(b/s)$ is the largest value of φ on *this* interval, and hence

$$\left|\Phi\left(\frac{b}{s}\right) - \Phi(b)\right| \le \left|\frac{b}{s} - b\right| \varphi\left(\frac{b}{s}\right) = |s-1| \left|\frac{b}{s}\right| \varphi\left(\frac{b}{s}\right) \le |s-1|.$$

If $\frac{1}{2} \le s \le 1$ then $\varphi(b)$ is the largest value of φ on this interval, and hence

$$\left|\Phi\left(\frac{b}{s}\right) - \Phi(b)\right| \le \left|\frac{b}{s} - b\right| \varphi(b) = \frac{|s-1|}{s}\,|b|\,\varphi(b) \le \frac{|s-1|}{s} \le 2\,|s-1|. \quad \blacksquare$$

Theorem 2 Let $S = X_1 + \cdots + X_n$ where $X_1, \ldots, X_n$ are independent random variables with finite variances. Assume that $ES = 0$ and $ES^2 = 1$. For real $c > 0$ let

$$g(c) = \sum_{k=1}^{n} E[d_c(X_k)]^2.$$

If for a particular c we have $g(c) \le c^3$ then for that c and every real b

$$|P[S < b] - \Phi(b)| \le 9c^{1/4}. \quad \blacksquare$$

In Theorem 2 the number $g(c) \ge 0$ is the sum of the second moments of the difference variables $d_c(X_k)$. If each $EX_k = 0$ and c converges to zero, then $E[d_c(X_k)]^2$ converges to $EX_k^2 = \operatorname{Var} X_k$, so $g(0+) = 1$. Thus if c starts at 0 and increases then g starts at $g(0+) = 1$ and decreases, so that for some $c \le 1$ we will satisfy $g(c) \le c^3$ and deduce the error bound $9c^{1/4}$. But this must occur for very small c if $9c^{1/4}$ is to be small. For example, if $g(10^{-4}) \le 10^{-12}$ then we can deduce the error bound $9(10^{-4})^{1/4} = .9$. Similarly if $g(10^{-8}) \le 10^{-24}$ then we can deduce the error bound .09. This may happen if almost all the contribution to each EX_k^2 comes from the interval $|x| \le 10^{-8}$, in which case n must be large since $ES^2 = 1$. Thus Theorem 2 says that S

is approximately normally distributed if g decreases from $g(0+)$ to almost the value zero almost immediately.

Proof of Theorem 2 If $c > 1$ then $9c^{1/4} \geq 1$, so the theorem is trivially true. Therefore we may assume that $0 < c \leq 1$ as well as $g(c) \leq c^3$. Let

$$S' = \sum_{k=1}^{n} t_c(X_k) \qquad \text{and} \qquad S'' = \sum_{k=1}^{n} d_c(X_k)$$

so that $S = S' + S''$. Let $(s')^2 = \text{Var } S'$. Then Theorem 1 yields

$$|P[S < b] - \Phi(b)| \leq |P[S < b] - P[S' < b]| + |P[S' < b] - \Phi(b)|$$

$$\leq |P[S < b] - P[S' < b]| + 5c^{1/4} + |ES'| + 2\,|s' - 1|.$$

We will show that the right-hand side is $c^{1/4} + 5c^{1/4} + c^{1/4} + 2c^{1/4}$ or smaller. For the first term we can use

$$P[S < b] = P[S < b, S' < b] + P[S < b, S' \geq b]$$

and

$$P[S' < b] = P[S < b, S' < b] + P[S \geq b, S' < b]$$

so that

$$|P[S < b] - P[S' < b]| \leq P[S < b, S' \geq b] + P[S \geq b, S' < b].$$

But the two events on the right-hand side are disjoint subsets of $[S \neq S']$, which is a subset of $\bigcup_{k=1}^{n} [|X_k| > c]$. Therefore

$$|P[S < b] - P[S' < b]| \leq \sum_{k=1}^{n} P[|X_k| > c] \leq \frac{1}{c^2} g(c) \leq c \leq c^{1/4}$$

where the second inequality follows from Lemma 2*c* (Sec. 20-1) with $r = 2$.

For the $|ES'|$ term note that $0 = ES = ES' + ES''$, so using Lemma 2*d* (Sec. 20-1) with $s = 1$ and $r = 2$, we get

$$|ES'| = |-ES''| \leq \sum_{k=1}^{n} E\,|d_c(X_k)| \leq \frac{1}{c} g(c) \leq c^2 \leq c^{1/4}.$$

To complete the proof of the theorem we need only show that $(s' - 1)^2 \leq \sqrt{c}$. We first make the *observation* that $S = S' + S''$ implies that

$$(\sigma_S - \sigma_{S'})^2 \leq \sigma_{S''}^2.$$

To prove this use

$$(\sigma_S - \sigma_{S'})^2 = \sigma_S^2 - 2\sigma_S\sigma_{S'} + \sigma_{S'}^2$$

and

$$\sigma_{S''}^2 = \text{Var }(S - S') = \sigma_S^2 - 2\,\text{Cov }(S,S') + \sigma_{S'}^2$$

so that the desired inequality reduces to $\text{Cov }(S,S') \leq \sigma_S\sigma_{S'}$, which is true by Theorem 3*h* (Sec. 11-3) [this observation is also implied by Exercise 10 (Sec. 11-4)]. Using the observation, we have

$$(1 - s')^2 \le \text{Var } S'' = \sum_{k=1}^{n} \text{Var } d_c(X_k) \le g(c) \le c^3 \le \sqrt{c}. \quad \blacksquare$$

Theorem 3 For each n let k_n be a positive integer and let $S_n = X_{n,1} + X_{n,2} + \cdots + X_{n,k_n}$ where $X_{n,1}, X_{n,2}, \ldots, X_{n,k_n}$ are k_n independent random variables with finite variances. Let $s_n{}^2 = \text{Var } S_n > 0$ and $S_n^* = (S_n - ES_n)/s_n$. For each real $c > 0$ let

$$g_n(c) = \frac{1}{s_n{}^2} \sum_{j=1}^{k_n} Ed^2_{cs_n}(X_{n,j} - EX_{n,j}).$$

If for each $c > 0$ we have $g_n(c) \to 0$ as $n \to \infty$ then for every real b

$$P[S_n^* < b] \to \Phi(b) \qquad \text{as } n \to \infty. \quad \blacksquare$$

Proof Apply Theorem 2 to S_n^* and use

$$d_c\left(\frac{X_{n,j} - EX_{n,j}}{s_n}\right) = \frac{1}{s_n} d_{cs_n}(X_{n,j} - EX_{n,j})$$

to obtain

$$|P[S_n^* < b] - \Phi(b)| \le 9c^{1/4} \qquad \text{if } g_n(c) \le c^3.$$

But $g_n(c)$ goes to zero as n goes to infinity. Therefore regardless of how small we select the number $9c^{1/4}$ we know that it will be a valid bound for all large enough n. ■

The fact that convergence of g_n to zero is sufficient for the CLT is due to Lindeberg. Feller showed[1] that this condition is also *essentially* necessary by proving a converse obtained by replacing the last sentence of Theorem 3 by the following:

If as n goes to infinity

$$P[S_n^* < b] \to \Phi(b) \qquad \text{for every real } b$$

and if

$$\frac{1}{s_n{}^2} \max \{\text{Var } X_{n,1}, \ldots, \text{Var } X_{n,k_n}\} \to 0$$

then for each $c > 0$ we have $g_n(c) \to 0$.[1]

If $f\colon R \to R$ satisfies $f(0) = 0$ then

$$f(x) = f(t_c(x)) + f(d_c(x)) \qquad \text{for all real } x$$

since this reduces to

$$f(x) = \begin{cases} f(x) + f(0) & \text{if } |x| \le c \\ f(0) + f(x) & \text{if } |x| > c. \end{cases}$$

[1] For a proof see W. Feller, "An Introduction to Probability Theory and Its Applications," vol. II, p. 492, John Wiley & Sons, Inc., New York, 1966, or M. Loève, "Probability Theory," 3d ed., p. 280, D. Van Nostrand Company Inc., Princeton, N.J., 1963.

Therefore $X^2 = d_c^2(X) + t_c^2(X)$, so that $g_n(c)$ of Theorem 3 may easily be expressed in terms of the truncated variables by $g_n(c) = 1 - h_n(c)$ where

$$h_n(c) = \frac{1}{s_n^2} \sum_{j=1}^{k_n} Et_{cs_n}^2(X_{n,j} - EX_{n,j}).$$

Clearly $g_n(c)$ goes to zero iff $h_n(c)$ goes to 1.

In particular if each $X_{n,j}$ has a continuous density with zero mean then

$$h_n(c) = \frac{1}{s_n^2} \sum_{j=1}^{k_n} \int_{-cs_n}^{cs_n} x^2 p_{X_{n,j}}(x)\, dx$$

$$g_n(c) = \frac{1}{s_n^2} \sum_{j=1}^{k_n} \left[\int_{-\infty}^{-cs_n} x^2 p_{X_{n,j}}(x)\, dx + \int_{cs_n}^{\infty} x^2 p_{X_{n,j}}(x)\, dx\right].$$

If $k_n = n$ and $X_{n,j} = X_j$ in Theorem 3 then we obtain Corollary 1.

Corollary 1 Let $S_n = X_1 + \cdots + X_n$ where $X_1, X_2, \ldots$ are independent random variables with finite variances. Let $s_n^2 = \text{Var } S_n > 0$ and $S_n^* = (S_n - ES_n)/s_n$. For each real $c > 0$ let

$$g_n(c) = \frac{1}{s_n^2} \sum_{j=1}^{n} Ed_{cs_n}^2(X_j - EX_j).$$

If for each real $c > 0$ we have $g_n(c) \to 0$ as $n \to \infty$ then for every real b

$$P[S_n^* < b] \to \Phi(b) \qquad \text{as } n \to \infty. \quad \blacksquare$$

Corollary 2 If $S_n = X_1 + \cdots + X_n$ where $X_1, X_2, \ldots$ are independent and identically distributed random variables with $0 < \sigma_1^2 = \text{Var } X_1 < \infty$ then for every real b

$$P\left[\frac{S_n - nEX_1}{\sigma_1\sqrt{n}} < b\right] \to \Phi(b) \qquad \text{as } n \to \infty. \quad \blacksquare$$

Proof From Corollary 1 we need only show that $g_n(c)$ goes to zero as n goes to infinity, where

$$g_n(c) = \frac{1}{n\sigma_1^2} nEd_{c\sigma_1\sqrt{n}}^2(X_1 - \mu_1).$$

Application of Lemma 2*a* (Sec. 20-1) completes the proof of Corollary 2. ■

Corollary 2 is the same as Theorem 1 (Sec. 18-3), which was proved by the use of characteristic functions.

20-3 THE STRONG LAW OF LARGE NUMBERS AND TRUNCATION

This section is devoted to a more detailed study of the strong law of large numbers. Lemma 1 relates the inequality version of Secs. 12-3 and 12-4 to the traditional asymptotic version. Then Theorems 1 and 2 state the asymptotic version for independent random variables having finite second moments. In Theorem 2 identical distributions are not assumed; this case specializes to Theorem 1. Lemmas 2 and 3 state general properties of the SLLN type of convergence. Truncated random variables first appear in Lemma 4, which is a special result needed, together with Theorem 2 and Lemmas 2 and 3, for the proof of Theorem 3. Theorem 3 is the SLLN for independent identically distributed random variables which need only have expectations.

If $Z_1, Z_2, \ldots$ is an *infinite* sequence of random variables, all defined on the *same* probability space, then we may seek the probability of the event

$$\{\omega: Z_n(\omega) \to 0 \text{ as } n \to \infty\}$$

that Z_n converges to zero. We say that Z_n **converges almost surely to zero** as n goes to infinity iff this event has probability 1. The notation $\xrightarrow{\text{a.s.}}$ is standard for this convergence, which is also called *convergence with probability* 1. Thus $Z_n \xrightarrow{\text{a.s.}} 0$ iff

$$P[Z_n \to 0 \text{ as } n \to \infty] = 1.$$

If d is a real number then $Z_n \xrightarrow{\text{a.s.}} d$ means that $(Z_n - d) \xrightarrow{\text{a.s.}} 0$. The following lemma exhibits two conditions, each of which is equivalent to $Z_n \xrightarrow{\text{a.s.}} 0$.

Lemma 1 Let $Z_1, Z_2, \ldots$ be an infinite sequence of random variables defined on the same probability space. Then the following three conditions are equivalent:

(*a*) $P\left[\lim_{n\to\infty} Z_n = 0\right] = 1.$

(*b*) For every $\epsilon > 0$

$$\lim_{n\to\infty}\left\{\lim_{m\to\infty} P\left(\bigcup_{k=n}^{m} [|Z_k| \geq \epsilon]\right)\right\} = 0.$$

(*c*) For every $\epsilon > 0$ there exists a sequence $a_{1,\epsilon}, a_{2,\epsilon}, a_{3,\epsilon}, \ldots$ of real numbers converging to zero and satisfying

$$P\left(\bigcup_{k=n}^{m} [|Z_k| \geq \epsilon]\right) \leq a_{n,\epsilon} \qquad \text{whenever } m \geq n. \quad \blacksquare$$

Proof We first deduce another condition equivalent to (*a*). Let A be the event that Z_n does not converge to zero, so that condition (*a*) becomes $P(A) = 0$. For every $\epsilon > 0$ define the event A_ϵ by $A_\epsilon = [|Z_k| \geq \epsilon$ for infinitely many $k]$. Then ω belongs to A iff there exists an integer m such that ω belongs to $A_{1/m}$, so that $A = \bigcup_{m=1}^{\infty} A_{1/m}$. If $P(A) = 0$ then every $P(A_{1/m}) = 0$, since every $A_{1/m}$ is a subset of A. If every $P(A_{1/m}) = 0$ then $P(A) \leq \sum_{m=1}^{\infty} P(A_{1/m}) = 0$. Thus condition (*a*) is equivalent to $P(A_{1/m}) = 0$ for every integer m. But if $P(A_{1/m}) = 0$ with $1/m \leq \epsilon$ then $A_{1/m}$ contains A_ϵ, so $P(A_\epsilon) = 0$. *Therefore* condition (*a*) is equivalent to: $P(A_\epsilon) = 0$ for every real $\epsilon > 0$.

To prove the equivalence of conditions (*a*) and (*b*) we now need only prove that we always have

$$P(A_\epsilon) = \lim_{n\to\infty}\left\{\lim_{m\to\infty} P\left(\bigcup_{k=n}^{m}[|Z_k| \geq \epsilon]\right)\right\}.$$

Now ω belongs to A_ϵ iff for every r there is some $k \geq r$ such that $|Z_k(\omega)| \geq \epsilon$. Therefore ω belongs to A_ϵ iff ω belongs to every B_r where

$$B_r = \bigcup_{k=r}^{\infty}[|Z_k| \geq \epsilon].$$

We do not exhibit the dependence of B_r on ϵ. Thus $A_\epsilon = \bigcap_{r=1}^{\infty} B_r$ so

$$P(A_\epsilon) = \lim_{n\to\infty} P\left(\bigcap_{r=1}^{n} B_r\right).$$

But B_r decreases as r increases, so $\bigcap_{r=1}^{n} B_r = B_n$. Hence $P(A_\epsilon) = \lim_{n\to\infty} P(B_n)$ where

$$P(B_n) = \lim_{m\to\infty} P\left(\bigcup_{k=n}^{m}[|Z_k| \geq \epsilon]\right).$$

The equivalence of conditions (*a*) and (*b*) is established.

If condition (*c*) holds let m, and then n, go to infinity to get condition (*b*). If condition (*b*) holds let

$$a_{n,\epsilon} = \lim_{m\to\infty} P\left(\bigcup_{k=n}^{m}[|Z_k| \geq \epsilon]\right).$$

to get condition (*c*). ■

Conditions (*b*) and (*c*) of Lemma 1 have the property that only events involving finitely many random variables need be considered. Typically in the past we have formulated problems in such a way that they involve a finite but arbitrarily large number of random variables. We seldom required that an infinite sequence of random variables all be defined on the same probability space. For example, we might specify some distribution for X_1 and then introduce independent identically distributed $X_1, \ldots, X_n$. This assumption *uniquely* and *consistently* determines the joint distribution of $X_1, \ldots, X_n$ for every $n = 1, 2, \ldots,$ as described more generally in Sec. 22-2. Thus, in this spirit, we may say that Z_n *converges almost surely to zero* iff for every $\epsilon > 0$

$$\lim_{n\to\infty}\left\{\lim_{m\to\infty} b_{n,m}\right\} = 0 \qquad \text{where } b_{n,m} = P\left(\bigcup_{k=n}^{m}[|Z_k| \geq \epsilon]\right).$$

If $n \leq n' \leq m' \leq m$ then obviously $0 \leq b_{n',m'} \leq b_{n,m} \leq 1$. Therefore $b_{n,m}$ increases with m, so $b_n = \lim_{m\to\infty} b_{n,m}$ always exists. Letting m go to infinity in $b_{n+1,m} \leq b_{n,m}$, we see that $b_{n+1} \leq b_n$ so that b_n decreases as n increases. *Therefore* $b = \lim_{n\to\infty} b_n$ *is always a* well-defined number; and we see that $Z_n \xrightarrow{\text{a.s.}} 0$, iff, $b = 0$ for all $\epsilon > 0$.

Theorem 1: An asymptotic version of the SLLN If $X_1, X_2, \ldots$ are independent and identically distributed random variables with $EX_1^2 < \infty$ then

$$\frac{1}{n}(X_1 + \cdots + X_n) \xrightarrow{\text{a.s.}} EX_1 \qquad \text{as } n \to \infty. \quad \blacksquare$$

Proof Let $a_{n,\epsilon} = 2(\text{Var } X_1)/\epsilon^2 n$ and note that Theorem 3 (Sec. 12-3) yields condition (*c*) of Lemma 1. ∎

Thus, with probability 1, the arithmetic average reward converges to the expected reward for one trial. In particular, with probability 1, the fraction of successes in Bernoulli trials converges to the probability of success on one trial.

Theorem 2: Kolmogorov's criterion Let $X_1, X_2, \ldots$ be an infinite sequence of independent random variables with each $EX_k^2 < \infty$. If

$$\sum_{k=1}^{\infty} \frac{\text{Var } X_k}{k^2} < \infty$$

then

$$\frac{1}{n}\sum_{k=1}^{n}(X_k - EX_k) \xrightarrow{\text{a.s.}} 0 \qquad \text{as } n \to \infty. \quad \blacksquare$$

Proof Theorem 1 (Sec. 12-4) with $\epsilon_k = k\epsilon > 0$ yields

$$P\left(\bigcup_{k=n}^{m}\left[\left|\frac{1}{k}\sum_{k=1}^{n}(X_k - EX_k)\right| \geq \epsilon\right]\right) \leq a_{n,\epsilon}$$

where

$$a_{n,\epsilon} = \frac{1}{\epsilon^2}\left[\left(\frac{1}{n^2}\sum_{k=1}^{n}\text{Var } X_k\right) + \left\{\sum_{k=n+1}^{\infty}\frac{\text{Var } X_k}{k^2}\right\}\right].$$

Clearly the term in { } goes to zero as n goes to infinity, and the Kronecker lemma of Exercise 2 (Sec. 20-1) shows that

$$\frac{1}{n^2}\sum_{i=1}^{n}\text{Var } X_i = \frac{1}{n^2}\sum_{k=1}^{n}k^2\left(\frac{\text{Var } X_k}{k^2}\right) \to 0 \qquad \text{as } n \to \infty.$$

Thus $a_{n,\epsilon}$ goes to zero as n goes to infinity. ∎

Example 1 To apply Theorem 2 we specify that $X_1, X_2, \ldots$ are independent and we specify the distribution of *every* X_k. Let

$$P[X_k = k^{.49}] = P[X_k = -k^{.49}] = .5$$

so that each X_k has zero mean and takes only two values. Then

$$\sum_{k=1}^{\infty}\frac{EX_k^2}{k^2} = \sum_{k=1}^{\infty}\frac{k^{.98}}{k^2} = \sum_{k=1}^{\infty}\frac{1}{k^{1.02}} < \infty$$

so $(1/n)(X_1 + \cdots + X_n) \xrightarrow{\text{a.s.}} 0$. Thus independent errors X_k can grow in magnitude $|X_k|$ almost as fast as $\sqrt{k}$ and still the arithmetic average error will converge to zero with probability 1 because of cancellations of large positive and negative errors. ∎

If $Z_1, Z_2, \ldots$ are all defined on the same probability space, and if real numbers $c_1, c_2, \ldots$ converge to zero, then the following two events are obviously equal:

$$\left[\lim_{n\to\infty} Z_n = 0\right] = \left[\lim_{n\to\infty} (Z_n + c_n) = 0\right]$$

yielding the following lemma.

Lemma 2 Let $Z_1, Z_2, \ldots$ be a sequence of random variables such that $Z_n \xrightarrow{\text{a.s.}} 0$. If $c_1, c_2, \ldots$ is a sequence of real numbers converging to zero then $Z_n + c_n \xrightarrow{\text{a.s.}} 0$. ■

Lemma 3 Let $X_1, Y_1, X_2, Y_2, \ldots$ be a sequence of random variables satisfying

$$\sum_{k=1}^{\infty} P[X_k \neq Y_k] < \infty.$$

Let $S_k = X_1 + \cdots + X_k$ and $T_k = Y_1 + \cdots + Y_k$. If $(1/k)T_k \xrightarrow{\text{a.s.}} 0$ then $(1/k)S_k \xrightarrow{\text{a.s.}} 0$. ■

Thus if the terms of the sequences $X_1, X_2, \ldots$ and $Y_1, Y_2, \ldots$ differ seldom enough, in the sense of the assumption in the lemma, then either both or neither of the arithmetic averages converges almost surely to zero.

Proof of Lemma 3 If

$$A = \{\omega\colon X_k(\omega) \neq Y_k(\omega) \text{ for infinitely many } k\}$$

then for each n the event A is a subset of

$$\bigcup_{k=n}^{\infty} [X_k \neq Y_k]$$

which, by assumption, has arbitrarily small probability for large n. Therefore $P(A) = 0$ so that

$$A^c = \{\omega\colon X_k(\omega) \neq Y_k(\omega) \text{ for only finitely many } k\}$$

has probability 1. If $(1/k)T_k(\omega)$ goes to zero as k goes to infinity, and if ω belongs to A^c then clearly $(1/k)S_k(\omega)$ goes to zero as k goes to infinity. Thus

$$\left[\lim_{n\to\infty} \frac{1}{n} T_n = 0\right] \cap A^c \text{ is a subset of } \left[\lim_{n\to\infty} \frac{1}{n} S_n\right].$$

If $(1/n)T_n \xrightarrow{\text{a.s.}} 0$ then the left-hand event has probability 1, and hence $(1/n)S_n \xrightarrow{\text{a.s.}} 0$. ■

The next lemma is a special truncation result needed in the proof of Theorem 3.

Lemma 4 If $E|X| < \infty$ then

$$\sum_{k=1}^{\infty} \frac{E[t_k(X)]^2}{k^2} \leq 2(1 + E|X|). \quad ■$$

Proof Applying Lemma 1 (Sec. 20-1) to $t_k^2(X)$ yields

$$\sum_{k=1}^{\infty} \frac{Et_k^2(X)}{k^2} \leq \sum_{k=1}^{\infty} \sum_{n=1}^{k} a_{k,n} \qquad \text{where } a_{k,n} = \frac{n^2}{k^2} P[n-1 < |X| \leq n].$$

The double sum above is the sum of the "row sums" in the triangular array of Fig. 1.

$$\begin{array}{ccccc} a_{11} & & & & \\ a_{21} & a_{22} & & & \\ a_{31} & a_{32} & a_{33} & & \\ \cdots & \cdots & \cdots & \cdots & \cdots \\ a_{k1} & a_{k2} & a_{k3} & \cdots & a_{kk} \\ \cdots & \cdots & \cdots & \cdots & \cdots \end{array}$$

(rows indexed by k downward, columns by n to the right)

Fig. 1

The first inequality in the following chain was shown in connection with Fig. 2 (Sec. 12-4):

$$\sum_{k=n}^{\infty} \frac{1}{k^2} = \frac{1}{n^2} + \sum_{k=n+1}^{\infty} \frac{1}{k^2} \leq \frac{1}{n^2} + \int_n^{\infty} \frac{dx}{x^2} = \frac{1}{n^2} + \frac{1}{n} \leq \frac{1}{n} + \frac{1}{n}.$$

Therefore the sum of the entries in the nth column of Fig. 1 is

$$n^2 P[n-1 < |X| \leq n] \sum_{k=n}^{\infty} \frac{1}{k^2} \leq 2nP[n-1 < |X| \leq n].$$

Adding these "column-sum" bounds over $n = 1, 2, \ldots$ yields

$$2 \sum_{n=1}^{\infty} nP[n-1 < |X| \leq n] \leq 2(1 + E|X|)$$

where we used Lemma 1 (Sec. 20-1). ■

Theorem 3: The SLLN for independent identically distributed random variables having expectations Let $X_1, X_2, \ldots$ be a sequence of independent and identically distributed random variables satisfying $E|X_1| < \infty$. If $S_k = X_1 + \cdots + X_k$ then $(1/k)S_k \xrightarrow{\text{a.s.}} EX_1$. ■

Proof Without loss of generality we may assume that $EX_1 = 0$. Let $T_k = Y_1 + \cdots + Y_k$ where $Y_i = t_i(X_i)$. Then $Y_1, Y_2, \ldots$ are independent, but in general they are not identically distributed, nor is EY_i zero. The X_i are identically distributed so

$$\sum_{k=1}^{\infty} P[X_k \neq Y_k] = \sum_{k=1}^{\infty} P[|X_k| > k] = \sum_{k=1}^{\infty} P[|X_1| > k].$$

The last sum above is finite by Lemma 1 (Sec. 20-1), since $E|X_1| < \infty$. Using Lemma 3 we need only show that $(1/k)T_k \xrightarrow{\text{a.s.}} 0$. Now $0 = EX_1 = Et_i(X_1) + Ed_i(X_1)$ so

$$\frac{1}{k} ET_k = \frac{1}{k} \sum_{i=1}^{k} Et_i(X_i) = \frac{1}{k} \sum_{i=1}^{k} Et_i(X_1) = \frac{-1}{k} \sum_{i=1}^{k} Ed_i(X_1).$$

By Lemma 2*a* (Sec. 20-1) we know that $Ed_i(X_1)$ goes to zero as i goes to infinity, and hence the Toeplitz lemma of Exercise 1 (Sec. 20-1) shows that $(1/k)ET_k$ goes to zero. But

$$\frac{1}{k}T_k = \frac{T_k - ET_k}{k} + \frac{ET_k}{k}$$

so from Lemma 2 we need only show that $(1/k)(T_k - ET_k) \xrightarrow{\text{a.s.}} 0$. From Theorem 2 we need only show that the following left-hand sum is finite:

$$\sum_{k=1}^{\infty} \frac{\text{Var } Y_k}{k^2} \leq \sum_{k=1}^{\infty} \frac{Et_k^2(X_k)}{k^2} = \sum_{k=1}^{\infty} \frac{Et_k^2(X_1)}{k^2}.$$

But the right-hand sum is finite from Lemma 4, which completes the proof. ■

EXERCISES

1. *The SLLN Implies the WLLN.* If $Z_1, Z_2, \ldots$ is a sequence of random variables we say that Z_n *converges to zero in probability* iff for all $\epsilon > 0$, $P[|Z_n| \geq \epsilon]$ converges to zero as n increases to infinity. The notation $Z_n \xrightarrow{P} 0$ is standard. If d is a real number then $Z_n \xrightarrow{P} d$ means that $Z_n - d \xrightarrow{P} 0$. In terms of distribution functions, we have $Z_n \xrightarrow{P} 0$ iff

$$\lim_{n\to\infty} F_{Z_n}(z) = \begin{cases} 0 & \text{for all } z < 0 \\ 1 & \text{for all } z > 0. \end{cases}$$

Thus this type of convergence depends only on the individual distributions of the Z_i, and not otherwise on their joint distributions. The conclusion of the WLLN of Theorem 1 (Sec. 20-1) can be stated in the form $(1/n)S_n \xrightarrow{P} EX_1$.

Let $Z_1, Z_2, \ldots$ be a sequence of random variables and assume that $Z_k \xrightarrow{\text{a.s.}} d$ where d is real number. Prove that $Z_k \xrightarrow{P} d$. HINT: $P[|Z_n| \geq \epsilon] \leq P\left(\bigcup_{k=n}^{m} [|Z_k| \geq \epsilon]\right)$.

COMMENT: Clearly, then, the SLLN of Theorem 3 immediately implies the WLLN of Theorem 1 (Sec. 20-1).

2. *A Converse to Theorem 3*

(*a*) Let $Z_1, Z_2, \ldots$ be independent random variables such that $Z_n \xrightarrow{\text{a.s.}} 0$. Let $\epsilon > 0$ and prove that $\sum_{n=1}^{\infty} P[|Z_n| \geq \epsilon] < \infty$. HINT: Apply Exercise 3*a* (Sec. 7-4).

(*b*) Let $X_1, X_2, \ldots$ be a sequence of independent identically distributed random variables. Let $S_k = X_1 + \cdots + X_k$ and assume that there is a real number d such that $(1/k)S_k \xrightarrow{\text{a.s.}} d$. Prove that $E|X_1| < \infty$. Naturally, then, $d = EX_1$ from Theorem 3. HINT:

$$Z_n = \frac{X_n}{n} = \frac{S_n}{n} - \frac{n-1}{n}\frac{S_{n-1}}{n-1} \xrightarrow{\text{a.s.}} 0.$$

3. *A Partial Converse to the Kolmogorov Criterion.* Let σ_k for $k = 1, 2, \ldots$ be any sequence of positive numbers such that $\sum_{k=1}^{\infty} (\sigma_k)^2/k^2 = \infty$. This exercise shows that there exists a sequence $X_1, X_2, \ldots$ of independent random variables satisfying $EX_k = 0$ and Var $X_k = (\sigma_k^2)$ but such that $(1/n)(X_1 + \cdots + X_n)$ does not converge almost surely to zero. Thus if the assumption of Theorem 2 is violated then the conclusion is also violated, *at least for some distributions.*

If $\sigma_k/k > 1$ define the distribution of X_k by

$$P[X_k = \sigma_k] = P[X_k = -\sigma_k] = \tfrac{1}{2}$$

so that $P[|X_k| \geq k] = 1$. If $\sigma_k/k \leq 1$ define the distribution of X_k by

$$P[X_k = -k] = P[X_k = k] = \frac{\sigma_k^2}{2k^2} \qquad \text{and} \qquad P[X_k = 0] = 1 - \frac{\sigma_k^2}{k^2}$$

so that $P[|X_k| \geq k] = \sigma_k^2/k^2$.

(*a*) Prove that $\sum_{k=1}^{\infty} P[|X_k| \geq k] = \infty$.

(*b*) Let $S_k = X_1 + \cdots + X_k$ and show that $\bigcup_{k=n+1}^{m} [|X_k| \geq k]$ is contained in $\bigcup_{k=n}^{m} [|S_k| \geq k/2]$.

HINT: $|X_k| \leq |S_k| + |S_{k-1}|$.

(*c*) For each n prove that

$$\lim_{m\to\infty} P\left(\bigcup_{k=n}^{m} \left[\left|\frac{1}{k} S_k\right| \geq \tfrac{1}{2}\right]\right) = 1$$

so that Lemma 1*b* implies that $(1/n)S_n$ cannot converge almost surely to zero.

21 Probability Spaces: A Rigorous Presentation

This chapter shows how Chap. 15 may be modified so as to agree with the theory which was systematized by Kolmogorov[1] in 1933 and has been the basis for most advanced probability research since then. To attain this goal it seems best just to start and carry a rigorous development of the theory fairly far along. This is easily done, since only brief comments are needed to justify many of our earlier proofs. Also there is little need to discuss the significance of theorems we have used so often. The more mathematical and technical results are placed in the lemmas, whose proofs may well be omitted on first reading. The theory is developed only through the introduction of expectation, since our goal is only to firmly bridge the gap between our presentation and the many excellent presentations of the advanced theory.

In Sec. 21-1 we modify the definition of a probability space by permitting $P(A)$ to be defined for only *some* subsets A of the sample space Ω. Henceforth only these subsets are called events. Sec. 21-2 introduces the special subsets of R^n which are normally used for events when $\Omega = R^n$. Section 21-3 introduces random variables, and functions of them, in such a way that our standard calculations never require $P(A)$ when A is not an event. Induced distributions, intermediate distributions, independence, etc. remain much as before. Section 21-4 introduces expectations and then closes with a few incidental comments on the further development of the theory.

[1] A. N. Kolmogorov, "Foundations of the Theory of Probability," Chelsea Publishing Company, New York, 1950, a translation of the original German monograph. The basis of this theory is measure theory, or the theory of integration over abstract spaces, for which there are a great many excellent modern texts. Many of these do not go into probability theory. There is a chapter on probability in the classic Paul R. Halmos, "Measure Theory," D. Van Nostrand Company, Inc., Princeton, N.J., 1950. The encyclopedic Michel Loève, "Probability Theory," 3d ed., D. Van Nostrand Company Inc., Princeton, N.J., 1963, contains a complete concise treatment of the necessary measure theory, as does the brief Jacques Neveu, "Mathematical Foundations of the Calculus of Probability," Holden-Day, Inc., San Francisco, 1965. Some measure theory appears in the classic J. L. Doob, "Stochastic Processes," John Wiley & Sons, Inc., New York, 1953, and also in Kai Lai Chung, "A Course in Probability Theory," Harcourt, Brace & World, Inc., New York, 1968, and William Feller, "An Introduction to Probability Theory and Its Applications," vol. II, John Wiley & Sons, Inc., New York, 1966, H. G. Tucker, "A Graduate Course in Probability," Academic Press, Inc., New York, 1967, and L. Breiman, "Probability," Addison-Wesley Publishing Company, Inc., Reading, Mass., 1968.

21-1 PROBABILITY SPACES

As explained early in Sec. 16-1, it is sometimes impossible to define $P(A)$ for every subset A of a sample space and still have P nicely behaved. Therefore we only require that $P(A)$ be defined for certain subsets A, and we call these subsets *events*. It is important that the collection of events be closed under complementation and countable unions and intersections. That is, use of "not" or countable use of "or" and "and" should not lead us out of the class of events. We first attach a name to such a collection $\mathscr{A}$ of subsets of a set Ω. Lemma 1 shows that a few of the desired properties of $\mathscr{A}$ imply the rest. We then define a probability space as requiring that P need only be defined on such a collection. Conditioned distributions and other elementary properties remain essentially unchanged.

A class $\mathscr{A}$ of subsets of a nonempty set Ω is called a **σ field** (read sigma field) of subsets of Ω iff it has the following properties:

1. φ belongs to $\mathscr{A}$, and Ω belongs to $\mathscr{A}$.
2. A^c belongs to $\mathscr{A}$ whenever A belongs to $\mathscr{A}$.
3. Both $\bigcup_{k=1}^{n} A_k$ and $\bigcap_{k=1}^{n} A_k$ belong to $\mathscr{A}$ whenever each of $A_1, \ldots, A_n$ belongs to $\mathscr{A}$.
4. Both $\bigcup_{k=1}^{\infty} A_k$ and $\bigcap_{k=1}^{\infty} A_k$ belong to $\mathscr{A}$ whenever each of $A_1, A_2, \ldots$ belongs to $\mathscr{A}$.

Lemma 1 Let $\mathscr{A}$ be a class of subsets of a nonempty set Ω. Suppose that $\varphi \in \mathscr{A}$, and $A^c \in \mathscr{A}$ whenever $A \in \mathscr{A}$. Also suppose that $\left(\bigcup_{k=1}^{\infty} A_k\right) \in \mathscr{A}$ whenever each of $A_1, A_2, \ldots \in \mathscr{A}$. Then $\mathscr{A}$ is a σ field. ■

Proof Clearly $\Omega = \varphi^c$ belongs to $\mathscr{A}$. If $A_1, A_2, \ldots$ each belong to $\mathscr{A}$ then so do $A_1{}^c, A_2{}^c, \ldots$, while

$$\bigcap_{k=1}^{\infty} A_k = \left(\bigcup_{k=1}^{\infty} A_k{}^c\right)^c$$

so the left-hand side belongs to $\mathscr{A}$. If $A_1, \ldots, A_n$ each belongs to $\mathscr{A}$ then so do $\bigcup_{k=1}^{n} A_k = \bigcup_{k=1}^{\infty} A_k$ where we let $A_k = \varphi$ for $k > n$, and $\bigcap_{k=1}^{n} A_k = \bigcap_{k=1}^{\infty} A_k$ where we let $A_k = \Omega$ for $k > n$. ■

A **probability space** $(\Omega, \mathscr{A}, P)$ consists of a nonempty set Ω, a σ field $\mathscr{A}$ of subsets of Ω, and a real-valued function P defined on $\mathscr{A}$ and satisfying the three conditions below. The set Ω is called the **sample space** and each member ω of Ω is called a **sample point.** Any subset A of Ω is called an **event** iff $A \in \mathscr{A}$, and in this case $P(A)$ is called the **probability of** A.

1. $P(\Omega) = 1$.
2. $P(A) \geq 0$ for every event A.
3. If $A_1, A_2, \ldots$ is any sequence of disjoint events then

$$P\left(\bigcup_{n=1}^{\infty} A_n\right) = \sum_{n=1}^{\infty} P(A_n).$$

The function P is called a **probability measure.** Thus a probability measure is a normed, nonnegative, countably additive set function defined on the σ field of all events.

Theorem 2 (Sec. 6-4) collected together a number of elementary properties of P. The theorem and its proof are valid for any probability space $(\Omega,\mathscr{A},P)$. The definition in Chap. 7 of conditional probability, $P(B \mid A)$, and the definitions and properties of independent events remain as before. Theorem 1 (Sec. 15-1) concerned P_A, and it is restated as Theorem 1 below. The earlier proof is valid.

Theorem 1 Let $(\Omega,\mathscr{A},P)$ be a probability space and A an event with $P(A) > 0$. If the function P_A is defined for every event B by

$$P_A(B) = P(B \mid A) = \frac{P(AB)}{P(A)}$$

then $(\Omega,\mathscr{A},P_A)$ is a probability space, which is said to be obtained by **conditioning** $(\Omega,\mathscr{A},P)$ **by** A. ■

Thus P is now defined only for events, and not for every subset. But our proofs used only operations which do not lead us outside the class of events, so the proofs still apply. In the discrete case it is possible, but not necessary, to let $\mathscr{A}$ be the class of all subsets of Ω. As will be clear from the next section, in order to preserve the same formalism in all cases, we usually let $\mathscr{A}$ be a smaller collection even in the discrete case.

21-2 BOREL SETS

Suppose that we have some set Ω which we plan to use as a sample space. Suppose that we also have some collection $\mathscr{C}$ of subsets of Ω, and we insist that every set C belonging to $\mathscr{C}$ is to be an event. If $\mathscr{C}$ does not happen to be a σ field then it is reasonable to enlarge $\mathscr{C}$ to some larger collection $\mathscr{A}$ of sets which is a σ field. Among such possible $\mathscr{A}$ we may be content, or required, to use the smallest such $\mathscr{A}$. Lemma 1 shows that a unique such smallest $\mathscr{A}$ always exists, and it is called the *σ field generated by $\mathscr{C}$*.

Lemma 2 concerns $\Omega = R^n$ and shows, roughly speaking, that if $\mathscr{C}$ is *any* rich enough collection of simple subsets of R^n then we always get the *same* σ field $\mathscr{B}_n$ generated by $\mathscr{C}$. We call $\mathscr{B}_n$ the *σ field of Borel sets in R^n*. Thus there is an essentially unique smallest σ field $\mathscr{B}_n$ of subsets of R^n which contains all subsets of practical interest. A constructive description of $\mathscr{B}_n$ simpler than its definition does not seem to exist. We do not prove the nonobvious fact that there are subsets of R^n which do not belong to $\mathscr{B}_n$. None such has ever been "explicitly" exhibited, although it can be proved that they exist. Usually $\mathscr{B}_n$ is used as the σ field of events when R^n is the sample space.

Lemma 3 shows that if two probability measures agree on a collection $\mathscr{C}$ of events which is closed under finite intersections then the two probability measures must also agree on the σ field generated by $\mathscr{C}$. Theorem 1 is an immediate consequence of Lemma 3, and it shows that if two probability measures agree on all closed rectangles in R^n then they must agree on $\mathscr{B}_n$. Thus if we assign a probability to each closed rectangle in R^n then there is at most one way to extend to a probability measure on all Borel sets. That is, the probabilities for closed rectangles determine uniquely the probability of every complicated Borel set.

The proof of Lemma 1 essentially defines the σ field $\sigma(\mathscr{C})$ generated by $\mathscr{C}$, to be the intersection of all σ fields containing $\mathscr{C}$. It easily follows that $\sigma(\mathscr{C})$ is the unique σ field contained in every σ field containing $\mathscr{C}$.

Lemma 1 Let $\mathscr{C}$ be a nonempty class of subsets of a nonempty set Ω. Then there is a *unique* σ field $\sigma(\mathscr{C})$ of subsets of Ω, called the **smallest σ field containing** $\mathscr{C}$, or the **σ field generated by** $\mathscr{C}$, which satisfies the following conditions:
(*a*) $\mathscr{C} \subset \sigma(\mathscr{C})$; that is, C belongs to $\sigma(\mathscr{C})$ whenever C belongs to $\mathscr{C}$.
(*b*) If $\mathscr{B}$ is any σ field of subsets of Ω such that $\mathscr{C} \subset \mathscr{B}$ then $\sigma(\mathscr{C}) \subset \mathscr{B}$. ■

Proof Let $\mathscr{B}_{\mathscr{C}}$ denote any σ field of subsets of Ω such that $\mathscr{C}$ is contained in $\mathscr{B}_{\mathscr{C}}$. For example, the class of all subsets of Ω is such a σ field. Define $\sigma(\mathscr{C})$ as follows: a subset A of Ω belongs to $\sigma(\mathscr{C})$ iff A belongs to *every* $\mathscr{B}_{\mathscr{C}}$. If C belongs to $\mathscr{C}$ then C belongs to every $\mathscr{B}_{\mathscr{C}}$, so C belongs to $\sigma(\mathscr{C})$. Hence $\mathscr{C}$ is contained in $\sigma(\mathscr{C})$.

We next show that $\sigma(\mathscr{C})$ is a σ field. Clearly φ belongs to every $\mathscr{B}_{\mathscr{C}}$, so φ belongs to $\sigma(\mathscr{C})$. If A belongs to $\sigma(\mathscr{C})$ then A belongs to every $\mathscr{B}_{\mathscr{C}}$, so A^c belongs to every $\mathscr{B}_{\mathscr{C}}$, and hence A^c belongs to $\sigma(\mathscr{C})$. If each of $A_1, A_2, \ldots$ belongs to $\sigma(\mathscr{C})$ then each of $A_1, A_2, \ldots$ belongs to every $\mathscr{B}_{\mathscr{C}}$, so $\bigcup_{k=1}^{\infty} A_k$ belongs to every $\mathscr{B}_{\mathscr{C}}$, and hence $\bigcup_{k=1}^{\infty} A_k$ belongs to $\sigma(\mathscr{C})$. Therefore Lemma 1 (Sec. 20-1) shows $\sigma(\mathscr{C})$ to be a σ field.

Take any $\mathscr{B}_{\mathscr{C}}$. If A belongs to $\sigma(\mathscr{C})$ then A belongs to $\mathscr{B}_{\mathscr{C}}$, so condition (*b*) is satisfied.

Let $\mathscr{A}$ be any σ field of subsets of Ω such that $\mathscr{C}$ is contained in $\mathscr{A}$ and $\mathscr{A}$ is contained in every $\mathscr{B}_{\mathscr{C}}$. Clearly $\mathscr{A}$ is contained in $\sigma(\mathscr{C})$, which is a $\mathscr{B}_{\mathscr{C}}$, and also $\sigma(\mathscr{C})$ is contained in $\mathscr{A}$ by condition (*b*), so $\mathscr{A} = \sigma(\mathscr{C})$. Thus $\sigma(\mathscr{C})$ is the *unique* σ field satisfying conditions (*a*) and (*b*). ■

A set A in R^n is called an **open set** in R^n iff for every x in A there is some open rectangle in R^n such that x belongs to the open rectangle and the open rectangle is contained in A. The rectangle definitions were stated in Sec. 15-1.

Lemma 2 Let each $\mathscr{C}_i$ be a class of subsets of R^n, defined as follows:
$\mathscr{C}_1 =$ the class of all closed rectangles
$\mathscr{C}_2 =$ the class of all open rectangles
$\mathscr{C}_3 =$ the class of all open sets
$\mathscr{C}_4 =$ the class of all sets of the form

$$\{(x_1, \ldots, x_n): x_i < b_i \text{ for all } i = 1, \ldots, n\}$$

where $b_1, \ldots, b_n$ can be any real numbers.

Then $\sigma(\mathscr{C}_1) = \sigma(\mathscr{C}_2) = \sigma(\mathscr{C}_3) = \sigma(\mathscr{C}_4)$. This σ field is denoted by $\mathscr{B}_n$ and is called the **σ field of Borel sets** in R^n. A subset of R^n is called a **Borel set** iff it belongs to $\mathscr{B}_n$. ■

Proof We first note an obvious fact which will be used in the proof. If $\mathscr{C}$ and $\mathscr{C}'$ are collections of subsets of Ω and $\mathscr{C} \subset \mathscr{C}'$ then $\sigma(\mathscr{C}) \subset \sigma(\mathscr{C}')$, because $\sigma(\mathscr{C}')$ is a σ field containing $\mathscr{C}$. More generally if every set in $\mathscr{C}$ can be obtained not necessarily as a set in $\mathscr{C}'$, but at least by using complementations and countable unions and intersections of sets in $\mathscr{C}'$, then $\mathscr{C} \subset \sigma(\mathscr{C}')$, so that $\sigma(\mathscr{C}) \subset \sigma(\mathscr{C}')$.

We first show that $\sigma(\mathscr{C}_1) = \sigma(\mathscr{C}_2)$. The union over $k = 1, 2, \ldots$ of the closed rectangles

$$\left\{(x_1, \ldots, x_n): a_i + \frac{1}{k} \le x_i \le b_i - \frac{1}{k} \text{ for } i = 1, \ldots, n\right\}$$

is the open rectangle determined by a and b, so the open rectangle belongs to $\sigma(\mathscr{C}_1)$. Thus $\mathscr{C}_2 \subset \sigma(\mathscr{C}_1)$ so $\sigma(\mathscr{C}_2) \subset \sigma(\mathscr{C}_1)$. The intersection over $k = 1, 2, \ldots$ of the open rectangles

$$\left\{(x_1, \ldots, x_n): a_i - \frac{1}{k} < x_i < b_i + \frac{1}{k} \text{ for } i = 1, \ldots, n\right\}$$

is the closed rectangle determined by a and b, and it must belong to $\sigma(\mathscr{C}_2)$. Thus $\mathscr{C}_1 \subset \sigma(\mathscr{C}_2)$, so $\sigma(\mathscr{C}_1) \subset \sigma(\mathscr{C}_2)$. Hence $\sigma(\mathscr{C}_1) = \sigma(\mathscr{C}_2)$.

We now show that $\sigma(\mathscr{C}_2) = \sigma(\mathscr{C}_4)$. Taking the union over $k = 1, 2, \ldots$ of the open rectangles

$$\{(x_1, \ldots, x_n): -k < x_i < b_i \text{ for } i = 1, \ldots, n\}$$

we obtain a set in $\mathscr{C}_4$. Thus $\mathscr{C}_4 \subset \sigma(\mathscr{C}_2)$, so $\sigma(\mathscr{C}_4) \subset \sigma(\mathscr{C}_2)$. Taking the union over $k = 1, 2, \ldots$ of the sets

$$\{(x_1, \ldots, x_n): x_1 < k, \ldots, x_{i-1} < k, x_i < b_i, x_{i+1} < k, \ldots, x_n < k\}$$

in $\mathscr{C}_4$, we obtain the set $\{(x_1, \ldots, x_n): x_i < b_i\}$ having a constraint on only one coordinate, and this set must belong to $\sigma(\mathscr{C}_4)$. Therefore its complement

$$\{(x_1, \ldots, x_n): x_i \ge b_i\}$$

belongs to $\sigma(\mathscr{C}_4)$. Intersecting two sets of the last two forms, we get one of the form

$$\{(x_1, \ldots, x_n): a_i \le x_i < b_i\}$$

which must belong to $\sigma(\mathscr{C}_4)$. Taking the union over $k = 1, 2, \ldots$ of

$$\left\{(x_1, \ldots, x_n): a_i + \frac{1}{k} \le x_i < b_i\right\}$$

we see that $\{(x_1, \ldots, x_n): a_i < x_i < b_i\}$ belongs to $\sigma(\mathscr{C}_4)$. The intersection of n such sets produces an open rectangle. Thus $\mathscr{C}_2 \subset \sigma(\mathscr{C}_4)$, so $\sigma(\mathscr{C}_2) \subset \sigma(\mathscr{C}_4)$. Hence $\sigma(\mathscr{C}_2) = \sigma(\mathscr{C}_4)$.

We now show that $\sigma(\mathscr{C}_2) = \sigma(\mathscr{C}_3)$. Clearly $\mathscr{C}_2 \subset \mathscr{C}_3$ so $\sigma(\mathscr{C}_2) \subset \sigma(\mathscr{C}_3)$. Let O be any open set in R^n. Consider the countable collection of all open rectangles all of whose end points a_i and b_i are rational numbers. Then O equals the union of the countably many such rational open rectangles which are contained in O. We used the obvious fact that if x belongs to O then x belongs to at least one such rational open rectangle contained in O. Thus O belongs to $\sigma(\mathscr{C}_2)$. Hence $\mathscr{C}_3 \subset \sigma(\mathscr{C}_2)$, so $\sigma(\mathscr{C}_3) \subset \sigma(\mathscr{C}_2)$. Therefore $\sigma(\mathscr{C}_3) = \sigma(\mathscr{C}_2)$. ■

Lemma 3 Let $(\Omega,\mathscr{A},P)$ and $(\Omega,\mathscr{A},Q)$ be probability spaces such that $P(C) = Q(C)$ whenever C belongs to $\mathscr{C}$, where $\mathscr{C}$ is some collection of events. Suppose that $(C \cap C')$ belongs to $\mathscr{C}$ whenever C and C' belong to $\mathscr{C}$. Then $P(B) = Q(B)$ whenever B belongs to $\sigma(\mathscr{C})$. ■

Proof We consider Ω, $\mathscr{A}$, and $\mathscr{C}$ to be fixed, but we use various P and Q satisfying the stated assumptions. If $(\Omega,\mathscr{A},P)$ and $(\Omega,\mathscr{A},Q)$ satisfy $P(C) = Q(C)$ whenever $C \in \mathscr{C}$ then we define

$$\mathscr{A}_{P,Q} = \{A\colon A \in \mathscr{A} \text{ and } P(A) = Q(A)\}$$

to be the class of all events on which P and Q agree. Clearly $\mathscr{C} \subset \mathscr{A}_{P,Q}$ and $\varphi \in \mathscr{A}_{P,Q}$. If $A \in \mathscr{A}_{P,Q}$ then $P(A^c) = 1 - P(A) = 1 - Q(A) = Q(A^c)$, so $A^c \in \mathscr{A}_{P,Q}$. If A_1, $A_2, \ldots$ are *disjoint* events each in $\mathscr{A}_{P,Q}$ then

$$P\left(\bigcup_{n=1}^{\infty} A_n\right) = \sum_{n=1}^{\infty} P(A_n) = \sum_{n=1}^{\infty} Q(A_n) = Q\left(\bigcup_{n=1}^{\infty} A_n\right)$$

so that $\left(\bigcup_{n=1}^{\infty} A_n\right) \in \mathscr{A}_{P,Q}$.

Suppose for a moment that $\mathscr{A}_{P,Q}$ is closed under finite intersections. Then if $A_1, A_2, \ldots$ are any events in $\mathscr{A}_{P,Q}$ then each $A_1^c \cdots A_{n-1}^c A_n$ is in $\mathscr{A}_{P,Q}$, so that $\bigcup_{n=1}^{\infty} A_n = \bigcup_{n=1}^{\infty} (A_1^c \cdots A_{n-1}^c A_n)$ can be written as a countable *disjoint* union of events in $\mathscr{A}_{P,Q}$, hence $\bigcup_{n=1}^{\infty} A_n \in \mathscr{A}_{P,Q}$. Then from Lemma 1 (Sec. 21-1) $\mathscr{A}_{P,Q}$ is a σ field containing $\mathscr{C}$, and hence $\mathscr{A}_{P,Q}$ contains $\sigma(\mathscr{C})$ and the lemma is proved. To actually prove the lemma we proceed in a less direct fashion than this.

Let $\mathscr{B}$ consist of those events which belong to every $\mathscr{A}_{P,Q}$. Thus an event B belongs to $\mathscr{B}$ iff *every* P and Q which agree on $\mathscr{C}$ also agree at B. Then $\varphi \in \mathscr{B}$ and $\mathscr{C} \subset \mathscr{B}$, and $\mathscr{B}$ is closed under complements and countable disjoint unions, because every $\mathscr{A}_{P,Q}$ has these properties. For example, if $A \in \mathscr{B}$ then A belongs to every $\mathscr{A}_{P,Q}$, so A^c belongs to every $\mathscr{A}_{P,Q}$; hence $A^c \in \mathscr{B}$. We will prove that $\mathscr{B}$ is closed under finite intersections. It then follows that $\mathscr{B}$ is a σ field containing $\mathscr{C}$, and hence $\mathscr{B}$ contains $\sigma(\mathscr{C})$, so that every $\mathscr{A}_{P,Q}$ contains $\sigma(\mathscr{C})$.

We first show that

$$(BC) \in \mathscr{B} \qquad \text{whenever } B \in \mathscr{B},\ C \in \mathscr{C}.$$

Select any $C \in \mathscr{C}$ and any P and Q agreeing on $\mathscr{C}$. We must show that

$$P(BC) = Q(BC) \qquad \text{whenever } B \in \mathscr{B}.$$

This is certainly true if $P(C) = Q(C) = 0$. If $P(C) = Q(C) > 0$ then we condition on the event C. Then

$$P_C(A) = \frac{P(AC)}{P(C)} = \frac{Q(AC)}{Q(C)} = Q_C(A)$$

holds whenever $A \in \mathscr{C}$, because $\mathscr{C}$ is closed under finite intersections. Hence P_C and Q_C agree on $\mathscr{C}$, so that P_C and Q_C must agree on $\mathscr{B}$, as was to be shown.

We now show that

$$BB' \in \mathscr{B} \qquad \text{whenever } B \in \mathscr{B},\ B' \in \mathscr{B}.$$

Select any $B \in \mathscr{B}$ and any P and Q agreeing on $\mathscr{C}$. We must show that

$$P(BB') = Q(BB') \qquad \text{whenever } B' \in \mathscr{B}.$$

This is certainly true if $0 = P(B) = Q(B)$. If $P(B) = Q(B) > 0$ then we condition on the event B. Then

$$P_B(A) = \frac{P(AB)}{P(B)} = \frac{Q(AB)}{Q(B)} = Q_B(A)$$

holds whenever $A \in \mathscr{C}$, by the preceding paragraph. Hence P_B and Q_B agree on $\mathscr{C}$, so that P_B and Q_B must agree on $\mathscr{B}$, as was to be shown. ■

Let $(R^n,\mathscr{B}_n,P)$ be a probability space, where $\mathscr{B}_n$ is the σ field of Borel sets. As in Chap. 17, the **distribution function** F: $R^n \to R$ **corresponding to** $(R^n,\mathscr{B}_n,P)$ is defined by

$$F(x_1, \ldots, x_n) = P\{(x_1', \ldots, x_n'): x_1' < x_1, \ldots, x_n' < x_n\}$$

for all reals $x_1, \ldots, x_n$. We now prove a generalization of Theorem 2 (Sec. 15-1), which was for the case $\mathscr{C}_1$ of closed rectangles.

Theorem 1 Let $(R^n,\mathscr{B}_n,P)$ and $(R^n,\mathscr{B}_n,Q)$ be probability spaces. Let $\mathscr{C}$ be any one of the four classes $\mathscr{C}_1$, $\mathscr{C}_2$, $\mathscr{C}_3$, and $\mathscr{C}_4$ defined in Lemma 2. If

$$P(C) = Q(C) \qquad \text{whenever } C \in \mathscr{C}$$

then $P = Q$. In particular if both probability spaces have the same distribution function then $P = Q$. ■

Proof Each of the $\mathscr{C}_i$ in Lemma 2 is closed under finite intersections and generates $\mathscr{B}_n$, so the result for each $\mathscr{C}_i$ follows from Lemma 3. Also P and Q have the same distribution function iff they agree on $\mathscr{C}_4$. ■

We now state a standard result whose proof can be found in advanced texts on probability theory or measure theory. If F: $R^n \to R$ is a distribution function on R^n, as defined in Sec. 17-3, then there is a probability space $(R^n,\mathscr{B}_n,P)$ such that F is the distribution function corresponding to $(R^n,\mathscr{B}_n,P)$. This fact, together with Theorem 1 and Sec. 17-3, shows that the correspondence which assigns "to each $(R^n,\mathscr{B}_n,P)$ its corresponding distribution function F on R^n" is indeed a one-to-one correspondence.

21-3 RANDOM VARIABLES AND INDEPENDENCE

This section carries our theoretical development up to expected values, which are left to the next section. Only some of the subsets of the sample space are events, so we define random vectors, and functions of them, in such a way that their corresponding "naturally interesting subsets of the sample space" will always be events. We immediately obtain Theorems 1 and 2 on induced and intermediate distributions. Then in Lemma 1 we collect together those basic properties of these functions needed to justify our previous manipulations and proofs.

A **random vector** f **with associated probability space** $(\Omega,\mathscr{A},P)$ is any function $f\colon \Omega \to R^n$ such that $[f \in B]$ is an event whenever B is a Borel set in R^n, that is, iff $f^{-1}(B) \in \mathscr{A}$ whenever $B \in \mathscr{B}_n$. If $n = 1$ then f may be called a **random variable.** Thus we apply the name "random variable" to a real-valued function defined on the sample space, iff it has the desired property that for every Borel set B in R the set $f^{-1}(B) = \{\omega : f(\omega) \in B\}$ is an event and so has a probability assigned to it. We have the following theorem on induced distributions, with the same proof as for Theorem 1 (Sec. 15-2).

Theorem 1 If $f\colon \Omega \to R^n$ is a random vector with associated probability space $(\Omega,\mathscr{A},P)$ and if P_f is defined for every Borel set B in R^n by $P_f(B) = P[f \in B]$ then $(R^n,\mathscr{B}_n,P_f)$ is a probability space, which is said to be **induced** by f with associated $(\Omega,\mathscr{A},P)$ ∎

Let $(\Omega,\mathscr{A},P_A)$ be obtained by conditioning $(\Omega,\mathscr{A},P)$ by the event A with $P(A) > 0$. Then the random vector $f\colon \Omega \to R^n$ with associated $(\Omega,\mathscr{A},P_A)$ induces a probability space $(R^n,\mathscr{B}_n,P_{f|A})$ called the **conditional distribution of** f **given** A. Therefore $P_{f|A}$ is defined by

$$P_{f|A}(B) = P_A[f \in B] = P[f \in B \mid A] = \frac{P([f \in B] \cap A)}{P(A)} \qquad \text{for every } B \in \mathscr{B}_n.$$

A function $g\colon R^n \to R^m$ is said to be a **Borel function** iff

$$g^{-1}(B) \in \mathscr{B}_n \qquad \text{whenever } B \in \mathscr{B}_m$$

that is, iff inverse images of Borel sets are always Borel sets. *The random vectors on a probability space* $(R^n,\mathscr{B}_n,P)$ *are just the Borel functions.*

Let I_A be the indicator function of a non-Borel subset A of R^n. If $\{1\}$ is the Borel set in R containing just the one real number 1 then the inverse image of $\{1\}$ under I_A is just A, so I_A is not a Borel function. Thus functions on R^n which are not Borel functions do indeed exist, although none has ever been "explicitly" exhibited.

If $f\colon \Omega \to R^n$ is a random vector with associated $(\Omega,\mathscr{A},P)$ and if $g\colon R^n \to R^m$ is a Borel function then $h = g(f)$ is a random vector with associated $(\Omega,\mathscr{A},P)$. This fundamental fact is true because if $B \in \mathscr{B}_m$ then $g^{-1}(B) \in \mathscr{B}_n$ so that

$$h^{-1}(B) = f^{-1}(g^{-1}(B))$$

is an event. Thus *Borel functions of random vectors are random vectors*, and hence the reasons for considering Borel functions are obvious.

We have the following obvious theorem, with the same proof as for Theorem 2 (Sec. 15-2).

Theorem 2: Intermediate distributions If f is a random vector with associated probability space $(\Omega,\mathscr{A},P)$ and if

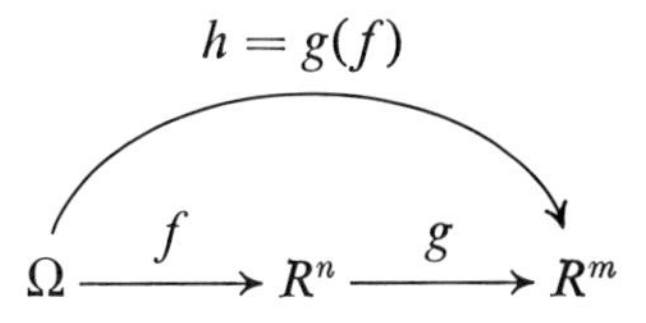

where g is a Borel function then $P_h = P_g$, where P_g is induced by g with associated $(R^n,\mathscr{B}_n,P_f)$. ■

If $(\Omega,\mathscr{A},P)$ is a probability space and f: $\Omega \to R^n$ is a function then by definition f is a random vector iff $f^{-1}(B) \in \mathscr{A}$ for every $B \in \mathscr{B}_n$. This definition does not involve P. Similarly the definition of a Borel function did not involve a P. We introduce the following definition in order to handle both cases simultaneously.

Let $\mathscr{A}$ be a σ field of subsets of a set Ω. A function f: $\Omega \to R^n$ is said to be $\mathscr{A}$ **measurable** iff $f^{-1}(B) \in \mathscr{A}$ whenever $B \in \mathscr{B}_n$, that is, iff all inverse images of Borel sets in R^n belong to $\mathscr{A}$.

Then f: $\Omega \to R^n$ is a random vector with associated probability space $(\Omega,\mathscr{A},P)$ iff f is $\mathscr{A}$ measurable. Also g: $R^n \to R^m$ is a Borel function iff g is $\mathscr{B}_n$ measurable.

Lemma 1 below says that if $\mathscr{A}$ is a σ field of subsets of Ω then f: $\Omega \to R^n$ is $\mathscr{A}$ measurable iff each of its real-valued components f_i in $f = (f_1, \ldots, f_n)$ is. Also to check for $\mathscr{A}$ measurability we need only take inverse images of closed rectangles, or open rectangles, etc. Furthermore the class of all $\mathscr{A}$ measurable functions into R^n has the fundamental property that it is closed under limits. The class of continuous functions, for example, does not have this property.

Lemma 1 Let $\mathscr{A}$ be a σ field of subsets of Ω.

(*a*) A function f: $\Omega \to R^n$ is $\mathscr{A}$ measurable iff each of its components f_i: $\Omega \to R$ is $\mathscr{A}$ measurable, where, as usual, $f = (f_1, \ldots, f_n)$.

(*b*) Let $\mathscr{C}$ be any one of the four classes $\mathscr{C}_1$, $\mathscr{C}_2$, $\mathscr{C}_3$, and $\mathscr{C}_4$ defined in Lemma 2 (Sec. 21-2). If a function f: $\Omega \to R^n$ satisfies

$$f^{-1}(C) \in \mathscr{A} \qquad \text{whenever } C \in \mathscr{C}$$

then f is $\mathscr{A}$ measurable.

(*c*) Suppose that each f_k: $\Omega \to R^n$ in the sequence $f_1, f_2, \ldots$ is $\mathscr{A}$ measurable. Suppose also that the sequence has a pointwise limit f: $\Omega \to R^n$; that is,

$$\lim_{k\to\infty} f_k(\omega) = f(\omega) \qquad \text{for all } \omega \text{ in } \Omega.$$

Then f is $\mathscr{A}$ measurable. ■

Proof Let

$$\mathscr{B} = \{B\colon B \subset R^n \text{ and } f^{-1}(B) \in \mathscr{A}\}$$

be the class of sets in R^n whose inverse images belong to $\mathscr{A}$. We show that $\mathscr{B}$ is a σ field.

Clearly $f^{-1}(\varphi) = \varphi$ which belongs to $\mathscr{A}$, so $\varphi \in \mathscr{B}$. We always have $f^{-1}(B^c) = [f^{-1}(B)]^c$. If $B \in \mathscr{B}$ then $f^{-1}(B) \in \mathscr{A}$, so $[f^{-1}(B)]^c \in \mathscr{A}$, so $B^c \in \mathscr{B}$. If each $B_n \in \mathscr{B}$ then each $f^{-1}(B_n) \in \mathscr{A}$, so

$$f^{-1}\left(\bigcup_{n=1}^{\infty} B_n\right) = \bigcup_{n=1}^{\infty} f^{-1}(B_n)$$

belongs to $\mathscr{A}$. Thus $\mathscr{B}$ is a σ field.

To prove part (*b*), if $f^{-1}(C) \in \mathscr{A}$ whenever $C \in \mathscr{C}$ then clearly $\mathscr{C} \subset \mathscr{B}$, so that $\mathscr{B}_n = \sigma(\mathscr{C}) \subset \mathscr{B}$ because $\mathscr{B}_n$ is the *smallest* σ field containing $\mathscr{C}$.

To prove half of part (*a*), we assume that $f\colon \Omega \to R^n$ is $\mathscr{A}$ measurable and show that each $f_i\colon \Omega \to R$ is $\mathscr{A}$ measurable. For convenience let $i = 1$. For each real x_1 the set

$$B_{x_1} = \{(x_1', \ldots, x_n') : x_1' < x_1\}$$

is a Borel set, so that

$$f^{-1}(B_{x_1}) = \{\omega : (f_1(\omega), \ldots, f_n(\omega)) \in B_{x_1}\} = \{\omega : f_1(\omega) < x_1\} \in \mathscr{A}.$$

Thus $f_1^{-1}\{x_1' : x_1' < x_1\}$ always belongs to $\mathscr{A}$, so part (*b*) applied to $\mathscr{C}_4$ shows that f_1 is $\mathscr{A}$ measurable.

To prove the other half of part (*a*) we assume that each f_i is $\mathscr{A}$ measurable. Then each $f_i^{-1}\{x_i' : x_i' < x_i\}$ belongs to $\mathscr{A}$, so that

$$f^{-1}\{(x_1', \ldots, x_n') : x_1' < x_1, \ldots, x_n' < x_n\} = \bigcap_{i=1}^{n} f_i^{-1}\{x_i' : x_i' < x_i\}$$

also belongs to $\mathscr{A}$, and hence part (*b*) shows that f is $\mathscr{A}$ measurable, thus proving part (*a*).

Turning to the proof of part (*c*), we may assume the real-valued case, $R^n = R$, since this special case, together with part (*a*), implies the general case. Assume that $f\colon \Omega \to R$ is the pointwise limit of $f_1, f_2, \ldots$. For any real x define the subset A_x of Ω by

$$A_x = \bigcup_{k=1}^{\infty} \bigcup_{m=1}^{\infty} \bigcap_{n=m}^{\infty} \left\{\omega : f_n(\omega) < x - \frac{1}{k}\right\}.$$

We first show that $\{\omega : f(\omega) < x\} = A_x$ for every real x. This equation asserts that any ω in Ω satisfies $f(\omega) < x$ *iff* there exist a k and an m such that

$$f_n(\omega) < x - \frac{1}{k} \qquad \text{whenever } n \geq m.$$

Fix x and take any ω belonging to $\{\omega : f(\omega) < x\}$. Select k large enough that $f(\omega) < x - (1/k)$. By assumption $f_n(\omega)$ converges to $f(\omega)$ as n goes to infinity, so

$$f_n(\omega) < x - \frac{1}{k} \qquad \text{for all large enough } n.$$

Therefore ω belongs to A_x.

Fix x and take any ω belonging to A_x. Then there exist a k and an m such that

$$f_n(\omega) < x - \frac{1}{k} \qquad \text{whenever } n \geq m.$$

Clearly, then, $f(\omega) = \lim f_n(\omega) \leq x - (1/k) < x$, so that $f(\omega) < x$. Therefore $\{\omega : f(\omega) < x\} = A_x$ for all real x.

Under the hypotheses of part (c) every set $\{\omega : f_n(\omega) < x - (1/k)\}$ belongs to $\mathscr{A}$. But A_x is obtained from these sets by performing, in succession, three operations which do not take us out of the σ field $\mathscr{A}$. Therefore $\{\omega : f(\omega) < x\} = A_x$ belongs to $\mathscr{A}$ for every real x, so part (b) implies that f is $\mathscr{A}$ measurable. ■

Lemma 1a shows that a function $X: \Omega \to R^n$ is a random vector with associated probability space $(\Omega,\mathscr{A},P)$ iff each of its components X_i in $X = (X_1, \ldots, X_n)$ is a random variable with associated $(\Omega,\mathscr{A},P)$. Lemma 1b shows that X is a random variable iff the set $[X < x]$ is an event, for every real x, that is, iff its distribution function $F_X: R \to R$ is defined. Lemma 1c shows that any pointwise limit of random vectors is also a random vector. Thus the collection of random variables, on a fixed probability space $(\Omega,\mathscr{A},P)$, is closed under pointwise limits.

Lemma 1a shows that $f: R^n \to R^m$ is a Borel function iff each component is a Borel function. Lemma 1b with $\mathscr{C} = \mathscr{C}_3$ shows that if $f: R^n \to R^m$ satisfies

$$f^{-1}(O) \in \mathscr{B}_n \qquad \text{whenever } O \text{ is an open set in } R^m$$

then f is a Borel function. As is well known, a function $f: R^n \to R^m$ is everywhere continuous iff the inverse image of every open set in R^m is an open set in R^n. Every open set in R^n is a Borel set in R^n. Therefore *every continuous function $f: R^n \to R^m$ is a Borel function*. For example, $f(x_1, \ldots, x_n) = x_1 + \cdots + x_n$ is continuous, so that if $X_1, \ldots, X_n$ are random variables then $(X_1, \ldots, X_n)$ is a random vector, and hence $f(X_1, \ldots, X_n) = X_1 + \cdots + X_n$ is a random variable. Lemma 1c shows that the class of Borel functions on R^n is closed under pointwise limits. Therefore every pointwise limit of continuous functions is a Borel function. The same is true for pointwise limits of "pointwise limits of continuous functions." And so on. Thus all functions $f: R^n \to R^m$ normally encountered in analysis are Borel functions, even the highly discontinuous ones.

Clearly $f(x,y) = x$ is a continuous function $f: R^2 \to R$. Thus if D_1 belongs to $\mathscr{B}_1$ then $f^{-1}(D_1) = D_1 \times R$ must belong to $\mathscr{B}_2$. The same is true for $R \times D_2$, and intersecting $D_1 \times R$ and $R \times D_2$ produces $D_1 \times D_2$. More generally *if D_i is a Borel set in R^{n_i} then $D_1 \times \cdots \times D_k$ is a Borel set in $R^{n_1 + \cdots + n_k}$*. Thus, roughly speaking, Borel restrictions on coordinate axes yield Borel sets.

The following theorem, with the same proof as for Theorem 3 (Sec. 15-2), shows that $P_{f,g}$ yields P_f.

Theorem 3 If $f: \Omega \to R^n$ and $g: \Omega \to R^m$ are random vectors with the same associated probability space $(\Omega,\mathscr{A},P)$ then

$$P_f(B) = P_{f,g}(B \times R^m) \qquad \text{for every Borel set } B \text{ in } R^n. \quad ■$$

Random vectors $f_1, \ldots, f_n$ with associated probability space $(\Omega,\mathscr{A},P)$ are said to be **independent** iff

$$P[f_1 \in B_1, \ldots, f_n \in B_n] = P[f_1 \in B_1]P[f_2 \in B_2] \cdots P[f_n \in B_n]$$

whenever the B_i are Borel subsets in R^{k_i}, where $f_i\colon \Omega \to R^{k_i}$. The immediate consequences remain as in Sec. 15-3. In particular we have the following theorem, with the same proof as for Theorem 1 (Sec. 15-3).

Theorem 4 If $f_1, \ldots, f_d$ are $d \geq 2$ independent random vectors with the same associated probability space $(\Omega,\mathscr{A},P)$, and if

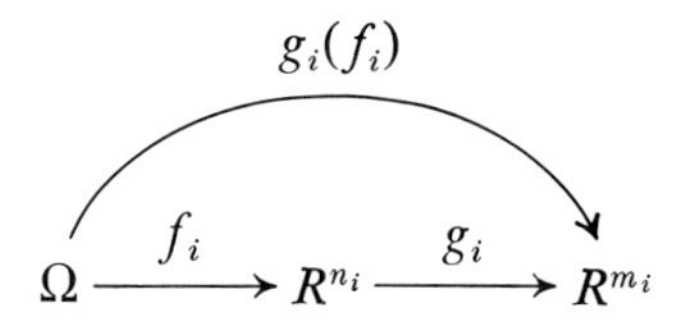

where $g_1, \ldots, g_d$ are Borel functions, then $g_1(f_1), \ldots, g_d(f_d)$ are d independent random vectors. ∎

As in Sec. 15-3, the independence of $f_1, \ldots, f_m, \ldots, f_n$ implies the independence of $(f_1, \ldots, f_m)$ and $(f_{m+1}, \ldots, f_n)$. Lemma 1 (Sec. 15-3) can be generalized as below to show that if A is independent of every event in a class $\mathscr{C}$, closed under finite intersections, then A is independent of every event in $\sigma(\mathscr{C})$.

Lemma 2 Let A be an event and let $\mathscr{C}$ be a class of events in a probability space $(\Omega,\mathscr{A},P)$. Suppose that $(C \cap C') \in \mathscr{C}$ whenever $C \in \mathscr{C}$ and $C' \in \mathscr{C}$. Also suppose that A and C are two independent events whenever $C \in \mathscr{C}$. Then A and C are two independent events whenever $C \in \sigma(\mathscr{C})$. ∎

Proof If $P(A) = 0$ then $P(AB) = 0 = P(A)P(B)$ for every event B, so the lemma is true. If $P(A) > 0$ then

$$P_A(B) = \frac{P(AB)}{P(A)} = P(B)$$

holds if $B \in \mathscr{C}$. Thus P_A and P agree on $\mathscr{C}$, and hence from Lemma 3 (Sec. 21-2) they agree on $\sigma(\mathscr{C})$. ∎

21-4 EXPECTATION

In this section we define EX and relate it to intermediate distributions in Theorem 1. The remainder of the section is devoted to miscellaneous comments concerning the further development of the theory.

If $(R,\mathscr{B}_1,P)$ is a probability space whose sample space is the real line, and whose events are the Borel sets, then the **mean** of P, as well as whether or not P has a mean, is defined in Theorem 1 (Sec. 15-4). A random variable X with associated probability space $(\Omega,\mathscr{A},P)$ induces a probability space $(R,\mathscr{B}_1,P_X)$. The **expectation** EX of a random variable X with associated probability space $(\Omega,\mathscr{A},P)$ is defined to be the mean of P_X, if P_X has a mean.

We immediately obtain the following theorem, with the same proof as for Theorem 2 (Sec. 15-4).

Theorem 1 If f is a random vector with associated probability space $(\Omega,\mathscr{A},P)$, and if

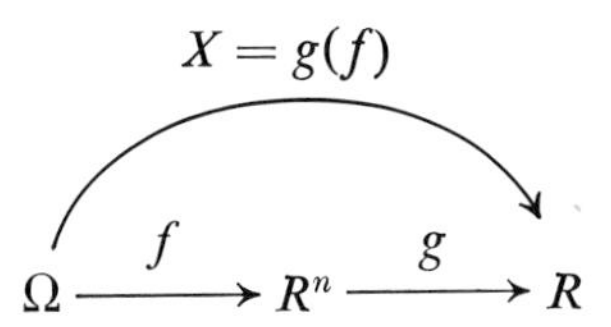

where g is a Borel function, then $EX = Eg$. That is, X with associated $(\Omega,\mathscr{A},P)$ and g with associated $(R^n,\mathscr{B}_n,P_f)$ have the same expectation, in the sense that if either expectation exists then both exist and are equal. ■

At this point we stop our formal rigorous development of the general theory. All the usual properties of expectation hold, including those in Theorem 3 (Sec. 11-3), the Chebychev inequality of Theorem 3 (Sec. 15-4), and the MGF Theorem 4 (Sec. 15-4). The remainder of this section consists of miscellaneous comments.

The following theorem, stated without proof, justifies the conclusion that EX_n converges to EX. We need only know that X_n converges to X, and then find some Z with $E|Z| < \infty$ such that $|Z|$ dominates each $|X_n|$.

Lebesgue dominated-convergence theorem Let Z, X, X_1, X_2, . . . be random variables with the same associated probability space $(\Omega,\mathscr{A},P)$. Suppose that X_n converges to X as n goes to infinity and that $|X_n| \le Z$ for all $n = 1, 2, \ldots$, and that $E|Z| < \infty$. Then EX_n converges to EX as n goes to infinity. ■

We next indicate how the dominated-convergence theorem may be used to deduce linearity of expectation. Assume that $|X| + |Y|$ has a finite expectation. Using the function g_n from the addendum to Sec. 15-4, we see that each of $|g_n(X)|$, $|g_n(Y)|$, and $|g_n(X) + g_n(Y)|$ is dominated by $|X| + |Y|$. Also $g_n(X)$ goes to X, $g_n(Y)$ goes to Y, and $g_n(X) + g_n(Y)$ goes to $X + Y$ as n goes to infinity, so we can apply the dominated-convergence theorem to each sequence. Therefore

$$E(X+Y) = \lim_{n\to\infty} E[g_n(X) + g_n(Y)] = \lim_{n\to\infty} [Eg_n(X) + Eg_n(Y)] = EX + EY$$

where the middle equality used the linearity of expectation in the discrete case.

The use of σ fields may be explained by the fact that often it is not possible to define a reasonably well behaved P for all subsets of the sample space. This paragraph indicates how σ fields are also widely used in a quite different, philosophically fundamental fashion. From Example 2 (Sec. 14-1), if $X\colon \Omega \to R$ is a random variable with associated probability space $(\Omega,\mathscr{A},P)$ and $g\colon \Omega \to R^n$ is a random vector with the same $(\Omega,\mathscr{A},P)$ then the function $f(\omega) = \mu_{X|g}(g(\omega))$, for all ω in Ω, may be called the *conditional expectation of X given g*. Thus f is a random variable which, among other things, satisfies the *condition* that it can be expressed as a Borel function of g. It can be shown that a function $f: \Omega \rightarrow R$ can be so expressed iff it is $\mathscr{A}_g$ measurable, where $\mathscr{A}_g = \{A: A \subset \Omega \text{ and } A = g^{-1}(B) \text{ for some } B \in \mathscr{B}_n\}$. Thus the condition that a number must be calculable from only certain restricted information, such as knowledge of the value of g, is translatable into a sample-space condition involving certain σ fields.

If in our definition of a probability space $(\Omega,\mathscr{A},P)$ we drop the requirement that $P(\Omega) = 1$, and we also let P take on the value ∞ then we have the definition of a **measure space**. In this case the symbol P is normally replaced by a symbol such as μ. For example, a special selection for μ_n gives us $(R^n,\mathscr{B}_n,\mu_n)$, where μ_n is volume (length if $n = 1$ and area if $n = 2$) or Lebesgue measure in R^n. The definitions and proofs concerning induced distributions, expectation, etc., can be generalized to measure spaces with very little extra effort. Then the resulting *integral of X with respect to μ* may be called the *expectation of X with associated* $(\Omega,\mathscr{A},\mu)$, in the special case when $\mu(\Omega) = 1$. Typically, then, the general theory of integration over measure spaces is developed and is then specialized to expectation for probability spaces. Then, of course, the theory includes continuous probability densities on R^n, which involve integration with respect to Lebesgue measure. Three theorems are considered to be especially important in the theory. One is a dominated-convergence theorem like the one stated above. Another is the *Fubini theorem*, involving interchanges of iterated integrals. The third is the *Radon-Nikodym theorem*, which asserts that certain functions can be expressed as indefinite integrals. This last theorem is fundamental for conditional expectations.

There are many ways to develop integration theory, such as the Daniell integral approach, which are quite different from the one just described and yet lead to essentially equivalent theories.

22
Introduction to Stochastic Processes

This chapter introduces some of the basic ideas in the extensive theory of stochastic processes. Stochastic processes can be thought of as generalizations of random vectors, so that it is not surprising that essentially all of the more important ideas and distributions from the earlier chapters reappear in this theory. Collections of probability densities will be used to define a probability *measure* on a sample space in which each sample point is essentially a function. In this context the advantages to emphasizing probability measures rather than probability densities should be particularly apparent. Section 22-1 provides some initial perspective, together with indications of the many broad areas of application.

In Sec. 22-2 we consider the description, construction, and some of the classifications of processes. The primary emphasis is on generalities concerning processes having stationary and independent increments. This class of processes includes the Poisson and Weiner processes, which are particularly important continuous-parameter processes. The Poisson process, together with some of its generalizations, will be studied in Chap. 23, where it is shown to unite the Poisson, exponential, and gamma densities. The Weiner process, based on normal densities, will be studied in Chap. 24. Thus Secs. 22-1 and 22-2 provide the general background for Chaps. 23 and 24, which may be read in either order.

Section 22-3 is devoted to the construction or description of processes in terms of consistent families of finite distributions, the resulting lack of uniqueness of certain probabilities associated with continuous-parameter processes, and the role of regularity assumptions in this connection. This section is relevant to the broad conceptual framework for stochastic processes, but it may be postponed until after there has been contact with explicit processes.

A rigorous development of the theory of stochastic processes presumes the rigorous foundation of probability theory, as introduced in Chap. 21. However, this does not prevent us from introducing, in this and the next two chapters, many fundamental concepts and obtaining a number of interesting results. Furthermore our conceptual framework requires essentially only technical modifications in order to make it rigorous. The Addendum to Sec. 22-3 indicates the flavor of such modifications.

22-1 INTRODUCTION

Except for the definition of stochastic processes, the discussions of this section are intuitive and motivational. Broad classes of applications are indicated, and processes are described pictorially. Attention is given to the initially confusing fact that a stochastic process is a function of two variables. Several trivial examples are used as illustrations, but our interest is primarily in processes which naturally involve infinitely many random variables.

Suppose that each performance of an experiment yields a real-valued function on the interval $0 \leq t \leq 1$. Three possible *outcomes*, or *realizations*, or *sample functions* are shown in Fig. 1*a*. Instead of the interval $0 \leq t \leq 1$ we could use any interval, or half-line, or the whole real line. Typical realizations might be like the two in Fig. 1*b*, or the two in Fig. 1*c*. In most applications t corresponds to time, but it may correspond to distance from a fixed point, or to volume, etc. Very roughly, a stochastic process is a mathematical model for an experiment which is such that each performance of the experiment yields such a function, and there is significant "randomness" in outcomes.

For interpretation, if we observe the function f then $f(t)$ might be the voltage at time t from an electrode on the scalp of a patient, so that the experiment produces the "brain wave" f. Or $f(t)$ might be the volume of water in a dam at time t, or the total rainfall up to time t, or the height of the water at time t above a particular point in a harbor during a storm. Each experiment might produce a waveform f on an oscilloscope. Or f may be a transmitted or received signal in a communication system, or the noise voltage waveform across a resistor, or the angular velocity record of the shaft of a motor in a control problem, or the airspeed graph from an airplane flight, or the learning curve for a subject. The theory of stochastic processes has been applied extensively in all the fields indicated, and many more. Very roughly, a stochastic process is a rule which assigns probabilities to realizations. That is, a "physical stochastic process" is idealized to a "mathematical stochastic process," which is nothing more than the assignment of probabilities to sets of realizations.

Suppose that at time 0 you look through a high-powered microscope and observe a very small particle at the origin of a three-dimensional coordinate system in a gas or liquid. If you plot the x coordinate of the particle as a function of time you may obtain an irregular graph like one of the two in Fig. 1*c*. This phenomenon was discovered by Robert Brown, an English botanist, in 1827. In 1905 Albert Einstein assumed continual bombardment of the particle by the molecules of the liquid and deduced that such motion should occur and with certain characteristics. Later this process and various generalizations of it were studied by physicists, and by mathematicians including N. Wiener and P. Lévy. It is variously called the Brownian motion process, or the Wiener-Lévy process, or the *Wiener process*. The process is particularly useful for the construction, or representation, of other processes. The Wiener process assigns probabilities to realizations by using the normal densities in a special way. "Typical realizations" appear as in Fig. 2 (Sec. 24-1), which depicts the result of a sampling experiment. That figure can be interpreted as having been obtained by observing a realization from a Wiener process and then plotting only the finitely many points shown. We study the Wiener process in less detail than Poisson type processes, which are considered next.

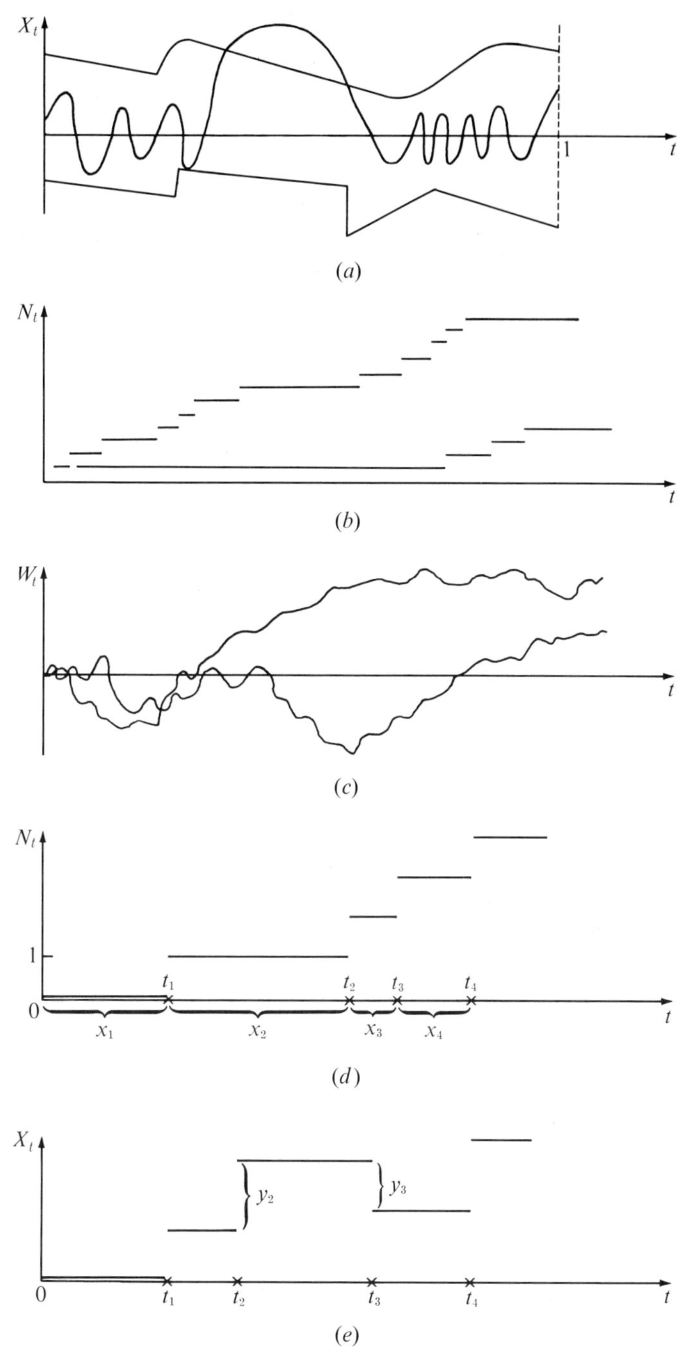

Fig. 1 (*a*) Three realizations from a process on $0 \le t \le 1$; (*b*) two realizations from a Poisson process on $t \ge 0$; (*c*) two realizations from a Wiener process on $t \ge 0$; (*d*) a counting function; (*e*) random jumps at random times.

If the ith airplane to arrive at an airport arrives at time t_i after time 0 then we call $0 < t_1 < t_2 < t_3 \cdots$ the *successive arrival times*. We assume that two airplanes never arrive at the same moment. As in Fig. 1*d*, we call $x_1, x_2, \ldots$ the *successive interarrival times*, where $x_i = t_i - t_{i-1}$ for $i \geq 2$ while we arbitrarily let $x_1 = t_1$. The realization from *one* performance of the experiment is assumed to consist of a sequence $0 < t_1 < t_2 < \cdots$, which can be depicted as a configuration of *crosses*, or *successes*, on the time axis, as in Fig. 1*d*, where there is a cross at each t_i. For any such outcome we can define a corresponding *counting function*, whose value at any $t \geq 0$ is the number of successes up to and including time t. The successive arrival times are intuitively natural, but the corresponding counting function is sometimes more convenient mathematically. Clearly either determines the other. In Chap. 23 the counting functions are used for realizations. One important method for assigning probabilities to counting functions, by using Poisson or exponential densities, yields the Poisson process. Realizations generated by sampling experiments are shown in Fig. 4 (Sec. 23-1) and Fig. 2 (Sec. 23-3). The Poisson process very roughly determines the arrival times "completely at random." For interpretation, t_i could be the time of arrival of the ith ship at a harbor, passenger at a terminal, egg in a hatchery (pure-birth process, no deaths), electron at the plate of a diode (shot noise), α particle at a radioactive counter, order for goods at a factory, customer at a service counter, etc. Similarly t_i may be the time of the ith death, or suicide (pure-death process, no births). Or t_i may be the time of initiation of the ith telephone call, or ith voltage impulse along a nerve cell. The Poisson process and its generalizations have been used extensively in all the applications indicated, and many more.

Figure 1*e* indicates a natural generalization of a Poisson process. Instead of a counting function we now use a function which has a jump of value y_i at time t_i, where y_i need not equal 1 but may be any real number. Now an outcome for one performance of the experiment consists of the two sequences $0 < t_1 < t_2 < \cdots$ and $y_1, y_2, \ldots$, in terms of which we can define the function of Fig. 1*e* to have at t the value which is the sum of the y_i over those i satisfying $t_i \leq t$. From this function we can reconstruct the two sequences, if all $y_i \neq 0$. Letting the y_i have values 1 and -1, we get "birth and death" processes, and we can obviously generalize to multiple births (or deaths, as in auto accidents) at one instant. Also y_i might be the size of the order received at time t_i, or the amount of life insurance carried by the person dying at time t_i.

Naturally in books on each of the areas indicated above, the Poisson process and its simplest properties are developed, interpreted, and applied, perhaps in terms of the language and motivation of that area. For obvious reasons we will *not* continually interpret everything in terms of a variety of such applications. The reader may well find it helpful, however, to occasionally make such interpretations himself in terms of applications like those indicated above.

Suppose that we have an experiment which produces realizations on the unit interval $0 \leq t \leq 1$. We perform the experiment, obtain and plot a realization, and then measure the height or value of the realization at $t = .5$. We thus obtain a real number. From n independent repetitions of the experiment we could obtain n real numbers in this way. It is reasonable to say that there is a random variable $X_{.5}$

associated with the experiment, and if the experiment is performed then the value of $X_{.5}$ for this performance is just the value at $t = .5$ of the realization obtained. More generally for every real t satisfying $0 \le t \le 1$ we can say that we have a random variable X_t, and if the experiment is performed then the value of X_t for this performance is just the value at t of the realization obtained. Thus, for example, if the experiment is performed once then the value obtained for the random vector $(X_0, X_{.1}, X_{.2}, \ldots, X_{.9}, X_1)$ can be portrayed by plotting the 11 values of the observed realization at $t = 0, .1, \ldots, .9, 1$. More generally if for *every* t we plot the value of X_t above t then we will just have a plot of the whole realization on the unit interval. Thus we are led to represent the process mathematically as a collection of random variables X_t, one for each t satisfying $0 \le t \le 1$, and all defined on the same probability space (Ω, P). Assume that P has somehow been given to us. Each X_t is a function $X_t: \Omega \to R$ defined on the sample space Ω. Thus we have a real-valued function of two variables having the value $X_t(\omega)$ at the argument t and ω. If we keep t fixed and consider $X_t(\omega)$ as a function of ω then we have the random variable X_t. If we keep ω fixed and consider $X_t(\omega)$ as a function of t then we have the realization obtained if the outcome is this ω. We will shortly elaborate on these interpretations.

As a premature observation, if realizations are to be continuous, or have at most a few discontinuities, then for s near to t we *usually* expect X_s to have a value near to X_t. Thus we anticipate that for most processes of interest, such as the Wiener or Poisson, that for every t

$$P[|X_s - X_t| \ge \epsilon] \to 0 \qquad \text{as } s \to t,$$

regardless of how small $\epsilon > 0$ may be. For both these processes if $0 \le s < t$ then X_s and $[X_t - X_s]$ are independent, so that we get X_t by adding to X_s an independent increment $[X_t - X_s]$ which has a normal 0 and $(t - s)\sigma^2$ distribution, or a Poisson distribution with mean $\lambda(t - s)$. Thus the closer s is to t the more chance there is for X_s to be close to X_t. We postpone, until the next two sections, consideration of general methods for assigning probabilities to sets of realizations.

Suppose that we have a container whose bottom is the unit square $T = \{(\alpha,\beta): 0 \le \alpha \le 1, 0 \le \beta \le 1\}$ in the $\alpha\beta$ plane, and the container is half filled with water. Suppose that the experiment consists of dropping a pebble into the water at time 0 and observing the whole irregular surface of the water at time 1. If the outcome of the experiment is ω then we can let $X_{(\alpha,\beta)}(\omega)$ represent the height of the water at time 1 above the point (α,β). For fixed ω if we consider $X_{(\alpha,\beta)}(\omega)$ as a function of (α,β) then we obtain the whole water surface at time 1 for this sample point ω. Thus a stochastic process can be used to describe random surfaces, etc. However, we will restrict our definitions and examples almost exclusively to stochastic processes whose realizations are real-valued functions defined on some set of real numbers of restricted form. We next make the appropriate definitions, interpret them, and examine several trivial processes.

A **stochastic process** consists of a nonempty set T, called the *parameter set*, a probability space (Ω, P), and the assignment of a random variable $X_t: \Omega \to R$ to each t in T. Thus a stochastic process is an indexed collection of random variables, X_t for $t \in T$, all having the same associated probability space. If T contains only a finite number of members, say the integers $1, 2, \ldots, n$, then the stochastic process is a random vector $(X_1, \ldots, X_n)$, or a random variable if $n = 1$. We call the stochastic process, X_t for $t \in T$, a **discrete-parameter process** iff T is some countable set of real numbers. In this case T is usually some set of integers, but it is sometimes useful to let T be the set of all rational numbers. A **continuous-parameter stochastic process** is one for which the parameter set T is some interval of real numbers. *In this connection* by an **interval** we mean either a finite or an infinite interval of real numbers, as defined below, where any sign $\leq$ may be replaced by the sign $<$. A **finite interval,** with end points $a < b$, is the set of x satisfying $a \leq x \leq b$. An **infinite interval** is either the set of all real numbers, or the set of x satisfying $x \leq a$, or the set of x satisfying $a \leq x$. Thus an interval has 0, 1, or 2 *end points.*

Henceforth by the term **stochastic process** *we always mean a discrete- or a continuous-parameter stochastic process, unless explicitly indicated otherwise. Thus the parameter set must be a countable set of real numbers, or an interval as defined above.*

A random variable is a real-valued function defined on the sample space Ω, so that a stochastic process is a real-valued function defined on the cartesian product $T \times \Omega$ with an associated (Ω,P). That is, to each (t,ω) we assign a real number $X_t(\omega)$. If we write X_t we think of t as fixed, so that X_t is a function on Ω. We may fix ω and obtain a function of t. If "X_t for $t \in T$" is a stochastic process with associated (Ω,P) then for any ω in Ω we define the function $r_\omega: T \to R$ by

$$r_\omega(t) = X_t(\omega) \qquad \text{for all } t \in T,$$

and we call r_ω a **realization** or **sample function** of the stochastic process. Thus a realization consists of the value assignment of *every* random variable for that sample point. Whenever we have a function of two variables, say $G: R^2 \to R$, then for each fixed value x of the first variable we have a function $G(x,\cdot)$ of the second variable, and for each fixed value y of the second variable we have a function $G(\cdot,y)$ of the first variable. A stochastic process is a function of two variables (t,ω), and instead of using the notations $X_t(\cdot)$ and $X_{\cdot}(\omega)$ we often prefer the notations X_t and r_ω. In addition, we also use the notation $X(t)$ in place of X_t. Thus, especially when the argument has a subscript, we may write $X(t_i)$ rather than X_{t_i}.

A stochastic process is a generalization of a random vector, or sequence of random variables, so we first examine these two special cases. If $(X_1, \ldots, X_n)$ is a random vector and we evaluate each random variable at the same sample point ω_1 then we obtain the realization

$$r_{\omega_1} = (X_1(\omega_1), X_2(\omega_1), \ldots, X_n(\omega_1))$$

pictured by the n crosses of Fig. 2*a*. Similarly we obtain the realization r_{ω_2} pictured by the n dots. Thus it suits us to consider and picture a vector $(x_1, \ldots, x_n)$ as a function defined on the integers $1, 2, \ldots, n$ and assigning the value x_i to the integer i. Figures 2 and 5 (Sec. 12-2) and Fig. 1 (Sec. 12-3) exhibit five realizations of random vectors. Naturally if $X_1, X_2, \ldots$ is a sequence of random variables then a particular realization $r_{\omega_1} = (X_1(\omega_1), X_2(\omega_1), \ldots)$ is a sequence of real numbers which we can depict as in Fig. 2*a* and consider to be a function defined on the integers $1, 2, \ldots$ and having the

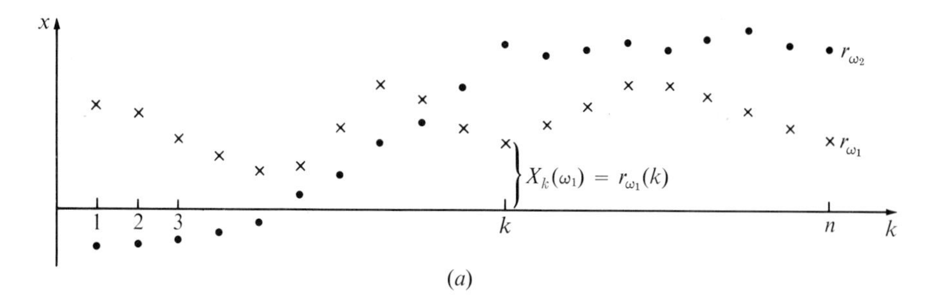

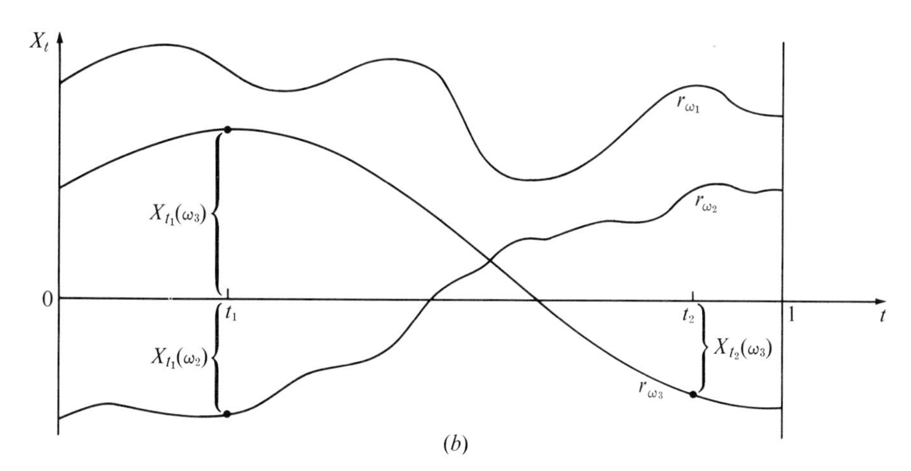

Fig. 2 (*a*) Two realizations from a random vector; (*b*) three realizations from a stochastic process.

value $r_{\omega_1}(k) = X_k(\omega_1)$ at k. Similarly if X_t is a stochastic process with the unit interval for parameter set then three possible realizations are shown in Fig. 2*b*. Thus if we fix t_1 then the value at the sample point ω of the random variable X_{t_1} is obtained by evaluating the realization r_ω at t_1.

Suppose that what we observe when an experiment is performed is the value of every X_t for $t \in T$. Then when the experiment is performed we observe a realization, and the collection of all realizations is just the collection of all observable outcomes. If T' is a proper subset of T then we can consider the new stochastic process, X_t for $t \in T'$. For example, T might be the real line while T' is a finite interval, or the set of all rational numbers. A realization from "X_t for $t \in T$" is defined on T, but if an experiment yields such a realization, and for some reason we happen to observe it only at all times in T', then this is equivalent to observing the whole corresponding realization from "X_t for $t \in T'$."

There are many ways to form new processes from a given process "X_t for $t \in T$." For example, the process "e^{X_t} for $t \in T$" has realization e^{r_ω} when the original process has realization r_ω. More generally for any function f: $R \to R$ we may introduce the process "$f(X_t)$ for $t \in T$."

If two stochastic processes X_t for $t \in T$ and Y_t for $t \in T$ have the same parameter set and the same probability space then we can combine them in various ways to obtain new processes. For example, the process $X_t e^{Y_t}$ for $t \in T$ assigns realization $r_\omega e^{s_\omega}$ to the sample point ω, where r_ω and s_ω are the realizations of X_t for $t \in T$ and Y_t for $t \in T$, respectively.

The study of stochastic processes is usually concerned with properties which involve many or all of the random variables simultaneously. For example, we may define the random variable Y by

$$Y(\omega) = \int_0^1 [X_t(\omega)]^2\,dt = \int_0^1 [r_\omega(t)]^2\,dt$$

so that Y is the mean-squared value of the realization obtained. In order to deal with Y it is usually not enough to consider just distributions of single X_t variables, or pairs, etc. Thus the use of the term "stochastic process" usually suggests that nontrivial joint distributions of many random variables are crucial to the problems being considered.

A stochastic process assigns a realization r_ω to each sample point ω. This defines a function X on Ω, where the value of X at ω is the whole realization r_ω. In many senses this is the most natural way to think about and to describe a stochastic process. Then a stochastic process induces a distribution on the space of possible realizations, using the same definition for an induced distribution as in the case of random vectors. By a Poisson process we mean any stochastic process which induces a certain distribution. Analogously we could say that a Poisson random variable is any random variable which induces a Poisson density. However, as explained in Sec. 22-3, there are certain complications associated with the concept of the distribution induced by a continuous-parameter process, and hence we usually avoid this concept.

Example 1 We easily construct a trivial continuous-parameter stochastic process as follows. Let $\Omega = \{\omega_1,\omega_2,\omega_3\}$ consist of precisely three sample points, and define a discrete density $p: \Omega \to R$ and random variables Y and Z as follows:

i	$p(\omega_i)$	$Y(\omega_i)$	$Z(\omega_i)$
1	.7	2	0
2	.2	−1	0
3	.1	1	1

For all real $t \geq 0$ define the random variable X_t by

$$X_t = e^{tY} \cos 2\pi Zt.$$

This continuous-parameter stochastic process "X_t for $t \geq 0$" has precisely three realizations, defined for all $t \geq 0$ by $X_t(\omega_1) = e^{2t}$, $X_t(\omega_2) = e^{-t}$, $X_t(\omega_3) = e^t \cos 2\pi t$ and depicted in Fig. 3. We obtain these three realizations with respective probabilities .7, .2, and .1. The three realizations could have been used as the three sample points. In constructing interesting complicated stochastic processes the realizations are often used as the sample points, hence the name "sample function."

The distribution of the random vector $(X_{.5},X_{1.5})$, for example, assigns probabilities .7, .2, and .1, respectively, to the three points (e,e^3), $(e^{-.5},e^{-1.5})$, and $(-e^{.5},-e^{1.5})$. With

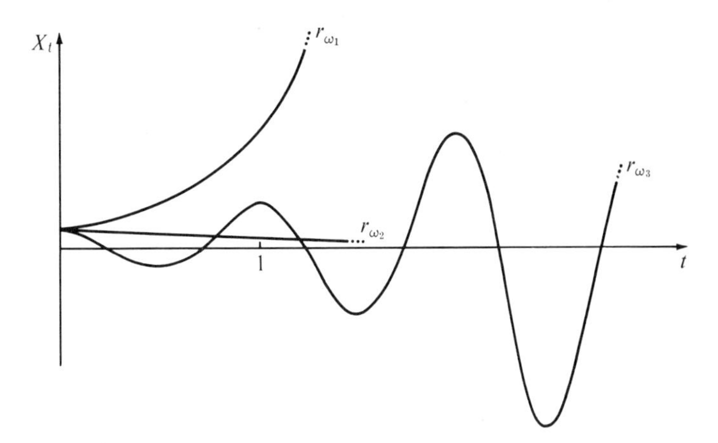

Fig. 3

probability .7 the random variable

$$Y = \int_0^1 (X_t)^2 \, dt$$

has the value

$$\int_0^1 (e^{2t})^2 \, dt = \frac{e^4 - 1}{4}. \quad \blacksquare$$

Example 2 We construct another trivial continuous-parameter stochastic process to further clarify the fact that we have a function of (t,ω). Let Ω be the real line and let P correspond to the normal 0 and 1 density φ. For any real $t \geq 0$ define the random variable X_t by "$X_t(\omega) = e^{\omega t}$ for all real ω." We then have a stochastic process "X_t for $t \geq 0$." If we consider $X_t(\omega)$ as a function of (t,ω) then we have the surface over the $t\omega$ half-plane, as indicated in Fig. 4.

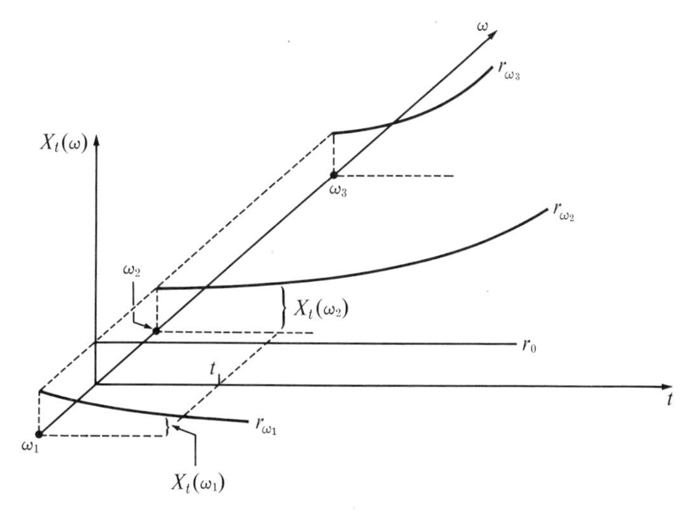

Fig. 4

Thus we select ω according to $\varphi(\omega)$ (not shown in Fig. 4) and we obtain the realization r_ω (four such are shown in Fig. 4). The distribution of $X_{2.3}$, for example, is the continuous density induced by $e^{2.3\omega}$ when ω has density φ. We may interpret $X_t(\omega)$ as the size of a population at time t when its randomly determined exponential growth constant is ω. A realization is a growth function.

Let $Y = \int_0^1 (X_t)^2\,dt$ so that if the outcome is ω then

$$Y(\omega) = \int_0^1 e^{2\omega t}\,dt = \frac{e^{2\omega} - 1}{2\omega}.$$

Thus Y has the density induced by $(e^{2\omega} - 1)/2\omega$ when ω has density φ. ■

Usually when we speak of a stochastic process we have in mind something more complicated than some "$X_t = f(t, Y_1, \ldots, Y_n)$ for $t \in T$," where the function $f\colon T \times R^n \to R$ and the distribution of the random vector $(Y_1, \ldots, Y_n)$ are "simply" definable in terms of finitely many elementary functions. Such $f(t, Y_1, \ldots, Y_n)$ are, however, of interest and arise naturally in many contexts. For example, the Y_i may be random initial conditions, or random errors in parameters, etc. However, our interest is primarily in processes which "naturally" involve infinitely many random variables. In particular, for our later processes (as opposed to Examples 1 and 2 above) a typical realization is still largely undetermined even after it has been observed at an arbitrarily large finite number of time points.

EXERCISES

1. A stochastic process "X_t for $t \in T$" is said to be **discrete valued** iff there exists some countable set C of real numbers such that $P[X_t \in C] = 1$ for all $t \in T$. Otherwise it is said to be **continuous valued**. Thus a process has either a discrete or continuous parameter, and it is either discrete or continuous valued; hence there are four types. For example, a Poisson process is a discrete-valued continuous-parameter process. Give an example of each of the other three types and sketch "typical realizations."

2. Define a stochastic process by

$$X_t = (1 + t)e^{-Yt}\cos Zt \qquad \text{for } t \geq 0$$

where Y and Z are random variables. We may interpret the process as a damped oscillation, where the damping constant $Y \geq 0$ and frequency Z are random.

(*a*) Sketch the three realizations corresponding to (y,z) having the values $(0,1)$ $(1,0)$, and $(.1,\pi/2)$.

(*b*) Express the random variable

$$W = \int_0^\infty X_t\,dt$$

as a rational function of Y and Z, not involving t.

(*c*) Show that $EW = \sqrt{\pi/2}$ if Y and Z are independent, Y is exponential $\lambda = 1$, and Z is normal 0 and 1. HINT: Ignore part (*b*).

(*d*) Let $V_t = |X_t|$ and for $(y,z) = (.1,\pi/2)$ sketch the realizations obtained from both the process "X_t for $t \geq 0$" and the process "V_t for $t \geq 0$."

22-2 CONSISTENT FAMILIES, AND PROCESSES HAVING STATIONARY INDEPENDENT INCREMENTS

The initial portion of this section is devoted to the basic method of describing a process by means of an infinite family of consistent distributions of random vectors. The major central portion of the section is devoted to the description of, and a few properties common to, stationary independent increment processes, in preparation for Chaps. 23 and 24. Stationary and Markov processes are mentioned near the end of the section. No attempt is made to systematically study the many important classes of Markov processes.

Let T be a countable set of real numbers, or an interval of real numbers. By a **family of finite distributions with parameter set** T we mean that for every positive integer n and for every n points $t_1 < t_2 < \cdots < t_n$ in T there has been assigned a probability space $(R^n, P_{t_1,\ldots,t_n})$ called an **nth-order distribution.** The *finite* refers to each joint distribution being for *finitely many* random variables. The family is said to be **consistent** iff for any one of the probability spaces for $n \geq 2$ in the family, if we consider random variables $X(t_1), \ldots, X(t_n)$ having this joint distribution, and from this joint distribution we obtain the joint distribution of some $n - 1$ of the random variables $X(t_1), \ldots, X(t_{i-1}), X(t_{i+1}), \ldots, X(t_n)$, as usual by use of Theorem 3 (Sec. 15-2), then this joint distribution on R^{n-1} is the same as the one assigned to $t_1, \ldots, t_{i-1}, t_{i+1}, \ldots, t_n$ by the given family. We assumed $t_1 < t_2$, since otherwise the family would contain both P_{t_1,t_2} and P_{t_2,t_1}, which would then be required to be related in an obvious fashion.

A stochastic process, $X(t)$ for $t \in T$, **generates** an obvious **family of finite distributions** with parameter set T; the distribution assigned to $t_1 < \cdots < t_n$ is the distribution induced by the random vector $(X(t_1), \ldots, X(t_n))$. The family is consistent because all the random variables have the same associated probability space. Conversely when proper attention is devoted to technicalities the following condition can be precisely stated and proved.

Kolmogorov consistency condition Let T be any interval or countable set of real numbers. For any consistent family of finite distributions with parameter set T there exists a stochastic process, X_t for $t \in T$, which generates the family. Furthermore the family uniquely determines the probability of any event which is definable in terms of countably many of the X_t random variables. ■

In other words, a stochastic process X_t for $t \in T$ can always be constructed so as to satisfy the following specifications. We specify for each $t \in T$ the first-order distribution (R,P_t) which we want X_t to have. Then for each pair $t_1 < t_2$ we specify the second-order distribution (R^2,P_{t_1,t_2}) which we want (X_{t_1},X_{t_2}) to have. Naturally the two marginals from the P_{t_1,t_2} must be the P_{t_1} and P_{t_2} specified earlier. Then we specify for each $t_1 < t_2 < t_3$ the joint distribution (R^3,P_{t_1,t_2,t_3}) which we desire for $(X_{t_1},X_{t_2},X_{t_3})$, being consistent, of course, with the three marginals P_{t_1,t_2}, P_{t_1,t_3}, and P_{t_2,t_3} specified earlier. And so on. Thus we can construct a stochastic process to agree with any rule which consistently specifies all joint distributions of all finite orders. Naturally each $P_{t_1,\ldots,t_n}$ may be given in terms of a discrete density, or a continuous density, or a distribution function on R^n.

The concept of a stochastic process is desirably rich, since practical applications give rise to varied and complicated relationships determining probabilities. In some applications high-order joint distributions cannot be simply determined in terms of low-order distributions, so this richness is desirable. From another point of view an undesirably large amount of data, or assumptions, is required to determine a stochastic process. However, many results can be proved to apply to every stochastic process satisfying a few simple assumptions. For example, the laws of large numbers and central-limit theorems of Chaps. 12 and 13 are valid for all stochastic processes which satisfy certain independence and moment assumptions. For many such results not even one distribution need be completely specified. At the other extreme, there are important explicit processes such as the Poisson and Wiener.

The stochastic process guaranteed by the consistency condition is normally constructed by letting the sample space Ω consist of all possible realizations, and using the consistent family to define probabilities. The next section describes this construction in greater detail. For most stochastic processes which we consider, the consistent nature of the definitions is intuitively apparent and fairly easy to prove. Our definitions of explicit processes are essentially equivalent to definitions of consistent families.

Let us briefly describe our orientation toward the consistency condition, although we devote little attention to these matters. Suppose that we decide to introduce a process which satisfies a specific verbal description: "its increments are to be independent, etc." Suppose also that we can completely convert this description into a description of a *unique* family which is to be generated by the process. If this family turns out to *not* be consistent then obviously no process can satisfy this verbal description, which must have been inherently self-contradictory in some sense. If the family turns out to be consistent then we know that the description makes sense in that there is a process satisfying the description. Suppose that we construct such a process and then calculate the probability of some event defined in terms of the random variables of the process: "at no time during the first hour will the angular displacement exceed . . . , etc." Will everyone else who constructs a process generating this same family neccessarily calculate the same probability for this event? In other words, does every event of practical interest necessarily have a unique probability assigned to it once we have made assumptions which unambiguously determine a consistent family? For discrete-parameter processes the answer is yes. For continuous-parameter processes the answer is yes *if* we restrict ourselves to processes whose realizations are regular. For example, in order to make the probabilities unique we require that Wiener process realizations be continuous functions. *Then* every *reasonable* description for an event can be converted into an *equivalent* description involving only the countably many random variables corresponding to rational time points, so that the consistency condition yields the desired uniqueness. The necessity for regularity assumptions is discussed in the next section.

Three stochastic processes, $X(t)$ for $t \in T$, $Y(s)$ for $s \in S$, $Z(h)$ for $h \in H$, all having the same associated probability space are said to be **independent stochastic processes** iff

$$(X(t_1), \ldots, X(t_m)) \qquad (Y(s_1), \ldots, Y(s_n)) \qquad (Z(h_1), \ldots, Z(h_r))$$

are three independent random vectors for every three finite sets of parameter points; and similarly for any finite collection of stochastic processes. The interpretation is that completely independent experiments yield the realizations for each process. That is, for two processes, any event (or more generally any random vector) defined in terms of the random variables of one process is independent of any event defined in terms of the random variables of the other process. For continuous-parameter processes the justification of this interpretation requires some qualifications concerning regularity of realizations, but we ignore these complications.

PROCESSES WHOSE INCREMENTS ARE STATIONARY AND INDEPENDENT

We now introduce the two definitions which are natural to a large class of stochastic processes including the Poisson and Wiener processes. It should not be assumed, however, that "most" stochastic processes of interest belong to this class.

A stochastic process X_t for $t \in T$ is said to have **stationary** or **homogeneous increments** iff $X_t - X_s$ and $X_{t'} - X_{s'}$ are identically distributed whenever $s < t$ and $s' < t'$ all belong to T, and $t - s = t' - s'$. That is, the distribution of the *increment random variable* $X_t - X_s$ depends on s and t only through their difference $t - s$. If we let $h = t - s > 0$ then the distribution of $X_{s+h} - X_s$ depends on h but not on s. Thus the distribution of the increase over every time span of 3 minutes is the same, and similarly for other time spans. If X_t for $t \in T$ has stationary increments then so does the process X_t for $t \in T_1$, for any subset T_1 of T.

A stochastic process $X(t)$ for $t \in T$ is said to have **independent increments** iff for every $n \geq 3$ points $t_1 < t_2 < \cdots < t_n$ in T the increment random variables

$$X(t_2) - X(t_1), X(t_3) - X(t_2), \ldots, X(t_n) - X(t_{n-1})$$

are independent. If $n = 4$ then independent increments imply that the two extreme increment variables

$$X(t_2) - X(t_1) \qquad \text{and} \qquad X(t_4) - X(t_3) \qquad \text{where } t_1 < t_2 < t_3 < t_4$$

are necessarily independent. Thus in general the independence applies to any finite number of "not necessarily adjacent" finite intervals which are disjoint, except possibly for end points. Also functions of independent random variables are independent, so that $X(t_2) - X(t_1)$ may be replaced by its negative $X(t_1) - X(t_2)$ without destroying independence. If X_t for $t \in T$ has independent increments then so does X_t for $t \in T_1$, for any subset T_1 of T.

Naturally a **stationary-independent-increment process** is one whose increments are both stationary and independent. Many such processes are constructed in Sec. 23-4.

Let $X(t)$ for $t \geq 0$ be a stationary-independent-increment process satisfying $X(0) = 0$. Fix real numbers $b > 0$, τ_0, c, d, e and for each real $\tau \geq \tau_0$ define a random variable

$$Y_\tau = d + e\tau + cX(b(\tau - \tau_0)).$$

Then the process Y_τ for $\tau \geq \tau_0$ starts at time τ_0 with the constant value $Y_{\tau_0} = d + e\tau_0$. For $s \geq 0$ and $h > 0$ the new process has increment

$$Y_{\tau_0+s+h} - Y_{\tau_0+s} = eh + c[X(bs + bh) - X(bs)]$$

whose distribution depends on the two indices only through their difference h. Similarly independence of increments is also easily demonstrated. *Thus* Y_τ for $\tau \geq \tau_0$ has stationary independent increments. If $0 = \tau_0 = d = e$ then we have just applied scale changes to the two axes, for every realization. For example, if t is in seconds and $X(t)$ is in feet then $Y_\tau = 12X(60\tau)$ for $\tau \geq 0$, is essentially the same process, but τ is in minutes and Y_τ is in inches.

We first consider discrete-parameter stationary-independent-increment processes. These most clearly reveal the nature of such processes. Fix any distribution for Z_1 and let $Z_1, Z_2, \ldots$ be independent and identically distributed. Let $S_n = Z_1 + \cdots + Z_n$. Then the sequence $S_1, S_2, S_3, \ldots$ clearly has stationary independent increments. When expressed in the form $[S_2 - S_1]$ and $[S_4 - S_2]$ these two increments both involve the same random variable S_2 in their definition, suggesting that the increments may be dependent. But of course

$$[S_2 - S_1] = Z_2 \qquad \text{and} \qquad [S_4 - S_2] = Z_3 + Z_4$$

so that *they can be expressed* as functions of the independent random vectors Z_2 and (Z_3,Z_4); hence they are independent. If Z_0 is *any* random variable, for example $Z_0 = Z_3 + (Z_1)^2$, then the sequence $Z_0, Z_0 + S_1, Z_0 + S_2, Z_0 + S_3, \ldots$ has stationary independent increments, because the addition of Z_0 to each term has no effect on increments. Conversely if $S_0, S_1, S_2, \ldots$ is *any* stationary-independent-increment sequence then it has the preceding form, because we can express S_n as

$$S_n = [S_n - S_{n-1}] + [S_{n-1} - S_{n-2}] + \cdots + [S_1 - S_0] + S_0$$

where the increments in square brackets must be independent and identically distributed. In order to eliminate the complication associated with S_0 we usually assume that $S_0 = 0$. We could assume only that $P[S_0 = 0] = 1$, but there is no loss of generality in assuming that S_0 has only the value 0. Thus a stationary-independent-increment sequence $0 = S_0, S_1, S_2, \ldots$ is nothing more than the sequence of partial sums of independent random variables each having the same distribution as S_1. The distribution for S_1 is completely arbitrary, and it determines uniquely the consistent family generated by the process.

The expression "stationary independent increments" is usually used only for continuous-parameter processes. For reasons like those in the discrete case, we usually "start" the process at $t = 0$ by requiring $X_0 = 0$, so we normally use $t \geq 0$ for parameter set. The parameter set can always be extended to the whole real line, in which case realizations from such a Poisson process, for example, appear as in Fig. 3 (Sec. 23-3). This extension is sometimes convenient, although the origin continues to play a unique role since $X_0 = 0$.

If "X_t for $t \geq 0$ with $X_0 = 0$" has stationary independent increments, and $\tau > 0$, then $0 = X_0, X_\tau, X_{2\tau}, X_{3\tau}, \ldots$ is clearly a stationary-independent-increment sequence. This fact should make the probabilistic structure of such processes fairly apparent. To *define* such a process we do not have available to us a first nontrivial random variable, in the same way S_1 was available in the discrete case. A satisfactory approach is to use the collection of all desired first-order distributions P_t for $t > 0$ in order to define the family of all finite joint distributions. Naturally these first-order distributions must satisfy some conditions, which we now describe.

A family of probability spaces (R,P_t), one for each real $t > 0$, is called a **semigroup** iff for every $s > 0$ and $t > 0$ the given P_{s+t} equals the distribution of the sum of two independent random variables having distributions P_s and P_t.

Let $X(t)$ for $t \geq 0$ be any stationary independent increment process for which $X(0) = 0$. Each random variable $X(t)$ induces a distribution (R,P_t). The family of these for all $t > 0$ is a semigroup because

$$X(t + s) = [X(t + s) - X(s)] + [X(s) - X(0)]$$

so that the distribution of $X(t + s)$ is that of the sum of two independent random variables having distributions P_t and P_s. Thus each process of this kind generates a semigroup. Furthermore if $0 < t_1 < \cdots < t_n$ then, when $n = 3$, for example, the random vector $(X(t_1),X(t_2),X(t_3))$ can be written in the form

$$(X(t_1),\ X(t_1) + [X(t_2) - X(t_1)],\ X(t_1) + [X(t_2) - X(t_1)] + [X(t_3) - X(t_2)])$$

and so has the distribution of the consecutive sums of three independent random variables having the same distributions as $X(t_1)$, $X(t_2 - t_1)$, $X(t_3 - t_2)$, respectively. Clearly, then, the whole family of finite distributions can be calculated from the semigroup of first-order distribution.

Furthermore if we have any semigroup (R,P_t) for $t > 0$ then we can use it as above to define a family of finite distributions with parameter set $t > 0$. This family is easily shown to be consistent, and hence there is a process generating this family. Such processes are completely defined by their first-order distributions, which can be *any* semigroup. We summarize below, and also exhibit another property.

Theorem I Let X_t for $t \geq 0$ be a stationary-independent-increment continuous-parameter stochastic process satisfying $X_0 = 0$. Then the distributions, P_{X_t} for $t > 0$, form a semigroup which we say **is generated by** the process.

Furthermore for any semigroup there exists such a process which generates this semigroup. Also two such processes generate the same semigroup iff they generate the same family of finite distributions.

In addition if $E(X_t)^2$ is bounded on $0 \leq t \leq 1$ then

$$EX_t = tEX_1 \qquad \text{and} \qquad \operatorname{Var} X_t = t \operatorname{Var} X_1 \qquad \text{for all } t \geq 0. \quad ∎$$

We indicate briefly how the consistent family which determines such a process is defined. Let any semigroup (R,P_t) for $t > 0$ be given, and use it for the first-order distributions. If $0 < t_1 < t_2$ then (R^2,P_{t_1,t_2}) is defined to be the distribution induced by $(Z_1, Z_1 + Z_2)$ when Z_1 and Z_2 are independent and have distributions P_{t_1} and $P_{t_2-t_1}$, respectively. Then the two marginals from P_{t_1,t_2} are just the distributions of Z_1 and $Z_1 + Z_2$, which are P_{t_1} and P_{t_2} by the semigroup property. Thus the second-order distributions are consistent with the first-order distributions. The obvious definition for higher-order distributions and the remainder of the consistency proof are left to Exercise 2.

All exercises except 1 and 4 may be done before reading the remainder of this section.

For the proof of the last part of Theorem 1 let X_t for $t \geq 0$ be a stationary-independent-increment process with $X_0 = 0$. We have stationary increments, so that if $f(t) = EX_t$ and $t > 0$, $s > 0$ then

$$f(t + s) = E([X_{t+s} - X_s] + X_s) = f(t) + f(s)$$

and hence $f(t) = f(1)t$ from Lemma 1 (Sec. 16-1). The same technique can be applied to the variance, because we have stationary *and* independent increments. Thus we have the asserted linearity in t for both the mean and the variance.

For any process such as that in Theorem 1, Chebychev's inequality yields

$$P[|X_t - tEX_1| < \delta\sqrt{t \operatorname{Var} X_1}] \geq 1 - \frac{1}{\delta^2} \qquad \text{for every } t > 0,\ \delta > 0.$$

In particular for every $t > 0$ the probability is at least .75 that X_t will have some value in the interval $tEX_1 \pm 2\sqrt{t \operatorname{Var} X_1}$ within two standard deviations of its mean. Thus it is helpful to plot the mean tEX_1 as a function of t, as well as the two curves $tEX_1 \pm \sqrt{t \operatorname{Var} X_1}$ one standard deviation on either side of the mean. This is done in Fig. 2 (Sec. 23-3) for the Poisson process and in Fig. 2 (Sec. 24-1) for the Wiener process. For these two processes the next paragraph adds to the interpretation of these curves.

We prematurely mention that a Poisson (Wiener) process X_t for $t \geq 0$ satisfying $X_0 = 0$, with parameter $\lambda > 0$ $(\sigma^2 > 0)$, is a stationary-independent-increment process satisfying certain regularity conditions on realizations, and such that X_t induces a Poisson (normal) density with mean λt (mean 0 and variance $\sigma^2 t$). In both cases the semigroup property is easily checked. For the Wiener process, Table 1 (Sec. 13-1) of the normal distribution yields

$$P[|X_t| < \sigma\sqrt{t}] = .6826 \qquad \text{and} \qquad P[|X_t| < 2\sigma\sqrt{t}] = .9544 \qquad \text{for all } t > 0.$$

Thus for any $t > 0$ about 68 percent of the realizations should at this t be within $\pm\sigma\sqrt{t}$. From Example 6 (Sec. 13-2) we know that the Poisson distribution is approximated by the normal, as the mean grows, and hence for a Poisson process

$$P[|N_t - \lambda t| < \sqrt{\lambda t}] \doteq .6826$$

if λt is much larger than 1.

STATIONARY PROCESSES AND MARKOV PROCESSES

We next introduce two basic definitions for stochastic processes and relate the second definition to stationary-independent-increment processes. A stochastic process $X(t)$ for $t \in T$ is said to be a **stationary stochastic process** iff for every $n \geq 1$ the two random vectors

$$(X(t_1), \ldots, X(t_n)) \qquad \text{and} \qquad (X(t_1 + s), \ldots, X(t_n + s))$$

are identically distributed whenever $t_1 < \cdots < t_n$ and $t_1 + s < \cdots < t_n + s$ are in T. In other words, the joint distribution of $X(t_1 + s), \ldots, X(t_n + s)$ does not depend on s. That is, joint distributions are invariant to translation. If the parameter set T is the real line then the intuitive meaning of stationarity is that every translation of any particular realization is as likely as is the particular realization. For a stationary process the distribution of $(X(s + h), X(s))$, and hence of $X(s + h) - X(s)$, does not depend on s. Thus a stationary process has stationary increments, but the converse is not true. In particular for a stationary process every X_t must have the same distribution, which is certainly not true for a stationary-independent-increment process. We will find it possible to transform the Poisson and Wiener processes into stationary processes.

A stochastic process $X(t)$ for $t \in T$ is called a **Markov process** iff for any $n \geq 3$ points $t_1 < \cdots < t_n$ in T and reals $x_1, \ldots, x_n$

$$P[X(t_n) < x_n \mid X(t_{n-1}) = x_{n-1}, \ldots, X(t_2) = x_2, X(t_1) = x_1] = P[X(t_n) < x_n \mid X(t_{n-1}) = x_{n-1}].$$

Such a process is said to have the *Markov property* above. Thus the conditional distribution function for any one of the variables, given the value of some one earlier variable *and* the values of finitely many still earlier variables, does not depend on the still earlier variables. In other words, if we are given $[X(t_{n-1}) = x_{n-1}]$ then the conditional distribution of any future variable is not affected by observing variables earlier than time t_{n-1}. For such a process the future behavior does not depend on which history led to the present condition $X(t_{n-1}) = x_{n-1}$. If we have only partial knowledge of the present condition, say $[X(t_{n-1}) < 3]$, we may not be able to ignore earlier observations, which might give additional information concerning the present condition. If X_t for $t \in T$ is Markov and T' is a subset of T then X_t for $t \in T'$ is certainly Markov.

If $X(t)$ for $t \geq 0$ satisfies $X(0) = 0$ and has independent increments then it is Markov, because $X(t_1), \ldots, X(t_n)$ are just consecutive sums of independent random variables.

A stationary-independent-increment process X_t for $t \geq 0$ with $X_0 = 0$ is a special kind of Markov process; namely, if $0 < s < t$ then the conditional distribution of $X_t = X_s + [X_t - X_s]$, given $[X_s = x]$, is just the distribution induced by $x + X_{t-s}$. Thus it depends on s and t only through their difference $t - s$ *and* its dependence on x is only through a translation of the conditional distribution by x. More intuitively, as indicated in Fig. 1, if the event $[X_s = x]$ occurs then, regardless of the realization up to time s which happened to yield this event, the future increment process $X_t - X_s$ for $t \geq s$ starts at the new origin (s,x) by precisely the same probability law by which the original process started at $(0,0)$.

Let $X(t)$ for $t \in T$ be a process such that each $X(t)$ induces a discrete probability density. Clearly it is a Markov process iff

$$P[X(t_n) = x_n \mid X(t_{n-1}) = x_{n-1}, \ldots, X(t_1) = x_1] = P[X(t_n) = x_n \mid X(t_{n-1}) = x_{n-1}]$$

whenever the first conditioning event has positive probability. Thus in this case we need not use distribution functions, and we need only condition on events having positive

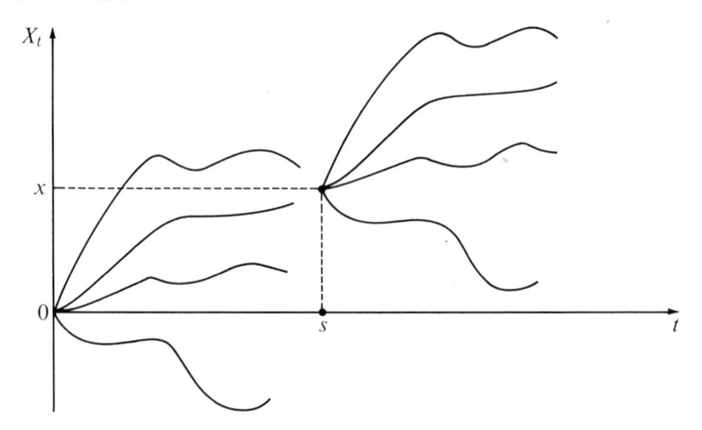

Fig. 1 Typical realizations starting at (0,0) and (s,x).

probability. In general, however, the Markov property requires conditioning on events having zero probability, and this can be very delicate, as explained after Example 3 (Sec. 16-6). However, we will only so condition when we have "nice" continuous densities, as for the Wiener process.

SOME FACTS ABOUT STATIONARY-INDEPENDENT-INCREMENT PROCESSES

In order to give a general perspective we mention a few facts about stationary-independent-increment processes.[1] As defined in Sec. 24-1, the Wiener processes, W_t for $t \geq 0$ with $W_0 = 0$, have stationary independent increments, satisfy $EW_t = 0$ for all $t \geq 0$, and have only *continuous* realizations. It is a fact that no other processes have these properties. Thus only the normal-density semigroup yields continuous realizations. In particular, realizations from the processes determined by the *continuous* density semigroups of Exercises 5*a* and *b* have jumps.

More generally a process "X_t for $t \geq 0$ with $X_0 = 0$" obviously has independent increments and only continuous realizations if it is defined by

$$X_t = f(t) + W(m(t)) \qquad \text{for } t \geq 0$$

where f and m are continuous functions with $f(0) = m(0) = 0$, m is nondecreasing, and $W(\tau)$ for $\tau \geq 0$ with $W(0) = 0$ is a Wiener process. By analogy with Sec. 23-5 we might call such a process a *nonhomogeneous Wiener process*. It is a fact that all processes "Y_t for $t \geq 0$ with $Y_0 = 0$" having independent increments and only continuous realizations can be constructed in this way.

We next indicate the generality of the class of stationary-independent-increment processes. Clearly X_t for $t \geq 0$ with $X_0 = 0$ has stationary independent increments if it is defined by

$$X_t = \beta t + N_t + W_t \qquad \text{for } t \geq 0$$

where N_t for $t \geq 0$, and W_t for $t \geq 0$ are two independent processes each having stationary independent increments. Let N_t for $t \geq 0$ be a compound Poisson process

[1] For precise statements and proofs see J. L. Doob, "Stochastic Processes," chap. 8, John Wiley & Sons, Inc., New York, 1953.

and let W_t for $t \geq 0$ be a Wiener process. It is a fact that in a certain sense every stationary-independent-increment process "Y_t for $t \geq 0$ with $Y_0 = 0$" can be approximated by such "X_t for $t \geq 0$" processes.

EXERCISES

1. *The Future Impact of Knowledge of the Past, for a Stationary-independent-increment Process,* $X(t)$ for $t \geq 0$ with $X(0) = 0$. In order to condition on events having positive probability assume that each X_t induces a discrete density. Let $0 \leq s_1 < s_2 < \cdots < s_k < s < s + t_1 < \cdots < s + t_n$ and prove the following equation:

$$P[X(s + t_1) - x = x_1, \ldots, X(s + t_n) - x = x_n \mid X(s) = x, X(s_1) = y_1, \ldots, X(s_k) = y_k] \\ = P[X(t_1) = x_1, \ldots, X(t_n) = x_n]$$

Thus, as in Fig. 1, if we are given $[X(s) = x]$ and we are given earlier information, then the future process "$X(s + t) - x$ for $t \geq 0$" transformed to start at (0,0), has the same joint distributions as the original process.

2. *Completing the Proof of Theorem 1.* Let (R,P_t) for $t > 0$ be any semigroup. Define a family of finite distributions with parameter set $t > 0$ as follows. For any $n \geq 1$ points $0 < t_1 < t_2 < \cdots < t_n$ let $(R^n,P_{t_1,\ldots,t_n})$ be the distribution induced by

$$(Z_1, Z_1 + Z_2, Z_1 + Z_2 + Z_3, \ldots, Z_1 + \cdots + Z_n)$$

where $Z_1, \ldots, Z_n$ are independent, Z_1 has distribution P_{t_1}, and Z_i has distribution $P_{t_i - t_{i-1}}$ for $i \geq 2$.

(*a*) Show that the marginal distribution for the second two coordinates of (R^3,P_{t_1,t_2,t_3}) is the already defined P_{t_2,t_3}.

(*b*) Sketch a proof of the fact that the family is consistent.

(*c*) From the Kolmogorov consistency condition there is a process X_t for $t > 0$ which generates this family. Define $X_0 = 0$ and show that the process X_t for $t \geq 0$ has stationary independent increments.

3. *Determination of the Whole Semigroup by the Distributions for t Near to Zero.* A semigroup on an interval of length $\delta > 0$ is a family of probability spaces (R,P_t), one for each t satisfying $0 < t \leq \delta$, such that whenever $0 < s < s + t \leq \delta$ then P_{s+t} equals the distribution induced by the sum of two independent random variables having respective distributions P_s and P_t. Obviously if we have a semigroup and throw away all distributions for t greater than some extremely small $\delta > 0$ then we are left with a semigroup on an interval of length δ.

(*a*) Given any semigroup (R,P_t) on an interval $0 < t \leq \delta > 0$, prove that *there is a unique extension* to a semigroup on the half-line $t > 0$.

(*b*) Prove that if "X_t for $0 \leq t \leq \delta > 0$ with $X_0 = 0$" is any stationary-independent increment process then there is a stationary independent increment process "Y_t for $t \geq 0$ with $Y_0 = 0$" such that X_t for $0 \leq t \leq \delta$ and Y_t for $0 \leq t \leq \delta$ generate the same family. Roughly speaking, we can extend such a process from an interval to the half-line $t \geq 0$.

4. *A Stationary Process Can Be Constructed by Randomizing the Phase of Any Periodic Function.* Let f: $R \to R$ be periodic with period 1; that is, $f(t) = f(1 + t)$ for all real t. For example, $f(t) = \cos^2 4\pi t + \sin (6\pi t + 3)$. Let Y have a uniform density on $0 < y < 1$ and let $X_t = f(t + Y)$ for all real t.

(*a*) If $f(t) = \sin (2\pi t + 3)$ find EX_t and $E(X_t)^2$.

(*b*) Show that in general, "X_t for t real" is a stationary stochastic process. It is sufficient to give a generalizable proof that $P[X_{s+h} < \epsilon, X_{t+h} < \delta]$ is independent of h.

5. Show that we obtain a semigroup if for $t > 0$ we define P_t to correspond to the following:

(*a*) Gamma density $\gamma_{\lambda,t}$ with λ fixed

(*b*) Density induced by tV, where V has a standardized Cauchy density

(*c*) Compound Poisson density with parameter $\lambda = t$, as defined at the beginning of Example 4 (Sec. 14-2); that is, each p_{X_1} determines a semigroup. A generalization of this fact arises naturally in Sec. 23-4.

6. An orbiting satellite expends W_i watts of energy in transmitting data, essentially instantaneously, at time t_i when it passes its control station for the ith time. The infinite sequence $0 < t_1 < t_2 < \cdots \to \infty$ of transmission times is fixed in advance. Let S_t be the total energy expended up to time t in this way; that is, S_t is the sum of those W_i for which $t_i \le t$. Assume that $W_1, W_2, \ldots$ are independent random variables. Does the process S_t for $t \ge 0$ necessarily have independent increments? Stationary increments? Stationary increments *if* we assume that $t_i = i$ for $i = 1, 2, \ldots$ and $W_1, W_2, \ldots$ are independent and identically distributed?

7. Let $W_1, W_2, \ldots$ be a sequence of independent random variables for which each $P[|W_i| \le 1] = 1$. Let $t_1, t_2, \ldots$ be some infinite sequence of distinct positive real numbers. For $t > 0$ let S_t be the sum of those $W_i/2^i$ for which $t_i \le t$. Explain why this definition makes sense even if $t_1, t_2, \ldots$ is an enumeration of the positive rationals. Note that S_t for $t \ge 0$ has independent increments.

COMMENT: A process X_t for $t \in T$ is said to have a *fixed point of discontinuity* at t iff there is positive probability for obtaining a realization having a discontinuity at t. The process "S_t for $t \ge 0$" above has a fixed point of discontinuity at t iff t equals some t_i with $P[W_i = 0] < 1$. It can be shown[1] that any independent increment X_t for $t \ge 0$ can be represented in the form

$$X_t = f(t) + S_t + Y_t$$

where S_t may be a slight generalization of the above process but "S_t for $t \ge 0$" is independent of the process "Y_t for $t \ge 0$," which has no fixed points of discontinuity and actually has some desirable continuity properties, although its realizations may have jumps.

22-3 CONSTRUCTION AND UNIQUENESS OF PROCESSES

This section first describes the theoretical technique for construction of a stochastic process from a given consistent family. Discussion and examples then illustrate that many events of interest, described in terms of realizations from a *continuous*-parameter process, do not have their probabilities uniquely determined by the family. The family is, however, often the basis for calculations. We then see how this difficulty can be overcome by requiring that realizations be regular functions.

If T is a nonempty set let R^T denote the **collection of all real-valued functions defined on** T; that is, an object f belongs to R^T iff it is a function f: $T \to R$. If we interpret R^n as an abbreviation for $R^{\{1,2,\ldots,n\}}$ then a member of R^n is a real-valued function on $\{1, 2, \ldots, n\}$, that is, a vector $(x_1, \ldots, x_n)$, just as before. In other words, R^T is the collection of all possible realization for all stochastic processes "X_t for $t \in T$" which have T for parameter set. Note that even for a random vector X: $\Omega \to R^n$, with

[1] See Doob, *ibid.*, p. 417.

probability 1 a realization may necessarily belong to a small subset of R^n, but this subset depends on the distributions involved—Bernoulli, exponential, etc.

A **tube**,[1] or a tube of functions, in R^T consists of those functions $f: T \to R$ satisfying

$$a_i \leq f(t_i) \leq b_i \qquad \text{for } i = 1, \ldots, n.$$

Thus a tube is defined in terms of $n \geq 1$ members $t_1 < t_2 < \cdots < t_n$ of T and $2n$ reals $a_i \leq b_i$. As pictured in Fig. 1, the tube consists of all the functions satisfying the

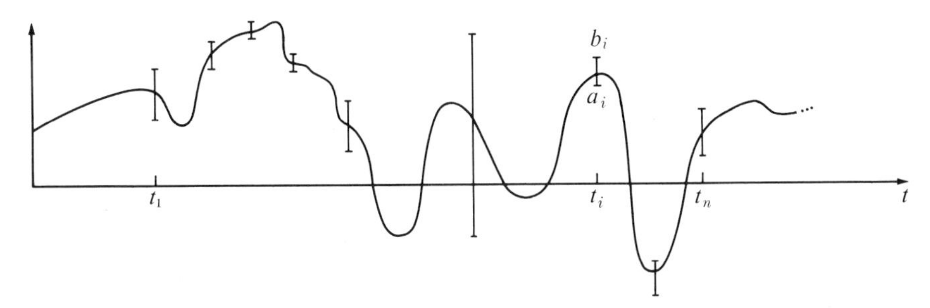

Fig. 1 One of the functions in a tube of functions.

finite-interval constraints at the n points. If $R^T = R^n$ then the tube is just a closed rectangle in R^n. Note that the complement of the tube consists of those f such that

$$f(t_i) < a_i \qquad \text{or} \qquad f(t_i) > b_i \qquad \text{for some } i = 1, \ldots, n.$$

Hence only the same t_i are involved in describing the complement.

If $X(t)$ for $t \in T$ is a stochastic process then the event

$$[a_1 \leq X(t_1) \leq b_1, \ldots, a_n \leq X(t_n) \leq b_n]$$
$$= \{\omega: a_1 \leq r_\omega(t_1) \leq b_1, \ldots, a_n \leq r_\omega(t_n) \leq b_n\}$$

can be described in terms of realizations, as shown. Thus this event consists of the sample points whose realizations belong to the tube of Fig. 1. Therefore the probability of obtaining a realization belonging to the tube of Fig. 1 is nothing more than the probability of the corresponding closed rectangle in R^n as calculated from the distribution of the random vector $(X(t_1), \ldots, X(t_n))$, which is in turn determined by the probabilities of closed rectangles in R^n. Clearly, then, if two stochastic processes have the same parameter set then they generate the same family iff both processes assign the same probability to every tube.

Assume that we are given a consistent family with parameter set T. We construct a stochastic process generating the family as follows. Let the sample space be $\Omega = R^T$. For any $t \in T$ define the function $X_t: \Omega \to R$ as follows: the value assigned by X_t

[1] *Tube* is not standard terminology. In relation to standard terminology, a tube is a special case of a **product cylinder** in R^T, which consists of all $f: T \to R$ satisfying $f(t_i) \in A_i$ for $i = 1, \ldots, n$, where $n \geq 1$, $t_i \in T$, $A_i \subset R$ are arbitrary.

to an f in Ω is just $f(t)$; that is, if realization f: $T \to R$ is obtained then the value of X_t is just the value $f(t)$ of this realization at t. Then to the tube

$$[a_1 \leq X_{t_1} \leq b_1, \ldots, a_n \leq X_{t_n} \leq b_n]$$

in R^T we assign the probability of the corresponding closed rectangle in R^n, as specified by $P_{t_1,\ldots,t_n}$ from the consistent family. Thus the family assigns consistent probabilities to tubes in $\Omega = R^T$, and in such a way that any random vector $(X_{t_1},\ldots,X_{t_n})$ has the distribution specified by the family.

Now if we take some very complicated event in R^T we can attempt to approximate the event by a countable collection of events, each of which is defined in terms of only finitely many random variables X_t and hence has a probability assigned to it by the family. We can then attempt to assign a probability to the complicated event by taking the limit of appropriate probabilities calculated from such approximations. The proof of the Kolmogorov consistency condition shows that this program can be carried out for a large collection of complicated events, and the resulting assignment is indeed a probability measure. The result is stated rigorously in the addendum, using the terminology of Chap. 21.

We now show that by specifying a family we are uniquely specifying the probability of many events which are more complicated than tubes. Suppose that we have a consistent family of finite distributions with an infinite parameter set T. We know that there exists a process generating this family. Let X_t for $t \in T$ with associated (Ω,P) be such a process, and let $Y(t)$ for $t \in T$ with associated (Ω',P') be another such process. For an *infinite sequence* of interval constraints we can use decreasing events, each of whose probabilities is determined by this same family, to get

$$\begin{aligned} P[a_i \leq X(t_i) \leq b_i \text{ for } i = 1, 2, \ldots] &= \lim_{n\to\infty} P[a_i \leq X(t_i) \leq b_i \text{ for } i = 1, \ldots, n] \\ &= \lim_{n\to\infty} P'[a_i \leq Y(t_i) \leq b_i \text{ for } i = 1, , \ldots n] \\ &= P'[a_i \leq Y(t_i) \leq b_i \text{ for } i = 1, 2, \ldots]. \end{aligned}$$

Therefore both processes assign the same probability to the event that some realization f: $T \to R$ will be obtained which satisfies

$$a_i \leq f(t_i) \leq b_i \qquad \text{for all } i = 1, 2, \ldots .$$

Having obtained agreement for such events, we can similarly use complements, countable disjoint unions, etc., of events for which agreement has already been obtained, to obtain agreement for still more complicated events. It should appear reasonable that agreement can be proved for essentially every event defined in terms of countably many of the random variables. In particular for a discrete-parameter process the family uniquely determines all probabilities of interest. This is *not* the case for continuous-parameter processes, so we now consider the phenomenon which leads to this lack of uniqueness for continuous-parameter processes.

UNIQUENESS FOR CONTINUOUS PARAMETER PROCESSES

For every $t \in T$ let A_t be an event and assume that $P(A_t) = 0$ for all $t \in T$. If T is countable then naturally

$$P\left(\bigcup_{t \in T} A_t\right) \le \sum_{t \in T} P(A_t) = 0.$$

That is, the probability is 0 for the occurrence of at least one of *a countable collection* of zero-probability events. No such deduction can be generally valid if T is uncountable. For example, if X has a normal 0 and 1 density then $P[X = t] = 0$ for all real t, since $[X = t]$ is the event that X has precisely the value t. Therefore, as above, if T is the set of all rational numbers then $P\left(\bigcup_{t \in T} [X = t]\right) = 0$, but if T is the set of all positive numbers then

$$P\left(\bigcup_{t \in T} [X = t]\right) = P[X > 0] = \tfrac{1}{2}.$$

Thus the probability of an uncountable union of disjoint events is not uniquely determined merely by the specification of the probability of each of the events.

For a continuous-parameter process let $g: R \to R$ be a fixed function and consider the event

$$[X_t \le g(t) \text{ for all real } t \text{ satisfying } 0 \le t \le 1] = \bigcap_{0 \le t \le 1} [X_t \le g(t)],$$

as well as its complement

$$[X_t > g(t) \text{ for at least one } t \text{ satisfying } 0 \le t \le 1] = \bigcup_{0 \le t \le 1} [X_t > g(t)].$$

For example, g might be the function e^{3t}. Thus we are interested in the event that the realization satisfy a certain constraint over the whole interval $0 \le t \le 1$. For example, if $g \equiv 4.2$ we have the event

$$\{\omega: r_\omega(t) \le 4.2 \text{ for all real } t \text{ satisfying } 0 \le t \le 1\}.$$

A casual review of the various fields of application indicated at the beginning of this chapter should convince the reader that events of this kind are of considerable practical importance. But these events involve uncountably many random variables. It turns out that two processes can generate the same family but assign different probabilities to such events. We next construct some examples which exhibit this lack of uniqueness.

By the **trivial zero process** with parameter set T we mean the process V_t for $t \in T$ which has only the single identically zero realization on T. Thus $V_t(\omega) = 0$ for all t and ω, so that

$$P[V_t = 0 \text{ for all } t \in T] = 1.$$

If the experiment is performed then the observed realization is zero for *all* t, not for all except one or a few t.

We call any stochastic process X_t for $t \in T$ a **zero process** iff $P[X_t = 0] = 1$ for all $t \in T$. If $T' = \{t_1, t_2, \ldots\}$ is any countable subset of T then

$$P[X_t \neq 0 \text{ for at least one } t \in T'] = P\left(\bigcup_{n=1}^{\infty} [X_{t_n} \neq 0]\right) \leq \sum_{n=1}^{\infty} P[X_{t_n} \neq 0] = 0$$

so that for the complement we have

$$P[X_t = 0 \text{ for all } t \in T'] = 1.$$

In particular this shows that a process is a zero process iff it generates the same family of finite distributions as does the trivial zero process. *If* a zero process happens to satisfy

$$P[X_t = 0 \text{ for all } t \in T] = 1,$$

as it must if T is countable for example, then there is no loss of generality in replacing the process by the trivial zero process. We next exhibit zero processes which assign different probabilities, to events of interest, from those assigned by the trivial zero process, even though all zero processes generate the same family. We will not usually use such processes, but they do arise naturally and can cause confusion if not properly related to the general conceptual framework.

Example 1 For the sake of explicitness let Y have an exponential density with mean 1. More generally we could use any continuous distribution function on R. Define the process X_t for $t \geq 0$ as follows. Each X_t has only the values 0 and 1. Let X_t have the value 1 iff $[Y = t]$ occurs. If $[Y = 2.3]$ occurs, for example, then $X_{2.3}$ has the value 1, while every other X_t has the value 0. Thus the process just tells us the value of Y. Three of its realizations are shown in Fig. 2. Thus y is selected according to the exponential density, and if $y = 2.3$ occurs then we observe the realization $r_{2.3}$ given by

$$r_{2.3}(2.3) = 1 \qquad \text{and} \qquad r_{2.3}(t) = 0 \qquad \text{for all } 0 \leq t \neq 2.3.$$

Obviously X_t for $t \geq 0$ is a zero process, since

$$P[X_t = 0] = P\,[Y \neq t] = 1 \qquad \text{for all } t \geq 0.$$

However, *no* realization is identically zero, so that

$$P[X_t = 0 \text{ for all } t \geq 0] = 0 \neq 1.$$

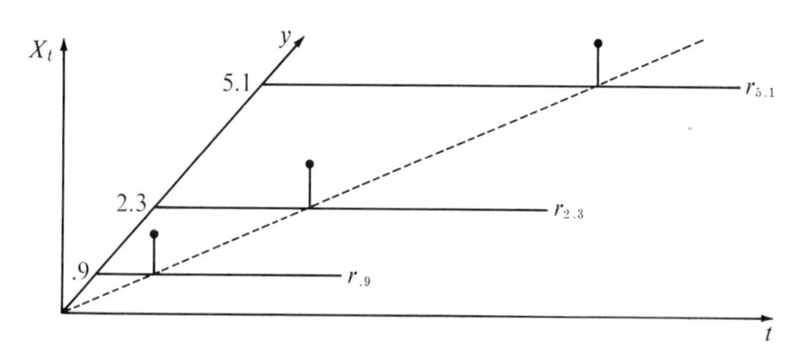

Fig. 2

If we observe this process at only a countable set of times T' then $P[Y \in T'] = 0$, so with probability 1 we will observe *the zero realization on* T' and hence not learn the value of Y.

We can generalize the construction as follows. Let Z_t have the value t if $[Y = t]$ occurs; otherwise Z_t has the value 0. Thus if $y = 2.3$ is selected then we observe the realization which is identically zero on $t \geq 0$, except that it has the value 2.3 at $t = 2.3$. Clearly Z_t for $t \geq 0$ is a zero process. However,

$$P[0 \leq Z_t \leq 2 \text{ for all } t \geq 0] = P[0 \leq Y \leq 2] = \int_0^2 e^{-y}\, dy.$$

For a more interesting example let Fig. 3 represent a realization from a Poisson process. Section 23-3 will justify the following self-contained description of Fig. 3. Place crosses at the points $T_k = X_1 + \cdots + X_k$ for $k = 1, 2, \ldots$, where $X_1, X_2, \ldots$ are independent and

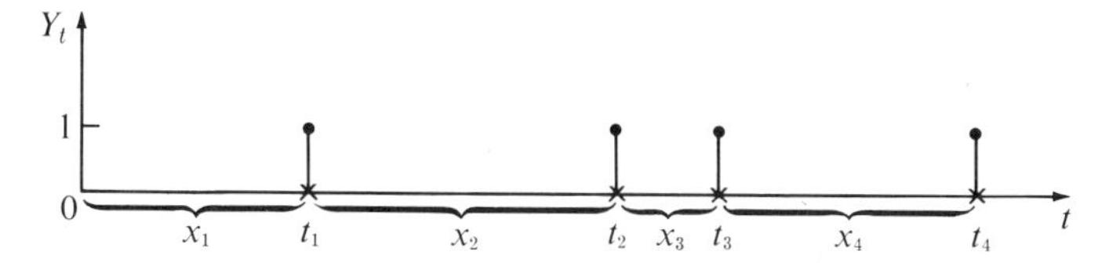

Fig. 3 A realization from a Poisson process.

each is exponential with mean $1/\lambda$. We define for each $t \geq 0$ a random variable Y_t to have the value 1 or 0, depending on whether there is or is not a cross at t. Then Fig. 3 depicts a realization from the zero process Y_t for $t \geq 0$ which describes the Poisson process as well as does the more common counting function process, and in some senses even more naturally. If we replace this process by one generating the same family but having only regular realizations, as defined below, then we essentially throw the process away and use the single identically zero realization.

For later reference let us complete this example by further use of processes from later chapters. If we add a function like that of Fig. 3 to some continuous function we get a function which has denumerably many discontinuities. Let W_t for $t \geq 0$ be a Wiener process, as defined in Sec. 24-1 to have continuous realizations, let Y_t for $t \geq 0$ be as above, and let the two processes be independent. Then "$W_t + Y_t$ for $t \geq 0$" generates the same family as "W_t for $t \geq 0$," but its realizations have infinitely many discontinuities. Thus a process may generate the Wiener process family but have only discontinuous realizations. Similarly if Y_t for $t \geq 0$ is as above and is independent of a new Poisson process "N_t for $t \geq 0$" described by counting random variables, then the process "$N_t + Y_t$ for $t \geq 0$" generates the same family as "N_t for $t \geq 0$," but its realizations are not counting functions, as defined in Sec. 23-1. ■

A function f: $T \to R$ on an interval T is said to be **regular** iff it has at most finitely many points of discontinuity in every finite interval; furthermore if t is a point of discontinuity, and t is not an end point of T then f is required to have left- and right-hand limits at t, and $f(t)$ must be equal to one of these limits. Also f must have a one-sided limit at every end point. Note that if f is a continuous function and if its value is changed at just one interior point of T, as was done in Example 1, then the modified function is not regular. Clearly if f is a regular function on T then its restriction[1]

[1] In general if $A \subset B$ and f: $B \to C$ then by the **restriction of** f **to** A we naturally mean the function g: $A \to C$ defined by $g(a) = f(a)$, for all a in A.

to a subinterval T_1 of T is then a regular function on T_1. For example, if

$$f(t) = \begin{cases} 0 & \text{if } 0 \leq t < 1 \\ 1 & \text{if } 1 \leq t \end{cases}$$

then its restriction to $0 \leq t \leq 1$ is a regular function because we permit a discontinuity at an end point.

If "X_t for $t \in T$" is a continuous-parameter process and if ω is a sample point then the realization r_ω may be a regular function on T, in which case we call it a **regular realization.** If every realization is regular then we say that the **process has only regular realizations.** Clearly if X_t for $t \in T$ is such a process then so is X_t for $t \in T_1$, for any subinterval T_1 of T, so the following theorem may be applied to subintervals.

For continuous-parameter processes the following theorem, with its restrictions on both the processes and the events, is a weak but sometimes adequate substitute for the basic property that two *discrete-parameter* processes which generate the same family must necessarily assign the same probability to all events of interest. The theorem, and its proof, are meant only to illustrate the way in which a regularity assumption can yield uniqueness.

Theorem I If two continuous-parameter stochastic processes generate the same family of finite distributions with parameter set T, and *have only regular realizations*, then both processes assign the same probability to every event of one of the forms

$$[f(t) \leq X_t \text{ for all } t \in T] \qquad [X_t \leq g(t) \text{ for all } t \in T]$$
$$[f(t) \leq X_t \leq g(t) \text{ for all } t \in T]$$

where f: $T \to R$ and g: $T \to R$ are continuous functions on T. ■

The terminology employed in the theorem may need clarification. Let X_t for $t \in T$ be a process with associated probability space (Ω,P), and let Y_t for $t \in T$ be a process with associated (Ω',P'). To say that both processes assign the same probability to every event of the form $[f(t) \leq X_t$ for all $t \in T]$ means that

$$P\{\omega : f(t) \leq X_t(\omega) \text{ for all } t \in T\} = P'\{\omega' : f(t) \leq Y_t(\omega') \text{ for all } t \in T\}.$$

These are two different events, perhaps even in different probability spaces. However, for any f: $T \to R$ we have essentially given a description which selects a unique event for any process with parameter set T, and two events selected according to this description are supposed to have the same probability.

We now indicate the basic idea behind the proof of the theorem. The event

$$[f(t) \leq X_t \leq g(t) \text{ for all } t \in T]$$

occurs iff we obtain a realization in the shaded band shown in Fig. 4. It should be fairly clear that if *a regular function*, like the one exhibited, satisfies this constraint whenever t is a rational number then it must satisfy the constraint for all t. In other words, when there are only regular realizations then the event above is equal to an

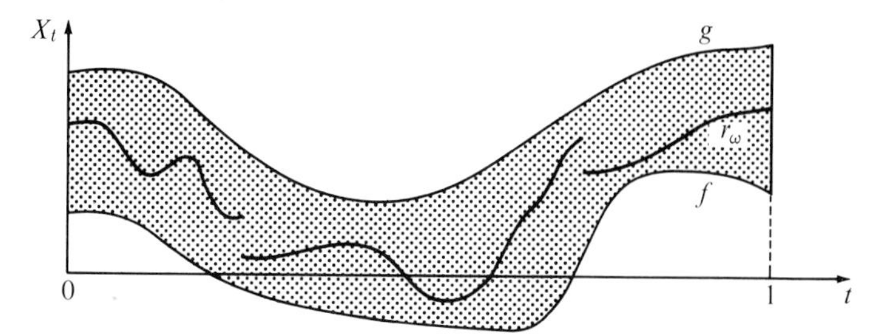

Fig. 4 A regular realization r_ω satisfying $f \le r_\omega \le g$.

event defined in terms of only countably many t, and to such events the family assigns a unique probability.

Proof of Theorem I We consider only the two-sided constraint. Let X_t for $t \in T$ be a continuous-parameter process and let

$$A = [f(t) \le X_t \le g(t) \text{ for all } t \in T] = \{\omega : f(t) \le r_\omega(t) \le g(t) \text{ for all } t \in T\}.$$

Let T' consist of all rational numbers in T, plus the end points of T if they are irrational. Let

$$B = [f(t) \le X_t \le g(t) \text{ for all } t \in T'] = \{\omega : f(t) \le r_\omega(t) \le g(t) \text{ for all } t \in T'\}$$

be the less constrained event that the realization be in the band at least for all of the countably many times in T'. Since A is a subset of B we have $P(A) \le P(B)$. [Example 1 shows that $P(A) < P(B)$ can occur, with $f \equiv -.5$ and $g \equiv .5$.] Assume that a particular ω in B happens to yield a regular realization r_ω. Let t be any number in T but not in T'. We can find a sequence $t_1 < t_2 < \cdots$ of rationals increasing to t. Then letting n go to infinity in

$$f(t_n) \le r_\omega(t_n) \le g(t_n)$$

yields

$$f(t) \le r_\omega(t - 0) \le g(t).$$

We can similarly take a limit along rationals decreasing to t, so that the definition of regularity then yields $f(t) \le r_\omega(t) \le g(t)$. Therefore any regular realization r_ω from B must also belong to A.

If the process has *only* regular realizations then $A = B$, so $P(A) = P(B)$. More generally if C is the event that we obtain a regular realization then $CA = CB$, so that if $P(C) = 1$ then $P(A) = P(CA) = P(CB) = P(B)$, as before.

If two processes generate the same family then they must assign the same probability to B, and hence also to A if they have only regular realizations. ■

For any consistent family of interest to us there exists a process generating the family and having only regular realizations. Often this will be apparent because we can construct such a process, as is done in the two characterizations for the Poisson process in Sec. 23-3.

Unless explicitly indicated otherwise, any continuous-parameter stochastic process which we consider is assumed to have only regular realizations. For such processes the family assigns a unique probability to events like those in Theorem 1, and to many other events of interest. In the future we will often assume the uniqueness without any attempt at justification.

The reader will probably at some time encounter mention of Doob's concept of *separability*, which is generalization of the foregoing. A process having only regular realizations is necessarily a separable process. Separable processes are more general in that they may have nonregular realizations, and it can be shown that for every consistent family there is a separable process generating it. However, the realizations are still sufficiently constrained to permit a reduction to countably many random variables, as was done above, yielding Theorem 1 with "have only regular realizations" replaced by "are separable." Sometimes a nonseparable process is replaced by a *separable version*, that is, by a separable process generating the same family. This certainly seems to be appropriate if, as often happens, the two processes assign the same probability to every event of interest. However, Example 1 shows that sometimes such a replacement is not appropriate. The trivial zero process with the single identically zero realization can be considered to be a version having only regular realizations, and hence a "separable version" of *every* zero process.

ADDENDUM ON THE KOLMOGOROV CONSISTENCY CONDITION

The consistency condition is stated in the rigorous terminology of Chap. 21, and then for continuous-parameter processes a characterization is given for the class of events whose probabilities are uniquely determined by the family of finite distributions.

Let T be any interval or countable set of real numbers and let $\mathscr{B}_T$ be the smallest σ field containing all tubes in R^T. The *Kolmogorov consistency condition* can be stated as follows: for any consistent family of finite distributions with parameter set T *there exists a unique* P on $\mathscr{B}_T$ such that $(R^T, \mathscr{B}_T, P)$ is a probability space, such that the stochastic process "X_t for $t \in T$" generates the given family, where, as before, the value of X_t at f: $T \to R$ is just $f(t)$.

Actually Lemma 3 (Sec. 21-2) immediately yields the uniqueness of P. If T is countable then every subset of R^T of any practical interest belongs to $\mathscr{B}_T$.

Assume that T is uncountable. Let $\mathscr{C}$ be the collection of all subsets of R^T which are definable in terms of only countably many t in T. We show that $\mathscr{B}_T$ is a subset of $\mathscr{C}$. More precisely let $\mathscr{C}$ consist of those subsets C of R^T satisfying the condition that there exists some countable subset T_1 of T and some subset C_1 of R^{T_1} such that any f: $T \to R$ belongs to C iff its restriction to T_1 belongs to C_1. In other words, such a C is defined by arbitrary constraints involving only the countably many t in T_1. It is easily seen that $\mathscr{C}$ is a σ field containing all tubes in R^T, and hence $\mathscr{B}_T$ is contained in $\mathscr{C}$, as was asserted.

If in the definition of $\mathscr{C}$ we require that C_1 belong to $\mathscr{B}_{T_1}$ then $\mathscr{C}$ is still a σ field, and $\mathscr{C}$ is easily shown to equal $\mathscr{B}_T$. Thus a subset B of R^T belongs to $\mathscr{B}_T$ iff there exists some countable subset T_1 of T and some subset B_1 of $\mathscr{B}_{T_1}$ such that any f: $T \to R$ belongs to B iff its restriction to T_1 belongs to B_1.

We restate the uniqueness in terms of induced distributions. If X_t for $t \in T$ is a stochastic process with associated $(\Omega,\mathscr{A},P)$, and B is any subset of R^T then we define $P_X(B)$ by

$$P_X(B) = P\{\omega : r_\omega \in B\},$$

whenever $\{\omega : r_\omega \in B\}$ belongs to $\mathscr{A}$. This set is easily shown to belong to $\mathscr{A}$ whenever B is in $\mathscr{B}_T$. We thus obtain the induced probability space $(R^T,\mathscr{B}_T,P_X)$, where R^T contains all possible realizations. Then the uniqueness assertion states that if two processes, X_t for $t \in T$ and Y_t for $t \in T$, generate the same family then $P_X = P_Y$ on $\mathscr{B}_T$.

EXERCISE

1. The technique used to prove Theorem 1 can be used to show that the family of finite distributions uniquely determines the probability of many other events involving uncountably many random variables. As an example prove the following fact, which applies in particular to every process which generates the same family as a Poisson process: Let "N_t for $t \geq 0$ with $N_0 = 0$" be any stochastic process having only regular realizations and satisfying the condition

$$P[N_t - N_s \geq 0] = 1 \qquad \text{whenever } 0 \leq s < t.$$

(This is clearly a condition on the family.) Then $P(A) = 1$ if A is the set of sample points whose corresponding realizations are nondecreasing functions on $t \geq 0$. HINT: Let T' be the set of all pairs (s,t) of nonnegative rational numbers with $s < t$.

23
The Poisson Process and Some of Its Generalizations

Almost all the processes considered in this chapter are continuous-parameter processes with $t \geq 0$ for parameter set and with all realizations jump functions, as depicted in Fig. 3 (Sec. 23-1), so that there is a jump of size y_i at time t_i while $x_i = t_i - t_{i-1}$ is the ith interarrival time. For a pure-birth process, such as a Poisson process, every $y_i = 1$, so that all jumps are of the same size, and hence the process essentially just generates the arrival times $0 < t_1 < t_2 < \cdots$.

In Sec. 23-1 approximating experiments suggest the definition and basic characteristics of a Poisson process. Selecting a realization from a Poisson process is sometimes described as selecting an infinite sequence of points purely at random on $t \geq 0$, since the probability of obtaining a point in any short interval is proportional to the length of the interval, and the distributions of the number of points in any two equal-length intervals is the same, and we have independence for the numbers of points in disjoint intervals. Theorem 1 (Sec. 23-2) shows that *only* the Poisson process can have these qualitative characteristics. Section 23-3 shows that Poisson processes are also characterized by the fact that the interarrival times are independent and all have the same exponential density. Exponential densities permeate this chapter, essentially because of their unique lack of memory as shown in Example 4 (Sec. 16-1). Section 23-3 also relates the Poisson process to the uniform continuous densities.

The compound Poisson processes of Sec. 23-4 are constructed by selecting arrival times t_i by a Poisson process, while the arbitrary jump sizes y_i are selected independently from a fixed distribution.

For the nonhomogeneous Poisson processes of Sec. 23-5 all $y_i = 1$, but now the distribution of the number of arrivals in a 1 hr period may be different for different 1-hr periods.

For the Markov pure-birth processes of Sec. 23-6 the distribution of the waiting time for the next birth is determined by the present population size but is otherwise unaffected by the time required to attain this size. In many applications large populations may be expected to yield close births. The Markov property follows from the exponential distributions of the independent interarrival times, which are identically distributed only for a Poisson process. For the Markov birth and death processes of

Sec. 23-7 not only the next interarrival time, but also the chance for its being a birth $y_i = 1$, rather than a death $y_i = -1$, may depend on the present population size.

23-1 INTRODUCTION

In this section a sequence of approximating experiments is used to suggest the definition, intuitive nature, and some of the basic properties of a Poisson process. A sampling experiment further illustrates the nature of realizations. Then the process is transformed so as to display the stationarity of the sequence of successive arrival times.

Let λ be a fixed positive real number. For each integer $m > \lambda$ we describe an experiment. The Poisson process with parameter λ may be thought of as an experiment which is the limit of these experiments as m approaches infinity. The mth experiment consists of an infinite sequence of Bernoulli trials with probability λ/m for a success on each trial. For the mth experiment suppose that the ith Bernoulli trial, for $i = 1, 2, \ldots,$ takes place at time i/m, and we place a cross at the point i/m on the time axis iff the ith trial results in a success. Figure 1 shows the initial portion up to $t = 1$ for one possible outcome for the mth experiment. Thus there are m trials per unit interval, so that

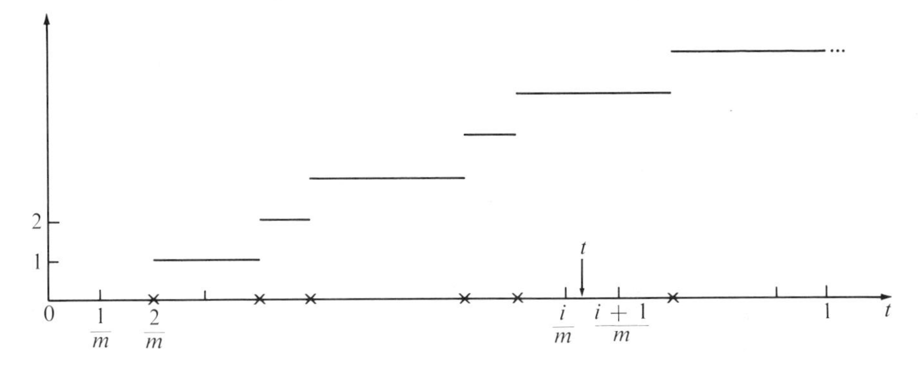

Fig. 1 A realization from the mth experiment.

the expected number of successes per unit interval is $m(\lambda/m) = \lambda$, which is the same for every experiment. The counting random variable N_t^m for the mth experiment is the number of successes up to and including time t. The superscript m refers to the mth experiment and is not to be interpreted as an mth power. The value of N_t^m for any outcome is just the value at t of the corresponding counting function for that outcome. As in Fig. 1, either the configuration of crosses or the corresponding counting function on $t \geq 0$ describes the outcome equally well, so we call either of them a possible realization for the mth experiment. A realization, or outcome, for the limit experiment is some configuration of crosses, or the corresponding counting function, where crosses can occur at any time instant and many can occur in a short interval, but the expected number of them per unit interval is λ. By tradition, it is often said that an "event" occurs at t iff there is a cross at t. Then a Poisson process describes "random events in time." We avoid this use of the word "event."

We now define our approximating experiments more explicitly. Let $\lambda > 0$ be fixed, and for each integer $m > \lambda$ let $Z_1^m, Z_2^m, \ldots$ be the Bernoulli trials. Thus $Z_1^m, Z_2^m, \ldots$ are independent and

$$P[Z_i^m = 1] = \frac{\lambda}{m} = 1 - P[Z_i^m = 0] \qquad \text{for } i = 1, 2, \ldots.$$

We will not consider anything like $Z_1^m + Z_1^{m+1}$, and hence only joint distributions within each experiment need be specified. For each real $t \geq 0$ we define the random variable N_t^m as follows: Let $N_t^m = 0$ if $0 \leq t < 1/m$, so that each of these random variables is the constant zero. If $t \geq 1/m$ then, as in Fig. 1, there is a unique integer $i \geq 1$ such that $i/m \leq t < (i+1)/m$, and we let

$$N_t^m = Z_1^m + \cdots + Z_i^m = N_{i/m}^m.$$

Hence

$$Z_i^m = N_{i/m}^m - N_{(i-1)/m}^m.$$

Thus the Z_i^m determine the N_t^m, and vice versa.

For each m we have defined a stochastic process "N_t^m for $t \geq 0$." If $0 \leq s < t$ then the increment random variable $N_t^m - N_s^m$ is the number of successes after time s and up to and including time t. Trials in disjoint time intervals are independent; hence the stochastic process "N_t^m for $t \geq 0$" has independent increments. It does not have stationary increments, except in the limit as m goes to infinity. For example, if $t - s < 1/m$ then, depending on s, there may or may not be a trial in this interval. For fixed $0 \leq s < t$ clearly $N_t^m - N_s^m$ is, for large enough m, the sum of about $(t-s)m$ Bernoulli trials with probability λ/m for a success on each trial. Thus the number of trials increases with m, while the expected number of successes remains at about $(t-s)\lambda$. From Sec. 8-3 we know that the density for $N_t^m - N_s^m$ converges to a Poisson density, with mean $(t-s)\lambda$ depending only on the length of the interval. We shortly define a Poisson process to be one having the properties just indicated. Before doing so we first exhibit an approximating experiment graphically and then take note of two fundamental properties of approximating experiments. These properties reappear in Sec. 23-3.

For graphical portrayal consider the $m = 3$ experiment, restricted to $0 \leq t \leq 1.2$ so there are Bernoulli trials at times $\frac{1}{3}$, $\frac{2}{3}$, and 1. The $8 = 2^3$ counting functions are shown in Fig. 2. Eight sample points are arbitrarily associated with the integers 1, 2, ..., 8 of one axis, while the t axis is the other axis of a plane, and values of counting functions are along the third axis. To the left of each sample point is its probability when $\lambda = 1$, so that the probability of obtaining a success on any Bernoulli trial is $\frac{1}{3}$. The eight sample points could be the eight triples of 0's and 1's, or they could be the eight counting functions. In any case, for each of the eight possible values of ω we have a corresponding counting function. Also for each t we have a random variable N_t^3. For example, the eight values of $N_{.5}^3$ are shown as dots. We now turn our attention to two properties of approximating experiments.

If X_1^m is the first interarrival time, that is, the time at which the first success occurs, then

$$P[X_1^m > t] = P[N_t^m = 0] \doteq \left(1 - \frac{\lambda}{m}\right)^{mt} \to e^{-\lambda t} \qquad \text{as } m \to \infty.$$

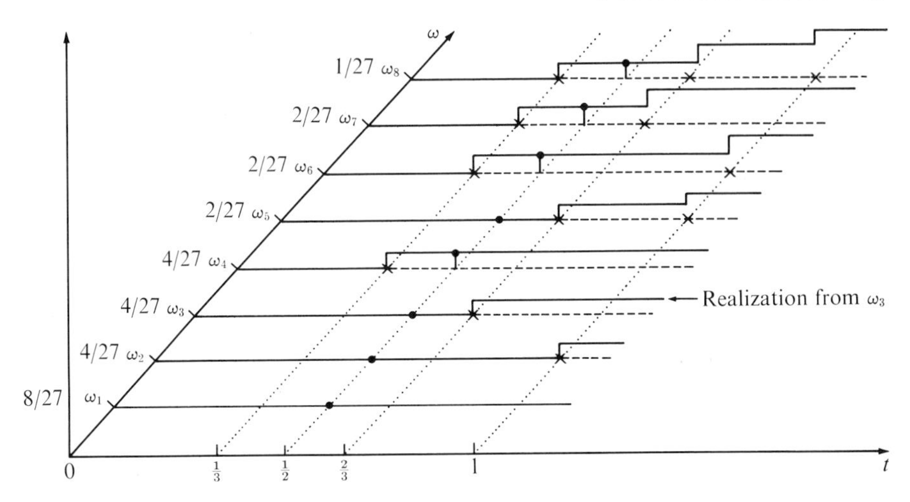

Fig. 2 The eight realizations from, "N_t^3 for $0 \leq t \leq 1.2$."

In other words, for fixed t the density for N_t^m converges to a Poisson density with mean λt. This density assigns probability $e^{-\lambda t}$ to the integer 0, and this implies that in the limit as m goes to infinity the first interarrival time has distribution function $1 - e^{-\lambda t}$. Also it is intuitively obvious that our approximating experiments have independent and identically distributed interarrival times, as was proved near the end of Sec. 8-4. Therefore we correctly anticipate that a Poisson process has the fundamental property that its interarrival times are independent and each is exponentially distributed with mean $1/\lambda$.

Suppose that mt is large and we learn that $[N_t^m = 3]$ occurred. Then the nature of our approximating experiments suggests that the locations of these three successes have equal likelihood of being at any of the about mt points in this interval. The Poisson process will exhibit this same fundamental property.

THE POISSON PROCESS

We wish to precisely define the class of jump functions, such as the one of Fig. 3, having a jump of size y_i, possibly negative, at the jump point t_i. Assume that we have

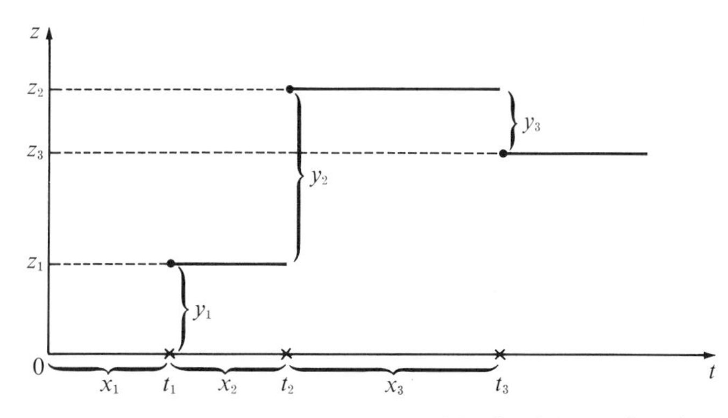

Fig. 3 A jump function.

two infinite sequences of real numbers $x_1, x_2, \ldots$ with each $x_i > 0$ and $y_1, y_2, \ldots$ with each $y_i \neq 0$. We call $x_1, x_2, \ldots$ the sequence of successive **interarrival times.** Each of these sequences determines its corresponding sequence of partial sums, and vice versa, where

$$t_n = x_1 + \cdots + x_n \qquad \text{and} \qquad z_n = y_1 + \cdots + y_n.$$

We call $0 < t_1 < t_2 < \cdots$ the sequence of successive **arrival times,** or jump points, and we require t_n to converge to infinity as n does. Then we define f for all $t \geq 0$ by

$$f(t) = \begin{cases} 0 & \text{if } 0 \leq t < t_1 \\ z_n & \text{if } t_n \leq t < t_{n+1}. \end{cases}$$

Thus $f(0) = 0$ and f is continuous from the right, jumps by amount y_i at t_i, and is constant between jump points. Naturally if we start with only finitely many jump points $t_1, \ldots, t_N$ then we let $f(t) = z_N$ if $t \geq t_N$. We call a real-valued function f on $t \geq 0$ a **jump function** iff it is constructable as above, with either a finite or infinite sequence of jump points, or is identically zero (no jumps). A **counting function** on $t \geq 0$ is any jump function for which every y_i is a positive integer. For interpretation, there are y_i successes, or simultaneous births, at time t_i. A **unit counting function** is any jump function for which every $y_i = 1$. Thus every jump is a unit jump, so only one success can occur at any one instant of time. The identically zero function on $t \geq 0$ is assumed to belong to all three classes of functions. Note that every jump function determines a corresponding unit jump function which makes unit jumps at exactly the same jump points.

A **Poisson process** "N_t for $t \geq 0$ with $N_0 = 0$ and parameter λ" is any stationary-independent-increment process such that for each $t > 0$ the random variable N_t induces a Poisson density with mean λt, and every realization for the process is a counting function. The parameter λ, which may be any positive real number, is called the **intensity** or **mean rate.** Thus if $0 \leq s < t$ then

$$P[N_t - N_s = k] = \frac{e^{-\lambda(t-s)}[\lambda(t-s)]^k}{k!} \qquad \text{for } k = 0, 1, \ldots.$$

The use of $P[N_t - N_s = 0] = e^{-\lambda(t-s)}$ is especially frequent. The next section shows that every realization is a *unit* counting function having *infinitely* many jump points.

The distributions induced by N_t for $t > 0$ form a semigroup; hence there exists a stationary-independent-increment process generating this semigroup. To prove the semigroup property we need only recall that if X and Y are independent and have Poisson densities with means λs and λt then $X + Y$ has a Poisson density with mean $\lambda(s + t)$. Section 23-3 shows that the restriction to counting-function realizations can be met.

If we were to omit the requirement that every realization for the process be a counting function then Example 1 (Sec. 22-3) shows that all realizations might be more irregular than jump functions. For continuous-parameter processes the joint distributions, such as that for $N(t_1), \ldots, N(t_n)$, do *not* determine the nature of the realizations, so a regularity assumption is imposed on realizations.

If $N(t)$ for $t \geq 0$ is a Poisson process with mean rate 1 then $N(\lambda\tau)$ for $\tau \geq 0$ is easily seen to be a Poisson process with mean rate $\lambda > 0$. Thus the parameter λ corresponds to a scale change in time. We do not usually apply a scale change to ordinates, $cN(t)$, since we would lose Poisson densities and counting functions.

Example 1 We perform a sampling experiment in order to obtain a typical realization from a Poisson process, at least as it would appear if observed at only finitely many equispaced time points. Let $N(t)$ for $t \geq 0$ be a Poisson process with mean rate $\lambda = 1$, so that $EN(t) = t$. We wish to generate the values of a realization at the times $0, \tau, 2\tau, \ldots, M\tau$ for $M = 400$. For $k = 1, 2, \ldots$ we have

$$N(k\tau) = \sum_{i=1}^{k} [N(i\tau) - N((i-1)\tau)]$$

where the increments in square brackets are independent and each is Poisson with mean τ. Let $\tau = .02$ so that only rarely will an increment value exceed 1. Then $M\tau = 8$, so we "expect" about eight successes in the interval $0 \leq t \leq 8$. We evaluate as follows:

$$P[N(\tau) = 0] = e^{-.02} \doteq .9802$$

$$P[N(\tau) = 1] = .02e^{-.02} \doteq .0196$$

$$P[N(\tau) \geq 2] \doteq .0002$$

We next generate 400 independent observations from this Poisson density with mean .02. Let $x_1, x_2, \ldots, x_M$ be the $M = 400$ different 4-tuples of decimal digits in the random number table of Sec. 12-2, listed column by column. We may write $x_1 = .2952$, $x_2 = .4167, \ldots, x_M = .9806$ and consider x_i to be a point in the unit interval. Define the transformation of x_i into y_i as follows:

If $0 \leq x_i < .9802$ then $y_i = 0$.

If $.9802 \leq x_i < .9998$ then $y_i = 1$.

If $x_i = .9998$ or $x_i = .9999$ then $y_i \geq 2$?

Then $x_1, \ldots, x_M$ generates $y_1, \ldots, y_M$. Fortunately an awkward undefined $y_i \geq 2$ did not occur. All but five of the y_i are zero: $y_i = 1$ when $i = 137, 341, 347, 366, 400$. We may consider $y_1, \ldots, y_M$ to be $M = 400$ independent observations from (an approximation to) the Poisson density with mean .02.

We interpret y_i as the observed value of the increment $[N(i\tau) - N((i-1)\tau)]$ from a sample function of the Poisson process. Figure 4 was obtained by plotting the value $y_1 + \cdots + y_k$ above the point $k\tau$ and then drawing a continuous line through each horizontal row of dots, since a Poisson process generates counting functions. Thus we may assume

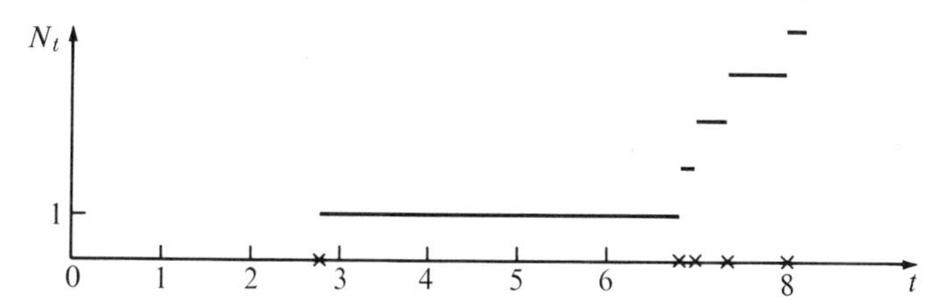

Fig. 4 The realization from a sampling experiment.

that we have observed the realization at *every* t in the interval $0 \le t \le 8$, except for the five subintervals where the successes occurred.

The following procedure seems like a reasonable way to more precisely locate the five successes. The procedure is justified by Theorem 2 (Sec. 23-3). We select a point "at random" from each of the five subintervals and assume that the five successes occurred at the five points selected. In other words, we perform a sampling experiment in order to obtain independent observations $u_1, \ldots, u_5$ from (an approximation to) the uniform continuous density on $0 < s < .02$. Then we locate the first success at $136\tau + u_1$; we follow a similar procedure for the other four successes. More explicitly if we obtain 803 from a *new* table of random numbers, not the already overused table, then we let $u_1 = (.02)(.803)$, etc. We thus locate the successes at multiples of 2×10^{-5} in the interval $0 \le t \le 8$.

Repeated independent performances of this sampling experiment, including the refinement of the last paragraph, should yield "fair" approximations to what in theory would be obtained from repeated independent observations of the Poisson process on $0 \le t \le 8$. That is, the chances, or relative frequencies, for the various configurations of successes should be "close" to theoretical. The idealized sampling experiment, with no numerical approximations and utilizing a (never-to-exist) "table of random real numbers from the unit interval," generates realizations from a Poisson process precisely. ■

Example 2 Let N_t for $t \ge 0$ be a Poisson process. If f is any real-valued function defined on $n = 0, 1, 2, \ldots$ then we can define a new stochastic process, $f(N_t)$ for $t \ge 0$. In particular if $f(n) = (-1)^n$ we obtain the process "$(-1)^{N_t}$ for $t \ge 0$," which takes only the two values ± 1. In Fig. 5 the dashed curve is a realization from the Poisson process and the solid curve is the

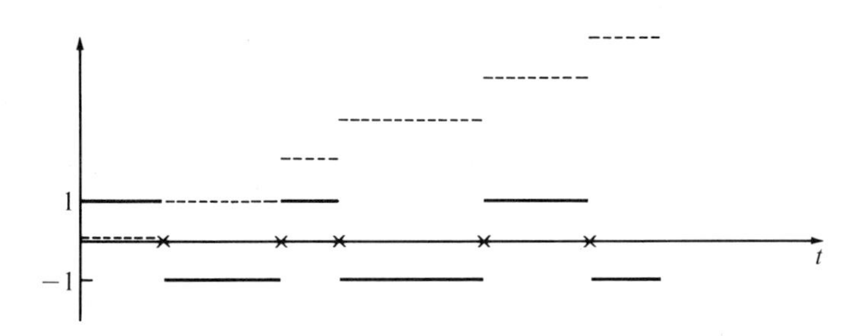

Fig. 5 The random telegraph signal.

corresponding realization from the new process. For symmetry we may start with $+1$ or -1 depending on a fair coin; that is, we multiply the whole realization by a random plus or minus sign.

More precisely we introduce a random variable X_0, independent of the Poisson process and having distribution $P[X_0 = 1] = P[X_0 = -1] = .5$, and then let

$$X_t = X_0(-1)^{N_t} \qquad \text{for all } t \ge 0.$$

This process is sometimes referred to as a *random telegraph signal*. Exercise 5 shows that this process is stationary. Speaking intuitively, every left translation $\bar{r}_\omega$ by real $t_0 > 0$ of a particular realization r_ω,

$$\bar{r}_\omega(t) = \begin{cases} r_\omega(t_0 + t) & \text{if } t \ge 0 \\ 0 & \text{if } t < 0 \end{cases}$$

is as likely as is the particular realization. That is, the distribution of the "success sequence" from a Poisson process is stationary. The random telegraph signal displays this stationarity; hence for some purposes it is more natural than the counting process. The random telegraph signal does not have independent increments. For example, it obviously cannot increase by $+2$ over both of two adjacent intervals, and hence

$$P[X_2 - X_1 = 2, X_3 - X_2 = 2] = 0 < P[X_2 - X_1 = 2]P[X_3 - X_2 = 2]. \quad \blacksquare$$

EXERCISES

1. Electrons are emitted from the cathode of a diode according to a mean rate of λ per second. Thus we expect about one electron to be emitted over a span of $1/\lambda$ sec. Find the distribution of the number of the first n time intervals $(k - 1)/\lambda \le t < k/\lambda$, where $k = 1, \ldots, n$, for which at least two electrons are emitted, assuming a Poisson process.

2. If $0 < s < t$ and it is learned that exactly $n \ge 1$ suicides occurred up to time t, according to a Poisson process, show that the conditional distribution of the number of suicides up to time s is binomial with parameters n and s/t independent of λ.

3. (*a*) Find the density induced by the first arrival time $T_1 = X_1$ in a Poisson process.
(*b*) If failures occur according to a Poisson process with mean rate $\lambda = .2$ per week what is the expected waiting time until the first failure?

4. (*a*) For a Poisson process show that for large t an observation of $(1/t)N_t$ can be used as a reasonable estimate for the mean rate λ. More precisely show that for all $\epsilon > 0$

$$P\left[\left|\frac{N_t}{t} - \lambda\right| \ge \epsilon\right] \to 0 \qquad \text{as } t \to \infty.$$

(*b*) Generalize this to stationary independent increment processes like those in Theorem 1 (Sec. 22-2).
(*c*) For a Poisson process find $\lim_{t \to \infty} P[N_t \le \lambda t + t^{1/3}]$.

5. Let N_t for $t \ge 0$ with $N_0 = 0$ be any stationary independent increment process having only integer values. Thus

$$\sum_{n=-\infty}^{\infty} P[N_t = n] = 1 \qquad \text{for all } t \ge 0.$$

Let $X_t = X_0(-1)^{N_t}$ where X_0 is a random variable independent of the process and having distribution $P[X_0 = -1] = P[X_0 = 1] = .5$. Prove that X_t for $t \ge 0$ is a stationary stochastic process.

6. Let X_t for $t \ge 0$ and Y_t for $t \ge 0$ be two independent Poisson processes with respective mean rates λ_1 and λ_2. For interpretation, we might have accident processes in two cities, or particles from two sources, or arrivals and departures, etc.
(*a*) Show that $X_t + Y_t$ for $t \ge 0$ is a Poisson process with mean rate $\lambda_1 + \lambda_2$.
(*b*) Show that the conditional distribution of X_t, given $[X_t + Y_t = n]$, is binomial with parameters $\lambda_1/(\lambda_1 + \lambda_2)$ and n, independent of $t > 0$.
(*c*) Let Z be the number of arrivals from "X_t for $t \ge 0$" which occur prior to the first arrival from "Y_t for $t \ge 0$." Show that Z has a geometric density

$$P[Z = k] = \frac{\lambda_2}{\lambda_1 + \lambda_2}\left(\frac{\lambda_1}{\lambda_1 + \lambda_2}\right)^k \qquad \text{for } k = 0, 1, \ldots.$$

(*d*) Private and commercial airplanes land at an airport according to independent Poisson processes with respective mean rates of $\lambda_1 = 10$ and $\lambda_2 = 5$ planes per hour. What is the expected number of private landings prior to the first commercial landing? In a particular day 300 planes landed. For that day what is the expected number of private landings?

(*e*) Show that $X_t - Y_t$ for $t \geq 0$ has stationary independent increments.

(*f*) If $0 < b < \lambda_1 - \lambda_2$ show that

$$P[X_t - Y_t > bt] \to 1 \qquad \text{as } t \to \infty.$$

7. In Sec. 23-4 it is shown that if births occur according to a Poisson process, and the sex of each individual is determined by an independent fair-coin (or biased-coin) toss, then male births form a Poisson process; that is, random selection from a Poisson process yields a Poisson process. The random selection may be interpreted as whether or not a suicide attempt is reported, or a nuclear particle is detected, etc. This exercise shows that if every second birth is male then male births do *not* form a Poisson process; that is, deterministic selection destroys most of the nice properties.

Let N_t for $t \geq 0$ be a Poisson process and let $X_t = f(N_t)$, where

$$f(2n) = f(2n + 1) = n \qquad \text{for } n = 0, 1, \ldots$$

Thus X_t for $t \geq 0$ is the counting process which counts only arrivals $t_2, t_4, t_6, \ldots$ from the Poisson process. Let $0 < t - r < t < t + h$.

(*a*) Show that

$$P[X_{t+h} - X_t = 0] = e^{-\lambda h} + \frac{1 + e^{-2\lambda t}}{2} e^{-\lambda h} \lambda h.$$

Thus X_t for $t \geq 0$ does *not* have stationary increments. Note that for large t the chance for counting an arrival is about $\frac{1}{2}$. HINT: To find $P[N_t$ is even$]$ expand $e^{-\lambda t}$ in a power series.

(*b*) Find $P[X_{t+h} - X_t = 0 \mid X_t = 0]$ and note that X_t for $t \geq 0$ does *not* have independent increments.

(*c*) Show that

$$P[X_{t+h} = 1 \mid X_t = 1] < P[N_h = 0 \text{ or } 1],$$

$$\lim_{\substack{r \to 0 \\ r > 0}} P[X_{t+h} = 1 \mid X_t = 1, X_{t-r} = 0] = P[N_h = 0 \text{ or } 1].$$

Thus $f(N_t)$ for $t \geq 0$ is *not* a Markov process. This is an example of the basic fact that a non-one-to-one function of a Markov process may not be a Markov process.

8. This exercise demonstrates for our approximating experiments a property analogous to that of Theorem 2 (Sec. 23-3) for Poisson processes. The mth approximating experiment, when considered only on the unit interval, consists of m Bernoulli trials occurring at times $1/m, 2/m, \ldots, 1$, with the ith trial having probability λ/m for generating a cross at i/m. Let $1 \leq n \leq m$ and suppose that we are given that $[N_1{}^m = n]$ occurred, so that exactly n crosses were obtained at some n unknown locations in the interval. There are $\binom{m}{n}$ distinct subsets, with each subset consisting of n of the m time points. Hence there are $\binom{m}{n}$ possible configurations of crosses. Let $0 < t_1 < \cdots < t_n \leq 1$ be some particular n of these m time points. Let A be the event that the crosses occur at precisely these n time points.

(*a*) Show that

$$P(A \mid N_1{}^m = n) = \frac{1}{\binom{m}{n}}.$$

This is independent of λ. Clearly, then, each of the $\binom{m}{n}$ possible configurations of crosses is equally likely. This fact was also observed in Exercise 7*a* (Sec. 8-2). Thus if we learn that $[N_1^m = n]$ occurred then the n unknown locations are essentially "at random." In some sense this joint conditional distribution should be the same as that obtained by using random selections. There are two complications in making this precise. The first is that random selections may yield two crosses at the same point. Let $U_1, \ldots, U_n$ be independent random variables each uniformly distributed over the m time points $1/m$, $2/m, \ldots, 1$. Let D_n be the event that $U_1, \ldots, U_n$ all have different values.

(*b*) Find $P(D_n)$ and show that for fixed n

$$P(D_n) \to 1 \qquad \text{as } m \to \infty.$$

To illustrate the second complication suppose that $n = 2$ and we place crosses at U_1 and U_2. Then we obtain the same configuration of crosses if $(u_1,u_2) = (3/m,7/m)$ or $(u_1,u_2) = (7/m,3/m)$. That is, different values for the random vector $(U_1, \ldots, U_n)$ result in the same configuration of crosses.

(*c*) Prove that

$$P[(U_1, \ldots, U_n) \text{ equals some permutation of } (t_1, \ldots, t_n) \mid D_n] = \frac{1}{\binom{m}{n}}.$$

(*d*) On the basis of this exercise describe a sampling experiment which should generate a realization from the process N_t^m for $0 \le t \le 1$. Assume that m is much larger than λ. Select an observation n from a Poisson density with mean λ; and so on.

23-2 THE UNIQUENESS OF THE POISSON PROCESS

The first part of this section is devoted to a variety of simple properties possessed by Poisson processes. In particular we examine probabilities related to short intervals. We also show that realizations have only unit jumps, and there must be an infinite sequence of these. Then Theorem 1 shows that the Poisson process is the unique process possessing its qualitative characteristics.

For every $t > 0$ we have $P[\text{a jump at } t] = 0$ because

$$\begin{aligned} P[\text{a jump at } t] &\le P\left[N\left(t + \frac{1}{n}\right) - N\left(t - \frac{1}{n}\right) \ge 1\right] \\ &= 1 - P\left[N\left(t + \frac{1}{n}\right) - N\left(t - \frac{1}{n}\right) = 0\right] \\ &= 1 - e^{-2\lambda/n} \to 0 \qquad \text{as } n \to \infty. \end{aligned}$$

Thus for every time t the probability is 0 that we will have a jump at precisely that t. Therefore for every *countable* set T' the probability is 0 that we will have a jump at at least one t in T'. Because of this we can sometimes be casual about the inclusion or exclusion of end points of intervals.

The probability of a unit increase (precisely one arrival) over a short interval is approximately proportional to the length of the interval. That is, if $0 \le s < s + t$

then $P[N_{s+t} - N_s = 1] \doteq \lambda t$ for small enough t, since

$$\frac{1}{\lambda t} P[N_{s+t} - N_s = 1] = \frac{1}{\lambda t} e^{-\lambda t} \lambda t \to 1 \qquad \text{as } t \to 0.$$

This is partial justification for the use of infinitesimals, $P[N_{s+dt} - N_s = 1] = \lambda\, dt$. Thus for λt small enough we are almost certain to see no increase,

$$P[N_{s+t} - N_s = 0] = e^{-\lambda t} \doteq 1$$

and the probability of an increase of 1 is approximately λt.

We next note that in several senses there is negligible probability of an increase of more than 1, over a short enough interval. Since we have stationary increments, for notational convenience we often consider $N_t - N_0 = N_t$ rather than $N_{s+t} - N_s$. We show that if t goes to zero then

$$P[N_t = 1 \mid N_t \geq 1] = \frac{P[N_t = 1]}{P[N_t \geq 1]} \to 1$$

$$P[N_t \geq 2 \mid N_t \geq 1] = \frac{P[N_t \geq 2]}{P[N_t \geq 1]} \to 0$$

$$\frac{P[N_t \geq 2]}{P[N_t = 1]} \to 0.$$

The first two limits are equivalent, since

$$P[N_t = 1 \mid N_t \geq 1] + P[N_t \geq 2 \mid N_t \geq 1] = 1.$$

The first and third limits are equivalent, since

$$\frac{P[N_t = 1]}{P[N_t \geq 1]} = \frac{P[N_t = 1]}{P[N_t = 1] + P[N_t \geq 2]} = \frac{1}{1 + P[N_t \geq 2]/P[N_t = 1]}.$$

Thus there are many natural *and equivalent* ways in which to assert that for short intervals the probability of an increase of two or more is negligible in comparison to the probability of an increase of one. For a Poisson process the first limit is proved by

$$\frac{e^{-\lambda t}\, \lambda t}{1 - e^{-\lambda t}} = \frac{\lambda t}{e^{\lambda t} - 1} \to 1 \qquad \text{as } t \to 0$$

since $e^x = 1 + x + x^2/2! + \cdots$.

We next show that all realizations for a Poisson process are *unit* counting functions. There will be a jump of more than unit size in the interval $0 \leq t \leq 1$ iff there is such a jump in at least one of the n disjoint subintervals of length $1/n$. But the probability of a union is at most the sum of the probabilities, and we have stationary increments; hence

$$P[\text{at least one jump} \geq 2 \text{ in } 0 \leq t \leq 1] \leq nP\left[N\left(\frac{1}{n}\right) \geq 2\right]$$

$$= \frac{1 - e^{-\lambda/n}}{1/n} \frac{P[N(1/n) \geq 2]}{P[N(1/n) \geq 1]} \to 0 \qquad \text{as } n \to \infty$$

since the two factors converge to λ and 0, respectively. Now if

$$A_k = [\text{at least one jump} \geq 2 \text{ in } k - 1 \leq t \leq k]$$

then $P(A_k) = 0$ from stationary increments, and hence $P\left(\bigcup_{k=1}^{\infty} A_k\right) = 0$. Thus, if we neglect an event of probability 0, as we often do, then every realization from a Poisson process is a *unit* counting function.

There must be an infinite number of successes on $t \geq 0$, because if B_k is the event that there are at most k successes in the whole interval $t \geq 0$ then for every time $t = n$ we have

$$P(B_k) \leq P[N_n \leq k] = \sum_{i=0}^{k} P[N_n = i] \to 0 \qquad \text{as } n \to \infty$$

since each of the $k + 1$ terms goes to zero. Thus $P(B_k) = 0$ for each k, and hence $P\left(\bigcup_{k=0}^{\infty} B_k\right) = 0$. Therefore a realization determines an *infinite* sequence of arrival times $0 < t_1 < t_2 < \cdots$; and by assigning the value t_k to this realization, we define a random variable T_k. Thus the *infinite* sequence $0 < T_1 < T_2 < \cdots$ of successive, arrival-time random variables is well defined. The same is true for the successive interarrival times $X_1, X_2, \ldots,$ where $X_n = T_n - T_{n-1}$. Note that if $t > 0$ then

$$P[\text{finitely many successes up to time } t] \geq P[N_t \leq n] \to 1 \qquad \text{as } n \to \infty$$

so there are only finitely many successes in every finite interval. In other words, our probability assumptions are consistent with our assumption that t_n goes to infinity in the definition of a counting function.

THE UNIQUENESS THEOREM

We now motivate the uniqueness theorem. In Sec. 23-4 it is shown that a stationary-independent-increment process is obtained if jump functions are selected as follows. We select the jump points $0 < t_1 < t_2 < \cdots$ according to a Poisson process and select the jump values $y_1, y_2, \ldots$ independently according to a fixed distribution function $F: R \to R$. We may interpret jumps as births or deaths if F happens to be concentrated on the integers. For example, we might let $P[Y_1 = k] = .6, .3, .1$ for $k = 1, 2, 3$ in order to permit multiple births but no deaths. Thus the class of stationary-independent-increment processes which have counting-function realizations is extremely rich. The following theorem shows that in this class only the Poisson processes have *unit* counting-function realizations. Therefore only the Poisson processes satisfy the following qualitative description. A finite integer number of births occur in any finite interval, but no multiple births occur, and the distribution of the number of births in an interval depends only on the length of the interval, while there is independence for disjoint intervals. The assumption $P[N_1 = 0] < 1$ is made in order to rule out the case $P[N_t = 0] = 1$ for all $t > 0$.

Theorem 1 Let N_t for $t \geq 0$ be a stationary-independent-increment stochastic process, all of whose realizations are counting functions and for which $P[N_1 = 0] < 1$. Then with probability 1, all its realizations are *unit* counting functions iff

$$P[N_t = 1 \mid N_t \geq 1] = \frac{P[N_t = 1]}{P[N_t \geq 1]} \to 1 \qquad \text{as } t \to 0,$$

and this is the case iff it is a Poisson process. ■

Proof Let N_t for $t \geq 0$ be any stationary-independent-increment counting-function process satisfying $P[N_1 = 0] < 1$. For all $t > 0$ let $g(t) = P[N_t = 0]$. If $t > 0$ and $s > 0$ then

$$g(t+s) = P[N_{t+s} = 0] = P[N_t = 0]P[N_{t+s} - N_t = 0] = g(t)g(s)$$

so that Lemma 2 in the Addendum to Sec. 16-1 shows that $g(t) = [g(1)]^t$ for all $t > 0$. By assumption $0 \leq g(1) < 1$. If $g(1) = 0$ then $g(t) = 0$ for all $t > 0$, so that $P[N_t \geq 1] = 1$ for all $t > 0$. Therefore for every integer n

$$P[N_1 \geq n] \geq P\left(\bigcap_{i=1}^{n}\left[N\left(\frac{i}{n}\right) - N\left(\frac{i-1}{n}\right) \geq 1\right]\right) = \left\{P\left[N\left(\frac{1}{n}\right) \geq 1\right]\right\}^n = 1$$

contradicting the fact that N_1 has a probability distribution. Therefore $0 < g(1) < 1$, so that we can define $\lambda > 0$ by $g(1) = P[N_1 = 0] = e^{-\lambda}$, and we have

$$P[N_t = 0] = e^{-\lambda t} \qquad \text{for all } t \geq 0.$$

If $P[N_t = 1 \mid N_t \geq 1]$ converges to 1 as t goes to zero then, using the earlier proof for the Poisson process, we see that all realizations must be unit counting functions. Therefore to prove the theorem we need only show that if all realizations are unit counting functions then each N_t has the desired Poisson distribution.

For the remainder of the proof let $t > 0$ and $k \geq 1$ be fixed. Let B be the event that there are exactly k jumps in the interval from 0 to t. We show, without assuming unit jumps, that

$$P(B) = \frac{e^{-\lambda t}(\lambda t)^k}{k!}\,.$$

Clearly if a process has only unit jumps then $B = [N_t = k]$, so the proof of the theorem will be complete.

We first subdivide the interval from 0 to t into n equal-length disjoint subintervals. Let C_n be the event that at least one of the n subintervals contains two or more jumps. Then C_{2^i} for $i = 1, 2, \ldots$ is a decreasing sequence of events, since in going from i to $i + 1$ we divide each subinterval in half. But the intersection of this sequence of events is empty. That is, a counting function has only a finite number of jumps in the interval 0 to t, so a counting function will not contribute to C_n when $1/n$ is less than the distance between the closest pair of its jumps in the interval from 0 to t. Therefore

$$P(C_n) \to 0 \qquad \text{when } n = 2^i \text{ and } i \to \infty.$$

Let A_n be the event that exactly k of the n subintervals contain *at least one jump*, and the remaining $n - k$ contain no jumps. Then

$$P(B) = P(BA_n) + P(BA_n{}^c) \qquad \text{for every } n.$$

But BA_n occurs iff there are exactly k jumps with each in a different one of the n

subintervals. Clearly, then, BA_n^c is a subset of C_n, and hence its probability goes to zero as $n = 2^i$ goes to infinity. Therefore

$$P(B) = \lim_{n \to \infty} P(BA_n).$$

But similarly B^cA_n is a subset of C_n, so that its probability goes to zero as $n = 2^i$ goes to infinity, and hence

$$P(B) = \lim_{n \to \infty} [P(BA_n) + P(B^cA_n)] = \lim_{n \to \infty} P(A_n).$$

Now for $P(A_n)$ we just have n Bernoulli trials with probability $1 - e^{-\lambda t/n}$ for heads (at least one jump in the subinterval) on each trial. Hence

$$P(A_n) = \binom{n}{k}(1 - e^{-\lambda t/n})^k (e^{-\lambda t/n})^{n-k}$$

$$= \frac{e^{-\lambda t}(\lambda t)^k}{k!}\left[\frac{e^{\lambda t/n} - 1}{\lambda t/n}\right]^k \left\{\frac{n(n-1)\cdots(n-k+1)}{n^k}\right\}$$

where both [ ] and { } converge to 1 as n goes to infinity. [This is a special case of Exercise 9*c* (Sec. 8-3).] ■

EXERCISES

1. *The Poisson Process via Differential Equations.* A traditional technique for deducing the probability law for a desired type of stochastic process is to use natural qualitative assumptions to derive differential equations whose solutions yield the probability law. This technique has great generality. It is illustrated here for the Poisson process with $\lambda > 0$. Suppose that we seek some process "N_t for $t \geq 0$" having stationary independent increments and counting-function realizations. These conditions arise naturally in considering "random events in time." We assume that as h goes to zero we have

$$\frac{1}{\lambda h} P[N_h = 1] \to 1 \qquad \text{and} \qquad \frac{1}{h} P[N_h \geq 2] \to 0.$$

That is, for short intervals the chance for one success is proportional to the length of the interval, and the chance for more than one success is negligible in comparison.

For $0 \leq t < t + h$ we have

$$P[N_{t+h} = 0] = P[N_t = 0, N_{t+h} - N_t = 0]$$

$$= P[N_t = 0]P[N_h = 0] = P[N_t = 0](1 - P[N_h \geq 1]).$$

Rearranging and dividing by h, we get

$$\frac{P[N_{t+h} = 0] - P[N_t = 0]}{h} = -P[N_t = 0]\frac{P[N_h = 1] + P[N_h \geq 2]}{h}.$$

For each $n = 0, 1, \ldots$ we define a function f_n on $t \geq 0$ by $f_n(t) = P[N_t = n]$. For fixed t if we let h go to zero then our assumptions imply that f_0 has a derivative satisfying

$$f_0'(t) = -\lambda f_0(t).$$

The unique solution on $t \geq 0$ for this differential equation, subject to the initial condition $f_0(0) = 1$, is $f_0(t) = e^{-\lambda t}$. This same technique can be used to find f_1, then f_2, and so on.
(*a*) Show for $t \geq 0$ and $n \geq 1$ that f_n satisfies the differential equation

$$f_n'(t) = -\lambda f_n(t) + \lambda f_{n-1}(t).$$

For $n \geq 1$ we have initial condition $f_n(0) = 0$.
(*b*) Check that for $n = 1$ the solution is $f_1(t) = \lambda t e^{-\lambda t}$.
(*c*) Prove by induction that

$$f_n(t) = \frac{e^{-\lambda t}(\lambda t)^n}{n!} \qquad \text{for } t \geq 0,\ n = 0, 1, \ldots$$

is the solution to this set of differential equations.

2. Show that there is zero probability for each complete outcome from a Poisson process "N_t for $t \geq 0$." That is, for every real-valued function f defined on $t \geq 0$ we have $P\{\omega: r_\omega = f\} = 0$. Note that this is not so for the process "N_t for $0 \leq t \leq 1$" and exhibit for this process all those complete outcomes which have positive probability.

3. *More on Multiple Births.* Let N_t for $t \geq 0$ be a stationary-independent-increment process all of whose realizations are counting functions and for which $P[N_1 = 0] < 1$. Jumps are *not* assumed to be of unit size. The proof of Theorem 1 showed that

$$P[N_t = 0] = e^{-\lambda t} \qquad \text{for } t \geq 0$$

where $\lambda > 0$. More generally, although not needed for this exercise, it was shown that for any $t > 0$ and $k = 0, 1, \ldots$ that

$$P[\text{exactly } k \text{ jumps in the interval from 0 to } t] = \frac{e^{-\lambda t}(\lambda t)^k}{k!}.$$

Prove that there is a number λ_1 satisfying $0 \leq \lambda_1 \leq \lambda$ and such that

$$P[N_t = 1] = \lambda_1 t e^{-\lambda t} \qquad \text{for all } t \geq 0.$$

Note that Theorem 1 then implies that the process is a Poisson process iff $\lambda_1 = \lambda$. HINT: Show that Lemma 1 in the Addendum to Sec. 16-1 applies to $f(t) = e^{\lambda t}P[N_t = 1]$ by considering

$$P[N_{t+s} = 1] = P[N_t = 0, N_{t+s} - N_t = 1] + P[N_t = 1, N_{t+s} - N_t = 0].$$

23-3 TWO CHARACTERIZATIONS OF A POISSON PROCESS

Theorem 1 in this section shows that interarrival times are independent and each has an exponential density with mean $1/\lambda$. This fact completely characterizes a Poisson process, and it provides a transparent method for generating it. Theorem 2 shows that selecting n points at random from an interval is equivalent, from the standpoint of distributions, to observing the unknown locations of the n arrivals, if we learn that there were exactly n arrivals in the interval. This fact also essentially characterizes the Poisson process. Thus these two results relate the Poisson process in a transparent fashion to Poisson, exponential, gamma, and uniform densities. Then the Poisson process is extended to the whole real line, and the waiting-time paradox is examined.

We now show, as stated in the following theorem, that for a Poisson process "N_t for $t \geq 0$" the successive interarrival times $X_1, X_2, \ldots$ are independent and each

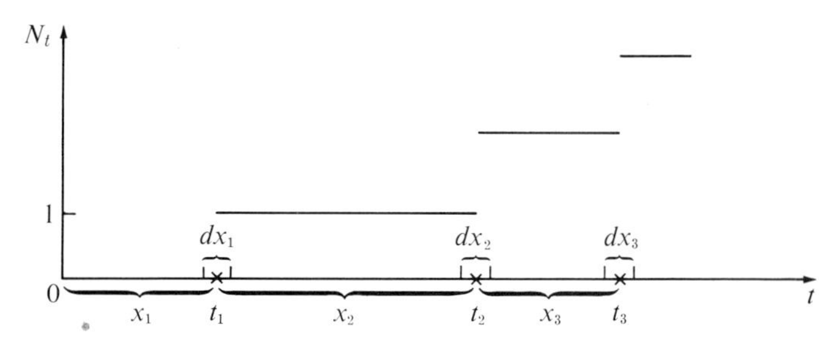

Fig. 1

has an exponential density with mean $1/\lambda$. Given positive numbers $x_1, \ldots, x_n$, let $t_1, \ldots, t_n$ be their partial sums $t_k = x_1 + \cdots + x_k$. In order to obtain the configuration of successes shown in Fig. 1 we must have a success in the interval between t_i and $t_i + dx_i$, and this has probability

$$P[N(t_i + dx_i) - N(t_i) = 1] \doteq \lambda\, dx_i$$

Also we must not have a success in the interval between $t_{i-1} + dx_{i-1}$ and t_i, and this has probability

$$P[N(t_i) - N(t_{i-1} + dx_{i-1}) = 0] \doteq e^{-\lambda x_i}.$$

Clearly, then, this configuration has probability

$$P[x_1 < X_1 < x_1 + dx_1, \ldots, x_n < X_n < x_n + dx_n] = \prod_{i=1}^{n} e^{-\lambda x^i} \lambda\, dx_i$$

as was asserted. Therefore if airplanes arrive according to a Poisson process with $\lambda = .2$ planes per minute then their successive interarrival times are independent and each is exponentially distributed with a mean spacing of $1/\lambda = 5$ minutes between planes.

Thus the arrival time $T_n = X_1 + \cdots + X_n$ is the sum of n independent random variables each of which is exponential with mean $1/\lambda$. Therefore Example 2 (Sec. 16-4) shows that T_n has a gamma density. We can derive this density more directly from the Poisson process: obviously the two events $[T_n \le t]$ and $[N_t \ge n]$ are equal, and $P[T_n = t] = 0$, so that we obtain

$$P[T_n < t] = P[N_t \ge n] = 1 - P[N_t < n] = 1 - \sum_{k=0}^{n-1} \frac{e^{-\lambda t}(\lambda t)^k}{k!}$$

for the distribution function for T_n. Differentiating and cancelling terms yields the asserted density for T_n. Even more simply, if we use infinitesimals then the gamma density can always be recalled, or rederived, by the transparent manipulation

$$P[t < T_n \le t + dt] = P[N_t = n - 1, N_{t+dt} - N_t = 1]$$

$$= \frac{e^{-\lambda t}(\lambda t)^{n-1}}{(n-1)!} \lambda\, dt\, e^{-\lambda\, dt} \doteq \frac{e^{-\lambda t}(\lambda t)^{n-1}}{(n-1)!} \lambda\, dt$$

where the sign $\doteq$ requires that $dt \doteq 0$. We summarize in the following theorem.

Theorem 1 If N_t for $t \geq 0$ is a Poisson process with mean rate $\lambda > 0$ then its successive interarrival times $X_1, X_2, \ldots$ are independent and each induces an exponential density with mean $1/\lambda$ and variance $1/\lambda^2$. The nth arrival time $T_n = X_1 + \cdots + X_n$ satisfies $P[T_n < t] = P[N_t \geq n]$ and induces the gamma density

$$p_{T_n}(t) = \gamma_{\lambda,n}(t) = \frac{e^{-\lambda t}(\lambda t)^{n-1}}{(n-1)!}\lambda \qquad \text{for all } t > 0. \quad \blacksquare$$

Suppose that we perform the following *idealized sampling experiment*. We select $x_1, x_2, \ldots$ independently from an exponential density with mean 1, place crosses at the points $t_1 = x_1$, $t_2 = x_1 + x_2$, $t_3 = x_1 + x_2 + x_3, \ldots$ of the time axis, and then construct the corresponding unit counting function. It is intuitively obvious, and easily proved below, that this idealized experiment generates realizations according to a Poisson process with $\lambda = 1$. With only 25 x_i and only four decimal places, Table 1 and corresponding Fig. 2 exhibit the outcome of such a sampling experiment. These $x_1, \ldots, x_{25}$ are interpretable as 25 rounded-off independent observations from the exponential density with mean 1. We used the first 25 entries from a table of 10^5 such, instead of obtaining x_i from $x_i = -\log z_i$ where the z_i are supposed to be independent observations from an approximation to the uniform density over $0 < z < 1$. Let us now precisely state this basic method for constructing and thinking about Poisson processes.

Characterization 1: Interarrival-time characterization of a Poisson process Let $X_1, X_2, \ldots$ be an infinite sequence of independent random variables each inducing an exponential density with mean $1/\lambda > 0$. Define a unit-counting-function stochastic process "N_t for $t \geq 0$ with $N_0 = 0$" to have these successive interarrival times; that is, *for each* $t \geq 0$ we define the random variable N_t to have the value 0 on the event $[X_1 > t]$ and to have the integer value $k \geq 1$ on the event

$$[X_1 + \cdots + X_k \leq t < X_1 + \cdots + X_{k+1}].$$

In other words, N_t is the number of partial sums which are t or smaller. Then N_t for $t \geq 0$ is a Poisson process with mean rate λ. $\blacksquare$

We describe a simple method for justifying this characterization. Suppose that we have *some* process "N_t for $t \geq 0$" all of whose realizations are *unit* counting functions having an *infinite* sequence of jump points. Let X_i and T_i be the interarrival and arrival times, respectively. Then any one of the following uniquely determines the other two:

1. The distribution of $(X_1, \ldots, X_n)$ for every $n \geq 1$
2. The distribution of $(T_1, \ldots, T_n)$ for every $n \geq 1$
3. The family of finite distributions generated by "N_t for $t \geq 0$"; that is, the distribution of $(N_{t_1}, \ldots, N_{t_n})$ for every finite $0 < t_1 < \cdots < t_n$.

The equivalence of properties 1 and 2 is immediate from the equations

$$\begin{aligned} T_1 &= X_1 & X_1 &= T_1 \\ T_2 &= X_1 + X_2 & X_2 &= T_2 - T_1 \\ &\cdots\cdots\cdots\cdots & &\cdots\cdots\cdots\cdots \\ T_n &= X_1 + \cdots + X_n & X_n &= T_n - T_{n-1}. \end{aligned}$$

Table I

n	x_n	$t_n = \sum_{i=1}^{n} x_i$	n	x_n	$t_n = \sum_{i=1}^{n} x_i$
1	2.7070	2.7070	14	.0208	13.7087
2	.3143	3.0213	15	.7720	14.4807
3	.1727	3.1940	16	4.2728	18.7535
4	.0213	3.2153	17	.3598	19.1133
5	.5928	3.8081	18	1.8020	20.9153
6	.5367	4.3448	19	.3378	21.2531
7	4.1464	8.4912	20	.1327	21.3858
8	.9065	9.3977	21	.4996	21.8854
9	2.7173	12.1150	22	1.3079	23.1933
10	1.1490	13.2640	23	1.4689	24.6622
11	.1245	13.3885	24	1.6647	26.3269
12	.1621	13.5506	25	1.6101	27.9370
13	.1373	13.6879			

SOURCE: C. E. Clark and B. W. Holz, "Exponentially Distributed Random Numbers," The Johns Hopkins Press, Baltimore, Md., 1960.

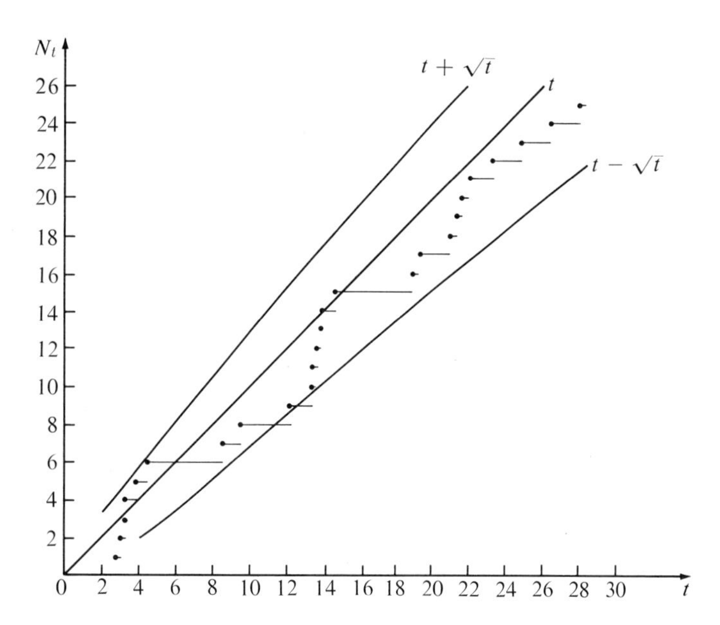

Fig. 2 A sampling experiment using the random exponential interarrival times from Table 1.

The equivalence of properties 2 and 3 is obtained by taking finite intersections of events of the form

$$[N_t = k] = [T_k \le t < T_{k+1}] \qquad \text{or} \qquad [T_k \le t] = [N_t \ge k].$$

Now a Poisson process *and* the process constructed in Characterization 1 both assign the same distribution to $(X_1, \ldots, X_n)$ for every $n \ge 1$, and hence both processes generate the same family. Therefore the process constructed in Characterization 1 is a Poisson process.

As a special case of the Kolmogorov criterion, we know that for any sequence $F_1, F_2, \ldots$ of distribution functions on R it is possible to construct a probability space and independent random variables $X_1, X_2, \ldots$ all defined on this *same* probability space and such that X_n induces F_n for $n = 1, 2, \ldots$. Therefore it is possible to construct the process described in Characterization 1. Hence continuous-parameter Poisson processes *having unit-counting-function realizations* can be simply constructed from a sequence of independent and identically exponentially distributed random variables.

Obviously we can select *any* distribution for $X_1 > 0$ and construct a unit-counting-function process "N_t for $t \ge 0$" having independent and identically distributed successive interarrival times $X_1, X_2, \ldots$. If N_t for $t \ge 0$ happens to have stationary independent increments, and $P[X_1 < 1] > 0$, then Theorem 1 (Sec. 23-2) shows that N_t for $t \ge 0$ is a Poisson process, and hence X_1 must have an exponential density. This suggests the unique importance of exponentially distributed interarrival times. In Sec. 23-6, by using independent interarrival times having different *exponential* densities we are able to construct important *Markov* processes.

THE INTERVALWISE CHARACTERIZATION

We turn to the second characterization for a Poisson process N_t for $t \ge 0$. If $0 \le r < s$ then $N_s - N_r$ has a Poisson density with mean $\lambda(s - r)$. If we learn that $[N_s - N_r = n]$ occurred, then the nature of the Poisson process suggests that the conditional distribution of the locations of these n successes should be the same as for n independent observations from the uniform density on the interval $s < t < r$, which density is independent of λ. This idea was used for $n = 1$ in the sampling experiment of Example 1 (Sec. 23-1). It yields the following informally stated characterization, which is justified in Exercise 4.

Characterization 2: Intervalwise characterization of a Poisson process Select an integer n according to a Poisson density with mean λ. If $n \ge 1$ then we select n numbers $u_1, u_2, \ldots, u_n$ independently from the uniform density on the unit interval $0 \le t < 1$ and place a cross at each of these n points. We perform this subexperiment independently on each of the intervals $i - 1 \le t < i$ for $i = 1, 2, \ldots$. The whole experiment yields a realization which is a configuration of crosses, or its corresponding unit counting function. This experiment generates the Poisson process on $t \ge 0$ with parameter λ. ■

One aspect of this method for generating the Poisson process deserves particular attention. The order $0 < t_1 < t_2 \cdots$ in which crosses appear in the final configuration is not necessarily the same as the order in which they were generated. That is, the cross closest to the origin may not correspond to the first observation from the first uniform density. If $n = 2$ then *both* the outcomes $(u_1,u_2) = (.2,.7)$ and $(u_1,u_2) = (.7,.2)$ yield the *same* configuration of crosses in the unit interval. In other words, the cross closest to the origin is located at the minimum of $u_1, \ldots, u_n$, and not necessarily at u_1. In general we obtain n crosses at the points $0 < t_1 < \cdots < t_n < 1$ iff $(u_1, \ldots, u_n)$ equals one of the $n!$ permutations of $(t_1, \ldots, t_n)$. Thus the configuration of crosses in the unit interval is the ordered vector determined by $(u_1, \ldots, u_n)$, as defined in Example 11 (Sec. 17-2).

The following theorem shows that the Poisson process possesses the property which suggests Characterization 2.

Theorem 2 Let N_t for $t \geq 0$ be a Poisson process with successive arrival times $T_1, T_2, \ldots$. For any $t > 0$ and any $n = 1, 2, \ldots$ the conditional distribution of $(T_1, \ldots, T_n)$, given the event $[N_t = n]$, is the same as the distribution of the ordered vector corresponding to $(U_1, \ldots, U_n)$ where $U_1, \ldots, U_n$ are independent and each has a uniform density over the interval from 0 to t. ■

Proof The probability of no success in an interval of length τ_i is $e^{-\lambda\tau_i}$. Hence the probability of no successes in any of $n + 1$ disjoint intervals is

$$e^{-\lambda\tau_1}e^{-\lambda\tau_2}\cdots e^{\lambda\tau_{n+1}} = \exp\left(-\lambda\sum_{i=1}^{n+1}\tau_i\right)$$

If $0 < t_1 < \cdots < t_n < t$ then by letting τ_i be the length of the ith no-success subinterval we obtain

$$P[t_1 < T_1 < t_1 + dt_1, \ldots, t_n < T_n < t_n + dt_n \mid N_t = n]$$

$$= \frac{e^{-\lambda t}\,\lambda\,dt_1\,\lambda\,dt_2\cdots\lambda\,dt_n}{P[N_t = n]} = n!\,\frac{dt_1}{t}\frac{dt_2}{t}\cdots\frac{dt_n}{t}$$

$$= n!\,P[t_1 < U_1 < t_1 + dt_1, \ldots, t_n < U_n < t_n + dt_n]$$

which is the asserted density. ■

Exercises 1 to 5, and 8 and 9 may be done before reading the remainder of this section.

POISSON PROCESSES HAVING THE SET OF ALL REAL NUMBERS FOR PARAMETER SET

We next extend the Poisson process to the whole real line and then note and examine a paradoxical property of Poisson processes. A Poisson process on the whole real line is defined by replacing "$t \geq 0$" by "t real" in our previous definition. In particular $N_0 = 0$, and the realizations appear as in Fig. 3. The process has stationary increments; hence if $t < 0$ then the increment $N_0 - N_t = -N_t$ must have a Poisson density with mean $\lambda(0 - t)$. Thus if $t < 0$ then $-N_t$ is interpreted as the number of successes in the interval between t and the origin. We call $f: R \to R$ a counting function iff both $f(t)$ and $-f(-t)$ are counting functions on $t \geq 0$. For any reals $r < s$ it is irrelevant whether $N_s - N_r$ is interpreted as the number of successes in $r < t < s$, or in $r \leq t \leq s$, since P[success at t] $= 0$ for all t. Similarly it is irrelevant whether a realization is continuous from the left or the right at a particular t. Nor is there any harm in never having a

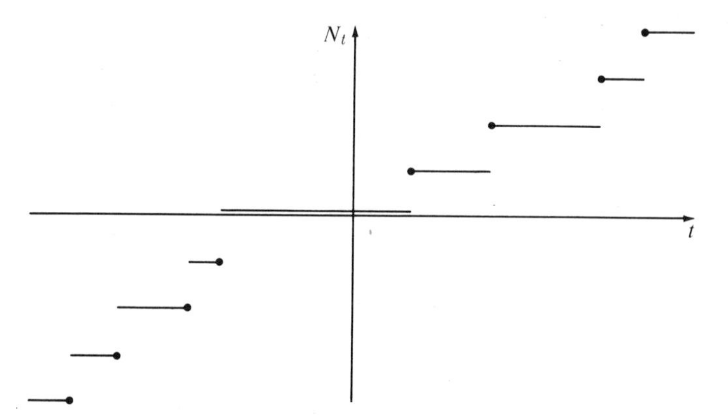

Fig. 3 A realization from a Poisson process on the whole real line.

success at the unique point $t = 0$. If "N_t for t real" is such a process then its restriction "N_t for $t \geq 0$" is obviously a Poisson process as previously defined.

More generally we can extend to the whole real line any stationary independent increment "X_t for $t \geq 0$ with $X_0 = 0$." Let Y_t for $t \geq 0$ be another process independent of "X_t for $t \geq 0$" but generating the same family of finite distributions. If $t < 0$ we define $X(t) = -Y(-t)$; that is, we take an independent version of "X_t for $t \geq 0$" and run it backward in time, with opposite values so as to preserve the increment distributions. Then "X_t for t real" is easily shown to have stationary independent increments. Also it is a Markov process. Exercise 10 shows how the conditional distributions are now affected by the unique role of the origin $X_0 = 0$.

Let "N_t for t real" be a Poisson process. Clearly the distribution of successes is the same near $t = 0$ as it is near any other t. This "stationarity" is indicated by using approximating experiments with Bernoulli trials at times $0, \pm 1/m, \pm 2/m, \ldots$. If X_0 is independent of the process and has distribution $P[X_0 = 1] = P[X_0 = -1] = .5$ then the random telegraph signal $X_t = X_0(-1)^{N_t}$ for t real is easily shown to be a stationary process. Thus for *this* "X_t for t real" process the origin does not have a unique role.

THE WAITING-TIME PARADOX

The Poisson process on the whole real line can be generated intervalwise just as for the case $t \geq 0$. We can also generate its two halves independently by using interarrival times. Roughly speaking, let $x_1, x_1', x_2, x_2', \ldots$ be independent observations from the exponential density with mean $1/\lambda$, and place crosses on the t axis at the points

$$t_n = x_1 + \cdots + x_n \qquad \text{and} \qquad t_n' = -(x_1' + \cdots + x_n') \qquad \text{for } n = 1, 2, \ldots$$

However, note that $x_1' + x_1$ with mean $2/\lambda$ is the distance between the two successes embracing the origin. This does *not* mean that successes are distributed differently near $t = 0$ than near any other t. We next examine this important *waiting-time paradox*. The paradox appears in simpler form for the parameter set "t real" than for "$t \geq 0$."

Let "N_t for t real" be a Poisson process, and select and fix some real t, for example, $t = 0$. Let $S_t(R_t)$ be the positive *spent* (*residual*) *waiting time* between t and the closest success before (after) t. That is, you arrive at the airport at time t; the last plane arrived $S_t > 0$ minutes ago, and the next plane arrives in $R_t > 0$ minutes. If $s > 0$ and $r > 0$ then

$$P[S_t > s, R_t > r] = P[N_t - N_{t-s} = 0, N_{t+r} - N_t = 0] = e^{-\lambda s}e^{-\lambda r}.$$

Thus S_t and R_t are independent and each is exponential with mean $1/\lambda$. Hence $S_t + R_t$, having expectation $2/\lambda$, is the interarrival time between the two arrivals embracing time t, for every t. Thus if the mean rate is $\lambda = .2$ planes per minute then each of the successive interarrival times $X_1, X_2, \ldots$ after $t = 0$ has expectation $1/\lambda = 5$ minutes between planes, but "on the average" the next plane after 3 P.M. (or any other fixed time) arrives about 10 minutes after the preceding plane. For the "N_t for $t \geq 0$" process the same phenomenon occurs for all large $t > 0$, but with the complication $0 < S_t \leq t$. *Roughly speaking*, there is more time between successive arrival pairs that are far apart, and hence most time instants are to be found between such pairs.

More generally let some rule select some particular one of the successive arrival times $0 < t_1 < t_2 < \cdots$ for each realization, and let z be the time between this and the preceding arrival. Let $\lambda = .2$. If the rule is to always select t_2 then $Z = X_2$ and $EZ = 5$. If the rule is to select the first t_n with $t_n - t_{n-1} \geq 20$ then $EZ \geq 20$. That is, we start at $t = 0$ and wait for the first 20-minute gap, and then select the next arrival. If the rule is always to select the first plane after time $t = 10^6$ then $EZ = 10$. The last two rules usually result in larger z than does the first.

EXERCISES

1. *Gamma Interarrival Times.* Sketch a modified version of Fig. 2 which can legitimately be interpreted as obtained from a sampling experiment which generates realizations from a process having independent interarrival times, each of which has a gamma density with parameters $\lambda = 1$ and $\nu = 2$. Note that it was shown in Exercise 7 (Sec. 23-1) that this process does *not* have stationary or independent increments, nor is it Markov. Sketch a version for parameters $\lambda = 1$ and $\nu = 5$.

2. Customers arrive according to a Poisson process with mean rate λ and mean spacing $1/\lambda$. Over a long period what percentage of interarrival times do you anticipate will exceed $1/\lambda$? $2/\lambda$?

3. Let X and Y be independent with exponential densities having means $1/\lambda_1$ and $1/\lambda_2$.
(*a*) Find the distribution of $Z = \min\{X, Y\}$.
(*b*) Relate this result to the two independent Poisson processes of Exercise 6*a* (Sec. 23-1).

4. *Justification of Characterization 2*
(*a*) Let crosses be placed in the unit interval $0 \leq t < 1$, as in Characterization 2. For fixed $0 < s < 1$ let X and Y be the number of crosses in the intervals $0 < t < s$ and $s < t < 1$, respectively. Prove that X and Y are independent and have Poisson densities with means λs and $\lambda(1 - s)$, respectively.
(*b*) Justify Characterization 2; that is, show that if the process is constructed as described then it has stationary independent Poisson increments.

5. Show that $[N_t = n]$ can be replaced by $[t \leq T_{n+1} \leq t + dt]$ in the statement of Theorem 2.

6. The traffic-control officer informs you that during the daylight hours planes arrive essentially according to a Poisson process with a mean rate of $\lambda = .1$ planes per minute, or a mean spacing of $1/\lambda = 10$ minutes. As a check you plan to examine the records for the past 365 days and then calculate the average spacing between the two planes which arrived just before and just after noon. What number should this calculation yield, approximately?

7. For a Poisson process "N_t for $t \geq 0$," if $t > 0$ define the spent waiting time S_t to be t if $[N_t = 0]$ occurs, with the residual waiting time R_t defined as before. Show that S_t and R_t are independent, and find and plot their distribution functions.

8. *Aftereffects from Only the Last Impulse.* A noise impulse occurs at time $t = 0$, and later impulses occur at Poisson process times with mean rate λ. Each impulse triggers an electronic circuit which charges a capacitor to 1 volt, and the voltage then decreases exponentially as e^{-t} until the next impulse occurs. If V_t is the voltage at time t then a realization from "V_t for $t \geq 0$" appears as in Fig. 4. Let

$$Z_n = \int_0^{T_n} V_t \, dt$$

be the integrated voltage up to the time of the nth impulse occurring after $t = 0$. Find EZ and Var Z_n.

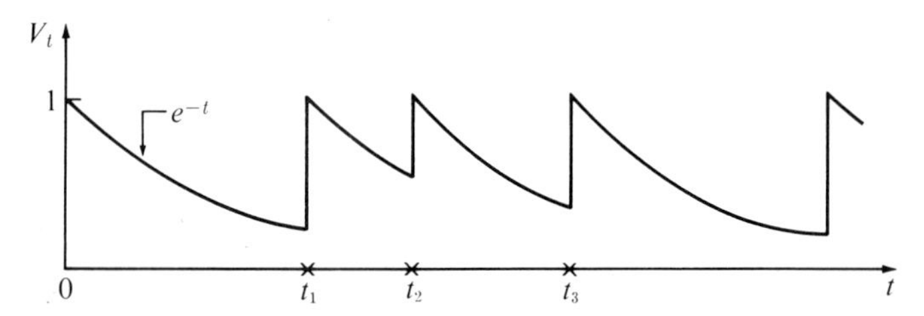

Fig. 4 A capacitor discharges through a resistor but is recharged at Poisson times.

9. *Cumulative Aftereffects.* You deposit $Z_n > 0$ dollars in your bank account at time T_n so that this deposit contributes $Z_n e^{\alpha h}$ dollars to your account h days later, where α determines the interest rate. Deposit times are a Poisson process with mean rate λ, and deposits Z_1, $Z_2, \ldots$ are independent (of each other and of the Poisson process) and identically distributed, with $EZ_1 < \infty$. Your balance at time $t > 0$, assuming $B_0 = 0$, is

$$B_t = \sum_{k=1}^{N_t} Z_k e^{\alpha(t-T_k)}.$$

Figure 5 depicts a realization from B_t for $t \geq 0$. Show that

$$EB_t = \lambda(EZ_1)\frac{e^{\alpha t} - 1}{\alpha}.$$

10. *A Stationary Independent Increment Process on the Whole Real Line Is a Markov Process with Conditional Distributions Affected by the Origin.* In order to condition on events having positive probability, assume that all random variables in this exercise induce discrete densities, and assume positive probability for every conditioning event below. Let $S_k = Z_1 + \cdots + Z_k$ where $Z_1, \ldots, Z_n$ are $n \geq 3$ independent random variables.

(*a*) Prove that

$$P[S_n = s_n \mid S_{n-1} = s_{n-1}, \ldots, S_2 = s_2, S_1 = s_1] = P[S_n = s_n \mid S_{n-1} = s_{n-1}]$$
$$= P[Z_n = s_n - s_{n-1}].$$

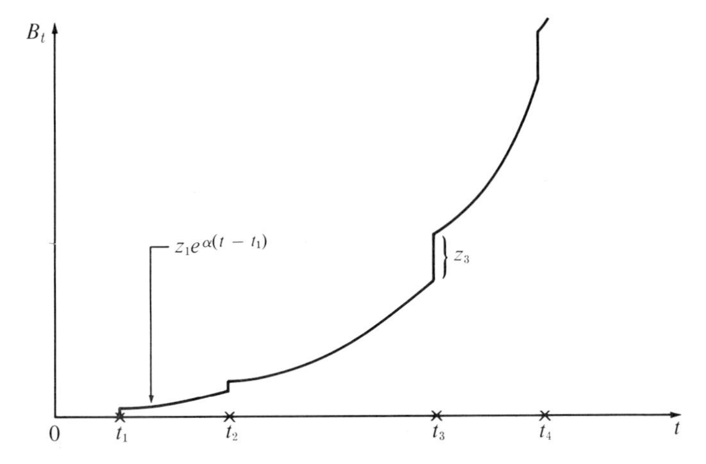

Fig. 5 B_t is the bank balance at time t due to random deposits $Z_1, Z_2, \ldots$ at Poisson times.

(*b*) Prove that

$$P[S_1 = s_1 \mid S_2 = s_2, S_3 = s_3, \ldots, S_n = s_n] = P[S_1 = s_1 \mid S_2 = s_2].$$

(*c*) Let the Z_i be Bernoulli trials with $P[Z_i = 1] = p$ where $0 < p < 1$. Find $P[S_1 = s_1 \mid S_2 = s_2]$ for $s_1 = 0, 1$ and $s_2 = 0, 1, 2$. Note that part (*b*) cannot have a further reduction as in part (*a*).

(*d*) Let "X_t for t real with $X_0 = 0$" have stationary independent increments, and let $t_1 < \cdots < t_n$ be any $n \geq 3$ real numbers. There are three natural cases, depending on whether the origin is inside or to the left or right of the interval between t_{n-1} and t_n. Prove that

$$P[X(t_n) = x_n \mid X(t_{n-1}) = x_{n-1}, \ldots, X(t_1) = x_1] = P[X(t_n) = x_n \mid X(t_{n-1}) = x_{n-1}]$$

and furthermore show that this reduces to

$$P[X(t_n) = x_n] \qquad \text{if } t_{n-1} \leq 0 \leq t_n$$

$$P[X(t_n - t_{n-1}) = x_n - x_{n-1}] \qquad \text{if } 0 < t_{n-1} \;.$$

(*e*) For a Poisson process, if $t < s < 0$ show that the conditional distribution of $-N_s$, given $[-N_t = n]$, is binomial n and s/t.

11. *The Waiting-time Paradox Must Arise for Arbitrary Arrivals*

(*a*) For positive interarrival times $x_1, x_2, \ldots, x_n$ let $t_k = x_1 + \cdots + x_k$ for $k = 1, \ldots, n$ be the arrival times. For $0 \leq t < t_n$, as in Fig. 6, define f by

$$f(t) = x_k \qquad \text{if } t_{k-1} \leq t < t_k$$

so that $f(t)$ is the distance between the closest pair of crosses embracing t. The area under f and over the interval from t_{k-1} to t_k is x_k^2. Let

$$d_n = \frac{1}{t_n}\int_0^{t_n} f(t)\,dt = \frac{1}{t_n}\sum_{k=1}^{n} x_k^2$$

be the time average of $f(t)$. That is, d_n is the expected "distance between the embracing

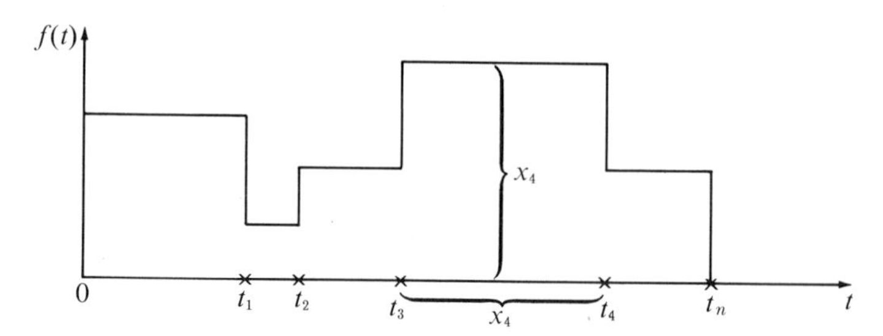

Fig. 6

pair of crosses" for a point selected at random between 0 and t_n. Let $\bar{x} = (x_1 + \cdots + x_n)/n = (1/n)t_n$ be the average interarrival time *when* each pair of adjacent crosses is selected with equal chance for each pair. Show that

$$d_n = \bar{x} + \frac{1}{\bar{x}}\left\{\frac{1}{n}\sum_{k=1}^{n}(x_k - \bar{x})^2\right\}$$

In particular if the x_k are not all equal then $d_n > \bar{x}$. Thus the average point is embraced by a pair of crosses which are farther apart than the average adjacent pair.

(*b*) We make the arrivals random and interpret the random variable D_n in the same way as above. Let $X_1 > 0$ have a finite fourth moment and note that

$$\frac{EX_1^2}{EX_1} = \frac{\mu^2 + \operatorname{Var} X_1}{\mu} = \mu + \frac{\sigma^2}{\mu} \qquad \text{where } \mu = EX_1 \text{ and } \sigma^2 = \operatorname{Var} X_1.$$

Let the interarrival times $X_1, X_2, \ldots$ be independent and identically distributed. Let

$$D_n = \frac{1}{T_n}\sum_{k=1}^{n} X_k^2 = \frac{(1/n)\sum_{k=1}^{n} X_k^2}{(1/n)\sum_{k=1}^{n} X_k}$$

Show that for every $\epsilon > 0$

$$P\left[\left|D_n - \left(\mu + \frac{\sigma^2}{\mu}\right)\right| < \epsilon\right] \to 1 \qquad \text{as } n \to \infty.$$

Thus for large n we expect that D_n will be near $\mu + \sigma^2/\mu$. For the exponential density $\sigma^2 = \mu^2$, so this becomes 2μ, in agreement with the waiting-time paradox. HINT: The intersection of two events is almost certain if each is.

(*c*) We now assume that arrivals occur at integer times, and essentially find the distribution of $f(t)$ of part (*a*) when t is uniform and the arrivals are random. More precisely, let the interarrival times $X_1, X_2, \ldots$ be independent and identically distributed with

$$P[X_1 = 1] + P[X_1 = 2] + \cdots + P[X_1 = r] = 1.$$

For each n let Y_n be the distance between the embracing pair of crosses when a point is selected at random from the interval from 0 to T_n. Fix an integer k satisfying $1 \le k \le r$ and prove that

$$\lim_{n\to\infty} P[Y_n = k] = \frac{k}{EX_1} P[X_1 = k].$$

HINT: Let M_n be the number of $X_1, \ldots, X_n$ which have the value k. Clearly if $[M_n = m, T_n = t]$ occurs then fraction km/t of the interval from 0 to t consists of points whose embracing crosses have distance k. First show that

$$P[Y_n = k] = kE\left(\frac{M_n}{T_n}\right).$$

COMMENT: Naturally it then follows that

$$\lim_{n\to\infty} EY_n = \frac{EX_1^{\,2}}{EX_1} = \mu + \frac{\sigma^2}{\mu}.$$

The random variable D_n of part (*b*) is the conditional expectation of Y_n given $X_1, \ldots, X_n$, that is, $D_n = E(Y_n \mid X_1, \ldots, X_n)$. Part (*c*) suggests that if X_1 has continuous density p_{X_1} then Y_n should, when n goes to infinity, have continuous density $(y/\mu)p_{X_1}(y)$, which becomes $\lambda^2 ye^{-\lambda y}$ when X_1 is exponential. This is the same density as found for $S_t + R_t$ for a Poisson process.

23-4 COMPOUND POISSON PROCESSES

This section is devoted to stationary-independent-increment processes which make jumps at times determined by a Poisson process, while the jump values are independent and identically distributed according to an arbitrary distribution. In particular we find that random selections from a Poisson process yield a new Poisson process.

The set of all real numbers could be used for the parameter set, but for simplicity we use $t \geq 0$ in both this and the next section. Let N_t for $t \geq 0$ be a Poisson process with mean rate λ, and let $F: R \rightarrow R$ be any distribution function which is not concentrated on the one point 0. Let "V_t for $t \geq 0$ with $V_0 = 0$" be the process having only jump function realizations and constructed by letting the Poisson process determine the jump times, with the jump values determined independently and each according to F. Such a process is called a **compound Poisson process,** or if F is concentrated on the nonnegative integers it may be called a **generalized Poisson process.**

We may construct the process explicitly as follows. Let $X_1, Y_1, X_2, Y_2, \ldots$ be independent, where each X_i is exponential with mean $1/\lambda$ and each Y_i has F for distribution function. Then N_t is defined in terms of $X_1, X_2, \ldots$ as in Characterization 1 (Sec. 23-3), and for $t > 0$ we let

$$V_t = \sum_{i=0}^{N_t} Y_i$$

In other words, let $Z_0 = 0$ and $Z_n = Y_1 + \cdots + Y_n$ for $n \geq 1$, and then for each $t > 0$ define V_t to equal Z_n on the event $[N_t = n]$. Clearly realizations are like Fig. 3 (Sec. 23-1), where jump functions were defined. For interpretation, Y_i might be the rainfall from the ith storm, or the number of fish in the ith school of fish.

Thus V_t is the sum of a random number N_t of random variables. This section is essentially independent of Secs. 14-2 and 15-5 where such sums are also considered. In the preceding construction any distribution satisfying $P[X_1 > 0] = 1$ could be used for X_1, but we used the exponential density in order to obtain stationary independent increments.

The distribution of each V_t is determined by F and λt as follows:

$$P[V_t < v] = \sum_{n=0}^{\infty} [V_t < v, N_t = n] = \sum_{n=0}^{\infty} P[Z_n < v, N_t = n]$$

$$= \sum_{n=0}^{\infty} P[Z_n < v]P[N_t = n] \qquad \text{for all real } v$$

where F determines the distribution of each of $Z_1, Z_2, \ldots$.

We now prove, as is intuitively obvious, that any compound Poisson process V_t for $t \geq 0$ has stationary independent increments. To show stationarity of increments let $0 \leq s < t$ so that

$$P[V_t - V_s < v] = \sum_{n=0}^{\infty} \sum_{k=0}^{\infty} P[V_t - V_s < v, N_t - N_s = n, N_s = k].$$

But we have widespread independence and the Y_i are identically distributed; hence the general term in the summation equals

$$P[Y_{k+1} + \cdots + Y_{k+n} < v, N_t - N_s = n, N_s = k]$$
$$= P[Z_n < v]P[N_{t-s} = n]P[N_s = k].$$

Summing over k and then n yields $P[V_{t-s} < v]$, so we have stationary increments.

We next prove that the increments are independent. If $0 \leq r < s < t$ then

$$P[V_s - V_r < v, V_t - V_s < v'] = \sum_{k=0}^{\infty} \sum_{n=0}^{\infty} \sum_{j=0}^{\infty} [\text{general term}]$$

where the general term is

$$P[V_s - V_r < v, V_t - V_s < v', N_s - N_r = k, N_t - N_s = n, N_r = j]$$
$$= P[Z_k < v]P[Z_n < v']P[N_{s-r} = k]P[N_{t-s} = n]P[N_r = j].$$

Summing first over j leaves a product of two sums which reduce to

$$P[V_{s-r} < v]P[V_{t-s} < v']$$

The proof generalizes to more than two intervals, so we have independent increments.

Example I If $0 < p \leq 1$ and if F is Bernoulli

$$P[Y_1 = 1] = p = 1 - P[Y_1 = 0]$$

then we say that V_t for $t \geq 0$, constructed as above, is obtained by *random selection from the Poisson process "N_t for $t \geq 0$."* For example, N_t for $t \geq 0$ might represent radioactive emissions or airplane arrivals, while V_t for $t \geq 0$ represents recorded emissions or landing accidents. We show that random selection with probability p from a Poisson process with mean rate λ yields a Poisson process with mean rate λp.

Naturally for the process of this example, as well as in every case where $P[Y_1 = 0] > 0$, the jumps of V_t for $t \geq 0$ occur only at *some* of the Poisson times, so all Poisson times cannot be calculated from V_t for $t \geq 0$. Nevertheless V_t for $t \geq 0$ is a stationary-independent-increment process having unit-counting-function realizations. Thus we need only find $P[V_t = k]$. Certainly $P[Z_n = k] = 0$ if $n < k$; hence, letting $q = 1 - p$, we have

$$P[V_t = k] = \sum_{n=k}^{\infty} \binom{n}{k} p^k q^{n-k} \frac{e^{-\lambda t}(\lambda t)^n}{n!}$$

$$= \frac{e^{-\lambda pt}(\lambda pt)^k}{k!} \sum_{n=k}^{\infty} \frac{e^{-\lambda qt}(\lambda qt)^{n-k}}{(n-k)!}$$

where the last summation equals 1, yielding the stated result.

Instead of the preceding proof, we could use Theorem 1 (Sec. 23-2) to conclude that V_t for $t \geq 0$ is a Poisson process. Its mean rate can be obtained from Theorem 3 (Sec. 15-5) by

$$EV_t = (EY_1)(EN_t) = p\lambda t$$

Thus if suicide attempts are describable by a Poisson process with $\lambda = 8$ attempts per month and there is an independent 25 percent chance of death for each attempt then suicides are describable by a Poisson process with a mean rate of two suicides per month. Note that if precisely every fourth attempt must result in death then, as in Exercise 7 (Sec. 23-1), suicides would *not* follow a Poisson process. In particular, the time between the nth and $(n + 1)$st suicide would, from Theorem 1 (Sec. 23-3), have the gamma density $\gamma_{\lambda,4}$ rather than an exponential density. ■

***Example 2** For $0 < p < 1$ and $p + q = 1$ let Y_1 have the logarithmic discrete probability density defined by

$$l(n) = \frac{q^n}{n \log(1/p)} \qquad \text{for } n = 1, 2, \ldots.$$

Let V_t for $t \geq 0$ be constructed as described above, with this distribution for Y_1. Exercise 10 (Sec. 14-2) shows that if $t > 0$ then V_t has a negative binomial density with parameters $r > 0$ and p, where $r = \lambda t/\log(1/p)$. Thus V_t has mean rq/p, variance rq/p^2, probability-generating function $[p/(1 - qz)]^r$, and density

$$P[V_t = k] = \frac{(r + k - 1)(r + k - 2) \cdots (r + 1)r}{k!} p^r q^k \qquad \text{for } k = 0, 1, \ldots.$$

If for a particular t the corresponding r happens to be a positive *integer* then, of course, V_t has a Pascal density with parameters r and p. ■

EXERCISES

1. Let V_t for $t \geq 0$ be the compound Poisson process determined by P_{Y_1} and the Poisson λ process N_t for $t \geq 0$. For $s \geq 0$ and $t \geq 0$ show that

$$EV_t = \lambda t EY_1 \qquad \text{Var } V_t = \lambda t E{Y_1}^2 \qquad \text{Cov }(V_s, V_t) = (\min\{s,t\})\lambda E{Y_1}^2$$

2. A pebble at the bottom of a river does not move except for an occasional sudden movement or jump along the direction toward the ocean. These independent jumps have a normal distribution with $\mu = \sigma = 1$ inch, so occasionally the pebble happens to roll away from the ocean. The jumps occur at Poisson times with a mean rate of one per hour.

(*a*) Let V_t be the number of inches the pebble has moved toward the ocean in t hours, where $V_0 = 0$. Show that $P[|V_{100} - 100| \geq 40]$ is small.

(*b*) Let W_t for $t \geq 0$ be the number of jumps up to time t which are toward rather than away from the ocean. Explain briefly why W_t for $t \geq 0$ is a Poisson process, and find its mean rate. In other words, note that deterministic selection according to the jump values of a compound Poisson process is equivalent to random selection from the underlying Poisson process.

3. *Linear Combinations of Independent Poisson Processes.* Suppose that 40 percent of all automobile accidents involve only one car and the remaining 60 percent involve exactly two cars. In one approach to setting up a mathematical model we might assume that accidents occur according to a Poisson process with a mean rate of λ accidents per month, and then define the number V_t of cars involved in accidents as in this section, using $P[Y_1 = 1] = .4$ and $P[Y_1 = 2] = .6$. That is, accidents occur according to a Poisson process, and a biased coin determines independently for each accident whether one or two cars is involved. Instead of using a compound Poisson process, another approach would be to let $V_t = N_1(t) + 2N_2(t)$, where $N_1(t)$ and $N_2(t)$ for $t \geq 0$ are two independent Poisson processes with mean rates $.4\lambda$ and $.6\lambda$. That is, the single and double accident processes are independent, and V_t is the number of cars involved in accidents from either process. This exercise shows that these two approaches are equivalent.

Let $a_1, \ldots, a_k$ be $k \geq 2$ distinct nonzero real numbers and let $P[Y_1 = a_i] = p_i > 0$, where $p_1 + \cdots + p_k = 1$. Let N_t for $t \geq 0$ be a Poisson λ process and let V_t for $t \geq 0$ be the corresponding compound Poisson process using P_{Y_1} as above. For $t > 0$ let $N_i(t)$ be the number of jumps of value a_i for the V_t process which occur prior to time t. Then we have the obvious identity

$$V_t = a_1 N_1(t) + \cdots + a_k N_k(t) \qquad \text{for all } t \geq 0.$$

Furthermore $N_i(t)$ for $t \geq 0$ is a Poisson process with mean rate λp_i, since it can obviously be described as obtained by random selection from the N_t for $t \geq 0$ process.

Show that for every $t \geq 0$ the k random variables $N_1(t), \ldots, N_k(t)$ are independent.

COMMENT: The proof can be generalized to show that the k *processes* are independent. Also Y_1 can be permitted to have a denumerable number of values.

4. Let V_t for $t \geq 0$ be the compound Poisson process determined by P_{Y_1} and the Poisson λ process N_t for $t \geq 0$.

(*a*) For all real r and for all real $t \geq 0$ show that

$$Ee^{irV_t} = \exp\{\lambda t(Ee^{irY_1} - 1)\}.$$

Thus, in a fashion similar to Theorem 1 (Sec. 14-2), the characteristic function for V_t is obtained by replacing z in the probability-generating function $Ez^{N_t} = e^{\lambda t(z-1)}$ for N_t by the characteristic function for Y_1.

(*b*) Let Y_1 be normal 0 and 1 and note that V_t does not have a normal density.

(*c*) Let $P[Y_1 = 1] = p = 1 - P[Y_1 = 0]$ and note that V_t has a Poisson density.

23-5 NONHOMOGENEOUS POISSON PROCESSES

Let $M(s)$ be the number of vehicles which pass a fixed point on a highway between midnight and s hours after midnight. In the vicinity of 8 A.M. we might expect that the increment process $M(8 + s) - M(8)$ is approximately a Poisson process with a mean rate of 10^4 vehicles per hour. In the vicinity of 10 A.M. we might expect that $M(10 + s) - M(10)$ is approximately a Poisson process with a mean rate of 10^3 vehicles per hour. This section is devoted to independent-increment Markov processes which in the vicinity of any time s appear to be approximately a Poisson process, with mean rate depending on s.

If we observe the nth vehicle pass at time s then our waiting time for the next vehicle has a conditional distribution which is independent of n but does depend on s, which is the sum of the first n interarrival times. Thus interarrival times are not independent.

Let $m(s) = EM(s)$ be the expected number of vehicles up to time s. To each time s we can correspond a fictitious "intrinsic" time $t = m(s)$; that is, time $t = 5$ is the time at which, on the average, five vehicles will have passed. If we measure time by this fictitious scale then we expect our reparameterized process to be a Poisson process. Therefore we will define our new processes to be just those obtainable by a natural reparameterization of a Poisson process.

Let $N(t)$ for $t \geq 0$ with $N_0 = 0$ be a Poisson process with mean rate or intensity $\lambda = 1$. Let m be any real-valued function defined on $s \geq 0$, having a positive derivative $m'(s)$ everywhere, and satisfying $m(0) = 0$, with $m(s)$ converging to infinity as s does. For each $s \geq 0$ define the random variable $M(s)$ by $M(s) = N(m(s))$. Then we call the process "$M(s)$ for $s \geq 0$," which obviously has unit-counting-function realizations, a **Poisson process with intensity function** m', and we call it a **nonhomogeneous Poisson process** if m' is not a constant. Thus "$M(m^{-1}(t)) = N(t)$ for $t \geq 0$" is a Poisson $\lambda = 1$ process. If $m(s) = \lambda s$ then $M(s)$ for $s \geq 0$ is a Poisson process; that is, its intensity function is the constant $m' = \lambda$. Naturally $M(s)$ has a Poisson density with mean $m(s) = EM(s)$ for any $s > 0$. Thus each realization r_ω from "N_t for $t \geq 0$" determines a corresponding realization $\bar{r}_\omega$ from "$M(s)$ for $s \geq 0$" as follows. For any $s \geq 0$ we find the corresponding $t = m(s)$ and let $\bar{r}_\omega(s) = r_\omega(t)$.

Figure 1 shows a realization from the Poisson process, along with the corresponding transformed realization which is produced by the nonhomogeneous Poisson process. Thus $0 < s_1 < s_2 < \cdots$ are the successive arrivals for the $M(s)$ for $s \geq 0$ process, and $t_n = m(s_n)$ is the nth Poisson arrival.

It is immediate that $M(s)$ for $s \geq 0$ is Markov and has independent increments, since $N(t)$ for $t \geq 0$ possesses these properties. For example, if $0 \leq q < r < s$ then

$$\begin{aligned} P[M(s) = k \mid M(r) = j, M(q) = i] &\\ = P[N(m(s)) = k \mid N(m(r)) = j, N(m(q)) = i] &\\ = P[N(m(s)) = k \mid N(m(r)) = j] = P[M(s) = k \mid M(r) = j]. & \end{aligned}$$

An analogous proof yields the independence of increments.

For small $\Delta > 0$ the probability is approximately $m'(s)\Delta$ that the process $M(s)$ for $s \geq 0$ will have an arrival in the interval between s and $s + \Delta$ since

$$\begin{aligned} \frac{P[M(s+\Delta) - M(s) = 1]}{m'(s)\Delta} &= \frac{P[N(m(s+\Delta)) - N(m(s)) = 1]}{m'(s)\Delta} \\ &= \frac{m(s+\Delta) - m(s)}{m'(s)\Delta} e^{-[m(s+\Delta)-m(s)]} \to 1 \quad \text{as } \Delta \to 0. \end{aligned}$$

Thus locally near s, arrivals occur as for a Poisson process with mean rate or intensity $m'(s)$.

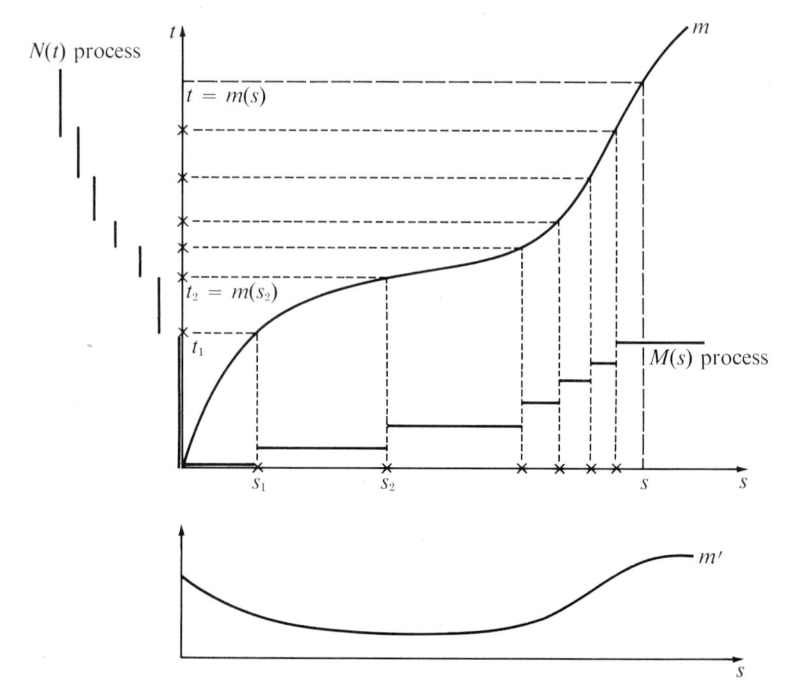

Fig. 1 The nonhomogeneous Poisson process, $M(s) = N(m(s))$ for $s \geq 0$, with intensity function m'.

Let $0 < S_1 < S_2 < \cdots$ be the successive arrival times of $M(s)$ for $s \geq 0$, so that $T_n = m(S_n)$ is the nth arrival of $N(t)$ for $t \geq 0$. For the first arrival we have

$$P[S_1 > s] = P[M(s) = 0] = P[N(m(s)) = 0] = e^{-m(s)}$$

and differentiating the distribution function $1 - e^{-m(s)}$ yields the continuous density

$$p_{S_1}(s) = e^{-m(s)}m'(s) \qquad \text{for all } s > 0.$$

As an example of many analogous computations, we generalize the above and now find, for any $n \geq 1$, the conditional distribution of S_n given $[S_{n-1} = s]$. Using the obvious relationships indicated in Fig. 2, and the independent exponentially distributed Poisson interarrival times $X_1, X_2, \ldots$, we find for all $0 \leq s \leq r$ that

$$\begin{aligned} P[S_n < r \mid S_{n-1} = s] &= P[X_n < m(r) - m(s) \mid T_{n-1} = m(s)] \\ &= P[X_n < m(r) - m(s)] = 1 - e^{-[m(r)-m(s)]}. \end{aligned}$$

Differentiating with respect to r yields

$$p_{S_n|S_{n-1}=s}(r) = e^{-[m(r)-m(s)]}m'(r).$$

Also if $n \geq 2$ then $Y_n = S_n - S_{n-1}$ is the nth interarrival time which has conditional distribution function

$$P[Y_n < y \mid S_{n-1} = s] = P[S_n < s + y \mid S_{n-1} = s] = 1 - e^{-[m(s+y)-m(s)]}$$

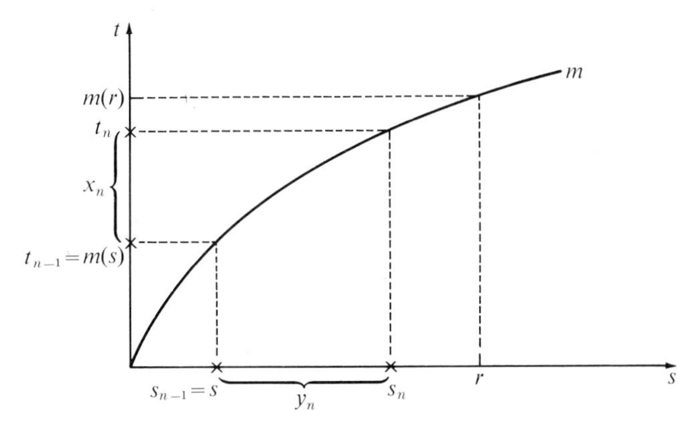

Fig. 2

depending on s; hence interarrival times are dependent for a nonhomogeneous Poisson process.

In Sec. 23-3 we saw that a Poisson process could be generated (1) by independent exponential interarrival times or (2) by random arrivals in an interval once the number of arrivals in the interval is known. The arrivals for $M(s)$ for $s \geq 0$ can then be taken as $m^{-1}(t_1), m^{-1}(t_2), \ldots$. Naturally $M(s)$ for $s \geq 0$ can be generated directly without reference to $N(t)$ for $t \geq 0$. Method 2 is essentially left as an exercise. For method 1 we select s_1 from p_{S_1} above, and then select s_2 from $p_{S_2|S_1=s_1}$ above, etc. Thus "$M(s)$ for $s \geq 0$ with $M_0 = 0$" can be thought of as described by the family of conditional densities

$$p(r \mid s) = e^{-[m(r)-m(s)]}m'(r) \qquad \text{for } 0 \leq s \leq r$$

which determine the next success r, given that the last success was at s.

EXERCISES

1. For a nonhomogeneous Poisson process with nth arrival time S_n and nth interarrival time Y_n show that if $n \geq 2$ then for all $s > 0$ and $y > 0$ we have

$$p_{S_{n-1}}(s) = \frac{e^{-m(s)}[m(s)]^{n-2}}{(n-2)!} m'(s)$$

$$p_{Y_n|S_{n-1}=s}(y) = e^{-[m(s+y)-m(s)]}m'(s+y)$$

$$p_{Y_n}(y) = \int_0^\infty \frac{e^{-m(s+y)}m(s)^{n-2}}{(n-2)!} m'(s+y)m'(s)\, ds$$

2. Prove the following generalization of Theorem 2 (Sec. 23-3) for the successive arrival times $S_1, S_2, \ldots$ of a nonhomogeneous Poisson process $M(s)$ for $s \geq 0$ with intensity function m': For any $s > 0$ and any $n = 1, 2, \ldots$ the conditional distribution of $(S_1, \ldots, S_n)$, given the event $[M(s) = n]$, is the same as the distribution of the ordered vector corresponding to $(U_1, \ldots, U_n)$ where $U_1, \ldots, U_n$ are independent and each has distribution function $m(u)/m(s)$ for $0 \leq u \leq s$.

3. Let $N(t)$ for $t \geq 0$ with $N_0 = 0$ be a Poisson λ process. Sketch typical realizations for the following processes:
(*a*) $M_s = N(s/(1-s))$ for $0 \leq s < 1$.
(*b*) $M_s = N(1-e^{-s})$ for $s \geq 0$. What is the expected total number of arrivals for this process?

4. *Compound Nonhomogeneous Poisson Processes.* Let $M(s) = N(m(s))$ for $s \geq 0$ be a nonhomogeneous Poisson process, where the Poisson process $N(t)$ for $t \geq 0$ has $\lambda = 1$. Let $V(t)$ for $t \geq 0$ be the compound Poisson process having independent jump values according to a distribution P_{Y_1} and occurring at the Poisson times $N(t)$ for $t \geq 0$. Let

$$W(s) = V(m(s)) = \sum_{i=0}^{N(m(s))} Y_i$$

so that $W(s)$ for $s \geq 0$ has independent jumps determined by P_{Y_1} but at times determined by $M(s)$ for $s \geq 0$. State the generalization from "$V(t)$ for $t \geq 0$" to "$W(t)$ for $t \geq 0$" of the following:
(*a*) Exercise 1 (Sec. 23-4)
(*b*) Exercise 3 (Sec. 23-4)
(*c*) Exercise 4*a* (Sec. 23-4)

23-6 PURE-BIRTH PROCESSES

This section is devoted to continuous-time Markov processes which have births, but no deaths or multiple births. We assume the infinitesimal relations

$$P[\text{a birth between } t \text{ and } t + dt \mid n \text{ births between times 0 and } t] = \lambda_n \, dt$$

for $n = 0, 1, \ldots$. Such a process is determined by its birth-rate sequence $\lambda_0, \lambda_1, \ldots$ of positive numbers. Thus the chance for a birth or arrival during the next short interval after time t is permitted to depend on the number of births between 0 and t, but not on t, so we say that the transition probabilities are stationary. If all $\lambda_n = \lambda$ we have a Poisson process. In general the increments are neither stationary nor independent. An interesting natural case is that of a linear birthrate $\lambda_n = (n + \nu)\lambda$, given $\nu \geq 1$ initiators at time 0, so that the chance of a birth is proportional to the present population size. For this case with $\nu = 1$, Example 1 exhibits the distribution of the population size at time t and Example 2 describes a transparent method for generating realizations up to time t.

A standard technique for Markov processes is to make infinitesimal assumptions such as the ones above and then derive differential equations whose solutions yield the unique probability law which must describe such a process. Instead of taking this approach, we relate such processes to an obvious generalization of Characterization 1 (Sec. 23-3) for Poisson processes. More explicitly we show that for such a process the interarrival times $X_1, X_2, \ldots$ are independent and X_{n+1} has an exponential density with mean $1/\lambda_n$. Naturally we can then use these properties in order to construct such processes, and then the lack of memory for exponential densities yields the Markov property and the desired infinitesimal relations.

If $\lambda_n \, dt$ in the infinitesimal relation is replaced by $\lambda_{n+1} \, dt$ then the nth interarrival time X_n would have mean $1/\lambda_n$. However, this notational advantage disappears in the next section, so we adhere to the tradition that λ_n is the parameter which determines the time to the next birth if we are presently in state n. Then the $(n+1)$st interarrival time X_{n+1} has mean $1/\lambda_n$.

We call a process "N_t for $t \geq 0$ with $N_0 = 0$" a **pure-birth Markov process** with birth-rate sequence $\lambda_0, \lambda_1, \ldots,$ each $\lambda_n > 0$, iff it is Markov, has unit-counting-function realizations, satisfies the infinitesimal relations

$$\lim_{\substack{\Delta \to 0 \\ \Delta > 0}} \frac{1}{\Delta} P[N_{t+\Delta} - N_t = 1 \mid N_t = n] = \lambda_n \qquad \text{for all } t > 0, n = 0, 1, \ldots$$

and has **stationary-transition probabilities;** that is, $P[N_{t+r} = k \mid N_t = n]$ may depend on r, n, k but not on t. Here $r > 0$ and $0 \leq n \leq k$, and if $n = 0$ we can let $t = 0$. We may interpret N_t as the number of births between times 0 and t.

Actually the stationarity of the transition probabilities is implied by the other assumptions, since λ_n is not permitted to depend on t. However its assumption simplifies the derivation of the interarrival-time distributions. We first show that for such a process the successive interarrival times $X_1, X_2, \ldots$ are independent and X_{n+1} has an exponential density with mean $1/\lambda_n$. We then show that if $\lambda_0, \lambda_1, \ldots$ is a sequence of positive numbers then there exists such a process with this sequence as its birth-rate sequence iff $\sum_{n=0}^{\infty} 1/\lambda_n = \infty$. We then derive a recursion relation for $P[N_t = n]$, after which Examples 1 and 2 discuss the case of a linear birth rate.

A PURE-BIRTH MARKOV PROCESS MUST HAVE INDEPENDENT EXPONENTIALLY DISTRIBUTED INTERARRIVAL TIMES

Assume that we are given a Markov pure-birth process. We deduce its interarrival-time properties. We have stationary transitions so $P[N_{t+r} = 0 \mid N_t = 0]$ must equal its value at $t = 0$, hence

$$P[N_{t+r} = 0 \mid N_t = 0] = P[N_r = 0].$$

In other words,

$$P[X_1 > t + r \mid X_1 > t] = P[X_1 > r] \qquad \text{for all } t > 0, r > 0.$$

If X_1 has an exponential density with mean $1/\lambda$ then this condition is satisfied, since

$$\frac{e^{-\lambda(t+r)}}{e^{-\lambda t}} = e^{-\lambda r}.$$

Furthermore we saw in Example 4 (Sec. 16-1) that only an exponential distribution can have this unique lack of memory. In addition the first infinitesimal relation is satisfied iff $\lambda = \lambda_0$ because

$$P[N_{t+dt} = 1 \mid N_t = 0] = P[t < X_1 < t + dt \mid X_1 > t] = \frac{\lambda e^{-\lambda t}\, dt}{e^{-\lambda t}} = \lambda\, dt.$$

Thus X_1 must be exponential with mean $1/\lambda_0$.

Proceeding by induction, we assume that the interarrival times $X_1, \ldots, X_n$ are independent and X_i for $1 \leq i \leq n$ is exponential with mean $1/\lambda_{i-1}$. For positive $x_1, \ldots, x_n$ let $t_i = x_1 + \cdots + x_i$ and let $r > 0$. By assumption we have a Markov process with stationary transition probabilities, so that we can add earlier constraints to a conditioning event $[N(t) = n]$, and thus in effect assume a particular past history. Therefore the probability

$$P[N(t_n + r) = n \mid N(t_n) = n]$$
$$= P[N(t_n + r) = n \mid N(t_i) = i,\ N(t_i - dx_i) = i - 1 \text{ for } i = 1, \ldots, n]$$
$$= P[X_{n+1} > r \mid x_n - dx_n < X_n < x_n, \ldots, x_1 - dx_1 < X_1 < x_1]$$

may depend on n and r, but it cannot depend on $x_1, \ldots, x_n$. But this is true for each $r > 0$, and hence X_{n+1} is independent of $(X_1, \ldots, X_n)$, so $X_1, \ldots, X_{n+1}$ are independent. In particular we have

$$P[N(t + r) = n \mid N(t) = n] = P[X_{n+1} > r] \qquad \text{for all } t > 0, r > 0$$

which is used for the last equality below. To show that X_{n+1} must have the zero-memory property we note that X_{n+1} is independent of $T_n = X_1 + \cdots + X_n$; hence we may assume a particular value for T_n and use the Markov property to get

$$P[X_{n+1} > t + r \mid X_{n+1} > t] = P[X_{n+1} > t + r \mid X_{n+1} > t,\ \tau - d\tau < T_n < \tau]$$
$$= P[N(\tau + t + r) = n \mid N(\tau + t) = n,\ N(\tau) = n,\ N(\tau - d\tau) = n - 1]$$
$$= P[N(\tau + t + r) = n \mid N(\tau + t) = n]$$
$$= P[X_{n+1} > r]$$

so X_{n+1} must be exponential. Also X_{n+1} must have mean $1/\lambda_n$ in order to satisfy the infinitesimal relation

$$P[N(t + dt) - N(t) = 1 \mid N(t) = n]$$
$$= P[N(t + dt) - N(t) = 1 \mid N(t) = n,\ N(t - dt) = n - 1]$$
$$= P[0 < X_{n+1} < dt \mid t - dt < T_n < t]$$
$$= P[0 < X_{n+1} < dt] = \lambda_n\, dt.$$

Thus the interarrival times for such a process have been shown to be independent and exponential, as was asserted.

PERMISSIBLE BIRTH-RATE SEQUENCES

We next characterize those sequences which can arise as birth-rate sequences. If we are given a sequence $\lambda_0, \lambda_1, \ldots$ of positive numbers, we may attempt to construct a corresponding process. Let $X_1, X_2, \ldots$ be independent with X_{n+1} exponential with mean $1/\lambda_n$. Let $T_n = X_1 + \cdots + X_n$ so that the infinite sequence $T_1, T_2, \ldots$ of arrival times is well defined and a realization from this sequence consists of an infinite sequence $0 < t_1 < t_2 < \cdots$. Now for each $t > 0$ let N_t be the number of arrivals up to and including time t. That is, N_t is zero on the event $[X_1 > t]$ and has the value n on the event $[T_n \le t < T_{n+1}]$. For this definition to make sense there must, with probability 1, be only finitely many arrivals prior to time t; otherwise N_t is undefined on an event having positive probability.

If we assume that $ET_n = \sum_{k=0}^{n-1} 1/\lambda_k$ goes to infinity as n does then our construction can be carried out, since we can use Lemma 1 from the Addendum for the second inequality below to obtain

$$P[\text{finitely many arrivals prior to time } t]$$

$$\geq P[T_n > t] \geq 1 - 2\sqrt{\frac{t}{ET_n}} \to 1 \qquad \text{as } n \to \infty.$$

Thus if $\sum_{n=0}^{\infty} 1/\lambda_n = \infty$ then we can construct a process N_t for $t \geq 0$ having unit-counting-function realizations and independent interarrival times $X_1, X_2, \ldots,$ with X_{n+1} being exponential with mean $1/\lambda_n$. We shall see shortly that this process is a pure-birth Markov process with this birth-rate sequence.

Exercise 2 shows that a sequence $\lambda_0, \lambda_1, \ldots$ cannot be a birth-rate sequence if it has a finite sum. As an example of this phenomenon let our arrival-time sequence $T_1, T_2, \ldots$ be constructed as above with $\lambda_n = 2^n$ so that

$$ET_n = \sum_{k=0}^{n-1} \frac{1}{\lambda_k} < \sum_{k=0}^{\infty} \frac{1}{\lambda_k} = 2 \qquad \text{for all } n = 1, 2, \ldots.$$

Hence successive arrivals usually occur *extremely* close to each other. It should not be surprising that, as shown in the aforementioned exercise, the probability is at least $\frac{2}{3}$ that *all* arrivals occur before time 4. Thus if N_4 is the number of arrivals prior to time 4 then $\frac{2}{3} \leq P[N_4 = \infty] < 1$; however, we do not permit a random variable to take infinity for a value. Of course, N_4 can take finite values, since $P[N_t = 0] = P[T_1 > t] > 0$ for every t, so the first birth may be very late.

Exercise 2 may be done before reading the remainder of this section.

THE CONSTRUCTED PROCESS IS A PURE-BIRTH MARKOV PROCESS

We next show that if $\sum_{n=0}^{\infty} 1/\lambda_n = \infty$ then N_t for $t \geq 0$ is Markov if it is constructed as above with independent exponential interarrival times. Let L and R be the indicated events to the left and right of time t in Fig. 1, so that

$$L = [t_i < T_i < t_i + dt_i \text{ for } i = 1, \ldots, n] \cap [N_t = n]$$

and R is the event that the next two births after time t occur at about s_1 and $s_1 + s_2$ time units after t. For simplicity we consider only the first two births after t. Clearly

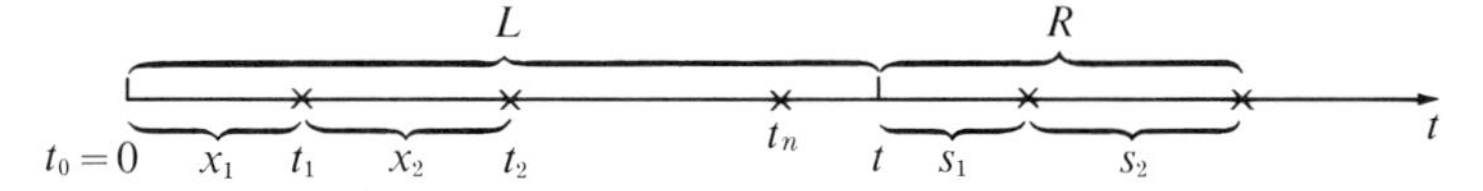

Fig. 1

$L \cap R$ is the event that the first $n + 2$ births occur as in Fig. 1, but we have independent exponential interarrival times, so $P(L \cap R)$ equals

$$\left(\prod_{i=1}^{n} e^{-\lambda_{i-1}(t_i - t_{i-1})}\lambda_{i-1}\, dt_i\right) e^{-\lambda_n(t-t_n+s_1)}\lambda_n\, ds_1\, e^{-\lambda_{n+1}s_2}\lambda_{n+1}\, ds_2.$$

In the definition of L we could replace $[N_t = n]$ by $[X_{n+1} > t - t_n]$, where $P[X_{n+1} > t - t_n] = e^{-\lambda_n(t-t_n)}$, so the above expression can be written as

$$P(L \cap R) = P(L)\{e^{-\lambda_n s_1}\lambda_n\, ds_1\, e^{-\lambda_{n+1}s_2}\lambda_{n+1}\, ds_2\}. \tag{1}$$

This basic equation implies that $P[R \mid L]$ equals the term in $\{\ \}$. Therefore if we are given all the arrival times prior to time t and leading to the event $[N_t = n]$ then the conditional joint distribution of s_1 and s_2 is just the unconditional distribution of X_{n+1} and X_{n+2}. *Thus the future evolution, given the past, is quite transparent.*

To deduce the Markov property explicitly we let

$$A = [N_{h_1} = k_1, \ldots, N_{h_m} = k_m, N_t = n]$$

where $0 < h_1 < \cdots < h_m < t$, $0 \le k_1 \le \cdots \le k_m \le n$, and $m \ge 0$. Summing, or integrating, equation (1) over all disjoint events L contained in A yields

$$P(A \cap R) = P(A)\{\ \}.$$

Hence $P(R \mid A) = \{\ \}$ independently of t, where the term in braces is as in equation (1), for any such A, including $A = [N_t = n]$. Summing $P(R \mid A) = P(R \mid N_t = n) = \{\ \}$ over all disjoint R contained in $[N_t = n, N_{t+r} = n + 1]$ yields

$$P[N_{t+r} = n + 1 \mid A] = P[N_{t+r} = n + 1 \mid N_t = n]$$

where neither expression depends on t. This proof obviously generalizes to yield the Markov property. Furthermore from the fact that $P[R \mid N_t = n]$ equals the term in braces we see that if we are given $[N_t = n]$ then the residual waiting time to the next birth is exponential with mean $1/\lambda_n$, so we have the desired infinitesimal relations. Thus we have shown that our process, as constructed from independent exponential interarrival times, is the desired Markov pure-birth process.

THE DISTRIBUTION OF N_t

We next derive the basic *recursion relation*

$$P[N_t = n] = \int_0^t e^{-(t-\tau)\lambda_n} P[N_\tau = n - 1]\lambda_{n-1}\, d\tau \qquad \text{for all } t \ge 0, n \ge 1.$$

The Markov property yields

$$P[\tau < T_n < \tau + d\tau] = P[N_\tau = n - 1]\lambda_{n-1}\, d\tau$$

but $P[N_t = n]$ is the integral with respect to $d\tau$, over the range from 0 to t, of

$$P[X_{n+1} > t - \tau, \tau < T_n < \tau + d\tau] = P[X_{n+1} > t - \tau]P[\tau < T_n < \tau + d\tau]$$

which proves the recursion relation.

If we are given the birth-rate sequence we can substitute

$$P[N_\tau = 0] = P[X_1 > \tau] = e^{-\tau\lambda_0}$$

into the integrand of the recursion relation and "perform the integration" to obtain $P[N_t = 1]$ for all $t \geq 0$. We substitute this into the integrand to obtain $P[N_t = 2]$ for all $t \geq 0$, etc. We thus obtain $P[N_t = n]$ for all n and t.

Example 1 If there are $\nu \geq 1$ initiators and N_t is the number of births between times 0 and t then $M_t = \nu + N_t$ is the total population size at time t, so for a linear birth rate we let

$$\begin{aligned}(\nu + n)\lambda\, dt &= P[M_{t+dt} - M_t = 1 \mid M_t = \nu + n] \\ &= P[N_{t+dt} - N_t = 1 \mid N_t = n] = \lambda_n\, dt\end{aligned}$$

and hence we define $\lambda_n = (n + \nu)\lambda$ for $n = 0, 1, \ldots$, where $\lambda > 0$.

In particular for $\nu = 1$ initiator and a linear birth rate we have $\lambda_n = (n + 1)\lambda$ for $n = 0, 1, \ldots$, so the recursion relation yields

$$P[N_t = 1] = e^{-2\lambda t} \int_0^t e^{2\lambda\tau} e^{-\lambda\tau} \lambda\, d\tau = e^{-2\lambda t}(e^{\lambda t} - 1) = e^{-\lambda t}(1 - e^{-\lambda t}).$$

A similar result holds for $P[N_t = 2]$, and we quickly guess that

$$P[N_t = n] = e^{-\lambda t}(1 - e^{-\lambda t})^n \qquad \text{for } t > 0,\ n = 0, 1, \ldots.$$

For a proof by induction we assume the formula for some $n - 1 \geq 0$, and the recursion relation yields

$$e^{\lambda t(n+1)} P[N_t = n] = \int_0^t n\lambda e^{\lambda\tau}(e^{\lambda\tau} - 1)\, d\tau.$$

But then the integral must equal $f(t) = (e^{\lambda t} - 1)^n$, since $f(0) = 0$ and f' equals the integrand. The resulting equation yields the desired formula for $P[N_t = n]$, which completes the proof. Of course, $P[M_t = n + 1] = P[N_t = n]$. ■

Example 2 We exhibit a revealing alternate experiment equivalent to the original experiment of observing the one initiator $\nu = 1$ linear birth-rate process "N_t for $t \geq 0$ with $N_0 = 0$" on the interval between 0 and t, where $t > 0$ is fixed. The alternate experiment may be described as follows. Let $Z_1, Z_2, \ldots$ be independent and each have density

$$p(z) = \lambda e^{-\lambda(t-z)} \qquad \text{for } z < t.$$

As in Fig. 2, this is the exponential density starting at t and going backward. We place a cross at Z_1 in the interval from 0 to t, then a cross at Z_2 in the interval from 0 to t, etc. until

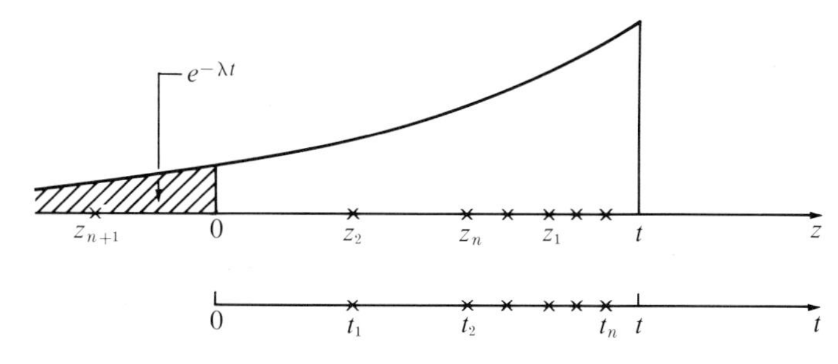

Fig. 2 Density used for alternate experiment.

we first obtain a negative number, at which trial the experiment stops, and the negative number is ignored.

Take any $0 < t_1 < \cdots < t_n < t$ and let A be the event that the original experiment yields exactly n successes, with the kth success between t_k and $t_k + dt_k$. Let B be the event satisfying the same description, but for this alternate experiment. We must show that $P(A) = P(B)$. For the original experiment the interarrival times are independent and the kth has density $k\lambda e^{-k\lambda x}$. Therefore if $t_0 = 0$ then

$$P(A) = \left(\prod_{k=1}^{n} k\lambda e^{-k\lambda(t_k - t_{k-1})}\, dt_k\right) e^{-(n+1)\lambda(t-t_n)} = e^{-\lambda t}\left\{n! \prod_{k=1}^{n} \lambda e^{-\lambda(t-t_k)}\, dt_k\right\}$$

where we used the easily checked equation

$$(t_1 - t_0) + 2(t_2 - t_1) + 3(t_3 - t_2) + \cdots + n(t_n - t_{n-1}) - (n+1)t_n = -\sum_{k=1}^{n} t_k$$

But then clearly $P(A) = P(B)$, because, roughly speaking, the term in braces above is the probability that $(Z_1, \ldots, Z_n)$ equals one of the $n!$ permutations of $(t_1, \ldots, t_n)$ while $e^{-\lambda t} = P[Z_{n+1} < 0]$.

This alternate experiment is quite transparent, and both experiments assign the same probability to every configuration of successes. In particular

$$P[0 < Z_1, 0 < Z_2, \ldots, 0 < Z_n, Z_{n+1} < 0] = (1 - e^{-\lambda t})^n e^{-\lambda t}$$

is the probability that the alternate experiment yields exactly n successes, and this, of course, equals $P[N_t = n]$, as found in Example 1. Thus this geometric density can be interpreted as arising from Bernoulli trials in the same way in which the geometric density was originally introduced in Sec. 8-4. ■

ADDENDUM

This addendum is devoted to the statement and proof of the following lemma, with a few comments preceding the proof.

Lemma 1 If $T = X_1 + \cdots + X_n$ is the sum of some finite number of independent exponentially distributed random variables, which need not have the same density, then

$$P[T \le t] \le 2\sqrt{\frac{t}{ET}} \qquad \text{for all } t > 0. \quad ■$$

If we know that some nonnegative random variable $T \geq 0$ has a large expectation $ET \geq 10^6$ then we would usually anticipate that $P[T \leq 1]$ will be small. Of course this need not be true, even if T were assumed to be the sum of a large number of independent and identically distributed random variables $T = X_1 + \cdots + X_n$. For example, $P[T = 0] = .9$ if $n = 10^6$ and the density for X_1 is concentrated on the two points 0 and α, with

$$P[X_1 = 0] = (.9)^{10^{-6}}$$

and $\alpha > 0$ is so large that $EX_1 \geq 1$. However, the lemma permits the desired conclusion if each X_i is assumed to have an exponential density, since then, as shown by the proof, either some one X_j has a large enough mean to assure the result or else n is large enough to apply the weak law of large numbers. An advantage of formalizing the result as above is that in the next section it is desirable to have such a bound which depends on the particular sequence $EX_1, \ldots, EX_n$ of means only through their sum, since the particular sequence will depend on the random birth-death sequence.

Proof of Lemma 1 If X_i has mean μ_i we let m be the largest of the means $\mu_1, \ldots, \mu_n$. Then for some j we know that X_j has density $(1/m)e^{-x/m}$ for $x > 0$. This density is always less than $1/m$, yielding the second inequality in

$$P[T < t] \le P[X_j < t] \le \frac{t}{m}.$$

If $m \ge (\frac{1}{2})\sqrt{tET}$ the conclusion of the lemma follows. This takes care of the case when at least one of the X_i has a large mean.

We now assume that $m \le (\frac{1}{2})\sqrt{tET}$. In this case

$$\text{Var } T = \sum_{i=1}^{n} \mu_i^2 \le \sum_{i=1}^{n} \mu_i m = mET \le \frac{ET}{2}\sqrt{tET}$$

which is used for the last inequality below. For the first inequality below we assume that $t \le ET/2$, since otherwise the bound of the lemma exceeds 1. Chebychev's inequality then yields

$$P[T \le t] \le P\left[T \le \frac{ET}{2}\right] \le P\left[|T - ET| \ge \frac{ET}{2}\right] \le \frac{\text{Var } T}{(ET/2)^2} \le 2\sqrt{\frac{t}{ET}} \quad \blacksquare$$

EXERCISES

1. For the one-initiator linear birth rate $\lambda_n = (n+1)\lambda$ Markov pure-birth process, as considered in Examples 1 and 2, show that if $0 < s < t$ then the distribution of the number of births between s and t is given by

$$P[N_t - N_s = n] = \frac{e^{-\lambda t}(1 - e^{-\lambda(t-s)})^n}{(1 - e^{-\lambda(t-s)} + e^{-\lambda t})^{n+1}} \qquad \text{for } n = 0, 1, \ldots.$$

2. Let $T_n = X_1 + \cdots + X_n$ where $X_1, X_2, \ldots$ are independent and X_{n+1} is exponential with mean $1/\lambda_n$ for $n = 0, 1, \ldots$. Assume that $\mu = \sum_{n=0}^{\infty} 1/\lambda_n$ is finite. Prove that the probability is 0 that $T_1, T_2, \ldots$ will converge to infinity. That is, with probability 1 the sequence of successive arrival times must converge to a *finite* limit. This limit depends, of course, on the realization, and this limit may be large, since $P[T_1 > r]$ is positive for every real r.

HINT: Clearly $\sigma^2 = \sum_{n=0}^{\infty} 1/\lambda_n^2$ is finite, since $1/\lambda_n$ converges to zero; hence $1/\lambda_n^2 \le 1/\lambda_n \le 1$ for all large enough n. For real r let

$$B_r = [T_n < r \text{ for all } n = 1, 2, \ldots].$$

It is sufficient to show that

$$P(B_r) \ge 1 - \frac{\sigma^2}{(r - \mu)^2} \qquad \text{for all } r > \mu$$

since $B_1, B_2, \ldots$ is an increasing sequence of events, and hence

$$P\left(\bigcup_{r=1}^{\infty} B_r\right) = \lim_{r\to\infty} P(B_r) \geq \lim_{r\to\infty} \left\{1 - \frac{\sigma^2}{(r-\mu)^2}\right\} = 1.$$

3. Let N_t for $t \geq 0$ with $N_0 = 0$ be a linear birth-rate $\lambda_n = (n + \nu)\lambda$ Markov pure-birth process with $\nu \geq 1$ initiators, so that $M_t = \nu + N_t$ is the total population size at time t.
(*a*) Use the recursion formula to prove that

$$P[N_t = n] = \binom{\nu + n - 1}{n}(e^{-\lambda t})^{\nu}(1 - e^{-\lambda t})^n \qquad \text{for all } \nu \geq 1, n \geq 0, t > 0.$$

Thus N_t has a Pascal density with parameters ν and $e^{-\lambda t}$. This is reasonable from the earlier result for $\nu = 1$, since each of the ν initiators generates an independent geometrically distributed number of offspring. In particular

$$EN_t = \nu\frac{1 - e^{-\lambda t}}{e^{-\lambda t}} = \nu(e^{\lambda t} - 1)$$

$$\text{Var } N_t = \nu\frac{1 - e^{-\lambda t}}{e^{-2\lambda t}} = \nu e^{\lambda t}(e^{\lambda t} - 1)$$

(*b*) Generalize the alternate experiment of Example 2 to an arbitrary number $\nu \geq 1$ of initiators. Show that the alternate experiment is equivalent to the original experiment, and that the alternate experiment makes part (*a*) obvious.
(*c*) Let $H_\tau = N_{m(\tau)}$ where $m(0) = 0$, $m'(\tau) > 0$ for all $\tau \geq 0$, and $m(\tau)$ goes to infinity as τ does. Thus H_τ for $\tau \geq 0$ is a nonhomogeneous linear birth-rate pure-birth Markov process. Exhibit the infinitesimal relations for "H_τ for $\tau \geq 0$" and find $P[H_\tau = n]$, EH_τ, and Var H_τ.

23-7 BIRTH-AND-DEATH PROCESSES

This section contains a brief introduction to the extensive theory of continuous-time Markov birth-and-death processes. It is assumed that no multiple births or deaths occur, so that, as in Fig. 1, all realizations of "N_t for $t \geq 0$ with $N_0 = 0$" are jump functions, with every jump being either $+1$ for a birth or -1 for a death. We still call the jump points the arrival times, so each arrival is a birth or death. If ν is the

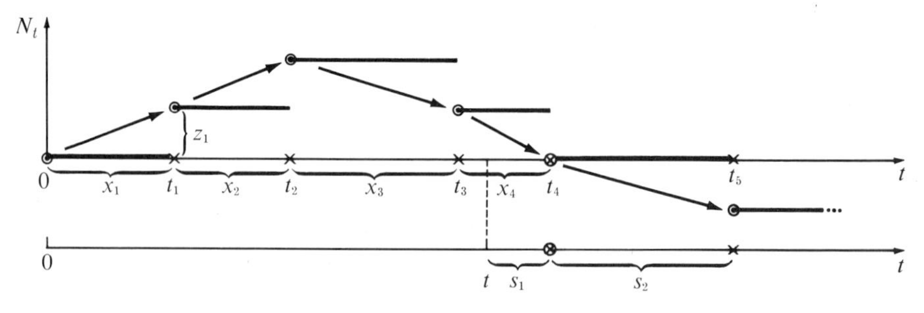

Fig. 1 A realization from a birth and death process.

population size at time 0 then $M_t = \nu + N_t$ is the population size at time t; hence N_t is interpreted as the increase in population during the first t time units. In general we do not rule out processes for which N_t may take arbitrarily large negative values, since, for example, we might assume an infinite initial population, or we might interpret N_t for $t \geq 0$ as a random walk. That is, a man, or a molecule, starts at the origin of the real line at time 0, and then at random times the man takes a unit step in the plus or minus direction, so that N_t is his location on the real line at time $t \geq 0$.

Given a set of parameters, we will define a birth-and-death process by using *conditionally* exponentially distributed interarrival times, in such a way that our construction reduces to the construction of the preceding section when no deaths are permitted. We assume that the parameters satisfy a certain condition which is sufficient, but not necessary, to guarantee that every finite interval has only finitely many arrivals, and then show that we do indeed have a Markov process. Our derived infinitesimal relations are often used as basic initial assumptions, leading to differential equations whose solutions are then used to construct a Markov process which can then be shown to have the structure we use as our definition. The modest goal of this section is to describe the structure of such processes, as exhibited in our definition.

Let $0 \leq p_n \leq 1$ and $\beta_n > 0$ for $n = 0, \pm 1, \pm 2, \ldots$ be two doubly infinite sequences of real numbers, and for convenience let $q_n = 1 - p_n$. Assume that

$$\sum_{n=0}^{\infty} \frac{1}{\beta_n} = \infty \qquad \text{and} \qquad \sum_{n=0}^{\infty} \frac{1}{\beta_{-n}} = \infty.$$

We will construct a *birth-and-death Markov process* "N_t for $t \geq 0$ with $N_0 = 0$" in terms of these parameters.

INTERPRETATIONS FOR THE PARAMETERS

If we are given $[N_t = n]$, so that our *state at time t is n*, then p_n is the probability that the next arrival is a birth, $\beta_n\,dt$ is the probability that there will be an arrival between t and $t + dt$, and $p_n\beta_n\,dt$ is the probability that there is a birth between t and $t + dt$. Thus if we are given $[N_t = n]$ then the waiting time to the next arrival, as well as the chance for its being a birth, may depend on the present population size n, but not on the time t, or on any additional information concerning the number, times, and types (birth or death) of arrivals to date. Note that the subscript n in p_n and β_n need not have a close relationship to the number of arrivals to date, which in some cases might be much larger than n. If $1 = p_0 = p_1 = \cdots$ then we have a pure-birth process, so if $[N_t = n]$ then n is both the state at time t and the number of arrivals to date, in which case β_n is the λ_n of the preceding section; and we need not introduce parameters p_n and β_n for $n < 0$, since these states will never be reached.

Traditionally a state ν is called a **reflecting barrier** if $\nu \geq 0$ and $p_\nu = 0$, or if $\nu \leq 0$ and $q_\nu = 0$. If $\beta_\nu = 0$ then ν is called an **absorbing state.** However, in our development we insist that all $\beta_\nu > 0$. In order to obtain an absorbing state ν we may then introduce a new process which agrees with the original process up to the first time, if any, that the original process reaches state ν, after which time the new process forever remains in state ν.

For each n we will have the following three infinitesimal relations, with the third relation being obtained by adding the first two:

$$P[N_{t+dt} - N_t = 1 \mid N_t = n] = p_n\beta_n \, dt = \lambda_n \, dt$$

$$P[N_{t+dt} - N_t = -1 \mid N_t = n] = q_n\beta_n \, dt = \mu_n \, dt$$

$$P[|N_{t+dt} - N_t| = 1 \mid N_t = n] = \beta_n \, dt = (\lambda_n + \mu_n) \, dt$$

The parameters λ_n and μ_n as they appear above are traditional, and in terms of these as initial data, we could introduce $\beta_n = \lambda_n + \mu_n > 0$, $p_n = \lambda_n/\beta_n$, and $q_n = \mu_n/\beta_n$.

CONSTRUCTION OF THE PROCESS

Given the p_n and β_n parameters, we proceed to construct a corresponding process. Let $0 = X_0 = T_0 = Z_0$, and for $k = 1, 2, \ldots$ interpret X_k, T_k, and Z_k as in Fig. 1, so that X_k is the kth interarrival time, T_k is the time of the kth arrival, and Z_k is the state immediately after the kth arrival. The arrows between circled points indicate the transitions, for $k = 1, 2, \ldots$, between the point (T_{k-1}, Z_{k-1}) of plane to the point (T_k, Z_k). These transitions are completely determined by all the transitions (X_{k-1}, Z_{k-1}) to (X_k, Z_k), which are the transitions which we now describe.

We assume that *if we are given* $[Z_{k-1} = n]$ then X_k and Z_k are independent, with X_k being exponential with mean $1/\beta_n$ while Z_k may take the value $n + 1$ or $n - 1$ with probability p_n or q_n, respectively. *Furthermore*, this conditional distribution is unchanged if we are also given additional information concerning $X_1, \ldots, X_{k-1}$ and $Z_1, \ldots, Z_{k-2}$. Therefore if we are given the complete evolution of the process up to (t_{k-1}, z_{k-1}) then the conditional distribution of (X_k, Z_k) depends on this evolution only through the value z_{k-1} as described, so we select the waiting time and type of the next arrival independently of each other.

Proceeding more formally, we define $(0,0) = (X_0, Z_0), (X_1, Z_1), \ldots$ to be a Markov sequence with transition distributions

$$P[X_k > x, Z_k = n + 1 \mid X_{k-1} = x_{k-1}, Z_{k-1} = n] = p_n e^{-x\beta_n}$$

$$P[X_k > x, Z_k = n - 1 \mid X_{k-1} = x_{k-1}, Z_{k-1} = n] = q_n e^{-x\beta_n}$$

independent of $k \geq 1$ and $x_{k-1} > 0$. Naturally $x \geq 0$ while n is an integer. Introducing the function

$$b(n + 1 \mid n) = p_n \qquad b(n - 1 \mid n) = q_n \qquad b(z \mid n) = 0 \qquad \text{if } |z - n| \neq 1$$

to describe the population size transitions, we may write this as

$$P[X_k > x_k, Z_k = z_k \mid X_{k-1} = x_{k-1}, Z_{k-1} = z_{k-1}] = b(z_k \mid z_{k-1}) e^{-x_k \beta_{z_{k-1}}}.$$

For an alternate but equivalent definition we may define for each $m \geq 1$ the joint distribution of $X_1, \ldots, X_m, Z_1, \ldots, Z_m$ by

$$P[X_k > x_k, Z_k = z_k \text{ for } k = 1, \ldots, m] = \prod_{k=1}^{m} b(z_k \mid z_{k-1}) e^{-x_k \beta_{z_{k-1}}}$$

where $z_0 = 0$. Dividing this by

$$P[Z_1 = z_1, \ldots, Z_m = z_m] = \prod_{k=1}^{m} b(z_k \mid z_{k-1})$$

which was obtained by letting all $x_k = 0$, we see that the joint distribution may be described as follows. The population size sequence $0 = Z_0, Z_1, \ldots$ is a Markov sequence with transitions according to the function b, which does not depend on the β_n parameters, and *if* we are given $[Z_1 = z_1, \ldots, Z_m = z_m]$ *then* $X_1, \ldots, X_m$ are independent, with X_k being exponential with mean $1/\beta_{z_{k-1}}$. In infinitesimal form this joint distribution becomes

$$P[x_k < X_k < x_k + dx_k, Z_k = z_k \text{ for } k = 1, \ldots, m]$$

$$= \prod_{k=1}^{m} b(z_k \mid z_{k-1})\beta_{z_{k-1}}\, dx_k e^{-x_k\beta_{z_{k-1}}}$$

and dividing this by the same expression for $m - 1$ shows that our alternate definition makes (X_0,Z_0), (X_1,Z_1), ... a Markov sequence having the transition distributions used in the earlier definition.

Example 1 We pause in our construction to note in this example that some questions naturally involve only the sequence $0 = Z_0, Z_1, Z_2, \ldots$. Let the type of arrival be *completely* unaffected by the population increase, so that $Z_n = Y_1 + \cdots + Y_n$ where $Y_1, Y_2, \ldots$ are independent and each has distribution

$$P[Y_1 = 1] = p \qquad P[Y_1 = -1] = q \qquad p + q = 1.$$

Clearly Z_n may take the value $-n$ if $q > 0$. We interpret Z_n as the population increase due to the first n arrivals. Let $B = \bigcup_{n=1}^{\infty} [Z_n = -1]$ be the event that the *initial* population size is decreased by one unit at least once during the whole evolution. Exercise 7*b* (Sec. 14-3) shows that $P(B) = 1$ if $p \le \frac{1}{2}$ while $P(B) = q/p$ if $p > \frac{1}{2}$. Thus if $p = \frac{3}{4}$ then the probability is $\frac{1}{3}$ that at some time we will have fewer people than we had initially. Naturally, then, $P(B)$ equals the probability of eventual extinction for a one-initiator population whose births and deaths occur as above as long as the population increase Z_n is nonnegative, but becomes extinct if some Z_n ever equals -1. ■

The sequence $(T_k,Z_k) = (X_1 + \cdots + X_k, Z_k)$ for $k = 0, 1, \ldots$ is now well defined, and realizations can be imagined to be plotted as in Fig. 1. For each $t \ge 0$ we naturally define $N_t = Z_i$ on the event $[T_i \le t < T_{i+1}]$, as indicated by Fig. 1. That is, for any sample point ω we find the unique integer i such that $T_i(\omega) \le t < T_{i+1}(\omega)$ and then let $N_t(\omega) = Z_i(\omega)$. If $T_k(\omega)$ goes to infinity as k does then this definition makes sense for this ω and *every* $t \ge 0$. In the next paragraph the set of these ω is shown to have probability 1 if

$$\sum_{n=0}^{\infty} \frac{1}{\beta_n} = \infty = \sum_{n=0}^{\infty} \frac{1}{\beta_{-n}}$$

so that the birth-and-death process "N_t for $t \ge 0$ with $N_0 = 0$" has now been defined. This sufficient restriction for constructability is a restriction on only the β_n, so that if it is satisfied we may use any p_n. If this condition is violated then, as can be seen from Exercise 2 (Sec. 23-6), there are parameters p_n such that the construction cannot be carried out (for example, use $p_n = 1$ for all $n \geq 0$, or use $p_n = 0$ for all $n \leq 0$) although for others (for example, $p_0 = 1$ and $p_5 = 0$) it can. We do not consider more general conditions, on the *two* parameter sequences, which guarantee constructability.

PROOF THAT $T_k \to \infty$

Let us now show that the probability is zero for the event that T_k does not converge to infinity. This event equals $\bigcup_{t=1}^{\infty} [T_k \leq t \text{ for all } k = 1, 2, \ldots]$, so we need only show that $P[T_k \leq t \text{ for all } k = 1, 2, \ldots] = 0$, for each t. But $[T_k \leq t \text{ for all } k = 1, 2, \ldots]$ is, for each k, contained in $[T_k \leq t]$, and hence we need only show that $P[T_k \leq t]$ goes to zero as k goes to infinity. The conditional density of T_k, given

$$[Z_0 = z_0, \ldots, Z_{k-1} = z_{k-1}],$$

has expectation

$$\left\{\frac{1}{\beta_{z_0}} + \cdots + \frac{1}{\beta_{z_{k-1}}}\right\} \geq a_k$$

where a_k is defined to be the minimum of $\{\ \}$ over all permissible $(z_0, \ldots, z_{k-1})$, namely, those $0 = z_0, \ldots, z_{k-1}$ for which each $|z_i - z_{i-1}| = 1$, since $P[Z_0 = z_0, \ldots, Z_{k-1} = z_{k-1}]$ may be positive only for such $(z_0, \ldots, z_{k-1})$. Lemma 1 in the Addendum to Sec. 23-6 yields the inequality below,

$$P[T_k \leq t] = \sum P[T_k \leq t \mid Z_0 = z_0, \ldots, Z_{k-1} = z_{k-1}]P[Z_0 = z_0, \ldots, Z_{k-1} = z_{k-1}]$$

$$\leq \sum 2\sqrt{\frac{t}{a_k}}\, P[Z_0 = z_0, \ldots, Z_{k-1} = z_{k-1}] = 2\sqrt{\frac{t}{a_k}}$$

where the summation is over all permissible $(z_0, \ldots, z_{k-1})$. Thus we need only show that if $\sum_{n=0}^{\infty} (1/\beta_n) = \infty = \sum_{n=0}^{\infty} (1/\beta_{-n})$ then a_k goes to infinity as k does. If, for example, larger populations result, on the average, in closer births,

$$\cdots \geq \frac{1}{\beta_{-1}} \geq \frac{1}{\beta_0} \geq \frac{1}{\beta_1} \geq \cdots$$

then clearly $(z_0, \ldots, z_{k-1}) = (0, 1, \ldots, k-1)$ yields the minimum $a_k = \sum_{n=0}^{k-1} 1/\beta_n$, which does go to infinity. The general case is left to Exercise 2.

THE CONSTRUCTED PROCESS IS MARKOV

We now give a crude but fairly transparent proof of the fact that all our constructed birth-and-death processes N_t for $t \geq 0$ are Markov processes. Certain events are made specific in order to avoid notational complications. We fix $t > 0$ and let L be any event which completely describes the evolution of the process up to time t, and for which $[N_t = 1]$ occurs. For example, as in Fig. 1, we may let L be the event that there are precisely three arrivals prior to t, with the ith arrival between t_i and $t_i + dx_i$ for $i = 1, 2, 3$, and these are birth, birth, and death, respectively, in which case

$$P(L) = (p_0 e^{-\beta_0 x_1}\beta_0\, dx_1)(p_1 e^{-\beta_1 x_2}\beta_1\, dx_2)(q_2 e^{-\beta_2 x_3}\beta_2\, dx_3)e^{-\beta_1(t-t_3)}.$$

Let R be the event that the next two arrivals after time t are deaths at the times shown in Fig. 1. So that the description of R will not involve the number of arrivals prior to time t, we introduce s_1 and s_2 as in Fig. 1. Using the obvious but fundamental identity

$$e^{-\beta_1(t-t_3+s_1)} = e^{-\beta_1(t-t_3)}e^{-\beta_1 s_1}$$

we see that

$$P(L \cap R) = P(L)\{(q_1 e^{-\beta_1 s_1}\beta_1\, ds_1)(q_0 e^{-\beta_0 s_2}\beta_0\, ds_2)\}. \tag{1}$$

Clearly the fact that $P(R \mid L)$ equals the term in braces has an interpretation analogous to the interpretation made in the preceding section. Summing, or integrating, equation (1) over all such disjoint L yields

$$P[N_t = 1, R] = P[N_t = 1]\{\ \}$$

and hence

$$P[R \mid N_t = 1] = \{\ \}$$

where the term in braces is as in equation (1). Similarly if $0 < s < t$ then summing equation (1) over all disjoint L contained in $[N_s = 5, N_t = 1]$ yields

$$P[N_s = 5, N_t = 1, R] = P[N_s = 5, N_t = 1]\{\ \}.$$

Therefore for this and still more general R, we have

$$P[R \mid N_t = 1, N_s = 5] = P[R \mid N_t = 1].$$

If $h > 0$ and we sum this last equation over all R contained in $[N_{t+h} - N_t = 1]$ then we obtain

$$P[N_{t+h} = 2 \mid N_t = 1, N_s = 5] = P[N_{t+h} = 2 \mid N_t = 1].$$

These calculations clearly generalize, so we have shown that any N_t for $t \geq 0$, constructed as above, is a Markov process.

We may imitate the foregoing, using for R the event that the first arrival after time t is a death between $t + s$ and $t + s + ds$, in which case we obtain

$$P[R \mid N_t = n] = q_n\beta_n e^{-\beta_n s}\, ds \qquad \text{for all } s > 0.$$

Thus if we are given $[N_t = n]$ then the residual waiting time to the next arrival is exponential with mean $1/\beta_n$ and the type of arrival is independent of this waiting time and has distribution p_n, q_n. In particular, this result immediately implies the previously stated infinitesimal relations.

EXERCISES

1. *Expected Population Size for Linear Growth with Immigration.* Let M_t be the total population size at time $t \geq 0$ for a Markov birth-and-death process having initial size

$M_0 = \nu \geq 0$. For real $\lambda \geq 0$, $\mu \geq 0$, and $\beta_0 > 0$ let the infinitesimal relations be

$$P[M_{t+dt} - M_t = 1 \mid M_t = n] = (n\lambda + \beta_0)\,dt$$

$$P[M_{t+dt} - M_t = -1 \mid M_t = n] = (n\mu)\,dt \qquad \text{for } n = 0, 1, 2, \ldots.$$

Note that $\nu \geq 0$, and if $n = 0$ then the next arrival must be a birth, so that $M_t \geq 0$ for all t.

(*a*) Let $m(t) = EM_t$ and show that $m(t)$ must satisfy the differential equation

$$\frac{dm(t)}{dt} = \beta_0 + (\lambda - \mu)m(t) \qquad \text{with } m(0) = \nu$$

HINT: Express $P[M_{t+dt} = n]$ in terms of the three events $M_t = n - 1, n, n + 1$, then multiply by n and form $\sum_{n=1}^{\infty}$.

(*b*) Check that the solution is given by

$$m(t) = \begin{cases} \beta_0 t + \nu & \text{if } \lambda = \mu \\ \nu e^{(\lambda-\mu)t} + \dfrac{\beta_0}{\lambda - \mu}(e^{(\lambda-\mu)t} - 1) & \text{if } \lambda \neq \mu \end{cases}$$

and note that $m(t) > 0$ if $t > 0$, for all cases.

(*c*) A new lake starts without any fish $\nu = 0$ but receives $\beta_0 = 1$ fish per hour from incoming streams. Evaluate $m(t)$ for $t = 10^4$ hours, or about 1 year, for the three values 0, 10^{-3}, -10^{-3} of the parameter $\lambda - \mu$, where $\lambda - \mu$ clearly reflects the favorableness of the impact of a growing fish population on its environment. Which of the three values permits an eventually stable population?

2. *Completing the Proof That $P[T_k \leq t]$ Converges to Zero as k goes to Infinity.* Let β_n for $n = 0, \pm 1, \pm 2, \ldots$ be any doubly infinite sequence of positive real numbers such that $\sum_{n=0}^{\infty} 1/\beta_n = \sum_{n=0}^{\infty} 1/\beta_{-n} = \infty$, and let a_k be the minimum of

$$\frac{1}{\beta_{z_0}} + \frac{1}{\beta_{z_1}} + \cdots + \frac{1}{\beta_{z_{k-1}}}$$

over all permissible $(z_0, \ldots, z_{k-1})$, namely, those for which $z_0 = 0$ and $|z_i - z_{i-1}| = 1$ for all $i = 1, \ldots, k - 1$. Prove that a_k converges to infinity as k does.

It should be kept in mind that the $1/\beta_n$ may be very irregular. For example, we might have

n	$\cdots$	-5	-4	-3	-2	-1	0	1	2	3	4	$\cdots$
$\dfrac{1}{\beta_n}$	$\cdots$	1	10^{-4}	5	10	100	1	10	10^{-3}	10^5	10^{-6}	

24
The Wiener Process

In Sec. 24-1 a sequence of approximating experiments, and a sampling experiment, are used to suggest the definition and nature of the Wiener process. After this intuitive introduction, Sec. 24-2 is devoted to some of the simpler properties of a Wiener process. Sec. 24-3 introduces covariance functions and normal processes, and the Wiener process is transformed to exhibit a stationary normal process. This section has been kept brief, partly because there are so many books devoted to its extensions and applications in such fields as communication and control theory, economics, etc. Section 24-3 does not presume Sec. 24-2.

24-1 INTRODUCTION

We now describe a sequence of approximating experiments which suggest the definition and properties of a Wiener process. We fix the parameter $\sigma^2 > 0$, and for each positive integer m we define an experiment. The Wiener process on $t \geq 0$ with parameter σ^2 may be thought of as an experiment which is the limit of these experiments as m goes to infinity. We now describe the mth experiment. At each time instant i/m for $i = 1, 2, \ldots$ we select one of the two numbers $\pm\sigma/\sqrt{m}$ according to a fair-coin toss, so that the variance for one trial is σ^2/m and the mean is zero. If $z_1, z_2, \ldots$ are the numbers so selected we plot z_i above the point i/m on the t axis, as shown in Fig. 1*a*. We let $w_{i/m} = z_1 + \cdots + z_i$ and plot this above i/m, as in Fig. 1*b*, and connect these points by straight lines. The resulting polygonal line is the realization for this performance of the mth experiment. The variance of its height at $t = i/m$ is $i(\sigma^2/m) = t\sigma^2$, independently of m. The Wiener process can be thought of as the limit of these experiments as m approaches infinity. The Wiener process has continuous nonpolygonal functions for realizations. In terms of the Brownian motion of one of the three xyz coordinates of a particle in a liquid, we may think of the particle as being struck by a molecule of the liquid at time i/m and consequently moving by $\pm\sigma/\sqrt{m}$ along this coordinate during the next time interval of length $1/m$.

We now define the approximating experiment more explicitly. We fix $\sigma^2 > 0$ and let $Y_1, Y_2, \ldots$ be independent, each with distribution

$$P[Y_i = 1] = P[Y_i = -1] = .5.$$

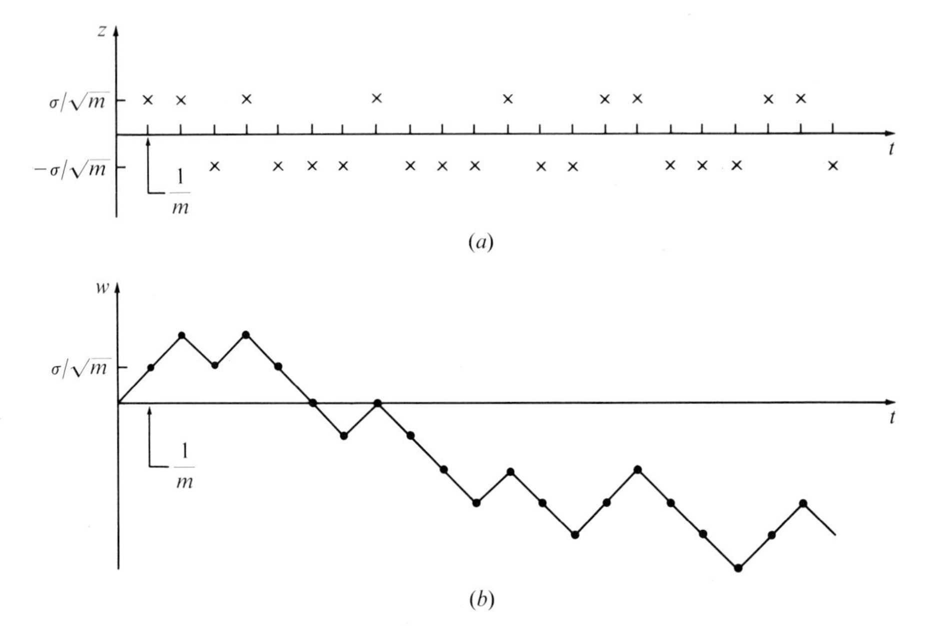

Fig. 1 (*a*) A graphed outcome for $z_1, z_2, \ldots$ for the mth experiment; (*b*) a realization from the mth experiment.

For each m and i let

$$W_{i/m}^{m} = \frac{\sigma}{\sqrt{m}} Y_1 + \frac{\sigma}{\sqrt{m}} Y_2 + \cdots + \frac{\sigma}{\sqrt{m}} Y_i$$

$$W_t^{\,m} = W_{i/m}^{m} + (mt - i)\frac{\sigma}{\sqrt{m}} Y_{i+1} \qquad \text{for } \frac{i}{m} \le t \le \frac{i+1}{m}.$$

The superscript m refers to the mth experiment and is not to be interpreted as an mth power. Thus Y_i is our ith fair-coin toss, and the previous z_i is the value of $(\sigma/\sqrt{m})Y_i$. Hence $W_t^{\,m}$ is the random height of the polygonal line at $t \ge 0$ for the mth experiment. We naturally define $W_0^{\,m} = 0$ for all m. The approximating experiments may be thought of as just different ways to plot the data from the one basic experiment of observing $Y_1, Y_2, \ldots$. Clearly $EY_i = 0$ and Var $Y_i = 1$, so $EW_t^{\,m}$ is always zero and if $t = i/m$ then

$$\text{Var } W_t^{\,m} = i\frac{\sigma^2}{m} = t\sigma^2.$$

Thus the coefficients for the Y_i were selected so that Var $W_1^{\,m} = \sigma^2$ for all m. That is, for every experiment, σ^2 is the variance of the height of the realization at time 1.

For each m we have defined a stochastic process "$W_t^{\,m}$ for $t \ge 0$." For intervals with end points of the form i/m we have stationary independent increments. Furthermore for fixed $t > 0$ and large m clearly $W_t^{\,m}$ is the sum of about mt variables, so the central-limit theorem in the form of Corollary 1*b* (Sec. 13-2) easily yields

$$\left| P\left[\frac{W_t^m}{\sigma\sqrt{t}} < b\right] - \Phi(b)\right| \le \frac{3}{(mt)^{1/8}} \qquad \text{for all real } b,$$

whenever mt is an integer. Thus for fixed $t > 0$ if m goes to infinity then the distribution of W_t^m converges to the normal distribution with mean 0 and variance $t\sigma^2$. To keep mt integer valued we could restrict m to have the form 2^r with integer r and first obtain the result for all t of the form $k/2^j$. We ignore these details in this motivational context. Naturally the CLT can be applied to other intervals $0 \le s < t$, where

$$\text{Var}\,(W_t^m - W_s^m) \to (t - s)\sigma^2 \qquad \text{as } m \to \infty$$

since there are about $(t - s)m$ trials in the interval. We have been led to the following definition.

DEFINITION OF A WIENER PROCESS

A **Wiener process** "W_t for $t \ge 0$ with $W_0 = 0$ and parameter $\sigma^2 > 0$" is any stationary independent increment process such that for each $t > 0$, W_t induces a normal density mean 0 and variance $t\sigma^2$, and every realization for the process is a continuous function. Thus if $0 \le s < t$ then $W_t - W_s$ has variance $(t - s)\sigma^2$, proportional to the length of the interval, and normal density

$$\frac{1}{\sqrt{2\pi(t-s)\sigma^2}}\, e^{-w^2/2(t-s)\sigma^2} \qquad \text{for all real } w.$$

By a **standard Wiener process** we mean "W_t for $t \ge 0$ with $W_0 = 0$ and $\sigma^2 = 1$." From such a process we easily obtain a more general Wiener process

$$X_t = \sigma W(t - t_0) + b \qquad \text{for } t \ge t_0$$

with parameter $\sigma^2 > 0$ and $X(t_0) = b$. Unless stated otherwise, we always assume that $t_0 = b = 0$.

The distributions induced by W_t for $t > 0$ form a semigroup, so we are consistent with this required property. For proof recall that if X and Y are independent and both have zero mean normal densities with variances $s\sigma^2$ and $t\sigma^2$ then $X + Y$ has a zero mean normal density with variance $(s + t)\sigma^2$.

Example 1 (Sec. 22-3) showed that if the requirement of continuous realizations for a Wiener process is dropped then there are processes satisfying the resulting definition but such that every realization has infinitely many discontinuities. As should be intuitively reasonable from the approximating experiments, Wiener processes can be shown to exist, but we do not prove this. That is, it is possible to construct a probability space with random variables W_t for $t \ge 0$ satisfying the stated definition, *including* the requirement of continuous realizations. It turns out that although the realizations are continuous, they are still so irregular that they have derivatives almost nowhere. For the Wiener process we do not have simple characterizations like the two in Sec. 23-3 for Poisson processes.

Let $X(\tau)$ for $\tau \geq 0$ be a standard Wiener process, so that $\text{Var}\, X(\tau) = \tau$. For $\sigma \neq 0$ if W_t for $t \geq 0$ is introduced by either of the definitions

$$W_t = X(t\sigma^2) \qquad \text{for all } t \geq 0$$

$$W_t = \sigma X(t) \qquad \text{for all } t \geq 0$$

then it is easily seen to be a Wiener process with parameter $\sigma^2 > 0$. Thus σ^2 corresponds to a scale change in time applied to every realization, or σ corresponds to a scale change in the values. Furthermore $-W_t$ for $t \geq 0$ is a Wiener process, so that, roughly speaking, every realization is as likely as is its reflection in the time axis. As we will see in the next section, this is a very useful property.

Table I $w_i = x_1 + \cdots + x_i$ **where** $x_1, \ldots, x_{100}$ **are random normal deviates**

i	x_i	w_i	i	x_i	w_i	i	x_i	w_i
1.	−1.276	−1.276	35.	−.099	3.919	69.	.170	12.805
2.	−.318	−1.594	36.	−.463	3.456	70.	.389	13.194
3.	−1.377	−2.971	37.	.503	3.959	71.	−.305	12.889
4.	2.334	−.637	38.	−.857	3.102	72.	−.321	12.268
5.	−1.136	−1.773	39.	−.122	2.980	73.	1.900	14.468
6.	.414	−1.359	40.	1.632	4.612	74.	−.778	13.690
7.	−.494	−1.853	41.	2.072	6.684	75.	.617	14.307
8.	1.048	−.805	42.	−.435	6.249	76.	−1.430	12.877
9.	.347	−.458	43.	.876	7.125	77.	.267	13.144
10.	.637	.179	44.	.833	7.958	78.	.978	14.122
11.	2.176	2.355	45.	−.891	7.067	79.	−1.235	12.887
12.	−1.185	1.170	46.	.644	7.711	80.	−.258	12.629
13.	.972	2.142	47.	.105	7.816	81.	.243	12.872
14.	1.210	3.352	48.	−1.192	6.624	82.	−.292	12.580
15.	2.647	5.999	49.	−.042	6.582	83.	−.505	12.075
16.	.398	6.397	50.	.498	7.080	84.	.397	12.472
17.	.846	7.243	51.	−1.329	5.751	85.	−.605	11.867
18.	.654	7.897	52.	1.284	7.035	86.	1.360	13.227
19.	.522	8.419	53.	.619	7.654	87.	.480	13.707
20.	−1.288	7.131	54.	.699	8.353	88.	−.027	13.680
21.	1.372	8.503	55.	.101	8.454	89.	−1.482	12.198
22.	.854	9.357	56.	−1.381	7.073	90.	−1.256	10.942
23.	−.148	9.209	57.	−.574	6.499	91.	−1.132	9.810
24.	−1.148	8.061	58.	.096	6.595	92.	−.780	9.030
25.	.348	8.409	59.	1.389	7.984	93.	−.859	8.171
26.	.284	8.693	60.	1.249	9.233	94.	.447	8.618
27.	−1.016	7.677	61.	.756	9.989	95.	.269	8.887
28.	1.603	9.280	62.	−.860	9.129	96.	.097	8.984
29.	−.190	9.090	63.	−.778	8.351	97.	−.686	8.298
30.	−.722	8.368	64.	.037	8.388	98.	.957	9.255
31.	−1.696	6.672	65.	2.619	11.007	99.	−.976	8.279
32.	−2.543	4.129	66.	−.420	10.587	100.	.274	8.553
33.	−.359	3.770	67.	1.048	11.635			
34.	.248	4.018	68.	1.000	12.635			

SOURCE: $x_1, \ldots, x_{100}$ are the first 100 random normal deviates from the Rand Corporation, "A Million Random Digits and 100,000 Normal Deviates," Free Press, New York.

A SAMPLING EXPERIMENT

Let $W(t)$ for $t \geq 0$ be a standard Wiener process. The remainder of this section is devoted to a sampling experiment in order to obtain a typical realization, at least as it would appear if observed only at the times $0, 1, \ldots, M = 100$. Let $x_1, \ldots, x_M$ be the $M = 100$ random normal deviates from Table 1. These are interpreted as having been obtained from M independent observations from a normal 0 and 1 density, with each observation recorded to three decimal places. For $k = 1, 2, \ldots$ we have

$$W(k) = \sum_{i=1}^{k} [W(i) - W(i-1)]$$

where the increments in square brackets are independent and each is normal 0 and 1. We interpret x_i as the observed value of $[W(i) - W(i-1)]$, and let $w_i = x_1 + \cdots + x_i$, and plot w_i against i in Fig. 2. Thus Fig. 2 can be interpreted as having been obtained by observing a realization at times $0, 1, \ldots, M$. Naturally $\sigma w_1, \ldots, \sigma w_M$ may be interpreted as having been observed at times $0, 1, \ldots, M$ from a Wiener process with parameter $\sigma^2 > 0$.

Exercise 2 shows how to "fill in" interpolating observations in Fig. 2. That is, we can use a new set of 100 random normal deviates to assign values at times $\frac{1}{2}$, $\frac{3}{2}, \ldots, 99 + \frac{1}{2}$; then we can use a new set of 200 to assign values to the time points midway between those to which values have already been assigned; and so on.

Suppose that we are interested in the "local" appearance of realization near the origin. We show that with proper scale factors on the two axes Fig. 2 can be interpreted as having been obtained locally. We fix any $d > 0$, say $d = 10^{-6}$, and define a new standard Wiener process $U(\tau)$ for $\tau \geq 0$ by

$$U(\tau) = \sqrt{d}\, W\left(\frac{\tau}{d}\right)$$

so that Var $U(\tau) = \tau$. Therefore if $w_1, \ldots, w_M$ are as in Fig. 2 then $\sqrt{d} w_1, \ldots, \sqrt{d} w_M$ may be interpreted as having been obtained by observing a standard Wiener process at times $d, 2d, \ldots, Md$. (This technique cannot be applied to a Poisson process, since we cannot apply a scale change to values, and more fundamentally most realizations are identically zero locally near $t = 0$.)

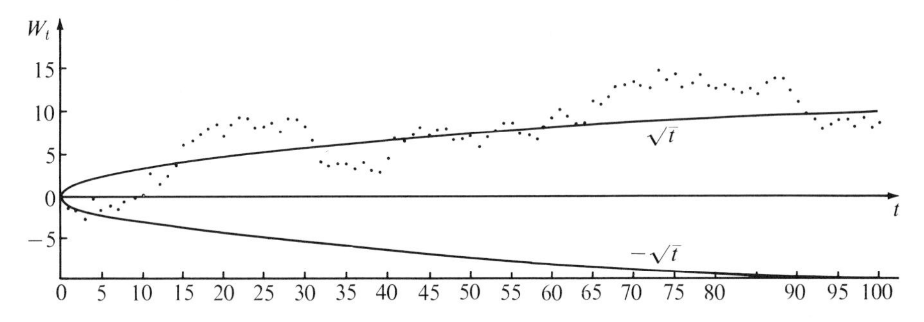

Fig. 2 A realization from a sampling experiment.

Think of Fig. 2 as obtained from a "typical" realization from a standard Wiener process over the interval $0 \le t \le 100$. Obviously this realization had to intersect the time axis at least once between times 9 and 10. But with proper scale factors we may interpret Fig. 2 as obtained from a realization from a standard Wiener process over the interval $0 \le t \le 10^{-10}$. This realization also intersects the time axis at least once close to $t = 0$. A little thought suggests, as justified in the next section, that typical realizations must intersect the time axis infinitely often near $t = 0$. Furthermore these points of intersection are not isolated. Such properties make it impossible to draw these realizations if, as opposed to Fig. 2, the "fine structure" is meant to be exhibited. Thus the reader should keep in mind that such drawings are inherently misleading, even though we find them quite helpful.

EXERCISES

1. (*a*) Let the random vector $(X_1, \ldots, X_n)$ have a continuous density and let $S_k = X_1 + \cdots + X_k$ be the sum of its first k components. Show that $S_1, \ldots, S_n$ have joint continuous density

$$p_{S_1,\ldots,S_n}(s_1, \ldots, s_n) = p_{X_1,\ldots,X_n}(s_1, s_2 - s_1, \ldots, s_n - s_{n-1})$$

for all real $s_1, \ldots, s_n$.
(*b*) Let $W(t)$ for $t \ge 0$ be a standard Wiener process and let $w_0 = 0$ and $0 = t_0 < t_1 < \cdots < t_n$. Show that the joint density for $W(t_1), \ldots, W(t_n)$ is given by

$$p_{W(t_1),\ldots,W(t_n)}(w_1, \ldots, w_n) = \prod_{i=1}^{n} \frac{1}{\sqrt{2\pi(t_i - t_{i-1})}} \exp\left\{-\sum_{i=1}^{n} \frac{(w_i - w_{i-1})^2}{2(t_i - t_{i-1})}\right\}.$$

2. Let W_t for $t \ge 0$ be a standard Wiener process and let $0 \le h < h + t$. We desire the conditional distribution of some intermediate value of the realization if we observe the realization only at the end points of an interval. Let $0 < p < 1$ and show that if we are given that $[W_h = a]$ and $[W_{h+t} = a + b]$ occurred then the conditional density of W_{h+pt} is normal with mean $a + pb$ and variance $p(1 - p)t$. Naturally $a = 0$ if $h = 0$. Thus as shown in Fig. 3

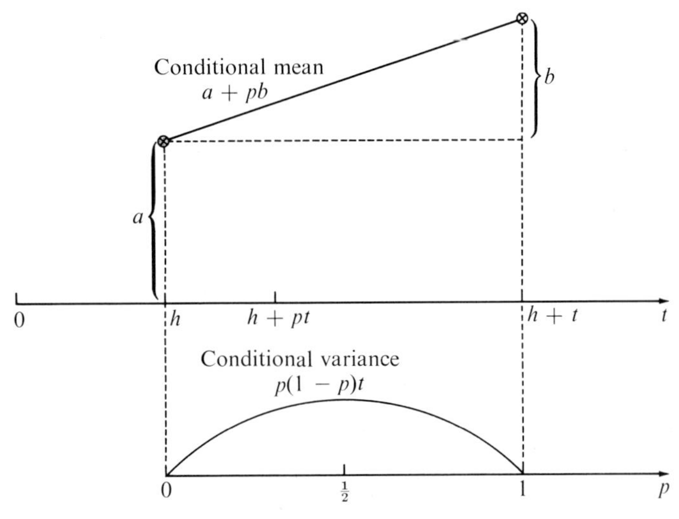

Fig. 3 Interpolating an intermediate point.

the conditional mean is obtained by linear interpolation between the two observed values, while the conditional variance does not depend on a or b, and it is largest at the midpoint of the interval.

3. From Exercise 2, if $0 < p < 1$ and we are given that $[W(t) = b]$ occurred then the conditional density of $W(pt)$ is normal with mean pb and variance $p(1 - p)t$. In the study of properties of typical realizations near the origin suppose that we plan a sampling experiment in order to generate $W(10^{-4n})$ in the order $n = 0, 1, 2, \ldots$. Let $w_0, w_1, \ldots$ represent the samples.

(a) Use Table 1 to generate an initial few samples.

(*b*) How frequently would you anticipate adjacent values to have opposite signs, that is, $w_i w_{i+1} < 0$?

24-2 SOME SIMPLE PROPERTIES OF WIENER PROCESSES

This section is devoted to some simple properties and random variables of Wiener processes. The main result is Theorem 1, which gives, for a fixed interval $0 \leq \tau \leq t$, the distribution of "the maximum of the random realization," depicted as m in Fig. 1. Example 2 derives the probability of obtaining a realization which intersects the time axis at least once in a specified interval. Example 3 derives the continuous density for "the first time to reach level λ," depicted as t_λ in Fig. 5. This density has properties similar to the Cauchy density.

For Examples 1 and 4 consider the continuous function f of Fig. 2. This function has the property that as $t > 0$ decreases to zero the function oscillates more and more rapidly between positive and negative values. Example 1 shows that Wiener process realizations have this same property of intersecting the time axis infinitely often near the origin. The upper and lower envelopes for f are $\pm\sqrt{t}$, which have derivatives of $\pm\infty$ at the origin, so that f does not have a right-hand derivative at $t = 0$. Example 4 shows that, similarly, Wiener process realizations do not have a right-hand derivative at $t = 0$, or at almost any $t > 0$.

As was mentioned earlier, Wiener process realizations do *not* intersect the time axis at isolated points like the function of Fig. 2. That is, if a typical realization intersects the axis at time t then it intersects the axis at infinitely many points near t. More precisely with probability 1 we will obtain a realization r_ω such that its set of zeros $\{t: r_\omega(t) = 0, t \geq 0\}$ is uncountable, has no isolated points, and has Lebesgue measure 0. The Cantor set of Exercise 7g (Sec. 17-1) has all these properties.

With regard to proofs Examples 1, 3 and 4 are based on Theorem 1 while Exercise 7 shows that Example 2 can also be derived from Theorem 1. Thus all of this section follows from Theorem 1, which is shortly "proved" by use of the reflection principle. This principle was used in the proof of Bertrand's ballot theorem in Example 6 (Sec. 2-3), but in our present context it can be considered only to *suggest* results, even though it is powerful and widely used for this purpose. For this reason the Addendum is devoted to a rigorous proof of Theorem 1, based on Lemma 1 in the Addendum. This lemma is a version of the reflection technique for sequences.[1]

[1] References, historical comments, and many deep results for Wiener processes and their generalizations may be found in the advanced monograph by Kiyosi Itô and Henry P. McKean, Jr., "Diffusion Processes and Their Sample Paths," Academic Press, Inc., New York, 1965. Many results go back to P. Lévy. For a discussion of a recent application of the Wiener process, see R. Hersh and R. J. Griego," Brownian motion and potential theory," *Scientific American*, March, 1969.

We next state and interpret Theorem 1 before using the reflection technique to indicate its truth.

Theorem 1 If W_t for $t \geq 0$ with $W_0 = 0$ is a Wiener process then

$$P[W_\tau \geq \lambda \text{ for at least one } \tau \text{ satisfying } 0 \leq \tau \leq t] = 2P[W_t \geq \lambda]$$

for all $\lambda \geq 0$, $t \geq 0$. ■

As indicated in Fig. 1, we are considering the event that we obtain a realization which reaches level λ at least once in the interval $0 \leq \tau \leq t$. Since W_t is normal, we

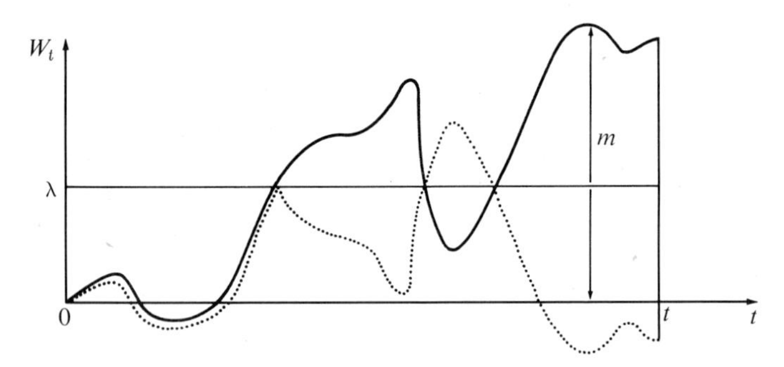

Fig. 1 The maximum m on the interval $0 \leq \tau \leq t$.

can easily evaluate the right-hand side of the conclusion of Theorem 1,

$$2P[W_t \geq \lambda] = 2\{1 - P[W_t < \lambda]\}$$

$$= 2\left\{1 - P\left[\frac{W_t}{\sigma\sqrt{t}} < \frac{\lambda}{\sigma\sqrt{t}}\right]\right\} = 2\left\{1 - \Phi\left(\frac{\lambda}{\sigma\sqrt{t}}\right)\right\}.$$

For example, if $\sigma^2 = t = 1$ then Table 1 (Sec. 13-1) for Φ yields

$$2P[W_1 \geq 1] = .3174 \qquad \text{and} \qquad 2P[W_1 \geq 2] = .0456.$$

Thus the probability is about .05 that the standard Wiener process will reach level 2 at some time during the first second. That is, in many independent experiments, with each experiment generating one realization, this event should occur in about 5 percent of the experiments.

For fixed $t \geq 0$ let the random variable M_t be the maximum value of the continuous realization on the interval from 0 to t. Thus M_t has the value m for the solid realization of Fig. 1. Clearly

$$[M_t \geq \lambda] = [W_\tau \geq \lambda \text{ for at least one } \tau \text{ satisfying } 0 \leq \tau \leq t].$$

Therefore we can restate the conclusion of Theorem 1 by the formula

$$P[M_t \geq \lambda] = 2P[W_t \geq \lambda] \qquad \text{for } t \geq 0, \lambda \geq 0.$$

Note that $2P[W_t \geq \lambda] = P[|W_t| \geq \lambda]$ because W_t has a symmetric density. Therefore the conclusion of Theorem 1 can also be stated as

$$P[M_t \geq \lambda] = P[|W_t| \geq \lambda] \qquad \text{for all } t \geq 0, \lambda \geq 0.$$

That is, M_t and $|W_t|$ are identically distributed. But W_t is normal 0 and $\sigma^2 t$, so the density for M_t is just twice this density, on the positive axis. Thus

$$p_{M_t}(\lambda) = \begin{cases} \dfrac{2}{\sigma\sqrt{2\pi t}} e^{-\lambda^2/2t\sigma^2} & \text{for } \lambda > 0 \\ 0 & \text{for } \lambda \leq 0. \end{cases}$$

USE OF THE REFLECTION PRINCIPLE TO INDICATE THE TRUTH OF THEOREM 1

Note that $[M_t \geq \lambda]$ is the union of the two disjoint events

$$[M_t \geq \lambda,\ W_t \geq \lambda] = [W_t \geq \lambda] \qquad \text{and} \qquad [M_t \geq \lambda,\ W_t < \lambda].$$

That is, $[M_t \geq \lambda]$ consists of those realizations which are λ or greater at time t, together with those realizations which are λ or greater at some time before t, but smaller than λ at time t. Thus the conclusion of Theorem 1 can be stated as

$$P[M_t \geq \lambda,\ W_t < \lambda] = P[W_t \geq \lambda].$$

For any realization belonging to $[W_t \geq \lambda]$, such as the solid curve in Fig. 1, find the first time point where the realization equals λ, and then for all later times reflect the realization in the horizontal line at level λ to obtain the dotted curve. We may think of these two curves as equally likely, since, roughly speaking, the process starts afresh when we first reach level λ, and the negative of a Wiener process is a Wiener process. This reflection technique establishes a one-to-one correspondence between

$$[M_t \geq \lambda,\ W_t < \lambda] \qquad \text{and} \qquad [W_t > \lambda].$$

Also the correspondence "preserves" probabilities, so these two events should have the same probability. Thus the reflection technique suggests Theorem 1. ∎

Example 1 In this example we use Theorem 1 to show that realizations from a Wiener process behave near the origin somewhat like the continuous function f of Fig. 2. Roughly speaking, initially a realization wanders indecisively across the time axis infinitely often before deciding to stay on one side of the time axis for a while. In particular it is *not* true that over some small enough interval near the origin that half the realizations remain above the time axis and the other half remain below. We say that r_ω **has a zero** at τ iff $r_\omega(\tau) = 0$. We show that realizations have infinitely many zeros near the origin. Examples 2 and 4 generalize this example.

Applying Theorem 1 to an increasing sequence of events, we see that if $t > 0$ then

$$P\left(\bigcup_{n=1}^{\infty}\left[M_t \geq \frac{1}{n}\right]\right) = \lim_{n\to\infty} P\left[M_t \geq \frac{1}{n}\right] = \lim_{n\to\infty} P\left[|W_t| \geq \frac{1}{n}\right] = 1$$

where the union on the left is just the event $[M_t > 0]$. Therefore

$$P[M_t > 0] = 1 \qquad \text{if } t > 0.$$

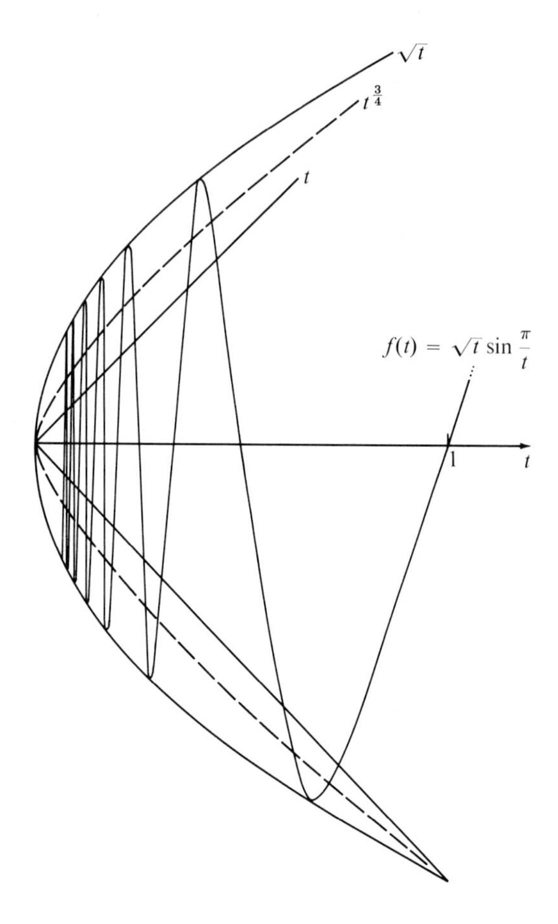

Fig. 2

Thus realizations must take at least one positive value in any arbitrarily small interval containing the origin.

We can apply this result to the Wiener process $-W_t$ for $t \geq 0$, and hence realizations must also take at least one negative value in any such interval. Therefore realizations must cross the time axis. That is, if $t > 0$ then

$$P[W_\tau = 0 \text{ for at least one } \tau \text{ satisfying } 0 < \tau < t]$$
$$= P\{\omega\colon r_\omega(\tau) = 0 \text{ for at least one } \tau \text{ satisfying } 0 < \tau < t] = 1.$$

In this result we let $t = 1/n$, so that for $n = 1, 2, \ldots$ we have a countable sequence of events whose intersection must have probability 1, since each event does. Thus with probability 1 we will obtain a realization which has at least one zero in *every* interval $0 < \tau < 1/n$ for $n = 1, 2, \ldots$. That is, there is probability 1 for the set of those ω whose corresponding realizations r_ω have the origin as a limit point of the set $\{\tau\colon r_\omega(\tau) = 0,\ \tau > 0\}$ of zeros of r_ω on the positive axis.

For *fixed* $s \geq 0$ the process "$W_{s+t} - W_s$ for $t \geq 0$" is a Wiener process, and hence the preceding results apply to it. Thus for each fixed $s \geq 0$, with probability 1 each realization will, for infinitely many times shortly after s, have the same value that it had at time s. ∎

Example 2 This example presumes a slight familiarity with the multivariate normal densities of Sec. 18-2 and illustrates the use of the reflection principle in deriving a formula for the probability of obtaining a realization having at least one zero in a specified interval. Exercise 7 derives the same formula rigorously from Theorem 1, without using Sec. 18-2. The formula asserts that if $0 \leq s \leq t$ then

$$P[W_\tau = 0 \text{ for at least one } \tau \text{ satisfying } s \leq \tau \leq t] = \frac{2}{\pi} \arccos \sqrt{\frac{s}{t}}$$

where W_t for $t \geq 0$ with $W_0 = 0$ is a Wiener process with parameter σ^2. Note that if $\delta = \sqrt{s/t}$ then $0 \leq \delta \leq 1$, and by $\arccos \delta$ we mean the unique angle $0 \leq \varphi \leq \pi/2$ for which $\cos \varphi = \delta$. The formula is obviously true if $s = 0$ or $s = t$, so we need only consider $0 < s < t$.

Note that if we fix t and let s go to zero then the formula has a limit of 1, consistent with Example 1. For interpretation, let $t = 2s$. Thus for any $s > 0$ it is as likely as not that we will obtain a realization which has at least one zero in the interval from s to $2s$. That is, we have a 50 percent chance of obtaining a realization which does not intersect the time axis in this interval. We turn to the proof.

All exercises except 2 to 4 may be done before reading the remainder of this section.

From Exercise 2 (Sec. 18-2) we know that if (Y_1, Y_2) has a bivariate normal density with correlation $-1 < \rho < 1$ and $EY_1 = EY_2 = 0$ then

$$P[Y_1 Y_2 \geq 0] = \tfrac{1}{2} + \frac{\arcsin \rho}{\pi}$$

where $-\pi/2 < \arcsin \rho < \pi/2$. As in Fig. 3, let $\varphi_1 = \arcsin \rho$, so that $\cos [(\pi/2) - \varphi_1] = \rho$; hence

$$\arccos \rho = \frac{\pi}{2} - \varphi_1 = \frac{\pi}{2} - \arcsin \rho$$

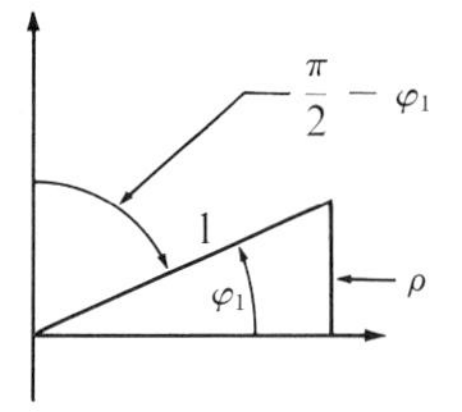

Fig. 3

and substituting this yields

$$P[Y_1 Y_2 \geq 0] = 1 - \frac{\arccos \rho}{\pi}$$

where $0 < \arccos \rho < \pi$. Also then

$$P[Y_1 Y_2 \leq 0] = \frac{\arccos \rho}{\pi}.$$

Clearly $[Y_1 Y_2 \geq 0]$ is the event that Y_1 and Y_2 have the same sign; that is, both are positive or both are negative. Similarly $[Y_1 Y_2 \leq 0]$ is the event that Y_1 and Y_2 have opposite signs. The event $[Y_1 Y_2 = 0]$ can be ignored, since its probability is zero. In particular if $\rho = 1/\sqrt{2}$

then $P[Y_1 Y_2 \geq 0] = \frac{3}{4}$. Thus 70 percent correlation implies a 75 percent chance that we will obtain a point (y_1, y_2) in the first or third quadrants of the plane.

If $0 < s < t$ then the random vector (W_s, W_t) has a bivariate normal density, since (W_s, W_t) can be obtained as follows by applying a nonsingular linear transformation to independent normal 0 and 1 random variables:

$$\begin{bmatrix} W_s \\ W_t \end{bmatrix} = \begin{bmatrix} \sigma\sqrt{s} & 0 \\ \sigma\sqrt{s} & \sigma\sqrt{t-s} \end{bmatrix} \begin{bmatrix} W_s/(\sigma\sqrt{s}) \\ (W_t - W_s)/(\sigma\sqrt{t-s}) \end{bmatrix}$$

Although we do not now need the result, clearly this proof generalizes to show that if $0 < t_1 < \cdots < t_n$ then $(W_{t_1}, \ldots, W_{t_n})$ has a multivariate normal density. Since $EW_t = 0$, we have

$$\begin{aligned} \text{Cov}(W_s, W_t) &= EW_s W_t = EW_s[W_s + (W_t - W_s)] \\ &= EW_s^2 + EW_s(W_t - W_s) = s\sigma^2 \end{aligned}$$

because W_s and $W_t - W_s$ are independent. For the correlation coefficient we get

$$\rho_{W_s, W_t} = \frac{\text{Cov}(W_s, W_t)}{[(\text{Var } W_s)(\text{Var } W_t)]^{1/2}} = \frac{s}{\sqrt{st}} = \sqrt{\frac{s}{t}}.$$

Thus if $0 < s < t$ then we have proved that

$$P[W_s W_t \leq 0] = \frac{1}{\pi} \arccos \sqrt{\frac{s}{t}}$$

Note that this expression does not have the factor 2 which appears in the original formula of this example.

In particular $P[W_s W_{2s} \leq 0] = \frac{1}{4}$ so that only 25 percent of the realizations are on different sides of the time axis at times 1 and 2. However, if $t = s\ 10^{10}$ then $P[W_s W_t \leq 0] \doteq P[W_s W_t \geq 0]$, so that after a long time it is as likely as not that the sign is opposite.

We use the reflection principle, as in Fig. 4, to complete the derivation. Let A be the event that there is at least one zero in the interval from s to t. Then $A = B \cup C$, where B and C are the disjoint events defined by

$$B = A \cap [W_s W_t \leq 0] = [W_s W_t \leq 0] \qquad C = A \cap [W_s W_t > 0].$$

That is, A consists of those realizations which have opposite signs at s and t, together with those which have the same sign at s and t but intersect the time axis at least once in the interval. Thus $P(A) = P(B) + P(C)$, where $P(B)$ was found above. We use the reflection

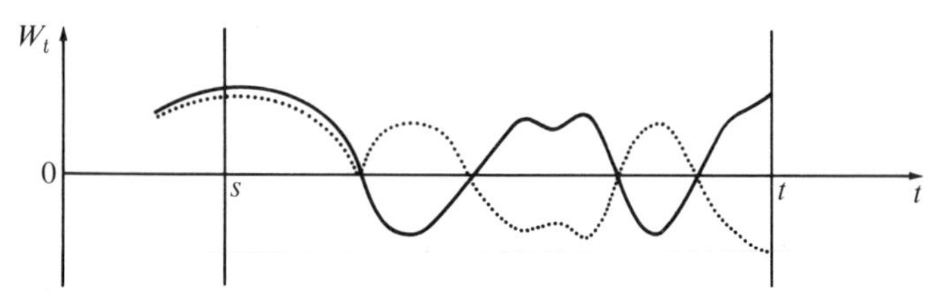

Fig. 4 Zeros in the interval $s \leq \tau \leq t$.

technique to justify $P(C) = P(B)$, for then $P(A) = 2P(B)$, as was asserted. Take any realization in C, such as the solid curve in Fig. 4; when it first intersects the time axis we reflect all later values in the time axis to obtain the "equally likely" dotted curve in B. This one-to-one correspondence between C and B correctly suggests that $P(C) = P(B)$. ■

Example 3 We now derive the continuous density induced by a random variable $T_1 > 0$ which arises in the standard Wiener process. The family of densities obtained from this density by scale changes is closed under independent additions, just as was shown to be true for Cauchy and normal densities in Sec. 16-5. However, $E\,|T_1| = \infty$, and this density is even more spread than the Cauchy density. The density induced by the arithmetic average of independent errors from this new density is *more* spread than is the density for one error. In this example W_t for $t \geq 0$ is a *standard Wiener process.*

Theorem 1 shows that with probability 1 each realization must eventually exceed any level $\lambda > 0$; that is,

$$P[M_t \geq \lambda] = P[|W_t| \geq \lambda] \to 1 \qquad \text{as } t \to \infty$$

because W_t has variance t. Therefore for each $\lambda > 0$ we may define the random variable T_λ to be the first time at which level λ is reached. For the realization of Fig. 5 the value of T_λ

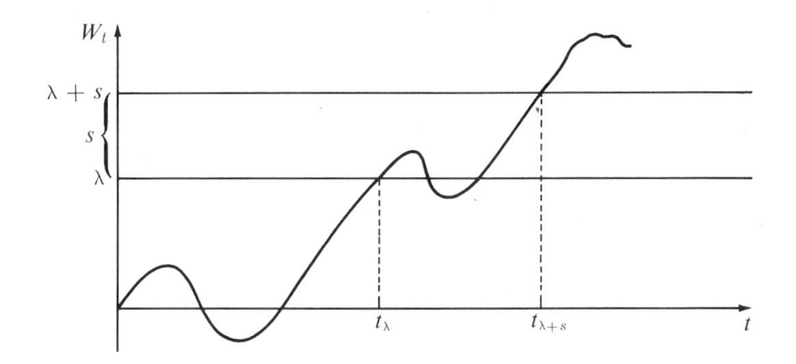

Fig. 5 The first time t_λ at which level λ is reached.

is t_λ. But for any $t > 0$ we obviously have $[T_\lambda \leq t] = [M_t \geq \lambda]$. That is, a realization reaches level λ on or before time t iff its maximum value up through time t is at least λ. Also $[T_\lambda = t]$ is a subset of $[M_t = \lambda]$, which has zero probability. Therefore the distribution function for T_λ is given for $t > 0$ by

$$F_{T_\lambda}(t) = P[T_\lambda < t] = P[T_\lambda \leq t] = 2P\left[\frac{W_t}{\sqrt{t}} \geq \frac{\lambda}{\sqrt{t}}\right] = 2\left[1 - \Phi\left(\frac{\lambda}{\sqrt{t}}\right)\right]$$

Differentiating, we obtain the continuous density

$$p_{T_\lambda}(t) = \frac{d}{dt} F_{T_\lambda}(t) = (-2)\frac{d}{dt}\int_{-\infty}^{\lambda/\sqrt{t}} \varphi(x)\,dx$$

$$= (-2)\,\varphi\left(\frac{\lambda}{\sqrt{t}}\right)\frac{d}{dt}\left(\frac{\lambda}{\sqrt{t}}\right) = \frac{\lambda}{\sqrt{2\pi t^3}}\,e^{-\lambda^2/2t} \qquad \text{for } t > 0$$

and $p_{T_\lambda}(t) = 0$ for $t \le 0$. For fixed λ if we consider this density as t goes to infinity, the exponential approaches 1, so that the density drops off as $t^{-3/2}$. Therefore the density does not have a finite expectation. It has a finite rth moment iff $0 < r < 1/2$.

If $\lambda^2 > 0$ the density induced by $S = \lambda^2 T_1$ is

$$p_S(s) = \frac{1}{\lambda^2} p_{T_1}\left(\frac{s}{\lambda^2}\right)$$

which is seen to equal p_{T_λ}. Thus T_λ and $\lambda^2 T_1$ are identically distributed, for every $\lambda > 0$.

As in Fig. 5, let $0 < \lambda < \lambda + s$ and consider the identity

$$T_{\lambda+s} = T_\lambda + [T_{\lambda+s} - T_\lambda].$$

If we reach level λ for the first time at time t_λ then the process essentially starts anew, independently of the past, with origin (t_λ,λ), and then eventually reaches a level s units above this origin. This suggests that the time $T_{\lambda+s} - T_\lambda$ between first reaching level λ and then first reaching level $\lambda + s$ must be independent of T_λ and have the same distribution as T_s. Therefore if Y_1 and Y_2 are independent with densities p_{T_λ} and p_{T_s} then $Y_1 + Y_2$ should have density $p_{T_{\lambda+s}}$. This may be proved directly by performing a difficult convolution. Thus the parameters just add under independent additions.

We may state the last result more generally as follows. If $Y_1, \ldots, Y_n$ are independent and Y_i has the same density as T_{λ_i} then $Y_1 + \cdots + Y_n$ has the same density as T_λ with $\lambda = \lambda_1 + \cdots + \lambda_n$. Using the earlier result on scale changes, if $X_1, \ldots, X_n$ are independent and each has the same density p_{T_1} then $\sum_{i=1}^n \lambda_i^2 X_i$ and $\left(\sum_{i=1}^n \lambda_i\right)^2 X_1$ are identically distributed. For $\lambda_i^2 = 1/n$ we see that $(1/n)(X_1 + \cdots + X_n)$ and nX_1 are identically distributed. Thus the arithmetic average of 100 independent observations is 100 times *more* spread than one observation. ■

Example 4. We show that Wiener process realizations do not have derivatives. That is, for every $s \ge 0$ the probability is *zero* for the set of realizations which have a right-hand derivative at s. We need only consider derivatives at the origin $s = 0$, since the result may then be applied to the Wiener process $W_{s+r} - W_s$ for $r \ge 0$. From Example 1, if there were a derivative at the origin it would have to equal zero, but we show that it does not exist.

Realizations behave, almost everywhere, somewhat like the behavior exhibited *only* at the origin by the continuous function f of Fig. 2. If we select two positive sequences, t_n and t_n', converging to zero and such that each $\sin(\pi/t_n) = 1$ and each $\sin(\pi/t_n') = -1$, then for the difference quotients for this f we find that as n goes to infinity

$$\frac{f(t_n) - f(0)}{t_n} = \frac{1}{\sqrt{t_n}} \to \infty \quad \text{and} \quad \frac{f(t_n') - f(0)}{t_n'} = \frac{-1}{\sqrt{t_n'}} \to -\infty.$$

Thus the limit of the difference quotient depends on the particular sequence t_n. Therefore f does not have a right-hand derivative at the origin, even if an infinite derivative is permitted. Clearly by properly selecting our approach to the origin we can make the difference quotient converge to *any* real number.

We start the proof by generalizing the function of Fig. 2. Suppose that h is some function on $t \ge 0$ satisfying $h(0) = 0$, and let $0 < \delta < 1$. Suppose there is a sequence $0 < t_n$

converging to zero, for which each $h(t_n) \geq (t_n)^\delta$. Then for the difference quotient we have

$$\frac{h(t_n) - h(0)}{t_n} \geq \left(\frac{1}{t_n}\right)^{1-\delta} \to \infty \qquad \text{as } n \to \infty.$$

Similarly suppose there is also a sequence $0 < t_n'$ converging to zero for which $h(t_n') \leq -(t_n')^\delta$. Then

$$\frac{h(t_n') - h(0)}{t_n'} \leq -\left(\frac{1}{t_n'}\right)^{1-\delta} \to -\infty \qquad \text{as } n \to \infty$$

so that h does not have a right-hand derivative at the origin. Thus we need only show that realizations will satisfy the conditions above. Now W_t is normal with standard deviation $\sigma\sqrt{t}$, so we often expect $r_\omega(t) \doteq \pm\sigma\sqrt{t}$. But $\sqrt{t} > t^{3/4}$ for $0 < t < 1$. Therefore we select $\delta = \frac{3}{4}$, actually any $.5 < \delta < 1$ will do, and show that with probability 1 we will obtain a realization having the properties assumed above for h.

Fix $.5 < \delta < 1$ and let

$$B_n = \left[W_\tau \geq \tau^\delta \text{ for at least one } \tau \text{ satisfying } 0 < \tau < \frac{1}{n}\right]$$

$$= \left\{\omega : r_\omega(\tau) \geq \tau^\delta \text{ for at least one } \tau \text{ satisfying } 0 < \tau < \frac{1}{n}\right\}.$$

To show that $P(B_n) = 1$ we first note that $[M_t \geq t^\delta]$ is a subset of B_n, for all t satisfying $0 < t < 1/n$. To prove this we assume that ω satisfies $M_t(\omega) \geq t^\delta$, so that realization r_ω reaches level t^δ at some point τ in the interval $0 < \tau \leq t$. Then $r_\omega(\tau) \geq t^\delta \geq \tau^\delta$ so that ω belongs to B_n. Therefore Theorem 1 yields

$$P(B_n) \geq P[M_t \geq t^\delta] = P[|W_t| \geq t^\delta] = P\left[\frac{|W_t|}{\sqrt{t}} \geq t^{\delta-.5}\right] \to 1 \qquad \text{as } t \to 0$$

since $W_t/\sqrt{t}$ is normal 0 and σ^2 while $t^{\delta-.5}$ goes to zero. Thus each $P(B_n) = 1$, and hence for their intersection $B = B_1 \cap B_2 \cap \cdots$ we also have $P(B) = 1$. Clearly if ω belongs to B then $r_\omega(\tau) \geq \tau^\delta$ for an infinite sequence of different $\tau > 0$ converging to zero. Naturally if

$$C_n = \left[W_\tau \leq -\tau^\delta \text{ for at least one } \tau \text{ satisfying } 0 < \tau < \frac{1}{n}\right]$$

then $P(C_n) = 1$, since $-W_t$ for $t \geq 0$ is a Wiener process. Therefore $P(C) = 1$, where $C = C_1 \cap C_2 \cap \cdots$. Thus $P(BC) = 1$, and clearly every realization from BC satisfies the properties specified earlier for the function h. ■

ADDENDUM

This addendum is devoted to the proof of Theorem 1. Lemma 1 is the heart of this proof. Early in the proof of Lemma 1 the lemma is interpreted as a reflection principle.[1]

Lemma 1 For $1 \leq k \leq n$ let $S_k = X_1 + \cdots + X_k$ where $X_1, \ldots, X_n$ are independent random variables, each with a symmetric distribution but not necessarily the same

[1] The proofs presented here essentially follow J. L. Doob, "Stochastic Processes," p. 392, John Wiley & Sons, Inc., New York, 1953.

distribution. That is, for each k the two random variables X_k and $-X_k$ are identically distributed. Fix $\lambda > 0$ and let

$$A = [S_k \geq \lambda \text{ for at least one } k \text{ satisfying } 1 \leq k \leq n]$$

$$= \bigcup_{k=1}^{n} [S_k \geq \lambda] = \left[\max_{1 \leq k \leq n} S_k \geq \lambda\right].$$

Then $P(A) \leq 2P[S_n \geq \lambda]$. Furthermore for every $\epsilon > 0$ we have

$$2P[S_n \geq \lambda + 2\epsilon] - \sum_{k=1}^{n} P[X_k \geq \epsilon] \leq P(A). \quad \blacksquare$$

Proof of Lemma 1 Certainly A equals the union of the two disjoint events $A[S_n \geq \lambda] = [S_n \geq \lambda]$ and $A[S_n < \lambda]$. That is, A consists of the event that the last sum is λ or greater, together with the event that the last sum is smaller than λ but one of the earlier sums is λ or more. Hence

$$P(A) = P[S_n \geq \lambda] + P(A[S_n < \lambda]).$$

Thus we need only prove the two inequalities

$$P[S_n \geq \lambda + 2\epsilon] - \sum_{k=1}^{n} P[X_k \geq \epsilon] \leq P(A[S_n < \lambda]) \leq P[S_n \geq \lambda]$$

since we can then add them to

$$P[S_n \geq \lambda + 2\epsilon] \leq P[S_n \geq \lambda] \leq P[S_n \geq \lambda]$$

in order to complete the proof.

Note that if $\epsilon \doteq 0 \doteq \sum_{k=1}^{n} P[X_k \geq \epsilon]$ then we will have shown that

$$P(A[S_n < \lambda]) \doteq P[S_n \geq \lambda]$$

which is a form of the reflection principle.

Let $A_k = [S_1 < \lambda, S_2 < \lambda, \ldots, S_{k-1} < \lambda, S_k \geq \lambda]$ be the event that S_k is the first partial sum to equal or exceed λ. Clearly $A_1, \ldots, A_n$ are disjoint, and their union equals A.

Very roughly, we show that for each k we have

$$P(A_k[S_n < \lambda]) \doteq P(A_k[S_n \geq \lambda]).$$

This is depicted as a reflection principle in Fig. 6. We obtain inequalities rather than an equality largely because A_k must use the constraint $S_k \geq \lambda$ instead of $S_k = \lambda$. In the proof of the second half of the lemma we essentially assume that $S_k \doteq \lambda$ and then bound the probability that this assumption is false.

Note that the sum of two independent symmetric random variables is also symmetric. That is, $-(X_1 + X_2) = (-X_1) + (-X_2)$ clearly has the same distribution as $X_1 + X_2$. Therefore if $1 \leq k < n$ then

$$S_n - S_k = X_{k+1} + \cdots + X_n$$

is symmetric and independent of any event defined in terms of $X_1, \ldots, X_k$.

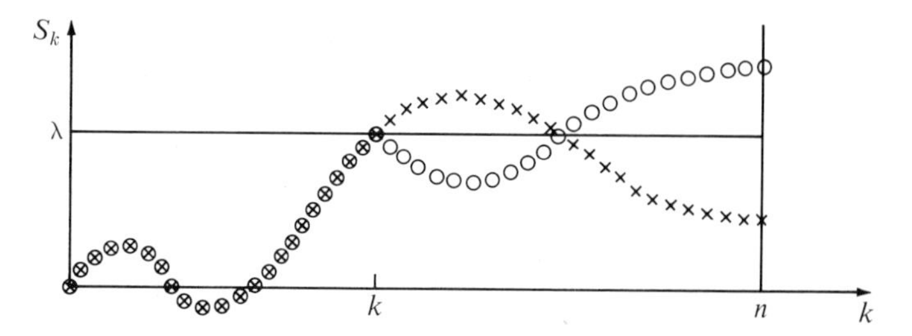

Fig. 6 The × curve $\in A_k[S_n < \lambda]$; the ○ curve $\in A_k[S_n \geq \lambda]$.

Let us now prove the first half of the lemma. In the following chain the two inequalities use the fact that $S_k(\omega) \geq \lambda$ whenever ω belongs to A_k:

$$\begin{aligned} P(A_k[S_n < \lambda]) \leq P(A_k[S_n < S_k]) &= P(A_k)P[S_n - S_k < 0] \\ &= P(A_k)P[S_n - S_k > 0] \\ &= P(A_k[S_n > S_k]) \leq P(A_k[S_n \geq \lambda]). \end{aligned}$$

Adding the two extremes over $k = 1, 2, \ldots, n - 1$ yields

$$P(A[S_n < \lambda]) \leq \sum_{k=1}^{n-1} P(A_k[S_n \geq \lambda]) \leq P[S_n \geq \lambda]$$

since $A_1[S_n \geq \lambda], \ldots, A_{n-1}[S_n \geq \lambda]$ are disjoint subsets of $[S_n \geq \lambda]$. *The first half of the lemma is proved.*

We now prove the second half of the lemma. Consider chain of inequalities

$$\begin{aligned} P(A_k[S_n \geq \lambda + 2\epsilon]) &\overset{1}{\leq} P(A_k[S_k < \lambda + \epsilon][S_n \geq \lambda + 2\epsilon]) + P[X_k \geq \epsilon] \\ &\overset{2}{\leq} P(A_k[S_k < \lambda + \epsilon][S_n > S_k + \epsilon]) + P[X_k \geq \epsilon] \\ &\overset{3}{=} P(A_k[S_k < \lambda + \epsilon])P[S_n - S_k > \epsilon] + P[X_k \geq \epsilon] \\ &\overset{4}{=} P(A_k[S_k < \lambda + \epsilon][S_n - S_k < -\epsilon]) + P[X_k \geq \epsilon] \\ &\overset{5}{\leq} P(A_k[S_n < \lambda]) + P[X_k \geq \epsilon]. \end{aligned}$$

For inequality 1 note that $A_k[S_k \geq \lambda + \epsilon][S_n \geq \lambda + 2\epsilon]$ is a subset of $[X_k \geq \epsilon]$. For inequality 2 if $S_n(\omega) \geq (\lambda + \epsilon) + \epsilon$ and $(\lambda + \epsilon) > S_k(\omega)$ then $S_n(\omega) > S_k(\omega) + \epsilon$. Equalities 3 and 4 used the independence of $S_n - S_k$ and $A_k[S_k < \lambda + \epsilon]$. Equality 4 also used $P[S_n - S_k > \epsilon] = P[S_n - S_k < -\epsilon]$ which follows from the symmetry of $S_n - S_k$. For inequality 5, just as for 2, we replaced $[S_n - S_k < -\epsilon]$ by $[S_n < \lambda]$ and then dropped $[S_k < \lambda + \epsilon]$ from the intersection. Summing the two extremes of the chain yields

$$\sum_{k=1}^{n-1} P(A_k[S_n \geq \lambda + 2\epsilon]) \leq P(A[S_n < \lambda]) + \sum_{k=1}^{n-1} P[X_k \geq \epsilon]$$

which is used for the second inequality below.

If $S_n(\omega) \geq \lambda + 2\epsilon$ then ω may belong to one of $A_1, \ldots, A_{n-1}$; but if it does not then $S_{n-1}(\omega) < \lambda$, and hence certainly $X_n(\omega) \geq \epsilon$. Therefore we have the first of the inequalities

$$P[S_n \geq \lambda + 2\epsilon] \leq \sum_{k=1}^{n-1} P(A_k[S_n \geq \lambda + 2\epsilon]) + P[X_n \geq \epsilon]$$

$$\leq P(A[S_n < \lambda]) + \sum_{k=1}^{n-1} P[X_k \geq \epsilon] + P[X_n \geq \epsilon].$$

The two extremes yield the second half of the lemma. ■

Proof of Theorem I Let W_t for $t \geq 0$ with $W_0 = 0$ be a Wiener process. The theorem is obviously true if $\lambda = 0$ or $t = 0$. Fix $\lambda > 0$ and $t > 0$. We must prove that

$$P[M_t \geq \lambda] = 2P[W_t \geq \lambda].$$

For integer m let

$$B_m = \left[W\left(\frac{kt}{2^m}\right) \geq \lambda \text{ for at least one integer } k \text{ satisfying } 1 \leq k \leq 2^m\right].$$

These discrete time points $t/2^m, 2t/2^m, \ldots, t$ for B_m form a subset of the time points for B_{m+1}, and hence B_m is a subset of B_{m+1}. We may express

$$W\left(\frac{kt}{2^m}\right) = \sum_{j=1}^{k}\left[W\left(\frac{jt}{2^m}\right) - W\left(\frac{(j-1)t}{2^m}\right)\right]$$

as a sum of independent symmetric normal increments; hence Lemma 1 yields

$$P(B_m) \leq 2P[W_t \geq \lambda]$$

and letting m go to infinity gives the same bound for $P(B)$, where $B = \bigcup_{m=1}^{\infty} B_m$. But realizations are continuous, so that $[M_t > \lambda]$ is a subset of B, and hence

$$P[M_t > \lambda] \leq 2P[W_t \geq \lambda].$$

Letting $\epsilon > 0$ go to zero in

$$P[M_t \geq \lambda] \leq P[M_t > \lambda - \epsilon] \leq 2P[W_t \geq \lambda - \epsilon]$$

yields $P[M_t \geq \lambda] \leq 2P[W_t \geq \lambda]$.

To prove the reverse inequality note that for every $\epsilon > 0$ and integer n

$$P[M_t \geq \lambda] \geq P\left[W\left(\frac{kt}{n}\right) \geq \lambda \text{ for some } k = 1, 2, \ldots, n\right]$$

$$\geq 2P[W_t \geq \lambda + 2\epsilon] - nP\left[W\left(\frac{t}{n}\right) \geq \epsilon\right].$$

The last inequality was obtained by applying Lemma 1 to S_k given by

$$W\left(\frac{kt}{n}\right) = \sum_{j=1}^{k}\left[W\left(\frac{jt}{n}\right) - W\left(\frac{(j-1)t}{n}\right)\right].$$

Exercise 3 (Sec. 13-1) showed that if X has the standard normal density φ and if $x \geq 1$ then $P[X \geq x] \leq \varphi(x)$. Letting $x_n = \epsilon\sqrt{n}/\sigma\sqrt{t}$ we have

$$nP\left[W\left(\frac{t}{n}\right) \geq \epsilon\right] = nP\left[\frac{W(t/n)}{\sigma\sqrt{t/n}} \geq x_n\right] \leq n\varphi(x_n)$$

and this bound goes to zero as n goes to infinity, since x_n goes to infinity and the exponential factor dominates. Therefore

$$P[M_t \geq \lambda] \geq 2P[W_t \geq \lambda + 2\epsilon]$$

and letting ϵ go to zero completes the proof of Theorem 1. ■

EXERCISES

1. For a standard Wiener process fix $t > 0$ and, as in Fig. 7, let the random variables L and R be the closest zeros to the left and right of t, respectively. Thus L is the largest zero not

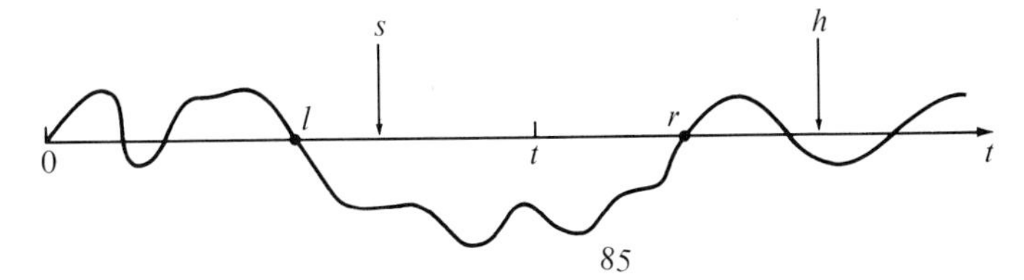

Fig. 7 The closest zeros, l and r, to t.

exceeding t while R is the smallest zero exceeding t. For $0 < s < t < h$ use the result of Example 2 to find the following:
(*a*) $P[L < s]$
(*b*) $P[R < h]$
(*c*) $P[L < s,\ R < h]$
2. Let X be normal 0 and 1 and show that $1/X^2$ has the same density as T_1 of Example 3.
3. In Example 3 it was shown by simple direct calculation that T_λ and $\lambda^2 T_1$ are identically distributed. This exercise shows how this same fact naturally arises within the context of a standard Wiener process. Justify the derivation

$$P[T_\lambda < t] = P[W(\tau) \geq \lambda \text{ at least once in } 0 < \tau < t]$$

$$= P\left[\frac{1}{\lambda} W(\lambda^2 s) \geq 1 \text{ at least once in } 0 < s < \frac{t}{\lambda^2}\right] = P\left[T_1 < \frac{t}{\lambda^2}\right].$$

4. Show that we obtain a semigroup if for $t > 0$ we define P_t to correspond to the density induced by $t^2 T_1$ where T_1 is as in Example 3.
5. *Realizations Do Not Have Isolated Roots.* Let W_t for $t \geq 0$ with $W_0 = 0$ be a Wiener process. It is a fact[1] that if $0 < s < t$ then

$$P[W_\tau = 0 \text{ for exactly one } \tau \text{ satisfying } s < \tau < t] = 0.$$

[1] Itô and McKean, *op. cit.* p. 29.

A realization r_ω has an **isolated root** at τ_0 iff $r_\omega(\tau_0) = 0$ and there is some $\epsilon > 0$ such that

$$r_\omega(\tau) \neq 0 \qquad \text{whenever } 0 < |\tau - \tau_0| < \epsilon.$$

Prove that the probability is 1 for the set of those ω whose realizations have no isolated roots.

6. *A. Hinčin's Local ($t \doteq 0$) and Global ($t \doteq \infty$) Laws of the Iterated Logarithm.* By iterated logarithm we mean log (log x), which is positive when $x > e$, since log (log e) $=$ log 1 $= 0$. Let $W(t)$ for $t \geq 0$ be a standard Wiener process. Define $X_0 = 0$ and $X_\tau = \tau W(1/\tau)$ for $\tau > 0$.

(*a*) Prove that X_τ for $\tau \geq 0$ has normal increments with $\text{Var}(X_t - X_s) = t - s$ whenever $0 \leq s < t$.

(*b*) Prove that X_τ for $\tau \geq 0$ is a standard Wiener process. Note that $\tau = 0, \infty$ corresponds to $1/\tau = \infty, 0$. (Results from Sec. 18-2 may be helpful.)

(*c*) For small $t \doteq 0$ realizations oscillate roughly between $\pm\sqrt{2t \log [\log (1/t)]}$. More precisely, considering just the "upper envelope," it is a fact that

$$\lim_{t \to 0} P[W(\tau) \geq (1 + \epsilon)\sqrt{2\tau \log [\log (1/\tau)]} \text{ for at least one } \tau \text{ satisfying } 0 < \tau \leq t] = \begin{cases} 0 & \text{if } \epsilon > 0 \\ 1 & \text{if } \epsilon < 0. \end{cases}$$

State precisely, as above, and prove that it then follows for large $t \doteq \infty$ that realizations roughly oscillate between $\pm\sqrt{2t \log (\log t)}$.

7. *Derivation for Example 2 by Using Theorem 1 Instead of the Reflection Principle.* As shown in Fig. 8, if W_s has the value $w < 0$ then we obtain a zero in the later interval iff $m \geq -w$. Let W_t for $t \geq 0$ with $W_0 = 0$ be a Wiener process and fix $0 < s < t$. Then $W_{s+h} - W_s$ for $h \geq 0$ is a Wiener process, so let M and M' be its maximum and minimum, respectively, over the interval $0 \leq h \leq t - s$. Clearly M and $-M'$ are identically distributed and each is independent of W_s. Justify each equality in the derivation below. For equation 4 we use Theorem 1 to introduce two random variables, X and Y, which are independent and each normal 0 and 1, so that their joint density is radially symmetric.

$$\begin{aligned} P[W_\tau &= 0 \text{ for at least one } \tau \text{ satisfying } s \leq \tau \leq t] \\ &\overset{1}{=} P[W_s < 0, M \geq -W_s] + P[W_s \geq 0, -M' \geq W_s] \\ &\overset{2}{=} P[W_s < 0, M \geq -W_s] + P[W_s \geq 0, M \geq W_s] \\ &\overset{3}{=} P[M \geq |W_s|] \\ &\overset{4}{=} P[\sigma\sqrt{t-s}\,|X| \geq \sigma\sqrt{s}\,|Y|] \\ &\overset{5}{=} P\left[\left|\frac{Y}{X}\right| \leq \frac{\sqrt{t-s}}{\sqrt{s}}\right] \\ &\overset{6}{=} \frac{4 \arccos \sqrt{s/t}}{2\pi} \end{aligned}$$

8. Show directly, without appeal to Theorem 1 or Example 4, that realizations cannot have *finite* right-hand derivatives at the origin. HINT: If r_ω has a finite right-hand derivative at the origin then $|r_\omega(t)/t|$ must be bounded and hence must be less than or equal to $1/t^{1/4}$ for small

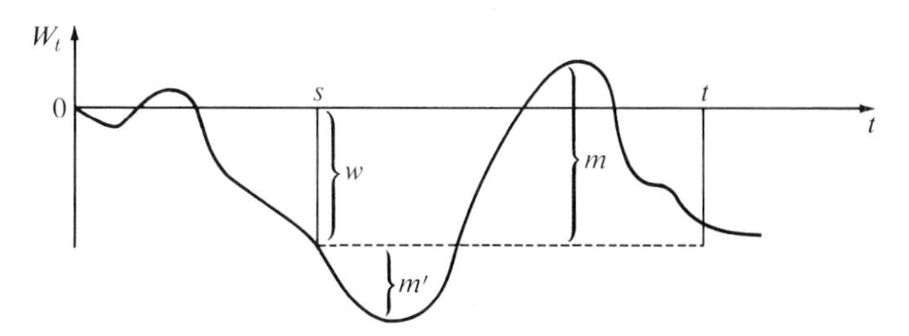

Fig. 8

enough t. Let

$$A_n = \left\{\omega : |r_\omega(t)| \le t^{3/4} \text{ for all } t \text{ satisfying } 0 < t < \frac{1}{n}\right\}.$$

9. If f is a continuous real-valued function on the unit interval $0 \le t \le 1$, $f(0) = 0$ and $f(1) \ne 0$ then it is easily proved that f has a last zero; that is, there is a unique τ, depending on f, and satisfying

$$0 \le \tau < 1 \qquad f(\tau) = 0 \qquad f(t) \ne 0 \qquad \text{whenever } \tau < t \le 1.$$

(For proof let τ be the supremum of the set of zeros, then use the continuity of f.) Therefore with probability 1 a Wiener process must produce a realization which has a last zero on the interval $0 \le t \le 1$. Reconcile this fact with Example 1.

24-3 COVARIANCE FUNCTIONS AND NORMAL PROCESSES

Many theoretical and practical problems naturally involve quite detailed assumptions or knowledge concerning probability distributions. However, only means, variances, and covariances are required for *linear* predictors, and for certain useful asymptotic results such as the laws of large numbers and central-limit theorems. Naturally, then, for certain problems involving stochastic processes we need only the covariance between each pair of random variables, and possibly the mean of each. Also sometimes additional information is lacking, or there may not be available any tractable mathematical theory capable of incorporating the additional information in a meaningful fashion.

At any rate, there exists an extensive, widely applied theory[1] based almost exclusively on covariance functions and their Fourier transforms. This severe restriction to use of only means and covariances is not a limitation when all joint densities are

[1] See, for example, H. Cramér and M. R. Leadbetter, "Stationary and Related Stochastic Processes," John Wiley & Sons, Inc., New York, 1967; W. B. Davenport, Jr., and W. L. Root, "Random Signals and Noise," McGraw-Hill Book Company, New York, 1958; A. Papoulis, "Probability, Random Variables, and Stochastic Processes," McGraw-Hill Book Company, New York, 1965; E. Parzen, "Stochastic Processes," Holden-Day, Inc., San Francisco, 1962, especially chap. 3; D. Sakrison, "Communication Theory: Transmission of Waveforms and Digital Information," John Wiley & Sons, Inc., New York, 1968; and A. M. Yaglom, "An Introduction to the Theory of Stationary Random Functions," trans. by R. S. Silverman, Prentice-Hall, Inc., Englewood Cliffs, N.J., 1962. This last book gives particular emphasis to extrapolation and filtering.

normal, and these more tractable processes are widely used and studied. The definitions of this section are basic to the aforementioned theory, although we make no attempt to properly introduce that theory.

If X_t for $t \in T$ is a stochastic process for which $E(X_t)^2 < \infty$ for all t in T then the function $m: T \to R$ defined by

$$m(t) = EX_t \qquad \text{for all } t \in T$$

is called its **mean-value function** and the function $K: T^2 \to R$ defined by

$$K(s,t) = \text{Cov}\,(X_s,X_t) = E(X_s - EX_s)(X_t - EX_t)$$

for all s and t in T is called its **covariance kernel**. Clearly $K(s,t) = K(t,s)$ and $K(t,t) = \text{Var}\, X_t$.

If "X_t for $t \geq 0$ with $X_0 = 0$" has independent increments and $0 \leq s \leq t$ then

$$\begin{aligned}\text{Cov}\,(X_s,X_t) &= \text{Cov}\,(X_s, X_s + X_t - X_s)\\ &= \text{Cov}\,(X_s,X_s) + \text{Cov}\,(X_s, X_t - X_s) = \text{Var}\, X_s\end{aligned}$$

since X_s and $X_t - X_s$ are independent and hence uncorrelated. Thus for any s and t, $\text{Cov}\,(X_s,X_t)$ equals the variance of the random variable with the smaller index,

$$K(s,t) = \text{Var}\, X_{\min\{s,t\}}.$$

Thus if "X_t for $t \geq 0$ with $X_0 = 0$" is a stationary-independent-increment process and $0 \leq s \leq t$ then Theorem 1 (Sec. 22-2) yields $K(s,t) = K(t,s) = s\,\text{Var}\, X_1$. Hence for *any* s and t we have

$$\begin{aligned}m(t) &= tEX_1 && \text{for all } t \geq 0\\ K(s,t) &= (\min\{s,t\})(\text{Var}\, X_1) && \text{for all } s \geq 0 \text{ and } t \geq 0.\end{aligned}$$

In particular $EX_1 = \text{Var}\, X_1 = \lambda$ for a Poisson process, and $EX_1 = 0$ and $\text{Var}\, X_1 = \sigma^2$ for a Wiener process. These processes are clearly not covariance stationary.

A process is said to be **covariance stationary** iff $K(s,t)$ is only a function of the difference $t - s$. That is, there is a real-valued function f called the **covariance function** such that $K(s,t) = f(t - s)$ for all s and t in T. This property is also referred to as *wide-sense* or *second-order stationary*. The mean EX_t may depend on t. Even if the mean-value function is constant, a covariance-stationary process need not be stationary. However, every stationary process, which has finite second moments, must obviously be covariance stationary, because *the distribution* of (X_s,X_t) depends only on $t - s$. For a covariance-stationary process

$$f(s - t) = \text{Cov}\,(X_s,X_t) = \text{Cov}\,(X_t,X_s) = f(t - s)$$

so that f must be an even function.

To prove that a process is covariance stationary we need only find a function f defined on the nonnegative reals and such that $K(s,t) = f(t - s)$ whenever $s \leq t$, because when $s > t$ then

$$K(s,t) = K(t,s) = f(s - t) = f(|t - s|)$$

so that $K(s,t) = f(|t - s|)$ is true for *all* s and t.

A stochastic process "$X(t)$ for $t \in T$" is called a **normal** or **gaussian process** iff for every $n \geq 1$ points $t_1 < \cdots < t_n$ in T the random vector $(X(t_1), \ldots, X(t_n))$ induces a multivariate normal distribution. In other words, the family which the process generates contains only multivariate normal distributions. But such a distribution is completely determined by its vector of means and its covariance matrix. Therefore the entire family of finite distributions generated by a normal process is completely determined by its mean-value function and covariance kernel.

A normal process is stationary iff its mean-value function is a constant and the process is covariance stationary. We need only observe that in this case the vector of means and the covariance matrix for $(X(t_1 + h), \ldots, X(t_n + h))$ do not depend on h, so the distribution does not depend on h.

A Wiener process W_t for $t \geq 0$ is a normal process because

$$W(t_k) = \sum_{i=1}^{k} \sigma\sqrt{t_i - t_{i-1}} \left\{\frac{W(t_i) - W(t_{i-1})}{\sigma\sqrt{t_i - t_{i-1}}}\right\}$$

so that $(W(t_1), \ldots, W(t_n))$ can be obtained by applying a linear transformation to the independent normal 0 and 1 variables in braces. This was done explicitly for $n = 2$ in Example 2 (Sec. 24-2). We next transform a Wiener process into a stationary normal process.

THE ORNSTEIN-UHLENBECK PROCESS

Let $W(\tau)$ for $\tau \geq 0$ be a standard Wiener process. Let

$$X_t = \frac{1}{\sqrt{g(t)}} W(\alpha g(t))$$

where $\alpha > 0$, and the function $g: R \to R$ is positive, strictly increasing and normalized to $g(0) = 1$. Clearly $EX_t = 0$ and Var $X_t = \alpha$ independently of t, since the multiplicative function $1/\sqrt{g(t)}$ was uniquely selected to ensure this. Clearly the new process is normal, since $(X(t_1), \ldots, X(t_n))$ is obtained by applying a diagonal linear transformation to $(W(\alpha g(t_1)), \ldots, W(\alpha g(t_n))$, which has a multivariate normal density. We wish to specialize g so as to make the new process stationary.

If $\tau > 0$ then $g(t) < g(t + \tau)$ so that

$$EX(t + \tau)X(t) = \frac{\text{Var } W(\alpha g(t))}{[g(t)g(t + \tau)]^{1/2}} = \alpha\left[\frac{g(t)}{g(t + \tau)}\right]^{1/2}$$

which we want to be independent of t, and so always equal to its value $\alpha/\sqrt{g(\tau)}$ at $t = 0$. We are thus led to the equation $g(t + \tau) = g(t)g(\tau)$, so that $g(t) = e^{2\beta t}$ must be an exponential function with some constant 2β. In this case the expression above becomes $\alpha e^{-\beta\tau}$ where $\tau > 0$.

To summarize, if $W(\tau)$ for $\tau \geq 0$ is a standard Wiener process then the process "$X(t)$ for t real" defined by

$$X(t) = e^{-\beta t} W(\alpha e^{2\beta t}) \qquad \text{for all real } t$$

is called the **Ornstein-Uhlenbeck process** with parameters $\alpha > 0$ and $\beta > 0$. It is a stationary normal process with zero mean-value function and covariance kernel

$$\text{Cov}\,(X(s),X(t)) = \alpha e^{-\beta|s-t|}$$

for all real s and t. All its realizations are continuous functions, since this property was assumed for Wiener processes.

Figure 1 shows a realization from the Wiener process and the corresponding transformed realization from the process "$Y(t)$ for t real," where $Y(t) = W(\alpha e^{2\beta t})$. If

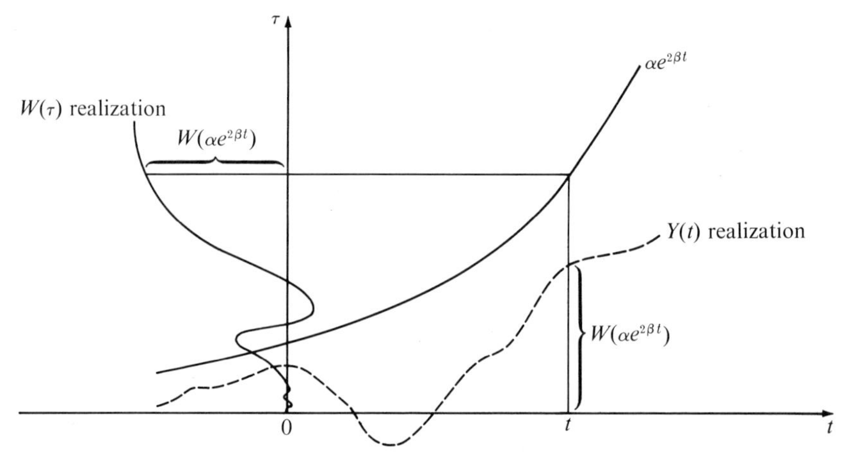

Fig. 1

this latter realization, which is shown as a dashed line, is multiplied by the function $e^{-\beta t}$, then we obtain the corresponding realization from "$X(t)$ for t real." Obviously realizations from $X(t)$ for t real must, as for a Wiener process, be continuous, non-differentiable, have no isolated roots, etc.

Exercise 2 shows that if $\alpha = 1$ and $\beta = 2\lambda$ then the Ornstein-Uhlenbeck process has the same mean-value function and covariance function as the random telegraph signal of Example 2 (Sec. 23-1). Thus drastically different processes can have the same covariance function.

EXERCISES

1. Let $X(t)$ for t real be an Ornstein-Uhlenbeck process with $\alpha = \beta = 1$. For real $r \leq h$ prove that

$$P[X(\tau) = 0 \text{ for at least one } \tau \text{ satisfying } r \leq \tau \leq h] = \frac{2}{\pi} \arccos e^{-(h-r)}.$$

2. In Example 2 (Sec. 23-1) the random telegraph stationary process X_t for $t \geq 0$ was defined by $X_t = X_0(-1)^{N_t}$ where N_t for $t \geq 0$ is a Poisson process and X_0 is an independent random variable with distribution $P[X_0 = 1] = P[X_0 = -1] = .5$. Each X_t has this same distribution, so $EX_t = 0$ for all $t \geq 0$. Show that its covariance function is

$$e^{-2\lambda|t|} \qquad \text{for } t \text{ real.}$$

In Sec. 23-3 the random telegraph signal was extended to the whole real line, and this extended process obviously has the same covariance function.

Appendixes

APPENDIX 1: SEVERAL FREQUENTLY USED FORMULAS

$$1 + 2 + 3 + \cdots + n = \frac{n(n+1)}{2} \qquad \text{for } n = 1, 2, \ldots \tag{1}$$

$$1 + q + q^2 + \cdots + q^{n-1} = \frac{1 - q^n}{1 - q} \qquad \text{for } q \neq 1,\ n = 1, 2, \ldots \tag{2}$$

$$1 + q + q^2 + \cdots = \frac{1}{1 - q} \qquad \text{for } -1 < q < 1 \tag{3}$$

$$\sum_{k=0}^{\infty} (k + r - 1) \cdots (k + 1) q^k = \sum_{k=0}^{\infty} \frac{(k + r - 1)!}{k!} q^k$$

$$= \frac{(r - 1)!}{(1 - q)^r} \qquad \text{for } -1 < q < 1,\ r = 1, 2, \ldots \tag{4}$$

$$e^x = \sum_{n=0}^{\infty} \frac{x^n}{n!} = 1 + x + \frac{x^2}{2!} + \frac{x^3}{3!} + \cdots \qquad \text{for } x \text{ real} \tag{5}$$

$$\lim_{n \to \infty} \left(1 + \frac{x}{n}\right)^n = e^x \qquad \text{for } x \text{ real} \tag{6}$$

$$1 - x \leq e^{-x} \qquad \text{for } x \text{ real} \tag{7}$$

$$\log x \leq x - 1 \qquad \text{for } x > 0 \tag{8}$$

$$(1 - x_1)(1 - x_2) \cdots (1 - x_N) \leq e^{-(x_1 + x_2 + \cdots + x_N)}$$

$$\text{if } x_1 \leq 1, \ldots, x_N \leq 1,\ N = 1, 2, \ldots \tag{9}$$

Proof of formula (1) Let S_n denote the sum of the first n positive integers, so that

$$S_n = 1 + 2 + 3 + \cdots + (n - 1) + n$$

$$S_n = n + (n - 1) + (n - 2) + \cdots + 2 + 1.$$

Then $2S_n = n(n + 1)$, since adding terms placed one above the other always yields $n + 1$, and there are n terms.

Proof of formula (2) Let $f_n(q)$ denote the sum of the first n terms in a geometric progression, so that

$$f_n(q) = 1 + q + q^2 + \cdots + q^{n-1}$$

and hence

$$qf_n(q) = q + q^2 + \cdots + q^{n-1} + q^n.$$

Subtracting the lower equation from the upper yields $f_n(q) - qf_n(q) = 1 - q^n$.

Proof of formula (3) If $-1 < q < 1$ let n go to infinity in formula (2), so that q^n goes to zero, yielding formula (3) for the sum of the *geometric series*.

Proof of formula (4) Differentiate both sides of formula (3) to get

$$\frac{d}{dq}\frac{1}{1-q} = \frac{1}{(1-q)^2} = 1 + 2q + 3q^2 + \cdots + (k+1)q^k + \cdots$$

which is formula (4) for $r = 2$. Differentiate this last equation to get

$$\frac{d^2}{dq^2}\frac{1}{1-q} = \frac{2}{(1-q)^3} = 2 + (3)(2)q + \cdots + (k+2)(k+1)q^k + \cdots$$

which is formula (4) for $r = 3$. The general formula (4) follows by induction. Also, as in Exercise 10 (Sec. 14-2), the binomial series yields formula (4).

Proof of formulas (5) and (6) These are standard facts from calculus. A generalization of formula (6) is proved in Exercise 9*b* (Sec. 8-3).

Proof of formula (7) The inequality is depicted in Fig. 1. Let $f(x) = e^{-x} - (1 - x)$ so $f(0) = 0$, $f'(0) = 0$, and $f''(x) = e^{-x} > 0$. Therefore f is convex and tangent to the x axis at the origin, and hence never below the x axis.

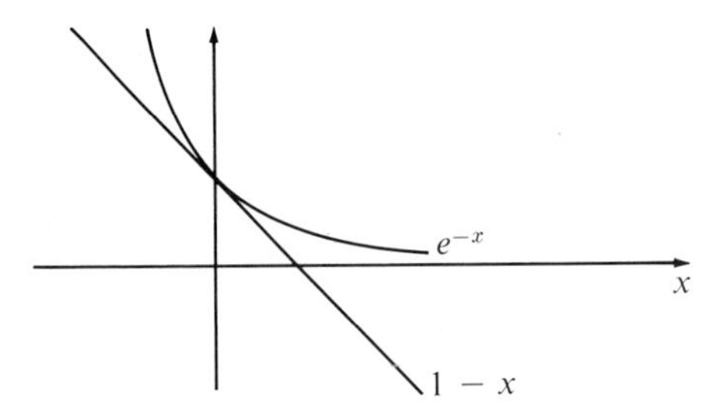

Fig. 1

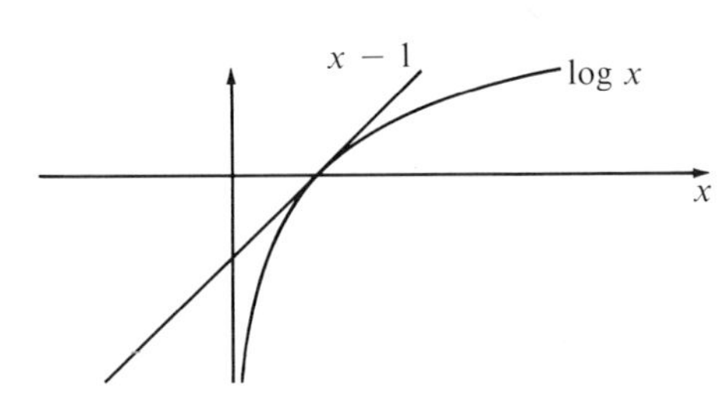

Fig. 2

Proof of formula (8) The inequality is depicted in Fig. 2. Let $y = 1 - x$ in formula (7) so $y \le e^{y-1}$. If $y > 0$ we can take logarithms and since log is an increasing function we get $\log y \le y - 1$.

Proof of formula (9) From formula (7) $1 - x_i \le e^{-x_i}$, and if $x_i \le 1$ then both sides are nonnegative, so these N inequalities can be multiplied.

APPENDIX 2: THE SUPREMUM

This appendix introduces the supremum, or least upper bound, of a set of real numbers. The supremum is used in the proof of the central-limit theorem of Chap. 19 to define a distance between distribution functions on R. Appendix 3 uses the supremum to prove the two theorems of Chap. 5 on absolute convergence.

We always assume the usual addition, subtraction, multiplication, division, and order properties of the rational numbers (ratios of integers) and the real numbers. The real number system is the essentially unique number system having the following property.

Property 1 Every bounded nondecreasing sequence of real numbers converges to a real number. That is, assume that $c_1 \le c_2 \le c_3 \le \cdots$ is any nondecreasing sequence of real numbers such that there is a real number b satisfying $c_n \le b$ for all n. Then there exists a real number c such that c_n converges to c as n goes to infinity. ■

We will shortly introduce Property 2 and show it to be equivalent to Property 1. We then define the supremum and collect its basic properties in Lemma 1. The missing details in this appendix are trivial exercises.

It is easy to construct a nondecreasing sequence of rational numbers which converge to $\sqrt{2}$. But Exercise 3 (Sec. 4-3) showed that $\sqrt{2}$ is not a rational number. Thus if Property 1′ is obtained by replacing the term "real" by "rational" *throughout* Property 1, then Property 1′ is false. Thus many sequences of rational numbers which obviously should be considered to be converging cannot have a rational number assigned as their limit. In a sense the aforementioned sequence of rational numbers is converging to a "hole" or "missing point" in the collection of rational numbers, and this "hole" may be "plugged" by the real number $\sqrt{2}$. Property 1 says that the real number system does not have any "holes."

We now indicate another way of saying that the real number system does not have any holes. Let c be a real number and consider the real number axis as cut into two pieces at the point c. Let C consist of all real $x \le c$, or all real $x < c$, so that C and C^c are the two pieces so obtained. Property 2 asserts that if we have some set C of reals which *appears* to be the left piece from such a cut then there is a real-number "cut point" which yields C. The rational numbers do not have this property since we can let C' be the set of all rational numbers less than $\sqrt{2}$; that is, C' consists of all negative rational numbers together with those nonnegative rationals whose square is less than 2. Then there is no rational number which might reasonably be called the cut point for C'.

A set C of real numbers is called a **cut set** iff $C \ne \varphi$, $C^c \ne \varphi$, and $x < y$ whenever $x \in C$ and $y \in C^c$. That is, both C and its complement C^c in the set of real numbers are nonempty, and every member of C is to the left of every member of C^c. Clearly a cut set *appears* to be the left piece of a cut. Also if $x' < x \in C$ then $x' \in C$, because $x' \in C^c$ would violate the definition of C. Thus C contains every number to the left of any number in C, and C^c contains every number to the right of any number in C^c.

Property 2 If C is any cut set then there is a unique real number c such that either

$$C = \{x : x \le c\} \qquad \text{or} \qquad C = \{x : x < c\}. \quad ■$$

Proof that Property 1 implies Property 2 Let C be a cut set. We construct a nondecreasing sequence which should converge to a real-number cut point for C. For any positive integer n let

$$R_n = \left\{x : x = \frac{j}{2^n} \text{ for some integer } j\right\} = \left\{0, \pm\frac{1}{2^n}, \pm\frac{2}{2^n}, \ldots\right\}.$$

There is some real number in C, and every smaller $j/2^n$ belongs to C, so that $CR_n \ne \varphi$. Also there is a real $b \in C^c$ and $x < b$ whenever $x \in C$. Thus CR_n is nonempty and bounded above. From the definition of R_n it is clear that there is a unique largest number c_n in CR_n. We have defined c_n for every n. Now $R_n \subset R_{n+1}$, so $CR_n \subset CR_{n+1}$, and hence $c_n \le c_{n+1}$. From Property 1 there is a real number c such that c_n converges to c as n goes to infinity.

To complete the proof we must show that $x < c$ implies $x \in C$, and that $c < x$ implies $x \notin C$. If $x < c$ then from $c_n \to c$ we know that $x < c_n \in C$ for large enough n, and hence $x \in C$. We always have $c_n \leq c$ and $[c_n + (1/2^n)] \in C^c$, so $[c + (1/2^n)] \in C^c$. If $c < x$ then $c + (1/2^n) < x$ for large enough n, so $x \in C^c$. ■

Proof that Property 2 implies Property 1 Let c_n be a nondecreasing sequence of real numbers which is bounded above. Let $C = \{x: x \leq c_n \text{ for some } n\}$. Clearly C is a cut set, so Property 2 says that there is a real number c such that $C = \{x: x \leq c\}$ or $\{x: x < c\}$. Each $c_n \in C$ so each $c_n \leq c$. If $\epsilon > 0$ then $(c - \epsilon) \in C$ so there exists a $c_n \geq c - \epsilon$. Therefore c_n converges to c. ■

The next paragraph contains some standard definitions. After that we define the supremum of a cut set and then generalize the definition.

Let A be a nonempty set of real numbers. A number b, which may or may not belong to A, is an **upper** (lower) **bound** for A iff $x \leq b$ ($x \geq b$) whenever $x \in A$. If b is an upper (lower) bound for A then clearly so is every b' greater (less) than b. The set A is said to be **bounded above** (below) iff it has at least one upper (lower) bound. The set A is said to be **bounded** iff there exists a real b such that $|x| \leq b$ whenever $x \in A$. That is, A is bounded both above and below. We say that A has, or achieves, a **maximum** (minimum) iff there is a real $M(m)$ such that $M \in A$ ($m \in A$) and $x \leq M$ ($x \geq m$) whenever $x \in A$. That is, A has an upper (lower) bound *which belongs to* A. Clearly if A has a maximum (minimum) it is unique.

If C is either of the cut sets $\{x: x \leq c\}$ or $\{x: x < c\}$ then the real number c is called the **supremum of** C, denoted by sup C. If $C = \{x: x \leq c\}$ then c achieves a maximum and sup C = max C. If $C = \{x: x < c\}$ then C does not have a maximum, and sup C does not belong to C, and $x <$ sup C for every $x \in C$. We often wish to refer to that point which determines the right end of C, regardless of whether or not C has a maximum, so we have named that point.

If C is one of the cut sets $\{x: x \leq c\}$ or $\{x: x < c\}$ then *in both cases* sup C may be described as the *smallest* number s such that $x \leq s$ whenever $x \in C$. That is, the set $\{x: x \geq c\}$ of all those real numbers which are upper bounds for the set C is the same in both cases, and $\{x: x \geq c\}$ *has a minimum* or least number. Thus sup C may be called the *least upper bound* for C, and it is often denoted by lub C. We now easily generalize these ideas.

Let A be an arbitrary nonempty set of real numbers which is bounded above. Often such a set A will arise in a very complicated fashion, and we may not know whether or not A has a maximum, and sometimes we do not really care. However, we frequently want to refer to that number which determines the right end of A. We may adjoin to A all numbers to the left of numbers in A, roughly by just filling in gaps when A is viewed as a subset of the real-number axis. More precisely, we form the set

$$C = \{x: x \leq a \text{ for some } a \in A\}$$

so that

$$C^c = \{x: x > a \text{ for all } a \in A\}.$$

Clearly A is a subset of C, which is a cut set. Naturally we think of the same point as determining the "right end" of both A and C, so we make the following definition: if A is any nonempty set of real numbers which is bounded above then

$$\textbf{sup}\, A = \sup \{x: x \leq a \text{ for some } a \in A\}.$$

The set $U = \{x: x \geq a \text{ for all } a \in A\}$ of all upper bounds for A contains C^c and sup C, but nothing else. Hence U has a minimum, so sup A can also be called the *least upper bound* for A, and it is often denoted by lub A. Thus sup A can be described as the smallest number s such that $a \leq s$ whenever $a \in A$. For example, if

$$A = \left\{x: x = -\frac{1}{n} \text{ for some positive integer } n\right\} = \left\{-\frac{1}{1}, -\frac{1}{2}, -\frac{1}{3}, \ldots\right\}$$

then lub A = sup A = sup $\{x: x < 0\} = 0$. Zero is certainly that point which determines the right end of A, even though zero does not belong to A, which does not have a maximum.

Let A be a nonempty set of real numbers which is bounded above. Then sup A is an upper bound for A, and no smaller number is an upper bound for A. That is, the number sup A satisfies the following two conditions:

1. $a \leq \sup A$ whenever $a \in A$.
2. For any $x < \sup A$ there exists an $a \in A$ such that $x < a$.

Clearly no other number can satisfy these two conditions, which are often used when working with suprema, and have the advantage of applying in either of the two cases when A does, or does not, achieve a maximum. We collect our results into the following lemma, which also names the analogous point which determines the "left end" of a set of real numbers.

Lemma 1 If A is any nonempty set of real numbers which is bounded above then the set $\{x: x \geq a \text{ for all } a \in A\}$ of all upper bounds for A has a minimum which is called the *supremum of A* or the *least upper bound for A* and is denoted by sup A or lub A. If A has a maximum then sup A = max A. Thus sup A is the unique real number satisfying the following two conditions:

(*a*) $a \leq \sup A$ whenever $a \in A$.
(*b*) For any $x < \sup A$ there exists an $a \in A$ such that $x < a$.

Similarly if A is any nonempty set of real numbers which is bounded below then the set $\{x: x \leq a \text{ for all } a \in A\}$ of all lower bounds for A has a maximum, which is called the **infimum of** A or the **greatest lower bound for** A, denoted by inf A or glb A. If A has a minimum then inf A = min A. Thus inf A is the unique real number satisfying the following two conditions:

(*c*) $a \geq \inf A$ whenever $a \in A$.
(*d*) For any $x > \inf A$ there exists an $a \in A$ such that $a < x$. ■

The following property of suprema is used often—as, for example, in Appendix 3. If $\varphi \neq A \subset B$ and B is bounded above then $\sup A \leq \sup B$. This is true because sup B is an upper bound for B and hence for A. Thus a larger set must have a larger supremum. More generally if A is nonempty, and if B is bounded above, and if for every $a \in A$

$$a \leq b \qquad \text{for some } b \in B$$

then $\sup A \leq \sup B$. Thus *if for every number in A there is one at least as large in B then*

$$\sup A \leq \sup B$$

because every upper bound for B must be an upper bound for A.

If $F: R \to R$ and $G: R \to R$ are two distribution functions, so that $|F - G| \leq 1$, then in Chap. 19 we define the ρ distance between F and G to be the real number

$$\sup \{y: y = |F(\alpha) - G(\alpha)| \text{ for some real } \alpha\}.$$

APPENDIX 3: SUMMABILITY

In this appendix we use the supremum concept from Appendix 2 to define the sum of a function $f: \Omega \to R$ over an "essentially countable" subset A of Ω. We require a condition similar to absolute convergence, but not depending on a particular ordering. Lemma 1 relates this sum to approximations using only finite summations. Lemma 2 and 3 prove countable additivity over disjoint sets. Lemma 4 shows that the same sum is obtained from every series constructed from a proper enumeration of A. We then easily prove the two theorems of Chap. 5 on absolutely convergent single and double sums. In a sense this appendix provides a rigorous general definition and notation for summations as used in the discrete case.

If $f: \Omega \to R$ is any real-valued function defined on some set Ω and if $J = \{\omega_1, \ldots, \omega_n\}$ is a finite set consisting of some number $n \geq 1$ of distinct members of Ω then we call $S_f(J) = f(\omega_1) + \cdots + f(\omega_n)$ the **sum of f over J**. We set $S_f(\varphi) = 0$. Clearly $0 \leq |S_f(J)| \leq S_{|f|}(J)$. We say that **$f$ is summable over a subset** A of Ω iff there is some real number b such that $S_{|f|}(J) \leq b$ whenever J is a finite subset of A: That is, there is some b which is an upper bound to the sum of the absolute values of f over *every* finite subset of A.

Obviously f is summable over A iff $|f|$ is summable over A. If f is summable over A then f is summable over every subset of A. Also f is summable over $A \cup B$ iff f is summable over A and over B.

Let

$$A^+ = \{\omega: \omega \in A \text{ and } f(\omega) > 0\} \qquad \text{and} \qquad A^- = \{\omega: \omega \in A \text{ and } f(\omega) < 0\}.$$

Clearly f is summable over A iff f is summable over

$$A^+ \cup A^- = \{\omega: \omega \in A \text{ and } f(\omega) \neq 0\}.$$

Hence f is summable over A iff f is summable over A^+ and over A^-.

Furthermore f is summable over A iff there is some real number b such that $|S_f(J)| \leq b$ whenever J is a finite subset of A. For proof note that we always have $|S_f(J)| \leq S_{|f|}(J)$, and equality holds whenever J is a finite subset of A^+, or of A^-.

If f is summable over A then the set $A^+ \cup A^-$ must be countable, because

$$A^+ \cup A^- = \bigcup_{n=1}^{\infty} A_n \qquad \text{where } A_n = \left\{\omega: \omega \in A \text{ and } |f(\omega)| \geq \frac{1}{n}\right\}$$

so we need only show that each A_n is finite. If $S_{|f|}(J) \leq b$ for all finite subsets J of A and if distinct $\omega_1, \ldots, \omega_k$ each belong to A_n then

$$k\frac{1}{n} \leq |f(\omega_1)| + \cdots + |f(\omega_k)| \leq b$$

so $k \leq nb$, and hence A_n must be finite.

If $f: \Omega \to R$ and $A \subset \Omega$ then we let $R_f(A) = \{s:$ there exists a finite subset J of A with $s = S_f(J)\}$. Thus $R_f(A)$ is the set of those real numbers which are obtainable as sums of f over finite subsets of A. Clearly $0 \in R_f(B) \subset R_f(A)$ whenever $B \subset A$. Also $R_f(A) = R_f(A^+ \cup A^-)$.

Clearly f is summable over A iff $R_{|f|}(A)$ is bounded above, iff $R_f(A^+)$ is bounded above and $R_f(A^-)$ is bounded below, iff $R_f(A)$ is bounded above and below. If $f: \Omega \to R$ is summable over $A \subset \Omega$ then we can make the following definitions:

$$S_f(A^+) = \sup R_f(A^+) = \sup R_f(A) \geq 0$$
$$S_f(A^-) = \inf R_f(A^-) = \inf R_f(A) \leq 0$$
$$S_f(A) = S_f(A^+) + S_f(A^-)$$

Clearly our new more general definition of $S_f(A)$ agrees with our old definition when A is a finite set. If f is summable over A then certainly $0 \leq S_{|f|}(B) \leq S_{|f|}(A)$ whenever $B \subset A$. Also $S_f(A) = S_f(A^+ \cup A^-)$, and it is easily seen that $|S_f(A)| \leq S_{|f|}(A)$.

Lemma I If $f: \Omega \to R$ is summable over $A \subset \Omega$ then for every $\epsilon > 0$ there is a finite subset J of A such that

$$|S_f(A) - S_f(B)| \leq \epsilon \qquad \text{whenever } J \subset B \subset A. \quad \blacksquare$$

The set B need not be finite, but as an extreme case we may let $B = J$. Thus there is a finite subset J with $S_f(J)$ close to $S_f(A)$, and every intermediate B has $S_f(B)$ close to $S_f(A)$.

Proof of Lemma I Take any $\epsilon > 0$. From Lemma 1 (Appendix 2) there is a finite subset J^+ of A^+ and there is a finite subset J^- of A^- such that

$$S_f(J^+) \geq \sup R_f(A^+) - \frac{\epsilon}{2} \qquad S_f(J^-) \leq \inf R_f(A^-) + \frac{\epsilon}{2}.$$

Let $J = J^+ \cup J^-$ and take any B satisfying $A \supset B \supset J$. Then

$$\sup R_f(A^+) \geq \sup R_f(B^+) \geq S_f(J^+) \geq \sup R_f(A^+) - \frac{\epsilon}{2}.$$

Therefore

$$|\sup R_f(A^+) - \sup R_f(B^+)| \leq \frac{\epsilon}{2}$$

and similarly

$$|\inf R_f(A^-) - \inf R_f(B^-)| \leq \frac{\epsilon}{2}$$

which together prove the lemma. $\blacksquare$

Lemma 2 If A and B are disjoint subsets of Ω and if $f: \Omega \to R$ is summable over $A \cup B$ then

$$S_f(A \cup B) = S_f(A) + S_f(B). \quad \blacksquare$$

Proof Take any $\epsilon > 0$ and let $C = A \cup B$. From Lemma 1 above there are finite subsets J'_C, J'_A, and J'_B of C, A, and B, respectively, such that each of

$$|S_f(C) - S_f(C')| \qquad |S_f(A) - S_f(A')| \qquad |S_f(B) - S_f(B')|$$

is $\epsilon/3$ or less whenever

$$J'_C \subset C' \subset C \qquad J'_A \subset A' \subset A \qquad J'_B \subset B' \subset B.$$

Let $J_A = J'_A \cup (J'_C \cap A)$ and $J_B = J'_B \cup (J'_C \cap B)$, so that $J_A \cup J_B$ contains J'_C. Then J_A and J_B are finite disjoint subsets of A and B, respectively, so that

$$S_f(J_A \cup J_B) = S_f(J_A) + S_f(J_B).$$

Adding and subtracting $S_f(J_A \cup J_B)$ inside the first absolute-value sign below yields

$$|S_f(A \cup B) - S_f(A) - S_f(B)| \leq |S_f(A \cup B) - S_f(J_A \cup J_B)| + |S_f(A) - S_f(J_A)|$$
$$+ |S_f(B) - S_f(J_B)| \leq \frac{\epsilon}{3} + \frac{\epsilon}{3} + \frac{\epsilon}{3} = \epsilon.$$

But this is true for each $\epsilon > 0$, so the lemma is proved. $\blacksquare$

If A_1, A_2, A_3 are disjoint then two applications of Lemma 2 yield

$$S_f((A_1 \cup A_2) \cup A_3) = S_f(A_1 \cup A_2) + S_f(A_3) = S_f(A_1) + S_f(A_2) + S_f(A_3).$$

Thus Lemma 2 extends to finite unions, and we now extend it to denumerable unions.

Lemma 3 Let $f\colon \Omega \to R$ and let $A = \bigcup_{n=1}^{\infty} A_n$ where $A_1, A_2, \ldots$ are disjoint subsets of Ω. Then f is summable over A, *iff*, f is summable over each A_n *and* the series $\sum_{n=1}^{\infty} S_{|f|}(A_n)$ is convergent. If f is summable over A then

$$S_f(A) = \sum_{n=1}^{\infty} S_f(A_n). \quad \blacksquare$$

Proof Assume that f is summable over each A_n and assume that the series $\sum_{n=1}^{\infty} S_{|f|}(A_n)$ is convergent. If J is any finite subset of A, and if M is the largest integer such that A_M contains at least one member of J, then $\bigcup_{n=1}^{M} A_n$ contains J, hence Lemma 2 yields

$$S_{|f|}(J) \le S_{|f|}\left(\bigcup_{n=1}^{M} A_n\right) = \sum_{n=1}^{M} S_{|f|}(A_n) \le \sum_{n=1}^{\infty} S_{|f|}(A_n)$$

so that f is summable over A.

If f is summable over A then Lemma 2 yields

$$\sum_{n=1}^{m} S_{|f|}(A_n) = S_{|f|}\left(\bigcup_{n=1}^{m} A_n\right) \le S_{|f|}(A) \qquad \text{for every } m,$$

so that the series is convergent.

Assume that f is summable over A. We must show that $\sum_{n=1}^{m} S_f(A_n)$ converges to $S_f(A)$ when m goes to infinity. Take any $\epsilon > 0$. Lemma 1 guarantees a finite subset J of A such that

$$|S_f(A) - S_f(B)| \le \epsilon \qquad \text{whenever } J \subset B \subset A.$$

Let M be the largest integer such that A_M contains at least one member of J. If $m \ge M$ then $J \subset \left(\bigcup_{n=1}^{m} A_n\right) \subset A$, so that

$$\left|S_f(A) - \sum_{n=1}^{m} S_f(A_n)\right| = \left|S_f(A) - S_f\left(\bigcup_{n=1}^{m} A_n\right)\right| \le \epsilon. \quad \blacksquare$$

Lemma 4 Let $f\colon \Omega \to R$ and let $\omega_1, \omega_2, \ldots$ be any sequence of *distinct* members of $A \subset \Omega$ such that every ω in A with $f(\omega) \ne 0$ appears in the sequence. Then f is summable over A iff the series $f(\omega_1) + f(\omega_2) + \cdots$ is absolutely convergent, and in this case the series converges to $S_f(A)$. $\blacksquare$

Proof Let A_n consist of the single point ω_n so that $S_f(A_n) = f(\omega_n)$. Applying Lemma 3 to $\bigcup_{n=1}^{\infty} A_n$ completes the proof. $\blacksquare$

Proof of Theorem I (Sec. 5-2) Let $a_1 + a_2 + \cdots$ be an infinite series. Let $\Omega = \{1, 2, \ldots\}$ and define $f\colon \Omega \to R$ by $f(n) = a_n$. Applying Lemma 4 to the enumeration $1, 2, \ldots$ shows that f is summable over Ω iff the series $a_1 + a_2 + \cdots$ is absolutely convergent, and in this case the series converges to $S_f(\Omega)$.

Assume that $a_1 + a_2 + \cdots$ is absolutely convergent, so f is summable over Ω, and let $b_1 + b_2 + \cdots$ be any reordering of $a_1 + a_2 + \cdots$. Then $b_n = a_{g(n)}$ where g is the reordering function, so every positive integer appears exactly once in the sequence $g(1), g(2), \ldots$. Lemma 4 shows that the series $b_1 + b_2 + \cdots = f(g(1)) + f(g(2)) + \cdots$ is absolutely convergent and converges to $S_f(\Omega)$. ■

Proof of Theorem 1 (Sec. 5-3) Let $\sum_{n=1}^{\infty} \sum_{m=1}^{\infty} a_{n,m}$ be a double series. Let Ω be the set of all ordered pairs (n,m) of positive integers, and define $f\colon \Omega \to R$ by $f(n,m) = a_{n,m}$. Clearly the double series is absolutely convergent iff f is summable over Ω. If the double series is absolutely convergent them Lemma 4 shows that its sum, as defined in Sec. 5-3, is just $S_f(\Omega)$.

Let $A_n = \{(n,1), (n,2), (n,3), \ldots\}$, so that $S_f(A_n) = a_{n1} + a_{n2} + \cdots$ is the nth row sum, if f is summable over A_n. Lemma 3 shows that f is summable over Ω iff f is summable over each A_n *and* the series $\sum_{n=1}^{\infty} S_{|f|}(A_n)$ is convergent. Also if f is summable over Ω then

$$S_f(\Omega) = \sum_{n=1}^{\infty} S_f(A_n) = \sum_{n=1}^{\infty}\left(\sum_{m=1}^{\infty} a_{nm}\right).$$

Similarly if $A_n = \{(1,n), (2,n), \ldots\}$ then we get the other iterated sum. ■

Next we examine briefly the application of this appendix to the discrete case. As in Chap. 6, we may define a discrete probability density as a function $p\colon \Omega \to R$ such that $p \geq 0$, p is summable over Ω, and $S_p(\Omega) = 1$. Then $\{\omega : p(\omega) > 0\}$ is countable, and the probability $P(A) = S_p(A)$ of any event $A \subset \Omega$ can be obtained from a series, as in Lemma 4, while Lemma 3 shows P to be countably additive. As in Chap. 11, a random variable $X\colon \Omega \to R$ induces a discrete probability density given by

$$p_X(x) = P[X = x] = S_p(\{\omega : X(\omega) = x\}) \qquad \text{for all real } x.$$

We can, for example, restate Theorem 1 (Sec. 11-3) as follows: the function $X(\omega)p(\omega)$ is summable over Ω iff the function $xp_X(x)$ is summable over R, and if this is the case the two sums are equal and EX is defined to equal this sum. For a proof let $x_1, x_2, \ldots$ be an enumeration of the set of real x for which $P[X = x] > 0$. Then Lemmas 3 and 4, together with the following manipulation, provide a proof,

$$S_{Xp}(\Omega) = \sum_{n=1}^{\infty} S_{Xp}([X = x_n]) = \sum_{n=1}^{\infty} x_n S_p([X = x_n]) = \sum_{n=1}^{\infty} x_n p_X(x_n) = S_{xp_X(x)}(R).$$

Note that if $f\colon \Omega \to R$ and $g\colon \Omega \to R$ and $A \subset \Omega$ and a and b are real numbers then we often (certainly when A is finite) have linearity,

$$S_{af+bg}(A) = aS_f(A) + bS_g(A).$$

In this appendix we have been more interested in using the same function with different sets. If the contrary were true then the notation $\sum_A f$ might be preferable, suggesting that we have something like an integral. Actually if in the general theory of integration, as briefly described in Sec. 21-4, we define the measure of every set $\{\omega\}$ consisting of exactly one point ω of Ω to be 1, $\mu(\{\omega\}) = 1$, then our $S_f(A)$ is just the integral with respect to μ of f over A.

Summaries

CHAPTER 1: SETS

A **set** is a collection of elements. The elements in A are also called **members** of A or **points** of A. Two sets are **equal** iff they both consist of the same elements. *Family* or *class* or *collection* are synonyms for "set" and are used in contexts such as "the collection $\{A,B,C,D\}$ of four sets." A set is said to be **finite** iff it has only a finite number of members; otherwise it is said to be **infinite.** Unless stated otherwise, the notation $\{x_1, x_2, \ldots, x_k\}$ for a finite set means that the x_i are all different, so that the set has k members. The notation $\{x: \pi(x)\}$ denotes the set of all x which make $\pi(x)$ true and is read "the set of all x such that $\pi(x)$." The notation $x \in A$ means that x **belongs to** the set A, and $x \notin A$ means that x **does not belong to** A. A set A is said to be **contained** in a set B or is called a **subset** of a set B iff every member of A also belongs to B. The notation $A \subset B$ means that A is a subset of B, that is, A is contained in B. A set A is called a **proper subset** of a set B iff $A \subset B$ and $A \neq B$. A sequence of sets $A_1, A_2, \ldots$ is said to be **increasing** iff each set is contained in the next one, that is, $A_1 \subset A_2 \subset A_3 \subset \cdots$. A sequence of sets is said to be **decreasing** iff each set is contained in the preceding one, that is, $A_1 \supset A_2 \supset A_3 \supset \cdots$. The set φ, which does not have any members, is called the **empty set.** Thus $A \neq \varphi$ means that A is **nonempty;** that is, A has at least one member. A collection of sets is said to be **disjoint** iff no object belongs to more than one set in the collection.

If A is a subset of Ω then the set $A^c = \{x: x \in \Omega \text{ and } x \notin A\}$ is called the **complement of A with respect to** Ω. If the set Ω is clear from the context then we may abbreviate to "the complement of A." Clearly $(A^c)^c = A$.

The **union** of a collection of sets is the set of all objects which belong to at least one set in the collection. Standard notations are shown below for the union of two, a finite number, and an infinite sequence of sets:

$$A \cup B$$

$$A_1 \cup A_2 \cup \cdots \cup A_N \qquad \bigcup_{n=1}^{N} A_n$$

$$A_1 \cup A_2 \cup \cdots \qquad \bigcup_{n=1}^{\infty} A_n$$

The **intersection** of a collection of sets is the set of all objects which belong to every set in the collection. Standard notations are shown below for the intersection of two, a finite number, and an infinite sequence of sets:

$$A \cap B \qquad AB$$

$$A_1 \cap A_2 \cap \cdots \cap A_N \qquad A_1A_2 \cdots A_N \qquad \bigcap_{n=1}^{N} A_n$$

$$A_1 \cap A_2 \cap \cdots \qquad A_1A_2 \cdots \qquad \bigcap_{n=1}^{\infty} A_n$$

An **identity** in the algebra of sets is an equality between sets which is always true regardless of which specific sets are used in the equation. For example, both $A \cup B = B \cup A$ and $(A^c)^c = A$ are identities. To prove that $C = D$ it is common to prove that $C \subset D$ and $C \supset D$. A set equation is proved to not be an identity if it is shown to be false for one example.

The *De Morgan laws* state that the complement of a union equals the intersection of the complements, and the complement of an intersection equals the union of the complements. All sets involved are assumed to be subsets of the set Ω with respect to which complements are taken. The two laws are shown below for two, a finite number, and an infinite sequence of sets:

$$(A \cup B)^c = A^c \cap B^c \qquad (A \cap B)^c = A^c \cup B^c$$

$$\left(\bigcup_{n=1}^{N} A_n\right)^c = \bigcap_{n=1}^{N} A_n^{\,c} \qquad \left(\bigcap_{n=1}^{N} A_n\right)^c = \bigcup_{n=1}^{N} A_n^{\,c}$$

$$\left(\bigcup_{n=1}^{\infty} A_n\right)^c = \bigcap_{n=1}^{\infty} A_n^{\,c} \qquad \left(\bigcap_{n=1}^{\infty} A_n\right)^c = \bigcup_{n=1}^{\infty} A_n^{\,c}$$

If $A_1, A_2, \ldots$ is *any* sequence of sets and $B_n = A_nA_{n-1}^cA_{n-2}^c \cdots A_2^{\,c}A_1^{\,c}$ then $B_1, B_2, \ldots$ is a sequence of *disjoint* sets, and $A_1 \cup A_2 \cup \cdots = B_1 \cup B_2 \cup \cdots$. Clearly B_n consists of those points of A_n which do not belong to any of the earlier sets $A_1, A_2, \ldots, A_{n-1}$.

CHAPTER 2: COMBINATORICS

If n is a positive integer then an **n-tuple** is an *ordered* collection $(x_1, x_2, \ldots, x_n)$ of n elements, where x_i is called the *ith coordinate*. Two tuples are **equal** iff they have the same number of coordinates and are equal coordinatewise.

The **cartesian product** of sets $A_1, A_2, \ldots, A_n$ is the set of all n-tuples $(x_1, x_2, \ldots, x_n)$ such that $x_i \in A_i$ for all $i = 1, 2, \ldots, n$. It is denoted by $A_1 \times A_2 \times \cdots \times A_n$ or by $\displaystyle\bigtimes_{k=1}^{n} A_k$. It may be denoted by A^n when all n sets $A_1, A_2, \ldots, A_n$ are equal to the same set A. Thus A^n is the set of all n-tuples with all coordinates belonging to the set A. We often ignore extra parentheses in cartesian products and make the convention that

$$(A \times B) \times C = A \times (B \times C) = A \times B \times C.$$

Unless stated otherwise, the letter R denotes **the set of all real numbers** and R^n denotes **the set of all n-tuples of real numbers**. A real number will often be referred to as a **real.**

The symbol $\doteq$ means **approximately equals** and does not have a precise definition. The **size** of a finite set is the number of members which belong to it.

Theorem 1 (Sec. 2-2) If $A_1, A_2, \ldots, A_n$ are finite sets then the size of $A_1 \times A_2 \times \cdots \times A_n$ equals the product of the sizes of $A_1, A_2, \ldots, A_n$. In particular

$$[\text{the size of } A^n] = [\text{the size of } A]^n. \quad \blacksquare$$

The Principle of Mathematical Induction For every positive integer n let $S(n)$ be a sentence which is either true or false; that is, $S(n)$ may be true for some values of n but false for other values of n. *If*

1. $S(1)$ is true
2. For every positive integer n, if $S(n)$ is true for that particular n then $S(n + 1)$ is true

then $S(n)$ is true for every positive integer. $\blacksquare$

Any set of size n has exactly 2^n subsets.

If $n \geq 1$ balls are drawn from an urn containing k balls labeled $1, 2, \ldots, k$ and if x_i is the number on the ball obtained from the ith draw, so that x_i is one of the integers $1, 2, \ldots, k$ then the n-tuple $(x_1, x_2, \ldots, x_n)$ obtained in this way is called the resulting **sample.**

From Theorem 1 (Sec. 2-2) there are exactly k^n different possible samples obtainable from n draws with replacement from an urn with k balls. By **with replacement** we mean that each ball is returned to the urn before the next ball is drawn, so that repeats are permitted, and the sample can be any member $(x_1, x_2, \ldots, x_n)$ of $\{1, 2 \ldots, k\}^n$.

If $1 \leq n \leq k$ then there are exactly $k(k - 1)(k - 2) \cdots (k - n + 1)$ different possible samples obtainable from n draws without replacement from an urn with k balls. By **without replacement** we mean that no ball is returned to the urn after it is drawn, so that repeats are not permitted, and the sample can be any member $(x_1, x_2, \ldots, x_n)$ of $\{1, 2, \ldots, k\}^n$ which satisfies $x_i \neq x_j$ whenever $i \neq j$. As usual, we define $0! = 1$ and $k! = k(k - 1)(k - 2) \cdots (2)(1)$ for any positive integer k, where $k!$ is called k-**factorial** Thus there are $k!$ samples obtainable from k draws without replacement from an urn with k balls, and these samples are just the $k!$ permutations of the integers $1, 2, \ldots, k$.

If A is a subset of a finite set Ω and we think of each member of Ω being in some sense equally likely then we define the number $P(A)$ by

$$P(A) = \frac{[\text{the size of } A]}{[\text{the size of } \Omega]}$$

and we call $P(A)$ the **probability of** A in Ω or, more briefly, the *probability of A*. Thus $P(A)$ is the fraction of the points in Ω which belong to A. Clearly $P(\Omega) = 1$ and $0 \leq P(A) \leq 1$. This is a temporary definition which is generalized and discussed in Chap. 6.

By $n \geq 1$ **fair-coin tosses** we mean that each of the 2^n members of $\{H,T\}^n$ is assumed to be equally likely.

If $1 \leq n \leq k$ and Ω is the set of all k^n samples $(x_1, x_2, \ldots, x_n)$ obtainable from n draws *with* replacement from an urn with k balls, and if A is the event that there are no repeats, then

$$P(A) = \frac{k(k-1)(k-2)\cdots(k-n+1)}{k^n}.$$

If k and n are any integers satisfying $0 \le k \le n$ we define the **binomial coefficient** $\binom{n}{k}$ by

$$\binom{n}{k} = \frac{n!}{[k!][(n-k)!]}.$$

Note that

$$\binom{n}{k} = \binom{n}{n-k} \qquad \binom{n}{0} = \binom{n}{n} = 1 \qquad \binom{n}{1} = \binom{n}{n-1} = n.$$

The basic *recursion formula for binomial coefficients*,

$$\binom{n+1}{k} = \binom{n}{k-1} + \binom{n}{k} \qquad \text{if } 1 \le k \le n,$$

justifies the *Pascal triangle* of Fig. 1 (Sec. 2-3), proves that the binomial coefficients are integers, and proves the *binomial theorem*: for any reals x and y and any positive integer n

$$(x+y)^n = \sum_{k=0}^{n} \binom{n}{k} x^k y^{n-k}.$$

Theorem I (Sec. 2-3) If k and n are integers satisfying $0 \le k \le n$ then there are exactly $\binom{n}{k}$ different n-tuples $(x_1, x_2, \ldots, x_n)$ which have k coordinates equal to 1, and the remaining $n - k$ coordinates equal to 0. ■

A set of size n has exactly 2^n subsets, and exactly $\binom{n}{k}$ of these are of size k, so that $2^n = \sum_{k=0}^{n} \binom{n}{k}$.

In Example 5 (Sec. 2-3) it is shown that an N-row Galton quincunx produces a binomial distribution with parameters N and $\frac{1}{2}$.

In Example 6 (Sec. 2-3) the following theorem is proved:

Bertrand's ballot theorem (1887) If n fair-coin tosses resulted in n_H heads and n_T tails with $n_H > n_T$ then the probability is

$$\frac{n_H}{n_H + n_T} - \frac{n_T}{n_H + n_T} = \frac{n_H - n_T}{n_H + n_T}$$

that for *every* $k = 1, 2, \ldots, n$ there were more heads than tails in the first k tosses; that is, the number of heads exceeded the number of tails throughout the whole sequence of $n = n_H + n_T$ tosses. ■

In Section 2-4 formulas (1), (2), (3) *below are derived.*

If n and k are positive integers then there are exactly

$$\binom{n+k-1}{n} = \binom{n+k-1}{k-1} \tag{1}$$

different k-tuples $(r_1, r_2, \ldots, r_k)$ of nonnegative integers which satisfy the equation $r_1 + r_2 + \cdots + r_k = n$.

If n and k are positive integers and $r_1, r_2, \ldots, r_k$ are nonnegative integers satisfying $r_1 + r_2 + \cdots + r_k = n$ then there are exactly

$$\frac{n!}{(r_1!)(r_2!)\cdots(r_k!)} \tag{2}$$

different n-tuples $x = (x_1, x_2, \ldots, x_n)$ such that for each $j = 1, 2, \ldots, k$ exactly r_j of the coordinates of x are equal to j. Expression (2) is called a **multinomial coefficient.**

If D balls are drawn *without replacement* from an urn containing R red balls and B black balls then the probability is

$$\frac{\binom{R}{r}\binom{B}{D-r}}{\binom{R+B}{D}} \tag{3}$$

that exactly r of the D balls drawn will be red. The nonnegative integers R, B, D, and r are subject to the constraints $D \leq R + B$, $r \leq D$, $r \leq R$, and $D - B \leq r$. Expression (3) when considered as a function of r is called the **hypergeometric distribution.**

CHAPTER 3: FUNCTIONS

A **function** is a set of ordered pairs with the property that no two different ordered pairs in the set have the same first coordinate. *Transformation*, *mapping*, and *correspondence* are among the many synonyms for "function." The **domain (range)** of a function is the set of all of the first (second) coordinates of ordered pairs which belong to the function. A function is said to be defined *on* its domain. Two functions are said to be **equal** iff they are equal as sets, that is, iff they both consist of the same set of ordered pairs. If (x,y) belongs to a function f then another notation for y is $f(x)$. We say that $f(x)$ is the **value** of f at x, or f carries x into $f(x)$, or $f(x)$ is the **image** of x under f, or $f(x)$ is the value of f at the argument x. Clearly $f = g$ iff both functions have the same domain and $f(x) = g(x)$ for every x in this domain.

In the remainder of this book the notation and verbal treatment of functions will usually be more casual than is indicated by the definition of a function as a set of ordered pairs. However, we will always try to be both precise and consistent with the definitions which have been made.

The **notation** $f\colon \mathscr{X} \to \mathscr{Y}$ means that f is a function whose domain *is* $\mathscr{X}$ and whose range *is contained in* $\mathscr{Y}$. We say that f is **on** $\mathscr{X}$ and **into** $\mathscr{Y}$.

A function is said to be a **function of n real variables** iff its domain is contained in R^n. A function is said to be **real valued** iff its range is contained in R. If there is a positive integer n such that the range of a function is contained in R^n then the function is said to be **vector valued.**

Let $f: \mathscr{X} \to R$ and $g: \mathscr{X} \to R$ be two real-valued functions with the same domain. By $f \le g$ we mean that $f(x) \le g(x)$ for all x in $\mathscr{X}$. By $|f| \le g$ we mean that $|f(x)| \le g(x)$ for all x in $\mathscr{X}$, and this is clearly equivalent to $-g(x) \le f(x) \le g(x)$ for all x in $\mathscr{X}$.

If $f: \mathscr{X} \to \mathscr{Y}$ and $g: \mathscr{Y} \to \mathscr{Z}$ then the **composition** of g applied after f is the function $h: \mathscr{X} \to \mathscr{Z}$ defined to have the value $h(x) = g(f(x))$ at any point x of $\mathscr{X}$. The notations $g(f)$ and $g \circ f$ are common for the function h, which is often introduced by the diagram

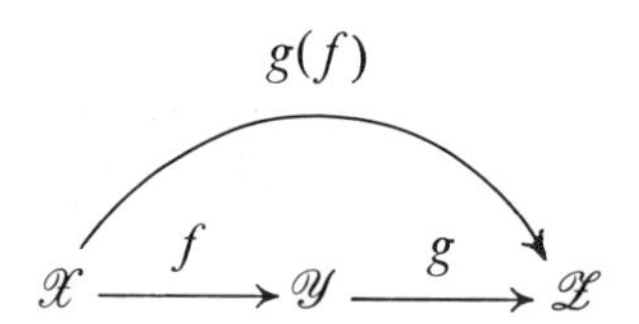

If $f: \mathscr{X} \to \mathscr{Y}$ and A is a subset of $\mathscr{Y}$ then the **inverse image of A under f** is the set of all x in $\mathscr{X}$ such that $f(x)$ belongs to A, and it is denoted by $f^{-1}(A)$. Thus $f^{-1}(A)$ is the set of all x which are carried into A by f. Hence f^{-1} as so defined has the class of all subsets of $\mathscr{Y}$ for its domain, and any member of its range is a subset of $\mathscr{X}$. Clearly $f^{-1}(\varphi) = \varphi$ and $f^{-1}(\mathscr{Y}) = \mathscr{X}$. Also f^{-1} preserves inclusion, disjointness, and unions. Thus if $A \subset B \subset \mathscr{Y}$ then $f^{-1}(A) \subset f^{-1}(B) \subset \mathscr{X}$. If A and B are disjoint subsets of $\mathscr{Y}$ then $f^{-1}(A)$ and $f^{-1}(B)$ are disjoint subsets of $\mathscr{X}$. The inverse image of a union equals the union of the inverse images. That is, $f^{-1}(A \cup B) = [f^{-1}(A)] \cup [f^{-1}(B)]$. If A is a subset of $\mathscr{Z}$ and

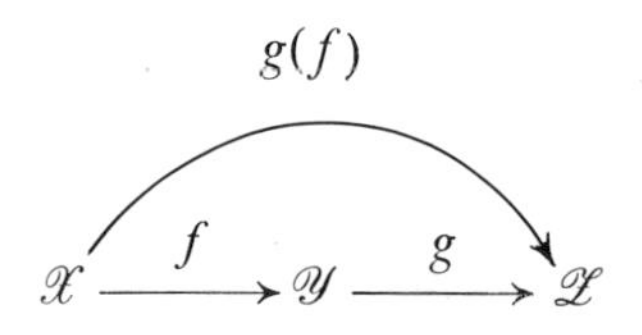

then the inverse image of A under the function $g(f)$ is the set $f^{-1}(g^{-1}(A))$.

If $f: \mathscr{X} \to \mathscr{Y}$ and $\mathscr{Y}$ actually equals the range of f then we say that f is **onto** $\mathscr{Y}$. A function $f: \mathscr{X} \to \mathscr{Y}$ is said to be **one-to-one** iff for every y in $\mathscr{Y}$ there is at most one x in $\mathscr{X}$ for which $y = f(x)$. A function $f: \mathscr{X} \to \mathscr{Y}$ is called a **one-to-one correspondence** between $\mathscr{X}$ and $\mathscr{Y}$ iff it is both one-to-one and onto.

If $f: \mathscr{X} \to \mathscr{Y}$ is a one-to-one correspondence between $\mathscr{X}$ and $\mathscr{Y}$ then the **inverse of f on $\mathscr{Y}$** is the function $f^{-1}: \mathscr{Y} \to \mathscr{X}$, defined by: the value of f^{-1} at y in $\mathscr{Y}$ is the unique x in $\mathscr{X}$ for which $y = f(x)$. Thus $f^{-1}(y) = x$ iff $y = f(x)$. Clearly f^{-1} is a one-to-one correspondence between $\mathscr{Y}$ *and* $\mathscr{X}$. Also $f^{-1}(f(x)) = x$ whenever x is in $\mathscr{X}$, and $f(f^{-1}(y)) = y$ whenever y is in $\mathscr{Y}$.

Thus if $f: \mathscr{X} \to \mathscr{Y}$ then the *inverse of f on* $\mathscr{Y}$ can be defined *only* when $f: \mathscr{X} \to \mathscr{Y}$ is a one-to-one correspondence between $\mathscr{X}$ and $\mathscr{Y}$, and this f^{-1} has *points* of $\mathscr{Y}$ as elements of its domain. However, the quite different f^{-1}, defined as the *inverse image of A under f*, can be defined for *every* $f: \mathscr{X} \to \mathscr{Y}$, and this f^{-1} has *subsets* of $\mathscr{Y}$ as elements of its domain.

CHAPTER 4: COUNTABILITY

Any two sets $\mathscr{X}$ and $\mathscr{Y}$ are said to have the same **cardinality** iff there exists a one-to-one correspondence $f: \mathscr{X} \to \mathscr{Y}$ between $\mathscr{X}$ and $\mathscr{Y}$. Having the same cardinality is interpreted as having the same number of members. The concept of cardinality agrees with the usual facts about the sizes of finite sets and gives us new insights into the sizes of infinite sets.

The property of having the same cardinality is reflexive, symmetric, and transitive. That is, for any sets $\mathscr{X}$, $\mathscr{Y}$, and $\mathscr{Z}$ we have the properties that $\mathscr{X}$ has the same cardinality as $\mathscr{X}$; if $\mathscr{X}$ has the same cardinality as $\mathscr{Y}$ then $\mathscr{Y}$ has the same cardinality as $\mathscr{X}$; and if $\mathscr{X}$ has the same cardinality as $\mathscr{Y}$ and if $\mathscr{Y}$ has the same cardinality as $\mathscr{Z}$ then $\mathscr{X}$ has the same cardinality as $\mathscr{Z}$.

The **identity function** on any set $\mathscr{X}$ is the function $f: \mathscr{X} \to \mathscr{X}$ defined by $f(x) = x$ for every x in $\mathscr{X}$. In particular the identity function shows that having the same cardinality is a reflexive property.

A set $\mathscr{X}$ is said to be **denumerable** iff there exists a one-to-one correspondence $f: I \to \mathscr{X}$ between the **set** $I = \{1, 2, \ldots\}$ **of positive integers** and $\mathscr{X}$, that is, iff I and $\mathscr{X}$ have the same cardinality. We may define x_n by $x_n = f(n)$, so that a denumerable set $\mathscr{X}$ is one which can be written as an infinite sequence $x_1, x_2, \ldots$ of members of $\mathscr{X}$ such that each member of $\mathscr{X}$ appears exactly once in the sequence. Such a function or sequence is called an **enumeration** of $\mathscr{X}$. Similarly a finite sequence $x_1, x_2, \ldots, x_n$ of members of a finite set $\mathscr{X}$ such that every member of $\mathscr{X}$ appears exactly once in the sequence is called an enumeration of $\mathscr{X}$.

Two denumerable sets have the same cardinality. Also the denumerable sets are the smallest infinite sets in the sense that any infinite set has a denumerable subset.

A set is said to be **countable** iff it is either finite or denumerable. Sometimes a countable set is called a **discrete set.** A set is said to be **uncountable** iff it is not countable. Thus an uncountable set is one which is infinite but not denumerable.

An infinite set can have the same cardinality as a proper subset of itself; for example, the set $\{\ldots, -1, 0, 1, \ldots\}$ of all integers, the set $I = \{1, 2, \ldots\}$ of all positive integers, and the set $\{2, 4, \ldots\}$ of all even integers are all denumerable and so have the same cardinality.

Theorem I (Sec. 4-2)

(*a*) A subset of a countable set is countable.
(*b*) If the domain of a function is countable its range is countable.
(*c*) The cartesian product of a finite number of countable sets is countable.
(*d*) The union of a finite or infinite sequence of countable sets is countable. ∎

A **rational number** is a real number which can be expressed as a ratio m/n of two integers. The set of all rationals is countable, even though it is dense in the real-number axis.

In Sec. 4-3 the interval $0 < x \leq 1$ of reals was proved to be uncountable.

CHAPTER 5: ABSOLUTELY CONVERGENT SERIES

A sequence $A_1, A_2, \ldots$ of real numbers is said to

1. **Converge** to the real number A iff for all $\epsilon > 0$ there is an N such that $|A_n - A| \leq \epsilon$ if $n \geq N$
2. **Converge to infinity** iff for every real L there is an N such that $A_n \geq L$ if $n \geq N$
3. **Converge to minus infinity** iff for every real L there is an N such that $A_n \leq L$ if $n \geq N$
4. Be **bounded** iff there is a number B such that $|A_n| \leq B$ for all n
5. Be **unbounded** iff it is not bounded
6. Be **nondecreasing (nonincreasing)** iff $A_n \leq A_{n+1}$ $(A_n \geq A_{n+1})$ for all n

A basic property of the set of real numbers is that a nondecreasing (nonincreasing) sequence of reals either converges to infinity (minus infinity) or is bounded and converges to a real. We do *not* consider ∞ and $-\infty$ to be real numbers.

If $a_1, a_2, \ldots$ is a sequence and $g: I \to I$ is a one-to-one correspondence between the set I of positive integers and itself, then $a_{g(1)}, a_{g(2)}, \ldots$ is called a **reordering** of the sequence $a_1, a_2, \ldots$.

The property of being a reordering is reflexive, symmetric, and transitive. That is, $a_1, a_2, \ldots$ is a reordering of itself; if $b_1, b_2, \ldots$ is a reordering of $a_1, a_2, \ldots$ then $a_1, a_2, \ldots$ is a reordering of $b_1, b_2, \ldots$; if $b_1, b_2, \ldots$ is a reordering of $a_1, a_2, \ldots$ and if $c_1, c_2, \ldots$ is a reordering of $b_1, b_2, \ldots$ then $c_1, c_2, \ldots$ is a reordering of $a_1, a_2, \ldots$.

Let $a_1 + a_2 + \cdots$ be an infinite series and let $A_1, A_2, \ldots$ be the corresponding sequence of partial sums, where $A_n = a_1 + a_2 + \cdots + a_n$. The *series* $a_1 + a_2 + \cdots$ is said to *converge* to the real number A (or to infinity, or to minus infinity) iff the sequence $A_1, A_2, \ldots$ converges to A (or to infinity, or to minus infinity).

The series $b_1 + b_2 + \cdots$ is said to be a *reordering* of the series $a_1 + a_2 + \cdots$ iff the sequence $b_1, b_2, \ldots$ is a reordering of the sequence $a_1, a_2, \ldots$.

A **series** $a_1 + a_2 + \cdots$ is said to be **absolutely convergent** iff there is a number B such that $|a_1| + |a_2| + \cdots + |a_n| \leq B$ for all n.

Theorem I (Sec. 5-2) Every reordering of an absolutely convergent series is absolutely convergent and converges to the same real number. ∎

A **double series** $\sum_{n=1}^{\infty} \sum_{m=1}^{\infty} a_{nm}$ is said to be **absolutely convergent** iff there is a number B such that $\sum_{n=1}^{N} \sum_{m=1}^{M} |a_{nm}| \leq B$ for all N and M. If an absolutely convergent double series is given an arbitrary ordering as a single-index series then the resulting series is absolutely convergent and the limiting real is the same for every ordering. This limiting sum is denoted by $\sum_{n=1}^{\infty} \sum_{m=1}^{\infty} a_{nm}$. The next theorem asserts that the sum of an absolutely convergent double series may also be obtained by iterated summation.

Theorem I (Sec. 5-3) The double series $\sum_{n=1}^{\infty} \sum_{m=1}^{\infty} a_{nm}$ is absolutely convergent iff for each n the nth row-sum series $\sum_{m=1}^{\infty} |a_{nm}|$ of absolute values is convergent *and* the series $\sum_{n=1}^{\infty} \left[\sum_{m=1}^{\infty} |a_{nm}| \right]$, consisting of the sum of these row-sums, is convergent. The same is true for the iterated sum obtained by summing column sums. If the double series is absolutely convergent then

$$\sum_{n=1}^{\infty} \sum_{m=1}^{\infty} a_{nm} = \sum_{n=1}^{\infty} \left[\sum_{m=1}^{\infty} a_{nm} \right] = \sum_{m=1}^{\infty} \left[\sum_{n=1}^{\infty} a_{nm} \right].$$

In particular if $a_{nm} = b_n c_m$, where $b_n c_m \neq 0$ for at least one pair of n and m, then the double series is absolutely convergent iff both $\sum_{n=1}^{\infty} b_n$ and $\sum_{m=1}^{\infty} c_m$ are, in which case

$$\sum_{n=1}^{\infty} \sum_{m=1}^{\infty} b_n c_m = \left[\sum_{n=1}^{\infty} b_n \right] \left[\sum_{m=1}^{\infty} c_m \right]. \quad \blacksquare$$

CHAPTER 6: DISCRETE PROBABILITY DENSITIES

A **discrete probability density** is a real-valued function $p: \Omega \to R$ defined on some set Ω and satisfying the following three conditions:

1. $p(\omega) \geq 0$ for all ω in Ω.
2. $\Omega_0 = \{\omega: p(\omega) > 0\}$ is a countable nonempty set.
3. If $\Omega_0 = \{\omega_1, \omega_2, \ldots, \omega_r\}$ then $p(\omega_1) + p(\omega_2) + \cdots + p(\omega_r) = 1$, and if $\Omega_0 = \{\omega_1, \omega_2, \ldots\}$ then $p(\omega_1) + p(\omega_2) + \cdots = 1$.

A discrete probability density can always be interpreted as a *random-wheel experiment*. If ω_i belongs to Ω_0 let $p_i = p(\omega_i) > 0$ and label the ith sector by ω_i and assign fraction p_i of the circumference to it.

If $p: \Omega \to R$ is a discrete probability density then the set Ω is called the **sample space,** an element ω of Ω is called a **sample point,** and each subset of Ω is called an **event.** If A is an event then the **probability** of A is the real number denoted by $P(A)$ and defined to be the sum of all of the positive values $p(\omega)$ for all of the ω in A. We say that p, or P, is **concentrated on** an event A iff $P(A) = 1$. A discrete probability density $p: \Omega \to R$ can be thought of as a discrete distribution of unit mass over Ω. Mass $p(\omega)$ is concentrated at the point ω, so that $P(A)$ is the total mass of the set A.

Our usual interpretation for a discrete probability density $p: \Omega \to R$ is as follows. An experiment is to be performed and the outcome is in some sense random or unpredictable. A sample point ω is a complete description of a conceivable outcome of the experiment in the sense that if the experiment is performed then its outcome must correspond to precisely one sample point. The collection Ω of all of these sample points is called the sample space. If A is an event we say that A has occurred iff the sample point describing the outcome of the experiment belongs to A. The probability that A will occur is the real number $P(A)$.

Clearly we always have $P(\varphi) = 0$ and $P(\Omega) = P(\Omega_0) = 1$ and $0 \leq P(A) \leq 1$ for every event A, and $P(\{\omega\}) = p(\omega)$ for every sample point ω. Also $P(A \cup B) = P(A) + P(B)$ whenever A and B are *disjoint* events.

If $e = (e_1, e_2, \ldots, e_n)$ is any n-tuple, ω is any object, and A is any set we let

$$m_e(\omega) = \frac{1}{n} \text{ [the number of coordinates of } e \text{ which equal } \omega]$$

$$M_e(A) = \frac{1}{n} \text{ [the number of coordinates of } e \text{ which belong to } A].$$

We call $m_e(\omega)$, and $M_e(A)$, the **relative frequency** of ω in e, and A in e, respectively. Relative frequencies can be used to describe predictions based on a model or to construct a model from experimental data. The relationship between M_e and m_e is similar to the relationship between P and p.

If $p: \Omega \to R$ is a discrete probability density and A is an event then we have defined the number $P(A)$, which we call the probability of A. Thus we have assigned a unique real number $P(A)$ to each event A. Therefore we have defined a real-valued function P whose domain is the class of events. This function P is called the **discrete probability measure** corresponding to the discrete probability density $p: \Omega \to R$. The function P is sometimes said to be a *set function* because each member of its domain is a set. The ordered pair (Ω, P), where Ω is the sample space and P is the discrete probability measure, is called the **discrete probability space** corresponding to $p: \Omega \to R$.

It is often more convenient to deal with P than with the corresponding density $p: \Omega \to R$. In nondiscrete cases there is always a P to work with, although there may not be a corresponding discrete density p. For such reasons, we often emphasize P more than p.

Theorem 1 (Sec. 6-4) If (Ω, P) is a discrete probability space then the following properties hold:

(*a*) $P(\Omega) = 1$.

(*b*) $P(A) \geq 0$ for every event A.

(*c*) If $A_1, A_2, \ldots$ is any sequence of disjoint events then

$$P\left(\bigcup_{n=1}^{\infty} A_n\right) = \sum_{n=1}^{\infty} P(A_n).$$

(*d*) The set $\Omega_0 = \{\omega: P(\{\omega\}) > 0\}$ is countable and $P(\Omega_0) = 1$. ■

Property (*c*) of this theorem, called the **countable additivity of** P, is of fundamental importance. In nondiscrete cases we will use various techniques to assign a probability $P(A)$ to each event A. In each case P will have properties (*a*) to (*c*) above and, these properties are essentially all that is needed to develop probability theory in the standard way. Theorems 1 and 3 (Sec. 6-4) show that the discrete case consists of those P which satisfy property (*d*) as well. In our treatment of the discrete case we use properties (*a*) to (*c*) freely, but we use property (*d*) only where necessary or when some simplification results. In this way we achieve greater generality, and emphasize a way of thinking which requires little change later.

Theorem 2 below collects together a number of simple properties of P, which are true in the most general case. The only properties of P used in the proof are that P assigns a real number $P(A) \geq 0$ to each event A, and $P(\Omega) = 1$, and P is countably additive.

Theorem 2 (Sec. 6-4) If (Ω,P) is a discrete probability space then the following properties hold:

(*a*) $P(\varphi) = 0$.

(*b*) $P(A_1 \cup A_2 \cup \cdots \cup A_N) = P(A_1) + P(A_2) + \cdots + P(A_N)$ for any disjoint events $A_1, A_2, \ldots, A_N$.

(*c*) $0 \leq P(A) \leq 1$ and $P(A) + P(A^c) = 1$ for any event A.

(*d*) $P(A) \leq P(B)$ whenever event A is a subset of event B.

(*e*) $P(A \cup B) = P(A) + P(B) - P(AB)$ for any events A and B.

(*f*) $P(A_1 \cup A_2 \cup \cdots \cup A_N) \leq P(A_1) + P(A_2) + \cdots + P(A_N)$ for any events $A_1, A_2, \ldots, A_N$.

(*g*) $P(A_1 \cup A_2 \cup \cdots) \leq P(A_1) + P(A_2) + \cdots$ for any events $A_1, A_2, \ldots$.

(*h*) If $A_1 \subset A_2 \subset \cdots$ is an increasing sequence of events then $P(A_1 \cup A_2 \cup \cdots) = \lim_{N\to\infty} P(A_N)$.

(*i*) If $A_1 \supset A_2 \supset \cdots$ is a decreasing sequence of events then $P(A_1 \cap A_2 \cap \cdots) = \lim_{N\to\infty} P(A_N)$. ∎

Henceforth we often use results from Theorem 2 (Sec. 6-4) without comment.

Theorem 3 (Sec. 6-4) Let every subset of a nonempty set Ω be called an event. Let P be *any* real-valued function whose domain is the class of all events, and for which properties (*a*) to (*d*) of Theorem 1 (Sec. 6-4) hold. Define the function $p: \Omega \to R$ by $p(\omega) = P(\{\omega\})$ for all ω in Ω. Then p is a discrete probability density, and the given function P is its corresponding discrete probability measure. ∎

CHAPTER 7: INDEPENDENT EVENTS

If (Ω,P) is a discrete probability space and if A and B are events with $P(A) > 0$ then $P(B \mid A)$ is defined by

$$P(B \mid A) = \frac{P(AB)}{P(A)} .$$

We read $P(B \mid A)$ as the **conditional probability of** B **given** A, or the probability of B given A.

If $p: \Omega \to R$ is a discrete probability density and A is an event with $P(A) > 0$ then the function $p_A: \Omega \to R$ defined by

$$p_A(\omega) = \begin{cases} \dfrac{p(\omega)}{P(A)} & \text{if } \omega \in A \\ 0 & \text{if } \omega \notin A \end{cases}$$

is called the **discrete probability density obtained by conditioning** p **by** or on, **the event** A. If P_A is the discrete probability measure corresponding to p_A then we call (Ω,P_A) the **discrete probability space obtained by conditioning** (Ω,P) **by the event** A. We always have

$$P_A(B) = P(B \mid A) = \frac{P(AB)}{P(A)} \qquad \text{for every event } B.$$

Theorem I (Sec. 7-2): Bayes' Theorem Let $B_1, B_2, \ldots, B_n$ be disjoint events for which each $P(B_i) > 0$. Let A be an event such that $P(A) > 0$ and such that A is contained in the union $B_1 \cup B_2 \cup \cdots \cup B_n$. Then

$$P(B_i \mid A) = \frac{P(A \mid B_i)P(B_i)}{\sum_{k=1}^{n} P(A \mid B_k)P(B_k)} \qquad \text{if } i = 1, 2, \ldots, n. \quad \blacksquare$$

At times $P(B_i)$ may be called the *a priori probability* of B_i while $P(B_i \mid A)$ is called the *a posteriori probability* of B_i given that A occurred.

Let (Ω,P) be a discrete probability space and let A and B be events. The events A and B are said to be **independent** iff $P(AB) = P(A)P(B)$. Clearly A and B are independent iff B and A are independent. We may say "A and B are **dependent**" instead of "A and B are not independent."

Independence of events is primarily a numerical fact about probabilities rather than a fact about the relationship between the sets A and B without regard to P. To emphasize this we may use **stochastically independent** or **statistically independent** rather than just independent. The adjectives "stochastically" and "statistically" may also be used later with independence of $n \geq 2$ events and independence of random variables.

If $0 < P(A) < 1$ then A is not independent of A.

If one of the four pairs (A,B) or (A^c,B) or (A,B^c) or (A^c,B^c) is an independent pair of events then all four are.

Let (Ω,P) be a discrete probability space and let $A_1, A_2, \ldots, A_n$ be a finite sequence of $n \geq 2$ events. The events $A_1, A_2, \ldots, A_n$ are said to be **pairwise independent** iff A_i and A_j are two independent events whenever $i \neq j$. In this case $P(A_i \mid A_j) = P(A_i)$, so that the probability of the occurrence of any one of the n events is unaffected by the knowledge of the occurrence of *any one* of the other $n - 1$ events. In discussing pairwise independence of events, or random variables, we never drop the adjective "pairwise." The notion of pairwise independence is used only rarely.

Mutual independence as defined below is used widely and is normally abbreviated to "independence." Mutual independence means, roughly, that the probability of the occurrence of any one of the events is unaffected by the knowledge of the occurrence of *any one or more* of the other events. For $n = 2$ events, pairwise and mutual independence are equivalent. Mutually independent events are necessarily pairwise independent, but $n \geq 3$ pairwise independent events may not be mutually independent. For the sake of simplicity we usually follow tradition and seldom mention pairwise independence, even though some later results which we state and prove under the assumption of mutual independence also hold, with the same proof, under the weaker assumption or pairwise independence of the events or random variables involved.

Let (Ω,P) be a discrete probability space and let $A_1, A_2, \ldots, A_n$ be a finite sequence of $n \geq 2$ events. The events $A_1, A_2, \ldots, A_n$ are said to be **mutually independent,** or **independent**, iff

$$P(A_{n_1}A_{n_2} \cdots A_{n_k}) = P(A_{n_1})P(A_{n_2}) \cdots P(A_{n_k})$$

for every one of the $2^n - n - 1$ subsequences of two or more events. We call $A_{n_1}, A_{n_2}, \ldots, A_{n_k}$ a **subsequence** of $A_1, A_2, \ldots, A_n$ iff the subscript integers satisfy $1 \leq n_1 < n_2 \cdots < n_k \leq n$.

Let $A_1, A_2, \ldots, A_n$ be a finite sequence of $n \geq 2$ independent events. Then any subsequence of two or more of them is independent. Also reordering, or complementing some of the events, does not destroy the independence. Furthermore $A_1A_2 \cdots A_m$ and $A_{m+1}A_{m+2} \cdots A_n$ are two independent events. More generally $B_1, B_2, \ldots, B_N$ are independent if each B_i is the intersection of some of the A_k, and if no A_k is used more than once.

CHAPTER 8: INDEPENDENT EXPERIMENTS

We use the term **distribution** to refer to any function, such as a density or measure, which determines the probability of every event.

Theorem 1 (Sec. 8-1) Let $p_i\colon F_i \to R$ for $i = 1, 2, \ldots, n$ be $n \geq 2$ discrete densities. Let $\Omega = F_1 \times F_2 \times \cdots \times F_n$ and let $p\colon \Omega \to R$ be defined by the product rule as

$$p(\alpha_1, \alpha_2, \ldots, \alpha_n) = p_1(\alpha_1)p_2(\alpha_2) \cdots p_n(\alpha_n) \qquad \text{for all } (\alpha_1, \alpha_2, \ldots, \alpha_n) \text{ in } \Omega.$$

Then $p\colon \Omega \to R$ *is a discrete density with corresponding discrete probability space* (Ω,P). *Also,* if

$$A_i(\alpha_i) = \{(\alpha_1', \alpha_2', \ldots, \alpha_n')\colon \alpha_i' = \alpha_i\}$$

is the event in Ω that the ith subexperiment yields α_i then $A_1(\alpha_1), \ldots, A_n(\alpha_n)$ are always independent, and $P[A_i(\alpha_i)] = p_i(\alpha_i)$ always holds. *Furthermore* p is the unique discrete density on Ω with these properties. ■

Thus an independent subexperiments model is obtained by using the cartesian product space with probabilities assigned by the product rule.

The **Bernoulli discrete probability density** $b_1\colon R \to R$ with parameter p is defined by

$$b_1(0) = q \qquad b_1(1) = p \qquad b_1(x) = 0 \qquad \text{if } x \neq 0,$$

where p is a real number satisfying $0 \leq p \leq 1$ and $q = 1 - p$. We will refer to n independent trials with probability p for a success on each trial as n **Bernoulli trials with parameter** p.

The **binomial discrete probability density** $b_n\colon R \to R$ with parameters n and p is defined by

$$b_n(k) = \binom{n}{k} p^k q^{n-k} \qquad \text{if } k = 0, 1, \ldots, n$$

$$b_n(x) = 0 \qquad \text{otherwise}$$

where n is a positive integer and p is a real number satisfying $0 \leq p \leq 1$ and $q = 1 - p$. There is a different density for each pair n and p. It was shown that *the probability of obtaining exactly k successes in n Bernoulli trials with parameter p, is given by the value* $b_n(k)$ *of the binomial density* b_n *with parameters n and p.* We call np the **mean number of successes** or the *expected number of successes.* It was shown that b_n is unimodal and has its maximum value $b_n(k)$ for k near to np.

The **Poisson discrete probability density** $p_\lambda \colon R \to R$ with parameter $\lambda > 0$ is defined by

$$p_\lambda(k) = \frac{e^{-\lambda}\lambda^k}{k!} \qquad \text{if } k = 0, 1, 2, \ldots$$

$$p_\lambda(x) = 0 \qquad \text{otherwise.}$$

The Poisson density with parameter λ *is close to every one of the binomial densities for which* $np = \lambda$ *and n is large enough.* More precisely, *for every fixed* $\lambda > 0$ *and every nonnegative integer k we have*

$$\lim_{n\to\infty} \binom{n}{k}\left(\frac{\lambda}{n}\right)^k\left(1 - \frac{\lambda}{n}\right)^{n-k} = \frac{e^{-\lambda}\lambda^k}{k!}.$$

Thus if $np = \lambda$ is fixed and n is large enough then the distribution of the number of successes on n Bernoulli trials with parameter p is approximately Poisson with parameter $\lambda = np$.

It was shown that p_λ is unimodal and has its maximum value $p_\lambda(k)$ for k near to λ.

Example 2 (Sec. 8-3) shows that the sum of two independent random variables each having a Poisson distribution with parameters λ_1 and λ_2, respectively, also has a Poisson distribution with parameter $\lambda_1 + \lambda_2$.

If r is a positive integer and p is a real number satisfying $0 < p < 1$ and $p + q = 1$ then the **Pascal discrete probability density** $p_r \colon R \to R$ with parameters r and p is defined by

$$p_r(k) = \binom{k + r - 1}{k} p^r q^k \qquad \text{if } k = 0, 1, 2, \ldots$$

$$p_r(x) = 0 \qquad \text{otherwise.}$$

The **geometric density** p_1 is the special case $r = 1$ of the Pascal density, so

$$p_1(k) = pq^k \qquad \text{if } k = 0, 1, 2, \ldots$$

$$p_1(x) = 0 \qquad \text{otherwise.}$$

The distribution of "the number of failures prior to the first success" is an infinite sequence of Bernoulli trials with parameter p, is geometric with parameter p. Actually, as shown in Sec. 8-4, an infinite sequence of Bernoulli trials does not fit within the discrete case.

For $n \geq k + r$ Bernoulli trials the probability is $p_r(k)$ that exactly k failures will occur prior to the rth success. Also $p_r(k)$ is the probability that $x_1 + x_2 + \cdots + x_r = k$, where $x_1, x_2, \ldots, x_r$ are obtained from r repeated independent spins of a random wheel corresponding to the geometric density p_1 with the same parameter p.

CHAPTER 9: THE MEAN AND VARIANCE

If $p: R \to R$ is a discrete probability density and if $\{x: p(x) > 0\} = \{x_1, x_2, \ldots, x_r\}$ is finite then the **mean** of p, or the first moment of p, is the real number μ defined by

$$\mu = x_1 p(x_1) + x_2 p(x_2) + \cdots + x_r p(x_r).$$

Thus if $p(x)$ is positive we form the product $xp(x)$ and add all such products. Similarly if $\{x: p(x) > 0\} = \{x_1, x_2, \ldots\}$ then the mean of p is defined by the series $\mu = x_1 p(x_1) + x_2 p(x_2) + \cdots$ if it is absolutely convergent, so that by Chap. 5 the sum does not depend on the order in which the terms are added. If the series is not absolutely convergent the mean is not defined. The mean of $p: R \to R$ can be interpreted as the center of gravity of the discrete mass distribution p on the real axis. In a large number of independent spins of the random wheel corresponding to p, the arithmetic average of the outcomes should be near to μ if the relative frequencies behave properly.

If $p: R \to R$ is concentrated on some interval then p has a mean and it is in this interval. If $p: R \to R$ is symmetric about some real number b,

$$p(b + x) = p(b - x) \qquad \text{for all real } x$$

and if p has a mean then it is equal to b. Figure 2 (Sec. 9-1) summarizes some of the results of Chap. 9 and is repeated here.

Name and parameters	Discrete probability density	Mean	Variance	Moment-generating function
Poisson $\lambda > 0$	$p_\lambda(k) = e^{-\lambda}\lambda^k/k!$ $k = 0, 1, \ldots$	λ	λ	$\exp\{\lambda(e^t - 1)\}$, $-\infty < t < \infty$
Bernoulli $0 < p < 1$	$b_1(0) = q,\ b_1(1) = p$	p	pq	$q + pe^t$, $-\infty < t < \infty$
Binomial $n = 1, 2, \ldots, 0 < p < 1$	$b_n(k) = \binom{n}{k} p^k q^{n-k}$ $k = 0, 1, \ldots, n$	np	npq	$(q + pe^t)^n$, $-\infty < t < \infty$
Geometric $0 < p < 1$	$p_1(k) = pq^k$ $k = 0, 1, \ldots$	q/p	q/p^2	$p/(1 - qe^t)$, $qe^t < 1$
Pascal $r = 1, 2, \ldots, 0 < p < 1$	$p_r(k) = \binom{r + k - 1}{k} p^r q^k$ $k = 0, 1, \ldots$	$r(q/p)$	$r(q/p^2)$	$[p/(1 - qe^t)]^r$, $qe^t < 1$

Fig. 2 In each case $q = 1 - p$ and $p(x) = 0$ except when x equals one of the values of k shown. The lower expression in the last column states the constraint on t needed for convergence. The Bernoulli density is the special case $n = 1$ of the binomial density. The geometric density is the special case $r = 1$ of the Pascal density.

If $p: R \to R$ is a discrete probability density and $\{x: p(x) > 0\} = \{x_1, x_2, \ldots\}$ then we define the **second moment** of p to equal $\sum_{i=1}^{\infty} x_i^2 p(x_i)$ if this series is convergent; otherwise we say that p does not have a second moment. It was shown that if p has a second moment then it has a mean, so we can define the **variance** of p by

$$\operatorname{Var} p = \sum_{i=1}^{\infty} (x_i - \mu)^2 p(x_i) = \left[\sum_{i=1}^{\infty} x_i^2 p(x_i)\right] - \mu^2$$

where μ is the mean of p, and these two expressions are easily shown to be equal. Since $0 \le \operatorname{Var} p$ we see that

$$[\text{the mean of } p]^2 \le [\text{the second moment of } p].$$

The **standard deviation** of p is defined to equal the *nonnegative* square root of the variance of p,

$$[\text{the standard deviation of } p] = [\operatorname{Var} p]^{1/2} \ge 0.$$

If $\{x: p(x) > 0\}$ contains two or more points then the variance of $p: R \to R$ is positive.

If $p_X: R \to R$ is a discrete probability density and $y = ax + b$, with $a \ne 0$, is a linear transformation then we define the **transformed discrete probability density** $p_Y: R \to R$ in any of the following three equivalent fashions:

1. $p_Y(y) = p_X(x)$ if x and y are related by $y = ax + b$.
2. $p_Y(ax + b) = p_X(x)$ for all real x.
3. $p_Y(y) = p_X((y - b)/a)$ for all real y.

We have

$$\mu_Y = a\mu_X + b \qquad \text{and} \qquad \sigma_Y^2 = a^2\sigma_X^2 \qquad \text{and} \qquad \sigma_Y = |a|\,\sigma_X$$

so that the mean is transformed in the same way as the observables, while the standard deviation is unaffected by a change of direction or a change of origin but is proportional to the expansion of the scale.

If $p_X: R \to R$ is a discrete probability density with a nonzero variance and if $x^* = (x - \mu_X)/\sigma_X$ then $\mu_{X^*} = 0$ and $\sigma_{X^*} = 1$, where $p_{X^*}: R \to R$ is defined by

$$p_{X^*}\left(\frac{x - \mu_X}{\sigma_X}\right) = p_X(x) \qquad \text{for all real } x$$

or equivalently by

$$p_{X^*}(x^*) = p_X((\sigma_X)x^* + \mu_X) \qquad \text{for all real } x^*.$$

We call p_{X^*} the **standardized** or **normalized discrete probability density** corresponding to p_X.

If $p: R \to R$ is a discrete probability density and if r is any positive real number then we define the r**th absolute moment** of p to equal

$$\sum_{j=1}^{\infty} |x_j|^r p(x_j) \qquad \text{where } \{x: p(x) > 0\} = \{x_1, x_2, \ldots\}$$

if this series converges; otherwise we say that p does not have an rth absolute moment, or the rth absolute moment of p is infinite. If $n = 1, 2, \ldots$ is a positive integer then we define the n**th moment** of p to equal

$$\sum_{j=1}^{\infty} (x_j)^n p(x_j)$$

if the series is absolutely convergent; otherwise we say that p does not have an nth moment.

Theorem 1 (Sec. 9-4) If a discrete probability density $p: R \to R$ has a finite sth absolute moment for some real $s > 0$ then it has a finite rth absolute moment for any real r satisfying $0 < r < s$, and furthermore

$$\left[\sum_{j=1}^{\infty} |x_j|^r p(x_j)\right]^{1/r} \leq \left[\sum_{j=1}^{\infty} |x_j|^s p(x_j)\right]^{1/s}. \quad \blacksquare$$

Throughout probability theory various statements and proofs of results such as the law of large numbers and the central-limit theorem require that several moments exist. The theorem above shows that it is necessary to assume only that the highest moment needed exists, since all lower moments must then also exist. Thus it is standard practice in the hypothesis of a theorem to state that a certain moment exists and then to use, without comment, lower moments in the remainder of the statement of the theorem and in its proof.

If $p: R \to R$ is a discrete probability density and $\{x: p(x) > 0\} = \{x_1, x_2, \ldots\}$ then the **moment-generating function (MGF)** of p is defined by

$$m(t) = \sum_{k=1}^{\infty} p(x_k) e^{tx_k}.$$

The MGF is defined for all real t for which the series converges. The MGF is positive valued and is always defined at $t = 0$, where $m(0) = 1$. For some densities the MGF is defined *only* at $t = 0$, and for such densities the MGF is useless.

The following theorem is not proved in this book.

Theorem 2 (Sec. 9-4) Let

$$m(t) = \sum_{k=1}^{\infty} e^{tx_k} p(x_k) \qquad \text{where } \{x: p(x) > 0\} = \{x_1, x_2, \ldots\}$$

be the MGF of a discrete probability density $p: R \to R$. Assume that $m(t_0)$ and $m(-t_0)$ are finite, where $t_0 > 0$. Then m is finite and has derivatives of all orders whenever $|t| < t_0$. Differentiation under the summation sign is permitted, so

$$m^{(n)}(t) = \frac{d^n}{dt^n} m(t) = \sum_{k=1}^{\infty} (x_k)^n e^{tx_k} p(x_k) \qquad \text{if } n = 1, 2, \ldots, |t| < t_0.$$

In particular all moments of p exist and $m^{(n)}(0)$ equals the nth moment, so

$$\mu = m^{(1)}(0) \qquad \sigma^2 = m^{(2)}(0) - [m^{(1)}(0)]^2.$$

Furthermore the power series

$$m(t) = \sum_{n=0}^{\infty} \frac{m^{(n)}(0)}{n!} t^n$$

converges absolutely for $|t| < t_0$.

If the MGFs m_1 and m_2 of discrete probability densities $p_1: R \to R$ and $p_2: R \to R$ are finite whenever $|t| \le t_0$, where $t_0 > 0$, then $p_1 = p_2$ iff $m_1(t) = m_2(t)$ for all $|t| \le t_0$. ■

CHAPTER 10: THE LAW OF LARGE NUMBERS FOR BERNOULLI TRIALS

Lemma 1 (Sec. 10-1) Let $p: R \to R$ be a discrete probability density and let A be any event. If a function $f: R \to R$ is nonnegative valued, $f \ge 0$, and satisfies $f(x) \ge 1$ whenever $x \in A$ then

$$P(A) \le \sum_{i=1}^{\infty} f(x_i)p(x_i) \qquad \text{where } \{x: p(x) > 0\} = \{x_1, x_2, \ldots\}$$

if the series has a finite sum. ■

Theorem 1 (Sec. 10-1) Chebychev's inequality Let $p: R \to R$ be a discrete probability density with mean μ and finite variance σ^2. If ϵ is any positive real then

$$1 - P\{x: |x - \mu| < \epsilon\} = P\{x: |x - \mu| \ge \epsilon\} \le \frac{\sigma^2}{\epsilon^2}. \quad ■$$

Theorem 2 (Sec. 10-1) Let $p: R \to R$ be a discrete probability density with mean μ and finite variance σ^2. If ϵ is any positive real then

$$1 - P\{x: x < \mu + \epsilon\} = P\{x: x \ge \mu + \epsilon\} \le \frac{\sigma^2}{\epsilon^2 + \sigma^2},$$

and

$$1 - P\{x: x > \mu - \epsilon\} = P\{x: x \le \mu - \epsilon\} \le \frac{\sigma^2}{\epsilon^2 + \sigma^2}. \quad ■$$

Theorem 1 (Sec. 10-2): The law of large numbers for Bernoulli trials Let P correspond to the binomial density $b_n: R \to R$ with parameters n and p. If ϵ is any positive real then

$$P\left\{x: \left|\frac{x}{n} - p\right| \ge \epsilon\right\} \le \frac{pq}{n\epsilon^2} \le \frac{1}{4\epsilon^2 n}. \quad ■$$

Theorem 1 (Sec. 10-3): The Weierstrass approximation theorem Let f be a continuous real-valued function defined on the real interval $0 \le p \le 1$ and let

$$B_n(p) = \sum_{k=0}^{n} f\left(\frac{k}{n}\right)\binom{n}{k} p^k (1-p)^{n-k} \qquad \text{for } 0 \le p \le 1$$

be the nth Bernstein polynomial for f. Then $B_n(p)$ converges to $f(p)$ as $n \to \infty$, and the convergence is uniform in p. That is, there is a sequence of constants $\delta_1, \delta_2, \ldots$, determined solely by f, which converge to zero and satisfy

$$|B_n(p) - f(p)| \leq \delta_n \qquad \text{for all } p,\ 0 \leq p \leq 1;\ \text{all } n = 1, 2, \ldots. \qquad \blacksquare$$

Corollary 1 (Sec. 10-3) For $0 \leq r < s \leq 1$ let the function $f_{r,s}$ be defined by

$$f_{r,s}(t) = \begin{cases} 0 & \text{if} \quad 0 \leq t \leq r \\ \dfrac{t-r}{s-r} & \text{if} \quad r \leq t \leq s \\ 1 & \text{if} \quad s \leq t \leq 1. \end{cases}$$

For any such r and s and for any positive integer n there is an nth-degree polynomial $F_n(t) = \sum_{k=0}^{n} a_{n,k} t^k$ with coefficients $a_{n,k}$ depending on r, s, and n such that

$$|f_{r,s}(t) - F_n(t)| \leq n^{-1/3}\left(1 + \frac{1}{s-r}\right) \qquad \text{for } 0 \leq t \leq 1. \qquad \blacksquare$$

The preceding corollary is used in Sec. 10-3 to prove the uniqueness of the Laplace transform. It is similarly used in Example 8 (Sec. 11-3) to prove that if two *different* probability distributions are both concentrated on the positive half-axis then their MGFs are different.

CHAPTER 11: RANDOM VARIABLES

Any function $f: \Omega \to F$ defined on the sample space of a discrete probability space (Ω,P) is said to be a **random point f with associated discrete probability space** (Ω,P), usually called the *random point f in F* or the *random point f*. If $F = R^n$ then f may be called a **random vector.** If $F = R$ then f may be called a **random variable.** Thus a random variable X with associated discrete probability space (Ω,P) is any real-valued function $X: \Omega \to R$.

If $f: \Omega \to F$ is a random point with associated discrete probability space (Ω,P) then the function $p_f: F \to R$ defined by

$$p_f(\alpha) = P[f = \alpha] \qquad \text{for each } \alpha \in F$$

is called the **discrete probability density of f**, or for f, or **induced by f**, or induced by f with associated (Ω,P). Thus $p_f(\alpha)$ is the probability that the random point will have the value α in F. If (F,P_f) is the discrete probability space corresponding to $p_f: F \to R$ then we say that (F,P_f) is induced by f.

Theorem 1 (Sec. 11-1) Let the discrete probability space (F,P_f) be induced by the random point $f: \Omega \to F$ with associated (Ω,P). Then

$$P_f(A) = P[f \in A]$$

for every subset A of F. $\blacksquare$

Thus if we want the probability that f will take on a value in A then we can obtain this same number from P_f or from P.

Theorem 2 (Sec. 11-1): Intermediate distributions If $f\colon \Omega \to F$ is a random point with associated density $p\colon \Omega \to R$ and if

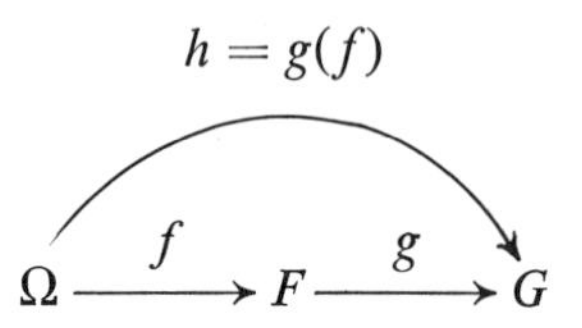

then $p_h = p_g$, where p_g is induced by g with associated p_f. ■

If $f_i\colon \Omega \to F_i$ for $i = 1, 2, \ldots, n$ are n random points with the same associated density $p\colon \Omega \to R$ then we can form the random point $f = (f_1, f_2, \ldots, f_n)$, where $f\colon \Omega \to F$ is into the cartesian product $F = F_1 \times F_2 \times \cdots \times F_n$ and is defined by

$$f(\omega) = (f_1(\omega), f_2(\omega), \ldots, f_n(\omega)) \qquad \text{for all } \omega \text{ in } \Omega.$$

The density induced by f is defined as usual by

$$p_f(\alpha) = P[f = \alpha] \qquad \text{for all } \alpha \text{ in } F$$

or equivalently by

$$p_f(\alpha_1, \alpha_2, \ldots, \alpha_n) = P[f_1 = \alpha_1, \ldots, f_n = \alpha_n].$$

Often the notation $p_{f_1, \ldots, f_n}$ is used instead of p_f, and this density is called the **joint density** of $f_1, \ldots, f_n$.

Theorem 3 (Sec. 11-1) If $f_1\colon \Omega \to F_1$ and $f_2\colon \Omega \to F_2$ are two random points with the same associated discrete probability density $p\colon \Omega \to R$ then

$$p_{f_1}(\alpha_1) = \sum_{\alpha_2} p_{f_1,f_2}(\alpha_1, \alpha_2) \qquad \text{for each } \alpha_1 \in F_1. \quad ■$$

A common application of this theorem is when $f_1 = (X_1, X_2, \ldots, X_m)$ and $f_2 = (X_{m+1}, X_{m+2}, \ldots, X_n)$ are random vectors, in which case

$$p_{X_1, \ldots, X_m}(x_1, \ldots, x_m) = \sum_{x_{m+1}, \ldots, x_n} p_{X_1, \ldots, X_n}(x_1, \ldots, x_m, \ldots, x_n)$$

for all $(x_1, \ldots, x_m)$ in R^m. Thus a joint density is obtained from a larger joint density by summing out the extra coordinates.

If $f_1\colon \Omega \to F_1$ and $f_2\colon \Omega \to F_2$ are two random points such that the sets $\{\alpha\colon p_{f_1}(\alpha) > 0\}$ and $\{\beta\colon p_{f_2}(\beta) > 0\}$ are both finite then the joint density p_{f_1,f_2} can be exhibited in the standard tabular form of Table 1 (Sec. 11-1). The table has a marginal column on the right for p_{f_1} and a marginal row on the bottom for p_{f_2}. Sometimes p_{f_1} is called the **marginal density of** f_1.

Random points $f_1, f_2, \ldots, f_n$ into the same set F are said to be **identically distributed** iff $p_{f_1} = p_{f_2} = \cdots = p_{f_n}$. The f_i need not all have the same associated discrete probability density; however, they must all be into the same set F which is the domain of every p_{f_i}.

If $f: \Omega \to F$ is a random point with associated discrete probability space (Ω, P) and A is an event with $P(A) > 0$ then the function $p_{f|A}: F \to R$ defined by

$$p_{f|A}(\alpha) = P_A[f = \alpha] = P[f = \alpha \mid A] = \frac{P([f = \alpha] \cap A)}{P(A)} \qquad \text{for each } \alpha \in F$$

is called the **conditional discrete probability density of f given A**. Thus $p_{f|A}(\alpha)$ is the probability that f will have the value α if we are given that A has occurred.

If $\Omega = F_1 \times F_2 \times \cdots \times F_n$ is the cartesian product of $n \geq 2$ sets and $1 \leq i \leq n$ then the ith **coordinate function** on Ω is the function $f_i: \Omega \to F_i$ defined by

$$f_i(\alpha_1, \alpha_2, \ldots, \alpha_n) = \alpha_i \qquad \text{for all } (\alpha_1, \alpha_2, \ldots, \alpha_n) \in \Omega.$$

Unless otherwise stated, it is always assumed that all of the random points in a discussion have the same associated discrete probability density. Sample points will often be suppressed in the notation. These are standard conventions in probability theory.

Random points $f_1, f_2, \ldots, f_n$ with $n \geq 2$ are said to be **independent,** or mutually independent, iff

$$P[f_1 \in A_1, \ldots, f_n \in A_n] = P[f_1 \in A_1] \cdots P[f_n \in A_n]$$

or equivalently

$$P_{f_1, \ldots, f_n}(A_1 \times \cdots \times A_n) = P_{f_1}(A_1) \cdots P_{f_n}(A_n)$$

for all subsets A_i of F_i, where $f_i: \Omega \to F_i$. At times we may use the terms **stochastically independent** or **statistically independent** rather than just "independent." Dependent means not independent. Clearly $f_1, \ldots, f_n$ are independent iff the events

$$[f_1 \in A_1], \ldots, [f_n \in A_n]$$

are *always* independent. Reordering or taking a subsequence of two or more does not destroy independence.

As stated below, functions of different independent random points, or subexperiments, must be independent.

Theorem 1 (Sec. 11-2) If $f_1, f_2, \ldots, f_n$ are $n \geq 2$ independent random points and

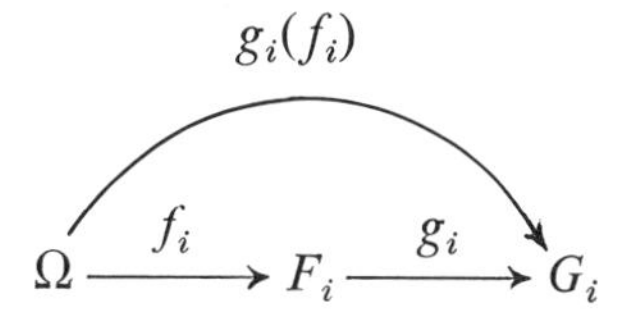

then $g_1(f_1), g_2(f_2), \ldots, g_n(f_n)$ are n independent random points. ■

It has been shown that *random points $f_1, f_2, \ldots, f_n$ are independent iff their joint density equals the product rule applied to the marginals*, $p_{f_1, \ldots, f_n} = p_{f_1} p_{f_2} \cdots p_{f_n}$. The joint-probability table for two independent random points has any two rows, or columns, proportional to each other. Also every entry in the table must be positive. Knowledge of the probabilistic behavior of each of several *independent* random points completely determines their unique probabilistic relationship by the product rule; that is, under independence the marginals determine the joint table.

If $f_1, f_2, \ldots, f_m, f_{m+1}, \ldots, f_n$ *are independent random points then the two random points* $(f_1, \ldots, f_m)$ *and* $(f_{m+1}, \ldots, f_n)$ *are independent.* We could break $f_1, \ldots, f_n$ into more than two disjoint parts and still have independence. Thus probabilities about some of them are never affected by knowledge concerning the others,

$$P[(f_1, \ldots, f_m) \in A \mid (f_{m+1}, \ldots, f_n) \in B] = P[(f_1, \ldots, f_m) \in A]$$

whenever $P[(f_{m+1}, \ldots, f_n) \in B] > 0$.

Random points $f_1, f_2, \ldots, f_n$ with $n \geq 2$ are said to be **pairwise independent** iff f_i and f_j are two independent random points whenever $i \neq j$. In this case knowledge about any *one* f_j does not alter the probabilities about any *one* other f_i,

$$P[f_i \in A_i \mid f_j \in A_j] = P[f_i \in A_i] \qquad \text{if} \qquad P[f_j \in A_j] > 0,\ i \neq j.$$

In discussing pairwise independence we never drop the adjective "pairwise" because pairwise independent random points may not be mutually independent.

Theorem 1 (Sec. 8-1) shows how to construct a model of independent subexperiments. It is restated below in the terminology of Chap. 11.

Theorem 2 (Sec. 11-2) Let $p_i: F_i \to R$ be a discrete probability density for each $i = 1, 2, \ldots, n$, where $n \geq 2$. Let $\Omega = F_1 \times F_2 \times \cdots \times F_n$ and let p be defined by

$$p(\alpha_1, \alpha_2, \ldots, \alpha_n) = p_1(\alpha_1)p_2(\alpha_2) \cdots p_n(\alpha_n) \qquad \text{for all } (\alpha_1, \ldots, \alpha_n) \in \Omega.$$

Then $p: \Omega \to R$ is a discrete probability density, and the coordinate functions $f_1, f_2, \ldots, f_n$ on Ω are independent, and $p_{f_i} = p_i$ for $i = 1, 2, \ldots, n$. ■

If A is a subset of Ω then the function $I_A: \Omega \to R$ defined by

$$I_A(\omega) = \begin{cases} 1 & \text{if } \omega \in A \\ 0 & \text{if } \omega \notin A \end{cases}$$

is called the **indicator** of A, or the indicator of A on Ω. Events $A_1, \ldots, A_n$ are independent iff the random variables $I_{A_1}, \ldots, I_{A_n}$ are independent.

As shown in Example 3 (Sec. 11-2), the sum of $d \geq 2$ independent random variables is binomial with parameters $n = n_1 + \cdots + n_d$ and p if the ith is binomial n_i and p. To prove this introduce n Bernoulli trials $X_1, \ldots, X_n$ and define Y_i by

$$S = \underbrace{X_1 + \cdots + X_{n_1}}_{Y_1} + \underbrace{X_{n_1+1} + \cdots + X_{n_1+n_2}}_{Y_2} + \cdots + X_n.$$

so that S is binomial n and p while $S = Y_1 + \cdots + Y_d$ where $Y_1, \ldots, Y_d$ are independent and Y_i is binomial n_i and p. Similarly by letting X_i be geometric with parameter p we see that the sum of $d \geq 2$ independent random variables is Pascal $n = n_1 + \cdots + n_d$ and p if the ith is Pascal n_i and p. This technique is based on the fact that if two different random points, f and f', with possibly different associated densities or experiments, happen to have the same induced densities $p_f = p_{f'}$ then so do functions of them, $p_{g(f)} = p_{g(f')}$. In particular if $Y_1, Y_2, \ldots, Y_d$ are independent random variables then their joint density is determined by their marginals,

$$p_f = p_{Y_1,\ldots,Y_d} = p_{Y_1}p_{Y_2}\cdots p_{Y_d}$$

and therefore the density induced by any function of them, such as

$$g(f) = g(Y_1, Y_2, \ldots, Y_d) = Y_1 + Y_2 + \cdots + Y_d,$$

is uniquely determined by the marginals $p_{Y_1}, p_{Y_2}, \ldots, p_{Y_d}$, so that in calculating $p_{g(f)}$ we can utilize *any* experiment having independent random variables with these marginals.

If X is a random variable with induced density p_X then the mean of p_X is denoted by EX and called the **expectation of** X, or the expected value of X, or the mean value of X, so that

$$EX = \sum_{i=1}^{\infty} x_i p_X(x_i) \qquad \text{where } \{x : p_X(x) > 0\} = \{x_1, x_2, \ldots\}$$

if the series is absolutely convergent. Instead of saying that X has an expectation, we often write $E|X| < \infty$. If the series is not absolutely convergent we say that X does not have an expectation, or EX does not exist. Identically distributed random variables have the same expectation. If I_A is the indicator of the event A then $EI_A = P(A)$. If c is a real number then $Ec = c$. Furthermore $E|X| = 0$ iff $P[X = 0] = 1$.

Theorem 1 (Sec. 11-3) Let X be a random variable with associated density $p: \Omega \to R$ and let $\{\omega : p(\omega) > 0\} = \{\omega_1, \omega_2, \ldots\}$ and $\{x : p_X(x) > 0\} = \{x_1, x_2, \ldots\}$. Then

$$\sum_{j=1}^{\infty} X(\omega_j)p(\omega_j) = \sum_{n=1}^{\infty} x_n p_X(x_n)$$

in the sense that if either series is absolutely convergent then so is the other, and equality holds. ■

Theorem 2 (Sec. 11-3) If

$$X = g(f)$$

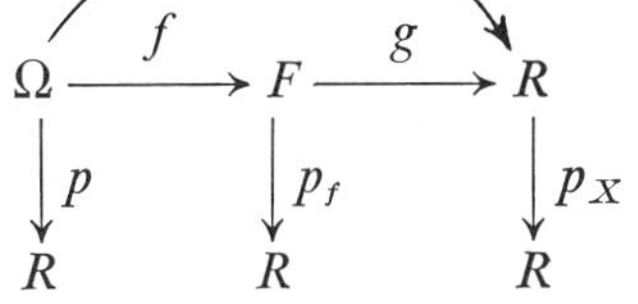

$$\begin{array}{ccccc} \Omega & \xrightarrow{f} & F & \xrightarrow{g} & R \\ \downarrow p & & \downarrow p_f & & \downarrow p_X \\ R & & R & & R \end{array}$$

and

$\omega_1, \omega_2, \ldots$ is an enumeration of $\{\omega : p(\omega) > 0\}$

$\alpha_1, \alpha_2, \ldots$ is an enumeration of $\{\alpha : p_f(\alpha) > 0\}$

$x_1, x_2, \ldots$ is an enumeration of $\{x : p_X(x) > 0\}$

then

$$\sum_{j=1}^{\infty} g(f(\omega_j))p(\omega_j) = \sum_{i=1}^{\infty} g(\alpha_i)p_f(\alpha_i) = \sum_{n=1}^{\infty} x_n p_X(x_n)$$

in the sense that if any one of the three series is absolutely convergent then all three are, and the equalities hold. ■

The conclusion of the last theorem might be written as

$$E_p X = E_{p_f} g = E_{p_X} x$$

where x in $E_{p_X} x$ stands for the identity function on R. In each case the calculation of the expectation is by the same rule: take the product of the random variable and its associated density and sum all nonzero values. In particular if $F = R^n$, so that $f = (X_1, \ldots, X_n)$ is a random vector, then

$$\begin{aligned} Eg(X_1, \ldots, X_n) &= \sum_{\omega} g(X_1(\omega), \ldots, X_n(\omega))p(\omega) \\ &= \sum_{x_1, \ldots, x_n} g(x_1, \ldots, x_n)p_{X_1, \ldots, X_n}(x_1, \ldots, x_n) \\ &= \sum_{x} x p_{g(X_1, \ldots, X_n)}(x). \end{aligned}$$

For any positive real r the **rth absolute moment** of X is the expectation of $Y = |X|^r$ as given by any of the series

$$E|X|^r = \sum_{j=1}^{\infty} |X(\omega_j)|^r p(\omega_j) = \sum_{n=1}^{\infty} |x_n|^r p_X(x_n) = \sum_{i=1}^{\infty} y_i p_{|X|^r}(y_i)$$

if absolutely convergent; otherwise we say that X does not have an rth absolute moment. For any $n = 1, 2, \ldots$ the **nth moment** of X is EX^n, and it exists iff X has an nth absolute moment. If $EX^2 < \infty$ we define the **variance** of X by $\operatorname{Var} X = E[X - EX]^2$ and we have

$$\operatorname{Var} X = E(X^2) - (EX)^2 \qquad [\text{the **standard deviation** of } X] = \sqrt{\operatorname{Var} X} \geq 0$$

$$E(aX + b) = a(EX) + b \qquad \operatorname{Var}(aX + b) = a^2(\operatorname{Var} X).$$

If X has mean μ_X and variance $\sigma_X^2 > 0$ then $X^* = (X - \mu_X)/\sigma_X$ is called the standardized or **normalized random variable** corresponding to X, and we have $EX^* = 0$ and $\operatorname{Var} X^* = 1$.

The following theorem is true in all cases in probability theory; if the hypotheses of a part are true then any expectations appearing in the conclusion exist.

Theorem 3 (Sec. 11-3): Properties of expectation Let X, Y, and $X_1, \ldots, X_n$ be random variables.

(*a*) If $E|Y| < \infty$ and $|X| \leq |Y|$ then $E|X| \leq E|Y|$.

(*b*) If $E|X| < \infty$ and a is a real then $EaX = aEX$ and $|EX| \leq E|X|$.

(*c*) If each $E|X_i| < \infty$ then $E\left(\sum_{i=1}^{n} X_i\right) = \sum_{i=1}^{n} EX_i$.

(*d*) If each $E|X_i| < \infty$ and $X_1, \ldots, X_n$ are independent then

$$E\left(\prod_{i=1}^{n} X_i\right) = \prod_{i=1}^{n} EX_i.$$

(*e*) If $E|X| < \infty$ and $E|Y| < \infty$ and $X \leq Y$ then $EX \leq EY$.
(*f*) If $E|X|^s < \infty$ and $0 < r < s$ then $(E|X|^r)^{1/r} \leq (E|X|^s)^{1/s}$.
(*g*) If $0 < r$ and each $E|X_i|^r < \infty$ then $E\left|\sum_{i=1}^{n} X_i\right|^r < \infty$.
(*h*) If $EX^2 < \infty$ and $EY^2 < \infty$ then $(EXY)^2 \leq (EX^2)(EY^2)$.
(*i*) If each $EX_i^2 < \infty$ and $X_1, \ldots, X_n$ are independent then

$$\operatorname{Var}\left(\sum_{i=1}^{n} X_i\right) = \sum_{i=1}^{n} \operatorname{Var} X_i. \quad \blacksquare$$

The basic property

$$E\sum_{i=1}^{n} a_i X_i = \sum_{i=1}^{n} a_i EX_i$$

is referred to as the *linearity of expectation*.

The **moment-generating function (MGF) of a random variable** X is defined by

$$m(t) = Ee^{tX} = \sum_{x} e^{tx} p_X(x)$$

If $X_1, \ldots, X_n$ are independent random variables then so are $e^{tX_1}, \ldots, e^{tX_n}$, so the MGF of a sum of independent random variables is equal to the product of their MGFs.

If $S_n = \sum_{i=1}^{n} X_i$ where $X_1, \ldots, X_n$ are independent and identically distributed, then

$$ES_n = nEX_1 \qquad \text{if } E|X_1| < \infty$$

$$\operatorname{Var} S_n = n \operatorname{Var} X_1 \qquad \text{if } E(X_1)^2 < \infty$$

$$m_{S_n}(t) = [m_{X_1}(t)]^n \qquad \text{if } m_{X_1}(t) = Ee^{tX_1} < \infty$$

For example, if $X_1, \ldots, X_n$ are Bernoulli trials with parameter p then we easily obtain $EX_1 = p$, $\operatorname{Var} X_1 = pq$, and $m_{X_1}(t) = q + pe^t$. In Sec. 8-2 it is shown that p_{S_n} is binomial n and p, so the equations above immediately yield the mean, variance, and MGF of the binomial n and p density. Similarly if X_1 has a geometric density with parameter p then we know that $EX_1 = q/p$, $\operatorname{Var} X_1 = q/p^2$, $m_{X_1}(t) = p/(1 - qe^t)$, and p_{S_n} is Pascal n and p. Therefore the equations above yield the mean, variance, and MGF of the Pascal density.

Example 8 (Sec. 11-3) *proves* that the distribution of a random variable X is completely determined by its moments if X is bounded, or by its MGF if X is non-negative. The results and proofs are valid in general, not just in the discrete case. Thus random variables X and Y must be identically distributed if $0 \leq X \leq 1$ and $0 \leq Y \leq 1$ and

$$EX^n = EY^n \qquad \text{for all } n = 1, 2, \ldots,$$

or if $0 \leq X$ and $0 \leq Y$ and

$$Ee^{tX} = Ee^{tY} \qquad \text{for all real } t \leq 0.$$

If X and Y have finite variances then the **covariance** of X and Y is given by

$$\text{Cov}\,(X,Y) = E[(X - EX)(Y - EY)] = E(XY) - (EX)(EY).$$

We have $\text{Cov}\,(X,Y) = \text{Cov}\,(Y,X)$ and $\text{Cov}\,(X,X) = \text{Var}\,X$ and $\text{Var}\,(X + Y) = \text{Var}\,X + \text{Var}\,Y + 2\,\text{Cov}\,(X,Y)$. We say that $X_1, \ldots, X_n$ are **uncorrelated** iff $\text{Cov}\,(X_i,X_j) = 0$ whenever $i \neq j$. If $X_1, \ldots, X_n$ are independent then they are uncorrelated, but the converse is false. If each $EX_i^2 < \infty$ then

$$\text{Var}\left(\sum_{i=1}^{n} X_i\right) = \sum_{i=1}^{n}\sum_{j=1}^{n} \text{Cov}\,(X_i,X_j) = \sum_{k=1}^{n} \text{Var}\,X_k + 2 \sum_{1 \leq i < j \leq n} \text{Cov}\,(X_i,X_j).$$

If X and Y have finite nonzero variances then their **correlation coefficient** is

$$\rho_{X,Y} = \text{Cov}\,(X^*,Y^*) = E\left(\frac{X - \mu_X}{\sigma_X}\right)\left(\frac{Y - \mu_Y}{\sigma_Y}\right) = \frac{\text{Cov}\,(X,Y)}{\sigma_X\sigma_Y}.$$

We have $\rho_{X,Y} = \rho_{Y,X}$ and $\rho_{X,X} = 1$; and $\rho_{X,Y} = 0$ iff $\text{Cov}\,(X,Y) = 0$, iff X and Y are uncorrelated. Also $-1 \leq \rho_{X,Y} \leq 1$.

The minimum over a and b of the expected squared error $E[Y - (aX + b)]^2$ is achieved by the unique linear predictor

$$Y_p = \mu_Y + \sigma_Y \rho \frac{X - \mu_X}{\sigma_X} \qquad \rho = \rho_{X,Y}$$

of Y in terms of X, and it yields

$$E[Y - Y_p]^2 = \sigma_Y^2(1 - \rho^2).$$

The line $y_p = \mu_Y + \sigma_Y\rho(x - \mu_X)/\sigma_X$ is called the **least squares regression line** of, or for, Y on X. These results indicate that the correlation coefficient ρ can be interpreted as a standardized measure of the degree $|\rho|$ and direction $\pm$ of the *proportionality* of the statistical dependence between two random variables.

CHAPTER 12: THE LAW OF LARGE NUMBERS

Theorem 1 (Sec. 12-1): A WLLN for independent identically distributed random variables Let $X_1, X_2, \ldots, X_n$ be independent and identically distributed random variables with $EX_1^2 < \infty$, and let $S_n = X_1 + X_2 + \cdots + X_n$. If $\epsilon > 0$ then

$$1 - P\left[EX_1 - \epsilon < \frac{1}{n} S_n < EX_1 + \epsilon\right] = P\left[\left|\frac{1}{n} S_n - EX_1\right| \geq \epsilon\right] \leq \frac{\text{Var}\,X_1}{\epsilon^2 n}. \qquad \blacksquare$$

This theorem implies that

$$P\left[EX_1 - \epsilon < \frac{1}{n} S_n < EX_1 + \epsilon\right] \geq 1 - \frac{\operatorname{Var} X_1}{\epsilon^2 n} \to 1 \qquad \text{as } n \to \infty.$$

That is, regardless of how small a tolerance interval $EX_1 \pm \epsilon$ we insist on, if n is large enough then we can be quite certain that the value of S_n/n will be inside the tolerance interval.

Theorem 2 (Sec. 12-1): A WLLN for independent random variables Let $X_1, X_2, \ldots, X_n$ be independent random variables with all $EX_i^2 < \infty$ and let $S_n = X_1 + X_2 + \cdots + X_n$ and $m_n = E(1/n)S_n = (1/n)(EX_1 + EX_2 + \cdots + EX_n)$. If $\epsilon > 0$ then

$$P\left[\left|\frac{1}{n} S_n - m_n\right| \geq \epsilon\right] \leq \frac{1}{\epsilon^2 n^2} (\operatorname{Var} X_1 + \operatorname{Var} X_2 + \cdots + \operatorname{Var} X_n). \quad \blacksquare$$

The most important case of this theorem is when the variances are all bounded, $\operatorname{Var} X_i \leq A$, by a constant which does not depend on n, so that

$$P\left[\left|\frac{1}{n} S_n - m_n\right| \geq \epsilon\right] \leq \frac{A}{\epsilon^2 n} \to 0 \qquad \text{as } n \to \infty.$$

Theorem 3 (Sec. 12-1) Let $Y_1, Y_2, \ldots, Y_n$ be independent and identically distributed random variables with $P[Y_1 > 0] = 1$ and $E(\log_b Y_1)^2 < \infty$, where $b > 1$ is any base. If $\epsilon > 0$ then

$$P[b^{n(E \log_b Y_1 - \epsilon)} < Y_1 Y_2 \cdots Y_n < b^{n(E \log_b Y_1 + \epsilon)}] \geq 1 - \frac{\operatorname{Var} (\log_b Y_1)}{n\epsilon^2}. \quad \blacksquare$$

It was shown that if $E|Y| < \infty$ and $P[Y > 0] = 1$ then $e^{E \log Y} \leq EY$, so that the true growth $e^{nE \log Y_1}$ of the product $Y_1 Y_2 \cdots Y_n$ is always less than the wrong guess $(EY_1)^n$.

The WLLN says that under certain conditions if n is large enough then we can be quite sure that $(1/n)S_n$ will be close to EX_1. The SLLN says that under certain conditions if n is large enough then we can be quite sure that every one of S_n/n, $S_{n+1}/(n + 1)$, $S_{n+2}/(n + 2)$, . . . will close to EX_1.

Theorem 1 (Sec. 12-3): An SLLN for independent identically distributed random variables Let $X_1, X_2, \ldots, X_n, \ldots, X_m$ be independent and identically distributed random variables with $EX_1^2 < \infty$, and let $S_k = X_1 + X_2 + \cdots + X_k$. If $\epsilon > 0$ then

$$1 - P\left[\left|\frac{1}{n} S_n - EX_1\right| < \epsilon, \left|\frac{1}{n+1} S_{n+1} - EX_1\right| < \epsilon, \ldots, \left|\frac{1}{m} S_m - EX_1\right| < \epsilon\right]$$

$$= P\left[\left|\frac{1}{k} S_k - EX_1\right| \geq \epsilon \text{ for at least one } k \text{ satisfying } n \leq k \leq m\right] \leq \frac{2 \operatorname{Var} X_1}{\epsilon^2 n}. \quad \blacksquare$$

This theorem, when specialized to Bernoulli trials, immediately yields the following theorem.

Theorem 2 (Sec. 12-3): An SLLN for Bernoulli trials If $X_1, X_2, \ldots, X_n, \ldots, X_m$ are m Bernoulli trials with parameter p, and if $S_k = X_1 + X_2 + \cdots + X_k$ and $\epsilon > 0$ then

$$P\left[p - \epsilon < \frac{1}{k} S_k < p + \epsilon \text{ whenever } n \le k \le m\right] \ge 1 - \frac{2pq}{\epsilon^2 n} \ge 1 - \frac{1}{2\epsilon^2 n}. \quad \blacksquare$$

Thus the fraction of successes must become *and remain* close to p.

Theorem 1 (Sec. 12-3) is an immediate consequence of the following theorem.

Theorem 1 (Sec. 12-4): The Kolmogorov-Hájek-Rényi (KHR) inequality Let $X_1, X_2, \ldots, X_n, \ldots, X_m$ be independent random variables, with all $EX_j^2 < \infty$, and let $S_k = X_1 + X_2 + \cdots + X_k$. If $0 < \epsilon_n \le \epsilon_{n+1} \le \cdots \le \epsilon_m$ then

$$P[|S_k - ES_k| \ge \epsilon_k \text{ for at least one } k \text{ satisfying } n \le k \le m]$$
$$\le \frac{1}{(\epsilon_n)^2} \sum_{i=1}^{n} \text{Var } X_i + \sum_{k=n+1}^{m} \frac{\text{Var } X_k}{(\epsilon_k)^2}. \quad \blacksquare$$

CHAPTER 13: THE CENTRAL-LIMIT THEOREM

For any real μ and any positive real σ we define the **normal continuous probability density $\varphi_{\mu,\sigma}: R \to R$ with mean μ and standard deviation** σ, or variance σ^2, by

$$\varphi_{\mu,\sigma}(x) = \frac{1}{\sigma\sqrt{2\pi}} e^{-(x-\mu)^2/2\sigma^2} \qquad \text{for all real } x.$$

By analogy with the discrete case, the following equations justify the multiplicative constant and the names of the parameters:

$$\int_{-\infty}^{\infty} \varphi_{\mu,\sigma}(x)\,dx = 1 \qquad \int_{-\infty}^{\infty} x\varphi_{\mu,\sigma}(x)\,dx = \mu \qquad \int_{-\infty}^{\infty} (x-\mu)^2\varphi_{\mu,\sigma}(x)\,dx = \sigma^2.$$

We let φ denote the normal density $\varphi_{0,1}$ with zero mean and unit variance, and define $\Phi: R \to R$ by $\Phi(b) = \int_{-\infty}^{b} \varphi(x)\,dx$. We call Φ the **cumulative normal.** The values $\Phi(b)$ are tabulated for some nonnegative b in Table 1 (Sec. 13-1). The identity $\Phi(-b) = 1 - \Phi(b)$ can be used for negative arguments. By making the transformation $y = (x - \mu)/\sigma$ we see that

$$\int_a^b \varphi_{\mu,\sigma}(x)\,dx = \int_{(a-\mu)/\sigma}^{(b-\mu)/\sigma} \varphi(y)\,dy = \Phi\left(\frac{b-\mu}{\sigma}\right) - \Phi\left(\frac{a-\mu}{\sigma}\right) \qquad \text{if } a < b,$$

so that areas under any $\varphi_{\mu,\sigma}$ can be obtained from Table 1 (Sec. 13-1).

Theorem 1 (Sec. 13-2) Let $S_n = X_1 + \cdots + X_n$ where $X_1, \ldots, X_n$ are independent random variables with each $E|X_k|^3 < \infty$. Assume that $s_n^2 = \text{Var } S_n > 0$ and let

$S_n^* = (S_n - ES_n)/s_n$. Then for every real b

$$|P[S_n^* < b] - \Phi(b)| \leq 3\left\{\sum_{k=1}^{n} E\left|\frac{X_k - EX_k}{s_n}\right|^3\right\}^{1/4}. \quad \blacksquare$$

Corollary 1 (Sec. 13-2) Let $S_n = X_1 + \cdots + X_n$ where $X_1, \ldots, X_n$ are independent and identically distributed random variables satisfying $E|X_1|^3 < \infty$ and $\sigma_{X_1}^2 = \text{Var } X_1 > 0$. If $S_n^* = (S_n - ES_n)/\sigma_{X_1}\sqrt{n}$ then for every real b

(*a*) $\lim_{n \to \infty} P[S_n^* < b] = \Phi(b)$

(*b*) $|P[S_n^* < b] - \Phi(b)| \leq 3\left\{\dfrac{E|X_1^*|^3}{\sqrt{n}}\right\}^{1/4}$

where $X_1^* = (X_1 - EX_1)/\sigma_{X_1}$. $\blacksquare$

It is shown in Sec. 20-2 that Corollary 1*a* holds if the assumption $E|X_1|^3 < \infty$ is relaxed to $EX_1^2 < \infty$. Thus the asymptotic CLT applies to any distribution for X_1 for which $EX_1^2 < \infty$.

Corollary 2 (Sec. 13-2) Let $S_n = X_1 + \cdots + X_n$, where $X_1, \ldots, X_n$ are independent random variables with each $E|X_k|^3 < \infty$. Assume that $s_n^2 = \text{Var } S_n > 0$ and let $S_n^* = (S_n - ES_n)/s_n$. Assume that M is a constant such that

$$E|X_k - EX_k|^3 \leq M \text{ Var } X_k \qquad \text{whenever } 1 \leq k \leq n.$$

Then for every real b

$$|P[S_n^* < b] - \Phi(b)| \leq 3\left(\frac{M}{s_n}\right)^{1/4}. \quad \blacksquare$$

In this corollary we can, as shown in Example 8 (Sec. 13-2), let $M = 2c$ if c is a constant such that each $P[|X_k| \leq c] = 1$. Thus the CLT always applies to the sum of independent *uniformly* bounded (that is, c does not depend on n) random variables, whenever the variance of the sum goes to infinity.

If for all real b we have $P[S_n^* < b] \to \Phi(b)$ as n goes to infinity then for all real b we have $P[S_n^* = b]$ converging to zero as n increases to infinity. That is, if the distribution of S_n^* converges to the normal distribution then *in a limiting sense* each real b has zero probability, so the inclusion of end points does not affect the limiting value.

If X_λ has a Poisson distribution with parameter λ much larger than 1 then X_λ^* is approximately normally distributed; that is, as shown in Example 6 (Sec. 13-2), if λ goes to infinity then

$$P[\lambda + a\sqrt{\lambda} < X_\lambda < \lambda + b\sqrt{\lambda}] = P[a < X_\lambda^* < b] \to \Phi(b) - \Phi(a) \quad \text{if } a < b.$$

CHAPTER 14: CONDITIONAL EXPECTATIONS AND BRANCHING PROCESSES

If $f: \Omega \to F$ and $g: \Omega \to G$ are random points and if β belongs to G with $p_g(\beta) > 0$ then the discrete density defined on F by

$$p_{f|g=\beta}(\alpha) = p_{f|g}(\alpha \mid \beta) = \frac{p_{f,g}(\alpha,\beta)}{p_g(\beta)} \qquad \text{for all } \alpha \in F$$

is called the **conditional density of f given the event** $[g = \beta]$. Then

$$p_f(\alpha) = \sum_\beta p_{f,g}(\alpha,\beta) = \sum_\beta p_{f|g}(\alpha \mid \beta)p_g(\beta)$$

is said to be a **mixture** of the densities $p_{f|g=\beta}$ according to p_g, or p_f is said to be obtained by **randomizing the parameter** β in $p_{f|g}(\alpha \mid \beta)$ according to p_g.

If X is a random variable with $E\,|X| < \infty$, g is a random point, and $p_g(\beta) > 0$ then the mean of $p_{X|g=\beta}$ is denoted by $\mu_{X|g=\beta}$ or $E[X \mid g = \beta]$. The function $\mu_{X|g} = E(X \mid g)$ is called the **conditional expectation of X given g**, or the **regression function of X on g**. If $EX^2 < \infty$ then the variance of $p_{X|g=\beta}$ is denoted by $\sigma^2_{X|g=\beta}$ or Var $[X \mid g = \beta]$. The function $\sigma^2_{X|g} = \text{Var}\,(X \mid g)$ is called the **conditional variance of X given g**.

The linearity of conditional expectation

$$E\left(\sum_{i=1}^n a_i X_i \mid g\right) = \sum_{i=1}^n a_i E(X_i \mid g)$$

is valid whenever each $E\,|X_i| < \infty$. Also $E(c \mid g) = c$. If X and g are independent then

$$E(X \mid g) = EX \qquad \text{if } E\,|X| < \infty$$

$$\text{Var}\,(X \mid g) = \text{Var}\,X \qquad \text{if } EX^2 < \infty.$$

Theorem 1 (Sec. 14-1) Let X and g be a random variable and a random point with the same associated discrete probability space, so that p_X can be expressed as a mixture

$$p_X(x) = \sum_\beta p_{X|g=\beta}(x)p_g(\beta).$$

Then p_X has a mean iff "each $p_{X|g=\beta}$ with $p_g(\beta) > 0$ has a mean and the series below converges absolutely," in which case

$$EX = \sum_\beta \mu_{X|g}(\beta)p_g(\beta) = \sum_\beta E[X \mid g = \beta]p_g(\beta).$$

Also p_X has a variance iff "each $p_{X|g=\beta}$ with $p_g(\beta) > 0$ has a variance and both series below converge," in which case

$$\text{Var}\,X = \sum_\beta \sigma^2_{X|g}(\beta)p_g(\beta) + \sum_\beta [\mu_{X|g}(\beta) - \mu_X]^2 p_g(\beta). \quad \blacksquare$$

The number

$$\sum_X \sum_\beta [x - h(\beta)]^2 p_{X,g}(x,\beta)$$

is minimized by the unique, for β satisfying $p_g(\beta) > 0$, function $h = \mu_{X|g}$. Therefore $\mu_{X|g}$ is the unique predictor of X, given g, which achieves the minimum expected squared error of $\sum_\beta \sigma^2_{X|g}(\beta)p_g$. The above theorem shows that this quantity is less than the expected squared error of Var X for the best constant predictor EX, by an amount equal to the variance of the predictor $\mu_{X|g}$.

As explained in Example 2 (Sec. 14-1), the random variable which is defined at a sample point ω by $\mu_{X|g}(g(\omega))$ is also often called the *conditional expectation of X given g*.

If $N, X_1, X_2, \ldots, X_m$ are $m + 1$ random variables such that $P[N = 0$ or $N = 1$ or $\cdots$ or $N = m] = 1$ then the random variable S_N defined by

$$S_N(\omega) = \begin{cases} 0 & \text{if } N(\omega) = 0 \\ X_1(\omega) + X_2(\omega) + \cdots + X_{N(\omega)}(\omega) & \text{otherwise} \end{cases}$$

is *the sum of a random number of random variables*. If $N, X_1, X_2, \ldots, X_m$ are independent then

$$p_{S_N}(s) = \sum_{n=0}^{m} p_{S_n}(s)p_N(n)$$

where $p_{S_0}(0) = 1$ and $S_n = X_1 + \cdots + X_n$ for $1 \leq n \leq m$. If $p_N(n) > 0$ for infinitely many n then, as in the theorem below, p_{S_N} is defined by this mixture and given the same interpretation.

If $P[X \geq 0] = 1$ then

$$G_X(z) = Ez^X = \sum_x z^x p_X(x) \qquad \text{for } 0 \leq z \leq 1$$

is called the **probability-generating function for** X, or p_X. We let $G_X(0) = G_X(0+) = p_X(0)$, while $G_X(1) = 1$ and $0 \leq G_X(z) \leq 1$ if $0 \leq z \leq 1$. G_X determines p_X. If $X_1, \ldots, X_n$ are independent then

$$G_{X_1 + \cdots + X_n} = G_{X_1}G_{X_2} \cdots G_{X_n}$$

which reduces to $[G_{X_1}]^n$ when they are identically distributed.

Theorem 1 (Sec. 14-2): The sum $S_N = X_1 + \cdots + X_N$ of a random number N of random variables Let p_N and p_{X_1} be discrete probability densities on R such that $\sum_{n=0}^{\infty} p_N(n) = 1$. Define the discrete density p_{S_0} by $p_{S_0}(0) = 1$, and for $n = 1, 2, \ldots$ let p_{S_n} be the density of $S_n = X_1 + \cdots + X_n$ where $X_1, \ldots, X_n$ are independent and each has density p_{X_1}. Define the discrete density $p_{S_N}: R \to R$ by

$$p_{S_N}(s) = \sum_{n=0}^{\infty} p_{S_n}(s)p_N(n).$$

Then p_{S_N} has a mean iff both p_N and p_{X_1} do, and in this case

$$ES_N = (EX_1)(EN).$$

Also p_{S_N} has a variance iff both p_N and p_{X_1} do, and in this case

$$\text{Var } S_N = (EN)(\text{Var } X_1) + (EX_1)^2(\text{Var } N).$$

Furthermore if p_{X_1} is concentrated on the nonnegative reals and G_N, G_{X_1}, and G_{S_N} are the probability-generating functions for p_N, p_{X_1}, and p_{S_N}, respectively, then

$$G_{S_N}(z) = G_N(G_{X_1}(z)) \qquad \text{for } 0 \leq z \leq 1. \quad \blacksquare$$

Compound Poisson and infinitely divisible densities are introduced and studied in Examples 4 and 6 (Sec. 14-2).

Let $p: R \to R$ be a discrete density concentrated on $0, 1, 2, \ldots$ and satisfying $0 < p(0) < 1$. We may interpret $p(k)$ as the probability that a parent will have k offspring. The sequence of densities $p_{W_1}, p_{W_2}, \ldots$ defined below is referred to as the **branching process determined by** p, where W_n is the size of the nth generation. Let $p_{W_1} = p$ so that the 0th generation consists of one initiator. If p_{W_n} has been defined let $p_{W_{n+1}}$ be the density induced, as in Theorem 1 (Sec. 14-2), by

$$W_{n+1} = X_1 + X_2 + \cdots + X_{W_n}$$

where W_n and the X_i are independent, W_n has density p_{W_n}, and each X_i has density p.

Theorem 1 (Sec. 14-3): Fundamental theorem for branching processes Let the discrete density $p: R \to R$ satisfy $0 < p(0) < 1$ and $\sum_{k=0}^{\infty} p(k) = 1$, and have probability-generating function G. Let $p = p_{W_1}, p_{W_2}, \ldots$ be the branching process determined by p, so that $p_{W_n}(0) \le p_{W_{n+1}}(0)$ and the limit $\beta = \lim_{n \to \infty} p_{W_n}(0)$ is interpretable as the **probability of eventual extinction.** Let $\mu = \sum_{k=0}^{\infty} kp(k)$ be the expected number of offspring per parent. If $\mu \le 1$ then $\beta = 1$, and if $\mu > 1$ then β is the unique real number satisfying $0 < \beta < 1$ and $\beta = G(\beta)$. ■

CHAPTER 15: PROBABILITY SPACES

In this chapter the structure of the general theory is developed in a fashion applicable to discrete and continuous densities and to all other cases. Much of the remainder of the book is not rigorous, primarily because a rigorous development valid for nondiscrete cases requires that $P(A)$ be defined for only some but not all subsets A of the sample space, and that some but not all real-valued functions $X: \Omega \to R$ defined on the sample space should be called random variables. Except in Chap. 21, we normally ignore these two complications.

A **probability space** (Ω, P) consists of a nonempty set Ω and a function P. The set Ω is called the **sample space,** every point ω in Ω is called a **sample point,** and every subset A of Ω is called an **event.** P is a real-valued function defined on the collection of all events, and if A is an event then $P(A)$ is called the **probability of** A. The function P is called a **probability measure** and is required to satisfy the following three conditions:

1. $P(\Omega) = 1$.
2. $P(A) \ge 0$ for every event A.
3. If $A_1, A_2, \ldots$ is any sequence of disjoint events then

$$P\left(\bigcup_{n=1}^{\infty} A_n\right) = \sum_{n=1}^{\infty} P(A_n).$$

Theorem 3 (Sec. 6-4) is valid in general.

The definition of conditional probability remains the same,

$$P(B \mid A) = \frac{P(AB)}{P(A)} \qquad \text{if } P(A) > 0$$

as do the statement and proof of Bayes' Theorem 1 (Sec. 7-2) and the definition, proofs, and properties of independent events given in Chap. 7.

Theorem 1 (Sec. 15-1) Let (Ω,P) be a probability space and A an event with $P(A) > 0$. If the function P_A is defined for every event B by

$$P_A(B) = P(B \mid A) = \frac{P(AB)}{P(A)}$$

then (Ω,P_A) is a probability space, which is said to be obtained by **conditioning** (Ω,P) **by** A. ■

If $a = (a_1, \ldots, a_n)$ and $b = (b_1, \ldots, b_n)$ are two points of R^n then the set

$$R_{a,b} = \{(x_1, \ldots, x_n): a_i \leq x_i \leq b_i \text{ for all } i = 1, 2, \ldots, n\}$$

is called a **closed rectangle** in R^n. The word "closed" refers to the fact that we used only $\leq$. If we replace every $\leq$ by $<$ then the set is called an **open rectangle** in R^n. If some or all of the $\leq$ are replaced by $<$ then we still call the set a **rectangle** in R^n.

Theorem 2 (Sec. 15-1) If (R^n,P) and (R^n,P') are probability spaces for which $P(R_{a,b}) = P'(R_{a,b})$ for every closed rectangle $R_{a,b}$ in R^n then $P = P'$. ■

Any function $f: \Omega \to R^n$ defined on the sample space of a probability space (Ω,P) is said to be a **random vector** f **with associated probability space** (Ω,P). If $n = 1$ then f may be called a **random variable.** At times we permit f to be defined only on an event having probability 1.

Theorem 1 (Sec. 15-2) If $f: \Omega \to R^n$ is a random vector with associated probability space (Ω,P) and if P_f is defined for every subset D of R^n by

$$P_f(D) = P[f \in D] = P\{\omega : f(\omega) \in D\} = P[f^{-1}(D)]$$

then (R^n,P_f) is a probability space, which is said to be **induced by** $f: \Omega \to R^n$ with associated (Ω,P). ■

Random vectors $f_1, \ldots, f_k$ into the same R^n are said to be **identically distributed** iff $P_{f_1} = \cdots = P_{f_k}$.

If $f = (f_1, \ldots, f_n)$, where $f_1, \ldots, f_n$ are random vectors with the same associated probability space (Ω,P) then, as before, we may write $P_f = P_{f_1, \ldots, f_n}$ and call it the **joint distribution** of $f_1, \ldots, f_n$.

Let (Ω,P_A) be obtained by conditioning the probability space (Ω,P) by the event A with $P(A) > 0$. Then the *random vector* $f: \Omega \to R^n$ *with associated* (Ω,P_A) induces a probability space $(R^n,P_{f|A})$, called the **conditional distribution of** f **given** A. Therefore $P_{f|A}$ is defined by

$$P_{f|A}(D) = P_A[f \in D] = P[f \in D \mid A] = \frac{P([f \in D] \cap A)}{P(A)}$$

for every subset D of R^n. That is, $P_{f|A}(D)$ is the conditional probability that f will belong to D if we are given that A has occurred.

Theorem 2 (Sec. 15-2): Intermediate distributions If (Ω,P) is a probability space and if

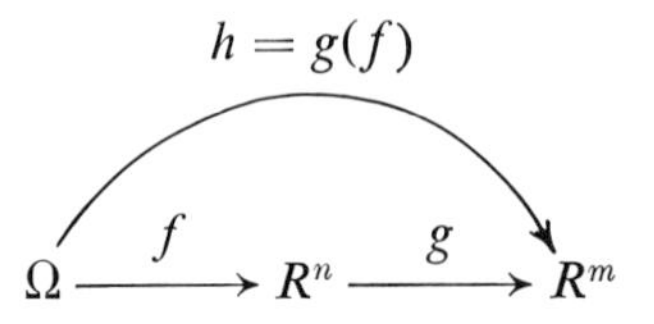

then $P_h = P_g$, where P_g is induced by g with associated (R^n,P_f). ■

Theorem 3 (Sec. 15-2) If $f\colon \Omega \to R^n$ and $g\colon \Omega \to R^m$ are random vectors with the same associated probability space (Ω,P) then

$$P_f(D) = P_{f,g}(D \times R^m) \qquad \text{for every subset } D \text{ of } R^n. \quad ■$$

As before, $n \geq 2$ random vectors $f_1, \ldots, f_n$ having the same associated probability space (Ω,P) are said to be **independent** iff

$$P[f_1 \in D_1, \ldots, f_n \in D_n] = P[f_1 \in D_1]P[f_2 \in D_2] \cdots P[f_n \in D_n]$$

for all subsets D_i of R^{k_i} where $f_i\colon \Omega \to R^{k_i}$. As before, taking a subsequence of two or more, or reordering, does not destroy independence, so that $f_1, \ldots, f_n$ are independent iff the events $[f_1 \in D_1], \ldots, [f_n \in D_n]$ are always independent. As before, events $A_1, \ldots, A_n$ are independent iff the random variables $I_{A_1}, \ldots, I_{A_n}$ are independent, where I_{A_i} is the indicator of A_i on Ω. As before, $n \geq 2$ random vectors $f_1, \ldots, f_n$ are **pairwise independent** iff f_i and f_j are two independent random vectors whenever $i \neq j$.

It is always possible to construct an (Ω,P) with independent random vectors with specified marginals $P_{f_1}, \ldots, P_{f_d}$. The marginals $P_{f_1}, \ldots, P_{f_d}$ of *independent* random vectors determine their joint distribution $P_f = P_{f_1,\ldots,f_d}$ and hence the distribution of any $g(f)$. We can find $P_{g(f)}$ directly from P_f and g, or within the context of a convenient experiment generating P_f.

Theorem 1 (Sec. 15-3) If $f_1, \ldots, f_d$ are $d \geq 2$ independent random vectors with the same associated probability space (Ω,P), and if

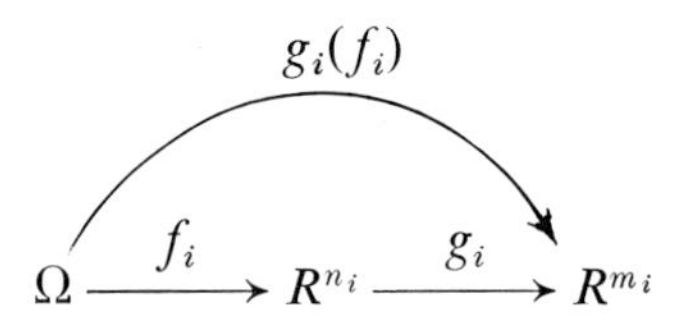

then $g_1(f_1), g_2(f_2), \ldots, g_d(f_d)$ are d independent random vectors. ■

Theorem 2 (Sec. 15-3) If $f_1, \ldots, f_m, \ldots, f_n$ are $n \geq 3$ independent random vectors with associated probability space (Ω,P) then $(f_1, \ldots, f_m)$ and $(f_{m+1}, \ldots, f_n)$ are two independent random vectors. ■

The proof of this theorem utilizes the following two lemmas.

Lemma 1 (Sec. 15-3) Let $f\colon \Omega \to R^n$ be a random vector and A an event such that A and $[f \in D]$ are two independent events whenever D is a closed rectangle in R^n. Then A and $[f \in D]$ are two independent events whenever D is any subset of R^n. ∎

Lemma 2 (Sec. 15-3) Random vectors $f_1, \ldots, f_d$ are independent if

$$P[f_1 \in R_1, \ldots, f_d \in R_d] = P[f_1 \in R_1]P[f_2 \in R_2] \cdots P[f_d \in R_d]$$

for all closed rectangles R_i of R^{n_i} where $f_i\colon \Omega \to R^{n_i}$. ∎

Theorem 1 (Sec. 15-4) Let (R,P) be a probability space with the real line for sample space. For each $n = 1, 2, \ldots$ define A_n^+ and A_n^- by

$$A_n^+ = \sum_{j=1}^{M_n} \frac{j}{2^n} P\left\{x\colon \frac{j}{2^n} \le x < \frac{j+1}{2^n}\right\}$$

$$A_n^- = \sum_{j=1}^{M_n} \frac{-j}{2^n} P\left\{x\colon -\frac{j+1}{2^n} < x \le -\frac{j}{2^n}\right\}$$

where $M_n = n2^n - 1$. Then $0 \le A_1^+ \le A_2^+ \le \cdots$ and $0 \ge A_1^- \ge A_2^- \ge \cdots$. Also the two sequences A_n^+ and A_n^- both converge to finite limits A^+ and A^- iff the series

$$\sum_{k=1}^{\infty} kP\{x\colon k - 1 \le |x| < k\}$$

has a finite sum B. We say that P has a mean iff this is the case, and we define

$$\textbf{the mean of } P = A^+ + A^-$$

and it satisfies

$$|\text{the mean of } P| \le A^+ - A^- \le B. \quad ∎$$

The **expectation** EX of the random variable $X\colon \Omega \to R$ with associated probability space (Ω,P) is defined to be the mean of P_X, if P_X has a mean; otherwise we say that X does not have an expectation. The Addendum to Sec. 15-4 shows that our definition of EX may be interpreted as the limit of the expectations of special discrete approximations to X.

The following properties of expectation are almost immediate. Identically distributed random variables, $P_X = P_Y$, have the same expectation, $EX = EY$. If a random variable is essentially a constant, $P[X = c] = 1$, then $EX = c$. If I_A is the indicator function of A on Ω then $EI_A = P(A)$. If there is a number B such that $P[|X| > B] = 0$ then X has an expectation and $-B \le EX \le B$. Also $E|X| = 0$ iff $P[|X| = 0] = 1$. Furthermore X has an expectation iff $|X|$ does, and this is denoted by $E|X| < \infty$.

Theorem 2 (Sec. 15-4) If (Ω,P) is a probability space and if

$$X = g(f)$$

$$\Omega \xrightarrow{\ f\ } R^n \xrightarrow{\ g\ } R$$

then $EX = Eg$. That is, the expected value of $X: \Omega \to R$ with associated (Ω,P) equals the expected value of g with associated (R^n,P_f) in the sense that if either expectation exists then both exist and are equal. ■

The general properties of expectation in the summary of Chap. 11 hold in general. This includes Theorem 3 (Sec. 11-3) and the facts about EX^n, $\sigma_X{}^2$, σ_X, X^*, MGFs, Cov (X,Y), ρ, and least-squares regression.

If r is a positive real and if $E|X|^r$ is finite then we call it the **rth absolute moment** of X. The **moment-generating function** (MGF) of a random variable X is defined by $m(t) = Ee^{tX}$ for any real t for which the positive-valued random variable e^{tX} has a finite expectation.

Theorem 3 (Sec. 15-4): Chebychev's inequality If $EX^2 < \infty$ and ϵ is any positive real then

$$P[|X - EX| \geq \epsilon] \leq \frac{\text{Var } X}{\epsilon^2}. \quad ■$$

Theorem 4 (Sec. 15-4) Let $m(t) = Ee^{tX}$ be the MGF of a random variable X. Assume that $m(t_0)$ and $m(-t_0)$ are finite, where $t_0 > 0$. Then m is finite and has derivatives of all orders whenever $|t| < t_0$. Differentiation under the expectation sign is permitted, so

$$m^{(n)}(t) = \frac{d^n}{dt^n} m(t) = EX^n e^{tX} \qquad \text{if } n = 1, 2, \ldots, |t| < t_0.$$

In particular all moments of X exist and $m^{(n)}(0)$ equals the nth moment, so

$$EX = m^{(1)}(0) \qquad \text{and} \qquad \text{Var } X = m^{(2)}(0) - [m^{(1)}(0)]^2.$$

Furthermore the power series

$$m(t) = \sum_{n=0}^{\infty} \frac{m^{(n)}(0)}{n!} t^n$$

converges absolutely for $|t| < t_0$.

If Ee^{tX} and Ee^{tY} are finite whenever $|t| \leq t_0$ with $t_0 > 0$ then

$$P_X = P_Y \text{ iff } Ee^{tX} = Ee^{tY} \text{ for all } |t| \leq t_0. \quad ■$$

Let $f: \Omega \to R^n$ and $g: \Omega \to R^m$ be random vectors with the same associated probability space (Ω,P). Assume that g induces a discrete probability density, although f need not. If β is a point for which $P[g = \beta] = p_g(\beta) > 0$ then the **conditional distribution of f given** $[g = \beta]$ is, as in Sec. 15-2, the distribution of f with associated $(\Omega,P_{[g=\beta]})$; that is,

$$P_{f|g=\beta}(D) = P[f \in D \mid g = \beta] = \frac{P[f \in D, g = \beta]}{p_g(\beta)}$$

for every subset D of R^n. Therefore

$$P_f(D) = \sum_{\beta} P[f \in D, g = \beta] = \sum_{\beta} p_g(\beta) P_{f|g=\beta}(D)$$

so that the distribution P_f of f is a **mixture**, according to discrete density p_g, of the conditional distributions $P_{f|g=\beta}$, and we write

$$P_f = \sum_\beta p_g(\beta) P_{f|g=\beta}.$$

We may say that P_f is obtained by **randomizing the parameter** β in $P_{f|g=\beta}$.

Theorem 1 (Sec. 15-5) Let $p\colon R^m \to R$ be a discrete density, and for each β for which $p(\beta) > 0$ let (R^n, P_β) be a probability space. Then there is a probability space (Ω, P) and two random vectors $g\colon \Omega \to R^m$ and $f\colon \Omega \to R^n$ such that $p_g = p$ and $P_{f|g=\beta} = P_\beta$ whenever $p_g(\beta) > 0$. Therefore

$$P_f = \sum_\beta p_g(\beta) P_{f|g=\beta} = \sum_\beta p(\beta) P_\beta \quad \blacksquare$$

Let X be a random variable and g a random vector with the same associated probability space (Ω, P). Assume that g induces a discrete density. If $E|X| < \infty$ and $p_g(\beta) > 0$ we introduce the notation

$$\begin{aligned}\mu_{X|g=\beta} = \mu_{X|g}(\beta) = E[X \mid g = \beta] &= [\text{the mean of } P_{X|g=\beta}] \\ &= [\text{the expectation of } X \text{ with associated } (\Omega, P_{[g=\beta]})].\end{aligned}$$

We call $\mu_{X|g}(\beta)$ the **conditional expectation of** X **given** $g = \beta$. The function $\mu_{X|g}$ is called the **conditional expectation of** X **given** g, or the **regression function of** X **on** g. Clearly $E(c \mid g) = c$, and if X and g are independent then $E(X \mid g) = EX$. Similarly if $EX^2 < \infty$ we introduce, for $p_g(\beta) > 0$, the notation

$$\begin{aligned}\sigma^2_{X|g=\beta} = \sigma^2_{X|g}(\beta) = \text{Var}\,[X \mid g = \beta] &= [\text{the variance of } P_{X|g=\beta}] \\ &= [\text{the variance of } X \text{ with associated } (\Omega, P_{[g=\beta]})].\end{aligned}$$

The function $\sigma^2_{X|g}$ is called the **conditional variance of** X **given** g. Clearly if X and g are independent then $\text{Var}\,[X \mid g] = \text{Var}\, X$.

The linearity of conditional expectation,

$$E\left(\sum_{i=1}^n a_i X_i \mid g\right) = \sum_{i=1}^n a_i E(X_i \mid g),$$

is valid whenever each $E|X_i| < \infty$.

Theorem 2 (Sec. 15-5) Let X be a random variable and g a random vector with the same associated probability space. Assume that g induces a discrete density, so that P_X can be expressed as a mixture

$$P_X = \sum_\beta p_g(\beta) P_{X|g=\beta}.$$

Then P_X has a mean iff "each $P_{X|g=\beta}$ with $p_g(\beta) > 0$ has a mean and the series below converges absolutely," in which case

$$EX = \sum_\beta \mu_{X|g}(\beta) p_g(\beta) = \sum_\beta E[X \mid g = \beta] p_g(\beta).$$

Also P_X has a variance iff "each $P_{X|g=\beta}$ with $p_g(\beta) > 0$ has a variance and both series below converge," in which case

$$\operatorname{Var} X = \sum_{\beta} \sigma^2_{X|g}(\beta)p_g(\beta) + \sum_{\beta} [\mu_{X|g}(\beta) - \mu_X]^2 p_g(\beta). \quad \blacksquare$$

As in Example 2 (Sec. 14-1), the function defined for all ω in Ω by $\mu_{X|g}(g(\omega))$ is referred to in many books on probability theory as the *conditional expectation of X given g*.

If P_X is concentrated on the nonnegative reals then we define the **probability-generating function** for P_X by

$$G_X(z) = Ez^X \qquad \text{for } 0 \leq z \leq 1$$

where X has distribution P_X. Then $G_X(0) = G_X(0+) = P[X = 0]$ and $0 \leq G_X(z) \leq 1$ and $G_X(1) = 1$.

Theorem 3 (Sec. 15-5): The sum $S_N = X_1 + \cdots + X_N$ of a random number N of random variables Let (R,P_{X_1}) be a probability space and $p_N: R \to R$ be a discrete density for which $\sum_{n=0}^{\infty} p_N(n) = 1$. Let $P_{S_0}(\{0\}) = 1$, and for each $n = 1, 2, \ldots$ let P_{S_n} be induced by $S_n = X_1 + \cdots + X_n$ where $X_1, \ldots, X_n$ are independent and each has distribution P_{X_1}. Define the probability space (R,P_{S_N}) by the mixture

$$P_{S_N} = \sum_{n=0}^{\infty} p_N(n)P_{S_n}$$

Let random variables X_1, N, and S_N have distributions P_{X_1}, P_N, and P_{S_N}, respectively. Then S_N has an expectation iff both N and X_1 do, and in this case

$$ES_N = (EX_1)(EN).$$

Also S_N has a variance iff both N and X_1 do, and in this case

$$\operatorname{Var} S_N = (EN)(\operatorname{Var} X_1) + (EX_1)^2(\operatorname{Var} N).$$

Furthermore if P_{X_1} is concentrated on the nonnegative reals then the probability-generating functions satisfy

$$G_{S_N}(z) = G_N(G_{X_1}(z)) \qquad \text{for } 0 \leq z \leq 1. \quad \blacksquare$$

CHAPTER 16: CONTINUOUS PROBABILITY DENSITIES

A **continuous probability density** is a function $p: R^n \to R$ satisfying

$$p(x_1, \ldots, x_n) \geq 0 \qquad \text{for all } (x_1, \ldots, x_n) \in R^n$$

and

$$\int_{-\infty}^{\infty} \cdots \int_{-\infty}^{\infty} p(x_1, \ldots, x_n)\, dx_1\, dx_2 \cdots dx_n = 1.$$

The **probability space** (R^n,P) corresponding to $p\colon R^n \to R$ is defined by setting

$$P(A) = \int_A \cdots \int p(x_1, \ldots, x_n)\, dx_1\, dx_2 \cdots dx_n$$

for every event A in the sample space R^n. Thus $p\colon R^n \to R$ is nonnegative, and $P(A)$ equals the integral of p over A, and $P(R^n) = 1$. The adjective "continuous" in "continuous probability density" means that integrals are to be employed, rather than sums as in the discrete case, and the density is "continuous enough" for the integrals to be defined. Two continuous densities, $p\colon R^n \to R$ and $p'\colon R^n \to R$, are said to be **equivalent** iff $P = P'$, and this is denoted by $p \equiv p'$.

If $p\colon R^n \to R$ is a continuous density and A is an event with $P(A) > 0$ then the continuous density $p_A\colon R^n \to R$ defined by

$$p_A(x_1, \ldots, x_n) = \begin{cases} \dfrac{1}{P(A)} p(x_1, \ldots, x_n) & \text{if } (x_1, \ldots, x_n) \in A \\ 0 & \text{if } (x_1, \ldots, x_n) \notin A \end{cases}$$

is said to be obtained by **conditioning p on the event A.** If (R^n,P_A) corresponds to $p_A\colon R^n \to R$ then $P_A(B) = P(B \mid A)$ for every event B.

For a continuous density $p\colon R \to R$ we make the following definitions:

1. The **mean** of p is $\mu = \int_{-\infty}^{\infty} xp(x)\, dx$,
2. The **nth moment** of p is $\int_{-\infty}^{\infty} x^n p(x)\, dx$ for $n = 1, 2, \ldots,$
3. The **rth absolute moment** of p is $\int_{-\infty}^{\infty} |x|^r p(x)\, dx$ for real $r > 0$,
4. The **variance** of p is $\int_{-\infty}^{\infty} (x - \mu)^2 p(x)\, dx = \int_{-\infty}^{\infty} x^2 p(x)\, dx - \mu^2$,
5. The **standard deviation** of p is [the variance of $p]^{1/2} \geq 0$.

For the nth moment to be defined we require that the nth absolute moment be finite. If p has a second moment it has a mean, and hence a variance. If $p(x) = 0$ whenever $|x| > B$ then p has a mean μ, and it satisfies $|\mu| \leq B$. If p has a mean μ and is symmetric about some real b then $\mu = b$.

Theorem 1 (Sec. 16-1) If a continuous density $p\colon R \to R$ has a finite sth absolute moment for some real $s > 0$ then it has a finite rth absolute moment for any real r satisfying $0 < r < s$, and furthermore

$$\left[\int_{-\infty}^{\infty} |x|^r p(x)\, dx\right]^{1/r} \leq \left[\int_{-\infty}^{\infty} |x|^s p(x)\, dx\right]^{1/s}. \quad \blacksquare$$

The **moment generating function** of a continuous density $p\colon R \to R$ is

$$m(t) = \int_{-\infty}^{\infty} e^{tx} p(x)\, dx$$

defined for all t for which the integral is finite. If $p(x) = 0$ whenever $|x| > B$ then $m(t) \leq e^{|t|B}$ is defined for all real t.

The obvious analog of Theorem 2 (Sec. 9-4) on MGFs is valid for continuous densities, where now

$$m^{(n)}(t) = \frac{d^n}{dt^n} m(t) = \int_{-\infty}^{\infty} x^n e^{tx} p(x)\, dx \qquad \text{if } n = 1, 2, \ldots, |t| < t_0.$$

In the last conclusion of that theorem $p_1 = p_2$ becomes $p_1 \equiv p_2$ for continuous densities.

Example 6 (Sec. 16-1) shows how to "continuously calibrate" a random wheel so that it will correspond to a given continuous density $p: R \to R$.

Let (Ω,P) correspond to the unit interval $0 \leq x \leq 1$ with a uniform density. Example 7 (Sec. 16-1) shows that for each fixed p, with $0 < p < 1$, it is possible to construct on this same (Ω,P) an *infinite* sequence $X_1, X_2, \ldots$ of Bernoulli trials with parameter p.

Table I

Name, parameters	Continuous probability density	Mean, variance	MGF
Uniform $a < b$	$p(x) = \begin{cases} \dfrac{1}{b-a} & \text{if } a < x < b \\ 0 & \text{otherwise} \end{cases}$	$\dfrac{a+b}{2}$ $\dfrac{(b-a)^2}{12}$	$\dfrac{e^{tb} - e^{ta}}{t(b-a)}$ $-\infty < t < \infty$
Exponential $\lambda > 0$	$p(x) = \begin{cases} \lambda e^{-\lambda x} & \text{if } x > 0 \\ 0 & \text{if } x \leq 0 \end{cases}$	$\dfrac{1}{\lambda}$ $\dfrac{1}{\lambda^2}$	$\dfrac{1}{1-(t/\lambda)}$ $t < \lambda$
Normal μ $\sigma > 0$	$\varphi_{\mu,\sigma}(x) = \dfrac{1}{\sigma\sqrt{2\pi}} e^{-(x-\mu)^2/2\sigma^2}$	μ σ^2	$e^{\mu t + \sigma^2 t^2/2}$ $-\infty < t < \infty$
Lognormal μ $\sigma^2 > 0$	$p(x) = \dfrac{1}{x\sigma\sqrt{2\pi}} e^{-(1/2\sigma^2)[(\log x)-\mu]^2}$	$e^{\mu+(\sigma^2/2)}$ $e^{2\mu+2\sigma^2}(e^{\sigma^2} - 1)$	... ...
Gamma $\lambda > 0$ $\nu > 0$	$\gamma_{\lambda,\nu}(x) = \dfrac{\lambda^\nu}{\Gamma(\nu)} x^{\nu-1} e^{-\lambda x}$	$\dfrac{\nu}{\lambda}$ $\dfrac{\nu}{\lambda^2}$	$\left[\dfrac{1}{1-(t/\lambda)}\right]^\nu$ $t < \lambda$

The lower expression in the last column states the constraint on t needed for convergence.
The exponential density is the special case $\gamma_{\lambda,1}$ of the gamma density.
The lognormal density with parameters μ and σ^2 is induced by e^X when X is normal μ and σ^2.

If $p: R \to R$ is an exponential density with mean $1/\lambda$ and $A_r = \{x: x \geq r\}$ then $P(A_r) = e^{-\lambda r}$ for all $r \geq 0$. This yields the *lack-of-memory property*,

$$P(A_{r+t} \mid A_r) = P(A_t) \qquad \text{for all real } r \geq 0,\ t \geq 0.$$

Example 4 (Sec. 16-1) shows that the exponential densities are the only distributions having this property. More precisely if (R,P) is any probability space for which $P(A_1) > 0$ and

$$P(A_{r+t}) = P(A_r)P(A_t) \qquad \text{for all real } r \geq 0,\ t \geq 0$$

then $P(A_r) = e^{-\lambda r}$ for all real $r \geq 0$, where λ is determined by $P(A_1) = e^{-\lambda}$. Thus any such P corresponds to an exponential density. The proof of this result was based on the uniqueness of the solutions to the functional equations satisfied by cx and b^x, as proved in the addendum to Sec. 16-1 and stated below.

Lemma 1 (Sec. 16-1) Let f be a real-valued function defined for all positive reals and bounded on the unit interval

$$|f(x)| \leq M \qquad \text{whenever } 0 < x \leq 1.$$

If f satisfies the equation

$$f(x + y) = f(x) + f(y) \qquad \text{whenever } x > 0,\ y > 0$$

then $f(x) = f(1)x$ for all $x > 0$. ■

Lemma 2 (Sec. 16-1) Let g be a real-valued function defined for all positive reals and bounded on the unit interval

$$|g(x)| \leq M \qquad \text{whenever } 0 < x \leq 1.$$

If g satisfies the equation

$$g(x + y) = g(x)g(y) \qquad \text{whenever } x > 0,\ y > 0$$

then $g(1) \geq 0$ and $g(x) = [g(1)]^x$ for all $x > 0$. ■

Let (R^n,P_f) be induced by random vector $f: \Omega \to R^n$ with associated probability space (Ω,P). If (R^n,P_f) happens to correspond to a continuous density $p_f: R^n \to R$ then we call p_f the **continuous density induced by** f **with associated** (Ω,P). If random vectors $f_1, \ldots, f_k$ into the same R^n all induce continuous densities then they are **identically distributed** iff $p_{f_1} \equiv \cdots \equiv p_{f_k}$. If $f = (f_1, \ldots, f_k)$, where each f_i is a random vector, induces a continuous density $p_f = p_{f_1,\ldots,f_k}$ then we may call it the **joint density of** $f_1, \ldots, f_k$.

Theorem 1 (Sec 16-2) Let $f: \Omega \to R^n$ and $g: \Omega \to R^m$ be random vectors with the same associated probability space (Ω,P). If the random vector $(f,g): \Omega \to R^{n+m}$ induces a continuous density $p_{f,g}: R^{n+m} \to R$ then f induces a continuous density obtainable by integrating out the extra coordinates in $p_{f,g}$,

$$p_f(x_1, \ldots, x_n) = \int_{-\infty}^{\infty} \cdots \int_{-\infty}^{\infty} p_{f,g}(x_1, \ldots, x_n, \ldots, x_{n+m})\, dx_{n+1} \cdots dx_{n+m}. \quad ■$$

Let X be a random variable with a continuous density p_X and let $Y = g(X)$, where $g\colon R \to R$ has a continuous positive, or negative, derivative everywhere. Then Y has a continuous density which is zero off the range of g while

$$p_Y(y) = \frac{1}{|dg(x)/dx|} p_X(x) \qquad \text{where } y = g(x)$$

if y belongs to the range of g.

If $Y = aX + b$ with $a \neq 0$ and if X has a continuous density then Y has a continuous density given by

$$p_Y(y) = \frac{1}{|a|} p_X\left(\frac{y-b}{a}\right) \qquad \text{for all real } y$$

and, as always, $\mu_Y = a\mu_X$ and $\sigma_Y^2 = a^2\sigma_X^2$. In particular if $X^* = (X - \mu_X)/\sigma_X$ then X^* is **normalized** since $\mu_{X^*} = 0$ and $\sigma_{X^*} = 1$. Also if X is normal and $Y = aX + b$ with $a \neq 0$ then Y is normal.

If X is normal μ and σ^2 and $Y = e^X$ then p_Y is called the **lognormal density** with parameters μ and σ^2. If $Z = cY^a$ where $c > 0$ and $a \neq 0$, then

$$Z = c(e^X)^a = e^{aX + \log c}$$

so Z is lognormal $a\mu + \log c$ and $a^2\sigma^2$.

Let X be a random variable with a continuous density p_X such that $Y = g(X)$, with $g\colon R \to R$, induces a continuous density p_Y. If dg/dx is "reasonably continuous" and if for a particular y there are a finite number $N_y \geq 1$ of solutions $y = g(x_1) = \cdots = g(x_{N_y})$ to the equation $y = g(x)$, and if each $g'(x_i) \neq 0$, then

$$p_Y(y) = \sum_{i=1}^{N_y} \frac{1}{\left|\left[\dfrac{dg}{dx}\right]_{x=x_i}\right|} p_X(x_i)$$

If X and Y are independent random variables having continuous densities then $Z = X + Y$ has a continuous density obtainable by **convolving** p_X and p_Y; that is,

$$p_Z(z) = \int_{-\infty}^{\infty} p_X(x)p_Y(z-x)\,dx = \int_{-\infty}^{\infty} p_X(z-y)p_Y(y)\,dy \qquad \text{for all real } z.$$

If $p_X(x) = 0 = p_Y(y)$ whenever $x < 0$ and $y < 0$ then $\int_{-\infty}^{\infty}$ may be replaced by $\int_0^z$. We sometimes denote convolution by the sign $*$, so $p_Z = p_X * p_Y$. Convolution is commutative and associative.

The following theorem asserts that if a candidate can be found for the density induced by g with associated p_f, such that the candidate yields probabilities which are consistent with p_f, then the candidate is indeed p_g.

Theorem I (Sec. 16-3) Assume that the random vector $f\colon \Omega \to R^n$ with associated probability space (Ω, P) induces a continuous density $p_f\colon R^n \to R$, and that

$$h = g(f)$$

$$\Omega \xrightarrow{f} R^n \xrightarrow{g} R^m.$$

If a continuous density $p': R^m \to R$ can be found for which

$$\int_{R_{a,b}} \cdots \int p'(y_1, \ldots, y_m)\, dy_1 \cdots dy_m = \int_{g^{-1}(R_{a,b})} \cdots \int p_f(x_1, \ldots, x_n)\, dx_1 \cdots dx_n$$

holds for every closed rectangle $R_{a,b}$ in R^m, then h and g induce continuous densities and $p_h \equiv p_g \equiv p'$. ■

Using an infinitesimal rectangle of volume dv' in this theorem yields

$$p_g(y)\, dv' = \int_{g^{-1}(R_{y,y+dy})} \cdots \int p_f(x_1, \ldots, x_n)\, dx_1 \cdots dx_n.$$

This formula is used in Examples 1 to 3 (Sec. 16-3), where $n > m = 1$, and where for a typical y the equation $y = g(x)$ is satisfied by infinitely many x. If $n = m$, and $y = g(x)$ has only one solution, then the above reduces to

$$p_g(y) = \frac{1}{dv'/dv} p_f(x) \qquad \text{where } y = g(x)$$

and where dv'/dv is the local scale factor by which small volumes near x are multiplied when transformed by g. [If $y = g(x)$ has a finite number $N_y \geq 1$ of solutions then $p_g(y)$ is the sum of N_y terms like the above.] In general dv'/dv is the absolute value of the *jacobian* of g evaluated at x. If $Y = TX + b$ where T is a nonsingular $n \times n$ matrix of real numbers then $dv'/dv = |\det T|$, so

$$p_Y(y) = \frac{1}{|\det T|} p_X(T^{-1}(y - b)) = |\det T^{-1}| p_X(T^{-1}(y - b)) \qquad \text{for all } y \in R^n.$$

In the continuous-density case, of course, we have all the general results of Sec. 15-3 for independent random vectors. Furthermore the following theorem shows that independence is equivalent to the joint density being given by the product rule. In particular this shows that it is always possible to construct independent random vectors having arbitrarily specified continuous marginal densities.

Theorem 1 (Sec. 16-4) Let $f_1, \ldots, f_k$ where $f_i: \Omega \to R^{n_i}$ be random vectors with the same associated probability space (Ω, P) and let $n = n_1 + \cdots + n_k$.

(*a*) If $f_1, \ldots, f_k$ induce a joint continuous density on R^n then from Theorem 1 (Sec. 16-2) each f_i induces a continuous density. In this case $f_1, \ldots, f_k$ are independent iff the joint density is equivalent to the density

$$p_{f_1}(\alpha_1) p_{f_2}(\alpha_2) \cdots p_{f_k}(\alpha_k) \qquad \text{where each } \alpha_i \in R^{n_i},$$

obtained from the product rule applied to the marginals.

(*b*) If $f_1, \ldots, f_k$ are independent and each induces a continuous density $p_{f_i}: R^{n_i} \to R$ then they have a joint continuous density, given as in part (*a*). ■

If $X_1, \ldots, X_n$ are independent and each is normally distributed then so is their sum $X_1 + \cdots + X_n$. If $Y_1, \ldots, Y_n$ are independent and each Y_i is lognormal μ_i and σ_i^2 then their product $Y_1 Y_2 \cdots Y_n$ is lognormal $\sum_{i=1}^{n} \mu_i$ and $\sum_{i=1}^{n} \sigma_i^2$.

The **gamma function** Γ is defined by

$$\Gamma(\nu) = \int_0^\infty t^{\nu-1}e^{-t}\,dt \qquad \text{for all real } \nu > 0$$

and satisfies the recursion relation

$$\Gamma(\nu) = (\nu - 1)\Gamma(\nu - 1) \qquad \text{for all real } \nu > 1.$$

Therefore $\Gamma(\nu) = (\nu - 1)!$ if $\nu = 1, 2, 3, \ldots$, while $\Gamma(\tfrac{1}{2}) = \sqrt{\pi}$. The gamma densities $\gamma_{\lambda,\nu}$ are defined in a preceding table. They satisfy

$$\gamma_{\lambda,\nu} * \gamma_{\lambda,\mu} = \gamma_{\lambda,\nu+\mu} \qquad \text{if } \nu > 0,\ \mu > 0.$$

Assume that $X_1, \ldots, X_n$ are independent and identically distributed. If X_1 is exponential $\gamma_{\lambda,1}$ with mean $1/\lambda$ then $X_1 + \cdots + X_n$ is $\gamma_{\lambda,n}$. If X_1 is normal 0 and 1 then $(X_1)^2$ is $\gamma_{1/2,1/2}$, so $X_1{}^2 + \cdots + X_n{}^2$ is $\gamma_{1/2,n/2}$, which is called the **χ-square density** with n degrees of freedom. *Student's t* and the *Snedecor F* distributions are derived in the exercises of Sec. 16-4.

The following theorem shows that the special definitions for means of densities are consistent with the general approach. The next theorem shows that $Ef(Y)$ can be calculated from p_Y just as in the discrete case. Thus an expectation can always be calculated from the standard formula in terms of *any* discrete or continuous density which exists.

Theorem 1 (Sec. 16-5) Let (R,P) be a probability space and let the mean of P be defined as in Theorem 1 (Sec. 15-4). If P corresponds to a discrete density $p: R \to R$ which is positive only at $x_1, x_2, \ldots$ then the mean of P exists iff

$$\sum_{i=1}^\infty |x_i| p(x_i) < \infty$$

and if this is the case then [the mean of P] $= \sum_{i=1}^\infty x_i p(x_i)$. If P corresponds to a continuous density $p: R \to R$ then the mean of P exists iff

$$\int_{-\infty}^\infty |x|\, p(x)\, dx < \infty$$

and if this is the case then [the mean of P] $= \int_{-\infty}^\infty x p(x)\, dx$. ∎

Theorem 2 (Sec. 16-5): The expectation, as calculated from an intermediate density If the random variable $X = f(Y_1, \ldots, Y_k)$ is a function of the random vector $(Y_1, \ldots, Y_k)$ which has a continuous density,

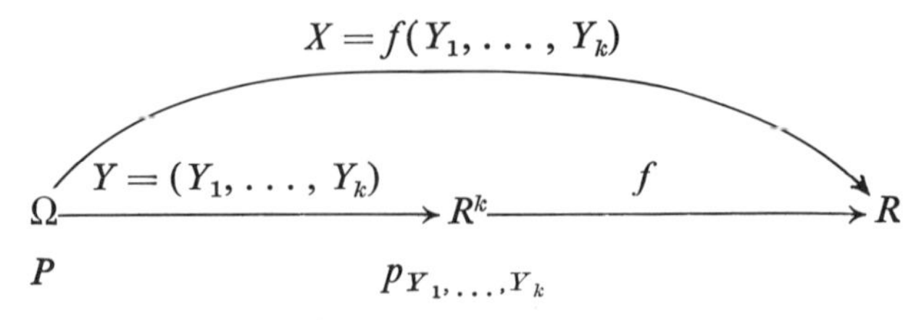

then X has an expectation iff

$$\int_{-\infty}^{\infty} \cdots \int_{-\infty}^{\infty} |f(y_1, \ldots, y_k)| \, p_{Y_1,\ldots,Y_k}(y_1, \ldots, y_k) \, dy_1 \cdots dy_k < \infty$$

and if this is the case then

$$EX = Ef(Y_1, \ldots, Y_k)$$

$$= \int_{-\infty}^{\infty} \cdots \int_{-\infty}^{\infty} f(y_1, \ldots, y_k) p_{Y_1,\ldots,Y_k}(y_1, \ldots, y_k) \, dy_1 \ldots dy_k. \quad \blacksquare$$

Uniqueness of a property of normal distributions Let random variables X and Y be independent and each have the same distribution P_X such that $EX = 0$ and $EX^2 = 1$. Assume that $aX + bY$ induces this same distribution P_X whenever $a^2 + b^2 = 1$. Then P_X corresponds to the normal density φ. $\blacksquare$

The stated uniqueness is related to and proved by the central-limit theorem. Exercise 2 (Sec. 16-5) shows that the second sentence above may be replaced by: Assume that $(X + Y)/\sqrt{2}$ induces this same distribution P_X. If the assumption of finite second moments is dropped then besides the normal density there are many, like the Cauchy density below, which by translations and scale changes generate a family closed under independent additions.

The continuous density

$$p_Z(z) = \frac{1}{\pi} \frac{\lambda}{\lambda^2 + (z - \mu)^2} \qquad \text{for all real } z$$

is called the **Cauchy density** with parameters μ and $\lambda > 0$. If p_V is Cauchy $\mu = 0$ and $\lambda = 1$ then we call it **standardized** in which case $Z = \lambda V + \mu$ has the density p_Z above if $\lambda > 0$. Also if $a \neq 0$ then $aZ + b$ is Cauchy $a\mu + b$ and $|a| \lambda$. Clearly p_Z is symmetric about μ, although Z does *not* have an expectation, $E|Z| = \infty$. It was shown that $P[|Z - \mu| < \lambda] = .5$. Hence μ and λ may be called the *location* and *scale parameters*, respectively. V has a standard Cauchy density if $V = \tan U$, where U has a uniform density over $-\pi/2 < u < \pi/2$, or if $V = Y/X$, where X and Y are independent and each is normal 0 and σ^2.

If $Z_1, \ldots, Z_n$ are independent and Z_i is Cauchy μ_i and λ_i then $\sum_{i=1}^{n} Z_i$ is Cauchy $\sum_{i=1}^{n} \mu_i$ and $\sum_{i=1}^{n} \lambda_i$. Therefore $(1/n)(V_1 + \cdots + V_n)$ has a Cauchy 0 and 1 density if $V_1, \ldots, V_n$ are independent and each is Cauchy 0 and 1. Thus the law of large numbers does not apply to Cauchy densities.

As in Sec. 16-6, we consider conditional distributions and conditional expectations when there are densities, discrete or continuous. Let $f: \Omega \to R^n$ and $g: \Omega \to R^m$ be random vectors. Assume first that g induces a discrete density p_g so that, as in Sec. 15-5, P_f is a mixture,

$$P_f(D) = \sum_{\beta} p_g(\beta) P_{f|g=\beta}(D) \qquad \text{for all subsets } D \text{ of } R^n.$$

If each $P_{f|g=\beta}$ for $p_g(\beta) > 0$ corresponds to a discrete density we are back in the discrete case. If each $P_{f|g=\beta}$ for $p_g(\beta) > 0$ corresponds to a continuous density then f has continuous density

$$p_f(\alpha) = \sum_{\beta} p_g(\beta) p_{f|g=\beta}(\alpha) \qquad \text{for all } \alpha = (\alpha_1, \ldots, \alpha_n) \in R^n.$$

Otherwise P_f corresponds to a mixture of a discrete and a continuous density. We say that (R^n, P) corresponds to a **mixture of a discrete and a continuous density** iff there is a discrete density $p_d\colon R^n \to R$ and a continuous density $p_c\colon R^n \to R$ and a real $0 < d < 1$ such that P is obtained from $dp_d + (1 - d)p_c$ by

$$P(D) = d \sum_{\alpha \in D} p_d(\alpha) + (1 - d) \int_D p_c(\alpha)\, d\alpha$$

for each subset D of R^n. If $f = X$ happens to be a random variable then we can calculate conditional means and variances, and Theorem 2 (Sec. 15-5) applies to all cases.

We now turn to mixtures according to a continuous density. If $\bar{p}\colon R^m \to R$ is any continuous density and if (R^n, P_β) is a probability space for each β for which $\bar{p}(\beta) > 0$ then we can define the probability space (R^n, P) by

$$P(D) = \int_{R^m} P_\beta(D) \bar{p}(\beta)\, d\beta \qquad \text{for all subsets } D \text{ of } R^n.$$

If $n = 1$ then the mean and variance of P have expressions analogous to those of Theorem 2 (Sec. 15-5). As before, P may correspond to a mixture of a discrete and a continuous density. If each P_β for $\bar{p}(\beta) > 0$ corresponds to a discrete density p_β concentrated on the *same* countable subset C of R^n then P corresponds to the discrete density $p\colon R^n \to R$ obtained by taking the same mixture,

$$p(\alpha) = \int_{R^m} p_\beta(\alpha) \bar{p}(\beta)\, d\beta.$$

If each P_β corresponds to a continuous density then this same expression gives the continuous density corresponding to P. In this case we may interpret $p_\beta(\alpha)\bar{p}(\beta)$ as a joint continuous density, and we now examine this important case.

If $f\colon \Omega \to R^n$ and $g\colon \Omega \to R^m$ are random vectors having a continuous joint density and if $p_g(\beta) > 0$ then the continuous density defined on R^n by

$$p_{f|g}(\alpha \mid \beta) = p_{f|g=\beta}(\alpha) = \frac{p_{f,g}(\alpha,\beta)}{p_g(\beta)} \qquad \text{for all } \alpha \in R^n$$

is called the **conditional density of** f **given** $g = \beta$. Clearly

$$p_f(\alpha) = \int_{R^m} p_{f,g}(\alpha,\beta)\, d\beta = \int_{R^m} p_{f|g=\beta}(\alpha) p_g(\beta)\, d\beta$$

so that p_f is the mixture, determined by p_g, of these conditional densities.

If a random variable $X: \Omega \to R$ and a random vector $g: \Omega \to R^m$ have a continuous joint density $p_{X,g}$ and if $E|X| < \infty$ and $p_g(\beta) > 0$ then

$$\mu_{X|g=\beta} = \mu_{X|g}(\beta) = E[X \mid g = \beta] = [\text{the mean of } p_{X|g=\beta}]$$

$$= \int_{-\infty}^{\infty} x p_{X|g=\beta}(x)\, dx = \frac{1}{p_g(\beta)} \int_{-\infty}^{\infty} x p_{X,g}(x,\beta)\, dx$$

is called the **conditional expectation of** X **given** $g = \beta$. The function $\mu_{X|g}$ is called the **conditional expectation of** X **given** g, or the **regression function of** X **on** g. We have

$$EX = \int_{R^m} \mu_{X|g}(\beta) p_g(\beta)\, d\beta$$

so once again EX equals the expectation of its conditional expectation.

Similarly if $EX^2 < \infty$ and $p_g(\beta) > 0$ we let

$$\sigma^2_{X|g=\beta} = \sigma^2_{X|g}(\beta) = \text{Var}\,[X \mid g = \beta] = [\text{the variance of } p_{X|g=\beta}]$$

and call the function $\sigma^2_{X|g}$ the **conditional variance of** X **given** g. As in Theorem 2 (Sec. 15-5), we easily see that if X has a variance then

$$\text{Var}\, X = \int_{R^m} \sigma^2_{X|g}(\beta) p_g(\beta)\, d\beta + \int_{R^m} [\mu_{X|g}(\beta) - EX]^2 p_g(\beta)\, d\beta.$$

The interpretations in Sec. 14-1 in terms of predictions of X, given g, still apply. As in Example 2 (Sec. 14-1), the function defined by

$$\mu_{X|g}(g(\omega)) \qquad \text{for all } \omega \in \Omega$$

is also often called the *conditional expectation of* X *given* g.

CHAPTER 17: DISTRIBUTION FUNCTIONS

If (R,P) is a probability space then the function $F: R \to R$ defined for all real x by $F(x) = P\{x' : x' < x\}$ is called the **distribution function corresponding to** (R,P). The **distribution function of**, or induced by, **a random variable** X is the one corresponding to its induced P_X,

$$F_X(x) = P[X < x] = P_X\{x' : x' < x\} \qquad \text{for all real } x.$$

We call a distribution function $F: R \to R$ corresponding to (R,P) **discrete (absolutely continuous)** iff P corresponds to a discrete (continuous) density,

$$F(x) = \sum_{x' < x} p(x') \qquad F(x) = \int_{-\infty}^{x} p(x')\, dx'$$

and in the latter case $dF/dx = p$, at least at points where p is continuous. Naturally $F: R \to R$ is said to be **continuous** iff it is continuous at every real number.

A function $F: R \to R$ is a **distribution function on** R iff the following four conditions hold:

1. $\lim_{x\to\infty} F(x) = 1$.
2. $\lim_{x\to-\infty} F(x) = 0$.
3. $\lim_{\substack{\epsilon\to 0\\ \epsilon>0}} F(x-\epsilon) = F(x)$ for each real x.
4. $F(x_1) \le F(x_2)$ whenever $x_1 < x_2$.

Thus $F: R \to R$ is a distribution function on R iff it converges to 1 to the right, converges to zero to the left, is continuous from the left at every point, and is nondecreasing. It can be shown that a function $F: R \to R$ is a distribution function on R iff there is a probability space (R,P) such that F is the distribution function corresponding to (R,P). Also $(R,P_1) = (R,P_2)$ iff $F_1 = F_2$. Thus the correspondence between probability spaces (R,P) and distribution functions on R is a one-to-one correspondence.

If $F: R \to R$ corresponds to (R,P) then the difference between the right- and left-hand limits of F at $x = a$ is called the **jump** of F at $x = a$, and it equals the probability $P(\{a\})$ of that one point. If it is zero (positive) then a is called a **point of continuity (discontinuity)** of F. Clearly

$$P\{x: a \le x < b\} = F(b) - F(a) \qquad \text{and} \qquad P\{x: x \ge a\} = 1 - F(a)$$

and, as in Fig. 2 (Sec. 17-1), the probability of every interval can similarly be found from F.

A set A of reals is said to have **zero length** (Lebesgue measure) iff for every $\epsilon > 0$ there is a finite or infinite sequence $I_n = \{x: a_n \le x \le b_n\}$ of intervals with $a_n \le b_n$, such that A is a subset of $\bigcup_{n=1}^{\infty} I_n$ and $\sum_{n=1}^{\infty} (b_n - a_n) < \epsilon$. That is, A has zero length iff we can cover it by a sequence of intervals the sum of whose lengths is arbitrarily small. If $A = \{x_1, x_2, \ldots\}$ is countable we can let $a_n = b_n = x_n$, so that A has zero length. Example 4 (Sec. 17-1) as well as Exercise 5 (Sec. 4-3) exhibit uncountable sets which have zero length. A distribution function $F: R \to R$ is said to be **continuous singular** iff it is continuous and there is a set A of reals of zero length such that $P(A) = 1$, where P is the probability measure corresponding to F. That is, continuous singular F are those corresponding to probability measures which assign zero probability to every point but are nevertheless concentrated on a set of zero length. Such a P cannot correspond to any discrete or continuous probability density, or mixture of such.

The following decomposition is not proved in this book, although the lemma below, which is proved, yields part of the proof of the decomposition.

Basic decomposition of a distribution function on R Every distribution function $F: R \to R$ has a representation $F = \beta_d F_d + \beta_{ac} F_{ac} + \beta_{cs} F_{cs}$ where $\beta_d + \beta_{ac} + \beta_{cs} = 1$ and each $\beta \ge 0$. The F_d, F_{ac}, and F_{cs} are, respectively, discrete, absolutely continuous, and continuous-singular distribution functions on R. Furthermore the representation is unique in the sense that the numbers β_d, β_{ac}, and β_{cs} are unique, and each distribution function with a positive coefficient is unique. ■

Lemma 1 (Sec. 17-1) Every distribution function $F: R \to R$ has a representation $F = \beta_d F_d + (1 - \beta_d)F_c$ where $0 \le \beta_d \le 1$ and where F_d and F_c are, respectively, discrete and continuous distribution functions on R. Furthermore the representation is unique in the sense that β_d is unique, and each distribution function with a positive coefficient is unique. ■

We now relate distribution functions, and mixtures of them, to expectations. The **mean (variance)** of a distribution function $F: R \to R$ is just the mean (variance) of its corresponding (R,P), if this mean (variance) exists. The mean of P was defined in Theorem 1 (Sec. 15-4) in terms of sequences A_n^+ and A_n^- which used expressions such as

$$P\left\{x: \frac{j}{2^n} \le x < \frac{j+1}{2^n}\right\} = F\left(\frac{j+1}{2^n}\right) - F\left(\frac{j}{2^n}\right)$$

so that A_n^+ and A_n^- can easily be expressed in terms of F rather than P. At times it will be convenient to think of A_n^+ and A_n^- as so expressed in terms of F. Also EX was defined to equal the mean of P_X, so EX may be thought of as defined directly in terms of F_X or P_X or P.

If $F_1, \ldots, F_n$ are distribution functions on R and $b_1, \ldots, b_n$ are nonnegative real numbers whose sum is 1 then the **mixture** $F = \sum_{i=1}^{n} b_i F_i$ is easily seen to be a distribution function on R. Let F and F_i correspond to (R,P) and (R,P_i), respectively, so that $P = \sum_{i=1}^{n} b_i P_i$ equals the mixture of the P_i. If P_i has mean μ_i and variance σ_i^2 then the mean and variance of P, or F, are given by

$$\mu = \sum_{i=1}^{n} b_i \mu_i \qquad \sigma^2 = \sum_{i=1}^{n} \sigma_i^2 b_i + \sum_{i=1}^{n} (\mu_i - \mu)^2 b_i.$$

Thus the mean of the mixture $F = \sum_{i=1}^{n} b_i F_i$ equals the same mixture of the means, while the variance of F exceeds the same mixture of the variances by an amount interpretable as the variance of the means.

Let $X_1, X_2, \ldots$ be an infinite sequence of Bernoulli trials with

$$p = P[X_i = 1] = 1 - P[X_i = 0].$$

Let ρ be a real number satisfying $0 < \rho < 1$ and let $F_Y: R \to R$ be the distribution function induced by

$$Y = \sum_{n=1}^{\infty} X_n \rho^n.$$

Example 4 (Sec. 17-1) shows that F_Y is continuous singular whenever $0 < p < 1$ and $0 < \rho < \frac{1}{2}$. Such an F_Y is constant over each of the many intervals which do not belong to the range of Y. Exercise 8 (Sec. 17-1) shows that if $\rho = \frac{1}{2}$, $0 < p < 1$, and $p \ne \frac{1}{2}$ then F_Y is continuous singular *and* strictly increasing over $0 \le y \le 1$, where, in essence, Y has the digits in its binary expansion selected according to independent biased-coin tosses. If $\rho = p = \frac{1}{2}$ then we have fair-coin tosses, and $F_Y(y) = y$ for $0 \le y \le 1$, so that P_Y corresponds to length in $0 \le y \le 1$. If

$0 < p < 1$ and $0 < \rho < 1$ then the usual manipulations for expectations of *finite* sums of random variables are valid for Y, so EY and Var Y are easily obtained in this way rather than from distribution functions. Such Y fit within our theory, and in particular the laws of large numbers and central-limit theorems can be applied to sums of independent random variables, each distributed like any one of the Y above.

If $Y = g(X)$ where $g: R \to R$ is continuous and strictly increasing then, as shown in Sec. 17-2,

$$F_Y(y) = F_X(g^{-1}(y)) \qquad \text{for any } y \text{ in the range of } g.$$

For example, if $Y = e^X$ then by direct manipulation

$$F_Y(y) = P[e^X < y] = P[X < \log y] = F_X(\log y) \qquad \text{for any } y > 0.$$

This manipulation is valid for *every* F_X and is so straightforward that distribution functions are often used for such $g(X)$ even when X has a density. Section 17-2 finds distribution functions for e^X, X^n, $aX + b$, and $X + Y$.

Example 7 (Sec. 17-2) shows that if $F: R \to R$ is *any* distribution function on R and if Y has a continuous uniform density over $0 < y < 1$ then there is a function $h: R \to R$ such that $h(Y)$ induces F. In simple cases h is just F^{-1}. Thus every distribution function is induced by the random variables on the unit-interval sample space with length for probability. A random-number table (0, 1, . . . , 9 independent and equally likely) may be used to obtain what are supposed to be independent samples $y_1, y_2, \ldots, y_n$ from the uniform distribution on $0/10^k, 1/10^k, 2/10^k, \ldots, (10^k - 1)/10^k$. Therefore $h(y_1), \ldots, h(y_n)$ may be used as independent samples from an approximation to F.

If X has a continuous distribution function $F_X: R \to R$ then the random variable $Y = F_X(X)$ induces a uniform continuous density on $0 < y < 1$. Thus, as in Example 8 (Sec. 17-2), we let $Y = g(X)$ where g is the distribution function of X. This transformation of X is called the **probability-integral transformation,** because when X has a continuous density then F_X is its indefinite integral.

For any $n \geq 1$ define a function $f_n: R^n \to R^n$ as follows. For any reals $v_1 \leq v_2 \leq \cdots \leq v_n$ if $(x_1, \ldots, x_n)$ is any permutation of $(v_1, \ldots, v_n)$ then let $f_n(x_1, \ldots, x_n) = (v_1, \ldots, v_n)$. Thus $f_n(x_1, \ldots, x_n)$ is just the vector of arguments placed in nondecreasing order. If $(X_1, \ldots, X_n)$ is any random vector then we call the random vector

$$(V_1, \ldots, V_n) = f_n(X_1, \ldots, X_n)$$

the **ordered vector determined by,** or **corresponding to** $(X_1, \ldots, X_n)$. Usually V_k is called the **order statistic** of rank k from $X_1, \ldots, X_n$. Naturally V_k may be described as the kth smallest of the random numbers $X_1, \ldots, X_n$. Each V_k depends on every $X_1, \ldots, X_n$ so sometimes the notation $V_k^{(n)}$ is used.

If $X_1, \ldots, X_n$ are independent and each has the same continuous density $p: R \to R$ then the corresponding ordered vector $(V_1, \ldots, V_n)$ has a continuous density on R^n given by

$$p_{V_1,\ldots,V_n}(v_1, \ldots, v_n) = \begin{cases} n!\, p(v_1)p(v_2)\cdots p(v_n) & \text{if } v_1 < v_2 < \cdots < v_n \\ 0 & \text{otherwise.} \end{cases}$$

Roughly speaking, we take the probability that $(X_1, \ldots, X_n) = (v_1, \ldots, v_n)$ and then multiply by $n!$ because the same $(v_1, \ldots, v_n)$ is obtained, with the same probability, if $(X_1, \ldots, X_n)$ should equal any one of the $n!$ permutations of $(v_1, \ldots, v_n)$. If we think of selecting x_i at time i and then placing a cross at x_i, then $(v_1, \ldots, v_n)$ may be interpreted as the final resulting configuration of crosses, *without* regard to the time order of selection.

As in Sec. 17-3 the **distribution function** $F_X: R^n \to R$ **for** or induced by **a random vector** $X = (X_1, \ldots, X_n)$ with associated probability space (Ω,P) is defined by

$$F_{X_1,\ldots,X_n}(x_1, \ldots, x_n) = P[X_1 < x_1, \ldots, X_n < x_n]$$

for all reals $x_1, \ldots, x_n$. Naturally if (R^n,P) is any probability space with sample space R^n then its corresponding distribution function $F: R^n \to R$ is defined by

$$F(x_1, \ldots, x_n) = P\{(x_1', \ldots, x_n'): x_1' < x_1, \ldots, x_n' < x_n\}.$$

Therefore $F_{X_1,\ldots,X_n}$ is also describable as the distribution function corresponding to the induced $P_{X_1,\ldots,X_n}$. The distribution function for a subsequence of $X_1, \ldots, X_n$ can be obtained by letting the extra coordinates in $F_{X_1,\ldots,X_n}$ converge to infinity.

If $X = (X_1, \ldots, X_n)$ induces a continuous density p_X then F_X can be obtained from p_X by integration. Then F_X is continuous and p_X can be obtained from F_X by differentiation,

$$p_{X_1,\ldots,X_n}(x_1, \ldots, x_n) = \frac{\partial^n F_{X_1,\ldots,X_n}(x_1, \ldots, x_n)}{\partial x_1\, \partial x_2 \cdots \partial x_n}$$

at least at points where the density is continuous.

Theorem 1 (Sec. 17-3) If a random vector $X = (X_1, X_2, \ldots, X_n)$ induces distribution function $F: R^n \to R$ then

$$P[x_1^{(1)} \le X_1 < x_1^{(0)}, x_2^{(1)} \le X_2 < x_2^{(0)}, \ldots, x_n^{(1)} \le X_n < x_n^{(0)}] = \Delta_F(x^{(1)},x^{(0)})$$

for any $x^{(1)} = (x_1^{(1)}, \ldots, x_n^{(1)})$ and $x^{(0)} = (x_1^{(0)}, \ldots, x_n^{(0)})$ if $x^{(1)}$ is coordinatewise less than $x^{(0)}$, that is, if $x_i^{(1)} < x_i^{(0)}$ for all i. The function $\Delta_F: R^{2n} \to R$ is defined by

$$\Delta_F(x^{(1)},x^{(0)}) = \sum_{\epsilon_1=0}^{1} \sum_{\epsilon_2=0}^{1} \cdots \sum_{\epsilon_n=0}^{1} (-1)^{\epsilon_1+\cdots+\epsilon_n} F(x_1^{(\epsilon_1)}, x_2^{(\epsilon_2)}, \ldots, x_n^{(\epsilon_n)}). \quad \blacksquare$$

This theorem asserts that the probability of the exhibited event equals $\Delta_F(x^{(1)},x^{(0)})$, which is defined to equal the sum of the 2^n terms

$$\pm F(x_1^{(\epsilon_1)}, \ldots, x_n^{(\epsilon_n)})$$

where the superscripts can be 0's and 1's. The sign is positive or negative, depending on whether an even or odd number of superscripts are 1, where 1's correspond to smaller coordinates.

Random vectors are identically distributed iff their distribution functions are equal. They are independent iff their joint distribution function satisfies the product rule. A discontinuity in F may be caused by a point (or by a line parallel to a coordinate axis) having positive probability.

A function $F: R^n \to R$ is called a **distribution function on R^n** iff the following four conditions hold:

1. $\lim_{x \to \infty} F(x, \ldots, x) = 1$.
2. $\lim_{x_i \to -\infty} F(x_1, \ldots, x_n) = 0$ for all i and $x_1, \ldots, x_{i-1}, x_{i+1}, \ldots, x_n$.
3. $F(x'_1, \ldots, x'_n)$ converges to $F(x_1, \ldots, x_n)$ whenever each x'_i converges to x_i, where $x'_i < x_i$ is always satisfied.
4. $\Delta_F(x^{(1)}, x^{(0)}) \geq 0$ whenever $x^{(1)} = (x_1^{(1)}, \ldots, x_n^{(1)})$ is coordinatewise less than $x^{(0)} = (x_1^{(0)}, \ldots, x_n^{(0)})$.

The correspondence assigning $F: R^n \to R$ to (R^n, P) is a one-to-one correspondence between the class of all probability spaces (R^n, P) and the class of all distribution functions on R^n.

If the indicator function of an event can be expressed as a linear combination of indicator functions of other, not necessarily disjoint, events then the linearity of expectation yields the following theorem, since $EI_B = P(B)$. An "elementary" proof of this theorem appears in Sec. 17-4.

Theorem 1 (Sec. 17-4) If events B and $B_1, \ldots, B_n$ in a probability space (Ω, P) satisfy

$$I_B = \sum_{i=1}^{n} b_i I_{B_i}$$

where $b_1, \ldots, b_n$ are reals, then

$$P(B) = \sum_{i=1}^{n} b_i P(B_i). \quad \blacksquare$$

Simple algebraic manipulations with indicators, together with the theorem above, comprise what we call the *method of indicators*. The method is used to prove the following theorem, where A is the event that all larger but no smaller events occur.

Theorem 2 (Sec. 17-4) Let $A_1^{(1)}, A_2^{(1)}, \ldots, A_n^{(1)}$ and $A_1^{(0)}, A_2^{(0)}, \ldots, A_n^{(0)}$ be $2n$ events such that $A_i^{(1)}$ is a subset of $A_i^{(0)}$ for each i. If

$$A = \bigcap_{i=1}^{n} [A_i^{(0)} (A_i^{(1)})^c]$$

then

$$P(A) = \sum_{\epsilon_1=0}^{1} \sum_{\epsilon_2=0}^{1} \cdots \sum_{\epsilon_n=0}^{1} (-1)^{\epsilon_1 + \cdots + \epsilon_n} P(A_1^{(\epsilon_1)} A_2^{(\epsilon_2)} \cdots A_n^{(\epsilon_n)}). \quad \blacksquare$$

Specializing this theorem to $B_i = A_i^{(1)}$ with every $A_i^{(0)} = \Omega$ yields the next theorem, and specializing to

$$A_i^{(0)} = [X_i < x_i^{(0)}] \qquad \text{and} \qquad A_i^{(1)} = [X_i < x_i^{(1)}]$$

proves Theorem 1 (Sec. 17-3).

Theorem 3 (Sec. 17-4) For any events $B_1, \ldots, B_n$ we have

$$P\left(\bigcup_{j=1}^{n} B_j\right) = \sum_{j=1}^{n} (-1)^{j-1} S_j = S_1 - S_2 + S_3 - \cdots + (-1)^{n-1} S_n$$

where

$$S_1 = \sum_{i=1}^{n} P(B_i) \qquad S_2 = \sum_{\substack{i=1 \\ i<k}}^{n} \sum_{k=1}^{n} P(B_i B_k) \qquad \cdots \qquad S_n = P(B_1 \cdots B_n). \quad \blacksquare$$

The number S_j can be described as equal to the sum of the probabilities of the $\binom{n}{j}$ events which can be written as an intersection,

$$B_{i_1} B_{i_2} \cdots B_{i_j} \qquad \text{where } 1 \le i_1 < i_2 < \cdots < i_j \le n$$

of j events, using each of $B_1, \ldots, B_n$ at most once.

CHAPTER 18: MULTIVARIATE NORMAL DENSITIES AND CHARACTERISTIC FUNCTIONS

A $k \times n$ **random matrix** X has random variable X_{ij} for its (i,j) entry in its ith row and jth column, and the transpose X' has X_{ji} for its (i,j) entry. If each $E|X_{ij}| < \infty$ then we define EX to be the matrix of real numbers whose (i,j) entry is EX_{ij}. Most often we have random vectors, so $n = 1$ or $k = 1$,

$$X = \begin{bmatrix} X_1 \\ \cdot \\ \cdot \\ \cdot \\ X_k \end{bmatrix} \qquad EX = \begin{bmatrix} EX_1 \\ \cdot \\ \cdot \\ \cdot \\ EX_k \end{bmatrix} \qquad \begin{matrix} X' = (X_1, \ldots, X_k) \\ \\ \\ EX' = (EX_1, \ldots, EX_k) \end{matrix}$$

If X and Y are random $k \times n$ matrices then so is $aX + bY$ for real a and b. It has $aX_{ij} + bY_{ij}$ for its (i,j) entry. We obviously have linearity of expectation

$$E(aX + bY) = aEX + bEY$$

if each of X and Y has an expectation. If A is an $r \times k$ matrix of real numbers and X is a random $k \times n$ matrix then $E(AX) = A(EX)$. More generally

$$E(AXB + CYD) = A(EX)B + C(EY)D$$

for six matrices of proper dimensions, where only X and Y are random.

If $X' = (X_1, \ldots, X_k)$ is a random vector for which $EX_i^2 < \infty$ for $i = 1, \ldots, k$ then the **covariance matrix** for X', or for X, is the $k \times k$ matrix C whose (i,j) entry is the real number $\operatorname{Cov}(X_i, X_j)$. Clearly C is symmetric, $c_{ij} = c_{ji}$, with diagonal entries $c_{ii} = \operatorname{Var}(X_i) \ge 0$. Also

$$C = E[(X - EX)(X - EX)']$$

If $X_1, \ldots, X_k$ are independent then C is diagonal. If $X_1, \ldots, X_k$ are independent and each has unit variance then the covariance matrix is just the identity matrix. For linear combinations of components we have

$$0 \le \text{Var}\left(\sum_{i=1}^{k} t_i X_i\right) = t'Ct \qquad \text{where } t = \begin{bmatrix} t_1 \\ \cdot \\ \cdot \\ \cdot \\ t_k \end{bmatrix}.$$

Therefore the vector $EX' = (EX_1, \ldots, EX_k)$ of means together with the covariance matrix are precisely what is needed in order to find the mean and variance of every linear combination of components.

The fact that $t'Ct$ is nonnegative for all t leads to the ideas of positive-definiteness and degeneracy, which are related by the next lemma.

If C is *any* $k \times k$ matrix of real numbers then its associated **quadratic form** is the function $Q: R^k \to R$ defined by

$$Q(t_1, \ldots, t_k) = t'Ct = \sum_{i=1}^{k} \sum_{j=1}^{k} c_{ij} t_i t_j.$$

The correspondence between real *symmetric* $k \times k$ matrices and their quadratic forms on R^k is a one-to-one correspondence. We say that C and Q are **nonnegative** iff $Q \ge 0$ on R^k. Clearly $Q(0, \ldots, 0) = 0$. We say that C and Q are **positive definite** iff $Q(t_1, \ldots, t_k) > 0$ whenever at least one $t_i \ne 0$; that is, $t'Ct > 0$ whenever $t \ne 0$. A symmetric positive-definite matrix must be nonsingular, and its inverse is also symmetric positive definite.

A *random variable* W, or the distribution induced by W, is said to be **degenerate** iff its distribution is concentrated on one point so that $P[W = EW] = 1$. Equivalently $\text{Var } W = 0$; that is, W is essentially a constant. A *random vector* $X' = (X_1, \ldots, X_k)$ is said to be **degenerate** iff there exist real numbers $t_1, \ldots, t_k$ not all zero, such that $t_1X_1 + \cdots + t_kX_k$ is a degenerate random variable, that is, has zero variance. This is the case iff one of $X_1, \ldots, X_k$ can be expressed as a constant plus a linear combination of the others, so that the distribution induced by X is concentrated on a space of dimension $k - 1$ or smaller.

Lemma 1 (Sec. 18-1) Let $X' = (X_1, \ldots, X_k)$ be a random vector for which $EX_j^2 < \infty$ for $j = 1, \ldots, k$ and let C be its covariance matrix. Then

$$0 \le \text{Var}\left(\sum_{i=1}^{k} t_i X_i\right) = t'Ct \qquad \text{where } t = \begin{bmatrix} t_1 \\ \cdot \\ \cdot \\ \cdot \\ t_k \end{bmatrix}$$

so that C is nonnegative. Furthermore the following three conditions are equivalent:

(*a*) X is nondegenerate.

(*b*) C is positive definite.

(*c*) C is nonsingular. ■

Assume that we know the joint distribution of $k + 1$ random variables Y and $X_1, \ldots, X_k$. We desire some linear combination $Y_p = \beta_0 + \beta_1 X_1 + \cdots + \beta_k X_k$ of the X_i which can be used to predict Y with minimum expected squared error. Section 11-4 was devoted to $k = 1$. The following theorem shows that such Y_p are characterized by the property that the resulting prediction error $[Y - Y_p]$ has zero expectation and is uncorrelated with each of the variables $X_1, \ldots, X_k$ used for prediction. These conditions are easily expressed as linear equations with constants which are covariances. Thus for a solution to this problem only the means and covariances of $Y, X_1, \ldots, X_k$ need be known.

Theorem 1 (Sec. 18-1) Assume that the random variables Y and $X_1, \ldots, X_k$ with $k \geq 1$ have finite second moments. For *fixed* real $\beta_0, \ldots, \beta_k$ let

$$Y_p = \beta_0 + \sum_{i=1}^{k} \beta_i X_i.$$

Then the following three conditions are equivalent:
(*a*) Y_p is a best linear predictor for Y; that is,

$$E[Y - Y_p]^2 \leq E\left[Y - \left(b_0 + \sum_{i=1}^{k} b_i X_i\right)\right]^2 \qquad \text{for all real } b_0, \ldots, b_k.$$

(*b*) $E[Y - Y_p] = 0$ and $E[Y - Y_p]X_j = 0$ for $j = 1, \ldots, k$.
(*c*) $E[Y - Y_p] = 0$ and

$$C \begin{bmatrix} \beta_1 \\ \cdot \\ \cdot \\ \cdot \\ \beta_k \end{bmatrix} = \begin{bmatrix} \text{Cov}(X_1, Y) \\ \cdot \\ \cdot \\ \cdot \\ \text{Cov}(X_k, Y) \end{bmatrix}$$

where C is the covariance matrix for $(X_1, \ldots, X_k)$.
Furthermore there is a Y_p satisfying these conditions. Also this Y_p is unique iff C is nonsingular. ■

The following lemma shows that all symmetric positive-definite matrices arise in a particularly transparent fashion. Assume that S is some nonsingular $k \times k$ matrix and let $t \neq 0$ be some k vector. If $v = St$ then $v \neq 0$, so the square of the length of v is positive,

$$0 < \sum_{i=1}^{k} (v_i)^2 = v'v = (St)'(St) = t'S'St \qquad \text{for all } t \neq 0.$$

Thus for any nonsingular S the matrix $S'S$ is symmetric, since it equals its transpose, and positive definite. The following lemma shows that every symmetric positive-definite matrix arises in this way. Furthermore S may be restricted to be triangular.

Lemma 2 (Sec. 18-1) A real symmetric $k \times k$ matrix C is positive definite iff there exists a nonsingular real upper-triangular $k \times k$ matrix

$$S = \begin{bmatrix} s_{11} & s_{12} & s_{13} & \cdots & s_{1k} \\ 0 & s_{22} & s_{23} & \cdots & s_{2k} \\ 0 & 0 & s_{33} & \cdots & s_{3k} \\ \cdots & \cdots & \cdots & \cdots & \cdots \\ 0 & 0 & \cdots & \cdots & s_{kk} \end{bmatrix} \qquad s_{ii} > 0,\ s_{ij} = 0 \text{ if } i > j$$

such that $C = S'S$. ■

If $X' = (X_1, \ldots, X_k)$ is a random vector with covariance matrix C, and T is a constant $k \times k$ matrix then the covariance matrix for TX is TCT'. In particular if C is the identity matrix then TX has TT' for covariance matrix, and hence $S'X$ has $S'S$ for covariance matrix. But the preceding lemma shows that every symmetric positive-definite matrix can be expressed as $S'S$ and hence must arise as a covariance matrix, as stated next.

Lemma 3 (Sec. 18-1) If C is any real symmetric positive-definite $k \times k$ matrix then there exists a nonsingular real lower-triangular matrix T such $C = TT'$. Therefore C is the covariance matrix of TX where the components of $X' = (X_1, \ldots, X_k)$ are independent and each is normal 0 and 1. ■

We introduce normal densities on R^k, as in Sec. 18-2. Let $X' = (X_1, \ldots, X_k)$ where $X_1, \ldots, X_k$ are independent and each is normal 0 and 1. A continuous density on R^k is called a **normal density** iff there is a nonsingular $k \times k$ matrix T with real entries, and a vector μ in R^k such that this density is induced by $Y = TX + \mu$. Clearly $\mu = EY$ and the covariance matrix $C = TT'$ for Y is symmetric positive definite, as is C^{-1}, and $(\det C)(\det C^{-1}) = 1$. We find that

$$p_Y(y) = \frac{\sqrt{\det C^{-1}}}{(2\pi)^{k/2}} e^{-(1/2)(y-\mu)'C^{-1}(y-\mu)} \qquad \text{for all } y \in R^k.$$

Clearly two normal densities on R^k are equal iff they have the same vector of means and the same covariance matrix. Thus we have a one-to-one correspondence between the family of normal densities on R^k and the set of pairs (μ, C), where μ is any k vector and C is any symmetric $k \times k$ positive-definite matrix.

Let $Y' = (Y_1, \ldots, Y_k)$ have a normal density. Then so does any vector obtained by permuting its coordinates. If $EY = \mu$ then clearly there exists a nonsingular T such that $T^{-1}(Y - \mu)$ has independent normal 0 and 1 coordinates. Each component of Y has a normal density, as does every linear combination $t'Y$ with $t \neq 0$. If U is a nonsingular $k \times k$ matrix then UY has a normal density. Every marginal density from Y is normal. Also $Y_1, \ldots, Y_k$ are independent iff they are pairwise uncorrelated.

If $Y' = (Y_1, \ldots, Y_k)$ has a normal density then the conditional density for Y_1, given $[Y_2 = y_2, \ldots, Y_k = y_k]$ is normal, the conditional variance is a constant, and the conditional mean $f(y_2, \ldots, y_k)$ is a constant plus a linear combination of $y_2, \ldots, y_k$. Therefore *no* predictor $g(Y_2, \ldots, Y_k)$ for Y_1 in terms of $Y_2, \ldots, Y_k$ can achieve a smaller expected squared error than the best linear predictor $f(Y_2, \ldots, Y_k)$, which is of the kind considered in Theorem 1 (Sec. 18-1).

Exercise 2 (Sec. 18-2) shows that if Y_1 and Y_2 have a bivariate normal density with $EY_1 = EY_2 = 0$ and correlation coefficient ρ then

$$P[Y_1 \geq 0, Y_2 \geq 0] = \tfrac{1}{4} + \frac{\arcsin \rho}{2\pi}$$

and hence $P[Y_1 Y_2 \geq 0]$ is twice this.

The **MGF of a random vector** $Y' = (Y_1, \ldots, Y_k)$ is the function

$$m_Y(t) = Ee^{t'Y}$$

defined for all t in R^k for which the expectation exists. Once again the MGF determines the distribution, in the same sense as before. More precisely two random vectors $X' = (X_1, \ldots, X_k)$ and $Y' = (Y_1, \ldots, Y_k)$ are identically distributed if there exists a real $t_0 > 0$ such that $Ee^{t'X}$ and $Ee^{t'Y}$ are finite and equal whenever each $|t_i| \leq t_0$ where $t' = (t_1, \ldots, t_k)$.

If $Y' = (Y_1, \ldots, Y_k)$ has a normal density with $EY = \mu$ and covariance matrix C then for fixed $t' = (t_1, \ldots, t_k) \neq 0$ the random variable $t'Y$ has a normal density; hence Y has MGF

$$Ee^{t'Y} = e^{E(t'Y)+(\frac{1}{2})\mathrm{Var}(t'Y)} = e^{t'\mu+(\frac{1}{2})t'Ct} \qquad \text{for all } t \in R^k.$$

As in Sec. 18-3, we next introduce complex-valued random variables and then specialize to characteristic functions. We continue to reserve the name "random variable" for real-valued functions. Let C denote the **set of all complex numbers** $z = (x,y) = x + iy$.

Summary of complex-valued random variables If X and Y are random variables then $Z = (X,Y) = X + iY$ is a **complex-valued random variable**, so $Z: \Omega \to C$ has an associated probability space (Ω,P) and $|Z| = (X^2 + Y^2)^{1/2}$ is a random variable. By the distribution induced by Z we mean the distribution induced by the random vector (X,Y). We say that Z has an **expectation** iff $E|Z| < \infty$, or equivalently iff $E|X| < \infty$ and $E|Y| < \infty$. In this case we let $EZ = EX + iEY$ and we have $|EZ| \leq E|Z|$.

If $c_1, \ldots, c_n$ are complex numbers, and $Z_1, \ldots, Z_n$ are complex-valued random variables, and each $E|Z_k| < \infty$ then the following sum has an expectation and we have linearity,

$$E\left(\sum_{k=1}^{n} c_k Z_k\right) = \sum_{k=1}^{n} c_k EZ_k$$

If $Z_1, \ldots, Z_n$ with $Z_k = X_k + iY_k$ are n **independent** complex-valued random variables, by which we mean that $(X_1, Y_1), \ldots, (X_n, Y_n)$ are n independent random vectors, and if each Z_k has an expectation then so does their product and we have

$$E\left(\prod_{k=1}^{n} Z_k\right) = \prod_{k=1}^{n} EZ_k \quad \blacksquare$$

If X is a random variable then the function $\varphi: R \to C$ defined for all real t by

$$\varphi(t) = Ee^{itX} = E(\cos tX) + iE \sin tX$$

is called the **characteristic function** of X, or of its induced distribution or distribution function. Clearly $\varphi(0) = 1$, and $|e^{itX}| = 1$ so that $|\varphi(t)| \le 1$ for all t. Thus *every* random variable has a characteristic function defined for *all real t. These, and only these, functions are called characteristic functions.*

Every characteristic function is uniformly continuous. Two random variables are identically distributed iff they have the same characteristic function. If $X_1, \ldots, X_n$ are independent random variables then

$$\varphi_{X_1+\cdots+X_n}(t) = \varphi_{X_1}(t) \cdots \varphi_{X_n}(t) \qquad \text{for all real } t.$$

For real a and b we have

$$\varphi_{aX+b}(t) = e^{itb}\varphi_X(ta) \qquad \text{for all real } t.$$

The characteristic function for $-X$ is the complex conjugate $\bar{\varphi}_X$ of that for X. Thus a characteristic function has only real values, $\varphi_X = \bar{\varphi}_X$, iff X and $-X$ are identically distributed, that is, iff X has a **symmetric distribution**, in which case $\varphi_X(t) = E(\cos tX)$.

For all our standard discrete and continuous densities which have finite MGFs in some interval containing the origin, we may obtain φ by $\varphi(t) = m(it)$; that is, we may take the closed form found for the MGF and just replace t by it. In particular this procedure can be justified for all densities in Fig. 2 (Sec. 9-1), which is reproduced in the summary for Chap. 9, and in Table 1 of the summary for Chap. 16.

If Z has a Cauchy density with parameters μ and $\lambda > 0$ then

$$p_Z(z) = \frac{1}{\pi}\frac{\lambda}{\lambda^2 + (z - \mu)^2} \qquad \text{for all real } z$$

$$\varphi_Z(t) = Ee^{itZ} = e^{it\mu - \lambda|t|} \qquad \text{for all real } t.$$

It immediately follows that if $Z_1, \ldots, Z_n$ are independent and Z_k is Cauchy μ_k and $\lambda_k > 0$ then $Z_1 + \cdots + Z_n$ is Cauchy $\sum_{k=1}^{n} \mu_k$ and $\sum_{k=1}^{n} \lambda_k$. This fact was used and interpreted in Sec. 16-5.

We next examine derivatives of characteristic functions and the related polynomial approximations which enable us to use the continuity theorem, for which a special case is stated first.

Summary of differentiability of characteristic functions Let X be a random variable with characteristic function φ. Let integers k and m satisfy $1 \le k \le m$, and assume that $E|X|^m < \infty$. Then φ has a continuous kth derivative obtainable by differentiating under the expectation sign,

$$\varphi^{(k)}(t) = E(iX)^k e^{itX} \qquad \text{for all real } t.$$

Hence

$$\varphi^{(k)}(0) = i^k EX^k.$$

Furthermore $r_m(t)$ goes to zero as t goes to zero, if $r_m: R \to C$ is defined by the equation

$$\begin{aligned}\varphi(t) &= \sum_{k=0}^{m-1} \frac{\varphi^{(k)}(0)}{k!} t^k + \frac{t^m}{m!} [\varphi^{(m)}(0) + i^m r_m(t)] \\ &= \sum_{k=0}^{m-1} \frac{EX^k}{k!} (it)^k + \frac{(it)^m}{m!} [EX^m + r_m(t)].\end{aligned}$$

For notational convenience we have used $\varphi^{(0)}(0) = EX^0 = 1$. In particular for $m = 2$ we see that if $EX^2 < \infty$, $EX = 0$, and r is defined by

$$Ee^{itX} = 1 - \frac{t^2}{2} [EX^2 + r(t)]$$

then $r(t)$ goes to zero as t goes to zero. ■

Continuity theorem for convergence to the standardized normal distribution function Φ If $F_1, F_2, \ldots$ is a sequence of distribution functions on R, with corresponding characteristic functions $\varphi_1, \varphi_2, \ldots$ then

$$\lim_{n\to\infty} F_n(x) = \Phi(x) \qquad \text{for all real } x$$

iff

$$\lim_{n\to\infty} \varphi_n(t) = e^{-t^2/2} \qquad \text{for all real } t. \quad ■$$

P. Lévy's continuity theorem Let $F_1, F_2 \cdots$ be a sequence of distribution functions on R^k and let $\varphi_1, \varphi_2, \ldots$ be the corresponding characteristic functions. If there is a distribution function F on R^k such that

$$\lim_{n\to\infty} F_n(x) = F(x) \qquad \text{for every } x \text{ at which } F \text{ is continuous,} \tag{*}$$

and if φ is the characteristic function for F then

$$\lim_{n\to\infty} \varphi_n(t) = \varphi(t) \qquad \text{for all } t \in R^k.$$

Conversely, if there is some function $g: R^k \to C$ which is continuous at the origin of R^k and such that

$$\lim_{n\to\infty} \varphi_n(t) = g(t) \qquad \text{for all } t \in R^k$$

then there is a distribution function F on R^k such that (*) above holds, and hence g must be the characteristic function for F. ■

The following lemma is easily obtained from the quadratic approximation to characteristic functions. This lemma, together with the continuity theorem, yields the following central-limit theorem [which is the same as Corollary 2 (Sec. 20-2)], where only the second moment is required to be finite.

Lemma 1 (Sec. 18-3) If the random variables $W_1, W_2, \ldots$ are independent and have the same distribution satisfying $EW_1^2 < \infty$ and $EW_1 = 0$ then

$$E \exp\left\{\frac{i(W_1 + \cdots + W_n)}{\sqrt{n}}\right\} \to \exp\left\{-\frac{EW_1^2}{2}\right\} \qquad \text{as } n \to \infty. \quad \blacksquare$$

Theorem 1 (Sec. 18-3): A CLT If $S_n = X_1 + \cdots + X_n$ where $X_1, X_2, \ldots$ are independent and identically distributed random variables with $0 < \sigma_1^2 = \text{Var } X_1 < \infty$, then for every real b

$$P\left[\frac{S_n - nEX_1}{\sigma_1\sqrt{n}} < b\right] \to \Phi(b) \qquad \text{as } n \to \infty. \quad \blacksquare$$

The following generalization to this theorem is proved by using characteristic functions of random *vectors*.

Theorem 1 (Sec. 18-4): A vector CLT Let $Y' = (Y_1, \ldots, Y_k)$ be a random vector which induces a nondegenerate distribution on R^k. Assume that $EY_m^2 < \infty$ for $m = 1, \ldots, k$ and let $\mu = EY$. Let $S_n = X_1 + \cdots + X_n$ where $X_1, X_2, \ldots$ are independent random vectors each having the same distribution as Y. Then, for each point of R^k, the distribution function for $(S_n - n\mu)/\sqrt{n}$ converges, as n goes to infinity, to the distribution function for the normal density with zero means and the same covariance matrix as Y. $\blacksquare$

The **characteristic function of a random vector** $(Y_1, \ldots, Y_k)$ is the function $\varphi: R^k \to C$ defined by

$$\varphi(t_1, \ldots, t_k) = Ee^{i(t_1Y_1 + \cdots + t_kY_k)} \qquad \text{for all real } t_1, \ldots, t_k.$$

Every random vector $(Y_1, \ldots, Y_k)$ has a characteristic function φ defined on all of R^k, and φ is uniformly continuous, $|\varphi| \leq 1$, $\varphi(0, \ldots, 0) = 1$. Letting appropriate $t_i = 0$ in this joint characteristic function yields the characteristic function of any marginal. It is a fact that two random vectors into R^k are identically distributed iff they have the same characteristic function. Clearly, then, random variables $Y_1, \ldots, Y_k$ are independent iff

$$\varphi_{Y_1, \ldots, Y_k}(t_1, \ldots, t_k) = \varphi_{Y_1}(t_1) \cdots \varphi_{Y_k}(t_k) \qquad \text{for all real } t_1, \ldots, t_k.$$

If $Y' = (Y_1, \ldots, Y_k)$ has a normal density with mean $\mu' = (\mu_1, \ldots, \mu_k)$ and covariance matrix C then for fixed $t' = (t_1, \ldots, t_k) \neq 0$ the random variable $t'Y$ has a normal density, and hence Y has characteristic function

$$Ee^{i(t'Y)} = e^{iE(t'Y) - (1/2)\text{Var}(t'Y)} = e^{it'\mu - (1/2)t'Ct}.$$

CHAPTER 19: PROOF OF THE CENTRAL-LIMIT THEOREM

This chapter is devoted to a proof of Theorem 4, which is a restatement of the central-limit Theorem 1 (Sec. 13-2). For two random variables U and V let $\rho(U,V)$ equal, essentially, the largest vertical distance between their distribution functions. More precisely

$$\rho(U,V) = \sup_{\alpha} |P[U < \alpha] - P[V < \alpha]| = \sup_{\alpha} |F_U(\alpha) - F_V(\alpha)|$$

where we use the supremum concept from Appendix 2. The conclusion of the CLT can be stated in terms of the distance ρ. However, ρ lacks certain properties desirable for a proof of the CLT. Therefore we introduce a distance ρ_ϵ, depending on a real parameter $\epsilon > 0$ at our disposal,

$$\rho_\epsilon(U,V) = \sup_{\alpha} \left| EG\left(\frac{U-\alpha}{\epsilon}\right) - EG\left(\frac{V-\alpha}{\epsilon}\right) \right|$$

where the fairly smooth function $G: R \to R$ is defined by

$$G(x) = \begin{cases} 0 & \text{if } x \leq 0 \\ 140\displaystyle\int_0^x t^3(1-t)^3\,dt & \text{if } 0 < x < 1 \\ 1 & \text{if } x \geq 1. \end{cases}$$

The function $G((u - \alpha)/\epsilon)$ of u has the value 0 or 1, depending on whether $u \leq \alpha$ or $u \geq \alpha + \epsilon$, so that, very roughly,

$$EG\left(\frac{U-\alpha}{\epsilon}\right) \doteq P[U \geq \alpha]$$

if ϵ if small. Therefore, as shown by the next theorem, small ϵ and small $\rho_\epsilon(U,V)$ can guarantee a small $\rho(U,V)$, at least when V is normal.

Theorem I Let V be normally distributed with variance σ^2. If U is any random variable and if $\epsilon > 0$ then

$$\rho(U,V) \leq \frac{\epsilon}{2\sigma} + \rho_\epsilon(U,V). \quad \blacksquare$$

For any real β we have $\rho_\epsilon(U + \beta, V + \beta) = \rho_\epsilon(U,V)$, as is apparent from the definition of ρ_ϵ. This can be modified, by the replacement of β by an independent Z, to $\rho_\epsilon(U + Z, V + Z) \leq \rho_\epsilon(U,V)$. Also G has a continuous bounded third derivative on R, so we can use Taylor's theorem with remainder to obtain the following lemma, which yields the next theorem.

Lemma I Let U, V, Z be independent random variables such that U and V have finite third absolute moments and satisfy $EU = EV$ and $EU^2 = EV^2$. Then

$$|EG(U + Z) - EG(V + Z)| \leq 10[E\,|U|^3 + E\,|V|^3]. \quad \blacksquare$$

Theorem 2 Let U, V, Z be independent random variables such that U and V have finite third absolute moments and satisfy $EU = EV$ and $EU^2 = EV^2$. If $\epsilon > 0$ then

$$\rho_\epsilon(U + Z, V + Z) \le \frac{10}{\epsilon^3}\,[E\,|U|^3 + E\,|V|^3]. \quad \blacksquare$$

Now ρ_ϵ satisfies the triangle inequality, so that for any $W_0, W_1, \ldots, W_n$ the distance $\rho_\epsilon(W_0, W_n)$ between the two extreme random variables is bounded by the sum of the distances between adjacent pairs,

$$\rho_\epsilon(W_0, W_n) \le \sum_{i=0}^{n-1} \rho(W_i, W_{i+1}).$$

Therefore if we are interested in $\rho_\epsilon(W_0, W_n)$ for W_0 and W_n as below, where we exhibit for $n = 3$, then we may introduce the proper intermediate W_i,

$$\begin{aligned} W_0 &= U_1 + U_2 + U_3 \\ W_1 &= V_1 + U_2 + U_3 \\ W_2 &= V_1 + V_2 + U_3 \\ W_3 &= V_1 + V_2 + V_3 \end{aligned}$$

so that each adjacent pair differs in only one component. Thus we go from $U_1 + \cdots + U_n$ to $V_1 + \cdots + V_n$ in n steps, changing only one component at each step, so that the impact of the kth step is small if $\rho_\epsilon(U_k, V_k)$ is small. Thus Theorem 2 yields Theorem 3, which says that the distributions of any two sums of independent matching random variables must be close to each other if sums of third absolute moments are small. Specializing $V_1, \ldots, V_n$ to be independent and normal, so that their sum is normal, then relating ρ_ϵ to ρ by Theorem 1, and then optimizing ϵ yields Theorem 4.

Theorem 3 Let $U_1, \ldots, U_n, V_1, \ldots, V_n$ be $2n$ independent random variables having finite third absolute moments and satisfying

$$EU_k = EV_k \qquad EU_k^{\,2} = EV_k^{\,2} \qquad \text{whenever } 1 \le k \le n.$$

If $\epsilon > 0$ then

$$\rho_\epsilon\left(\sum_{k=1}^{n} U_k, \sum_{k=1}^{n} V_k\right) \le \frac{10}{\epsilon^3} \sum_{k=1}^{n} [E\,|U_k|^3 + E\,|V_k|^3]. \quad \blacksquare$$

Theorem 4. The CLT Let $U = U_1 + \cdots + U_n$ where $U_1, \ldots, U_n$ are independent and each $EU_k = 0$ and $E\,|U_k|^3 < \infty$. If $EU^2 = 1$ and V is normal 0 and 1 then

$$\rho(U, V) \le 3B^{1/4} \qquad \text{where } B = \sum_{k=1}^{n} E\,|U_k|^3. \quad \blacksquare$$

To convert Theorem 4 to Theorem 1 (Sec. 13-2) let

$$U_k = \frac{X_k - EX_k}{\sqrt{\operatorname{Var} S_n}}.$$

CHAPTER 20: THE METHOD OF TRUNCATION

Our previous laws of large numbers and central-limit theorems presumed finite second and third absolute moments, respectively. This chapter introduces the method of truncation and uses it to prove laws of large numbers when only expectations need exist, and central-limit theorems when only variances need exist. Section 20-3 is devoted largely to a more detailed study of the SLLN.

The method of truncation may be motivated as follows. Assume an experiment involving only a finite collection of random variables of interest. A number m can be selected large enough that the probability is negligible that at least one of the variables will have a value exceeding m in absolute value. Therefore, practically speaking, we should be able to replace any X by an essentially equivalent X' which never exceeds m in absolute value. Such an X' has finite moments of all orders.

For any real $c > 0$ define the functions $t_c: R \to R$ and $d_c: R \to R$ by

$$t_c(x) = \begin{cases} x & \text{if } |x| \le c \\ 0 & \text{if } |x| > c \end{cases}$$

and

$$d_c(x) = \begin{cases} 0 & \text{if } |x| \le c \\ x & \text{if } |x| > c. \end{cases}$$

Thus $|t_c(x)| \le c$ and $t_c(x) + d_c(x) = x$ for all x. We call t_c the *truncation* at level c of the identity function on R, and d_c the *difference* between the identity function and t_c.

For any random variable X and any real $c > 0$ we call the random variable $t_c(X)$ the **truncation of X at c**. Thus $X = t_c(X) + d_c(X)$, so that X equals the sum of its truncation and difference variables.

Typically $t_c(X)$ and $d_c(X)$ are highly dependent. No moments of X and $t_c(X)$ need be equal. Also if $c_1 \ne c_2$ then $t_{c_1}(X)$ and $t_{c_2}(Y)$ need not be identically distributed even though X and Y are. However, truncation also has desirable properties. By definition $t_c(X)$ agrees with X on $[|X| \le c]$ and is zero off $[|X| \le c]$. Clearly $|t_c(X)| \le c$ so $t_c(X)$ has all moments finite, while X and $d_c(X)$ always have finite moments of the same orders. Furthermore if X and Y are independent (identically distributed) then $t_{c_1}(X)$ and $t_{c_2}(Y)$ are independent (identically distributed if $c_1 = c_2$).

The method of truncation as applied to the WLLN, for a typical example, consists of replacing the given random variables by appropriate truncations, showing that the difference variables are negligible in the sense that a WLLN for the truncated variables implies one for the original variables, and then proving a WLLN for the truncated variables.

Lemma 1 below shows that finiteness of moments of X depends only on the sequence of numbers $P[k - 1 < |X| \le k]$ for $k = 1, 2, \ldots$. Lemma 2 collects some of the properties of X, $t_c(X)$, and $d_c(X)$ which are implied by $E\,|X|^r < \infty$. The method of truncation then yields the WLLN of the following theorem.

Lemma 1 (Sec. 20-1) Let X be a random variable and r a positive real number. Then $E|X|^r < \infty$ iff $B_r < \infty$, where

$$B_r = \sum_{k=1}^{\infty} k^r P[k - 1 < |X| \leq k].$$

Furthermore if $E|X|^r < \infty$ then $E|X|^r \leq B_r$. In particular $E|X| < \infty$ iff $B_1 < \infty$, where

$$B_1 = \sum_{k=1}^{\infty} kP[k - 1 < |X| \leq k] = \sum_{n=0}^{\infty} P[|X| > n].$$

Furthermore if $E|X| < \infty$ then $B_1 - 1 \leq E|X| \leq B_1$. ■

Lemma 2 (Sec. 20-1) If $E|X|^r < \infty$, where $r > 0$, then the following are true:
(*a*) $E|d_c(X)|^r \to 0$ as $c \to \infty$.
(*b*) For any $s > 0$, $(1/c^s)E|t_c(X)|^{r+s} \to 0$ as $c \to \infty$.
(*c*) For any $c > 0$, $P[|X| > c] \leq (1/c^r)E|d_c(X)|^r$.
(*d*) If $0 < s < r$ and $c > 0$ then $E|d_c(X)|^s \leq (1/c^{r-s})\, E|d_c(X)|^r$. ■

Theorem 1 (Sec. 20-1): The WLLN for independent identically distributed random variables having expectations Let $X_1, X_2, \ldots$ be independent and identically distributed random variables satisfying $E|X_1| < \infty$. If $S_n = X_1 + \cdots + X_n$ and $\epsilon > 0$ then

$$P\left[\left|\frac{1}{n} S_n - EX_1\right| \geq \epsilon\right] \to 0 \qquad \text{as } n \to \infty. \quad ■$$

Section 20-2 applies truncation to the CLT. The CLT, proved in Chap. 19, yields the following very special case for random variables bounded by a constant c. If $S = X_1 + \cdots + X_n$ where $X_1, \ldots, X_n$ are independent, and each $EX_k = 0$ and $P[|X_k| \leq c] = 1$ then

$$\left|P[S^* < b] - \Phi(b)\right| < 4\left(\frac{c}{\sqrt{\operatorname{var} S}}\right)^{1/4} \qquad \text{for all real } b.$$

From this special case quite general central-limit theorems are deduced with comparative ease, making this section probably the most impressive application of truncation.

Truncation alters means and variances, so Theorem 1 below modifies the special case above so as to permit the mean and variance to be only near to zero and 1, respectively. Theorem 1, together with truncation, yields the key inequality of Theorem 2, from which the remaining asymptotic results are immediate. Theorem 3 is the CLT for general triangular arrays under the Lindeberg condition. It is specialized in Corollary 1 to consecutive sums of a sequence and in Corollary 2 to the CLT for independent and identically distributed random variables having a finite variance.

Theorem 1 (Sec. 20-2) Let $S = X_1 + \cdots + X_n$ where $X_1, \ldots, X_n$ are independent random variables with each $P[|X_k| \leq c] = 1$, where c is a constant. If $\mu = ES$ and $s^2 = \text{Var } S$ then for every real b

$$|P[S < b] - \Phi(b)| \leq 5c^{1/4} + |\mu| + 2\,|s - 1|. \quad \blacksquare$$

Theorem 2 (Sec. 20-2) Let $S = X_1 + \cdots + X_n$ where $X_1, \ldots, X_n$ are independent random variables with finite variances. Assume that $ES = 0$ and $ES^2 = 1$. For real $c > 0$ let

$$g(c) = \sum_{k=1}^{n} E[d_c(X_k)]^2.$$

If for a particular c we have $g(c) \leq c^3$ then for that c and every real b

$$|P[S < b] - \Phi(b)| \leq 9c^{1/4}. \quad \blacksquare$$

Theorem 3 (Sec. 20-2) For each n let k_n be a positive integer and let $S_n = X_{n,1} + X_{n,2} + \cdots + X_{n,k_n}$ where $X_{n,1}, X_{n,2}, \ldots, X_{n,k_n}$ are k_n independent random variables with finite variances. Let $s_n^2 = \text{Var } S_n > 0$ and $S_n^* = (S_n - ES_n)/s_n$. For each real $c > 0$ let

$$g_n(c) = \frac{1}{s_n^2} \sum_{j=1}^{k_n} Ed^2_{cs_n}(X_{n,j} - EX_{n,j}).$$

If for each $c > 0$ we have $g_n(c) \to 0$ as $n \to \infty$ then for every real b

$$P[S_n^* < b] \to \Phi(b) \qquad \text{as } n \to \infty. \quad \blacksquare$$

Corollary 1 (Sec. 20-2) Let $S_n = X_1 + \cdots + X_n$ where $X_1, X_2, \ldots$ are independent random variables with finite variances. Let $s_n^2 = \text{Var } S_n > 0$ and $S_n^* = (S_n - ES_n)/s_n$. For each real $c > 0$ let

$$g_n(c) = \frac{1}{s_n^2} \sum_{j=1}^{n} Ed^2_{cs_n}(X_j - EX_j).$$

If for each $c > 0$ we have $g_n(c) \to 0$ as $n \to \infty$ then for every real b

$$P[S_n^* < b] \to \Phi(b) \qquad \text{as } n \to \infty. \quad \blacksquare$$

Corollary 2 (Sec. 20-2) If $S_n = X_1 + \cdots + X_n$ where $X_1, X_2, \ldots$ are independent and identically distributed random variables with $0 < \sigma_1^2 = \text{Var } X_1 < \infty$, then for every real b

$$P\left[\frac{S_n - nEX_1}{\sigma_1\sqrt{n}} < b\right] \to \Phi(b) \qquad \text{as } n \to \infty. \quad \blacksquare$$

A more detailed study of the SLLN follows. Truncation appears only in Lemma 4 and the proof of Theorem 3.

Lemma 1 (Sec. 20-3) Let $Z_1, Z_2, \ldots$ be an infinite sequence of random variables defined on the same probability space. Then the following three conditions are equivalent:

(*a*) $P\left[\lim_{n\to\infty} Z_n = 0\right] = 1.$

(*b*) For every $\epsilon > 0$

$$\lim_{n\to\infty}\left[\lim_{m\to\infty} P\left(\bigcup_{k=n}^{m}[|Z_k| \geq \epsilon]\right)\right] = 0.$$

(*c*) For every $\epsilon > 0$ there exists a sequence $a_{1,\epsilon}, a_{2,\epsilon}, a_{3,\epsilon}, \ldots$ of real numbers converging to zero and satisfying

$$P\left(\bigcup_{k=n}^{m}[|Z_k| \geq \epsilon]\right) \leq a_{n,\epsilon} \qquad \text{whenever } m \geq n. \quad \blacksquare$$

We write $Z_n \xrightarrow{\text{a.s.}} 0$ and say that Z_n **converges almost surely to zero** iff one of the three equivalent conditions above holds. For real d we write $Z_n \xrightarrow{\text{a.s.}} d$ iff $(Z_n - d) \xrightarrow{\text{a.s.}} 0$. Condition (*a*) is the traditional definition, and it requires that all $Z_1, Z_2, \ldots$ be defined on the same probability space. Conditions (*b*) and (*c*) involve only joint distributions of finitely many of $Z_1, Z_2, \ldots$, but all of these must always be defined uniquely and consistently, as in Sec. 22-2, for $Z_n \xrightarrow{\text{a.s.}} 0$ to make sense. The double limit of condition (*b*) then always exists, and by definition it equals zero iff $Z_n \xrightarrow{\text{a.s.}} 0$. In Chap. 12 we state the SLLN essentially in the form of condition (*c*). The results of Chap. 12 yield the next two theorems.

Theorem 1 (Sec. 20-3): An asymptotic version of the SLLN If $X_1, X_2, \ldots$ are independent and identically distributed random variables with $EX_1^2 < \infty$ then

$$\frac{1}{n}(X_1 + \cdots + X_n) \xrightarrow{\text{a.s.}} EX_1 \qquad \text{as } n \to \infty. \quad \blacksquare$$

Theorem 2 (Sec. 20-3): Kolmogorov's criterion Let $X_1, X_2, \ldots$ be an infinite sequence of independent random variables with each $EX_k^2 < \infty$. If

$$\sum_{k=1}^{\infty} \frac{\operatorname{Var} X_k}{k^2} < \infty$$

then

$$\frac{1}{n}\sum_{k=1}^{n}(X_k - EX_k) \xrightarrow{\text{a.s.}} 0 \qquad \text{as } n \to \infty. \quad \blacksquare$$

Lemma 2 below is completely natural intuitively. Lemma 3 says that if X_k and Y_k differ with small enough probability for most k, then either neither or both of their arithmetic averages converge almost surely to zero.

Lemma 2 (Sec. 20-3) Let $Z_1, Z_2, \ldots$ be a sequence of random variables such that $Z_n \xrightarrow{\text{a.s.}} 0$. If $c_1, c_2, \ldots$ is a sequence of real numbers converging to zero then $Z_n + c_n \xrightarrow{\text{a.s.}} 0$. ■

Lemma 3 (Sec. 20-3) Let $X_1, Y_1, X_2, Y_2, \ldots$ be a sequence of random variables satisfying

$$\sum_{k=1}^{\infty} P[X_k \neq Y_k] < \infty.$$

Let $S_k = X_1 + \cdots + X_k$ and $T_k = Y_1 + \cdots + Y_k$. If $(1/k)\,T_k \xrightarrow{\text{a.s.}} 0$ then $(1/k)S_k \xrightarrow{\text{a.s.}} 0$. ■

The next lemma is a special truncation result which, together with the preceding results, yields Theorem 3.

Lemma 4 (Sec. 20-3) If $E|X| < \infty$ then

$$\sum_{k=1}^{\infty} \frac{E[t_k(X)]^2}{k^2} \leq 2[1 + E\,|X|]. \quad ■$$

Theorem 3 (Sec. 20-3): The SLLN for independent identically distributed random variables having expectations Let $X_1, X_2, \ldots$ be a sequence of independent and identically distributed random variables satisfying $E\,|X_1| < \infty$. If $S_k = X_1 + \cdots + X_k$ then $(1/k)S_k \xrightarrow{\text{a.s.}} EX_1$. ■

CHAPTER 21: PROBABILITY SPACES: A RIGOROUS PRESENTATION

This chapter shows how Chap. 15 may be modified so as to agree with the theory which was systematized by Kolmogorov in 1933 and has been the basis for most advanced probability research since then. We develop the theory rigorously, through the introduction of expectation and then stop, since our goal is only to firmly bridge the gap to the advanced theory.

A class $\mathscr{A}$ of subsets of a nonempty set Ω is called a **σ field**, read "sigma field," of subsets of Ω iff it has the following properties:

1. φ belongs to $\mathscr{A}$, and Ω belongs to $\mathscr{A}$.
2. A^c belongs to $\mathscr{A}$ whenever A belongs to $\mathscr{A}$.
3. Both $\bigcup_{k=1}^{n} A_k$ and $\bigcap_{k=1}^{n} A_k$ belong to $\mathscr{A}$ whenever each of $A_1, \ldots, A_n$ belongs to $\mathscr{A}$.
4. Both $\bigcup_{k=1}^{\infty} A_k$ and $\bigcap_{k=1}^{\infty} A_k$ belong to $\mathscr{A}$ whenever each of $A_1, A_2, \ldots$ belongs to $\mathscr{A}$.

A **probability space** $(\Omega, \mathscr{A}, P)$ consists of a nonempty set Ω, a σ field $\mathscr{A}$ of subset of Ω, and a real-valued function P, called a **probability measure,** defined on $\mathscr{A}$ and satisfying the three conditions below. The set Ω is called the **sample space** and each member ω of Ω is called a **sample point.** Any subset A of Ω is called an **event** iff $A \in \mathscr{A}$, and in this case $P(A)$ is called the **probability** of A.

1. $P(\Omega) = 1$.
2. $P(A) \geq 0$ for every event A.
3. If $A_1, A_2, \ldots$ is any sequence of disjoint events then

$$P\left(\bigcup_{n=1}^{\infty} A_n\right) = \sum_{n=1}^{\infty} P(A_n).$$

Thus P is now defined only for events, and not for every subset. But our proofs used only operations which do not lead us outside the class of events, so the proofs still apply. In the discrete case it is possible, but not necessary, to let $\mathscr{A}$ be the class of all subsets of Ω.

In proving that a class of sets is a σ field it is often convenient to use the following lemma, which asserts that a class of sets is a σ field iff it contains the empty set, and is closed under complements and countable unions.

Lemma 1 (Sec. 21-1) Let $\mathscr{A}$ be a class of subsets of a nonempty set Ω. Suppose that $\varphi \in \mathscr{A}$, and $A^c \in \mathscr{A}$ whenever $A \in \mathscr{A}$. Also suppose that $\left(\bigcup_{k=1}^{\infty} A_k\right) \in \mathscr{A}$ whenever each of $A_1, A_2, \ldots \in \mathscr{A}$. Then $\mathscr{A}$ is a σ field. ∎

The elementary properties of P in Theorem 2 (Sec. 6-4) remain unchanged. So do the definitions in Chap. 7 of $P(B \mid A)$ and independent events, and their elementary properties. Theorem 1 (Sec. 15-1) is essentially restated below.

Theorem 1 (Sec. 21-1) Let $(\Omega,\mathscr{A},P)$ be a probability space and A an event with $P(A) > 0$. If the function P_A is defined for every event B by

$$P_A(B) = P(B \mid A) = \frac{P(AB)}{P(A)}$$

then $(\Omega,\mathscr{A},P_A)$ is a probability space, which is said to be obtained by **conditioning** $(\Omega,\mathscr{A},P)$ by A. ∎

The next lemma essentially defines the σ field generated by a class $\mathscr{C}$ of subsets to be the intersection of all σ fields containing $\mathscr{C}$. It is then shown that many reasonably rich collections of reasonable subsets of R^n generate the same important σ field of Borel sets.

Lemma 1 (Sec. 21-2) Let $\mathscr{C}$ be a nonempty class of subsets of a nonempty set Ω. Then there is a *unique* σ field $\sigma(\mathscr{C})$ of subsets of Ω, called the **smallest σ field containing** $\mathscr{C}$, or **the σ field generated by** $\mathscr{C}$, which satisfies the following conditions:
(*a*) $\mathscr{C} \subset \sigma(\mathscr{C})$; that is, C belongs to $\sigma(\mathscr{C})$ whenever C belongs to $\mathscr{C}$.
(*b*) If $\mathscr{B}$ is any σ field of subsets of Ω such that $\mathscr{C} \subset \mathscr{B}$ then $\sigma(\mathscr{C}) \subset \mathscr{B}$. ∎

A set A in R^n is called an **open set** in R^n iff for every $x \in A$ there is some open rectangle in R^n such that x belongs to the open rectangle, and the open rectangle is contained in A. The rectangle definitions are made in Sec. 15-1.

Lemma 2 (Sec. 21-2) Let each $\mathscr{C}_i$ be a class of subsets of R^n, defined as follows:

$\mathscr{C}_1 =$ the class of all closed rectangles
$\mathscr{C}_2 =$ the class of all open rectangles
$\mathscr{C}_3 =$ the class of all open sets
$\mathscr{C}_4 =$ the class of all sets of the form

$$\{(x_1, \ldots, x_n): x_i < b_i \text{ for all } i = 1, \ldots, n\}$$

where $b_1, \ldots, b_n$ can be any real numbers.

Then $\sigma(\mathscr{C}_1) = \sigma(\mathscr{C}_2) = \sigma(\mathscr{C}_3) = \sigma(\mathscr{C}_4)$. This σ field is denoted by $\mathscr{B}_n$ and is called the **σ field of Borel sets** in R^n. A subset of R^n is called a **Borel set** iff it belongs to $\mathscr{B}_n$. ■

It can be proved that there exist subsets of R^n which are not Borel sets, but none has ever been explicitly exhibited.

Lemma 3 shows that if two probability measures agree on a collection $\mathscr{C}$ of events which is closed under finite intersections then the two probability measures must also agree on the σ field generated by $\mathscr{C}$. Theorem 1 is an immediate consequence of Lemma 3, and it shows in particular that if two probability measures agree on all closed rectangles in R^n then they must agree on $\mathscr{B}_n$. Thus if we assign a probability to each closed rectangle in R^n then there is at most one way to extend to a probability measure on all Borel sets. That is, the probabilities for closed rectangles determine uniquely the probability of every complicated Borel set.

Lemma 3 (Sec. 21-2) Let $(\Omega,\mathscr{A},P)$ and $(\Omega,\mathscr{A},Q)$ be probability spaces such that

$$P(C) = Q(C) \qquad \text{whenever } C \in \mathscr{C}$$

where $\mathscr{C}$ is some collection of events. Suppose that

$$(C \cap C') \in \mathscr{C} \qquad \text{whenever } C \in \mathscr{C},\ C' \in \mathscr{C}.$$

Then $P(B) = Q(B)$ whenever $B \in \sigma(\mathscr{C})$. ■

Let $(R^n,\mathscr{B}_n,P)$ be a probability space, where $\mathscr{B}_n$ is the σ field of Borel sets. As in Chap. 17, the **distribution function** F: $R^n \to R$ corresponding to $(R^n,\mathscr{B}_n,P)$ is defined by

$$F(x_1, \ldots, x_n) = P\{(x_1', \ldots, x_n'): x_1' < x_1, \ldots, x_n' < x_n\}$$

for all reals $x_1, \ldots, x_n$.

Theorem 1 (Sec. 21-2) Let $(R^n,\mathscr{B}_n,P)$ and $(R^n,\mathscr{B}_n,Q)$ be probability spaces. Let $\mathscr{C}$ be any one of the four classes $\mathscr{C}_1$, $\mathscr{C}_2$, $\mathscr{C}_3$ and $\mathscr{C}_4$ defined in Lemma 2. If

$$P(C) = Q(C) \qquad \text{whenever } C \in \mathscr{C}$$

then $P = Q$. In particular if both probability spaces have the same distribution function then $P = Q$. ■

We now state a standard result whose proof can be found in advanced texts on probability theory or measure theory. If F: $R^n \to R$ is a distribution function on R^n, as defined in Sec 17-3, then there is a probability space $(R^n,\mathscr{B}_n,P)$ such that F is the distribution function corresponding to $(R^n,\mathscr{B}_n,P)$. This fact, together with Theorem 1

(Sec. 21-2), and Sec. 17-3, shows that the correspondence which to each $(R^n,\mathscr{B}_n,P)$ assigns its corresponding distribution function F on R^n is indeed a one-to-one correspondence.

A **random vector** f with associated probability space $(\Omega,\mathscr{A},P)$ is any function $f\colon \Omega \to R^n$ such that $[f \in B]$ is an event whenever B is a Borel set in R^n, that is, iff $f^{-1}(B) \in \mathscr{A}$ whenever $B \in \mathscr{B}_n$. If $n = 1$ then f may be called a **random variable.** Thus we apply the name "random variable" to a real-valued function defined on the sample space iff it has the desired property that for every Borel set B in R the set $f^{-1}(B) = \{\omega : f(\omega) \in B\}$ is an event, and so has a probability assigned to it. We have the following version of Theorem 1 (Sec. 15-2) on induced distributions.

Theorem 1 (Sec. 21-3) If $f\colon \Omega \to R^n$ is a random vector with associated probability space $(\Omega,\mathscr{A},P)$ and if P_f is defined for every Borel set B in R^n by $P_f(B) = P[f \in B]$ then $(R^n,\mathscr{B}_n,P_f)$ is a probability space, which is said to be **induced** by f with associated $(\Omega,\mathscr{A},P)$. ∎

Let $(\Omega,\mathscr{A},P_A)$ be obtained by conditioning $(\Omega,\mathscr{A},P)$ by the event A with $P(A) > 0$. Then the random vector $f\colon \Omega \to R^n$ with associated $(\Omega,\mathscr{A},P_A)$ induces a probability space $(R^n,\mathscr{B}_n,P_{f|A})$ called the **conditional distribution of** f **given** A. Therefore $P_{f|A}$ is defined by

$$P_{f|A}(B) = P_A[f \in B] = P[f \in B \mid A] = \frac{P([f \in B] \cap A)}{P(A)} \qquad \text{for every } B \in \mathscr{B}_n.$$

A function $g\colon R^n \to R^m$ is said to be a **Borel function** iff

$$g^{-1}(B) \in \mathscr{B}_n \qquad \text{whenever } B \in \mathscr{B}_m.$$

That is, iff inverse images of Borel sets are always Borel sets. The random vectors on a probability space $(R^n,\mathscr{B}_n,P)$ are just the Borel functions.

It can be proved that there exist functions $f\colon R^n \to R^m$ which are not Borel functions, but none has ever been explicitly exhibited.

If $f\colon \Omega \to R^n$ is a random vector with associated $(\Omega,\mathscr{A},P)$ and if $g\colon R^n \to R^m$ is a Borel function then $h = g(f)$ is a random vector, with associated $(\Omega,\mathscr{A},P)$. *Thus Borel functions of random vectors are random vectors*, and hence the reason for considering Borel functions is obvious. We have the following obvious version of Theorem 2 (Sec. 15-2).

Theorem 2 (Sec. 21-3): Intermediate distributions If f is a random vector with associated probability space $(\Omega,\mathscr{A},P)$ and if

$$h = g(f)$$

$$\Omega \xrightarrow{\ f\ } R^n \xrightarrow{\ g\ } R^m$$

where g is a Borel function, then $P_h = P_g$, where P_g is induced by g with associated $(R^n,\mathscr{B}_n,P_f)$. ∎

If $(\Omega,\mathscr{A},P)$ is a probability space and $f: \Omega \to R^n$ is a function then by definition f is a random vector iff $f^{-1}(B) \in \mathscr{A}$ for every $B \in \mathscr{B}_n$. This definition does not involve P. Similarly the definition of a Borel function did not involve a P. We introduce the following definition in order to handle both cases simultaneously.

Let $\mathscr{A}$ be a σ field of subsets of a set Ω. A function $f: \Omega \to R^n$ is said to be $\mathscr{A}$ **measurable** iff $f^{-1}(B) \in \mathscr{A}$ whenever $B \in \mathscr{B}_n$, that is, iff all inverse images of Borel sets in R^n belong to $\mathscr{A}$. Then $f: \Omega \to R^n$ is a random vector with associated probability space $(\Omega,\mathscr{A},P)$ iff f is $\mathscr{A}$ measurable. Also $g: R^n \to R^m$ is a Borel function iff g is $\mathscr{B}_n$ measurable.

Lemma 1 below says that if $\mathscr{A}$ is a σ field of subsets of Ω then $f: \Omega \to R^n$ is $\mathscr{A}$ measurable iff each of its real-valued components f_i in $f = (f_i, \ldots, f_n)$ is. Also to check for $\mathscr{A}$ measurability we need only take inverse images of closed rectangles, or open rectangles, etc. Furthermore the class of all $\mathscr{A}$-measurable functions into R^n has the fundamental property that it is closed under limits. The class of continuous functions, for example, does not have this property.

Lemma 1 (Sec. 21-3) Let $\mathscr{A}$ be a σ field of subsets of Ω.

(*a*) A function $f: \Omega \to R^n$ is $\mathscr{A}$ measurable iff each of its components $f_i: \Omega \to R$ is $\mathscr{A}$ measurable, where, as usual, $f = (f_1, \ldots, f_n)$.

(*b*) Let $\mathscr{C}$ be any one of the four classes $\mathscr{C}_1$, $\mathscr{C}_2$, $\mathscr{C}_3$, and $\mathscr{C}_4$ defined in Lemma 2 (Sec. 21-2). If a function $f: \Omega \to R^n$ satisfies

$$f^{-1}(C) \in \mathscr{A} \qquad \text{whenever } C \in \mathscr{C}$$

then f is $\mathscr{A}$ measurable.

(*c*) Suppose that each $f_k: \Omega \to R^n$ in the sequence $f_1, f_2, \ldots$ is $\mathscr{A}$ measurable. Suppose also that the sequence has a pointwise limit $f: \Omega \to R^n$; that is,

$$\lim_{k\to\infty} f_k(\omega) = f(\omega) \qquad \text{for all } \omega \in \Omega.$$

Then f is $\mathscr{A}$ measurable. ■

Lemma 1*a* shows that a function $X: \Omega \to R^n$ is a random vector with associated probability space $(\Omega,\mathscr{A},P)$ iff each of its components X_i in $X = (X_1, \ldots, X_n)$ is a random variable with associated $(\Omega,\mathscr{A},P)$. Lemma 1*b* shows that X is a random variable iff the set $[X < x]$ is an event, for every real x, that is, iff its distribution function $F_X: R \to R$ is defined. Lemma 1*c* shows that any pointwise limit of random vectors is also a random vector. Thus the collection of random variables, on a fixed probability space $(\Omega,\mathscr{A},P)$, is closed under pointwise limits.

Lemma 1*a* shows that $f: R^n \to R^m$ is a Borel function iff each of its real-valued components f_i in $f = (f_1, \ldots, f_m)$ is a Borel function. Lemma 1*b* for open sets $\mathscr{C}_3$ shows that every continuous function $f: R^n \to R^m$ is a Borel function. Lemma 1*c* then shows that limits of continuous functions, and limits of such limits, etc., are also Borel functions. Thus all functions $f: R^n \to R^m$ normally encountered in analysis, including very discontinuous ones, are Borel functions.

If D_i is a Borel subset of R^{n_i} then $D_1 \times \cdots \times D_k$ is a Borel subset of $R^{n_1+\cdots+n_k}$. We have the following version of Theorem 3 (Sec. 15-2).

Theorem 3 (Sec. 21-3) If $f: \Omega \to R^n$ and $g: \Omega \to R^m$ are random vectors with the same associated probability space $(\Omega,\mathscr{A},P)$ then

$$P_f(B) = P_{f,g}(B \times R^m) \qquad \text{for every Borel set } B \in R^n. \quad \blacksquare$$

Random vectors $f_1, \ldots, f_n$ with associated probability space $(\Omega,\mathscr{A},P)$ are said to be **independent** iff

$$P[f_1 \in B_1, \ldots, f_n \in B_n] = P[f_1 \in B_1]P[f_2 \in B_2] \cdots P[f_n \in B_n]$$

whenever the B_i are Borel subsets in R^{k_i} where $f_i: \Omega \to R^{k_i}$. The immediate consequences remain as in Sec. 15-3. In particular we have the following version of Theorem 1 (Sec. 15-3).

Theorem 4 (Sec. 21-3) If $f_1, \ldots, f_d$ are $d \geq 2$ independent random vectors with the same associated probability space $(\Omega,\mathscr{A},P)$, and if

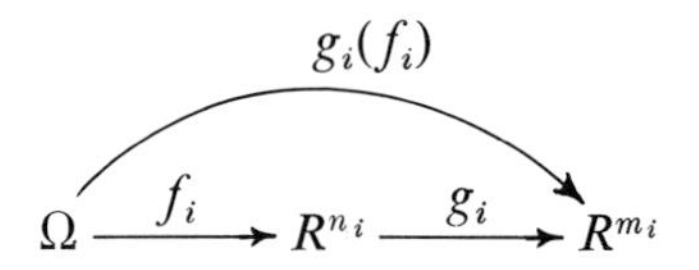

where $g_1, \ldots, g_d$ are Borel functions, then $g_1(f_1), \ldots, g_d(f_d)$ are d independent random vectors. $\blacksquare$

As in Sec. 15-3, the independence of $f_1, \ldots, f_m, \ldots, f_n$ implies the independence of $(f_1, \ldots, f_m)$ and $(f_{m+1}, \ldots, f_n)$. Lemma 1 (Sec. 15-3) can be generalized as below to show that if A is independent of every event in a class $\mathscr{C}$, closed under finite intersections, then A is independent of every event in $\sigma(\mathscr{C})$.

Lemma 2 (Sec. 21-3) Let A be an event and let $\mathscr{C}$ be a class of events in a probability space $(\Omega,\mathscr{A},P)$. Suppose that $(C \cap C') \in \mathscr{C}$ whenever $C \in \mathscr{C}$ and $C' \in \mathscr{C}$. Also suppose that A and C are two independent events whenever $C \in \mathscr{C}$. Then A and C are two independent events whenever $C \in \sigma(\mathscr{C})$. $\blacksquare$

If $(R,\mathscr{B}_1,P)$ is a probability space whose sample space is the real line and whose events are the Borel sets then the **mean** of P, as well as whether or not P has a mean, is defined in Theorem 1 (Sec. 15-4). A random variable X with associated probability space $(\Omega,\mathscr{A},P)$ induces a probability space $(R,\mathscr{B}_1,P_X)$. The **expectation** EX of a random variable X with associated probability space $(\Omega,\mathscr{A},P)$ is defined to be the mean of P_X, if P_X has a mean.

We immediately obtain the following version of Theorem 2 (Sec. 15-4).

Theorem 1 (Sec. 21-4) If f is a random vector with associated probability space $(\Omega,\mathscr{A},P)$ and if

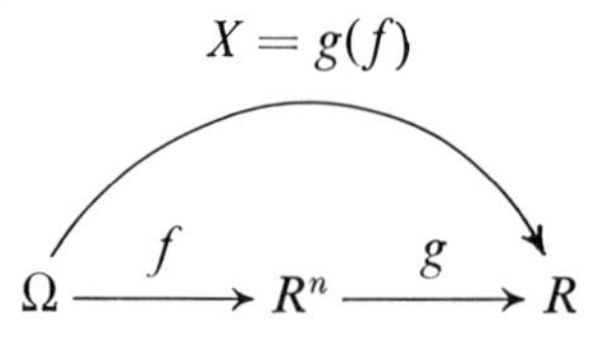

where g is a Borel function, then $EX = Eg$. That is, X with associated $(\Omega,\mathscr{A},P)$ and g with associated $(R^n,\mathscr{B}_n,P_f)$ have the same expectation, in the sense that if either expectation exists then both exist and are equal. ∎

The Addendum to Sec. 15-4 shows that expectations can be (and usually are) equivalently defined in terms of discrete approximations to X as follows. Let X be a random variable with associated probability space $(\Omega,\mathscr{A},P)$. For each n define the finitely-many valued random variable X_n by

$$X_n(\omega) = \begin{cases} 0 & \text{if } |X(\omega)| \geq n \\ \dfrac{j}{2^n} & \text{if } \dfrac{j}{2^n} \leq X(\omega) < \dfrac{j+1}{2^n} \\ -\dfrac{j}{2^n} & \text{if } -\dfrac{j+1}{2^n} < X(\omega) \leq -\dfrac{j}{2^n} \end{cases}$$

where $j = 0, 1, \ldots, M_n = n2^n - 1$. Let EX_n be defined as in the discrete case, or equivalently by $EX_n = A_n^+ + A_n^-$ where

$$A_n^+ = \sum_{j=1}^{M_n} \frac{j}{2^n} P\left[\frac{j}{2^n} \leq X < \frac{j+1}{2^n}\right]$$

$$A_n^- = \sum_{j=1}^{M_n} \frac{-j}{2^n} P\left[-\frac{j+1}{2^n} < X \leq -\frac{j}{2^n}\right].$$

Then $E|X_n| = A_n^+ - A_n^-$ is nondecreasing as n increases, and it converges to a finite limit iff

$$\sum_{k=1}^{\infty} kP[k-1 \leq |X| < k]$$

has a finite sum B. We say that X *has an expectation* iff this is the case. Then X has an expectation iff $|X|$ does, in which case EX_n has a limit which we define to be *the expectation of* X,

$$EX = \lim_{n\to\infty} EX_n$$

and we have $|EX| \leq E|X| \leq B$.

The usual properties of expectation can be proved within this rigorous framework, in particular the properties in Theorem 3 (Sec. 11-3), the Chebychev inequality of Theorem 3 (Sec. 15-4), and the MGF Theorem 4 (Sec. 15-4).

CHAPTER 22: INTRODUCTION TO STOCHASTIC PROCESSES

A **stochastic process** consists of a nonempty set T, called the *parameter set*, and a probability space (Ω,P), and the assignment of a random variable $X_t\colon \Omega \to R$ to each t in T. Thus a stochastic process is an indexed collection of random variables, X_t for $t \in T$, all having the same associated probability space. If T contains only a finite number of members, say the integers $1, 2, \ldots, n$, then the stochastic process is a random vector $(X_1, \ldots, X_n)$, or a random variable if $n = 1$. We call the stochastic process, X_t for $t \in T$, a **discrete-parameter process** iff T is some countable set of real numbers. In this case T is usually some set of integers, but it is sometimes useful to let T be the set of all rational numbers. A **continuous-parameter stochastic process** is one for which the parameter set T is some interval of real numbers. By an **interval** we mean either a finite or an infinite interval of real numbers, as defined below, where any $\leq$ may be replaced by $<$. A **finite interval** with end points $a < b$ is the set of x satisfying $a < x < b$. An **infinite interval** is either the set of all real numbers, or the set of x satisfying $x < a$, or the set of x satisfying $a < x$. Thus an interval has 0, 1, or 2 *end points*.

Henceforth by the term **stochastic process** *we always mean a discrete- or continuous-parameter stochastic process, unless explicitly indicated otherwise. Thus the parameter set must be a countable set of real numbers, or an interval as defined above.*

If X_t for $t \in T$ is a stochastic process with associated (Ω, P) then for any ω in Ω we define the function $r_\omega: T \rightarrow R$ by

$$r_\omega(t) = X_t(\omega) \qquad \text{for all } t \in T,$$

and we call r_ω a **realization** or **sample function** of the stochastic process.

We next consider, as in Sec. 22-2, a standard method for defining stochastic processes. Let T be a countable set of real numbers, or an interval of real numbers. By a **family of finite distributions with parameter set** T we mean that for every positive integer n and for every n points $t_1 < t_2 < \cdots < t_n$ in T there has been assigned a probability space $(R^n, P_{t_1, \ldots, t_n})$ called an n**th-order distribution.** The *finite* refers to the joint distributions of *finitely many* random variables. The family is said to be **consistent** iff for any one of the probability spaces for $n \geq 2$ in the family, if we consider random variables $X(t_1), \ldots, X(t_n)$ having this joint distribution, and from this joint distribution we obtain the joint distribution of some $n - 1$ of the random variables $X(t_1), \ldots, X(t_{i-1}), X(t_{i+1}), \ldots, X(t_n)$, as usual by use of Theorem 3 (Sec. 15-2), then this joint distribution on R^{n-1} is the same as the one assigned to $t_1, \ldots, t_{i-1}, t_{i+1}, \ldots, t_n$ by the given family. We assume $t_1 < t_2$, since otherwise the family would contain both P_{t_1, t_2} and P_{t_2, t_1} which would then be required to be related in an obvious fashion.

A stochastic process $X(t)$ for $t \in T$ **generates** an obvious **family of finite distributions** with parameter set T; the distribution assigned to $t_1 < \cdots < t_n$ is the distribution induced by the random vector $(X(t_1), \ldots, X(t_n))$. The family is consistent because all the random variables have the same associated probability space. Conversely, when proper attention is devoted to technicalities the following condition can be precisely stated and proved.

Kolmogorov consistency condition Let T be any interval or countable set of real numbers. For any consistent family of finite distributions with parameter set T there exists a stochastic process, X_t for $t \in T$, which generates the family. Furthermore the family uniquely determines the probability of any event which is definable in terms of countably many of the X_t random variables. ∎

The stochastic process guaranteed by the consistency condition is normally constructed by letting the sample space Ω consist of all possible realizations, and using the consistent family to define probabilities. Section 22-3 describes this construction in greater detail. For most stochastic processes which we consider, the consistent nature of the definitions is intuitively apparent and fairly easy to prove. Our definitions of explicit processes are essentially equivalent to definitions of consistent families.

Three stochastic processes, $X(t)$ for $t \in T$, $Y(s)$ for $s \in S$, $Z(h)$ for $h \in H$, all having the same associated probability space are said to be **independent** iff

$$(X(t_1), \ldots, X(t_m)) \qquad (Y(s_1), \ldots, Y(s_n)) \qquad (Z(h_1), \ldots, Z(h_r))$$

are three independent random vectors for every three finite sets of parameter points. Similarly for any finite collection of stochastic processes. The interpretation is that completely independent experiments yield the realizations for each process.

A stochastic process X_t for $t \in T$ is said to have **stationary** or **homogeneous increments** iff $X_t - X_s$ and $X_{t'} - X_{s'}$ are identically distributed whenever $s < t$ and $s' < t'$ all belong to T and $t - s = t' - s'$. That is, the distribution of the *increment random variable* $X_t - X_s$ depends on s and t only through their difference $t - s$. If X_t for $t \in T$ has stationary increments then so does the process X_t for $t \in T_1$, for any subset T_1 of T.

A stochastic process $X(t)$ for $t \in T$ is said to have **independent increments** iff for every $n \geq 3$ points $t_1 < t_2 < \cdots < t_n$ in T the increment random variables

$$X(t_2) - X(t_1), X(t_3) - X(t_2), \ldots, X(t_n) - X(t_{n-1})$$

are independent. In general the independence applies to any finite number of not necessarily adjacent finite intervals which are disjoint, except possibly for end points. If X_t for $t \in T$ has independent increments then so does X_t for $t \in T_1$, for any subset T_1 of T.

Naturally a **stationary-independent-increment process** is one whose increments are both stationary and independent.

Let $X(t)$ for $t \geq 0$ be a stationary-independent-increment process satisfying $X(0) = 0$. Fix real numbers $b > 0$, τ_0, c, d, e and for each real $\tau \geq \tau_0$ define a random variable

$$Y_\tau = d + e\tau + cX(b(\tau - \tau_0)).$$

Then the process Y_τ for $\tau \geq \tau_0$ starts at time τ_0 with the constant value $Y_{\tau_0} = d + e\tau_0$ and the process is easily seen to have stationary independent increments.

If a sequence $0 = S_0, S_1, S_2, \ldots$ of random variables has stationary independent increments then from the expression

$$S_n = [S_n - S_{n-1}] + [S_{n-1} - S_{n-2}] + \cdots + [S_1 - S_0]$$

it is clear that we just have the sequence of partial sums of independent random variables each having the distribution of $S_1 - S_0 = S_1$. Each distribution for S_1 essentially defines such a sequence.

If "X_t for $t \geq 0$ with $X_0 = 0$" is a stationary-independent-increment process, and $\tau > 0$, then $0 = X_0, X_\tau, X_{2\tau}, X_{3\tau}, \ldots$ is clearly a stationary-independent-increment sequence. This fact should make the structure of such processes intuitively apparent. To *define* such a process we do not have a first random variable, like S_1 above, so instead we use the whole collection of desired first-order distributions. These distributions must at least be a semigroup, and as stated in the next theorem this is all that is required.

A family of probability spaces (R,P_t), one for each real $t > 0$, is called a **semigroup** iff for every $s > 0$ and $t > 0$ the given P_{s+t} equals the distribution of the sum of two independent random variables having distributions P_s and P_t.

Theorem I (Sec. 22-2) Let X_t for $t \geq 0$ be a stationary-independent-increment continuous parameter stochastic process satisfying $X_0 = 0$. Then the distributions P_{X_t} for $t > 0$ form a semigroup, which we say is **generated** by the process. Furthermore for any semigroup there exists such a process which generates this semigroup. Also two such processes generate the same semigroup iff they generate the same family of finite distributions. In addition if $E(X_t)^2$ is bounded on $0 \leq t \leq 1$ then

$$EX_t = tEX_1 \qquad \text{Var } X_t = t \text{ Var } X_1 \qquad \text{for all } t \geq 0. \quad \blacksquare$$

Chebychev's inequality applied to a process as in the above theorem yields

$$P[|X_t - tEX_1| < \delta\sqrt{t \text{ Var } X_1}] \geq 1 - \frac{1}{\delta^2} \qquad \text{for every } t > 0,\ \delta > 0.$$

For this reason we sometimes plot the mean tEX_1 as a function of t, and the two curves $tEX_1 \pm \sqrt{t \text{ Var } X_1}$ one standard deviation on either side of the mean.

A stochastic process $X(t)$ for $t \in T$ is said to be **stationary** iff for every $n \geq 1$ the two random vectors

$$(X(t_1), \ldots, X(t_n)) \qquad (X(t_1 + s), \ldots, X(t_n + s))$$

are identically distributed whenever $t_1 < \cdots < t_n$ and $t_1 + s < \cdots < t_n + s$ are in T. That is, joint distributions are invariant to translation. For a stationary process the distribution of $(X(s + h), X(s))$, and hence of $X(s + h) - X(s)$, does not depend on s. Thus a stationary process has stationary increments, but the converse is not true. We will find it possible to transform the Poisson and Weiner processes into stationary processes.

A stochastic process $X(t)$ for $t \in T$ is called a **Markov process** iff for any $n \geq 3$ points $t_1 < \cdots < t_n$ in T and reals $x_1, \ldots, x_n$

$$P[X(t_n) < x_n \mid X(t_{n-1}) = x_{n-1}, \ldots, X(t_2) = x_2, X(t_1) = x_1] = P[X(t_n) < x_n \mid X(t_{n-1}) = x_{n-1}].$$

Such a process is said to have the *Markov property* above. For such a process the future behavior does not depend on which history led to the present condition $X(t_{n-1}) = x_{n-1}$. If we have only partial knowledge of the present condition, say $[X(t_{n-1}) < 3]$, we may not be able to ignore earlier observations, which might give additional information concerning the present condition. If X_t for $t \in T$ is Markov and T' is a subset of T then X_t for $t \in T'$ is certainly Markov.

Let $X(t)$ for $t \in T$ be a process such that each $X(t)$ induces a discrete probability density. Clearly it is a Markov process iff

$$P[X(t_n) = x_n \mid X(t_{n-1}) = x_{n-1}, \ldots, X(t_1) = x_1] \\ = P[X(t_n) = x_n \mid X(t_{n-1}) = x_{n-1}]$$

whenever the first conditioning event has positive probability. Thus in this case we need not use distribution functions, and we need only condition on events having positive probability. In general, however, the Markov property requires conditioning on events having zero probability, and this can be very delicate,

If $X(t)$ for $t \geq 0$ satisfies $X(0) = 0$ and has independent increments then it is Markov, because $X(t_1), \ldots, X(t_n)$ are just consecutive sums of independent random variables.

A stationary-independent-increment process, X_t for $t \geq 0$ with $X_0 = 0$, is a special kind of Markov process; namely, if $0 < s < t$ then the conditional distribution of $X_t = X_s + [X_t - X_s]$, given $[X_s = x]$, is just the distribution induced by $x + X_{t-s}$. More intuitively if the event $[X_s = x]$ occurs then the future increment process, $X_t - X_s$ for $t \geq s$, starts at the new origin (s,x) by precisely the same probability law by which the original process started at $(0,0)$.

As in Sec. 22-3, we next describe the construction of a stochastic process from a given consistent family. If T is a nonempty set let R^T denote the **collection of all real-valued functions defined on** T; that is, an object f belongs to R^T iff it is a function $f\colon T \to R$. A **tube,** or a tube of functions, in R^T consists of those functions $f\colon T \to R$ satisfying

$$a_i \leq f(t_i) \leq b_i \qquad \text{for } i = 1, \ldots, n.$$

(A *tube*, which is not standard terminology, is a special case of a *product cylinder in* R^T, which is standard terminology.) Thus a tube is defined in terms of $n \geq 1$ members $t_1 < t_2 < \cdots < t_n$ of T and $2n$ reals $a_i \leq b_i$.

If $X(t)$ for $t \in T$ is a stochastic process then, as exhibited below, the event

$$[a_1 \leq X(t_1) \leq b_1, \ldots, a_n \leq X(t_n) \leq b_n] \\ = \{\omega\colon a_1 \leq r_\omega(t_1) \leq b_1, \ldots, a_n \leq r_\omega(t_n) \leq b_n\}.$$

can be described, in terms of realizations, as consisting of those sample points whose realizations belong to the corresponding tube in R^T. Clearly, then, two stochastic processes generate the same family of finite distributions iff both processes assign the same probability to every tube in R^T.

Assume that we are given a consistent family with parameter set T. To construct a stochastic process generating the family we let the sample space be $\Omega = R^T$. Then for any t in T define the function $X_t\colon \Omega \to R$ as follows: the value assigned by X_t to an f in Ω is just $f(t)$; that is, if realization $f\colon T \to R$ is obtained then the value of X_t is just the value $f(t)$ of this realization at t. Then to the tube $[a_1 \leq X_{t_1} \leq b_1, \ldots, a_n \leq X_{t_n} \leq b_n]$ in R^T we assign the probability of the corresponding closed rectangle in R^n, as specified by $P_{t_1,\ldots,t_n}$ from the consistent family. Thus the family assigns consistent probabilities to tubes in $\Omega = R^T$, and in such a way that any random vector $(X_{t_1}, \ldots, X_{t_n})$ has the distribution specified by the family.

We then assign probabilities to complicated events by taking the limit of appropriate probabilities calculated from approximations using tubes, etc. The proof of the Kolmogorov consistency condition shows that this program can be carried out for a large collection of complicated events, and the resulting assignment is indeed a probability measure. The result is stated rigorously in the Addendum to Sec. 22-3, in the terminology of Chap. 21.

For discrete-parameter processes the family uniquely determines all probabilities of interest. This is *not* the case for continuous-parameter processes. In particular we call X_t for $t \in T$ a **zero process** iff $P[X_t = 0] = 1$ for all $t \in T$. Clearly all zero processes with the same parameter set T generate the same family. However, Example 1 (Sec. 22-3) exhibits many different zero processes on $t \geq 0$, and they do not all assign the same probabilities to events of interest. In particular a Poisson process has a natural representation as a zero process. These difficulties with continuous-parameter processes can largely be eliminated if we restrict ourselves to processes all of whose realizations are regular functions.

A function $f\colon T \to R$ on an interval T is said to be **regular** iff it has at most finitely many points of discontinuity in every finite interval; furthermore if t is a point of discontinuity and t is not an end point of T then f is required to have left- and right-hand limits at t, and $f(t)$ must be equal to one of these limits. Also, f must have a one-sided limit at any end point. The restriction of such an f to a subinterval is also regular.

If X_t for $t \in T$ is a continuous-parameter process and if ω is a sample point then the realization r_ω may be a regular function on T, in which case we call it a **regular realization.** If every realization is regular then we say that the process **has only regular realizations.** Clearly if X_t for $t \in T$ is such a process then so is X_t for $t \in T_1$, for any subinterval T_1 of T, so the following theorem may be applied to subintervals.

The following theorem and its proof are meant only to illustrate the way in which a regularity assumption can yield uniqueness.

Theorem 1 (Sec. 22-3) If two continuous-parameter stochastic processes generate the same family of finite distributions with parameter set T and *have only regular realizations* then both processes assign the same probability to every event of one of the forms

$$[f(t) \leq X_t \text{ for all } t \in T] \qquad [X_t \leq g(t) \text{ for all } t \in T]$$

$$[f(t) \leq X_t \leq g(t) \text{ for all } t \in T]$$

where $f\colon T \to R$ and $g\colon T \to R$ are continuous functions on T. ■

Unless explicitly indicated otherwise, any continuous-parameter stochastic process which we consider is assumed to have only regular realizations. For such processes the family assigns a unique probability to events like those in the preceding theorem, and to many other events of interest. We often assume the uniqueness without any attempt at justification.

Doob's concept of *separability* is a generalization of the foregoing. A process having only regular realizations is necessarily a separable process. The preceding theorem can be extended to this more general class of processes; that is, "have only regular realizations" may be replaced by "are separable."

CHAPTER 23: THE POISSON PROCESS AND SOME OF ITS GENERALIZATIONS

Almost all the processes considered in this chapter are continuous-parameter processes with $t \geq 0$ for parameter set and with all realizations jump functions as shown in Fig. 3 (Sec. 23-1), so that there is a jump of size y_i at arrival time t_i and $x_i = t_i - t_{i-1}$ is the *ith* interarrival time.

For a pure-birth process, such as a Poisson process, every $y_i = 1$, so that all jumps are of the same size, and hence the process essentially just generates the arrival times $0 < t_1 < t_2 < \cdots$. Selecting a realization from a Poisson process is sometimes described as selecting an infinite sequence of points purely at random on $t \geq 0$. Theorem 1 (Sec. 23-2) shows that *only* the Poisson process can have its qualitative characteristics. Poisson processes are also characterized by the fact that the interarrival times are independent and all have the same exponential density. Exponential densities permeate this chapter, essentially because of their unique lack of memory, as shown in Example 4 (Sec. 16-1). The Poisson process is also related to the uniform continuous densities.

Compound Poisson processes are constructed by selecting arrival times t_i by a Poisson process, while the arbitrary jump sizes y_i are selected independently from a fixed distribution.

For nonhomogeneous Poisson processes all $y_i = 1$, but now the distribution of the number of arrivals in a 1-hour period may be different for different 1-hour periods.

For the Markov pure-birth processes the distribution of the waiting time for the next birth is determined by the present population size, but is otherwise unaffected by the time required to attain this size. The Markov property follows from the exponential distributions of the independent interarrival times, which are identically distributed only for a Poisson process. For the Markov birth-and-death processes not only the next interarrival time, but also the chance for its being a birth $y_i = 1$, rather than a death $y_i = -1$, may depend on the present population size.

To define the realizations of interest we assume that we have two infinite sequences of real numbers, $x_1, x_2, \ldots$ with each $x_i > 0$ and $y_1, y_2, \ldots$ with each $y_i \neq 0$. We call $x_1, x_2, \ldots$ the sequence of successive **interarrival times.** Each of these sequences determines its corresponding sequence of partial sums, and vice versa, where

$$t_n = x_1 + \cdots + x_n \qquad \text{and} \qquad z_n = y_1 + \cdots + y_n.$$

We call $0 < t_1 < t_2 < \cdots$ the sequence of successive **arrival times,** or jump points, and we require t_n to go to infinity as n does. Then we define f for all $t \geq 0$ by

$$f(t) = \begin{cases} 0 & \text{if } 0 \leq t < t_1 \\ z_n & \text{if } t_n \leq t < t_{n+1}. \end{cases}$$

Thus $f(0) = 0$, and f is continuous from the right, jumps by amount y_i at t_i, and is constant between jump points. Naturally if we start with only finitely many jump points $t_1, \ldots, t_N$ we let $f(t) = z_N$ if $t \geq t_N$. We call a real-valued function f on $t \geq 0$ a **jump function** iff it is constructable as above with either a finite or infinite sequence of jump points, or is identically zero (no jumps). A **counting function** on $t \geq 0$ is any jump function for which every y_i is a positive interger. A **unit counting function** is any jump function for which every $y_i = 1$. The identically zero function on $t \geq 0$ is assumed to belong to all three classes of functions.

A **Poisson process**, N_t for $t \geq 0$ with $N_0 = 0$ and parameter λ, is any stationary-independent-increment process such that for each $t > 0$ the random variable N_t induces a Poisson density with mean λt, and every realization for the process is a counting function. The parameter λ, which may be any positive real number, is called the **intensity** or **mean rate.** Thus if $0 \leq s < t$ then

$$P[N_t - N_s = k] = \frac{e^{-\lambda(t-s)}[\lambda(t-s)]^k}{k!} \qquad \text{for } k = 0, 1, \ldots .$$

The use of $P[N_t - N_s = 0] = e^{-\lambda(t-s)}$ is especially frequent. In Sec. 23-2 it is shown that every realization is a *unit* counting function having *infinitely* many jump points. Also for every time t there is zero probability for a jump at t. The simplest way to show that Poisson processes exist is to directly construct them as in Characterization 1 (Sec. 23-3).

If $N(t)$ for $t \geq 0$ is a Poisson process with mean rate 1 then $N(\lambda\tau)$ for $\tau \geq 0$ is easily seen to be a Poisson process with mean rate $\lambda > 0$. Thus the parameter λ corresponds to a scale change in time.

The Poisson λ process, as defined above, is motivated in Sec. 23-1 as a "limit experiment" if each experiment consists of infinitely many Bernoulli trials at equispaced time points, and if the number of Bernoulli trials per unit interval goes to infinity while the expected number of successes per unit interval is kept constant at λ. These approximating experiments suggest the intuitive nature of a Poisson process, and they also suggest Characterizations 1 and 2 (Sec. 23-3).

If $P[X_0 = 1] = P[X_0 = -1] = .5$, where X_0 is independent of the Poisson process N_t for $t \geq 0$ then

$$X_t = X_0(-1)^{N_t} \qquad \text{for all } t \geq 0$$

is a stationary process, which is sometimes called a *random telegraph signal.*

It is easily seen that for a Poisson process both

$$\frac{1}{\lambda t} P[N_{s+t} - N_s = 1] \qquad \text{and} \qquad P[N_{s+t} - N_s = 1 \mid N_{s+t} - N_s \geq 1]$$

converge to 1 as t converges to zero, and both have obvious interpretations.

The following theorem shows that among the many stationary-independent-increment counting-function processes, such as the compound Poisson processes of Sec. 23-4, only the Poisson processes have only unit jumps and hence satisfy the following qualitative description. A finite integer number of births occur in any finite interval, but no multiple births occur, and the distribution of the number of births in an interval depends only on the length of the interval, while there is independence for disjoint intervals. The assumption $P[N_1 = 0] < 1$ is made in order to rule out the case $P[N_t = 0] = 1$ for all $t > 0$.

Theorem 1 (Sec. 23-2) Let N_t for $t \geq 0$ be a stationary-independent-increment stochastic process all of whose realizations are counting functions and for which $P[N_1 = 0] < 1$. Then with probability 1, all its realizations are *unit* counting functions iff

$$P[N_t = 1 \mid N_t \geq 1] = \frac{P[N_t = 1]}{P[N_t \geq 1]} \to 1 \qquad \text{as } t \to 0,$$

and this is the case iff it is a Poisson process. ■

The following theorem states a property of Poisson processes. But the property completely describes a process; hence we obtain Characterization 1, which can be thought of as a method for generation of a Poisson process, or as an equivalent definition. The same is true for the next theorem and its associated characterization.

Theorem 1 (Sec. 23-3) If N_t for $t \geq 0$ is a Poisson process with mean rate $\lambda > 0$ then its successive interarrival times $X_1, X_2, \ldots$ are independent and each induces an exponential density with mean $1/\lambda$ and variance $1/\lambda^2$. The nth arrival time $T_n = X_1 + \cdots + X_n$ satisfies $P[T_n < t] = P[N_t \geq n]$ and induces the gamma density

$$p_{T_n}(t) = \gamma_{\lambda,n}(t) = \frac{e^{-\lambda t}(\lambda t)^{n-1}}{(n-1)!}\lambda \qquad \text{for all } t > 0. \quad ■$$

Characterization 1 (Sec. 23-3): Interarrival-time characterization of a Poisson process Let $X_1, X_2, \ldots$ be an infinite sequence of independent random variables each inducing an exponential density with mean $1/\lambda > 0$. Define a unit-counting-function stochastic process, N_t for $t \geq 0$ with $N_0 = 0$, to have these successive interarrival times. That is, for each $t \geq 0$ define the random variable N_t to have the value 0 on the event $[X_1 > t]$, and to have the integer value $k \geq 1$ on the event

$$[X_1 + \cdots + X_k \leq t < X_1 + \cdots + X_{k+1}].$$

In other words, N_t is the number of partial sums which are t or smaller. Then N_t for $t \geq 0$ is a Poisson process with mean rate λ. ■

Theorem 2 (Sec. 23-3) Let N_t for $t \geq 0$ be a Poisson process with successive arrival times $T_1, T_2, \ldots$. For any $t > 0$ and any $n = 1, 2, \ldots$ the conditional distribution of $(T_1, \ldots, T_n)$, given the event $[N_t = n]$, is the same as the distribution of the ordered vector corresponding to $(U_1, \ldots, U_n)$, where $U_1, \ldots, U_n$ are independent and each has a uniform density over the interval from 0 to t. ■

Characterization 2 (Sec. 23-3): Intervalwise characterization of a Poisson process Select an integer n according to a Poisson density with mean λ. If $n \geq 1$ then select n numbers $u_1, u_2, \ldots, u_n$ independently from the uniform density on the unit interval $0 \leq t < 1$, and place a cross at each of these n points. Perform this same subexperiment independently on each of the intervals $i - 1 \leq t < i$ for $i = 1, 2, \ldots$. The whole experiment yields a realization which is a configuration of crosses, or its corresponding unit counting function. This experiment generates the Poisson process on $t \geq 0$ with parameter λ. ■

A Poisson process can be extended to "N_t *for t real* with $N_0 = 0$," in which case $-N_t$ is the number of successes between $t < 0$ and 0. The corresponding random telegraph signal is a stationary process on the whole real line. Also if $S_t(R_t)$ is the positive spent (residual) waiting time between t and the closest arrival before (after) t then S_t and R_t are independent and each is exponential with mean $1/\lambda$; hence we have the *waiting-time paradox*. That is, the interarrival time $S_t + R_t$ between the closest pair of arrivals embracing time t has expectation $2/\lambda$ and is not exponentially distributed.

As in Sec. 23-4, let N_t for $t \geq 0$ be a Poisson λ process and let $F: R \to R$ be any distribution function which is not concentrated on the one point zero. Let V_t for $t \geq 0$ with $V_0 = 0$ be the process having only jump-function realizations, and constructed by letting the Poisson process determine the jump times, while the jump values are determined independently and each according to F. Such a process is called a **compound Poisson process,** or if F is concentrated on the nonnegative integers it may be called a **generalized Poisson process.** Every compound Poisson process has stationary independent increments. If $0 < p \leq 1$ and F assigns probability p and $1 - p$ to 1 and 0, respectively, then we say that V_t for $t \geq 0$ is obtained by *random selection with probability p from the Poisson process*, in which case V_t for $t \geq 0$ is a Poisson process with mean rate λp.

As in Sec. 23-5, let $N(t)$ for $t \geq 0$ with $N_0 = 0$ be a Poisson process with intensity $\lambda = 1$. Let m be any real-valued function defined on $s \geq 0$, having a positive derivative $m'(s)$ everywhere, and satisfying $m(0) = 0$, with $m(s)$ approaching infinity as s does. For each $s \geq 0$ define the random variable $M(s)$ by $M(s) = N(m(s))$. Then we call the process $M(s)$ for $s \geq 0$, which obviously has unit-counting-function realizations, a **Poisson process with intensity function** m'; we call it a **nonhomogeneous Poisson process** if m' is not a constant. Thus $M(m^{-1}(t)) = N(t)$ for $t \geq 0$ is a Poisson $\lambda = 1$ process. If $m(s) = \lambda s$ then $M(s)$ for $s \geq 0$ is a Poisson process; that is, its intensity function is the constant $m' = \lambda$. Naturally $M(s)$ has a Poisson density with mean $m(s) = EM(s)$ for any $s > 0$. Clearly $M(s)$ for $s \geq 0$ is Markov and has independent increments, since $N(t)$ for $t \geq 0$ possesses these properties. If S_n is the nth arrival time for "$M(s)$ for $s \geq 0$" then $T_n = m(S_n)$ is the nth arrival time for "$N(t)$ for $t \geq 0$." For small $\Delta > 0$ the probability is approximately $m'(s)\Delta$ that the process "$M(s)$ for $s \geq 0$" will have an arrival in the interval between $s \geq 0$ and $s + \Delta$. The process "$M(s)$ for $s \geq 0$" can be thought of as generated by the Poisson process, or by the sequential construction that the next arrival time r should be selected according to the density

$$p(r \mid s) = e^{-[m(r)-m(s)]}m'(r) \qquad \text{for } 0 \leq s \leq r$$

if the last arrival was at time s. Clearly the interarrival times are not independent.

As in Sec. 23-6, we call a process N_t for $t \geq 0$ with $N_0 = 0$ a **pure-birth Markov process** with birthrate sequence $\lambda_0, \lambda_1, \ldots$, with each $\lambda_n > 0$, iff it has unit-counting-function realizations, satisfies the infinitesimal relations

$$\lim_{\substack{\Delta \to 0 \\ \Delta > 0}} \frac{1}{\Delta} P[N_{t+\Delta} - N_t = 1 \mid N_t = n] = \lambda_n \qquad \text{for all } t > 0,\ n = 0, 1, \ldots$$

and has **stationary transition probabilities;** that is, $P[N_{t+r} = k \mid N_t = n]$ may depend on $r > 0$ and $0 \le n \le k$, but not on t. We may interpret N_t as the number of births between times 0 and t. Such processes are characterized by the fact that the successive interarrival times $X_1, X_2, \ldots$ are independent, and X_{n+1} has an exponential density with mean $1/\lambda_n$. A sequence $\lambda_0, \lambda_1, \ldots$ of positive numbers can be a birth-rate sequence for such a process iff $\sum_{n=0}^{\infty} 1/\lambda_n = \infty$, since the following lemma then guarantees that $T_n = X_1 + \cdots + X_n$ goes to infinity as n does.

Lemma 1 (Sec. 23-6) If $T = X_1 + \cdots + X_n$ is the sum of some finite number of independent exponentially distributed random variables, which need not have the same density, then

$$P[T \le t] \le 2\sqrt{\frac{t}{ET}} \qquad \text{for all } t \ge 0. \quad \blacksquare$$

The first-order distributions for a pure-birth process can be found from the *recursion relation*

$$P[N_t = n] = \int_0^t e^{-(t-\tau)\lambda_n} P[N_\tau = n - 1]\lambda_{n-1}\, d\tau \qquad \text{for all } t \ge 0,\ n \ge 1.$$

Examples 1 and 2 (Sec. 23-6) are devoted to linear birth rates with $\nu \ge 1$ initiators, so that the population size at time t is $M_t = \nu + N_t$, and if we are given that $[M_t = \nu + n] = [N_t = n]$ has occurred then the chance for a birth between t and $t + dt$ is $\lambda_n\, dt = (n + \nu)\lambda\, dt$, which is proportional to the present population size.

As in Sec. 23-7, let $0 \le p_n \le 1$ and $\beta_n > 0$ for $n = 0, \pm 1, \pm 2, \ldots$ be two doubly infinite sequences of real numbers, and let $q_n = 1 - p_n$. In order to guarantee a finite number of arrivals in each finite interval we assume that

$$\sum_{n=0}^{\infty} \frac{1}{\beta_n} = \infty = \sum_{n=0}^{\infty} \frac{1}{\beta_{-n}}.$$

Then it is possible to construct a Markov process N_t for $t \ge 0$, all of whose realizations are jump functions having jumps of $+1$ (birth) or -1 (death), and which satisfies the following infinitesimal relations:

$$P[N_{t+dt} - N_t = 1 \mid N_t = n] = p_n\beta_n\, dt = \lambda_n\, dt$$

$$P[N_{t+dt} - N_t = -1 \mid N_t = n] = q_n\beta_n\, dt = \mu_n\, dt$$

$$P[|N_{t+dt} - N_t| = 1 \mid N_t = n] = \beta_n\, dt = (\lambda_n + \mu_n)\, dt$$

The parameters λ_n and μ_n are traditional, and in terms of these we could introduce $\beta_n = \lambda_n + \mu_n > 0$, $p_n = \lambda_n/\beta_n$ and $q_n = \mu_n/\beta_n$.

The construction is roughly as follows. Let X_k be the kth interarrival time, T_k the kth arrival time, and Z_k the state immediately after the kth arrival. If we are given the complete evolution of the process up to the $(k - 1)$st arrival t_{k-1} and state z_{k-1}, then the conditional distribution of (X_k,Z_k) depends on this evolution only through the state, say $z_{k-1} = n$, and also X_k and Z_k are conditionally independent, with X_k being exponential with mean $1/\beta_n$ while Z_k takes the values $n + 1$ or $n - 1$ with probabilities p_n or q_n, respectively.

CHAPTER 24: THE WIENER PROCESS

A **Wiener process** "W_t for $t \geq 0$ with $W_0 = 0$ and parameter $\sigma^2 > 0$" is any stationary-independent-increment process such that for each $t > 0$, W_t induces a normal density with mean 0 and variance $t\sigma^2$, and every realization for the process is a continuous function. Thus if $0 \leq s < t$ then $W_t - W_s$ has variance $(t - s)\sigma^2$, proportional to the length of the interval, and normal density

$$\frac{1}{\sqrt{2\pi(t - s)\sigma^2}} e^{-w^2/2(t-s)\sigma^2} \qquad \text{for all real } w.$$

By a **standard Wiener process** we mean W_t for $t \geq 0$ with $W_0 = 0$ and $\sigma^2 = 1$.

The Weiner process with parameter $\sigma^2 > 0$ can be interpreted as the limit, as m goes to infinity, for a sequence of experiments, one for each positive integer m. For the mth experiment we let each of $z_1, z_2, \ldots$ be one of $\pm\sigma/\sqrt{m}$ according to a fair-coin toss; then we plot $z_1 + \cdots + z_i$ above the point $t = i/m$ and connect these points by a polygonal line to obtain the realization from this performance of the mth experiment. Thus the variance of the value plotted above $t = 1$ is σ^2 for every experiment, and the central-limit theorem leads to the normal distribution. These experiments clearly suggest stationary and independent increments.

For a Wiener process the distributions induced by the individual W_t for $t > 0$ obviously form a semigroup. Therefore Theorem 1 (Sec. 22-2) implies that there is a process generating this semigroup. However, we do not prove the fact that such a process can be constructed so as to have continuous realizations.

Let $X(\tau)$ for $\tau \geq 0$ be a standard Wiener process, so that $\text{Var } X(\tau) = \tau$. For $\sigma \neq 0$ if W_t for $t \geq 0$ is introduced by either of the definitions

$$W_t = X(t\sigma^2) \qquad \text{for all } t \geq 0$$

$$W_t = \sigma X(t) \qquad \text{for all } t \geq 0$$

then it is easily seen to be a Wiener process with parameter $\sigma^2 > 0$. Thus σ^2 corresponds to a scale change in time being applied to every realization, or σ corresponds to a scale change in the values.

All the results in Sec. 24-2 follow directly from the following theorem, suggested by use of the reflection principle, and proved in the Addendum to Sec. 24-2.

Theorem 1 (Sec. 24-2) If W_t for $t \geq 0$ with $W_0 = 0$ is a Wiener process then

$$P[W_\tau \geq \lambda \text{ for at least one } \tau \text{ satisfying } 0 \leq \tau \leq t] = 2P[W_t \geq \lambda]$$

for all $\lambda \geq 0$ and $t \geq 0$. ■

Let M_t be the maximum of the random realization on the interval $0 \leq \tau \leq t$. If $\lambda \geq 0$ then

$$[M_t \geq \lambda] = [W_\tau \geq \lambda \text{ for at least one } \tau \text{ satisfying } 0 \leq \tau \leq t].$$

Hence the theorem states that if $\lambda \geq 0$ and $t \geq 0$ then

$$P[M_t \geq \lambda] = 2P[W_t \geq \lambda] = P[|W_t| \geq \lambda].$$

Thus *for a Wiener process* the rather esoteric random variable M_t has a simple density, namely, that of the absolute value of a normal 0 and $t\sigma^2$ variable.

Example 1 (Sec. 24-2) shows that for any fixed $s \geq 0$, with probability 1, each Wiener process realization will, for infinitely many times shortly after s, have the same value that it had at time s.

Example 2 (Sec. 24-2) derives the probability for obtaining a realization having at least one zero in a specified interval,

$$P[W_\tau = 0 \text{ for at least one } \tau \text{ satisfying } s \leq \tau \leq t] = \frac{2}{\pi} \arccos \sqrt{\frac{s}{t}} \qquad \text{if } 0 \leq s \leq t$$

where W_t for $t \geq 0$ with $W_0 = 0$ is a Wiener process with parameter σ^2.

Example 4 (Sec. 24-2) shows that for each $s \geq 0$ the probability is zero for the set of Wiener process realizations which have a right-hand derivative at s.

For fixed $\lambda > 0$ let T_λ be the first time that a standard Wiener process reaches level λ. Obviously if $t > 0$ then $[T_\lambda \leq t] = [M_t \geq \lambda]$, so the distribution function and density induced by T_λ are easily found. It is easily shown that if Z is normal 0 and 1 then $(\lambda/Z)^2$ and T_λ are identically distributed; also $E\,|T_\lambda| = \infty$. As argued in Example 3 (Sec. 24-2), if $X_1, \ldots, X_n$ are independent and each has the same distribution as T_1 then

$$\frac{1}{n}(X_1 + \cdots + X_n) \qquad \text{and} \qquad nX_1$$

are identically distributed. Thus the arithmetic average of 100 independent observations is 100 times *more* spread than one observation.

Theorem 1 (Sec. 24-2) is deduced from the following lemma, which is interpretable as a reflection technique for sequences.

Lemma 1 (Sec. 24-2) For $1 \leq k \leq n$ let $S_k = X_1 + \cdots + X_k$ where $X_1, \ldots, X_n$ are independent random variables, each having a symmetric distribution but not necessarily the same distribution. That is, for each k the two random variables X_k and $-X_k$ are identically distributed. Fix $\lambda > 0$ and let

$$A = [S_k \geq \lambda \text{ for at least one } k \text{ satisfying } 1 \leq k \leq n]$$

$$= \bigcup_{k=1}^{n} [S_k \geq \lambda] = \left[\max_{1 \leq k \leq n} S_k \geq \lambda\right].$$

Then $P(A) \leq 2P[S_n \geq \lambda]$. Furthermore for every $\epsilon > 0$ we have

$$2P[S_n \geq \lambda + 2\epsilon] - \sum_{k=1}^{n} P[X_k \geq \epsilon] \leq P(A). \quad \blacksquare$$

Section 24-3 contains only the briefest of introductions to covariance functions. If X_t for $t \in T$ is a stochastic process for which $E(X_t)^2 < \infty$ for all t in T then the function m: $T \to R$ defined by

$$m(t) = EX_t \qquad \text{for all } t \in T$$

is called its **mean-value function,** and the function K: $T^2 \to R$ defined by

$$K(s,t) = \text{Cov}\,(X_s,X_t) = E(X_s - EX_s)(X_t - EX_t) \qquad \text{for all } s,t \in T$$

is called its **covariance kernel.** Clearly $K(s,t) = K(t,s)$ and $K(t,t) = \text{Var}\,X_t$. If X_t for $t \geq 0$ with $X_0 = 0$ has independent increments then

$$K(s,t) = \text{Var}\,X_{\min\{s,t\}} \qquad \text{for all } s \geq 0,\ t \geq 0$$

and if it also has stationary increments then

$$m(t) = tEX_1 \qquad \text{for all } t \geq 0$$

$$K(s,t) = (\min\{s,t\})(\text{Var}\,X_1) \qquad \text{for all } s \geq 0 \text{ and } t \geq 0.$$

A process is said to be **covariance stationary** iff $K(s,t)$ is only a function of the difference $t - s$. That is, there is a real-valued function f, called the **covariance function,** such that $K(s,t) = f(t - s)$ for all s and t in T. This property is also referred to as *wide-sense* or *second-order* stationary. The mean EX_t may depend on t. Even if the mean-value function is constant, a covariance-stationary process need not be stationary. However, every stationary process which has finite second moments must obviously be covariance stationary. For a covariance-stationary process

$$f(s - t) = \text{Cov}\,(X_s,X_t) = \text{Cov}\,(X_t,X_s) = f(t - s)$$

so that f must be an even function.

A stochastic process "$X(t)$ for $t \in T$" is called a **normal** or **gaussian process** iff for every $n \geq 1$ points $t_1 < \cdots < t_n$ in T the random vector $(X(t_1), \ldots, X(t_n))$ induces a multivariate normal distribution. In other words, the family which the process generates contains only multivariate normal distributions. But such a distribution is completely determined by its vector of means and its covariance matrix. Therefore the entire family of finite distributions generated by a normal process is completely determined by its mean-value function and covariance kernel. Every Wiener process W_t for $t \geq 0$ with $W_0 = 0$ is a normal process. A normal process is stationary iff its mean-value function is a constant and the process is covariance stationary. We need only observe that in this case the vector of means and the covariance matrix for $(X(t_1 + h), \ldots, X(t_n + h))$ do not depend on h, so the distribution does not depend on h.

If $W(\tau)$ for $\tau \geq 0$ is a standard Wiener process then the process "$X(t)$ *for t real*," defined by

$$X(t) = e^{-\beta t} W(\alpha e^{2\beta t}) \qquad \text{for all real } t,$$

is called the **Ornstein-Uhlenbeck process** with parameters $\alpha > 0$ and $\beta > 0$. It is a stationary normal process with zero mean-value function and covariance kernel

$$\text{Cov}\,(X(s),X(t)) = \alpha e^{-\beta|s-t|} \qquad \text{for all real } s,t.$$

All its realizations are continuous functions, since this property was assumed for Wiener processes.

Exercise 2 (Sec. 24-3) shows that if $\alpha = 1$ and $\beta = 2\lambda$ then the Ornstein-Uhlenbeck process has the same mean-value function and covariance function as the random telegraph signal of Example 2 (Sec. 23-1). Thus drastically different processes can have the same covariance function.

Selected Answers and Solutions

SECTION 1-3

1. (*a*) $B = \{1,5,6\}$. (*b*) No, $AB = \{1\}$. (*c*) $C = \{1,7,6\}$. (*d*) $A \cup B = \{1,2,4,6,5\}$. (*e*) $A \cup (B \cap C^c) = \{1,2,4\} \cup (\{1,6,5\} \cap \{2,3,4,5\}) = \{1,2,4\} \cup \{5\} = \{1,2,4,5\}$ = those who are either married or are nontall males.
(*f*) Yes, $AC = \{1\}$. $AC \subset B$ means that everyone in this population who is tall and married is also male.
3. $A = A_1$ = unit disk including boundary. B = origin of the plane. C = punctured unit disk, that is, unit disk including boundary but excluding the origin.
4. See Fig. 1 (Sec. 2-2).

SECTION 1-4

1. (*b*), (*j*), and (*p*) are not identities. If $\Omega = \{1,2\}$ then $\varphi = \Omega \cap \varphi \neq \Omega = \{1,2\}$ shows that (*b*) is not an identity.
3. Let $A = A_1 \cup A_2 \cup \cdots$, where $A_1 \supset A_2 \supset \cdots$. If $\omega \in A$ then $\omega \in A_n$ for some n, but $A_1 \supset \cdots \supset A_n$, so $\omega \in A_1$. Thus $A \subset A_1$, but clearly $A \supset A_1$, so $A = A_1$.

Let $A = A_1 \cap A_2 \cap \cdots$, where $A_1 \subset A_2 \subset \cdots$. We wish to prove that $A = A_1$. If $\omega \in A$ then ω belongs to every A_n so in particular $\omega \in A_1$. Thus $A \subset A_1$. If $\omega \in A_1$ then ω belongs to every A_n since the sets are increasing; therefore $\omega \in A$. Thus $A \supset A_1$; hence $A = A_1$.

SECTION 2-1

1. (α,α), $(\gamma,3.6)$
$(3,4,8,3)$, $(3,4,1,3)$
$(-1,-1,-1,0,1,2,-1)$, $(0,-1,-1,0,2,2,-1)$
$(0,0,0,8,-2,6,7,85.72)$, $(\sqrt{2},5,1,1,1,-3,7,8)$
3. By definition $(x,y) \in (A \times B)$ iff $x \in A$ and $y \in B$. Thus if (x,y) and (x',y') belong to $(A \times B)$ then x and x' belong to A while y and y' belong to B, so (x,y') and (x',y) belong to $(A \times B)$. However, a circular disk has two points $(0,r)$ and $(r,0)$ belonging to it, while (r,r) does not.

SECTION 2-2

1. 133^{1000}. **2.** 2^{100}. **3.** (*a*) 10^{11}. (*b*) 10^{220}.
4 (*b*) $S(1)$ reduces to $1 = (\frac{1}{6})(1)(2)(3)$, which is true. If $S(n)$ is true for a particular n then by adding $(n+1)^2$ to both sides we get $1 + 4 + \cdots + n^2 + (n+1)^2 = (\frac{1}{6})n(n+1)(2n+1) + (n+1)^2 = (\frac{1}{6})[n+1][(n+1)+1][2(n+1)+1]$, where we used simple manipulations to get the second equality. Thus $S(n+1)$ is true. Therefore $S(n)$ is true for all n
(*c*) $S(1)$ is true. If $S(n)$ is true, we multiply both sides by $1 - y$, which is nonnegative, so that

$$(1-y)^{n+1} \geq (1-ny)(1-y) = 1 - (n+1)y + ny^2 \geq 1 - (n+1)y,$$

since $ny^2 \geq 0$. Thus $S(n+1)$ is true. Therefore $S(n)$ is true for all n.
9. (*a*) [The number of members of Ω whose first two coordinates are $H]/2^{100} = 2^{98}/2^{100} = \frac{1}{4}$.
11. $(12)(11)(10)(9)(8)(7)/12^6 = 385/1728 \doteq .22$.

SECTION 2-3

1. $\binom{52}{13}$. **2.** $\binom{10}{4} = 210$. **8.** $\binom{12}{8}$ See Sec. 2-4. **9.** .711. **10.** Only for (*e*).

SECTION 2-4

1. $\binom{n+k-1}{n} = \binom{6+3-1}{10} = 28$. **3.** (*a*) 3^{10}, $\binom{10+3-1}{10}$. (*b*) $\dfrac{10!}{2!\,4!\,4!}$.

5. $\dfrac{n!}{r_1!\,r_2!\cdots r_k!}$

SECTION 3-1

1. (*a*) Yes. (*b*) Yes. (*c*) Yes. (*d*) No.
2. (*a*), (*d*), (*e*), and (*f*) are functions.

SECTION 3-3

1. Yes, $\cos x^4$ is the value of both at x.
2. $f(h(x)) = e^{2\cos x}$, $h(f(x)) = \cos e^{2x}$, and $g(f(x),h(x)) = [\sin e^{2x}]^2 + [\cos(\cos x)]^2$.

$$g(f(\pi/2),h(\pi/2)) = [\sin e^{\pi}]^2 + 1$$

4. $|f| \leq g$ and $\sin f \leq g$ are true; $f^2 \leq g$ is false.

SECTION 3-4

1. Let A and B be two disjoint subsets of the plane. Then $f^{-1}(A)$ is the set of those people (or more precisely, their integer labels) having (height,weight) pairs in A. The two sets of people, $f^{-1}(A)$ and $f^{-1}(B)$, are disjoint.

SECTION 3-5

1.

Function	Onto R	One-to-One
$\sin x$	No	No
e^x	No	Yes
$3x^3 + 2x^2$	Yes	No
x^3	Yes	Yes

2. A one-to-one correspondence $f\colon \mathscr{X} \to \mathscr{Y}$ need not only be one-to-one; *it also must be onto* $\mathscr{Y}$.
3. (*b*) $f(x) = x + 5$ for all $x \in \mathscr{X}$.

SECTION 4-2

2. Let R' be the set of all rationals and let R^+ be the set of all positive rationals. Then $R' \times R' \times R^+$ is countable, and each member (x,y,r) of this set corresponds to a rational open disk, and the correspondence is a one-to-one correspondence.
7. (*a*) Yes. (*b*) No.

8. Let $f(x,y) = \left(\dfrac{1}{x} - \dfrac{1}{1-x}, \dfrac{1}{y} - \dfrac{1}{1-y}\right)$ for $0 < x < 1$, $0 < y < 1$. For fixed $x = 1/3$, this function transforms the vertical segment $\{(1/3, y)\colon 0 < y < 1\}$ one-to-one onto the vertical line $\{(3/2, z)\colon -\infty < z < \infty\}$. By varying x we transform the segments, one by one, onto the lines.

SECTION 4-3

1. Let $\mathscr{X} = \mathscr{X}_1 \times \mathscr{X}_2 \times \cdots$, where $\{0, 1, \ldots, 9\} = \mathscr{X}_1 = \mathscr{X}_2 = \cdots$. Let $\mathscr{X}'$ consist of those members $(x_1, x_2, \ldots)$ of $\mathscr{X}$ which have an infinite number of nonzero coordinates. Then $(x_1, x_2, \ldots) \to .x_1x_2\cdots$ is a one-to-one correspondence between $\mathscr{X}'$ and the set of reals in the unit interval when represented by their decimal expansions. Thus $\mathscr{X}'$ is uncountable, so $\mathscr{X}$ is uncountable by Theorem 1*a* (Sec. 4-2).
2. Yes.
3. If the set T of transcendentals were countable then from Theorem 1*d* (Sec. 4-2) the set $A \cup T$ would be countable. But $A \cup T = R$ is uncountable, and hence T must be uncountable.

SECTION 5-1

1. (1) $0, \frac{1}{2}, \frac{2}{3}, \ldots, 1 - \frac{1}{n}, \ldots$; (2) $8.3, 8.3, 8.3, \ldots$; (3) $1, \frac{1}{2}, \frac{1}{3}, \ldots, \frac{1}{n}, \ldots$;

(4) $-1, \frac{1}{2}, -\frac{1}{3}, \ldots, (-1)^n \frac{1}{n}, \ldots$; (5) $-1, 1, -1, \ldots, (-1)^n, \ldots$; (6) $1, 2, 3, \ldots, n, \ldots$;

(7) $0, 3, 2, 5, \ldots, n + (-1)^n, \ldots$; (8) $-1, -2, -3, \ldots, -n, \ldots$;

(9) $0, -3, -2, -5, \ldots, -[n + (-1)^n], \ldots$; (10) $-1, 2, -3, 4, \ldots, (-1)^n n, \ldots$.

SECTION 5-3

1. $\sum_{n=0}^{N} \sum_{m=0}^{M} |a|^n |b|^m = \left(\sum_{n=0}^{N} |a|^n\right)\left(\sum_{m=0}^{M} |b|^m\right) = \frac{1 - |a|^{N+1}}{1 - |a|} \frac{1 - |b|^{M+1}}{1 - |b|} \leq \frac{1}{1 - |a|} \frac{1}{1 - |b|}$, so it is absolutely convergent. From Theorem 1 we get $\left(\sum_{n=0}^{\infty} a^n\right)\left(\sum_{m=0}^{\infty} b^m\right) = \frac{1}{1 - a} \frac{1}{1 - b}$.

2. We calculate, and then equate, the sum of the row sums, and the sum of the column sums. The absolute convergence is easily verified.

$$\sum_{n=1}^{\infty}\left(\sum_{m=1}^{\infty} a_{nm}\right) = \sum_{n=1}^{\infty} (\rho^{n-1} + \rho^n + \rho^{n+1} + \cdots) = \sum_{n=1}^{\infty} \rho^{n-1}(1 + \rho + \rho^2 + \cdots)$$

$$= \sum_{n=1}^{\infty} \frac{\rho^{n-1}}{1-\rho} = \frac{1}{(1-\rho)^2} \text{ while } \sum_{m=1}^{\infty}\left(\sum_{n=1}^{\infty} a_{nm}\right) = \sum_{m=1}^{\infty} m\rho^{m-1}.$$

SECTION 6-1

1. (*a*) 600. (*b*) 400. (*c*) 300. (*d*) 800.

SECTION 6-2

1. (*a*) $A = \{\omega_8, \omega_{12}, \omega_{14}, \omega_{15}, \omega_{16}\}$, $P(A) = .32$.
(*b*) $.33 = P(A \cup B) = P(A) + P(B) - P(AB) = .32 + .21 - .20 = .33$.

4. (*c*) .34. **5.** (*c*) $P(B) = \frac{x}{x + y}$. **6.** Yes.

SECTION 6-4

1. $P(\varphi) = 0$, $P\{\omega_1\} = P\{\omega_2\} = P\{\omega_3\} = .3$, $P(\omega_4) = .1$

$P\{\omega_1,\omega_2\} = P\{\omega_1,\omega_3\} = P\{\omega_2,\omega_3\} = .6$

$P\{\omega_1,\omega_4\} = P\{\omega_2,\omega_4\} = P\{\omega_3,\omega_4\} = .4$

$P\{\omega_2,\omega_3,\omega_4\} = P\{\omega_1,\omega_3,\omega_4\} = P\{\omega_1,\omega_2,\omega_4\} = .7$

$P\{\omega_1,\omega_2,\omega_3\} = .9$, $P(\Omega) = 1$

4. (*a*) $1 - P(A_1A_2 \cdots A_N) \overset{1}{=} P([A_1A_2 \cdots A_N]^c) \overset{2}{=} P(A_1{}^c \cup A_2{}^c \cup \cdots \cup A_N{}^c) \overset{3}{\leq} P(A_1{}^c) + P(A_2{}^c) + \cdots + P(A_N{}^c)$, where $\overset{2}{=}$ is by DeMorgan's law Example 4 (Sec. 1-4), and $\overset{3}{\leq}$ is by Theorem 2*f*.

6. (*a*) $P(A) = \int_0^2 (\int_0^\infty e^{-x-y}\,dy)\,dx = 1 - e^{-2} = .865.$

7. (*a*) $P(B) = P\{x: .7 < x < .8\} = .1.$ (*b*) $P(C) = P\{x: .05 < x < .06 \text{ or } .15 < x < .16 \text{ or } \cdots \text{ or } .95 < x < .96\} = 10(.01) = .1.$

8. (*a*) .6.

SECTION 7-1

1. Extending the formula of Example 1 to four events yields $P(A_4A_3A_2A_1) = P(A_4 \mid A_3A_2A_1) P(A_3A_2A_1) = P(A_4 \mid A_3A_2A_1)P(A_3 \mid A_2A_1)P(A_2 \mid A_1)P(A_1)$. Thus the answer is .084.

3. (*a*) .32.

(*b*) Let A be the set of those $\alpha = (x_1, \ldots, x_{100})$ such that 32 of the x_i are H and the remaining 68 are T. Then $P(A) = \binom{100}{32}(1/2^n)$, so that

$$p_A(\alpha) = \begin{cases} 0 & \text{if } \alpha \notin A \\ \dfrac{p(\alpha)}{P(A)} = \dfrac{1/2^n}{\binom{100}{32}\frac{1}{2^n}} = \dfrac{1}{\binom{100}{32}} & \text{if } \alpha \in A \end{cases}$$

Thus the conditional density p_A is zero off A and uniform on A, so every 32 head outcome is equally likely.

4. (*a*) $P(B) = .44 > .392 = P(B \mid A).$

5. (*a*) $P(A_1 \mid A_2) = .72.$ (*b*) $P(A_1 \mid A_2 \cup A_3) = .743.$

SECTION 7-2

3. (*a*) $P(C_3 \mid F_1F_2) = .267.$ (*b*) Replace β because . . . .

SECTION 7-4

5. (*d*) 9/14.

6. We denote the two definitions by "independence" and "alternate independence." Assume that $A_1, \ldots, A_n$ are independent. Then we saw that $A_1, \ldots, A_{n-1}, A_n{}^c$ are independent, and hence

$$P(A_1 \cdots A_{n-1}A_n{}^c) = P(A_1) \cdots P(A_{n-1})P(A_n{}^c).$$

This is one of the 2^n equations required for alternate independence. The other equations can obviously be derived similarly. Therefore $A_1, \ldots, A_n$ must be alternate independent.

Now assume that $A_1, \ldots, A_n$ are alternate independent. We first show that any subsequence of two or more of them is also alternate independent. Note that

$$\begin{aligned} P(A_1 \cdots A_{n-1}) &= P(A_1 \cdots A_{n-1}A_n) + P(A_1 \cdots A_{n-1}A_n{}^c) \\ &= P(A_1) \cdots P(A_{n-1})P(A_n) + P(A_1) \cdots P(A_{n-1})P(A_n{}^c) \\ &= P(A_1) \cdots P(A_{n-1})[P(A_n) + P(A_n{}^c)] = P(A_1) \cdots P(A_{n-1}). \end{aligned}$$

Replacing A_1 by $A_1{}^c$ in this proof yields $P(A_1{}^cA_2 \cdots A_{n-1}) = P(A_1{}^c)P(A_2) \cdots P(A_{n-1})$. More generally we may replace any subset of the $A_1, \ldots, A_{n-1}$ by their complements and obtain the corresponding equation. Therefore $A_1, \ldots, A_{n-1}$ are alternate independent. But alternate independence is not destroyed by reordering, and hence it is not destroyed by dropping any one of $A_1, \ldots, A_n$. Clearly, then, any subsequence of two or more from $A_1, \ldots, A_n$ is also alternate independent. For example, A_2, A_5, A_6 are alternate independent; hence $P(A_2A_5A_6) = P(A_2)P(A_5)P(A_6)$. This is one of the $2^n - n - 1$ equations which must be satisfied in order to assure that $A_1, \ldots, A_n$ are independent. The remaining equations are derived similarly. Hence $A_1, \ldots, A_n$ are independent.

SECTION 8-2

7(c) $^{12}/_{100}$. The answer remains unchanged if the probability .8 for a downward movement is changed.

SECTION 8-3

2. If n is large enough then, to a fair approximation, we can replace each spin of the p_λ wheel by the sum of the outcomes from n independent spins of the wheel in Fig. 2. This yields the sum from $3n$ independent spins of Fig. 2, and the binomial density can be approximated by the Poisson density with parameter $(3n)(\lambda/n) = 3\lambda$. Both approximations improve as n approaches infinity.
5. $1 - (1 - 10^{-4})^{1,000}$. The Poisson density p_λ with $\lambda = np = .1$ approximates this probability by $1 - p_\lambda(0) = 1 - e^{-.1}$.
9. (*a*) Replacing x by $-x$ in inequality (7) of Appendix 1 yields $1 + x \leq e^x$. We must have $g(x) \leq 1 + x$ if $-.2 \leq x \leq .5$, since g is convex upward here, and tangent to $1 + x$ at $x = 0$. But then $g'(x) \leq 0$ for $x \geq .5$ assures that $g(x) \leq g(.5) \leq 1 + .5 \leq 1 + x$ for $x \geq .5$.
(*b*) In $g(x) \leq 1 + x \leq e^x$ replace x by y/n and take the nth power of each of the three expressions.

10. (*a*) $|(1+z)^n - 1| = \left|\sum_{k=1}^{n}\binom{n}{k}z^k\right| \leq \sum_{k=1}^{n}\binom{n}{k}|z|^k = (1+|z|)^n - 1$. Letting $z = \delta_n/n$ and applying

Exercise 9*b* yields $\lim_{n\to\infty}\left|\left(1+\frac{\delta_n}{n}\right)^n - 1\right| \leq \lim_{n\to\infty}\left(1+\frac{|\delta_n|}{n}\right)^n - 1 = e^0 - 1 = 0.$

(*b*) The exhibited definition of δ_n is equivalent to $\delta_n = (c_n - b)/[1 + (b/n)]$ so δ_n goes to zero as n goes to infinity. Therefore $[1 + (c_n/n)]^n = [1 + (b/n)]^n[1 + (\delta_n/n)]^n$ converges to e^b as n goes to infinity, from Exercises 9*b* and 10*a*.

SECTION 8-4

2. $\sum_{k=k_0}^{\infty}\binom{9+k}{k}(.995)^k(.005)^{10}$ where $k_0 = 10{,}001$.

4. Let p be the Poisson density with mean λ. If $0 \leq r < r + k$ then

$$p(r+k) = p(r)\frac{p(r+1)}{p(r)}\frac{p(r+2)}{p(r+1)}\cdots\frac{p(r+k)}{p(r+k-1)}$$

Each of the last k factors satisfies $\frac{p(r+i)}{p(r+i-1)} = \frac{\lambda}{r+i} \leq \frac{\lambda}{r+1}$ so that $p(r+k) \leq \left(\frac{\lambda}{r+1}\right)^k p(r)$ for $k = 0, 1, \ldots$. If $\lambda < r + 1$ then summing over k yields $\sum_{k=0}^{\infty} p(r+k) \leq \frac{p(r)}{1 - [\lambda/(r+1)]}$, which can be written as $\frac{p(r)}{\sum_{k=0}^{\infty} p(r+k)} \geq 1 - \frac{\lambda}{r+1}$. Now let r go infinity.

5. (*b*) Let m be the largest integer satisfying $m \leq (r-1)q/p$. If $(r-1)q/p$ is an integer, so that m equals it, then $p_r(0) < p_r(1) < \cdots < p_r(m-1) = p_r(m) > p_r(m+1) > \cdots$, and if $(r-1)q/p$ is not an integer then the only equals sign in the chain becomes the sign $<$.

SECTION 9-1

6. The sum of the r sector labels obtained from r repeated independent spins from a geometric p wheel has a Pascal r and p density.

9. $(1/n)\sum_{i=1}^{n} e_i \geq r2^r m_e(r2^r)$ which converges to r as n goes to infinity.

SECTION 9-2

1. (*a*) 1.56. (*b*) 1.6, 2.64.

4. (*a*) Each $(x_i - \mu)^2 \leq B^2$, so that $\sigma^2 = \sum_{i=1}^{\infty} (x_i - \mu)^2 p(x_i) \leq B^2 \sum_{i=1}^{\infty} p(x_i) = B^2$, and hence $\sigma \leq B$.

(*b*) If $p(-2) = p(2) = .5$ then $\mu = 0$ and $\sigma^2 = 4$, and we can let $B = 2$.

SECTION 9-3

1. $\mu_{Y_n} = ½$ and $\sigma_{Y_n} = 1/(2\sqrt{n})$.

5. Not Poisson, since $p_{X^*}(-\sqrt{3.2}) > 0$.

SECTION 9-4

3. (*c*) $t = 0$.

4. $tx_k \leq |tx_k| \leq |t|\, B$, so that $\sum_{k=1}^{\infty} e^{tx_k}p(x_k) \leq \sum_{k=1}^{\infty} e^{|t|B}p(x_k) = e^{|t|B}$.

5. (*a*) $m(t) = e^{0t}q + e^t p = q + pe^t$

$m'(t) = pe^t,\ \mu = m'(0) = p$

$m''(t) = pe^t,\ \sigma^2 = m''(0) - [m'(0)]^2 = p - p^2 = pq$

6. $m_2'(t) = n[m_1(t)]^{n-1}m_1'(t)$; hence $\mu_2 = m_2'(0) = n(1)^{n-1}m_1'(0) = n\mu_1$. Similarly for $\sigma_2{}^2 = n\sigma_1{}^2$.

7. (*a*) $m(t) = \sum_{k=0}^{n} e^{tk}\binom{n}{k} p^k q^{n-k} = \sum_{k=0}^{n} \binom{n}{k}(pe^t)^k q^{n-k} = (q + pe^t)^n$ by the binomial theorem. Hence Exercises 5*a* and 6 imply that the binomial n and p density has mean np and variance npq.

10. From Theorem 1 we need only show that for every positive integer n that p has a finite nth absolute moment. The exhibited inequality yields $\frac{|t_0 x_k|^n}{n!} \leq e^{t_0 x_k} + e^{-t_0 x_k}$. Multiplying by $p(x_k)$ and summing over k yields $\sum_{k=1}^{\infty} |x_k|^n p(x_k) \leq \frac{n!}{|t_0|^n}[m(t_0) + m(-t_0)]$.

SECTION 10-1

1. (*a*) .1 or smaller. **2.** (*a*) .836 or greater. **3.** (*a*) .171 or smaller.

5. In Lemma 1 let $f(x) = |(x - \mu)/\epsilon|^r$.

8. (*a*) The function $f(x) = [(x - \mu)/\delta\sigma]^2$ in the proof of Theorem 1 satisfies $f(\mu) = 0$ and $f(\mu \pm \delta\sigma) = 1$. Nowhere else does it take the values 0 or 1.

(*b*) On the three points 0, $-\delta$, and δ let $p_\delta(0) = 1 - (1/\delta^2)$ and $p_\delta(\pm\delta) = 1/2\delta^2$, while $p_\delta(x) = 0$ otherwise. This density has mean 0 and variance 1 and satisfies $h_{p_\delta}(\delta) = P_\delta\{x\colon |x - \mu| \geq \delta\sigma\} = 1/\delta^2$.

SECTION 10-2

1. The probability is at least .9984.

SECTION 10-3

2. $1/(1 - p)$ on $0 \leq p < 1$.

SECTION 11-1

1. (*a*) $P[R \geq 1.5] = P\{\omega_2,\omega_4,\omega_6,\omega_{10},\omega_{12}\} = .1 + .1 + .05 + .1 + .05 = .4$
$P[R \geq 1.5] = p_R(2) + p_R(3) = .3 + .1 = .4$

M \ m	0	1	2	p_M
2	0	.5	.1	.6
3	.1	.3	0	.4
p_m	.1	.8	.1	1

SECTION 11-2

4. (*d*) Let X_i be the number of zeros between the $(i-1)$st and ith ones. Then, as shown in Sec. 8-4, the random variables $X_1, X_2, \ldots$ are independent and each induces the geometric density with parameter p, so let

$$Y_1 = X_1 + \cdots + X_{r_1}$$
$$Y_2 = X_{r_1+1} + \cdots + X_{r_1+r_2}$$
$$\cdots\cdots\cdots\cdots\cdots\cdots$$

5. Assume that $A_1, \ldots, A_n$ are independent according to the alternate definition of Exercise 6 (Sec. 7-4). We must show that the equation $P[I_{A_1} = x_1, \ldots, I_{A_n} = x_n] = P[I_{A_1} = x_1] \cdots P[I_{A_n} = x_n]$ holds for all real $x_1, \ldots, x_n$. If $0 \neq x_i \neq 1$ for some i then $[I_{A_i} = x_i]$ is empty, so the equation becomes $0 = 0$. Otherwise every x_i is either 0 or 1, so the equation becomes one of the 2^n equations assumed to be true. For example, if $n = 4$, $x_1 = x_2 = x_4 = 1$, and $x_3 = 0$ then the equation becomes $P[A_1A_2A_3^cA_4] = P[A_1]P[A_2]P[A_3^c]P[A_4]$.

SECTION 11-3

1. $EX_1 = 4.1$ and $EY_1 = 1.4$. **2.** $Eg(X) \doteq (10)(2/3) - (5)(1/3) = 5$.
3. (*a*) Essentially by definition we have

$$P[X_2 = 0 \mid X_1 = 0] = q \qquad P[X_2 = 1 \mid X_1 = 0] = p$$
$$P[X_2 = 0 \mid X_1 = 1] = (3/7)p \qquad P[X_2 = 1 \mid X_1 = 1] = 1 - (3/7)p.$$

Therefore $P[X_2 = 0, X_1 = 0] = P[X_2 = 0 \mid X_1 = 0]P[X_1 = 0] = q(.3)$, with similar results for the other three entries.

7. (*a*) $$E\left[(X_2 - 3)^2\left(\cos\frac{\pi}{4}X_3\right)e^{2X_1}\right] = [E(X_2 - 3)^2]\left[E\left(\cos\frac{\pi}{4}X_3\right)\right][Ee^{2X_1}]$$
$$= [9q + 4p][q][q + pe^2].$$

12. $P[X \leq Y] = P[(X,Y) \in D] = P_{X,Y}(D)$, so that $P[X \leq Y] = 1$ iff $p_{X,Y}$ is concentrated on D. Of course, $X \leq Y$ means $X(\omega) \leq Y(\omega)$ for *all* sample points ω, not just for those having positive probability; hence $X \leq Y$ certainly implies $P[X \leq Y] = 1$.

13. (*e*) $a\sum_x g(x) \sum_y p_{X,Y}(x,y) + b \sum_y h(y) \sum_x p_{X,Y}(x,y)$.

SECTION 11-4

2. $$\rho_{X_1,X_2} = \frac{.7 - p}{.7}$$

5. (*a*) None. (*b*) The lower-right-hand-corner entry must be changed to False. For example, Y and X of Exercise 3 are uncorrelated, but Y and X^2 are not.

10. We need only show that the inequalities hold between the squares of the three expressions. Using $\sigma^2_{X+Y} = \sigma_X{}^2 + \sigma_Y{}^2 + 2\,\text{Cov}\,(X,Y)$, the inequalities reduce to $-\sigma_X\sigma_Y \le \text{Cov}\,(X,Y) \le \sigma_X\sigma_Y$, which is equivalent to $-1 \le \rho \le 1$ and hence true.
12. (*c*) $\rho(A,B) = -\rho(A,B^c) = -\rho(A^c,B) = \rho(A^c,B^c)$

SECTION 12-1

1. (*a*) $p_{V_1}(0) = p_{V_1}(1) = .3$, $p_{V_1}(4) = .4$, $EV_1 = 1.9$, $\sigma_{V_1}{}^2 = 3.09$, and $\sigma_{V_1} = 1.76$. We select a symmetric subinterval about ES_n and apply Chebychev's inequality. For $n = 100$,

$$P[.9 < (1/n)S_n < 2.5] = P[90 < S_n < 250] = P[-100 < S_n - 190 < 60]$$

$$\ge P[-60 < S_n - 190 < 60] \ge 1 - \frac{(100)(3.09)}{(60)^2} = .914.$$

(*b*) .0109 or less.
(*c*) We easily show that this probability is .034 or less, while binomial tables show that it is approximately 10^{-5}.
5. (*a*) For pairwise-uncorrelated random variables the variance of a sum equals the sum of the variances hence the same proofs apply.
7. .97 or more.

SECTION 12-2

1. (*a*) $P(A) \le .196$.

SECTION 12-3

1. (*a*) For Example 2 (Sec. 12-1) the probability of the event $[|(1/k)(X_1{}^2 + \cdots + X_k{}^2) - 10.2| \ge \epsilon$ for at least one k satisfying $n \le k \le m]$ is at most $(2)(110)/\epsilon^2 n$. This is true for all m.
(*e*) .94 or more.
2. (*a*) Let $Y_1, Y_2, \ldots, Y_n, \ldots, Y_m$ be independent and identically distributed random variables with $P[Y_1 > 0] = 1$ and $E(\log_b Y_1)^2 < \infty$, where $b > 1$ is any base. If $\epsilon > 0$ then

$$P[b^{k[E(\log_b Y_1) - \epsilon]} < Y_1 Y_2 \cdots Y_k < b^{k[E(\log_b Y_1) + \epsilon]} \text{ for all } k \text{ with } n \le k \le m]$$

$$\ge 1 - \{2[\text{Var}\,(\log_b Y_1)]/n\epsilon^2\}.$$

For Proof, if we take logarithms we can write the exhibited event as follows, so that Theorem 1 (Sec. 12-3) can be applied to $X_1, \ldots, X_m$, where $X_i = \log_b Y_i$:

$$\left[E(\log_b Y_1) - \epsilon < (1/k)\sum_{i=1}^{k} \log_b Y_i < E(\log_b Y_1) + \epsilon\right].$$

3. .9 or more.

SECTION 13-1

1. (*a*) .9104.
2. If k is odd the integrand is an odd function. If $k = 2n$ we differentiate equation (2) a total of n times with respect to t and then let $t = 1$. A simpler derivation is suggested by Exercise 8 (Sec. 16-1).

SECTION 13-2

3. $P[Z_n = b] \le P[b - \epsilon \le Z_n < b + \epsilon] = P[Z_n < b + \epsilon] - P[Z_n < b - \epsilon]$ which converges to

$$\Phi(b + \epsilon) - \Phi(b - \epsilon) = \int_{b-\epsilon}^{b+\epsilon} \varphi(x)\,dx \le 2\epsilon/\sqrt{2\pi}$$

since $\varphi(x) \le 1/\sqrt{2\pi}$ for all x. This is true for every $\epsilon > 0$, so $P[Z_n = b]$ must converge to zero.

4. (*c*) The statement is disproved by the Poisson example exhibited in this exercise. Thus it is not possible to find a sequence, A_n converging to zero as n goes to infinity, of error-bounding *constants* for the CLT that is valid for this whole class of distributions. The A_n must depend on some property of the distribution, such as the third absolute moment.

7. When λ converges to zero the first term goes to zero, the second term goes to 1, and the third term goes to zero since it is bounded by $\lambda\left\{\sum_{k=2}^{\infty} k^3\lambda^{k-2}\right\}$, where the series in braces is convergent for $\lambda < 1$, by the ratio test, and decreases as λ decreases.

SECTION 14-1

4. $\mu_{Y|X}(x) = f(x)$ for all x with $p_X(x) > 0$, so this predictor has no error. Exercise 6*a* (Sec. 11-4) showed that for some f the best *linear* predictor may just be the constant μ_X and so have Var Y for its expected squared error.

6. (*c*) The regression function has a strictly smaller expected squared error than the linear regression function iff they differ on at least one β with $p_g(\beta) > 0$.

7. (*a*) The predictor $\mu_{X|g}$ and its conditional variance $\sigma^2_{X|g}$ are unaffected. However, p_X and the expected squared error may well be affected. For example, p_g might assign more probability to those β for which $\sigma^2_{X|g}(\beta)$ is small.

SECTION 14-2

1. (*b*) $P[S_N = 0] = .196, \ldots.$ **5.** $e^{-.8}[.1 + (.2)^2/2 + (.1)^3/6]$.

SECTION 14-3

7. (*b*) The conditional distribution of W_{n+1}, given $[W_n = k]$, is that of

$$k + Y_1' + \cdots + Y_k' = (1 + Y_1') + \cdots + (1 + Y_k') = X_1 + \cdots + X_k$$

where $P[X_k = 2] = p = 1 - P[X_k = 0]$. Thus W_{n+1} has the distribution of $X_1 + \cdots + X_{W_n}$; hence part (*a*) applies.

SECTION 15-1

3. (*b*) $P[T < 1] = 1 - e^{-1}$, $P[T \le 1] = 1.2 - e^{-t}$.

SECTION 15-2

1. (*a*) No. (*b*) Yes.

3. Apply Theorem 2 to

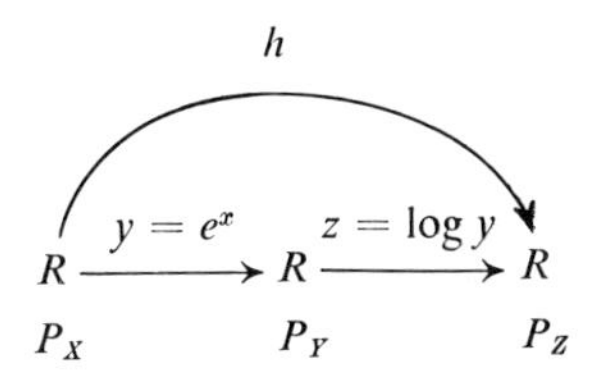

where the composite function h is the identity function on R.

5. $P[X \in D] = P_X(D) = P_{X,Y}[D \times R] \doteq .3$.

SECTION 15-3

2. If $A = [h \in D_h]$ and $D_f = g^{-1}(D_h)$ then we also have $A = [f \in D_f]$. But $0 < P(A) < 1$, so

$$P[h \in D_h, f \in D_f] = P(A) \neq P(A)P(A) = P[h \in D_h]P[f \in D_f].$$

SECTION 16-1

10. (*a*) 1/12. (*b*) 0. (*c*) 11/12.

SECTION 16-2

2. (*b*) $X^* = (X - 1/\lambda)/(1/\lambda) = \lambda X - 1$, $p_{X^*}(x^*) = \begin{cases} e^{-(x^*+1)} & \text{if } x^* + 1 > 0 \\ 0 & \text{otherwise.} \end{cases}$

5. (*c*) $p_F(y) = \begin{cases} 0 & \text{if } y \geq 0 \\ \dfrac{1}{y^{3/2}\sigma\sqrt{2\pi}} e^{-1/(2\sigma^2 y)} & \text{if } y > 0. \end{cases}$

SECTION 16-3

9. $p_{Y_1,Y_2}(y_1,y_2) = y_1 e^{-y_1}/(1 + y_2)^2$ if $y_1 > 0$ and $y_2 > 0$.

SECTION 16-4

4. (*a*) .726. (*b*) Yes. **6.** (*b*) .973.
7. (*a*) p_{S_2} is a triangular density,

$$p_{S_2}(s) = \begin{cases} s & \text{if } 0 \leq s \leq 1 \\ 2 - s & \text{if } 1 \leq s \leq 2. \end{cases}$$

(*b*) p_{S_3} is beginning to look like a normal density,

$$p_{S_3}(s) = \begin{cases} s^2/2 & \text{if } 0 \leq s \leq 1 \\ 3/4 - (s - 3/2)^2 & \text{if } 1 \leq s \leq 2 \\ (s - 3)^2/2 & \text{if } 2 \leq s \leq 3. \end{cases}$$

A general expression for p_{S_n} appears in Feller, vol. II, p. 27.† Naturally $p_{S_n}(s) > 0$ iff $0 < s < n$. A different $(n - 1)$st-degree polynomial represents p_{S_n} in each of the n unit intervals. Also p_{S_n} has a continuous derivative of order $n - 2$, everywhere.
11. The gamma density with parameters $1/m\sigma^2$ and $3n/2$.
12. (*b*) 2, 1, $2^2/3$, $1/2!$, $2^3/[(5)(3)]$, $1/3!$, $2^4/[(7)(5)(3)]$, since

$$C_n = \begin{cases} \dfrac{1}{(n/2)!} & \text{for } n \text{ even} \\ \dfrac{1}{[n/2][(n/2) - 1] \cdots [3/2][1/2]} & \text{for } n \text{ odd.} \end{cases}$$

SECTION 16-5

1. $EY = 0$, $\text{Var } Y = 3$.
6. (*a*) $EW^r = E(X^r/Y^r) = (EX^r)E(1/Y^r)$. From Sec. 16-4 $EX^r = \Gamma(\nu + r)/\Gamma(\nu)$ if $r > \nu$. The same formula yields $EY^{-r} = \Gamma(\mu - r)/\Gamma(\mu)$ if $-r > -\mu$.
7. (*e*) Apply Example 5 (Sec. 16-2) to $W_\nu = M^\nu e^{-S_\nu}$ where S_ν has the gamma density with parameters 1 and ν.
14. .773. **16.** (*b*) .023. (*d*) 1/16. **17.** (*d*) .0062.

SECTION 16-6

4. (*b*) The conditional densities $p_{X|Y=y}$ are not concentrated on the *same* countable set for all y with $p_Y(y) > 0$.
5. The resulting density is geometric with parameter $\lambda'/(1 + \lambda')$.

† "An Introduction to Probability Theory and Its Applications," John Wiley & Sons, Inc., New York, 1966.

SECTION 17-1

5. (*b*) Follows from $G_X(x-0) = P[X < x]$ and $G_X(x+0) = P[X \le 0]$ which are easily proved.
7. (*b*), (*c*), (*d*) *Comments.* As indicated in Fig. 4 each $C_{x_1,\ldots,x_k}$ is a subset of an interval of length $\rho^{k+1}/(1-\rho)$. The end points of $C_{x_1,\ldots,x_k}$ depend on ρ, but the values of F_Y at these two end points do not depend on ρ, and neither does the ordering between all these end points. Parts (*b*) and (*c*) yield two different calculations for the value of F_Y at the left end point of any $C_{x_1,\ldots,x_k}$. For example, if $k = 3$ and $(x_1,x_2,x_3) = (1,0,1)$ so $y = \rho + \rho^3$, then for part (*b*) we sum the probabilities of (0,0,0), (0,0,1), (0,1,0), (0,1,1), (1,0,0) to get $q^3 + 3q^2p + qp^2$, while part (*c*) gives the alternate calculation $F_Y(y) = P[X_1 = 0] + P[X_1 = 1, X_2 = 0, X_3 = 0] = q + pq^2$, and these two expressions are easily shown to be equal.

SECTION 17-2

4. Case 3, because $P[X = a] > 0$ implies that $P[Y = g(a)] \ge P[X = a] > 0$.

SECTION 17-3

2. If X, Y, Z are independent then Y, Z are independent so $F_{Y,Z}(y,z) = F_Y(y)F_Z(z)$; hence $F_{X,Y,Z}(x,y,z) = F_X(x)F_Y(y)F_Z(z) = F_X(x)F_{Y,Z}(y,z)$ for all x, y, z; therefore X and (Y,Z) are independent.
4. (*a*) If the joint distribution function F_X of $X = (X_1, \ldots, X_n)$ satisfies the product rule then Theorem 1 (Sec. 17-3) yields

$$
\begin{aligned}
P[x_1^{(1)} \le X_1 < x_1^{(0)}, \ldots, x_n^{(1)} \le X_n < x_n^{(0)}] &= \Delta_{F_X}(x^{(1)},x^{(0)}) \\
&= \left[\sum_{\epsilon_1=0}^{1} (-1)^{\epsilon_1} F_X(x_1^{(\epsilon_1)})\right] \cdots \left[\sum_{\epsilon_n=0}^{1} (-1)^{\epsilon_n} F_{X_n}(x_n^{(\epsilon_n)})\right] \\
&= [F_{X_1}(x_1^{(0)}) - F_{X_1}(x_1^{(1)})] \cdots [F_{X_n}(x_n^{(0)}) - F_{X_n}(x_n^{(1)})] \\
&= P[x_1^{(1)} \le X_1 < x_1^{(0)}] \cdots P[x_n^{(1)} \le X_n < x_n^{(0)}]
\end{aligned}
$$

from which we can deduce the same equation with every $<$ replaced by $\le$. Hence Lemma 2 (Sec. 15-3) shows that $X_1, \ldots, X_n$ are independent.

SECTION 18-1

4. Let T be diagonal with $t_{ii} = \delta_i$ so that the ith coordinate of Tt is $\delta_i t_i$. If T is singular then there exists $t \ne 0$ with $Tt = 0$, but for some i we have $t_i \ne 0$, hence δ_i must be zero. Conversely if $\delta_i = 0$ let t have all zero coordinates except for $t_i = 1$ so that $Tt = 0$ with $t \ne 0$; hence T is singular.

SECTION 18-2

5. Construct T by

$$
TX + \mu = \begin{bmatrix} \begin{matrix}\text{matrix for}\\(Y_1, \ldots, Y_r)\end{matrix} & \text{all zeros} \\ \text{all zeros} & \begin{matrix}\text{matrix for}\\(Y_{r+1}, \ldots, Y_k)\end{matrix} \end{bmatrix} \begin{bmatrix} X_1 \\ \vdots \\ X_r \\ \vdots \\ X_k \end{bmatrix} + \begin{bmatrix} \mu_1 \\ \vdots \\ \mu_r \\ \vdots \\ \mu_k \end{bmatrix}
$$

Clearly the normal density induced by $TX + \mu$ has the same mean and covariance matrix as p_Y; hence the densities are equal. But the first r coordinates of $TX + \mu$, considered as a random r vector, is a function of $(X_1, \ldots, X_r)$ while the last $k - r$ coordinates of $TX + \mu$ is a function of $(X_{r+1}, \ldots, X_k)$.

6. We need only show that Y is independent of $(X_2 - Y, \ldots, X_k - Y)$. But $(Y, X_2 - Y, \ldots, X_k - Y)$ has a normal density so from Exercise 5 we need only show that Y is uncorrelated with each $X_i - Y$ for $i = 2, \ldots, k$: $EY[X_i - Y] = EYX_i - (1/k) = E(X_i/k)X_i - (1/k) = 0$.

SECTION 18-3

7. (*b*)
$$\varphi_X(t) = \begin{cases} \left[(1-|t|) + \left(1 - \frac{|t|}{2}\right)\right] \Big/ 2 = 1 - (3/4)\,|t| & \text{if } |t| \le 1 \\ \left[1 - \frac{|t|}{2}\right] \Big/ 2 & \text{if } 1 \le t \le 2 \\ 0 & \text{otherwise.} \end{cases}$$

Thus $\varphi_X(t) = \varphi_Y(t)$ if $|t| \le 1$.
(*d*) Clearly $\varphi_X(t)\varphi_Z(t) = \varphi_Y(t)\varphi_Z(t)$ for all real t since $\varphi_X(t) = \varphi_Y(t)$ if $|t| \le 1$, while $\varphi_Z(t) = 0$ if $|t| > 1$.

SECTION 18-4

5. (*d*) $\rho(Z,Z^3) = \sqrt{3/5}$, $1/4 + \arcsin \sqrt{3/5} = .391$.

CHAPTER 19

8. Assume that F is a continuous distribution function. Given an integer M let x_k satisfy $F(x_k) = k/M$ for $k = 1, \ldots, (M-1)$. For any distribution function G let δ be the largest of $|G(x_k) - F(x_k)|$ for $k = 1, \ldots, (M-1)$. If x satisfies $x_k < x < x_{k+1}$ then $k/M - \delta \le G(x_k) \le G(x) \le G(x_{k+1}) < (k+1)/M + \delta$ but then using $k/M \le F(x) \le k/M + 1/M$ on the extremes yields

$$F(x) - (1/M) - \delta \le G(x) \le F(x) + (1/M) + \delta.$$

Thus we easily see (even if $x < x_1$ or $x > x_{M-1}$) that $|G(x) - F(x)| \le (1/M) + \delta$ for all x. Therefore for every X_n and X if F_X is continuous then

$$\rho(X_n,X) \le \frac{1}{M} + [\max\{|F_{X_n}(x_1) - F_X(x_1)|, \ldots, |F_{X_n}(x_{M-1}) - F_X(x_{M-1})|\}].$$

Assume that $F_{X_n}(x) \to F_X(x)$ for all real x, as $n \to \infty$. Fix M and let $n \to \infty$, then $[\ ] \to 0$ so the limit of $\rho(X_n,X)$ is at most $1/M$. But this is true for each M hence $\rho(X_n,X) \to 0$ as $n \to \infty$.

To see that the continuity of F_X is essential, note that if $P[X = 0] = 1$ and $P[X_n = 1/n] = 1$ then for each real x we have $F_n(x) \to F(x)$ as $n \to \infty$, while $\rho(X_n,X) = 1$ for all n.

SECTION 20-3

2. (*a*) Fix $\epsilon > 0$. If $\sum_{k=1}^{\infty} P[|Z_k| \ge \epsilon] = \infty$ then $\sum_{k=n}^{\infty} P[|Z_k| \ge \epsilon] = \infty$ for every n. Then Exercise 3*a* (Sec. 7-4) implies that $P\left(\bigcup_{k=n}^{m} [|Z_k| \ge \epsilon]\right) \ge \left\{1 - \exp\left(-\sum_{k=n}^{m} P[|Z_k| \ge \epsilon]\right)\right\} \to 1$ as $m \to \infty$. But this is true for all n so Z_n cannot $\xrightarrow{\text{a.s.}} 0$.
(*b*) Since $(1/n)X_n \xrightarrow{\text{a.s.}} 0$ we may apply part (*a*) with $\epsilon = 1$ to get

$$\infty > \sum_{n=1}^{\infty} P[|X_n| \ge n] = \sum_{n=1}^{\infty} P[|X_1| \ge n]$$

where the equality follows from the identical distributions. But now Lemma 1 (Sec. 20-1) implies that $E|X_1| < \infty$.

SECTION 22-2

2. (*a*) The second two coordinates have the distribution induced by $([Z_1 + Z_2], [Z_1 + Z_2] + Z_3)$ where, by the semigroup property, $[Z_1 + Z_2]$ has distribution $P_{t_1+(t_2-t_1)} = P_{t_2}$ and Z_3 has distribution $P_{t_3-t_2}$. But $[Z_1 + Z_2]$ and Z_3 are independent so this induced distribution is just P_{t_1,t_2} by the definition of P_{t_1,t_2}.
3. (*a*) Repeated application of the semigroup requirement implies that for $k = 1, 2, \ldots$ the only candidate for $P_{k\delta}$ is the distribution of $Z_1 + \cdots + Z_k$ where $Z_1, \ldots, Z_k$ are independent and each has distribution P_δ. Similarly if we have an intermediate point $s = k\delta + b$, $0 < b < \delta$, then we must define P_s to be the distribution of $Z + W$ where Z and W are independent and have distributions $P_{k\delta}$ and P_b. Thus there is at most one possible extension. This extension should now be shown to be a semigroup.
4. (*a*) $EX_t = 0$ and $E(X_t)^2 = \frac{1}{2}$ for all t.

SECTION 23-1

4. (c) $\frac{1}{2}$. **7.** (*b*) $e^{-\lambda h}[1 + (\lambda h)/(1 + \lambda t)]$.
8. (*d*) Then select n independent observations $u_1, \ldots, u_n$ from a uniform distribution on the m points $1/m, 2/m, \ldots, 1$ and place crosses at $u_1, \ldots, u_n$. This assumes that $u_1, \ldots, u_n$ are distinct. If not, then repeat this part of the experiment.

SECTION 23-3

1. If $t_1, t_2, \ldots, t_{25}$ are the arrival times in Fig. 2, then construct the counting function having arrivals $t_2, t_4, t_6, \ldots, t_{24}$ and only these. That is, $X_1 + X_2, X_3 + X_4, \ldots, X_{23} + X_{24}$ are independent and each has a $\gamma_{1,2}$ density. For $\gamma_{1,5}$ use the arrivals $t_5, t_{10}, t_{15}, t_{20}, t_{25}$ and only these.
6. 20 minutes, by the waiting-time paradox.
8. $EZ_n = n/(\lambda + 1)$ and $\operatorname{Var} Z_n = n\lambda/[(\lambda + 2)(\lambda + 1)^2]$.

11. (*c*)
$$\begin{aligned} P[Y_n = k] &= \sum_{m,t} P[Y_n = k, M_n = m, T_n = t] \\ &= \sum_{m,t} P[Y_n = k \mid M_n = m, T_n = t] P[M_n = m, T_n = t] \\ &= \sum_{m,t} \frac{km}{t} P[M_n = m, T_n = t] = kE\left\{\frac{M_n}{T_n}\right\}. \end{aligned}$$

For large n, with high probability, $(1/n)M_n$ will be near $P[X_1 = k]$ while T_n/n will be near EX_1 so that the distribution of the ratio M_n/T_n will be concentrated near $P[X_1 = k]/EX_1$. Furthermore, $0 \le M_n/T_n \le 1$ so that the small probability not near $P[X_1 = k]/EX_1$ cannot have much effect on the mean, hence $E\left\{\dfrac{M_n}{T_n}\right\} \to \dfrac{P[X_1 = k]}{EX_1}$ as $n \to \infty$.

SECTION 23-6

3. (*c*)
$$\begin{aligned} P[H(\tau + d\tau) - H(\tau) = 1 \mid H(\tau) = n] &= P[N(m(\tau + d\tau)) - N(m(\tau)) = 1 \mid N(m(\tau)) = n] \\ &= (n + \nu)\lambda m'(\tau)\, d\tau. \end{aligned}$$

$$P[H_\tau = n] = \binom{\nu + n - 1}{n} e^{-\lambda m(\tau)\nu}(1 - e^{-\lambda m(\tau)})^n.$$

$$EH_\tau = \nu(e^{\lambda m(\tau)} - 1).$$

$$\operatorname{Var} H_\tau = \nu e^{\lambda m(\tau)}(e^{\lambda m(\tau)} - 1).$$

SECTION 23-7

2. Obviously $a_1 \leq a_2 \leq \cdots$ so we need only show that for every number M, no matter how large, there is a k such that $a_k \geq M$. Given M select m large enough so that

$$\sum_{n=0}^{m} \frac{1}{\beta_n} \geq M \quad \text{and} \quad \sum_{n=0}^{m} \frac{1}{\beta_{-n}} \geq M \tag{a}$$

then select k large enough so that

$$k \min \left\{\frac{1}{\beta_{-m}}, \ldots, \frac{1}{\beta_0}, \ldots, \frac{1}{\beta_m}\right\} \geq M. \tag{b}$$

Take any permissible $(z_0, \ldots, z_{k-1})$. We must show that

$$\frac{1}{\beta_{z_0}} + \frac{1}{\beta_{z_1}} + \cdots + \frac{1}{\beta_{z_{k-1}}} \geq M. \tag{c}$$

If for some i we have $m \leq z_i$ then every one of the integers $0, 1, \ldots, m$ must appear as a coordinate of $(z_0, \ldots, z_i)$, hence (*c*) follows from (*a*). If for some i we have $z_i \leq -m$ then similarly (*c*) follows from (*a*). For the remaining case we have $-m \leq z_i \leq m$ for $i = 0, \ldots, k-1$ so (*c*) follows from (*b*).

SECTION 24-1

3. (*a*) $-1.276, -.331 \times 10^{-2}, -1.38 \times 10^{-4}, \ldots$. (*b*) About half of the time because . . . .

SECTION 24-2

1. (*a*) $P[L < s] = (2/\pi) \arcsin \sqrt{s/t}$.

8. For fixed n if $0 < t < 1/n$ then $P(A_n) \leq P[|W_t| \leq t^{3/4}] = P[|W_t^*| \leq t^{1/4}] \to 0$ as $t \to 0$ because W_t^* is normal 0, 1. Therefore $P(A_n) = 0$ so $P\left(\bigcup_{n=1}^{\infty} A_n\right) = 0$.

Index

The definitions of the terms in the table below can have various forms, depending on the context. The pages in the summaries that contain the definitions, and the corresponding chapter numbers, are indicated for each group of terms. (In addition to this listing, Chap. 21 also contains rigorous definitions of many of these terms.)

The index itself does not contain those general mathematical terms (concerning sets, functions, countability, and convergence) which are concisely collected and defined in the summaries of Chaps. 1, 3, 4, and 5.

A table of popular discrete probability densities can be found on pp. 130, 653; a table of popular continuous probability densities can be found on p. 678.

Summary locations for the definitions of several common terms

Term	Chapter				
	6	11	14	15	16
Event, probability measure, probability space, sample space	647, 648 (discrete case)			670 (general case)	
Conditional distribution of f given A, identical distribution, induced distribution, joint distribution, random variable, random vector		657–659 (discrete case)		671	679 (continuous case)
Mutual and pairwise independence of random vectors		659, 660		672	681
Expectation, mean, MGF, moments, standard deviation, standardized variable, variance (These first appear in Chap. 9 and pp. 653–655, prior to the introduction of random variables.)		661–663		673, 674	677, 680, 682
Conditional distribution of f given $[g = \beta]$, conditional expectation and variance, mixture, randomize a parameter, regression function			668	674, 675	684, 685

Where appropriate, a text page number is immediately followed by the corresponding summary page number (summaries begin on p. 639). For example, the definition of a "Binomial coefficient" appears on text p. 20 and again on p. 642 of the summary.